ENVIRONMENTAL SCIENCE AND TECHNOLOGY

A Wiley-Interscience Series of Texts and Monographs

Edited by JERALD L. SCHNOOR, *University of Iowa*
ALEXANDER ZEHNDER, *Swiss Federal Institute for Water Resources and Water Pollution Control*

PHYSIOCHEMICAL PROCESSES FOR WATER QUALITY CONTROL
Walter J. Weber, Jr., Editor

pH AND pION CONTROL IN PROCESS AND WASTE STREAMS
F. G. Shinskey

AQUATIC POLLUTION: An Introductory Text
Edward A. Laws

INDOOR AIR POLLUTION: Characterization, Prediction, and Control
Richard A. Wadden and Peter A. Scheff

PRINCIPLES OF ANIMAL EXTRAPOLATION
Edward J. Calabrese

SYSTEMS ECOLOGY: An Introduction
Howard T. Odum

INTEGRATED MANAGEMENT OF INSECT PESTS OF POME AND STONE FRUITS
B. A. Croft and S. C. Hoyt, Editors

WATER RESOURCES: Distribution, Use and Management
John R. Mather

ECOGENETICS: Genetic Variation in Susceptibility to Environmental Agents
Edward J. Calabrese

GROUNDWATER POLLUTION MICROBIOLOGY
Gabriel Bitton and Charles P. Gerba, Editors

CHEMISTRY AND ECOTOXICOLOGY OF POLLUTION
Des W. Connell and Gregory J. Miller

SALINITY TOLERANCE IN PLANTS: Strategies for Crop Improvement
Richard C. Staples and Gary H. Toenniessen, Editors

ECOLOGY, IMPACT ASSESSMENT, AND ENVIRONMENTAL PLANNING
Walter E. Westman

CHEMICAL PROCESSES IN LAKES
Werner Stumm, Editor

INTEGRATED PEST MANAGEMENT IN PINE-BARK BEETLE ECOSYSTEMS
William E. Waters, Ronald W. Stark, and David L. Wood, Editors

PALEOCLIMATE ANALYSIS AND MODELING
Alan D. Hecht, Editor

BLACK CARBON IN THE ENVIRONMENT: Properties and Distribution
E. D. Goldberg

GROUND WATER QUALITY
C. H. Ward, W. Giger, and P. L. McCarty, Editors

TOXIC SUSCEPTIBILITY: Male/Female Differences
Edward J. Calabrese

ENERGY AND RESOURCE QUALITY: The Ecology of the Economic Process
Charles A. S. Hall, Cutler J. Cleveland, and Robert Kaufmann

AGE AND SUSCEPTIBILITY TO TOXIC SUBSTANCES
Edward J. Calabrese

ECOLOGICAL THEORY AND INTEGRATED PEST MANAGEMENT PRACTICE
Marcos Kogan, Editor

AQUATIC SURFACE CHEMISTRY: Chemical Processes at the Particle
Water Interface
Werner Stumm, Editor

RADON AND ITS DECAY PRODUCTS IN INDOOR AIR
William W. Nazaroff and Anthony V. Nero, Jr., Editors

PLANT STRESS–INSECT INTERACTIONS
E. A. Heinrichs, Editor

INTEGRATED PEST MANAGEMENT SYSTEMS AND COTTON PRODUCTION
Ray Frisbie, Kamal El-Zik, and L. Ted Wilson, Editors

ECOLOGICAL ENGINEERING: An Introduction to Ecotechnology
William J. Mitsch and Sven Erik Jorgensen, Editors

ARTHROPOD BIOLOGICAL CONTROL AGENTS AND PESTICIDES
Brian A. Croft

AQUATIC CHEMICAL KINETICS: Reaction Rates of Processes in Natural Waters
Werner Stumm, Editor

GENERAL ENERGETICS: Energy in the Biosphere and Civilization
Vaclav Smil

FATE OF PESTICIDES AND CHEMICALS IN THE ENVIRONMENT
J. L. Schnoor, Editor

ENVIRONMENTAL ENGINEERING AND SANITATION, Fourth Edition
Joseph A. Salvato

AQUATIC POLLUTION: An Introductory Text, Second Edition
Edward A. Laws

AQUATIC POLLUTION

AQUATIC POLLUTION

An Introductory Text

Second Edition

EDWARD A. LAWS
University of Hawaii
Honolulu, Hawaii

An Interscience Publication
JOHN WILEY & SONS, INC.
New York • Chichester • Brisbane • Toronto • Singapore

Library of Congress Cataloging in Publication Data:
Laws, Edward A., 1945–
    Aquatic pollution : an introductory text / Edward A. Laws. — 2nd
ed.
      p.   cm. — (Environmental science and technology)
    Includes index.
    ISBN 0-471-53457-9. — ISBN 0-471-58883-0 (pbk.)
    1. Water—Pollution.   I. Title.   II. Series.
    TD420.L387   1993   92-26501
    628.1′68—dc20   CIP

Printed in the United States of America

10 9 8 7 6 5 4 3 2 1

PREFACE

The decade of the 1970s was a period of environmental awareness and increasing appreciation of the fact that the quantities and types of waste materials discharged to the environment by human society were creating serious pollution problems that could be solved only by reducing the quantities of waste and/or identifying more satisfactory means of disposal. Concern over water pollution was a major issue at that time, a concern reflected in the United States by the passage of federal legislation such as the Safe Drinking Water Act of 1974 and the Clean Water Act of 1977. There have been some success stories, cases in point being the reversal of cultural eutrophication in Lake Washington following diversion of sewage during the 1960s and the recovery of the southern California brown pelican after the 1972 ban on DDT use in the United States. In other cases, however, efforts to reduce or eliminate water pollution have met with mixed results. Eutrophication in Lake Erie, for example, continues to be a problem, despite the fact that literally billions of dollars have been spent to reduce nutrient inputs to the lake. Over \$7 billion have been spent to clean up contaminated aquifers in the United States since the Comprehensive Environment Response, Compensation, and Liability Act was passed in 1980, but according to some authorities the cleanup program has little success to show for the money that has been spent.

The record leaves little doubt that an effective approach to solving water pollution problems will require policy makers and environmentalists to be knowledgeable about scientific issues and cognizant of the efficacy and limitations of corrective measures. It is unrealistic, however, to expect that many of these persons will have PhDs in environmental science, toxicology, oceanography, ecology, or some related field of natural science. PhDs can always be consulted for their opinions on specific environmental issues. It would be desirable, however, for environmentalists and persons involved in establishing water quality policy to be aware of fundamental ecological and toxicological principles relevant to water pollution and of the issues and considerations that influence decision making in specific cases. Indeed such knowledge is desirable for the general public, whose votes and opinions weigh heavily in environmental decision making.

Aquatic Pollution is intended to provide that knowledge. It was written to be a comprehensive introduction to the subject of water pollution, the intended audience being college undergraduates. This second edition, written a little more than 10 years after publication of the first edition, is in part simply an update on the status of particular water pollution problems. As in the first edition, numerous case studies have been used to illustrate concepts and principles discussed in the text. Chapters 1–3 introduce the reader to basic ecological principles relevant to water pollution. Chapter 8 likewise introduces basic principles of toxicology. The remaining chapters are concerned with particular types of water pollution, in most cases involving an introduction to concepts and issues associated with the particular type of pollution followed by one or more case studies. This second edition includes three topics not covered in the first edition, namely acid rain (Chapter 15), groundwater pollution (Chapter 16), and plastics in the sea (Chapter 17). These topics have been the cause of increasing concern in the past decade and certainly deserve a place in a textbook of this kind. Covering all the material in the second edition requires roughly two quarters of classroom time. Some judicious selection of material will probably be necessary if the book is to be used in a one-semester course.

As in the case of the first edition, I am indebted to numerous persons who made available information to me that was not readily available in the literature or helped to answer questions. I would particularly like to thank the staff of the science and technology and government documents sections of Hamilton Library at the University of Hawaii. Their assistance was invaluable. I am also indebted to the Hawaii Institute of Geophysics publications staff, who helped with the preparation of many of the illustrations in the book. Others whose help deserves mention include Dr. Peter Betzer, Dr. W. T. Edmondson, Dr. Paul LaRock, Mr. Jack Huizingh, Dr. Roger Fujioka, Mr. Clifford Henry, Dr. Daniel Anderson, Dr. Jota Kanda, Dr. Nancy Yamaguchi, Dr. John Bardach, Dr. Fred Mackenzie, Dr. James Galloway, Dr. Roland Wollast, and Ms. Christine Curick. I would especially like to thank my wife, Stephanie, and son, Ryan, who provided support and encouragement throughout the preparation of this second edition.

EDWARD A. LAWS

Honolulu, Hawaii
April, 1992

SERIES PREFACE
Environmental Science and Technology

The Environmental Science and Technology Series of Monographs, Textbooks, and Advances is devoted to the study of the quality of the environment and to the technology of its conservation. Environmental science therefore relates to the chemical, physical, and biological changes in the environment through contamination or modification, to the physical nature and biological behavior of air, water, soil, food, and waste as they are affected by man's agricultural, industrial, and social activities, and to the application of science and technology to the control and improvement of environmental quality.

The deterioration of environmental quality, which began when man first collected into villages and utilized fire, has existed as a serious problem under the ever-increasing impacts of exponentially increasing population and of industrializing society. Environmental contamination of air, water, soil, and food has become a threat to the continued existence of many plant and animal communities of the ecosystem and may ultimately threaten the very survival of the human race.

It seems clear that if we are to preserve for future generations some semblance of the biological order of the world of the past and hope to improve on the deteriorating standards of urban, suburban, and rural public health, environmental science and technology must quickly come to play a dominant role in designing our social and industrial structure for tomorrow. Scientifically rigorous criteria of environmental quality must be developed. Based in part on these criteria, realistic standards must be established and our technological progress must be tailored to meet them. It is obvious that civilization will continue to require increasing amounts of fuel, transportation, industrial chemicals, fertilizers, pesticides, and countless other products; and that it will continue to produce waste products of all descriptions. What is urgently needed is a total systems approach to modern civilization through which the pooled talents of scientists and engineers, in cooperation with social scientists and the medical profession, can be focused on the development of order and equilibrium in the presently disparate segments of the human environment. Most of the skills and tools that are needed are already in existence. We surely have a right to hope a technology that has created such manifold environmental problems is also capable of solving them. It is our hope that this Series in Environmental Sciences and Technology will not only serve to make this challenge more explicit to the established professionals, but that it also will help to stimulate the student toward the career opportunities in this vital area.

JERALD L. SCHNOOR
ALEXANDER ZEHNDER

CONTENTS

1

FUNDAMENTAL CONCEPTS

The introduction of pollutants into aquatic systems is a perturbation that can set off a complicated series of biological and chemical reactions. In order to understand how and why these reactions occur, it is first necessary to understand something about the interactions that occur in the system in the absence of such perturbations. For example, if a toxic metal such as mercury were discharged into the ocean, it is quite possible that its presence in the water would affect adversely the photosynthetic rates of marine plants in the discharge area. This reduction of photosynthetic rates might be the primary and most direct effect of the pollutant, but it is unlikely that the impact would end there. To the extent that photosynthetic rates were lowered, the food supply of plant-eating animals would be reduced, and their biomass and production rates would also be lowered. Furthermore, if the mercury were absorbed by or adsorbed to the plants, plant-eating animals would probably assimilate some of the mercury and become stressed by the presence of the mercury in their tissues. Thus the marine plant eaters could be adversely affected both by a reduction in their food supply and by the presence of the mercury in their bodies. Using the same logic, it is easy to imagine how animals who preyed on the plant eaters could be affected through similar mechanisms, and how feeding relationships could ultimately spread the mercury to every organism in the water. Obviously some understanding of the feeding relationships in a natural aquatic system is necessary to appreciate and anticipate the effects of such pollutants.

Now let us suppose that the mercury discharges cease. Will the system recover and return to its original condition? Perhaps, but not necessarily. The stability of natural systems to perturbations such as pollutant discharges is a fundamental area of study in systems analysis and a critical consideration in the understanding of pollutant effects. The fact that a natural system is in equilibrium by no means guarantees that the system will return to the original state following a perturbation. To cite a popular example, had a very small meteor struck the earth 65 million years ago, it is possible that a few dinosaurs might have been killed or injured. However, the condition of the dinosaur population would have returned very likely to normal

1

within a short time through natural processes. It is now generally agreed, however, that the extinction of all the dinosaurs was probably caused by a very large meteor that struck the earth about 65 million years ago. Conditions on the earth for a period of time following that event are believed to have been incompatible with the survival of dinosaurs, the result being that the system did not return to its pre-event status. Perturbations caused by pollution can have similar consequences. Some present-day biological systems are believed to be much more sensitive than others to perturbations, and our ability to quantify this sensitivity requires some basic understanding of perturbation theory. In summary, there are some fundamental ecological concepts that we need to appreciate before we can address intelligently problems caused by aquatic pollution.

SIMPLE FOOD CHAIN THEORY

All animals require food. Food may be burned (respired) to provide energy or incorporated into the animal's body in the form of proteins, fats, carbohydrates, and other compounds to provide essential structural or metabolic components. Plants are by far the most important producers of food in most aquatic systems, although certain bacteria may be significant producers in some parts of the deep sea (Jannasch and Wirsen, 1977). Plants utilize sunlight as an energy source to manufacture organic compounds from carbon dioxide, water, and various inorganic nutrients in a process called *photosynthesis*. For example, a simplified equation describing the manufacture of glucose may be written

$$\text{energy} \quad + \quad 6CO_2 \quad + \quad 6H_2O \quad \underset{\text{respiration}}{\overset{\text{photosynthesis}}{\rightleftharpoons}} \quad C_6H_{12}O_6 \quad + \quad 6O_2$$

$$\underset{\text{carbon dioxide}}{} \quad \underset{\text{water}}{} \qquad\qquad \underset{\text{glucose}}{} \quad \underset{\text{oxygen}}{}$$

In this case glucose is the organic compound, the adjective organic meaning that the compound is found in organisms. If the reaction proceeds from left to right, the energy source is sunlight. Part of this energy is stored chemically in the glucose molecule. If the glucose is then oxidized by burning it with oxygen, the reaction proceeds from right to left, and the energy stored in the glucose is released. Some of that energy is made available to the organism mediating the respiratory process and is used to perform various metabolic functions. It is common practice to use either organic carbon or its associated chemical energy content as a metric for food supply, 1 gram (g) of organic carbon being associated with an energy content of 8–11 kilocalories (kcal). All animals have the ability to transform organic compounds from one form to another and hence to convert their food into the compounds they require. However, only plants and certain bacteria have the ability to manufacture organic, high-energy compounds from inorganic, low-energy constituents, and it is this transformation that is referred to as *primary production*. If the energy needed to drive the transformation comes from light, the process is called photosynthesis. If the energy is obtained from chemical reactions involving inorganic compounds, the process is called *chemosynthesis*. Only certain types of bacteria are capable of mediating the latter process. All living organisms depend either directly or indirectly on primary producers as a source of food. Organisms that can produce most or all of the substances they need from inorganic compounds are called *photoautotrophs* or *chemoautotrophs*, depending on whether the energy needed to effect the conversion comes from light or the reactions of inorganic chemicals, respectively. Organisms that lack autotrophic capabilities are called *heterotrophs*. Plants are autotrophs, and animals are heterotrophs. Most bacteria are heterotrophs, although some bacteria do have well-developed photoautotrophic or chemoautotrophic capabilities.

Food produced initially by a plant may be consumed by a plant-eating heterotroph, or herbivore. The production of herbivore biomass due to the metabolism of plant biomass by the herbivore is secondary production. The herbivore may in turn be eaten by another heterotroph, or primary carnivore, which converts part of the herbivore biomass into primary carnivore biomass. The production of primary carnivore biomass from herbivore biomass is *tertiary production*. The primary carnivore may in turn be eaten by another heterotroph, or secondary carnivore, which may in turn be eaten by a tertiary carnivore, and so forth. Ecologists refer to such a system of successive food transfers as a *food chain*. Each component of the food chain is called a *trophic level*. In the example given, plants would make up the first trophic level, herbivores the second trophic level, primary carnivores the third trophic level, and so forth. Such a food chain is depicted schematically in Figure 1.1.

In most aquatic systems the transfer of food from one trophic level to the next is believed to occur with an efficiency of only about 20%. In other words, the rate at which food is ingested by a trophic level is about five times greater than the rate at which food is passed on to the next trophic level. This efficiency is referred to as an *ecological efficiency*, or more specifically as a *trophic level intake efficiency* (Odum, 1971, p. 76). Ecological efficiencies are generally low because much of the food ingested by a trophic level is either respired to provide energy or excreted because it cannot be incorporated into new trophic level biomass. However, ecological efficiencies are also reduced when, for example, an organism dies from disease or a female fish releases her eggs into the water. Eggs occupy a trophic level that is always lower than that of the organism that produced them.

Ecological Pyramids

Because ecological efficiencies are only about 20% in aquatic systems, the flux of food from one trophic level to the next steadily decreases as one moves up the food chain. The result is that the primary production rate is likely to exceed greatly the production of top level carnivores, the magnitude of the discrepancy depending on the number of trophic levels in the food chain. Ryther (1969) has estimated that there are roughly six trophic levels in typical open ocean marine food chains. On the

```
                      ⌠ secondary carnivores  (trophic level 4)
                      │           │
    heterotrophs     ⟨  primary carnivores    (trophic level 3)
                      │           │
                      ⌊ herbivores             (trophic level 2)
                                  │
    autotrophs  ⟷  plants                      (trophic level 1)
                                  │
                      inorganic nutrients
```

Figure 1.1 Diagram of a food chain through trophic level four.

other hand, some coastal and upwelling areas may have food chains with as few as three trophic levels. This difference stems in part from the fact that the primary producers in open ocean systems are dominated by very small microscopic plants called *phytoplankton*, whereas in coastal and upwelling areas the individual phytoplankton cells tend to be larger, and the cells tend to form chains and gelatinous masses. In the coastal and upwelling areas the primary producers can therefore be efficiently grazed by rather large herbivorous crustaceans such as copepods or even small fish. However, in the open ocean most of the phytoplankton are much too small to be consumed by crustaceans and small fish, and several intermediate trophic levels therefore separate these two categories of organisms. Regardless of the length of the food chain, the steady decrease in the flux of food to higher and higher trophic levels results usually in a decrease in the biomass of organisms on successively higher trophic levels. Thus, if one were to represent the biomass of each trophic level by a bar whose length was proportional to the biomass of organisms in the trophic level and if one were to lay these bars on top of each other, the resulting figure would look qualitatively like Figure 1.2. Arranged in this way the bars of trophic level biomass form a pyramid, often referred to as an *ecological pyramid*.

There are two caveats relative to the issue of ecological pyramids. First, although Figure 1.2 correctly depicts the average distribution of biomasses in a food chain, it is quite possible in non-steady-state systems for the biomass distribution in two or more trophic levels to become temporarily inverted. In other words, trophic level biomass increases rather than decreases with increasing trophic level number. For example, in temperate oceans and lakes, a so-called bloom of plant biomass may occur in the spring as the water temperature and average daily solar insolation increase. This plant bloom generally does not occur at a time when the herbivore biomass is large, but the herbivore biomass begins to increase rapidly shortly thereafter in direct response to the increase in herbivore food. Typically herbivore grazing reduces the plant biomass to a low level, whereas herbivore biomass peaks and then declines. The fall in herbivore biomass is caused by both the decrease in herbivore food and grazing pressure by primary carnivores. Figure 1.3 shows qualitatively how plant and herbivore biomass may vary with time during this period.

A system in which the herbivore biomass is greater than the plant biomass for a short period following the plant bloom is shown in Figure 1.3. Such a condition may exist for a short time in many aquatic systems that are subject to large-scale seasonal

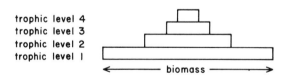

Figure 1.2 Trophic level biomass through trophic level four in a hypothetical food chain.

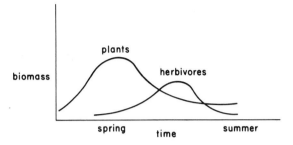

Figure 1.3 Biomass of plants and herbivores during spring and early summer in a hypotheti-cal temperate aquatic ecosystem.

cycles. During this period the first two trophic level biomasses form a so-called inverted pyramid, because the second trophic level biomass is greater than that of the first. This situation lasts for only a short time, and the average distribution of biomass is similar to Figure 1.2. The logical arguments that lead us to expect a nor-mal pyramid of biomass do not necessarily apply in a non-steady-state system, because over short time intervals predators may consume more food than prey are producing and hence reduce the prey biomass to a low level. Obviously this situa-tion cannot persist for long; otherwise the predators would destroy their food sup-ply. Hence on the average one does expect to see a normal pyramid of biomass.

A second point about ecological pyramids concerns the size of organisms mak-ing up each trophic level. In general one expects predators to be larger than prey, and hence higher trophic level organisms should be larger than lower trophic level organisms. This expectation is generally fulfilled, although there are certainly exceptions to the rule. For example, animals that hunt in groups or packs, such as wolves or killer whales, may kill organisms larger than themselves. However, pred-ators are usually larger than their prey, and as a result the numbers of organisms on successively higher trophic levels decrease even more rapidly than the total bio-mass. Although it is generally true that large organisms consume more food than small organisms, it is also generally true that large organisms consume less food *per unit biomass* than do small organisms. The relationship between organism size and metabolic rate is such that, if two organisms differ in weight by a factor of 10,000, the larger organism can be expected to consume only 10% as much food per unit body weight as the smaller organism. In other words, the larger organism would consume about 1000 times as much food as the smaller organism, or 1000/10,000 = 1/10 as much food per unit body weight.

Now consider a case in which the size of individual organisms on two successive trophic levels differs by a factor of 10,000 and the ecological transfer efficiency between the trophic levels is 20%. In this case a steady-state situation might exist in which the total biomass of the second trophic level was twice that of the first; although the second trophic level received only 20% as much food as the first trophic level, the second trophic level would need only 10% as much food to support a given amount of biomass as the first trophic level. Thus the logical arguments that lead us to expect an ecological pyramid of biomass need not apply to food chains in which the size of organisms on successive trophic levels differs greatly, because these arguments implicitly assumed the food requirements per unit biomass of all trophic levels to be identical. The fact that normal ecological pyramids of biomass are found in most natural aquatic food chains (e.g., Odum, 1971, p. 80; Sheldon et al., 1972) indicates that differences in organism size on successive trophic levels are not sufficiently great to invert the pyramids. Nevertheless, the difference in successive trophic level biomasses is often less than the factor of 5 which would be expected to result from transfer efficiencies of 20% if all organisms required the same amount of food per unit biomass. Thus organism size differences tend to reduce, but not eliminate, the effect of low ecological transfer efficiencies on trophic level biomass structure.

Recycling and the Microbial Loop

The food chain we have discussed up to this point is called the *grazing food chain*, because the second and higher trophic levels consist of predators who graze upon prey. The first trophic level of the grazing food chain is customarily considered to be occupied by the primary producers. A very important companion of the grazing food chain in any healthy aquatic system is the detritus food chain. The first trophic level in the detritus food chain is the nonliving organic matter produced by living organisms. This nonliving organic matter may exist either as particles or as dissolved organic substances and is referred to as *detritus*. The detritus provides food for a category of organisms called *detritivores*, a designation that includes both bacteria and certain metazoans. Bacteria have no mouth parts, and hence strictly speaking must feed entirely on dissolved organics. However, by exuding enzymes they are able to solubilize and hence make use of particulate material as well. Metazoan detritivores such as benthic worms feed primarily on particulate detritus. Because detritivores are living organisms, they respire and excrete organics, just as do the members of the grazing food chain. The organic compounds excreted by detritivores may be utilized very likely as food by other detritivores, and as a result only the most refractory organic compounds accumulate in the system. Most of the organic matter initially synthesized by the primary producers is ultimately respired, either by organisms in the grazing food chain or by detritivores. The detritivores themselves are consumed by animals or protozoans, and in this way some of the organic carbon excreted by the grazing food chain is recycled back into the grazing food chain. The process is illustrated schematically in Figure 1.4. The portion of the detritus food chain involving dissolved organics, bacteria, and protozoans is often referred to as the *microbial loop* and is believed to account for much of the degradation of detritus in aquatic systems.

It is apparent from Figure 1.4 that the grazing food chain and the detritus food chain are interconnected and do not function independently of each other. The interaction between the two food chains is a mutualistic one, i.e., favorable to both and obligatory. The grazing food chain benefits the detritus food chain by excreting much of the organic matter needed by the detritivores for food; the detritus food chain benefits the grazing food chain by removing potentially toxic waste products excreted by both food chains. An approximate balance between the anabolism and catabolism of organic matter is essential to the maintenance of a stable aquatic ecosystem. In a system in which primary production on the average exceeds respiration, organic matter in the form of either plant or animal biomass or detritus will

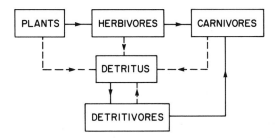

Figure 1.4 Box model of the grazing and detritus food chains and the interactions between the two food chains. Solid lines represent feeding relationships. Dashed lines represent excretion.

accumulate in the system. Eventually the whole system may fill up with organic sediments. In fact exactly this process does occur, although often at a very slow rate, in most freshwater habitats and in some marine basins. This gradual accumulation of organic debris results in part from the fact that some detritus is rather refractory and not broken down efficiently by detritivores. On the other hand, if respiration exceeds primary production, then it is clear that a net consumption of biomass is occurring within the system. Such a system cannot persist unless subsidized by an external input of organic compounds, as, for example, from stream runoff.

It is important to realize that the waste products resulting from respiration and excretion are utilized by primary producers and detritivores, respectively, to create living biomass. For example, carbon dioxide, which is a direct product of respiration, is the source of carbon for primary production, and ammonia (as ammonium ions), which is excreted by many aquatic organisms, can be directly assimilated by primary producers as a source of nitrogen for the production of proteins and nucleic acids. Waste products can be, and often are, toxic to the organisms that produced them. However, in a well-balanced ecosystem, waste products never reach high concentrations because they are constantly being utilized as a source of food by other organisms in the system. Detritivores play a crucial recycling role in aquatic systems by consuming organic wastes and converting them to inorganic forms that are utilized by primary producers. The grazing food chain utilizes the organic matter synthesized by the primary producers and releases part of it in the form of detritus, which in turn provides the food for the detritus food chain.

Due to this internal recycling, there is a tendency for both organic and inorganic compounds to accumulate in aquatic systems. Inorganic carbon can of course escape as carbon dioxide to the atmosphere, and inorganic nitrogen may similarly escape as ammonia or N_2. However, under normal circumstances the latter escape routes are not very efficient for nitrogen, and removal of organic compounds and essential nutrients by means of washout rarely occurs with 100% efficiency. The accumulation of refractory organic debris in the sediments and buildup of organic matter and nutrient concentrations in the water column are natural processes in most aquatic systems. Associated with these phenomena is an increase in primary production and respiration and a decrease in the depth of the system caused by sediment accumulation. The whole process is referred to as *eutrophication*, which eventually causes most lakes to fill up with sediments after a time of perhaps hundreds, thousands, or even tens of thousands of years. Sediments do accumulate at the bottom of the ocean, but the sediments are removed by tectonic processes at subduction zones at rates that approximately balance their rate of formation. Obviously there is no danger that the oceans will fill up with sediments. However, some regions of the ocean are much more productive than others, and this fact directly reflects the relative efficiency with which essential nutrients are recycled by the grazing and detritus food chain in different parts of the ocean.

Any unnatural acceleration of the eutrophication process due to the activities of humans is called *cultural eutrophication*. Cultural eutrophication could be caused, for example, by the discharging of sewage containing a high concentration of nutrients and organic matter. Instances of cultural eutrophication constitute one of the most common and widespread examples of water pollution problems. We will explore a few of these examples in detail in Chapter 4.

Food Chain Magnification

Respiration and excretion obviously play a critical role in controlling the flux of organic and inorganic materials between the grazing and detritus food chains. However, from the standpoint of water pollution, respiration and excretion are also important in determining the movement of pollutants both between and within these same food chains. If the pollutant is biodegradable, it may of course be catabolized and rendered harmless. However, if the pollutant is nonbiodegradable, it may be passed from prey to predator and in this way be spread throughout the grazing food chain. If some of the pollutant is excreted, then it may spread to the detritus food chain as well. One of the most important applications of food chain theory to water pollution problems has been the effort to explain how these transfers of the pollutant between food chains and trophic levels affect the concentration of the pollutant in the organisms. In cases where it has been possible to examine in some detail the distribution of pollutant concentrations among the trophic levels in a simple food chain, results have sometimes indicated a steady increase in concentration with increasing trophic level number. Table 1.1 shows concentrations of the pesticide DDT (plus the closely related compounds DDD and DDE) in the water and in various organisms taken from a Long Island, New York, salt marsh. The residue concentrations increase steadily from the plankton to the small fish to the larger fish and finally to the fish-eating birds. The total concentration factor from plankton to fish-eating birds is roughly 600. Observations such as this one led some scientists to believe that a common mechanism or explanation might underlie similar observations of increasing pollutant concentrations at higher trophic levels in some food chains; a phenomenon that they termed *food chain concentration* or *biological magnification*.

A logical explanation for biological magnification is forthcoming from food chain theory if you assume that certain pollutants ingested with an organism's food are not respired or excreted as effectively as is the remainder of the food. DDT would seem to be a likely candidate for such a pollutant because it is resistant to biological breakdown and tends to be stored in an organism's fatty tissues rather than being directly excreted with other waste materials that the organism is unable to utilize. Consider for example a case in which the ecological transfer efficiency of food between trophic levels is 20%, but in which the transfer efficiency of DDT, caused by

Table 1.1   DDT Residues in Organisms Taken from a Long Island Salt Marsh

Organism	DDT residues (ppm)[a]
Water	0.00005
Plankton	0.04
Silverside minnow	0.23
Sheephead minnow	0.94
Pickerel (predatory fish)	1.33
Needlefish (predatory fish)	2.07
Heron (feeds on small animals)	3.57
Tern (feeds on small animals)	3.91
Herring gull (scavenger)	6.00
Fish hawk (osprey) egg	13.8
Merganser (fish-eating duck)	22.8
Cormorant (feeds on larger fish)	26.4

[a] Parts per million (ppm) of total residues, DDT + DDD + DDE (all of which are toxic), on a wet weight, whole organism basis.

Source. Woodwell et al. (1967).

its resistance to respiration and excretion, is 60%. As a result the steady-state concentration of DDT in a predator will be about three times greater than the DDT concentration in its food, because the predator retains three times (60% versus 20%) as much of the DDT that it eats as it does of the remaining food. If this process is repeated through four trophic level transfers, the concentration of DDT in the fifth trophic level will be $3^4 = 81$ times greater than the DDT concentration in the first trophic level.

Although this reasoning is logical enough, the logicality of the reasoning by no means guarantees that biological magnification of the sort described is responsible for observations such as those in Table 1.1. Pollutant concentration trends of exactly this sort may be produced by mechanisms very different from food chain magnification. Only carefully designed experiments can sort out the possible causes of such concentration trends. In Chapter 10 we will have a chance to examine one such experiment in some detail. The point here is not to argue the pros and cons of the theory of food chain magnification, but rather to show how some knowledge about the characteristics of food chains can lead to a logical hypothesis regarding pollutant effects, which is worth testing with further study. Logical thinking and hypothesis testing of this sort, based on sound ecological principles, represent the best means for studying and solving water pollution problems.

FOOD WEBS

Now that we have developed a simple food chain model as a conceptual basis for examining ecological problems, it is best to back up a step and remind ourselves that the world is not really so simple. The feeding behavior of a great many animals is such that they cannot be assigned to a unique trophic level. Some animals, such as shrimp, will eat almost anything they can swallow, including plants, detritus, and other animals. Obviously they cannot be assigned to one or even a few trophic levels. Other organisms may feed on one trophic level as juveniles, a second trophic level at a later developmental stage, and a third trophic level as adults. In such a case you would have to treat each developmental stage of the species as a different organism in order to make unique trophic level assignments. Certainly the feeding habits of many organisms are influenced by the kind and quantity of various food types available to them. Figure 1.5 indicates the complex feeding relationships found in a small stream community.

The pattern of lines indicating the feeding relationships in such a system form a sort of web, and the feeding pattern has therefore come to be known as a *food web*. Certainly a food web depicts a more complex system than that represented by a food chain. One may therefore ask whether any of the implications deduced from food chain theory are relevant to a world that seems to be much more complex. The answer to this question is yes for the following reasons.

First, even though it is true that many organisms feed on more than one trophic level, it is also true that many organisms show a preference for one kind of food or another and can be reasonably defined as belonging to primarily one or two trophic levels. Although there are exceptions to this rule, the exceptions are not so numerous as to warrant discarding the concept.

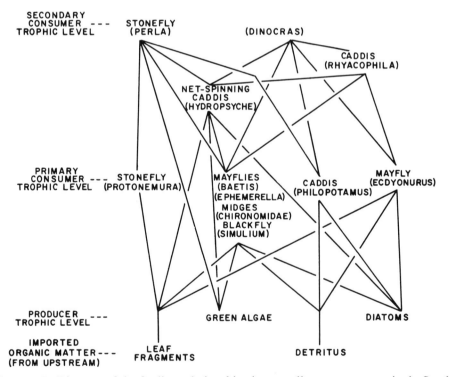

Figure 1.5 Diagram of the feeding relationships in a small stream community in South Wales. (Redrawn from Jones, 1949.)

Second, and perhaps more important, is the fact that some implications of food chain theory do not even require one to assign specific organisms to specific trophic levels. The idea that a large percentage of a prey will be either respired or excreted by the predator is valid regardless of the identity of the prey and predator. Thus food in an abstract sense is produced by plants and is passed through a series of feeding transfers from one trophic level to the next, with roughly 80% of the food's being respired or excreted at each transfer. Therefore, on the average the biomass of food that has passed through m such transfers is likely to be much less than the biomass of food that has passed through $m - 1$ transfers. If you think of trophic levels in terms of a series of food transfers and not necessarily in terms of particular organisms eating particular organisms, the problem of assigning each organism to a unique trophic level disappears.

Despite this second point, it is in fact the case that the effect of pollutants on particular organisms is discussed sometimes with reference to an organism's position in the food chain. The relevance of food chain theory to such discussions is based on the first of the two previous arguments; namely, that many organisms may be reasonably assigned to one or two trophic levels. For example, it is reasonable to assume that a pelican does not eat microscopic plants and animals, bacteria, protozoans, or even small fish; nor does a pelican eat whales, dolphins, or sea lions. Rather, a pelican eats fish of the size (approximately a few tens of centimeters) that can be conveniently scooped up in its mouth. Thus a pelican would be assigned to roughly the fourth trophic level in a grazing food chain. Such reasoning, while not flawless, is not likely to be wildly in error. Nevertheless, whenever food chain arguments are invoked, one should keep in mind that the theory is not without its shortcomings and that the feeding relationships of some organisms can be extremely complex.

Unstructured Food Webs

To illustrate some of the controversy surrounding the use of food chain models in pollution studies, we consider an analysis performed by John Isaacs (1972) concerning the concentrations of cesium in the muscle tissue of certain marine fish. A study performed by Young (1970) had revealed that the ratio of cesium to potassium of several species of fish in the Salton Sea in southeastern California increased by about a factor of three between successive trophic levels in the simple grazing food chain which seemed to characterize the feeding relationships in that ecosystem. However, when similar fish were analyzed from the Gulf of California, the cesium/potassium (Cs/K) ratio was found to be very similar in all the fish and about 16 times higher than the Cs/K ratio in the algae. This observation led Young (1970) to speculate that the results from the Gulf of California might reflect the existence of a complex food web in which biological magnification would not adhere to the predictions of simple food chain theory.

Following up on this hypothesis, Isaacs (1972) developed what he referred to as an unstructured food web, which is illustrated diagrammatically in Figure 1.6. Isaacs (1973) contrasted his model with the classical structured food web models in which heterotrophs were considered to occupy relatively well-defined positions within a finite sequence of trophic steps. In Isaacs' unstructured food web model, each moiety of food was considered to pass through an infinite number of steps and conversions, and at each such step/conversion was considered to be transformed into living biomass, detritus, or nonrecoverable material (e.g., CO_2). This approach completely avoided the assignment of particular organisms to particular trophic levels. In fact, since a given heterotroph might consist of an assemblage of organic moieties that had passed through a number of transformations ranging from some small integer to infinity, the position of a given organism in Isaacs' food web would in general be spread out over an infinite number of food web elements.

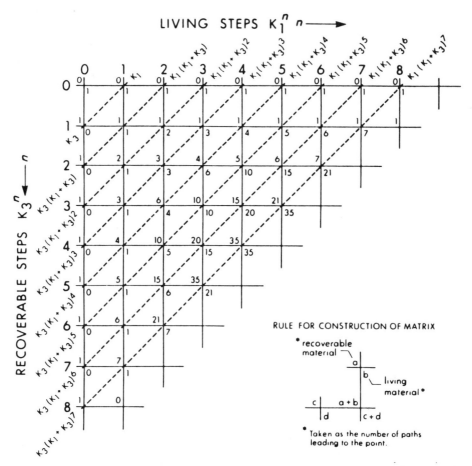

Figure 1.6   Isaacs' unstructured food web matrix. Each diagonal line passes through points representing biomass which has undergone an equal number of transformations since being introduced at point (0,0). The number of transformations associated with each diagonal line is indicated by the number at both ends of the diagonal. The expressions along the top of the matrix are the total biomass of living organisms along each diagonal in the steady state. The expressions at the left have a similar meaning for the detritus. The number of possible pathways leading to living biomass at a particular lattice point is indicated by a number below and to the right of the lattice point. The number of pathways leading to detritus at a particular lattice point is indicated by a number above and to the left of the lattice point. Redrawn from Isaacs (1972).

In his original model Isaacs (1972) assumed that food was converted to living tissue with an efficiency K_1, to irretrievable forms with an efficiency K_2, and to detritus with an efficiency K_3. If these efficiencies are expressed in fractional form, then $K_1 + K_2 + K_3 = 1$. In Figure 1.6 new food in the form of plant (algal) biomass is introduced at the upper left-hand corner of the diagram. This point is designated (0,0), i.e., row 0 and column 0. The flow of food is either to the right or down. Flow to the right produces living biomass; flow down produces detritus. A critical assumption of the original model was that the time interval between steps in the matrix was the same for all steps. If the system is in steady state and if the biomass at point (0,0) is unity, then the biomass at (0,1) (i.e., row 0 and column 1) is K_1 and the biomass at (1,0) (i.e., row 1 and column 0) is K_3.

Proceeding with the analysis, we note that the diagonal lines in Figure 1.6 pass through points, each of which represents biomass that has gone through the same number of steps since being introduced at point (0,0). With the exception of points lying along row 0 and column 0, each point in the matrix represents a combination of both living biomass and detritus. For example, living biomass at point (1,1) is produced by a transformation of biomass from point (1,0); detritus at point (1,1) is produced by transforming biomass from point (0,1). The total number of pathways leading to the production of living biomass and detritus at a given point is easily counted, and the method of calculation is indicated in Figure 1.6. If the numbers of pathways leading to detritus and living biomass at point $(n - 1, m)$ are a and b, respectively, then the number of pathways leading to detritus at point (n, m) is $a + b$. Similarly, if the numbers of pathways leading to detritus and living biomass at point $(n, m - 1)$ are c and d, respectively, then the number of pathways leading to living biomass at point (n, m) is $c + d$.

Given the assumption of steady state, it is possible to calculate the amount of living biomass and detritus at each point in the matrix relative to the amount at point (0,0) by remembering that the efficiency of living biomass production is K_1, and the efficiency of detritus production is K_3. The sum of all the living biomass and of all the detritus along each diagonal line forms a simple geometric series, as indicated in Figure 1.6, and one can therefore easily calculate the amount of living biomass and detritus in the system. If the algal biomass at point (0,0) is M_a, then the total heterotrophic biomass M_h in the system is given by the equation

$$M_h = M_a K_1 / K_2 \tag{1.1}$$

and the total amount of detritus M_d in the system is given by the equation

$$M_d = M_a K_3 / K_2 \tag{1.2}$$

Although Isaacs' model does not require the assignment of particular organisms to particular trophic levels, it is nevertheless possible to derive expressions for the potential biomass of certain types of organisms. These expressions provide useful predictive tools, however, only to the extent that the feeding niche of no other type of organism in the system overlaps with the type of organism in question. For example, if all the organisms in a system could be classified as being either herbivores, carnivores, or detritivores, there would obviously be no overlap in any of the feeding niches. However, if the organisms in the system included herbivores and omnivores, or herbivores and particle feeders, then there would be overlap in the feeding niches and Isaacs' expressions for potential biomass would not be strictly applicable to these types of organisms. The expressions Isaacs (1972) provided included the following:

$$\text{potential biomas of herbivores} \quad = \quad K_1 M_a \tag{1.3}$$

$$\text{potential biomass of carnivores} \quad = \quad K_1 M_h \ = \ M_a K_1^2 / K_2 \tag{1.4}$$

$$\text{potential biomass of detritivores} \quad = \quad K_1 M_d \ = \ K_1 K_3 M_a / K_2 \tag{1.5}$$

$$\text{potential biomass of omnivores} \quad = \quad M_h \ = \ M_a K_1 / K_2 \tag{1.6}$$

Now let's see how Isaacs' model can be applied to the issue of biological magnification. Suppose that the coefficients K_1, K_2, and K_3 apply to cesium and that the coefficients K_1', K_2', and K_3' apply to potassium. Then according to equation 1.6 the Cs/K ratio in the potential biomass of omnivores is related to the Cs/K ratio in M_a by the expression $K_1 K_2' / (K_1' K_2)$. Similar expressions can be derived for the Cs/K ratios in other groups of organisms. Using this approach, Isaacs (1972) was able to show that the predictions of his unstructured food web model were consistent with the distribution of cesium and potassium in fish from the Gulf of California.

Fair enough, but is it really necessary to abandon food chain theory to explain Young's (1970) observations? The answer is no. In fact, Laws (1986) has shown that all of the equations utilized by Isaacs can be derived from the very simple box model shown in Figure 1.7. The model in this case is the simplest possible representation of the combined grazing and detritus food chains. Not only is the box model in Figure 1.7 much easier to analyze mathematically than the unstructured food web, but use of the box model allows one to relax several of the unrealistic restrictions which Isaacs (1972) imposed on his unstructured food web. In particular, it no longer becomes necessary to assume that all conversions to living biomass and all conversions to detritus occur with the same efficiency, and it is unnecessary to assume that the turnover times of autotrophs, heterotrophs, and detritus are identical.

The conclusion is that the distribution of cesium and potassium in fish from the Gulf of California can be explained with the use of a box model that depicts in the simplest possible way the grazing and detritus food chains and the interactions between the two. This conclusion is undoubtedly comforting to anyone who finds

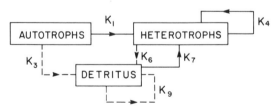

Figure 1.7 Simplest possible box model of the grazing and detritus food chains and the interactions between them. The K's are defined in Table 1.2.

the mathematical analysis of the unstructured food web confusing. Two very good rules of thumb in ecological modeling are the following:

1. Clearly define the question to be addressed with the model.
2. Make the model no more complicated than necessary to answer the question.

In this case at least, it was unnecessary to abandon food chain theory in order to explain the observations of Young (1970). It was, however, necessary to represent, at least in a crude way, the interactions within and between the grazing and detritus food chains.

Box Model Analysis

For those with an interest in modeling, the following analysis of the box model in Figure 1.7 is provided. The symbols used in the equations are defined in Table 1.2. The following three equations define the rate of change of the biomass of autotrophs, heterotrophs, and detritus.

$$(d/dt)M_a = [P - (K_1 + K_2 + K_3)M_a]/T \tag{1.7}$$

$$(d/dt)M_h = [K_1M_a + K_4M_h + K_7M_d - (K_4 + K_5 + K_6)M_h]/T \tag{1.8}$$

$$(d/dt)M_d = [K_3M_a + K_6M_h + K_9M_d - (K_7 + K_8 + K_9)M_d]/T \tag{1.9}$$

In the steady state, the rates of change and hence the right-hand sides of each equation are zero. The time constant T is arbitrary, and if it is chosen to be equal to the turnover time of the autotrophs, then $K_1 + K_2 + K_3 = 1$. In that case $M_a = P$, and with a little algebraic manipulation one can show that

Table 1.2   Definitions of Symbols used in Equations 1.7–1.9

Symbol	Definition
K_1	fraction of autotrophic biomass converted to heterotrophs in time T
K_2	fraction of autotrophic biomass respired in time T
K_3	fraction of autotrophic biomass converted to detritus in time T
K_4	fraction of heterotrophic biomass converted to heterotrophs in time T
K_5	fraction of heterotrophic biomass respired in time T
K_6	fraction of heterotrophic biomass converted to detritus in time T
K_7	fraction of detritus converted to heterotrophs in time T
K_8	fraction of detritus respired in time T
K_9	fraction of detritus converted to detritus in time T
M_a	biomass of autotrophs
M_h	biomass of heterotrophs
M_d	biomass of detritus
P	amount of autotrophic biomass produced in time T
T	arbitrary time constant

$$M_h = M_a[K_1(K_7 + K_8) + K_3K_7]/[K_6K_8 + K_5(K_7 + K_8)] \qquad (1.10)$$

$$M_d = M_a[K_3(K_5 + K_6) + K_1K_6]/[K_6K_8 + K_5(K_7 + K_8)] \qquad (1.11)$$

Had we assumed, as did Isaacs, that the turnover times were the same for all elements of biomass in the system and that the conversion efficiencies to living biomass and detritus were the same for all transformations in the system, then $K_1 = K_4 = K_7$, $K_2 = K_5 = K_8$, and $K_3 = K_6 = K_9$. It is straightforward to show that equations 1.10 and 1.11 reduce to equations 1.1 and 1.2 when these assumptions are made.

At equilibrium the right-hand side of equation 1.8 may be rearranged to show that

$$M_h = K_1M_a/(K_4 + K_5 + K_6) + K_4M_h/(K_4 + K_5 + K_6) + \\ K_7M_d/(K_4 + K_5 + K_6) \qquad (1.12)$$

The first, second, and third terms on the right-hand side of equation 1.12 represent the contribution to the heterotropic biomass of herbivores, carnivores, and detritivores, respectively. These terms are therefore the potential biomasses of herbivores, carnivores, and detritivores, respectively. The terms become identical to the right-hand sides of equations 1.3, 1.4, and 1.5, respectively, if one makes the simplifying assumptions noted above.

FOOD WEBS AND ECOSYSTEM STABILITY

The complexity of aquatic food webs is associated with another controversy that is particularly relevant to the problem of water pollution. The controversy concerns the stability of ecosystems. To what extent will the biological community be able to resist change when stressed by pollution, and will the system return to its original state if the pollution stress is removed?

In the 1950s and 1960s, a number of publications appeared in the literature suggesting that the complex interactions between organisms in a food web tended to stabilize the biological system. MacArthur (1955) for example suggested that community stability might be roughly proportional to the logarithm of the number of links in the food web, an hypothesis which unfortunately has sometimes been accorded the status of a theorem. The hypothesis is based on conclusions derived from information theory, in which it is shown that such a logarithm provides a measure of the degree of organization or complexity. One then argues intuitively that the greater the number of links and pathways in the food web, the greater the ability of the system to damp down perturbations and hence the greater the chance that the system can absorb environmental shocks without falling apart. Examples of stable and very complex ecosystems include tropical rain forests and coral reefs.

A very elegant and understandable treatment of this issue has been provided by Robert May (1974). The somewhat counterintuitive result of May's analysis is that greater food web complexity *per se* does not impart greater stability to the ecosystem. In fact, the more complex the linkages in the food web, the more unstable the system is likely to become. As noted by May (1974, p. 75), "The greater the size and connectance of a web, the larger the number of characteristic modes of oscillation it possesses: since in general each mode is as likely to be unstable as to be stable (unless the increased complexity is of a highly special kind), the addition of more and more modes simply increases the chance for the total web to be unstable." The fact that tropical rain forests and coral reefs are highly complex systems may therefore be due more to the fact that these ecosystems exist in stable environments than to some inherent stabilizing characteristics conferred by the complex links in their food webs. May (1974) is careful not to rule out the latter possibility and notes that some very special and mathematically atypical sorts of complexity may enhance ecosystem stability, but that it would be unwise to assume that as a general rule food web complexity implies stability. In fact, "If there is a generalization, it could be that stability [of the environment] permits complexity" (May 1974, p. 76).

Please note that the foregoing discussion of stability and complexity pertains to complexity of food web linkages only. Other types of complexity may confer stability on ecosystems. For example, habitat complexity may provide refuges that allow some members of the biological community to survive adverse conditions. Similarly, complexity or diversity of the gene pool in a species may result in a subset of the population that is unusually resistant to a particular stress and hence better able to survive the impact of certain types of pollution. There are many types of complexity, and certainly not all of them are inherently destabilizing. However, it appears from May's (1974) analysis that as a general rule increased complexity in food web linkages does not increase ecosystem stability.

REFERENCES

Isaacs, J. D. 1972. Unstructured marine food webs and 'pollutant analogues'. *Fish. Bull.*, **70**, 1053–1059.

Isaacs, J. D. 1973. Potential trophic biomasses and trace-substance concentration in unstructured marine food webs. *Mar. Biol.*, **22**, 97–104.

Jannasch, H. W., and C. O. Wirsen. 1977. Microbial life in the deep sea. *Sci. Amer.*, **236**(6), 42–52.

Jones, J. R. E. 1949. A further ecological study of a calcareous stream in the 'Black Mountain' district of South Wales. *J. Anim. Ecol.*, **18**, 142–159.

Laws, E. A. 1986. Unstructured food webs: a model for aquaculture. *Aquacul. Eng.*, *5*, 161–170.

MacArthur, R. H. 1955. Fluctuations of animal populations, and a measure of community stability. *Ecology*, **36**, 533–536.

May, R. M. 1974. *Stability and Complexity in Model Ecosystems*, 2nd ed. Princeton University Press, Princeton. 265 pp.

Odum, E. P. 1971. *Fundamentals of Ecology*, 3rd ed. Saunders, Philadelphia. 574 pp.

Odum, H. T. 1957. Trophic structure and productivity of Silver Springs, Florida. *Ecol. Monogr.*, **27**, 55–112.

Ryther, J. H. 1969. Photosynthesis and fish production in the sea. *Science*, **166**, 72–76.

Sheldon, R. W., A. Prakash, and W. H. Sutcliffe, Jr. 1972. The size distribution of particles in the ocean. *Limnol. Oceanogr.*, **17**, 327–340.

Woodwell, G. M., C. F. Wurster, and P. A. Isaacson. 1967. DDT residues in an east coast estuary: A case of biological concentration of a persistent insecticide. *Science*, **156**, 821–824.

Young, D. R. 1970. The distribution of cesium, rubidium, and potassium in the quasi-marine ecosystem of the Salton Sea. Ph.D. dissertation. Scripps Inst. of Oceanogr., University of California San Diego, La Jolla.

2

PHOTOSYNTHESIS

As the organisms primarily responsible for the synthesis of organic matter from inorganic constituents, plants occupy a uniquely important position in aquatic food chains. Therefore a study of the factors that control photosynthetic rates is a logical place to begin examining how changes in the environment may be expected to influence aquatic systems.

Aquatic plants are for the most part very different from the terrestrial plants with which most people are familiar. There are relatively few trees or grasses in aquatic systems. Along coastlines mangrove and cypress trees are, to be sure, aquatic organisms, and benthic algae such as kelp are somewhat analogous to terrestrial grasses or shrubs. In shallow lakes or along coastlines where the water is sufficiently shallow for light to effectively penetrate to the bottom, benthic algae may be the most important primary producers in an aquatic ecosystem. However, since plants need light to carry out photosynthesis and since water absorbs light, the depth range within which rooted aquatic plants can survive is obviously limited. In systems where there is a high concentration of particulate materials suspended in the water, water transparency is further reduced, because such particles scatter and absorb light. A similar effect is produced by certain dissolved organic substances such as tannic and humic acids, which also absorb light. In the clearest ocean water, only about 1% of surface light is transmitted to a depth of 100 m, whereas in clear coastal waters the 1% light level is typically reached at a depth of only 20 m. Since most aquatic plants cannot effectively photosynthesize at light intensities much less than 1% of surface light, the foregoing observations set the limit of the euphotic zone (the region of the water column within which plants can photosynthesize) at a few tens of meters in clear coastal waters and at roughly 150 m or at most 200 m in the clearest open ocean water. These depth limits would obviously have to be reduced in waters containing significant concentrations of particulate or dissolved organic materials. Obviously benthic algae and/or rooted aquatic plants are not to be found in parts of aquatic systems where the bottom lies below the euphotic zone. Since the depth below about 93% of the surface area of the world's oceans is greater than 180 m

(Ryther, 1969), it is clear that benthic plants contribute nothing to primary production in most parts of the ocean. Furthermore many lakes are sufficiently deep and/ or turbid as to prevent the development of benthic plants except in the immediate vicinity of the shoreline.

From the previous discussion, it is clear that aquatic plants found in the surface waters of the open ocean or of deep lakes must have the ability to float or drift about in the euphotic zone and must be able to derive all the essential nutrients for photosynthesis directly from the water. Seaweeds such as the *Sargassum* weed provide one example of such planktonic (drifting) plants. However, most such plants are microscopic or nearly so in size and are frequently unicellular. These tiny plants are called *phytoplankton* and may range in size from microscopic cells a few microns in diameter to "giant" cells visible to the unaided eye with a diameter of as much as 2 mm. Some species of phytoplankton form colonies or long chains of cells that are visible to the unaided eye, although the individual cells are not. It is perhaps surprising to realize that such tiny organisms are responsible for the vast majority of aquatic primary production and that top-level carnivores such as sharks and whales depend almost entirely, although indirectly, on such minute organisms as a source of food. However, despite their essential role as primary producers in virtually all aquatic food chains, phytoplankton may create serious problems, particularly when their concentrations in the water exceed certain limits. For example, red tides, which are an unwelcome but recurrent event in a number of coastal areas throughout the world, are caused by a population explosion of certain species of a class of phytoplankton called *dinoflagellates*. At such times the concentration of these dinoflagellates often becomes sufficiently great to give the water a reddish color over distances of as much as several kilometers. Unfortunately these dinoflagellates release neurotoxins into the water. During red tides these toxin concentrations may become high enough to kill great numbers of small fish, some of which frequently wash up on the beach to rot. Furthermore, the toxins may be concentrated by shellfish (which are seemingly unaffected) and subsequently poison humans who eat the shellfish. In severe cases this poisoning may result in paralysis or even death. Although some phytoplankton population explosions are completely natural phenomena, there is no doubt that in many cases human activities have largely caused or greatly exacerbated conditions that lead to undersirably high phytoplankton concentrations. To avoid and/or correct conditions that stimulate excessive phytoplankton growth, it is obviously essential to first understand the factors that normally control phytoplankton growth rates, a problem to which we now turn our attention.

LIGHT LIMITATION OF PHOTOSYNTHESIS

There is no doubt that light is the most important factor limiting photosynthetic rates in the world's oceans. Over 95% of the ocean's volume lies below the euphotic zone and is hence unable to support plant life. In addition many lakes are sufficiently deep that the euphotic zone makes up only a small fraction of the lake's volume, and even shallow lakes may show pronounced seasonal patterns in plant production that are undoubtedly influenced in part by changes in insolation. Figure 2.1 shows the variation of light intensity with depth in a hypothetical body of water.

Only visible light is graphed, because the plant pigments that absorb light are effective in absorbing only certain parts of the visible light spectrum. For example, chlorophyll absorbs predominantly blue and red light. In fact not all parts of the visible light spectrum are equally useful for photosynthesis, but the total intensity of visible light does provide a convenient, if approximate, measure of the amount of light available for photosynthesis.

As indicated in Figure 2.1 light intensity drops off with depth in the water column in approximately an exponential fashion, that is, at a rate proportional to the intensity of the light. Photosynthetic rates tend to become saturated at high light intensities and may be inhibited by the ultraviolet (UV) light in direct sunlight. However, UV light is rapidly attenuated by water, and at depths where the irradiance is less than about 30% of surface values, photosynthetic rates are almost directly proportional to light intensity. Thus a graph of photosynthetic rate versus depth in a body of water might appear qualitatively very similar to the solid curve in Figure 2.2. Therefore, to the extent that light limits photosynthesis, phytoplankton growth rates should be highest near the surface, where light intensities are close to optimal. However, most phytoplankton cannot position themselves in the water column, because they have very little locomotive ability, and for the most part are moved about by turbulence and currents, both in the horizontal and vertical directions. At the surface, winds blowing over the water generate waves and turbulence that keep the water column well mixed and the phytoplankton concentration correspondingly uniform. If the depth of this mixed layer is less than the depth of the euphotic zone, the gross production of biomass by the mixed layer phytoplankton population will exceed phytoplankton biomass losses to respiration, and the resultant net primary production will provide food for heterotrophs and/or increase the standing stock of phytoplankton. If the depth of the mixed layer equals a certain depth known as the *critical depth*, gross production in the mixed layer will be exactly balanced by plant and animal respiration, the result being that the net production of organic matter in the mixed layer equals zero (Smetacek and Passow, 1990). The net production of organic matter by the plant and animal community is referred to as *net community production*. If the mixed layer is deeper or shallower than the critical depth, net community production in the mixed layer will be negative or positive, respectively. Obviously the former situation cannot persist indefinitely in the absence of allochthonous (external) inputs of organic matter, or there would be no organic matter left. The dashed curve in Figure 2.2 shows qualitatively the relationship between net community production and depth in the mixed layer of a hypothetical body of water on a day when the mixed layer extends below the critical depth. Net community production is positive in the upper part of the water column

Figure 2.1   Variation of visible light intensity with depth in a hypothetical body of water.

and negative in the lower part of the water column. The areas of positive and negative net community production balance each other above the critical depth, so that on the average there is zero net community production in the water column above the critical depth.

Certain kinds of pollution associated with human activities may reduce the depth of the euphotic zone by increasing the turbidity of the water. The result is a decrease in net community production and sometimes a dramatic change in the composition of the phytoplankton community. For example, sediment runoff from construction sites may greatly diminish water clarity and therefore decrease the amount of light available for phytoplankton. In such cases the phytoplankton population may become dominated by *cyanobacteria* (also referred to as *blue-green algae*), many species of which are able to maintain themselves near the surface of the water by means of special gas-filled vacuoles that give the plants a slight positive buoyancy. These cyanobacteria may form floating surface scums or mats that are aesthetically objectionable, particularly if they wash ashore and rot along the water's edge. Furthermore, the cyanobacteria may be unpalatable to the natural community of herbivorous zooplankton, and through this and other feeding relationships the initial change in the composition of the phytoplankton community may impact the entire biological community.

NUTRIENT LIMITATION OF PHOTOSYNTHESIS

Excessive stimulation of algal production by the addition of essential nutrients to the euphotic zone is more often the cause of algal pollution problems than changes in water clarity. In fact, even in cases where cyanobacteria come to dominate a phytoplankton community, the principal cause of change is usually a shift in the relative input of certain nutrients to the euphotic zone rather than increased turbidity. The sensitivity of phytoplankton production and/or composition to changes in nutrient inputs arises because phytoplankton biomass in the euphotic zone is often controlled largely by the availability of certain nutrients that, in addition to

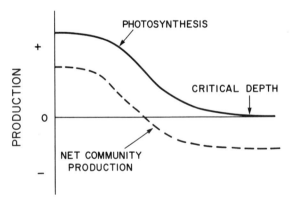

Figure 2.2   Relationship between photosynthetic rate (solid curve), net community production (dashed curve), and depth in a well-mixed body of water. The areas of positive and negative net community production balance each other exactly above the critical depth.

light, are essential for growth. If the supply of these nutrients is increased to a nutrient-limited phytoplankton community, an increase in production and biomass roughly proportional to the increase in nutrient supply can be expected. Obviously no immediate increase in phytoplankton production would occur if such nutrient additions were made to the water column below the euphotic zone (the so-called aphotic zone), where light is inadequate to support photosynthesis even if nutrients are abundant. However, production in the surface waters would subsequently increase if aphotic zone water were later mixed into the euphotic zone, a process that occurs regularly, albeit sometimes slowly, in all aquatic systems.

Usually only one or a few nutrients are limiting to the production of phytoplankton biomass in a given system at a given time, but the identification of these nutrients has proven to be a controversial problem. Table 2.1 lists chemical elements that are known to be essential for plants. Macronutrient elements are required in relatively large amounts compared to micronutrients, although for some elements the distinction is not clear cut. Of the macronutrients, carbon, oxygen, and hydrogen are required in the largest amounts, since they are essential components of organic compounds such as carbohydrates, fats, and proteins. Nitrogen and phosphorus are required in somewhat smaller amounts, with the atomic ratio of nitrogen to phosphorus averaging about 16 (3–30 range) in phytoplankton and 30 (10–70 range) in macroalgae (Ryther and Dunstan, 1971; Atkinson and Smith, 1983). Nitrogen is an essential component of proteins, nucleic acids, and certain pigments (e.g., chlorophyll), and phosphorus is required to produce phospholipids and sugar phosphate bonds in molecules such as adenosine triphosphate (ATP). These high-energy phosphate bonds provide a convenient and essential storage unit for small amounts of energy derived primarily from the stepwise catabolism of organic molecules and may be used at any time by the organism as an energy source for metabolic processes. The remaining macronutrient elements are required in even smaller amounts, and their concentrations in both marine and freshwaters are more than adequate to supply the nutritional needs of aquatic plants. The micronutrients or trace elements are required in the smallest amounts, in many cases functioning as catalysts to speed up metabolic reactions. Such catalysts are not consumed or altered by the reactions they mediate, and therefore only small amounts are required by an organism.

Realizing the relative requirements of plants for the essential nutrients listed in Table 2.1, one may ask which of these elements are most likely to limit phytoplankton biomass. Carbon, hydrogen, and oxygen are needed in the largest amounts, but these elements are readily obtained from H_2O (hydrogen) and CO_2 (carbon and oxygen). Obviously there is no lack of H_2O in an aquatic environment. Carbon dioxide is a gas that is found in the atmosphere and dissolves in water, reaching at equilibrium a concentration proportional to its concentration in the atmosphere. The chemistry of the oceans is such that there is invariably an abundance of CO_2 to support photosynthesis. On the other hand, the CO_2 concentration in some freshwater lakes is extremely low, and it has sometimes been argued that photosynthesis in such systems might be limited by a lack of CO_2. However, Schindler (1974) has convincingly shown that the exchange of CO_2 between the atmosphere and the water is sufficiently rapid to provide adequate CO_2 for the development of large phytoplankton blooms over a time period of no more than a few weeks. In other words the atmosphere acts as a CO_2 reservior for the mixed layer

Table 2.1   Essential Macro- and Micronutrient Elements for Plants

Essential macronutrient elements	Symbol	Essential micronutrient or trace elements	Symbol
Oxygen	O	Iron	Fe
Carbon	C	Manganese	Mn
Nitrogen	N	Copper	Cu
Hydrogen	H	Zinc	Zn
Phosphorus	P	Boron	B
Sulfur	S	Silicon	Si
Potassium	K	Molybdenum	Mo
Magnesium	Mg	Chlorine	Cl
Calcium	Ca	Vanadium	V
		Cobalt	Co
		Sodium	Na

Source. Odum (1971) p. 127.

of an aquatic system, and the flux of CO_2 from the atmosphere into the water may easily provide the CO_2 needed for photosynthesis, even if the ambient CO_2 concentration in the water is low. We will have a chance to examine more closely how Schindler arrived at this conclusion later in this chapter. For the moment suffice it to say that CO_2 appears to limit phytoplankton biomass in few if any natural aquatic systems.

Table 2.2 lists the average concentrations of most of the remaining macro- and micronutrient elements in typical river water and seawater. Sulfur (as sulfate, SO_4^{2-}), potassium, magnesium, calcium, chlorine, and sodium are the elements that make up the principal salts in seawater. Their concentrations in seawater range from 10 millimolar (mM) to 0.56 molar (M). There is no evidence that a lack of these elements ever limits photosynthesis in the sea. The concentrations of the same elements in river water are much lower and range from ~40 to 400 µM. Again, however, there is no evidence that such concentrations are limiting to photosynthesis. The micronutrient elements are for the most part found at much lower concentrations than any of the macronutrients. Although it is true that most of the micronutrients are required in very small amounts, there is evidence that the availability of certain micronutrients may limit algal biomass in some aquatic systems or at least limit the biomass of certain species of algae. For example, the class of phytoplankton known as diatoms build elaborate skeletons of silica that enclose the rest of the cell. Consequently, diatoms require much more silicate than other classes of phytoplankton, and there is fairly convincing evidence that changes in silicate concentration may affect the abundance of certain diatoms in freshwater systems (Hutchinson, 1967, pp. 446–455). There is little evidence that silicate is limiting the photosynthetic rates of marine algae, although the distribution of marine diatoms may be influenced by silicate availability. Silicate concentrations in ocean surface waters can easily drop as low as 5 µM. The greatest abundance of marine diatoms is found in the Antarctic Seas, where there is also an unusually high concentration of silicate. Molybdenum (Mo) plays a role in the formation of the enzyme nitrate reductase, and hence is required for the assimilation of nitrogen in the form of nitrate (NO_3^-). Molybdenum is also required for nitrogen fixation, the process by which certain plants and bacteria convert atmospheric nitrogen gas (N_2) into a form that can be used in primary production. There is evidence that the availability of Mo limits nitrate uptake and photosynthetic rates in Castle Lake, California (Axler, Gersber, and Goldman, 1980), and evidence presented by Howarth and Cole (1985) indicates that Mo availability limits nitrogen fixation rates in the ocean. In the latter case limitation may result from competitive interference with Mo uptake by sulfate, which has an effective radius and charge distribution nearly identical to those of molybdate. Similar competitive interference is presumably insignificant in fresh water, because the sulfate concentration in fresh water is less than 1% of the sulfate concentration in seawater. Howarth and Cole's (1985) hypothesis has been challenged, however, by Paulsen, Paerl, and Bishop (1991), who could find no evidence of Mo limitation of nitrogen fixation in North Carolina coastal waters.

Nitrogen and phosphorus, the two essential elements listed in Table 2.1 not yet discussed, are found in seawater below the euphotic zone in concentrations of ~20–40 µM and 1.3–2.5 µM, respectively. Typical aphotic zone concentrations of these elements in freshwater systems may be two or three times lower, but can vary greatly from one system to another. However, when there is adequate light to support photosynthesis, it is not unusual for the concentrations of inorganic nitrogen and phosphorus in the euphotic of both freshwater and marine systems to be in the 50–100 nanomolar (nM) range (Edmondson 1972; Smith, Kimmerer, and Walsh, 1986). These concentrations are several orders of magnitude smaller than the average concentrations of the other macronutrients listed in Table 2.1; and of those macronutrients, phytoplankton require N and P in much larger amounts

Table 2.2 Average Concentrations of Selected Elements in River Water and Seawater (S = 35 °/oo)[a]

Element	River Water	Seawater
B	1.67	416
Ca	332	10,300
Cl	226	546,000
Co	0.003	0.00002
Cu	0.024	0.004
Fe	0.716	0.001
K	38	10,200
Mg	128	53,200
Mn	0.149	0.0005
Mo	0.005	0.11
Na	391	468,000
S	116	28,200
Si	178	100
V	0.02	0.03
Zn	0.459	0.006

[a] Concentration units are µM.

Source. For seawater, Bruland (1983). For freshwater, Martin and Whitfield (1983) with exception of S, Cl, and Na, which were taken from Riley and Chester (1971).

than S, K, Mg, or Ca. Thus, based on observed concentrations, N and P are the most likely of the macronutrients to be limiting photosynthesis. However, because many of the micronutrients are present in extremely low concentrations in the euphotic zone, it is impossible to tell, based simply on measured nutrient concentrations, whether N, P, or one of the micronutrients is limiting photosynthetic rates. As a result phytoplankton ecologists have resorted to a bioassay type of experiment to determine which nutrient or nutrients are limiting phytoplankton production. These bioassay experiments are commonly referred to as nutrient enrichment experiments.

Nutrient Enrichment Experiments

The usual approach in a nutrient enrichment experiment is to fill a series of clear flasks with the water to be assayed and enrich some of the flasks with various nutrients to see whether these nutrient additions have any effect on phytoplankton production in the flasks. In some cases the water is filtered first to remove the natural phytoplankton and other organisms and then inoculated with a monoculture of a particular phytoplankton species. In other cases the water is not filtered, so that the natural phytoplankton community becomes the test population. Following enrichment the flasks are incubated under appropriate light and temperature conditions, and the response of the phytoplankton is monitored for a period of time in both the enriched flask and a control flask, which receives no nutrient additions. The response of the phytoplankton is monitored by normally one of two methods. In the first method the phytoplankton biomass is determined, usually in terms of cell counts or chlorophyll concentration, and the effect of a given enrichment is measured in terms of the difference in phytoplankton biomass between the enriched and the control flask. Generally, such a bioassay is based on the yield (i.e., biomass) in the enriched flasks. In the second method the actual rate of photosynthesis in each flask is measured, usually after waiting a fixed time interval (approximately one day to a week) after the nutrient enrichments. There are pros and cons to both methods. If, as is usually the case, the test population consists of the natural phytoplankton community, one often finds that the composition of the community in the flask changes after several days of incubation. This change results because some species evidently do not grow well under artificial conditions, and because nutrient enrichments may not stimulate all species equally. Because the biomass in the enriched flasks generally peaks as much as several weeks following enrichment, one can argue that yield type experiments may misrepresent the nutrient limitation characteristics of the natural phytoplankton assemblage. On the other hand relative photosynthetic rates in enriched flasks may vary greatly with time following enrichment, because there may be a lag in the response of phytoplankton to certain enrichments, whereas other enrichments produce a rapid increase in production that subsequently declines. For example, Menzel, Hulbert, and Ryther (1963) were initially led to believe that Sargasso Sea water was iron-limited, because after a few days, incubation photosynthetic rates in flasks enriched with N, P, and Fe were substantially higher than in flasks that received only N and P additions. However, they later discovered that after about a week's incubation flasks enriched with only N and P showed photosynthetic rates just as high as the flasks enriched with N, P and Fe had shown after a few days. They concluded that N and P were the principal limiting nutrients, and that addition of Fe simply speeded up the response of the phytoplankton to N and P additions. The effect of the added Fe is perhaps not surprising, since Menzel, Hulbert, and Ryther (1963) added N in the form of nitrate, and iron is required for the reduction of nitrate to a form that can be utilized in the photosynthetic production of organic nitrogen compounds. This example illustrates why nutrient enrichment experiments must be interpreted cautiously and carefully to avoid jumping to unwarranted conclusions.

Figure 2.3 illustrates the methodology and interpretation of so-called single nutrient and multiple nutrient enrichment experiments. In a single nutrient enrichment experiment, a series of experimental flasks receives an enrichment with only one nutrient. In the multiple nutrient enrichment experiment, a series of experimental flasks is enriched with all essential plant nutrients except one. In both cases the productivity or yield in the experimental flasks is compared to that of the control flasks.

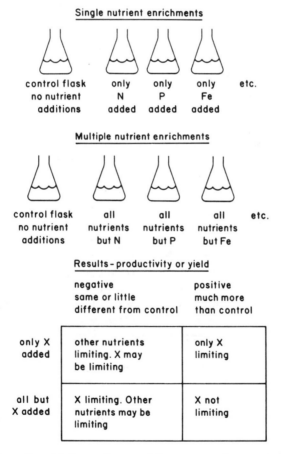

Figure 2.3   Single and multiple nutrient enrichment experiments and interpretation of results.

The rationale behind the interpretation of the results is as follows: If the water being studied contains an abundance of all essential nutrients except one, then addition of this single nutrient should greatly stimulate production. If addition of a single nutrient does not greatly stimulate production, then some other nutrient(s) is (are) limiting. Since several nutrients may be simultaneously limiting (i.e., there is very little of each of several nutrients in the water to support additional growth), it is impossible to say in the case where production is not stimulated whether the single nutrient added was one of the limiting nutrients or not. We only know that there is at least one limiting nutrient other than the one tested.

In the case of multiple nutrient enrichments, production little different from the control flask indicates that the nutrient omitted is limiting, since all other nutrients were added in the enrichment. It is impossible to tell without doing additional experiments whether other nutrients are simultaneously limiting. However, if production is much greater than that of the control flask, then the omitted nutrient must not have been limiting. Figure 2.3 summarizes the possible results and conclusions to be reached from the two types of nutrient enrichment experiments.

Unfortunately the simple interpretability of the four outcomes indicated at the bottom of Figure 2.3 is not always encountered in practice. One obvious complicating factor is the variable composition of phytoplankton. For example, we previously noted that the N/P ratio by atoms in marine phytoplankton could vary between roughly 3 and 30. Hence a population initially with a N/P ratio of 10 might respond to N enrichment by assimilating additional N until the N/P ratio reached 20. However, the same population might respond to P enrichment by assimilating additional P until the N/P ratio was reduced to 5. In either case the additional nutrient assimilation would probably result in an increase in photosynthetic rate and/or yield. Applying the simple logic in Figure 2.3 appropriate to a single nutrient enrichment, you might conclude in the first case (positive response to N addition) that only N was limiting, and in the second case (positive response to P addition) that only P was limiting. These two conclusions are obviously inconsistent with one another.

Fortunately, changes in productivity and/or yield associated with shifts in cellular composition are unlikely to be large. For example, it is unlikely that yield as measured by changes in cell numbers or chlorophyll would change by more than a factor of two in response to shifts in N/P ratios. On the other hand changes in yield by an order of magnitude or more are not uncommon in cases where a single nutrient is added to water that contains all other nutrients in abundance. It is therefore best to consider only large (e.g., order of magnitude) changes in yield or productivity as clear evidence of a positive response to nutrient addition, and Figure 2.3 defines accordingly a positive response as a productivity or yield *much more* than the control.

We have now seen that nutrient enrichment experiments may be complicated by the following factors.

1. Possible changes in the species composition of the culture in the enrichment flask with time.
2. Variability over time in the response of the same population to different nutrient additions.
3. Changes in cellular composition in response to nutrient additions.

Such complications need not be serious problems if the enrichment experiments are designed carefully and if the enrichment flasks are monitored frequently. In fact such complications may provide useful insights for the experimentalist (e.g., Menzel, Hulbert, and Ryther, 1963). Keeping these observations in mind, let us examine several cases in which nutrient enrichment experiments have been used to study nutrient limitation questions.

Long Island Bays

The release of organic waste from duck farms began to cause noticeable water pollution problems in certain bays and tributary streams along the southern side of Long Island as the Long Island duckling industry developed during the 1940s. The affected area is shown in Figure 2.4. Most apparent among these problems was the development of massive phytoplankton blooms in the waters of the tributary streams along which the duck farms were located. These blooms extended downstream from the duck farms and into the bays along the coast, where the demise of the once productive oyster and hard-shell clam fishery closely coincided with the development of the phytoplankton blooms (Ryther and Dunstan, 1971). Oysters, which feed by filtering water through their gills, were unable to feed or respire effectively, because their gills became coated with the phytoplankton cells (Wagner, 1971). Although clams thrived in the phytoplankton-rich waters, the clams were so contaminated with bacteria that they were often commercially unusable. Studies by the Woods Hole Oceanographic Institution between 1950 and 1955 showed that the algal concentrations in the bay water dropped off steadily with increasing distance from the mouths of the principal tributary streams in Moriches Bay, in a manner similar to what would be expected from dilution by coastal water. In other words there was little indication that the dense phytoplankton populations in the bays were actively growing, but rather that they had simply been washed into the bays from tributary streams and subsequently dispersed by tidal currents.

Nutrient analyses performed on the bay and tributary stream waters showed that phosphate and phytoplankton concentrations were closely correlated, with the maximum phosphate concentrations in Moriches Bay of about 7 µM dropping off to about 0.25 µM at the eastern and western ends of the affected area. Similar

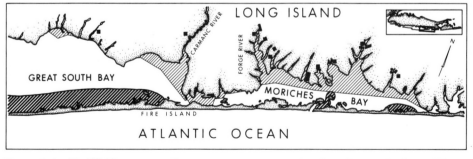

Figure 2.4 Shellfishing areas affected by duck farm wastes along the south shore of Long Island. Solid blocks along tributary streams indicate duck farms in 1966. Hatched regions indicate shellfishing areas. Lightly hatched shellfishing areas were closed to shellfishing due to pathogen contamination. [Redrawn from U.S. Dept. of the Interior (1966).]

analyses for nitrogen in the form of nitrate (NO_3^-), nitrite (NO_2^-), ammonium (NH_4^+), and uric acid (ducks excrete uric acid) revealed virtually no detectable nitrogen in any of these forms except in the immediate vicinity of the duck farms. It was tentatively concluded that nitrogen was the principal nutrient limiting phytoplankton production in the study area, that nitrogenous wastes from the duck farms were stimulating the massive algal blooms in the tributary streams, and that phosphate, being present in excess in the duck farm wastes, simply acted as a tracer of the water and phytoplankton from the tributary streams.

To see whether nitrogen was in fact the nutrient limiting phytoplankton biomass in the system, a series of nutrient enrichment experiments was performed on water samples taken from Moriches Bay, Great South Bay, and the Forge River, a Moriches Bay tributary on which several duck farms were located. Each water sample was filtered and then poured into three flasks, one of which served as a control and received no nutrient additions. A second flask was enriched with N, and the third flask with P. No other nutrients were added. The experimental setup therefore conformed to the single nutrient enrichment design of Figure 2.3. All three flasks at each station were inoculated with a pure culture of the phytoplanker *Nannochloris atomus*, the dominant algal species in the phytoplankton blooms, and were incubated for one week, at which time the number of cells in each flask was counted. Figure 2.5 shows the results of the experiments. Cell counts in the N-enriched flasks were an order of magnitude or more greater than counts in the control flasks after 1-week incubations. Virtually no change in cell numbers occurred in the P-enriched flasks. Counts in the control flasks increased by factors of about 2–4, indicating that the inoculum possessed a moderate potential for additional growth even without additional nutrients. The order-of-magnitude difference in cell numbers between the N-enriched flasks and the controls indicates that phytoplankton biomass in the system was limited by nitrogen, and only by nitrogen.

Canadian Experimental Lakes

During the 1960s and early 1970s, the Canadian government authorized a series of nutrient enrichment experiments on whole lakes in the Experimental Lakes Area of northeastern Ontario. The nutrient additions were made directly to the lakes rather than to water samples taken from the lakes to make the experiments as realistic as possible. Of particular concern was the importance of CO_2 exchange between the air and water. In laboratory flasks no mechanism such as stirring or shaking is usually provided to simulate mixing effects and encourage gaseous exchange with the atmosphere. Thus water taken from a lake with a low CO_2 content might be CO_2 limited if incubated in a flask in the laboratory, whereas in the real world CO_2 exchange with the atmosphere, stimulated by wind-generated turbulence and mixing, might be more than adequate to provide the phytoplankton population with CO_2.

As a test of the effectiveness of CO_2 exchange with the atmosphere, Lake 227, a lake having an extremely low dissolved CO_2 content, was enriched with N and P. Within weeks a massive bloom of phytoplankton had developed, with algal concentrations roughly 100 times greater than in other lakes of the area (Schindler, 1974). Gas exchange studies revealed that some of the CO_2 to support the phytoplankton bloom had come from the atmosphere. Considering the results of this experiment

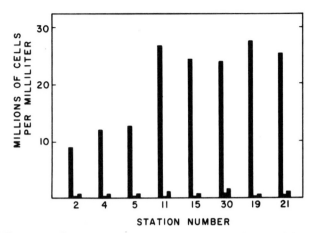

Figure 2.5 Cell counts after a one-week incubation in nutrient enrichment flasks at indicated stations. At each station, the left-hand, middle, and right-hand vertical bars represent the N-enriched flask, the P-enriched flask, and the control flask, respectively. Stations 2, 4, and 5 were located in Great South Bay; station 11 was located between Great SouthBay and Moriches Bay; stations 15, 19, and 21 were located in Moriches Bay; and station 30 was located in the Forges River. [Data from Ryther and Dunstan (1971).

and the fact that most lakes (and the oceans) have a much higher CO_2 content than Lake 227, Schindler (1974) concluded that CO_2 probably does not limit phytoplankton biomass in most bodies of water.

To determine the nutrient limiting photosynthetic rates in the experimental lakes, multiple nutrient enrichments were made to a second lake, Lake 226. This lake consisted of two similar basins separated by a shallow neck. A divider consisting of a sea curtain made from vinyl reinforced with nylon was sealed into the sediments and fastened to the bedrock of the shallow neck (Schindler, 1974). The two basins were therefore separated physically. Beginning in May, 1973, additions of carbon and nitrogen were made to one basin, and carbon, nitrogen, and phosphorus to the other basin. The phytoplankton population in the basin that received only carbon and nitrogen additions was little different from the population before nutrient additions, but a massive algal bloom developed in the basin that received added phosphorus. This experiment clearly showed phosphorus to be a limiting nutrient in the lake.

In a third series of experiments, phosphorus, nitrogen, and carbon additions were made in 1971 and 1972 to Lake 304. As expected a large phytoplankton crop developed in response to the nutrient additions. In 1973, although additions of nitrogen and carbon continued, no phosphorus was added. As a result the phytoplankton population declined dramatically, reaching typical pre-1971 levels. Table 2.3 shows the mean annual phytoplankton biomass in Lake 304 as measured by the mean chlorophyll *a* concentration in the water. Most of the phosphorus added to the lake in 1971 and 1972 was evidently trapped either in the form of chemical precipitates or in microbial biomass on the bottom of the lake and was hence

Table 2.3   Mean Chlorophyll *a* Concentrations in Lake 304

Year	Chlorophyll *a* (mg m^{-3})
1969	6.5
1970	11.0
1971	21.5
1972	24.3
1973	8.9

Source. Data derived from figures reported by Schindler (1974).

unavailable to support a phytoplankton bloom in 1973. This experiment showed convincingly not only that phosphorus was limiting photosynthetic rates in Lake 304, but also that little of the phosphorus added to the lake in one year was recycled in the next year. This latter result suggested that lakes having undesirably high phytoplankton populations might be effectively treated by reducing phosphorus inputs. In Chapter 4 we will have a chance to see how well this idea has worked in practice.

Nitrogen Versus Phosphorus Limitation

Ryther and Dunstan (1971) found convincing evidence that only nitrogen limited photosynthetic rates in the Long Island bays, whereas Schindler (1974) found phosphorus to be limiting in the Canadian experimental lakes. These results are rather typical of findings reported in freshwater and marine systems; that is, in freshwater systems, phosphorus is often found to be the principal limiting nutrient, whereas in marine systems, nitrogen is most commonly found to be limiting (e.g., Thomas, 1969). Why would phosphorus tend to be limiting in freshwater systems and nitrogen in marine systems?

A number of papers have been written on this subject, and Howarth (1988) has given an excellent review and discussion of the issues. Which nutrient limits photosynthetic rates depends on the N/P ratio of the external inputs to the system and on the biogeochemical processes that alter the availability of N and P within the system. Domestic sewage typically has a N/P ratio of about 9 by atoms (Weibel 1969), which is about one-third to one-half the values typically associated with benthic algae and phytoplankton, respectively. The use of polyphosphates as a component of laundry detergents undoubtedly accounts in part for the low N/P ratio in domestic wastes, but it is clear from Ryther and Dunstan's (1971) work in the Long Island bays that wastewater from the duck farms also contained a low N/P ratio relative to phytoplankton needs. A consequence of the low N/P ratio in domestic wastewater and sewage is that photosynthetic rates in systems enriched artificially with such water almost invariably become nitrogen limited. Consequently Ryther and Dunstan's (1971) results cannot be attributed so much to the fact that the systems contained seawater but rather to the fact that the water was enriched with wastewater having a low N/P ratio relative to the needs of the phytoplankton.

The biogeochemical processes that occur within many freshwater systems tend to create phosphorus limitation. One of the most important of such processes is nitrogen fixation, of which many species of cyanobacteria are capable. As a result several experimental lakes deliberately enriched by Schindler (1977) with fertilizer containing a low N/P ratio developed large populations of cyanobacteria. The algae were able to make up for the nitrogen deficiency in the fertilizer by fixing atmospheric nitrogen. Thus a mechanism exists for providing needed nitrogen from the atmosphere (nitrogen fixation) even if nitrogen is limiting in the external inputs to a lake. The atmosphere thus serves as a source both of carbon (by means of CO_2 exchange at the air–water interface) and of nitrogen. Phosphorus has no such atmospheric reservoir, however, and all essential phosphorus must come from external inputs or from recycling within a lake.

In most parts of the oceans, on the other hand, N-fixation by cyanobacteria appears to be of very little importance. The principal reason appears to be the lower availability of iron and perhaps molybdenum in the ocean compared to most fresh waters (Howarth, Marino, and Cole, 1988b). Both metals are required for nitrogen fixation. The concentration of Fe in the ocean is orders of magnitude lower than the Fe concentration in typical fresh waters (Table 2.2). The idea that competition with sulfate lowers the availability of molybdate in marine waters is being debated (Paulsen, Paerl, and Bishop, 1991). Although N-fixing cyanobacteria are abundant in some regions of the ocean such as the Baltic Sea and Harvey-Peel Inlet in Australia, such systems are the exceptions to the rule in the marine environment (Howarth et al. 1988a).

An important factor in the phosphorus limitation question is the fact that the phosphate ion, PO_4^{3-}, forms insoluble compounds with several positive ions, including aluminum (Al^{3+}), calcium (Ca^{2+}), and iron (Fe^{3+}). In fresh water the most important of these chemical precipitates is frequently ferric phosphate, $FePO_4$. This precipitate will sink to the bottom of a lake, effectively trapping phosphate in the sediments. Such chemical precipitation may have been partly responsible for the rapid recovery of Lake 304 in 1973. If the oxygen concentration in the water is very low, however, ferric iron, Fe^{3+}, is spontaneously converted to ferrous iron, Fe^{2+}. Ferrous iron does not form insoluble precipitates with phosphate. Thus if the bottom waters of a lake become anoxic, phosphate trapped in the sediments may be released and circulate back into the water. In his work on the Canadian experimental lakes, however, Schindler found that in some cases phosphate was efficiently trapped in the sediments even though the bottom waters were anoxic for periods of several months. He reported that uptake of the phosphate by microorganisms in the bottom waters and subsequent sedimentation explained the failure of the phosphate to return to the water column. Nevertheless, the bottom waters of many lakes do remain oxidizing throughout the year, and in such cases the formation and sedimentation of ferric phosphate may be a significant mechanism for removing phosphate from the water column. Since nitrogen does not form insoluble chemical precipitates, there is no mechanism analogous to inorganic iron precipitation to trap nitrogen in the sediments of either marine or freshwater systems.

Since the average depth of the oceans is roughly 4 kilometers (km), particulate materials that sink out of the surface mixed layer are essentially lost from the euphotic zone. Upwelling of water from below the nutricline does occur at a slow rate in almost all parts of the ocean and at a rapid rate in certain "upwelling" areas; but with the exception of shallow coastal regions, recycling of nutrients from below the nutricline and from the sediments is a very slow process in most parts of the ocean. Thus recycling of nutrients *within the surface mixed layer* takes on special importance in most marine food chains. The principal mechanisms for recycling nutrients within the mixed layer are direct excretion (e.g., zooplankton excrete ammonia) and regeneration from detritus. Evidence to date suggests that both animal excretion and regeneration of nutrients from detritus tend to create N-limited conditions. Nutrient release from particulate detritus must be rapid if the nutrients are to be recycled before the particles sink out of the euphotic zone. Phosphorus seems to be released more rapidly from detritus than N, presumably because phosphate ester bonds are more easily cleaved than the covalent bonds of organic nitrogen (Howarth, 1988). As a result, fecal material and sedimenting detritus tend to be enriched in N relative to P (Knauer, Martin, and Bruland, 1979; Lehman, 1984). For example, the studies of Knauer, Martin, and Bruland (1979) revealed that material that settled into sediment traps near the base of the euphotic zone in the North Pacific subtropical gyre contained N and P in an atomic ratio of 29, which is about twice the ratio of 16 typically associated with oceanic particulate matter (Copin-Montegut and Copin-Montegut, 1983). Corresponding ratios in coastal areas were 22 and 27 under upwelling and nonupwelling conditions, respectively. With respect to excretion, studies summarized by Lehman (1984) indicate that the soluble compounds released by zooplankton are enriched in P relative to N, and studies by Le Borgne (1982) clearly show that the net growth efficiencies of zooplankton are higher for N than P, i.e., the zooplankton excrete a higher percentage of the P than of the N in the food they assimilate. Hence biological processes that occur within the surface mixed layer tend to create N-limited conditions, and this fact in part accounts for the tendency of open ocean systems to be N-limited.

Denitrification is the process by which nitrate nitrogen is reduced in a series of steps to N_2, the gas which makes up about 78% of the earth's atmosphere. The conversion is mediated by certain species of bacteria that utilize the oxygen from the nitrate ion to oxidize organic matter under anoxic or nearly anoxic conditions. Denitrification obviously tends to create N-limited conditions, because N_2 is unavailable to aquatic plants other than nitrogen-fixing cyanobacteria. Because of the requirement for low oxygen concentrations, denitrification is restricted to only certain portions of aquatic habitats. Sediments are usually anoxic below the surface layer, particularly when the overlying water column is highly productive. According to Seitzinger (1988, p. 702), "During the mineralization of organic matter in sediments, a major portion of the mineralized nitrogen is lost from the ecosystem via denitrification" and, "The loss of nitrogen via denitrification exceeds the input of nitrogen via N_2 fixation in almost all river, lake, and coastal marine ecosystems in which both processes have been measured."

Nor is denitrification confined to the sediments. Denitrification may occur in the water column when the oxygen concentrations drops below ~0.2 grams per cubic meter ($g\,m^{-3}$), a condition which is quite common in some lakes and certain parts of the ocean. Furthermore, even though the concentration of oxygen in a bulk water sample may be well above 0.2 $g\,m^{-3}$, it is possible that very low O_2 concentrations exist in microzones associated with particles. Such microzones may be sites of denitrification. Studies summarized by Hattori (1983) indicate that the low-oxygen intermediate waters of the eastern tropical Pacific Ocean account for a major fraction of the denitrification that occurs in the water column of the ocean, and that substantial amounts of denitrification also occur in the low-oxygen intermediate waters of the Arabian Sea, the bottom waters of the southwest African shelf, and perhaps in the Bay of Bengal. Following their formation, oceanic bottom waters do not return to the surface again for times of typically 500–1000 years. It is reasonable to expect that denitrification would cause these subsurface waters to become depleted in nitrogen. The denitrification of bottom and intermediate waters is probably the major factor that causes subsequently upwelled marine surface waters to be N limited.

The general picture that emerges from this analysis is that freshwater systems tend to be P limited because of the precipitation and burial of inorganic phosphate compounds in the sediments, particularly in the form of ferric phosphates. Ferric phosphate precipitation is of much less consequence in the marine P cycle because of the very low Fe concentrations in seawater. The tendency of marine waters to be N limited reflects the low iron concentration in seawater, the fact that P is recycled more efficiently than N by biological processes in the euphotic zone, and the long time during which denitrification may act to deplete subsurface waters of fixed nitrogen. While these generalizations are useful for understanding the role of nutrients in limiting photosynthetic rates in aquatic systems, it is important to realize that not all lakes and rivers are P limited, nor are all marine waters N limited. The magnitude and elemental composition of external nutrient inputs and the relative importance of various biogeochemical processes occurring within the system all combine to determine which nutrient or nutrients limit photosynthetic rates. The relative importance of these inputs and processes will vary from one system to another. One of the best examples of an exception to the general picture is the apparent limitation of photosynthetic rates in offshore waters of the Antarctic and northeast Pacific subarctic oceans by iron. In those areas the concentrations of phosphate and nitrate in the euphotic zone are well above the levels associated with nutrient limitation, and nutrient enrichment studies conducted by Martin and Fitzwater (1988) indicate that Fe is the single nutrient limiting photosynthetic rates. The explanation seems to be that the Fe/N ratio of intermediate depth ocean waters is low compared to the needs of phytoplankton (Martin and Gordon 1988); and in the absence of external inputs of Fe, these waters would become Fe limited when upwelled into the euphotic zone. In coastal waters the input of Fe from land runoff and release from sediments is apparently more than adequate to supply the needed iron, and in the large subtropical gyres, which account for about 40% of the ocean's surface area, upwelling is so slow that atmospheric fallout of Fe from dust and rainfall is sufficient to provide the Fe required for photosynthesis. However, in the subarctic offshore waters of the Antarctic and northeast Pacific Oceans, atmospheric inputs and lateral transport of Fe are insufficient to keep pace with the upward movement of N into the euphotic zone, and as a result phytoplankton strip the water of Fe long before the supply of N is exhausted.

REFERENCES

Atkinson, M., and S. V. Smith. 1983. C:N:P ratios of benthic marine plants. *Limnol. Oceanogr.*, **28**, 568–574.

Axler, R. P., R. M. Gersber, and C. R. Goldman. 1980. Stimulation of nitrate uptake and photosynthesis by molybdenum in Castle Lake, California. *Can. J. Fish. Aquat. Sci.*, **37**, 707–712.

Bruland, K. W. 1983. Trace elements in seawater. In J. P. Riley and R. Chester (Eds.), *Chemical Oceanography*. Vol. 8. Academic Press, New York. pp. 157–220.

Copin-Montegut, C., and G. Copin-Montegut. 1983. Stoichiometry of carbon, nitrogen, and phosphorus in marine articulate matter. *Deep-Sea Res.*, **30**, 31–46.

Edmondson, W. T. 1972. Nutrients and phytoplankton in Lake Washington. In G. E. Likens (Ed.), *Nutrients and Eutrophication*. American Society of Limnology and Oceanography, Lawrence, Kansas. pp. 172–193.

Hattori, A. 1983. Denitrification and dissimilatory nitrate reduction. In E. J. Carpenter and D. G. Capone (Eds.), *Nitrogen in the Marine Environment*. Academic Press, New York. pp. 191–232.

Howarth, R. W. 1988. Nutrient limitation of net primary production in marine ecosystems. *Annu. Rev. Ecol.*, **19**, 89–110.

Howarth, R. W., and J. J. Cole. 1985. Molybdenum availability, nitrogen limitation, and phytoplankton growth in natural waters. *Science*, **229**, 653–655.

Howarth, R. W., R. Marino, J. Lane, and J. J. Cole. 1988a. Nitrogen fixation in freshwater, estuarine, and marine ecosystems. 1. Rates and importance. *Limnol. Oceanogr.*, **33**, 669–687.

Howarth, R. W., R. Marino, and J. J. Cole. 1988b. Nitrogen fixation in freshwater, estuarine, and marine ecosystems. 2. Biogeochemical controls. *Limnol. Oceanogr.*, **33**, 688–701.

Hutchinson, G. E. 1967. *A Treatise on Limnology*. Vol. 2. Wiley, New York. 1115 pp.

Knauer, G. A., J. H. Martin, and K. W. Bruland. 1979. Fluxes of particulate carbon, nitrogen, and phosphorus in the upper water column of the northeast Pacific. *Deep-Sea Res.*, **26A**, 97–108.

Le Borgne, R. 1982. Zooplankton production in the eastern tropical Atlantic Ocean: net growth efficiency and P:B in terms of carbon, nitrogen, and phosphorus. *Limnol. Oceanogr.*, **27**, 681–698.

Lehman, J. T. 1984. Grazing, nutrient release, and their impacts on the structure of phytoplankton communities. In D. G. Meyers and J. R. Strickland (Eds.), *Trophic Interactions Within Aquatic Ecosystems*. Westview, Boulder, Colo. pp. 49–72.

Martin, J. H., and S. E. Fitzwater. 1988. Iron deficiency limits phytoplankton growth in the north-east Pacific subarctic. *Nature*, **331**, 341–343.

Martin, J. H., and R. M. Gordon. 1988. Northeast Pacific iron distributions in relation to phytoplankton productivity. *Deep-Sea Res.*, **35**, 177–196.

Martin, J.-M., and M. Whitfield. 1983. The significance of the river input of chemical elements to the ocean. In C. S. Wong (Ed.), *Trace Metals in Sea Water*. Proc. NATO Adv. Res. Inst., March 30–April 1, 1981, Erice, Italy. Plenum, New York. pp. 265–296.

Menzel, D. W., E. M. Hulbert, and J. H. Ryther. 1963. The effects of enriching Sargasso Sea water on the production and species composition of the phytoplankton. *Deep-Sea Res.*, **10**, 209–219.

Odum, E. P. 1971. *Fundamentals of Ecology*. 3rd ed. Saunders, PA. 574 pp.

Paulsen, D. M., H. W. Paerl, and P. E. Bishop. 1991. Evidence that molybdenum-dependent nitrogen fixation is not limited by high sulfate concentrations in marine environments. *Limnol. Oceanogr.*, **36**, 1325–1334.

Riley, J. P., and R. Chester. 1971. *Introduction to Marine Chemistry*. Academic Press, New York. 465 pp.

Ryther, J. H., and W. M. Dunstan. 1971. Nitrogen, phosphorus, and eutrophication in the coastal marine environment. *Science*, **171**, 1008–1013.

Schindler, D. W. 1974. Eutrophication and recovery in experimental lakes: Implications for lake management. *Science*, **184**, 897–899.

Schindler, D. W. 1977. Evolution of phosphorus limitation in lakes. *Science*, **195**, 260–262.

Seitzinger, S. P. 1988. Denitrification in freshwater and coastal marine ecosystems: Ecological and geochemical significance. *Limnol. Oceanogr.*, **33**, 702–724.

Smetacek, V., and U. Passow. 1990. Spring bloom initiation and Sverdrup's critical-depth model. *Limnol. Oceanogr.*, **35**, 228–234.

Smith, S. V., W. J. Kimmerer, and T. W. Walsh. 1986. Vertical flux and biogeochemical turnover regulate nutrient limitation of net organic production in the North Pacific Gyre. *Limnol. Oceanogr.*, **31**, 161–167.

Thomas, W. H. 1969. Phytoplankton nutrient enrichment experiments off Baja California and in the eastern equatorial Pacific Ocean. *J. Fish. Res. Bd. Can.*, **26**, 1133–1145.

U.S. Department of the Interior. 1966. *Proceedings, Conference in the Matter of Pollution of the Navigable Waters of Moriches Bay and the Eastern Section of Great South Bay and their Tributaries*, Patchogue, New York. September. 496 pp.

Wagner, R. H. 1971. *Environment and Man*. Norton, New York. 491 pp.

Weibel, S. R. 1969. Urban drainage as a factor in eutrophication. In *Eutrophication: Causes, Consequences, Correctives*. National Academy of Sciences, Washington, D.C. pp. 383–403.

3

PHYSICAL FACTORS AFFECTING PRODUCTION

Although the availability of light and nutrients most directly influences photo-synthetic rates, purely physical processes such as upwelling, vertical mixing, and currents may affect greatly the availability of light and nutrients to aquatic plants. This chapter examines the physical properties of water relevant to these effects. Seasonal production cycles and the susceptibility of aquatic systems to cultural eutrophication stress are then examined using the physical and chemical concepts developed in the first two chapters. Estuaries are singled out for special considera-tion because of their importance in the life history of many aquatic organisms and because of their susceptibility to pollution stress.

PHYSICAL PROPERTIES OF WATER

One of the most peculiar properties of pure water is the fact that its density max-imum occurs at a temperature (4°C) above the freezing point (0°C). Thus either heating or cooling pure water initially at a temperature of 4°C causes its density to decrease. On the other hand the density of typical seawater with a salinity of 35 parts per thousand ($^0/oo$) steadily increases as the water is cooled, no matter what the ini-tial temperature is. What is there about pure water that causes it to have a density maximum at 4°C?

First of all water is, chemically speaking, a polar molecule. The fact that the oxygen atom in the water molecule is more electronegative than the hydrogen atoms causes the distribution of negative electrons in the molecule to be shifted slightly toward the oxygen atom. As a result of this shift, the oxygen end of the molecule has a small negative charge, and the side with the two hydrogen atoms has a small positive charge. Figure 3.1 shows a water molecule in its equilibrium configuration and the polarization of charge associated with that structure. Note

Figure 3.1 Simple chemical representation of a water molecule. Lines between the
oxygen atom (O) and each hydrogen atom (H) represent electron-pair bonds which
hold the atoms in the molecule together. Positive and negative signs indicate partial
polarization of charge.

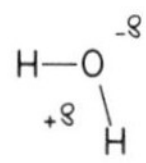

$$H\!-\!\overset{-\delta}{O} \\ {}^{+\delta}\backslash{}_{H}$$

that the molecule is not linear. In the gaseous phase the H-O-H angle is about 105°.
In the liquid and solid phases, individual water molecules interact with one
another. Since positive and negative charges attract each other, the molecules tend
to become oriented with their positive and negative ends close together. Figure 3.2
shows the hexagonal arrangement of water molecules that is believed to be one of
the principal structural forms of ice. The hexagonal pattern of the molecules
actually extends in three dimensions, although only a single planar hexagonal unit
is shown in Figure 3.2.

The bridge between two adjacent oxygen atoms formed by the hydrogen atom
attached to one of the oxygen atoms is called a *hydrogen bond*, which is indicated by
dashed lines in Figure 3.2. Hydrogen bonds are caused by the attraction of a
positively charged hydrogen atom for a negatively charged atom, in this case
oxygen. Although the strength of hydrogen bonds is relatively small compared to
that of normal electron-pair bonds, it is the hydrogen bonds that largely serve to
orient the water molecules in ice into structures such as that shown in Figure 3.2. An
important feature of this type of structure is its high degree of porosity. When ice
melts, some of the hydrogen bonds in the ice crystal are broken, and this breakage
causes some of the structural units to collapse totally or partially. As these structural
units collapse the density of the water increases, because some of the gaps in the
original ice structure are filled. Thus the density of liquid water at 0°C is greater than
that of ice at 0°C, and ice therefore floats on water. As the temperature of the water is
further raised above 0°C, more hydrogen bonds are broken, and the structural units
collapse even more. As a result the density of the water is increased further as more
gaps in the structure are filled in with smaller collapsed units or individual molecules. However, the increase in temperature is accompanied by another effect that
tends to reduce the density. As the water is heated the individual molecules and
structural units move about at faster speeds, just as the translational speed of the
atoms or molecules in a gas increases as the temperature of the gas increases. The
higher speed of the molecules and structural units in the liquid water causes the
water to expand slightly and become less dense, just as a gas expands and becomes
less dense when it is heated. The latter phenonenon is routinely utilized by hot air
balloon enthusiasts. As pure water is warmed from 0°C to 4°C the density of the
water increases, because the effect of the collapsing structural units on the water's
density more than offsets the reduction in density caused by the increased translational speed of the water molecules and structural units. However, the latter effect
becomes more important above 4°C, and as a result the density of water steadily
decreases as the temperature of the water rises above 4°C. Hence the maximum density of pure water occurs at 4°C.

Substances dissolved in water disrupt to a certain degree the arrangement of the
structural units. If the dissolved substances are ionic or polar themselves, water
molecules become oriented about them with positive and negative ends in close
proximity. Figure 3.3 shows how water molecules would be expected to orient themselves around a positive sodium ion and a negative chlorine ion.

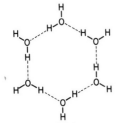

Figure 3.2   Hexagonal arrangement of water molecules in ice. Each hydrogen atom is joined by an electron-pair bond to the nearest oxygen atom and by a hydrogen bond to the next closest oxygen atom.

The concentration of dissolved ions in seawater with a salinity greater than about 24.7 $^0/oo$ is high enough to significantly disrupt the porous structure of pure water. As a result changes in structure associated with heating or cooling are never great enough to override the effect on the water's density of changes in the translational speed of the molecules and structural units in the seawater. Thus the density of seawater with a salinity greater than 24.7 $^0/oo$ steadily decreases as the water is warmed above the freezing point.

WATER COLUMN STABILITY AND OVERTURNING

A column of water is resistant to vertical mixing (i.e., stable) if the density of the water increases steadily with increasing depth. In other words, the densest water is found on the bottom and the least dense water is on the top. If the densest water is on the top, the water column is top heavy and will spontaneously mix vertically if disturbed slightly. In a freshwater system the temperature in a stable water column will decrease steadily with depth if the temperature of the water is everywhere greater than 4°C. In the ocean the temperature in a stable water column will almost always decrease with increasing depth. However, since changes in salinity also affect density (the higher the salinity, the higher the density), it is possible that in a stable water column the temperature may increase with depth if the salinity also increases, and if the effect of the salinity increase on the density more than offsets the effect of the temperature increase. The warm salt brines at the bottom of the Red Sea, for example, remain at the bottom of the water column because their high salinity makes them more dense than the cooler overlying waters. In most parts of the ocean, however, the temperature, or more correctly the potential temperature, decreases steadily with increasing depth. Figure 3.4 shows the characteristic variation of temperature with depth in a hypothetical body of water. The water column is being heated from above by radiant energy, and hence the temperature is highest at the surface. Waves and turbulence generated by winds have created a region of almost constant temperature near the surface. This region is called the *mixed layer* or, in strictly limnological work, the *epilimnion*. The depth of the mixed layer may vary greatly, depending on the strength of the winds and the stability of the water column. The mixed layer in a lake may be only a meter or so deep during parts of the summer, but perhaps a hundred or more meters deep at certain times during the fall or winter when the water column is destabilized. The region of relatively rapid temperature change below the mixed layer is called the *thermocline*, or in limnological work, the *metalimnion*. Below the thermocline is a region of relatively constant temperature

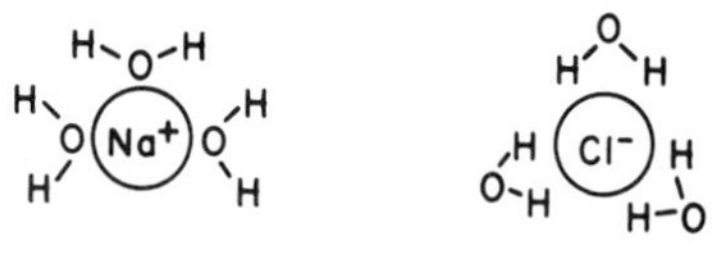

Figure 3.3 Qualitative arrangement of water molecules around a positive sodium (Na^+) ion and a negative chlorine (Cl^-) ion. Note that the negative (oxygen) end of the water molecule is closest to the positive Na^+, while the positive (hydrogen) end of the water molecule is closest to the negative Cl^-.

referred to as the *hypolimnion* by limnologists. Oceanographers have coined no special name for the corresponding region of the ocean. The decrease of temperature with depth shown in Figure 3.4 would stabilize any water column in the ocean and would stabilize any freshwater column if the temperature were everywhere $\geq 4°C$. Figure 3.5 shows the variation of temperature with depth in a stable freshwater system in which the temperature is everywhere $\leq 4°C$. In this case the water column is being cooled from above, so that the lowest temperature is found at the surface. The water column is nevertheless stable because the density of freshwater increases with increasing temperature at temperatures below $4°C$.

Now let us consider what happens to the temperature structure in a freshwater lake located in a temperate climate as the temperature of the atmosphere and the amount of radiant heating change seasonally. Figure 3.6 shows the sequence of temperature profiles that might be observed in the lake from the middle of summer until the middle of winter. As the atmosphere cools in the fall, heat fluxes from the surface waters of the warmer lake into the atmosphere, causing the temperature of the surface water to drop. As the surface water cools, it becomes denser and sinks. Consequently the mixed layer becomes deeper and the thermocline begins to break down as shown. After sufficient cooling the water column becomes isothermal, and any further cooling of the surface water causes the water column to mix from top to bottom. This period of downward mixing of surface waters caused by surface cooling is called the *fall overturn*. If the temperature of the atmosphere becomes sufficiently low, the lake water will be cooled and will remain isothermal until the temperature of the water reaches $4°C$. At that point any further cooling of the surface water causes the water column to stratify, with the coldest water being at the top and the warmest water on the bottom. If the lake eventually freezes, the water temperature near the surface will be $0°C$, and the temperature of the bottom water will be somewhere between $0°C$ and $4°C$.

Between the end of winter and the end of summer the sequence of temperature profiles follows a pattern similar to the mirror image of Figure 3.6. In the early spring, warming of the surface water causes it to become denser and hence induces downward mixing. The water column becomes isothermal at some temperature $\leq$ $4°C$ and mixes to the bottom until the temperature reaches $4°C$. This period of water column mixing is called the *spring overturn*. Once the water temperature has reached $4°C$ any further warming at the surface causes the water column to stratify. The highest temperature is now found at the surface and the lowest temperature at the bottom. Thus any further warming at the surface leads to a typical summer temperature profile such as that in Figure 3.4.

Bodies of water that have two overturning periods per year separated by periods of stratification are called *dimictic* (twice-mixing). Most lakes in temperate climates in which the summer water temperature is $> 4°C$ and the winter water temperature

< 4°C are dimictic. Bodies of water that overturn only once per year are termed *monomictic* (once-mixing). Warm lakes in which the winter water temperature never drops below 4°C may be monomictic, as may be cold lakes in which the summer water temperature never rises above 4°C. In the former case overturning occurs during the winter while the surface waters are being cooled; in the latter case overturning occurs during the summer when the surface waters are being warmed. Temperate marine waters with a salinity > 24.7 ‰ never thermally stratify during the winter while the surface water is being cooled because the density of the water increases steadily all the way to the freezing point. In such waters overturning may occur throughout the winter, and the mixed layer at such times may be hundreds of meters deep.

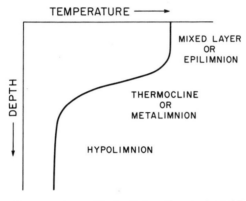

Figure 3.4   Variation of temperature with depth in a thermally stable marine water column or in a stable freshwater column in which the temperature is everywhere $\geq$ 4°C.

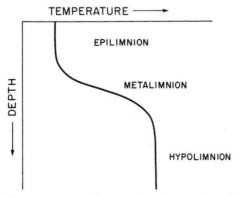

Figure 3.5   Variation of temperature with depth in a thermally stable freshwater system in which the temperature is everywhere $\leq$ 4°C.

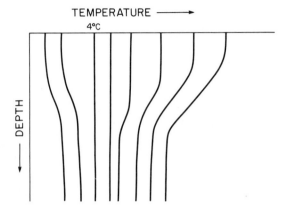

Figure 3.6   Temperature profiles in a hypothetical freshwater lake in a temperate climate from the end of summer (right-hand side) until the end of winter (left-hand side).

The Importance of Overturning

Overturning of the water column serves two highly important functions. First, downward mixing of oxygen-rich surface waters below the thermocline introduces oxygen into the bottom waters of aquatic systems. Without this mechanism of oxygen recharge the hypolimnions of many lakes would become anoxic, since the simple downward diffusion of gases through a stratified water column is quite slow. For the hypolimnion to remain oxygenated, the respiratory consumption of oxygen by organisms living below the thermocline must be sufficiently slow so as not to consume all the oxygen between overturning periods.

The oxygen concentration in the bottom waters of the deep ocean, at depths of several kilometers or more, is almost everywhere unaffected by winter overturning in the waters directly above, since overturning in the oceans rarely occurs to depths as great as several kilometers. Exceptions to this rule are found in the Atlantic Ocean near Greenland and near Antarctica in the Weddel Sea, where surface water temperatures drop as low as $1-2°C$ below zero during the winter. The formation of sea ice at such times leaves the surrounding waters enriched in dissolved salts, because sea ice has a relatively low salt content. The high salinity and very low temperature of these surface waters causes them to sink (with some mixing) all the way to the bottom. These areas are the only parts of the ocean where bottom waters are formed. Deep-ocean currents transport these bottom waters to all the major ocean basins, where mixing and very slow upwelling gradually bring the water back to the surface after a period of approximately 500–1000 years. Despite the long residence time of these bottom waters, only a very few areas of the ocean are anoxic at the bottom. The respiratory consumption of oxygen by organisms in the bottom waters of the ocean is very slow due to the cold temperatures (see discussion of Q_{10} in Chapter 8) and low rate of food supply. Consequently the bottom waters remain oxygenated despite being out of contact with the atmosphere for hundreds of years. The only exceptions to this rule are found in places such as the Black Sea and Cariaco Trench where the fallout of organic matter is exceptionally high and/or the bottom waters are unusually stagnant.

The second important function of overturning is the recharging of the surface waters with nutrients. Inorganic nutrients regenerated by excretion or detrital decay below the mixed layer diffuse at only a slow rate upward through the thermocline. Thus in a lake at the end of the summer stratification period one frequently finds very high nutrient concentrations in the hypolimnion and very low nutrient concentrations in the epilimnion. The low epilimnetic nutrient concentrations of course reflect uptake by the phytoplankton community. The breakdown of the thermocline and the mixing of epilimnetic and hypolimnetic waters mixes regenerated nutrients into the surface waters. A similar process occurs in the sea. In aquatic systems where overturning is a weak phenomenon, biomass in the surface waters is low throughout the year due to the lack of an efficient mechanism for bringing nutrients from below the thermocline into the euphotic zone. The large oceanic subtropical gyres are illustrative of such systems.

SEASONAL PRODUCTION CYCLES

With this information one may explain, at least in a qualitative way, the seasonal production cycle found in many temperate aquatic systems. Table 3.1 lists nutrient, light, and photosynthetic characteristics during each of the four seasons in a hypothetical temperate system. During the summer the water column is highly stratified with a shallow mixed layer whose depth is well above the critical depth. However, the concentration of nutrients in the mixed layer is quite low, and since no efficient mechanism exists for returning nutrients trapped below the thermocline to the mixed layer, photosynthetic rates are also low.

As the surface waters begin to cool in the fall, vertical mixing brings nutrient-rich waters into the euphotic zone. The surface light intensity is decreasing at this time, and the increasing depth of the mixed layer further reduces the average light intensity in the mixed layer. However, the bottom of the mixed layer is well above the critical depth, at least initially. With abundant nutrients and adequate light to support photosynthesis, a phytoplankton bloom occurs frequently at this time. The bloom is terminated when the declining surface light intensity and increasing mixed layer

Table 3.1   Nutrient, Light, and Photosynthetic Characteristics During Each of the Four Seasons in a Typical Temperate Aquatic System in Which the Mixed Layer Does Not Extend to the Bottom

Season	Nutrients	Light	Photosynthesis
Summer	Low	High	Low and limited by lack of nutrients
Autumn	Increasing	Decreasing	An autumn bloom occurs and is terminated by decreasing light
Winter	High	Low	Low and limited by lack of light
Spring	Decreasing	Increasing	Vernal bloom occurs and is terminated by decreasing nutrient concentrations and grazing

depth cause the bottom of the mixed layer to descend below the critical depth, or at least become close enough to the critical depth that net community production in the mixed layer is quite low.

During the winter, nutrient levels in the mixed layer remain high, even if the water column stratifies, because the lack of light prevents any significant phytoplankton uptake of nutrients. Photosynthetic rates during this time are severely limited by the lack of light in the deep mixed layer.

As the surface waters are warmed during the spring, the water column begins to stratify, and the depth of the mixed layer is therefore reduced. At the same time the average surface light intensity is increasing, and the shallow winter critical depth begins to deepen. The combination of increasing surface light intensity and decreasing mixed layer depth obviously increases the light available to phytoplankton in the mixed layer. By utilizing the high mixed layer nutrient concentrations left over from the fall and perhaps winter mixing periods, the phytoplankton population begins to multiply, and a bloom develops when the mixed layer extends no deeper than the upper region of the euphotic zone, where photosynthetic rates are uniformly high (Smetacek and Passow, 1990). This spring phytoplankton bloom is usually terminated by a combination of herbivore grazing and the exhaustion of nutrient reserves in the mixed layer by the phytoplankton population. Production during the rest of the summer remains at a low level, because much of the nutrient reserve used to set off the spring bloom has been temporarily lost below the thermocline as detritus or is tied up in higher trophic level biomass.

TROPHIC STATUS

This admittedly simplified picture of photosynthetic seasonality in a temperate aquatic system underlines the importance of vertical mixing processes and water column stratification in determining the availability of light and nutrients for photosynthesis. Vertical mixing stimulates the recycling of nutrients from deep water, but at the same time reduces the amount of light available to the phytoplankton by increasing the mixed layer depth. How then do aquatic systems become highly productive (i.e., eutrophic) for more than the short periods of time typical of spring or fall blooms? The answer is that in shallow systems the mixed layer may extend all the way or most of the way to the bottom during much of the year. In such systems recycling of nutrients is highly efficient, because there is little or no part of the system into which detritus may sink and regenerated nutrients become trapped. Because the system is shallow, it is impossible for the mixed layer to extend to great depths, and in fact the entire water column may be in the euphotic zone during much of the year. By contrast, in a very deep system recycling of nutrients from detritus that has fallen far below the euphotic zone is extremely inefficient. In fact, in most parts of the ocean and in some deep lakes overturning of the water column may never extend to the bottom. Furthermore, in deep systems the mixed layer may extend to great depths during overturning periods, so that production is brought to a halt due to the lack of light in the mixed layer. Thus, barring unusual circumstances (e.g., sewage disposal, upwelling), deep aquatic systems are inherently less productive than shallow aquatic systems. Recalling that the eutrophication process involves a gradual reduction in the depth of the aquatic system, it should not be surprising to learn that annual production in a system undergoing eutrophication is an accelerating function of time; i.e., the rate of increase of production is small at first but becomes progressively larger over the course of years as the system both accumulates nutrients and recycles them more efficiently. Relatively deep, unproductive systems are often referred to as being *oligotrophic* (few nutrients), whereas highly productive, usually shallow systems are called *eutrophic* (many nutrients or nourishing). The term *mesotrophic* is sometimes applied to systems with intermediate characteristics. One should not get the impression that all aquatic systems are initially oligotrophic. A lake, for example, may be shallow from its inception and therefore tend toward an initial high rate of production. The present trophic status of a body of water is determined both by its original characteristics and by its history since formation. In all cases, however, the natural tendency of aquatic systems is to become shallower and more productive; that is, to undergo eutrophication.

SUSCEPTIBILITY OF SYSTEMS TO OXYGEN DEPLETION

The aphotic zone of a body of water will become anoxic if the consumption of oxygen by biological or chemical processes exceeds the rate of resupply by vertical mixing and diffusion. Since virtually all aquatic organisms require oxygen for respiration, it is generally considered desirable for all parts of the water column to remain oxygenated. From our discussion of trophic status it is obvious that depth is an important determinant of productivity in an aquatic system, and depth obviously influences the percentage of the water column which is impacted by vertical mixing and overturning. Consequently, depth plays an important role in determining the susceptibility of an aquatic system to oxygen depletion.

Deep, oligotrophic systems are the least likely of aquatic systems to develop oxygen depletion problems. This conclusion is based on the following two considerations:

1. With the exception of allochthonous (external) inputs such as stream runoff, virtually all the organic carbon that is metabolized by aphotic zone organisms must have been produced in the euphotic zone. Consequently, the respiratory activity of aphotic zone organisms is related directly to the amount of productivity in the euphotic zone. Since oligotrophic systems are by definition unproductive systems, the amount of food available to aphotic zone organisms is small, so that the numbers of these organisms and their overall respiratory rate are also small.

2. In a deep aquatic system the volume of water below the thermocline is large relative to the volume of the mixed layer. Therefore the total amount of oxygen potentially available for respiration in a deep system is quite large relative to the small part of the system that is utilized for production. Thus the productivity of the surface waters would have to be exceptionally high in order for respiration to significantly reduce the average deep-water oxygen concentration. If the deep waters are reoxygenated regularly by means of overturning, it is virtually impossible for a deep oligotrophic system to become anoxic at any depth.

Shallow systems that are mixed to the bottom at all times obviously do not develop seasonal oxygen depletion problems. If there is a sufficiently vigorous mixing of the water column and exchange of gases with the atmosphere, the oxygen concentration in such systems is likely to remain near the saturation level at all times. If the wind dies down, however, so that oxygen exchange with the atmosphere is sluggish, the oxygen concentration in a highly productive, shallow system may drop to almost zero within a few days or even a few hours. The lowest concentrations are of course observed at night, when there is no photosynthetic production of oxygen. The shallow western basin of Lake Erie, with an average depth of only about 7.4 m, provides a good example of a shallow, productive system that may develop serious oxygen depletion problems after several days of calm weather. Aquaculture ponds, which are typically only about 1.0 m deep and receive large inputs of allochthonous organic matter in the form of feeds, are illustrative of the more extreme cases in which the respiratory activity of organisms in the system may strip the water of oxygen within a few hours on a calm night.

Seasonal oxygen depletion problems are undoubtedly most common in bodies of water having somewhat intermediate depths; that is, shallow enough to be highly productive, yet deep enough so that the mixed layer does not extend to the bottom except during overturning periods. In lakes of this sort the hypolimnion may be rather small compared to the epilimnion. Hence even a moderate amount of production in the epilimnion may result in enough food being consumed and respired in the hypolimnion to reduce the hypolimnetic oxygen concentration to virtually zero between overturning events. The central basin of Lake Erie, with a mean depth of 18.5 m, is a good example of such a system. The disappearance from Lake Erie's central basin of certain fish species that normally inhabit cold, deep waters and that function efficiently only when oxygen concentrations are near saturation levels is undoubtedly explained in part by the periodic low oxygen concentrations that characterize the central basin's hypolimnion. We will have a chance to study conditions in Lake Erie in more detail in Chapter 4. For the moment suffice it to say that the development of low oxygen concentrations below the thermocline in any aquatic system is frequently associated with an undesirable change in the type and abundance of organisms living in the water.

Large-scale fish kills, in some cases involving hundreds of thousands or even millions of fish, are probably the most dramatic and most highly publicized results of oxygen depletion problems associated with eutrophication. Seasonal fluctuations in oxygen levels are not likely to be associated with the sudden killing of large numbers of organisms, because seasonal declines in oxygen concentration are gradual. In such cases organisms are more likely to be eliminated from the system due to their inability to function efficiently (e.g., to escape predators, obtain food, or reproduce) than to suffocation. It is possible, however, in extremely productive systems, for oxygen levels to fluctuate from saturating or supersaturating conditions during the day to virtually zero at night. The aforementioned aquaculture ponds are examples of systems with the potential for this sort of behavior, but similar problems may develop in more natural systems seriously impacted by eutrophication. If the concentration of phytoplankton in the water is very dense, one can expect that the abundance of herbivores, primary carnivores, and higher trophic level organisms will also be high, since organisms are naturally attracted to a source of food. If this situation should develop in a fairly open system, and if the oxygen level does drop dangerously low at night, all motile organisms will rapidly try to leave the area, and for the most part will be successful as long as there are wide avenues of escape. Large-scale kills of aquatic organisms do, however, sometimes occur in bodies of water that have only restricted escape routes. In such an isolated and highly eutrophic body of water, large numbers of organisms may be attracted by the abundance of food during the day when the oxygen concentration is high. At night, however, the respiration of all these organisms may consume the oxygen in the water, and organisms that are unable to find their way out of the system will suffocate. A system must be highly eutrophic for such a situation to develop, but there is no doubt that large-scale kills of the sort described do occur from time to time in some systems.

ESTUARIES—A SPECIAL CASE

Estuaries are defined as semienclosed coastal bodies of water having a free connection with the open ocean, and within which seawater is measurably diluted by freshwater derived from land drainage (Lauff, 1967). Estuaries may be formed by the drowning of river valleys (Chesapeake Bay), by glacial scouring (Puget Sound), by the formation of barrier islands or sand spits (Pamlico Sound), or by tectonic processes (San Francisco Bay).

Regardless of their mode of formation, estuaries tend to be highly productive systems because of the nature of the estuarine circulation pattern that characterizes these systems. Since freshwater is less dense than saltwater of comparable temperature, there is a natural tendency in estuarine systems for the freshwater from land runoff to flow from the head to the mouth of the estuary along the surface, whereas seawater moves in and out with the tides along the bottom. Figure 3.7 depicts the general pattern of water movement in an estuary. As the curved arrows in Figure 3.7 indicate, there is invariably some upward mixing of seawater into the freshwater, so that some of the seawater that enters the estuary near the bottom flows back out near the surface. As a result there is a net outflow of water (freshwater mixed with some saltwater) at the mouth of the estuary in the upper water column, and a net inflow of seawater in the lower water column. If the flux of freshwater into the estuary at the head is large compared to the tidal in-and-out flux of seawater, there is generally a sharp demarcation between the freshwater at the top of the water column and the saltwater below. Due to the mixing of saltwater and freshwater, this sharp transition region gradually blurs as one approaches the mouth of the estuary. Such an estuary is commonly referred to as a *salt wedge estuary*, because of the shape of the saltwater wedge in the lower part of the water column when the estuary is viewed in profile (Figure 3.8).

If the in-and-out flux of the tides is large compared to the flux of freshwater, then freshwater and seawater tend to be mixed together thoroughly throughout the estuary, except of course in the immediate vicinity of the head. Such an estuary is commonly referred to as a *well-mixed estuary*. Undoubtedly many estuaries are best classified as having circulation patterns intermediate between those of typical salt

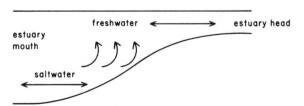

Figure 3.7   Simplified estuarine circulation pattern.

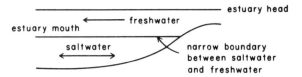

Figure 3.8 Profile of a salt wedge estuary.

wedge and well-mixed estuaries. The important point to bear in mind, however, is that there is a net outflow of water near the surface and a net inflow near the bottom at the mouth of all estuaries, regardless of whether the details of the circulation pattern correspond most closely to those of a salt-wedge, well-mixed, or intermediate type situation.

Because there is a net inflow and upward mixing of seawater at the bottom of an estuary, detritus that has sunk out of the mixed layer at the surface and regenerated nutrients from the deeper water are constantly being carried back into the estuary and mixed up into the surface waters. Plankton that drift out of the estuary on the surface current and that subsequently sink offshore or are eaten and then excreted offshore tend to be swept back into the estuary by the net influx of bottom water. Figure 3.9 depicts the cycling of nutrients and organic matter in an estuary as influenced by the estuarine circulation pattern. Thus the physical circulation pattern in estuaries provides a natural mechanism for recycling food and inorganic nutrients, and thereby maintains a high level of production in the system. Unfortunately pollutants introduced into an estuary tend to be recycled by the same circulation mechanism. From this standpoint alone estuaries are one of the last places one would choose to discharge pollutants, since the pollutants will not be conveniently washed out to sea and dispersed. Rather they will tend to be recycled repeatedly within the estuarine system. Admittedly there will be leakage to the open ocean, and some pollutants will tend to be trapped in the sediments rather than recycled in the water column. However, as a general rule the estuarine circulation pattern can be expected to exacerbate the impact of pollutants discharged into the estuary. Estuaries are susceptible particularly to problems associated with cultural eutrophication, because the estuarine circulation pattern tends to recycle discharged nutrients and hence magnify their impact on production and biomass.

Estuaries are naturally eutrophic systems because of the high efficiency with which the estuarine circulation pattern recycles nutrients. Because of their high productivity, estuaries such as those found along the Gulf Coast of the United States account for some important coastal fisheries. Furthermore, estuaries serve as breeding and/or nursery grounds for many organisms that one usually associates with the open ocean such as sharks and whales. Finally, estuaries are traversed by a number of migratory species that breed either in freshwater (e.g., salmon) or in saltwater (e.g., the American eel). Contamination of estuaries by any sort of pollution may therefore have much graver consequences for aquatic systems than would perhaps be apparent from a casual examination of the abundance and kinds of organisms present in the estuary at any one time. Many large population and industrial centers have developed adjacent to estuaries because of the easy access to both the ocean and to inland river systems for water transportation. Unfortunately the wastes from these large population/industrial centers have often been discharged carelessly into

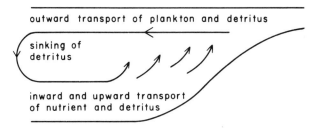

Figure 3.9 Cycling of nutrients and organic matter in a typical estuary.

estuarine waters, with little or no awareness of the biological importance of the estuary or of the tendency of pollutants to be recycled within the system. The following example is a particularly good illustration of the role of circulation in determining the impact of pollutant discharges on an estuarine system.

A CASE STUDY—ESCAMBIA BAY

Escambia Bay is a well-mixed estuary on the Gulf of Mexico near Pensacola, Florida. The principal stream flowing into the bay is the Escambia River, although several other streams also contribute freshwater input. Figure 3.10 shows the principal features of the bay.

During the 1950s Escambia Bay was an important commercial fishing area, with brown and white shrimp, oysters, and menhaden among the important species taken from its waters. By 1970, however, commercial fishing had disappeared from the bay, along with the benthic grass beds that had contributed both food and shelter for many of the organisms that had previously lived there. During this roughly 20-year period a number of industries had become established or expanded their operations around Escambia Bay, including a container plant, an electric power plant, and several chemical companies. All of these industries discharged wastes of one sort or another into the bay, including toxic chemicals such as polychlorinated biphenyls (Monsanto), sodium thiocyanate and acrylonitrile (American Cyanamid), heated wastewater (Gulf Power), and various organic wastes whose decomposition by microorganisms consumed oxygen in the bay.

The Louisville and Nashville (L&N) railroad trestle across the bay (Figure 3.10) was supported by wooden pilings that had to be periodically replaced. However, no attempt was made to remove the old, slowly rotting pilings from beneath the trestle when replacement pilings were driven in at nearby positions. As a result this transect across the bay gradually became cluttered with many wooden pilings, whose presence restricted tidal flushing and caused the northern part of the bay to become somewhat isolated from the rest of the system and relatively stagnant. The construction of an interstate (I-10) highway bridge across the bay adjacent to the railroad bridge further restricted circulation. Fill for the I-10 construction was taken from Mulatto Bayou, an extension of the bay on the eastern side and north of the railroad and highway bridges (Figure 3.10). Dredging for I-10 fill produced a number of deep holes in the bottom of the bayou. The circulation in Mulatto Bayou was sluggish even before this dredging and was further reduced by the restrictions to tidal mixing

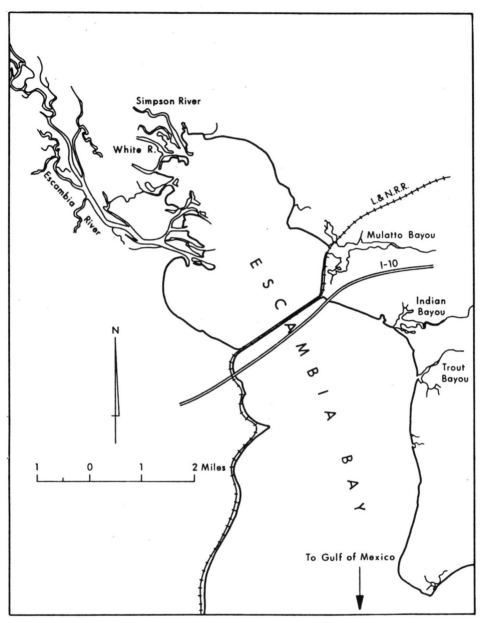

Figure 3.10 Escambia Bay and adjacent land areas.

Florida Wins Dubious Title: Nation's No. 1 Fish-Killer

By CHRISTOPHER CUBBISON
Of The Times Staff

Escambia Bay Was The Main Culprit In Boosting The State's Ascendancy To The Fish-Kill Crown

For the schools of unaware fish that swim through the inland and coastal waters of the United States, Florida is the most lethal spot to live.

Pollution from pesticides, fertilizers, manure, mining operations, improperly treated sewage and other sources killed an estimated 31.6-million Florida fish in 1971, making the state the nation's No. 1 fish-killer from pollution.

STATISTICS compiled by the U.S. Environmental Protection Agency (EPA) show that Florida waters killed almost twice as many fish from pollution as the next leading state, Texas, with a record 16.2-million fish reported killed in 1971, the last full year for which figures are available.

Florida now has led the nation in fish killed by pollution in two of the last three years. The national total of 74.3-million fish killed was 81 per cent higher than the previous high of 41-million reported in 1969 (federal records on fish kills date back to 1960). The 860 pollution-caused fish-kills were about a third higher than in 1970.

The overall figures do not represent all the fish killed by pollution, of course, because many small kills go unreported.

And the figures do not indicate whether the record total is due to better reporting by a concerned public or to greater fish-kills.

Escambia Bay and environs, often considered the most polluted waterway system in the state, was the main culprit in Florida's ascendancy to fish-kill supremacy.

Of 62 pollution-caused fish kills reported in Florida in 1971, 41, or exactly two-thirds, occurred in or around Escambia Bay.

The figures, significantly, do not include the millions of fish killed by the Tampa Bay area's long siege of Red Tide in the summer of 1971. Red Tide is caused by a natural organism, not from pollution.

OHIO HAD more than twice as many reported kills as Florida – 134 – but 12 massive Florida kills of over 1-million each gave the Sunshine State a huge margin over Ohio's 1971 total fish kill of 1.2-million.

All 12 massive Florida kills occurred in the Escambia Bay waterway system, including America's largest pollution-caused kill of 1971: 5.5-million on Aug. 22, 1971. None of the big kills could be traced to a single source.

In addition to the municipal sewage that pours into the Escambia Bay system, paper mills and other heavy industries contribute heavily to

(See FISH-KILL, 4-B)

(See FISH-KILL, 4-B)

Fish-Kills

From 1-B

the toxic environment for fish in the waters around Pensacola.

The largest kill outside the Escambia area occurred in Banana Lake in Lakeland, where an estimated 273,000 fish died March 29, 1971 from an overdose of sewage.

ELSEWHERE, two kills took place each in Avon Park, Sebring, Jacksonville and Lake Thonotosassa in Hillsborough County, where the largest pollution-caused U.S. fish-kill on record took place in 1969 – a staggering 26.5-million-victim debacle blamed on the discharge of wastes from food-processing plants.

Nationally, municipal sewage systems were the leading killer of fish in 1971. Almost 30 per cent of the fish killed nationally by pollution in 1971 literally suffocated in sewage.

Sewage was the cause of 19.6 per cent of the total fish kills in the southeast region, which includes Florida.

In Florida as the rest of the nation, more fish died in coastal waters than in lakes and rivers, with 44 of 62 kills occurring along the shore.

THE NUMBER of fish killed in estuarine waters like Escambia Bay rose from 44 to 77 per cent from 1970 to 1971 in the nation, a fact that "could be of great national concern since estuaries serve as nursery grounds for many species of marine fish," the report states.

However, the report adds the large increase results from a series of massive kills principally in Escambia Bay and Galveston Bay in Texas and says that "interpretation as a national trend, therefore, is not in order."

Figure 3.11 Article that appeared in the February 21, 1973 edition of the St. Petersburg Times. Reproduced with permission from the St. Petersburg Times.

in the northern bay caused by the railroad and highway bridges. Not surprisingly circulation in the deep holes dredged for the I-10 fill was extremely poor, and the water in the bottom of these holes frequently became anoxic. This condition was exacerbated by the construction of "dream homes" along Mulatto Bayou. The runoff from exposed ground on construction sites contained high nutrient concentrations that stimulated the growth of phytoplankton in the bayou. Dredging of canals to serve the dream homes created additional pockets of water that were subject to very poor circulation. Mulatto Bayou became a highly eutrophic system, characterized by poor circulation and extreme fluctuations in oxygen levels.

As a result of these conditions, 12 fish kills involving over one million fish each occurred in and around Mulatto Bayou during 1971 alone (Figure 3.11). One kill involved over 5.5 million fish (Cubbison, 1973). These kills resulted from fish's literally suffocating in the water during the night due to the lack of oxygen. The oxygen-consuming organic wastes discharged into the bay both from industries and from sewer outfalls, the runoff of nutrients into the bayou from construction sites, the poor circulation created in the northern bay and in Mulatto Bayou in particular, and the additional stress imposed on organisms by the discharge of toxic wastes undoubtedly all contributed to the situation that led to the fish kills. A spokesman for one of the industries suggested another explanation for the fish kills. He suggested that the fish might have committed suicide, since the low nighttime oxygen levels in Mulatto Bayou would surely have caused a normal fish to avoid the bayou at night.

It is unlikely, however, that the millions of fish that died in Escambia Bay were trying to commit suicide, nor is it likely that many of them swam into Mulatto Bayou at night. The fish were undoubtedly attracted to the bay by the abundance of food and probably swam into Mulatto Bayou during the day when oxygen levels in the euphotic zone were high. It is not difficult to imagine from an examination of Figure 3.10 that fish might have found it difficult to quickly escape from Mulatto Bayou when the oxygen concentration rapidly dropped at night. The dredging of boat canals (not shown in Figure 3.10) along the bayou created additional blind alleys, which compounded the escape problem. There is little doubt that the fish kills that occurred in Escambia Bay could have been largely prevented had some understanding of basic ecological principles been applied in the planning of developments in and around the bay.

REFERENCES

Cubbison, C. 1973. Florida Wins Dubious Title: Nation's No. 1 Fish Killer. St. Petersburg Times, Feb. 21.

Lauff, G. H., Ed. 1967. *Estuaries*. American Association for the Advancement of Science. Washington, D.C., 757 pp.

Smetacek, V., and U. Passow. 1990. Spring bloom initiation and Sverdrup's critical-depth model. *Limnol. Oceanogr.*, **35**, 228–234.

4

CULTURAL EUTROPHICATION— CASE STUDIES

This chapter is devoted to an in-depth study of three examples of cultural eutrophication. The examples include two freshwater lakes, Lake Washington and Lake Erie, and an estuary, Kaneohe Bay. Before beginning our study of these cases, it is appropriate to review some of the points covered in the first three chapters. In particular, what is cultural eutrophication, and what sort of problems may be created or intensified by cultural eutrophication?

Eutrophication is a natural process that occurs in virtually all bodies of water. The gradual accumulation of nutrients and organic biomass accompanied by increased levels of production and a decrease in the average depth of the water column caused by sediment accumulation constitute the natural eutrophication process. *Cultural eutrophication* is simply the anthropogenic acceleration of eutrophication. This anthropogenic acceleration is often brought about by discharges of organic wastes and/or nutrients. What sort of problems does cultural eutrophication cause?

Actually, cultural eutrophication may be beneficial to many aquatic systems. Indeed the deliberate fertilization of ponds or similar enclosed systems is a basic technique used in aquaculture to produce large crops of fish or shellfish. There is certainly nothing inherently wrong with stimulating production. Cultural eutrophication may create problems, however, if the system of interest is not properly managed and, in particular, if the increased production levels and associated effects are incompatible with other uses of the system that are considered more important. The following effects are frequently encountered in cases of problem cultural eutrophication:

1. Species associated with eutrophic systems are sometimes less desirable than species characteristic of oligotrophic systems. Cyanobacteria, which are frequently associated with organic nutrient enrichment, are a class of organisms frequently associated with undesirable water quality conditions. Problems created by large

cyanobacterial populations were discussed in Chapter 2. In Lake Erie the types of fish that have come to dominate the polluted areas of the lake are commercially less valuable than the species that previously inhabited the same areas. In Kaneohe Bay, a virtually worthless form of algae grew over and destroyed once-healthy coral reefs in some of the polluted sectors of the bay. From the human standpoint such changes in the biota of polluted systems are certainly undesirable. Although cultural eutrophication almost invariably increases the productivity of a system, the value (monetary, aesthetic, scientific, etc.) of the organisms produced frequently declines.

2. Oxygen concentrations in highly eutrophic systems generally fluctuate over a much wider range than is the case in oligotrophic or mesotrophic systems. Since it is impossible for some organisms to function efficiently unless the oxygen concentration in the water is near saturation, such organisms are often absent from eutrophic environments. Low oxygen concentrations may be seasonal phenomena, as in the hypolimnions of eutrophic lakes, or a noctural phenomenon in the water column of highly productive systems. Fish kills involving as many as several million fish are in some cases attributed largely to a rapid drop in oxygen concentrations during the night in relatively confined and highly eutrophic bodies of water such as Mulatto Bayou. Large-scale fish kills and the elimination of desirable species as a result of oxygen depletion may constitute a serious eutrophication problem in some aquatic systems.

3. Excessive amounts of phytoplankton and/or macroscopic plants in the water create aesthetic problems and reduce the value of the body of water as a recreational resource. From a purely aesthetic standpoint, crystal clear water characteristic of highly oligotrophic systems is most attractive for swimming or boating. High phytoplankton concentrations cause the water to appear turbid and aesthetically unappealing. Macroscopic plants can cover literally the entire surface of eutrophic ponds or lakes and consequently make the water almost totally unfit for swimming or boating. In such highly eutrophic systems the large plant biomass is often not entirely grazed off by herbivores. The death and decomposition of large amounts of plant biomass may create a foul smell and highly unaesthetic appearance. Such occurrences are clearly undesirable from the standpoint of recreational use of the body of water. It should be noted here that the various potential uses of a body of water, such as flood control, recreation, fish production, and irrigation, frequently conflict with one another. The high concentration of phytoplankton in a well-managed catfish pond, for example, would hardly be conducive to use of the pond for swimming; nor would the periodic raising and lowering of the water level in a reservoir used for flood control be desirable from the standpoint of recreational boating. The importance attached to the pros and cons of some eutrophication effects depends to a certain degree on the intended use of the body of water.

4. Because of the high concentration of organisms in a eutrophic system, competition for resources and predator pressure is often severe. This high degree of competition and predation combined with sometimes severe chemical and/or physical stresses make the struggle for survival in eutrophic systems particularly keen. As a result the diversity of organisms is frequently much lower in eutrophic than in oligotrophic systems. In other words, a smaller number of kinds of organisms are able to survive in eutrophic systems. For certain purposes, of course, it is desirable for the organisms in a system to be dominated by only a few species. Aquaculturists

routinely design and manage systems that produce one or only a few species of fish or shellfish. Such systems, although highly productive, are rather monotonous and uninteresting from the standpoint of the naturalist–scientist or nature lover.

CASE STUDY 1. LAKE WASHINGTON

Lake Washington is located in Washington state adjacent to Seattle and about 10 km east of Puget Sound (Figure 4.1). The lake has an approximate 88 km^2 area and a 33 m mean depth. Inflow comes primarily from the Cedar River to the south and from the Sammamish River to the north. Outflow occurs via two parallel locks to Puget Sound. The residence time of water in the lake (i.e., the ratio of the lake volume to the inflow rate of fresh water) is about 3.4 years. The lake is monomictic with a minimum recorded winter water temperature of 4.6°C. Overturning generally occurs from the end of November until April or May.

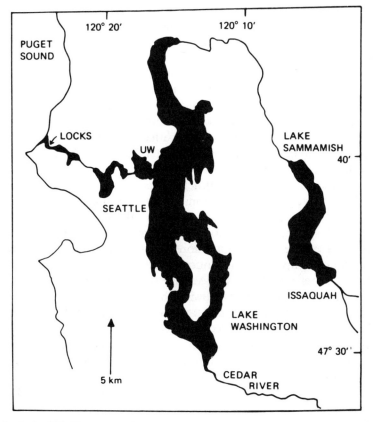

Figure 4.1   Lake Washington and surrounding area. [Redrawn from Edmondson (1969). Reproduced from *Eutrophication: Causes, Consequences, Correctives* (1969, p. 134) with permission from the National Academy of Sciences, Washington, D.C.]

History of Eutrophication

During the 1920s Lake Washington received increasing amounts of raw sewage from the growing Seattle metropolitan area. However, concern over public health problems caused this sewage to be diverted to Puget Sound by about 1936. The lake was used as a source of drinking water by some communities until as late as 1965. Beginning in 1941 a series of 10 secondary sewage treatment plants serving the Seattle metropolitan area was built with outfalls entering Lake Washington. As of 1941 the population served by these outfalls numbered about 10,000, but by 1957 the population had risen to 64,300 (Edmondson, 1972). In 1959 an additional sewage treatment plant was built with its outfall entering the Sammamish River. As of 1960 treated sewage from a population of about 70,000 people was being discharged into Lake Washington or its tributaries.

In addition to the treated sewage discharged into Lake Washington, additional sewage from the outflow of improperly operating septic tanks entered the lake from tributary streams. The actual amount of this septic tank overflow is not known accurately, but in 1957 it was estimated to be equivalent to the treated sewage from a population of about 12,000 persons (Edmondson, 1972). Finally sewage periodically entered Lake Washington from the combined sewer system of the city of Seattle. A combined sewer system consists of a single system of pipes for handling both street runoff and sanitary sewage. During dry weather all flow in the system consists of sanitary sewage, which in the case of Seattle was given secondary treatment before being discharged into Lake Washington. Secondary treatment, which is discussed in more detail in Chapter 6, usually consists of removal of at least 85% of the settleable and floatable solids and potentially oxygen-consuming wastes, and disinfection by means of chlorination. During storms the flow in a combined sewer system consists of both sanitary sewage and street runoff. If the street runoff is small the entire flow may be handled by the sewage treatment plant, but during heavy storms a portion of the combined flow is simply diverted from the treatment plant and discharged directly into the receiving body of water. This flow obviously contains raw sewage, both from the normal operation of the sanitary sewer system and from the so-called stranded filth (Weibel, 1969), defined as sewage that is left in the pipes during normal low-flow operating conditions and that is scoured out during storm-flow conditions. The amount of raw sewage entering Lake Washington from the Seattle combined sewer system during storms was estimated to be equivalent to the continuous input of sewage from a population of about 4500 persons in 1957 (Edmondson, 1972). Summing up the various sources of sewage entering Lake Washington as of roughly 1960, one concludes that sewage from a population of about 85,000–90,000 persons was entering the lake, with about 15–20% of this sewage being partially decomposed sewage from septic tanks and combined sewer overflow and the remainder being secondarily treated sewage.

The effect of the steadily increasing enrichment of Lake Washington with sewage is a bit hard to document in quantitative terms, since only two significant studies had been performed on the lake prior to 1959, the first by Scheffer and Robinson (1939) in 1933 and the second by Comita and Anderson (1959) in 1950. Both studies were of a general limnological nature and were not directly concerned with the question of pollution. However, as a result of these studies some information on water transparency, nutrient concentrations, and the abundance and characteristics of the phytoplankton community is available prior to the extensive studies by Dr. W. T. Edmondson and colleagues from approximately 1955 to 1985.

Scheffer and Robinson (1939) characterized the lake as "distinctly oligotrophic," although they noted several species of cyanobacteria and observed that *Anabaena lemmermanni*, a cyanobacterium, could be "abundant in early summer and very common again in fall." By 1950, when the studies of Comita and Anderson (1959) were performed, the lake had begun to show some definite signs of nutrient enrichment. For example, figures given by Edmondson (1969) indicate that winter phosphate concentrations were about 50% higher in 1950 than in 1933. In June of 1955 a bloom of the cyanobacterium *Oscillatoria rubescens*, a species frequently associated with sewage pollution, was noticed in the lake, and a series of studies on water quality conditions in the lake was begun shortly thereafter (Edmondson, 1969).

During this time public concern over conditions in Lake Washington had been growing, primarily as a result of the obvious reduction in summer water clarity that accompanied the steady increase in phytoplankton abundance during the 1950s. The concern over water clarity was based primarily on aesthetic considerations, since the lake was used extensively for swimming and boating activities by local residents. One simple measure of water clarity is the Secchi disk transparency or simply the Secchi depth. A Secchi depth measurement is made by lowering a white disk into the water until it just disappears from sight. The disk is then lowered into the water a bit more and then raised until it just becomes visible. The mean of the two depths, disappearance depth and reappearance depth, is taken to be the Secchi depth. In limnological and oceanographic work the diameter of a Secchi disk is usually 20 cm and 30 cm, respectively. Obviously one would expect a Secchi disk to be more visible at greater depths in clear water than in turbid water. Figure 4.2C shows the results of summer (July–August) Secchi depth measurements made in Lake Washington from 1950 to 1979. The mean Secchi depth in 1950 was 3.7 m, but had declined to about 2 m by 1955 and had dropped to 1 m by 1963. Thus almost a fourfold reduction in water clarity had occurred from 1950 to 1963. Suffice it to say that water with a Secchi depth of 1 m is not particularly enticing to swimmers.

Figure 4.2D shows that during the same 13-year period the average summer chlorophyll *a* concentration increased from less than 3 mg m^{-3} to almost 35 mg m^{-3}, and winter phosphate levels increased from 14 to 55 mg m^{-3}. It is reasonable to conclude from these graphs and from what we have learned in the first three chapters that the high winter phosphate concentrations were assimilated by the phytoplankton during the spring and summer, so that the concentration of phytoplankton was roughly proportional to the concentration of phosphate in the winter. The fact that the chlorophyll *a* concentration increased almost 13-fold while winter phosphate levels increased only fourfold probably reflects adaptation on the part of the phytoplankton community to the reduced water clarity. As the irradiance experienced by phytoplankton cells is reduced, the cells tend to increase their content of chlorophyll *a* and other light-absorbing pigments so as to capture as much of the available light as possible (Laws and Bannister, 1980). Certainly much of the decrease in summer water clarity can be attributed to the high summer phytoplankton population. A liter of water with a chlorophyll *a* concentration of 35 mg m^{-3} is visibly green. Thus the postulated sequence of events is as follows:

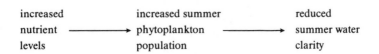

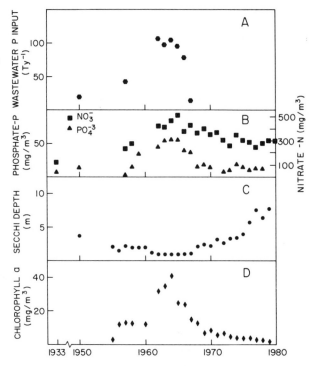

Figure 4.2   Changes in Lake Washington between 1933 and 1980. (A) Annual loading of P to the lake. (B) Annual means of phosphate-P and nitrate-N. (C) Mean Secchi depth during July and August. (D) Mean Chlorophyll *a* in top 10 m during July and August. [Source of data: Edmondson and Lehman (1981).]

Following similar but more quantitative and detailed reasoning, Seattle voted in 1958 to spend $121 million to develop an effective sewage disposal system for the entire area (Edmondson, 1969). Although some of this money was spent to stop discharges of raw sewage into Puget Sound and for other improvements, the majority of the funds was spent to improve the water quality of Lake Washington. The plan developed by the Municipality of Metropolitan Seattle (METRO) called for diverting all of the secondary sewage discharged into Lake Washington to Puget Sound, where tidal mixing and dilution were expected to prevent the development of objectionably high phytoplankton populations. Even though the vote for this project took place in 1958, the first diversion did not occur until February 20, 1963, and the last sewage treatment plant diversion did not occur until February, 1968.

Effects of Sewage Diversion

The effects of the sewage diversion on the lake can be seen from the data graphed in Figures 4.2–4.4. By the early 1970s summer chlorophyll *a* concentrations had dropped to about 7 mg m^{-3} and had further declined to about 3 mg m^{-3} by the end of the decade. In 1950 less than 20% of the July–August phytoplankton crop consisted of cyanobacteria (Comita and Anderson, 1959), but by 1962 this figure had increased to 96% (Edmondson, 1966) and remained at approximately that level until

1973 (Figure 4.3). The contribution of cyanobacteria to the phytoplankton population declined rapidly in the next few years, and by the end of the decade was comparable to or less than the levels observed in 1950. Total phosphorus concentrations dropped from about 70 mg m^{-3} between 1963 and 1964 to roughly 17 mg m^{-3} by 1971 and in subsequent years have remained at that level, which is virtually identical to the concentration measured in 1933. Secchi depths increased from a low of 1 m just prior to sewage diversion to over 2 m between 1968 and 1970 and to an average of 3.4 m between 1971 and 1975. Thus after a period of about 5–10 years the lake in many respects had returned to a condition similar to what existed before eutrophication became a serious problem. The concentration of phosphorus had been reduced to a level similar to that reported in 1933, and both the abundance of cyanobacteria and Secchi depths were comparable to values observed in 1950. Considering the fact that the residence time of water in the lake is about 3.4 years, the time course of this response seems reasonable.

In 1976, however, the July–August Secchi depth increased to 5.7 m and has averaged over 7 m in subsequent years. This dramatic increase in water clarity was apparently caused by an increase in the abundance of certain species of *Daphnia*, commonly known as *water fleas*. The *Daphnia* are herbivorous, and, "Phytoplankton becomes relatively scarce at times of *Daphnia* abundance" (Infante and Edmondson, 1985, p. 161). The cause of the increase in *Daphnia* abundance is somewhat problematic, but, "Because of the timing, there is no reason to think that the success of *Daphnia* can be attributed directly to cleaning up the lake" (Edmondson and Litt, 1982, p. 284). At least two events seem to have been responsible for the increase of *Daphnia* since 1976. The first event was a rapid decline in the abundance of the large carnivorous zooplankter *Neomysis mercedis* in the mid-1960s. The decline of *Neomysis* does not appear to have been related to diversion of the sewer outfalls, but may have been caused by an increase in the abundance of the long-fin smelt, which is known to feed in large part on *Neomysis* in Lake Washington. In any case the abundance of *Neomysis*, a *Daphnia* predator, would have made establishment of a sizeable *Daphnia* population difficult prior to the mid-1960s. The second important event was the decline of filamentous cyanobacteria of the genus *Oscillatoria* beginning in approximately 1973 (Figure 4.3). *Daphnia* cannot digest *Oscillatoria*, but the cyanobacteria accumulate in *Daphnia*'s feeding mechanism and must be ejected periodically. Inevitably, some edible algae are simultaneously ejected, and the act of rejection is metabolically costly (Edmondson and Litt, 1982). The presence of *Oscillatoria* therefore interferes with the ability of *Daphnia* to feed and grow. The increase of *Daphnia* is closely correlated with the decline of *Oscillatoria* and other cyanobacteria in the mid-1970s.

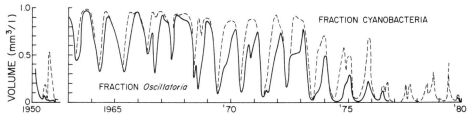

Figure 4.3   Fraction of total phytoplankton in Lake Washington composed of cyanobacteria (---) and the genus Oscillatoria. [Redrawn from Edmondson and Litt (1982).]

The rapid decline of the phosphate concentration in Lake Washington as a result of the sewage diversion provides an interesting contrast to the much less dramatic change in nitrate levels (Figure 4.2). During the winter overturning period the concentration of both nutrients in the lake is high, a characteristic of monomictic temperate lakes. The rapid decline of both elements during the summer stratification period attests to the efficiency with which phytoplankton can take up nutrients when there is adequate light to support photosynthesis. The winter increase in phosphate is clearly much less pronounced in 1969 than in 1963, although still substantially higher than the winter buildup in 1933. The 1969 winter nitrate concentration, however, was quite comparable to the winter nitrate level in 1963, and by 1980 winter nitrate-N concentrations were still averaging over 300 mg m^{-3}. What is the explanation for the relatively modest response of the winter nitrate concentrations to sewage diversion?

The explanation is that the sewage being discharged into Lake Washington contributed a minor percentage of the nitrogen input to the system. This condition is explained by the fact that the sewage delivered N and P in a ratio of about 7 by atoms, whereas the N:P ratio in streams entering the lake was about 43. In the year (1962) before the sewage diversions began, sewage effluent contributed about 72% of the total P input to Lake Washington but only 35% of the total N (Edmondson and Lehman, 1981). The average winter phosphate and nitrate concentrations in Lake Washington from 1975 to 1980 were in fact 72% and 36%, respectively, lower than the corresponding 1962–1963 values. Thus the changes in the nutrient concentrations of the lake are in fact quite consistent with estimates of the contribution of the sewage effluent to nutrient loading just prior to the initiation of the diversion program.

Nutrient Limitation

A good picture of the nutrient limitation situation in Lake Washington may be obtained from an examination of phosphate and nitrate levels in the lake during the development of spring blooms. As the high winter nutrient concentrations are assimilated by the rapidly growing phytoplankton, the concentration of both nitrate and phosphate drops to a low level (Figure 4.4). Figure 4.5 shows plots of nitrate versus phosphate during the spring bloom period in each of 5 years. In 1933 and 1967 phosphate and nitrate were exhausted almost simultaneously, and the values therefore extrapolate to a point close to the origin. In 1968 and 1969 phosphate was exhausted first, and the values therefore extrapolate to a point on the positive nitrate axis. However, in 1962 just before the sewage diversion began, nitrate was exhausted first, and the results therefore extrapolate to a point on the positive phosphate axis. The steady shift of the lines across Figure 4.5 from right to left during the period 1962–1969 may be interpreted as a shift in the system from nitrate limitation during sewage pollution to phosphate limitation following sewage diversion. The lake evidently became nitrate limited during the period of sewage pollution due to the low N:P ratios in the sewage and phosphorus limited following sewage diversion because of the high N:P ratio in the stream inputs. Recall that phytoplankton utilize N and P in ratios ranging from roughly 3 to 30 by atoms and that a ratio of about 16 is associated with optimal growth conditions. Judging from Figure 4.5 Lake Washington was more clearly phosphate limited in 1968 and 1969 than in 1933. This observation is consistent with the fact that the ratio of winter nitrate and phosphate concentrations increased from 37 by atoms in 1933 to 48 in 1969.

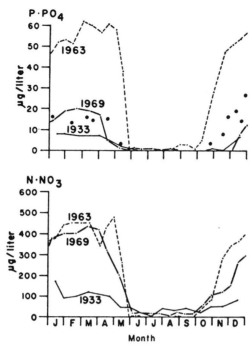

Figure 4.4 Seasonal changes in phosphate and nitrate in Lake Washington surface water during 1933, 1963, and 1969. Isolated dots are taken from 1950. [Redrawn from Edmondson, (1972).]

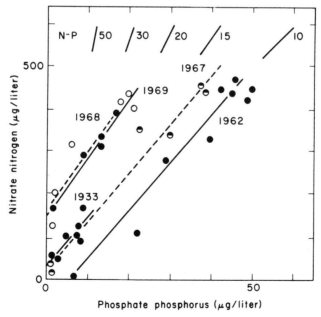

Figure 4.5 Correlation between surface values of phosphate and nitrate during the spring phytoplankton bloom when the concentrations of nutrients are decreasing. The lines were fit to the data by eye. Nitrogen:phosphorus ratios are shown by the numbered lines radiating from the origin. [Redrawn from Edmondson (1970). Copyright © 1970 by the American Association for the Advancement of Science.]

Oxygen Depletion

Despite the high productivity levels in Lake Washington during the period of sewage enrichment, severe oxygen depletion problems never occurred, primarily because of the large volume of the hypolimnion relative to the euphotic zone. The Secchi depths reported for the lake during the summer suggest that the euphotic zone probably did not extend much deeper than about 3–6 m between 1950 and 1970, whereas the hypolimnion extended from roughly 10 m to the bottom (mean depth = 33 m). Nevertheless, a steady decline in the hypolimnetic oxygen concentration was apparent following the winter overturning period even in 1933 (Figure 4.6). This decline was of course due to the respiratory activity of organisms in the hypolimnion and the fact that a stratified water column permits very little downward movement of oxygen from the surface waters. The decline in oxygen concentration in 1950 was quite comparable to the decline in 1933, but by 1955 a noticeable increase in the rate of oxygen consumption was apparent (Figure 4.6). Table 4.1 shows the mean oxygen concentration late in the summer in the hypolimnion of Lake Washington for 5 different years. The rather steady decline in late summer hypolimnetic oxygen levels during the period of rapid cultural eutrophication in the 1950s is evident from this table. The data suggest that the hypolimnion of Lake Washington would have become anoxic in the late summer had cultural eutrophication continued at the same rate for approximately another 20 years.

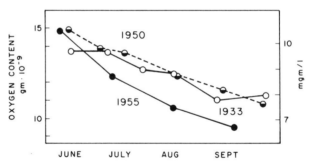

Figure 4.6   Total content and mean concentration of oxygen in the hypolimnion of Lake Washington during three summers. The volume of the hypolimnion is 1.41×10^9 m³. [Redrawn from Edmondson (1956).]

Table 4.1   Mean Hypolimnetic Oxygen Concentration Late in the Summer in Lake Washington

Year	Mean oxygen concentration (mg/m³)
1933	8.0
1950	8.1
1955	6.9
1957	4.6
1958	5.1

Source. Edmondson (1961).

Water Clarity

Since concern over water clarity was the principal motivation for diverting the sewage from Lake Washington, it seems appropriate to examine this issue in some detail. Figure 4.7 shows the empirical relationship between summer chlorophyll *a* concentrations and Secchi depths based on data summarized by Edmondson and Lehman (1981). It is apparent from this figure that Secchi depths in Lake Washington are very insensitive to chlorophyll *a* concentrations greater than about 15 mg m^{-3} and highly sensitive to chlorophyll *a* concentrations in the approximate 1–10 mg m^{-3} range. Based on Figure 4.7 it seems fair to say that a reduction of summer chlorophyll *a* concentrations by a factor of two, for example, from prediversion values would have produced no discernable effect on water clarity in Lake Washington. Reduction of the chlorophyll *a* concentrations by at least a factor of 3–4 was necessary to produce a significant impact on water clarity. Just prior to the first sewage diversion, the total N to total P ratio in all inputs to Lake Washington was about 15, which is very close to the ideal ratio for phytoplankton. It was therefore reasonable to assume, at least as a first approximation, that the reduction in phytoplankton biomass caused by a diversion of the sewer outfalls would be directly proportional to the reduction in P loading. Since wastewater accounted for 72% of the P loading at that time, the expected reduction would be a factor of $1/(1 - 0.72) = 3.6$, which is comparable to the estimated minimum reduction needed to significantly improve water clarity. The fact that the actual reduction in chlorophyll *a* has been closer to a factor of 10 reflects changes in the composition of the zooplankton community (i.e., increased abundance of *Daphnia*) and (probably) a reduction in the chlorophyll *a* content of the phytoplankton in response to the improved water clarity. The former effect was not anticipated; the latter could have been predicted, at least in a qualitative way. In any case it was important at the time the sewage diversions were being considered that knowledgeable and respected scientists were able to say with some confidence that diverting the sewage would indeed improve water quality. Otherwise it would have been difficult to justify spending over $120 million to finance the project.

Theoretical Predictions

Predicting how an aquatic system such as Lake Washington will respond to a change in its nutrient loading rate is a problem that limnologists have discussed and debated for many years. Among the theoretical models most commonly applied to such problems are those developed by Vollenweider (1968, 1969, 1975, 1976). Pertinent equations can be derived by considering the simple box model depicted in Figure 4.8. If it is assumed that the system is in steady state, then the rate of nutrient addition to the lake must be balanced by losses due to sedimentation and washout. It is assumed that per unit time a fraction f_w of the phosphorus in the lake washes out and that a fraction f_s sinks to the bottom and is buried in the sediments. The decrease in phosphorus concentration P per unit time due to washout and sedimentation is therefore $(f_w + f_s)P$. The loading rate L is commonly expressed per unit of lake surface area, and the increase in phosphorus concentration per unit time due to external inputs is therefore L/Z, where Z is the average depth of the lake. In the steady state $L/Z = (f_w + f_s)P$ or $P = L/[Z(f_w + f_s)]$. One therefore predicts that the average concentration of phosphorus in the lake will be directly proportional to the loading rate and inversely proportional to the depth of the lake and the fractional

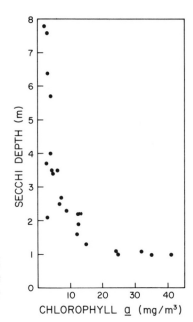

Figure 4.7 Relationship between Secchi depth and mean July–August chlorophyll in upper 10 m of the water column. [Source of data: Edmondson and Lehman (1981).]

loss rate of phosphorus to sedimentation and washout. Pertinent data for Lake Washington are summarized in Table 4.2. Vollenweider (1976) defined somewhat arbitrarily the transitions from oligotrophy to mesotrophy and mesotrophy to eutrophy as occurring at total P concentrations of 10 and 20 mg m^{-3}, respectively. By these standards, Lake Washington was clearly eutrophic prior to the diversion of the sewer outfalls, but became mesotrophic during the postdiversion period.

Toward the end of the Lake Washington sewage diversion effort, METRO appropriated an additional $3 million to initiate a similar wastewater diversion program for Lake Sammamish, which had begun to develop cultural eutrophication problems. The sewage effluent from the city of Issaquah and the waste from a dairy were both diverted in 1968. The lake was monitored from 1964 to 1966 to obtain baseline data and from 1969 to 1975 to determine the rate and extent of its recovery. Results are summarized in Table 4.2.

The response of Lake Sammamish was not as dramatic as that of Lake Washington, and an examination of the P loading rates indicates one of the reasons for the difference. The sewage effluent and dairy plant wastewater accounted for only about 35% of the P loading of Lake Sammamish in 1968. In the case of Lake Washington sewage effluent accounted for approximately 70% of the external P inputs at the time the diversions began. Vollenweider's model indicates that Lake Sammamish was on the borderline between mesotrophy and eutrophy even after the diversion program. By 1975 the total P concentration in Lake Sammamish had declined from a prediversion mean of 33 mg m^{-3} to a value of 19.8 mg m^{-3} (Welch et al., 1980), the latter figure being virtually identical to Vollenweider's transition level of 20 mg m^{-3}.

From an aesthetic standpoint, Lake Sammamish showed very little response to the wastewater diversion. The effect most obvious to nearby residents was a decline by almost a factor of two in the percentage of the phytoplankton accounted for by cyanobacteria. As a result scums of cyanobacteria are no longer observed along the shoreline in late autumn (Welch et al., 1980). However, chlorophyll *a* concentrations and Secchi depths were almost unchanged by the diversion, and continued to average about 6.6 mg m^{-3} and 3.3 m, respectively. Hence despite the fact that postdiversion total P concentrations were comparable in Lake Washington and Lake Sammamish, the postdiversion phytoplankton population was substantially higher and water clarity correspondingly lower in Lake Sammamish. Furthermore, the hypolimnion continued to become anoxic for several months prior to the fall overturn.

Several considerations explain why Lake Sammamish did not respond as dramatically as Lake Washington to wastewater diversion. First, of course, is the fact that wastewater accounted for only about one-third of the P loading of Lake Sammamish. Second is the fact that Lake Sammamish is only about half as deep as Lake Washington. Although prediversion summer chlorophyll *a* concentrations were 5–6 times higher in the euphotic zone of Lake Washington, the hypolimnion of Lake Washington never became anoxic. The greater susceptibility of the hypolimnion of Lake Sammamish to anoxia is undoubtedly related to its smaller size. Once hypolimnetic waters become anoxic, it is no longer possible to trap phosphorus in the sediments with iron, and large amounts of P therefore may be released to the overlying waters. The transition from oxic to anoxic conditions in the hypolimnion of a lake, even for only a few weeks, is therefore a crucial event in the eutrophication process. It is a transition that is not easily reversed.

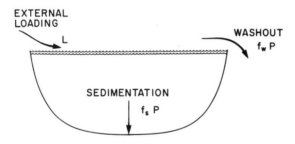

Figure 4.8   Vollenweider model of phosphorus fluxes in a lake.

Table 4.2   Characteristics of Lake Washington and Lake Sammamish relevant to Vollenweider's Model of Lake Eutrophication

	Washington	Sammamish
Area (km^2)	87.6	19.8
Z (m)	32.9	17.7
f_s (y^{-1})	0.83	1.2
f_w (y^{-1})	0.29	0.56
L (mesotrophic → eutrophic) (g m^{-2} y^{-1})	0.74	0.70
L (oligotrophic → mesotrophic) (g m^{-2} y^{-1})	0.37	0.35
L (prediversion) (g m^{-2} y^{-1})	1.71	1.02
L (postdiversion) (g m^{-2} y^{-1})	0.64	0.67
P (prediversion) (g m^{-3})	65	33
P (postdiversion) (g m^{-3})	17.4	19.8

CASE STUDY 2. LAKE ERIE

Lake Erie has been turned into a cesspool. RENE DUBOIS. [Cited by Cy Adler (1973) in *Ecological Fantasies*, p. 111]

Lake Erie represents the first large-scale warning that we are in danger of destroying the habitability of the Earth. Mankind is in an environmental crisis, and Lake Erie constitutes the biggest warning. BARRY COMMONER [Cited by Cy Adler (1973) in *Ecological Fantasies*, p. 111]

You see, Lake Erie has died. The lake can no longer support organisms that require clean, oxygen-rich water. Much of this shallow body of water is a stinking mess—more reminiscent of a septic tank than the beautiful lake it once was . . . No one in his right mind would eat a Lake Erie fish today. PAUL EHRLICH (1968, in *The Population Bomb*, p. 39)

Once clear and filled with valuable fish and game, Lake Erie today is the embodiment of all that can go wrong in an aquatic system. The more desirable species of fish have disappeared; the once clear water is filled with excessive numbers of microorganisms; mats of filamentous algae at times cover whole square miles of lake surface; swimming is impossible in many places because of the quantity of untreated sewage in the water and the decaying vegetation covering the beaches; oil scums often cover harbors and coves, making boating and water sports unpleasant. RICHARD WAGNER (1974, in *Environment and Man*, pp. 128–129)

Despite some pollution, it should be obvious . . . that Lake Erie, with its enormous fish populations, its low bacterial count, and enormous quantities of clean, potable water, is far from dead. Millions of Americans living near the lake swim, fish and boat in Erie's waters each year. CY ADLER (1973, in *Ecological Fantasies*, p. 119)

The preceding statements indicate that a certain amount of disagreement exists regarding the quality of Lake Erie's waters. Let us therefore begin this discussion by trying to make an intelligent evaluation of the condition of Lake Erie and of the changes that have occurred as a result of human activities.

Lake Erie is a glacially scoured depression that was filled with water from glacial runoff about 12,000 years ago. With an area of 25,820 km², it is the thirteenth largest lake in the world by area. Its 18.5 m mean depth, however, makes it the shallowest of the Great Lakes. The western, central, and eastern basins of the lake are somewhat arbitrarily defined by lines drawn between Point Pelee (Bar Point) and Sandusky, Ohio, and between Long Point and Erie, Pennsylvania (Figure 4.9). The western basin is the shallowest basin, with a mean depth of about 7.4 m. The central and eastern basins have mean depths of 18.5 m and 24.4 m, respectively. The major inflow to Lake Erie comes from Lake Huron via the St. Clair and Detroit rivers, and the outflow is of course to Lake Ontario via Niagra Falls and the Welland Canal. The residence time of water in the lake is about 2.5 years. Lake Erie is dimictic and invariably freezes over in the winter.

The various ecological stresses that have been imposed on the biota of Lake Erie over approximately the last 200 years cannot all be classified as cultural eutrophication, although there is no doubt that cultural eutrophication has had a significant impact on the system. Unfortunately one sometimes gets the impression from newspaper articles or similar news sources that removing phosphates from detergents (for example) would solve all of Lake Erie's problems. The following summary

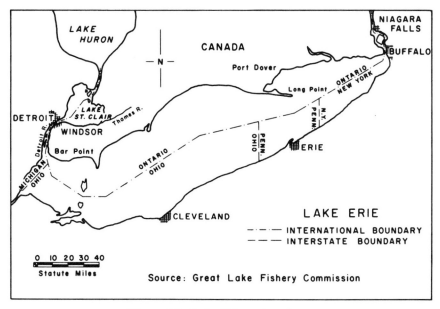

Figure 4.9 Lake Erie and environs.

of ecological stresses imposed on the lake will hopefully give you a better perspective on the complex problems that have been created as a result of human activities in the Lake Erie watershed.

The Destruction of Fish Spawning and Nursery Grounds as a Result of Land Use Modifications

As of the eighteenth century, when the human population in the Lake Erie watershed was less than 0.1% of the present population, the land in the watershed supported large forests, primarily beech-birch, maple-hemlock, and oak-hickory associations (Regier and Hartman, 1973). Interspersed between the forests were savannahs of grass and wild oats. The lake was bordered in many areas by marshes, the largest being the Great Black Swamp, which covered nearly 10,000 km² at the southwest corner of the lake. Because of the extensive vegetative cover, soil erosion was minimal, and streams flowing into the lake were generally clear. By roughly 1870 most of the forests had been cleared, the grasslands burned, and some of the swamps drained. Much of this modified land was given over to farming. Clay and silt eroded from exposed land and washed into streams and nearshore lake areas ruined the spawning grounds of many fish such as the walleye and lake whitefish. As a result of the increased sediment runoff, water quality in nursery marshes and bays declined, as did the biomass of aquatic vegetation, which provided food and shelter for the juvenile fish in these areas. By 1900 nearly all of the swamps, which had provided important nursery and spawning areas for the fish, had been drained. Furthermore, hundreds of mill dams prevented or impeded fish such as walleye and sturgeon from reaching their traditional river spawning areas. Thus the reproductive success of numerous species of fish was greatly reduced as a result of human activities that eliminated or seriously degraded the quality of traditional nursery and spawning areas.

The Depletion of Fish Stocks Due to Overfishing

Although other factors undoubtedly contributed to the decline of certain species of fish in Lake Erie, the reduction of population numbers in many cases is strongly correlated with a known increase in fishing effort. Indeed, even in the absence of other ecological stresses, it seems likely that the numbers of certain valuable fish species would have been greatly reduced as a result of overfishing alone. The problem of regulating fishing efforts has been compounded by the fact that four states (Michigan, Ohio, Pennsylvania, and New York) and one Canadian province (Ontario) have jurisdiction in determining fishing practices. Unfortunately fish are mobile creatures, and do not recognize state or national boundaries. As a result the fish resources of Lake Erie are to a certain degree held in joint or common possession by several parties. It is too often the case that when valuable resources are held in common ownership, the owners find it difficult to agree on a rational policy of utilization, the result being that an every-man-for-himself attitude prevails. Although relations between Canada and the United States are generally good, the so-called tragedy of the commons (Hardin, 1968) has undoubtedly influenced fishing policies in Lake Erie. The following is a summary of the changes that have occurred in the commercial fishing picture since roughly 1820.

The commercial fishery did not begin to develop around Lake Erie until after the War of 1812. Until that time basically subsistence fishing was practiced in the tributary streams and nearshore waters of the lake. The failure of any significant commercial fishery to develop until roughly 1820 can be attributed to the following factors:

1. A lack of sophisticated methods for catching fish.
2. A lack of reliable refrigeration methods to preserve fish for market.
3. An inadequate transportation system for getting fish to market.

During the period from roughly 1820 until 1890 a number of economic and technological changes resulted in the steady growth of the Lake Erie fishery at the rate of about 20% per year.

The development of reliable refrigeration methods during the midnineteenth century eliminated the need for smoking and salt curing as a means of preservation and facilitated the transportation of fish without spoilage over greater distances to market. During the 1820s and 1830s the opening of canals greatly expanded the potential range of markets available to Lake Erie fishermen. The Erie Canal joining Buffalo with the Hudson River at Albany was completed in 1825; the Welland Canal connecting Lake Erie and Lake Ontario was opened in 1829; and the Erie–Ohio Canal joining the Ohio River with the Maumee River upstream from Toledo was completed in 1832. In the latter half of the nineteenth century the development of an extensive railroad network eliminated the dependence of the fishery on canal transportation, reduced the transit time to markets, and further expanded the range of potential markets.

Prior to 1850 most Lake Erie fishing was done with hooks, seine nets, or small stationary gear such as traps and weirs. During the 1850s fishermen began to use gill nets and pound nets in nearshore waters. During the next 20 years the use of these nets was extended further offshore, and by 1880 fishing gear was used throughout the lake (Regier and Hartmen, 1973). In 1905 the use of a large and very efficient gill net, the *bull net*, was first introduced in the Lake Erie fishery. Due to its great efficiency the bull net became highly popular, but because of its nonselectivity the bull net was eventually outlawed during the 1930s. The efficiency of gill nets was greatly increased in the early 1950s by the substitution of nylon for the traditional cotton or linen as a netting material. Furthermore, since nylon is much less subject to attack by microorganisms than cotton or linen, it was no longer necessary for fishermen to rack and dry their nets every few days. As a result the amount of fishing time per unit of gear was greatly increased.

Over the last 100 years, the annual commercial catch of fish from Lake Erie has remained remarkably constant at roughly 22 million kilograms per year, a figure that has equaled about 75% of the commercial catch from the other Great Lakes combined. However, the principal kinds of fish caught and their commercial value has varied greatly with time. In the early nineteenth century the lake was inhabited by large numbers of the generally preferred species of food and game fish. In nearshore waters these species included smallmouth and largemouth bass, muskellunge, northern pike, and channel catfish. The important offshore species included lake herring (cisco), blue pike, lake whitefish, lake sturgeon, walleye, sauger, freshwater drum (sheepshead), white bass, and (primarily in the eastern end of the lake) lake trout. Of these species, the lake sturgeon and lake trout were among the first to suffer drastic population declines. The sturgeon was not valued commercially until the 1860s, when the knowledge of how to smoke it, render its oil, and make caviar from its eggs and isinglass from its air bladder was acquired from a European immigrant (Regier and Hartman, 1973). Prior to that time the lake sturgeon had been considered a nuisance fish, since its large size and external bony armor enabled it to tear nets set for smaller fish. To combat the lake sturgeon Lake Erie fishermen had devised larger mesh and stronger nets specifically designed for its capture. However, they made no attempt to utilize the lake sturgeon they caught, but simply disposed of the fish, often by burning piles of them on the beach. The effort to catch lake sturgeon intensified during the 1860s and 1870s after the fish's commercial value became known. Unfortunately the lake sturgeon is a slow-growing fish, which becomes sexually mature only at an age of 15–25 years, and is rather easily taken with most types of fishing gear. As Figure 4.10 shows, the catch of lake sturgeon declined steadily after 1880 and by 1960 had reached a commercially insignificant level. How much of this decline can be attributed to overfishing and how much to other causes is problematic. The construction of mill dams, for example, interfered with the ability of the sturgeon to reach river spawning grounds. However, there is no doubt that the slow growth rate and long maturation period of the lake sturgeon combined with its vulnerability to most types of fishing gear made it a particularly susceptible species to overfishing effects.

As the lake sturgeon population declined after 1880, fishing pressure on other species intensified. As Figure 4.10 shows, the catch of lake trout began to drop precipitously in about 1880, and despite some pronounced peaks due to good year classes, the lake trout catch had become insignificant by 1930–1940. Although it is

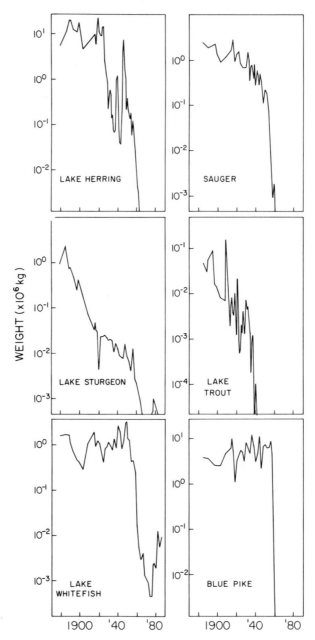

Figure 4.10 Commercial catch of six species of fish whose populations suffered serious declines in Lake Erie. [Source of data: International Board of Inquiry (1943) and Baldwin et al. (1979).]

again impossible to evaluate accurately the impact of fishing pressure on the lake trout population, it seems likely that the irregular decline in annual catches between roughly 1880 and 1935 was due in no small part to overfishing. However, the elimination of native lake trout from Lake Erie after about 1940 was brought about almost certainly by deterioration of the environment as well as by overfishing.

During the period between 1890 and 1910, U.S. sports fishermen succeeded in having a number of species declared off limits to commercial fishermen. In Lake Erie the restricted species included smallmouth and largemouth bass, northern pike, and muskellunge. However, similar restrictions were not adopted by the Canadian government, apparently because the large number of lakes in Ontario provided an adequate resource for Canadian anglers. Nevertheless, the U.S. restrictions provided these species with at least some refuge from commercial fishing pressure.

Figures 4.10 shows annual commercial catch statistics for several species of fish that suffered serious population declines. The whitefish, sauger, and blue pike catches all show precipitous declines beginning around 1960. Recalling the substitution of nylon for cotton or linen in the manufacture of gill nets during the early 1950s, one is tempted to attribute these rapid declines to overfishing pressure associated with the use of the more efficient nets. However, cultural eutrophication effects as well as other forms of pollution had already become serious problems in Lake Erie, and it therefore seems unwise to attribute these declines entirely or even primarily to overfishing. A more realistic assessment might conclude that the increase in fishing pressure during this time may have triggered the rapid decline of populations that were already severely stressed from pollution and the elimination of many spawning and nursery grounds. The decline in the lake herring (cisco) catch began in about 1920 and continued rather steadily thereafter, with the exception of a few good years around 1945. It seems likely that some of this decline has been due to overfishing, although a quantitative assessment is again impossible.

The Creation of Anoxic Bottom Water Conditions Due to Cultural Eutrophication

As the shallowest of the Great Lakes, Lake Erie has always been the most productive by far and, at the same time, the lake most likely to be adversely affected by cultural eutrophication. The human population in its watershed rose from about three million in 1900 to almost 15 million by 1980, a more rapid increase than in any other Great Lakes watershed. The major cities in the watershed are Detroit, Cleveland, and Erie. The sewage and other wastes from these population centers have been routinely discharged into Lake Erie. Between 1930 and 1965 the average nitrogen and phosphorus content of Lake Erie's water increased by about a factor of 3, as did the average biomass of phytoplankton. The concentration of all major ions in the water increased significantly during the first half of the twentieth century, with chloride and sulfate, which are conspicuous components of wastewater, showing the greatest increases (Beeton, 1965). Although the average biomass of phytoplankton in the lake increased by a factor of 3, the rate of phytoplankton production increased by roughly a factor of 20 over the same time period, a consequence primarily of longer duration of the spring and fall blooms (Regier and Hartman, 1973). The increase of biomass and productivity of the surface waters was reflected by an increase in the fallout of detritus into the hypolimnion and a concomitant increase in the respiration rate of the detritus food chain. Much of the increased oxygen demand was apparently associated with detritus that settled to the bottom of the lake and was decomposed by microorganisms living on or in the sediments. As of 1953, 28 days of calm weather were required for oxygen levels in the shallow western basin to drop below 3 ppm.[1] By 1963 only 5 days of similar weather were required to produce the same effect (Beeton, 1969). Thus the sediment oxygen demand had increased by over a factor of 5 during this 10-year period. Since the shallow western basin is usually mixed to the bottom, oxygen depletion of the water column is not generally a problem except during occasional periods of calm weather when the water column stratifies. However, the western basin was at one time an important breeding area for many fish. The female fish would typically deposit her fertilized eggs in the upper layer of sediment, where they would incubate prior to hatching. For the eggs to survive the oxygen concentration in the immediate vicinity of the eggs must remain high. The high oxygen demand of the western basin's sediments has been partially responsible for making this area poorly suited as a breeding ground. In addition the accumulation of fine-grained clay and silt deposits on the bottom produced a compact sediment that inhibited the circulation of oxygenated water within the sediment surface layer. Both factors—the high sediment oxygen demand and the low porosity of the sediments—created a sediment type that was virtually useless for fish reproduction.

The effect of nutrient enrichment on hypolimnetic oxygen levels has been undoubtedly most pronounced in Lake Erie's central basin, which is deep enough to stratify but has only a small hypolimnion. The average thickness of the central basin's hypolimnion is about 4.7 m (International Joint Commission (IJC), 1985). Low hypolimnetic oxygen concentrations were first noted in the central basin as early as 1929, and the oxygen demand of the hypolimnetic waters increased by about a factor of 2 between 1930 and 1970. By the late 1950s large areas of the hypolimnion were becoming anoxic for as much as several weeks during mid- and late summer (Regier and Hartman, 1973). Lake trout, lake herring, lake whitefish, and blue pike are all typical cold-water fish that inhabit the hypolimnion during the summer and require oxygen concentrations at or near saturation to function effectively. Such fish were excluded from progressively larger areas of the central basin during the summer as low hypolimnetic oxygen conditions became more and more widespread and of longer duration. The eastern basin of the lake provided some refuge for these species, both because it received lower nutrient inputs and because its greater depth gave it a larger hypolimnion. However, by the 1960s hypolimnetic oxygen levels in the eastern basin were found to be as low as 40–50% of saturation (about 4–5 ppm), a value approaching the critical level for many cold-water fish species (Beeton, 1965). It seems likely that the decline in commercial catches of cold-water fish such as the blue pike, lake whitefish, and cisco was due in part to the development of low oxygen concentrations in first the central and later the eastern basin of Lake Erie as a result of cultural eutrophication.

---

[1]The temperature of the bottom water in the western basin is about 19°C during the summer, and the saturation oxygen concentration in freshwater is about 8.6 ppm at that temperature. Concentrations below 3–4 ppm are considered dangerously low for fish by the Environmental Protection Agency (EPA, 1986).

The Disposal of Toxic Wastes

About 500 toxic compounds have been identified in the Great Lakes. Some of these compounds have been discharged directly into the water; others have been introduced via stream runoff or fallout from the atmosphere. Persistent biocides, toxic metals, and exotic organic chemicals have all been found in the flesh of Lake Erie fish. The impact of these substances on the health of the fish population has not been thoroughly studied, but undoubtedly the effect has not been beneficial.

The toxic substances of greatest concern have been polychlorinated biphenyls (PCBs) and mercury. PCBs are no longer manufactured or used in the United States, but they are highly persistent and were used for many years in a variety of industrial applications. Their toxicological properties are discussed in some detail in Chapter 10. The 1978 Great Lakes Water Quality Agreement between the United States and Canada set an objective criterion of 0.1 ppm PCBs (whole organism, wet weight) for fish. However, a number of species of fish from the Great Lakes contain PCB concentrations that exceed this limit. In Lake Erie carp, catfish, sheepshead, walleye, and muskellunge were all identified as being contaminated highly by PCBs in a 1980 Great Lakes Basin Commission report, and consumptive warnings have been established for carp and catfish (Simmons, 1984).

Mercury is a highly toxic metal that can cause genetic abnormalities and damage to the brain, kidneys and liver (see Chapter 12). In 1970 the fishery in Lake Saint Clair was closed because of excessive levels of mercury in the fish, especially walleye; and with the exception of perch, similar action was taken for the commercial fishery in Lake Erie. Since that time efforts to reduce mercury discharges from industrial operations have caused the mercury concentrations in the fish to drop below federal guidelines. Although a number of other toxic metals have been detected in fish taken from the Great Lakes, according to Shear (1984, p. 34), "The levels of metals . . . . in whole fish samples are generally extremely low, well below any guidelines established for human consumption of fish flesh."

Other pollutants introduced into Lake Erie include a wide variety of organic compounds discharged either directly into the lake or into tributary streams. For example, during 1969 more than 1000 barrels of oil and grease per day were discharged into the Detroit River, the principal tributary of Lake Erie (IJC, 1970), and in 1973 Regier and Hartman reported that the Detroit River transported 6 billion liters of industrial and domestic waste water per day into the western basin of the lake. The Cuyahoga River at Cleveland has carried such a high concentration of oil and other flammable industrial wastes that it actually caught fire in 1936, 1952, and 1969. The June 22, 1969 incident burned two railroad bridges to such an extent that the bridges were unusable. Runoff from farms as well as from lawns and backyard gardens have contributed persistent pesticides that have been detected both in fish and fish-eating birds (IJC, 1987). The concentrations of these pesticides have been generally decreasing since the mid-1970s as a result of a greater awareness of their environmental impact and the use of alternative approaches to pest management.

Contamination of Nearshore Areas with Sewage Wastes

Sewage, in some cases with inadequate treatment, has been discharged into Lake Erie or its tributaries for many years. Untreated sewage has also found its way into the lake as a result of overflows from combined sewer systems during storm runoff.

As a result bacterial counts in nearshore waters, particularly near large cities, have been found frequently to exceed limits considered safe for swimming. Although beaches have been closed by public health authorities, people have continued to use the beaches despite posted warnings. Even though bacterial counts in offshore waters are rather uniformly low (see Cy Adler's quotation at the beginning of the Lake Erie case study), swimmers are interested in using nearshore rather than off-shore waters. Since sewage is discharged at nearshore outfalls as a matter of convenience, it is the nearshore swimming areas that have become contaminated. Cleveland has for many years drawn its water from an intake located 5 km offshore, and the bacterial counts in this water have usually been low enough to satisfy drinking water standards. Thus even though it is correct to say that bacterial counts in most of Lake Erie's waters are low and that the water is potable, this generalization does not apply to nearshore areas that have become polluted by sewage discharges from nearby outfalls.

In addition to discharging pathogens into the water, sewage outfalls have introduced large amounts of nutrients, particularly nitrogen and phosphorus. The nutrients from the sewage have stimulated phytoplankton production greatly, and the fallout of the organic matter produced by the phytoplankton has been a major cause of the hypolimnetic oxygen depletion problem. The increased phytoplankton concentrations also created an aesthetic problem that greatly reduced the recreational value of certain areas of the lake. The principal aesthetic problem was the formation of dense surface mats of algae, which have a tendency to drift ashore and rot on the beaches. The most offensive algal genus in this respect has been the green alga *Cladophora*, a macroscopic species that generally grows best attached to the bottom in shallow water. Although apparently *Cladophora* has inhabited Lake Erie for many years, its biomass did not begin to reach nuisance proportions until the 1950s. By the late 1950s accumulations of *Cladophora* up to 0.5 m thick were being reported along the shoreline in some areas, and when these shoreline accumulations decomposed, local residents sometimes had to move out of their homes because of the odors produced by the decay process (IJC, 1975, p. 7). In addition to the *Cladophora* problem, cyanobacteria, which tend to form surface mats, began to dominate the fall phytoplankton bloom by about 1970 (Regier and Hartmen, 1973). The formation of these dense accumulations of algae was clearly undesirable from the standpoint of human recreational use of the lake.

Perhaps surprisingly, water transparency in Lake Erie has been comparable to that of Lake Washington in the first few years after sewage diversion, Secchi depths in the central and eastern basins of the lake falling usually in the 3–7 m range (IJC, 1985). The mean Secchi depth in the shallow western basin of only 1.5 m (IJC, 1985) is most likely a consequence of the generally high concentration of particulate material stirred up from the bottom (Beeton, 1969).

Remedial Efforts

A massive cleanup of Lake Erie, initiated largely as the result of an agreement signed by Prime Minister Trudeau and President Nixon in 1972, has been underway for some time now. The cost of this program has run into literally billions of dollars. What has this cleanup accomplished to date, and realistically what can the program ultimately be expected to achieve? Is it possible that the lake will be restored to a condition comparable to what existed in the early 1800s?

Toxic Substances

The technology is available to remove from wastewater or even better to recycle most of the toxic substances that have been discharged into Lake Erie. In some cases the solution has been to eliminate the need for the toxic chemical, either by finding a suitable nontoxic substitute or by changing to an alternative technology that does not require the use of the toxic substance. In many cases regulations regarding the discharge of toxic wastes have been established, and industries have been forced either to comply or shut down (IJC, 1977, p. 40).

Figures 4.11–4.13 indicate some of the results that have been achieved as a result of the efforts to eliminate or at least reduce the discharge of toxic substances to Lake Erie. Mercury pollution in the Great Lakes has been associated primarily with fish taken from Lake St. Clair and the western basin of Lake Erie. The principal cause of the contamination was a chlor-alkali plant operated at Sarnia, Ontario, by the Dow Chemical Company (see Chapter 12). Elimination of these discharges resulted in a rapid decline of the mercury content of Lake St. Clair fish (Figure 4.11). A similar pattern has been observed in fish sampled from Lake Erie. For example, mercury levels in the tissue of Lake Erie white bass and walleye declined from 1.19 to 0.21 ppm and from 0.55 to 0.31 ppm, respectively, between 1971 and 1977 (IJC, 1977).

Voluntary restrictions on the manufacture of PCBs in the United States went into effect in 1971, and all production of PCBs in the United States was voluntarily terminated in 1977. An Environmental Protection Agency ban on the manufacture, processing, and distribution of PCBs in the United States took effect in July of 1979. The result of these actions on the PCB levels in Lake Erie fish and fish-eating birds is shown in Figure 4.12. With the exception of lake trout, the pattern of decline is obvious in these figures, but the target goal of 100 ppb had not been reached in some fish as much as eight years after PCB production ceased. Furthermore, the PCB concentrations in lake trout increased dramatically from 1984 to 1987, although according to the IJC (1989, p. 88), "Limited sample size for lake trout precludes any analysis of significance." Concentrations of DDT, which was banned in the United States as of January, 1973, showed a general pattern of decline during the subsequent decade and were well below the target goal of 1.0 ppm in Lake Erie fish by 1983 (Figure 4.13).

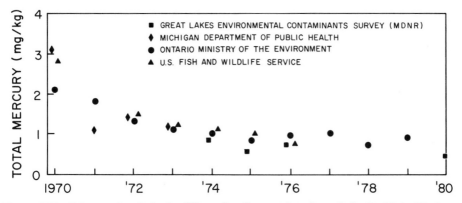

Figure 4.11 Mercury levels in the fillets of walleyes taken from Lake St. Clair. [Redrawn from IJC (1981).]

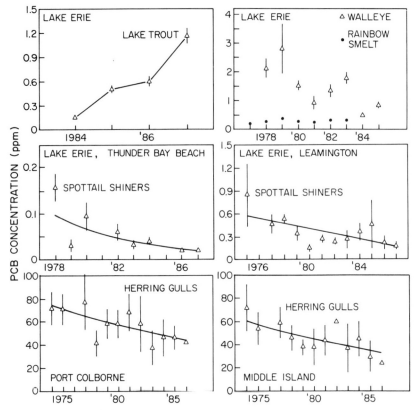

Figure 4.12   PCB levels in Lake Erie fish and herring gulls. [Redrawn from IJC (1989).]

Eutrophication

The effort to reduce nutrient inputs to Lake Erie has been ongoing since approximately 1970. Controls have been focused on phosphorus, which is believed to be the primary limiting nutrient in the lake. Figure 4.14 shows the record of phosphorus loading to Lake Erie from 1967 to 1986. The 1978 revisions to the 1972 Great Lakes Water Quality Agreement between the United States and Canada called for annual P loading of the lake to be reduced to 11,000 tonnes, and this target had been very nearly achieved by 1981. Initial efforts to control P inputs were focused on point sources such as municipal and industrial outfalls. The agreement required that all such point sources that discharged more than 1 million gallons per day (mgd) reduce the P content of their effluent to no more than 1.0 ppm. The annual P loading from these point sources was expected to be about 2500 tonnes if the P concentration in the effluent were 1.0 ppm. The actual P loading rate from point sources dropped below 2500 tonnes per year in 1982, although at that time the effluent from some sources was still above the 1.0 ppm goal.

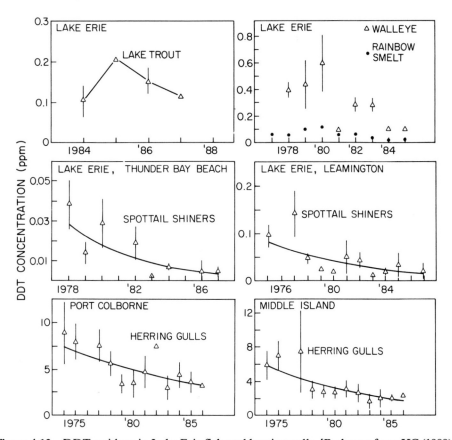

Figure 4.13 DDT residues in Lake Erie fish and herring gulls. [Redrawn from IJC (1989).]

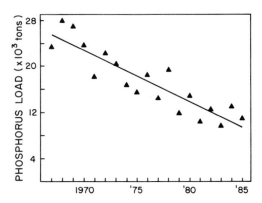

Figure 4.14 Total phosphorus loading of the central basin of Lake Erie. [Redrawn from IJC (1987).]

At the present time about 75% of the P loading to Lake Erie is from nonpoint sources such as agricultural runoff. The focus in recent years has therefore shifted from control of point sources to control of nonpoint sources. In fact, calculations have shown that even if all major point sources achieved the 1.0 ppm P concentration goal, the total loading of P to Lake Erie would still be 13,000 tonnes per year, or 2000 tonnes per year above the goal established by the 1978 revisions. The 1983 Phosphorus Load Reduction Supplement to Annex 3 of the 1978 Agreement therefore called for an additional reduction of 2000 tonnes per year in the P loading of Lake Erie. About 70% of the additional reduction was to be achieved by erosion control, conservation tillage, and improved animal waste management in the area of Ohio west of Cleveland (IJC, 1987).

The impact of the P loading reduction is illustrated in part in Figures 4.15–4.17. Both P and chlorophyll *a* concentrations in the lake have declined since the latter half of the 1970s. Furthermore, diatoms and green algae rather than cyanobacteria are now the dominant forms of phytoplankton in the central portion of the lake (IJC, 1985, 1987). The oxygen depletion rate in the hypolimnion of the central basin peaked around 1980 and declined precipitously between 1986 and 1988 (Figure 4.17). Portions of the central basin hypolimnion become anoxic during the late summer if the oxygen depletion rates exceeds approximately 3.0 parts per million per month (Hartman, 1973), and in 1985 a period of late summer anoxia caused some 2000 tonnes of P to be released from sediments in the central basin (IJC, 1987). This internal recycling of P from the sediments has undoubtedly frustrated efforts to improve the oxygen regime in the central basin through control of allochthonous P inputs. In 1988 oxygen depletion rates in the hypolimnion of the central basin dropped substantially below 3.0 ppm per month for the first time in approximately 30 years, but it will probably be necessary for these rates to decline to about 2.0 ppm per month before the hypolimnetic waters of the central basin can provide a suitable habitat for cold stenotherms throughout the year. For such fish the oxygen concentration must remain above 3–4 ppm.

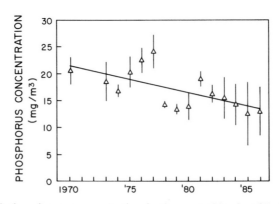

Figure 4.15 Total phosphorus concentration in the central basin of Lake Erie. [Redrawn from IJC (1987).]

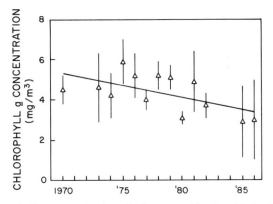

Figure 4.16   Chlorophyll *a* concentrations in the central basin of Lake Erie. [Redrawn from IJC (1987).]

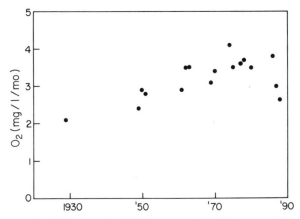

Figure 4.17   Oxygen depletion rates in the hypolimnion of the central basin of Lake Erie. [Source of data: IJC (1987, 1989).]

Prospects for Lake Erie

Problems associated with the introduction of toxic substances into Lake Erie certainly appear to have diminished. For example, in the early 1970s the reproductive success of herring gulls in Lake Erie was reduced to nearly zero, apparently due to exposure to PCBs and other organochlorine substances (see Chapter 10). Between 1969 and 1972 legislation was enacted to restrict or ban the use of PCBs, mercury, and the organochlorine pesticides dieldrin, heptachlor, DDT, and mirex within the Great Lakes basin. Virtually all use of DDT in the United States ceased in January, 1973; production of mirex in the Great Lakes basin ended in 1976; and the manufacture of PCBs was voluntarily terminated by the Monsanto Corporation in 1977. In the latter half of the 1970s both reproduction levels and the number of adult herring gulls increased (Sonzogni and Swain, 1984).

The increase of the herring gull population and the decrease of contaminant levels in fish (Figures 4.12–4.13) have been in the desired direction, but PCB levels in fish such as walleye and lake trout remain disturbingly high (Figure 4.12). Streams such as the Maumee River, Black River, Cuyahoga River, and Ashtabula River are still significant sources of toxic organic chemicals, heavy metals, oil, and grease. Pollution control measures have improved water quality in these streams, but it is doubtful whether water quality in the former three will ever meet all of the objectives set by the 1978 Great Lakes Water Quality Agreement (IJC, 1982).

The effort to reduce phosphorus inputs to Lake Erie has obviously produced some positive results, but the hypolimnetic oxygen demand of the central basin remains a cause for concern. The rates reported in 1987 and 1988 were comparable to the rates observed around 1950, when the development of low oxygen conditions in the hypolimnion during the late summer was considered serious enough to have adversely impacted a number of species of fish. Reduction of nonpoint source P inputs will be more difficult and expensive than point source P control, but almost 75% of existing P inputs come from nonpoint sources. In the past regeneration of P from sediments during periods of anoxia in the hypolimnion of the central basin has frustrated efforts to reduce P inputs to the lake. A dramatic change in P cycling and hypolimnetic water quality can be anticipated if the hypolimnetic oxygen depletion rate in the central basin can be kept below 3.0 ppm per month. However, further reductions in allochthonous P inputs will clearly be needed to keep the oxygen concentrations above 3–4 ppm at all times.

Between 1972 and 1979 New York, Indiana, Michigan, Minnesota, and Wisconsin all passed laws limiting the P content of laundry detergents to 0.5%, and Canada imposed a 2.2% limit in 1973. Legislation that became effective in 1990 limited the P content of laundry detergents to 0.5% P in the counties of Ohio and Pennsylvania located in the Lake Erie watershed, a move long sought by environmentalists. Prior to these restrictions, detergent phosphates had accounted for about 50% of the P in municipal wastewater. In its 1989 report, the IJC recommended that Canada further reduce its detergent phosphate limit to 0.5%, because, "A further reduction in Canada would allow municipal treatment plants which remove phosphorus to achieve lower effluent concentrations with the same effort. In addition, a lower phosphorus limit would lower the effluent concentrations from those municipal and private plants that do not remove phosphorus, as well as reduce the phosphorus content in wastewater from combined sewer overflows and treatment plant bypasses" (IJC, 1989, p. 35).

It is thought provoking to realize that the P concentrations of 10–15 mg m^{-3} in Lake Erie today are less than the post sewage diversion P concentrations of 17 mg m^{-3} in Lake Washington (Table 4.2 and Figure 4.15). If 10 and 20 mg P m^{-3} are taken to be the concentrations associated with the oligotrophic/mesotrophic and mesotrophic/eutrophic transitions (Hern et al., 1981), then the central basin of Lake Erie today is mesotrophic. It seems problematic, however, whether the very low late-summer oxygen concentrations in the central basin are a condition to be associated with mesotrophy. It is certainly possible that the transition concentrations of 10 and 20 mg P m^{-3} are not the same for all lakes and may be smaller in a relatively shallow lake (e.g., the central basin of Lake Erie) than is the case in Lake Washington, which is almost 80% deeper and has a hypolimnion almost five times thicker (23 m vs. 4.7 m).

Contamination of nearshore waters with pathogens continues to be a problem in Lake Erie, but the severity of the problem is unclear. The aim of remedial efforts has been to protect persons involved in water contact sports, but in its 1987 report, the IJC noted that most jurisdictional beach monitoring programs involve several agencies, a fact that made it "nearly impossible to routinely report on basinwide or lakewide microbiologically related problems" (IJC, 1987, p. 123). Furthermore, water quality standards were not uniform either between or within jurisdictions, and more than one monitoring scheme and classification system was employed. There is even now uncertainty as to the appropriateness of the standard fecal coliform test as an assay for the presence of sewage-derived pathogens (see Chapter 7).

The technology is certainly available for sewage treatment plants to remove virtually all the pathogens from raw sewage. It seems likely, therefore, that the pathogen problem reflects nonpoint source pollution, the most important sources probably being combined sewer system overflows and treatment plant bypasses. Correcting this problem will not be accomplished easily or cheaply. Construction and operation of facilities capable of treating storm water runoff is an enormous task (see Chapter 5). The most cost-effective solution to the combined sewer system overflow problem is probably construction of separate sewer systems, but this task is also a major undertaking. Whether such decisions are made will undoubtedly depend very much on the public's perception of the environmental and public health costs associated with allowing untreated sewage to enter Lake Erie.

The fish population picture is a rather sobering one. The native populations of blue pike, sauger, and lake trout have been virtually eliminated from the lake; sturgeon, lake herring, lake whitefish, and muskellunge are found only in small numbers; and the walleye and northern pike populations have been greatly reduced. Human impact on the lake's natural fish community has taken four principal forms: overfishing, degradation of water quality, change or loss of habitat, and introduction of exotic species. Overfishing is a theoretically manageable problem, but implementation requires the cooperation and understanding of all concerned parties. To the extent that the United States and Canada remain good neighbors and recognize the errors of the past, there is reason to hope that Lake Erie fisheries will be managed in a more intelligent manner than was the case during the nineteenth and first half of the twentieth century. A principal goal of the phosphorus control program has been the restoration of year-round aerobic conditions to the hypolimnion of the central basin. It has become clear that achieving this goal will require some control of nonpoint source pollution, a difficult task. However, the present control program may be close to achieving its objective. The recent decisions by Ohio and Pennsylvania to limit laundry detergent P levels to 0.5% will certainly help, but further control over nonpoint source P will be necessary to prevent late summer oxygen concentrations from dropping below 3–4 ppm.

However, even if aerobic conditions return to the hypolimnion and sensible fishing practices are followed, it is highly unlikely that the pre-1800 distribution of fish will be restored in the lake. First, there is virtually no chance that many of the nursery and spawning grounds that were destroyed for various land developments or have been covered with silt or dammed off will be restored. Indeed, it is unlikely that many people would welcome the return of swamps with their attendant mosquito populations. Secondly, a number of exotic species, introduced in many cases from the ballast water of ocean-going ships, have greatly altered the natural biological community of the Great Lakes. Of particular concern in Lake Erie are the alewife, rainbow smelt, and sea lamprey. The alewife is zooplanktivorous, and its consumption of herbivorous zooplankton may have contributed to the development of nuisance algal blooms in the lake (IJC, 1977). Rainbow smelt, which first appeared in Lake Erie in 1931, have increased greatly in numbers and are now the most abundant pelagic fish. Like the alewife, rainbow smelt are largely zooplanktivorous, but yearling and older rainbow smelt feed to a certain degree on the young of other fish such as cisco, sauger, blue pike, and lake whitefish. Any attempt to reintroduce large numbers of these fish (e.g., by seeding with fingerlings from hatcheries) would be greatly inhibited probably by the now established large population of rainbow smelt. Other now abundant species such as the freshwater drum, carp, and goldfish would undoubtedly offer similar resistance to restocking attempts. The sea lamprey is a marine organism that reproduces and hatches in freshwater streams. However, in some parts of the world it has adapted to a completely freshwater life cycle and is an old inhabitant of the St. Lawrence River and Lake Ontario. The lamprey is an eel-like fish with a suckerlike mouth and sharp teeth which attaches itself to other fish, rasps a hole in the fish's body, and sucks the blood and body juices. A sea lamprey can kill a delicate fish such as the lake trout in as little as 4 hours. Niagra Falls blocked the access of sea lampreys to the upper Great Lakes until the Welland Ship Canal was completed in 1829. The lamprey was evidently slow to take advantage of this access route, however, because no lampreys were reported in Lake Erie until 1921 (Applegate and Moffett, 1955). The lamprey then moved on into Lake Huron and Lake Michigan, where it decimated the lake trout populations during the 1940s. During this time the sea lamprey never became a serious problem in Lake Erie, evidently because water quality conditions in the lake did not favor its propagation. However, as conditions in Lake Erie improved during the 1970s and 1980s, the population of sea lampreys in the lake began to increase. Even then, however, it was not considered a serious problem, because its predation was spread over a large number of species, no one of which was perceived as seriously impacted by its presence. Lamprey predation did become an issue, however, when attempts were made to introduce lake trout into the eastern basin of the lake. These stocking efforts were impeded seriously by the lamprey, because lake trout seem to be the preferred prey of the sea lamprey in the Great Lakes. The problem was brought under control during the latter half of the 1980s by applying chemical treatments to tributary streams where the lamprey offspring spend the first 4 years of their life as larva in the sediments.

As far as the fish communities are concerned, the purpose of the restoration program has not been to reestablish the historical fish populations, but rather to sensibly manage the valued species that remain (e.g., yellow perch, white bass, walleye) and (perhaps) to introduce other valuable species that would be able to coexist with the present populations. Pacific salmon were introduced into the lake in 1968 to satisfy the needs of U.S. sports fishermen, but the impact on U.S. sportsfishing has been small, since the salmon apparently migrate during the summer into Canadian waters, where the temperature is coldest and there is adequate oxygen. Both rainbow trout and lake trout are also being stocked into the lake, although in the latter case the stocking attempts have been of much significance only in the eastern basin. Ohio commercial fishermen are prohibited currently from catching blue pike, northern pike, sauger, sturgeon, whitefish, trout, walleye, and salmon in Lake Erie. Almost all the commercial catch of whitefish, smelt, walleye, white bass, and yellow perch is taken by Canadian fishermen. Fishing restrictions in some cases have been imposed to protect the interests of sportsfishermen, but in other cases are aimed simply at maximizing the probability that the population will increase. The walleye population, for example, increased dramatically during the 1970s and is now a major target of U.S. sportsfishermen.

Figure 4.18 shows the temporal pattern in the commercial catch of the six species of fish that presently contribute the most to the commercial catch in Lake Erie. Of these six species the yellow perch, carp, white bass, and walleye are warm water species and hence little impacted by water quality conditions in the hypolimnion during late summer. Despite the collapse of cold water fish such as lake trout, lake herring, and white fish (Figure 4.10), it is obvious from Figure 4.19 that the total catch by weight has remained fairly constant since the latter part of the nineteenth century. The statement by Ehrlich at the beginning of this section that, "Lake Erie has died" is therefore rather an overstatement of the case. However, it is obvious from an examination of Table 4.3 that there have been major changes in the composition of the commercial catch. While the weight of the catch has been relatively constant, the decline in species such as sauger and blue pike has reduced greatly its monetary value.

CASE STUDY 3. KANEOHE BAY

The example of Kaneohe Bay provides an informative comparison with the previous two cases of Lake Washington and Lake Erie. Kaneohe Bay was polluted with sewage for a period of about 30 years, but this stress was removed in 1978. Because the residence time of water in the bay is only a few weeks, the response of the system to this change has in some respects been much more rapid than in the case of Lake Washington or Lake Erie. Nevertheless, as will become apparent in the following discussion, years were required for the bay to recover from some effects of the sewage enrichment, a sobering realization when one considers that the bay is a relatively well mixed and well flushed system compared to Lake Washington or Lake Erie.

Physical Setting

Kaneohe Bay is a subtropical embayment on the northeast side of the island of Oahu in the Hawaiian Islands (Figure 4.20). The bay has an area of about 46 km² and a mean depth of about 6 m, although the average depth in the southeast sector is about 12–13 m. The salinity of the bay's waters normally falls in the 33–35‰ range, and water temperature varies seasonally between roughly 20 and 27°C (Bathen, 1968). A barrier reef that restricts circulation with the ocean extends along much of the bay's mouth, with two narrow channels, the Ship Channel and the Sampan Channel, providing access to the bay for boats. The southeast sector of the bay is relatively isolated physically from the rest of the bay by Coconut Island and a system of shallow reefs. As a result, circulation in the southeast sector is more sluggish than in the rest of the bay. However, the residence time of water in the southeast sec-

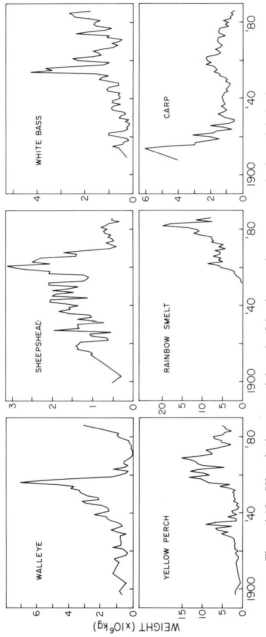

Figure 4.18 Historical commercial catch of the six most important species by weight in the present commercial fish catch from Lake Erie. [Source of data: Baldwin et al. (1979) and subsequent catch summaries from the Great Lakes Fishery Commission.]

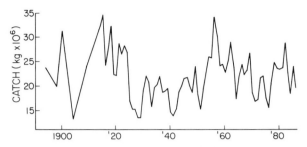

Figure 4.19 Total commercial fish catch from Lake Erie. [Source of data: Baldwin et al. (1979) and subsequent catch summaries from the Great Lakes Fishery Commission.]

Table 4.3 **Importance (by Weight) of Species to Commercial Fish Catch in Lake Erie in Selected Years (weight expressed in millions of kg)**

	1885		1918		1952		1986	
Rank	Species	wt	Species	wt	Species	wt	Species	wt
1	cisco	11.5	cisco	22.1	blue pike	6.4	smelt	7.9
2	sauger	2.5	carp	2.2	walleye	3.3	yellow perch	5.0
3	sturgeon	2.4	yellow perch	1.4	sheepshead	2.1	walleye	3.1
4	whitefish	1.7	whitefish	1.2	yellow perch	1.8	white bass	1.8
5	yellow perch	0.7	sheepshead	1.4	carp	1.5	carp	0.5
6	northern pike	0.1	sauger	1.0	whitefish	1.3	sheepshead	0.5
	others	8.0	others	2.9	others	3.1	others	0.9
	total	26.8		32.2		19.4		19.7

tor is only about 24 days, and in the rest of the bay about 12 days (Sunn et al., 1975). These short residence times are due largely to tidal flushing through the channels and over the barrier reef.

The Kaneohe Bay watershed is only slightly larger in area than the bay itself and is bounded on its landward side by a series of steep cliffs rising 500–850 m above sea level. Runoff from the watershed normally drains into the bay by way of 11 small streams, but during heavy rains the streams may overflow their banks, and water may flow into the bay from virtually all shoreline areas (Banner, 1968). The northeast trade winds blow directly against the front of the bay about 70% of the year, and much rain is precipitated from the moist air as it rises against the cliffs at the landward margin of the watershed. The average annual rainfall in the watershed is estimated to be about 230 cm (Cox and Fan, 1973).

The Coral Reefs

Concern over pollution in Kaneohe Bay centered around the deterioration of the coral reef community. Although historical information on the reef community is rather sketchy, some idea of the characteristics of the coral reefs prior to their demise can be had by putting together bits and pieces of information. In a popular article written in 1915, MacKay commented that, "Probably no other one spot in the Territory of Hawaii can show such a wonderful variety of corals as the waters of Kaneohe Bay . . ." He commented in particular about the beauty of the reefs in the southeastern part of the bay, the so-called coral gardens that were viewed by tourists from glass-bottom boats. In 1928, Edmondson noted that, "Kaneohe Bay, on the windward coast, is recognized as one of the most favorable localities for the development of shallow water corals. Nearly all the reef-forming genera known in Hawaiian waters are represented . . . and many species grow luxuriantly." In 1946 Edmondson again commented that, "Kaneohe Bay harbors a large population of marine animals. Here is one of the best exhibitions of living corals to be seen about the islands. . . ."

Banner (1974) provided a more detailed description of the coral reefs. As of 1951 he reported that by far the dominant coral species in the lagoon area was *Porites compressa* vaughn, with *Montipora verrucosa* (Lamark) of secondary importance. In areas exposed to greater wave action, the most abundant species included *Pocillopora meandrina* var. *nobilis* Verrill, *P. ligulata* dana, and the sturdier forms of *Porites*, such as *P. lobata* (Dana). Roughly 20 years later, Maragos (1972) reported a drastic change

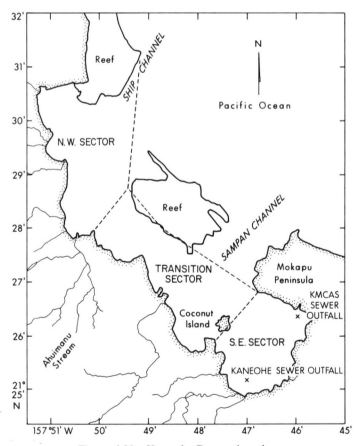

Figure 4.20 Kaneohe Bay and environs.

in the condition of the coral reefs. Based on a number of quantitative transects he estimated that in the southeast sector 99.9% of the original coral reefs had died or been destroyed, 87% had died or been destroyed in the transition sector, and 26% in the northwest sector. What caused this severe decline in a once-thriving coral reef community?

Urbanization of the Watershed

Most of the adverse effects on the coral reef system in Kaneohe Bay can be traced to problems associated with urbanization. To be sure some corals were removed outright by dredging activities associated with construction of the Marine Corps Air Station on Mokapu Peninsula from 1939 to 1941, and to a lesser extent by private individuals seeking access for their boats across the reef flat. However, Banner (1974) has commented that, "By 1951 the reefs in most sections of the bay were again thriving, and some of the reefs cut to 3 m had a recolonization of usual bay corals." Evidently had the environment of the bay remained favorable to the growth of corals, the recolonization of dredged areas noted by Banner in 1951 presumably would have continued. However, changes in the ecosystem caused by urbanization brought this recovery to a halt.

Changes in Land Runoff

Until roughly 1940 the Kaneohe Bay watershed was used primarily for agricultural purposes. The region's native Hawaiian population, which numbered only about 5000 in 1830, cultivated taro as the staple of their diet (Chave and Maragos, 1973). The taro was grown in a system of terraces, into which stream runoff was diverted to provide irrigation water. Because most of the watershed's low floodplains were once covered with taro patches, runoff of sediment as well as overland water runoff into the bay was greatly reduced, as water diverted into the taro patches tended to percolate into the ground, depositing its sediment load on the taro patches. Overland runoff that reached the bay was of reduced intensity, and as a result had less erosive capability and transported less sediment than would otherwise have been the case. In short, there is little reason to believe that the agricultural practices of the native Hawaiians, who may have lived in the watershed for as long as 1500 years, significantly increased the rate of land erosion and concomitant siltation of Kaneohe Bay. In fact, their extensive taro patch irrigation system may have reduced the rate of sediment runoff.

In the early nineteenth century a variety of foreigners began to arrive in Hawaii following the early explorers. The diseases they brought with them devastated the native Hawaiian population, and by 1860 the population of the Kaneohe Bay region had dropped to less than 2000 (Chave and Maragos, 1973). During the latter half of the nineteenth century the growing Hawaiian sugar cane industry began to import large numbers of Chinese, Japanese, and other ethnic groups to work in the cane fields, but the population of the Kaneohe Bay region did not reach 5000 again until about 1940. During the influx of these nationalities some changes in agricultural practices took place in the watershed. Rice replaced taro in some of the terraces, and there were small-scale commercial efforts to cultivate coffee, bananas, sugar cane, pineapple, and perhaps oranges (Cox and Fan, 1973). However, the major agricultural development affecting stream runoff appears to have been the utilization of the drier lands in the northern part of the bay's watershed for grazing cattle, sheep, horses, and goats. In some cases overgrazing of these lands left persistent patches of bare ground. These patches, usually consisting of red volcanic clay, would, "Dry to pavement-like hardness during dry weather and would wash away in heavy rains" (Banner, 1974, p. 687). During this time Agassiz (1889) noted that some corals were killed, evidently by sedimentation in nearshore reef areas. However, the areal extent of this damage appears to have been small.

In 1913 a series of partial stream diversions was begun in the northern watershed to provide irrigation water for leeward Oahu. The tapping of several streams near their headwaters for this purpose reduced the annual stream runoff into the bay by about 42% (Chave and Maragos, 1973). At least some of the potential increase in sediment runoff from the effects of overgrazing and the reduction in the number of taro patches was counterbalanced undoubtedly by the reduced flux of stream runoff. Thus sedimentation and freshwater runoff appear to have had only a limited effect on the coral reefs prior to at least 1925.

Since that time several important changes in the watershed have greatly increased the runoff problem. The most significant of these changes has been the urbanization of the town of Kaneohe. The rapid rise in population began in approximately 1940, and has continued at a rate of about 5.3% per year since then. By 1990 the population of the watershed numbered about 70,000. Most of this population increase was confined to the southeastern sector's watershed, in which the town of Kaneohe is located.

Urbanization is inevitably associated with two processes that greatly increase runoff problems. First, the clearing of land for the construction of roads and buildings exposes, albeit temporarily, much bare soil to erosion from rainwater. In the case of Kaneohe, rather weak land clearing ordinances were passed to alleviate this problem, but in practice even these weak laws were poorly enforced. Second, roads, parking lots, rooftops, and other such surfaces are impervious to water. Rainfall that lands on these areas has no chance to sink into the ground, as would be the case if the land were covered with vegetation. The percentage of land cover by impervious surfaces in business districts may approach 100% and in residential areas may be easily 50%. Thus the amount of overland runoff from urban areas is much greater than in comparable rural areas (see Chapter 5). Furthermore, runoff from impervious surfaces is more rapid than from land covered with vegetation, since there is little to impede the flow of water over parking lots, streets, sidewalks, or rooftops. As a result both the total runoff and the intensity of runoff are greater from urban areas than from rural areas. Because the erosive power of water increases with the intensity of water flow and because there is more total runoff in urban areas, exposed land is eroded away more rapidly in urban than in rural areas.

Freshwater runoff can adversely affect coral reefs because of both the stress of reduced salinity and the smothering effects of excessive sedimentation. Corals can tolerate salinities in the approximate range 27–40‰ (Shepard, 1973), and although salinities in Kaneohe Bay are generally well within this range, salinities of the surface waters may drop to much lower levels after particularly heavy rains. For example, on May 2–3, 1965, approximately 50–60 cm of rain fell on the Kaneohe Bay watershed, and 5 days later surface water salinities were still only 23–24‰ (Banner, 1968). As a result of this storm, corals in parts of the bay were killed to a depth of as much as 1.5 m. Although torrential rains of this sort have undoubtedly occurred in the past in Kaneohe Bay, the effects of such rains have been exacerbated by the urbanization of the watershed, because the high percentage of impervious surfaces in urban areas results in a greater volume of freshwater runoff and in a more rapid delivery of runoff waters to the bay.

With regard to sedimentation it is noteworthy that Charles Darwin (1842) attributed the breaks in fringing reefs off the mouths of streams to the effects of sedimentation and not to the effect of reduced salinity. He reasoned that the freshwater would spread out on the surface, whereas the sediment particles would settle down on the corals. However, Dr. Robert Johannes (quoted by Banner and Bailey, 1970) has noted that, "The temporary production of turbid waters due to man's activities appears not to affect the surface reef community seriously unless siltation is so great that the biota is fairly evenly coated with sediments," but that, "The exposure of reefs to brackish, silt-laden water associated with flood runoff seems to be a major cause of reef destruction historically." Certainly the combined stress of reduced salinity and heavy siltation may prove lethal in some cases where either stress taken separately would not be lethal.

There have been no direct measurements of sedimentation rates in Kaneohe Bay, but several indirect measurements provide some insight into the problem. Based on old bathymetric charts of the bay, Roy (1970) concluded that there had been very little change in the mean depth of the bay between 1882 and 1927, but a 1976 bathymetric survey conducted by Hollett (1977) indicated that the depth of the lagoonal area of the bay had decreased by about 1.0 m between 1927 and 1976. The shoaling was most pronounced in the southeast sector of the bay, where the decrease in depth amounted to 1.6 m. According to Hollett (1977), only about 27% of the bay's sediments were terrigenous, the remainder consisting of reef carbonate detritus (63%) and dredge spoils (11%). Hollett (1977) concluded that shoaling rates in the bay had increased substantially since 1927 and attributed the increase to higher stream sediment loads caused by urbanization and to extensive dredging and disposal activities. Based on water samples taken from Kamooalii Stream (the major stream contributing runoff to the southeast sector) during runoff from a storm of about 6–7 cm of rainfall, Fan (1973) calculated that the stream discharged about 8250 tonnes of sediment into the bay. The observed accumulation of terrigenous sediments on the bottom of Kaneohe Bay can easily be accounted for if sediment loadings of the magnitude calculated by Fan (1973) occur approximately 8–9 times per year. Given the present average depth of the bay (6 m) and the rate of sediment accumulation estimated by Hollett (1977) from his bathymetric studies, the bay will completely fill up the sediments in about another 300 years.

Sewage Disposal

The first sewage treatment plant built in the Kaneohe Bay watershed was constructed to serve the Kaneohe Marine Corps Air Station (KMCAS) on Mokapu peninsula. The outfall from this plant was located near one corner of the southeastern sector of the bay as indicated in Figure 4.20. Originally constructed as a primary

---

[1] Primary treatment involves removal of solid waste from the raw sewage. Secondary treatment includes removal of dissolved organics in addition to solid materials. Tertiary treatment involves nutrient removal in addition to standard secondary treatment.

treatment facility, the KMCAS plant was upgraded to secondary treatment in 1972.[1] As of 1972 it was discharging about 4.1×10^3 m^3 of sewage per day into the bay. During dry weather, about 25% of this discharge was diverted to water the military golf course (Banner, 1974).

Until 1963 sewage from the nonmilitary population of the watershed was handled by private cesspools or septic tanks. In 1963, however, the Kaneohe municipal sewage treatment plant, a secondary treatment facility serving the population of Kaneohe Town, was put into operation. Its outfall, also located in the southeast sector of the bay, is indicated in Figure 4.20. As of 1972 the Kaneohe municipal plant was discharging about 11.5×10^3 m^3 per day of treated sewage at this outfall. In 1970 a third sewage treatment plant serving a housing development in the Kahaluu valley began discharging effluent into Ahuimanu Stream, which flows into the northwest sector of the bay. The plant was designed as a tertiary treatment facility. As of 1972 it was discharging about 0.5×10^3 m^3 per day of effluent into Ahuimanu Stream. The remainder of the watershed is still served by cesspools.

As was the case with Lake Washington, the principal concern over sewage disposal into Kaneohe Bay was centered around the nutrient enrichment problem. According to figures given by Sunn et al. (1975), about 74% of the nitrogen and 75% of the phosphorus that entered the bay came directly from the Kaneohe municipal and KMCAS sewage treatment plants, with the remainder attributed to stream runoff. The nutrient enrichment problem was aggravated by the fact that the sewage was discharged into the most stagnant part of the bay, so that mixing with low-nutrient open ocean water was minimized, and the residence time of the discharged water was maximal. From a pollution standpoint, the southeast sector was the worst part of the bay to choose for sewage disposal. The decisions to locate the outfalls in the southeast sector were clearly made on the basis of convenience and expense rather than on the basis of ecological considerations.

Effects of Sewage Disposal. The sewage discharges into Kaneohe Bay affected the coral reef community in basically three ways. The first effect resulted from a reduction in water clarity caused by the increased populations of phytoplankton, particularly in the southeast sector. Hermatypic corals such as those found in Kaneohe Bay contain symbiotic algae, and these algae require light to carry out photosynthesis. The algal symbionts benefit the corals by producing and excreting organic compounds used by the corals as a major source of food. Furthermore, removal of the symbiotic algae has been shown to markedly reduce coral calcification rates (Shepard, 1973). A series of coral transplant experiments performed by Maragos (1972) showed that when healthy corals were transplanted to the southeast sector of the bay the transplants invariably died, and even when still alive grew slowly and erratically. The failure of these transplants appears to have been caused in large part by the low water clarity in the southeast sector of the bay.

The second effect of the sewage discharges was to create conditions favorable to the growth of filter feeding organisms such as sponges and zoanthids, which then overgrew many of the former coral reefs. Figure 4.21 illustrates the food chain leading to these filter feeding organisms. Had the principal signal perceived by the natural coral reef community been an elevation of inorganic nutrient concentrations, the effect on the reefs would have been a stimulation of photosynthesis

(Kinsey and Domm 1974), as indicated in Figure 4.21. However, because the effluent was not released directly onto the reefs, much of the nutrient content of the wastewater was assimilated by phytoplankton. Although nutrient concentrations in the water flowing over the reefs were elevated by the sewage discharges, the principal signal perceived by the reefs was an increase in the concentration of plankton rather than of nutrients. Hermatypic corals are also filter feeders, but appear to derive most of their nutrition from their algal symbionts rather than from plankton. As as result dense populations of plankton create a condition more favorable to the growth of noncoral filter feeders than to corals. Metabolic studies performed on a fringing reef in the southeast sector of Kaneohe Bay reflect the takeover of the reef by organisms other than corals. These results are summarized in Table 4.4. Photosynthesis and calcification on the southeast sector reef were both about 25% less than the rates characteristic of a typical coral reef, a result reflecting the reduced amount of light available to the reef community and the concomitant reduction in the activity of calcareous algae, which are among the principal calcifiers on healthy coral reefs. Respiration, on the other hand, was about 85% higher than that of a typical coral reef, a consequence of the greater abundance of plankton in the water column and hence the greater supply of food for benthic filter feeders. The more than twofold imbalance between photosynthesis and respiration is very uncharacteristic of a healthy coral reef. The behavior of the southeast sector fringing reef during the period of sewage enrichment contrasts sharply with the metabolic behavior of a reef enriched with inorganic nutrients. In the study of Kinsey and Domm (1974) for example, enrichment of an experimental reef with inorganic nutrients produced no change in respiration rates, but caused a 27% increase in photosynthetic rates.

The third adverse effect of the sewage discharges on the coral reefs was the stimulation of the alga *Dictyosphaeria cavernosa*, commonly known as bubble algae. This alga may establish itself within a coral head at the base of a frond and then grow outward, eventually enveloping the coral head and killing the coral. In the northwest sector of the bay, bubble algae were grazed by certain fish and other organisms (Banner, 1974) and consequently were prevented from enveloping coral heads. However, in the middle sector of the bay, the low abundance of *D. cavernosa* grazers allowed it to envelop and kill numerous coral heads. Maragos (1972)

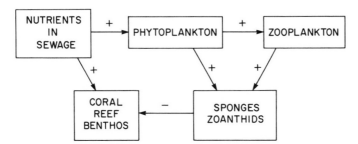

Figure 4.21 Effect of nutrients in sewage on production of organic matter by organisms in Kaneohe Bay. Positive effects are indicated by a +; negative effects by a −. The direct effect of nutrients on production of organic matter by the coral reef benthic community is positive, but indirect effects, transmitted through the food chain, are negative.

Table 4.4   Metabolic Characteristics of Reef Flat Benthic Communities

	Production (P) (g C m^{-2} d^{-1})	Respiration (R) (g C m^{-2} d^{-1})	P/R	Calcification kg CaCO$_3$ m^{-2} y^{-1}
Typical coral reef	6.9	6.8	1.0	4.3
SE Kaneohe Bay reef	5.2	12.7	0.4	3.1

estimated that about 24% of the corals in the lagoon portion of the bay had been killed by this mechanism. A study reported by Smith et al. (1981) showed that bubble algae grew poorly, if at all, in offshore water unless the water was artificially enriched with nutrients. Thus it seems likely that the proliferation of bubble algae in Kaneohe Bay in large part resulted from the elevated nutrient concentrations in the water caused by sewage enrichment.

Response to Sewage Diversion

Between December 1977 and June 1978 both the Kaneohe Municipal and KMCAS sewer outfalls were diverted to a new outfall seaward of Mokapu peninsula. Since the residence time of water in the bay is only about 2 weeks, it was expected that characteristics of the water column would rapidly respond to this reduction in nutrient inputs. Table 4.5 summarizes mean water column parameters in the southeast and northwest sectors of the bay before and after the sewage diversions.

Phytoplankton populations as measured by chlorophyll *a* decreased by about 45% in the southeast sector, whereas water clarity as measured by Secchi depth increased by 30%. The effect of the reduced plankton concentrations on the populations of benthic filter feeders was apparent within a matter of weeks, as colonies of these organisms began to rapidly disappear from previously colonized areas. Brock and Smith (1983) documented the decrease of these organisms by sampling at approximately 60-day intervals from June, 1976 to August, 1979. They discovered that the hard substratum benthic communities near the municipal sewer outfall were dominated by filter- and suspension-feeding organisms. Following the sewage diversion, the abundance of these organisms was 2.5–10 times lower than the prediversion abundance on a fringing reef in the southeast sector of the bay.

The increase in water clarity associated with the reduction in plankton created conditions more favorable to the growth of corals in all parts of the bay, and the decline of inorganic nutrient concentrations by a factor of two or more created conditions less favorable to the growth of bubble algae. The response of the corals to the sewage diversion was expected to be much slower than that of the plankton community, because corals grow slowly even under optimal conditions (Shepard, 1973). However, studies by Evans, Maragos, and Holthus (1986) and Dettelbach (1990) revealed that the response of the corals was quite dramatic in the 6 years following the sewage diversion. The results are summarized in Figure 4.22. On the average live coral coverage in the lagoon almost doubled between 1971 and 1983. There was relatively little additional change in coral coverage between 1983 and 1990. Coverage by *D. cavernosa* declined by a factor of 4 between 1971 and 1983, but by 1990 *D. cavernosa* abundance at depths of 3 m or less was comparable to the values reported in

Table 4.5   Mean Water Column Parameters in Kaneohe Bay During Prediversion and Postdiversion Conditions[a]

	Southeast sector		Remainder of bay	
	Prediversion	Postdiversion	Prediversion	Postdiversion
Chl a (mg/m^3)	2.0	1.1	0.67	0.53
Secchi depth (m)	4.9	6.4	7.1	7.9
Inorganic N (mg/m^3)	26	10	23	11
Inorganic P (mg/m^3)	14	5	6.5	2.5

[a]*Source.* Laws and Redalje (1982).

1971. The change in *D. cavernosa* coverage between 1983 and 1990 did not appear to be related to any obvious source of nutrients. Discharges from the Ahuimanu sewage treatment plant were diverted to the Mokapu Point ocean outfall in 1986, and the areas with the heaviest *D. cavernosa* coverage in 1990 were offshore regions not subject to the immediate impact of land runoff.

Prospects for Kaneohe Bay

The condition of the coral reefs in the central part of Kaneohe Bay only six years after the diversion of the sewer outfalls from the southeast sector was a very encouraging discovery. Because corals grow slowly, the complete recovery process was expected to take 10–20 years. In the southeast sector, which was most heavily impacted by the sewage discharges, coral coverage at monitoring stations increased from virtually zero in 1971 to approximately 10% in 1983 and to 20% in 1990. *D. cavernosa*, on the other hand, has never become established in the southeast sector. The fact that coral coverage in the other parts of the bay did not continue to increase between 1983 and 1990 may be related to the increase in *D. cavernosa* outside the southeast sector.

The resurgence of corals but not of *D. cavernosa* in the southeast sector is a curious phenomenon. The amount of sediment that has washed into this sector has left in many areas a thick deposit of mud on the bottom. If these sediments are in the least stirred up, visibility near the bottom is reduced to virtually zero, so that, "In diving, one would literally swim into the bottom" (Banner, 1974). Coral larvae require a hard substrate for attachment, and such muddy substrates are therefore very unlikely places for corals to become established. Coral coverage has increased dramatically in the southeast sector, indicating that despite the prevalence of muddy sediments there are some hard substrates available for attachment. Why bubble algae have never become established in the southeast sector remains something of a mystery, but may be related to the physical circulation regime. Water movement in the southeast sector is somewhat sluggish, and Dettelbach's (1990) study revealed that the greatest coverage by bubble algae in the rest of the bay occurred near channel mouths where water movement was relatively brisk. The flow of water over and around benthic algae can greatly increase the effeciency of nutrient delivery to the plants and hence stimulate their growth.

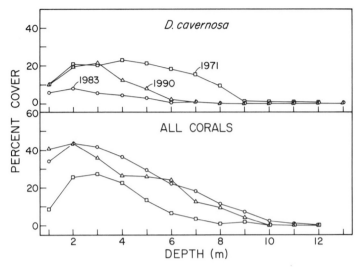

Figure 4.22   Percent cover of bottom by corals and *Dictyosphaeria cavernosa* along experimental transects in Kaneohe Bay. [Redrawn from Dettelbach (1990).]

With respect to the issue of nutrient limitation, it is noteworthy that stream runoff can introduce large amounts of nutrients as well as sediments into a body of water, and after heavy rains the perturbation caused by these nutrient inputs can be quite substantial. For example, from November, 1987 to January, 1988 rainfall in the southeast sector watershed averaged 40 cm per month. The nutrients delivered to the bay in the runoff associated with these heavy rains stimulated a phytoplankton bloom that produced chlorophyll *a* concentrations in excess of 40 mg m^{-3} in the southeast sector, almost four times the highest concentration measured during the year immediately preceding the sewage diversion (Taguchi and Laws, 1989). The amount of water and nutrients delivered to the southeast sector as a result of these heavy rains would undoubtedly have been much less if the watershed were undeveloped rather than heavily urbanized. If urbanization continues to spread from Kaneohe northwestward through the watershed, it is quite probable that the fluxes of water, sediment, and nutrients to the central and northwest sectors of the bay will increase. Such developments would undoubtedly have an adverse impact on the coral reefs.

The principal use of Kaneohe Bay is for recreation, primarily sailing and water skiing in the lagoon. The local skipjack tuna fishery uses the bay as a source of nehu, a small planktivorous anchovy used as baitfish. The diversion of sewage from the bay has been associated with a reduction in the abundance of plankton and hence of nehu. Public health problems associated with sewage disposal have been noted. At least one aquaculture operation was forced to move from the bay due to contamination of its shellfish with sewage pathogens, and water skiers and swimmers have suffered at times from infections that may have been caused by sewage-introduced organisms. Thus diversion of sewage has been beneficial from the standpoint of some but not all important uses of Kaneohe bay.

Although the coral reefs have benefited from the sewage diversion, there is little indication that concern over the reefs will be adequate to prompt public authorities to design and implement an effective plan to control land runoff. The local government's attitude (based on conversations by the author with public officials) is to control runoff passively, i.e., through enforcement of land clearing and grading ordinances. Unfortunately the present land clearing and grading ordinance is extremely weak due to the large number of exceptions that are written into the law, and the ordinance itself is ineffectively enforced due in part to the lack of an adequate staff to monitor land clearing and grading operations. In any case it is questionable whether strict control of land clearing and grading operations would in itself be adequate to prevent deterioration of the reefs if the watershed continues to urbanize, since the high percentage of impervious surfaces in urban areas increases the amount of overland runoff, regardless of the status of construction operations.

The long-term prospects for the coral reefs will probably be determined in large part by their perceived value to the state of Hawaii versus the value of proposed developments in the watershed. To the state government the value of the reefs as a recreational resource and tourist attraction will be the most compelling reasons to protect them. Careful planning and control of land use in the watershed will be necessary to provide that protection.

REFERENCES

Adler, C. A. 1973. *Ecological Fantasies*. Green Eagle Press, New York. 350 pp.

Agassiz, A. 1889. The coral reefs of the Hawaiian Islands. *Bull. Mus. Comp. Zool. Harvard College*, **17**(3), 121–170.

Applegate, V. C., and J. W. Moffett. 1955. The sea lamprey. *Sci. Amer.*, 192(4), 36–41.

Baldwin, N. S., R. W. Saalfeld, M. A. Ross, and H. J. Buettner. 1979. Commercial fish production in the Great Lakes 1867–1977. Great Lakes Fishery Commission Tech. Rept. No. 3. Ann Arbor, MI. 187 pp.

Banner, A. H. 1968. A fresh-water 'kill' on the coral reefs of Hawaii. Hawaii Inst. Mar. Biol. Tech. Rep. **15**. 29 pp.

Banner, A. H. 1974. Kaneohe Bay, Hawaii: Urban pollution and a coral reef ecosystem. Proc. Second Int. Coral Reef Symp. pp. 685–702.

Banner, A. H., and J. H. Bailey. 1970. The effects of urban pollution upon a coral reef system. Hawaii Inst. Mar. Biol. Tech. Rep. **25**. Oahu, HI. 66 pp.

Bathen, K. H. 1968. A descriptive study of the physical oceanography of Kaneohe Bay, Oahu, Hawaii. Hawaii Inst. Mar. Biol. Tech. Rep. **14**. 353 pp.

Beeton, A. M. 1965. Eutrophication of the St. Lawrence Great Lakes. *Limnol. Oceanogr.*, **10**, 240–254.

Beeton, A. M. 1969. Changes in the environment and biota of the Great Lakes. In *Eutrophication: Causes, Consequences, Correctives*. National Academy of Sciences, Washington, D.C., pp. 150–187.

Brock, R. E., and S. V. Smith. 1983. Response of coral reef cryptofaunal communities to food and space. *Coral Reefs*, **1**, 179–183.

Chave, E. H., and J. E. Maragos. 1973. A historical sketch of the Kaneohe Bay region. In *Atlas of Kaneohe Bay: A Reef Ecosystem under Stress*. University of Hawaii Sea Grant, Oahu, HI. pp. 9–14.

Comita, G. W., and G. C. Anderson. 1959. The seasonal development of a population of *Diaptomus ashlandi* marsh, and related phytoplankton cycles in Lake Washington. *Limnol. Oceanogr.*, **4**, 37–52.

Cox, D. C., and P. F. Fan. 1973. The Kaneohe Area. In *Estuarine Pollution in the State of Hawaii*. Vol. 2. *Kaneohe Bay Study*. University of Hawaii Water Resources Res. Center Tech. Rept. No. 31. pp. 7–25.

Darwin, C. 1842. *The Structure and Distribution of Coral Reefs*. Smith, Elder, London. 214 pp. Reprinted in 1962 by University of California Press, Berkeley–Los Angeles, CA.

Dettelbach, A. 1990. Community Structure and Change in the Coral Reefs of Kaneohe Bay: Change in Percent Cover of *D. cavernosa* as an Indicator of Community-structuring forces. Unpublished manuscript. University of Hawaii Research Experience for Undergraduates.

Edmondson, C. H. 1928. Ecology of a Hawaiian coral reef. *B. P. Bishop Mus. Bull.*, **45**, 1–64.

Edmondson, C. H. 1946. *Reef and shore fauna of Hawaii*. 2nd ed. B. P. Bishop Mus. Sp. Publ. 22. 381 pp.

Edmondson, W. T. 1956. Artificial eutrophication of Lake Washington. *Limnol. Oceanogr.*, **1**, 47–53.

Edmondson, W. T. 1961. Changes in Lake Washington following an increase in the nutrient income. *Verh. Internat. Verein. Limnol.*, **14**, 167–175.

Edmondson, W. T. 1966. Changes in the oxygen deficit of Lake Washington. *Verh. Internat. Verein. Limnol.*, **16**, 153–158.

Edmondson, W. T. 1969. Eutrophication in North America. In *Eutrophication: Causes, Consequences, Correctives*, National Academy of Sciences, Washington, D.C. pp. 124–149.

Edmondson, W. T. 1970. Phosphorus, nitrogen, and algae in Lake Washington after diversion of sewage. *Science*, **169**, 690–691.

Edmondson, W. T. 1972. Nutrients and phytoplankton in Lake Washington. In Likens, G. E. (Ed.), *Nutrients and Eutrophication*. American Society of Limnology and Oceanography, Lawrence, Kans. pp. 172–193.

Edmondson, W. T., and J. T. Lehman. 1981. The effects of changes in the nutrient income on the condition of Lake Washington. *Limnol. Oceanogr.*, **26**, 1–29.

Edmondson, W. T., and A. H. Litt. 1982. *Daphnia* in Lake Washington. *Limnol. Oceanogr.*, **27**, 272–293.

Ehrlich, P. 1968. *The Population Bomb*. Ballantine, New York. 223 pp.

Environmental Protection Agency. 1986. *Quality Criteria for Water*. EPA 440/5-86-001. Washington, D.C.

Evans, C. W., J. E. Maragos, and P. F. Holthus. 1986. Reef corals in Kaneohe Bay. Six years before and after termination of sewage discharges (Oahu, Hawaiian Archipelago). In Jokiel, P. J., R. H. Richmond, and R. A. Rogers (Eds.), *Coral Reef Population Biology*, Hawaii Institute Mar. Biol. Tech. Rept. No. 37. pp. 76–90.

Fan, P. F. 1973. Sedimentation. In *Estuarine Pollution in the State of Hawaii*. Vol. 2. Kaneohe Bay Study. Water Resources Res. Center Tech. Rept. No. 31. University of Hawaii. pp. 229–265.

Hardin, G. 1968. The tragedy of the commons. *Science*, **162**, 1243–1245.

Hartman, W. L. 1973. Effects of exploitation, environmental changes, and new species of the fish habitats and resources of Lake Erie. Great Lakes Fishery Commission Tech. Rept. No. 22. Ann Arbor, MI. 43 pp.

Hern, S. C., V. W. Lambou, L. R. Williams, and W. D. Taylor. 1981. Modifications of models predicting trophic state of lakes: adjustment of models to account for the biological manifestations of nutrients. U. S. Environmental Protection Agency. Project Summary. EPA-600/S3-81-001. U. S. Government Printing Office.

Hollett, K. J. 1977. Shoaling of Kaneohe Bay, Oahu, Hawaii, in the period 1927 to 1976, based on bathymetric, sedimentological, and geographical studies. M.S. Thesis. University of Hawaii, Honolulu. 145 pp.

Infante, A., and W. T. Edmondson. 1985. Edible phytoplankton and herbivorous zooplankton in Lake Washington. *Arch. Hydrobiol. Beih.*, **21**, 161–171.

International Board of Inquiry for the Great Lakes Fisheries. 1943. Report and Supplement. U.S. Government Printing Office, Washington, D.C. 213 pp.

International Joint Commission. 1970. *Pollution of Lake Erie, Lake Ontario and the international section of the St. Lawrence River.* Information Canada, Ottawa. 105 pp.

International Joint Commission. 1975. *Cladophora in the Great Lakes.* Great Lakes Research Advisory Board, Windsor, Ontario. 179 pp.

International Joint Commission. 1977. *Great Lakes Water Quality.* 1977 Annual Report. Great Lakes Water Quality Board. 89 pp.

International Joint Commission. 1982. *1982 Report on Great Lakes Water Quality.* Windsor, Ontario. 153 pp.

International Joint Commission. 1985. *A Review of Trends in Lake Erie Water Quality with Emphasis on the 1978–1979 Intensive Survey.* Report to the Surveillance Work Group. Windsor, Ontario. 129 pp.

International Joint Commission. 1987. *1987 Report on Great Lakes Water Quality.* Windsor, Ontario. 236 pp.

International Joint Commission. 1989. *1989 Report on Great Lakes Water Quality.* Windsor, Ontario. 128 pp.

Kinsey, D. W. 1979. Carbon turnover and accumulation by coral reefs. Ph.D. dissertation. University of Hawaii at Manoa. Oahu, HI. 248 pp.

Kinsey, D. W., and A. Domm. 1974. Effects of fertilization on a coral reef environment— primary production studies. *Proc. Second Int. Coral Reef Symp.*, **1**, 49–66.

Laws, E. A., and T. T. Bannister. 1980. Nutrient- and light-limited growth of *Thalassiosira fluviatilis* in continuous culture, with implications for phytoplankton growth in the oceans. *Limnol. Oceanogr.*, **25**, 457–473.

Laws, E. A., and D. G. Redalje. 1982. Sewage diversion effects on the water column of a tropical estuary. *Mar. Environ. Res.*, **6**, 265–279.

Mackay, A. I. 1915. Corals in Kaneohe Bay. In *Hawaiian Almanac and Annual for 1916.* B. P. Bishop Museum, Honolulu, HI. pp. 136–139.

Maragos, J. E. 1972. A study of the ecology of Hawaiian reef corals. Ph.D. dissertation. University of Hawaii. 290 pp.

Regier, H. A., and W. L. Hartman. 1973. Lake Erie's fish community: 150 years of cultural stress. *Science*, **180**, 1248–1255.

Roy, K. L. 1970. Changes in bathymetric configuration, Kaneohe Bay, Oahu. 1882–1969. University of Hawaii, Hawaii Inst. of Geophysics Rept. No. 70-15. 226 pp.

Scheffer, V. B., and R. J. Robinson. 1939. A limnological study of Lake Washington. *Ecol. Monogr.*, **9**, 95–143.

Shear, H. 1984. Contaminants Research and Surveillance—A Biological Approach. In Nriagu, J. O., and M. S. Simmons (Eds.), *Toxic Contaminants in the Great Lakes.* Wiley-Interscience, New York. pp. 31–51.

Shepard, F. P. 1973. *Submarine Geology*. 3rd ed. Harper and Row, New York. 517 pp.

Simmons, M. S. 1984. PCB contamination in the Great Lakes. In Nriagu, J. O. and M. S. Simmons (Eds.), *Toxic Contaminants in the Great Lakes*. Wiley-Interscience, New York, pp. 287–309.

Smith, S. V., W. J. Kimmerer, E. A. Laws, R. E. Brock, and T. W. Walsh. 1981. Kaneohe Bay sewage diversion experiment: Perspectives on ecosystem responses to nutritional perturbation. *Pac. Sci.*, **35**, 279–395.

Sonzogni, W. C., and W. R. Swain. 1984. Perspectives on human health concerns from Great Lakes contaminants. In Nriagu, J. O., and M. S. Simmons (Eds.), *Toxic Contaminants in the Great Lakes*. Wiley-Interscience, New York. pp. 1–29.

Sunn, Low, Tom, and Hara, Inc. 1975. Draft Kaneohe Bay data evaluation report.

Taguchi, S., and E. A. Laws. 1989. Biomass and compositional characteristics of Kaneohe Bay phytoplankton inferred from regression analysis. *Pac. Sci.*, **43**, 316–331.

Vollenweider, R. A. 1968. *Water management research. Scientific fundamentals of the eutrophication of lakes and flowing waters with particular reference to nitrogen and phosphorus as factors in eutrophication*. Org. for Econ. Cooperation and Development Tech. Rept. DAS/CSI/68.27, 159 pp. Revised 1971. Paris, France.

Vollenweider, R. A. 1969. Moglichkeiten und Grenzen elementarer modelle der Stoffbilanz von Seen. *Arch. Hydrobiol.*, **66**, 1–36.

Vollenweider, R. A. 1975. Input-Output models. *Schweiz. Z. Hydrologie*, **37**, 53–84.

Vollenweider, R. A. 1976. Advances in defining critical loading levels for phosphorus in lake eutrophication. *Mem. Ist. Ital. Idrobiol.*, **33**, 53–83.

Wagner, R. H. 1974. *Environment and Man*. Norton, New York. 528 pp.

Weibel, S. R. 1969. Urban drainage as a factor in eutrophication. In *Eutrophication: Causes, Consequences, Correctives*. National Academy of Sciences, Washington, D.C. pp. 383–403.

Welch, E. B., C. A. Rock, R. C. Howe, and M. A. Perkins. 1980. Lake Sammamish response to wastewater diversion and increasing urban runoff. *Water Res.*, **14**, 821–828.

5

URBAN RUNOFF

In the case of Kaneohe Bay we saw that urban runoff may carry large amounts of sediment into a receiving body of water, reducing water clarity and modifying the benthic substrate. Unfortunately sediments are not the only objectionable substances that are found in urban runoff. Urban runoff may also contain high concentrations of nutrients, oxygen-consuming wastes, pathogens, and toxic substances such as pesticides, heavy metals, and oil. As a result urban runoff is in some ways just as serious a problem as sewage pollution. There is no doubt, however, that land runoff from other than urban areas may create significant water pollution problems as well. For example, runoff from feedlots contains high nutrient and bacterial concentrations, and runoff from agricultural land may contain high concentrations of pesticides and/or metals and nutrients. Urban runoff is therefore best considered as one example of a variety of land drainage problems that come under the classification of nonpoint source pollution. Such nonpoint source pollution problems have been characterized by Peterson et al. (1985) as among the most pervasive, persistent and diverse water quality problems facing the United States. Because about 88% of the population of the United States is expected to be living in urban or suburban areas by the year 2000 (EPA, 1983), the nonpoint source pollution problems associated with urban runoff are of particular interest. Furthermore, the principles associated with understanding and solving urban runoff problems are in many cases highly relevant to other types of land drainage problems as well.

DEFINITIONS

Before going any further in our discussion of urban runoff, it is best to define a few terms that invariably arise in a discussion of land drainage problems. Urban runoff may contain high concentrations of sediments. Some of these sediments may consist of relatively large particles such as pebbles and gravel that are too heavy to be suspended in the water column under normal flow conditions. Instead these larger

particles along with some of the smaller particles such as sand and silt are transported along the bottom of the stream or storm sewer. The flux of particles along a
stream bed is referred to as the stream's *bed load*. Normally the majority of the sediment transported by a stream consists of small particles such as sand, silt, and clay,
many of which remain suspended in the water column rather than being transported along the stream bed. The stream's suspended load is determined from the
concentration of suspended solids (SS) in the water by simply filtering a known
volume of the water and weighing the filter before and after filtering. The difference
in weight is the weight of suspended solids in the given volume of water. Normally
most of the material retained on the filter consists of inorganic sand, silt, and clay
particles, although it is possible that in some cases significant amounts of biological
material such as leaf fragments may also contribute. The total sediment load
transported by a stream or storm sewer is defined as the sum of the bed load and suspended load.

What about the oxygen-consuming wastes in the runoff? How does one measure
the oxygen-consuming potential of a sample of water? There are two commonly
used metrics, one called the *biochemical oxygen demand* (BOD) and the other called
the *chemical oxygen demand* (COD). BOD is defined as the amount of oxygen consumed in a 300 ml sample of water incubated in a stoppered bottle in the dark at
20°C for 5 days. Since the sample is incubated in the dark, there is no possibility for
photosynthesis to occur, so the oxygen concentration must either remain constant
or decline. A decrease in oxygen concentration may be due to the respiratory activity
of organisms and/or the simple chemical oxidation of substances that are unstable
in the presence of oxygen, such as ferrous iron or ammonia. Because the decline in
oxygen concentration may be caused by both biological and strictly chemical processes, BOD should be understood to refer to biochemical oxygen demand, as
opposed to simply biological oxygen demand. The COD is the amount of oxygen
consumed when the substances in the water are oxidized by a strong chemical oxi-
dant. The COD is measured by refluxing the water sample in a mixture of chromic
and sulfuric acid for a period of two hours. This oxidation procedure almost
invariably results in a larger consumption of oxygen than the standard BOD test
because many organic substances, which because of their refractory nature are not
immediately available as food to aquatic microbes (e.g., cellulose), are readily
oxidized by a boiling mixture of chromic and sulfuric acids. On the other hand,
some compounds such as acetic acid, which are utilized by aquatic microbes, are
not efficiently oxidized in the standard COD test. Because of the variability in the
organic composition of different wastewaters, the ratio of COD to BOD may vary
widely from one type of wastewater to another. Obviously both the BOD and COD
are arbitrary measures of oxygen-consuming potential. The value of the tests lies in
the fact that from experience one can usually estimate the impact of a particular
wastewater on the oxygen content of the receiving water from empirical correlations
between either BOD or COD and the impact of the wastewater.

COMPOSITION OF URBAN RUNOFF

With this introduction, let's turn our attention to Table 5.1, which lists typical concentrations of SS, BOD, COD, nitrogen, and phosphorus in urban runoff, raw
sewage, and rainfall. The numbers for urban runoff and rainfall are taken from

studies conducted in Cincinnati, Ohio, with the urban runoff numbers coming from a residential-light commercial area (Weibel, 1969). The numbers are therefore not to be considered a national or worldwide norm, because the characteristics of both urban runoff and rainfall can vary greatly from one time and place to another. For example, Abernathy (1981) has reported that urban runoff may contain COD as high as 1000 mg/l, SS up to 2000 mg/l, and total phosphorus as high as 15 mg/l. However, it is probably fair to say that the order of magnitude of the urban runoff and rainfall numbers in Table 5.1 is representative of many areas. The characteristics of domestic raw sewage are much less variable than those of urban runoff and rainfall, and the numbers for raw sewage are typical of the raw sewage entering many sewage treatment plants in the United States.

Although urban runoff obviously contains water derived from rainfall, it is clear from Table 5.1 that the concentration of most pollutants in urban runoff is much higher than in rainfall. The implication is either that rainfall is a minor source of the pollutants or that the urban watershed retains rainwater more efficiently than the pollutants present in the rainwater. The latter explanation seems to account in part for the elevated concentrations of COD and nutrients in urban runoff, but there is no doubt that the watershed itself is a significant source of SS, and at least in commercial areas, of COD and nutrients as well (Randall et al., 1981).

Federal law requires sewage treatment plants to remove at least 85% of the BOD and SS from raw sewage and to keep the BOD and SS of the plant effluent below 30 mg/l. Given this information, one concludes from Table 5.1 that urban runoff may contain BOD and SS at concentrations comparable to or much higher than the concentrations in treated sewage. The concentrations of nitrogen and phosphorus in urban runoff are usually much lower than the concentrations found in either raw sewage or treated sewage, but are still several times higher than the N and P concentrations in rainfall. The contribution of the watershed to the SS, BOD, COD, N, and P in urban runoff reflect the results of soil erosion, the leaching of nutrients from exposed soils and detritus, and the transportation by the runoff of various accumulated wastes in the watershed during dry weather.

The daily discharge of sewage from a sewage treatment plant is a very predictable number, but urban runoff, which is much influenced by rainfall, is irregular and highly unpredictable (weathermen notwithstanding). Acknowledging that the concentrations of certain substances in urban runoff are of the same order of magnitude as the concentrations of the same substances in sewage, one may therefore ask

Table 5.1   Average Concentrations of Various Components of Urban Runoff, Raw Sewage, and Rainfall (in mg/l)

	Urban runoff	Raw sewage	Rainfall
SS	227	200	13
COD	111	350	16
BOD	17	200	no data
Total N	3.1	40	1.3
Total PO$_4$ (as P)	0.36	10	0.08

Source. Weibel (1969).

whether urban runoff might not create pollution problems similar to those created by sewage. To answer this question, one must have some idea of the total flux of substances delivered to a body of water by urban runoff versus sewage. Thus not only concentrations but flow rates must be taken into account. Table 5.2 lists fluxes of SS, BOD, COD, N, and P delivered by raw sewage and urban runoff in the Cincinnati watershed studied by Weibel (1969). Recalling that sewage treatment should remove at least 85% of the SS and BOD from raw sewage, one concludes from Table 5.2 that urban runoff could contribute about half as much BOD and 6–7 times as much SS to the receiving body of water as treated sewage. Most sewage treatment plants remove only about 30% or so of the N and P in raw sewage, and based on Table 5.2 one therefore concludes that in general treated sewage would be a far more important source of nutrient loading than urban runoff. Whether the nutrient loading from urban runoff would be sufficient to create eutrophication problems would depend on the characteristics of the receiving body of water. Recalling that the winter phosphate-P concentrations in Lake Washington at the height of eutrophication were only about 0.06 mg/l, whereas the phosphate-P concentrations in the urban runoff characterized by Weibel (1969) were 0.36 mg/l, the conclusion is that the nutrient levels in urban runoff can be more than enough to cause eutrophication problems if the runoff is not sufficiently diluted by the receiving body of water.

TYPES OF SEWER SYSTEMS

Given this state of affairs, you may say logically, "What should be done with urban runoff?" For many years the response of most communities in the United States was to do little or nothing. Until approximately 1980 flooding was considered to be the principal danger associated with urban runoff, and systems intended to handle urban runoff were designed to minimize the potential damage from flooding. Little or no attention was paid to the pollution potential of the runoff. However, as discharges from point sources of pollution were reduced during the 1970s, it became apparent that significant water pollution problems were being caused by nonpoint sources, including urban runoff. Lake Erie is a case in point. Another example that may have caught the attention of the federal government is the Occoquan Reservoir

Table 5.2   Fluxes (kg km^{-2} yr^{-1}) of Various Constituents from Residential-Light Commercial Watershed in Cincinnati, Ohio Estimated by Weibel (1969)

	Urban Runoff	Raw Sewage[a]
SS	64,050	68,250
BOD	4,725	68,250
COD	31,000	119,400
Total N	875	13,650
Total PO$_4$ (as P)	105	3,500

[a] The population density of Cincinnati at the time of the study was about 2500 persons per km^2.

in northern Virginia, which supplies drinking water for about 650,000 persons in the Virginia suburbs of Washington, D.C. Water quality in the reservoir began to deteriorate noticeably during the 1960s, and a program of point source pollution control in the watershed was initiated. Despite progress in reducing the point sources of pollution, water quality in the reservoir continued to decline, and extensive monitoring revealed that urban runoff was the cause of the problem. Urbanization of the watershed combined with the successful effort to reduce point source pollution had created a situation in which urban runoff was the primary source of the pollutants in the reservoir. This and similar incidents made it clear to public authorities that urban runoff could cause serious water quality problems. The extent of the problem was apparent in a 1983 EPA report to Congress, which summarized the results of 30 studies conducted under the auspices of its Nationwide Urban Runoff Program.

The crux of the problem was and still is the fact that in many communities urban runoff is simply routed to the nearest convenient watercourse and discharged without treatment. The conduit for routing the water from streets and parking lots to the discharge point is called the *storm sewer*. In most cases this system of pipes is separate and distinct from the sanitary sewer system, which routes domestic and other point source wastewater to the sewage treatment plant. According to the EPA (1983) about 46% of the U. S. population was served by separate storm and sanitary sewer systems in 1980. This figure is expected to increase to about 70% by the year 2000. Separate sewer systems obviously present no opportunity for treatment of urban runoff. However, it is much more difficult to design a system to treat urban runoff than sanitary sewage, because during storms the former system must be able to handle a potentially enormous volume of runoff in a short time. The flux of sanitary sewage is more predictable and less variable, although leaks in the sanitary sewer system can lead to substantial increases in flow during storms due to infiltration of the sewer lines by groundwater.

The wastewater from approximately 20% of the U. S. population is handled by so-called combined sewer systems, in which a single system of pipes transports both the urban runoff and sanitary sewage. During dry weather flow, the sanitary sewage is routed to an interceptor point, where it is diverted to the sewage treatment plant rather than flowing to the stormwater outlet. The interceptor system is deliberately designed to divert 1.5–5 times the normal dry weather sewage flow to the treatment plant. Hence runoff from small storms is given the same treatment as the sanitary sewage. When the flow in the pipes exceeds the capacity of the interceptor system, however, the additional flow is simply routed to the stormwater outlet. Stormwater runoff can amount easily to 50–200 times the dry weather flow of sanitary sewage (Weibel, 1969). Since the sanitary sewage and stormwater are mixed in the combined sewer system, almost all the sanitary sewage is discharged untreated with the stormwater during moderate to heavy rains. According to Weibel (1969), if the contribution of so-called stranded filth[1] is ignored, the amount of raw sewage that escapes untreated during storms amounts to about 3% of the total annual flux of raw

---

[1]Stranded filth is sanitary sewage that is deposited in the pipes during dry weather flow and that is scoured out by storm flows.

sewage. The advantages of a combined sewer system are lower cost and treatment of runoff of small storms. The major disadvantage is that a combined system allows raw sewage to escape when stormwater flow is greater than 1.5–5 times dry weather flow. The negative consequences of allowing 3% of the raw sewage to escape untreated are considered generally to outweigh the positive effects resulting from the treatment of runoff from small storms.

CORRECTIVES MEASURES

Urban runoff has been viewed not illogically by some people as a misplaced resource. If runoff is intercepted from impervious surfaces such as rooftops and parking lots, it is potentially usable for a variety of purposes, including cooling, watering lawns and gardens, and recharging groundwater. There is no doubt that water is put to a great many uses that do not require it to be potable (drinkable). If the quality of runoff from a particular area is inconsistent with some uses, it is possible that only a moderate amount of treatment would be required to sufficiently upgrade the quality of the runoff. Such considerations have been little applied in the United States, but may receive more attention in the future, particularly in areas where water conservation is critical (see Chapter 16). The following discussion summarizes a few approaches that have been proposed or implemented for dealing with urban runoff.

Use of Settling Basins

A settling basin is simply a large basin through which runoff is routed. In some cases the settling basin is constructed by merely deepening and greatly widening the channel of an existing stream near its mouth. In other cases the settling basin is an excavated pit or trench located adjacent to a parking lot, shopping mall, housing development, or highway. In the past such basins have been designed primarily to reduce the intensity of runoff from storms, and are referred to as detention or retention basins. If some or most of the water intercepted by these basins is percolated into the ground, they are referred to as *infiltration* or *recharge basins*. In all cases the speed of the runoff water is reduced because of the enlarged cross-sectional area of the channel across the settling basin. Because of the reduced speed of the water as it crosses the basin, some of the sediments sink to the bottom before the water flows through the exit. Since many of the pollutants in urban runoff are associated with suspended solids, settling of the suspended solids will usually remove a significant portion of the BOD, nutrients, hydrocarbons, metals, and pesticides from the water. During dry-flow conditions the accumulated sediments are periodically dredged out and trucked away to be used for landfill.

In practice such settling basins sometimes have little impact on pollutant concentrations in urban runoff. During dry-flow conditions, for example, nutrient concentrations in the water column of the settling basin may reach high levels due to recycling from accumulated sediment, and during even a small storm these nutrients can be rapidly flushed from the basin. Although detention basins may serve a useful purpose in reducing the intensity of storm runoff, the residence time of water in the basins during storms is often too short to allow a significant portion of the suspended solids to settle out. In fact sediment accumulated on the bottom during dry-flow conditions is more than likely to be resuspended and overflow with the runoff water. Furthermore a significant fraction (~25% or more) of the SS may consist of silt and clay size particles (Harriss and Turner, 1974) that stay in suspension much longer than the residence time of water in the settling basin. Such small particles are not amenable to removal by settling even in large settling basins unless chemical flocculants are added to speed up the settling process. Evans et al. (1968), for example, found that completely quiescent settling for less than one hour removed negligible amounts of SS, BOD, and nutrients from urban runoff, and Whipple and Hunter (1980) found that even 16 hours of settling removed only about 50% of the SS and phosphorus and less than 40% of the BOD from urban runoff. Extending the settling time to 32 hours increased the removal efficiency of the SS to almost 70%, but there was little improvement in the removal of phosphorus and BOD (Whipple and Hunter, 1980). Since conditions during storm runoff in a settling basin are far from quiescent and the retention time of storm runoff in typical settling basins is no more than a few hours, it should be obvious why settling basins have sometimes been ineffective in handling urban runoff problems.

Ground Recharge Basins

If land area is available, and if the soil is sufficiently porous, it is possible to route urban runoff to recharge basins where the water has a chance to percolate into the soil rather than running off the land. Such basins may be classified as either dry or wet, the latter containing a permanent pool of water below the level of stormwater containment. In areas where the water table is sufficiently low, percolation of runoff may also be accomplished with the use of dry wells, which are pits or trenches in porous soil backfilled with rock. A system of recharge basins has been used for many years in certain parts of Long Island, for example, where the sandy subsoil is favorable to rapid percolation. Some of the recharge basins are actually underground pits; others are fenced-off open basins. Runoff from roads, construction sites, and some industrial operations is routed to the recharge basins (Weibel, 1969). From time to time recharge basins must be dredged to remove accumulated sediments, but in principle it is preferable to locate the basins so that runoff from impervious surfaces has had little chance to traverse erodable soils; hence the sediment load is minimized. From this standpoint a system of many small pits is clearly preferable to one large pit.

An important concern with the use of recharge basins is the possibility that pollutants in the runoff will contaminate groundwater supplies. This possibility is minimal to the extent that the pollutants are associated with suspended solids, since percolation of the water through the soil removes virtually 100% of the suspended solids. Pathogens derived from animal wastes are also removed with virtually 100% efficiency by the percolation process (see Chapter 6). Groundwater may become contaminated, however, with certain dissolved substances that are not adsorbed effectively by soil particles. Whether the presence of such substances in urban runoff precludes percolation of the water depends on the concentrations of the substances in the runoff and the extent to which they are diluted by the water in the underground aquifer. In the case of Long Island, for example, cadmium concentrations in the groundwater near an aircraft manufacturing company were found to be over 100 times acceptable levels during the early 1950s (Chapter 12). The company responsible had been using cadmium to anodize aluminum and other metals and was disposing of its wastewater in a recharge basin. While runoff from streets and parking lots is unlikely to contain pollutant concentrations comparable to those in industrial wastewaters, one should be aware that some pollutants are not effectively removed by percolation, and that the possibility of groundwater contamination does exist. Some studies or estimates of the characteristics of runoff from existing or proposed urban sites may therefore be useful in determining whether groundwater recharge basins are a reasonable mechanism for treating the runoff.

Storage in Underflow Tunnels or Tanks

Overflow from combined sewer systems is sometimes diverted into underflow tunnels or tanks until it can be treated. The tunnels or tanks are placed underground, and the overflow from storm events is pumped out during dry weather and routed to a treatment plant. If the capacity of the underground reservoir system is large enough, the runoff from virtually all storms can be treated in this way. The lower the capacity of the reservoir system, the more frequent will be the release of untreated runoff.

Berlin, Germany, for example, has a system of underground tanks to handle overflow from the combined sewer system that serves a third of West Berlin. The system is able to handle small overflows completely. In the case of large overflows, water passes through the tank system, receives nominal sedimentation, and is discharged to local streams. With the system in operation about 85% of the total waste handled by the combined sewer system receives treatment. Before the tanks were installed only 55% of the waste was treated (Cohrs, 1962).

The most ambitious treatment system for combined sewer overflow is undoubtedly the tunnel and reservoir plan (TARP) developed by the City of Chicago. TARP consists of a system of huge tunnels, reservoirs, pumping stations, and treatment facilities. The system was designed in two phases, the first intended primarily for pollution control and the second aimed at flood control from large storms. Phase I includes approximately 175 km of tunnels, into which combined sewer overflow is routed. The tunnels are 3–13 m in diameter and are located 45–90 m below street level. Runoff is pumped out of the underflow tunnel system to the treatment plants during dry weather (Figure 5.1). Sediment and sludge are removed from the tunnels with a floating hydraulic dredge and pumped to the same treatment plants. The treated wastewater is pumped into the ground through deep groundwater recharge wells located near the treatment plants. When completed, the Phase I system will be capable of handling about 85% of the pollution caused by combined sewer overflows from an area of 970 km^2. Phase II, which has not yet been built, would extend the tunnel system to a total length of 210 km and would add 3 storage reservoirs with a combined capacity of 59 million m^3. The principal reservoir, with a capacity of 40 million m^3, will be located in the McCook-Summit area. Two smaller reservoirs will be located near O'Hare Airport and at the southern extension of the Calumet system (Figure 5.1). When completed, TARP is expected to eliminate over 99% of the release of SS and BOD from combined sewer overflows in the Chicago area. Phase I was sufficiently complete to begin operating in 1985, and by 1990 approximately 81 km of the tunnel system had been completed and another 40 km were under construction. The cost of the work completed to date or underway is $1.7 billion, with 75% of the bill paid by the EPA and the remainder by local authorities. The total cost of Phase I is expected to be $2.5 billion, with operating costs of about $9 million per year. The Corps of Engineers will pay 75% of the cost of Phase II. Construction of the entire TARP system is expected to cost $4.0 billion, with annual operating costs of $13.6 million (City of Chicago, 1972; Di Vita, 1990).

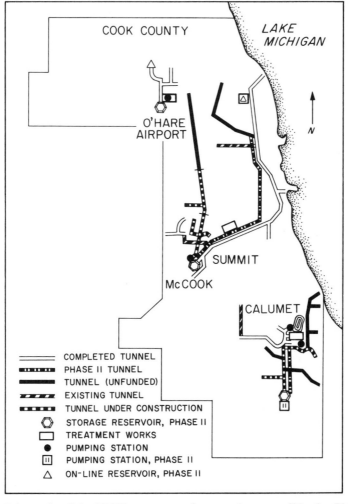

Figure 5.1 Schematic of the TARP to store and treat urban runoff in the Chicago area. (*Source*: Metropolitan Water Reclamation District of Greater Chicago Engineering Department.)

Minimizing Runoff

Urbanization is a process that invariably increases the percentage of impervious surfaces in a watershed. As a result less rainwater percolates into the ground and more flows overland. For example, it has been estimated that if a well-managed wooded area is converted to quarter-acre housing, the runoff from a storm of 5 cm would increase from 6 to 18 mm (Whipple et al., 1983). With careful planning, however, it is possible to reduce the amount of overland water flow. For example, one can intercept runoff from rooftops and percolate it directly into the ground with the use of dry wells or similar devices. Where practical, the use of porous paving material for sidewalks, parking lots, and streets can significantly reduce runoff from these surfaces. The usual impervious concrete or asphalt may be replaced with gravel or crushed rock. In some cases an even better solution is to use modular pavement, which consists of impervious material, usually concrete, interspersed with void areas filled with pervious materials such as sod, sand, or gravel. In suburban areas where the slope of the land is not too great, the traditional curb and gutter system may be replaced with gently sloping grassed channels or swales. In the past relatively little use has been made of such preventive techniques, but as awareness of the problems caused by urban runoff increases, these methods will likely receive more attention, particularly in the planning of new developments.

Summary

The foregoing examples give some idea of the variety of methods that have been used for dealing with urban runoff. There are basically two approaches to the problem. The first is to minimize the amount of overland runoff by intercepting the water as close to the source as possible. This goal is achieved most effectively by using pervious paving materials wherever practical, but may also be accomplished by intercepting runoff in small detention or recharge basins before it has had a chance to traverse exposed soils. The principal pollutant in urban runoff is almost always suspended sediment. The amount of soil erosion at construction sites can be minimized by planning developments so that as little land as possible is laid bare at one time and building interceptor basins around the construction site to permit percolation of runoff water. This first approach requires a relatively large number of individually small preventive measures and a good deal of planning by public authorities and cooperation by developers.

The second approach is to intercept the runoff in the storm sewer system and to give it some form of treatment before it is released. This approach is exemplified by the Berlin underflow tank system and by TARP. The principal advantage of the first system is that soil erosion is minimized, and because of its low SS concentration, the runoff may be percolated without treatment or utilized for purposes that do not require water of potable quality. The principal advantage of the second system is that its operation and maintenance fall usually under the authority of a single government entity. Consequently its performance is more likely to be monitored and maintained than is the performance of the many component parts of the first system, which are likely to be separately owned and operated by numerous persons and agencies. The principal disadvantage of the second system is its high cost. The price tag of course reflects the fact that the system must be able to handle very large volumes of water in a relatively short time. The $4.0 billion cost of TARP, for example, works out to about $520 per resident of the Chicago metropolitan area. Similar projects for the control of runoff into Kaneohe Bay have been estimated to cost $100–$1300 per Oahu resident (Gray and Lau, 1973). Obviously a rather aroused public is required to finance such projects. The fact that the EPA financed 75% of Phase I of TARP undoubtedly enhanced the image of the project in the eyes of local taxpayers.

A CASE STUDY—LAKE JACKSON, FLORIDA

Lake Jackson is a shallow lake (mean depth 2–3 m) with an area of about 20 km^2 located just north of Tallahassee, Florida (Figure 5.2). Because of the shallowness of Lake Jackson, almost all of its benthos lies in the euphotic zone. As a result benthic algae and grasses contribute substantially to the overall primary production of the lake. There is no doubt that the lake is highly productive, but this eutrophic condition does not appear by itself to have led to any undesirable effects. Water clarity in most parts of Lake Jackson is good, with Secchi depths of about 2–3 m where it is possible to make a Secchi measurement.[1] Until the early 1970s there were no reports of nuisance algal blooms involving either benthic algae or phytoplankton in any part of the lake. During the 1960s Lake Jackson gained national prominence due to the large number of trophy-size largemouth bass that were taken from its waters. It has been undoubtedly a valuable recreational resource to the city of Tallahassee, since it is used extensively for fishing, pleasure boating, and picnicking at parks along its shores.

Unfortunately, the northward expansion of Tallahassee during the 1960s led to increasing urbanization of the lake's southern watershed, particularly the watersheds draining into Megginnis Arm and Ford's Arm (Figure 5.2). Associated with this urbanization was the construction in the southern watershed of two shopping malls in 1968 and 1969 and an interstate highway link in 1972. In the early 1970s the effect of this urbanization became apparent to those using the lake, as large amounts of fine-grained sediment washed into both Megginnis Arm and Ford's Arm after every rain. Fortunately domestic sewage from northern Tallahassee was pumped to a sewage treatment plant south of the city rather than being discharged into Lake Jackson, but the nutrient loading from urban runoff alone was sufficient to create water quality problems in the lake. Massive blooms of rooted aquatic plants began to appear during the summer in both southern arms of the lake, creating an unsightly appearance and impeding boat traffic. In addition, cyanobacterial blooms characteristic of hypereutrophic waters began to occur in the same areas. As a result of these observations, the Florida Game and Freshwater Fish Commission initiated a study of Lake Jackson to determine how serious the eutrophication problem had become and to decide what corrective measures needed to be taken. This study lasted for three years, from July, 1971 to June, 1974.

---

[1] The bottom is visible from the surface in many locations.

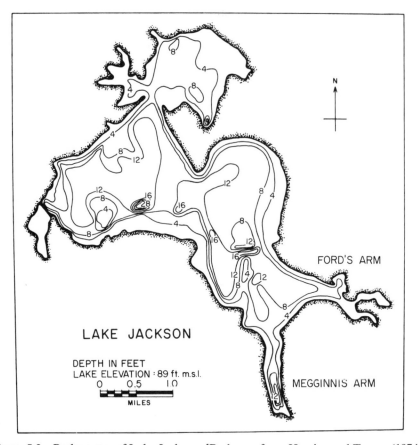

Figure 5.2 Bathymetry of Lake Jackson. [Redrawn from Harriss and Turner (1974).]

As a result of this study, the deterioration of water quality in the southern part of the lake caused by the effects of urban runoff from a rapidly developing watershed was documented in quantitative terms. Table 5.3 summarizes a few of the results reported in the study. The concentration of SS and the rate of primary production at the inner Megginnis Arm station were about 20 and 4.3 times greater, respectively, than the same parameters at the north midlake station. The mean Secchi depth of 0.5 m at the inner Megginnis Arm station reflects the combined effects of a high algal biomass and high concentration of SS, and may be compared to the mean Secchi depth of about 1 m in Lake Washington at the height of eutrophication from 1963 to 1965 and to the mean Secchi depth of 2.9 m at the north midlake station in Lake Jackson.

To quantify the effects of urban runoff on water quality in the southern part of Lake Jackson, a study was initiated comparing the runoff characteristics of two sub-basins of the lake's watershed. The two subbasins, which we will henceforth refer to as the forested and urban watershed, were of comparable area, topography, and soil

Table 5.3   Mean Values of Selected Parameters Over a Three-year Period in Lake Jackson

	Inner Megginnis Arm	North Midlake[a]
Secchi depth (m)	0.52	2.9
SS (mg/1)	61.3	4.0
Primary Production $(mg\ C\ m^{-3}\ hr^{-1})$	139	32.4

[a] Deep hole in central part of lake (see Figure 5.2).

Source. Harriss and Turner (1974).

type.[2] The principal difference between the two was the use of the land. The forested watershed, which drained into the northern part of the lake, consisted in large part of an old estate with limited public access. About 52% of the forested watershed was covered by a mixed pine-hardwood forest, with the remainder being in light agricultural use or in old fields. The urban watershed, which drained into the head of Megginnis Arm, was used primarily for residential housing (67%). About 13% of the land was occupied by commercial developments, 12% was forested, and 9% was used for agricultural purposes or was in old fields (Harriss and Turner, 1974). Figure 5.3 depicts the experimental watersheds and their relationship to Lake Jackson.

During the course of the watershed study (August, 1973–April, 1974) water samples were taken by automatic sampling devices from the streams draining the experimental watersheds. During storms, the urban watershed stream was sampled at 20- or 30-min intervals, and the forested watershed stream was sampled at 1-hour intervals. Stream flow rates were measured by the U.S. Geological Survey. During the study period 13 storms were sampled in the urban watershed, and eight storms were sampled in the forested watershed.

Table 5.4 lists mean concentrations of selected constituents of runoff from the urban and forested watersheds based on this sampling. In the urban watershed, SS and nutrient concentrations were both significantly higher in storm runoff than in baseflow. In the forested watershed SS concentrations were over twice as high in storm runoff than in baseflow, but nutrient concentrations were little different. The most dramatic comparison in Table 5.4 is the 30-fold increase in SS concentrations in urban stormwater runoff versus urban baseflow. This large difference is undoubtedly attributable to the many areas of exposed land that had been cleared for building construction purposes and for the construction of an interstate highway (I-10). In the case of the I-10 construction, the contractor had signed an agreement to clear only small areas of land at a time, but in fact after construction had begun the entire right-of-way for the highway was bulldozed all at once, leaving an enormous area of bare land that contributed large amounts of sediment to storm runoff for many months before highway construction was completed and the

[2] area: 6.3 km²-forested, 7.2 km²-urban; topography: 4.4% average slope-forested, 4.2% average slope-urban; soil type: sandy loam, well drained over loamy subsoil.

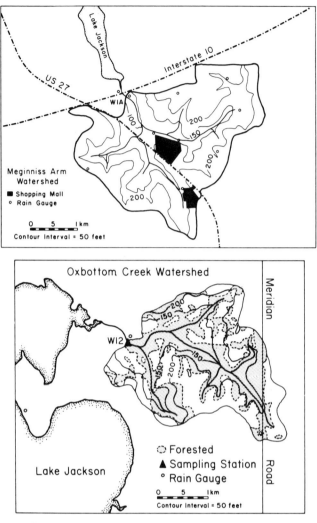

Figure 5.3 Maps of the urban watershed (Megginnis arm watershed) and the forested watershed (Oxbottom Creek watershed). [Redrawn from Harriss and Turner (1974).]

ground finally sodded. The failure of local authorities to enforce the signed contract or to penalize in any way the construction company is another example (recall Kaneohe Bay) of the futility of antipollution laws that are not enforced properly.

In this particular case the Department of Transportation, which had contracted for the work, was persuaded to install settling basins downstream of the I-10 construction in an attempt to reduce sediment loading to Megginnis Arm. As we noted earlier, such sediment traps are futile in moderate or heavy storms, since the water flushes through the traps so rapidly that there is virtually no chance for settling, and in fact accumulated sediments may be scoured out. It is true that the traps in this case retained some of the larger size sediment particles that washed into them during baseflow or during light storms. As a result it was necessary to periodically dredge the traps. Incredibly, the sediment dredged from the traps was deposited initially next to the stream on the downstream side of the trap, where it was well situated to wash back into the stream with the next storm and be transported down to Megginnis Arm. This situation was corrected later after notification of local authorities.

Table 5.5 lists annual discharges of water, SS, and nutrients into Lake Jackson from the urban and forested watersheds estimated from sampling during the study period. The discharge of water was about 1.5 times greater from the urban watershed than from the forested watershed. This difference was caused undoubtedly by the relatively large area of the urban watershed covered by impervious surfaces. As a result a greater percentage of the rainfall that fell on the urban watershed ran off rather than percolating into the ground. The discharge of SS was almost 60 times greater from the urban watershed than from the forested watershed, a result caused by both the greater amount of water runoff and the higher SS concentration in the water from the urban watershed. Inorganic P loading from the urban watershed was only 20% greater than from the forested watershed, but inorganic nitrogen loading from the urban watershed was about 150% higher. The greater sensitivity of inorganic N loading to urbanization probably reflects the fact that nitrate N, unlike phosphate P, does not bind to soil particles. As a result nitrate is more likely to be flushed from exposed soils than phosphate.

Finally, Table 5.6 compares the importance of storm runoff versus baseflow in the materials balance of Lake Jackson. The comparison has been made by calculating the number of days of baseflow that would be required to deliver to the lake the amount of the given substance delivered by an average storm. The importance of storm events in the materials balance of the lake is apparent from Table

Table 5.4 Mean Concentrations of SS, Inorganic Nitrogen, and Inorganic Phosphate in Dry Weather Runoff and Storm Runoff from the Forested and Urban Watersheds (in mg/l)

	Baseflow (dry weather)		Stormwater	
	Urban	Forested	Urban	Forested
SS	10	12	299	34
Inorganic N	0.17	0.10	0.29	0.12
Inorganic P	0.06	0.13	0.12	0.10

Source. Turner and Burton (1975).

Table 5.5 Rainfall and Estimated Annual Loadings of Water, SS, and Nutrients from the Urban and Forested Watersheds

	Urban	Forested
Rainfall (cm yr^{-1})	249	254
Total runoff (cm yr^{-1})	48.3	31.2
SS (kg ha^{-1} yr^{-1})	2376	40
Inorganic P (kg ha^{-1} yr^{-1})	0.18	0.15
Inorganic N (kg ha^{-1} yr^{-1})	0.41	0.16

Source. Turner et al. (1977).

5.6. If an average storm occurs about once per week, it is clear from Table 5.6 that storm runoff loadings are of at least comparable importance and in most cases of greater importance than baseflow loadings for all the constituents considered. Therefore any realistic attempt to obtain a materials balance for such a system, whether the watershed be forested or urban, must be based on sampling during both storms and dry weather. It is not adequate to sample only when the sun is shining. The most dramatic figure in Table 5.6 is the number of years of sediment baseflow loading to which an average storm event was equivalent in the urban watershed. This study illustrates dramatically the tremendous perturbation to the sediment loading characteristics of a watershed that can be caused by urbanization and land clearing.

A bit more can be learned about the characteristics of urban runoff by examining the time course of pollutant concentrations and runoff during storm events. Figure 5.4, for example, shows water discharge and the concentrations of SS, silicate, and dissolved inorganic P in the runoff from both the urban and forested watersheds during a storm on November 21, 1973. The rain itself was characterized by two intensity peaks that are mimicked by the water discharge from the urban watershed. This close correspondence between rainfall intensity and stream discharge is characteristic of watersheds covered by a large percentage of impervious surfaces. Runoff from such watersheds is rapid. In this case the runoff from the urban watershed was almost complete 6 hours after the beginning of the storm, whereas storm runoff from the forested watershed was barely underway by that time. The time lag in water discharge from the forested watershed reflects the fact that much of the early rainfall was absorbed by vegetation and soil, and the flow of overland runoff was delayed and impeded by the vegetative cover. In fact much of the early runoff from the forested watershed consisted probably of subsurface flow in the form of percolating groundwater rather than overland flow. This subsurface flow transports no sediment and creates no water pollution problems. Stream discharge from the forested watershed in no way reflects the two-peaked intensity pattern evident in the urban runoff, because the runoff has been delayed and the rainfall intensity characteristics consequently smeared out or integrated in the actual stream discharge. The relative importance of subsurface flow is much smaller in the case of the urban watershed than the forested watershed due to the high percentage of impervious surfaces in urban areas.

It is clear from Figure 5.4 that the concentration of all three constituents shown in the figure tends to peak prior to the peak in stream discharge. This phenomenon is referred to as the *first flush effect* and arises because nutrients and particulate

Table 5.6 Days of Baseflow Required to Deliver as Much of the Indicated Constituent as was Delivered by an Average Storm During the Lake Jackson Watershed Study

	Urban Watershed	Forested Watershed
Water	19.3	6.2
SS	3859 (10.6 yr)	21
Inorganic N	30.0	8.9
Inorganic P	21.8	5.8

Source. Turner and Burton (1975).

materials that have accumulated on the ground since the last storm wash into the storm sewer system with the early runoff. After this first flush, concentrations tend to drop, because most of the easily eroded or flushed materials are gone already. This explanation is somewhat more applicable to silicate and phosphate than SS, since the peak concentration of the former two occurred clearly before the peak in water discharge. SS concentrations, on the other hand, were at or very near their maximum at approximately the same time as the peak in water discharge. The concentration of SS was thus more closely correlated with stream flow than the concentration of the two dissolved constituents. This result should not be particularly surprising considering that sediment erosion is closely correlated with flow intensity, whereas simple chemical dissolution is not. The greater intensity of runoff from the urban watershed is apparent in Figure 5.4. The peak discharge from the urban watershed was about ten times greater than the peak discharge from the forested watershed. This difference reflects both the greater volume of water that ran

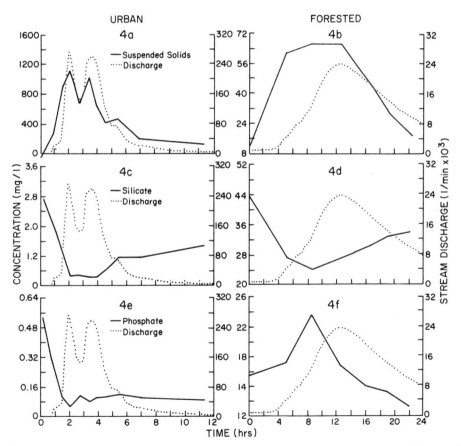

Figure 5.4 Comparison of stream discharge and concentrations of SS (a,b), silicate (c,d), and phosphate (e,f) versus time in the two experimental watersheds for the storm of November 21, 1973. Note differences in horizontal (time) and vertical (discharge and concentration) scales between watersheds. [Redrawn from Harriss and Turner (1974).]

off the urban watershed and the fact that the runoff was confined to a shorter time interval. The greater intensity of runoff from the urban watershed means that the urban runoff has a much greater erosive potential than the runoff from the forested watershed.

The first flush phenomenon has potential utility in the treatment of urban runoff if in fact a large percentage of the discharge of certain pollutants is associated with a relatively small percentage of the total runoff. In other words, if the early runoff from a storm is grossly contaminated and if subsequent runoff is relatively clean, then one may be able to eliminate much of the discharge of the pollutant from the watershed by retaining and treating only the first portion of the runoff. For example, data summarized by Whipple et al. (1983) suggest that about 60% of the efflux of extractable zinc and lead from urban watersheds is associated with only the first 20% of the runoff (Figure 5.5). The same study indicates that the first flush effect is much less dramatic for total N and P and virtually nonexistent for total dissolved P (Figure 5.5). The magnitude of the first flush effect obviously depends on the nature of the pollutant, but will also depend on the time since the previous storm, the degree and type of developments in the watershed, and perhaps a number of other factors. When the data shown in Figure 5.4, for example, are plotted in the same manner as the data in Figure 5.5, it becomes apparent that in the case of this particular storm the percent efflux of SS exceeded the percent of water runoff by at most ~10% at any time (Figure 5.6). Furthermore, the percent efflux of silicate lagged behind the percent of water runoff, because silicate concentrations tended to increase after the peak in water runoff. Thus while it is generally true that the early runoff from a storm contains the highest concentration of many contaminants, it is by no means the case that retention and treatment of only the early runoff can be consistently relied upon to eliminate a major portion of the pollution caused by urban runoff. Treatment of the early runoff is certainly a step in the right direction, but it is only a step.

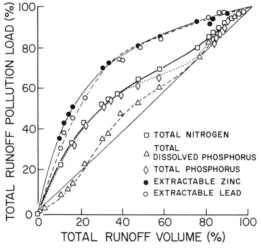

Figure 5.5   Percent total runoff of certain pollutants versus percent total runoff of water in storms. [Redrawn from Whipple et al. (1983).]

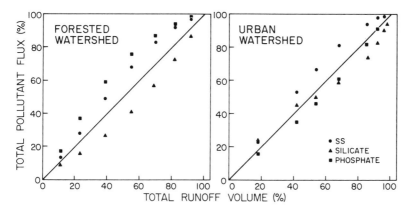

Figure 5.6   Percent total runoff of SS and nutrients versus percent total runoff of water calculated from the data in Figure 5.4.

Correctives

As a result of the 3-year study of Lake Jackson, Harriss and Turner (1974) made the following recommendations to local authorities to minimize the damaging effects of urban runoff on the lake:

1. Land clearing, site preparation, and utility installations be confined insofar as possible to the dry season of the year, when the possibility of heavy rains' washing away large amounts of exposed sediment would be minimal.[3]

2. Developments requiring drastic disturbances of the subsoil such as deep grading for level parking lots and slab foundations be prohibited in the watershed, or where permitted, that extensive stabilization measures be required.

3. Developers be required to provide facilities to percolate any additional storm runoff generated by land use change, where disposal of such additional runoff would otherwise be to the primary drainage system in the watershed.

4. Impact zoning principles be adopted for all future land development in the watershed. In other words, the impact of certain types of real estate developments on the environment be considered in planning the growth of the city.

5. Early storm runoff from large shopping malls and other heavily used, paved parking surfaces be percolated or treated before disposal, since such runoff was found to be grossly contaminated with heavy metals, hydrocarbons, and nutrients.

6. Porous paving materials be considered for the paving of storm drainage ditches, road shoulders, residential driveways, roadways, and parking areas near the lake and in any areas where the benefits caused by reduced runoff from the use of such porous paving materials be judged to exceed any possible liabilities.

---

[3]Rainfall in the Tallahassee area is highly seasonal.

7. Extant and future sedimentation and nutrient control ordinances be effectively enforced, as was not the case with the I-10 land clearing.

8. The effect of small and large ponds, including settling basins, on nutrient and SS loading to the lake from selected watersheds be determined, in the hopes that certain types of ponds, including, for example, ponds containing marsh plants or similar vegetation or a system of artificial baffles to retard flow, might prove effective in reducing stream inputs of SS and nutrients.

Most of these recommendations are nothing more than common sense measures that any municipality should consider if urban runoff were perceived to be a problem. The last suggestion is a bit more extreme. As it turns out, however, water quality conditions in Lake Jackson became of such concern that the City of Tallahassee decided finally to implement the last suggestion. A facility to treat urban runoff into the head of Megginnis Arm was completed in 1983 and is shown schematically in Figure 5.7. Runoff enters initially a detention pond with a capacity of 163 million liters. The water then passes through a sand filter with a surface area of 1.8 ha. The filtrate from the sand filter is routed to a triple box culvert running underneath I-10, and from there enters a diversion impoundment, where it is joined with direct runoff from I-10. The combined runoff then enters a 2.5 ha artificial marsh. The marsh has an average depth of 0.5 m and is divided into three cells. The first, second, and third cells are planted with the macrophytes *Thypha*, *Scirpus*, and *Pontederia*, respectively.

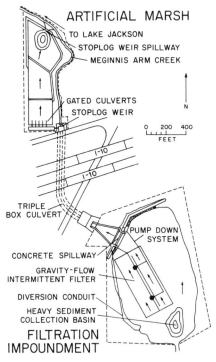

Figure 5.7 Schematic of the Lake Jackson Stormwater Treatment Facility. [Redrawn from LaRock (1988).]

The detention basin is intended to settle out suspended solids, and with a capacity of 163 million liters it can contain all the runoff from small to moderate storms. The sand filter is intended to remove particulate materials which do not settle out in the detention basin. Much of the material collected by the sand filter consists of silt and clay size particles that settle very slowly even under quiescent conditions. The artificial marsh is intended primarily to remove dissolved nutrients, which are expected to be assimilated by the aquatic macrophytes and associated epiphytes. A concrete spillway built into the detention pond dam and a stoplog weir at the diversion impoundment leading into the marsh permit runoff that exceeds the capacity of the system components to bypass the sand filter and marsh, respectively. The performance of the treatment facility was studied from 1983 to 1987 by LaRock (1988), and as might be expected the ability of the system to treat the runoff from a particular storm was found to depend very much on the amount of runoff and the antecedent weather conditions. It is sufficient for our purposes to examine the performance of the system only during the storm of June 11, 1985, because this analysis reveals clearly both the strengths and weaknesses of the treatment facility.

The storm in question actually consisted of three storm events lasting nearly one week. The total amount of runoff entering the head of the detention basin was 315 million liters. Of that amount, 155 million liters passed through the detention basin and sand filter, 108 million liters flowed over the detention basin dam and bypassed the sand filter, and 52 million liters remained in the detention basin. The $155 + 108 = 263$ million liters that flowed through the triple box culvert were joined in the diversion impoundment by an additional 1080 million liters of runoff from I-10. Of the $263 + 1080 = 1343$ million liters of water that passed through the diversion impoundment, 223 million liters were actually routed through the artificial marsh, the remaining $1343 - 223 = 1120$ million liters being bypassed to Megginnis Arm.

The facility did a rather good job of removing pollutants from the water that it treated (Table 5.7). The combination of the detention basin and sand filter reduced SS and total P concentrations by at least 97% and total N concentrations by 92%. The artificial marsh reduced SS concentrations by about 73% and total N and P concentrations by 36% and 60%, respectively. It is obvious that the system is capable of effecting a dramatic reduction in SS and nutrient concentrations when not overloaded.

Table 5.7   Percentage Change in Flux of Various Constituents Through Detention Basin/Sand Filter and Artificial Marsh During Storm of June 11, 1985

	Detention Basin/Sand Filter	Artificial Marsh
Inorganic solids	−99.0	−76.4
Organic solids	−98.1	−69.1
Total N	−91.6	−36.4
Total P	−97.0	−60.2
Calcium	91.7	−5.3
Magnesium	221.1	−22.1

Source. LaRock (1988).

One obvious problem with the system is the fact that a large percentage of the runoff from I-10 received no treatment at all, and much of the remaining I-10 runoff passed only through the artificial marsh. The magnitude of the runoff from I-10 was evidently not appreciated at the time the facility was built. As the drainage system is designed, a large percentage of the runoff from I-10 cannot reach the head of the detention basin and instead is routed directly to the diversion chamber. However, even if all the I-10 runoff were routed to the head of the detention pond, it is obvious, in this case at least, that most of the water would have bypassed both the sand filter and artificial marsh. This observation underscores the fact that urban runoff treatment systems must have the capacity to retain very large volumes of water if they are to give adequate treatment to the runoff from large storms. One major flaw in the Lake Jackson stormwater treatment facility is the fact that there exists no adequate mechanism for removing water from the detention basin between storm events. As a result the capacity of the detention basin is sometimes much less than 163 million liters. The detention basin actually contains a dewatering system that pumps water out of the detention basin and sprays it over the surface of the sand filter when the water level in the detention basin drops below the level of the sand filter. However, use of this dewatering system was discontinued after only a few weeks of use due to the high electrical costs associated with pumping the water. Subsequently, the storage capacity of the detention basin has been regulated by operating manually the drain valve in the spillway. This procedure, however, causes the water to bypass the sand filter and in practice has not provided adequate storage capacity for subsequent storms. Some bypass of the sand filter has occurred during 60% of all storm events (LaRock, 1988).

A second problem associated with the inefficient removal of water from the detention basin has been the accumulation of ammonia in the water that remains in the detention basin after storms. The ammonia accumulates as the result of the breakdown of organic matter trapped in the detention basin. The design of the sand filter is such that lateral water movement can occur from the detention basin through the sides of the filter. Ammonia-rich water therefore enters the sand filter and percolates down into the limestone that underlies the sand. As the water level is lowered, air enters the sand filter, and the ammonia is oxidized to nitric acid by chemosynthetic bacteria. The nitric acid is then neutralized by the calcium and magnesium ions in the limestone. The result is dissolution of the limestone and export of significant amounts of calcium and magnesium with the water that leaves the sand filter (Table 5.7). If this situation is allowed to continue, the sand filter will disappear someday into a sink hole.

A third problem, and one which was anticipated, is clogging of the sand filter. The combination of the detention basin and sand filter does an excellent job of removing SS (Table 5.7), and much of the particulate material retained by the sand filter consists of clay and silt size particles, which are too small to be settled efficiently in the detention basin. As the clay and silt accumulate on the sand filter, its effective porosity is reduced greatly, and the flow rate of water through the filter is slowed. According to LaRock (1988), the flow rate of water through the sand filter was reduced by about 70% during the first 3 years of operation of the sand filter. This problem could be corrected by replacing periodically roughly the upper 25 cm of the material in the sand filter. Serious clogging of the sand filter should not occur if a reasonable schedule of maintenance is followed.

The performance of the Lake Jackson stormwater treatment facility illustrates both the potential and problems of systems designed to treat urban runoff. The capacity of the system must be large to handle the amount of water that flows off urban areas during severe storms. It is often difficult to identify sufficient areas of land for treating urban runoff in existing urban settings. The problem is solved undoubtedly most economically if the treatment facilities are incorporated at the planning stage into the design of future urban developments. Because public awareness of the problems caused by urban runoff is a relatively recent phenomenon, most storm sewer systems have not been designed in a way that easily accommodates large treatment facilities. Chicago has solved the problem with its underflow tunnel system, but at considerable expense.

The Lake Jackson facility does an excellent job of treating the water that passes through it. Flaws in the design, however, and inadequate maintenance have resulted in the bypass of at least some runoff during about 60% of storm events. The artificial marsh is a creative idea, but at higher latitudes the biological uptake of nutrients by marsh plants would cease during the winter months. It seems fair to say that one important component of any plan to minimize pollution from urban runoff should be to minimize both the amount of runoff that must be treated and the concentration of pollutants in the water. In this regard, the recommendations made by Harriss and Turner (1974) should be given serious consideration.

REFERENCES

Abernathy, A. R. 1981. *Oxygen-consuming organics in nonpoint source runoff.* Environmental Protection Agency. EPA-600/S3-81-033.

City of Chicago. 1972. *Development of a flood and pollution control plan for the Chicagoland area.* 111 pp.

Cohrs, A. 1962. Storm water tanks in the combined sewerage system of Berlin. *Gas Wasserfach.*, **103**, 947–952.

Di Vita, L. R. 1990. *TARP Status Report as of 9/14/90.* Metropolitan Water Reclamation District of Greater Chicago.

Environmental Protection Agency. 1983. *The 1982 Needs Survey. Conveyance, Treatment, and Control of Municipal Wastewater Combined Sewer Overflows, and Stormwater Runoff.* Summary of Technical Data. EPA/43019-83-002.

Evans, F. L., E. E. Geldreich, S. R. Weibel, and G. G. Robeck. 1968. Treatment of urban stormwater runoff. *J. Water Pollut. Control Fed.*, **40**, R162-R170.

Gray, B., and L. S. Lau. 1973. Alternative methods for sewage disposal. In *Estuarine Pollution in the State of Hawaii.* Vol. 2. University of Hawaii Water Resources Research Center Tech. Rep. No. 31. pp. 343–435.

Harriss, R. C., and R. R. Turner. 1974. *Job Completion Report, Lake Jackson Investigations.* Florida State University Marine Laboratory, Tallahassee, Florida. 231 pp.

LaRock, P. 1988. *Evaluation of the Lake Jackson Stormwater Treatment Facility.* Final Report to Florida Dept. of Environmental Regulation. 246 pp.

Peterson, S. A., W. E. Miller, J. C. Greene, and C. A. Callahan. 1985. Use of bioassays to determine potential toxicity effects of environmental pollutants. In *Perspectives on Nonpoint Source Pollution.* Environmental Protection Agency, EPA 440/5-85-001. pp. 38–45.

Randall, C. W., T. J. Grizzard, D. R. Helsel, and D. M. Griffin, Jr. 1981. A comparison of pollutant mass loading in precipitation and runoff in urban areas. In *Proceedings, 2nd International Conference on Urban Storm Drainage*, IAWPR, University of Illinois, Urbana. pp. 2, 29–38.

Turner, R. R., T. M. Burton, and R. C. Harriss. 1977. Lake Jackson watershed study. In D. L. Correll (Ed.), *Watershed Research in Eastern North America. A Workshop to Compare Results*. Vol. 1. Chesapeake Bay Center for Environmental Studies. Edgewater, MD. pp. 19–32, 211–224, 323–342, and 471–485.

Weibel, S. R. 1969. Urban drainage as a factor in eutrophication. In *Eutrophication: Causes, Consequences, Correctives*. National Academy of Sciences, Washington, D.C. pp. 383–403.

Whipple, W., N. S. Grigg, T. Grizzard, C. W. Randall, R. P. Shubinski, and L. Scott Tucker. 1983. *Stormwater management in urbanizing areas*. Prentice-Hall, NJ. 234 pp.

Whipple, W., and J. V. Hunter. 1980. *Settleability of Urban Runoff Pollution*. Rep. Water Resources Research Institute, Rutgers University, New Brunswick, NJ.

6

SEWAGE TREATMENT

The toilet area shall be outside the camp. Each man must have a spade as part of his equipment; after every bowel movement he must dig a hole with the spade and cover the excrement. (Deuteronomy 23: 12,13)

Recognition of the need to properly dispose of sewage has existed for a long time. However, in the United States the first comprehensive sanitary sewer system was not constructed until 1855 (in Chicago), and the first sewage treatment facility was not completed until 1886 (in New York City). Problems associated with inadequate treatment or improper disposal of sewage led to the proliferation of sewage treatment plants (STPs) in the United States during the twentieth century, and by the year 2000 it is estimated that the wastewater generated by about 90% of the U.S. population will be processed by municipal STPs. Since 1972 U.S. governmental agencies have spent more than $72 billion upgrading the nation's sewage systems (Marinelli, 1990). At the present time there are approximately 15,500 municipal STPs in the United States, and they treat roughly 40 billion m^3 of raw sewage per year.

As we have seen in the previous few chapters, the discharge of sewage, whether treated or not, can create serious water pollution problems in the receiving body of water. The high concentrations of suspended solids and nutrients as well as the biological oxygen demand of raw sewage give it great potential for causing cultural eutrophication problems. The fact that raw sewage also contains a high concentration of pathogens is cause for concern from a public health standpoint as well. Standard sewage treatment removes most of the SS, BOD, and pathogens from the raw sewage, but less than half the nitrogen and phosphorus. Hence even treated sewage can be an environmental problem from the standpoint of nutrient enrichment. In this chapter we focus on primarily the methods used to remove SS and BOD from wastewater, but we also consider methods for dealing with the nutrients, either by removing them or putting them to practical use. We consider in detail the problem of sewage pathogens in Chapter 7.

PRIMARY, SECONDARY, AND TERTIARY TREATMENT

Sewage treatment plants differ greatly in the details of their operations, in part because the characteristics of the sewage that they treat may also differ greatly. Although the characteristics of domestic sewage are rather well defined, the composition of industrial sewage is highly variable. For example, some industrial wastes may contain toxic organic substances, pesticides, or heavy metals. As pollution laws become more strict, the release of such substances by industries will undoubtedly decline, but in the meantime sewage plants that receive effluent from such operations must be able to handle such substances as well as the usual domestic wastes. It is therefore perhaps surprising that the basic design of most STPs is so similar.

Figure 6.1 illustrates the basic features of a secondary STP. When the raw sewage first arrives at the STP, it usually passes through a screening device to remove large objects that have somehow gotten into the sewer pipes and might damage components of the STP if they are allowed to pass through. The screening device may consist of nothing more than a series of metal bars spaced 5–10 cm apart. The trash that accumulates on these racks is removed periodically and deposited in a landfill. An alternative to the trash rack is a mechanical device called a *comminuter*, which pulverizes any solid objects in the raw sewage. Once the sewage has passed through the screening device or comminuter, it flows into a so called grit chamber, where particles of roughly sand size and larger settle out. Removal of this coarse grit is intended to protect the pumps in the rest of the system. With large objects and grit thus removed, the sewage is pumped to a primary settling tank or primary clarifier. Much of the flow through the rest of the system is by gravity.

The primary clarifier is nothing more than a large quiescent tank in which settleable or floatable solids are removed from the sewage. The tank may be circular or rectangular. In the former case the sewage is introduced at the center and slowly flows to the edge, where it exits under a baffle. In the latter case the sewage is introduced at one end and exits at the other. Solid materials that float to the surface during the residence time of the sewage in the tank[1] are removed by a mechanical skimming device that pushes floating objects to a hopper from which they are pumped to the anaerobic digester. The bottom of the tank is inclined, and the sludge that accumulates on the bottom is moved by a similar mechanical device toward the deepest part of the tank, where it is drawn off and pumped to the anaerobic digester. The material skimmed from the surface and pumped from the bottom of the primary clarifier is called the *primary sludge*. Treatment involving no more than removal of this primary sludge is termed *primary sewage treatment*.

If the STP is a secondary treatment plant, the effluent from the primary clarifier flows into a second tank where biological processes are used to remove much of the BOD in the effluent. The basic idea is to allow organisms living in the tank to consume the organic substances in the primary effluent. In this process the consumer organisms grow and multiply, but at the same time a large percentage of the organic substances are respired. As a result, the amount of potentially oxidizable material in the waste is greatly reduced. We will describe this phase of the treatment in detail later on. The effluent from this treatment process consists in part of the organisms

---

[1]The residence time is usually a few hours.

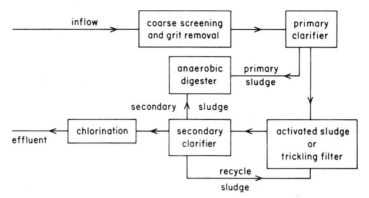

Figure 6.1   Diagram of flow through a secondary sewage treatment plant.

that have been feeding off the organic wastes. Most of the biomass of these organisms can be removed by running the effluent into a secondary settling tank or clarifier. The organisms' biomass settles to the bottom in the form of a floc or sludge and is removed in a manner similar to that employed in the primary clarifier. Normally no attempt is made to skim the surface of the secondary clarifier, since little if any of the secondary sludge floats to the surface. A portion of the secondary sludge is pumped to the anaerobic digester, while the remainder is recycled.

The entire treatment process just described is called *secondary sewage treatment*. According to federal law, the treatment process should remove at least 85% of the SS and BOD from the raw sewage and reduce the SS and BOD concentrations to less than 30 mg/l. Primary sewage treatment typically removes 40–75% of the SS and any BOD associated with the SS. BOD removal efficiencies associated with primary treatment are usually in the 30–40% range. Regardless of whether the STP is a primary or secondary facility, the liquid effluent is usually not released to the environment until it has been disinfected, usually by chlorination, to reduce the concentration of pathogens to an acceptably low level. With this overview of a typical secondary STP, let's take a closer look at several of the steps in the treatment process.

Secondary Treatment for BOD Removal

By far the most commonly used methods for removing BOD at the secondary level are the so-called activated sludge and trickling filter processes. Both methods use primarily biological means to reduce BOD.

Trickling Filters

A trickling filter usually consists of a cylindrical tank 2–3 m in depth filled with rocks. The rocks have a diameter of typically 2–10 cm and should be loosely packed and as nearly spherical in shape as possible to facilitate the downward flow of

---

[2]For example, if the plant is being forced to treat more sewage than it was designed to handle.

sewage and the upward flow of air through the system. The effluent from the primary clarifier is sprinkled over the surface of the bed of rocks by one or more rotating arms. Usually a given area of the bed is sprinkled every ~30 seconds, but faster application rates are possible and sometimes necessary.[2]

As the sewage trickles down through the bed of rocks, the dissolved and suspended organic substances in the water are ingested by bacteria and fungi that grow on the surface of the rocks in a glutinous film. The bacteria and fungi are in turn fed on by a variety of higher trophic level organisms, including protozoa, rotifers, nematodes, worms, and insects. Since the composition of the sewage changes (due to the feeding of the bacteria and fungi) as it trickles from the top to the bottom of the trickling filter, there is a tendency for the species composition of the food chain to change from top to bottom. Because the biological community living in the trickling filter consists of several trophic levels, the organic substances assimilated from the sewage by the bacteria and fungi may pass through several food chain transfers, and a large percentage of the organics may therefore be respired. Typically about 70% of the organic wastes in the primary clarifier effluent are oxidized in the trickling filter. Of the remainder, 15–20% is converted to organism biomass. The film of organic biomass that develops on the trickling filter rocks periodically sloughs off and is carried by the water to the bottom of the filter, where it is drained off with the liquid waste. Part of the trickling filter effluent is routed to the secondary clarifier, where settling of the sludge removes additional BOD from the water. However, part of the effluent is normally recycled back to the trickling filter to maintain the population of decomposer organisms in the filter and to further reduce the BOD.

Bacteria grow by simple cell division, and given adequate food their population numbers increase exponentially with time. Biomass production during this so-called logarithmic phase of growth is relatively efficient, i.e., respiration losses are relatively small. The operator of a STP, however, is not interested in producing large numbers of bacteria. Instead, the operator is interested in having the bacteria and other organisms burn up as much of their food as possible, or in other words, to grow inefficiently. As the concentration of bacterial food in the water declines below a certain level, both the efficiency of bacterial biomass production and the rate of biomass production begin to decline. Realizing this fact, the STP operator attempts to operate his trickling filter so that the organisms in the system are growing rapidly enough to consume most of the organic substances in the water but slowly enough to keep biomass production efficiency as low as possible. In a trickling filter, the concentration of food is obviously highest at the top of the filter and lowest at the bottom. Therefore the microorganisms near the top are growing rather rapidly and efficiently, whereas organisms near the bottom are growing more slowly and oxidizing much of their food. The STP operator can control to a degree the growth rate and growth efficiency of the organisms in the system by adjusting the amount of trickling filter effluent that is recycled. He is constrained of course by the fact that the net outflow of water from the trickling filter (gross minus recycled flow) must equal the inflow of water to the plant.

Activated Sludge

An activated sludge system usually consists of a long rectangular tank or series of tanks 3–5 m deep. The effluent from the primary clarifier is introduced at one end of

---

[2]For example, if the plant is being forced to treat more sewage than it was designed to handle.

the tank and exits at the other end. The residence time of the water in the system is usually about 4–8 hours. During its passage through the tank(s), the sewage is mixed vigorously from below with bubbles of air that keep the contents of the tank in a state of great turbulence. This vigorous mixing and aeration are necessary to prevent the oxygen concentration in the water from dropping below 2–3 mg/l. Otherwise the respiratory activity of the organisms in the tank is likely to be slowed.

Because of the extreme buffeting that goes on in an activated sludge tank, only very small aquatic organisms are suited to the environment. Bacteria are the principal decomposer organisms in the tank, with fungi being of strictly secondary importance unless the system is not operating properly. Protozoan ciliates and flagellates, small rotifers, and nematode worms usually complete the food chain, although other classes of organisms such as insects may occasionally become established. The bacteria, protozoans, and other organisms in the system are found primarily aggregated together in flocculent masses that are called collectively the activated sludge. Because the food chain in an activated sludge system is usually shorter than the food chain in a trickling filter, a smaller percentage of the organics in the inflow is oxidized and a greater percentage is incorporated into sludge biomass. Typically about 50% of the organics from the primary clarifier effluent are oxidized in the tank, and the volume of secondary sludge may be as much as twice that produced by a trickling filter. This secondary sludge is of course removed from the effluent in the secondary clarifier. A small part of the secondary sludge is usually pumped to the anaerobic digester; the remainder is recycled to the activated sludge tank(s). This recycling of sludge is essential with the activated sludge process, since the primary clarifier effluent normally produces only about 5% of the sludge required for its treatment (Warren, 1971, p. 358). As is the case with a trickling filter, the operator of an activated sludge STP may control to a certain extent the age and hence the biomass production efficiency of the sludge by adjusting the proportion of the sludge that is recycled. Although one pass through the activated sludge system may take only 4–8 hours, recycling increases the effective age of the sludge to 3–4 days. Manipulations of the system are of course constrained by the total volume of sewage per unit time that the plant must handle and by economic factors. For example, additional recycling of sludge will normally require additional aeration and therefore increase the cost of treatment. The objective of the treatment process is of course to oxidize as much of the waste organics as possible and therefore minimize the amount of secondary sludge that must be handled by the anaerobic digester.

Pros and Cons of Trickling Filter and Activated Sludge Treatment

Table 6.1 provides a comparison of the efficiency with which various sewage treatment systems remove SS and BOD. As is evident in Table 6.1, efficient trickling filter and activated sludge STPs are both capable of achieving the 85% removal of BOD and SS required by federal law. What, then, are the relative merits of these two types of treatment processes?

There is virtually no difference in the efficiency with which chlorination kills pathogens in the effluent from the two types of treatment. However, activated sludge plants tend to achieve higher SS and BOD removals than trickling filter plants. It is noteworthy that the difference between 90 and 95% BOD removal is a factor of 2 in the amount of BOD released by the plant. Thus activated sludge plants tend to be significantly more successful at accomplishing what they are designed to do, namely to remove SS and BOD. However, the trickling filter produces less second-

Table 6.1   Percent Removal of BOD and SS Using Various Treatment Methods

Process	% BOD Removal	% SS Removal
Septic tank	25–65	40–75
Primary treatment	30–40	40–75
Primary + trickling filter	80–90 or more	80–90 or more
Primary + activated sludge	85–95 or more	85–95 or more

Source. Klein (1966).

ary sludge than an activated sludge system, and trickling filter biological communities seem to be less perturbed by sudden changes in the character of the sewage than are activated sludge communities. This greater resilience of the trickling filter communities probably reflects the greater variety of organisms that inhabit trickling filters. For example, if certain bacterial species in a trickling filter are killed suddenly due to a change in the temperature or pH of the sewage, other bacterial species already present in the trickling filter community may quickly move in to fill the vacant niche. However, in an activated sludge system there are likely to be fewer types of decomposer organisms, and if some of these are killed there are relatively fewer species to move in and take their place. Because of the greater resilience of trickling filter communities to perturbations and the costs associated with aerating activated sludge tanks, trickling filters are generally less expensive to operate than activated sludge systems.[3] On the other hand, trickling filters require more land area to build and are more costly to construct. Furthermore, trickling filters tend to attract flies[4] and to have more of an odor problem than activated sludge systems.

Although the foregoing discussion makes it clear that there are pros and cons to both trickling filter and activated sludge systems, present federal regulations mandating 85% removal of BOD and SS have tended to make activated sludge systems more popular, because it is generally easier to achieve 85% BOD and SS removal with an activated sludge system than with a trickling filter. Trickling filters may be preferred, however, in cases where large fluctuations in sewage characteristics are anticipated. For example, one might prefer to build a trickling filter plant to treat sewage containing wastewater from industrial operations where accidents or shutdowns might lead to a sudden change in the characteristics of the sewage. On the other hand, sewage that is largely of domestic origin has rather predictable and constant characteristics. Judging from experience, fluctuations in the characteristics of domestic sewage usually do not create unmanageable problems in activated sludge STPs.

The Anaerobic Digester

What happens to the sludge that is collected from the primary and secondary clarifiers? This organic biomass is putrescible and therefore cannot be simply deposited at a convenient landfill. Although it is theoretically possible to com-

---

[3]In recent years, however, more efficient mechanisms for aerating activated sludge tanks have been developed. The strategy has been to reduce the size of the air bubbles and hence increase the contact area between air and water. The result has been roughly a twofold reduction in aeration costs.
[4]Recall that insects are one component of the trickling filter community.

pletely oxidize this sludge by burning it in a furnace, a minority of STPs in the United States incinerate their sludge. In the past some STPs have disposed of their sludge in the ocean, but this practice came to an end when federal legislation prohibiting ocean dumping of sludge was enacted in 1981 (Clark et al., 1984). The principal alternatives to incineration are either deposition in a sanitary landfill or some form of land application. In either case the putrescible nature of the sludge must be eliminated before it is disposed of. This task is accomplished by pumping the sludge to a closed, usually cylindrical tank where it provides food for a special class of microorganisms that carry on their metabolic activities in the absence of oxygen, i.e., anaerobically. The products of the anaerobic digester are digested sludge, a supernatant fluid, and gases. The digested sludge is relatively stable and inoffensive compared to the primary and secondary sludges. Furthermore, its volume is only about one-third that of the primary and secondary sludges. This reduction in volume is achieved both by the conversion of some of the organics in the raw sludge to gases, primarily methane, by anaerobic catabolism and by dewatering the sludge to reduce its moisture content. The raw sludge typically has a moisture content of 94–99% by weight, and most of its volume is due to water.

The anaerobic digester is invariably heated to speed up the metabolic rate of the microbes. Operating temperatures are usually in the 27–35°C range. The methane gas produced in the digester is more than adequate to provide fuel to heat the digester to the desired temperature. The excess methane and other gases produced in the digester are burned as waste products. The supernatant fluid from the digester retains some objectionable properties (e.g., odor) and is usually recycled through the secondary treatment process. Thus the only product of the digester that requires disposal is the stabilized sludge, which is either trucked off to a sanitary landfill or applied to the land as an organic fertilizer and soil conditioner.

The availability of suitable land to accommodate the sludge produced by the anaerobic digester can be a limiting factor in the use of this technique to treat the primary and secondary sludge. Under such conditions, the STP may resort to burning as a means of disposal. Because of its high water content, the sludge must first be dewatered, and the organic matter is then burned in a furnace. A small amount of inorganic ash remains, but the volume of this ash is far less than the amount of stabilized sludge produced by anaerobic digestion. About 20–25% of STPs in the United States dispose of their sludge in this way (Sopper and Kerr, 1981; Sopper and Seaker, 1984).

Tertiary Treatment

Tertiary treatment is, strictly speaking, any additional processing of secondary treatment effluents undertaken to further improve their quality. In some countries tertiary treatment means additional removal of SS and BOD, but in the United States tertiary treatment usually means nutrient removal (Ellis, 1980). From some of our previous case studies, it is clear that the nutrient concentrations in raw sewage are sufficiently high to stimulate objectionable algal growth in receiving waters, unless of course the effluent is greatly diluted in the receiving system. To get some idea of the nutrient problem, let's compare the amount of oxygen required to oxidize the algal biomass that could be produced using the nitrogen in raw sewage versus the typical BOD of 200 mg/l found in raw sewage. Recall that sewage usually has a low

N/P ratio relative to the needs of phytoplankton. Taking the total N of raw sewage to be 40 mg/l and assuming the oxidation of typical phytoplankton biomass to proceed via the equation (Riley and Chester, 1971, p. 173)

$$(CH_2O)_{106}(NH_3)_{16}H_3PO_4 \ + \ 138 \, O_2 \ \rightarrow \ 106 \, CO_2 \ + \ 122 \, H_2O \ + \ 16 \, NH_3 \ + \\ H_3PO_4$$

one concludes that it would take

$$\frac{40 \text{ mg N/l}}{14 \text{ mg N/mg atom N}} \quad \frac{138 \text{ mmoles O}_2}{16 \text{ mg atom N}} \quad \frac{32 \text{ mg O}_2}{\text{mmole O}_2} \ = \ 789 \text{ mg/l O}_2$$

to oxidize the phytoplankton biomass produced from the nitrogen in raw sewage. If a typical secondary sewage treatment plant removes 30% of the total N from the raw sewage and discharges the remaining 70% in the form of ammonia, and if the ammonia is subsequently assimilated by phytoplankton and converted into algal biomass, the decomposition of that algal biomass would be expected to consume $(0.7)(789) = 552$ mg/l O_2, which is about 2.8 times the BOD of the raw sewage. The depletion of oxygen in the hypolimnion of Lake Erie's central basin has been attributed largely to this sort of effect; that is, the loading of a system with nutrients that are incorporated into organic biomass whose ultimate decomposition consumes oxygen.

It should be clear from our discussion of primary and secondary sewage treatment that neither process is aimed directly at removing nutrients. In fact the oxidation of the organic substances in sewage releases nutrients rather than removing them. However, the sludge that is removed in the primary and secondary clarifier does contain nutrients. The nutrients are bound primarily in the organic biomass that makes up most of the sludge. If STPs operated so as to try to maximize BOD removal by way of sludge production rather than by way of oxidation, then nutrient removal would be more efficient. However, since finding suitable sites to dispose of sludge is often a problem, STPs prefer to remove as much BOD as possible through oxidation. Therefore nutrient removal efficiencies are generally rather low. According to Rohlich and Uttormark (1972), primary treatment plants and conventional activated sludge plants remove 5–15% and 30–50%, respectively, of the nitrogen and phosphorus from raw sewage. Under special conditions phosphorus removal efficiencies as high as 70–90% have been reported in some STPs, but nitrogen removal using conventional secondary treatment methods is rarely more than 30–40% efficient.

Phosphorus Removal

What methods are available to an STP operator for removing nitrogen and phosphorus? P removal is generally easier and cheaper than N removal, primarily because inorganic P forms insoluble compounds with calcium, aluminum, and iron under appropriate conditions, whereas inorganic N compounds are all highly soluble. A common procedure for removing phosphate consists of simply adding lime (CaO), aluminum sulfate (alum), sodium aluminate, or ferric salts to the effluent from the secondary clarifier. The phosphate precipitate that forms in the effluent must then be allowed to settle out before the effluent is discharged. Since phosphate is of commercial value in fertilizer, the precipitate may be chemically processed to recover the phosphate. Since the solubility of the phosphate precipitates is pH-dependent, some adjustment of effluent pH may be necessary before addition of the cation (Al^{3+}, Ca^{2+}, or Fe^{3+}) and again after separation of the precipitate to bring the pH of the final effluent back to a suitable value. In the case of Ca^{2+}, for example, the pH must be raised above 10 (highly basic) to obtain phosphate removals in the 90–95% range. Phosphate removal efficiencies by this chemical precipitation technique are typically in the 88–95% range, regardless of whether aluminum, iron, or calcium is used as the precipitating agent.

A variation on the strict chemical precipitation technique is to add the precipitating agent at some stage in the primary or secondary treatment process. The chemical precipitate is then collected along with the primary or secondary sludge. Addition of alum or coagulating chemicals to the primary clarifier promotes the removal of organic substances as well as phosphate (Rohlich and Uttormark, 1972). However, this approach obviously produces a larger volume of sludge and also requires greater chemical additions.

Several other tertiary techniques for removing phosphate have been tested. Those tertiary methods that have achieved P removal in excess of 90% are listed in Table 6.2 along with their estimated costs. The popularity of chemical precipitation as a tertiary treatment method for P removal obviously stems from its low cost and the fact that its removal efficiency compares favorably with that of the other methods. The actual cost depends very much on the method used and on the degree of P removal. Miller (1972), for example, has estimated the cost of tertiary treatment achieving 85% P removal by chemical addition to be only about 15–30% greater than conventional secondary treatment, and Lee and Jones (1986) indicate that 90% P removal can be achieved at a cost of less than one penny per person per day at STPs serving more than 10,000 persons. However, Miller (1972) has estimated that 95% P removal by means of unspecified polishing steps following secondary treatment would raise the total treatment cost by roughly a factor of 3. Although experience and additional research may well reduce the cost of P removal from sewage, it appears that the additional cost of achieving P removal of ~95% will probably continue to be approximately the same as the cost of secondary treatment alone.

Table 6.2 Removal Efficiencies and Estimated Costs of Various Treatment Techniques for Removing P from Sewage

Technique	% Removal efficiency	Estimated cost[a] ($/million l)
Chemical precipitation	88–95	9–65
Sorption	90–98	37–65
Chemical precipitation with filtration	95–98	65–83
Ion exchange	86–98	157–277
Reverse osmosis	65–95	231–370
Distillation	90–98	370–925

[a]In general, the larger the STP, the smaller the cost of P removal per million liters treated.

Source. Eliassen and Tchobanoglous (1969). Costs were multiplied by a factor of 3.5 to correct for inflation (Anonymous, 1989).

Nitrogen Removal

There are a variety of possible ways to remove nitrogen from sewage. Although secondary treatment involves vigorous oxidation of the organic wastes in sewage, most of the inorganic nitrogen in the effluent from the secondary clarifier is in the form of ammonium (NH_4^+) rather than oxidized forms such as nitrate (NO_3^-). The ammonium ion, however, can be converted to ammonia gas, NH_3, by shedding a proton (H^+). Therefore a common method for removing N is to convert NH_4^+ to NH_3 and then drive off the NH_3 gas by vigorous aeration. The equilibrium between NH_4^+ and NH_3 is strongly pH dependent, and at neutral pH over 99% of an ammonium-ammonia mixture is in the form of NH_4^+ at equilibrium. However, if the pH is raised above 11 over 98% of the mixture is NH_3 at equilibrium. The procedure in the so-called ammonia stripping process is therefore to artificially raise the pH of the secondary effluent to 11 or more, vigorously aerate the effluent, and then bring the pH back down to a suitable level. Since P removal by means of Ca^{2+} addition also requires a high pH, it is possible to combine the two procedures in the same step. A disadvantage of the ammonia stripping technique is the high solubility of NH_3 in water. As a result, a very large volume of air is required to strip the NH_3 from a small volume of effluent. About 3000 m^3 of air are required to strip 95% of the NH_3 from 1 m^3 of effluent (Rohlich and Uttormark, 1972). Furthermore, ammonia stripping efficiency is reduced in cold weather, because the solubility of NH_3 in water increases as the temperature decreases. To deal with this problem, special counter-current, low-pressure air flow stripping towers have been designed to remove the NH_3 from the effluent. Such a system has achieved about 90% NH_3 removal at a STP near Lake Tahoe (Slechta and Culp, 1967).

A second technique for removing nitrogen is to convert the nitrogen to nitrogen gas, N_2. There are two common procedures for accomplishing this task, one biological and the other chemical. In the biological procedure the inorganic N is first oxidized to nitrate by prolonged and vigorous aeration. The nitrate is then converted to N_2 by a process called *denitrification*, which proceeds in an anaerobic environment and is mediated by a special class of microorganisms that are able to utilize the oxygen from NO_3^- to oxidize organic substrates. Usually the organic content of secondary effluent is sufficiently low that an additional carbon source such as acetone, alcohol, or acetic acid must be added to provide sufficient food for the denitrifying microbes. In practice some NO_3^- is converted to nitrous oxide gas (N_2O), but both N_2O and N_2 may be stripped by aeration.

The chemical procedure is called *breakpoint chlorination* and involves addition of chlorine to the effluent until the total dissolved residual chlorine reaches a minimum (the breakpoint), and virtually all the ammonia N has disappeared. The overall chemical reaction is

$$3\,Cl_2 + 2\,NH_3 \rightarrow N_2 + 6\,HCl$$

The efficiency of the process is pH dependent, and production of nitrate and chloramines may become significant if reaction conditions are not properly maintained. Ammonia removal efficiencies of 95–99% are achievable with breakpoint chlorination if the pH is maintained in the 6.5–7.5 range (Pressley, Bishop, and Roan, 1972), but computer control and sophisticated monitoring of reaction conditions are necessary if the process is to work efficiently (Fertik and Sharpe, 1980).

Nitrogen removal from sewage has been tried using a variety of other techniques, including ion exchange, electrodialysis, reverse osmosis, electrochemical treatment, and distillation. Table 6.3 lists those methods which have achieved over 90% N removal. Based on its low cost and excellent N removal efficiency, ammonia stripping is the most attractive of the tertiary treatment methods. In fact its estimated cost is little different from that for P removal by chemical precipitation; and as pointed out earlier, the two methods may be conveniently combined if Ca^{2+} is used to precipitate the phosphate. Miller (1972) has estimated the cost of an STP that removes 85% of the N and virtually 100% of the P from sewage by a so-called multistage biological process to be about twice the cost of conventional secondary treatment. Thus removal of a large percentage of the N and P from sewage is possible if one is willing to pay the price.

Cost of Conventional Sewage Treatment

How high is the cost of conventional sewage treatment? Table 6.4 lists estimated costs of building and operating STPs providing various degrees of sewage treatment. The ratio of sewage flow to the number of persons served by a STP is about 400 liters per person per day. The operating cost of producing potable water from the 60 and 400 million-liters-per-day (mld) plants would be about 20 and 12 cents per person per day, respectively. If the construction of the plant were financed by selling 20-year municipal bonds at 10% interest, the cost of building the tertiary-potable water plants would be 17 and 8 cents per person per day for the 60 and 400 mld plants, respectively. This latter calculation assumes that the cost of financing the plant is divided equally among 150,000 taxpayers (= 60 mld/400 liters per day) and 1 million taxpayers (= 400 mld/400 liters per day) in the case of the 60 mld and 400 mld plants, respectively. Thus the total cost (capital plus operating cost) of a tertiary STP producing potable water would be 37 and 21 cents per person per day for the 60 mld and 400 mld plants, respectively. These results, along with similar total cost calculations for the other methods of sewage treatment, are listed in Table 6.5.

Table 6.3   N Removal Efficiencies and Estimated Costs of Various Tertiary Techniques for Removing N from Sewage

Technique	% N removal	Estimated cost[a] ($/million liters)
Ammonia stripping	80–98	8–23
Anaerobic denitrification	60–95	23–28
Ion exchange	80–92	157–277
Reverse osmosis	65–95	231–370
Distillation	90–98	370–925
Breakpoint chlorination	95–99	714–1615

[a] In general, the larger the STP, the lower the N removal cost per unit volume of sewage.

Source. With the exception of breakpoint chlorination, all numbers were taken from Eliassen and Tchobanoglous (1969), with costs multiplied by a factor of 3.5 to correct for inflation (Anonymous, 1989). The efficiency for breakpoint chlorination was taken from Pressley (1972). The costs for breakpoint chlorination reflect only the cost of the chlorine, which was assumed to lie in the range $1–$2.20/kg.

Table 6.4   Estimated Capital Cost and Operating Cost of a 60 Million Liter per Day (mld) and 400 mld STP as of 1990

	Capital cost (millions of dollars)		Operating cost (cents/1000 l)	
	60 mld	400 mld	60 mld	400 mld
Primary	8.1	35.1	4.8	3
Secondary	16.6	74	10	7.7
Tertiary-nutrient removal	27	111	21	16
Tertiary-potable water	93	285	50	31

Source. Stephen and Weinberger (1968). Costs were multiplied by a factor of 3.5 to adjust for inflation (Anonymous, 1989).

Table 6.5   Estimated Total Cost to the Taxpayer of the STPs Listed in Table 6.4

	Total cost to the taxpayer (cents/person/day)	
	60 mld plant	400 mld plant
Primary	3	2
Secondary	7	5
Tertiary-nutrient removal	13	9
Tertiary-potable water	37	20

From the numbers in Table 6.5 it is apparent that each additional step in the sewage treatment process roughly doubles the cost of treatment. The cost to the taxpayer of tertiary sewage treatment for nutrient removal, however, is still only about 10–15 cents per person per day.

LAND APPLICATION OF SEWAGE

Until now our discussion of sewage treatment has implicitly taken the attitude that sewage effluent is an undesirable, problem-creating substance. A little thought will convince one, however, that sewage effluent is, at least potentially, a valuable resource. This conclusion stems from the following reasoning:

1. The high nutrient concentrations in sewage obviously stimulate plant growth. Although excessive plant growth in aquatic systems is undesirable, sewage could be applied at a controlled rate to fields or forests to stimulate the growth of valued crops, just as high-nutrient fertilizer is presently applied to achieve the same purpose. Use of sewage to irrigate crops would eliminate the need to discharge the sewage into aquatic systems that might be adversely affected by the high nutrient levels in the sewage, reduce the dependence of the crop grower on synthetic fertilizers, and allow part of the sewage irrigation water to percolate down through the soil into the underground water table and thus to recharge groundwater supplies. This last observation is directly related to point 2 below.

2. In some parts of the world, freshwater is in scarce supply or is rapidly becoming scarce. Discharging sewage at controlled rates to land areas rather than to the nearest convenient watercourse could be used as a mechanism for recharging groundwater, since under suitable conditions a large percentage of the effluent could be expected to percolate into the underground water table.

Thus sewage effluent is potentially useful both for growing crops and for recharging groundwater. However, several problems associated with such applications immediately suggest themselves.

1. The effluent from a conventional secondary STP invariably contains some pathogens. Obviously no one would want to eat crops or drink groundwater that had become contaminated with sewage pathogens.
2. High concentrations of nitrate (NO_3^-) in drinking water may create health problems of two sorts. Nitrate breaks down in saliva and in the digestive tract of humans and animals into nitrite (NO_2^-). Nitrite may subsequently combine with amines to form nitrosamines, a class of compounds shown to cause cancer in laboratory animals (Smith, 1978; Tannenbaum et al., 1978). In addition, nitrite may be absorbed into the blood, where it converts hemoglobin to methemoglobin. The latter pigment is unable to transport oxygen, and in severe cases of methemoglobinemia the victim may literally suffocate.

   EPA water quality standards with respect to nitrate have been based largely on the methemoglobinemia problem. The disease, which is confined largely to infants less than 3 months old, is fatal in about 7–8% of reported cases (EPA, 1986). Approximately 2000 cases of infant methemoglobinemia have been reported in Europe and North America since 1945. Older children and adults are less affected because (a) the stomach pH of infants is higher than that of adults[5] and (b) infant gastrointestinal illnesses may permit reduction of nitrate to nitrite to occur in the intestinal tract.

   Methemoglobinemia is associated with nitrate–nitrogen concentrations in excess of 10 mg/l, although it is known that many infants have consumed water containing higher nitrate concentrations without developing the disease. To be on the safe side, the EPA has established an upper limit of 10 mg/l nitrate–nitrogen for public water supplies. Since the total nitrogen concentration in raw sewage is about 40 mg/l and since only about 30–40% of this nitrogen is removed by conventional secondary treatment, it is possible that continuous recharging of groundwater with sewage effluent could produce dangerously high nitrate levels in the groundwater.
3. Continuous irrigation of land with water containing dissolved salts can lead to an increase in the salt content of the soil, because some of the irrigation water inevitably evaporates, leaving its salt content behind. If the accumulation of salt is sufficiently great, the soil may become unfit for growing crops. This problem would be most severe in arid climates where evaporation rates are high. In addition to the dissolved salts that are abundant in aquatic systems, sewage effluent may also contain small amounts of industrial wastes,

---

[5]This higher pH permits the growth of bacteria that reduce nitrate to nitrite.

including heavy metals such as mercury, lead, copper, cadmium, and zinc. Accumulation of elements in the soil (e.g., copper and zinc) could poison the soil for plant growth, and uptake of elements such as lead, mercury, and cadmium could render crops unfit for human consumption. A gradual deterioration in soil quality due to the accumulation of salts and heavy metals might be difficult to detect for many years, since crop production would be stimulated at the same time due to the high nutrient concentrations in the effluent. Once soil quality had become sufficiently degraded to noticeably affect crop production, it might be difficult to restore soil quality, at least in a reasonably short time.

4. Spraying of sewage for irrigation purposes obviously introduces many small droplets of water into the air. The smaller of these droplets may be transported for considerable distances by the wind. Studies conducted at conventional STPs have shown that aerosols injected into the air from trickling filter or activated sludge operations may transport viable pathogens as much as several hundred meters downwind from STPs (Hickey and Reist, 1975). Spray irrigation systems would undoubtedly inject many more aerosols into the air than do trickling filters or activated sludge tanks. Thus spray irrigation operations utilizing sewage effluent might create significant health problems for persons living immediately downwind from the irrigation systems.

Results of Spray Irrigation Studies

Realizing the potential benefits as well as the potential problems of land application of sewage effluent, a number of scientists have set out to determine whether land application of sewage effluent is in fact a safe and more beneficial means of disposing of sewage effluent than the traditional method of discharging the effluent into the nearest body of water. A very thorough study was carried out at Pennsylvania State University beginning in 1963 (Kardos, 1970; Kardos et al., 1974). Interest in land application of sewage was stimulated both by the results of that study and by the passage in 1972 of the Clean Water Act, which proposed a zero discharge concept that encouraged the philosophy of reuse and recovery. A long-term study of sewage spray irrigation began near Tallahassee, Florida, in 1971 (Payne and Overman, 1987; Allhands and Overman, 1989), and reports summarizing work carried out at numerous other locations around the United States began to appear during the 1970s and 1980s (Gilbert et al., 1976; Hershaft and Truett, 1981; Majeti and Clark, 1981; Page et al., 1984; Shuval, Fattal, and Wax, 1984). As of 1978 the treated sewage from a population of about 9 million persons was being applied to the land (Hershaft and Truett, 1981). Virtually all the studies conducted to date have led to the same conclusion, namely that land application of sewage effluent is a safe and practical way to dispose of sewage wastewater if reasonable precautions are exercised. The principal limiting factors are the amount of land required and accumulation of nitrate in groundwater supplies. For our purposes it is sufficient to examine in some detail the results of the Pennsylvania State study and then to briefly summarize the results of some of the more recent work.

The Pennsylvania State studies were carried out near State College, Pennsylvania, where a series of experimental land plots were established on which various crops were grown using either recommended amounts of commercial fertilizer to stimulate growth (the control plot) or various application rates of secondary sewage effluent. The effluent was applied to the experimental crops from April to November each year for a period of 5 years beginning in 1963. Other plots consisting of mixed hardwood trees and grassy areas were sprinkled throughout the year. Table 6.6 compares some of the crop yields from the control and experimental plots in 1965. Irrigation with 2.5 cm/week of effluent roughly doubled the yields of alfalfa, hay, corn, and oats, and increased the yield of corn silage by about 26%. Increasing the irrigation rate further to 5 cm/wk produced little significant additional change in the crop yields.

The results of sprinkling forested areas with sewage effluent were somewhat mixed. The growth of red pines was actually retarded by the irrigation, perhaps because red pines grow better in dry soil[6], and perhaps because of the relatively high boron content of the sewage effluent. Boron is a common component in detergents; and boron, though an essential plant nutrient, may be toxic to plants if its concentration is in the approximate 0.3–0.4 mg/l or higher range (Allen, 1973). However, other trees responded favorably to effluent sprinkling, although in most cases the growth response was rather small (Kardos, 1970).

To determine the effect of the spray irrigation program on groundwater quality, a number of sampling wells were dug to various depths and at various locations in and around the experimental area. There was no evidence of fecal pathogens in the groundwater. Similar results have been reported by Bouwer, Lance, and Riggs (1974), Gilbert et al. (1976), and Hershaft and Truett (1981). Percolation of water through roughly 1 m of well-aerated soil appears to be an extremely efficient mechanism for removing sewage pathogens. Such pathogens simply do not survive well in soil. According to Allen (1973, p. 39), "Almost all die within two weeks." Thus sewage irrigation appears to pose little threat to groundwater quality as far as pathogen contamination is concerned.

The effect of sewage irrigation on the nutrient concentrations of the groundwater presents a somewhat different picture. Phosphorus removal by the soil-vegetation complex was generally quite efficient, although not as impressive as the removal of pathogens. On the plots that received 5 cm of effluent per week throughout the year, phosphorus removal was about 92% at the 120 cm depth level during the third year of application. After the first year or so most of the phosphorus removal in the forested areas was evidently due to adsorption of the

Table 6.6   Crop Yields from Control Plots (Commercial Fertilizer) and Plots Sprayed with Secondary Sewage Effluent

	Control plot	2.5 cm/wk effluent	5 cm/wk effluent
Alfalfa hay, tonnes/ha	5.2	10.5	12.1
Corn silage, tonnes/ha	6.9	8.7	9.6
Corn, bushels/ha	156	283	274
Oats, bushels/ha	117	198	179

Source. Kardos (1970).

[6]Control plots in the forested areas received neither fertilizer nor well water.

phosphorus to soil particles, since much of the phosphorus taken up by the vegetation during the spring and summer was returned to the soil in the fall by way of dead leaves and other litter. In the plots where crops were grown, phosphorus removal was more efficient, because much of the P taken up by the crops was permanently removed when the crops were harvested. By the third year of the study P removal by way of harvested crops ranged from 22% in corn silage irrigated with 5 cm/wk to 63% in red clover irrigated with 2.5 cm/wk (Kardos, 1970). With the additional P removal due to adsorption to soil particles, P removal at the 120-cm-depth level was 99% in the plots where crops were grown.

Nitrogen removal by the soil-vegetation complex was not as efficient as P removal, primarily because nitrogen in the form of nitrate does not effectively adsorb to soil particles.[7] In fact, soil water nitrate levels in a white-spruce area sprayed throughout the year with 5 cm/wk of sewage effluent were actually higher by the third year of the study than the nitrate levels in the effluent. This increase in nitrate was presumably caused by the recycling of N utilized by vegetation during the spring and summer and then returned by way of litter in the fall. The other forested plots (red pine, oak) that received similar sprinkling did remove, "a large percentage of the nitrate added to the soil by the effluent" (Kardos 1970, p. 15). However, on forested plots that were sprinkled with 10 cm/wk of effluent, soil water nitrate levels approached the EPA guideline for potable water of 10 mg/l nitrate-N.

Nitrogen removal in the plots planted with crops was of course more efficient because of the harvesting of the crops. In fact calculated removal efficiencies based on the N content of the crops sometimes exceeded 100%, presumably because the crops were utilizing N already in the soil in addition to that added with the effluent. Calculated N removal efficiencies were 60–90% for wheat, 65–127% for corn (grain), and 105–210% for corn silage (ear, stalk, and leaves). After 3 years of irrigation soil water samples at a depth of 120 cm in the crop plots were found to have a nitrate N concentration of about 6 mg/l, and groundwater samples from an unspecified depth had a nitrate N concentration of 3 mg/l. The highest soil water nitrate levels were usually found in the early spring, when the ground had thawed but when the plant root systems were not yet fully active. Kardos (1970) points out that perennial grasses are most effective in removing nitrogen from the soil in the early spring, because their root systems are already fully established, whereas the roots of annual grasses and most crops are just beginning to develop. At latitudes where there is a long period of the year when plant root systems are inactive, it may be necessary to set aside additional land areas for sewage irrigation during the winter to prevent the buildup of unacceptably high soil nitrate levels by late winter or early spring. Such measures would obviously not be necessary in tropical latitudes where root systems are active throughout the year.

Results reported from the Tallahassee study have been qualitatively similar to the Pennsylvania State results, but have also revealed some interesting temporal trends. The Tallahassee work has involved irrigation with secondary treated effluent to approximately 700 hectares (ha) of sandy soil south of the city of Tallahassee. Since 1981 application rates have been 4.5–6.0 cm/wk. Corn, soybeans, and coastal Bermuda grass have all been grown on the experimental plots. The soil pH, which was initially acidic (5.3), increased to an apparently steady-state value of 7.0 (neutral) in

---

[7]NH_4^+ does effectively adsorb to soil particles, but NH_4^+ is spontaneously oxidized to nitrate in the presence of even small amounts of oxygen.

the upper 1.5 m after about 6–7 years of irrigation, and had reached a value of 6.0 at a depth of 7.6 m (Allhands and Overman, 1989). The organic content of roughly the upper 50 cm of the soil had increased by about a factor of 2 after 8 years of irrigation and did not appear to have reached a steady-state value. Virtually all the P in the effluent was removed in the upper 1.0 m of soil during the first 6–7 years of irrigation. The P enrichment of the soil surface layer continued to increase throughout the study period, and the P front was moving down through the soil at about 10–15 cm per year. Nitrogen removal was less impressive, in part because supplemental N was added to both the corn and coastal bermuda grass plots during the growing season. Fertilizer was added because percolation of the sewage effluent through the soil was so rapid that the crop roots were unable to obtain adequate N from the sewage effluent. The corn and coastal bermuda grass removed 25% and 47%, respectively, of the total applied N (fertilizer N plus sewage effluent N). Allhands and Overman (1989) felt these percentages could have been increased to 39% and 71%, respectively, by fine tuning the fertilizer and sewage effluent application schedules. Because of the low N removal efficiencies, nitrate concentrations in groundwater increased steadily during the roughly 8-year study period, and had reached an average of about 18 mg/l by 1988. According to Allhands and Overman (1989), this trend could have been reversed by reducing the fertilizer application rate, but presumably at some sacrifice in crop production. Incidentally, corn yields generally increased during the study period and averaged about 320 bushels/ha from 1984 to 1988.

The buildup of salts in the soil because of repeated irrigation with sewage was evidently not a problem in either the Pennsylvania State or Tallahassee studies. However, it was obvious to scientists in the former case that harvesting of crops was in most cases removing only a small percentage of the sewage salts. Kardos (1970) states that in 1965 an alfalfa plot removed 118% of the irrigation water potassium, 19% of the calcium, 11% of the magnesium, and 0.4% of the sodium. These figures suggest that significant accumulations of calcium, magnesium, and sodium might be expected in alfalfa fields that are irrigated year after year with sewage effluent. In the case of the Tallahassee study calcium and magnesium concentrations were as much as tenfold or more higher in the upper roughly 1 m of sewage-irrigated plots after 6–7 years of irrigation, but sodium and potassium concentrations were relatively unchanged. Heukelekian (1957) reported that salt accumulation was not a problem in sewage-irrigated fields in Israel, because rainfall tended to leach out accumulated salts. The irrigation systems used in Israel, however, typically called for application rates of 50–75 cm of effluent per year, a somewhat lower rate than the 2.5–6.0 cm/wk application rates used in the Pennsylvania State and Tallahassee studies. Whether the gradual accumulation of salts in the soil would eventually create a problem is certainly a legitimate question. Rodhe (1962), for example, reported that soils at sewage farms that had been in operation for over 100 years near both Paris and Berlin had shown marked decreases in productivity, and that the cause had been traced to an accumulation of copper and zinc in the soils. Hershaft and Truett (1981), however, reported no evidence of metal accumulation in soils at six sewage irrigation projects in the western United States, one of which had been in operation for over 30 years. None of the six facilities received industrial wastewater with a high concentration of metals. It seems fair to say that the long-term accumulation of metals and salts in sewage-irrigated soils is a potential problem and that careful monitoring of the characteristics of both the sewage effluent and the soil and groundwater is advisable so that corrective action can be taken if necessary.

Until now we have discussed harvesting of crops from irrigated plots as a means of removing nutrients, but it should be obvious that such crops are potentially useful in themselves as sources, for example, of food, fuel, and fibers. If the crops are to be used for food, obviously one would be concerned over possible contamination with pathogens. There is also the possibility that such crops might concentrate toxic substances such as heavy metals from the sewage effluent.

As far as pathogens are concerned, it is important to keep in mind that enteric pathogens do not survive well in soil, and proper cooking will kill virtually any pathogen. Furthermore, there is no evidence that germs can be assimilated through a plant's root system and transported, for example, to fruits and seeds. A number of studies with sewage sludge (see later discussion) have clearly indicated that plants can pick up toxic substances from the soil (Sopper and Kerr, 1981; Dowdy et al., 1984), but such effects are not always observed (Robson and Sommers, 1982), and it appears that uptake can be minimized through plant breeding (Hinesly et al., 1984). The best safeguard against contamination with toxic substances is probably common sense and careful monitoring of the crops. For example, it would be unwise to plan on growing food crops using sewage irrigation water that is known to receive large amounts of industrial wastes and has been shown to contain high concentrations of mercury, lead, or some other toxic substance. Sewage that is largely of domestic origin may not pose such a problem.

In the United States, the EPA has specifically recommended against the use of untreated sewage for irrigation or land disposal (EPA, 1972), the principal concern being the high concentration of pathogens in raw sewage. Primary treated sewage is recommended for crop irrigation if the crops are not for human consumption, and secondary effluent is recommended on crops for human consumption if the crops are canned or similarly processed before sale (EPA, 1972). These recommendations are not binding, however, and in some states it is legal to use secondary sewage effluent to irrigate food crops, regardless of how the crops are to be processed, as long as the effluent satisfies certain criteria (Donovan and Bates, 1980). In Israel, where water conservation is critical and where sewage irrigation is common, it is acceptable to use primary treated sewage to irrigate both vegetables and fruits destined for human consumption, with the proviso that the vegetables must be cooked and the fruits peeled before eating, or in the case of fruits that are not peeled before eating, that irrigation be stopped 1 month before harvest (Heukelekian, 1957). At the present time Israel recycles the sewage from about two-thirds of its population, primarily for agricultural use. Israel's goal is to recycle 80% of its wastewater (Shuval et al., 1984). In the United States land application of sewage wastewater accounts for only 5–10% of the effluent produced by municipal sewage treatment plants (Hershaft and Truett, 1981).

Until now we have discussed the pathogen problem from the standpoint of contamination of groundwater and crops. However, studies by Shuval et al. (1984) have shown that sprinkler irrigation results in aerosolization of about 0.1–1.0% of the effluent, and pathogens may be dispersed as much as several hundred meters in the resultant aerosols (Majeti and Clark, 1981). This aerial transport of pathogens could create public health problems for persons living or working downwind from the spray irrigation operation.

Perhaps the most noteworthy study of this problem was carried out by Shuval et al. (1984) in Israel. During a 4-year period they compared the incidence of 12 illnesses associated with enteric pathogens at 11 kibbutzim (cooperative agricultural settlements) where lagoon-treated sewage effluent was used to irrigate crops during 2 of the 4 years. The incidence of an additional group of 12 illnesses not associated with enteric pathogens was monitored to check for any changes in the health of the roughly 3000 people in the study unrelated to the sewage spray irrigation. No effort was made to disinfect the sewage effluent, and based on the reported coliform and enteric virus concentrations, the effluent was grossly contaminated with pathogens. Shuval et al. (1984) found that the incidence of enteric diseases was 32–112% higher among children less than 5 years old during the periods of wastewater irrigation, but the corresponding excess risk in other age groups appeared to be no more than 10%. Shuval et al. (1984) concluded that any health problems associated with the spray irrigation system would have been probably negligible if the sewage had been disinfected using standard procedures.

A somewhat different approach to land application of sewage effluent was taken at the Flushing Meadows project, where the Salt River bed west of Phoenix, Arizona, was used for ground recharge of secondary sewage effluent (Bouwer et al, 1974). The objective of this project was to apply sewage effluent at as high a rate as possible to experimental plots on the river bed without degrading soil quality, while at the same time producing groundwater that could subsequently be used for unrestricted irrigation and recreation. The application rates averaged about 90 m/yr, or about 30–35 times the application rates used in the Pennsylvania State and Tallahassee studies. The effluent was applied to the plots by simply flooding the plots to various depths with effluent and then allowing the water to percolate into the ground. The soil was allowed to dry for variable periods of time between floodings. Some experimental plots contained no vegetation, but others were planted with various grasses or rice. No attempt was made to harvest the vegetation. The following is a brief summary of the experimental results.

1. Percolation of the effluent through the sand and gravel of the river bed removed virtually 100% of the SS, BOD, and fecal coliform bacteria.

2. With alternate flooding and drying periods of 2 weeks each, it was possible to remove about 30% of the N from the effluent. The mechanism appeared to be adsorption of NH_4^+ to soil particles during flooding, conversion of the NH_4^+ to NO_3^- during subsequent drying periods as the soil became partially aerobic, and finally denitrification in anaerobic parts of the soil during the drying period. This sequence of steps is very sensitive to oxygen levels. The conversion of NH_4^+ to NO_3^- requires oxygen, but the conversion of NO_3^- to N_2 gas can occur only in the almost complete absence of oxygen. Despite this problem laboratory studies indicated that N removal efficiencies as high as 90% might be achieved by further modifications in the system.

3. Phosphate removal of about 90% was achieved after about 100 m of underground travel. The mechanism of removal presumably involved a combination of precipitation and adsorption to soil particles.

4. Copper and zinc concentrations were reduced by about 80% and mercury concentrations by about 35%, but cadmium and lead concentrations were virtually unchanged. None of these metals was present in the renovated water at concentrations that would be objectionable for irrigation or recreational use.

Limiting Factors

The conclusion that emerges from the studies of land application of sewage effluent is that the two most important factors limiting the feasability of this mode of disposal are the land area requirements and the accumulation of nitrate in groundwater. If the effluent is applied at a rate of 5 cm/wk, for example, and if the per capita sewage flow is about 400 l/day, then a city with a population of 100,000 persons would need about 5.6 km² to dispose of its sewage effluent. Additional land areas would probably be needed in places where plant root systems are inactive during the winter. Large areas of land suitable for irrigation are apt to be scarce near large population centers, nor is the cost of such land likely to be cheap. The Flushing Meadows project showed that much higher application rates are possible if the soil is sufficiently permeable, but had the soil consisted of clay rather than sand, it would have been impossible to sustain such high irrigation rates.

One of the most serious limiting factors in the land application of sewage, regardless of whether the effluent is used to grow crops or not, is the accumulation of nitrate in groundwater. The Pennsylvania State study made it clear that under appropriate conditions a major portion of the nitrogen in the effluent may be assimilated by crops and removed at harvest. In the case of the Tallahassee work, however, the percolation rate of the effluent through the soil was so rapid that the crop root systems were unable to remove more than a minor portion of the nitrogen. Furthermore, crop root systems will obviously assimilate none of the nitrogen in temperate latitudes during the winter months. An alternative to growing crops to remove nitrogen is the approach used at the Flushing Meadows project, in which high average application rates and alternate flooding and drying periods resulted in N removal via denitrification. Regardless of which approach is used, it is clear that careful planning and monitoring of sampling wells are necessary if groundwater nitrate levels are to be maintained below 10 mg/l.

Use of Sewage Sludge

In addition to the liquid effluent from a STP, the solid waste (sludge) may also be put to good use in land application. Typical sludge from municipal STPs contains about 3.2% N, 1.8% P, and 0.3% K (Abron-Robinson et al., 1981). Because of its low K content the sludge is not an ideal fertilizer, but it is an excellent source of N and P. Because the nutrients are bound in organic matter, they are slowly released following land application, and experience has shown that a single application of sludge can provide the N and P requirements of terrestrial plants for as much as 3–5 years (Sopper and Seaker, 1984). At the present time STPs in the United States generate about 8.5 million tons of sludge per year (Marinelli, 1990). If this sludge were used as fertilizer to supply the nitrogen needs of nonleguminous plants, an application rate of about 25 tons per hectare per year would be required (Lewicke, 1972). Thus the fertilizer N requirements of about 0.34 million ha of cropland could be supplied with sewage sludge. The market value of the sludge as a source of N, P, and K would be about $200 million per year (Abron-Robinson et al., 1981). While these figures may sound impressive, all the sewage sludge produced in the United States could supply only 1–2% of the annual fertilizer N required for crop production in the United States (Dowdy et al., 1984). Furthermore, municipalities would have to bear the cost of transporting the sludge if land application were to be cost competitive with commercial fertilizer (Abron-Robinson et al., 1981).

Therefore, sewage sludge is not going to change significantly United States reliance on commercial fertilizer for growing crops. Several factors, however, do make land application of sewage sludge very appealing. First, legislation prohibiting ocean dumping of sludge was enacted in 1981 (Clark et al., 1984). Second, in many parts of the country sanitary landfills for disposal of sludge have been or are becoming rapidly filled. Third, incinerators used to burn sludge have been phased out in some parts of the country due to air pollution problems (Sopper and Kerr, 1981). Finally, strip mining in a total of 31 states has laid bare approximately 2.0 million ha of land, less than half of which has been properly reclaimed. About 0.25 million ha of land are disturbed each year as a result of strip mining activity (Sopper and Kerr, 1981). The Federal Surface Mining Control and Reclamation Act of 1977 requires backfilling and restoration of strip-mined land and maintenance for periods of either 5 or 10 years, depending on whether annual rainfall in the area is greater or less than 66 cm, respectively. Restoring strip-mined land has proven not to be an easy task, but research has shown that the job is greatly facilitated by applying sewage sludge to the land.

Thus there is considerable motivation for applying sludge to the land, and during the 1980s a number of studies addressed the feasibility of this method of disposal. The concerns were similar to those associated with land application of the liquid effluent: pathogens, nitrate in groundwater, and heavy metals. The studies conducted to date, some of which have lasted as long as 5–15 years, have indicated no evidence of contamination of groundwater with fecal pathogens or nitrate (Sopper and Seaker, 1984; Hinesly et al., 1984). According to Page et al. (1984) groundwater monitoring for nitrate is unnecessary where sludge N application rates do not exceed fertilizer N recommendations. The lack of a nitrate problem is probably associated with the slow release of the N from the organic matter in the sludge. Crop yields on land fertilized with sludge often exceed those obtained with commercial fertilizer (Hinesly et al., 1984; Sopper and Seaker, 1984). The lack of a pathogen problem in groundwater is not surprising considering the results obtained in spray irrigation studies, but reasonable precautions to avoid pathogen exposure must be exercised by persons directly involved in applying the sludge to the land (Clark et al., 1984). Heavy metals are sometimes found at much higher concentrations in municipal sludge than in typical agricultural soils, and the uptake of these metals by crops and/or their migration into groundwater appear to be the most serious factors limiting land application of sludge. Dowdy et al. (1984), for example, reported that cadmium and zinc concentrations were elevated in corn silage produced on sludge-fertilized land, but there was no evidence of elevated metal concentrations in the tissues of animals fed the corn silage. Sopper and Kerr (1981) found that metal concentrations were higher than usual in vegetation grown on strip-mined land restored with sewage sludge, but felt that the metal concentrations were below problem levels. However, lead concentrations in groundwater exceeded standards for potable water, and in a later study Sopper and Seaker (1984) reported that both lead and chromium concentrations exceeded periodically drinking water standards in groundwater below sludge-fertilized land.

The most dramatic success of land application of sewage sludge has been undoubtedly the restoration of strip-mined land. Establishing vegetation on such land is extremely difficult due to the lack of nutrients, low organic matter content, low pH, low water retention, and toxic levels of metals in the soil (Sopper and Kerr, 1981). The acidity of the soil results from the oxidation of sulfides such as pyrite

(FeS_2) to produce sulfuric acid when the sulfides are exposed to air and water. The acidic nature of the soil mobilizes metals and makes retention of certain essential nutrients difficult. Restoration of the land is accomplished by first grading and tilling. Lime is then applied to temporarily neutralize the acid. Sludge is then applied at typically 100–250 tonnes/ha and the ground seeded with vegetation. Plants then grow rapidly, and the soil pH stabilizes at a neutral value. Vegetative cover generally increases over the course of the next few years. In several documented cases strip-mined land which had resisted previous attempts at restoration recovered dramatically following application of sewage sludge (Sopper and Kerr, 1981; Sopper and Seaker, 1984; Hinkle, 1984). In the study of Hinkle (1984) progress was slowed by several years of drought, and there was no improvement in the acidity of a stream draining the disturbed area. About 90% vegetative cover, however, was established on the land after 7 years, and heavy metal concentrations generally declined in the drainage stream.

The consensus that emerges from these studies is that land application of sewage sludge is a very practical means of disposal. This conclusion undoubtedly accounts in part for the fact that 15–20% of municipal sewage sludge is now applied to the land. The chief cause for concern with this approach would appear to be the accumulation of toxic metals in the soil, groundwater, or vegetation. To date, however, there has been no evidence of a problem with phytotoxicity, and to the extent that crops are not to be marketed for food, there is little reason for concern over contamination of terrestrial food chains (Abron-Robinson et al., 1981). Furthermore, there is no evidence that animals fed crops grown on sludge-amended soils accumulate objectionable concentrations of heavy metals (Dowdy et al., 1984; Hinesly et al., 1984). Groundwater contamination with heavy metals is perhaps the most sensitive problem associated with land application of sewage sludge, and it clearly seems advisable to monitor groundwater in areas where sludge is being applied. Adjustments of sludge application rates may well be sufficient to keep groundwater metal concentrations within acceptable limits. The use of sewage sludge to restore strip-mined land is probably the most beneficial use of the sludge, and given the present and future anticipated extent of strip-mining activity, it seems likely that a large percentage of the sludge generated in the United States could be used for this purpose for many years.

UNCONVENTIONAL SEWAGE TREATMENT

There are several alternatives to conventional sewage treatment as it has been described in this chapter. For example, one can forego biological systems for removing BOD and rely instead on largely physical and chemical methods to treat the sewage. Kreissl and Lewis (1981), for example, describe a treatment process involving chemical clarification with hydrated lime, nitrification of the effluent from the clarifier, removal of additional particulate material with a dual-media filter, and removal of dissolved organics with activated carbon columns. The complete system achieved 97%, 93%, 80%, and 32% removal of SS, BOD, phosphate, and total N, respectively. The normal operation and maintenance of the plant, however, was more time consuming and complex than that of conventional treatment, and the manpower required for the hybrid system was about three times that of an extended aeration system designed to achieve comparable removals of SS and BOD. One advantage of the physical/chemical system over conventional secondary treatment was the much greater removal of phosphate, but it does not appear that physical/chemical treatment is any more cost effective than advanced biological treatment for tertiary nutrient removal (U.S. Army, 1972). The chief appeal of physical/chemical treatment systems would appear to be that they can be packaged in small units and brought on line rapidly, characteristics that may give them special appeal under certain conditions (Kreissl and Lewis, 1981).

Another alternative to conventional treatment is natural systems of microorganisms and higher plants. There are basically three types of such systems. The first consists of artificial marshes analogous to the marsh used by the City of Tallahassee to treat runoff into Lake Jackson. The second involves artificial rock marshes, in which wastewater flows laterally underground through a bed of rocks within which plants are rooted. The third is sometimes referred to as *solar aquatics*, and involves growing microorganisms, plants, and in some cases animals in artificial enclosures illuminated with natural sunlight. At the present time there are an estimated 150 artificial marsh and solar aquatic systems treating municipal sewage in the United States.

Probably the best known artificial marsh sewage treatment systems in the United States is operated by the town of Arcata, California. Sewage from the town's 15,000 residents is first routed to a primary treatment facility, and then to 20 ha of oxidation ponds containing microalgae. The effluent from the oxidation ponds is routed to two 1.0 ha artificial marshes planted with bulrush and cattails, and then into 18 ha of additional marshes constructed in part by the California Coastal Conservancy to help restore fish, shellfish, waterfowl, and other wildlife in the area. The residence time of water in the treatment system is about 2 months (Stewart, 1990), and by the time the effluent flows into Humboldt Bay it is cleaner than the water in the bay.

Rock marsh systems are variations on the more common artificial marshes, and are capable of producing effluent that satisfies secondary treatment standards. Wastewater is routed through lined trenches filled with rocks. Wastes in the sewage are broken down by microorganisms growing on the rocks and assimilated by plants rooted in the rocks. One of the largest rock marsh systems in the United States is operated by the town of Denham Springs, Louisiana. Wastewater from the town's 20,000 residents is treated partially in two 16-ha ponds, and then routed through three 2-ha rock marshes. The rock marshes are planted primarily with ornamental flowers and are designed to treat 11 mld. Compared to the cost of a secondary STP, the rock marshes saved the town of Denham Springs about $1 million in construction costs and are saving about $60,000 per year in maintenance costs (Marinelli, 1990).

Solar aquatic treatment systems have been pioneered by scientists at the New Alchemy Institute on Cape Cod, and frequently involve growing plants and animals in greenhouse environments. Microalgae are grown in vertically oriented cylindrical tanks; macrophytes are grown hydroponically. Fish and aquatic invertebrates are often grown in the microalgal tanks. Solar aquatic systems involve more manipulation and control of the environment than artificial marshes and are capable of achieving much higher wastewater treatment rates per unit area. In fact, solar aquatic systems require no more land area than conventional sewage treatment and can provide advanced wastewater treatment at about two-thirds the cost of conventional secondary treatment (Marinelli, 1990).

Whether artificial marsh and solar aquatic sewage treatment systems are the wave of the future remains to be seen. They are certainly less expensive to build and operate than conventional treatment systems. For example, the Arcata treatment system cost only a little over $0.5 million to build, or about $33 per capita. Much larger secondary treatment plants, which should be cheaper based on economics of scale, actually cost 2–3 times as much per capita to build.[8] Furthermore, the artificial marsh and solar aquatic systems produce little or no sludge, achieve a fair degree of tertiary nutrient removal, and have even been shown capable of removing pollutants such as heavy metals, pesticides, and industrial toxins. The artificial marsh systems, however, require considerably more area than a conventional STP. The 2-month residence time of wastewater in the Arcata artificial marsh, for example, is several orders of magnitude longer than the residence time of water in a conventional STP. Solar aquatic systems require about the same amount of land as conventional STPs, but at this point in time solar aquatic systems are less thoroughly tested than the other systems, and a different kind of expertise is needed to operate them. However, despite some uncertainties and limitations, the interest these nonconventional forms of treatment has generated in recent years suggests that they will be given careful consideration in the future and may one day largely replace conventional sewage treatment systems.

DETERGENT PHOSPHATES

The use of phosphates in laundry and dishwashing detergents is pertinent to our discussion of the treatment and use of sewage, because historically as much as 50% or more of the phosphorus in municipal wastewaters has come from the phosphates used in these detergents. Phosphorus in the form of sodium tripolyphosphate has been used as a component of laundry and dishwashing detergents ever since detergents were introduced on the market in the 1930s. Initially the amount of phosphorus released to the environment as a result of detergent use was quite small, because the public was rather slow to switch from the traditional soap flakes or soap powders for cleaning clothes and dishes. However, detergents had largely replaced soap products for such cleaning purposes by the 1950s, and by 1971 it was estimated that about 30–40% of the P entering the aquatic environment was coming from the phosphate in the wastewater from detergent washing operations (Grundy, 1971). Since there is reason to believe that many freshwater systems are phosphorus limited (Schindler, 1974), it occurred to many ecologists that removal of phosphorus from detergents might be an easy and cheap way to reduce or perhaps eliminate cultural eutrophication problems in some freshwater systems. As a result, pressure was brought on detergent manufacturers to find a substitute for sodium tripolyphosphate in their products, and a number of new detergent products containing no phosphorus rapidly appeared on the market. The president of Proctor and Gamble

---

[8]Given the figures in Table 6.4, the cost of a 60-mld secondary treatment plant is about $16.6 million. Assuming that this plant services $60 \times 10^6 / 400 = 150,000$ persons, the per capita cost is $111. The corresponding per capita cost of a 400-mld plant is $74 per capita.

was led to comment, "We recognize that the public wants phosphates out of laundry detergents, and we intend to take them out. Our job is to make certain that we remove them as rapidly as we can do so in a thoroughly reasonable manner. This we are doing" (Morgens, 1970). A glance at the detergent section in many grocery stores in the United States today will reveal that many detergents do contain little or no phosphorus, but some products contain P in concentrations as high as 11%.[9] After all the pressure to remove P, why do some detergents still contain appreciable amounts of phosphorus?

To answer this question, let us understand what phosphates are doing in detergents in the first place. A detergent contains two important components, a surfactant and a builder, which together often account for over half the weight of the product. A filler, typically sodium sulfate, and moisture account for most of the remaining weight of the detergent (Layman, 1984). Surfactants consist primarily of alkylsulfonates, fatty alcohol ethoxylates, and fatty alcohol sulfates. All of these compounds are bipolar, and because of this characteristic are able to penetrate between dirt and fabric. One side of the surfactant molecule is attracted to the fabric; the other side is attracted to the dirt. Surfactants also help to lift off grease by lowering surface tension. The principal purpose of the builder is to soften the water by removing calcium and magnesium ions. Otherwise these ions combine with the surfactant to form a gummy precipitate or curd that cannot be effectively removed with the rinse water. The curd settles onto the clothes, forming a dull film, and may build up in the washer, ultimately clogging the washing machine. Sodium tripolyphosphate acts as a builder by chemically sequestering calcium and magnesium so that they do not have a chance to combine with the surfactant. Phosphates also peptize and help suspend certain kinds of dirt, aid in germ killing, buffer the pH of the water to aid in the removal of fatty soils, and increase the effectiveness of the surfactant by breaking up surfactant polymeric units.

If sodium tripolyphosphate were simply removed from detergents without substituting another builder, it would still be possible to get a clean wash under any of the following conditions.

1. About 10 times as much detergent was added to the washing machine (Hammond, 1971).
2. A water softening agent such as sodium carbonate was added to the washing machine before both the wash and rinse cycles.[10]
3. A water softener had been used to remove the calcium and magnesium from the water before use of the water in the washing machine.
4. The washload contained a naturally low concentration of magnesium and calcium.

---

[9]A product containing 11% P in the form of sodium tripolyphosphate contains about 44% sodium tripolyphosphate.
[10]Carbonate forms a granular precipitate with calcium and magnesium ions, but this precipitate settles harmlessly to the bottom of the washing machine rather than gumming up the clothes.

How soft must water be before a surfactant alone (e.g., soap) will do an adequate job of washing clothes? According to Duthie (1972), about 3g of P are required to neutralize the hardness contributed by perspiration, soil, food, and previous rinsing of the clothes. If any calcium and magnesium are present in the water, then additional P will be required. Figure 6.2 shows the approximate amount of P that would be required to neutralize all the hardness in a washload as a function of water hardness. The implication of Figure 6.2 is that a certain amount of P, or of some other builder, is required to neutralize hardness even in water with virtually no calcium and magnesium. There are, however, alternatives to the use of detergents containing a high percentage of phosphate builder.

1. *Reduce or eliminate phosphate.* Obviously elimination of P would make detergents comparable to soap in cleaning ability. Any reduction in P content would presumably affect the quality of the wash in hard water areas if the recommended detergent addition left a significant amount of unneutralized hardness. You could compensate, of course, by adding more detergent, but at a proportionate additional cost. Furthermore, adding more detergent would increase the amount of P discharged with the wastewater. Detergent manufacturers might also produce regional detergents with P levels suited to the hardness of the water in each area, but the use of such regional detergents would not reduce the input of P to aquatic environments in hard-water areas.[11]

2. *Go back to soap.* Soap, which is made from soluble salts of fatty acids, does a fine job of cleaning clothes in soft water, in spite of hardness from perspiration, soil, food, and so forth. However, soap does form a gummy precipitate with calcium and magnesium ions, just as do detergent surfactants. Accustomed as people are to "whiter-than-white" clothes, most persons would not be satisfied with the cleaning job done by soaps in hard water.

3. *Substitute some other builder for all or most of the sodium tripolyphosphate in detergents.* A number of alternative builders have been tested and marketed in detergents since the early 1970s. The principal alternative builders on the market are sodium citrate, nitrilotriacetate (NTA), zeolites such as sodium aluminosilicate, and sodium carbonate. Sodium citrate and NTA are sequestering builders, and in this respect are similar to sodium tripolyphosphate. Zeolites soften water through ion exchange. Sodium carbonate softens water by precipitating calcium and magnesium ions.

During the late 1960s it appeared that much of the sodium tripolyphosphate in detergents would be replaced by NTA, but this alternative, upon which detergent manufacturers placed great hopes, and in which they invested much money, was squelched by the U.S. Surgeon General on December 18, 1970, as a result of tests performed by the National Institute of Environmental Health Science (NIEHS).

---

[11]A common but somewhat arbitrary scale of water hardness is the following classification scheme from Sawyer (1960):

$CaCO_3$ equiv. (mg l^{-1})	Description
0–75	soft
75–150	moderately hard
150–300	hard
>300	very hard

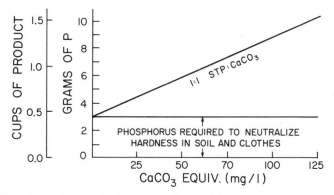

Figure 6.2   Phosphorus (p) required to neutralize total hardness from both washwater and clothes in a 64-l-capacity washer. Cups of product are determined assuming the detergent contains 8.7% P. [Redrawn from Duthie, (1972).]

The tests indicated that NTA interacted with heavy metals such as cadmium and mercury in such a way as to increase the transmission of these metals across the placental barrier to the fetus, thus increasing the likelihood of birth defects. Ruckelshaus and Steinfeld (1970) admitted that the doses of NTA and toxic metals used in the NIEHS animal tests were "considerably higher than would ordinarily be encountered by the human population", and that, "there is no evidence at this time to indicate that anyone has been or is being harmed by the combination of NTA with metals in the environment." Furthermore, NTA is normally degraded in sewage treatment plants (Swisher, Crutchfield, and Caldwell, 1967) and in aerobic aquatic systems, and therefore would not be present to interact with the metals anyway. However, it is not clear how effectively NTA may be degraded in anaerobic systems, so that it is possible, for example, that disposal of anaerobically digested sewage sludges might contaminate certain systems with NTA. Proctor and Gamble did some limited test-marketing of NTA as a builder for detergents in Indiana and New York, but concern over a possible legal ban on NTA caused the company not to pursue the matter further.

In contrast to the United States, both Canada and Finland have permitted the substitution of NTA for sodium tripolyphosphate in detergents, and in Canada about 125,000 tons of detergents containing an average of 15% NTA were purchased in 1975 (IJC, 1977, p. 16). Task forces appointed by the International Joint Commission concluded that there was no evidence that NTA had any teratogenic or mutagenic potential either alone or in combination with heavy metals (IJC, 1977, pp. 16–17). NTA appears to cause cancer of the urinary tract of rats and mice given large doses of NTA over their lifetimes, but the doses of NTA that produced these effects were on the order of a million times higher than levels reported in Canadian drinking water supplies. On the basis of the rat and mice studies, the task force estimated the maximum risk for the environmental doses found by the Canadian monitoring program to be within an order of one possible incident of tumor formation due to NTA in humans per two million population. Based on these estimates and other studies on the possible ecological effects of NTA, the International Joint Commission concluded that, "On the basis of health hazard there is no reasonable cause for restricting the use of NTA as a replacement for phosphate in detergents in the Great Lakes basin" (IJC, 1977, p. 3).

Despite this recommendation, and in spite of the fact that NTA is currently used for various industrial purposes in the United States to the tune of about 5000 tons annually (IJC, 1977, p. 16), U.S. authorities have been reluctant to approve its use in detergents. The EPA, for example, has said only that it sees no reason to ban NTA in detergents, a statement that hardly constitutes an endorsement. This state of affairs, combined with the generally anti-NTA attitude of labor unions and consumer groups, has caused detergent manufacturers in the United States and many other countries to look for other substitutes for sodium tripolyphosphate. Zeolites appear to be the most common alternative. In Japan, for example, over 90% of detergents are built with zeolites. Sodium carbonate and sodium citrate, though used in some detergents in the United States, have been less popular alternatives.

Impact of Detergent Phosphate Reductions

The impact of detergent phosphate reductions lies primarily in two areas, one financial and the other ecological. First of all, what has been the financial cost of eliminating or reducing the P content in many laundry detergents? Given the fact that detergent phosphates never accounted for more than 50% of the P in municipal wastewater, and that the wastewater from most households is routed through sewage treatment plants in the United States, would it not make more sense to remove the P at the sewage treatment plant? The answer is that where P enrichment of receiving waters is perceived to be a problem, it certainly makes sense to remove the P from raw sewage. This approach, for example, is being taken in the Lake Erie watershed, and in Sweden the wastewater from about 80% of the urban population is given tertiary treatment to remove nutrients (Layman, 1984). However, one should keep in mind that marine waters receive about half the sewage effluent from the United States population (Ryther and Dunstan, 1971), and a substantial body of data now indicates that nitrogen rather than phosphorus is the principal limiting nutrient in many marine waters. Given this state of affairs, it might be more cost effective to remove the nitrogen from the sewage than the phosphorus.

If P removal is considered to be a cost-effective way to control cultural eutrophication, then P removal from point source discharges may be the most cost-effective way to deal with the problem. However, the lower the P concentration in the raw sewage, the easier it will be for tertiary treatment to achieve a specified level of P removal. Ecologically the most effective strategy is to remove the P from detergents and also provide tertiary P removal at sewage treatment plants. P removal from detergents alone is ecologically a much less effective strategy, since on the average detergents never accounted for more than 30–40% of P inputs to aquatic systems in the United States. Lee and Jones (1986), for example, have argued that removal of P from laundry detergents will have no discernable effect on water quality unless detergent P accounts for at least 20% of the phosphorus load to a body of water.

From a strictly financial standpoint, P removal at sewage treatment plants is more costly than P removal from laundry detergents. Judging from the figures in Table 6.4, the incremental cost of upgrading a conventional secondary sewage treatment plant to remove nutrients is about $15–$20 per person per year. On the other hand, the cost of substituting NTA for sodium tripolyphosphate in laundry detergents would increase the cost to the consumer by about 40 cents per kg.[12] At the present time the worldwide average per capita use of detergents amounts to about 5 kg/yr (Layman, 1984). Hence the incremental cost to the consumer would be about $2 per person per year.

While this comparison makes detergent P removal seem much more cost effective than P removal at the sewage treatment plant, one should keep in mind that detergents never accounted for more than 50% of the P in municipal wastewater. Thus complete removal of P from detergents would reduce P inputs from sewage treatment plants by at most a factor of 2. According to Lee and Jones (1986), the cost of removing 90% of the P from municipal wastewaters by tertiary sewage treatment would be no more than about $3.50 per person per year. The implication is that if one is willing to back off from the goal of virtually 100% P removal and accept a 90% reduction, the cost per unit P removed is about the same for tertiary sewage treatment and detergent P removal. Land application of sewage effluent, particularly if used in conjunction with the growing of marketable crops, would offer a third alternative that would be very cost competitive with both detergent P removal and conventional tertiary treatment.

Judging from the case studies presented in this book, it seems unlikely that P removal from laundry detergents by itself would effect much improvement in the trophic status of many receiving waters. A 30–40% reduction in the P inputs to Lake Sammamish, Lake Washington, and Lake Erie would have produced little improvement in water quality. On the other hand, there is at least one example of an aquatic system that did respond rapidly and favorably to limitations on the P concentrations in detergents. On July 1, 1971, the city of Syracuse, New York, enacted legislation requiring that detergent P concentrations not exceed 8.7%. The state of New York passed similar legislation in January, 1972 (Murphy, 1973) and then further reduced the P limit to 0.5% in July, 1973. Table 6.7 shows average concentrations of inorganic P in the epilimnion and hypolimnion of Onondaga Lake (near Syracuse) during the 1970–1972 period. The inorganic P concentration declined by about 80% as a result of the 8.7% restriction on detergent phosphates. Murphy (1973) also reports that the total P concentration in the lake declined by about 57%, from 1.74 to 0.74 mg/l. Furthermore, in the 5 years prior to 1971, the fall phytoplankton population in Onondaga Lake had been dominated by a species of cyanobacterium, but in 1972 green algae dominated the phytoplankton crop throughout the summer and fall. Thus clear-cut and desirable changes in the nutrient characteristics of the lake as well as in the phytoplankton population occurred at the same time as the imposition of the 8.7% limitation on detergent phosphate levels. However, the Onondaga Lake watershed was heavily urbanized, and about 90% of the inorganic P loading to the lake could be accounted for by municipal discharges and combined sewer outflows, whereas only 8–9% of the loading was estimated to be natural (Murphy, 1973). Thus the Onondaga Lake case may well be an atypical example in the sense that an unusually high percentage of the P inputs to the lake were coming from detergent phosphates.

[12]Based on Hammond (1971) and an inflation factor of 3.5 (Anonymous, 1989).

Table 6.7   Mean Concentrations of Inorganic P in the Epilimnion and Hypolimnion of Onondaga Lake

Period of observation	Inorganic P (mg/l)	
	Epilimnion	Hypolimnion
1/1/70–6/30/71	0.73	0.81
7/1/71–12/31/71	0.22	0.58
1/1/72–12/31/72	0.11	0.19

Source. Murphy (1973).

The net result of all the commotion over detergent phosphates has been that many detergents now contain at most 0.5% P, and almost all contain less than 8.7% P. However, with a few exceptions, cultural eutrophication problems in most aquatic systems are unlikely to be solved by removal of detergent phosphates alone. Tertiary sewage treatment or land application of sewage will be needed to eliminate most of the point source inputs, and in some cases, such as Lake Erie, control of nonpoint source inputs will probably be necessary as well.

REFERENCES

Abron-Robinson, L. A., C. Lue-hing, E. J. Martin, and D. W. Lake. 1981. *Production of nonfood-chain crops with sewage sludge*. EPA-600/S2-199. Cincinnati, OH.

Allen, J. 1973. Sewage farming. *Environment*, **15**(3), 36–41.

Allhands, M. N., and A. R. Overman. 1989. *Effects of municipal effluent irrigation on agricultural production and environmental quality*. Report for Wastewater Operations Water & Sewer Department of Tallahassee. 377 pp.

Anonymous. 1989. Treatment plants: cost hikes back on track. *Engineering News-Record*, **223**(24), 81.

Bouwer, H., J. C. Lance, and M. S. Riggs. 1974. High-rate land treatment. *J. Water Pollut. Contrl. Fed.*, **46**, 834–859.

Clark, C. S., H. S. Bjornson, C. C. Linnemann, Jr., and P. S. Gartside. 1984. *Evaluation of health risks associated with wastewater treatment and sludge composting*. EPA-600/S1-84-014. Research Triangle Park, N.C.

Donovan, J. F., and J. E. Bates. 1980. *Guidelines for water reuse*. EPA-600/8-80-036. Cincinnati, OH.

Dowdy, R. H., R. D. Goodrich, W. E. Larson, B. J. Bray, and D. E. Pamp. 1984. *Effects of sewage sludge on corn silage and animal products*. EPA-600/S2-84-075. Cincinnati, OH.

Duthie, K. V. 1972. Detergents: Nutrient considerations and total assessment. In G. E. Likens (Ed.), *Nutrients and Eutrophication*. American Society of Limnology and Oceanography. Lawrence, Kans. pp. 205–216.

Eliassen, R., and G. Tchobanoglous. 1969. Removal of nitrogen and phosphorus from waste water. *Environ. Sci. Technol.*, **3**, 536–541.

Ellis, K. V. 1980. The tertiary treatment of sewages. *Effluent and Water Treatment J.*, **20**(9), 422–430; **20**(11), 527–537.

Environmental Protection Agency. 1972. *Water Quality Criteria*. EPA-R3-73-003. Washington, D.C. 594 pp.

Environmental Protection Agency. 1986. *Quality Criteria for Water*. EPA 440/5-86-001. Washington, D.C.

Fertik, H. A., and R. Sharpe. 1980. Optimizing the computer control of breakpoint chlorination. *ISA Trans.*, **19**(3), 3–17.

Gilbert, R. G., R. C. Rice, H. Bouwer, C. P. Gerba, C. Wallis, and J. L. Melnick. 1976. Wastewater renovation and reuse: virus removal by soil filtration. *Science*, **192**, 1004–1005.

Grundy, R. D. 1971. Strategies for control of man-made eutrophication. *Environ. Sci. Tech.*, **5**, 1184–1190.

Hammond, A. L. 1971. Phosphate replacements: Problems with the washday miracle. *Science*, **172**, 361–363.

Hershaft, A., and J. B. Truett. 1981. *Long-term effects of slow-rate land application of municipal wastewater*. EPA-600/S7-81-152. Washington, D.C.

Heukelekian, H. 1957. Utilization of sewage for crop irrigation in Israel. *J. Water Pollut. Contrl. Fed.*, **29**, 868–874.

Hickey, J. L. S., and P. C. Reist. 1975. Health significance of airborne microorganisms from wastewater treatment processes. *J. Water Pollut. Contrl. Fed.*, **47**, 2741–2773.

Hinesly, T. D., L. G. Hansen, D. J. Bray, and K. E. Redborg. 1984. *Long-term use of sewage sludge on agricultural and disturbed lands*. EPA-600/S2-84-128. Cincinnati, OH.

Hinkle, K. R. 1984. *Reclamation of toxic mine waste utilizing sewage sludge: Contrary creek demonstration project*, addendum report. EPA-600/S2-84-016. Cincinnati, OH.

International Joint Commission. 1977. *Annual Report of the Research Advisory Board*. 45 pp.

Kardos, L. T. 1970. A new prospect. *Environment*, **12**(2), 10–27.

Kardos, L. T., W. E. Sopper, E. A. Myers, R. R. Parizek, and J. B. Nesbitt. 1974. *Renovation of secondary effluent for reuse as a water resource*. EPA-660/2-74-016. Washington, D.C.

Klein, L. 1966. *River Pollution. 3. Control*. Butterworth, Washington, D.C. 484 pp.

Kreissl, J. F., and R. F. Lewis. 1981. *Demonstration physical chemical sewage treatment plant utilizing biological nitrification*. EPA-600/S2-81-173. Cincinnati, OH.

Layman, P. L. 1984. Brisk detergent activity changes picture for chemical suppliers. *Chem. and Eng. News*, Jan 23, pp. 17–20, 31–49.

Lee, G. F., and R. A. Jones. 1986. Detergent phosphate bans and eutrophication. *Environ. Sci. Technol.*, **20**, 330–331.

Lewicke, C. K. 1972. Recycling sludge and sewage effluent by land disposal. *Environ. Sci. Technol.*, **6**, 871–873.

Majeti, V. A., and C. S. Clark. 1981. *Potential health effects from viable emissions and toxins associated with wastewater treatment plants and land application sites*. EPA-600/S1-81-006. Cincinnati, OH.

Marinelli, J. 1990. After the flush. The next generation. *Garbage,* Jan/Feb, pp. 24–35.

Miller, S. S. 1972. Debugging physical-chemical treatment. *Environ. Sci. Technol.*, **6**, 984–985.

Morgens, H. J. 1970. The issue of phosphates in detergents. A statement to shareholders of Proctor & Gamble Company. October 13.

Murphy, C. B., Jr. 1973. Effect of restricted use of phosphate-based detergents on Onondaga Lake. *Science*, **182**, 379–381.

Page, A. L., T. L. Gleason, III, J. E. Smith, Jr., I. K. Iskandar, and L. E. Sommers. 1984. *Utilization of municipal wastewater and sludge on land: Proceedings of the 1983 workshop*. EP-600/S9-84-003. Washington, D.C.

Payne, J. F., and A. R. Overman. 1987. Performance and long-term effects of a wastewater spray irrigation system in Tallahassee, Florida. Report for Wastewater Division Underground Utilities City of Tallahassee, 329 pp.

Pressley, T. A., D. F. Bishop, and S. G. Roan. 1972. Ammonia-nitrogen removal by breakpoint chlorination. *Environ. Sci. Technol.*, **6**, 622–628.

Riley, J. P., and R. Chester. 1971. *Introduction to Marine Chemistry*. Academic Press, New York. 465 pp.

Robson, C. M., and L. E. Sommers. 1982. *Spreading lagooned sewage sludge on farmland: a case history*. EPA-600/S2-82-019. Cincinnati, OH.

Rodhe, G. 1962. The effects of trace elements on the exhaustion of sewage irrigated land. *Water Pollution Abstracts*, **36**, 421(2063).

Rohlich, G. A., and P. D. Uttormark. 1972. Wastewater treatment and eutrophication. In G. E. Likens (Ed.), *Nutrients and Eutrophication*. American Society of Limnology and Oceanography, Lawrence, Kansas. pp. 231–245.

Ruckelhaus, W. D., and J. L. Steinfeld. 1970. Statement on NTA. Joint press release of the Environmental Protection Agency and Surgeon General, December 18.

Ryther, J. H., and W. M. Dunstan. 1971. Nitrogen, phosphorus, and eutrophication in the coastal marine environment. *Science*, **171**, 1008–1013.

Sawyer, C. N. 1960. *Chemistry for Sanitary Engineers*. McGraw-Hill, New York. 367 pp.

Schindler, D. W. 1974. Eutrophication and recovery in experimental lakes: Implications for lake management. *Science*, **164**, 897–899.

Shuval, H. I., B. Fattal, and Y. Wax. 1984. *Retrospective epidemiological study of disease associated with wastewater utilization*. EPA-600/S1-84-006. Research Triangle, N.C.

Slechta, A. F., and G. L. Culp. 1967. Water reclamation studies at the south Lake Tahoe public utility district. *J. Water Pollut. Contrl. Fed.*, **39**, 787–813.

Smith, R. J. 1978. Ever so cautiously, the FDA moves toward a ban on nitrites. *Science*, **201**, 887–891.

Sopper, W. E., and S. N. Kerr. 1981. *Revegetating strip-mined land with municipal sewage sludge*. EPA-600/S2-81-182. Cincinnati, OH.

Sopper, W. E., and E. M. Seaker. 1984. *Strip mine reclamation with municipal sludge*. EPA-600/S2-84-035. Cincinnati, OH.

Stewart, D. 1990. Flushed with pride in Arcata, California. *Smithsonian,* **21**, (1), 174–180.

Swisher, R. D., M. M. Crutchfield, and D. W. Caldwell. 1967. Biodegradation of nitrilotriacetate in activated sludge. *Environ. Sci. Technol.*, **10**, 820–827.

Tannenbaum, S. R., D. Fett, V. R. Young, P. D. Lund, and W. R. Bruce. 1978. Nitrite and nitrate are formed by endogenous synthesis in the human intestine. *Science*, **200**, 1487–1489.

U. S. Army Corps of Engineers. 1972. *Regional Wastewater Management Systems for the Chicago Metropolitan Area*. Summary report and technical appendix. Office of the Chief of Engineers, Dept. of the Army, Washington, D.C.

Warren, C. E. 1971. *Biology and Water Pollution Control*. Saunders, Philadelphia. 434 pp.

7

PATHOGENS IN NATURAL WATERS

A wide variety of human pathogens may be found in the excrement from humans as well as from other animals. Most human pathogens can be classified as either viruses, protozoans, helminths (intestinal worms), or bacteria. Both raw sanitary sewage and land runoff contain pathogenic organisms, and virtually every sizeable body of water contains some pathogens. However, it does not follow that bathing in a stream, drinking water from a lake, or swimming in the ocean will automatically cause you to become ill. You must first come in contact with pathogens, the pathogens must gain entry into the body, and the dose of pathogens must be sufficiently great to overcome the body's natural defense mechanisms. Under special circumstances an infection can develop from a single virus, protozoan, or helminth, but the minimum infective dose for bacteria is between 100 and 100 million, depending on the species (Majeti and Clark, 1981).

It follows that neither recreational waters nor public water supplies need be absolutely free of pathogens to be nominally safe, but the higher the concentration of pathogens, the greater the probability that health problems will develop. The fact that the concentrations of pathogens in recreational waters and public water supplies have sometimes been well above safe limits is well documented in the historical record of waterborne disease outbreaks. Even in an advanced nation such as the United States, where sanitary procedures and waste disposal practices are far superior to those used in second- and third-world nations, waterborne disease occurrences are common. Lippy and Waltrip (1985), for example, have estimated that between 1946 and 1980 there were over 1300 waterborne disease outbreaks in the United States, and that these outbreaks caused approximately 350,000 cases of illness.

It is certainly possible to treat water so that the concentration of pathogens is reduced to a safe level, but careful treatment of drinking water and monitoring of recreational water quality has become common only in the last 100 years or so, due in no small part to the recognition and acceptance of the germ theory of disease as developed by Louis Pasteur and Robert Koch during the nineteenth century. There remain, however, practical and scientific limitations on the extent to which water is treated or monitored, even in advanced nations. In their analysis of waterborne disease outbreaks in the United States, for example, Lippy and Waltrip (1985, p. 74) commented that, "The glaring deficiencies were that disinfection was not in place where it was needed and not properly operated where it was in place." In this chapter we examine the nature of the waterborne pathogen problem and the methodologies available for treatment and monitoring.

SOURCES OF PATHOGENS

Pathogens found in human excrement, whether urine or fecal material, come from persons who are presently infected by the disease organism. That such persons are infected by a pathogen does not necessarily imply that they would feel or show any signs of disease. In other words, a person infected by a pathogen may be either symptomatic or asymptomatic. Some persons, following infection by a particular disease organism, may harbor the pathogen in their bodies for the rest of their lives without showing any ill effects. Such persons are referred to as *permanent carriers* of the pathogen. In most populations the percentage of persons who are permanent carriers of a particular pathogen is quite small. For example, roughly 2–4% of the persons who recover from typhoid fever become permanent carriers of the bacterium *Salmonella typhi*, which causes the disease (Frobisher, Sommermeyer, and Fuerst, 1969). In some cases a permanent carrier may be completely unaware that he has ever had the disease caused by the organism he carries, since some infections may be so mild as to pass almost without notice.

In the case of some pathogens there are no truly permanent carriers, but persons who are presently sick or recovering from a disease may excrete enormous numbers of the causative disease organisms in their feces or urine. Such persons are termed *temporary carriers* of the pathogen. Animals of course also become infected by organisms, some pathogenic to humans. Thus animals may be carriers of human pathogens just as are humans. For this reason both sanitary sewage and land runoff may contaminate water supplies with human pathogens.

TYPES OF PATHOGENS

Table 7.1 lists the causative agents of waterborne disease outbreaks in the United States between 1946 and 1980. An *outbreak* is here defined as an incident in which at least two cases of acute infectious disease or one case of an acute intoxicating illness[1] occurred and originated from a common source. In over half the reported cases, the cause of the outbreak was unknown. In the case of microbiological agents, two factors confound the problem of identifying the causative agent. First, the incubation period preceding overt symptoms of illness ranges from about 1 day to several weeks. As a result, outbreaks are normally not recognized less than 1 to 2 weeks following exposure. In the meantime the causative agent may have died off or

---

[1]Infectious diseases are associated with microbiological agents. Intoxicating illnesses are caused by chemical agents.

been flushed out of the water. Second, the isolation, culture, and identification of microbiological agents from natural waters is not a trivial task. The presence of viruses is particularly difficult to document, because they are too small to be seen under a light microscope and are not self-propagating organisms. Because viruses are obligatory intracellular parasites, they can be isolated and cultured only in the presence of living cells in the form of live animals, embryonated eggs or, more commonly, *in vitro* cell cultures. The presence of viruses is evidenced by the characteristic lesions which appear in the culture when the virus multiplies. Most identifications of microbiological agents have been based on samples of stools or blood from infected persons (Lippy and Waltrip, 1985).

Bacterial Pathogens

Approximately 20% of the reported outbreaks and illnesses associated with waterborne diseases were attributed to bacterial pathogens belonging to the genera *Salmonella* and *Shigella*. The diseases associated with these two bacterial genera are commonly referred to as *salmonellosis* and *shigellosis*, respectively. Salmonellosis usually involves acute gastroenteritis with diarrhea and stomach cramps. Fever, nausea, and vomiting may also occur. Under normal conditions, about 1–4% of human beings excrete *Salmonellae*. The excretion percentage is much higher of course during salmonellosis outbreaks. Over several hundred serotypes of *Salmonellae* are known to be pathogenic to humans. Typhoid fever, which is caused by *Salmonella typhi*, is probably the best known of the salmonelloses. *S. typhi* and other

Table 7.1   Causative Agents of Waterborne Disease Outbreaks in the United States Between 1946 and 1980

Cause	% Total outbreaks	% Total cases of illness
Bacterial		
Campylobacter	0.3	2.5
Pasteurella	0.3	0.004
Leptospira	0.15	0.006
Escherichia coli	0.7	0.8
Shigella	9.1	8.7
Salmonella	11.1	12.4
Viral		
Hepatitis A	10.1	1.5
Parvoviruslike	1.5	2.1
Polio	0.15	0.01
Protozoans		
Entamoeba	0.9	0.05
Giardia	6.3	13.1
Chemical		
Inorganic	4.3	0.6
Organic	3.1	1.8
Unknown	52.1	56.4

Source. Lippy and Waltrip (1985).

Salmonellae apparently pass through the stomach, but may multiply in the intestines. Subsequently, *S. typhi* appears in the blood and may establish itself in the periosteum (membranes covering bones), liver, gallbladder, bone marrow, spleen, or kidneys. It may cause meningitis or pneumonia. In cases of death it is usually these latter complications that are responsible for death rather than the original intestinal infection (Frobisher, Sommermeyer, and Fuerst, 1969). Between 1920 and 1945 an average of 37 deaths per year due to waterborne diseases were reported in the United States, and most were caused by typhoid fever. Since that time the number of deaths associated with waterborne diseases has averaged only one per year, and the number of typhoid fever cases has declined to an average of about two per million persons (Figure 7.1).

Shigellae are bacillus-type bacteria, and shigellosis is therefore sometimes called *bacillary dysentery*. The disease is spread by way of the feces of a carrier, whereas *S. typhi* is excreted both in feces and in urine. The percentage of persons excreting *Shigellae* is about 0.3–3%. Unlike *Salmonellae*, *Shigellae* are not commonly found in animals.

Survival of *Salmonellae* in water is variable and is enhanced by low temperatures and high nutrient levels. For example, *Salmonellae* discharged into the Red River at Fargo, North Dakota, and Moorhead, Minnesota, during January were detected 120 km downstream, 4 days flow time from the point of discharge (Geldreich, 1972). *S. typhi* can survive for several weeks in river water and 39 days in ice cream (Frobisher, Sommermeyer, and Fuerst, 1969). *Shigellae*, on the other hand, do not survive for long periods either in feces or in sewage. Like *Salmonellae*, their survival in aquatic systems is enhanced by low temperatures. With the exception of typhoid fever, the reported incidence of salmonellosis has been increasing in recent years, and the disease is now about twice as common as shigellosis in the United States (Figure 7.2).

Several other types of waterborne bacterial pathogens are worthy of mention, either for historical or other reasons. *Escherichia coli* is found in the intestines of all warm-blooded animals, including humans. In fact, human feces may consist of as much as 5–50% *E. coli*. Certain serotypes of *E. coli*, referred to as enteropathogenic *E. coli*, are pathogenic in the sense that they may cause diarrhea. Infant children are particularly susceptible to this form of diarrhea. Most instances of enteropathogenic *E. coli* infections have involved drinking contaminated water. In aquatic systems contaminated with human feces, the proportion of enteropathogenic *E. coli* is probably less than 1%, and the percentage of persons excreting enteropathogenic *E. coli* is no more than 1–10% (Geldreich, 1972). The survival time of *E. coli* in water is

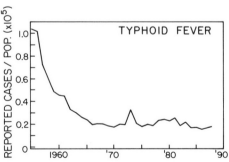

Figure 7.1 Rate of occurrence of typhoid fever in the United States from 1955 to 1988. [Redrawn from Morbidity and Mortality Weekly Report *37* (54), 1989.]

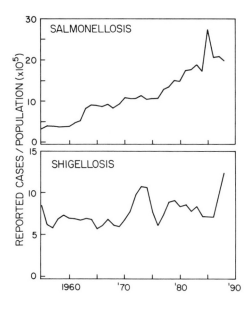

Figure 7.2 Rate of occurrence of salmonellosis (excluding typhoid fever) and shigellosis in the United States from 1955 to 1988. [Redrawn from Morbidity and Mortality Weekly Report *37* (54), 1989.]

influenced by a great many factors and is usually much shorter in seawater than in freshwater (Fujioka et al., 1981). Multiplication of *E. coli* in aquatic systems is believed to be rare, but may occur in heavily polluted warm water containing high concentrations of bacterial nutrients. In most cases, the presence of *E. coli* in aquatic systems can be taken as evidence of recent fecal pollution.

Campylobacteriosis is among the most common bacterial infections of humans throughout the world. The disease may be caused by any of several members of the bacterial genus *Campylobacter*. Although in the United States campylobacteriosis outbreaks are much less frequent than epidemics of shigellosis and salmonellasis, in most parts of the world *Campylobacter* is more commonly isolated from fecal specimens of diarrheal patients than either *Salmonella* or *Shigella*. The number of reported *Campylobacter* infections in England exceeds those of *Salmonella* and *Shigella* combined (Blaser, 1990). The most common symptoms of the disease are diarrhea, malaise, fever, and abdominal pains. Symptoms may last from 1 day to a week or longer. In the United States the peak incidence of campylobacteriosis is in children under 1 year old and persons in the 15–29-year-old age group. The disease is an important cause of acute diarrhea suffered by travelers who visit developing areas. In endemic regions, *Campylobacter* infections are frequently symptomatic in early life but tend to be asymptomatic later on (Blaser, 1990). The implication is that the body develops a certain resistance to the symptoms of the disease as a result of repeated infections.

Cholera is a serious and acute intestinal disease caused by the bacterium *Vibrio cholera*. The disease is characterized by diarrhea, vomiting, suppression of urine, rapid dehydration, fall of blood pressure, subnormal temperature, and complete collapse. Death may occur within a few hours of onset unless treatment is started in time. The treatment, which involves infusion of water and electrolytes into the veins, produces a rapid relief from the symptoms of the disease (Frobisher, Sommermeyer, and Fuerst, 1969). For example, in a recent case in Louisiana, a 67-

year-old woman suffering from cholera was admitted to the intensive care unit of a hospital with hypotension and bradycardia. She was resuscitated after administration of approximately 22 liters of fluids over a 24-hour period (Gergatz and McFarland, 1989).

Cholera is transmitted via contaminated feces. Roughly 2–10% or more of the human population are healthy carriers, and convalescent carriers may shed vibrios intermittently for as much as 4–15 months. Outbreaks of cholera have usually been associated with contaminated water supplies. The disease apparently first appeared in the Orient, the earliest recorded account of an epidemic being a 1563 medical report from India. Cholera did not spread to other parts of the world until the nineteenth century, when it first appeared in Europe. European epidemics were brought under control eventually through the protection and treatment of public water supplies. In the United States only about two cases per year have been reported in recent years (Craun, 1984, p. 249). Cholera epidemics, however, are quite common in other parts of the world. Approximately two epidemics per year occur in the endemic area of Bangladesh (Islam, Drasar, and Bradley, 1990), and during the latter 1980s outbreaks were reported in Thailand, India, Singapore, West Africa, and Tanzania (Tabtieng et al., 1989; Murthy et al., 1990; Fule et al., 1990; Goh et al., 1990; St. Louis et al., 1990; Killewo, Amsi, and Mhalu, 1989). A 1985 epidemic in Bangladesh resulted in 12,194 registered cholera cases and 51 deaths (Siddique et al., 1989), and a 1991 outbreak in Peru infected over 14,000 people and claimed at least 90 lives (Anonymous, 1991a). Because of the severity of the symptoms associated with cholera and the potential of the disease to be fatal, outbreaks of cholera can have very serious consequences. For the same reasons, there is speculation that *Vibrio cholera* might be used as a biological warfare agent by governments that do not subscribe to the accords of the 1972 Bacteriological (Biological) and Toxin Weapons Convention.

Leptospirosis and tularemia are diseases caused by bacteria of the genus *Leptospira* and the bacterium *Pasteurella tularensis*, respectively. In both cases the bacteria enter the blood stream through skin abrasions or mucus membranes. Leptospiral bacteria may cause acute infections involving the kidneys, liver, and central nervous system. Tularemia is characterized by both chills and fever, swollen lymph nodes, and a general prostrate condition. Both pathogens tend to be transmitted to humans by animals. *Leptospira* are excreted in urine but are not normally found in feces. Transmission to humans usually involves contact with water in which infected animals have urinated. In the United States leptospirosis outbreaks are confined almost exclusively to the summer recreational period and are associated with swimming in polluted waters. Leptospirosis is often severe and may be fatal. However, only about 60 cases per year are reported in the United States (Figure 7.3). Most cases of tularemia occur in the hunting season as a result of contact with wood ticks or infected animals. However, the disease may also be spread through water contaminated with the urine, feces, or dead bodies of infected animals, and a number of outbreaks of tularemia have been associated with drinking contaminated water (Geldreich, 1972). Reported cases of tularemia in the United States declined rapidly from 1950 to 1970 and have averaged about 100–150 cases per year since 1970 (Figure 7.3).

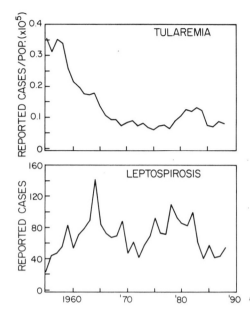

Figure 7.3 Rate of occurrence of leptospirosis and tularemia in the United States from 1955 to 1988. [Redrawn from Morbidity and Mortality Weekly Report *37* (54), 1989.]

Protozoan Pathogens

Numerous protozoan species normally inhabit the intestinal tract of warm-blooded animals, including humans. Many of these protozoan species are completely harmless to humans and are routinely found in the feces of both healthy and sick persons. However, some protozoans are pathogenic. In the United States the most important of the enteropathogenic protozoans is *Giardia lamblia*, which causes a disease known as *giardiasis*. The most common symptom of the disease is diarrhea, but giardiasis may also be associated with weakness, weight loss, abdominal cramps, nausea, greasy stools, abdominal distention, flatulence, vomiting, belching, and fever. *Giardia* was first described in 1681 by Anthony van Leeuwenhoek, incidentally from a sample of his own stools. Intestinal disorders caused by *G. lamblia* were not documented in the United States until 1966 (AWWA, 1985), but of the waterborne disease outbreaks where a causative agent has been identified, *G. lamblia* has accounted for more illnesses in recent years than any other single pathogen (Table 7.1). According to Hill (1990), roughly 4% of persons in the United States are carriers, but rates may be as high as 16% in certain areas.

    G. lamblia occurs in two forms, a free-living or trophozoite form and an encapsulated cyst form. Both forms may be excreted by an infected organism, but the cysts are more resistant and tend to be the dominant forms in natural waters. Furthermore, the trophozoite stage is not resistant to the initial stages of digestion, and the cyst form is therefore believed to be the cause of infections. Once the cysts reach the small intestine they undergo an excystation process and are transformed into the active trophozoite state, which reproduces and infects the host.

    Giardia cysts are associated primarily with surface waters, and have been reported in groundwater only in cases involving sewage contamination (AWWA, 1985). The survival of the cysts in water is enhanced by cold temperatures, and the cold environment of mountain streams may allow them to remain viable for up to 2 months. Almost half the giardiasis outbreaks in the United States have occurred in Colorado, and virtually all the outbreaks in the United States have been reported in mountaneous areas. In the Rocky Mountains, the frequent use of easily contaminated surface waters has undoubtedly contributed to the prevalence of the disease. Although most mammals are susceptible to *G. lamblia* infection, beavers, because of their aquatic habitat, are believed to be significant sources of *G. lamblia* in surface waters. Because *G. lamblia* cysts are resistant to chlorination, filtration appears to be the most practical and effective method of eliminating *G. lamblia* from public water supplies (Ongerth, Riggs, and Cook, 1989).

The only other protozoan associated with a waterborne disease of much consequence in the United States is *Entamoeba histolytica*, which causes a disease of the large intestine called *amebiasis*. The disease syndrome may range from mild abdominal discomfort involving diarrhea alternating with constipation to a more severe chronic dysentery referred to as *amebic dysentery*. Using tissue-dissolving enzymes, *E. histolytica* burrows into the intestinal lining and occasionally ruptures the intestine. The organism is eliminated in the feces, often in the form of cysts that may remain viable for many days in water. Geldreich (1972) reports human carrier rates of about 10–17% of the general population. The concentration of cysts in sewage is usually quite low, typically 1–5 per l, and these low densities are greatly reduced in receiving bodies of water by dilution, settling, and death of the cysts. Incidents of amebiasis have generally involved defects in plumbing systems, such as cross connections between sewer lines and water supply pipes, or leaky sewer and/or water lines.

Viral Pathogens

Over 120 different viral pathogens are excreted in human feces and urine. Table 7.2 lists the major categories of these viruses and their associated diseases. According to the data in Table 7.1, viral pathogens accounted for about 12% of waterborne disease outbreaks in the United States between 1946 and 1980, but because of the difficulty in detecting and identifying viruses, there is reason to suspect that a very significant portion of the unexplained waterborne disease outbreaks may also have been caused by viruses. There is circumstantial evidence to support this viewpoint. For example, a study of infant diarrheal cases in Houston, Texas, between 1964 and 1967 revealed that Group A coxsackie viruses were present 3.7 times more often in diarrheal than in nondiarrheal infants (Geldreich, 1972). The Norwalk virus was unknown until 1969, when it was detected after an epidemic of gastroenteritis in Norwalk, Ohio (Kapikian, 1972). Kaplan et al. (1982) reviewed 74 outbreaks of non-bacterial gastroenterities investigated by the Centers for Disease Control and attributed 31 of them to the Norwalk virus. However, only 13 of the 31 cases involved water pollution. A practical consideration is the fact that viruses are more resistant to standard water treatment procedures than the commonly used indicator bacteria (see later discussion). Consequently public water supplies and recreational waters that appear safe based on standard assays may in fact contain dangerous concentrations of viruses. Based on these as well as other considerations, Cabelli (1983) was led to conclude that the cases of gastroenteritis he documented in a study of swimming-related illnesses in the United States and Egypt were probably caused by human rotaviruses and/or parvolike viruses, although no attempt was made to isolate and identify the etiologic agent(s).

In the United States the waterborne disease most frequently linked with a known virus has been infectious hepatitis, also known as hepatitis A. The pathogen is excreted in feces of infected persons. The disease is associated with a marked inflammation of the liver. Symptoms include malaise, transient fever, myalgia, nausea, anorexia, and abdominal pain. The fact that this virus accounts for such a high percentage of the viruses linked to waterborne disease outbreaks probably reflects its unusually high resistance to chlorination. The issue of resistance was dramatically illustrated during an outbreak of infectious hepatitis that occurred in Delhi, India, between 1955 and 1956. In that outbreak over 20,000 clinical cases of infectious hepatitis were diagnosed. The cause of the outbreak was contamination of the public water supply with raw sewage. Although the water was treated by filtration and chlorination, the treatment process was evidently inadequate to kill

Table 7.2   Human Enteric Viruses that May Be Present in Water

Virus group	Number of types	Disease or symptom
Enteroviruses		
Poliovirus	3	Paralysis, meningitis, fever
Echovirus	34	Meningitis, respiratory disease, rash, fever, gastroenteritis
Coxsackievirus A	24	Herpangina, respiratory disease, meningitis, fever, hand, foot and mouth disease
Coxsackievirus B	6	Myocarditis, congenital heart anomalies, rash, fever, meningitis, respiratory disease, pleurodynia
New enteroviruses types 68–71	4	Meningitis, encephalitis, respiratory disease, rash, acute hemorrhagic conjunctivitis, fever
Hepatitis A (enterovirus 72)	1	Infectious hepatitis
Norwalk virus	2	Epidemic vomiting and diarrhea, fever
Rotavirus	4	Gastroenteritis, diarrhea
Reovirus	3	Not clearly established
Adenovirus	>30	Respiratory disease, conjunctivitis, gastroenteritis
Parvovirus		
Adeno-associated virus	3	Associated with respiratory disease of children, but etiology not clearly established
?	?	Acute infectious nonbacterial gastroenteritis

Source. Rao and Melnick (1986) and R. Fujioka (personal communication).

off sufficient numbers of the hepatitis A viruses. However, the fact that there was no change in the incidence of typhoid fever or dysentery during the infectious hepatitis epidemic suggests that the treatment process had been adequate to reduce the concentration of other pathogens to an acceptable level. Indian officials stated that during the period of contamination *E. coli* counts in the treated water did not exceed 2 counts per 100 ml (Dennis, 1959), a level which the Indian officials considered acceptable for drinking water. According to Dennis (1959), analysis of the raw water entering the treatment plant during November, 1955 indicated that as much as 50% of the water was sewage during the peak of contamination. Indian officials were aware that the raw water was contaminated and raised greatly the level of residual chlorine in the treated water in order to reduce the pathogen concentrations to a safe level (Fox, 1976). They evidently succeeded in the case of all pathogens except hepatitis A, an indication of its unusually high resistance to chlorination. The incident emphasizes the problem of using the concentration of a single indicator organism (*E. coli.*) as a criterion of water quality. The survival time in water and susceptibility of different pathogens to standard water treatment methods varies greatly.

The other category of viruses that contributed more than 1% to the number of waterborne disease outbreaks in the United States between 1946 and 1980 was the parvoviruslike viruses, which are very small DNA viruses. Most members of this group infect lower animals such as rats, cats, and dogs, but at least two types of parvoviruslike viruses are known to be of human origin. The adeno-associated viruses are associated with childhood respiratory disease, and have been recovered in the feces of infected children. A second category of parvoviruslike viruses is associated with acute infectious nonbacterial gastroenteritis (Fox, 1976).

Poliomyelitis is not a significant health problem in the United States today, but historically it has taken on much greater significance. In recent years there have been fewer than 10 reported cases per year of poliomyelitis in the United States (Figure 7.4). During the early 1950s the frequency of occurrence was over 1000 times greater. The disease is marked by inflammation of nerve cells in the anterior horns of the spinal cord and is characterized by fever, motor paralysis, and atrophy of skeletal muscles, often with permanent disability and deformity. Poliomyelitis is transmitted primarily by person-to-person contact, and between 1946 and 1980 only a single outbreak of poliomyelitis was attributed to waterborne transmission. In that case, a 1952 epidemic involving defects in the water distribution system in Huskerville, Nebraska, a total of 16 persons developed poliomyelitis. As indicated in Figure 7.4, the incidence of clinical poliomyelitis has declined greatly since the introduction of the Salk vaccine (1954–1955) and Sabin vaccine (1961–1962). The Sabin vaccine has proven particularly effective, because it involves the oral administration of live, but nonvirulent, polio viruses. The live viruses replicate in the gut of vaccinated persons for several weeks, and as a result individuals who have been vaccinated spread the virus among their contacts, who become similarly immunized. The Salk vaccine does not provide as strong or long-lasting immunity as the Sabin vaccine, and because it involves the administration of inactive viruses, there is no chance for immunity to be transferred from vaccinated to nonvaccinated persons. The negative aspect of the Sabin vaccine is that it occasionally causes a case of clinical poliomyelitis.

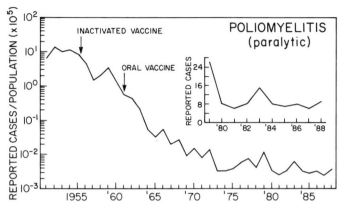

Figure 7.4   Rate of occurrence of poliomyelitis in the United States from 1951 to 1988. [Redrawn from Morbidity and Mortality Weekly Report *37* (54), 1989.]

In 1985 the Pan American Health Organization began an effort to eradicate polio from the Americas. The immunization program was so successful that during the first 4 months of 1991 not a single case of polio due to the wild virus was reported among the 700 million inhabitants of North, Central, and South America (Anonymous, 1991b). The World Health Organization has a similar immunization program underway in Asia and Africa and hopes to eradicate the disease by the year 2000. If that effort is successful, polio may join smallpox as the second major disease to be eradicated worldwide by immunization.

The problem of clinical polio cases caused by immunization unfortunately still remains. The 5–10 cases of poliomyelitis reported in the United States each year are all caused by the Sabin vaccine. Recently the United States government has been considering the use of a so-called enhanced inactivated polio vaccine (E-IPV), which is similar to the original Salk vaccine, but rivals the Sabin vaccine in effectiveness and, because the viruses are inactivated, cannot cause poliomyelitis. It is unlikely, however, that there will be a complete switch to E-IPV, because such a switch would eliminate the benefits derived from transfer of immunization to unvaccinated members of the population. In some disadvantaged areas of the United States, less than 50% of young children are vaccinated against poliomyelitis, in contrast to the national average of 95% (Roberts, 1988). Since the greatest probability of developing poliomyelitis is associated with the first administration of the Sabin vaccine, the compromise solution will probably be to administer the E-IPV vaccine the first few times, for example at age 2, 4, and 6 months, followed by the Sabin vaccine at age 18 months and upon entry to elementary school (Roberts, 1988).

Helminths

There are a number of parasitic intestinal worms that may be found in sewage. Most of them enter the human body through the mouth, although a few gain entry through the skin. Contamination of drinking water with these worms can be prevented effectively with modern water treatment methods. However, swimming or wading in sewage-polluted waters, or use of raw sewage to fertilize crops, may also lead to outbreaks of intestinal worm infections. Such incidents are of very little public health significance in the United States today, but they are much more important in some other parts of the world.

Perhaps the best known helminth is the beef tapeworm *Taenia saginata*. The adult tapeworm lives in the intestinal tract and may discharge as many as one million eggs per day in the feces of an infected person. These eggs are frequently found in sewage, but usually at low concentrations on the order of 1 or 2 eggs per 100 ml. The eggs remain viable for hundreds of days if kept cool and moist. Humans occasionally become infected by swallowing water containing the eggs. Symptoms of tapeworm infection include abdominal pain, digestive disturbance, and weight loss. Geldreich (1972) cited a 50% incidence of beef tapeworm infection in East Africa, but the percentage of carriers in the United States is less than 1%.

Ascariasis is an infection of the small intestine caused by the round worm *Ascaris lumbricoides*. The disease produces variable symptoms and is more common among young children. In the intestine the worm may produce as many as 200,000 embryonated eggs per day. After being excreted with the feces, the eggs may remain viable in water or soil for as much as several months. *A. lumbricoides* has been found in 2% of the feces of sewage plant workers and in 16% of the feces of farm workers who were involved in cultivating sewage-irrigated crops (Geldreich, 1972).

Certain species of the blood fluke genus *Schistosoma* spend part of their life cycle in humans, where they may cause a debilitating infection known as schistosomiasis. The worm enters the body through the skin and from there may be transported by the circulatory system to the lungs, liver, and intestines. The organism is excreted in both urine and fecal material. Completion of the organism's life cycle requires it to infect snails, and a given species of *Schistosoma* infects only one species of snail (Mahmoud, 1990). It is impossible for the infection to be transmitted in the United States because of the absence of the appropriate snail intermediate hosts. Nevertheless, it is estimated that as many as 400,000 persons in the United States are infected with *Schistosoma* (Mahmoud, 1990). Most of these persons are immigrants who were infected before entering the United States. The disease is endemic to certain areas in Africa, South America, and Asia. Worldwide it is estimated that about 200 million persons are carriers of *Schistosoma*.

Penetration of the skin by the larval *Schistosomae* may cause a skin rash known as swimmers itch. The more serious symptoms of the disease, which generally do not develop until 4–8 weeks later, include fever, chills, sweating, and headaches. Chronic cases are associated with fatigue, abdominal pain, and intermittent diarrhea or dysentery. The liver and intestines are the organs most commonly affected.

TESTS FOR PATHOGENS

The preceding discussion suggests one of the serious problems associated with testing for pathogens in water supplies. There are simply too many pathogens to make testing for all or even most of them practical. Since the survival of different enteric pathogens in natural waters and their resistance to standard water treatment methods vary greatly from one species to another, pathogen concentrations are likely to show considerable temporal and spatial variations that are correlated poorly with each other. Consequently monitoring the concentration of only one or a few pathogens would not necessarily provide a good measure of disease probability. The cost of setting up and operating a water quality laboratory capable of testing municipal water supplies and sewage treatment plant effluents for a large variety of organisms would, however, be prohibitive for most communities. Viral pathogens pose a particularly difficult detection problem because they are invisible under the light microscope and can be propagated only in living cells. Furthermore, although it is possible to estimate the concentrations of many viruses using sophisticated culture methods, present techniques are capable of demonstrating only the presence or absence of the infectious hepatitis virus and the parvolike viruses, which are among the most important of the enteric viral pathogens (Cabelli, 1983). Therefore, the general approach to checking water for pathogens has been to look for an indicator organism whose presence could be taken as evidence of sewage pollution. Ideally such an indicator organism should satisfy the following criteria.

1. The organism is proportional in abundance to the number of pathogens in the water and survives at least as long as the pathogens outside the intestinal tract.
2. The organism is easily detectable and present in greater quantity than any potential pathogen.
3. The organism is absent from the aquatic environment unless the water has been polluted with sewage or animal excrement.

Because human enteric pathogens are likely to be found in the intestinal tracts of only a small percentage of the population, the concentration of none of these pathogens is particularly high in sewage under normal circumstances. Because the concentrations in the sewage are likely to be reduced greatly in receiving waters by dilution, such pathogens would not make good indicator organisms of sewage pollution. What is needed is an indicator organism that is likely to be detected by a scientist who takes a sample from only a small part of a natural body of water. Such an indicator organism must be present consistently in high concentrations in raw sewage.

The search for a suitable indicator organism has not met with unqualified success. In the first half of the twentieth century many water quality standards were written in terms of the concentration of coliform bacteria, because it was believed that all coliform bacteria were of fecal origin. We now know, however, that the total coliform population includes four genera from the family *Enterobacteriaceae* and some species from the genus *Aeromonas*, and of the various coliform species only *Escherichia coli* of the family *Enterobacteriaceae* is found exclusively in feces (Cabelli, 1983). The use of total coliforms as an indicator of fecal pathogens received a severe jolt in 1941 when the Illinois Supreme Court overruled a lower court conviction of the director of the Illinois department of public welfare for failing to take action to correct contamination of the drinking water supply in a state hospital. The contamination resulted in a typhoid fever epidemic that produced many deaths. Contamination was suggested by the numbers of total coliforms in the water. However in *People vs. Bowen* (1941) 376 Ill. 317, 33 N.E. (2d) 587, the court ruled that:

> It appears from the record that coli aerogenes or colon bacillus may be friendly or inimical, and that the mere presence of the colon bacillus in water proves exactly nothing as far as typhoid fever is concerned. The tests seem to have been made by a method of broth fermentation, and determined nothing more than the presence or absence of some kind of colon bacillus. It further appears that this type of bacillus is present in the air one breathes, in milk, on fruits and practically everywhere.

> Despite this ruling, federal and many state water quality guidelines were still written in terms of total coliforms during the early 1970s. For example, the 1972 EPA water quality criteria state that a well-operated water treatment plant, "Can be expected to meet a value of 1 total coliform per 100 ml with proper chlorination practice" (EPA, 1972, p. 58).

As realization that total coliform counts could be a very misleading indicator of fecal pollution, guidelines were rewritten in terms of fecal coliforms. The fecal coliforms are operationally distinguished from other coliforms on the basis of their ability to grow on a defined medium at 44.5°C. The test for total coliforms can be conducted using the same growth medium, but the incubation temperature is 35°C (Wolf, 1972). Unfortunately the assay for fecal coliforms does not select exclusively for *E. coli*, but also includes thermotolerant members of the *Enterobacteriaceae* genus *Klebsiella*. *Klebsiella* is found infrequently in human feces, and when present is usually a minor portion of the coliform population (Cabelli, 1983). Furthermore, there are substantial nonfecal sources of *Klebsiella*, including the thermotolerant biotype. Hence the designation "fecal" coliform is something of a misnomer. Nevertheless, the use of fecal coliform counts was regarded as a significant improvement over total coliform counts as an indicator of water quality, and fecal coliform counts largely replaced or at least supplemented total coliform counts in water quality criteria during the 1970s. For example, the 1972 EPA guidelines for public water supplies stated that the geometric mean fecal coliform and total coliform counts should not exceed 2000/100 ml and 20,000/100 ml, respectively (EPA, 1972, p. 58).

The empirical evidence for setting these standards was surprisingly slim. At the time the 1972 EPA water quality criteria were established, the only data available from freshwater stream pollution studies on the correlation between pathogen occurrence and fecal coliform concentrations were for *Salmonella*. The data are summarized in Table 7.3. Taken at face value, Table 7.3 indicates that most water supplies containing the public water supply standard of 2000 fecal coliforms per 100 ml would be contaminated with *Salmonella*, and although treatment processes were expected to reduce the fecal coliform counts to 1 or less per 100 ml, it is unclear from the table whether even 1 count per 100 ml is a safe level. Indeed, the 1955–1956 infectious hepatitis epidemic in Delhi, India, showed dramatically that coliform counts of $< 2/100$ ml can be associated with a very unsafe level of at least one enteric pathogen.

Realizing the paucity of empirical evidence to support their recommended water quality standards with respect to fecal coliforms, the Environmental Protection Agency authorized a systematic study of both freshwater and marine recreational

Table 7.3   Fecal Coliform Abundance and *Salmonella* Occurrence

Fecal coliform density range (counts/100 ml)	% of samples containing *Salmonella*
1–200	31.7
201–1000	83
1001–2000	88.5
Over 2000	97.6

Source. EPA (1972, p. 57).

waters. The study of marine waters lasted from 1972 to 1978 and involved beaches in New York City, Boston, and Lake Pontchartrain in the United States and Alexandria, Egypt. The freshwater studies lasted from 1978 to 1982 and were conducted at Keystone Lake, Oklahoma, and Lake Erie near Erie, Pennsylvania. In both the marine and freshwater studies weekend beachgoers were interrogated 7–10 days after their visits to the beach to determine whether they had experienced any health problems that might be related to enteric pathogens in the water. The beachgoers were divided into a swimming and nonswimming group, depending on whether they had immersed their heads in the water while swimming, and the difference in the incidence of gastrointestinal (GI) illness symptoms between the swimmers and nonswimmers was correlated with the abundance of various possible indicator organisms in the water. The GI illness symptoms included vomiting, diarrhea, stomachache, and nausea.

The results of the study proved quite revealing. Not unexpectedly, there was a rather poor correlation between the incidence of GI illness among swimmers and total coliform concentrations. Surprisingly, there was an even poorer correlation between fecal coliform concentrations and GI illnesses. Of the concentrations of indicator species tested in the study, the concentration of enterococci bacteria was by far best correlated with the frequency of GI illnesses in the study of marine bathing beaches (Figure 7.5). Although *E. coli* counts were correlated positively with GI illnesses, the correlation coefficient was not significant statistically (Figure 7.5). In the freshwater beach studies, the occurrence of GI illnesses again showed no significant correlation with fecal coliforms concentrations, but did correlate significantly with the abundances of both enterococci and *E. coli.*

An important, and perhaps surprising, implication of the study was that fecal coliform concentrations are almost worthless as indicators of the probable occurrence of GI illnesses associated with waterborne pathogens. Evidently for natural waters at least, the presence of thermotolerant *Klebsiella* genera in the fecal coliform assay thoroughly confounds the implications of the fecal coliform count regarding public health. The poor correlation between *E. coli* counts and GI illnesses in marine waters probably reflects differences in the survival rates of *E. coli* and the etiological agent(s) responsible for causing the health problems. With respect to this point, it has been reported that enterococci survive better than *E. coli* in salt water (Cabelli, 1983). However, enterococci are found usually at lower densities in human feces and sewage effluents than *E. coli* (Dufour, 1984), and this fact may compromise their usefulness as indicator organisms in marine waters (see following discussion). Nevertheless, their concentrations were rather well correlated with the incidence of GI illnesses in both the marine and freshwater studies.

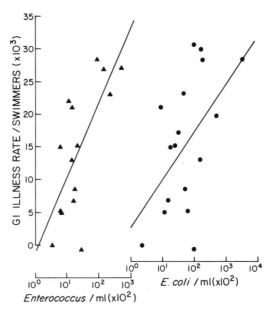

Figure 7.5 Empirical relationships between gastrointestinal illness rates of swimmers at marine beaches and the concentrations of enterococcus (left) and *E. coli* (right). [*Source.* Cabelli et al. (1983).]

It was obvious from the 1972–1982 studies that the relative survival rates of enterococci and the etiologic agent(s) responsible for the GI illnesses differed in marine and fresh waters. This fact is apparent when the regression lines relating the frequency of GI illnesses and enterococci concentrations are plotted on the same graph (Figure 7.6). The concentration of enterococci is lower in marine than in fresh waters at the same incidence of GI illnesses. The cause of this offset is believed to be the fact that enterococci die off more rapidly in marine than freshwaters, whereas the die-off rate of the etiolotic agent(s) is similar in both water types (Dufour, 1984).

Given the regression lines in Figure 7.6, what is an acceptable concentration of enterococci in bathing waters? The answer to this question has proven to be controversial. The average swimming-associated illness rates recorded in the 1972–1982 EPA studies were 15 and 6 GI illnesses per 1000 swimmers in marine waters and freshwaters, respectively. The associated geometric mean enterococcus concentrations were 25 and 20/100 ml, respectively. On the assumption that 6 GI illnesses per 1000 swimmers represented an acceptable illness rate, the EPA (1984) proposed that the criterion for fresh recreational waters be set at 20 enterococci per 100 ml. Based on the regression line in Figure 7.6, 6 GI illnesses per 1000 swimmers in marine waters would be associated with 3 enterococci per 100 ml. Rationalizing that comparable public health standards should be applied to both marine and fresh waters, the EPA (1984) proposed a criterion of 3 enterococci per 100 ml for marine waters. A number of public health officials, however, objected to the marine criterion, in part because laboratory methods for quantifying enterococci at such low densities are not dependable. In response to public comments, the EPA (1986) revised the marine and freshwater criteria to 35 and 33 enterococci per 100 ml, respectively. The rationale was that the fecal coliform criterion recommended previously for both marine and fresh recreational waters had been 200/100 ml and that based on the data collected in the 1972–1982 study, these concentrations of enterococci would be associated with a fecal coliform concentration of 200/100 ml in marine and fresh waters, respectively. These enterococci concentrations correspond to GI illness rates of about 16 and 8 per 1000 swimmers in marine and freshwaters, respectively. However, some public health officials consider 16 GI illnesses per 1000 swimmers unacceptably high. Furthermore, since the EPA study revealed no significant correlation between fecal coliform counts and GI illnesses in either marine or fresh waters, it is unclear why the fact that states had historically accepted a fecal coliform count of 200/100 ml in recreational waters should have any bearing on the water quality standard with respect to enterococci. It seems fair to say that these standards will continue to be debated and discussed in the years ahead.

TREATMENT OF PUBLIC WATER SUPPLIES

Surface waters are treated frequently to remove objectionable substances if they are to be used for drinking purposes. Groundwater is much less likely to require treatment, although aquifers do become contaminated sometimes and require treatment prior to use (see Chapter 16). Often the treatment is designed primarily to insure that the treated water does not contain a dangerous concentration of pathogens. However, the treatment process may also involve removal of suspended solids or dissolved organics, and in some cases the water is softened by precipitating out calcium and magnesium ions. The actual treatment process may differ greatly from one municipality to another, depending on the characteristics of the public water supply. Where treatment does occur, one or more of the following three steps is usually a part of the process.

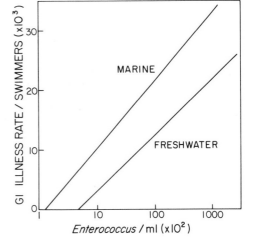

Figure 7.6 Empirical relationships between gastrointestinal illness rates of swimmers at marine and freshwater beaches and the concentrations of enterococcus bacteria. [*Source.* Cabelli (1983) and Dufour (1984).]

Removal of Suspended Solids

Perhaps the simplest form of water treatment is removal of suspended solids. This can be accomplished by merely drawing the water from the downstream end of a large reservoir where the residence time of the water is weeks or months. This long residence time in a quiescent system allows much of the suspended particulates to settle to the bottom. An alternative is to route the water through a special sedimentation basin where *alum*, a compound containing aluminum and sulfate ions, is added to abet sedimentation of particulate and colloidal substances. In most natural waters, the alum forms a positively charged precipitate that attracts colloidal substances, which are negatively charged at normal pH levels. The combination of positive and negatively charged substances tends to conglomerate into a larger and larger composite particle or floc. The flocculation process is speeded up by slow agitation. The large floc particles gradually settle out, taking with them a large portion of the bacterial population. The efficiency of bacterial removal in the sedimentation step can vary greatly, but a well-designed flocculation and sedimentation system may remove 90–95% of the bacterial population.

Filtration

In some cases sedimentation and flocculation do not remove a sufficiently large percentage of objectionable substances from the water. In such cases a filtration step, either alone or following sedimentation/flocculation, is a part of the water treatment process. Filtration is necessary, for example, to remove most of the *Giardia lamblia* cysts from raw surface waters. The filtration unit consists of layers of sand built up over layers of gravel or some other coarse material. The filter bed is typically several meters deep (Figure 7.7). The system is usually referred to simply as

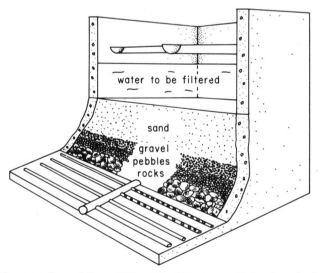

Figure 7.7   Diagram of a rapid sand filter. The filter is usually backwashed every 1–2 days. The two troughs above the sand carry away wash water when the filter is backwashed. [Redrawn from Frobisher et al. (1969).]

a *sand filter*, although the designation *dual-media filter* is sometimes applied to this or similar filtration systems. Particulate matter strained out on the surface of the sand provides food for microorganisms, which grow to form a film over the sand surface. This biological filter helps to strain out both colloidal and dissolved organic matter. In the early days of water treatment there was usually no treatment step preceding filtration. As a result the water was percolated rather slowly through the filters in order to bring about effective treatment. Bacterial removal by these slow sand filters could be as high as 99.99% (McKinney, 1962). With modern pretreatment methods, it is possible to run water through the sand filters at a much higher rate without impairing water quality. Ongerth, Riggs, and Cook (1989), for example, indicate that a well-functioning flocculation/filtration system can reduce the concentration of *Giardia lamblia* cysts by a factor of 10–100.

Chlorination

In many cases surface waters destined for drinking are chlorinated to kill pathogens, and of course sewage effluent is usually disinfected by the same procedure. When chlorine gas (Cl_2) is added to water, the gas reacts with the water to form hydrochloric acid (HCl) and hypochlorous acid (HOCl) according to the reaction

$$Cl_2 + H_2O \rightarrow HCl + HOCl$$

The HOCl is actually the chemical agent primarily responsible for killing microorganisms. It oxidizes key reducing enzyme systems and prevents normal respiration, and at high chlorine concentrations it denatures sufficient proteins to completely destroy cells. Viruses are generally more resistant to chlorination than bacteria, because killing viruses requires denaturation of the viral nucleoprotein. About twice the free chlorine residual is required to kill viruses as is needed to kill bacteria (McKinney, 1962). If the raw water supply contains suspended particles, chlorination following rather than preceding removal of these particles increases the effectiveness of the chlorine kill, because a large percentage of the pathogens are removed in the sedimentation and/or filtration step(s), and because most of the particles that might react otherwise with chlorine and/or provide refuges for pathogens have been removed.

Normally sufficient chlorine is added to the water to maintain a residual chlorine concentration in the water until it reaches the user. This residual chlorine level prevents reestablishment of pathogen populations in the water while the water stands in storage. If the residual chlorine level is too high, however, customers are likely to complain about the water's taste. Certainly no one expects his/her tap water to taste as if it had been drawn from a swimming pool. The degree of chlorination therefore represents something of a compromise between a high degree of microorganism kill on the one hand and water taste and treatment cost considerations on the other hand.

Alternatives to Chlorination

Chlorination is certainly not the only mechanism for killing waterborne pathogens, and in recent years much interest has developed in identifying suitable alternatives to chlorination for this purpose. The principal reasons are the potential formation of carcinogenic chlorinated organic compounds during the treatment process, the resistance of certain pathogens such as viruses and *Giardia lamblia* to chlorination, and the toxicity of residual chlorine to nontarget aquatic organisms (Scheible and Bassell, 1981). Possible alternatives include ultraviolet radiation, ozone, and chlorine dioxide, and a number of studies during the 1980s addressed the feasability of these methods (Fluegge, Metcalf, and Wallis, 1981; Roberts et al., 1981; Stover, Jarnis, and Long, 1981; Carlson et al., 1985; Scheible, Casey, and Forndran, 1986). Neither ultraviolet light nor ozone is associated with the formation of chlorinated organics, and in studies with sewage treatment plant effluent, disinfection with chlorine dioxide (ClO_2) produced less than 10% as many chlorinated organics as Cl_2 (Roberts et al., 1981).

With respect to pathogen-killing efficiency, ClO_2 inactivates more effectively viruses than does chlorine, but at least for typical sewage treatment plant effluent the cost of ClO_2 treatment would be 2–5 times as great as disinfection with Cl_2 (Roberts et al., 1981). The studies of Stover, Jarnis, and Long (1981) showed that satisfactory reduction of total coliform concentrations in treated sewage effluent could be achieved with ozone, but Scheible and Bassell's (1981) cost analysis indicates that disinfection of secondary sewage treatment plant effluent with ozone would cost at least twice as much as disinfection with Cl_2. For sewage treatment plants that handle flow rates of 40–400 million liters per day, ultraviolet light can reduce fecal coliform concentrations to less than 200/100 ml at a cost that is only about 25% higher than chlorination, and for small plants that handle on the order of 4 million liters per day, ultraviolet light disinfection may actually be 30% cheaper than chlorination (Scheible and Bassell, 1981). *Giardia lamblia* cysts do, however, survive ultraviolet light treatment substantially better than *E. coli*. Based on the studies of Scheible and Bassell (1981), Scheible et al. (1986), and Carlson et al. (1985), ultraviolet light doses that reduce fecal coliform concentrations by a factor of 10^3–10^4 would reduce the concentration of *Giardia lamblia* cysts by only a factor of 5. In a study of virus inactivation, Fluegge, Metcalf, and Wallis (1981) found that chlorine, ozone, and ultraviolet light treatments of sewage effluent all produced comparable reductions in virus isolation rates, the median reduction being 75%. At the present time, therefore, it does not seem that either ultraviolet light or ozone offers any advantage over chlorine in terms of killing pathogens. However, ultraviolet light treatment is more-or-less cost competitive with chlorination and is not associated with the production of chlorinated organics. Ultraviolet light treatment may therefore become more common as new water and wastewater treatment plants are built and old ones are upgraded.

REFERENCES

American Water Works Association. 1985. *Giardia lamblia in Water Supplies—Detection, Occurrence, and Removal*. American Water Works Association, Denver, CO. 242 pp.

Anonymous. 1991a. Life in the time of cholera. *Time*, **137**(8), 61.

Anonymous. 1991b. Polio in the Americas. *Harvard Health Letter*, **16**(8), 7.

Blaser, M. J. 1990. *Campylobacter* species. In G. L. Mandell, R. G. Douglas, Jr., and J. E. Bennett (Eds.), *Principles and Practice of Infectious Diseases*. 3rd ed. Churchill Livingstone, New York. pp. 1649–1658.

Cabelli, V. J. 1983. *Health Effects Criteria for Marine Recreational Waters*. EPA-600/1-80-031. Research Triangle Park, NC. 98 pp.

Carlson, D. A., R. W. Seabloom, F. B. Dewalle, T. F. Wetzler, J. Engeset, R. Butler, S. Wangsuphachart, and S. Wang. 1985. *Ultraviolet Disinfection of Water for Small Water Supplies*. EPA/600/S2-85/092. Cincinnati, OH.

Craun, G. F. 1984. Waterborne outbreaks of giardiasis. In S. L. Erlandsen and E. A. Meyer (Eds.), *Giardia and Giardiasis*. Plenum, New York. pp. 243–261.

Dennis, J. M. 1959. Infectious hepatitis epidemic in Delhi, India. *J. Amer. Water Works Assoc.*, **51**, 1288–1296.

Dufour, A. P. 1984. *Health Effects Criteria for Fresh Recreational Waters*. EPA-600/1-84-004. Research Triangle Park, N.C. 33 pp.

Environmental Protection Agency. 1972. *Water Quality Criteria*. EPA-R3-73-033. Washington, D.C. 594 pp.

Environmental Protection Agency. 1984. Water quality criteria: Request for comments. *Federal Register*, **49**, 21987–21988.

Environmental Protection Agency. 1986. Bacteriological ambient water quality criteria: Availability. *Federal Register*, **51**, 8012–8016.

Fluegge, R. A., T. G. Metcalf, and C. Wallis. 1981. *Virus inactivation in wastewater effluents by chlorine, ozone, and ultraviolet light*. EPA-600/S2-81-088. Cincinnati, OH.

Fox, J. P. 1976. Human-associated viruses in water. In G. Berg, H. L. Bodily, E. H. Lennette, J. L. Melnick, and T. G. Metcalf (Eds.), *Viruses in Water*. Am. Public Health Assoc. Washington, D.C. pp. 39–49.

Frobisher, M., L. Sommermeyer, and R. Fuerst. 1969. *Microbiology in Health and Disease*. 12th ed. Saunders, PA. 549 pp.

Fujioka, R. S., H. H. Hashimoto, E. B. Siwak, and R. H. F. Young. 1981. Effect of sunlight on survival of indicator bacteria in seawater. *Appl. Environ. Microbiol.*, **41**, 690–696.

Fule, R. P., R. M. Power, S. Menon, S. H. Basutkar, and A. M. Saoji. 1990. Cholera epidemic in Solapur during July-August, 1988. *Indian J. Med. Res.*, **91**, 24–26.

Geldreich, E. E. 1972. Water-borne pathogens. In R. Mitchel (Ed.), *Water Pollution Microbiology*. Wiley-Interscience, New York. pp. 207–241.

Gergatz, S. J., and L. M. McFarland. 1989. Cholera on the Louisiana Gulf coast: Historical notes and case report. *J. Louisiana State Med. Soc.*, **141**(110), 29–34.

Goh, K. T., S. H. Teo, S. Lam, and M. K. Ling. 1990. Person-to-person transmission of cholera in a psychiatric hospital. *J. Infect.*, **20**(3), 193–200.

Hill, D. R. 1990. Giardia lamblia. In G. L. Mandell, R. G. Douglas, Jr., and J. E. Bennett (Eds.), *Principles and Practice of Infectious Diseases*. 3rd ed. Churchill Livingstone, New York. pp. 2110–2115.

Islam, M. S., B. S. Drasar, and D. J. Bradley. 1990. Long-term persistence of toxigenic *Vibrio cholera* 01 in the mucilaginous sheath of a blue-green alga, *Anabaena variabilis. J. Trop. Med. Hyg.*, **93**(2), 133–139.

Kapikian, A. Z., R. G. Wyatt, R. Dolin, T. S. Thornhill, A. R. Kalica, and R. M. Chanock. 1972. Visualization by immune electron microscopy of a 27 nm particle associated with acute infectious nonbacterial gastroenteritis. *J. Virol.*, **10**, 1075–1081.

Kaplan, J. E., G. W. Gary, R. C. Baron, N. Singh, L. B. Schonberger, R. Feldman, and H. B. Greenberg. 1982. Epidemiology of Norwalk gastroenteritis and the role of Norwalk virus in outbreaks of acute nonbacterial gastroenteritis. *Ann. of Int. Med.*, **96**, 756–761.

Killewo, J. Z., D. M. Amsi, and F. S. Mhalu. 1989. An investigation of a cholera epidemic in Butiama village of the Mara. *J. Diarrhoeal Dis. Res.*, **7**(1-2), 13–17.

Lippy, E. C., and S. C. Waltrip. 1985. Waterborne disease outbreaks—1946–1980: A thirty-five year perspective. In *Giardia lamblia in Water Supplies—Detection, Occurrence, and Removal*. Am. Water Works Assoc. Denver, CO. pp. 67–74.

Mahmoud, A. F. 1990. Schistosomiasis. In G. L. Mandell, R. G. Douglas, Jr., and J. E. Bennett (Eds.), *Principles and Practice of Infectious Diseases*. 3rd ed. Churchill Livingstone, New York. pp. 2145–2151.

Majeti, V. A., and C. S. Clark. 1981. *Potential Health Effects From Viable Emissions and Toxins Associated with Wastewater Treatment Plants and Land Application Sites*. EPA-600/S1-81-006. Cincinnati, OH.

McKinney, R. E. 1962. *Microbiology for Sanitary Engineers*. McGraw-Hill, New York. 293 pp.

Murthy, G. V., A. Goswami, S. Narayanan, and S. Amar. 1990. Effect of educational intervention on defaecation habits in an Indian urban slum. *J. Trop. Med. Hyg.*, **93**(3), 189–193.

Ongerth, J. E., J. Riggs, and J. Cook. 1989. *A Study of Water Treatment Practices for the Removal of Giardia lamblia cysts*. Am. Water Works Assoc. Denver, CO. 56 pp.

Rao, V. C., and J. L. Melnick. 1986. *Environmental Virology*. Am. Soc. *Microbiol.*

Roberts, L. 1988. Change in polio strategy? *Science*, **240**, 1145.

Roberts, P. V., E. Marco Aieta, J. D. Berg, and B. M. Chow. 1981. *Chlorine Dioxide for Wastewater Disinfection: a Feasability Evaluation*. EPA-600/S2-81-092. Cincinnati, OH.

Scheible, O. K., and C. D. Bassell. 1981. *Ultraviolet Disinfection of a Secondary Wastewater Treatment Plant Effluent*. EPA-600/S2-81-152. Cincinnati, OH.

Scheible, O. K., M. C. Casey, and A. Forndran. 1986. *Ultraviolet Disinfection of Wastewaters From Secondary Effluent and Combined Sewer Overflows*. EPA/600/S2-86/005. Cincinnati, OH.

Siddique, A. K., Q. Islam, K. Akram, Y. Mazumder, A. Mitra, and A. Eusof. 1989. Cholera epidemics and natural disasters; where is the link? *Trop. Geogr. Med.*, **41**(4), 377–382.

St. Louis, M. E., J. D. Porter, A. Helal, K. Drame, N. Hargrett-Bean, J. G. Wells, and R. V. Tauxe. 1990. Epidemic cholera in west Africa: The role of food handling and high risk. *Am. J. Epidemial.*, **131**(4), 719–728.

Stover, E. L., R. N. Jarnis, and J. P. Long. 1981. *High-level Ozone Disinfection of Municipal Wastewater Effluents*. EPA-600/S2-81-040. Cincinnati, OH.

Tabtieng, R. S., S. Wattanasri, P. Echeverria, J. Seriwatana, L. Bodhidatta, A. Chatkaeomorakot, and B. Rowe. 1989. An epidemic of *Vibrio cholerae* el tor Inaba resistant to several antibiotics with a conjugative group C plasmid coding for type II dihydrofolate reductase in Thailand. *Am. J. Trop. Med. Hyg.*, **41**(6), 680–686.

Wolf, H. W. 1972. The coliform count as a measure of water quality. In R. Mitchell (Ed.), *Water Pollution Microbiology*. Wiley-Interscience, New York. pp. 207–241.

8

TOXICOLOGY

We assumed [10 or 15 years ago] we knew what the bad pollutants are, at what levels they cause adverse environmental or public health threats. We [assumed] we knew how to drive them down to a no-effect level, how to measure them, how to control them at a reasonable cost. All you needed was an enforcement presence that was sufficiently strong. [Except for the need for enforcement,] all those assumptions were wrong. [Statement by Environmental Protection Agency administrator William Ruckelhaus, quoted by Sun (1985)].

Toxicology is the quantitative study of the effects of harmful substances or stressful conditions on organisms. This rather broad field is broken down into three major divisions: economic, forensic, and environmental toxicology. *Economic toxicology* is concerned with the deliberate use of toxic chemicals to produce harmful effects on target organisms such as bacteria, parasites, and insects. The obvious applications of economic toxicology are in medicine, agriculture, and forestry management, where it is frequently desirable to eliminate or at least control the numbers of various infectious organisms, parasites, and pests (see Chapter 10). *Forensic toxicology* is concerned with the medical and legal aspects of the adverse effects of harmful chemicals and stressful conditions on humans. The medical aspects of forensic toxicology concern the diagnosis and treatment of the adverse effects of toxic chemicals and stressful conditions. The legal aspects concern the cause and effect relationships between exposure to harmful chemicals or conditions and the effect of this exposure on human health. *Environmental toxicology* is concerned with the incidental exposure of plants and animals, including humans, to pollutant chemicals and unnatural environmental stresses. Environmental toxicology is related directly to the subject of water pollution, because it is environmental toxicological studies that reveal the quantitative relationships between, for example, the concentration of chemicals found in the water or in aquatic organisms and the effect of these chemicals on aquatic organisms and on persons who drink the water or consume the organisms.

The remaining chapters of this book are concerned with substances or conditions which exert some form of toxic stress on organisms. The etiology of the effects is somewhat different from that of the consequences of cultural eutrophication, which is a problem associated with the overloading of natural aquatic systems with potentially useful substances. This chapter concerns the theoretical and practical techniques that have been used to study toxic effects of pollutants and to ultimately establish water quality standards.

TOXICITY TESTS

In a general sense toxicity tests may be assigned to one of two categories. The first category involves studies of the general overall effects of a compound or stress on an organism. Such overall effects are described often as being either acute or chronic. Acutely toxic effects become apparent over a relatively short time interval, usually no more than a few days. Chronic effects become apparent only after a relatively long time interval, usually 10% of an organism's life span or longer. Acute toxicity is usually, but not always, lethal. Chronic toxicity involves a lingering or continuous stimulus and may be lethal or sublethal.

The second category of toxicity tests involves experiments designed to evaluate in detail specific kinds of toxicity. The tests may be designed, for example, to study the tendency of a toxicant to cause abnormal development of the fetus (teratogenic tests), to affect the reproductive capacity of an organism, to cause mutations, to produce tumors, to cause cancer, to affect the photosynthetic rates of plants, and so forth. These specific tests are necessary because many of the important effects of toxicants on organisms, particularly at the sublethal level, do not become apparent in standard tests to evaluate overall effects.

Sublethal Effects

Concentrations of a toxic substance that are insufficient to kill organisms outright may nevertheless have a devastating effect on the population of the same organism over a sufficient period of time as a result of sublethal effects. Most sublethal effects may be classified as modifications or interferences with reproduction, development or growth, and behavior. The aim of setting water quality standards is to eliminate or at least minimize the sublethal effects of pollutants. Unfortunately the task of determining the level of stress that causes no significant sublethal effects has proven much more difficult to determine than was once imagined, in part because of the large number of potentially toxic substances and because of the large number of possible sublethal effects. In some cases simply designing an experiment to study certain types of sublethal effects is no trivial matter. A few examples illustrate the variety and complexity of sublethal effects.

Reproduction
Obviously any pollutant that interferes with or blocks the reproduction of a species may completely eliminate that species without having any apparent effect on the adult members of the population. For example, dichlorodiphenyldichloroethane (DDE), which is a metabolite of dichlorodiphenyltrichloroethane (DDT), was

implicated as the cause of the reproductive failure of certain fish-eating birds during the 1960s and 1970s. Certain toxicological tests designed to examine the specific effects of DDE on the reproduction of various birds showed convincingly that high levels of DDE in the feed of female birds lowered the level of estrogen in the birds and caused them to lay thin-shelled eggs. Table 8.1 illustrates the results of one such study.

In this particular experiment, which lasted over a period of 2 years, test ducks began receiving feed containing either 10 or 40 parts per million (ppm) DDE several weeks before the onset of the first laying season. A control group of ducks received feed containing no added DDE. The ducks fed 10 or 40 ppm DDE produced 49–80% fewer 14-day ducklings per hen than the control group and produced eggshells that were about 10% thinner than the control eggshells.

That fish-eating birds might have encountered food containing as high as 10 ppm DDE during the 1960s is confirmed by the data in Table 8.2. In 1969 northern anchovies, which account for about 92% of the food consumed by brown pelicans in the southern California bight (Anderson and Gress, 1983), were found to contain an average of 3.24 ppm DDE on a fresh weight basis. On a dry weight basis this concentration is equivalent to about 9.7 ppm and is therefore virtually identical to the concentration of 10 ppm dry weight which Table 8.1 indicates reduced the reproduction of mallard ducks by 50–80%. The sharp drop in anchovy DDE content between 1969 and 1970 coincides with the time that the Montrose Chemical Company, located in Los Angeles County and the sole manufacturer of DDT in the United States, began disposing of its liquid wastes in a sanitary landfill rather than discharging the wastes into the municipal sewer system.[1] The DDE content of the anchovies then remained constant for about 3 years, but dropped sharply between 1972 and 1973 when the EPA ban on DDT use in the United States went into effect.[2] The EPA ban resulted in no small part from the examination of toxicological results such as those shown in Table 8.1, which demonstrated convincingly the adverse effects of DDE on the reproduction of aquatic birds.

The correlation between DDT use and the reproductive success of the southern California brown pelican is illustrated in Table 8.3. In 1969 and 1970 only 9 fledglings were produced from a total of 1852 nesting attempts, and the thickness of intact eggshells averaged about 0.40 mm. In 1971 and 1972, following the diversion of Montrose's liquid waste to a sanitary landfill, reproduction improved to 249 fledglings out of 1161 nesting attempts, and eggshell thickness increased to about 0.45 mm. From 1973 to 1975, following the ban on DDT use in the United States, reproductive success improved further to 1575 fledglings out of 2175 nesting attempts, and intact eggshell thickness averaged 0.50 mm. The most recent data have shown almost no evidence of crushed eggs due to thin eggshells, although Anderson, Gress, and Maise (1982) noted that the thickness of the shells was still 19% below normal in 1978, over 5 years after the DDT ban went into effect.

The number of brown pelicans in the southern California bight colonies increased dramatically during the 1970s, in large part due to the increase in reproductive success of the adult birds. However, in recent years the fledgling rates of

---

[1] Use of the sanitary landfill began in April, 1970.
[2] The ban became effective in January, 1973.

these colonies have averaged only about 75–85% of the rates observed at similar colonies in the Gulf of California (Anderson and Gress, 1983), and according to the calculations of Anderson and Gress (1983), perhaps 25% of the adult population in 1980 was the result of immigration from colonies in southern Baja California or the Gulf of California. However, had the disastrous reproductive failures in the years immediately preceding the EPA's ban on DDT been allowed to continue for much longer, the southern California brown pelican would have very likely become extinct. Although some of the population decline during the late 1960s may have been due to adult mortality from direct poisoning (Anderson and Gress, 1983), the major cause of the decline was reproductive failure. Fortunately, careful toxicological studies of the sublethal effects of DDE on bird reproduction had been carried out and were presented to the EPA at the time of the 1972 DDT hearings. These studies did much to influence the EPA decision to ban DDT use in the United States.

In recent years reproductive rates of the southern California bight brown pelicans have been strongly influenced by fluctuations in the local availability of anchovies (Anderson et al., 1980; Anderson et al., 1982), a natural phenomenon. However, the fact that fledgling rates in these colonies are still below average rates observed at colonies outside the southern California bight suggests that the former colonies may still be under some unnatural stress, perhaps from polychlorinated biphenyls (see Chapter 10).

Table 8.1 Reproduction and Eggshell Thickness Data from Penned Mallard Ducks Maintained for Two Seasons on Feed Containing DDT

DDE added to feed (ppm)	No. of 14-day ducklings per hen (% of control results)		Shell thickness (% of control results)	
	Year 1	Year 2	Year 1	Year 2
10	51	24	92	89
40	35	20	87	86

Source. Heath et al. (1969).

Table 8.2 Geometric Mean Residues of DDE in Anchovies and Brown Pelican Eggs Off the Southern California Coast

Year	Anchovy whole bodies (ppm, fresh weight)	Intact brown pelican eggs (ppm, lipid weight)
1969	3.24	853
1970	0.84	—
1971	0.87	—
1972	0.74	221
1973	0.18	175
1974	0.12	97
1975[a]	—	113

[a] Data for 1975 for Anacapa and Santa Cruz islands only.

Source. Anderson et al. (1975, 1977).

Table 8.3   Recent History of Brown Pelicans Breeding Off the Coast of Southern California and Northwestern California[a]

Year	No. of nests built	No. of young fledged total	per nest	Intact eggshell thickness (mm)	Anchovy abundance
1969	1125	4	0.004	0.402	140
1970	727	5	0.007	0.393	70
1971	650	42	0.065	0.460	80
1972	511	207	0.405	0.438	195
1973	597	134	0.225	0.510	275
1974	1286	1185	0.922	0.482	355
1975	292	256	0.88	0.51	—
1976	417	279	0.67	—	—
1977	76	39	0.51	—	—
1978	210	37	0.18	—	—
1979	1258	980	0.78	—	—
1980	2244	1515	0.68	—	—
1981	2946	1805	0.61	—	—
1982	1862	1175	0.63	—	—
1983	1877	1159	0.62	—	—
1984	628	530	0.84	—	—
1985	6194	7902	1.28	—	—
1986	7349	4601	0.63	—	—
1987	7167	4898	0.68	—	—
1988	2878	2500	0.87	—	—
1989	5959	3500	0.59	—	—
1990	2400	—	—	—	—

[a]Data pertain to California brown pelicans nesting in the Anacapa Island area and Santa Barbara Island area.

Source. Anderson et al. (1975, 1977) and Gustafson (1990).

Development and Growth

Interference with the normal physiological processes of an organism may adversely affect its ability to grow and develop without directly killing the organism. Nevertheless, because of the pressure of competition and predation, a population of organisms whose development or growth has been retarded may be rapidly eliminated from an ecosystem. For example, studies have shown that exposure of certain species of marine phytoplankton to mercury concentrations as low as 0.5 parts per billion (ppb) can cause more than a 50% reduction in photosynthesis. Figure 8.1 illustrates some of the experimental results obtained with the diatom *Nitzschia delicatissima*. If the presence of a toxin in the water significantly reduces the photosynthetic rate of a particular species, it is possible that the species will be grazed to extinction by herbivores without being killed directly by the toxin. Similarly, phytoplankton division rates and/or photosynthetic rates have been found to be reduced at polychlorinated biphenyl (PCB) concentrations as low as 0.1 to 10 ppb (Mosser et al., 1972a; Fisher and Wurster, 1973; Harding and Phillips, 1978). The effect of PCBs on phytoplankton differs markedly between species, some species being much more sensitive than others. As a result, the species composition of a phytoplankton community may be altered greatly due to the presence of PCBs in the water at concentrations as low as 0.1 ppb (Mosser et al., 1972b; Fisher et al., 1974; Fisher, 1975). Studies conducted by O'Connors et al. (1978) in a natural estuarine marsh showed that exposure to as little as 1.0 ppb PCBs reduced phytoplankton production and caused a shift in the size structure of the phytoplankton community toward smaller cells. This shift in the mean cell size of the phytoplankton could effectively increase the number of trophic levels between the primary producers and commercially useful fish (Ryther, 1969) and/or could divert more production toward various gelatinous predators such as jellyfish (Greve and Parsons, 1977). Thus, the sublethal effects of a toxin on organisms at the base of the food chain could have a profound effect on the structure of the entire food chain and could affect significantly the productivity of the system with respect to commercially useful organisms.

One of the best documented effects of a stress on the development and growth of aquatic organisms is that of temperature. All organisms grow most efficiently over a limited temperature range. Temperatures above or below the optimum range lead to reduced growth rates or to reduced growth efficiency. Figure 8.2 illustrates in a qualitative way how temperature can influence the growth rate and efficiency of an organism. In the figure A_c is the amount of food consumed by the organism, and A_g is the amount of that food converted into biomass (growth). A_w is the amount of food that is excreted (waste), while A_a, A_d, and A_a represent the amounts of food that are respired to support the activity of the organism (A_a), the degradation and metabolism of consumed food (A_d), and standard or basal metabolism (A_s). It is clear from Figure 8.2 that there is some temperature at which growth, A_g, is a maximum, because the rate of food consumption, A_c, peaks at a certain temperature, whereas the basal respiration rate, A_s, rises exponentially with temperature. This dependence of A_c and A_s on temperature is characteristic of the behavior that might be expected from a poikilothermic (cold-blooded) organism. The amount of food converted to biomass is positively correlated with temperature only as long as the difference between A_c and A_w increases more rapidly than A_s.

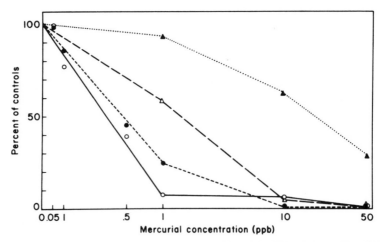

Figure 8.1 Photosynthesis by the marine diatom *Nitzschia delicatissima* after 24 hours of exposure to the following mercurials: solid triangles, diphenyl mercury; open triangles, phenylmercuric acetate; solid circles, methylmercury dicyandiamide; open circles, **MEMMI** (a compound containing methylmercury). [Redrawn from Harriss et al. (1970). Copyright © 1970 by the American Association for the Advancement of Science.]

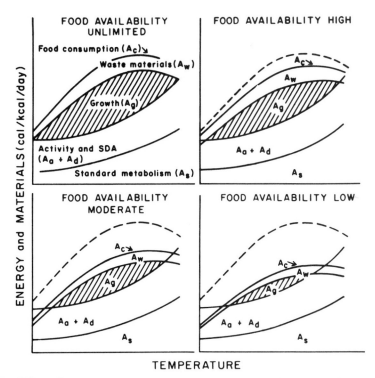

Figure 8.2   Effect of temperature and food availability on food consumption (A_c) and the partitioning of consumed food for various purposes by a poikilothermic animal. The curves are qualitative and purely theoretical. [Redrawn from Warren (1971).]

Figure 8.2 illustrates one of the difficult problems encountered in converting toxicologic information into water quality standards. In all cases shown, it is clear that a small deviation in temperature from the optimal value results in some decrease in growth, but this effect by itself will not be lethal. However, the effect of a change in temperature on growth is dependent obviously on the availability of food to the organism. A change of temperature that would cause only a modest reduction in growth when the food supply is unlimited could be lethal when food availability is low. Furthermore, although a modest reduction in growth might not by itself be lethal, in a highly competitive system, even a small reduction in growth could lead to the elimination of a species through the combined effects of competition and predation. Similarly, the ecological impact of a given temperature stress might be far greater in a system in which the organisms were already subjected to other forms of stress, such as the presence of toxic chemicals, than in a system in which no other stresses existed. Thus, translating apparently straightforward toxicological information into water quality standards can be quite difficult.

Behavior
Modification in behavior patterns involves a wide variety of effects, such as alterations in migratory behavior, learning ability, feeding behavior, predator avoidance, and so forth. A few examples illustrate some of the kinds of problems that pollutants may cause as a result of behavioral modification.

Salmon migrate from the ocean back to the freshwater stream where they were born in order to spawn. Although the mechanism that these fish use to find their way from the open ocean back to the general vicinity of their home stream is not well understood, experiments have shown clearly that salmon use their sense of smell to guide their migration upstream once they have entered fresh water. The chemical substances that guide this migration have been shown to include certain volatile organic compounds (Hasler and Larsen, 1971). Unfortunately, a number of streams once characterized by large salmon runs have become seriously polluted with industrial wastes, particularly in the northeastern United States. In many of these streams salmon runs are now no longer observed or involve insignificant numbers of fish. There is good reason to believe that the presence in these streams of industrial wastes, including a variety of organic substances, has so altered the characteristic small of the water that the fish are no longer able to identify their home stream. Thus, the presence of chemicals in the water at concentrations well below lethal levels may completely disrupt the migratory patterns of the salmon and, through this behavioral modification, effectively eliminate the species.

Seasonal changes in water temperature can lead to a variety of natural behavioral responses in aquatic organisms. These responses include the onset of reproductive cycles and migratory behavior. The discharge of heated wastewater from large electric power plants can completely disrupt these temperature cues and therefore interfere seriously with natural behavior patterns. For example, menhaden found in the coastal waters of New Jersey during the summer months normally migrate to warmer waters off the North Carolina coast during the winter. At Barnegat Bay, New Jersey, the Jersey Central Power and Light Company operates a nuclear power plant that discharges cooling water at a temperature about 14°C above ambient. During the winter, large numbers of menhaden are found often in the waters near the Oyster Creek outfall, either because they are naturally attracted to the warm water in the winter or because, finding themselves in the warmer waters near the outfall, they do not receive the proper temperature cue to migrate south. In this case the migratory behavior of the menhaden is completely altered by the presence of the heated effluent water, which by itself has no adverse physiological effect on the fish. In fact, during the winter months the temperature of the effluent water is undoubtedly much closer to the optimum temperature for the menhaden than is the temperature of the surrounding water. Unfortunately, the water temperature near the outfall can drop quickly back to ambient levels if the power plant is shut down. Such a shut-down occurred on January 28, 1972, and as a result 100,000–200,000 menhaden died of thermal shock (Clark and Brownell, 1973). In a similar incident in Long Island Sound, approximately 10,000 bluefish that had been attracted to the heated plume from the Northport power plant died from thermal shock when winds and tidal currents caused the plume to suddenly shift position on January 17, 1972 (Silverman, 1972). Similar incidents can be anticipated wherever the heated wastewater from power plants attracts aquatic organisms during the colder months of the year.

Acute/chronic Ratios

Although water quality standards are intended to be criteria that will provide reasonable protection against sublethal and chronic stresses, in many cases there is simply not enough information on sublethal and chronic toxicity effects alone to warrant setting a standard. In such cases an acute toxicity standard is established based on an adequate amount of short-term, usually lethal toxicity tests, and an average acute/chronic toxicity ratio is then calculated based on a smaller amount of information. Finally, a chronic toxicity standard is established by dividing the acute toxicity standard by the acute/chronic ratio. The rationale for this procedure is that for a given pollutant the acute/chronic ratio is likely to be more constant between species than is the chronic or sublethal stress level itself. The ratio of concentrations that cause acute and chronic stresses has therefore proven to be an important parameter in establishing water pollution standards.

An experiment performed by Weir and Hine (1970) on conditioned goldfish suggests how information on the two types of toxicity might be used in setting water quality criteria. The experiments concerned the effects of certain toxic metals on the fishes' learning ability, a particularly subtle form of sublethal stress. The goldfish were conditioned by teaching them to respond when a light was turned on in one side or the other of the aquarium in which they were kept. If the goldfish remained in the half of the aquarium with the lighted bulb for more than 3 seconds, they received an electrical shock. If they swam to the darkened side of the aquarium, they received no shock. The goldfish were trained to respond to the light until the percentage of goldfish responding correctly varied by no more than 5% for 3 consecutive days. Once trained in this way, the goldfish responded correctly about 69% of the time.

After this training period, the goldfish were put into tanks containing various concentrations of either arsenic, lead, mercury, or selenium for a total of 48 hours. After 24 hours and again after 48 hour, the goldfish were retested to check their ability to respond correctly to the light. After the 48-hours exposure period, the goldfish were returned to clean water and were retested every 24 hours until their percentage of correct responses varied by no more than 5% for 2 consecutive days. If this final score was found to be significantly lower than the score the fish had made prior to exposure to the metal, then the metal was concluded to have adversely affected the fishes' learning ability.

Table 8.4 lists the lowest concentration of each metal that significantly impaired the learning ability of the goldfish. For comparative purposes, the lethal concentration of each metal is listed in the second column of the table, and the third column gives the lethal/sublethal ratio. A metal concentration was considered to be lethal if 50% of the goldfish died during a 7-day exposure to the given concentration. The most dramatic aspect of the data in Table 8.4 is the large difference between the lethal and sublethal concentrations. The sublethal concentrations are 48 to 1570 times smaller than the corresponding lethal concentrations. Since the exposure in the sublethal experiments lasted only 48 hours, the impairment of the fish's learning ability was, strictly speaking, the result of acute toxicity. However, longer duration experiments with various organisms and pollutants have shown that chronic stresses are often detected at similarly low concentrations, i.e., an order of magnitude or more smaller than lethal, acutely toxic concentrations.

ACUTE TOXICITY DETERMINATION

There is a large amount of information available on the lethal effects of pollutants on aquatic organisms. This arises in part because experiments designed to test the lethality of a substance are relatively easy to perform. On the other hand, there are a

Table 8.4 Lowest Metal Concentration Found to Impair the Learning Ability of Conditioned Goldfish

Metal	Sublethal concentration (ppm)	Lethal concentration (ppm)[a]	Lethal/sublethal ratio
Arsenic	0.1	32	320
Lead	0.07	110	1570
Mercury	0.003	0.82	273
Selenium	0.25	12	48

[a]7-day TLm.

Source. Weir and Hine (1970).

large number of potentially serious sublethal effects, and experiments designed to examine these effects are often more difficult to design and more time consuming to perform than lethal toxicity experiments.

The principal concerns in lethal toxicity tests are the natural variability among the members of a species in tolerance to the given stress and the relationship between the duration of exposure and the effect of the stress. We consider first the problem of natural variability.

Median Survival Times

Because of the natural variability among members of a given species, any sample of that population can be expected to contain individuals that differ in their tolerance to a given stress. How then can the response of the population to the stress be quantified? Should you use the response of the most sensitive individual or the least sensitive individual, or should you use the mean or median response? In the United States at least, the tolerance of a particular species to a stress is based on the median response of a sample of individuals of the given species. Whether one considers this policy conservative or liberal depends to a certain extent on one's frame of mind. For example, suppose we were interested in knowing how long a particular species could withstand a given stress before the stress became fatal. An industrialist might argue that the time required to kill all the members of the species should be termed the *survival time*, and that the median survival time was unduly conservative because half the members of the species were still alive at the end of the experiment. The environmentalist might argue that the time during which all members of the species survived should be termed the survival time, and that the median survival time was unduly liberal because half the members of the species were dead at the end of the experiment. However, one should keep in mind that median survival times do not themselves become the basis for long-term water quality standards. Ultimately, it is chronic toxicity studies that determine the long-term water quality standards. Furthermore, median survival times are far more reproducible from one subgroup of a species to the next than are survival times of the most sensitive and least sensitive members of a subgroup.

Figure 8.3 illustrates how median survival times are calculated. The data are taken from a study of the toxicity of low oxygen levels to juvenile brook trout (Shepard, 1955). The trout were conditioned for a period of time in water containing 10.5 mg/l dissolved oxygen and were then placed in experimental aquaria having lower concentrations of dissolved oxygen. The survival of the trout was then monitored over a period of 2000 minutes. As each fish died, it was removed from its aquarium and its time of death recorded. Ten fish were studied at each oxygen level. The results were plotted on semilog paper, and the data fit with straight lines. At the two highest oxygen levels, over half the fish survived for the entire 2000 minutes, and no effort was therefore made to estimate median survival times at these oxygen concentrations. However, at the 8 lowest oxygen levels, over half the fish died during the study period, and estimates of median survival time were made from the intercept of the regression lines fit to the data with the horizontal line at 50% mortality. This graphical technique of determining median survival times is common in acute toxicity studies.

Incipient Lethal Levels

Given the relationship between median survival time and pollutant concentration, it is possible to calculate lethal concentrations, but now explicit recognition must be taken of the relationship between survival time and degree of exposure. There are basically two approaches. In one approach a graph is made of median survival time versus toxicant concentration, and the graph is extrapolated to time infinity. The hypothetical concentration of the toxicant corresponding to an infinite median survival time is called the *incipient lethal level*. Figure 8.4 illustrates how incipient lethal levels are calculated. The data are taken from a study of the toxicity of zinc to rainbow trout (Lloyd, 1960). Median survival times were calculated as a function of zinc concentration at three different levels of water hardness. Note that both axes are plotted on a logarithmic scale. At water hardness levels of 50 and 320 mg/l, the data show distinct curvature, and a function of the

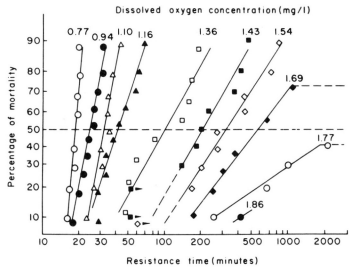

Figure 8.3 Percentage mortality versus resistance time for juvenile brook trout exposed to the indicated concentrations of dissolved oxygen after being conditioned to 10.5 mg/l oxygen. [Redrawn from Shepard (1955), in J. Fish. Res. Bd. Canada.]

form $(C - C_i)^n(T - T_i) = K$ was fit to the data. In this equation C is the concentration of zinc corresponding to median survival time T. As T approaches infinity, C approaches C_i, and C_i is therefore interpreted as the incipient lethal level. The values of C_i estimated from this curve fitting were 3.5 and 1.5 mg/l at water hardness levels of 320 and 50 mg/l, respectively. At a water hardness of 12 mg/l, it was impossible to estimate an incipient lethal level because the data were linear over the range of concentrations studied. All that can be said is that the incipient lethal level at 12 mg/l hardness is less than 0.5 mg/l zinc.

It should be obvious that the incipient lethal level is in a sense a hypothetical concept, because no organism survives for an infinite time, regardless of the level of stress. However, the incipient lethal level provides a standard measure of the concentration that produces no acute effects on survival, and it is apparent from Figure 8.4 that at water hardness levels of 50 and 320 mg/l the concentration of zinc corresponding to a median survival time in excess of about 2000 minutes is very insensitive to survival time. In other words, at a water hardness of 320 mg/l, the zinc concentration corresponding to a median survival time of 2000 minutes and the zinc concentration corresponding to a hypothetical median survival time of infinity are nearly the same. This latter observation leads to the second method for estimating lethal concentrations.

Median Tolerance Levels

Rather than extrapolating median survival times to infinity, you may determine instead the toxicant concentration corresponding to a standard survival time. In most cases this survival time is either 48 or 96 hours (2880 or 5760 min). The toxicant

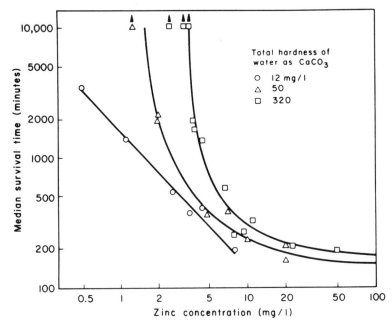

Figure 8.4   Median survival time of rainbow trout as a function of zinc concentration at three levels of water hardness. [Redrawn from Lloyd (1960).]

concentration corresponding to this median survival time is called the *median tolerance limit* or TLm. In some cases the abbreviations TL50 (tolerance limit corresponding to 50% survival) or LC50 (lethal concentration corresponding to 50% mortality) are used to designate the same parameter. In a more general sense the abbreviation EC50 is often used to designate the concentration that produces a specified effect on 50% of the organisms after a certain period of exposure. In all cases the symbol should be preceded by a designation of the associated period of time. Thus, a 96-hr TLm would be the concentration of a toxicant associated with 50% survival after 96 hr.

Figure 8.5 illustrates the calculations of a 96-hr TLm. The data are taken from a study of the toxicity of copper to bluegill fish (Trama, 1954). The fish were divided into two groups and exposed to various concentrations of either CuSO$_4$ or CuCl$_2$ for 96 hours. Percentage survival was then plotted against time. The two data sets were very similar, and a smooth curve was therefore drawn through the combined results. The intersection of this curve with the horizontal line at 50% survival corresponds to the TLm, which in this case was 0.74 mg/l.

From a scientific standpoint both incipient lethal levels and median tolerance limits are arbitrary measures of lethal concentrations, and there is often little reason to prefer one over the other when setting water quality standards. This fact is apparent in the case of Figure 8.4, where it is obvious that 96-hr TLm's would be virtually identical to incipient lethal levels at water hardness levels of 50 and 320 mg/l. At a water hardness of 12 mg/l, neither an incipient lethal level nor a 96-hour TLm can be estimated, because the data are linear, and in no case did half the fish survive for 96 hours. From a practical standpoint, however, TLm's are easier to determine because a detailed time series (e.g., Figure 8.3) is not required. In determining a TLm

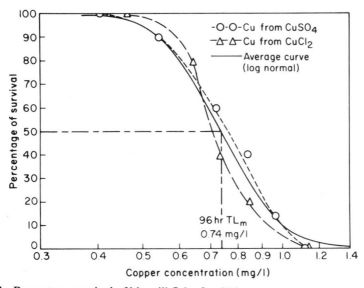

Figure 8.5   Percentage survival of bluegill fish after 96 hours versus copper concentration. The smooth curve is a log-normal curve adjusted to give the best fit to the combined CuSO$_4$ and CuCl$_2$ data. The intersection of this curve with the horizontal line at 50% survival gives the median tolerance limit (TLm) after 96 hours, 0.74 mg/l. [Redrawn from Trama (1954), with permission from Notulae Naturae of the Academy of Natural Sciences of Philadelphia.]

it is only necessary to record survival at a single time, usually at either 48 or 96 hours. Calculation of incipient lethal levels does require careful time course studies and, in addition, requires extrapolation of the results to time infinity, a procedure that in some cases may not be straightforward. Perhaps in part because of these considerations, present EPA water quality criteria are written in terms of TLm's rather than incipient lethal levels.

EPA Water Quality Guidelines

The rationale and procedures for determining water quality guidelines in the United States are described in guidelines promulgated by the EPA (1980a, 1986). In establishing these guidelines, the EPA has not attempted to suggest standards that would make the water safe for every organism. The rationale behind the guidelines is summarized as follows

> Aquatic communities can tolerate some stress and occasional adverse effects on a few species, and so total protection of all the species all of the time is not necessary. Rather, the Guidelines attempt to provide a reasonable and adequate amount of protection with only a small possibility of considerable overprotection or underprotection. Within these constraints, it seems appropriate to err on the side of overprotection. (EPA, 1980a, p. 79342)

Within the context of this statement, it is now useful to examine the minimum data base that the EPA feels is needed for purposes of establishing acutely toxic concentrations. The rationale behind establishing a minimum data base is that, "Results of acute and chronic toxicity tests with a reasonable number and variety of aquatic animals are necessary so that data available for tested species can be considered a useful indication of the sensitivities of the numerous untested species" (EPA, 1980a, p. 79343).

The minimum data base differs for freshwater and marine organisms and is stipulated as follows in the 1986 EPA guidelines. The criterion in freshwater must be based on acute tests with freshwater animals in at least eight different families, including all of the following categories: 1) salmonid fish, 2) nonsalmonid fish, 3) a third vertebrate family, 4) planktonic crustaceans, 5) benthic crustaceans, 6) insects, 7) a family not included among vertebrates or insects, and 8) a family in any order of insect or any phylum not already represented. For marine organisms, the criterion must be based on acute tests with saltwater animals in at least eight different families subject to the following five constraints: 1) at least two different vertebrate families are included, 2) at least one species is from a family not included among the vertebrates and insects, 3) either the Mysidae or Penaeidae family or both are included, 4) there are representatives from at least three other families not included among the vertebrates, and 5) at least one other family is represented.

If studies have been carried out on an appropriate number and variety of organisms, it may be possible to establish what the EPA refers to as the final acute value of the given toxicant, as long as the duration of exposure to the toxicant was appropriate. The guidelines require that results be based on 48–96-hour exposures, the recommended duration depending on the type of organism and the nature of the physiological response. In some cases acute values are based on TLm's, but EC50 values are used for effects such as decreased shell deposition in oysters. For each species for which satisfactory studies with a particular toxicant have been performed, the acute concentration is calculated as the geometric mean of the reported values. When available, the results of flow-through tests in which the toxicant concentrations were measured are used to determine the geometric mean. When no such data exist, the geometric mean is calculated from all available acute values, including flow-through tests in which the toxicant concentration was not measured and static and renewal tests in which the initial total toxicant concentration was specified.

Once mean acute concentrations have been determined for the appropriate number and variety of species, the species are grouped by genera, and genus mean acute values are calculated as the geometric mean of all the species mean acute values for each genus. The genus mean acute values are then ranked from lowest to highest, and a graphical technique is used to estimate the concentration of the toxicant that would exert an acutely toxic effect on no more than 5% of the species. The genus ranks are converted to cumulative probabilities by dividing each rank by $N + 1$, where N is the number of genera in the list. The four genera with cumulative probabilities closest to 0.05 are then used to determine the final acute value. A plot is made of the logarithm of the genus mean acute values for the four genera against the square root of the cumulative probability, and a model II geometric mean least squares regression line is fit to the data. The concentration of the toxicant corresponding to a cumulative probability of 0.05 on the regression line is taken to be the final acute value.

Table 8.5 and Figure 8.6 illustrate how this procedure would be used to establish a final acute value for the pesticide dieldrin in saltwater at the time the EPA water quality criteria for dieldrin were promulgated. In this case acute values were available for a total of 21 species, the range of values being about a factor of 70. The least sensitive species was the grass shrimp *Palaemonetes vulgaris*, with a TLm of 50.0 ppb, and the most sensitive species was the pink shrimp *Penaeus duorarum*, with a TLm of 0.7 ppb. Table 8.5 includes species from a total of 19 genera, and the cumulative probabilities associated with the four most sensitive genera are therefore 0.05, 0.10, 0.15, and 0.20. The genus mean acute values and the least squares regression line are shown in Figure 8.6. The regression line actually gives an excellent fit to the entire data set, although it was fit to only the four lowest genus mean acute values. According to the regression line, the dieldrin concentration corresponding to a cumulative probability of 5% is 0.63 ppb, and according to the 1986 EPA guidelines this concentration would therefore become the final acute value for dieldrin in saltwater. Based on the data available for dieldrin, it is expected that if a large random sample of saltwater genera were taken, only 5% would have TLm values less than 0.63 ppb. In 1980, when the dieldrin criteria were being determined (EPA, 1980b), the EPA was using a somewhat different procedure for estimating the final acute value, and the final acute value determined at that time was 0.71 ppb. According to the regression line in Figure 8.6, this concentration would be associated with a cumulative probability of 6%. The difference does not seem worth arguing about.

CHRONIC TOXICITY DETERMINATION

In some respects chronic toxicity studies are conducted in the same way as acute toxicity studies, but the former require much longer periods of time. Ideally studies

Table 8.5   Dieldrin Genus Mean Acute Values in Saltwater

Species	Species mean acute value (ppm)	Genus mean acute value (ppm)
Sphaeroides maculatus, northern puffer	34.0	34.0
Crassostrea virginica, eastern oyster	31.2	31.2
Mugil cephalus, striped mullet	23.0	23.0
Palaemonetes vulgaris, grass shrimp	50.0	20.7
Palaemonetes pugio, grass shrimp	8.6	
Morone saxatilis, striped bass	19.7	19.7
Pagurus longicarpus, hermit crab	18.0	18.0
Gasterosteus aculatus, threespine stickleback	14.2	14.2
Palaemon macrodactylus, Korean shrimp	10.8	10.8
Cyprinodon variegatus, sheepshead minnow	10.0	10.0
Crangon septemspinosa, sand shrimp	7.0	7.0
Fundulus heteroclitus, mummichog	8.9	6.7
Fundulus majalis, striped killifish	5.0	
Thalassoma bifasciatum, bluehead	6.0	6.0
Menidia menidia, Atlantic silverside	5.0	5.0
Mysidopsis bahia, mysid shrimp	4.5	4.5
Micrometrus minimus, dwarf perch	3.5	3.5
Cyamatogaster aggregata, shiner perch	2.3	2.3
Oncorhynchus tshawytscha, chinook salmon	1.5	1.5
Anguilla rostrata, American eel	0.9	0.9
Penaeus duorarum, pink shrimp	0.7	0.7

Source. EPA (1980b).

would last throughout the life cycle of the organism, and where such studies are feasible, this approach is the one recommended by the EPA (1980a). Partial life cycle studies are acceptable for fish that require more than a year to reach sexual maturity, and early life stage toxicity tests may be used to establish water quality criteria if complete or partial life cycle tests with a given species are unavailable. In the latter case exposure should begin shortly after fertilization and should last 28 to 32 days through early juvenile development, or 60 days posthatch for salmonids (EPA, 1986). According to the EPA guidelines (EPA, 1986), these chronic toxicity studies should be conducted in flow-through systems, and toxicant concentrations must be measured in the test solutions. In the case of daphnids, renewal systems may be substituted for flow-through systems. Obviously such tests involve considerably more time and effort than the 48–96-hour bioassays conducted for acute toxicity determination.

Calculation of the Final Chronic Value

If data on a sufficient number and diversity of organisms are available, a final chronic value for a particular toxicant may be calculated in the same way that final acute values are determined. However, since there are many effects that might be examined with respect to chronic toxicity and since species will invariably be more sensitive with respect to some kinds of effects than others, the particular effects that were studied can influence to a certain extent the mean chronic value for the species and genus. For example, Table 8.6 lists the values of four pollutants that have been shown to affect reproduction and the survival of adult *Daphnia magna* in chronic toxicity studies (Adema, 1978). The difference in the chronic values is as much as a factor of 10. Obviously you would reach rather different conclusions if you estimated chronic levels of these pollutants for *D. magna* solely on the basis of reproduction or solely on the basis of adult survival. The EPA practice of using the geometric mean of the reported values would result in the mean chronic values in the last column of Table 8.6 if indeed these were the only data available on the chronic toxicity of the given pollutants to *D. magna*. It is noteworthy that the mean chronic value for 3,4 dichloroaniline is over 3 times the concentration reported to adversely affect reproduction.

In practice there are seldom sufficient data to allow a direct graphical estimation of the toxicant concentration that would exert a chronic stress on no more than 5% of the species in the system. However, a final chronic value may be calculated from the final acute value and acute/chronic ratios if the latter are available for at least three species and 1) at least one of the species is a fish, 2) at least one is an invertebrate, and 3) at least one is an acutely sensitive freshwater species or saltwater species when the ratio is being used to establish freshwater or marine criteria, respectively.

Table 8.6   Chronic Toxicity of Four Compounds to *Daphnia magna*

Compound	Chronic values (ppb)		
	Survival of adults	Reproduction	Mean chronic value
1,1,2-Trichloroethene	32,000	18,000	24,000
Dieldrin	100	32	57
Pentachlorophenol	180	320	240
3,4-Dicloroaniline	56	5.6	18

Source. Adema (1978).

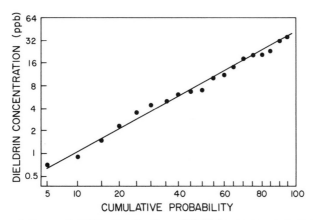

Figure 8.6   Cumulative probability distribution of dieldrin genus mean acute values based on the data in Table 8.5. The straight line is a least squares fit to the four lowest data points.

According to the methodology recommended by the EPA (1986), the final acute/ chronic ratio may be calculated in one of four ways. If there is no major trend apparent among the acute/chronic ratios for the different species and the species mean acute/chronic ratios lie within a factor of 10 for a number of species, the final acute/chronic ratio is the geometric mean of all the species acute/chronic ratios available for both freshwater and saltwater species. If the species mean acute/ chronic ratios seem to be correlated with the species mean acute values, the final acute/chronic ratio should be taken to be the acute/chronic ratio of species whose acute values lie close to the final acute value. In the case of acute tests conducted on metals and possibly other substances with embryos and larvae of barnacles, bivalve molluscs, sea urchins, lobsters, crabs, shrimp, and abalones, the acute/chronic ratio is assumed to be 2. The rationale is that chronic tests are very difficult to conduct with such species, and the sensitivities of embryos and larvae would likely determine the results of life-cycle tests. Assuming the acute/chronic ratio to be 2 causes the final chronic value to equal the criterion maximum concentration (see later discussion). Finally, if the most appropriate species mean acute/chronic ratios are less than 2, acclimation to the stress probably occurred during the chronic test (see later discussion). In such cases, the final acute/chronic ratio is assumed to be 2. The final chronic value for the toxicant is then calculated by dividing the final acute value by the final acute/chronic ratio.

An example of the calculation of a final acute/chronic ratio is shown in Table 8.7 for the pesticide dieldrin. Chronic values for this pesticide were available for only four species in 1980 when the guidelines for dieldrin were established (EPA, 1980b), and it was therefore necessary to use acute/chronic ratios to establish the final chronic value. Acute toxicity values were available in only three of the four cases where chronic effects were studied, but the three species satisfied the criteria for calculating acute/chronic ratios in both freshwater and saltwater. Since the acute/ chronic ratios for the three species differed by less than a factor of 2, it was appropriate to calculate the final acute/chronic ratio for dieldrin by taking the geometric mean of the three ratios, which is $[(11)(9.1)(6.2)]^{1/3} = 8.5$. Final acute values for dieldrin in freshwater and saltwater were 2.5 and 0.71 ppb, respectively. Hence the final chronic values for dieldrin in freshwater and saltwater were $2.5/8.5 = 0.29$ ppb and $0.71/8.5 = 0.084$ ppb, respectively.

TOXICITY TO PLANTS

The EPA guidelines for calculating acute and chronic concentrations of a toxicant clearly pertain to aquatic animals. Nevertheless, aquatic plants may be affected adversely by the presence of toxicants in the water, and sometimes at exceedingly low concentrations (e.g., Figure 8.1). According to EPA guidelines (EPA, 1986), a substance is toxic to a plant at a given concentration if the growth of the plant is decreased in a 96-hour or longer experiment with an alga or a chronic test with a vascular plant. Growth may be measured in a variety of ways, including photosynthetic rate, change in dry weight, or change in chlorophyll concentration. Table 8.8, for example, provides a summary of data on the toxicity of dieldrin to aquatic plants. In this particular case, the concentrations that reduced plant growth were substantially higher than both the final acute and final chronic values for animals and therefore had no influence on the final water quality guidelines for dieldrin. However, because plants are the chief producers of organic matter in most aquatic food chains, it is critical that the effects of toxicants on aquatic plants be assessed. Reductions in photosynthetic rates and/or changes in the composition of the plant community may cause repercussions throughout the entire food web. The final plant value is the lowest concentration of the toxicant that reduces plant growth in an appropriate experiment. Based on the data in Table 8.8, the final plant value for dieldrin was set at 100 ppb in freshwater and 950 ppb in saltwater (EPA, 1980b).

RESIDUE VALUES

Up to this point we have considered the direct adverse effects on plants and animals caused by toxicants present in the water. However, once these toxicants are incorporated into the organisms, there is obviously a potential for other kinds of adverse effects caused by the transfer of the toxicants through the food web. One obvious concern is that the toxicants may be transferred to humans as the result of consumption of contaminated fish or shellfish. The rationale in the 1986 EPA guidelines behind establishing a final residue value was actually twofold; first, to "prevent concentrations in commercially or recreationally important aquatic species from affecting marketability because of exceedence of applicable FDA action levels" and second, to "protect wildlife, including fishes and birds, that consume aquatic organisms from demonstrated unacceptable effects" (EPA, 1986). The residue value, which is the maximum permissible concentration in the water based on the foregoing two considerations, is calculated by dividing the maximum permissible tissue concentration by an appropriate bioconcentration factor (BCF), where the

Table 8.7   Acute/chronic Ratios for Dieldrin Toxicity to Three Species of Aquatic Animals

Species	Acute value (ppb)	Chronic value (ppb)	Ratio
Salmo gairdneri, rainbow trout	2.5	0.22	11
Poecilia reticulata, guppy	4.1	0.45	9.1
Mysidopsis bahia, mysid shrimp	4.5	0.73	6.2

Source. EPA (1980b).

Table 8.8   Concentrations of Dieldrin that Reduce Growth by Aquatic Plants

Species	Effects	Concentration (ppb)
Scenedesmus quadricaudata	22% reduction in biomass in 10 days	100
Navicula seminulum	50% reduction in growth in 5 days	12,800
Wolffla papulifera	Reduced population growth in 12 days	10,000
Agmenellum quadruplication	Reduced growth rate	950

Source. EPA (1980b).

BCF is the quotient "of the concentration of a material in one or more tissues of an aquatic organism divided by the average concentration in the solution in which the organism has been living (EPA, 1986). The maximum permissible tissue concentration is the lower of 1) a FDA action level for fish oil or the edible portion of fish or shellfish and 2) the tissue concentration that results in "a maximum acceptable dietary intake based on observations on survival, growth or reproduction in a chronic wildlife feeding study or a long-term wildlife field study" (EPA, 1986).

Strictly speaking, BCFs pertain only to net uptake directly from water, and hence must be measured in the laboratory. In the field, however, biological magnification can of course occur by way of the food chain, and hence the effective biological accumulation factor (BAF), which relates the concentration of a substance in an organism to the concentration in the water, may be substantially higher in the field than a laboratory-measured BCF. Unfortunately very few acceptable BAFs have been measured, because it is necessary to make enough measurements of the concentration of the material in water to show that it was reasonably constant for a long enough period of time over the range of territory inhabited by the organisms. Hence final residue values are calculated on the basis of reported BCFs.

Special considerations apply to lipid-soluble toxicants, because the BCF can be expected to increase more or less in direct proportion to the percentage of fat in an organism. Consequently the percentage of lipids must be measured in the tissue for which the BCF is calculated. A normalized BCF is then calculated by dividing the unnormalized BCF by the percentage of lipids. In effect this normalized BCF is the BCF that one would expect to measure if the tissue contained 1% lipids. The geometric mean of all normalized BCFs is then calculated using data from both freshwater and saltwater organisms. The residue value equals the quotient of the maximum permissible tissue concentration and the product of the mean normalized BCF and the percentage of lipids appropriate to the particular kind of tissue. The appropriate percentage of lipids is 11% and 10% for the edible portions of freshwater fish and saltwater fish, respectively, and 100% for fish oil. The freshwater figure of 11% is typical of lake trout and chinook salmon. The saltwater figure of 10% is typical of Atlantic herring. The edible portions of these species have the highest fat content of any important consumed species. The residue value is based on these species in order to ensure that FDA guidelines are satisfied for all edible fish species. In the case of chronic feeding studies with wildlife, the appropriate percentage of lipids is "that of an aquatic species or group of aquatic species which constitute a major portion of the diet of the wildlife species" (EPA, 1986).

The following example illustrates how the final residue value was determined for dieldrin, a lipid soluble substance. Table 8.9 lists the BCF values reported for dieldrin in various freshwater and saltwater species. Only two studies actually reported the percentage of lipids. The geometric mean of the normalized BCF values for these two species is $[(1160/1)(2300/1.1)]^{1/2} = 1557$. Since the FDA action level for dieldrin in fish, shellfish, and fish oil is 0.3 ppm, residue values would be calculated as follows.

> freshwater fish
> $(0.3 \text{ ppm})/[(1557)(11)]$ $= 0.018 \text{ ppb}$
>
> saltwater fish
> $(0.3 \text{ ppm})/[(1557)(10)]$ $= 0.019 \text{ ppb}$

fish oil
$$(0.3 \text{ ppm})/[(1557)(100)] = 0.0019 \text{ ppb}$$

Two additional residue values may be calculated based on the BCFs for dieldrin in the edible portions of fish and shellfish. The highest BCF for the edible portion of a freshwater fish is 2993 for the channel catfish. Therefore, the corresponding residue value would be 0.3 ppm/2993 = 0.10 ppb. The highest BCF value for the edible portion of a saltwater species is 8000 for the eastern oyster, and the corresponding residue value would be 0.3 ppm/8000 = 0.038 ppb.

Only two aquatic wildlife feeding studies have been performed with dieldrin, and both involved rainbow trout. The studies showed that a dieldrin concentration of 0.36 ppm in the diet of rainbow trout altered the fish's ability to detoxify ammonia and metabolize phenylalanine. Although the lipid content of the diet was not reported in either study, it is obvious that the lipid content did not exceed 100%. A minimum residue value based on these two feeding studies would be 0.36 ppm/ [(1557)(100)] = 0.0023 ppb. This figure is still larger than the residue value of 0.0019 ppb based on the fish oil action level. In fact, the smallest of any of the residue values is 0.0019 ppb, and that figure therefore became the final dieldrin residue value for both freshwater and saltwater.

It is important to realize that residue values calculated in this way may not be low enough. First, the calculations are based on BCFs rather than BAFs. Second, since a residue value calculated from an FDA action level will probably result in an average concentration in the edible portion of a fatty species that is at the action level, "Some individual organisms, and possibly some species, will have residue concentrations higher than the mean value" (EPA, 1986). Finally, some chronic feeding studies,

Table 8.9   Bioconcentration Values in Freshwater Species of Dieldrin on a Wet Weight Basis

Species	Tissue	% Lipid	BCF
Alga, *Scenedesmus obliquus*	—	—	128
Community dominated by the alga *Tribonema minus*	—	—	5558
Community of algae including *Stigeoclonium subsecundum, Synedra ulna, Epithemia sorex, Cocconeis placentula var. euglypta*, and *Nitzschia* sp.	—	—	3188
Cladoceran, *Daphnia magna*	—	—	1395
Freshwater mussel, *Lampsilis siliquoidea*	Whole body	1	1160
Steelhead trout (newly hatched alevin), *Salmo gairdneri*	Whole animal	—	3225
Lake trout (yearling), *Salvelinus namaycush*	Whole body	—	68286
Channel catfish, *Ictalurus punctatus*	Dorsal muscle	—	2385
Channel catfish, *Ictalusus punctatus*	Dorsal muscle	—	2993
Guppy, *Poecilia reticulata*	Whole animal	—	12708
Guppy, *Poecilia reticulata*	Whole animal	—	28408
Eastern oyster, *Crassostrea virginica*	Edible tissue	—	8000
Crab, *Leptodinus floridanus*	Whole body	—	400
Sailfin molly, *Poecilia latipinna*	Edible tissue	—	4867
Spot, *Leiostomus xanthurus*	Whole body	1.1	2300

Source. EPA (1980b).

including the two cited here for dieldrin, and long-term field studies with wildlife identify concentrations that cause adverse effects but do not identify concentrations which do not cause adverse effects.

THE TWO-NUMBER CRITERION

In the United States, EPA water quality guidelines with respect to toxic substances are based on a two-number criterion (EPA, 1986). The first number is the 1-hour average concentration that is not to be exceeded more than once every 3 years. This number is referred to as the *criterion maximum concentration*. The second number is the 4-day average concentration that is not to be exceeded more than once every 3 years. This number is referred to as the *criterion continuous concentration*. The rationale for using a two-number criterion is that organisms can tolerate brief exposures to toxicant concentrations substantially higher than can be tolerated over a longer period. This fact is evident in Figure 8.4. There is invariably a negative correlation between the concentration of a toxicant that causes a particular adverse effect and the duration of exposure. The recommended exceedence frequency of 3 years is the EPA's best scientific judgment of the average amount of time it would take an unstressed system to recover from a pollution event in which exposure to a toxicant exceeded the criterion.

Under the present system, the two numbers in the criterion are calculated from the final acute value, the final chronic value, the final plant value, and the final residue value. The four values for dieldrin in freshwater and saltwater are shown in Table 8.10. The criterion maximum concentration is equated to half the final acute value. The criterion continuous concentration is the smallest of the final chronic value, the final plant value, and the final residue value.

COMPLICATING FACTORS

Interactions with Harmless Substances or Conditions

The toxicity of a particular substance to aquatic organisms may sometimes depend strongly on the concentrations of other substances or on environmental conditions that are harmless by themselves. For example, in aquatic systems a continuous

Table 8.10   The Four Toxicant Concentrations Used to Establish Water Quality Guidelines for Dieldrin in Freshwater and Saltwater

	Value in freshwater (ppb)	Value in saltwater (ppb)
Final acute value	2.5	0.71
Final chronic value	0.29	0.084
Final plant value	100	950
Final residue value	0.0019	0.0019
Criterion maximum concentration	1.25	0.355
Criterion continuous concentration	0.0019	0.0019

interconversion takes place between ammonium ions (NH_4^+) and unionized ammonia (NH_3). The equation describing this interchange can be written $NH_3 + H^+ = NH_4^+$. Although NH_4^+ is nontoxic, NH_3 is highly toxic. The NH_3 96-hr median tolerance limits for a variety of freshwater fish lie in the range 0.08–4.6 ppm (EPA, 1986). The equilibrium distribution of NH_3 and NH_4^+ is obviously a function of pH, and as noted already (Chapter 6) the equilibrium shifts to NH_3 as the pH is raised. At steady state the concentrations of NH_3 and NH_4^+ are equal at a pH of 9.3, and the $[NH_3]/[NH_4^+]$ ratio increases by a factor of 10 whenever the pH increases by one unit. Thus water that was virtually harmless at a pH of 7 might become extremely toxic if the pH were raised to 9. Because of this fact, it is of relatively little use in water quality work to assay merely for $NH_3 + NH_4^+$ (by standard procedures) without simultaneously measuring the pH. EPA water quality criteria for ammonia are written in terms of NH_3 rather than $NH_3 + NH_4^+$ (EPA, 1986).

Temperature is another factor that can greatly influence the toxicity of a substance without itself causing any significant stress. In general terms the performance of an organism is positively correlated with temperature below the optimum temperature for the organism and then rapidly drops as the temperature increases further (Figure 8.7). The rate of increase of the respiration rate or metabolic rate of an organism is usually expressed in terms of the organism's Q_{10}, which is the change in respiration rate or metabolic rate per 10°C increase in temperature. For temperature conformers (i.e., cold-blooded organisms) metabolic rates typically increase by a factor of 2 or 3 for every 10°C rise in temperature as long as the temperature is not too close to the lethal temperature for the species. In other words, typical Q_{10} values for cold-blooded organisms are 2–3. Obviously any increase in respiration rate implies an increased demand for oxygen, and as a result, the TLm for oxygen can be expected to be positively correlated with temperature. This fact in part underlies the tendency of fish kills caused by inadequate oxygen concentrations to occur during the summer when water temperatures are highest. An increase in the respiration rate of a fish obviously implies an increase in the rate at which water is pumped over the gills. As a result, the intensity of exposure of the fish to any toxic substance in the water is increased. Therefore, the effects of a given toxicant concentration on the organism may become apparent more rapidly in warm water than in cold water.

Figure 8.4 provides another clear example of the interaction between otherwise harmless water quality characteristics and toxicant effects. The toxicity of zinc to rainbow trout is clearly correlated in a negative way with water hardness. Similar correlations between metal toxicity and water hardness have been noted for a variety of toxic metals. The mechanism responsible for the correlation is not well understood, and considering the wide range of metals for which the effect has been observed, it is unlikely that exactly the same mechanism is operative in all cases. Water hardness, which is a measure of the concentration of calcium and magnesium ions in the water, tends to be correlated with a number of other water quality characteristics such as pH and alkalinity. It is possible, therefore, that at least in some cases the observed negative correlation between metal toxicity and water hardness actually reflects a cause-and-effect relationship with one of these other covariates.

Incorporation into Water Quality Guidelines
If there is enough relevant information, it is possible to incorporate these interactions between toxicant effects and water quality characteristics into water quality

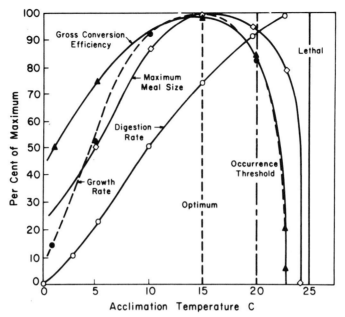

Figure 8.7 Performance of Sockeye salmon (*Oncorhynchus nerka*) as a function of acclimation temperature. [Redrawn after Brett (1971).]

criteria. In the United States, for example, the criterion maximum concentration and the criterion continuous concentration are sometimes expressed as explicit functions of certain water quality characteristics. In the case of NH_3, for example, toxicity has been shown to be negatively correlated with pH and, below 20–25°C, with temperature. Therefore, the EPA water quality criteria for NH_3 are expressed in terms of mathematical equations that relate the criterion maximum and continuous concentrations of NH_3 to temperature and pH (EPA, 1986).

In the case of heavy metals, the criteria are often expressed as functions of water hardness. Figure 8.8 shows some of the data that were used to determine the freshwater criterion maximum concentration in the case of cadmium (EPA, 1984). Pertinent data relating cadmium toxicity to water hardness were available for five species, and the logarithms of the cadmium 96-hr TLm values were plotted against water hardness. The slopes of the five regression lines fit to the data were similar, the average being 1.13. The implication is that cadmium 96-hr TLm values can be expected to be proportional to water hardness raised to the 1.13 power. Freshwater 96-hour TLm data reported in the literature were then normalized to the same water hardness by assuming this dependence on water hardness, and the normalized 96-hr TLm values were then plotted as in Figure 8.6. The normalized final acute value was then calculated in the manner described previously. However, this procedure resulted in a normalized final acute value greater than the normalized species mean acute values of four salmonids, which were the most sensitive species among the 50 species included in the study. Since the salmonids are a commercially and recreationally important group of fish, the normalized final acute value was arbitrarily set to the normalized species mean acute value of rainbow trout, the most sensitive salmonid for which results of flow-through tests were available. The normalized freshwater criterion maximum concentration was then set equal to half this normalized final acute value. The criterion maximum concentration is then determined from the equation

$$\text{criterion maximum concentration (ppb)} = 0.022\, H^{1.13}$$

where H is water hardness in mg/liter.

An analogous equation for the final chronic value may be calculated by simply dividing the equation for the final acute value by the final acute/chronic ratio. However, if there is evidence that there is a difference in the functional dependence of chronic toxicity and acute toxicity on water quality characteristics such as temperature and hardness, then the final chronic equation may be determined independently of the final acute equation. In the case of cadmium, for example, chronic toxicity appears to be less sensitive to water hardness than acute toxicity, and a final freshwater chronic equation was therefore developed solely from chronic toxicity studies performed with a total of 16 freshwater species. The final chronic equation for cadmium in freshwater is

$$\text{final chronic value (ppb)} = 0.0305\, H^{0.785}$$

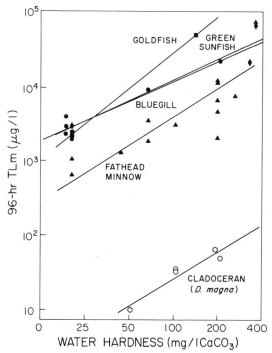

Figure 8.8   Cadmium freshwater TLm values as a function of water hardness for five species. [Data from EPA (1984).]

Conditioning and Acclimation

The toxicity of a particular stress to an aquatic organism can depend very much on the previous life history of the organism. This fact reflects the ability of many organisms to acclimate or adapt to changes in the environment and in particular to the presence of toxic substances in the water. Figure 8.9 provides a good example of the effects of adaptation on the toxic effects of temperature stress. The data are taken from a study of both lethal and sublethal temperature stresses on young sockeye salmon. Two features of the graph are of particular interest. First, sublethal effects on spawning and growth become apparent outside a much narrower temperature range than is the case for lethal effects. For example, the fish do not begin to die outright until the temperature exceeds 21–25°C or falls below 5–7°C. These observations simply illustrate and reemphasize the importance of distinguishing sublethal effects from lethal effects. Second, the range of temperatures within which the fish can effectively carry out certain activities is shifted to higher temperatures as the acclimation temperature is increased. For example, a young sockeye salmon acclimated to a temperature of 20°C would evidently die if placed in water at 4°C, but a similar fish acclimated at 10°C would not die if placed in the same water. There is, however, a limit to the adaptive capabilities of aquatic organisms. For example, it is apparent from Figure 8.9 that young sockeye salmon cannot spawn outside the approximate temperature range 5–14°C, regardless of the acclimation temperature.

In some cases exposure of an organism to a stress may actually increase the sensitivity of the organism to that stress, but in many cases prolonged exposure to a sublethal level of stress may actually desensitize or acclimate the organism to the stress.

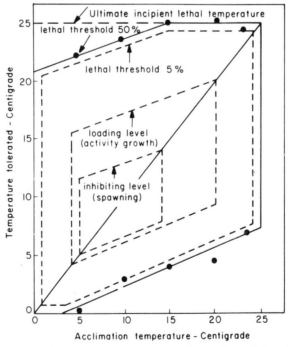

Figure 8.9 Temperature tolerance zones for young sockeye salmon (*Oncorhynchus nerka*) as a function of acclimation temperature. [Redrawn from Brett (1960).]

For example, many fish are capable of acclimation to low oxygen levels as well as to high and low temperatures. Some fish have been shown to acclimate to ammonia, cyanide, pH, phenol, synthetic detergents, and zinc (Degens et al., 1950; Bucksteeg, 'Thiele, and Stoltzel, 1955; Neil, 1957; Lloyd, 1960; Lemke and Mount, 1963; Jordan and Lloyd, 1964; Edward and Brown, 1967; Lloyd and Orr, 1969). In organisms such as bacteria that can multiply rapidly, adaptation may actually involve genetic changes. Mutant strains may by chance be more resistant to a particular stress than the other members of the population and over time will gradually dominate the population. Genetic selection of this sort is believed to have been responsible for the emergence of pesticide-resistant strains of about 200 different insect pests. Among organisms that reproduce more slowly, genetic selection would take much longer, and acclimation in the short term must therefore involve other mechanisms such as biochemical, physiological, or behavioral adjustments. The processes involved in these adjustments are in general not well understood, but probably involve changes in the functioning of hormones and/or enzymes and the response characteristics of the organism's nervous system.

In a general sense organisms that adapt to a stress can be classified as being either "conformers" or "regulators" (Warren, 1971). For example, cold-blooded animals are conformers with respect to temperature, because their internal body temperature is correlated strongly with the temperature of the environment. Warm-blooded animals are regulators with respect to temperature, because their internal temperature is relatively independent of the temperature of the environment. Of course, for both conformers and regulators there is an optimum internal state at which the organism functions most effectively. Acclimation among regulators usually produces a change in the range of the external variable within which the organism can effectively control its internal state, but does not change the optimum state of the organism. Acclimation among conformers may involve a change in the organism's optimal internal state and usually involves a change in the range of internal variability within which the organism can function efficiently.

Should the possible effects of acclimation be considered when setting water quality guidelines? The attitude of the U.S. Environmental Protection Agency is that acclimation effects should in general not be considered and that toxicity data obtained with acclimated organisms should not be used in deriving water quality guidelines. The rationale for this attitude is that, "Acclimated organisms are the exception rather than the norm. Rarely, if ever, can acclimation be depended on to protect organisms in a field situation because concentrations often fluctuate and motile organisms do not stay in one location very long" (EPA, 1980a, p. 79364). This attitude seems a reasonable one to the extent that acclimation in toxicology studies is avoidable and tends to desensitize organisms to a stress. However, conditioning may sometimes sensitize rather than desensitize an organism to a stress, and in the case of temperature stress, it is impossible to avoid the issue of acclimation in experimental studies. Certainly one should be aware of the potentially confounding effects of acclimation in toxicological work.

Interactions Between Toxic Substances

The discussion to this point has considered the effect of a toxicant as if it were the only stress-producing substance. What effect can be anticipated, however, if two or more toxic substances are present in the water at the same time? For example, according to the present EPA guidelines, the 4-day average concentrations of lead and zinc in saltwater should not exceed 5.6 and 86 ppm, respectively, more than once every 3 years. Now suppose that a particular body of saltwater is found to consistently contain 4 ppm of lead and 60 ppm of zinc. Would this water be likely to exert a chronic stress that would in some sense violate the intentions of the EPA guidelines?

The answer to this question is that the effect of two or more toxic substances on the organisms in the water will depend on the manner in which the toxicants interact. The following example illustrates the various ways in which two toxicants might interact. Suppose that the 96-hr TLm for a particular organism is 1.0 and 10 ppb for toxicants A and B, respectively. Now suppose that toxicants A and B are added to the water in a flow-through acute toxicity experiment so that the concentrations of A and B are in fact 1.0 and 10 ppb, respectively. The test organisms are placed in the water and their survival monitored for 96 hours. Any of the following outcomes is possible.

1. Exactly half the organisms are dead after 96 hours. This result is exactly what one would have expected if you had added 1.0 ppb A but no B to the water, or 10 ppb B but no A. In this case there has been *no interaction* between A and B.

2. Fewer than half the organisms are dead after 96 hours. In this case the toxicity of the mixture is apparently less than the toxicity of a solution containing only 1.0 ppb A or only 10 ppb B, and the interaction between A and B is said to be *antagonistic*.

3. If more than half the organisms are dead after 96 hr, there are several possible interpretations.

   a. If the percentage of dead organisms is exactly what would have been expected if the water contained only 2.0 ppb A or only 20 ppb B, then the interaction of A and B is said to be *strictly additive*. In other words, 1 ppb A produces the same toxicity as 10 ppb B, and when the two substances are present together the toxicity of the mixture can be calculated by assuming that every 10 ppb of B are equivalent to 1.0 ppb of A.

   b. If the fraction of organisms that are dead after 96 hr is greater than in case 3a, then the interaction is said to be *supra-additive*.

   c. If the percentage of organisms dead after 96 hr is greater than 50% but less than in case 3a, then the interaction is said to be *infra-additive*.

It is useful in discussing these interactions to speak in terms of toxic units rather than actual concentrations. In this hypothetical example, one acute toxicity unit of A would be 1.0 ppb of A, and one acute toxicity unit of B would be 10 ppb of B. The possible interactions between A and B may then be indicated graphically as in Figure 8.10.

How should these interactions be taken into account in setting water quality guidelines? One obvious practical problem is that there is not a great deal known about the interactions of toxic substances. In fact, a review of current EPA water quality criteria (EPA, 1987) shows that in many cases insufficient information is available to establish numerical guidelines for toxicants considered separately. The present guidelines make no allowance for interactions between toxicants. In

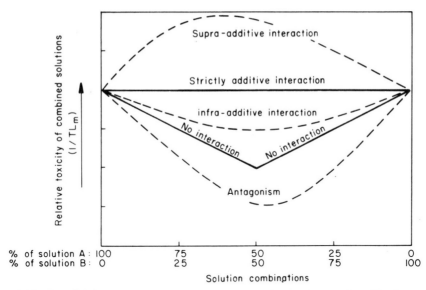

Figure 8.10   Possible interactions between two hypothetical toxicants, A and B. [Redrawn from Warren (1971).]

response to criticisms concerning this policy, the EPA has stated, "Synergism and antagonism are possible between numerous combination [sic] of two or more pollutants, and some data indicate that such interactions are not only species specific, but also vary with the ratios and absolute concentrations of the pollutants and the life stage of the species" (EPA, 1980a, p. 79358).

In other words, not much is known, the interactions may be very complicated, and given these facts it is very hard to estimate the toxicity of a solution containing more than one toxic substance. To the extent that the effects of two or more toxicants can be normalized by expressing concentrations in terms of toxicity units, Figure 8.10 shows that in two cases calculation of the toxicity of a mixture of toxicants is straightforward. First, if there is no interaction between the toxicants, then the toxicity of the mixture would be determined by the toxicant present in the greatest number of toxicity units. For example, if a solution contained 0.75 toxic units of A and 0.5 toxic units of B, then if A and B do not interact, the toxicity of the solution would be equal to that of a solution containing only 0.75 toxic units of A. Similarly, if the interaction were strictly additive, the toxicity of the mixture would be equal to that of a solution containing 1.25 toxic units. In other words, if the assumption is that there is no interaction or that the interactions are strictly additive, then the toxicity of a mixture of toxicants can be calculated easily. However, if the interactions are supra-additive, infra-additive, or antagonistic, then detailed information is needed about the nature of the interaction to calculate the toxicity of the mixture.

It is apparent from Figure 8.10 that the assumption of no interaction is conservative only if the true interactions are antagonistic. The assumption of no interaction would underestimate the toxicity if the true interactions were in any sense additive. The assumption of strictly additive interactions would appear to be conservative in all cases except the case in which the true interactions were supra-additive. This latter conclusion, however, is valid only to the extent that a given toxic effect can be uniquely associated with a certain number of toxicity units. Figure 8.10, for example, assumes implicitly that X toxicity units of A have the same effect as X toxicity units of B for all values of X. In other words, there is a universal curve that relates toxic effect to toxicity units. The normalization procedure ensures that the effects are the same when X = 1, but it does not follow automatically that the effects are the same for all values of X. For example, suppose that two acute toxicity units of A kill 75% of a given species in 96 hours and that two acute toxicity units of B kill 95% of the same species in 96 hours. Now suppose that 85% of the test organisms die after 96 hours of exposure in water containing one toxicity unit of both A and B. The interactions are clearly additive in some sense, since more than 50% of the test organisms died after 96 hours. However, the interactions appear supra-additive relative to toxicant A and infra-additive relative to toxicant B. Strictly speaking then, the assumption of strictly additive interactions cannot be applied unless it is certain that the curves relating toxic effects and toxicity units are the same or very similar for all toxicants of concern.

There is no doubt that some toxicants do not interact in a strictly additive manner. For example, the toxicity of cadmium is decreased by other metal ions, particularly low concentrations of copper and zinc (Parizek, 1957; Parizek et al., 1969; Ferm and Carpenter, 1967; Gunn et al., 1963a,b; Hill et al., 1963; Bunn and Matrone, 1966), and it is known that the cholinesterase-inhibiting pesticides EPN (O-ethyl-O-paranitrophenyl phenylphosphonothioate) and malathion interact in a supra-additive manner. The TLm of a mixture containing equal numbers of toxic units of EPN and malathion is only one-twelfth the TLm of the individual compounds (Frawley, 1965). The effect is due to the ability of EPN to inhibit the enzyme system that is responsible for detoxifying malathion after it has been converted to malaoxon. In contrast, chlorinated hydrocarbons such as DDT that induce some of the liver microsomal enzyme systems may interact antagonistically with chemicals that are detoxified by the same enzyme systems (e.g., hexobarbital). Although it is admittedly illogical to expect the toxicity of poisons with different toxicological properties and different concentration-response curves to be strictly additive, "Nevertheless, the method has been found to work empirically . . . If, for the present, the poisons can be regarded as agents producing stress . . . , each of which produces a degree of shock with resulting non-specific effects, it might be considered reasonable that summation of the overall stress is possible" (Brown, 1968).

Application of the strictly additive interaction assumption to water quality guidelines was made in the 1973 EPA water quality criteria, where it was noted, "The system of adding different toxicants in this way is based on the premise that their lethal actions are additive. Unlikely as it seems, this simple rule has been found to govern the combined lethal action of many pairs and mixtures of quite dissimilar toxicants, such as copper and ammonia, and zinc and phenol in the laboratory . . . The rule holds true in field studies" (Herbert, 1965; Sprague et al., 1965). "The method of addition is useful and reasonably accurate for predicting thresholds of lethal effects in mixtures" (EPA, 1973, p. 122).

This statement notwithstanding, the assumption of strictly additive interactions was abandoned in 1980 (EPA, 1980a) and replaced with the assumption of no interactions. Because the latter assumption causes one to underestimate the toxicity of a mixture whenever the interactions are in any sense additive, adoption of this policy seems rather inconsistent with the expressed intentions of the EPA, "to err on the side of overprotection" (EPA, 1980a, p. 79342). If a general guideline in establishing water quality criteria is to err on the side of overprotection and if detailed information on the interactions between combinations of toxicants is not available, then the assumption of strictly additive interactions would seem preferable to the assumption of no interactions at all.

PUBLIC HEALTH

In the previous discussion we have considered the use of toxicological studies for purposes of setting water quality criteria primarily for the protection of marine and freshwater organisms. An equally if not more important use of toxicology is to establish water quality criteria for the protection of the general public. In the United States at least, such criteria are established by different methods, depending on whether the toxicants are believed to be carcinogenic or not.

Noncarcinogenic Effects

In the case of noncarcinogenic substances, it is generally assumed that there is a threshold dose below which the substance exerts no adverse effects on humans. It is obvious that such a threshold must exist in the case of essential trace metals, which are required in small amounts by all organisms but become toxic if administered in large doses. In all cases, however, the existence and value of the threshold must be established based on experiments with animals or experience with humans. Water quality criteria are then established based on the estimated thresholds.

One can imagine several types of experimental or observational data that might be used to estimate the threshold doses for noncarcinogenic substances. First of all, there may have been unfortunate incidents in which humans were exposed to toxic concentrations of a particular substance. If the dose received by the people is known or can be estimated and the effect on the persons documented, then the data are useful for establishing an upper bound on the threshold dose. In some cases there may actually have been experimental feeding studies with human volunteers, and in such cases it may be possible to obtain an even better estimate of the threshold dose. In the absence of information directly related to humans, there may be data obtained from experimental feeding studies with animals. These data may be used to estimate human effects, but the uncertainty factor is greater in the absence of any direct observations on humans.

The duration of exposure must also be considered. Water quality criteria are based on an assumed lifetime exposure, but experimental observations, whether on humans or animals, do not always cover the lifetime of the organisms. If the only data available are from acute rather than chronic observational or experimental studies, then the uncertainty factor used in establishing water quality criteria must be increased.

Finally, you can imagine several general types of effects or responses associated with exposure to a toxicant. The first type of response would be no observed adverse effect. The dose associated with no observed adverse effect is referred to as the *no observed adverse effect level* (NOAEL) and is theoretically a lower bound to the threshold dose that causes some adverse effect. The second type of response would be an *observed adverse effect*, and the *lowest observed adverse effect level* (LOAEL) is theoretically an upper bound to the threshold dose. Adverse effects are defined by the U.S. EPA as, "Any effects which result in functional impairment and/or pathological lesions which may affect the performance of the whole organism, or which reduce an organism's ability to respond to an additional challenge" (EPA, 1980a, p. 79353). It is also possible that exposure to a toxicant will produce a response that is not an adverse effect, according to the previous definition. For example, rather than passing through the digestive tract and being excreted, the toxicant may be assimilated and then detoxified or broken down by the organism. The organism has therefore been affected by the toxicant to the extent that it has diverted some of its metabolic energy to detoxifying or breaking down the toxicant; but should this response be regarded as an adverse effect? The job of classifying such responses is obviously somewhat subjective, and has become increasingly difficult as more sophisticated testing protocols are developed and more subtle responses are identified. One result has been that some responses are classified simply as observed effects. One therefore finds in the EPA literature, for example, reference to *no observed effect levels* (NOELs) and *lowest observed effect levels* (LOELs). It is up to public health administrators, given information on these various types of responses or effects and after applying appropriate uncertainty factors, to determine the *acceptable dietary intake* (ADI). Table 8.11 lists the uncertainty factors used in the United States.

Once an ADI has been established, the criterion (concentration in water) for the toxicant is calculated from the equation

$$C = (ADI - DT - IN)/(2 + 0.0065R) \qquad (8.2)$$

where C is the criterion, DT is the estimated nonfish dietary intake, IN the estimated daily intake due to inhalation, and R the bioconcentration factor in liters/kg. Equation 8.2 assumes that an average person consumes 2 liters of water per day and con-

Table 8.11   Uncertainty Factors for Deriving Criteria for Threshold Effects of Toxicants from NOEL, NOAEL, LOEL, and/or LOAEL Data

Nature of experimental results	Uncertainty factor
Valid experimental results from studies on prolonged ingestion by humans with no indication of carcinogenicity	10
Experimental results of studies of human ingestion not available or scanty (e.g., acute exposure only) with valid results of long-term feeding studies on experimental animals, or in the absence of human studies, valid animal studies on one or more species; no indication of carcinogenicity	100
No long-term or acute human data; scanty results on experimental animals with no indication of carcinogenicity	1000
Additional judgemental uncertainty factor to be applied to LOAEL data	1–10

Source. National Academy of Sciences (1977).

sumes 6.5 g (= 0.0065 kg) of fish and shellfish per day. Adequate information on DT and IN may not always be available, in which case an assumption must be made concerning total exposure. Lack of information about or variability in DT and IN obviously can create a degree of uncertainty in the derived water quality guidelines.

Application to Cadmium

Application of the foregoing methodology is illustrated in the following example of cadmium. In this case there are a number of examples of humans who were exposed to levels which clearly caused adverse effects. In areas of Japan where itai-itai disease occurred (see Chapter 12), about 85% of the cadmium intake was due to the consumption of cadmium-contaminated rice (Muramatsu, 1974). The disease was associated with tubular proteinuria, the incidence of which rose above that observed in control populations when the cadmium concentration in the rice exceeded about 0.45 ppm (Nogawa, Ishizaki, and Kawano, 1978). Japanese in the area were known to consume about 0.43 kg of rice per day (Friberg et al., 1974). Therefore an adverse effect level for Japanese equals $(0.43)(0.45)/(0.85) = 0.23$ mg/day. For Americans and Western Europeans, who are about 32% larger than Japanese, the corresponding adverse effect level would be about 0.30 mg/day. Since only about 5% of ingested Cd is actually absorbed by the body (EPA, 1980c), the figure of 0.30 mg/day ingested corresponds to only about 15 µg/day absorbed.

Similar conclusions have been reached from other studies. For example, a study of workers who inhaled Cd dust concluded that effects began to appear when the concentration of Cd in the air equaled 21 µg m^{-3}. If the assumption is that the average person inhales about 10 m^3 of air per day, that lung retention of Cd is about 25% efficient (EPA, 1980c), and that the workers were exposed to the contaminated air 5 days per week, the calculated level of absorption corresponding to the appearance of effects was $(21)(10)(0.25)(5/7) = 38$ µg/day. Similarly, the Working Group of Experts for the Commission of European Communities has estimated that the threshold level of Cd exposure by ingestion is about 0.20 mg/day (Comm. Eur. Communities, 1978), which would correspond to an absorption of 10 µg/day.

Applying the uncertainty factor of 10 in Table 8.11 appropriate for studies on prolonged ingestion by humans, it would be concluded that an ADI (assuming 5% absorption efficiency) would be $10/[(10)(0.05)] = 20$ µg/day to $38/[(10)(0.05)] = 76$ µg/day for Cd. It is apparent from Table 8.12 that for the average American the rate of Cd intake in food exceeds the lower of these two figures and equals about 40% of the higher figure. It is also apparent that drinking water contributes a very minor fraction of the total Cd absorbed. In fact, the average drinking water in the United States contains only 1.3 ppb Cd. Only 3 of 969 community water supplies studied by the EPA contained Cd in excess of the existing drinking water standard of 10 ppb (EPA, 1980c).

How much Cd would an average person absorb from drinking water and consuming fish and shellfish if all water contained 10-ppb Cd? Bioconcentration factors (BCFs) for Cd in the edible parts of commercially important fish and shellfish are shown in Table 8.13. The BCFs are consistently higher for the 5 bivalve molluscs than for the other 2 species, the geometric mean BCFs being 763 and 11, respectively. Of the 6.5 g/day of fish and shellfish consumed by the average American, only 0.8 g is accounted for by bivalve molluscs. Finfish and other shellfish account for the other 5.7 g/day (EPA, 1980c). Therefore, it seems appropriate to weight the effective BCF according to the relative amounts of the different kinds of fish and shellfish, and the weighted BCF is therefore $[(0.8)(763) + (5.7)(11)]/6.5 = 104$. The daily intake of Cd from drinking water and the consumption of fish and shellfish if all water contained 10 ppb Cd would therefore be $10[2 + 0.0065(104)] = 26.8$ μg/day. Of this figure, only 25% comes from the consumption of fish and shellfish. For the average American then, the consumption of fish and shellfish makes a minor contribution to Cd intake. This conclusion may not apply, however, to persons who consume large amounts of bivalve molluscs.

In considering the effects of Cd on human health, the EPA (1980c) commented, "It appears that a water criterion needs to be no more stringent than the existing primary drinking water standard (10 μg/liter) to provide ample protection of human health." This conclusion seems debatable. If a drinking water supply actually contained 10 μg/l of Cd, the intake of Cd from drinking water alone would be 20 μg/day, which equals the lower bound of the ADIs for Cd estimated from the observed adverse effects of Cd on human health. If the estimated intake of 30 μg/day from food (Table 8.12) is added to this figure, the total is 2.5 times the lower bound. The EPA's conclusion seems to have been based largely on the fact that, "Water constitutes only a relatively minor portion of man's daily cadmium intake" (EPA, 1980c). This fact arises, however, not because average water contains 10 ppb cadmium, but because average water contains almost eight times less cadmium. Nevertheless, to this day 10 ppb remains the primary drinking water standard for Cd.

Table 8.12   Average Daily Intake and Absorption of Cadmium by Americans

Exposure source	Exposure	Cd intake (μg)	% Absorbed	Absorption (μg)
Air, ambient	0.03 μg m^{-3}	0.6	25	0.15
Air, smoking (one pack)	3.0 μg/pack	3.0	25	0.75
Food	—	30.0	5	1.50
Drinking water	1.3 μg/liter	2.6	5	0.13

Source. EPA (1980c).

Table 8.13   Cadmium Geometric Mean BCFs for Edible Parts of Consumed Aquatic Species

Species	Mean BCF
Bivalve molluscs	
Asiatic clam, *Corbicula fluminea*	2570
Blue mussel, *Mytilus edulls*	186
Bay scallop, *Argopecten irradians*	2040
Eastern oyster, *Crassostrea virginica*	1660
Soft-shell clam, *Mya arenaria*	160
Other fish and shellfish	
Brook trout, *Salmo gairdneri*	21.5
Green crab, *Carcinus maenas*	5.9

Source. EPA (1984).

Carcinogenic Effects

The whole issue of carcinogenic effects has become quite controversial because of the suggestion that faulty experimental and analytical methods may have been used to determine whether a substance is carcinogenic and, if so, the relationship between dose and response. Although in some cases evidence of carcinogenicity has been based on unintentional human exposure, in many cases toxicological studies relative to cancer have been conducted with rats or mice. In the United States the standard procedure has been to conduct initial feeding experiments with the test chemical, and on the basis of these preliminary experiments to determine the *maximum tolerated dose* (MTD). The MTD is the highest dose of the chemical that does not literally kill the animals, although if repeated over a 2-week period, the MTD, "usually leads to a noticeable but tolerated reduction in weight" (Abelson, 1987). Chronic feeding studies are then conducted at doses equal to or less than the MTD over the course of the animal's lifetime. The experimental doses vastly exceed those to which humans are likely to be exposed.

If an abnormal incidence of cancer occurs in the experimental animals compared to a control group that is not exposed to the chemical, the excess incidence of cancer is plotted as a function of dose and an analytical function is fit to the data. The analytical function has the form (Crump, 1984)

$$A(d) = 1 - \exp(-q_1 d - q_2 d^2 - \ldots - q_k d^k) \qquad (8.1)$$

where A(d) is the extra risk over background associated with a dose d, and the q_i are constants to be determined by a least squares curve-fitting procedure. An important characteristic of Equation 8.1 is the fact that at low doses it reduces to the form $A(d) = q_1 d$ and hence conforms to the assumption that there is no threshold dose below which the chemical exerts no carcinogenic effects. The levels of exposure for most humans are likely to fall in the range where, according to the model, A(d) can for all intents and purposes be assumed to be directly proportional to d. The cancer risks at these low dose levels represent considerable extrapolations of the experimental data.

So what's wrong with these studies? Actually this sort of approach was probably a reasonable one 10 or 20 years ago, because medical understanding of the etiology of cancer was relatively crude compared to the present time, and it undoubtedly seemed wise to error on the side of safety. As our understanding of the mechanisms that cause cancer has improved, however, this approach has drawn much criticism. Abelson (1990a), for example, has commented, "The standard carcinogen tests that use rodents are an obsolescent relic of the ignorance of past decades."

The problems with the rodent tests are severalfold. First, the rodents in question are pure strains that have been inbred for numerous generations. This procedure of inbreeding produces rats and mice with a more-or-less well-defined genotype, but such inbreeding often leads to genetic impairments. In the experimental mice, for example, there is a high natural incidence of liver tumors, and, "The usual response of these animals to massive doses of a chemical is to develop an even higher incidence of liver tumors" (Abelson, 1987). This increase in liver tumors among naturally tumorigenic mice is of doubtful relevance to humans, because primary liver cancer is rare in humans with the exception of alcoholics and persons who have suffered from hepatitis (Abelson, 1987).

A major criticism of the rodent studies is the fact that prolonged feeding at the MTD tends to cause mitogenesis (induced cell division), and it is now recognized that nongenotoxic chemicals such as saccharin can be carcinogens at high doses simply because they cause mitogenesis. The reason is that a dividing cell is much more at risk of mutating than a quiescent cell (Ames and Gold, 1990). Because there is believed to be a threshold dose below which such nonmutagenic toxins exert no effect on cell division, "at the low doses of most human exposures (where cell-killing and mitogenesis do not occur), the hazards [of nonmutagenic toxins] may be much lower than is commonly assumed and often will be zero" (Ames and Gold, 1990).

Given the above information, one should not be surprised to learn that about 58% of all chemicals tested chronically at the MTD are judged to be carcinogens. Although much more emphasis has been given to testing synthetic chemicals than naturally occurring compounds, the percentage of synthetic chemicals judged to be carcinogenic on the basis of these tests is not much different than the corresponding percentage of naturally occurring compounds, 61% versus 48%, respectively (Ames and Gold, 1990). Of the natural chemicals that have been tested, about two-thirds are natural pesticides produced by plants to defend themselves. According to Ames and Gold (1990), these natural pesticides account for over 99.99% of the pesticides consumed by persons in the United States, and based on the rodent bioassays, 27 out of 52, or about 52% of the natural pesticides tested are carcinogens. These 27 naturally occurring pesticides are found in 57 different foods, including apples, bananas, carrots, celery, coffee, lettuce, orange juice, peas, potatoes, and tomatoes (Abelson, 1990a). A cup of coffee, for example, contains about 10 mg of rodent carcinogens, primarily caffeic acid, catechol, furfural, hydrogen peroxide, and hydroquinone. The implication of these revelations is that the rodent bioassays have probably been very misleading concerning the carcinogenic threat associated with numerous chemicals, both synthetic and natural. Diets rich in fruits and vegetables, for example, tend to reduce the incidence of cancer in humans. This observation seems inconsistent with the results of the rodent bioassays.

Although not all scientists agree that the rodent bioassays have been misleading (Cogliano et al., 1991; Rall, 1991; Weinstein, 1991), there does seem to be a general consensus that a better understanding of the mechanisms that cause cancer is needed at this time. For example, we need to know the hormonal determinants of breast cancer, the viral determinants of cervical cancer, and the dietary determinants of stomach and colon cancer (Ames and Gold, 1990). Unleaded gasoline has been implicated as a carcinogen because it causes kidney tumors in male rats, but we now know that branched-chain hydrocarbons, which are key components of unleaded gas, interfere with the mechanism for excreting a low-molecular-weight protein by the male rat. Research conducted at the Chemical Industry Institute of Toxicology indicates that this interference may be the cause of the kidney cancer in male rats exposed to gasoline. A similar mechanism does not exist in female rats, in male or female mice, or in humans (Abelson, 1990b). Similarly, we now know that, "Saccharin's ability to induce bladder tumors in male rats is solely due to the proliferative effects that high doses have on the bladder lining" (Marx, 1990). It will clearly be impossible to make an informed judgment about the carcinogenic threat of various chemicals to humans until the mechanisms responsible for causing cancer in humans are more thoroughly understood.

COMMENTARY

At the present time there are approximately 70,000 chemicals in use or being distributed throughout the environment, and an additional 500–1000 are added each year (Postel, 1987). Although not all these chemicals are toxic, merely keeping up-to-date information on the possible toxicity of so many compounds is a prodigious task. The foregoing discussion has indicated how involved the process of setting water quality guidelines is for even a single chemical. Large amounts of the right kind of information are needed. In its 1987 update to *Quality Criteria for Water*, the EPA lists water quality criteria for approximately 100 toxic chemicals or stressful conditions. In only one-third of these cases was there sufficient information to determine a criterion maximum concentration and criterion continuous concentration. This fact dramatically illustrates the nature of the problem that confronts governmental agencies charged with the responsibility of controlling water pollution. On the one hand, there is an increasingly urgent demand to establish water quality guidelines for the large number of toxic chemicals that pollute both groundwater and surface water. On the other hand, there is a need to develop intelligent and legally defensible techniques for establishing those guidelines. Ultimately the government agencies must decide in a somewhat arbitrary manner how much of what kind of toxicologic information is necessary and sufficient to establish water quality criteria. Establishing the rules of the game is obviously a critical first step, and in the case of carcinogenic substances, it appears that there is considerable disagreement over what the rules should be. Even when the rules have been established, there often remain information gaps that must be filled before criteria can be recommended. Invariably disagreement will remain over whether the methodology required to establish the guidelines is too simple or too cumbersome, too strict or too lax, and too detailed or too ambiguous. There will remain differences of opinion as to what types of toxicological data are useful and acceptable and how best to utilize those data in developing guidelines. The wisdom and collaboration of many conscientious scientists and administrators will be needed if the legal efforts to control the quality of water is to keep pace with the growing threat of pollution.

REFERENCES

Abelson, P. H. 1987. Cancer phobia. *Science*, **237**, 473.

Abelson, P. H. 1990a. Incorporation of new science into risk assessment. *Science*, **250**, 1497.

Abelson, P. H. 1990b. Testing for carcinogens with rodents. *Science*, **249**, 1357.

Adema, D. M. M. 1978. *Daphnia magna* as a test animal in acute and chronic toxicity tests. *Hydrobiol*, **59**, 125–134.

Ames, B. N., and L. S. Gold. 1990. Too many rodent carcinogens: Mitogenesis increases mutagenesis. *Science*, **249**, 970–971.

Anderson, D. W., and F. Gress. 1983. Status of a northern population of California brown pelicans. *Condor*, **85**, 79–88.

Anderson, D. W., F. Gress, and K. F. Mais. 1982. Brown pelicans: influence of food supply on reproduction. *Oikos*, **39**, 23–31.

Anderson, D. W., F. Gress, K. Mais, and P. R. Kelly. 1980. Brown pelicans as anchovy stock indicators and their relationships to commercial fishing. *Calif. Coop. Oceanic Fish. Invest. Rep.*, **21**, 54–61.

Anderson, D. W., J. R. Jehl, R. W. Risebrough, L. A. Woods, L. R. Deweese, and W. G. Edgecomb. 1975. Brown pelicans: improved reproduction off the southern California coast. *Science*, **190**, 806–808.

Anderson, D. W., R. M. Jurek, and J. O. Keith. 1977. The status of brown pelicans at Anacapa Island in 1975. *Calif. Fish Game*, **63**, 4–10.

Brett, J. R. 1960. Thermal requirements of fish—three decades of study. In C. M. Tarzwell (Ed.), *Biological Problems of Water Pollution*. U.S. Dept. of Health, Education, and Welfare, Robert A. Taft Sanitary Engineering Center. Cincinnati, pp. 110–117.

Brown, V. M. 1968. The calculation of the acute toxicity of mixtures of poisons to rainbow trout. *Water Res.*, **2**, 723–733.

Bucksteeg, W., H. Thiele, and K. Stoltzel. 1955. Die Beeinflussung von Fischen durch Giftstoffe aus Abwassern. *Vom Wasser* (*Jahrburch fur Wasser-Chemie*), **22**, 194–211.

Bunn, C. R., and G. Matrone. 1966. *In vitro* interaction of cadmium, copper, zinc, and iron in the mouse and rat. *J. Nutr.*, **90**, 395–399.

Clark, J., and W. Brownell. 1973. *Electric Power Plants in the Coastal Zone: Environmental Issues*. American Littoral Society Special Publication No. 7, Highlands, NJ.

Cogliano, V. J., W. H. Farland, P. W. Preuss, J. A. Wiltse, L. R. Rhomberg, C. W. Chen, M. J. Mass, S. Nosnow, P. D. White, J. C. Parker, and S. M. Wuerthele. 1991. Carcinogens and human health: Part 3. *Science*, **251**, 606–607.

Commission of the European Communities. 1978. *Criteria (dose/effect relationships) for Cadmium*. Rep. Working Group of experts prepared for the Comm. Euro. Communities, Directorate-General for Social Affairs, Health and Safety Directorate. Pergamon Press, New York.

Crump, K. S. 1984. An improved procedure for low-dose carcinogenic risk assessment from animal data. *J. Env. Pathol. Toxicol.* **5**, 339–348.

Degens, P. N., H. van der Zee, J. D. Kommer, and A. H. Kamphuis. 1950. Synthetic detergents and sewage processing. Part 5. The effect of synthetic detergents on certain water fauna. *J. Inst. Sewage Purification*, **1950**(1), 63–68.

Edward, R. W., and V. M. Brown. 1967. Pollution and fisheries: a progress report. *Water Pollut. Contr.* (J. Inst. Water Pollut. Contr.), **66**, 63–78.

Environmental Protection Agency. 1973. *Water Quality Criteria 1972*. EPA/R3/73/033. Washington, D.C. 594 pp.

Environmental Protection Agency. 1980a. Water Quality Criteria Documents: Availability. *Fed. Reg.*, **45**(231), 79318–79379.

Environmental Protection Agency. 1980b. *Ambient Water Quality Criteria for Aldrin/dieldrin*. EPA 440/5-80-019. Washington, D.C.

Environmental Protection Agency. 1980c. *Ambient Water Quality Criteria for Cadmium*. EPA 440/5-80-025. Washington, D.C.

Environmental Protection Agency. 1984. *Ambient Water Quality Criteria for Cadmium—1984*. EPA 440/5-84-032. Washington, D.C.

Environmental Protection Agency. 1986. *Quality Criteria for Water*. EPA 440/5-86-001. Washington, D.C.

Environmental Protection Agency. 1987. *Update 2 to Quality Criteria for Water 1986*. May 1, 1987. Office of Water Regulations and Standards. Criteria and Standards Division. Washington, D.C.

Ferm, V. H., and S. J. Carpenter. 1967. Teratogenic effect of cadmium and its inhibition by zinc. *Nature*, **216**, 1123.

Fisher, N. S. 1975. Chlorinated hydrocarbon pollutants and photosynthesis of marine phytoplankton: A reassessment. *Science*, **189**, 463–464.

Fisher, N. S., E. J. Carpenter, C. C. Remsen, and C. F. Wurster. 1974. Effects of PCB on interspecific competition in natural and gnotobiotic phytoplankton communities in continuous and batch cultures. *Microbiol. Ecol.*, **1**, 39–50.

Fisher, N. S., and C. F. Wurster. 1973. Individual and combined effects of temperature and polychlorinated biphenyls on the growth of three species of phytoplankton. *Environ. Pollut.*, **5**, 205–212.

Frawley, J. P. 1965. Synergism and Antagonism. In C. D. Chichester (Ed.), *Research in Pesticides*. Academic Press, New York. pp. 69–83.

Friberg, L., M. Piscator, G. F. Nordberg, and T. Kjellstrom. 1974. Cadmium in the Environment. 2nd ed. Chemical Rubber Company Press. Cleveland, OH.

Greve, W., and T. R. Parsons. 1977. Photosynthesis and fish production: Hypothetical effects of climatic change and pollution. Helgol. Wiss. Meeresunters, **30**, 666–672.

Gunn, S. A., T. C. Gould, and W. A. D. Anderson, 1963a. Cadmium-induced interstitial cell tumors in rats and mice and their prevention by zinc. *J. Natl. Cancer Inst.*, **31**, 745–759.

Gunn, S. A., T. C. Gould, and W. A. D. Anderson. 1963b. The selective injurious response of testicular and epididymal blood vessels to cadmium and its prevention by zinc. *Am. J. Path.*, **42**, 685–702.

Gustafson, J. R. 1990. Five-year Status Report for California Brown Pelican. Calif. Dept. Fish and Game. Nongame Bird and Mammal Sec. Rep. 90-4 (draft).

Harding, L. W., Jr., and J. H. Phillips, Jr. 1978. Polychlorinated biphenyls (PCB) effects on marine phytoplankton photosynthesis and cell division. *Mar. Biol.*, **49**, 93–101.

Harriss, R. C., D. B. White, and R. B. MacFarlane. 1970. Mercury compounds reduce photosynthesis by plankton. *Science*, **170**, 736–737.

Hasler, A. D., and J. A. Larsen. 1971. The homing salmon. In J. R. Moore (Ed.), *Oceanography. Readings from Scientific American*. Freeman, San Francisco. pp. 253–256.

Heath, R. G., J. W. Spann, and J. F. Kreitzer. 1969. Marked DDE impairment on mallard reproduction in controlled studies. *Nature*, **224**, 47–48.

Herbert, D. W. M. 1965. Pollution and fisheries. In G. T. Goodman, R. W. Edwards, and J. M. Lambert (Eds.), *Ecology and the Industrial Society. Blackwell Scientific Publications*, Oxford. pp. 173–195.

Hill, C. H., G. Matrone, W. L. Payne, and C. W. Barber. 1963. *In vivo* interactions of cadmium with copper, zinc, and iron. *J. Nutr.*, **80**, 227–235.

Jordan, D. H. M., and R. Lloyd. 1964. The resistance of rainbow trout (Salmo Gairdnerii Richardson) and roach (Rutilus rutilus (L.)) to alkaline solutions. *Int. J. Air Water Pollut.*, **8**, 405–409.

Lemke, A. E., and D. I. Mount. 1963. Some effects of alkyl benzene sulfonate on the bluegill, *Lepomis macrochirus*. *Trans Am. Fish. Soc.*, **92**, 372–378.

Lloyd, D., and L. D. Orr. 1969. The diuretic response by rainbow trout to sublethal concentrations of ammonia. *Water Res.*, **3**, 335–344.

Lloyd, R. 1960. The toxicity of zinc sulfate to rainbow trout. *Annals Appl. Biol.*, **48**, 84–94.

Marx, J. 1990. Animal carcinogen testing challenged. *Science*, **250**, 743–745.

Mosser. J. L., N. S. Fisher, T. Teng, and C. F. Wurster. 1972a. Polychlorinated biphenyls: Toxicity to certain phytoplankters. *Science*, **175**, 191–192.

Mosser, J. L., N. S. Fisher, and C. F. Wurster. 1972b. Polychlorinated biphenyls and DDT alter species composition in mixed cultures of algae. *Science*, **176**, 533–535.

Muramatsu, S. 1974. Research about cadmium pollution in the Kakehashi River basin. *Jap. J. Pub. Health*, **21**, 299–308.

National Academy of Sciences. 1977. *Drinking Water and Health*. National Academy of Sciences, Washington, D.C.

Neil, J. H. 1957. Some effects of potassium cyanide on speckled trout (*Salvelinus fontinalis*). *Ontario Industrial Waste Conference*, **4**, 74–96.

Nogawa, K., A. Ishizaki, and S. Kawano. 1978. Statistical observations of the dose-response relationships of cadmium based on epidemiological studies in the Kakehashi River basin. *Environ. Res.*, **15**, 189–198.

O'Connors, H. B., Jr., C. F. Wurster, C. D. Powers, D. C. Biggs, and R. C. Rowland. 1978. Polychlorinated biphenyls may alter marine trophic pathways by reducing phytoplankton size and production. *Science*, **201**, 737–739.

Parizek, J. 1957. The destructive effect of cadmium ion on testicular tissue and its prevention by zinc. *J. Endocrin.*, **15**, 56–63.

Parizek, J., I. Benes, J. Kalousko, A. Babicky, and J. Lener. 1969. Metabolic interrelationships of trace elements. Effects of zinc salts on survival of rats intoxicated by cadmium. *Physl. Bohem.*, **18**, 89–94.

Postel, S. 1987. *Defusing the Toxics Threat: Controlling Pesticides and Industrial Waste*. Worldwatch Paper 79. Worldwatch Institute, Washington, D.C. 69 pp.

Rall, D. P. 1991. Carcinogens and human health: Part 2. *Science*, **251**, 10–11.

Ryther, J. H. 1969. Photosynthesis and fish production in the sea. *Science*, **166**, 72–76.

Shepard, M. P. 1955. Resistance and tolerance of young speckled trout (*Salvelinus fontinalis*) to oxygen lack, with special reference to low oxygen acclimation. *J. Fish. Res. Bd. Canada*, **12**, 387–446.

Silverman, M. J. 1972. Tragedy of Northport. *Underwater Naturalist*, **7**(2), 15–18.

Sprague, J. B., P. F. Elson, and R. L. Saunders. 1965. Sublethal copper-zinc pollution in a salmon river: A field and laboratory study. *Air Water Pollut.*, **9**(9), 531–543.

Sun, M. 1985. Legislative paralysis on the environment. *Science*, **227**, 496–497.

Trama, F. B. 1954. The acute toxicity of copper to the common bluegill (*Lepomis macrochirus* Rafinesque). *Acad. Nat. Sci. Phila.*, *Notulae Naturae*, **257**. 13 pp.

Warren, C. E. 1971. *Biology and Water Pollution*. W. B. Saunders. Philadelphia. 434 pp.

Weinstein, I. B. 1991. Mitogenesis is only one factor in carcinogenesis. *Science*, **251**, 387–388.

Weir, P. A., and C. H. Hine. 1970. Effects of various metals on behavior of conditioned goldfish. *Arch. Environ. Health*, **20**, 45–51.

9

INDUSTRIAL POLLUTION

Wastewaters discharged by industrial operations are in some cases among the worst sources of water pollution. Although the nature of the pollutants associated with these wastewaters differs greatly from one industry to another, in almost all cases the problems are caused by one or a combination of the following conditions in the wastewater.

1. high BOD
2. high concentration of suspended solids
3. presence of toxic substances

Indeed the BOD and SS in some industrial wastewaters may be a factor of 10 or more higher than the BOD and SS in raw sewage. It is not hard to imagine that severe oxygen depletion and/or turbidity and sedimentation problems could result from the discharge of such wastewater. The high concentration of toxicants in some industrial effluents can be lethal or at least stressful to organisms in the receiving waters, and of course may be concentrated by fish or shellfish to such an extent that they are unfit for human consumption. Efforts to reduce or eliminate the toxicant problem have tended to focus on three general types of solutions. The first is to eliminate the need for the toxicant in the manufacturing process. The second is to recover and recycle the toxicant before the wastewaters are released. The third is to remove the toxicant from the wastewater and dispose of it in an environmentally acceptable manner.

This chapter begins with a general examination of the problems caused by the release of BOD, SS, and toxic substances with industrial wastewaters, and the nature of corrective measures taken to alleviate these problems. This general discussion lays the groundwork for a more detailed examination of the wastewater problems associated with specific industries. The case studies that follow provide good examples of the types of problems that are caused by industrial wastewaters and illustrate rather well some of the approaches that have been tried, in all cases not successfully, to alleviate those problems. The remainder of the chapter is devoted to an in-depth study of two industries with significant wastewater problems, the Hawaiian sugar cane industry and the pulp and paper industry.

THE OXYGEN SAG

The wastewater from industrial operations such as pulp mills, sugar refineries, and some food processing plants may easily have a BOD as high as several thousand parts per million or more. By contrast, raw sewage, with a typical BOD of only about 200 ppm, seems like the nectar of the gods. If industries discharged only small amounts of such wastewater, the effect of these enormous BODs on receiving waters would be small in many cases, since the effluent would be greatly diluted in the receiving system. Unfortunately just the opposite is too often the case. For example, a 500-ton/day sulfite process pulp mill can be expected to produce about 100 million liters per day of wastewater with a BOD of about 1000 ppm (McCubbin, 1983). The discharge of wastewater alone from such a plant is equal to the volume of sanitary sewage produced by a city of about 250,000 persons and the BOD loading from a city of about 1.3 million persons.

If such large amounts of BOD are discharged into a system that is not mixed vigorously, there is an excellent chance that the oxygen concentrations will drop below levels necessary to maintain the natural aquatic community. Water initially saturated with oxygen at 20°C, for example, and containing 1000 ppm BOD would become completely anoxic in about 1 hour if placed in a closed container in the dark. Obviously such wastewater must be greatly diluted or the receiving waters vigorously aerated to prevent serious oxygen depletion problems.

Figure 9.1 illustrates qualitatively how the dissolved oxygen concentration may vary with distance in a river downstream from an industry discharging wastewater with a high BOD. The oxidation of the organic substances in the wastewater by microbes initially consumes oxygen faster than the exchange of oxygen between the stream and atmosphere can resupply oxygen. The result is that the oxygen concentration in the water drops rapidly downstream from the wastewater outfall. Due to the metabolic activity of the microbes, the concentration of oxidizable organic wastes steadily decreases with increasing distance from the outfall. Since the respiration rate of the microbes is correlated positively with the concentration of their food supply, the rate of microbial oxygen consumption also steadily decreases with distance form the outfall. The net influx of oxygen from the atmosphere to the stream is proportional to the difference between the oxygen concentration at 100% saturation and the actual concentration in the water. With increasing distance from the outfall, this net oxygen influx rate eventually exceeds the microbial respiration rate, and the oxygen concentration in the water begins to rise. At a sufficient distance downstream from the outfall, the oxygen concentration is back to virtually 100% saturation. The characteristic decline and subsequent rise in oxygen concentration downstream from an outfall discharging wastewater with a high BOD is usually referred to as an *oxygen sag*. The phenomenon was first treated theoretically by Streeter and Phelps (1925), who developed an analytical expression to describe the characteristics of the oxygen sag curve.

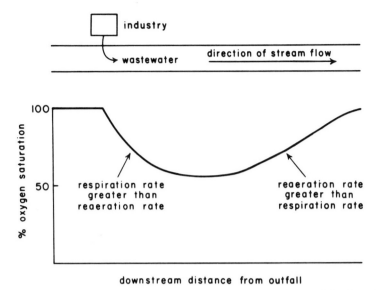

Figure 9.1   Qualitative variation of dissolved oxygen concentration resulting from BOD discharge into a stream. A characteristic oxygen sag curve.

Of particular importance are two variations of the oxygen sag curve illustrated in Figure 9.2. If the BOD loading from an outfall is sufficiently great, the oxygen concentration in a stream may drop to virtually zero and remain at that level for a considerable distance downstream from the outfall. In extreme cases the zone of anoxic water may extend many kilometers downstream before the combined effects of reaeration and declining respiration rates allow the oxygen concentration to increase again (Livingston, 1975). Such anoxic zones result in virtually a complete displacement of the stream's natural fauna. The second variation concerns the case in which two or more industries are discharging high BOD wastewater into a stream at locations along the stream sufficiently close to one another that the oxygen sag produced by one industry's wastes is still apparent at the point downstream where the next industry is discharging wastewater. As a result, the combined effect of two or more oxygen sags may have a much more deleterious effect on water quality than would any one of the contributing oxygen sags taken by itself. Obviously it is possible to reduce oxygen concentrations to virtually zero by a series of several relatively small but closely spaced BOD discharges along the course of a stream.

What can be done to solve the oxygen sag problem? Obviously one solution is to oxidize the organic matter in the wastewater before the water is discharged. This approach is of course the one adopted in conventional sewage treatment plants. If suspended solids are also a problem, they can also be removed with conventional treatment. The problem with using nothing more than conventional wastewater treatment methods to remove the BOD and SS from industrial wastewaters is twofold. First, conventional treatment may do a poor job of reducing the concentration of toxicants in the wastewater. Second, the concentrations of BOD and SS are sometimes so high that the cost of treatment becomes a serious obstacle. A more

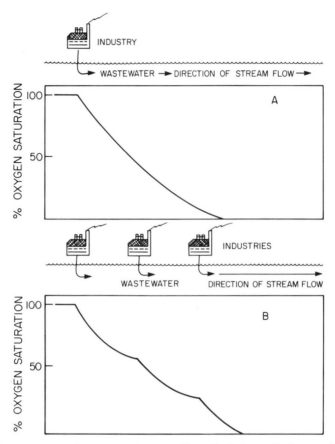

Figure 9.2 Oxygen sag leading to anoxia caused by a single large discharge of BOD (A) or a series of smaller but closely spaced BOD discharges (B).

practical alternative may be to reduce the amounts of toxic substances, BOD, and SS released from the industrial process in the first place. This objective may be accomplished by altering the industrial process or by recycling. These latter options may be not only more cost effective but also more environmentally appealing both in terms of resource conservation and pollution.

INNOVATIVE STRATEGIES FOR REDUCING INDUSTRIAL POLLUTION

How much does it cost to dispose properly of industrial wastes? This question is rather difficult to answer, in part because many industries have not disposed properly of their wastes in the past. In the United States about 95% of industrial hazardous waste is disposed of on the site where it is generated (Postel, 1987), and in a disturbing number of cases it would appear that the principal concern has been to simply get rid of the waste, without much concern over potential environmental impacts (see Chapter 16). Furthermore, in cases where environmentally acceptable disposal practices are followed, the costs do not always reflect manufacturing strategies designed to minimize waste production. Waste management costs at DuPont, the largest chemical producer in the United States, currently exceed $100 million per year (Postel, 1987). Companies like DuPont, realizing the high cost of waste disposal, have naturally begun to look for ways to minimize waste production. There now seems to be a general consensus that this approach is the real key to solving the industrial waste problem.

Table 9.1 summarizes a few of the innovative techniques that have been used by companies to minimize waste production. In the United States, the Minnesota Mining and Manufacturing Company (3M) has probably had the longest commitment to waste reduction of any major corporation. Its "Pollution Prevention Pays" program, initiated in 1975, has reportedly reduced waste production by a factor of 2 and saved 3M over $300 million. In some cases one of the most effective ways to reduce pollution is to recycle secondary materials rather than manufacturing products from virgin resources. Table 9.2 summarizes the results of recycling in the production of aluminum, steel, paper, and glass. Not only does recycling of manufactured products often substantially reduce water pollution, it may also greatly reduce energy costs, water use, and air pollution. The impact of such recycling

Table 9.1 Successful Innovative Techniques for Minimizing Industrial Waste

Company/location	Products	Strategy and Effect

Astra
    Sodertalje,
    Sweden
	Pharmaceuticals	Improved in-plant recycling and
		    substitution of water for solvents cut toxic
		    wastes by half

Borden Chemical
    California,
    US
	Resins;
	adhesives
		Altered rinsing and other operating
		    procedures cut organic chemicals in
		    wastewater by 93%. Sludge disposal costs
		    reduced by $49,000/yr

Cleo Wrap
    Tennessee,
    US
	Gift wrapping
	paper
		Substitution of water-based for solvent-
		    based ink virtually eliminated hazardous
		    waste, saving $35,000/yr

Duphar
    Amsterdam,
    The Netherlands
	Pesticides	New manufacturing process cut toxic waste
		    per unit of one chemical produced by
		    factor of 20

DuPont
    Barranquilla,
    Colombia
	Pesticides	New equipment to recover chemical used in
		    making a fungicide reclaims materials
		    valued at $50,000 annually; waste dis-
		    charges were cut 95%

DuPont
    Valencia,
    Venezuela
	Paint, finishes	New solvent recovery unit eliminated
		    disposal of solvent wastes, saving
		    $200,000/yr

3M
    Minnesota
    US
	Varied	Companywide, 12-year pollution prevention
		    effort has halved waste generation,
		    yielding total savings of $300 million.

Pioneer Metal
Finishing
    New Jersey
    US
	Electroplated
	metal
		New treatment system design cut water use
		    by 96% and sludge production by 20%;
		    annual net savings of $52,500; investment
		    paid back in three years

Source. Postel (1987).

Table 9.2 Environmental Benefits Derived from Substituting Recycled Materials for Virgin Resources

Percent reduction of:	Product			
	Aluminum	Steel	Paper	Glass
Energy use	90–97	47–74	23–74	4–32
Air pollution	95	85	74	20
Water pollution	97	76	35	—
Mining wastes	—	97	—	80
Water use	—	40	58	50

Source. Pollock (1987).

efforts on some industries has been substantial. For example, by 1983 the world's industrial market economies produced 30% of their raw steel in electric arc furnaces, which rely exclusively on scrap metal as a feedstock (Pollock, 1987). In the United States, more than half of the 300 billion aluminum cans sold since 1981 have been recycled, and the aluminum industry used 22% less energy to produce a kilogram of aluminum in 1984 than in 1972 (Pollock, 1987). Building a paper mill designed to use recycled paper rather than raw wood pulp is 50–80% cheaper and reduces the demand on valuable forest resources. According to Pollock (1987, p. 22), "Simply recovering the print run of a Sunday edition of the *New York Times* would leave 75,000 trees standing." In the United States approximately 200 paper mills now use recycled paper exclusively. These mills use less than half as much water to produce a given amount of paper product as do conventional pulp mills.

Although some progress towards reducing industrial waste has obviously been made, there is much room for improvement. Japan, which seems to have gone further than any major industrial country towards reducing and reusing industrial waste, recycled more than half its estimated 220 million tons of industrial waste in 1983 (Postel, 1987). In the United States, on the other hand, only 4% of hazardous waste was recycled in 1981. The U.S. Congressional Budget Office has estimated that 80% of waste solvents and 50% of the metals in industrial wastewater could be recovered (Postel, 1987). Such reductions would obviously do much to reduce the amount of toxicants released to the environment by some industries. As population pressures increase the demand for clean water, serious efforts to reduce the amount of water used by and polluted by industry will likely become more mandatory than optional. The following examples serve to illustrate this point.

THE HAWAIIAN SUGAR CANE INDUSTRY

The Hawaiian sugar cane industry was for many years one of the principal sources of income to the state of Hawaii. As a result of a reciprocity treaty signed between the United States and the Territory of Hawaii in 1876, the 2-cent-per-pound duty fee on raw sugar imported into the United States was waived on Hawaiian-produced raw sugar (HSPA, 1990). As a result of this competitive advantage over foreign sugar growers, the Hawaiian sugar cane industry expanded rapidly, and many land areas that otherwise would have been marginally profitable for cane growing were converted to sugar cane cultivation. Table 9.3 indicates the size of the Hawaiian sugar cane industry as measured by land area under cane cultivation and tons of raw sugar produced, and Table 9.4 shows its ranking relative to other sources of income to the state of Hawaii in 1988. The industry reached its peak in approximately 1968, when over 980 km^2 were planted in cane, and raw sugar production exceeded 1.1 million tonnes. Both the amount of land planted in cane and raw sugar production declined during the 1980s as a result of foreign competition and capture of the liquid sweetener market by high-fructose corn syrup. The roughly twofold increase in raw sugar production per unit area between 1908 and 1989 reflects primarily improvements in cane growing and harvesting and in sugar extraction techniques. Although the sugar cane industry in the state has declined somewhat in recent years, both the tourist industry and military operations have grown steadily throughout the twentieth century. As a result, the sugar cane industry now ranks a very poor third behind tourism and military operations in providing revenue to the state.

Until the early 1970s it was the custom of Hawaiian sugar cane companies to discharge their wastewater into the ocean. This wastewater consisted primarily of processing plant effluent, but also included excess irrigation water, called *tailwater*, and storm runoff from the cane fields. Although some companies treated their wastewater by means of settling basins or similar devices, other companies discharged their wastewater with no treatment whatsoever. During the cane harvesting season, roughly February through November, a cane processing mill would typically discharge about 20–40 million liters per day of wastewater into the ocean. As of 1966 there were a total of 26 sugar mills in the state, 6 on Kauai, 4 on Oahu, 4 on Maui, and 12 on Hawaii (Figure 9.3). Ultimately it became apparent to state officials that the discharges from the mills were creating conditions in the coastal ocean that were in direct conflict with the tourist promotional campaign, which described Hawaii's coastal waters as blue and sparkling clear. Because of the increasing importance of

Table 9.3 Area of Land Given Over to Sugar Cane Cultivation and Metric Tons of Raw Sugar Produced in 1908 and 1989

Year	Cane (km^2)	Raw sugar (tonnes)	Tonnes raw sugar/km^2
1908	816	495,085	607
1989	691	783,457	1,134

Source. Hawaii Sugar Planters Association (1990).

Table 9.4   Industries Ranked by Direct Income to the State of Hawaii in 1988

Industry	Income to the state of Hawaii
Tourism	$9.1 billion
Military	1.9 billion
Sugar cane	337 million
Diversified agriculture	257 million
Pineapple	247 million

Source. Hawaiian Sugar Planters Association (1990).

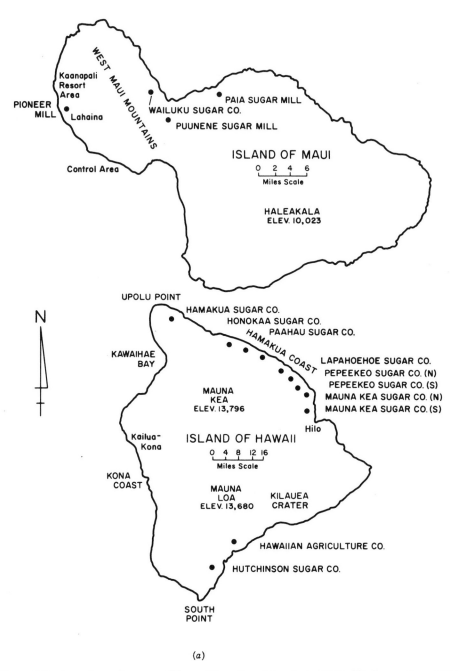

Figure 9.3 Location of sugar mills on Hawaiian Islands in 1966. [Redrawn from EPA (1971).]

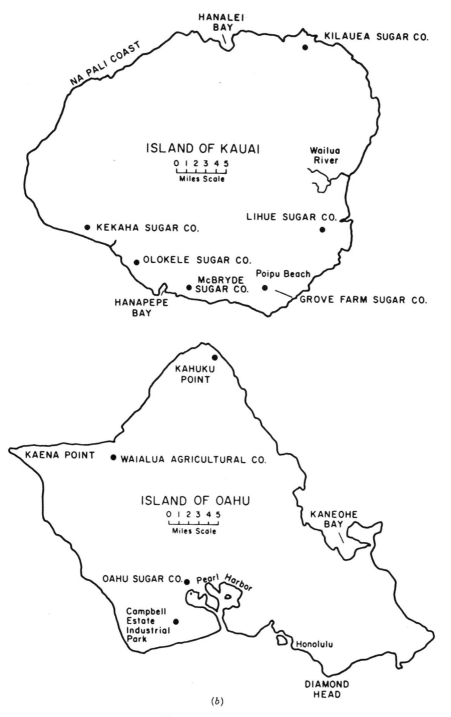

ISLAND OF KAUAI

0 1 2 3 4 5
Miles Scale

HANALEI BAY

NA PALI COAST

KILAUEA SUGAR CO.

Wailua River

LIHUE SUGAR CO.

KEKAHA SUGAR CO.

OLOKELE SUGAR CO.

McBRYDE SUGAR CO.

Poipu Beach

GROVE FARM SUGAR CO.

HANAPEPE BAY

ISLAND OF OAHU

0 1 2 3 4 5
Miles Scale

KAHUKU POINT

KAENA POINT

WAIALUA AGRICULTURAL CO.

KANEOHE BAY

OAHU SUGAR CO.

Pearl Harbor

Campbell Estate Industrial Park

Honolulu

DIAMOND HEAD

(b)

Figure 9.3   (Continued)

the tourist industry to the state, changes in the cane industry's waste disposal practices were needed. In June, 1963 a presidential board visited the state to survey the extent of the pollution problem. As a result of the recommendations of this board and the request of the Hawaii Department of Health, an in-depth study of the Hawaiian sugar cane industry wastewater problem was initiated by the Federal Water Pollution Control Administration (FWPCA).[1] The study, ultimately published by the Environmental Protection Agency (EPA), lasted from November, 1966 to September, 1968. We will henceforth refer to this study as the *EPA study*.

As a part of the EPA study, three sugar mills in the state were singled out for intensive study. These mills were the McBryde Sugar Co. mill on Kauai, the Pioneer Co. mill on Maui, and the Honokaa Sugar Co. mill on Hawaii. Of the three mills, the McBryde and Pioneer mills both subjected their wastewater to primary and secondary clarification before releasing it. The Honokaa wastewater received no treatment prior to discharge.

Table 9.5 summarizes the characteristics of the wastewater from the three mills as reported by the EPA (1971). The numbers in Table 9.5 are simply mean values of the given parameters measured at the three mills. The McBryde and Pioneer mill values were taken from the analyses of the washwater prior to clarification, so that the numbers would be comparable to the Honokaa values, since Honokaa wastewater received no treatment.

The concentrations of total N and total P in the mill wastewater were both comparable to the values for raw sewage (compare Table 5.1). The SS, BOD, and COD were much higher in the mill wastewater than in raw sewage. In fact, the SS concentration in the mill wastewater was over 30 times the SS concentration in raw sewage. Given the typical mill wastewater discharge rate of 20–40 million liters per day and assuming a per capita raw sewage discharge of 400 l/day, it is concluded that the loading of SS in the untreated wastewater from a single sugar cane mill would be equivalent to the loading of SS in the raw sewage from a population of 1.5–3.0 million persons. This calculation provides some feeling for the magnitude of the water pollution problem created by discharges of untreated cane mill wastes. Fecal coliform counts in the mill wastewater were about a factor of 1000 lower than fecal coliform counts in raw sewage (e.g., Gundersen, 1973), but far in excess of levels considered safe for water contact. Although the fecal coliform counts may in part have reflected the presence of nonfecal thermotolerant biotypes of *Klebsiella* (Chapter 7), the high counts raised the concern that the wastewater contained significant concentrations of fecal pathogens. What was the source of all the SS, BOD, nutrients, and fecal coliforms in the cane mill wastewater? To answer this question, you must understand something about the way sugar cane is grown, harvested, and processed to make raw sugar in Hawaii.

Sugar Cane Production: Field Operations

Fields in which cane is to be planted are first graded to control runoff and then deep-plowed to a depth of as much as 0.5 m. Short sections of freshly cut 8–10-month-old cane stalks are then planted in the furrows, given an initial dose of fertilizer, and covered with dirt. The planting is done by a mechanical planter rather

---

[1] The FWPCA became the Federal Water Quality Administration (FWQA) on April 30, 1970. The FWQA was absorbed by the Environmental Protection Agency (EPA) on December 2, 1970.

than manually. The cane is fertilized heavily during the first year of growth with nitrogen, phosphorus, and potassium. During the second year no fertilizer is applied, so that the cane will store sugar rather than produce additional plant foliage. In Hawaii, herbicides are used to control weeds, with 3–4 applications per crop being typical treatment rates. The heaviest applications are made during the first 6 months of growth.

Prior to 1876 sugar cane in Hawaii was grown only in those areas where there was adequate rainfall to support cane growth. After passage of the 1876 Reciprocity Treaty, the industry expanded rapidly its plantings into areas where irrigation was required, and at the present time about 60% of Hawaiian sugar cane fields are irrigated. The principal sources of pollution associated with field operations were the tailwater and stormwater runoff from the fields. The average concentration of selected constituents in the tailwater from five different fields on the McBryde plantation are listed in Table 9.6. These values are considerably lower than the mean values for mill wastewater, but still high enough to cause pollution problems. The SS concentration, for example, is over 4 times the typical SS concentration in raw sewage, and the fecal coliform count is still well above values considered safe for water contact.

Table 9.5   Mean Concentrations of Selected Constituents of the Wastewater from the McBryde, Pioneer, and Honokaa Mills and Typical Raw Sewage

Constituent	Mill wastewater[a]	Raw sewage
Fecal coliforms	53,000/100 ml	$\sim 10^6$/100 ml
Suspended solids	6900 ppm	200 ppm
Settleable solids	6600 ppm	—
BOD	755 ppm	200 ppm
COD	2033 ppm	350 ppm
Total N	39 ppm	40 ppm
Total P	15 ppm	10 ppm

[a] The values reported for the three mills have simply been averaged. In the cases of McBryde and Pioneer effluents, the values averaged pertain to the washwater prior to clarification.

Source. EPA (1971) and Weibel (1969).

Table 9.6   Mean Concentrations of Selected Constituents in Tailwater from the McBryde Plantation

Constituent	Concentration
Fecal coliforms	1934/100 ml
Suspended Solids	843 ppm
Settleable Solids	576 ppm
COD	91 ppm
Total N	10 ppm
Total P	5.5 ppm

Source. EPA (1971).

Sugar Cane Production: Harvesting

Prior to World War II, sugar cane in Hawaii was cut by hand and transported to the mill by flumes. After World War II, however, the cost of labor increased to such an extent that mechanical harvesting became more economical than manual harvesting. Unfortunately, conventional cane cutters proved generally unsatisfactory in Hawaii, due to the thickness of the cane growth, the rocky soil, and the furrowed and sometimes hilly terrain. A modified mechanical harvesting procedure has therefore been adopted. Just prior to harvesting, the fields are burned to remove excess foliage. The succulent cane stalk is virtually unaffected by this burning, but much of the relatively dry leafy material is removed. A conventional tractor with a modified push rake in front then snaps off the cane stalks near ground level. The root structure is left intact to allow growth of a second, third, or even fourth crop from shoots that grow from the old roots. The snapped-off stalks of cane are raked into windrows by bulldozers, and the cane is then lifted into 40-ton-load cane haulers by cranes equipped with special grabs. The cane haulers then transport the cane to the mill. This harvesting procedure accounts for many of the undesirable characteristics of cane mill wastewater, because rocks and dirt as well as fecal material[2] are invariably loaded onto the cane haulers along with the cane. This material must of course be removed at the mill.

Sugar Cane Production: Factory Operations

A flow diagram of factory operations in a typical sugar cane mill is shown in Figure 9.4. The sugar cane factory can be divided into basically three sectors, the cane preparation plant, the milling plant, and the boiling plant. The cane is brought from the field to the preparation plant, where the first step in processing consists of removal of the dirt and rocks. The cane is loaded onto a steeply inclined conveyer belt fitted with widely spaced prongs that hold the cane on the belt as it moves upward but hopefully allow rocks to roll back down to the bottom, where they are collected in a hopper. The rocks are trucked off to a landfill. Once over the top of the inclined plane, the cane travels through a floatation bath. The cane floats on the surface of the water while rocks sink to the bottom and some dirt is washed from the cane. Following the floatation bath, the cane is sprayed with jets of water to remove additional dirt. The cane stalks next pass over a series of oppositely spinning rollers that strip off any excess foliage. Finally a series of revolving knives cut the cane stalks into short segments in preparation for milling.

Wastewater from the cane preparation plant contains high concentrations of SS and fecal coliforms from the dirt washed from the cane. The preparation plant wastewater also contains the cane trash, which consists primarily of excess foliage and broken stalks.

In the milling plant, high pressure rollers, usually operated as a tandem of 3 per mill, squeeze the juice from the stalk segments. Usually a series of 3–5 mills are used to extract the juice. The stalk tissue that remains after this milling has the appearance of coarse sawdust and is called *bagasse*. In the past at least some of this bagasse was discharged with the mill wastewater into the ocean. At the present time, however, all the bagasse is burned as fuel in the plant, and on the island of Hawaii

---

[2]The fields are inhabited by cane rats, mongooses, and other animals.

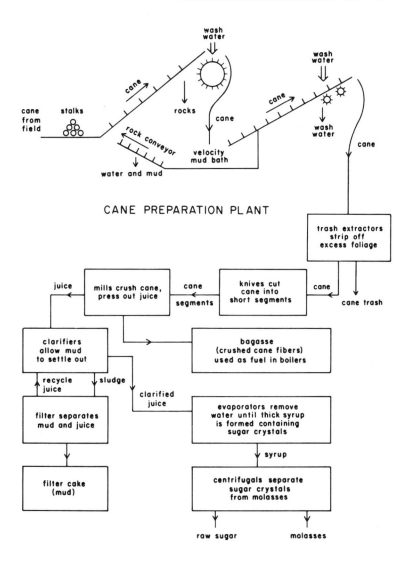

CANE PREPARATION PLANT

MILLING AND BOILING PLANT

Figure 9.4   Flow diagram of a typical sugar cane mill in Hawaii.

excess energy is sold to the local electric utility. Drying and pelletizing the bagasse further increase its energy yield; therefore, the bagasse has become an asset rather than a liability.

In the boiling plant, lime is added to the cane juice to help prevent fermentation and to aid in clarification. The juice is then pumped into a clarifier similar in design to the secondary clarifier in a sewage treatment plant. The supernatant juice taken from the top of the clarifier is pumped directly to the evaporators, while the muddy sludge from the bottom is pumped to a cylindrical filtration unit. Normally the juice that passes through the filtration unit is recycled to the clarifier. The sludge that is separated out is continuously scraped off the rotating filter and is referred to as *filter cake.*

In the evaporators the water in the clarified juice is gradually evaporated off until a thick syrup consisting of about 50% sugar remains. This thick syrup is then pumped to the vacuum pans, where a final evaporation step results in the precipitation of sugar crystals from the syrup. The mixture of sugar crystals and syrup is then centrifuged in a rotary filter. The crystals of sugar adhere to the inside of the spinning filter, while the liquid syrup passes through. The raw sugar is then scraped from the inside of the centrifuge. The syrup collected from the first vacuum pass is called *A molasses*. This syrup still has a high sugar content and is evaporated down further in a second vacuum pan to yield additional raw sugar. The syrup from the second centrifuge step is called *B molasses*, and has a rather low sugar content. Usually no attempt is made to obtain sugar from the B molasses. The chief waste products from the boiling mill are the filter cake from the clarification step and the heated wastewater from the boilers.

Survey of Water Pollution Problems

The 1966–1968 EPA study revealed that a variety of pollution problems were created when wastewater from cane factory operations was discharged to the ocean. Plumes of brown water discolored by sediment from the wastewater were visible along the coastline for as much as 3 km or more from the outfall. Secchi depths taken near the outfalls were sometimes no more than 5–10 cm. Total coliform counts exceeded 1000/100 ml for distances up to 3–5 km from mill outfalls. In some cases surfers were noted in the areas of high total coliform counts.

Mills that discharged cane trash and bagasse created an additional water pollution problem. The trash and bagasse tended to float on the surface in mats as much as 1.5 km long and 50 m wide. These mats were unsightly, created a hazard to navigation, interfered with fishing, and given an onshore wind would wash up along the shoreline, creating an unsightly mess. If the cane trash and bagasse did not wash up on the shore, it ultimately sank to the bottom.

Benthic surveys revealed that the accumulation of sediment and sludge from the mill discharges had killed or severely disrupted the natural coral reef communities up to as much as 1.5 km from the outfalls. Off the Honokaa mill outfall, sludge deposits as much as 3 m deep were found.[3] Killing of corals was due presumably to direct smothering by sludge and/or the high turbidity of the water near the outfalls. As was the case in Kaneohe Bay (Chapter 4), sponges and benthic algae had colonized many areas of dead coral.

Fish populations were reduced both in numbers and diversity near the outfalls. This change in the fish populations was believed to be caused in part by the turbidity of the water, since most of the local species use visual means for locating prey, and in part by the demise of the natural coral communities that normally would have provided food and shelter for the fish. Toxicity bioassays conducted with four local fish species revealed that Tilapia, Mullet, and Aholehole (flagtail fish) could survive in almost 100% mill wastewater for 96 hours if the wastewater was adequately aerated. However, Nehu (an anchovy) had a 48-hr TLm of about 6–10% mill wastewater.

---

[3] Recall that the Honokaa wastewater received no treatment.

Because of the wide variety of fish on a typical undisturbed coral reef, it seems safe to say that a more thorough study of acute toxicity effects on fish would have been required to determine whether water quality near the outfalls might have been lethal for a significant portion of the natural fish population.

Despite the high BOD and COD of the mill wastewater, oxygen levels in the ocean near the outfalls were virtually identical to oxygen concentrations in control areas. In both control areas and discharge areas, bottom water oxygen levels were typically 80–90% of saturation, whereas oxygen levels in the remainder of the water column were usually 90% of saturation or higher. These results indicate the effectiveness of ocean mixing processes in dispersing the effluent and in promoting oxygen exchange with the atmosphere. It is likely that cane mill effluent discharged into a confined body of water would create serious oxygen depletion problems (Officer and Ryther, 1977), but in Hawaii no mill wastewater discharges were made to estuaries with restricted circulation patterns. However, tailwater and stormwater runoff to the west loch of Pearl Harbor (Figure 9.3) may have created oxygen depletion problems in that system. No studies were made in Pearl Harbor as part of the EPA survey.

Response to the EPA Survey

The EPA study revealed that cane mill wastewater discharges to the ocean were creating conditions that violated eight different Hawaii state water quality guidelines. These guidelines pertained to the following conditions.

1. Objectionable sludge or bottom deposits (sediment, cane trash, bagasse);
2. Floating debris (cane trash, bagasse);
3. Objectionable color or turbidity (sediment);
4. Pathogens (suggested by presence of fecal coliforms);
5. Total nutrient concentrations (total N and total P in effluent).

To eliminate water pollution problems associated with cane mill operations, the EPA recommended that the following three policies should be followed:

1. No cane trash or bagasse should be discharged by any mills in the state;
2. Runoff of tailwater and stormwater should be minimized through the use of ponds and embankments for trapping water and through storage and reuse of irrigation water;
3. All wastewater discharged to the ocean should be treated to minimize the discharge of suspended solids.

Relevant to the second of these recommendations was a study of field runoff from the Waialua Sugar Company's fields on Oahu, which revealed that during most storms there was virtually no runoff from the fields (EPA, 1971). The fields were graded so that runoff was directed to ponds or against embankments at the end of the furrows. The fields could retain all the runoff from a 5-cm rain, and in one rain of 15 cm, only 1 cm of water ran off the fields. In fact, the Waialua cane fields retained significantly more rainfall than a control area of undeveloped land. Figure 9.5 shows the cumulative runoff from the Waialua cane fields and the control area during the 15-cm rain on February 1, 1969. Despite the much smaller total runoff from the cane fields, the higher concentration of suspended solids in the cane runoff resulted in a total discharge of SS from the cane fields that was little different than the SS discharged from the control area. The results do indicate, however, that the use of ponds and embankments for retaining field runoff can at least reduce runoff to the point where soil erosion from the fields is no worse than soil erosion from undeveloped land.

Hawaii followed through on the first two recommendations of the EPA panel. No cane mills in the state are allowed to discharge cane trash or bagasse, and all plantations must minimize field runoff through efforts similar to those practiced at Waialua. With respect to the third recommendation, the state was more stringent than the EPA panel. No sugar cane mills on the islands of Kauai, Oahu, and Maui are allowed to discharge any mill wastewater. On the island of Hawaii, an exemption to this restriction was granted to mills along the Hamakua coast, where pumping wastewater back to the fields was regarded as impractical because of the steep terrain and the fact that irrigation requirements are minimal or nonexistent due to the abundant natural rainfall, which averages 180–380 cm/yr along the Hamakua coast (Grigg, 1985). Wastewater from Hamakua coast cane mills is, however, treated with the use of hydroseparators and settling ponds. This treatment process reduced the discharge of sediment by roughly a factor of 30 between 1978 and 1982 (Grigg, 1985). The procedure is effective in removing much of the soil from the wastewater because of the high percentage of SS that is settleable (Table 9.5). On the other hand, the EPA study revealed that fecal coliform counts were not reduced to satisfactory levels through settling, and at the Pioneer mill fecal coliform counts actually increased by a factor of 10 between the mill washwater and the final settling pond effluent. This observation undoubtedly influenced the state's decision to require no wastewater discharges at most mills.

Present Status of the Industry

The response of Hawaii and the Hawaiian sugar cane industry to the problems caused by wastewater discharges associated with the industry illustrate rather well the nature of solutions to such problems and, to some extent, the practical limitations associated with implementing those solutions. The complete ban on discharges at most mills has led to a policy of recycling and reuse which has been beneficial to the industry. The bagasse, which was previously discharged to the ocean, is now burned as a fuel. The use of bagasse as a fuel reduces significantly the energy costs of the industry. Because Hamakua coast plantations do not pump water for irrigation purposes, their energy costs are substantially less than plantations that irrigate. As a result, they are able to generate and sell electricity to the Hawaiian Electric Company. These sales further reduce their operating costs. Electricity produced by burning bagasse presently accounts for about 10% of the state's electricity consumption.

Discharging topsoil and nutrients into the ocean is an obvious waste of a valuable resource. Along the Hamakua coast alone, the loss of topsoil amounted to 200,000–225,000 tonnes per year just prior to the implementation of measures to

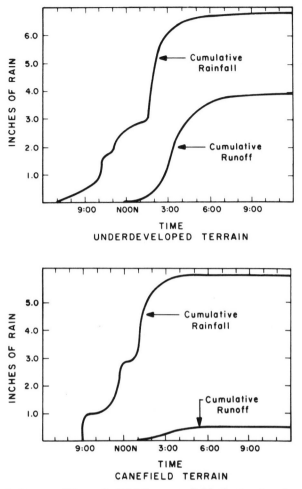

Figure 9.5 Cumulative runoff from the Waialua cane fields (bottom) and control area (top) versus cumulative rainfall for the storm of February 1, 1969. [Redrawn from EPA (1971).]

treat mill wastewater (Grigg, 1985). Recycling of mill wastewater and minimization of irrigation water runoff have largely eliminated such losses. The mill wastewater is now treated to separate to a certain extent the water from the sediment. The water is pumped back to the fields for irrigation. The topsoil is either returned to the fields, or in some cases applied to low elevation plots of land that would be otherwise unsuitable for cane cultivation. On Oahu, the Waialua Sugar Company has used the latter approach to expand its cane production acreage.

Irrigation with mill wastewater throughout the 2-year growth cycle of the cane reduces sugar yield, because the high nutrient levels in the wastewater cause the cane to store less sugar during the second year of growth.[4] Related to this problem have been the studies conducted by Lau et al. (1975) on the possibility of using

[4]Instead, more foliage is produced.

secondary sewage effluent to irrigate sugar cane in Hawaii. They found that irrigation with secondary sewage effluent throughout the 2-year cane growing period reduced sugar production by about 6% versus control fields that received standard fertilizer applications during the first year but no fertilizer during the second year. Sugar yields increased by about 6%, however, in fields irrigated with secondary sewage during the first year of growth and with groundwater or surface water during the second year of growth. Since the total N and total P concentrations in mill wastewater and raw sewage are comparable, it seems likely that a similar increase in sugar production might be achieved on fields irrigated with mill wastewater if the wastewater were applied during only the first year of growth. However, this approach would obviously require that about twice as much cane be irrigated with wastewater, and the additional cost of pumping the wastewater to higher elevation fields might more than offset the net revenue associated with the gain in sugar yield.

The industry experienced difficult times during the period from 1975 to 1985 due to the rapid expansion in the use of high-fructose corn syrup in the liquid sweetener market. The share of the U.S. sweetener market controlled by high-fructose corn syrup, however, has been stable since 1985. This competitive pressure forced an overall improvement in the economic efficiency of Hawaii's sugar cane industry, the real cost of production declining by about 25% from 1981 to 1986 (HSPA, 1990). One change of particular relevance to the issue of water pollution was a major conversion to drip rather than furrow irrigation on irrigated cane fields. Drip irrigation is now used on about 85% of irrigated fields. This change in methodology leads to much more efficient use of irrigation water and eliminates tailwater runoff.

Hawaiian sugar yields are among the highest in the world, about 13 tonnes per hectare per year. This figure compares to annual yields of 7.5, 6.4, and 4.2 tonnes ha^{-1} for raw sugar produced from sugar cane in Florida, Louisiana, and Texas, respectively. The profitability of the Hawaiian industry, however, is reduced by the high cost of labor in Hawaii. According to the HSPA (1990, p. 2), "Hawaii's cane field workers have the highest standard of living of any agricultural workers in the world, with daily earnings (including benefits) averaging $116.56 in 1989." High labor costs compromise the industry's competitive position with producers in countries such as India, Brazil, and Cuba, which are major sugar producers and where labor costs are much lower.

The profitability of the industry became a factor in the issue of water pollution during the 1980s, when an effort was made by the Hamakua coast mills to obtain a relaxation of the standards applied to their wastewater discharges. The Hamakua coast mills lost collectively about $40,000 in 1981 and 1982 from sugar production (Grigg, 1985), and although other sources of income including revenues from the sale of molasses and electricity largely offset this loss, it was argued that without some economic relief the plantations might be forced to close. Studies conducted by Grigg (1985) indicated that the approximately 30-fold reduction in suspended solids discharged by the mills had in fact produced only slight changes in the zones of measurable impact near the mill outfalls. The areas of coral cover, for example, had increased by only 4–8%. The EPA (1989), however, noted that the costs of wastewater treatment amounted to only 2–3% of operating costs for Hamakua coast mills, and that the tenuous financial position of the mills[5] would not be altered substantially

---

[5]In 1988 the mills showed a profit of $1.6 million, which was 2.5% of their operating costs.

by reducing the level of wastewater treatment. The EPA further noted that turbidity and concentrations of nitrate plus nitrite, copper, mercury, lead, arsenic, and manganese exceeded Hawaii water quality standards and EPA criteria within designated zones of mixing around one or another of the mill outfalls. The levels of copper and mercury exceeded EPA criterion maximum concentration limits. Based on these observations and other considerations, the EPA (1989) decided not to approve a reduction in the level of wastewater treatment for Hamakua coast mills.

THE PULP AND PAPER INDUSTRY

The pulp and paper industry in Canada is the largest single industrial employer (McCubbin, 1983) and in the United States is the largest discharger of conventional pollutants subject to national effluent standards (GAO, 1987). Annual pulp and paper production in the United States amounts to about 80 million tonnes, with a market value of $25–30 billion. Canadian production is about 25% of the U.S. figure. The industry is a major contributor to the U.S. economy, with major plants in 30 states, primarily in the southeast and western sections of the country. In Canada, pulp, paper, and forest products account for the largest dollar value of foreign exports, exceeding the value of such other exports as minerals, petroleum, and agricultural products (McCubbin, 1983). Obviously the pulp and paper industry is a major industry in North America.

As a general statement it is fair to say that the pulp and paper industry uses a great deal of water. The amount of water used varies greatly, however, depending on the pulping procedure and the desired characteristics of the pulp. Water requirements may be as low as $10 \text{ m}^3 \text{ tonne}^{-1}$ for dark, unbleached products such as cardboard and as high as $300 \text{ m}^3 \text{ tonne}^{-1}$ for high quality stationary paper (McCubbin, 1983). A water requirement of about 100 m^3 per tonne of product is probably a good working average for the industry as a whole. With an annual output of roughly 100 million tonnes, the United States and Canadian pulp and paper industries combined discharge about 10 km^3 of wastewater per year, which is about twice the annual flow of the Colorado River. The discharge of BOD associated with this wastewater is equal to about 75% of the BOD associated with the untreated sanitary sewage produced by the entire population of the United States and Canada. Obviously the task of adequately treating this wastewater before releasing it to the environment is a major undertaking. Since the characteristics of pulp and paper mill effluent depend very much on the procedure used to make the pulp and the nature of the paper products, it is appropriate here to examine in some detail how pulp and paper are produced.

Steps in the Production of Paper

A simplified diagram of typical pulp and paper processes is shown in Figure 9.6. There are basically four steps to the overall procedure. First, the bark is separated from the rest of the wood in the so-called woodroom. This debarking may be accomplished with the use of rotating drums, which use friction to remove the bark; hydraulic jets; or with mechanical debarkers, which use rotating knives to strip off the bark. The bark may account for as much as 15% of the weight of the raw wood. In the past the bark was regarded as waste, and its disposal was a major problem. In virtually all modern mills, however, the bark is utilized as a source of fuel in special

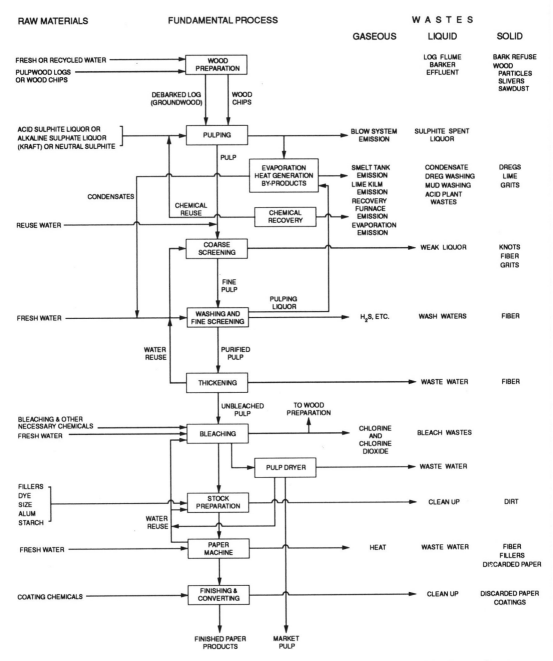

Figure 9.6 Simplified diagram of typical pulp and paper processes [Redrawn from McCubbin (1983).]

bark-burning boilers. The costs associated with preparing the bark and burning it in this manner are more than offset by the elimination of bark disposal costs and the reduced requirements of the mill for fossil fuels to provide energy. Stone ground-wood mills use the debarked logs directly to make pulp, but all other pulping procedures require that the logs be cut into small chips, generally 12–18 mm in size (McCubbin, 1983). The debarking and chipping operations may of course be bypassed to the extent that the mill receives wood chips from an external source.

The next step is the production of the pulp itself. This process consists of breaking down partially the cellulose and lignin that make up the cell walls of the wood tissue to produce a soft, spongy mass of pulp. Pulping may be accomplished by either mechanical or chemical means, or some combination of the two.

Once the pulp has been produced, it may be treated in a variety of ways to give the paper particular qualities. In many cases the pulp is bleached by chlorination. Most of the color in the pulp is due to the lignin in the wood, and the chlorine combines with the lignin to form chlorinated lignin compounds, which can be removed by dissolution in sodium hydroxide. Reclaimed paper or rag stock may be added to give the pulp a desired texture. Fillers such as clay, talc, and diatomite are usually added to give the paper increased opacity, brightness, and smoothness. In many cases the pulp is sized to provide wet strength, improved printability, and resistance of the paper surface to damage due to the pull of the ink in printing. Dyes may of course be added to impart color.

Finally the pulp goes to the paper machines, where it is formed into paper and dried. Although there are several different types of paper machines, all consist of a wet-end or forming section, a press section, and a dryer section. The pulp is formed into paper on a wire or plastic mesh. The press section removes much of the moisture from the paper and gives it a smoother surface. The dryer section removes remaining moisture and prepares the paper for reeling.

This general procedure is followed in virtually all pulp and paper mills. The chief factor that determines the quantity and characteristics of the mill's wastewater is the method used to produce the pulp. When the debarked wood enters the pulping department, it consists primarily of cellulosic fiber (45%), lignin (20–30%), and hemicellulose[6] (25%). Chemical pulping removes most of the lignin and hemicellulose; mechanical pulping allows much of the nonfibrous material to remain in the pulp.

Mechanical pulping is used mainly to produce pulp for newsprint. The earliest form of mechanical pulping was stone groundwood, a process in which logs were forced into contact with a revolving grindstone in the presence of water. The wastewater from this process contains a relatively small amount of dissolved organic substances leached from the wood and a great many small wood fibers. Until the mid-1970s newsprint was made from a mixture of about 75% groundwood pulp and 25%-chemical (sulfite) pulp. In recent years, however, an increasing percentage of mechanical pulp has been produced with refiners, which use rotating discs to break down wood chips into pulp. The pulp produced with refiners is stronger than groundwood pulp and superior for most purposes (McCubbin, 1983). A variation on this theme is the production of thermomechanical pulp (TMP), in which the

---

[6]Hemicellulose is a portion of the wood fiber consisting of sugarlike substances intimately associated with cellulose in the fiber wall.

wood chips are softened by steam under pressure and then defibered in a disc refiner. With the use of TMP, it is possible to reduce or even eliminate the need for chemical pulp in the production of newsprint. Both stone groundwood and TMP convert over 90% of the wood into pulp.

There are basically two ways of producing pulp by chemical means, the sulfite process and the sulfate or kraft process. Both methods consist of cooking the wood chips in a pressure cooker to which various chemicals have been added to help separate the wood fibers and break down the lignin. In most cases about 95% of the lignin is removed in chemical pulping, and only about 35–60% of the material in the wood chips is converted to pulp.

In the sulfite process sulfurous acid (H_2SO_3) is added along with calcium, magnesium, or some other base to cook the wood. The cooking solution is acidic. Lignin compounds are removed by sulphonation and hydrolysis reactions that form soluble lignosulphonates, and these account for roughly half the dissolved organic by-products in the cooking solution. Sugars derived from hemicellulose and cellulose, and sugarsulphonic and aldonic acids derived from the bisulphite substitution and oxidation of sugars account for most of the remaining dissolved organics in the spent sulfite liquor. While it is technically possible to recycle and burn much of the organic matter in the sulfite liquor, historically most sulfite mills have made no effort to recover these chemicals. The corrosive nature of the liquor and the lack of an economic incentive have been responsible primarily for this attitude. The dissolved organics leached from the wood chips and produced in the cooking process account for much of the BOD and toxicity of sulfite mill effluent. Typically, the BOD of sulfite mill effluent is about 1000 ppm if no effort is made to recover the organics produced in the cooking process (McCubbin, 1983).

In the kraft or sulfate pulping process, wood chips are cooked in a basic medium containing sodium hydroxide (NaOH) and sodium sulfide (Na_2S). The pulp produced by the kraft process has high strength properties relative to sulfite pulp, and carbohydrates that tend to degrade in the acidic sulfite medium are largely stable in the kraft process. Kraft mills are able to handle a greater variety of wood types than sulfite mills, because extracts of certain wood species that lead to problems with pitch in acid sulfite pulping are dissolved or dispersed in the basic kraft liquor (McCubbin, 1983). Historically, the chief disadvantage with the kraft process was the fact that a relatively large amount of residual condensed lignin remains in the pulp. This residual lignin gives kraft pulp a dark color that is much harder to remove by bleaching than the color associated with the lignin in sulfite pulp. Until 1946, in fact, most chemical pulping was done with the sulfite process, because it was impossible to produce bleached kraft pulp of comparable whiteness without impairing the strength of the product. In that year, however, two Canadians worked out a bleaching process using chlorine dioxide, which allowed the bleaching of kraft pulp to a whiteness which was previously thought impossible without loss of strength. Since that time the kraft process has become the preferred chemical pulping method for most purposes because of the difficulty of recovering extracted organics from spent sulfite liquor and the limited availability of wood species that could be pulped efficiently by the sulfite process.[7]

---

[7]Species pulped usually by the sulfite process are spruce, balsam, western hemlock, birch, and poplar.

From the standpoint of water pollution, a key feature of the kraft process is that most of the organic matter that is not converted to pulp is burned and about 95% of the cooking chemicals are recycled. The procedure is to evaporate down the cooking liquor to about 65% solids and then to burn the concentrated liquor in a recovery furnace (McCubbin, 1983). Combustion of the organic matter provides heat for steam generation, and under the conditions in the furnace sodium sulfate is reduced to Na_2S. In subsequent steps NaOH and CaO are recovered from calcium hydroxide and sodium carbonate by way of the overall reaction

$$Ca(OH)_2 + Na_2CO_3 \rightarrow 2NaOH + CaO + CO_2$$

Chemical recovery is essential economically for the kraft process. Because most of the organics that are not converted to pulp are burned, the BOD of kraft mill effluent is only 25–35% that of sulfite mill effluent. Kraft mill effluent may contain small amounts of hydrogen sulfide (H_2S) and various mercaptans, such as methyl mercaptan (CH_3SH), but these highly toxic compounds are unstable in the presence of oxygen, and under oxidizing conditions kraft mill effluent is about 100 times less toxic than sulfite mill effluent.

A hybrid of the standard mechanical and chemical pulping methods is semichemical pulping, in which wood chips are treated chemically first in a digester followed by a mechanical defibrating stage in a disc refiner. About 60–80% of the material in the wood chips can be converted to pulp with this procedure. The two major semichemical pulping methods are the *neutral sulfite* and *kraft semichemical techniques*. In the former case the cooking medium is neutralized, usually with the addition of sodium carbonate.

Objectionable Characteristics of Pulp and Paper Mill Effluent

Until roughly the mid-1960s, there was little effort to treat pulp and paper mill wastewater in order to reduce its undesirable characteristics. Since that time, growing environmental concern has caused the industry to make a serious effort to lessen the adverse impacts of its wastewater discharges both through modifications of the production process and by the installation of wastewater treatment systems. In this section we briefly review the undesirable characteristics of pulp and paper mill wastewater and the environmental impacts associated with its release.

Suspended Solids

Table 9.7 lists typical characteristics of pulp and paper mill effluents. The suspended solids concentrations range from about half to 2.5 times the SS of raw sanitary sewage. These suspended solids may exert directly toxic effects on certain organisms. Sprague and McLeese (1968b), for example, report that pulp and paper mill suspended solids may plug the gills of fish, leading in extreme cases to suffocation. Lower SS concentrations can produce sublethal and chronic stresses by the same mechanism.

Perhaps the principal adverse impact of pulp and paper mill suspended solids on aquatic communities occurs, however, when the SS settle to the bottom of the receiving system. Virtually all the pulp and paper SS consist of small chips of wood that

Table 9.7   Typical Characteristics of Pulp and Paper Mill Effluents

Pulping method	Flow (m³/tonne)	Suspended solids (ppm)	BOD (ppm)
Unbleached kraft	30	500	350
Bleached kraft	150	300	250
Newsprint mill (inc. sulfite)	50	400	1000
Sulfite without recovery	200	100	950
Neutral sulfite semichemical	30	350	500

Source. McCubbin (1983).

were not converted to pulp. Unless the bottom is scoured by currents, these small wood chips may accumulate to such a degree that the natural benthic community is destroyed completely. The impact of this destruction extends beyond the benthos, since many water column organisms depend on the benthos for food and shelter, and fish often utilize the benthos as a spawning and nursery area. When the natural benthos is covered by pulp mill sludge, it ceases to play its usual role in the aquatic ecosystem. Until wastewater treatment systems were installed at pulp and paper mills, the discharge of SS with the mill wastewater often had devastating effects on the benthic community in the receiving waters. Waldichuk (1962), for example, described the environmental impact of a pulp and paper mill that had been discharging wastewater at the head of Cousins Inlet, British Columbia, since 1913. The effluent from the mill consisted of a mixture of wastewater from the production of sulfite and kraft pulp, newsprint, and specialty paper. The SS in the wastewater had settled to the bottom of the inlet, due in part to the flocculating effect of saltwater on the solids. The decomposition of this precipitated sludge had produced anoxic conditions both in the sediments and in the bottom waters of the inlet. No living animals were found in the area of sludge deposits, and the odor of H_2S was apparent when a sediment sample was brought to the surface. Periodic dredging of this accumulated sludge to keep channels clear stirred the toxic H_2S into the water column. At least one fish kill was attributed to such dredging effects (Hourston and Herlinveaux, 1957).

Dissolved Organics
The dissolved organics in pulp and paper mill effluent create several types of problems. First, due to the presence of wood sugars and other readily metabolized organic substrates, the dissolved organics contribute a substantial amount of BOD. In fact, dissolved organics account for about 90% of a typical pulp and paper mill effluent's BOD, and this BOD can be as much as 5 times the BOD of raw sewage (Table 9.7). Hence there is considerable potential for anoxia to develop if the receiving waters are not well mixed.

Second, if the effluent is discharged into freshwater, objectionable growths of slime-forming fungi of the genus *Sphaerotilus* may develop. These slime fungi have been reported frequently downstream of sulfite pulp mills (Waldichuk, 1962). Overgrowth of fish eggs by this slime may reduce egg survival by as much as several orders of magnitude, apparently because the *Sphaerotilus* slime inhibits the exchange of gases between the egg and the surrounding water.

Some of the dissolved organics in pulp and paper mill effluent have surface-active properties that may lead to foaming in the receiving system, particularly if the water is turbulent. Although foaming is generally suppressed in saltwater, high concentrations of either kraft or sulfite effluent may produce considerable foaming even in seawater (Waldichuk, 1962). Foaming is of relatively little importance from a toxicological standpoint, but there is no doubt that water covered with foam is unappealing aesthetically. Surface-active agents may also retard the exchange of oxygen between the water and air, and in this way compound the oxygen depletion problem caused by the high BOD of the effluent. From the standpoint of wastewater treatment, one interesting benefit from the foaming caused by surface-active compounds is the restriction of heat exchange between the water and atmosphere. In Canada, for example, the foam blanket on lagoon-type secondary treatment systems helps prevent heat loss to the atmosphere during the winter months and is therefore important to the maintenance of an effective biological treatment process (McCubbin, 1983).

Both sulfite and kraft mill effluents are highly colored. Sulfite effluent has a characteristic amber color; kraft effluent is typically deep brown. These colors result primarily from the kinds and amounts of lignin-derived substances found in the wastewater. As with foaming, there is nothing particularly toxic about these colors or the compounds that produce them, but the appearance of such water is rather unappealing to most persons. The color does provide a sensitive tracer of the effluent.

Toxic Substances

Pulp and paper mill wastewater contains a variety of toxic chemicals that originate from mainly three sources: cooking chemicals and their byproducts, dissolved organics derived from the nonfibrous material broken down in chemical pulping, and chlorinated organics produced in the bleaching process. The most toxic cooking chemicals are the H_2S and mercaptans associated with kraft pulping, but these are fortunately unstable in the presence of oxygen. Acids and bases used in the pulping and bleaching operations create wastewaters that are often highly acidic or alkaline. Kraft pulping effluent is basic; sulfite process wastewater is acidic. Unless neutralized, these wastewaters can be toxic to aquatic organisms. For purposes of biological treatment, pH's in the 6.5–8.0 range are desirable (McCubbin, 1983).

The various dissolved organic substances in pulp and paper mill wastewater are the source of toxicity that is very difficult to eliminate. Toxicology studies have often involved simply diluting raw effluent with either freshwater or seawater and testing the effect of the diluted effluent on various organisms. In most cases no effort is made to determine the component or components of the effluent that are causing the toxic effects. For both fish and shellfish, such studies have indicated that the incipient lethal level of sulfite effluent is about one part per thousand.[8] Jones et al. (1956), for example, found that the 30-day TLm for chinook and coho salmon exposed to sulfite effluent was about 1–2 parts per thousand, and Wilbur (1969) reported that most oysters died within 2–29 days if exposed to sulfite effluent diluted with seawater to concentrations of 0.67–10 parts per thousand. Similar tox-

---

[8]In other words, 1 ml of effluent diluted to 1 liter.

icity studies indicate that kraft mill effluent is 10–100 times less toxic than sulfite mill effluent. Sprague and McLeese (1968a), for example, found that for juvenile Atlantic salmon the incipient lethal level of neutralized bleached kraft mill effluent was about 12–15% (120–150 parts per thousand), and that larval lobster survival was little affected by concentrations less than 10%. Galstoff et al. (1947) observed no lethal effects on oysters exposed for up to 30 days to 0.1 parts per thousand kraft mill effluent.

Sublethal effects have been observed unfortunately at concentrations several orders of magnitude lower than concentrations associated with lethal toxicity. Chronic effects on oyster growth and reproduction, for example, have been detected at sulfite effluent concentrations of 3–16 ppm. In other words, such effluent would have to be diluted by about a factor of a million to make it safe in terms of chronic stresses, at least for oysters. To dilute all the U.S. and Canadian pulp and paper mill effluent by even a factor of 200 would require more than all the freshwater which flows into the ocean from U.S. and Canadian river systems. Obviously dilution is not a practical answer to the toxicity problem.

One practical approach to detoxifying pulp and paper mill effluent has been suggested from experiments conducted by Sprague and McLeese (1968b). They stored fresh bleached kraft mill effluent (BKME) under quiescent conditions for periods of time ranging up to 2 weeks and then tested its toxicity on juvenile salmon and lobster larvae. During the storage period the oxygen level in the effluent dropped to 1–2 ppm due to microbial consumption of the organics in the effluent. No effort was made to aerate the effluent. After 2 weeks of storage, the BKME had lost virtually all of its acute toxicity for juvenile salmon. After 7 days exposure to 100%, 2-week-old BKME, no fish died. However, quiescent storage produced no significant change in the toxicity of the BKME toward lobster larvae. These results suggested to Sprague and McLeese that the acutely toxic effects on lobster larvae and juvenile salmon were being caused by different substances in the BKME. After 2 weeks, microbial activity had apparently degraded the components toxic to the juvenile salmon, but substances toxic to the lobster larvae remained in the effluent.

To speed up microbial degradation of the BKME, Sprague and McLeese (1968b) set up a biological oxidation system in which fresh BKME was seeded with a culture of microbes, primarily bacteria and protozoa, which had been conditioned to growing on BKME. The BKME was then stirred to effect aeration. During the first week of bio-oxidation the oxygen level in the BKME dropped to 1 ppm, an indication that aeration was not keeping pace with the microbial respiration rate. During the second week, however, the oxygen level recovered to 8 ppm. After 2 weeks of this treatment, the toxicity of the BKME to lobster larvae was tested. The median survival time of the lobster larvae in 100% BKME after 2 weeks of bio-oxidation was about 36 hours, compared to a median survival time of only about 2 hours in fresh BKME. If the concentration of the 2-week bio-oxidized BKME was diluted by about a factor of 2, the median survival time of the lobster larvae exceeded 7 days.[9]

---

[9] The survival experiments were terminated after 7 days, and at that time fewer than 50% of the lobster larvae had died.

Wastewater Treatment

By the mid-1960s it was becoming clear that simply discharging untreated pulp and paper mill wastewater was not acceptable environmentally, and that some form of wastewater treatment would be necessary. The issue became a legal matter in 1972 when both the U.S. and Canadian governments passed legislation that controlled the discharge of pollutants into aquatic systems by industrial operations. The U.S. Clean Water Act, for example, requires industrial facilities to have permits for effluent discharges, and permit limits specify the types and amounts of pollutants allowed in the effluent over time. The Canadian government has passed legislation specific to pulp and paper liquid effluent discharges (Environmental Protection Service, 1972).

The strategy adopted in the treatment of pulp and paper mill effluent is similar to that used in the conventional treatment of sanitary sewage, but with some important modifications to deal with the toxicity problem. Normally there is some form of pre-treatment to remove grit and debris and to bring the pH to an acceptable value, usually in the 6.5–8.0 range. The primary treatment process then consists of removal of suspended solids and is accomplished in a settling basin or clarifier similar to those used at conventional sewage treatment plants. Wastewaters that contain only a small concentration of suspended solids may bypass the primary clarifier and go directly to secondary treatment.[10] This strategy minimizes the amount of water that must be handled by the primary clarifier. Normally 80–95% of the settleable portion of the suspended solids is removed in the primary clarifier. Flocculants are some-times added to abet settling if a large fraction of the suspended solids is colloidal or very fine material that does not settle effectively by gravity. Roughly 10% of the BOD is removed along with the primary sludge. After dewatering, the sludge may be burned as a source of fuel; in some cases it is deposited in a landfill.

The secondary treatment step is intended to reduce both the BOD and toxicity of the effluent. Where land area is limited, some mills have opted for activated sludge type secondary treatment, because BOD reductions of 70–95% can be effected in a relatively small space. The problem with this type of treatment is that the residence time of water in the system is not long enough to allow microorganisms to detoxify the wastewater to an acceptable degree. As a result, many pulp and paper mills use bio-oxidation lagoons in which the residence time of the water is typically 4–10 days (McCubbin, 1983). The lagoons are aerated to ensure that there is an adequate con-centration of oxygen to support aerobic catabolism of the organics in the waste-water. In addition to being more efficient in reducing the toxicity of the effluent, bio-oxidation lagoons also produce very little sludge relative to activated sludge sys-tems. Much of the particulate material settles to the bottom of the lagoon and is con-sumed by auto-oxidation (McCubbin, 1983). The lagoons are also better able to absorb shock loads of concentrated effluent without appreciable change in treat-ment efficiency and are less costly to operate than activated sludge systems.

Tertiary treatment is possible of course to further reduce BOD and suspended solids concentrations, lower the toxicity of the effluent, or to ameliorate color, odor, and taste problems. Activated carbon absorption, for example, can be used to eliminate most of the dissolved organics. Such treatment is expensive, and at the present time is not deemed necessary at most mills.

---

[10]Bleach plant effluents from new kraft mills, for example, contain very few suspended solids.

A thought-provoking point made by McCubbin (1983) is that the design of the wastewater outfall can be of comparable importance to the treatment process in ameliorating the environmental impact of the effluent. It is important to try to dilute the wastewater as quickly as possible and to take steps to reduce or eliminate visible foam and color in the receiving system. A mill located near the mouth of a river, for example, might do well to build an outfall as much as several kilometers long to the ocean rather than discharging directly into the river.

A CASE STUDY—THE BUCKEYE CELLULOSE CORPORATION PULP AND PAPER MILL AT PERRY, FLORIDA

In 1954 the Buckeye Cellulose Corporation[11] began operating a kraft process pulp and paper mill at Perry, Florida, a small town in the Florida panhandle. The mill produces about 1250 tonnes per day of pulp and paper products. BKME from the mill is discharged at the rate of about 210 million liters per day into the Fenholloway River at a point about 20 km upstream from the mouth of the river (Figure 9.7). To accommodate this wastewater legally, the state of Florida declared the Fenholloway an "industrial river."

Prior to the operation of the pulp mill, the Fenholloway River had supported an abundant fish population, as evidenced by the presence of a fish camp near the present outfall. A survey of the Fenholloway River by Beck (1954) just prior to the opening of the pulp mill revealed that the river supported a healthy fauna of invertebrates and fish.

Unfortunately, the discharge of wastewater from the pulp mill displaced virtually all the natural fauna from the stream. Although this displacement in no sense violated the designation of the stream as an industrial river, the company's right to pollute was confined strictly to the Fenholloway River and did not extend into the Gulf of Mexico. As a result of preliminary studies conducted by several investigators in Apalachee Bay (Figure 9.7), it became apparent that the effects of the pulp mill pollution extended out into the bay and along the coast for as much as several kilometers. As a result of this discovery, the company initiated efforts to minimize the impact of its wastewater on the river and bay. In conjunction with this effort, an extensive 2-year study (1972–1974) of the bay and river was conducted by Dr. Robert Livingston and his students (Livingston, 1975; Zimmerman and Livingston, 1976). As a result, a rather clear picture of the impact of the pulp mill effluent on the Fenholloway River and Apalachee Bay is available.

In the study conducted by Livingston, the Econfina River and its estuary were used as a control system to be compared with the Fenholloway River and its estuary. The two streams are of similar length and discharge rate and drain the same swamp area. Oyster bars are found off the mouth of both rivers.

As a result of the mill discharges, water quality in the Econfina and Fenholloway rivers was found to be dramatically different. Below the mill outfall, oxygen concentrations at all depths in the Fenholloway River were virtually zero until the river flowed into Apalachee Bay. By comparison, oxygen concentrations in the Econfina River were about 6–8 ppm throughout the water column. The bottom of the

---

[11] The Buckeye Cellulose Corporation was a subsidiary of Proctor and Gamble. The company name was later changed to Proctor and Gamble Cellulose.

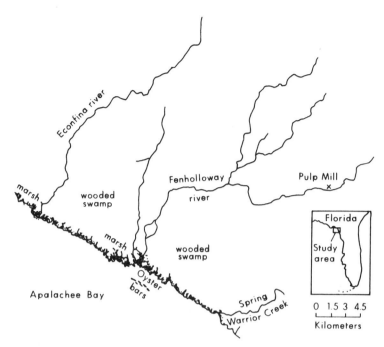

Figure 9.7 Apalachee Bay drainage area showing the Econfina and Fenholloway Rivers and the Buckeye Cellulose pulp mill. [Redrawn from Livingston (1975).]

Fenholloway River was covered with wood fibers and chips, and the odor of H_2S was evident in the air above the stream. Water clarity as measured by Secchi depth in the Fenholloway was significantly lower than in the Econfina due to the greater turbidity and color of the Fenholloway's water[12]. Lower Secchi depths associated with greater turbidity and water color were apparent also in the Fenholloway estuary as compared to the Econfina estuary.

Given this state of affairs, it is not surprising to learn that significant differences were found in the biological communities inhabiting the two systems. The anoxic conditions in the Fenholloway River precluded the existence of most natural inhabitants of the marsh ecosystem. The most obvious animals in the Fenholloway River were a few turtles and alligators. Sampling for fish in the Fenholloway River not surprisingly revealed no permanent fish assemblages, and a total of only three species were taken from the river during the study period. Most of these fish were topminnows, which could have moved into the river from adjacent marsh areas. By comparison, control areas yielded a total of thirty species of fish.

A study of fish populations was also conducted in marsh areas adjacent to the two rivers. A total of 48 species of fish were taken from the Econfina marsh area, versus only six species from Fenholloway marshes. In addition, over 80 times as many fish were taken from Econfina marsh areas as were taken from Fenholloway marsh areas. Although the Fenholloway marshes supported a seemingly healthy stand of the usual marsh vegetation, none of the usual marsh invertebrates (e.g., fiddler crabs) was found.

---

[12]Recall that pulp and paper mill effluent is strongly colored.

In the Fenholloway estuary several extensive fish kills were noticed during the study period. No such fish kills were observed anywhere in the Econfina system. The Econfina nearshore benthos was covered by extensive beds of benthic macrophytes, but these plants were largely absent from the Fenholloway nearshore area. Instead, the Fenholloway nearshore benthos was characterized by mud flats and only scattered clusters of benthic plants. No living oysters were found on the oyster bars off the mouth of the Fenholloway.

Moving away from the mouth of the Fenholloway, the effects of the pulp mill were apparent for the greatest distance in a southeasterly direction. The effects were evident in comparisons of both invertebrate and fish populations sampled at comparably positioned stations at various distances from the mouths of the Fenholloway and Econfina rivers. At the Fenholloway stations, significantly fewer numbers of invertebrates were taken in trawls (Hooks, 1973), and fish populations were significantly reduced both in terms of numbers of species and numbers of individuals taken each month. These effects were most acute in the area within 2–3 km from the mouth of the Fenholloway River.

In summary, the pulp and paper mill effluent had virtually eliminated all natural fauna from the Fenholloway River, had greatly reduced the biomass of plants and animals in the Fenholloway estuary, had eliminated many of the natural fauna from marsh areas adjacent to the Fenholloway, and had reduced significantly the numbers and kinds of organisms found in Apalachee Bay within a distance of roughly 2–3 km from the mouth of the Fenholloway. Within Apalachee Bay, the areal extent of the effluent's impact was determined by both wind-driven and tidal currents, and extended both west and southeast from the mouth of the Fenholloway.

In an effort to reduce the deleterious effects of its effluent on the Fenholloway River and Apalachee Bay, Buckeye Cellulose installed a primary clarifier to remove suspended solids and a bio-oxidation pond with a capacity of about 1.0 billion liters in January, 1974. Given the rate of effluent discharge from the mill, the residence time of the wastewater in the oxidation pond was expected to be about 5 days.

The effects of the wastewater treatment system during the first few years of its operation have been reported in a study by Livingston (1977). In the Fenholloway river itself, observations made within a month or 2 after initiation of water treatment showed that the accumulated wood fibers and chips had largely disappeared from the river bed and that natural species of fish and other animals (including porpoises) were beginning to reappear in the river (P. LaRock, personal communication). Livingston's (1977) study, conducted from 1974 through 1976, revealed that both water quality and turbidity had improved markedly in the Fenholloway River. Some recolonization by sea grasses was noted in the Fenholloway estuary, and in offshore areas there was an increase in the number of benthic macrophyte species, although the biomass of benthic macrophytes remained lower than in the Econfina control area. Fishes in Fenholloway marsh areas increased significantly both in numbers of species and in biomass, at times exceeding the number of species and individuals found in control areas. Fishes in nearshore Fenholloway coastal areas showed little change in biomass, but there was an increase in species numbers.

Toxicity studies conducted with pinfish showed that the fish suffered no mortality when placed in water containing up to 50% oxidation pond effluent, but the TLm for untreated mill effluent was about 10%. Growth and food conversion efficiency for these fish were affected adversely in water containing only 0.01–0.1% untreated effluent, but about 10 times as much oxidation pond effluent was required to produce the same effect.

Based on these reports, the conclusion is that the Fenholloway River system made a significant but incomplete recovery following installation of Buckeye Cellulose's initial wastewater treatment system. The treatment facility was upgraded in 1980 with the installation of an additional oxidation pond with a capacity of about 310 million liters between the primary clarifier and original oxidation pond. The overall system now removes 90% of the BOD and suspended solids from the mill's wastewater. Effluent BOD and SS concentrations are 19 and 27 ppm, respectively. The oxygen concentration in the treated wastewater averages 7.3 ppm, but the BOD in the effluent is sufficient to produce a considerable oxygen sag in the Fenholloway River. The minimum oxygen concentration downstream from the mill averages 0.7 ppm (C. Henry, personal communication).

In one sense the Buckeye Cellulose case study represents a worst-case scenario, because the natural flow of the Fenholloway River is small compared to the discharge from the mill. Even in the rainy season about 60% of the water in the river consists of mill effluent, and the river is dry upstream of the mill about 35% of the time. Thus dilution of the mill's wastewater by river water is in this case of minor significance in reducing the adverse effects of the effluent. The response of the Fenholloway River after January, 1974 is thus an encouraging sign that the harmful effects of pulp and paper mill effluents can be reduced greatly by a rather straight-forward treatment process.

COMMENTARY

Great progress has been made in reducing the adverse effects of pulp and paper mill effluents on aquatic systems since roughly 1970. The industry, whether by choice or by force, has taken the matter of wastewater treatment seriously. In the United States, for example, the industry spent $472 million on capital equipment to control water pollution and $1.8 billion to operate the equipment between 1979 and 1982 (General Accounting Office, 1987). In Canada, BOD and suspended solids discharged per tonne of pulp declined by factors of approximately 2 and 3, respectively, between 1969 and 1981 as a result of pollution control efforts.

Despite the success that has been achieved with secondary treatment, some troublesome issues remain outstanding. The use of chlorine to bleach pulp, for example, produces a wide variety of chlorinated organics, a number of which are known to contribute to the toxicity of pulp and paper mill effluent (Wallin and Condren, 1981). Organizations such as Greenpeace are now advocating the elimination of chlorine-bleached paper products by switching to oxygen-based bleaching methods or simply using unbleached paper products. Concern over this issue may produce the sort of economic incentives that will eliminate or greatly reduce chlorine bleaching by the industry. Some pulp and paper companies in Canada are estimated to have lost millions of dollars of business to competitors who use chlorine-free bleaching methods (Bohn, 1991). Another major issue in the pulp and paper industry is recycling. According to Pollock (1987), paper recycling programs in 9 of the world's 11 largest paper-consuming nations saved more than 4000 km^2 of trees in 1984; and the use of recycled paper rather than raw wood considerably reduces the pollution and demands for energy and water associated with the production of paper products (Table 9.2). Recognition of the environmental benefits associated with recycling paper products and of the problems caused possibly by chlorine bleaching has translated into some noteworthy consumer actions. The McDonald's fast food chain, for example, now uses carry-out bags made of 100% recycled brown paper and oxygen-bleached coffee filters in its restaurants. The wrap for the Big Mac is also made with unbleached paper (EDF, 1991). Such changes reflect an increasing public awareness of environmental issues. This awareness will lead very likely to even greater reliance on recycled paper and less use of chlorine bleaching in the future.

REFERENCES

Beck, W. M., Jr. 1954. A stream quality survey of the Fenholloway River. Report, Bureau of Sanitary Engineering, Florida State Board of Health, Jacksonville, FL. 3 pp. (unpublished).

Bohn, G. 1991. Minister says polluters risk losing sales. Vancouver Sun. June 4.

Environmental Defense Fund. 1991. A deeper shade of green: Cutting McDonald's waste. *EDF Letter*, **23**(3), 7.

Environmental Protection Agency. 1971. *Hawaii Sugar Industry Waste Study*. San Francisco, CA. 106 pp.

Environmental Protection Agency. 1989. *Report on the Evaluation of Wastewater Discharges from Raw Cane Sugar Mills on the Hilo-Hamakua Coast of the Island of Hawaii*. U.S., Washington, D.C.

EPS. 1972. *Pulp and Paper Effluent Regulations*. EPA 1-WP-72-1, Water Pollution Control Directorate, November, 1971.

Galstoff, P. S., W. A. Chipman, J. B. Engle, and H. N. Calderwood. 1947. Ecological and physiological studies of the effect of sulphate pulp mill wastes on oysters in the York River, Virginia. *Fish. Bull.*, **51**, 59–186.

General Accounting Office. 1987. *Water Pollution: Application of National Cleanup Standards to the Pulp and Paper Industry*. National Technical Information Service PB87-193231. U.S. Dept. of Commerce. Washington, D.C. 36 pp.

Grigg, R. W. 1985. *Hamakua Coast Sugar Mill Ocean Discharges Before and After EPA Compliance*. Univ. Hawaii Sea Grant Technical Rept. UNIHI-SEAGRANT-TR-85-02. U.H. Sea Grant College Program. Honolulu, HI. 25 pp.

Gundersen, K. R. 1973. "Microbiology." In *Estuarine Pollution in the State of Hawaii*. Vol. 2. Water Resources Research Center Technical Report No. 31. University of Hawaii at Manoa, Oahu, HI. pp. 235–341.

Hawaiian Sugar Planters Association. 1990. *Hawaiian Sugar Manual 1990*. Aiea, Oahu, HI. 27 pp.

Hooks, T. A. 1973. An analysis and comparison of the benthic invertebrate communities in the Fenholloway and Econfina estuaries of Apalachee Bay, Florida. M.S. thesis, Florida State University, Tallahassee.

Hourston, A. S., and R. H. Herlinveaux. 1957. A 'mass mortality' of fish in Alberni Harbour, British Columbia. Fish. Res. Bd. Can., Pac. Prog. Rep. 109.

Jones, B. F., C. E. Warren, C. E. Bond, and P. Doudoroff. 1956. Avoidance reactions of salmonid fishes to pulp mill effluents. *Sewage Indust. Wastes*, **28**, 1403–1413.

Lau, L. S., P. C. Ekern, P. C. S. Loh, R. H. F. Young, N. C. B. Burbank, and G. L. Dugan. 1975. *Recycling of sewage effluent by irrigation: A field study on Oahu*. Final progress report for August 1971 to June 1975. University of Hawaii Water Resources Research Center Technical Report No. 94. Honolulu, HI. 151 pp.

Livingston, R. J. 1975. Impact of kraft pulp-mill effluents on estuarine and coastal fishes in Apalachee Bay, Florida, USA. *Mar. Biol.*, **32**, 19–48.

Livingston, R. J. 1977. Recovery of the Fenholloway drainage system: Analysis of water quality, benthic macrophytes, and fishes in Apalachee Bay (1971–76) following implementation of a waste control program. Final report to Buckeye Cellulose Corp. 18 pp. + figures.

McCubbin, N. 1983. *The Basic Technology of the Pulp and Paper Industry and its Environmental Protection Practices*. EPA 6-EP-83-1. Environmental Protection Service. Environment Canada. 204 pp.

Officer, C. B., and J. H. Ryther. 1977. Secondary sewage treatment versus ocean outfalls: An assessment. *Science*, **197**, 1056–1060.

Pollock, C. 1987. *Mining Urban Wastes: The Potential for Recycling*. Worldwatch Paper 76. Worldwatch Institute. Washington, D.C. 58 pp.

Postel, S. 1987. *Defusing the Toxics Threat: Controlling Pesticides and Industrial Waste*. Worldwatch Paper 79. Worldwatch Institute. Washington, D.C. 69 pp.

Streeter, H. W., and E. B. Phelps. 1925. A study of the pollution and natural purification of the Ohio River. III. Factors concerned in the phenomena of oxidation and re-aeration. U.S. Public Health Service Bulletin 146.

Sprague, J. B., and D. W. McLeese. 1968a. Toxicity of kraft pulp mill effluent for larval and adult lobster, and juvenile salmon. *Water Res.*, **2**, 753–760.

Sprague, J. B., and D. W. McLeese. 1968b. Different toxic mechanisms in kraft pulp mill effluent for two aquatic animals. *Water Res.*, **2**, 761–765.

Waldichuk, M. 1962. Some water pollution problems connected with the disposal of pulp mill wastes. *Can. Fish Culturist*, **31**, 3–34.

Wallin, B. K., and A. J. Condren. 1981. *Fate of toxic and nonconventional pollutants in wastewater treatment systems within the pulp, paper, and paperboard industry*. Environmental Protection Agency. EPA-600/S2-81-158. Cincinnati, OH.

Wilber, C. G. 1969. *The Biological Aspects of Water Pollution*. Charles C. Thomas. Springfield, IL. 296 pp.

Weibel, S. R. 1969. Urban drainage as a factor in eutrophication. In *Eutrophication, Causes, Consequences, Correctives*. National Academy of Sciences, Washington, D.C. pp. 383–403.

Zimmerman, M. S., and R. J. Livingston. 1976. Effects of kraft-mill effluents on benthic macrophyte assemblages in a shallow-bay system (Apalachee Bay, North Florida, USA). *Mar. Biol.*, **34**, 297–312.

10

PESTICIDES

"I am deeply disturbed and sadly disillusioned by the actions of numerous educators. For years they have made almost a fetish of the 'scientific method' of investigation, and have stressed that method of analysis in their classes and in their writings. I had presumed that they also followed their own advice, but it is now obvious that often they do not.

Despite dozens of scientific studies proving that DDT is broken down rather quickly by the environment, many so-called 'ecologists' are heard telling the public that exactly the opposite is true.

Why do they do this, when it is certain that they will be later exposed as being either ignorant of the facts or being deliberately untruthful? Perhaps the reason is simply that they are either ignorant of the facts or untruthful . . .

If the emphasis in our classrooms and in the public media can be shifted toward more concern for the truth, no matter whose sacred cows get gored, then we'll be on the way back to the proper role of scientists in the American scene.

To begin with we must discount the erroneous beliefs that have been perpetuated by certain scientists and conservationists concerning DDT and attempt to inform the public as to the truth about those topics. DDT is *not* terribly persistent (under environmental conditions). People in the U.S. only ingest about 0.0005 parts per million of DDT [0.5 µg DDT per kg body weight] daily (and excrete all excess DDT from their body). A diet containing thousands of times the amount we ingest daily [still] does *not* produce cancer, sterility, mutations, or other undesirable effects, even in experimental mice, birds, and fish.

So-called "substitutes for DDT" are [usually] more toxic than DDT to most organisms, and are much more costly. Substitutes for DDT are exterminating honey bees, and we will not have them around to pollinate the crops and orchards and "set" the crops. DDT is not widespread in the environment . . . not even in the United States or in the states that have used it most heavily.

DDT has *increased* the number of birds in the United States, rather than causing declines, and the recent British government report stated that DDT has not been responsible for any decline of bird populations in that country, either. DDT probably causes wild birds and animals to be healthier if they are exposed to moderate doses of it, because it reduces tumors, inhibits cancer, . . . eliminates insect-transmitted diseases, and induces the production of enzymes that destroy harmful substances in the body." (Excerpt of a letter written by Dr. J. Gordon Edwards, Professor of Entomology, San Jose State College and published in the Feb. 12, 1974, edition of the Oregon State University Barometer. Italics and words in brackets were added later at Dr. Edwards' request.

Since 1945 chemical pesticides of one form or another have become a significant form of pest control throughout much of the world. Literally thousands of different commercial pesticides are available on the market. At the present time, however, there is a growing consensus that conventional chemical pesticides are by no means the panacea they were once considered to be, and that in many cases their indiscriminate use has created far more problems than it has solved. This realization has come gradually and not without controversy, as the quotation from Dr. Edwards suggests.

Undoubtedly the most notorious of the conventional chemical pesticides has been DDT, and this chapter is devoted in part to a case study of its use and misuse. DDT was chosen as an example because the impact of its use on the environment has been thoroughly studied, because the problems that have beset its use are in many ways typical of the adverse effects associated with pesticide use in general, and because the emotional nature of the debate over DDT provides a revealing picture of the way emotions and prejudices may cloud rational discussions of environmental issues.

DDT (*di*chloro*di*phenyl*tri*chloroethane) was first synthesized in 1877, but did not come into use as a pesticide until 1942. It was used on a broad scale during World War II by the U.S. Army, both to stop typhus fever epidemics that had broken out in Italy, and to eradicate malaria in the Mediterranean basin (Jukes, 1974). DDT was used extensively later in many tropical countries by the World Health Organization (WHO) to control diseases such as malaria, plague, typhus fever, yellow fever, sleeping sickness, and river blindness (Jukes, 1974). DDT has been used to control agricultural pests on many crops throughout the world,[1] and in forestry management to control pests such as the spruce budworm and the Dutch Elm disease vector (a beetle). Today, however, use of DDT in the United States is banned by the EPA, and its use in other countries is limited largely to malaria control. Similar fates have befallen the use of a number of other pesticides. What is it about some pesticides that causes them to rapidly lose popularity or literally be banned following a period of widespread use?

First, pesticides often affect nontarget species, i.e., organisms for which they are not intended. Determining the magnitude and weighing the importance of these side effects is a critical problem in judging the merits of pesticide use. A country with a serious public health problem, for example, may well feel that pesticide use to reduce the incidence of disease(s) is an acceptable policy even if certain species of wildlife are reduced greatly in number or even wiped out because of their susceptibility to pesticide poisoning.

---

[1] For example, cotton, tobacco, soybeans, and potatoes.

Second, target pests, particularly many species of insects, have shown a remarkable ability to develop resistance to pesticides. Since many pesticides are synthetic compounds that do not occur naturally in the environment, there has been no reason, until recently, for organisms with a natural resistance to pesticides to have any better chance of survival than otherwise similar organisms lacking such resistance. Obviously this picture has changed in recent years. The success of many pesticides, particularly many of those introduced since 1945, rests on the fact that pest populations have not been, at least initially, resistant. In other words, pest populations have not tended to be dominated by individuals that by chance were immune or resistant to the pesticide's effects. As pesticide use continues, pest populations tend to be dominated more and more by naturally resistant or immune individuals, since susceptible individuals are killed. Thus the pesticide user may find him- or herself switching constantly from one pesticide to another in order to keep one step ahead of the pest's genetic adaptability, or using larger and larger applications of the same pesticide in order to bring about the same reduction in pest numbers. These two problems, the killing of nontarget species and the development of resistance in target species, underlie many of the difficulties that have been encountered in pesticide use.

CLASSIFICATION OF PESTICIDES

Pesticides may be classified in a variety of ways, based on physical state, target species, purpose of application, or chemical nature. When classified according to target species, the most common pesticides may be defined broadly as being either insecticides, herbicides, or fungicides, depending on whether they are designed to kill insects, plants, or fungi, respectively. Of the total use associated with these three categories, roughly 70%, 20%, and 10% by weight are accounted for by herbicides, insecticides and fungicides in the United States (Pimentel et al., 1991).

When classified according to chemical nature, most synthetic chemical pesticides used since 1945 fall into one of the following categories: chlorinated organics, organophosphates, carbamates, and pyrethroids (Brattstein et al., 1986). In the United States, for example, the two most widely used herbicides are alachlor and atrazine, both of which are chlorinated organics (Postel, 1987). In malaria control, the most commonly used pesticides are DDT and malathion. The former is a chlorinated organic; the latter is an organic phosphate compound.

The chlorinated organics of course consist of chlorine atoms attached to organic moieties. They began to be marketed as pesticides during the early 1940s. Many are rather stable compounds that tend to accumulate in lipid tissue. Because of their persistence, organochlorine pesticides tend to be associated with biomagnification and food chain transfer problems. Some, such as DDT and toxaphene, are highly toxic to fish; some are also highly toxic to birds.

The chlorinated organics may be broken down into subgroups based on similarities in chemical structure. For example, formulas for DDT, DDD, DDE, and methoxychlor are shown in Figure 10.1. The similarity between these chlorinated hydrocarbons is obvious. DDE is not itself a pesticide, but is a relatively stable metabolite of DDT, and has been responsible for many of the adverse effects associated with DDT use. The chemical similarity of these compounds is of more than academic importance, since pest resistance often closely follows such group similarities (Metcalf, 1981; Erickson et al., 1985). Thus pests that have developed a resistance to DDT are likely to show resistance to DDD and methoxychlor as well. In the same vein, it is interesting to note that the chemically similar chlorinated organics chlordane, heptachlor, aldrin, and dieldrin (Figure 10.2) have all been banned by the EPA on the grounds that they are carcinogenic.[2] Historically, some of the most notorious synthetic chlorinated organic pesticides have been chlorinated hydrocarbons.

Organophosphorus pesticides began to appear on the market in approximately 1944. As their name implies, these compounds consist of one or more phosphate groups attached to an organic moiety. The chemical formulas for two such compounds, malathion and parathion, are shown in Figure 10.3. Organophosphorus pesticides are much less stable than the organochlorine compounds, and their use is therefore not associated with biomagnification and food chain transfer problems. However, they are general biocides that are toxic to nearly all animals. They are highly toxic to bees and to natural insect parasites and predators. Parathion and related compounds are also highly toxic to humans, and the majority of pesticide poisonings throughout the world are the result of the improper use of these organophosphorus pesticides. Others, such as malathion, are much less toxic to humans and are readily available in garden store formulations. The cost of pest control with organophosphorus compounds is usually much greater than the cost of comparable control with chlorinated organics, the difference being as much as a factor of 5 or more (Metcalf, 1981).

Carbamate pesticides did not become available until 1956. They are all derivatives of carbamic acid, NH_2CO_2H. Examples such as aldicarb and carbofuran are shown in Figure 10.4. Like the organophosphorus pesticides, carbamates are biodegradable and nonpersistent. They are not particularly toxic to fish, but they are highly toxic to birds and bees (Metcalf, 1981).

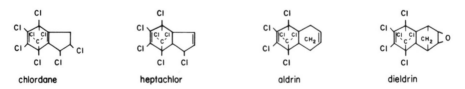

Figure 10.1 Structures of DDT (dichlorodiphenyltrichloroethane), DDD, (dichlorodiphenyldichloroethane), DDE (dichlorodiphenyldichloroethene), and methoxychlor.

Figure 10.2 Structures of chlordane, heptachlor, aldrin, and dieldrin.

[2]Use of chlordane and heptachlor is still permitted in use through subsurface ground insertion for termite control and the dipping of roots or tops of nonfood plants.

MALATHION   PARATHION

Figure 10.3 Structures of two organophosphorus pesticides, malathion and parathion.

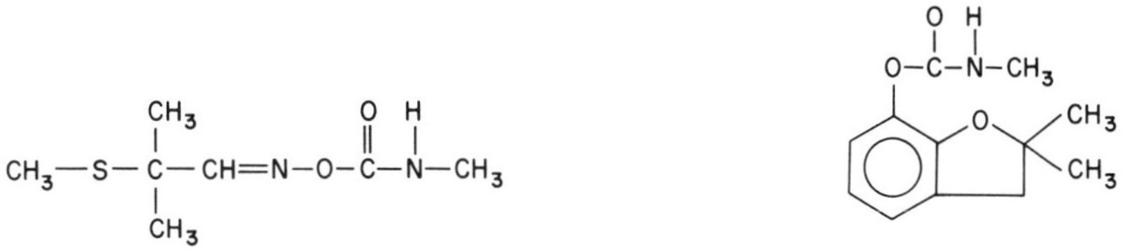

ALDICARB   CARBOFURAN

Figure 10.4 Structures of two carbamates, aldicarb and carbofuran.

The pyrethroid pesticides are synthetic relatives of the natural pyrethrin esters obtained from *Chrysanthemum* flowers. The rapid and selective insecticidal action of the natural pyrethrins combined with their high cost stimulated an effort to produce synthetic derivatives, the first of which was produced in 1949 (Metcalf, 1981). Examples of two such compounds, allethrin and dimethrin, are shown in Figure 10.5. Compared to the organochlorine, organophosphorus, and carbamate pesticides, the application rates of pyrethroids required to achieve a given level of pest control are lower, and in general the pyrethroids are less toxic to animals. The pyrethroids, however, are highly toxic to beneficial insects and are extremely toxic to fish (Metcalf, 1981).

Mode of Action

The mode of action of pesticides depends on both the pesticide and the pest. Herbicides, for example, kill weeds by interfering with growth, respiration, or photosynthesis. Growth inhibitors interfere with the natural function of plant hormones such as auxins and gibberellins. Herbicides can inhibit respiration and photosynthesis in a variety of ways, for example, by reacting with enzyme systems or uncoupling oxidative phosphorylation. Alachlor and atrazine both interfere with photosynthesis by inhibiting the Hill reaction in photosystem II (Plimmer, 1980).

ALLETHRIN   DIMETHRIN

Figure 10.5 Structures of two pyrethroids, allethrin and dimethrin.

Among the insecticides, it is known that many chlorinated organics affect the insect's nervous system. DDT, for example, affects peripheral sensory organs and causes violent trains of afferent impulses that lead to hyperactivity and convulsions. The paralysis and death that follow are believed to result from metabolic exhaustion or from elaboration of a naturally occurring neurotoxin (Metcalf, 1981). Chlordane and its relatives (Figure 10.2), however, affect the ganglia of the central nervous system rather than the peripheral nerves. Symptoms of poisoning include hypersensitivity, hyperactivity, convulsions, prostration, and death. Fish and shellfish are extremely sensitive to chlorinated hydrocarbons, but death comes from suffocation due to interference with oxygen uptake at the gills rather than from effects on the nervous system (Rudd, 1964). Chlorinated hydrocarbons tend to be stored and concentrated in an organism's fatty tissues and reach an equilibrium concentration that is determined by the balance between pesticide intake and excretion. They do not accumulate indefinitely. When pesticide exposure is removed, concentrations gradually return to normal.

Organophosphates and carbamates both inhibit the action of the enzyme cholinesterase, which normally hydrolyzes acetylcholine at nerve junctions in the body. If acetylcholine is not hydrolyzed, it accumulates at nerve junctions as a result of the passage of nerve impulses, and eventually blocks the transmission of these impulses to glands and muscles. Symptoms of poisoning in animals include lacrimation, vertigo, muscular weakness, tremors, and labored respiration (Rudd, 1964). Animals usually die from depression of activity in the respiratory control center of the brain (i.e., they suffocate). There is very little residual accumulation of organic phosphates in tissue, although additive effects are possible if exposure is repeated before cholinesterase levels have returned to normal.

Pyrethroids interfere with the transmission of nerve impulses, the sodium channel being their primary target site (Brattstein, 1986). They readily penetrate the insect cuticle, and it is believed that the effects on the insect are associated with a specific stereochemical interaction with a biological receptor. Paralyzed insects exhibit a characteristic vacuolization of the nerve tissues (Metcalf, 1981).

PESTICIDE USE

Most large-scale pesticide use is underaken for reasons of public health, agricultural production, or forestry management. In the following sections we review briefly the history of pesticide use in these three areas.

Public Health

When infectious microorganisms or parasites are transmitted to humans by means of other carrier animals, it is often easier to attack the host animal or vector than the microorganism itself. The mosquito provides an excellent example of such a vector. More than a dozen important diseases are transmitted by mosquitoes, including malaria, yellow fever, encephalitis, filariasis, and dengue fever. In the past DDT was the principal pesticide used to control such diseases, but DDT use is now confined almost exclusively to the control of malaria. The fact that other pesticides have been substituted for DDT in most disease control programs reflects some of the problems associated with pesticide use in general. Even in the case of malaria control, where

DDT is still used, the control program has been less than an unqualified success. The history of DDT use in malaria control is a good example of the benefits, costs, and limitations of pesticide use in general.

Use of DDT to Control Malaria

Anopheles mosquitoes carry the protozoan parasite that causes malaria. The parasite finds its way into the mosquito's salivary gland, from which it is injected into the next victim bitten by the mosquito. Once inside a human victim, the parasite reproduces in enormous numbers in the red blood cells, and in the case of the most malignant genus, often attacks the brain. Around 1900 there were about 250 million cases of malaria each year worldwide, with about 1% of the cases being fatal (WHO, 1988). Although most persons who become infected survive, the effects of the disease, ranging from lethargy to total incapacitation, are so widespread in the tropics that, "malaria has been blamed for impeding the development of entire nations" (Marshall, 1990). Survivors, often debilitated severely, are subject to reinfection, although persons do develop resistance as a result of repeated infections. Most persons who die from malaria are young children.

The mosquitoes that transmit the disease tend to rest on the interior walls of homes during the day and to attack sleeping humans at night. Effective control of malaria was achieved initially by simply spraying the interior walls of homes with DDT. To be effective when used in this manner, a pesticide must be persistent, since spray teams could not be expected to treat homes at frequent intervals. Also, the pesticide must be safe to humans.

Both the success and failure of DDT spraying for malaria control are rather well illustrated by the experience of Sri Lanka. There were about 2.8 million cases of malaria in Sri Lanka in 1946. DDT spraying for malaria control began the following year, and the incidence of malaria rapidly dropped to less than 40,000 clinical cases by 1954 (Figure 10.6). In the next few years the government reduced the extent and frequency of DDT spraying, presumably for economic reasons. The spraying interval was extended from 6 weeks to 6 months, and by 1955 the spraying was confined to development projects and settlements. The result was a resurgence of malaria cases in 1957, but the number of cases dropped rapidly in subsequent years with the recommencement of the spraying program. By 1963 there were only 17 reported cases of malaria in Sri Lanka, 11 of which were imported (Edirisinghe, 1988).

Perhaps feeling that malaria had been eradicated effectively from the country and perhaps partially in response to the publication of Rachel Carson's *Silent Spring* in 1962, the government of Sri Lanka reduced the DDT spraying program in 1963 and terminated spraying altogether in 1964. The result was a rapid increase in the incidence of malaria, until an epidemic of 2.5 million cases occurred during 1968 and 1969 (Jukes, 1974). Spraying with DDT was recommenced on a 4-month cycle in 1969, but tests conducted at that time indicated that the mosquito population was developing resistance to the pesticide. The spraying program, nevertheless, reduced the incidence of malaria to 130,000 cases by 1972. Tests conducted in 1974, however, indicated that resistance within the mosquito population had increased to an alarming level, and there were over 300,000 clinical cases of malaria that year. In response to this development and perhaps also as a result of the U.S. ban on DDT use in 1972, the government began substituting malathion for DDT in the spray control program in 1974, and DDT use was ultimately phased out entirely. Because malathion is less persistent than DDT, it must be applied more frequently than

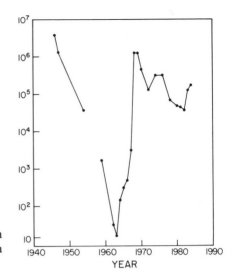

Figure 10.6 Incidence of malaria in Sri Lanka from 1946 to 1984. (*Source*: Jukes, 1974; Fonseka and Mendis, 1987; Edirisinghe, 1988.)

DDT to achieve equivalent pest control. Largely because of this fact, the malathion spraying program cost more per unit area than the DDT spraying program. Because of this increased cost, the government of Sri Lanka opted for less than total coverage spraying, the result being protection with malathion to only a portion of the population at risk (Edirisinghe, 1988). Since 1978 the incidence of clinical malaria cases in Sri Lanka has been on the order of 10^5 per year. This figure is roughly a 10-fold improvement over the million or more cases reported in 1946 and 1968, when there was no systematic spraying program to control the mosquito population, but it is sufficiently high to make malaria a significant public health problem in Sri Lanka.

There was euphoria in Sri Lanka in the early 1960s when malaria seemed to have been all but eradicated. This euphoria was short-lived. The impact of DDT on nontarget species, the development of resistance within the mosquito population, and the increased cost of alternative pesticides all conspired to reduce the effectiveness of the control program. This scenario is a reflection of the world situation as well. According to the WHO (1990), there are presently about 110 million clinical cases of malaria each year worldwide. This figure is certainly an improvement over the 250 million cases per year that occurred around 1900, but malaria is still a major public health problem in some parts of the world and particularly in Africa, where roughly 90% of the malaria cases occur. Treatment for the disease involves administration of quinine-related drugs, usually chloroquine, but the most lethal strain of the protozoan parasite that causes malaria, *Plasmodium falciparum*, is now resistant genetically to chloroquine (Marshall, 1990). In the late 1980s health authorities began switching to a new compound, mefloquine, but by 1990 there was evidence from Asia and Africa of resistance of the parasite to mefloquine as well (Marshall, 1990). Efforts to develop a vaccine for malaria have been unsuccessful (Hoffman et al., 1991).

It is fair to say that the use of pesticides such as DDT to prevent diseases such as malaria has saved tens of millions of lives and prevented hundreds of millions of illnesses since 1945. It is also fair to say that despite the use of pesticides in public health, some diseases such as malaria and river blindness remain major public health problems in some parts of the world (Walsh, 1986). The continued use of pesticides in these disease-control programs largely reflects our inability to identify a more satisfactory alternative. The most effective defense against yellow fever, for example, which is transmitted by mosquito vectors, is vaccination, and not pesticide spraying for mosquito control.

Agriculture

The use of pesticides in agriculture expanded greatly after World War II, in response both to the greater availability of pesticides and to concerted efforts to increase the world's food supply. Basic elements of the so-called Green Revolution have been the development of high-yield agricultural crops through genetic manipulations, increased use of water and fertilizers, and modifications of traditional agricultural practices to control crop pests (Walsh, 1991). For example, farmers reduced their use of mechanical tilling to control weeds and instead increased their use of chemical herbicides. This change in crop husbandry allowed them to plant row crops at more closely spaced intervals and hence increase yields per unit area. Similarly, control of insect pests through crop rotation was in some cases abandoned in favor of planting the same crop year after year. Such changes in agricultural practices allowed farmers to achieve greater yields of the desired crops, particularly rice and wheat, and turned chronic agricultural shortages into surpluses. There is no question that from the standpoint of crop production the Green Revolution has been a tremendous success.

This success, however, has not been achieved without some cost, and part of the cost has been a great increase in the use of chemical pesticides. In India, for example, pesticide use increased from about 2000 tons annually in the 1950s to more than 80,000 tons in the mid-1980s (Postel, 1987). DDT and benzene hexachloride (BHC), both of which have been banned in the United States and much of Europe, account for about 75% of pesticide use in India (Postel, 1987). The variety and extent of pesticide use in agriculture during the Green Revolution is reflected in Table 10.1, which lists chlorinated hydrocarbon insecticides used on agricultural crops during the 1960s. In the United States, pesticide use in agriculture continued to increase until approximately 1980, when annual use leveled off at roughly 350–400 thousand tons (Figure 10.7). Interestingly, the leveling off of agricultural pesticide use in the United States during the 1980s to a large extent reflects the substitution of more biologically effective pesticides for some of the earlier synthetic pesticides, such as those listed in Table 10.1. For example, essentially the same agricultural pest control that was achieved by applying DDT at 2 kg ha^{-1} in 1945 is now achieved by applying pyrethroids and aldicarb at 0.1 kg ha^{-1} and 0.05 kg ha^{-1}, respectively (Pimentel et al., 1991).

While it is true that agricultural production has increased greatly as a result of the Green Revolution, it is remarkable that the percentage of crops lost to pests has not declined as a result of the tremendous increase in pesticide use. For example, between 1942 and 1951 U.S. agricultural losses to pests were estimated to be about 31%. In 1974 and 1986 the percentage loss was estimated to be 33% and 37%, respectively (Pimentel et al., 1991). The fact that heavy reliance on pesticides has not reduced the percentage of crops lost to pests, and that in fact the percentage loss in the United States has increased, has suggested to many persons that such heavy use of pesticides may not be the most effective way to control crop losses to pests.

Table 10.1   Areas and Major Crops on Which Chlorinated Hydrocarbon Insecticides Are Used, Excluding US, Canada, Mainland China, and USSR

Crop	Areas of major significance in use	Insecticides used
Cotton	Mexico, Nicaragua, Egypt, Sudan Brazil, Guatemala Columbia, Australia, Turkey, Uganda, Ivory Coast, El Salvador	Various, but mainly DDT
Rice	Japan, India, Indonesia, Cambodia, Columbia, Spain, Venezuela, Taiwan, Brazil	BHC, endrin, DDT, toxaphene, aldrin
Other cereals (maize, small grains, sorghum, etc.)	India, United Kingdom, Mexico, Upper Volta, Niger, Turkey, France, Columbia, Chile, Spain, Japan, Argentina, Greece, East Africa	BHC, cyclodienes
Vegetables (excluding potatoes)	India, Japan, Mexico, Spain, Chile, United Kingdom, Thailand, South Africa	BHC, DDT
Onions, tomatoes, chillies, cabbage	Japan, Italy, Spain, France, Portugal	Cyclodienes
Potatoes and sweet potatoes	France, Spain, Brazil, Peru, Columbia, United Kingdom, Japan, Greece, Taiwan, Mexico, Australia	Cyclodienes, DDT
Sugar beets	Belgium, Italy, France, Spain, Greece, Chile, Turkey	Aldrin, heptachlor
Sugar cane	Mexico, Australia, India, Brazil, Ivory Coast, Taiwan, South Africa, Pakistan	BHC/lindane, endrin, dieldrin, aldrin, heptachlor
Tobacco	Mexico, Australia, South and East Africa, Japan, Italy, Columbia, Greece, Spain	DDT, BHC, cyclodienes
Oil seeds (sesame, soybeans, groundnuts, sunflower, etc.)	India, Japan, Argentina, Columbia, Nicaragua, Brazil, France, Venezuela	DDT, toxaphene
Crops in general	Asia and Africa	Dieldrin, BHC

Source. Ling et al. (1972).

Combined with this feeling is the realization that the present heavy use of pesticides in agriculture is associated with significant undesirable side effects. It is estimated, for example, that 400,000–2 million pesticide poisonings occur worldwide each year, most of them among farmers in developing countries. These poisonings result in 10,000–40,000 deaths each year (Postel, 1987). Chlorinated organics are not particularly dangerous to humans in this context, but some organophosphorus compounds such as parathion and methylparathion are acutely toxic and must be applied with great care. Unfortunately many third-world farmers lack the training and equipment needed to apply such pesticides safely. "A 1985 sur-

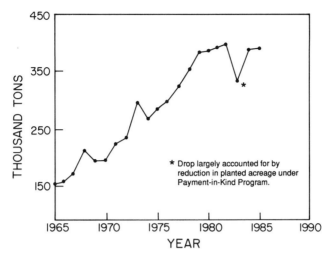

Figure 10.7   Pesticide use in agriculture in the United States from 1965 to 1985. [Redrawn from Postel (1987).]

vey in one county of the Brazilian state of Rio de Janeiro found that 6 out of 10 farmers using pesticides had suffered acute poisonings, two-thirds of them from organophosphates" (Postel, 1987). In addition to the overt poisoning of the individuals involved in applying the pesticides, there are the somewhat less obvious impacts on nontarget species, including humans. In a survey conducted in India's Punjab region, for example, samples of breast milk from 75 women were all found to contain residues of DDT and BHC, the average concentrations being over 20 times World Health Organization (WHO) guidelines (Postel, 1987). Similar studies of breast milk from Nicaraguan women have revealed DDT levels 45 times tolerance limits (Postel, 1987). Pesticides can obviously find their way into humans through the consumption of contaminated crops, but in some cases pesticides leach into groundwater and later appear in drinking water. Monitoring wells and groundwater supplies for pesticides is estimated to cost about $1.2 billion per year in the United States (Pimentel et al., 1991). The consensus now is that pesticides do have a place in the control of agricultural pests, but that the heavy reliance on pesticides that developed during the Green Revolution years is undesirable, and that if a full accounting of the costs and benefits of pesticide use in agriculture were made, a considerable scaling back of pesticide use would be needed. Alternatives do exist to the present large-scale use of pesticides in agriculture (see later discussion).

Forestry

Damage to forests from pests, mainly insects, is difficult to estimate. These losses include reduced growth rates, loss of reproductive ability, and direct mortality of trees, as well as general watershed damage[3] and loss of recreational value. In 1958 the U.S. Forest Service estimated timber losses from insects at 5 billion board-feet,

---

[3] For example, increased erosion due to loss of vegetative cover.

and growth losses were put at 3.6 billion board-feet (Rudd, 1964). The most serious forest insect pests in the United States are the gypsy moth, the tussock moth, the spruce budworm, and the Engelmann spruce beetle. An outbreak of spruce bud-worms in parts of Canada during the 1930s and 1940s affected over 400,000 km^2 and resulted in over 70% timber mortality in some areas (Rudd, 1964). The gypsy moth, which first appeared in the New World in 1869, has been steadily expanding its range, and is now found throughout much of the northeastern United States. In 1981 an outbreak of gypsy moths resulted in the defoliation of over 40,000 km^2 of timber (Marshall, 1981).

During the 1950s efforts to control such outbreaks involved widescale aerial applications of pesticides such as DDT. The adverse effects of such aerial spraying soon became apparent, however, as large numbers of nontarget species were frequently killed. A major and fundamental problem with such aerial spraying is the fact that only about 25–50% of the pesticide actually reaches the target area (Akesson and Yates, 1984). As a result, significant impacts on nontarget species are difficult to avoid. The consensus is that widescale aerial application of pesticides is not a satisfactory way to deal with insect pests in forestry management and that alternative controls must be sought. Satisfactory alternatives have not, however, been readily forthcoming (Evans, 1978; Marshall, 1982), and as a result synthetic pesticides continue to be used in forestry management, though on a much less grandiose scale. In the case of gypsy moth control, for example, "the campaign has tapered off to a sporadic application of pesticides now considered safe" (Marshall, 1981, p. 992).

PESTICIDE EFFECTS ON NONTARGET SPECIES

There is no doubt that the adverse effects of pesticides on nontarget species has been a major cause of both the public's and governments' disenchantment with the widescale use of pesticides. The adverse effects or their consequences include human pesticide poisonings, poisoning of agricultural livestock, the reduction of natural pest enemies, reduced pollination due to the poisoning of bees, losses of crops and trees, and fishery and wildlife losses. Pimentel et al. (1991) have estimated the cost of these adverse effects to be roughly $650 million per year in the United States. The following examples illustrate two of the numerous well-documented adverse effects of DDT use on nontarget species.

Forest Spraying with DDT to Control Spruce Budworms in New Brunswick, Canada

Beginning in 1952, large areas of forest in northern New Brunswick were sprayed each spring with DDT in an effort to control an outbreak of spruce budworms. The DDT was applied at about 50 kg km^{-2}, a typical dosage for control of many forest pests. Despite these efforts, the outbreak continued to spread, and by 1957 over 20,000 km^2 were sprayed, much for the second or third time. The Fisheries Research Board of Canada had been studying salmon populations in the major streams of the affected watershed for several years prior to the initiation of spraying. Therefore a reliable estimate of salmon populations before 1952 could be made. Continued monitoring of salmon populations after 1952 led to the conclusions that application of DDT at 50 kg km^{-2} produced the following results (Rudd, 1964).

1. Loss of 90% of under-yearling salmon.
2. Loss of 70% of salmon parr (juveniles over 1 year old).
3. Complete loss of aquatic insects.

Although aquatic insects began to reappear in affected streams about 3 weeks after spraying, the composition of the insect population was changed radically. The population was dominated by very small insects, which were too small to be satisfactorily utilized as food by the larger young salmon (Ide, 1956). The larger insects, which would normally have provided food for the juvenile salmon, did not reappear. As a result, the salmon fingerlings were forced to turn to alternate sources of food and fed largely on snails (Ide, 1956). Insect populations did not return to normal for 5–6 years after spraying. Since the coho salmon is an anadromous fish that normally spends 3 years in fresh water, repeated spraying in successive years would affect the same year class more than once. Simply restocking the affected streams with salmon would not allow salmon populations to recover until their food supply had also returned to normal.

As a result of this experience, a spraying program for blackheaded budworms on Vancouver Island was carried out in a different manner (Rudd, 1964). A special effort was made to avoid contact with streams by using the following techniques.

1. Not using streams as boundaries for spray plots, so that streams would not receive double spray doses.
2. Spraying parallel to major streams, but keeping one swath away.
3. Shutting off spray when streams were crossed.

Despite these precautions, the mortality of coho salmon fry approached 100% in four major streams, and ranged down to 10% in others (Rudd, 1964). These results point out the fact that streams may be contaminated with pesticides by way of surface runoff as well as by direct application. The contamination of streams during large-scale pesticide spraying operations in forestry management is an unavoidable problem that can obviously have serious consequences for aquatic species.

DDD Treatment to Control Gnat Populations on Clear Lake, California

Clear Lake is a rather shallow (maximum depth 10 m), highly productive lake located about 160 km north of San Francisco. The large number of fish in the lake has made it an attractive resort area. Unfortunately for the resort business, Clear Lake is normally populated during the summer months by an enormous number of gnats. Although the species is not blood-sucking, its numbers are so great as to make it a considerable nuisance, particularly at night when the gnats are attracted to lights. Between 1916 and 1941 several studies were done on the gnat population in an effort to discover a suitable control method, but no satisfactory means was found.

In 1946 laboratory experiments with DDT and DDD were begun. These experiments indicated that DDD would provide effective gnat control with less hazard to fish and other organisms in the lake than would DDT. On the basis of test applications in two nearby bodies of water, it was determined that a concentration of about 14 ppb DDD would provide adequate gnat control with minimal damage to the fish population (Hunt and Bischoff, 1960). Accordingly, a DDD application of 14 ppb was made to Clear Lake in September of 1949. This treatment resulted in about a 99% kill of gnat larvae, but by 1951 larvae began to reappear in significant numbers, and by September, 1954, a second treatment was needed. This second treatment was made at a higher concentration of 20 ppb, as was a third treatment in September, 1957. The rate of gnat larvae kill in the second treatment was again about 99%, but the third treatment was not as successful as the previous two. The lower success of the third treatment was attributed by some to adverse weather, but the more rapid recovery of the gnat population following the second versus the first treatment[4] suggests that the gnats may have been developing resistance (Rudd, 1964). In December, 1954, following the second treatment with DDD, 100 dead western grebes[5] were found on the lake (Hunt and Bischoff, 1960). Analysis of specimens by the Department of Fish and Game Disease Laboratory indicated no evidence of disease. Similarly, in December, 1957, 75 dead grebes were found, again with no evidence of disease. However, two sections of visceral fat taken from these birds were found to contain 1600 ppm DDD, 80,000 times higher than the DDD concentration applied to the lake (Hunt and Bischoff, 1960).

Although other explanations for the grebe deaths are possible, DDD poisoning is strongly implicated as the cause of death for the following reasons.

1. The decline in the grebe population corresponded with the period in which pesticide applications were made. Die-offs were not noted at other times.
2. There was no evidence of infectious diseases in dead grebes.
3. Clinical symptoms of poisoning (e.g., nervous tremors) were observed in some grebes on the lake.
4. Unusually high DDD concentrations were found in the fatty tissue of dead grebes.

In additional sampling of Clear Lake conducted by the California Department of Fish and Game, all fish and birds analyzed were found to contain DDD, and all parts of the lake were found to contain DDD-contaminated fish. Many of the fish taken from Clear Lake were found to contain DDD in their edible flesh at concentrations higher than the level of 5 ppm presently set by the Food and Drug Administration as the action level in marketed fish. Samples of flesh from largemouth bass and Sacramento blackfish hatched as long as 7–9 months after the last DDD application contained 22–25 ppm and 7–9 ppm DDD, respectively. Although it is impossible to say how the grebes came to have DDD levels as high as 1600 ppm in their fatty tissues[6], it seems probable that the major source of input was their food, since visceral fat in fish taken from the lake contained on the order of hundreds to several thousand parts per million DDD (Rudd, 1964). Upon the recommendation of county authorities, DDD applications to Clear Lake were discontinued after the 1957 treatment, and methyl parathion was substituted as a control agent. Since then the grebe population has recovered (Jukes, 1974).

---

[4]Three years versus 5 years, respectively.
[5]A species of bird.
[6]They could have absorbed DDD directly from the water, for example.

EXAGGERATED AND/OR ERRONEOUS CHARGES AGAINST PESTICIDE USE

In the previous two examples, there is little doubt that pesticide use harmed non-target species significantly. An understanding and appreciation of such problems is vital to the development of a sensible pesticide use program. Unfortunately, the literature on pesticides contains some examples of exaggerated or erroneous charges regarding the impact of pesticides on nontarget species. Some of these claims have come to be accepted as established facts, simply because they have been repeated and publicized so frequently. In the interests of scientific credibility, it is important to distinguish effects that can be attributed beyond reasonable doubt to a particular cause from effects whose cause is unclear. The following examples illustrate some of the controversial adverse effects that have been attributed to the use of DDT.

The Destruction of Speckled Sea Trout in the Laguna Madre, Texas

In 1971, the Panel on Monitoring Persistent Pesticides in the Marine Environment submitted a report to the National Academy of Sciences, which was widely publicized in the popular press. In one section, the panel reported that:

"In the speckled sea trout on the south Texas coast, DDT residues in the ripe eggs are about 8 ppm. This level may be compared with the residue of 5 ppm in freshwater trout that causes 100 percent failure in the development of sac fry or young fish. The evidence is presumptive for similar reproductive failure in the sea trout. Sea trout inventories in the Laguna Madre in Texas have shown a progressive decline from 30 fish per acre in 1964 to 0.2 fish per acre in 1969. It is significant that few juvenile fish have been observed there in recent years, although in less contaminated estuaries 100 miles away there is a normal distribution of sea trout year classes" (Goldberg et al., 1971, p. 10)

Table 10.2 is taken directly from the National Academy of Sciences report. Several major criticisms can be made of this report.

Table 10.2   Capture of Sea Trout in the Laguna Madre

Year	Number of sea trout captured per acre
1964	30
1965	25
1966	12
1967	—[a]
1968	2.7
1969	0.2

[a] No data exist for 1967 as hurricanes destroyed all the fishing gear. The decline in the fishery has been only in the juvenile trout, which eliminates the possibility of overfishing. The DDT residues in the gonads of the adult trout reach a maximum of 3 ppm prior to spawning.

Source. Reproduced from *Chlorinated Hydrocarbons in the Marine Environment* (1971), p. 10, with permission of the National Academy of Sciences, Washington, D.C.

1. The text of the report states that sea trout inventories declined from 30 fish per acre in 1964 to 0.2 fish per acre in 1969. In fact, these figures refer only to juvenile trout, as the footnote to the table indicates. There was no decline in adult populations. The figure of 0.2 juvenile trout per acre in 1969 is incorrect; the correct figure is 0.5.

2. Three hurricanes swept through the area during the study period, one in each of the years 1965, 1967, and 1968. The most destructive was probably hurricane Beulah, which occurred in 1967. Beulah flooded heavily all coastal areas, dropping 76 cm of rain and undoubtedly disrupted the shallow Laguna Madre ecosystem. The decline in juvenile sea trout is correlated strongly with the occurrence of these hurricanes. In 1970, when weather conditions were stable, juvenile trout numbers increased to 10 per acre and in 1971 reached 25–30 per acre. It is remarkable that the 1970 figures were not included in the NAS panel report, which was released in June, 1971.

3. Since pesticides are used for agricultural purposes in the coastal region around the Laguna Madre, it would not be surprising to find that the runoff from the three hurricanes had introduced unusually high pesticide concentrations into the estuary. However, the figure of 8 ppm DDT in ripe eggs is not a mean value, but rather the highest value found in any of the sea trout eggs at any time. Mean values were more in the 1–2 ppm range and declined in 1970 and 1971 as the hurricane effects diminished (Butler, 1971).

In summary, the decline in juvenile sea trout in the Laguna Madre from 1965 to 1969 appears to have been caused by hurricane damage and not by DDT effects. Juvenile sea trout populations had returned to normal by 1971, despite continued use of DDT in the watershed area.

DDT Reduces Photosynthesis by Marine Phytoplankton

The following excerpt from an article provides a good introduction to this issue.

"The end of the ocean came late in the summer of 1979, and it came even more rapidly than the biologists had expected. There had been signs for more than a decade, commencing with the discovery of 1968 that DDT slows down photosynthesis in marine plant life. It was announced in a short paper in the technical journal, *Science*, but to ecologists it smacked of doomsday. They knew that all life in the sea depends on photosynthesis, the chemical process by which green plants bind the sun's energy and make it available to living things. And they knew that DDT and similar chlorinated hydrocarbons had polluted the entire surface of the earth, including the sea." (Ehrlich, 1969)

The short paper referred to in the previous excerpt is an article by Wurster (1968), reporting the results of experiments on pure and mixed phytoplankton cultures to determine the effect of DDT exposure on photosynthesis. The DDT was added in concentrations ranging from 1 to 500 ppb to flasks containing the experimental cultures, the cultures incubated for approximately 1 day more, and the photosynthetic rate measured over the next 4–5 hours. The results of these experiments are shown graphically in Figure 10.8. The crucial point about these results is that the solubility limit of DDT in seawater is only 1.2 ppb. To achieve the concentrations

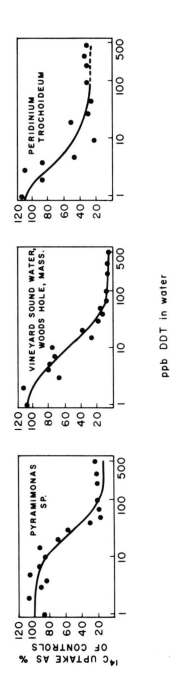

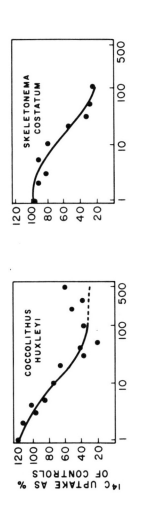

Figure 10.8 Photosynthetic rates of five cultures of marine phytoplankton as affected by DDT concentrations. [Redrawn from Wurster (1968). Copyright © 1968 by the American Association for the Advancement of Science.]

shown in the figures, it was necessary to dissolve the DDT in alcohol before adding it to the experimental flasks. One would never see concentrations this high in the ocean. Looking at the figures, it is concluded that even at its solubility limit, DDT has virtually no effect on phytoplankton photosynthesis. Subsequent experiments performed by Menzel, Anderson, and Randtke (1970), Moore and Harriss (1972), Mosser, Fisher, and Wurster (1972b), Luard (1973), and Fisher (1975) have been consistent with the results reported by Wurster (1968). No effects of DDT on marine phytoplankton photosynthesis have been reported at concentrations below the solubility limit, and in some cases no effects have been observed at concentrations 10 times or more the solubility limit.

On the other hand, it is reasonable to assume that the DDT dissolved in the water was not directly responsible for the effects observed by Wurster and others. Rather, the DDT *absorbed by or adsorbed to the phytoplankton* was responsible for the reduction in photosynthetic rates. The fact that the percentage reduction in photosynthetic rate and the concentration of phytoplankton cells are correlated negatively at the same total DDT concentration (Figure 10.9) is consistent with this hypothesis. In other words, the more DDT available per cell, the lower the photosynthetic rate. In the open ocean, where phytoplankton abundance in most places is quite low, it is quite probable therefore that DDT would depress photosynthetic rates at concentrations that were associated with no adverse effect on dense laboratory cultures. In order to circumvent this problem, Menzel et al. (1970) conducted their experiments at phytoplankton concentrations that they considered to be within the range of naturally occurring concentrations in oceanic surface waters, and Moore and Harriss' (1972) experiments were conducted with natural phytoplankton communities from St. George Sound in the northeastern Gulf of Mexico. In neither case were any adverse effects of DDT on photosynthesis apparent at DDT concentrations below the solubility limit of DDT in natural waters. The conclusion would seem to be that even at the low phytoplankton concentrations typical of many natural waters, DDT at concentrations below the solubility limit does not depress photosynthetic rates. There is no convincing evidence to show that DDT concentrations in the oceans have ever had a significant effect on primary production.

DDT Residues of 5 ppm (Wet Weight) in the Eggs of Freshwater Trout Result in 100% Mortality of Fry

The basis for this statement, which we saw earlier in the Laguna Madre sea trout discussion, stems from an article published in 1964 by Dr. G. E. Burdick and colleagues concerning the failure of the Lake George fish hatchery in New York state. A complete loss of fry at this hatchery was experienced in both 1955 and 1956. "The fry floated upside down on the surface, eventually sinking and dying. Symptoms appeared after absorption of the yolk sac when the fry were about ready to feed. Pathological examination failed to show the presence of any disease" (Burdick et al., 1964, p. 127). In 1956 and 1958, eggs from the Lake George hatchery were distributed to three other hatcheries as a check on hatchery conditions. There was no survival in 1957, and negligible survival in 1958, results that indicated that the problem was associated with the Lake George eggs and not due to water quality in the Lake George hatchery. Attempts to cross Lake George females with males from other lakes failed, though males from Lake George reproduced successfully with females from other lakes.

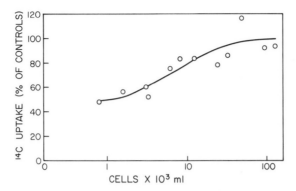

Figure 10.9 Effect of 10 ppb DDT on the photosynthetic rate of the marine diatom *Skeletonema costatum* as a function of cell concentration. [Redrawn from Wurster (1968). Copyright © 1968 by the American Association for the Advancement of Science.]

It was known that from 1951 to 1955 the conservation department had distributed over 3500 kg of DDT in the Lake George watershed, primarily for gypsy moth control. In 1956 and 1957 an additional 13,000 kg were distributed. Spraying and fogging with DDT for blackfly and mosquito control were also common in the watershed. Fish kills had been reported occasionally subsequent to such treatments, and a 1959 study reported that fish fry from Lake George contained measurable amounts of DDT. These observations suggested that DDT might be the cause of the fry mortality at the hatchery.

To check this hypothesis, female fish were collected from twelve different New York lakes from 1960 to 1962. It would appear from the numbering of the fish in Tables 2 and 3 of Burdick's paper that at least 95 fish were collected over this 3-year period, though only 61 fish are accounted for in the later analyses. Eggs were extracted from 51 of these females. Some of the eggs were incubated following usual procedures; the remainder were analyzed for DDT and DDE. The 5 ppm figure comes entirely from the 1960 results, and we will therefore concentrate on that data set.

In 1960, DDT concentrations were measured in the eggs of 20 fish, 4 from Fourth Lake, 3 each from Seneca, Saranac, Eighth, George, and Raquette lakes, and 1 from Lake Placid. Unfortunately, the wet weights of the eggs were not determined, and it was necessary to estimate wet weights based on 1960 dry weights and the mean water loss from dehydration (70%) determined from 1961 egg analyses. An attempt was then made to correlate fry survival with the DDT residues in the eggs. The result of this exercise is shown in Figure 10.10.

In Figure 10.10, % *syndrome* simply means the percentage of fry that died, exhibiting the previously described symptoms. The authors concluded from this figure that fry mortality is severe (over 50%) when DDT concentrations in the egg exceed 5 ppm. Several comments are in order regarding this conclusion.

1. Only 16 of the original 20 fish are accounted for. The results on 2 fish each from Lake Saranac and Eighth Lake are missing.
2. The experiments were terminated when mortality reached 50%. It is impossible to say what mortalities would have been had the experiments been allowed

to continue. There is clearly no basis for saying that 5 ppm DDT caused 100% mortality, since 100% mortality was never observed. If these results are accepted at face value, it might be argued that 5 ppm DDT caused at least 50% mortality, but that would be the strongest statement that could be made.

3. Establishing the breakpoint at 5 ppm DDT is completely arbitrary. The lowest DDT concentration in any of the eggs from Lake George and Fourth Lake was 10.10 ppm. Eggs from 1 Lake Raquette fish showed 15% mortality at a DDT level of 4.75 ppm, but a second fish from the same lake produced eggs with a DDT concentration of 5.81 ppm, yet there was no mortality. In short, the breakpoint for at least 50% mortality could be equally well established at 5 or 10 ppm DDT.

4. Only 3 of the fish are from Lake George, where 100% mortality of hatchery fry was observed. The lowest DDT concentration in the eggs from these fish was 13.55 ppm DDT.

5. The DDT analyses were made by the now-outmoded Schechter-Haller method, which fails to distinguish between DDT and a number of other compounds, including methoxychlor, DDD, chlorobenzylate, polychlorinated biphenyls (PCBs), and most aromatic compounds. After 1966 chromatographic methods completely replaced the Schechter-Haller method in pesticide analyses. After this change in analytical methodology, the edible flesh from over half the lake trout analyzed from Lake George was found to contain more than 5 ppm PCBs (Boyle, 1975). Therefore, it is possible that much of what Burdick et al. (1964) were measuring with the Schechter-Haller method was PCBs.

In summary, there is no doubt that 100% mortality of fish fry occurred at the Lake George fish hatchery in 1955 and 1956. Based on the work of Burdick et al. (1964), it can be said that the concentration of something (whatever the Schechter-Haller method was measuring) in fish eggs seems to correlate positively with mortality. Considering the rather small number of samples (16) involved, the statistical basis for this statement is a bit weak. This summary is clearly a far cry from the statement that 5 ppm DDT in fish eggs causes 100% mortality of the fry. In defense of Burdick

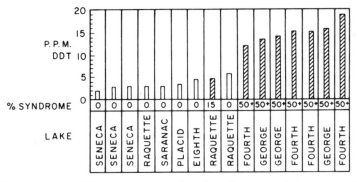

Figure 10.10   Total DDT of 1960 lake trout egg lots, based on wet weight, computed from dry weight and based on the average wet weight to dry weight ratio of 1961 eggs. Hatching indicates the occurrence of the syndrome. [Redrawn from Burdick et al. (1964). Reproduced with permission of the American Fisheries Society.]

et al. (1964), it should be pointed out that they never made this last assertion. Others, however, in reading their paper, have drawn this unjustified conclusion and have repeated it until it seems to have become an established fact. Indeed the NAS report (Goldberg et al., 1971) did not even bother to reference this assertion in its summary of DDT effects on fish.

DDT Causes Cancer

DDT is suspected of being a human carcinogen. The basis for this suspicion is the fact that DDT has been shown to cause liver tumors in mice. These results are of dubious relevance to human health, for reasons discussed in Chapter 8. In several other animal species, including rats, monkeys, hamsters, and dogs, no tumorigenic effects have been shown at doses less than 50 mg DDT/kg body weight per day (Ottoboni, 1972; EPA, 1980a). At doses higher than that level, evidence is equivocal for rats (EPA, 1980a). For comparison, average human DDT intake rates in the United States, primarily from the consumption of contaminated food, were about 0.0009 mg kg^{-1} in 1965 and 0.0004 mg kg^{-1} in 1970 (Duggan and Corneliussen, 1972).

There have been several studies of humans who ingested or were exposed to unusually high amounts of DDT. Ortelee (1958), for example, studied 40 men with extensive and prolonged occupational exposure to DDT in manufacturing or formulating plants. Exposure rates, which were estimated from observations on the job, ranged as high as 0.6 mg kg^{-1} d^{-1}. Physical, neurological, and laboratory findings were within normal ranges with the exception of minor skin irritations, and no correlations were found between DDT exposure and frequency and distribution of the few abnormalities seen. Laws (1967) found no evidence of adverse health effects and no cases of cancer in a group of 35 workers at the Montrose Chemical Company, which manufactured DDT exclusively. The workers had been ingesting and/or inhaling DDT for 9–19 years at rates estimated to be 0.05-0.26 mg kg^{-1} d^{-1}. Almeida et al. (1975) examined a group of men who had been involved for 6 or more years in spraying DDT as part of a malaria control program in Brazil. Although the spray workers' blood serum contained significantly higher DDT and DDE concentrations than the blood serum of a control group, there was no significant difference in the health of the two groups. Edmundson et al. (1969a,b) studied 154 persons with occupational exposure to DDT over a 2-year study period. No clinical effects related to DDT exposure were observed. Hayes, Dale, and Prinkle (1971) administered DDT at up to 0.51 mg kg^{-1} d^{-1} to a group of human volunteers for periods in excess of 600 days with absolutely no signs of pathological effects. DDT accumulated in the fatty tissues of the volunteers for a period of time until an equilibrium concentration was reached, with the equilibrium concentration being roughly proportional to the rate of DDT ingestion. When DDT feeding was terminated, tissue concentrations returned to normal.

Under certain conditions, DDT and related compounds actually appear to have anticarcinogenic properties. In fact, DDD has been used successfully as a chemotherapeutic in the treatment of adrenal carcinoma (Conney et al., 1967; Bledsoe et al., 1964; Sakauchi et al., 1969; Southren et al., 1966). In experiments with mice, Laws (1971) fed an experimental group 5.5 mg kg^{-1} d^{-1} DDT. At 2-4 week intervals, 6 mice each from a control group (no DDT feeding) and the experimental group were inoculated with cancer cells under the skin of the back. These transplants continued for 13 months. According to Laws, this transplant technique had, "produced a 100% take of tumor transplants over a ten-year period" (Laws, 1971). Not surprisingly, 100% of 87 control mice developed cancer and died after a mean time of 40 days following the tumor transplant. However, 8% of 89 DDT-fed mice did not develop cancer, and those that did lived significantly longer (mean = 56.5 days) after the transplant than did the controls. The results suggest that DDT may suppress or even block the growth of at least one kind of malignant tumor. The mechanism of this effect is unknown.

In summary, the idea that DDT causes cancer is based almost entirely on studies with mice fed DDT at rates 1000 times or more the rates of average human exposure in the United States during the 1960s. With the exception of some equivocal results from studies with rats fed even higher doses of DDT, there is absolutely no evidence that exposure to DDT causes cancer in other animals or humans. In fact, evidence suggests that under certain conditions DDT and DDD may actually have anticarcinogenic effects. The EPA, however, does not consider the results of the human studies an adequate basis for reaching conclusions regarding human carcinogenicity, because the sample sizes have been small and the studies short relative to the average human lifespan (EPA, 1980a).

Implications

Rational decisions about pesticides require an informed understanding of the benefits and costs associated with their use. DDT, for example, does have harmful effects on many nontarget species, and it is true that many target species have developed resistance to DDT. It is for these reasons that DDT use has been discontinued in many parts of the world. On the other hand, if DDT were a deadly carcinogen, it is very doubtful that governments would still authorize spraying the interior of homes with DDT to combat malaria. Problems associated with pesticide use in the past have often stemmed from an inadequate knowledge of the true environmental costs. It is important not to underestimate those costs; it is also important not to overestimate or exaggerate them. Synthetic chemical pesticides will probably continue to play a role in pest control for many years. An objective assessment of both the pros and cons of pesticide use will be necessary if that role is to be defined properly.

PESTICIDE PERSISTENCE IN THE BIOSPHERE AND FOOD CHAIN MAGNIFICATION

The persistence of pesticides in the environment is one of the important factors that determines their efficacy as pesticides and their impact on nontarget species. If pesticides degraded rapidly to harmless substances through natural processes there would be little opportunity for them to be spread throughout the food web through feeding relationships and their concentrations to increase through biological magnification. On the other hand, if pesticides degrade rapidly, repeated applications may be necessary to achieve the same level of pest control that could be obtained with one or a few applications of a more persistent pesticide. Thus the very characteristic that makes persistent pesticides desirable from the standpoint of pest control makes them undesirable from the standpoint of their impact on nontarget species.

The subject of pesticide persistence has been a controversial one, in part because the persistence of many pesticides depends very much on the surrounding environment. We again use DDT as an example. The half-life[7] of DDT sprayed on citrus crops is estimated to be about 50 days, but only 11–15 days on peaches and 7 days on alfalfa. Similarly, parathion has a half-life of about 78 days on citrus crops, 3–6 days on apples, and 2 days on alfalfa (Gunther and Jeppson, 1960). The situation is often very different if pesticides are washed into soils. Soil plots experimentally treated with 11.2 tonnes km^{-2} of DDT in 1947 had a residue of 3.16 tonnes km^{-2} in 1951 (Rudd, 1964). The implied half-life is about 2.2 years. Similar studies by Fleming and Maines (1953), Lichtenstein (1957), and MacPhee, Chisholm, and MacEachern (1960) suggest a half-life for DDT in soils of 6.6, 5.2, and 5.4 years, respectively. The persistence of pesticides in soils, especially halogenated hydrocarbons, has been of particular concern with respect to groundwater contamination. Persistent pesticides may migrate slowly down through the soil complex, eventually reaching the water table. In some cases the contamination does not become apparent until years after the first use of the pesticide and may continue for many years after use of the pesticide has been discontinued (Anonymous, 1981; Yoshishige, 1991). Such contamination can create very serious public health problems, because groundwater supplies are not decontaminated easily (see Chapter 16).

With respect to the marine environment, Wilson, Forester, and Knight (1970) reported that 92% of the DDT and its metabolites disappeared from a seawater sample stored in a sealed container that had been immersed in an outdoor flowing seawater tank for 38 days. This rate of disappearance would imply a half-life for DDT of only about 10 days under the experimental conditions. On the other hand, Clear Lake fish that had hatched as long as 7–9 months after the last DDD application to the lake contained 7–25 ppm DDD in their flesh, which obviously implies that DDD had persisted somewhere in the Clear Lake ecosystem for at least 7–9 months. Furthermore, intact brown pelican eggs taken off the southern California coast were found to contain 853 ppm (lipid weight basis) DDE in 1969 (Table 8.2), suggesting an impressive degree of environmental persistence, since the pelican eggs, the pelicans themselves, and the oceans for that matter were never sprayed with DDT. The DDT must have entered the ocean from land runoff or from fallout (e.g., rainfall), then entered the parent bird by way of the marine food chain or from direct contact with the water, and finally been transferred from the parent to the egg. Suffice it to say that this process must have taken longer than 10 days.

Some idea of the persistence of DDT and its metabolites in the marine environment may be gained from an examination of the historical record of total DDT residues in anchovies taken from southern California coastal waters. Values reported by Anderson et al. (1975) are given in Table 10.3. As noted in Chapter 8, the sharp drop in residues between 1969 and 1970 was probably due largely to the virtual elimination of DDT discharges from the Montrose Chemical Company outfall in

---

[7]The half-life is the time required for half the substance to disappear.

1970. The second sharp decline, between 1972 and 1973, may be attributed to the EPA's ban on DDT use in the United States, which went into effect in January, 1973. Since in both cases effects were evident in the anchovies within 1 year from the associated discharge reductions, it seems clear that the effective half-life of DDT and its metabolites in this system could not have been more than several months. However, the half-life must have been more than a few days; otherwise DDT residues would have been degraded long before they appeared in the anchovies.

Confusion over the persistence of DDT and other pesticides in aquatic systems sometimes arises because of the fact that the solubility of many pesticides in water is quite low and the dissolved concentrations therefore very small. For example, Harvey, Steinhauer, and Teal (1973) found the concentration of DDT residues in the North Atlantic Ocean to be less than one part per trillion, over a thousand times less than the solubility limit. Although such observations may be taken to imply a short residence time for DDT residues in the water, they do not necessarily imply that DDT residues have been degraded. It is entirely possible that most of the DDT residues in the water are present in the plankton and other living organisms rather than in the dissolved state. Indeed, we have seen that the concentration of DDT residues in some aquatic organisms may be quite high. The half-life of DDT residues in the fatty tissues of such organisms may be considerably longer than the residence time of DDT residues dissolved in the water.

The issue of pesticide persistence is closely associated with the controversy surrounding food chain magnification. The fact that some pesticides are stored primarily in fatty tissues, where they are metabolized or excreted at a rather slow rate, has suggested to ecologists that such pesticides might become greatly concentrated at higher trophic levels in a food chain. The high concentrations of pesticides sometimes found in the bodies of carnivorous fish and birds lend further support to this argument. A well-known paper by Woodwell, Wurster, and Isaacson (1967) has been cited often to support the idea of pesticide magnification through food chains. The Woodwell et al. paper deals with DDT residue concentrations measured in various animals from an extensive salt marsh on the south shore of Long Island, New York. The area selected was characterized by Woodwell, Wurster, and Isaacson as being "representative of relatively undisturbed marsh." Table 10.4 summarizes some of the findings of Woodwell et al. (1967).

Table 10.3   DDT Residues in Anchovies Taken Off the Southern California Coast

Year	Total DDT residues[a] in anchovies (ppm, fresh weight)
1969	4.27
1970	1.40
1971	1.34
1972	1.12
1973	0.29
1974	0.15

[a] DDT residues = DDT + DDE + DDD.

Source. Anderson et al. (1975).

Table 10.4   DDT Residues in Selected Organisms from a Long Island Salt Marsh

Organism	Concentration of DDT residues (ppm, wet weight)
Water	0.00005
Plankton	0.04
Silverside minnow	0.23
Sheephead minnow	0.94
Pickerel (predatory fish)	1.33
Needlefish (predatory fish)	2.07
Heron (feeds on small animals)	3.57
Tern (feeds on small animals)	3.91
Herring gull (scavenger)	6.00
Fish hawk (osprey) egg	13.8
Merganser (fish-eating duck)	22.8
Cormorant (feeds on larger fish)	26.4

Source. Woodwell et al. (1967).

The increase in pesticide concentrations at higher trophic levels is clearly evident in the table. The data are consistent with the effects expected from food chain magnification, but do not prove that this mechanism is responsible for the observations. It is possible, for example, that higher trophic level organisms simply absorb more DDT residues from the environment than do organisms further down the food chain.

In contrast to the Woodwell, Wurster, and Isaacson (1967) results, Harvey, Steinhauer, and Teal (1974) found no evidence of DDT residue magnification in Atlantic Ocean pelagic food chains. In fact, the highest DDT residue concentrations were found in the plankton, whereas flying fish, which feed largely on plankton, carried about 10 times *lower* DDT residues than the plankton. At higher trophic levels, sharks have typically been found to carry high concentrations of DDT residues; but barracuda, which are also top-level carnivores, contain about 100 times lower concentrations than sharks (Giam et al., 1972). Such observations led Harvey et al. (1974) to postulate that chlorinated hydrocarbon concentrations in aquatic organisms may largely reflect a simple physical–chemical equilibrium between chlorinated hydrocarbon concentrations in the water and the lipid tissues of the organisms. Concentration differences between aquatic organisms in the same environment might then reflect differences in the solubility of chlorinated hydrocarbons in the lipid tissues of the organisms and/or metabolic differences affecting the storage and breakdown or excretion of chlorinated hydrocarbons by the organisms.

At least one systematic study of the biological magnification question (Hamelink et al., 1971) has shown that some fish (largemouth bass, green sunfish, and pumpkinseed sunfish) may pick up DDT residues primarily from the water by adsorption on their gills rather than from their food. In one set of experiments, ponds were set up containing either algae, invertebrates, and fish or algae and fish only. In the latter case, the fish were fed laboratory-raised brine shrimp, uncontaminated with DDT. DDT was added to the sediments of both ponds, and appeared rapidly in the organisms in each system. There was no significant difference, however, between the DDT residue concentrations in the fish from the first pond, which ate invertebrates containing DDT residues, and in the fish from the second pond, which ate uncontaminated brine shrimp. Since the fish in the second pond obviously picked up their DDT residues entirely from the water, the implication is that the final concentration in the fish was independent of whether the DDT came solely from water or from a combination of water and food. Interestingly, there was almost a perfect correlation over the time of the experiment between the concentration of DDT residues in the invertebrates and the concentration of DDT residues in the water (Figure 10.11), an observation that suggests that the invertebrates obtained much of their DDT directly from the water rather than from the algae. Nevertheless, in all experimental ponds that contained algae, invertebrates, and fish, there was a steady increase in the concentration of DDT residues from the algae to the invertebrates to the fish, similar to the results found by Woodwell et al. (1967) in the Long Island salt marsh. Mean concentration factors (versus water) were 7.2×10^3 for the algae, 1.8×10^4 for the invertebrates, and 1.7×10^5 for the fish. The results of the Hamelink study suggest that effects similar to those attributed to food chain magnification may be caused by a very different mechanism, namely by the direct exchange of pollutants between an aquatic organism and the water, with the effective magnification determined by the metabolic characteristics of the organism. On the other hand, Hamelink's results do not rule out food chain transfers as a significant means of distributing pesticide residues within the biosphere, and indeed both mechanisms (direct exchanges between organism and environment and food chain transfers) may play an important role in some systems. Without conducting more detailed studies, however, it is impossible to say to what extent food chain effects and direct exchanges are responsible for observations such as those made by Woodwell et al. (1967). It seems probable, however, that the transfer of pesticides through food chains has been a major factor in some cases, such as the poisoning of raptorial and fish-eating birds (e.g., western grebes on Clear Lake, California), a subject to which we now turn our attention.

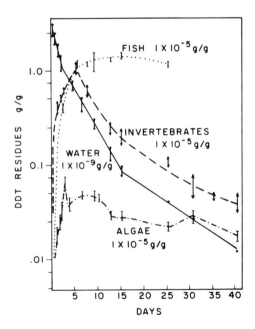

Figure 10.11 DDT residues in components of artificial ponds treated with 5 ppb DDT in 1966. [Redrawn from Hamelink et al. (1971). Reproduced with permission of the American Fisheries Society.]

PESTICIDE EFFECTS ON BIRDS

Pesticides, particularly some chlorinated organics, have been blamed as the cause of reproductive failure among certain birds of prey (raptorial birds). These effects, perhaps more than any other single factor, aroused public concern over the indiscriminate use of pesticides during the 1960s and 1970s. According to Peakall (1970b) and Peakall and Peakall (1973) reproductive failure was associated with the following three general types of effects.

1. A remarkable thinning of the eggshells and much breakage of the eggs that were laid.
2. Modification of parental behavior, leading to delayed breeding or failure to lay eggs altogether, improper incubation of eggs, and/or failure to produce more eggs after earlier clutches were lost.
3. High mortality of the embryos and among fledglings due to elevated pesticide levels in the offspring.

While acknowledging that some bird populations have been, at least temporarily, affected adversely by the unwise use of pesticides, Jukes (1974) has pointed out that some of the reports of declining bird populations are not borne out by bird census data,[8] and that in other cases population declines might have been caused by human harassment[9] or invasion of natural bird habitats. Peakall (1970b), however, reported that some brown pelican eggshells sampled off the California coast, "were so thin that the eggs could not be picked up without denting the shells," a condition that can hardly be considered healthy.

The allegations made against pesticides are based both on field studies and on laboratory experiments that involved controlled feeding of pesticides to birds. The former have been concerned mainly with the first of the above-mentioned effects and have attempted to show whether or not significant correlations existed between pesticide residues in bird eggs and egg weight or shell thickness. The latter experiments have tended to be of a wider scope, and have included studies of pesticide effects on numbers of eggs laid, percent eggshells cracked, mortality of fledglings, and on the activity of certain enzymes within the parent birds. The literature dealing with pesticide effects on birds is so voluminous that it would be impossible to survey all the relevant articles here. In the following discussion, however, we review some of the more important papers, pointing out areas where controversy has arisen. This discussion provides an excellent example of an issue on which supposedly knowledgeable scientists have sometimes reached very different conclusions, although their judgments have been based on essentially the same set of experimental facts and observations.

Field Observations

Changes either in eggshell weight or shell thickness of certain birds were reported both in England (Ratcliffe, 1970) and in the United States (Hickey and Anderson, 1968). The birds involved were peregrine falcons, sparrowhawks, and golden eagles

---

[8]e.g., the Audubon Christmas Bird Count.
[9]e.g., shooting bald eagles.

in England; and peregrine falcons, ospreys, and bald eagles in the United States. In England, Ratcliffe (1970) reported that the relative weight of peregrine falcon, sparrowhawk, and golden eagle eggs had declined by 19%, 17%, and 10% respectively, since 1946, and that since 1950 egg breakage had occurred with unprecedented frequency in the nests of these birds. Ratcliffe observed no effect, however, on the frequency of egg breakage or relative egg weight in a number of other species studied. His observations indicated that the problem was peculiar to certain species or that some species were much more susceptible than others. The onset of the high frequency of egg breakage followed by a few years the introduction of DDT and benzene hexachloride (BHC) use in England (1946–1948). Lockie et al. (1969) attributed the thinning of golden eagle eggshells to the use of dieldrin for dipping sheep[10]. Indeed a marked decline in the breeding success of English golden eagles did not begin until 1960, after the introduction of dieldrin (Peakall, 1970b).

Figures 10.12–10.13 show the results of two often-cited studies that relate DDE residues in eggs to eggshell thickness. In both cases, there appears to be a significant negative correlation between eggshell thickness and DDE residues. Actually no one believes that thin eggshells are caused by the pesticides in the eggs. The physiological processes of the female bird control the thickness of the eggshell. The pesticide residues in the egg, however, are commonly assumed to reflect the residues in the female bird, and the latter residues may affect the bird's physiological processes.

Results such as those in Figures 10.12–10.13 have been cited frequently as implicating DDE as a causative agent of eggshell thinning. According to Spitzer et al. (1978), eggshell thinning of about 15–20% appears to be a critical level associated with hatching failure due to cracking of eggs during incubation. Others, however, have criticized these field studies as being statistically meaningless. Hazeltine (1972), for example, pointed out that the data in Figure 10.13 consist of results obtained from three separate regions involving two subspecies of pelicans. He argued that if the overall correlation were significant, each region, "should show a similar trend in itself." If the data are broken down by region, only the results from the Florida study show a significant correlation between eggshell thickness and DDE levels.

Figure 10.12 Variation in shell thickness and DDE concentrations in the eggs of herring gulls in 1967. The eggs were taken off Block Island, RI; Green Island in Penobscot Bay, ME; Rogers City, Mich., on Lake Huron; near Knife River, MN, on Lake Superior; and the Sister Islands in Green Bay, WI. Some polychlorinated biphenyls probably also occurred in these eggs, but they have not been identified. [Redrawn from Hickey and Anderson (1968). Copyright © 1968 by the American Association for the Advancement of Science.]

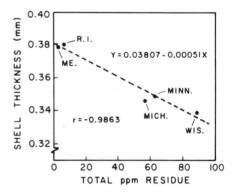

[10]Golden eagles feed on dead sheep.

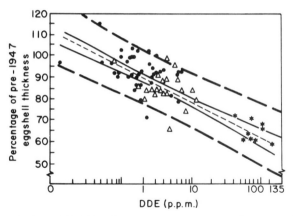

Figure 10.13 Association of DDE residues in 80 pelican eggs from Florida (solid circles), South Carolina (triangles), and California (stars) with the percentage of pre-1947 eggshell thickness. The regression line (narrow dashed line) is shown together with the 95% confidence limits for both the regression line (solid lines) and the individual ordinate values (thick dashed lines). [Redrawn from Blus et al. (1972b).]

The banning of DDT use in the United States at the end of 1972 provided scientists with an opportunity to study correlations between DDT residues (mainly DDE) and eggshell thickness in single geographic regions as residues in eggs declined. In 1969 the lipid tissue in southern California brown pelican eggs contained about 850 ppm DDE and intact eggshells averaged about 0.4 mm thick (Anderson et al., 1975). Between 1974 and 1975 DDE levels in the lipid tissue of the eggs had declined to about 100 ppm, and eggshell thickness had increased to 0.48–0.51 mm (Anderson et al., 1975, 1977). As noted in Chapter 8, the reproductive success of these birds increased dramatically as the DDE levels in the eggs declined. Between 1970 and 1976 the DDE concentrations in the eggs of Ospreys breeding in the Connecticut–Long Island area decreased fivefold, and mean eggshell thickness increased from about 0.41 to 0.45 mm (Spitzer et al., 1978). These results offer further support to the idea that DDE is associated with eggshell thinning in at least some species of birds. The existence of simple correlations in field data, however, by no means proves a cause and effect relationship. In order to examine the possible role of pesticides in eggshell thinning, as well as to look for possible effects on parental behavior and juvenile mortality, scientists have performed a number of laboratory experiments on bird populations under controlled conditions in which the only significant variable has been the degree of pesticide exposure. Such experiments hopefully eliminate much of the environmental "noise" caused by natural variations in environmental parameters other than pesticide exposure levels. Some of these experiments have been designed simply to determine whether exposure to pesticides influences the reproductive success of certain birds; other experiments have been aimed at elucidating the mechanism or mechanisms by which pesticides might affect reproduction.

Laboratory Studies

Why would we expect pesticides to interfere with the reproduction of birds? First of all, it is generally agreed that certain pesticides can upset the normal chemical balance in an organism by inducing the production of liver enzymes. Studies have shown that such enzymes can hydroxylate a number of substances, including the sex hormones estrogen, testosterone, and progesterone (Peakall, 1970b). The inference was then made that pesticides might be interfering with bird reproduction by upsetting the birds' hormone balance. Interference with a bird's sex hormones could affect parental behavior with respect to egg laying, incubation of eggs and/or rearing of hatchlings. With respect to eggshell thickness, it is known that estrogen promotes the storage in the bird's bone marrow of a large percentage of the calcium needed to produce the eggshell (Taylor, 1970). Carbonic anhydrase, an enzyme found in the oviduct of birds, is also believed to be important in eggshell formation (Pocker et al., 1971; Peakall, 1970b). Inhibition of this enzyme by sulfanilamide, for example, results in poorly calcified eggs (Gutowska and Mitchell, 1945). Thus pesticides could affect eggshell thickness by either depressing estrogen levels or by inhibiting carbonic anhydrase activity.

Peakall (1970a) performed a series of experiments with ringdoves to examine some of these hypotheses. He found that estrogen levels were depressed by 33% in the blood of birds given 10 ppm DDT in their feed for 3 weeks prior to breeding, and the DDT-fed birds were found to have stored only about 40% as much calcium in their bone marrow as had birds given uncontaminated feed. Furthermore, the time from mating to the laying of the first egg was 5 days longer (21.5 vs. 16.5 days) for the DDT-fed birds. Both the reduced amount of calcium deposited in the bone marrow and the lengthened time to egg laying may have resulted from the lower estrogen levels in the DDT-fed birds.

The possible role of DDT in causing birds to produce thin eggshells through inhibition of carbonic anhydrase is a somewhat controversial subject. Carbonic anhydrase is usually present in at least a 100-fold excess in biological systems (Maren, 1967), so that inhibition of enzyme activity by at least 99% would be needed to produce a significant physiological effect. Both Peakall (1970a) and Bitman et al. (1970) conducted experiments in which birds were exposed to either DDT and DDE, and then assayed the birds' blood or shell gland for carbonic anhydrase activity. They observed inhibition ranging from 20 to 59%. Inhibition by such amounts would not be likely to produce a noticeable physiological effect. Straus and Goldstein (1943), however, have noted that the degree of enzyme inhibition is often dependent on dilution effects, and since the method of sample preparation used by Peakall (1970a) and Bitman et al. (1970) involved dilution, there is a good chance that the inhibition they reported is an underestimation of the inhibition which may have occurred in vivo.

A number of other workers have attempted to observe carbonic anhydrase inhibition in vitro (Dvorchik et al., 1971; Pocker et al., 1971; Anderson and March, 1956). These experiments included studies of carbonic anhydrase from human red blood cells, from insects, and from cattle. The chemicals studied included DDT, DDE, and dieldrin. In no case was any chemical inhibition of carbonic anhydrase activity observed. Dvorchik et al. (1971) and Pocker et al. (1971) both reported that in tests with very high pesticide concentrations, some inhibition of activity occurred due to a physical precipitation of the enzyme with the pesticide. It seems unlikely, however, that this mechanism would underlie possible in vivo effects. Torda and

Wolff (1949) and Keller (1952) had reported previously inhibition of bovine and human red blood cell carbonic anhydrase by DDT, but later attempts (above) failed to reproduce these results. It now appears that the earlier results were in error.

In addition to the papers of Peakall (1970a) and Bitman et al. (1970), a number of other reports (Bitman et al., 1969; Porter and Wiemeyer, 1969; Heath et al., 1969; Wiemeyer and Porter, 1970; Cecil, Bitman, and Harris 1971) have been cited often as implicating DDT or DDE in the reproductive failure of birds. Unfortunately some of these studies have been conducted in a manner that makes unambiguous interpretation of the results difficult, and as a result the studies have been occasionally the object of some rather harsh criticism[11]. It is of interest to review those reports.

Bitman et al. (1969) fed Japanese quail a low calcium diet containing 100 ppm DDT. The quail produced eggs with about 4% less calcium than controls and had a lag in egg production analogous to the lag reported by Peakall (1970a) for ringdoves. The experiments were criticized by Edwards (1971) and others on the grounds that the birds were stressed artificially by the low calcium diet. In response to this criticism, Cecil et al. (1971) performed a second set of experiments in which Japanese quail were given an adequate amount of calcium and either 100 ppm DDT or DDE. The experimental birds produced over twice as many broken eggs. There was no significant difference, however, between the shell calcium content of the DDE-fed birds and controls, nor was there any significant difference in the eggshell thickness of the controls and either the DDT- or DDE-fed birds.

Porter and Wiemeyer (1969) fed combined doses of DDT and dieldrin to captive American sparrow hawks. They found a marked reduction in reproductive success, particularly among first-generation (yearling) hawks[12]. Eggshells produced by the yearling hawks were 15–17% thinner than controls. Unfortunately it is impossible to tell from these experiments whether DDT had anything to do with the results. Peakall (1970b) found dieldrin to be an even more powerful inducer of liver enzymes than DDT, and in England Lockie et al. (1969) found that golden eagle breeding success increased by 38% as soon as dieldrin sheep dips were banned. Edwards noted (1971), "Why set up experiments feeding only mixtures of DDT and dieldrin, then blame the DDT for all ill effects?" Wiemeyer and Porter (1970) reported later the results of a study on captive American kestrels, in which the experimental birds were given a diet containing 10 ppm DDE. No other pesticides were added to the feed. Eggshells produced during the first 7 weeks of feeding were virtually identical in thickness to control eggshells, but after 1 year of feeding the eggshells produced by the DDE-fed birds were about 10% thinner.

The laboratory study that demonstrated most convincingly that DDT residues in feed could affect bird reproduction adversely is probably that of Heath, Spann, and Kreitzer (1969), who conducted a 2-year study of the effects of feeding DDT, DDE, and DDD to mallard ducks. Several weeks before the onset of the first laying season, test ducks began receiving either 10 or 40 ppm DDE or DDD in their feed, or 2.5 or 10 ppm DDT. The results of the DDE feeding were discussed briefly in Chapter 8; a

---

[11]For example, Edwards (1971) and the letter to the *Barometer* quoted at the beginning of this chapter.

[12]In other words, the offspring of hawks fed DDT and dieldrin.

more complete summary of the results is shown in Table 10.5. Ducks fed 10 or 40 ppm DDE produced only 20–51% as many offspring as controls and produced eggshells about 10% thinner than controls. Ducks fed 10 or 40 ppm DDD produced only 44–57% as many offspring as controls, and although experimental eggshells were slightly thinner than controls, the difference was not statistically significant. DDT-fed ducks produced slightly fewer offspring than controls in the first study year, but surprisingly 23–37% more offspring in the second study year. Eggshell thinning for the DDT-fed birds was not statistically significant.

The fact that the DDT-fed ducks produced more offspring than controls during the second study year has been offered by some as evidence that DDT has little or no adverse effect on bird reproduction (Edwards, 1971). The results of this study, however, leave little doubt that DDE and DDD at concentrations as low as 10 ppm in feed can greatly reduce mallard duck reproduction. Since DDE is a rather stable metabolite of DDT, the inference can be made that release of DDT to the environment could interfere with mallard duck reproduction through the effects of the metabolite DDE. It is therefore noteworthy that over 3 times as much DDE as DDT was found in southern California anchovies by Anderson et al. (1975), despite the fact that DDE itself has never been used as a pesticide. The anchovy is a principal food source of southern California brown pelicans, and according to Anderson et al. (1975) the DDE in southern California anchovies in 1969 was about 3.2 ppm on a wet weight basis. This figure is equivalent to about 9.7 ppm on a dry weight basis, a concentration virtually identical to the 10 ppm dry weight concentration which Table 10.5 indicates reduced mallard duck reproduction by at least 50%.

Summary of Pesticide Effects on Birds

Having reviewed most of the better-known studies of pesticide effects on birds, what conclusions can we reach? First, the field studies of Blus et al. (1971, 1972a,b), Hickey and Anderson (1968), Anderson et al. (1975), and Spitzer et al. (1978) all show significant negative correlations between DDE residues in eggs and eggshell thickness. The latter two studies are particularly relevant, since the data were collected from birds in the same geographical area and were therefore not subject to possible interregional biases. Other studies have shown that dieldrin is probably more effective than DDE in interfering with bird reproduction.

Table 10.5   Reproduction and Eggshell Thickness Data from Penned Mallards Maintained for Two Seasons on Feed Containing DDE, DDD, or DDT[a]

| | | Pesticide added to feed (ppm) | | | | | |
| | | DDE | | DDD | | DDT | |
	Year	10	40	10	40	2.5	10
14-day ducklings per hen	1	51	35	57	51	97	89
	2	24	20	53	44	137	123
Shell thickness	1	92	87	97	97	—	—
	2	89	86	95	95	95	92

[a]Numbers are results as percent of control results.

Source. Heath et al. (1969).

Second, Peakall (1970a) has demonstrated convincingly that DDT can depress estrogen levels in the bird's blood and hence cause a delay in reproduction and a reduction in the amount of calcium stored in the bone marrow. Dieldrin was found to be even more potent than DDT in this respect.

Third, the evidence linking DDE to carbonic anhydrase inhibition is weak. Several studies have reported no effect at all in vitro. In the two cases where some inhibition has been reported (Peakall, 1970a; Bitman et al., 1970), the inhibition was almost certainly insufficient to produce a physiological effect. However, the issue is still open due to the difficulty of comparing in vitro and in vivo systems.

Fourth, at least one laboratory experiment (Heath, Spann, and Kreitzer, 1969) involving controlled feeding of DDE to birds has demonstrated that DDE has a significant effect on reproduction and eggshell thinning. The effects were evident at DDE concentrations of 10 ppm dry weight (about 3 ppm wet weight), a level that unfortunately could have been easily encountered in the natural food of some birds prior to the EPA ban on DDT use in 1972.

Finally, the recovery of raptorial birds such as golden eagles (Lockie, Ratcliffe, and Balharry, 1969), and brown pelicans (Anderson and Gress, 1983) following restrictions on the use of dieldrin and DDT, respectively, supports strongly the implications of the laboratory studies that these pesticides or their residues can have serious effects on bird populations when passed up through the food chain.

PEST RESISTANCE

Disenchantment with the widespread use of synthetic pesticides is often perceived as being largely the result of the adverse effects on nontarget species associated with pesticide use. Development of resistance in pest populations, however, has also been a major factor in the decline of pesticide popularity. In 1938, only seven insect and mite species were known to have acquired resistance to synthetic pesticides. By 1991, this figure had increased to 504 and included most of the world's major pests (Pimentel, 1991). Resistance to herbicides, which was virtually nonexistent before 1970, had become apparent in 273 weed species by 1991 (Pimentel, 1991).

In the case of insects, resistance, once acquired, appears to be more-or-less permanent. Resistance of Danish houseflies to the DDT and chlordane pesticide groups, for example, has persisted for more than 20 years, and according to Metcalf (1981, p. 472), "When these compounds were tried again they became useless within 2 months." A major problem with pest resistance is the fact that the effect of resistance genes is often not specific to a particular pesticide, but instead confers resistance to several or even many pesticides simultaneously. In northeastern Mexico, for example, the tobacco budworm, a major cotton pest, has developed resistance to every registered insecticide; and in Suffolk County, Long Island, the Colorado potato beetle is now resistant to all major insecticides registered for use on potatoes (Postel, 1987).

Mechanisms of Resistance

Understanding of the mechanisms responsible for the development of pest resistance does much to explain this behavior. The principal mechanisms involved appear to be enhancement of the capacity to metabolically detoxify pesticides and alterations in target sites that prevent pesticides from binding to them (Arntzen et al., 1982; Brattstein et al., 1986). Certain herbicides that interfere with photosystem II, for example, compete with quinone for binding to thylakoid membranes, and it has been proposed that herbicide resistance results when an alteration in the membrane reduces herbicide binding and hence allows quinone to bind even in the presence of the herbicide (Vermaas et al., 1983). Erickson et al. (1985) have identified three amino acid residues in the thylakoid membrane protein that can be independently altered to produce resistance to such herbicides.

In the case of insects, it is known that lipophilic insecticides are primarily detoxified by microsomal oxidases (Brattsten et al., 1986). With some exceptions, these enzymes convert the pesticides to polar metabolites that can be excreted. An important exception is the enzyme DDT dehydrochlorinase, which rapidly converts DDT to DDE (Metcalf, 1981). Considering the adverse effects of DDE on bird reproduction, it is ironic that DDE is noninsecticidal. Of the various microsomal oxidases, glutathione transferases are important in organophosphate detoxification (Montoyama and Dauterman, 1980), and carboxylesterases are often involved in pyrethroid resistance (Soderlund, Sanborn, and Lee, 1983).

Target site modification is believed to be less common than metabolic detoxification as a mechanism of insect resistance, but several types of target site modification have been documented among insects. So-called knockdown resistance is associated with target site modification in the sodium channel of the nervous system and was responsible for the rapid development of DDT resistance in houseflies (Winteringham, Loveday, and Harrison, 1951). The gene responsible for knockdown resistance also confers resistance to pyrethroids (Sawicki and Farnham, 1967). A second type of target site resistance is a modification in the synaptic acetylcholinesterase that renders it far less sensitive to inhibition by organophosphates and carbamates (Brattstein et al., 1986).

The Cost of Pest Resistance

The cost of pest resistance appears in two forms. The first is the obvious fact that certain pesticides or classes of pesticides become virtually worthless for controlling pests. DDT, for example, was initially used with great success to control the insect vectors of the pathogens causing malaria, typhus, yellow fever, and plague. Largely because of pest resistance, DDT is either no longer used to control these vectors or, in the case of malaria, used with much less success than initially. The second cost is quite simply the amount of money that must be spent to achieve a satisfactory level of pest control. Corn rootworm beetles, for example, were controlled from about 1954 to 1964 by soil applications of aldrin and heptachlor, but resistance to these insecticides became almost complete during the mid-1960s. A switch to organophosphates at that time was associated with almost a fivefold increase in pesticide cost per kg (Metcalf, 1981). Resistance of the Colorado potato beetle to insecticides has caused potato growers on Long Island to spray up to 10 times per season, and pesticide costs have risen as high as $700 ha^{-1} (Postel, 1987). These heavy pesticide applications have caused serious contamination of groundwater, the region's only source of drinking water (Dover and Croft, 1986; Holden, 1986). The pesticides being used currently to replace DDT in malaria control programs cost 5–20 times as much per capita, and as a result malaria control has become too costly for some developing countries. In addition to the previously dis-cussed case of Sri Lanka, major recrudescences of the disease have occurred in India and Pakistan (Metcalf, 1981).

ALTERNATIVES TO SYNTHETIC PESTICIDE USE

The foregoing discussion leads one to ask, "Isn't there a better way?" In many cases the answer is yes. In fact, there are several better ways. Most scientists now agree that widespread use of synthetic chemicals to control pests was a mistake, and in some cases even counterproductive. There is a general consensus now that reliance on such pesticides should be minimized, and that alternative pest control methods should be utilized to the maximum extent possible. This realization has stimulated much basic and applied research into alternative control strategies. In the following discussion we review the progress that has been made in this area.

Biological Control

The introduction of natural pest predators, competitors, or pathogens has in some cases been proven to be an effective means of controlling pest populations; the idea is not new. Postel (1987) claims that since the 1860s scientists have introduced roughly 300 organisms worldwide in biological control programs, but Holcomb (1970) maintains that almost 700 insect enemies alone have been introduced in such control efforts. Unfortunately many such introductions do not produce the desired results. Holcomb (1970), for example, states that less than 25% of the introduced insect enemies actually became established. Even if the pest enemy does become established, there is no guarantee that it will reduce the pest population to an accept-ably low level. Mongooses, for example, were introduced into Hawaii to control the rats that had taken up residence in sugar cane fields. The mongooses were quite suc-cessful at establishing themselves in the Islands, but because the cane rats are noc-turnal and the mongooses active primarily during the day, the introduction had virtually no effect on the rat population. When biological control works, however, the effects can be dramatic and far less costly than the use of synthetic pesticides. The following sections summarize some of the more noteworthy successes of biological control.

Natural Predators and Parasites

In 1868 an insect called the *cottony cushion scale (Icerya purchasi)* was introduced accidentally into California on acacia plants from Australia. The spread of the scale onto citrus trees threatened to destroy the California citrus industry. In 1886 ladybug beetles *(Rodalia carbinalis)*, a natural predator of the insect, were intro-duced in an effort to achieve biological control. The beetles reduced the pest popula-tion so effectively that no further effort to control the cottonly cushion scale has proven necessary. Similar use of a small parasitic wasp *(Aphytis melinus)* has helped to control California red scale on citrus orchards in southern California (Marx, 1977). A related wasp introduced into Florida citrus orchards to control the snow scale has saved as much as $10 million per year in reduced insecticide costs.

In Australia control of several species of prickly pear (genus *Opuntia*) has been achieved successfully by the accidental introduction of a moth (*Cactoblastis cac-*

torum) from Argentina. The prickly pear, which originated in the United States, had spread over 400,000 km² of Australian grazing land by 1925. In many areas the prickly pear had rendered the grazing land unusable, and in some areas the cactus was impenetrable (Gunn, 1976). The introduction of about 50 species of natural insect enemies of the prickly pear had proven largely unsuccessful. By 1930, however, following the accidental introduction of *C. cactorum*, the numbers of prickly pear had been so reduced that no further control measures were necessary. Deliberate introduction of the European beetle (*Chrysolina quadrigemina*) to control Klamath weed (*Hypericum perforatum*) on California range land in 1946 met with similar success. Roughly 8000 km² were cleared for grazing. About $750,000 was spent to locate and introduce the beetle; accumulated savings from the biological control have been estimated at more than $100 million (Postel, 1987).

The cassava is a major food crop of some 200 million Africans, and much of it is grown on small plots by subsistence farmers (Postel, 1987). Mealybugs, which first appeared in Zaire in 1973, had infiltrated a large portion of the cassava belt by 1982, and together with the spider mite, were causing cassava crop losses estimated at $2 billion per year. A pesticide program was ruled out because there was no infrastructure to deliver chemicals to the subsistence farmers. Searching in Latin America, the cassava's place of origin, turned up about 30 natural enemies of the mealybug, and one of these, a tiny wasp called *Epidinocarsis lopezi*, has produced remarkable results. By parasitizing the eggs of the mealybug, the wasp now effectively controls the population of the pest over 650,000 km² in 13 countries (Postel, 1987).

One of the principal insect pests of onions is the onion maggot, and without some form of pest control onion losses in the United States average about 40%. Pimentel et al. (1991), however, reported that losses could be reduced to 2–3% by simply raising cattle adjacent to the onion field and mulching the onions with straw. A parasitic wasp uses the maggots in the cattle manure as an alternate host, and the straw protects a beetle that preys on the onion maggot.

In China a parasitic wasp is used to control stem borers, a sugarcane pest, at one-third the cost of chemical control; and a fungus and parasitic wasp provide 80–90% control of a major corn pest (Postel, 1987). In Sri Lanka pest damage estimated at $11.3 million per year to coconuts has been prevented with the introduction of a parasite found and shipped for $32,250 in the early 1970s (Postel, 1987); and in Costa Rica the use of pesticides on bananas was simply stopped. Natural enemies reinvaded to control the banana pests.

One variation of the predator/parasite control strategy is to release large numbers of a natural enemy at a critical time of year. *Trichogramma*, a tiny wasp that parasitizes the eggs of certain butterflies and moths, is used extensively in this sort of control. The idea is to prevent the eggs from developing into crop-damaging caterpillars. *Trichogramma* is presently used to control moth pests on about 170,000 km² of cropland worldwide (Postel, 1987).

Pathogens and Natural Toxins

Insects are attacked by a great many pathogens, and about 450 viruses, 80 bacteria, 460 fungi, 250 protozoa, and 20 rickettsial diseases are effective natural enemies (Metcalf, 1981). A number of these are adaptable for use as insecticides, and in recent years much interest has developed in using these pathogens or toxins derived from them to control insect populations. Such natural pesticides tend to be highly specific to one or a few pests and are generally harmless to other animals. The bacterium *Bacillus popilliae*, for example, has been used to control the Japanese beetle (*Popilbia japonica*) on lawns and golf courses. The strategy has been to collect *B. popilliae* spores from infected larvae of the beetle and to mix the spores with talc to produce a standardized powder, which is then applied to the soil. *B. popilliae* causes a disease (milky disease) that effectively controls the beetle population under certain conditions, but the control is not so effective in rough grass. Parasites taken from Europe to Canada to control the European spruce sawfly (*Neodiprion sertifier*) apparently carried a virus that proved highly effective in controlling the sawfly population (Gunn, 1976). The virus has since been deliberately spread to other parts of Canada and the United States. The first insect virus to be developed commercially in the United States was the nuclear polyhedrosis virus (NPV) of the cotton bollworm. Since then NPV insecticides have been developed to control several other insect pests, including the corn earworm, the primary insect pest in sweet corn (Pimentel et al., 1991), and the gypsy moth (Marshall, 1981).

Undoubtedly the best known example of a bacterial control agent has been the use of the bacterium *Bacillus thuringiensis*. *B. thuringiensis* was first identified as an insecticide in 1927. The bacterium is lethal to insects, because it produces spores that release toxic proteins when eaten by insect larvae. It is nontoxic, however, to mammals and birds. *B. thuringiensis* can be used on a wide variety of agricultural crops to control insect pests, and spraying a mixture of spores and protein crystals on tree foliage has proven effective in controlling the gypsy moth (Moffatt, 1991). The bacterium has many subspecies, and each is effective against only a narrow range of insects belonging to the same order. This characteristic is desirable from the standpoint of impact on nontarget species, but it is undesirable from the perspective of companies contemplating commercialization. It is difficult to make money marketing a pesticide that affects only a very narrow range of pests. The specificity problem may well be overcome with genetic engineering techniques.

Most biological control methods developed to date have been aimed at insects, but in some cases biological control methods are available for weed control as well. The development of such alternatives is important particularly in countries such as the United States, where herbicides now account for a major fraction of pesticide use. Recently two bioherbicides that rely on a fungus as the biological agent have entered the U.S. market. DeVine, which is produced by Abbott Laboratories, is used to control the milkweed vine in Florida citrus groves. The weed, also known as the strangler vine, climbs a tree's trunk, covers the crown, and eventually smothers the tree (Postel, 1987). Treatment with DeVine has been found to give at least 5 years of protection from this pest. Use of Collego, a bioherbicide marketed by the Upjohn Company, has produced 90% control of northern jointvetch, a weed commonly found in Arkansas rice and soybean fields (Postel, 1987). Another form of biological control of weeds is the planting of cover crops that inhibit the germination or growth of weeds. This approach takes advantage of a phenomenon called *allelopathy*, the inhibition of one plant by another through the release of natural toxins. Leaving residues of rye, sorghum, wheat, or barley on a field, for example, can provide as much as 95% weed control for up to 2 months (Postel, 1987).

Genetic Control

Resistant Plants

Through genetic manipulations such as cross-breeding and selection of desirable mutant strains, it has been possible to develop crop strains that are wholly or largely resistant to predation or infection by certain pests. For example, the European wine industry was saved from ruin a century ago by the practice of grafting nonresistant but high-quality European vines onto American root stocks resistant to the attacking aphids *Phylloxera vastatix* (Gunn, 1976). Pest-resistant strains are available for a number of crops of agricultural importance in the United States, including alfalfa, beans, cole, corn, cotton, cucumbers, grapefruit, lemons, oranges, potatoes, rice, sweet potatoes, and wheat (Pimentel et al., 1991). In the case of wheat, for example, only about 7% of the hectarage is treated with insecticides, because host-plant resistance, crop rotations, and other agricultural practices have succeeded in minimizing insect damage. On the other hand, in the case of corn, which accounts for over 40% of agricultural insecticide use in the United States, farmers appear not to have taken full advantage of the availability of insect-resistant strains. Pimentel et al. (1991) estimate that the use of insecticides on corn in the United States could be reduced by 80% by rotating crops and planting corn resistant to the corn borer and chinch bug.

An important variation on the theme of resistant plants has been the use of genetic engineering to insert specific genes into the plant or into microorganisms that live in or on the plant. For example, the Rohm and Haas company has developed a genetically engineered strain of tobacco that contains a gene from the bacterium *Bacillus thuringiensis*. The gene triggers production of a *B. thuringiensis* protein, which is toxic to a broad spectrum of caterpillars that feed on plant leaves (Crawford, 1986). Similar insertions of *B. thuringiensis* genes into cotton, tomato, and potato plants have produced strains resistant to the cotton bollworm, tomato pinworm, and the Colorado potato beetle, respectively (Moffatt, 1991). A related strategy is to insert the *B. thuringiensis* toxin gene into other bacteria such as *Pseudomonas fluorescens*, which grows in roots, and *Clavibacter xyli*, which grows in the vascular tissue of corn and other plants. Plants infected with the genetically engineered bacteria receive a continuous supply of insect toxins to protect them from attackers (Moffatt, 1991).

Sterile Males

Releasing large numbers of sterile males into an insect population at the appropriate time may effectively reduce reproductive success, particularly if each female pest mates with only one male. One of the best known and certainly one of the most successful examples of this type of pest control has been the eradication of the screwworm fly population from the United States. The female screwworm fly lays her eggs in open wounds on warm-blooded animals. She can utilize an opening as small as a tick bite. The eggs hatch in about 1 day and release hundreds of larval worms that burrow into the wound and feed on the animal's flesh. A full-grown steer can be killed in a week if the worms enter a dehorning wound and get into the brain, and infestation of the umbilical cord of a newborn calf can lead to death within a few days. The fly has even been known to kill humans by entering the brain or lungs (Palea, 1990). Prior to the advent of the control program, livestock losses to screwworm flies in the United States were estimated to be $40 million per year (Carson, 1962).

During the late 1950s scientists at the U.S. Department of Agriculture began raising literally hundreds of millions of screwworm flies on artificial media and irradiating pupae with gamma rays just before they became adult flies. In this way the scientists produced vast numbers of sterile male screwworm flies, which they then released into infested areas. The sterile males compete with wild males for female mates, and since a female screwworm fly typically mates only once, the sterile male release program can reduce reproduction drastically as long as the wild male population is outnumbered greatly by the sterile males. In the case of the screwworm fly program, the ratio of sterile males to wild males was about 10:1 (Bush, Neck, and Kitto, 1976). The control program was initiated in Florida between 1958 and 1959 with remarkable success. The screwworm fly was eradicated completely at a total cost of $10 million. The focus of the program then shifted to other parts of the southern United States, and by 1962 the screwworm fly population in the United States was confined almost entirely to the state of Texas.

The effort to control the screwworm fly population in Texas was initiated in 1962, and at first met with remarkable success (Figure 10.14). The number of annual infestations dropped from almost 50,000 in 1962 to less than 500 between 1969 and 1971. There was a remarkable epizootic, however, from 1972 to 1976, during which screwworm fly infestations in Texas averaged over 30,000 per year. The cause of this epizootic has never been identified beyond reasonable doubt, but appears to have been due to a combination of factors, including local abiotic conditions, matters related to sterile fly production and release, seasonal effects, and general program execution (Krafsur, 1985). The success of the program returned in 1977, and the last screwworm fly infestations in the United States were reported in 1982. Since then the fly has been pushed south through Mexico, and is now confined to an area south of the Isthmus of Tehuantepec (Palea, 1990).

One of the chief problems with the sterile-male approach to insect control is getting enough sterile males into the population at the right time. In the Texas control program, for example, release rates averaged 1.0–9.2 million sterile males per day

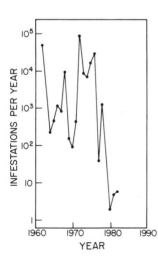

Figure 10.14 Incidence of screwworm fly infestations in Texas from 1962 to 1982. (*Source.* Krafsur, 1985.)

(Krafsur, 1985). Given that the density of this pest is only about 40–80 km^{-2} (Bush et al., 1976), you can imagine the problem of using sterile males to control a pest whose population density may be many orders of magnitude higher. For many insects the job of rearing and then releasing adequate numbers of sterile males to check the pest population is simply impossible. The technique has been used, however, with some success as part of a program to prevent fruit flies from becoming a serious problem in southern California (Anonymous, 1990).

Chemical Methods

Other than the synthetic pesticides, there are two other classes of chemical methods to control pests. The first method is the use of hormones or natural biochemicals that affect hormones to alter the growth and/or development processes of a pest and consequently render it unable to grow, feed, or reproduce. For example, two chemicals isolated from the East African medicinal plant, *Ajuga remota*, have been found to kill certain agricultural pests by inhibiting molting. The chemicals cause insect caterpillars to grow as many as three new head coverings or cuticles without shedding any old ones, and as a result their mouths become so deeply buried inside the coverings that the caterpillars are unable to eat and starve to death (Maugh, 1981). Unfortunately the cost of developing such products can be substantially more than the cost of developing a conventional pesticide (Holcomb, 1970); and in some cases the chemical must be applied at just the right time in the life cycle of the pest to be effective. Some such products are available for mosquito and aphid control, but it does not appear that this approach will become a major factor in the effort to control insect pests without the use of conventional synthetic pesticides.

The second class of nonconventional chemical control involves the use of pheromones, which are chemicals secreted by animals to influence other members of the same species. Pheromones may be used to control insect pests in basically three ways. First, pheromones may be used to attract insects into traps. This approach, for example, is used to control fire ants, rootworms on corn, the sweet potato weevil on sweet potatoes, and the grape berry moth on grapes (Maugh, 1981; Pimentel et al., 1991). In the case of corn, this strategy is so successful that conventional insecticide use is reduced by 99% (Paul, 1989). Second, pheromones can again be used in traps, but with the purpose of providing an indication of the abundance of the insect pest. This approach is used in so-called scouting, the idea being to determine when it is best to apply pesticide treatments. The use of scouting can reduce requirements dramatically for conventional pesticides by confining their use to those times when pest populations are approaching nuisance levels. It is clear from empirical studies that without the use of scouting conventional pesticides are often applied at times when their use is unnecessary (Pimentel et al., 1991). The third approach is to introduce sex-attractant pheromones into the air in amounts sufficient to confuse male insects, which are then unable to locate females for mating. This last approach has been used with some success, for example, in gypsy moth control. In laboratory studies the male gypsy moths become so confused by the smell of the pheromone that they, "have attempted copulation with chips of wood, vermiculite, and other small, inanimate objects" (Carson, 1962, p. 252), so long as the objects were suitably impregnated with the pheromone. This control method breaks down, however, if the pest population density is too great, because males encounter females an adequate number of times purely by chance.

Integrated Pest Management

Integrated pest management (IPM) has been defined by the United Nations Food and Agriculture Organization as, "a pest management system that, in the context of the associated environment and the population dynamics of the pest species, utilizes all suitable techniques and methods in as compatible a manner as possible and maintains the pest populations at levels below those causing economic injury" (Batra, 1982). What this statement refers to in practice is the attempt to minimize the use of conventional synthetic pesticides by using other means of pest control as much as possible and resorting to conventional pesticide applications in small doses if the pest population starts to get out of hand. The other methods used to control the pests may include one or more of the biological, genetic, or chemical control methods just discussed, and in the case of agricultural pests may also include the following:

1. Crop rotation to check the buildup of a pest specific to one crop.
2. Fallowing for a year to similarly retard the development of crop pests.
3. Use of ridge-till to facilitate weeding.
4. Choosing the date of planting so as to be able to harvest the crop before the pest population has had a chance to peak.
5. Use of rope-wick application of herbicides to increase the amount of herbicide that reaches the target weeds and hence reduce the amount of herbicide used.

Recent developments in cotton growing provide a good example of IPM. Before the advent of IPM, cotton accounted for about 45% of all insecticides applied to agricultural crops in the United States (Marx, 1977). In the 1950s and 1960s the boll weevil and cotton fleahopper were the key pests responsible for crop losses. Unfortunately the insecticides used to control these two pests also killed the natural enemies of the bollworm and tobacco budworm, which then became serious pests (Adkisson et al., 1982). Furthermore, both the bollworm and tobacco budworm began to develop remarkable resistance to conventional pesticides, and even though many fields were treated 15–20 times during the growing season, the crops were often severely damaged by insects. A strain of tobacco budworm began to emerge that was resistant to virtually all insecticides (Adkisson et al., 1982). It was obvious that some form of control involving minimal use of insecticides was needed.

The key element of the new control program has been the development of a new strain of cotton that matures quickly and can be harvested in late July or early August, about a month before other kinds of cotton. The population dynamics of the boll weevil is such that this short-season strain of cotton has been harvested before the boll weevil population becomes a serious problem. After harvesting, the cotton stalks are shredded and plowed under and insecticides applied to the harvested cotton fields. This procedure kills many diapausing boll weevils, and hence minimizes the number that emerge in the spring. Insecticides are applied again to the fields in the spring to reduce the number of surviving adults before they have a chance to reproduce. The springtime insecticide applications, which also control the cotton fleahopper, are limited and timed to have the least impact on the natural enemies of the bollworm and tobacco budworm (Adkisson et al., 1982). The overall strategy then is to let natural enemies control the bollworm and tobacco budworm and to minimize the amount of pesticides required to control the boll weevil and fleahopper by growing short-season cotton, plowing under the stalks after harvesting, and controlling carefully the amount and timing of pesticide applications in order to minimize harm to the natural enemies of the bollworm and tobacco budworm. Cotton grown in the short-season system requires only about 25% as much insecticide as conventionally grown cotton, and in addition only 25% as much nitrogen fertilizer and 50% as much irrigation water. The result is a substantial increase in profitability to the farmer (Table 10.6).

Experience with the use of IPM has been highly favorable. Not only is pesticide use reduced greatly, but in many cases profitability is increased substantially as well. Tables 10.6 and 10.7 summarize a few of the results. There is, however, considerable room for further improvement. In the United States for example, there is presently minimal use of IPM on corn, which now accounts for almost twice as much agricultural insecticide use as any other single crop. According to Pimentel et al. (1991), simply rotating corn with another high-value product such as soybeans and using strains of corn resistant to the corn borer and chinch bug would not only lower insect losses but would also reduce the use of insecticides on corn by 80%.

Another important area for improvement is the use of herbicides, which now account for almost 70% of agricultural pesticide use in the United States. In the past much of the development of alternative pest control techniques and implementation of IPM have focused on insect pests, but there are certainly techniques available to reduce the use of herbicides as well. In the United States, for example, corn accounts for over half of agricultural herbicide use, and there is good reason to

Table 10.6   Estimated Average Annual Economic Benefits from Use of IPM, Selected Cases, United States, Early 1980s

State	Crop	Increase in net returns to IPM users	
		Farm level (dollars/hectare)	Statewide (thousand dollars)
California	almonds	769	96,580
Georgia	peanuts	154	62,600
Indiana	corn	72	134,230
Kentucky	stored grain	<1	890
Massachusetts	apples	222	400
Mississippi	cotton	122	29,680
New York	apples	528	33,000
North Carolina	tobacco	6	780
Northwest[1]	alfalfa seed	132	2420
Texas	cotton	282	215,830
Virginia	soybeans	10	2570
Total			578,980

[1]Idaho, Nevada, Montana, Oregon, and Washington

Source. Postel (1987).

Table 10.7   Successful Applications of IPM in Agriculture

Country or region	Crop	Effect
Brazil	Soybeans	Pesticide use decreased 80–90% over 7 years
Jiangsu Province, China	Cotton	Pesticide use decreased 90%; pest control costs decreased 84%; yields increased
Orissa, India	Rice	Insecticide use cut by one-third to one-half
Nicaragua	Cotton	Early to midseventies effort cut insecticide use by one-third while yields increased
United States	Grain sorghum	Pesticide use reduced by 41% between 1971 and 1982
United States	Cotton	Pesticide use reduced by 75% between 1971 and 1982
United States	Peanuts	Pesticide use reduced by 81% between 1971 and 1982

Source. Postel (1987).

believe that crop rotation combined with mechanical weeding could reduce the use of herbicides on corn by about 60% (Pimentel et al., 1991). In the so-called ridge-till system, corn and other row crops are planted in ridges about 20–25 cm high. The ridges are spaced about 75 cm apart and are aligned perpendicular to the slope of the land in order to minimize soil erosion. Mechanical weeding is facilitated greatly by planting the crops in ridges, and reliance on herbicides can therefore be reduced greatly. Another important development in weed control is the use of rope-wick applicators rather than spraying to apply herbicides. In the rope-wick technique, a 10–13-cm-diameter pipe with small holes punched into it functions as the reservoir for a liquid pesticide. Segments of rope inserted into adjacent holes in the pipe serve as wicks that deliver the pesticide to the weeds as the applicator, positioned behind a tractor, is pulled between rows of crops. Use of the applicator allows for much more direct and efficient application of pesticides to weeds than spraying. In the case of soybeans, for example, use of rope-wick applicators has reduced herbicide use by about 90% and increased soybean yields by 51% compared to conventional treatments (Pimentel et al., 1991).

COMMENTARY

It should be clear from the foregoing discussion that the indiscriminate and widespread use of pesticides can and has created serious problems. The adverse effects of pesticides on nontarget species and the development of resistance in target species have both contributed to the decline in popularity of conventional synthetic pesticides and to the search for alternative means of pest control. Many of the alternative methods developed to date have focused on agricultural insect pests. Some remarkable success has been achieved with these alternatives, but it is clear that their potential has not been fully realized. We have a long way to go in terms of implementing alternative methods of controlling agricultural insect pests. There is also much to be done with respect to the identification and implementation of alternative methods of weed control in agriculture and pest control in general in forestry management and public health. In the United States, for example, Dutch Elm Disease, Chestnut Blight, and the Gypsy Moth remain serious forestry pests, and we have yet to identify a satisfactory way to control tropical diseases such as malaria and river blindness. Despite the progress that has been made in the development of alternatives to conventional synthetic pesticides, it therefore seems likely that synthetic pesticides will continue to be a significant component of many pest control programs in the foreseeable future. If we are to use such compounds wisely, it is important that we have a clear understanding of both the benefits and problems associated with their use.

Two examples serve to illustrate the dilemma faced by government officials who must decide whether or not to authorize the use of conventional synthetic pesticides. Following the ban on DDT use in the United States, the U.S. EPA granted requests to the states of Washington and Idaho and to the Forest Service to use DDT on the basis of economic emergency and no effective alternative to DDT being available (EPA, 1980a). In accord with this policy, permission was granted in June, 1974, to spray over 1700 km^2 of Douglas fir trees in Oregon with DDT to control an outbreak of tussock moths. The government evidently felt that an emergency existed and that there was no effective alternative to DDT spraying. The decision was made just 18 months after the ban on DDT use in the United States went into effect and was reached with a full awareness of the fact that large-scale killing of salmon and aquatic insects had accompanied forest spraying with DDT in parts of Canada. After evaluating the facts, the government was convinced, in this case at least, that the benefits of spraying outweighed the costs.

During the Vietnam War, the U.S. military used the herbicide Agent Orange extensively to defoliate areas of Vietnam. Unfortunately this herbicide turned out to be contaminated with dioxin.[13] Dioxin is formed primarily as a contaminant during the production of 2,4,5-trichlorophenol, which is the major chemical feedstock in the production of several herbicides, including 2,4,5-trichlorophenoxyacetic acid (2,4,5-T), 2,4,5-T esters, and 2-(2,4,5-trichlorophenoxy)propionic acid (Silvex). According to the EPA (1984), dioxin is one of the most toxic substances known to man. It affects the immune system in mammals and has been shown to be acnegenic, fetotoxic, teratogenic, mutagenic, and carcinogenic. Dioxin has caused birth defects in experimental animals at concentrations as low as 10–100 parts per trillion and is so toxic to humans that only 85 grams could kill the entire population of New York City (Costa, 1983). Following the Vietnam War, a number of military veterans experienced health problems that they attributed to their exposure to Agent Orange, which contained 2,4,5-T. The health problems included cancer, nerve, liver, and immune system disorders, birth defects, and chloracne. They filed a lawsuit against seven chemical companies that had produced Agent Orange for the U.S. military, and in 1984 won an out-of-court settlement of $180 million (Fox, 1984). Would the U.S. government use Agent Orange again? The answer is almost certainly no. The benefits do not outweigh the costs, and there are certainly alternatives available.

---

[13] Dioxins are, strictly speaking, any of 75 chlorinated dibenzo-para-dioxin compounds, but the dioxin of particular concern is 2,3,7,8-tetrachlorodibenzo-para-dioxin (2,3,7,8-TCDD), and we here refer to 2,3,7,8-TCDD simply as dioxin.

At this point it is appropriate to turn back to the beginning of this chapter and to reread the excerpt from Dr. Edwards' letter. In the light of what you have learned from reading this chapter, to what extent do you agree or disagree with the statements made by Dr. Edwards? The fact is that DDT continues to be used for malaria control and at times for other purposes because the benefits of its use, under certain conditions and for certain purposes, are still perceived to outweigh the costs. The same conclusion has not been reached about some other synthetic pesticides, Agent Orange being a case in point. If pesticides are to be used intelligently in the future, a realistic understanding of their effects, both good and bad, will be crucial to the decision-making process. Biased and extreme statements by antipesticide environmentalists or propesticide advocates do little to clarify a sometimes complex issue. Unfortunately such statements have a habit of creeping into the literature, particularly when the issues involved are controversial. If this fact has not yet become clear, the following verbal exchange, which took place at EPA hearings conducted on January 13, 1972 (Woodwell, 1972), should prove thought provoking. The two papers being discussed were published in *Science* by Dr. G. M. Woodwell (1964, 1967).

EPA Hearings, 13 January, 1972. Afternoon Session

MR. O'CONNOR: Q. *Doctor, in the introductory sentence in the abstract, the abstract reads, DDT residues in the soil of an extensive salt marsh on the south shore of Long Island averaged more than 13 pounds per acre, 15 kilograms per hectare. The maximum was 32 pounds per acre, 36 kilograms per hectare. What was the lowest—what were the lowest residues that you found?*

A. *Well, let's see. They're listed in the—in fact, the analyses are listed in Table 1 and where the samples were taken. We . . .*

Q. *Isn't it fact, Dr. Woodwell, that after you wrote this, or after you initially studied this salt marsh, that you continued your samplings, and that you found as a result of your continued samplings that you were getting around or less than one—an average of one pound per acre of DDT?*

A. *No, I wouldn't agree with that.*

Q. *Did you continue your samplings, Dr. Woodwell?*

A. *We have sampled the marsh subsequently and . . .*

Q. *What were your averages in . . .*

A. *—the conclusion we came to was that the marsh contains in the range of some pounds per acre of DDT.*

Q. *Tell me how many, Doctor.*

A. *I can't. It would be in the range of one to probably three pounds per acre.*

Q. *One to three pounds per acre?*

A. *Let me . . .*

Q. *That is the average that you now hold is the DDT residue in this marsh?*

A. *Let me clarify.*

Q. *Is that true, Doctor: Would you answer my question? Is it one to three pounds?*

A. *It is not true that the marsh now contains one to three pounds per acre.*

Q. *What was true at the time, Doctor? What was true at the time of your writing this article?*

A. *I would like to make a true statement. Would you like me to make a true statement or would you like me to simply say yes or no to what you say and which I consider untrue?*

EXAMINER SWEENEY: *Doctor, you're under oath so I imagine that every statement you make you realize must be true.*

DOCTOR WOODWELL: *That's correct, and I intend to speak the truth here.*

EXAMINER SWEENEY: *All right. Can you answer the question?*

MR. O'CONNOR: *Let me give you the question again, Dr. Woodwell.*

Q. *Isn't it a fact that since your initial sampling which resulted in this article which is Respondents'—Environmental Defense Fund's Exhibit 41, that you took additional samples of the salt marsh and concluded that the residues of DDT in that salt marsh were one to three pounds per acre?*

A. *It is true that we sampled the marsh subsequently and I believe that at the time of this sampling, the average residues in that marsh, obtainable from an extensive sampling of the marsh, would have revealed residues in the range of a few pounds per acre. It is also true that this sampling was deliberately biased in order to find the highest residues we could find, because at the time we did this study, we wondered whether we could find any residues in the marsh.*

Q. *What do you mean by "average of 13 pounds per acre," if what you're talking about is a bias? Is 13 pounds per acre, Doctor, an average or is it a bias?*

A. *It means the average of the samples we took in this marsh was 13 pounds per acre.*

Q. *That's not what this sentence says, Doctor. It says, "The DDT residues in the soil of an extensive salt marsh was 13 pounds per acre." Is that true or is it not?*

A. *That's true what that says. Based on this sampling, the average residues were 13 pounds per acre.*

Q. *Didn't you also find out later, Doctor, that one of the areas where you took your samples was an area of DDT dumping?*

A. *I can't say that I discovered that.*

Q. *Dr. Wurster perhaps?*

A. *I don't believe that he knows that either. I don't believe there's any evidence to that effect.*

Q. *Are you aware of the statement that Dr. Wurster made at the Wisconsin hearings—not the Wisconsin hearings, the Washington State hearings—where he said—and I'm quoting from the verbatim transcript of the proceedings, "we have since sampled that marsh much more extensively and we found that the average, the overall figure on the marsh is closer to one pound per acre. The discrepancy was caused by the fact that our initial sampling was in a convenient place, and this turned out to be a convenient place for the Mosquito Commissions's spray truck, too." Did you learn that after the fact, Doctor?*

A. *That is a true statement in my experience. I did not know that Dr. Wurster had said that, but that is a true . . .*

Q. *So one of the places of your sampling was a place where a dump truck had unloaded its DDT?*

A. *That is not what Dr. Wurster said.*

Q. *What did he say, then?*

A. *If I may read the record, I'd be glad to read it to you. Could I see . . .*

Q. *What is your interpretation of what we've just read, Doctor?*

A. *Could I read the paper that you've just introduced?*

Q. *Is what I just read to you unclear? Do you understand what Dr. Wurster said?*

A. *I would like to cite it directly.*

Q. *Well, I've just read it to you. Did you not understand that?*

MR. BUTLER: *Excuse me. The Doctor asked to see the paper to the question he answered. I think it would be reasonable if he could do so.*

EXAMINER SWEENEY: *The only point is, Mr. Butler, the Doctor has already answered the question, and then after he answered the question, all of a sudden, he apparently figures it wasn't read correctly. He answered the question. He said, "That is a true statement," and so forth. And later on, he decided for some reason or other, that he wanted to check it to see if it was. I will ask Mr. O'Connor to reread the question and we will then get the answer from the witness. Reread the question that you asked him, please.*

MR. O'CONNOR: *The statement which I read, Dr. Woodwell, was this:*

Q. *"We have since sampled that marsh much more extensively and we found that the average, the overall figure on the marsh is closer to one pound per acre. The discrepancy was caused by the fact that our initial sampling was in a convenient place and this turned out to be a convenient place for the Mosquito Commission's spray truck, too."*

A. *That's true. Now, you suggested that the spray truck had dumped there, and that is not true as far as I know.*

Q. *What does it mean by the statement that it was "a convenient place for the spray truck, too?"*

A. *My interpretation of what Dr. Wurster meant is that it was close to the road and the people who control mosquitoes tend to work most intensively close to the edge of the road, and the net result of that was that over a period of years they visited if very often and put in large quantities of DDT, perfectly straightforward . . .*

Q. *Doctor, have you ever published a retraction of this 13 pounds per acre or a qualification, or a further article which evidences and discloses the results of your further sampling which brings the average down to around one part per million.*

A. *I never felt that this was necessary.*

Q. *Isn't it a fact, Doctor, that 13 parts per acre—13 pounds per acre has been cited as a figure quite extensively in the literature on DDT, your figure here, which is used in this article, which is EDF Exhibit 41?*

A. *I am not sure that I would agree that it has been cited extensively. It has certainly been . . .*

Q. *You know that the figure "13 pounds per acre" has been cited in the literature, and you never took it upon yourself, Doctor, to write any kind of letter to the editor, a correction, further article, anything to clarify that error?*

A. *I don't consider that an error. It is not an error in the context of this paper. If I . . .*

Q. *How many samples did you take in this salt marsh, Doctor?*

A. *The sampling program is outlined in this paper, and . . .*

Q. *How many samples did you take, Doctor?*

A. *—the samples are listed . . .*

Q. *Was it six samples?*

A. *It looks as though there are six samples plus one in the bay bottom.*

Q. *A routine survey by a department such as the Department of Agriculture of a lake that has 2 to 300 acres to it, a marsh that has that number of acres to it, would be about 50 samplings, would it not?*

A. *It could be. It could be larger. You have to recognize that the objectives in this study were to determine whether we could find DDT in the marsh. We reported what we found.*

Q. *Didn't you also publish another article, Doctor, called "Persistence of DDT in Soils of Heavily Sprayed Forests?"*

EXAMINER SWEENEY: *Before you go on, Mr. O'Connor, I'm kind of confused here as to what samplings we're talking about. Let me ask the Doctor—now, with reference to Intervenor EDF Number 41, where in the abstract, in the beginning, they refer to 13 pounds per acre, it is your testimony that you continued the experiment or you renewed the experiment in this same salt marsh and determined that there was something other than 13 pounds per acre?*

DOCTOR WOODWELL: *We sampled more extensively over a period of a couple of years, a couple of years after that, and found that an estimate of average levels in that marsh based on a more extensive sampling, would give an average in the range of a pound or two pounds per acre with some spots rising to much higher levels and some spots having less.*

EXAMINER SWEENEY: *Is that difference of between 13 pounds and an average of one to three pounds a significant difference, in your opinion?*

DOCTOR WOODWELL: *Oh, yes.*

MR. O'CONNOR: *Mr. Sweeney, I move to strike this article as misleading.*

EXAMINER SWEENEY: *No, I don't think so. They have offered it as an exhibit and I think the testimony will show what this Doctor has said regarding this article. I think it should stay in.*

POLYCHLORINATED BIPHENYLS

Polychlorinated biphenyls (PCBs) are complex mixtures of chlorine substituted biphenyls (Figure 10.15). Although PCBs are not pesticides, their similarity in many ways to chlorinated hydrocarbon pesticides makes their inclusion in this chapter appropriate. PCBs are no longer produced in the United States, but they were manufactured by the Monsanto Corporation from 1929 to 1977. As of 1975, nine different PCB formulations were being produced by the company (Boyle, 1975), but all were sold under the trade name Askarel. Most PCB formulations are liquids or resins, although a few are powders. PCBs are exceptionally stable compounds that are nonflammable and highly resistant to strong reagents and heat. Destruction by burning requires a temperature of over 1300°C (Martin, 1977). Because of these characteristics, PCBs have been used in a variety of industrial applications. Prior to 1970 PCBs were used primarily in closed or semiclosed systems in electrical transformers, capacitors, heat transfer systems, and hydraulic fluids. They were used to a minor extent in paints, adhesives, caulking compounds, plasticizers, inks, lubricants, carbonless copy paper, and sealants, coatings, and dust control agents (EPA, 1979). Foreign imports of PCBs generally accounted for only a few percent of PCB use in the United States and were used primarily as plasticizers.

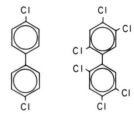

Figure 10.15 Two polychlorinated biphenyl compounds.

Problems with PCBs

PCBs have been found to display a degree of toxicity to certain organisms comparable to that displayed by some pesticides. Peakall (1970b), for example, found PCBs to be even more potent inducers of liver enzymes in birds than DDE, and Peakall and Peakall (1973) found that PCBs at 10 ppm in feed lowered reproduction significantly in ring doves. In fact, there is good evidence that PCBs interfere with the reproduction of a variety of organisms, including rodents, fish, fowl, and primates (Maugh, 1975). The reproduction of mink, for example, is completely halted if the PCB concentration in their food exceeds 5 ppm (Maugh, 1972). In experiments at the University of Wisconsin, eight female rhesus monkeys were given a diet containing 2.5 ppm of Aroclor 1248 (a PCB formulation) for 6 months. The females were then bred to healthy males. Three of the females resorbed their fetuses shortly after conception. The five babies born to the other females were all undersized, and two died while still nursing. The remaining three showed signs of being hyperactive (Boyle, 1975). The mothers' breast milk was found to contain 7–16 ppm PCBs on a lipid weight basis (EPA, 1980b). In experiments with chickens, hens were fed a diet containing 20 ppm of certain PCB formulations for 5 weeks. Only 8% of the fertilized eggs hatched, and many of the embryos exhibited teratogenic abnormalities. The effect on the hens, however, appeared to be specific to only certain PCB formulations (Maugh, 1972).

As was the case with chlorinated hydrocarbon pesticides, PCB concentrations in most parts of the ocean have always been quite low. Harvey et al. (1973), for example, reported mean PCB concentrations in the upper 200 m of the North Atlantic Ocean to be about 20 parts per trillion, although concentrations as high as 100 parts per trillion were sometimes measured (Harvey et al., 1974). The latter observation is significant, since Fisher et al. (1974) found that PCB concentrations as low as 100 parts per trillion could cause substantial disruption of phytoplankton communities grown in continuous culture.

As would be expected, PCB concentrations are much higher in the lipid tissue of aquatic organisms than in the water. Harvey et al. (1974), for example, reported PCB levels in North Atlantic plankton lipids as high as several parts per thousand in samples rich in phytoplankton. Whether PCB levels this high in phytoplankton lipids might be adversely affecting photosynthetic rates is unknown, because the appropriate laboratory studies have never been done.

PCB concentrations in marine fish have never been a serious problem, at least insofar as commercial catch utilization is concerned. PCB levels in the muscle tissue of fish such as cod, haddock, and halibut sampled by Harvey et al. (1974) ranged from 2 to 190 ppb, well below the present Food and Drug Administration (FDA) guideline of 2 ppm for edible fish flesh. These results are typical for marine fish. During the 1970s, however, PCB levels in some freshwater fish, particularly fatty species, were found to exceed 5 ppm. The contaminated fish included lake trout and coho and chinook salmon as well as anadromous species such as striped bass (Martin, 1977). Particularly contaminated freshwater systems included the Great Lakes, the Hudson River, and the Mississippi River.

With respect to human health, PCBs rank third in toxicity behind dioxins and furans when the most toxic isomer of each group is considered (Sun, 1983), but how dangerous PCBs in general are to human health is a bit unclear. In Yusho, Japan, 1291 persons became ill after eating rice bran oil contaminated with PCBs that had leaked from a heat exchanger. Twenty-nine of those persons died over a 2-month period in 1968, and many developed chloracne, a disfiguring acnelike disorder (Sun, 1983). The PCB concentrations in the rice oil averaged 2500 ppm, and persons affected by the sickness were estimated to have ingested about 2 g of PCBs. The rice oil, however, was later found to have contained 1.6–5 ppm of polychlorinated dibenzofurans, which are estimated to be several orders of magnitude more toxic than PCBs (Martin, 1977). Although many of the symptoms of illness displayed by persons who consumed the rice oil were typical of PCB poisoning, it is possible that the polychlorinated dibenzofurans contributed significantly to the health problems of the victims. One of the problems in evaluating the human health effects of PCBs is the fact that commercial PCB mixtures often do contain small amounts of polychlorinated dibenzofurans (EPA, 1980b).

A group of Michigan sport fishermen who had been consuming about 10 kg of PCB-contaminated fish per year over a 2-year period exhibited no adverse health effects, although PCB levels in their blood were higher than normal. The fishermen were estimated to have consumed about 46.5 mg of PCBs per person per year (Martin, 1977). These results are reminiscent of the experimental results of Hayes et al. (1971) in feeding DDT to human volunteers. However, in both cases it is possible that the PCB- or DDT-contaminated food might have induced long-term effects that did not become apparent during the course of the study.

There have been several studies of persons whose health was affected through occupational exposure to PCBs. Workers who used PCBs to make capacitors at two General Electric plants near Albany, New York, for example, complained of allergic dermatitis, nausea, dizziness, eye irritation, and asthmatic bronchitis (Maugh, 1975). Workers involved in testing dielectric fluids at a Westinghouse plant complained similarly of sore throats, skin rash, gastrointestinal disturbances, eye irritations, and headaches (EPA, 1980b). The blood PCB concentrations of the Westinghouse workers were substantially above the range of values reported for the general population (EPA, 1980b). Although the levels of PCBs to which these workers were exposed is a bit unclear, some of the same symptoms have been produced in rhesus monkeys fed a diet containing 300 ppm PCBs (Allen and Norback, 1973).

The issue of PCB pollution is particularly relevant to water pollution, because for most persons the principal source of exposure is the consumption of contaminated fish; and until the mid-1970s the concentrations of PCBs in the human diet were comparable to the levels that proved toxic to rhesus monkeys. A survey reported in 1977, for example, that the average PCB concentration in the breast milk of women in the United States was 1.8 ppm on a lipid weight basis (42 Federal Register, 17487), a figure that is disturbingly close to the range of 7–16 ppm which proved toxic to infant rhesus monkeys. During the first half of the 1970s the average daily intake of

PCBs by persons in the United States was about 10 µg (EPA, 1980b). It is noteworthy that if you actually consumed 6.5 g of fish and shellfish per day (Chapter 8) containing the present FDA tolerance level of 2 ppm PCBs, you would be consuming 13 µg of PCBs per day.

Persistence of PCBs

One of the chief concerns over PCBs has been their high degree of persistence in the environment. Because of this persistence, it was feared that serious environmental contamination with PCBs would be difficult to reverse in a short time, since the compounds would continue to cycle in the ecosystem for perhaps many years. Some indication of the persistence of PCBs is evident from the extensive sampling by Harvey et al. (1974) in the North Atlantic Ocean. Based on estimates of DDT and PCB production and loss rates to the environment, Harvey et al. (1973) estimated that the ratio of PCBs to DDT residues in the environment should be about 0.1. Harvey et al. (1974), however, found the ratio to be about 30 in the marine atmosphere, surface seawater, and plankton of the North Atlantic. The data from Lake Erie (Figure 4.12–4.13) indicate a ratio of 5–10. The implication is that PCBs are 50–300 times more persistent than DDT residues in the environment.

Realizing the potential seriousness of the PCB contamination problem, the Monsanto Chemical Corporation voluntarily restricted its manufacture of PCBs in 1971 to include only uses in electrical capacitors and transformers. The company reasoned that PCB leakage to the environment from capacitors and transformers would be much less of a problem than PCB leakage from such previous sources as discarded lubricants and hydraulic fluids. There is strong evidence that this restriction on PCB usage reduced greatly the extent of environmental loading with PCBs. Table 10.8 lists annual discharge rates of PCBs into the southern California bight from five major sewer outfalls in the Los Angeles–San Diego area (Young, McDermott, and Heeson, 1975). The discharge rate declined by a factor of at least 7.5 between 1972 and 1975.

Despite Monsanto's voluntary restriction on PCB production and despite the apparent reductions in discharge rates, the total amount of PCBs in the environment continued to increase during the first half of the 1970s (Maugh, 1975; Figure 10.16). As a result of this discovery and because of the demonstrated toxic effects of PCB formulations, Monsanto voluntarily terminated all manufacture of PCBs in 1977, and the EPA, under authorization of the Toxic Substances Control Act, banned the manufacture, processing, distribution in commerce, and use of PCBs in the United States effective July 2, 1979. The EPA ruling, however, allowed the continued use of PCBs in existing enclosed electrical equipment under controlled conditions. The lifetime of such equipment is approximately 20–30 years. Although it was felt initially that such use would not pose a significant environmental danger, several accidents involving electrical transformers heightened public concern and caused utility companies to accelerate the phase-out of PCB-containing equipment rather than simply waiting for it to wear out (Sun, 1983). The cost of the phase-out has not been cheap. Pacific Gas and Electric, for example, allocated $60 million in 1983 to replace 1000 large PCB transformers after one of them exploded and contaminated a high-rise office building in San Francisco. At the time it was estimated that there were 20,000–30,000 PCB-containing transformers in the United States (Sun, 1983).

Table 10.8   PCB Loading from a Total of Five Sewer Outfalls Discharging to the Southern California Bight

Year	Total PCB loading (kg/yr)
1972	>19,400
1973	>3400
1974	5400
1975	2600

Source. Young et al. (1975).

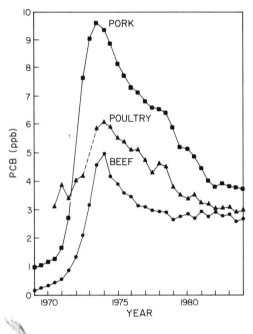

Figure 10.16   Average PCB levels in beef, pork, and poultry based on Canadian analyses from October, 1969–April, 1985. (*Source.* IJC, 1985.)

Ultimately most PCBs discharged to aquatic systems are presumably either degraded in the water column or buried in the sediments. Elder and Fowler (1977), for example, discovered that sinking of zooplankton fecal pellets may be the major mechanism for transporting PCBs to the ocean floor. Similar mechanisms are operative presumably in freshwater systems. Gradual degradation and burial has probably accounted for much of the decline of PCB concentrations in Great Lakes fish during the 1980s (Figure 4.12). It is noteworthy, however, that in 1985 several Beluga whales died in the St. Lawrence River, and autopsies revealed PCB levels that were utterly fantastic—575 ppm in the lipid tissue and as high as 1750 ppm in the milk (IJC, 1985). Such events underscore dramatically the persistence of PCBs and the tendency of such stable compounds to be concentrated through biological processes. It seems fair to say that the decision to terminate all use of PCBs and to phase out old PCB-containing equipment came none too soon.

REFERENCES

Adkisson, P. L., G. A. Niles, J. K. Walker, L. S. Bird, and H. B. Scott. 1982. Controlling cotton's insect pests: A new system. *Science*, **216**, 19–22.

Akesson, N. B., and W. E. Yates. 1984. Physical parameters affecting aircraft spray application. In W. Y. Garner and J. Harvey (Eds.), *Chemical and Biological Controls in Forestry*. Am. Chem. Soc. Ser. 238, Washington, D.C.

Allen, J. R., and D. H. Norback. 1973. Polychlorinated biphenyl- and triphenyl-induced gastric mucosal hyperplasia in primates. *Science*, **179**, 498–499.

Almeida, W. F., P. Pigati, R. Gaeta, and M. T. S. Ungaro. 1975. DDT residues in human blood serum in Brazil. In F. Coulston and F. Korte (Eds.), *Environmental Quality and Safety, Supplement Volume III: Pesticides*. Georg Thieme Publishers. Stuttgart.

Anderson, A. D., and R. March. 1956. Inhibitors of carbonic anhydrase in the American cockroach, *Periplaneta americana* (L.). *Can. J. Zool.*, **34**, 68–74.

Anderson, D. W., and F. Gress. 1983. Status of a northern population of California brown pelicans. *Condor*, **85**, 79–88.

Anderson, D. W., J. R. Jehl, R. W. Risebrough, L. A. Woods, L. R. Deweese, and W. G. Edgecomb. 1975. Brown pelicans: Improved reproduction off the southern California coast. *Science*, **190**, 806–808.

Anderson, D. W., R. M. Jurek, and J. O. Keith. 1977. The status of brown pelicans at Anacapa Island in 1975. *Calif. Fish Game*, **63**, 4–10.

Anonymous. 1981. DBCP poison found in California wells at levels above state's safety standards. Honolulu Advertiser, March 31, p. B-8.

Anonymous. 1990. Island fruit-fly funds OK'd. Honolulu Star-Bulletin. September 30.

Arntzen, C. J., K. Pfister, and K. Steinback. 1982. The mechanisms of triazine resistance; alterations in the chloroplast site of action. In. H. Lebaron and J. Gressel (Eds.), *Herbicide Resistance in Plants*. Wiley-Interscience. New York. pp. 185–214.

Batra, S. W. T. 1982. Biological control in agroecosystems. *Science*, **215**, 134–139.

Bitman, J., C. Cecil, S. J. Harris, and G. F. Fries. 1969. DDT induces a decrease in eggshell calcium. *Nature*, **224**, 44–46.

Bitman, J. H., C. Cecil, and G. F. Fries. 1970. DDT-induced inhibition of avian shell gland carbonic anhydrase: A mechanism for thin eggshells. *Science*, **168**, 594–595.

Bledsoe, T., D. P. Island, R. L. Ney, and G. W. Liddle. 1964. An effect of o,p'-DDD on the extra-adrenal metabolism of cortisol in man. *J. Clin. Endocr.*, **24**, 1303–1311.

Blus, L. J., R. G. Heath, C. D. Gish, A. A. Belisle, and R. M. Prouty. 1971. Eggshell thinning in the brown pelican: Implication of DDE. *Bioscience*, **21**, 1213–1215.

Blus, L. J., C. D. Gish, A. A. Belisle, and R. M. Prouty. 1972a. Logarithmic relationship of DDE residues to eggshell thinning. *Nature*, **235**, 376–377.

Blus, L. J., C. D. Gish, A. A. Belisle, and R. M. Prouty. 1972b. Further analysis of the logarithmic relationship of DDE residues to eggshell thinning. *Nature*, **240**, 164–166.

Boyle, R. H. 1975. The spreading menace of PCB. *Sports Illustrated*, Dec. 1, pp. 20–21.

Brattstein, L. B., C. W. Holyoke, Jr., J. R. Leeper, and K. F. Raffa. 1986. Insecticide resistance: Challenge to pest management and basic research. *Science*, **231**, 1255–1260.

Burdick, G. E., E. J. Harris, H. J. Dean, T. M. Walker, J. Skea, and D. Colby. 1964. The accumulation of DDT in lake trout and the effect on reproduction. *Trans. Am. Fish. Soc.*, **193**, 127–137.

Bush, G. L., R. W. Neck, and G. B. Kitto. 1976. Screwworm eradication: Inadvertent selection for noncompetitive ecotypes during mass rearing. *Science*, **193**, 491–493.

Butler, P. A. 1971. Testimony consolidated DDT hearings. Environmental Protection Agency, Washington, D.C.

Carson, R. 1962. *Silent Spring*. Fawcett, Greenwich, CT. 304 pp.

Cecil, H. C., J. Bitman, and S. J. Harris. 1971. Effects of dietary p,p'-DDT and p,p'-DDE on egg production and egg shell characteristics of Japanese quail receiving an adequate calcium diet. *Poultry Sci.*, **50**, 657–659.

Conney, A. H., R. M. Welch, R. Kuntzman, and J. J. Burns. 1967. Effects of pesticides on drug and steroid metabolism. *Clin. Pharmacol. Ther.*, **8**, 2–10.

Costa, P. 1983. Dioxin's deadliness breeds fear; regulation process complicated. Honolulu Star-Bulletin, HI. March 20.

Crawford, M. 1986. Test of tobacco containing bacterial gene approved. *Science*, **233**, 1147.

Dover, M. J., and B. A. Croft. 1986. Pesticide resistance and public policy. *Bioscience*, **36**, 78–85.

Duggan, R. E., and P. E. Corneliussen. 1972. Dietary intake of pesticide chemicals in the United States (III), June 1968–April 1970. *Pestic. Monitor. J.*, **5**, 331–341.

Dvorchick, B. H., M. Istin, and T. H. Maren. 1971. Does DDT inhibit carbonic anhydrase. *Science*, **172**, 728–729.

Edirisinghe, J. S. 1988. Historical references to malaria in Sri Lanka and some notable episodes up to present times. *Ceylon Med. J.*, **33**, 110–117.

Edmundson, W. F., J. E. Davies, M. Cranmer, and G. A. Nachman. 1969a. Levels of DDT and DDE in blood and DDA in urine of pesticide formulators following a single intensive exposure. *Industr. Med. Surg.*, **38**, 145–150.

Edmundson, W. F., J. E. Davies, G. A. Nachman, and R. L. Roeth. 1969b. p,p'-DDT and p,p'-DDE in blood samples of occupationally exposed workers. *Pub. Health Rep.*, **84**, 53–58.

Edwards, J. G. 1971. Effects of DDT. *Chem. Eng. News*, **49**, 6, 59.

Ehrlich, P. F. 1969. Eco-catastrophe. *Ramparts*, **8**(3), 24–28.

Elder, D. L., and S. W. Fowler. 1977. Polychlorinated biphenyls; Penetration into the deep ocean by zooplankton fecal pellet transport. *Science*, **197**, 459–461.

Environmental Protection Agency. 1979. EPA bans PCB manufacture; phases out uses. *Environmental News*. April 19. 3 pp.

Environmental Protection Agency. 1980a. *Ambient Water Quality Criteria for DDT*. EPA 440/5-80-038. Washington, D.C.

Environmental Protection Agency. 1980b. *Ambient Water Quality Criteria for Polychlorinated Biphenyls*. EPA 440/5-80-068. Washington, D.C.

Environmental Protection Agency. 1984. *Ambient Water Quality Criteria for 2,3,7,8-tetra-chlorodibenzo-p-dioxin*. EPA 440/5-84-007. Washington, D.C.

Erickson, J. M., M. Rahire, J-D Rochaix, and L. Mets. 1985. Herbicide resistance and cross-resistance: changes at three distinct sites in the herbicide-binding protein. *Science*, **228**, 204–207.

Evans, G. 1978. Dutch Elm Disease: Fighting a holding action. *TWA Ambassador Magazine*, August. pp. 70, 72.

Fisher, N. S. 1975. Chlorinated hydrocarbon pollutants and photosynthesis of marine phytoplankton: A reassessment. *Science*, **189**, 463–464.

Fisher, N. S., E. J. Carpenter, C. C. Remsen, and C. F. Wurster. 1974. Effects of PCB on interspecific competition in natural and gnotobiotic phytoplankton communities in continuous and batch cultures. *Microbial Ecol.*, **1**, 39–50.

Fleming, W. E., and W. W. Maines. 1953. Persistence of DDT in soils of the area infested by the Japanese beetle. *J. Econ. Entomol.*, **46**, 445–449.

Fonseka, J., and K. N. Mendis. 1987. A metropolitan hospital in a non-endemic area provides a sampling pool for epidemiological studies on vivax malaria in Sri Lanka. *Trans. Roy. Soc. Trop. Med. Hyg.*, **81**, 360–364.

Fox, J. L. 1984. Tentative agent orange settlement reached. *Science*, **224**, 849–850.

Giam, C. S., A. R. Hanks, R. L. Richardson, W. M. Sackett, and M. K. King. 1972. DDT, DDE and polychlorinated biphenyls in biota from the Gulf of Mexico and the Caribbean Sea— 1971. *Pestic. Monit. J.*, **6**, 139–143.

Goldberg, E. D., P. Butler, P. Meier, D. Menzel, G. Paulik, R. Risebrough, and L. F. Stickel. 1971. *Chlorinated hydrocarbons in the marine environment*. Report prepared by the panel monitoring persistent pesticides in the ocean, National Academy of Sciences, Washington, D.C. 42 pp.

Gunn, D. L. 1976. Alternatives to chemical pesticides. In D. L. Gunn and J. G. R. Stevens (Eds.), *Pesticides and Human Welfare*. Oxford University Press, Oxford. pp. 240–255.

Gunther, F. A., and L. R. Jeppson. 1960. *Modern insecticides and world food production*. Wiley, New York. 284 pp.

Gutowska, M. S., and C. A. Mitchell. 1945. Carbonic anhydrase in the calcification of the egg shell. *Poultry Sci.*, **24**, 159–167.

Hamelink, J. L., R. C. Waybrant, and R. C. Ball. 1971. A proposal: Exchange equilibria control the degree chlorinated hydrocarbons are biologically magnified in lentic environments. *Trans. Am. Fish. Soc.*, **100**, 207–214.

Harvey, G. R., W. G. Steinhauer, and J. M. Teal. 1973. Polychlorobiphenyls in North Atlantic ocean water. *Science*, **180**, 643–644.

Harvey, G. R., H. P. Miclas, V. T. Bowen, and W. G. Steinhauer. 1974. Observations on the distribution of chlorinated hydrocarbons in Atlantic ocean organisms. *J. Mar. Res.*, **32**, 103–118.

Hayes, W. J. Jr., W. E. Dale, and C. I. Prinkle. 1971. Evidence of safety of long-term high, oral doses of DDT for man. *Arch. Environ. Health*, **22**, 119–135.

Hazeltine, W. 1972. Disagreements on why brown pelican eggs are thin. *Nature*, **239**, 410–411.

Heath, R. G., J. W. Spann, and J. F. Kreitzer. 1969. Marked DDE impairment on mallard reproduction in controlled studies. *Nature*, **224**, 47–48.

Hickey, J. J., and D. W. Anderson. 1968. Chlorinated hydrocarbons and eggshell changes in raptorial and fish-eating birds. *Science*, **162**, 271–273.

Hoffman, S. L., V. Nussenzweig, J. C. Sadoff, and R. S. Nussenzweig. 1991. Progress toward malaria preerythrocytic vaccines. *Science*, **252**, 520–521.

Holcomb, R. W. 1970. Insect control: Alternatives to the use of conventional pesticides. *Science*, **168**, 456–458.

Holden, P. W. 1986. *Pesticides and Groundwater Quality: Issues in Four States*. National Academy Press. Washington, D.C. 124 pp.

Hunt. E. G., and A. I. Bischoff. 1960. Inimical effects on wildlife of periodic DDD applications to Clear Lake. *Calif. Fish & Game*, **46**, 91–106.

Ide, F. P. 1956. Effect of forest spraying with DDT on aquatic insects of salmon streams. *Trans. Am. Fish. Soc.*, **86**, 208–219.

IJC. 1985. *PCBs: A case study*. Proceedings of a workshop on Great Lakes research coordination. International Joint Commission, Windsor, Ontario. 113 pp.

Jukes, T. H. 1974. Insecticides in health, agriculture, and the environment. *Naturwissenschaften*, **61**, 6–16.

Keller, H. 1952. Die Bestimmung Kleinster Mengen DDT auf enzymanalytischem wege. *Naturwissenschaften*, **39**, 109.

Krafsur, E. S. 1985. Screwworm flies (*Diptera: Calliphoridae*): Analysis of sterile mating frequencies and covariates. *Bull. Entomol. Soc. Am.*, **31**, (3), 36–40.

Laws, E. R. 1967. Men with intensive occupational exposure to DDT. *Arch. Environ. Health*, **15**, 766–775.

Laws, E. R. 1971. Evidence of antitumorigenic effects of DDT. *Arch. Environ. Health*, **23**, 181–184.

Lichtenstein, E. P. 1957. DDT accumulation in midwestern orchard and crop soils treated since 1945. *J. Econ. Entomol.*, **50**, 545–547.

Ling, L., F. W. Whittemore, and E. E. Turtle. 1972. *Persistent Insecticides in Relation to the Environment and Other Unintended Effects*. Food and Agriculture Organization of the United Nations, Misc. Publ. No. 4, Rome, May.

Lockie, J. D., D. A. Ratcliffe, and R. Balharry. 1969. Breeding success and organochlorine residues in golden eagles in west Scotland. *J. Appl. Ecol.*, **6**, 381–389.

Luard, E. J. 1973. Sensitivity of *Dunaliella* and *Scenedesmus* (Chlorophyceae) to chlorinated hydrocarbons. *Phycologia*, **12**(1/2), 29–33.

MacPhee, A. W., D. Chisholm, and C. R. MacEachern. 1960. Breeding success and organochlorine residues in golden eagles in west Scotland. *J. Appl. Ecol.*, **6**, 381–389.

Maren, T. H. 1967. Carbonic anhydrase: Chemistry, physiology, and inhibition. *Physiol. Rev.*, **47**, 595–781.

Marshall, E. 1981. The summer of the gypsy moth. *Science*, **213**, 991–993.

Marshell, E. 1982. USDA retreats on gypsy moth front. *Science*, **216**, 716.

Marshall, E. 1990. Malaria research—what next. *Science*, **247**, 399–403.

Martin, R. G. 1977. PCBs—polychlorinated biphenyls. *Sport Fishing Institute Bulletin*, No. 288, Sept. pp. 1–3.

Marx, J. L. 1977. Applied ecology: Showing the way to better insect control. *Science*, **195**, 860–862.

Maugh, T. H., II. 1972. Polychlorinated biphenyls: Still prevalent, but less of a problem. *Science*, **178**, 388.

Maugh, T. H., II. 1975. Chemical pollutants: Polychlorinated biphenyls still a threat. *Science*, **190**, 1189.

Maugh, T. H., II. 1981. Starving in the midst of plenty. *Science*, **212**, 430.

Menzel, D. W., J. Anderson, and A. Randtke. 1970. Marine phytoplankton vary in their response to chlorinated hydrocarbons. *Science*, **167**, 1724–1726.

Metcalf, R. L. 1981. Insect control technology. In *Encyclopedia of Chemical Technology*, Volume 13. Wiley-Interscience, New York. pp. 413–485.

Moffatt, A. S. 1991. Research on biological pest control moves ahead. *Science*, **252**, 211–212.

Montoyama, N., and W. C. Dauterman. 1980. Glutathione S-transferases: Their role in the metabolism of organophosphorus insecticides. *Rev. Biochem. Toxicol.*, **2**, 49–69.

Moore, S. A., and R. C. Harriss. 1972. Effects of polychlorinated biphenyl on marine phytoplankton communities. *Nature*, **240**, 356–357.

Mosser, J. L., N. S. Fisher, and C. F. Wurster. 1972. Polychlorinated biphenyls and DDT alter species composition in mixed cultures of algae. *Science*, **176**, 533–535.

Ortelee, M. F. 1958. Study of men with prolonged intensive occupational exposure to DDT. *Am. Med. Ass. Arch. Ind. Health*, **18**, 433–440.

Ottoboni, A. 1972. DDT: The world has been doused with it for 25 years. With what results? *California's Health*, **27**(2), 1–2.

Palea, J. 1990. Libya gets unwelcome visitor from the west. *Science*, **249**, 117.

Paul, J. 1989. Getting tricky with rootworms. *Agrichem. Age*, **33**(3), 6, 25, 30.

Pimentel, D. 1991. Pesticide use. *Science*, **252**, 358.

Peakall, D. B. 1970a. p,p'-DDT: Effect on calcium metabolism and concentration of estradiol in the blood. *Science*, **168**, 592–594.

Peakall, D. B. 1970b. Pesticides and the reproduction of birds. *Sci. Amer.*, **222**(4), 73–78.

Peakall, D. B., and M. L. Peakall. 1973. Effect of a polychlorinated biphenyl on the reproduction of artificially and naturally incubated dove eggs. *J. Appl. Ecol.*, **10**, 863–868.

Pimentel, D., L. McLaughlin, A. Zepp, B. Lakitan, T. Kraus, P. Kleinmen, F. Vancini, W. J. Roach, E. Graap, W. S. Keeton, and G. Selig. 1991. Environmental and economic impacts of reducing U.S. agricultural pesticide use. In D. Pimentel and A. A. Hanson (Eds.), *CRC Handbook of Pest Management in Agriculture*. 2nd ed. Vol. I. CRC Press. Boca Raton. pp. 679–718.

Plimmer, J. R. 1980. Herbicides. In *Encyclopedia of Chemical Technology*. Vol. 12. Wiley-Interscience, New York. pp. 297–351.

Pocker, Y., W. M. Beug, and V. R. Ainardi. 1971. Carbonic anhydrase interaction with DDT, DDE, and dieldrin. *Science*, **174**, 1336–1338.

Porter, R. D., and S. N. Wiemeyer. 1969. Dieldrin and DDT: Effects on sparrow hawk eggshells and reproduction. *Science*, **168**, 199–200.

Postel, S. 1987. *Defusing the Toxics Threat: Controlling Pesticides and Industrial Waste*. Worldwatch Paper 79. Worldwatch Inst. Washington, D.C. 69 pp.

Ratcliffe, D. A. 1970. Changes attributable to pesticides in egg breakage frequency and eggshell thickness in some British birds. *J. Appl. Ecol.*, **7**, 67–107.

Rudd, R. L. 1964. *Pesticides and the Living Landscape*. University of Wisconsin Press. Madison. 320 pp.

Sakauchi, N., S. Kumaoka, T. Naruke, O. Abe, M. Kusama, and O. Takatani. 1969. A case of adrenocortical cancer treated with o,p'-DDD. *Endocr. Jap.*, **16**, 287–290.

Sawicki, R. M., and A. W. Farnham. 1967. Genetics of resistance to insecticides of the SKA strain of *Musca domestica* I. Location of the main factors responsible for the maintenance of high DDT-resistance in diazinon-selected SKA flies. *Entomol. Exp. Appl.*, **10**, 253–267.

Soderlund, D. M., J. R. Sanborn, and P. W. Lee. 1983. Metabolism of pyrethrin and pyrethroids in insects. *Prog. Pesticide Biochem. Toxicol.*, **3**, 401–435.

Southren, A. L., S. Tochimoto, L. Strom, A. Ratuschni, H. Ross, and G. Gordon. 1966. Remission in Cushing's syndrome with o,p'-DDD. *J. Clin. Endocr.*, **26**, 268–278.

Spitzer, P. R., R. W. Risebrough, W. Walker, R. Hernandez, A. Poole, D. Puleston, and I. C. Nisbet. 1978. Productivity of ospreys in Connecticut-Long Island increases as DDE residues decline. *Science*, **202**, 333–335.

Straus, O. H., and A. Goldstein. 1943. Zone behavior of enzymes. *J. Gen. Physiol.*, **26**, 559–585.

Sun, M. 1983. EPA, utilities grapple with PCB problems. *Science*, **222**, 32–33.

Taylor, T. G. 1970. How an eggshell is made. *Sci. Amer.*, **222**(3), 88–95.

Torda, C., and H. Wolff. 1949. Effects of convulsant and anticonvulsant agents on the activity of carbonic anhydrase. *J. Pharmacol. Exp. Ther.*, **95**, 444–447.

Vermaas, W., C. Arntzen, L. Q. Gu, and C. A. Yu. 1983. Interactions of herbicides and azidoquines at a photosystem II binding site in the thylakoid membrane. *Biochim. Biophys. Acta*, **723**, 266–275.

Walsh, J. 1986. River blindness: A gamble pays off. *Science*, **232**, 922–925.

Walsh, J. 1991. The greening of the green revolution. *Science*, **252**, 26.

Wiemeyer, S. N., and R. D. Porter. 1970. DDE thins eggshells of captive American kestrels. *Nature*, **227**, 737–738.

Wilson, A. J., Jr., J. Forester, and J. Knight. 1970. Chemical assays. In *U.S. Dept. of Interior Circular 335*, Gulf Breeze Lab., Florida, Prog. Rep. F. Y. 1969. pp. 18–20.

Winteringham, F. P. W., P. M. Loveday, and A. Harrison. 1951. Resistance of houseflies to DDT. *Nature*, **167**, 106–107.

Woodwell, G. M. 1972. Testimony, consolidated DDT hearings, Environmental Protection Agency. Washington, D.C.

Woodwell, G. M., and F. T. Martin. 1964. Persistence of DDT in soils of heavily sprayed forest stands. *Science*, **145**, 481–483.

Woodwell, G. M., C. F. Wurster, and P. A. Isaacson. 1967. DDT residues in an east coast estuary: A case of biological concentration of a persistent insecticide. *Science*, **156**, 821–824.

World Health Organization. 1988. Malaria control activities in the last 40 years. In *World Health Statistics Annual*. Geneva. pp. 28–29.

World Health Organization. 1990. World malaria situation 1988. *Bull. World Health Org.*, **68**(5), 667–673.

Wurster, C. F. 1968. DDT reduces photosynthesis by marine phytoplankton. *Science*, **159**, 1474–1475.

Yoshishige, J. 1991. DBCP in groundwater will be around for 'generations'. Honolulu Advertiser. May 8. p. A6.

Young, D. R., D. J. McDermott, and T. C. Heeson. 1975. *Polychlorinated biphenyl inputs to the southern California bight*. Southern California Coastal Water Research Project. Background paper prepared for the National Conference on Polychlorinated Biphenyls, 19–21 November, 1975. 50 pp.

11

THERMAL POLLUTION AND POWER PLANTS

Indian Point, on the Hudson River, is the site of a nuclear power station opened by Consolidated Edison (Con Ed), with great fanfare, in 1963. A lot of people don't like any nuclear power plants, and with some reason—the general idea is that utility company engineers are in a great rush to build them, while nobody really knows enough yet about what their effects might be—but there wasn't any really great outcry about Indian Point. Like all reactors so far, it is a thermal polluter—it raises the temperature of the river at its side by using river water as a coolant and then dumping the warm water back into the riverbed—but the Hudson is so fouled up anyway that many people have given up on it.

The only problem was that after the plant opened, somebody noticed a bunch of crows around a dump near the new power station—more crows than usual by far. When this continued somebody got in touch with the Long Island League of Saltwater Sportsmen, a group that—in marked contrast to a lot of sportsmen's groups—happened to retain a consulting biologist. His name is Dominick Pirone, and he went to take a look at the dump.

Although Con Ed didn't want him to look, he managed to see bulldozers at work, shoving dead bass into 12-foot-high piles, where lime was being dumped on them to hasten their decomposition, and he saw a line of trucks, bringing new loads of dead bass. Pirone decided to follow the trucks, and found himself at Indian Point.

"I saw and smelled," he said, "some 10,000 dead and dying fish under the dock."

The seven-degree rise in the temperature of the water, after the plant has sucked in Hudson River water and spat it back out again, is enough to attract spawning bass. They were trapped under the dock and ultimately suffocated. Intake pipes sucked them up into wire baskets, the baskets were dumped into the trucks—and before Con Ed, under pressure, put up a fine-mesh screen around the dock, 2,000,000 bass had been killed.

Con Ed did its best to kill the story. New York's representative Richard Ottinger (who would probably have leprosy by now if somebody on the Con Ed staff had the Evil Eye) described it this way:

> The story of the Indian Point fish kill is strangely obscure. There are reports of truckloads of fish carted away secretly; fish graveyards limed to hasten the destruction of evidence and guarded by Burns detectives to prevent witnesses' access to see the size of the kill. There are stories of pictures suppressed by state officials and state employees pressured into silence. Gene Marine (*America the Raped*, pp. 86–87)

Many industries and almost all nuclear and fossil fuel electric power plants discharge heated wastewater into aquatic systems. It is not hard to imagine some of the ways in which this heated effluent might adversely affect aquatic biota. In the summer and particularly in tropical climates, the ambient water temperature may already be a chronic stress to some organisms in the system. In such cases any further increase in temperature would only increase this stress and perhaps create lethal conditions for some organisms. On the other hand, heated effluent may often attract aquatic organisms during the winter in temperate latitudes when ambient water temperatures are lower than the preferred range of many species. As noted in Chapter 8, an abrupt plant shutdown during such times may produce a large-scale kill of aquatic organisms due to cold shock. Perhaps the most serious effect associated with thermal pollution, however, is the killing or stressing of organisms that are sucked into a plant's cooling water system and are either impinged on protective screens (see above) or, if they pass through, subjected to physical buffeting and possible chemical stresses as well as thermal shock.

POWER PLANT DESIGN

At the present time electric power plants account for roughly 75–80% of the thermal pollution in the United States. Industrial operations such as refineries, petrochemical plants, cokeries, and steel mills account for most of the remainder. We will therefore concentrate in this chapter on electric power plants, although much of what we say is relevant to industrial operations as well.

Figure 11.1 is a schematic diagram of a typical electric power plant's steam and cooling system. The plant generates electricity by first boiling liquid water to produce steam. The source of the heat may be either fossil fuel energy or nuclear energy. The expansion of the water during the transformation from the liquid to the gaseous state creates a pressure, which is used to drive a turbine, much as the burning of liquid gasoline to create vapor in a car is used to drive the pistons of the car. The turning of the turbine powers the generator, which produces electricity. The steam from the turbine is then condensed back to liquid water and recycled to the boiler. Condensation is achieved by running cooling water through a long coiled pipe exposed to the steam. Heat is transferred through the pipe from the hot steam to the cooling water until the steam condenses. The heated effluent from the condenser is the source of thermal pollution from such a power plant.

The increase in temperature of the condenser cooling water as it passes through the heat exchanger depends on the steam exhaust pressure and how fast the water is pumped. The higher the steam exhaust pressure, the higher the temperature at

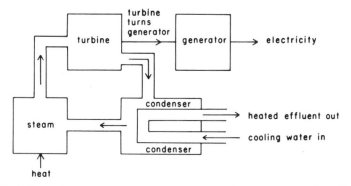

Figure 11.1   Schematic diagram of an electric power plant's steam and cooling system.

which steam condenses to water. In the United States virtually all electric power plants operate with a steam exhaust pressure of about 0.05 atmosphere, at which pressure the steam condenses at a temperature of about 33°C (Karkheck, Powell, and Beardsworth, 1977). At a given steam exhaust pressure, the temperature rise of the cooling water is correlated negatively with the flux of cooling water through the heat exchangers. Since organisms may be killed from physical buffeting as well as by thermal shock from the heated effluent, there is obviously some tradeoff between pumping cooling water rapidly to reduce the temperature rise and pumping slowly to reduce the number of organisms sucked into the cooling system.

The characteristics of nuclear and fossil fuel power plants are such that nuclear plants discharge usually about 45% more waste heat to the cooling water than do fossil fuel plants per unit of electricity generated (GESAMP, 1984). The difference is due both to the lower efficiency of nuclear plants (35% vs. 40%) and to the fact that fossil fuel plants discharge about 15% of their waste heat into the air. Nuclear plants release only about 3% of their waste heat to the atmosphere. Most power plants in the United States operate so that the effluent water is somewhere between 5 and 15°C above ambient. The rate at which cooling water must be pumped through the heat exchangers to achieve a given temperature rise will vary from one plant to another, but as a rough idea of the amount of cooling water required the GESAMP (1984) suggest that $2.6-5.2 \times 10^3$ m^3 d^{-1} are required per megawatt (MW) of electricity to limit the temperature increase to 10°C. One thousand MW power plants are by no means unusual, and power plants as large as 4000 MW exist in the United States. A 4000-MW power plant producing effluent 10°C above ambient requires $1.0-2.1 \times 10^7$ m^3 d^{-1} of cooling water. This rate of cooling water use equals about 75–150% of the flow of water in the Colorado River. Thus we are talking about large fluxes of water.

Water Quality Criteria

It is of interest to compare the increase in temperature of the cooling water from a power plant with federal water quality criteria. In its report to the Secretary of the Interior on April 1, 1968, the National Technical Advisory Committee made the recommendations shown in Table 11.1. The maximum recommended temperature elevations ranged from 0.8–2.8°C, and depended on the nature of the body of water and, in the case of estuaries, on the season of the year. The present EPA water quality criteria with respect to temperature were promulgated in 1976. In marine waters, the criteria specify that the maximum acceptable increase in the week'y average temperature resulting from artificial sources is 1.0°C during all seasons of the year. The criteria also specify that summer thermal maxima, which define the upper thermal limits for the communities of the discharge area, should be established on a site-specific basis. Recommended summer thermal maxima along the east coast of the United States, for example, are 27.8–29.4°C for the daily mean temperature and 30.6–32.2°C for the short-term maximum (EPA, 1976). The freshwater criteria are more complicated and site specific. They take into account the thermal tolerance of the most sensitive important species and requirements for successful migration, spawning, egg incubation, fry rearing, and other reproductive functions. The freshwater criteria also consider impacts resulting from cold shock when plants shut down during the cooler months of the year. The conclusion that obviously emerges from an examination of the criteria is that the typical cooling water temperature increase of 5°–15°C at U.S. power plants is well outside the range of acceptable temperature increases. To satisfy the criteria, the effluent water must either be cooled down before being discharged or else mixed vigorously and diluted with the receiving water in the zone of mixing.

Cooling Water System Characteristics

Two characteristics associated with cooling water systems should be mentioned at this time. First, since the cooling water is drawn invariably from a natural body of water, some effort must be made to prevent unwanted objects from being drawn through the cooling system. Figure 11.2 indicates the techniques commonly employed to keep debris out of the system. The water is drawn from under a 2–5 m deep baffle to prevent floating objects from entering the plant. The water then passes through a trash rack analogous to the trash rack in a sewage treatment plant. The trash rack consists of a grid of metal bars typically spaced about 7–8 cm apart. The water then passes through a screen with a mesh size of about 1 cm. The screen may be designed to rotate as in Figure 11.2 to facilitate periodic cleaning, or the screen may simply be removed for cleaning and replaced with another screen. Finally the water is pumped through the condenser tube manifold. This manifold may consist of as many as several tens of thousands of 2–3-cm-diameter tubes having a length of perhaps 15 m. All of these tubes are in direct contact with the exhaust steam from the turbines.

Table 11.1   Maximum elevations of monthly means of maximum daily water temperatures outside designated zones of mixing as recommended by the National Technical Advisory Committee in 1968

Area	Temperature elevation
lakes	3°F (1.7°C)
rivers	5°F (2.8°C)
estuaries	
summer	1.5°F (0.8°C)
other seasons	4°F (2.2°C)

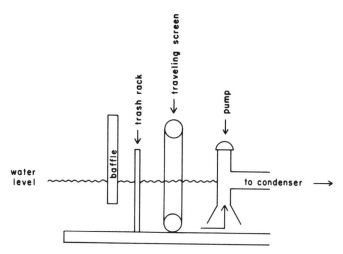

Figure 11.2   Diagram of the inflow system to the condenser cooling tubes in a typical electric power plant.

A second problem associated with the condenser system is the growth of fouling organisms on the interior walls of the condenser tubes. There are several approaches to solving the fouling organism problem.

1. The plant may simply introduce chlorine into the cooling water system, either continuously or intermittently.
2. The plant may backflush periodically the heated effluent so as to kill the fouling organisms by heat stress.
3. The plant may mechanically scour the tubes by releasing periodically a great many brushes or elastic spheres into the inflow water.

In some cases there is no need to go to any effort to prevent fouling, because natural mechanisms keep the condenser tubes clean. For example, the condenser tubes in the Hawaiian Electric power plant at Kahe Point on Oahu are kept free of fouling organisms by the scouring action of sand drawn in with the cooling water from the ocean. However, such circumstances are the exception rather than the rule in power plant operations. Most power plants use chlorine to control fouling. The chlorine may be introduced intermittently at a concentration of typically 12–15 mg/l, generally every 4–8 hours, or continuously at concentrations of 1–5 mg/l. Discharge concentrations are typically lower by a factor of 10–20 (Langford, 1983; GESAMP, 1984).

TOXIC EFFECTS OF EFFLUENT WATERS ON BIOTA

Direct kills of fish and other motile organisms due to the temperature of the effluent from power plants are fortunately rather uncommon. Data summarized by Langford (1983) indicate that about 2–3 voluntarily reported fish kills are associated with thermal discharges each year in the United States. Mass mortalities of Atlantic menhaden, for example, have been reported at three locations; on the Cape Cod

Canal (Fairbanks et al., 1968), at Millstone, Connecticut (IAEA, 1972), and at Northport on Long Island Sound (Young and Gibson, 1973). In the first two cases, the maximum temperature was only about 21°C, but the temperature rose abruptly by about 15°C, and the fish were unable to avoid the discharge. In the third case the water temperature was 37–38°C, almost certainly above the upper tolerance limit of Atlantic menhaden.

In most cases, however, motile organisms are apparently able to avoid power plant discharges by simply swimming away if water temperatures become too high. Furthermore, the tendency of the heated effluent plume to rise or remain on the surface provides a refuge in deeper water for organisms that prefer cooler temperatures. This effect may be minimized, however, if the power plant uses a diffuser discharge system to mix vigorously its effluent with the receiving waters, or if the receiving system is so shallow as to be mixed effectively to the bottom. In freshwater systems in the winter, if the temperature of the water is less than 4°C, the effluent plume may actually sink to the bottom, but under such conditions the temperature of the effluent water is likely to attract rather than repel most organisms.

Nonmotile benthic organisms are particularly susceptible to being killed by heated effluents if the effluents are mixed to the bottom. The incipient lethal temperature of most aquatic species lies within or below the 30–35°C range, so that in tropical climates where summer water temperatures may approach this limit naturally, a further temperature increase of only a few degrees centigrade may prove lethal to many organisms. Kills of shallow-water corals at Kahe Point, Oahu, Hawaii (Jokiel and Coles, 1974) and of virtually the entire benthos over a large area of Biscayne Bay, Florida (Zieman and Wood, 1975) provide two good examples of direct thermal kills of benthic organisms. Obviously there will be no direct effect on the benthos unless the effluent plume is mixed to the bottom.

In temperate climates during the winter, organisms may be attracted to the warmer waters from a power plant's discharge. Under such conditions, direct kills of fish and other motile organisms may occur due to cold shock if the plant shuts down suddenly. The fish kill that occurred in January, 1972 at the Oyster Creek power plant in Barnegat Bay, New Jersey (Chapter 8) illustrates this type of problem. In such cases motility is of no advantage to an organism, since virtually all the surrounding water is lethal. The impact of such temperature changes will of course depend both on the magnitude and the abruptness of the temperature change. In the case of the Oyster Creek power station, the effect of the shutdown was exacerbated by the fact that the temperature in the discharge canal fell from 22°C to 15°C over a period of 48 hours prior to the shutdown. The temperature dropped to 2°C in 6.5 hours when the plant shut down. As noted in Chapter 8, organisms can adapt to different temperatures, but in general such adaptation is accomplished more rapidly at higher temperatures than at lower temperatures (Clark, 1969).

In addition to kills resulting directly from temperature stress, aquatic organisms may be killed by the discharge of chlorine used to prevent fouling in the heat exchangers. As examples of such kills, Clark and Brownell (1973) cite the killing of "a number of menhaden" at the Cape Cod Canal Plant in Massachusetts in 1968 and the killing of 40,000 blue crabs at the Chalk Point Plant in Maryland. Several other comparable incidents are reported by Brungs (1973) and Truchan (1978). In most cases a kill results when fish swim into a discharge canal during a period of nonchlorination and are unable to escape when the chlorine concentration subsequently rises. Effluent chlorine concentrations during intermittent chlorination are typically 0.5–2.0 mg/l for periods of 20–30 minutes. For comparison, the EPA criterion maximum concentrations for chlorine are 0.013 and 0.019 mg/l in marine and freshwater, respectively (EPA, 1986). Kills resulting from intermittent chlorination, however, appear to be of minor importance compared to some of the other problems considered in this chapter.

Organisms that are sucked through a power plant's condenser tubes during intermittent chlorination are exposed to high concentrations of chlorine as well as thermal and physical stresses. In such cases, the old adage that you can only die once applies. In the case of fish eggs and larvae, for example, thermal stress and physical buffeting appear sufficient to account for observed mortality (GESAMP, 1984). Chlorine toxicity does, however, seem to play a role in the impact of entrainment on other types of organisms. The mortality of entrained zooplankton, for example, is correlated positively with chlorine concentration at chlorine concentrations above 0.25 mg/l (GESAMP, 1984).

There are several known instances of fish kills due to gas bubble disease associated with heated effluents from power plants (DeMont and Miller, 1971; Clark and Brownell, 1973; Schneider, 1980). The heated effluent is supersaturated with atmospheric gases, and excess gas taken into a fish's blood tends to bubble out, leading at first to disequilibrium and ultimately to death from embolism. Thousands of adult Atlantic menhaden were reportedly killed by gas bubble disease on April 9, 1973, near the discharge from the Pilgrim Power Station on Cape Cod Bay, Massachusetts. The incident was described in a Smithsonian Institution report as follows.

> There are preliminary indications that the menhaden are dying from a gas bubble disease which is caused by supersaturation of nitrogen in the water naturally. However, the water taken by the plant at a low temperature is increased roughly 26°–27°F so that when it comes out of the plant it is supersaturated.

> The fish have lesions and are hemorrhaging on their fins. They have numerous gas bubbles on their fins and in the membranes between the fin rays (Smithsonian Institution, 1973a).

In summarizing the lethal effects of plant discharges, it is worth noting that the absence of repeated kills in the vicinity of plant outfalls does not necessarily imply the absence of a continuing problem. Motile organisms may simply avoid the area, whereas nonmotile benthic types may be killed once and then never replaced. Thus the effect of the effluent would be to exclude certain organisms from the vicinity of the outfall. Actual kills occur primarily when the characteristics of the effluent suddenly change or when for some reason organisms are unable to escape as water quality deteriorates.

Sublethal Effects

Sublethal stresses of effluent water on biota can of course amount to nothing more than a mild exposure to the sort of stresses that are potentially lethal. Many chronic stresses associated with power plant effluent are, however, the result of complex interactions between a variety of factors. For example, an increase in temperature can be expected to increase the respiration rate of an aquatic organism, at least up to the point at which the temperature approaches the lethal level for the species. Thus the organism's demand for oxygen generally increases as the temperature is raised. Since the Q_{10} for many organisms is about 2, an increase of $10°C$ in the temperature of the receiving water can be expected to roughly double the respiration rate of the biological community. Increases in temperature, however, also reduce the solubility of oxygen in water. The relationship is such that a $10°C$ increase in temperature reduces the solubility of oxygen by about 20%. Thus an increase in temperature generally produces an increase in respiratory demand for oxygen but a decrease in the capacity of the water to dissolve oxygen. Finally, if the heated effluent plume is not mixed thoroughly with the receiving water and tends to rise to the surface, any thermal stratification of the water column will be intensified. The water column therefore becomes more stable and hence more difficult to mix. As a result, any effective exchange of oxygen between the atmosphere and subsurface water will be reduced. The heated effluent from a power plant thus exacerbates oxygen depletion problems by the following mechanisms.

1. Increasing the respiratory demand of aquatic organisms.
2. Reducing the solubility of oxygen in the water.
3. Stratifying or further stratifying the water column so that reoxygenation of subsurface water is inhibited.

Oxygen depletion problems are of course most likely to develop during the summer, when the ambient water temperature is naturally high and the water column stratified thermally.

Temperature increases invariably increase the rate of chemical reactions and therefore alter the rate of virtually all physiological processes. If the temperature becomes too high, the enzymes and hormones that catalyze and control biochemical reactions begin to break down.[1] As a result, metabolic processes begin to slow down and ultimately come to a halt if the temperature is raised sufficiently. Below this temperature range, increases in temperature speed up metabolic processes. Figure 8.7, for example, indicates the dependence of various metabolic processes on temperature for Sockeye salmon. Such curves are qualitatively characteristic of virtually all aquatic organisms.

Organisms cannot carry out metabolic functions effectively if the rates of biochemical processes are sufficiently slowed. The curves in Figure 8.7 suggest why there is both an upper and lower limit to the temperature range within which an organism can survive. As indicated in Figure 8.2, the temperature ranges for growth and reproduction are invariably narrower than the lethal temperature range. As a result, organisms may be eliminated from the thermal discharge area not by directly lethal effects, but simply because they are unable, for example, to grow efficiently or reproduce. In such cases, the area may be recolonized by species that are better suited to higher temperatures. Cyanobacteria, for example, are sometimes better able to survive in artificially warmed waters than are the natural phytoplankton communities of (usually) diatoms and/or green algae. Heated effluent from the Florida Power and Light power plant at Turkey Point, Biscayne Bay, Florida, is believed to have been largely responsible for the stimulation of cyanobacterial blooms in Biscayne Bay during the early 1970s (Clark and Brownell, 1973). Besides

---

[1] Both enzymes and hormones are composed of proteins whose complex structure is determined in part by hydrogen bonds, which are broken rather easily by thermal agitation.

tending to form unsightly algal scums on the surface, some species of cyanobacteria are known to release substances that are toxic to aquatic organisms as well as humans. Toxic substances released by cyanobacteria are believed to have been the cause of a series of fish and bird kills that occurred in Biscayne Bay during the winter from 1972 to 1973 (Smithsonian Institution, 1973b). Thus power plant discharges can lead both to the elimination of desirable species and possibly to their replacement by undesirable species. Cyanobacteria are a good example of the latter.

Temperature effects on reproduction of aquatic organisms are well known. Many species of fish and invertebrates initiate spawning activity at least partly in response to higher temperatures in the spring (Clark and Brownell, 1973). As a result, organisms attracted to thermal discharges during the winter may be induced to spawn earlier than usual in the spring, perhaps at a time when there is inadequate food to support their offspring. Early spawning of fish affected by thermal discharges has been reported for silver bream, tench, rudd, pike-perch, white suckers, large-mouth bass, and sauger (Langford, 1983). The reported advancements in spawning have been 1–5 weeks. In one interesting case involving a geothermally heated stream, rainbow trout changed their spawning period from spring to autumn, and thus avoided the hottest period of the year for hatching and fry development (Kaya, 1977).

It has been suggested that migratory patterns of fish may be disrupted by power plant discharges into rivers or estuaries that serve as migration routes for fish (Hawkes, 1969; Bush et al., 1974). The evidence in support of this hypothesis is, however, less than overwhelming. The salmon fishery in the River Severn in Great Britain was unaffected by the Ironbridge power station, which discharged thermal effluent into the river for over 45 years, and migrations of sockeye salmon, steelhead trout, and Coho salmon in the Columbia River were unaffected by thermal discharges from the Hanford nuclear facility in Washington (Langford, 1983). In the latter case most fish avoided the thermal plume in the hottest months of the year by migrating up the opposite side of the river, but the migration rates of fish that encountered the plume were similar to those that did not. Shad migrating up the Connecticut River similarly tend to avoid the discharge from the Haddam Neck power plant by migrating around or under the thermal plume (Clark and Brownell, 1973).

There is no doubt that elevated temperatures tend to increase the susceptibility of fish and shellfish to certain other stresses. As noted in Chapter 8, this interaction arises in part because increased temperatures lead to increased metabolic rates (at least below the lethal temperature range), which in turn require an increased pumping rate of water over the gills. As a result the organism's intensity of exposure to any toxic substance in the water is increased. There is good evidence, for example, that the toxicity of chlorine is increased at elevated temperatures (Cairns et al., 1978). At 0.1 ppm residual chlorine, for example, the median survival time of Chinook salmon decreases by a factor of 1000 as the temperature is raised from 25 to 30°C, and of pink salmon by a factor of 73 as the temperature is raised from 11 to 22°C (Langford, 1983). Furthermore, some diseases and pathogens are known to develop faster at higher temperatures, and thermal stress may make aquatic organisms more susceptible to infection (Clark and Brownell, 1973; Langford, 1983).

Commentary

The foregoing discussion of sublethal stresses has involved few concrete examples. Instead, the discussion has revolved around interactions that we believe could affect aquatic organisms and have been postulated to contribute to the impact of power plant effluents on aquatic biota. However, the gradual elimination of a species from a system due, for example, to reduced fecundity, greater susceptibility to disease, reduced growth efficiency, or other sublethal stresses is not likely to attract the sort of attention that is drawn to a fish kill. The effects of sublethal stresses generally do not lead to spectacular events. It is possible, however, that the greatest disruption to aquatic systems from power plant effluents may be caused by the continual exposure of organisms to sublethal stresses rather than the occasional killing of large numbers of organisms due to thermal shock, chlorination, or gas bubble disease. Langford (1983), for example, cites a number of examples of changes in the species composition of benthic invertebrate communities near thermal discharges. Not surprisingly, there is a tendency for the natural fauna to be replaced by more thermotolerant species.

A final comment concerns the siting of electric power plants. For reasons of convenience and economics, it has often been the case that power plants are located on estuaries. Such sitings are made because large population centers are frequently found adjacent to estuaries, and because the estuary obviously provides a source of cooling water for the power plant. From the standpoint of water pollution, however, estuaries are just about the last place where one would want to locate a power plant. Estuarine organisms are exposed often to a variety of natural stresses, including salinity variations, low oxygen levels due to high respiration rates, turbidity caused by the stirring up of bottom sediments and the high plankton concentrations in the water, and in some cases large temperature fluctuations due to the shallowness of the water and restricted circulation. Although indigenous estuarine species therefore tend to be, for example, euryhaline and eurythermal, there is of course a limit to the range of environmental parameters within which even these organisms can function effectively. Furthermore, as noted in Chapter 3, estuaries serve as important nursery and/or breeding grounds for a variety of aquatic organisms, and may serve as part of the travel route of migratory fish. Deterioration of estuarine water quality can therefore affect a great number of aquatic species, in fact a far greater number of species than might be apparent from a census of organisms in the estuary at any one time. Considering this fact and considering the magnitude of the stresses that are naturally imposed on estuarine organisms, any additional stress from thermal pollution is to be avoided if at all possible. With this introduction, we turn our attention to one of the most extreme examples of estuarine thermal pollution, the Florida Power and Light Company's Turkey Point power plant on Biscayne Bay, Florida.

A CASE STUDY—THE FLORIDA POWER AND LIGHT POWER PLANT AT TURKEY POINT

The Study Area

Biscayne Bay is a subtropical estuary located just south of Miami, Florida, near the lower tip of Florida (Figure 11.3). The bay has been formed by a series of barrier islands or keys about 14 km east of the Florida peninsula. The bay is bounded on the north and south by shoals, specifically Featherbed Bank on the north and Cutter Bank on the south. The bay is shallow, with a mean depth of only 1.5–1.8 m.

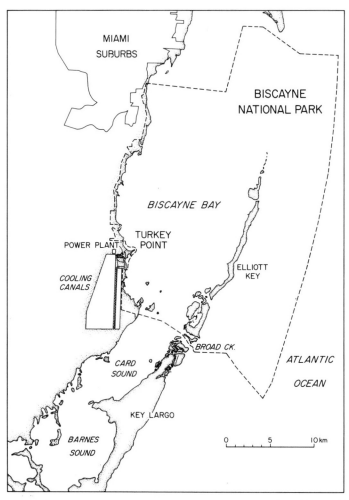

Figure 11.3   East Florida coast just south of Miami showing the location of Biscayne Bay, the area occupied by the Biscayne National Park, and the power plant at Turkey Point.

Biscayne Bay is a highly productive estuary and supports a large and diverse fauna of fish, shellfish, and invertebrates. At least 27 species of fish that are of importance for commercial, sport, or bait fishing are found in the bay's waters, including mullet, barracuda, mackerel, jack, snapper, grouper, bonefish, snook, and pompano (FWPCA, 1970). Furthermore, at least nine invertebrate species found in the bay are of major commercial importance, including sponges, queen conch, blue crab, stone crab, shrimp, and spring lobster. The bay also supports a wide variety of tropical birds, including some rare and endangered species. Most of the birds are wading and shore birds that are found in the shallows adjacent to the mangrove shoreline or near the mangrove islands in the southern portion of the bay. These birds include herons, pelicans, cormorants, egrets, white ibis, roseate spoonbills, frigate birds, waterfowl, gulls, and terns (FWPCA, 1970).

The food chain in Biscayne Bay is primarily a detritus food chain, the source of detritus being the extensive beds of turtle grass *Thalassia testudinum*, which cover much of the bay's benthos. The blades of turtle grass are not grazed directly to a significant degree, but when they break off or die their decomposition yields detritus, which forms the base of the detritus food chain. The turtle grass also serves as a refuge for the abundant benthic fauna and stabilizes the sediments against the erosive action of currents and turbulence.

Realizing the beauty and ecological significance of Biscayne Bay, the U.S. Congress authorized funds to set aside 420 km^2 of lower Biscayne Bay as the Biscayne National Monument on October 18, 1968. The Monument became a national park in 1980 and now includes an area of 700 km^2 (Figure 11.3). As stated by Congress, the original Monument was established to ". . . preserve and protect for the education, inspiration, recreation, and enjoyment of present and future generations a rare combination of terrestrial, marine, and amphibian life in a geographical setting of great natural beauty . . ." (U.S. Congress, 1968).

The Power Plant

In June of 1964 the Florida Power and Light Company (FPL) was granted a permit to build two oil-fired electric power generators at Turkey Point (Figure 11.3). The first of these units went into operation on April 22, 1967, and the second began operating on April 25, 1968. Cooling water for the condensers was drawn from Biscayne Bay at an intake just north of Turkey Point, and discharged back into the bay by way of a series of short canals (Figure 11.4) just south of Turkey Point. Each of the two units draws a total of 1.55×10^6 m^3 of cooling water per day. Under so-called normal full load, which corresponds to 720 MW of power from the two units combined, the temperature rise of the cooling water is 6.7°C, and at maximum capacity of 865 MW the temperature rise is 7.8°C. The actual temperature rises in 1969 exceeded 6.7°C about 5.5% and 16.4% of the time in July and August, respectively, the 2 hottest months of the year. During these 2 months in 1969 the company had been operating the Turkey Point plant under the policy that the Turkey Point plant would be the last plant in the South Florida network to go to full power and the first to come down from full power (FWPCA, 1970). This policy was adopted as the result of a massive fish kill which occurred in lower Biscayne Bay on June 25, 1969.

In early 1967, prior to the initial operation of the first oil-fired generator, FPL began construction of two additional generators at Turkey Point. These two additional units are nuclear powered, with a maximum power output of 760 MW each. Construction on the nuclear generators was completed in 1971 and 1972. When all four units are operating, the plant requires 10.4×10^6 m^3 d^{-1} of cooling water. Under normal full load of 2080 MW the temperature of the plant's cooling water is raised 7.9°C, and at maximum load of 2384 MW the temperature rise is 8.8°C (FWPCA, 1970).

Effects on Biota

The average maximum daily water temperature in Biscayne Bay is normally 30–31°C during the months of July and August (FWPCA, 1970). Since very few aquatic species are able to withstand temperatures in excess of 30–35°C for more than a short time, artificially raising the temperature of Biscayne Bay by only a few degrees centigrade could be expected to have a serious impact on the bay's biota.

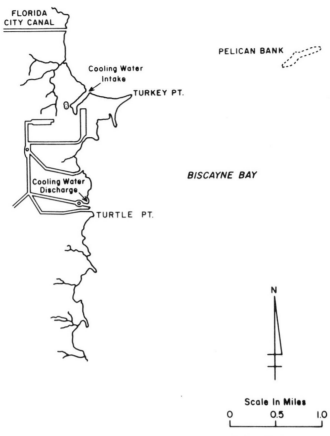

Figure 11.4   Detail of the original cooling canal system at Turkey Point power plant. Note discharge into Biscayne Bay at Turtle Point. [Redrawn from FWPCA (1970).]

The first adverse effect of the heated effluent from the power plant was detected in a survey of south Biscayne Bay during late July, 1968, about 3 months after the second oil-fired generator went on line. An area of barren sediment covering 8–12×10^3 m^2 was found off the mouth of the discharge canal. The turtle grass that flourished in similar areas of the bay was completely gone, and the denuded sediments "were covered with what appeared to be irregular mats of green or bluegreen microalgae" (FWPCA, 1970, p. 28). By mid-November, the area of denuded sediments had expanded to about 4×10^5 m^2. Unconsolidated sediments in the exposed area had been eroded seriously by currents, and scoured depressions were filling in with mangrove peat and dead turtle grass rhizomes. Noticeable damage to the benthos, and turtle grass in particular, was apparent over an additional 8×10^5 m^2 (FWPCA, 1970).

By May, 1969 a regrowth of turtle grass had occurred over much of the 8×10^5 m^2 area that had been adversely affected the previous summer and fall. In this area of turtle grass regrowth, a wide variety of benthic fauna was found, including shrimp, mollusks, immature crabs, sponges, corals, and small bottom fish. However, an area of 4×10^5 m^2 near the mouth of the discharge canal still remained devoid of the usual benthic biota. The sediments in this area were covered by colonies of green algae and cyanobacteria, with occasional patches of benthic diatoms. These microalgae provided virtually no protection to the sediments from erosive currents, and additional erosion was apparent from propellor tracks cut into sediments by motorboats (FWPCA, 1970).

Further deterioration of the benthos became apparent on June 5, 1969. The coral colonies within 350–550 m of the outfall had all died, and some dead corals were found as much as 750 m from the outfall. By June 19, other benthic organisms were beginning to disappear. The turtle grass at 550 m from the outfall was losing its green color and was coated with a brown gelatinous material, and some turtle grass was beginning to lose color and vigor at distances up to 900 m from the outfall. At 750 m from the outfall, 50–70% of the corals were dead.

A massive kill of organisms in the bay occurred approximately 1 week later. On June 25, 1969 the bay's water was unusually turbid and the water temperature concomitantly high.[2] Thousands of dead fish were found south of Turkey Point on June 26. A survey on June 27 revealed numerous dead pistol shrimp, blue crabs, spider crabs, stone crabs, mollusks, and algae. The dead animals were most numerous within 900 m of the outfall. A more extensive survey up to 1.4 km from the outfall on June 29 revealed that the bottom was littered with ". . . dead plants and animals, including sponges, spider crabs, blue crabs, corals, pistol shrimp, clams, snails, mussels, and several varieties of fish in addition to the wilted and browned algae" (FWPCA, 1970, p. 33).

Figure 11.5 indicates the area of biological damage that was apparent to scientists who surveyed the bay on June 27 and shortly thereafter. Virtually a complete kill of aquatic organisms occurred over an area of about 8.4×10^5 m^2 indicated by the light area in Figure 11.5. The shape of this area of acute effects follows closely the northeasterly flow of effluent water from the discharge canal. The total area of biological damage, somewhat arbitrarily defined in Figure 11.5, extended both north and south of the discharge canal and covered a total area of about 2.7 km^2. Subsequent temperature studies conducted during August, 1969, revealed that the power plant effluent was increasing the temperature of the bay water by at least 2.2°C over an area of 2.5 km^2, and that near the mouth of the discharge canal the temperature rise was about 4.4°C (FWPCA, 1970). Although some secondary treated sewage from the Homestead sewage treatment plant (a few kilometers west of Turkey Point) is discharged into the bay by way of a canal and "has caused a noticeably heavier growth of bottom vegetation near the mouth of the canal" (Florida State Board of Health, 1962), water quality studies in the bay conducted during January, 1970, revealed no indication of pollution as measured by standard chemical indicators.[3] This leads us to the conclusion that the June, 1969 kill was caused entirely by elevated temperatures resulting from the power plant's cooling water discharge.

Modifications

Water quality standards adopted by the state of Florida require that water temperature "shall not be increased so as to cause any damage or harm to the aquatic life or

[2] Turbid water absorbs more light energy.
[3] For example, BOD, dissolved oxygen, and nutrient concentrations.

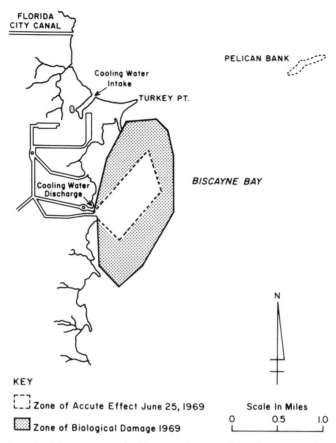

FLORIDA
CITY CANAL

PELICAN BANK

Cooling Water
Intake

TURKEY PT.

BISCAYNE BAY

Cooling Water
Discharge

N

KEY

┌─┐
└─┘ Zone of Accute Effect June 25, 1969

▓ Zone of Biological Damage 1969

Scale In Miles
0   0.5   1.0

Figure 11.5   Biological damage zone in Biscayne Bay resulting from conditions on June 25–27, 1969. [Redrawn from FWPCA (1970).]

vegetation of the receiving waters or interfere with any beneficial use assigned to such waters" (FWPCA, 1970, p. 15). Water uses assigned to Biscayne Bay include recreation and propagation and management of fish and wildlife. Clearly the deterioration of the benthos in 1968 and 1969 and the massive kill of fish and benthic organisms in late June of 1969 as a result of the discharge from the Turkey Point power plant directly violated this Florida law. Dade County, which includes Biscayne Bay, further required that water temperatures not be artificially raised above 35°C, and the present EPA water quality criteria stipulate that the short-term mean temperature will not exceed 32.2°C. During June of 1969, however, water temperature in Biscayne exceeded 35°C and 32.2°C during part of the day at distances up to 1.1 km and 1.4 km, respectively, from the discharge canal (FWPCA, 1970, Figure 12). For comparison, the maximum water temperature measured at any time during August (the hottest month of the year), 1969, at the cooling water intake was 31.1°C (FWPCA, 1970).

Considering the impact of the cooling water discharge on Biscayne Bay in 1969, realizing that the nuclear generators to be installed between 1971 and 1972 would increase the amount of cooling water discharge by a factor of about 3.4 and the temperature of the effluent by about an additional degree centigrade, and since lower Biscayne Bay had been declared a national monument in 1968, the state of Florida and FPL took steps to ameliorate the discharge problem before installation of the nuclear generators. The solution proposed initially was to discharge the effluent by way of a 9.7-km-long canal into Card Sound (Figure 11.3). The rationale behind this scheme was severalfold.

1. Card Sound was not a national monument.
2. The effluent water would have a chance to cool down by about $0.5°C$ during its passage through the canal (Bader and Roessler, 1971).
3. Card Sound is deeper (mean depth 2.7 m) than Biscayne Bay, and is more stable physically and chemically (FWPCA, 1970). Therefore, hopefully thermal discharges to Card Sound would have less of an impact on the system than the same discharges made to Biscayne Bay.
4. Studies of the mixing and flushing characteristics of Biscayne Bay had shown that effluent discharged at Turkey Point (Figure 11.4) tended to move northward and be recirculated to the intake canal. This recirculation produced a cumulative effect on the temperature rise of the water returning to the plant (Lyerly and Littlejohn, 1973). This problem would have been eliminated by discharging the water to Card Sound.

The initial FPL proposal also called for diluting the effluent up to 150% with Biscayne Bay water to reduce the temperature of the effluent. After months of hearings, this proposal was finally approved, and construction of the discharge canal begun. The project was abandoned, however, as the result of a Federal–State Conference on Biscayne Bay in February of 1970. Concern at this conference was expressed over possible damage to biota in Card Sound associated with the high velocity of cooling water and dilution water effluent that would be entering the sound (Lyerly and Littlejohn, 1973). Furthermore, studies conducted in Card Sound indicated that the water temperature in at least 35% of Card Sound would still be raised over $35°C$ during about 5% of the time in the summer if the diluted effluent were discharged to the sound as proposed (FWPCA, 1970).

As a result of these predictions, FPL adopted even more stringent measures for dealing with its effluent. The plan ultimately put into operation consists of an enormous system of 32 shallow cooling canals, each about 8.4 km long, through which cooling water from the plant flows before being recycled back in the intake system (Figure 11.3). The canal system provides about 15.6 km^2 of water surface area for heat exchange with the atmosphere, evaporation losses being made up by seepage from brakish groundwater underneath the canal system. The residence time of water in the canals is about 40 hours, during which time the water temperature drops to within $1-2°C$ of Biscayne Bay ambient temperature. According to Tucker (1977), "The canal system is a closed system. There has been no surface water makeup or blowdown since the system was closed in February 1973. There are no withdrawals [sic] from or discharges to Card Sound."

Commentary

From the standpoint of thermal pollution, the physical characteristics of Biscayne Bay make it an extremely poor choice as a place to discharge large quantities of heated water. In particular:

1. Circulation in the bay is poor; therefore, heated effluent is not flushed readily from the system (FWPCA, 1970).
2. The shallowness of Biscayne Bay makes the benthos in this system particularly susceptible to thermal pollution effects. Although some stratification of the water column is apparent even during stormy weather (FWPCA, 1970), this stratification is evidently inadequate to prevent the effluent's being mixed downward sufficiently to increase bottom water temperatures. The killing of benthic biota in 1968 and 1969 attests to this fact.
3. In 1968, the mean maximum daily water temperatures in the bay during the summer months of June, July, and August were 28.3, 29.8, and 30.7°C, respectively, outside the influence of the plume (FWPCA, 1970). These temperatures are close to the lethal tolerance limit of tropical marine organisms, and an increase of water temperature of only a few degrees centigrade would be expected to have serious consequences for the bay's biota. In short, the bay is thermally stressed naturally during the summer months.

Although the construction of a huge closed cooling canal system has eliminated the thermal pollution problem, it is not clear that this solution was the wisest choice environmentally. The construction of 15.6 km^2 of cooling canals destroyed a comparable area of mangroves, a loss whose ecological impact was of some consequence. During the 1 year that the cooling canal system was being dredged, effluents from the power plant were diluted with Biscayne Bay water and discharged both into Biscayne Bay by way of the original discharge canal and into Card Sound by way of a 9.7-km-long canal, or later only into Card Sound. According to Thorhaug et al. (1979), the discharge to Card Sound adversely affected an area of only 2–3×10^4 m^2. The much smaller impact on Card Sound was attributed to the greater depth of Card Sound, the fact that the effluent temperature never exceeded 34°C, and the fact that the canal mouth was constructed to direct effluent water upward, so that less of the effluent impinged directly on the bottom (Thorhaug et al., 1978). In retrospect, it seems probable that a less heroic cooling canal system combined with the release of makeup water to Card Sound could have largely eliminated the thermal pollution problem while sacrificing only a small area of mangroves.

CORRECTIVES

The cooling canal system used at Turkey Point represents one of several methods for dealing with thermal pollution. If the effluent can be cooled to a temperature close to ambient, then it may be practical to recycle the water back through the cooling system. A cooling system that involves more-or-less complete recycling of the cooling water is called a *closed cooling system*, as opposed to the open or once-through cooling systems, which use cooling water once and then discard it. Actually closed cooling systems are not closed completely. In many cases some water is lost to the atmosphere through evaporation in the process of cooling down the effluent, and in all cases some dilution of the effluent with fresh cooling water is needed to offset the buildup of corrosion products and other dissolved and particulate substances in the cooling water. However, the so-called makeup water requirements of a closed-cycle power plant are typically only 2–4% of the cooling water demand of a once-through power plant (Clark and Brownell, 1973). As a result, closed-cycle power plants withdraw 25–50 times less cooling water from the environment than do once-through plants. There is, of course, a continuum of possibilities between strictly closed-cycle and once-through power plants. In the case of Turkey Point, for example, a partially open system involving some discharge to Card Sound and less land area devoted to cooling canals might have been preferable to either the once-through or closed-cycle alternatives.

Cooling Canals

A closed-cycle cooling system may operate with nothing more sophisticated than a long channel or lake to allow the effluent water to exchange heat with the atmosphere. Such a system was installed ultimately at Turkey Point. Recommended design generally calls for a pond that is only a meter or so deep at the input end and about 15 m deep at the outlet. The cooling water to be recycled is drawn from about 9 m below the surface at the outlet (Clark, 1969). Such canals are designed usually so that the residence time of water in the cooling canal is about 2 weeks. The major disadvantage of such systems is the large areas of land required to provide adequate cooling. Typical 1000 MW fossil fuel and nuclear power plants require about 8 and 10 km^2 of pond surface area (Langford, 1983). The surface area required for the ponds can be reduced by as much as a factor of 20, however, if the canals are designed with spray devices that disperse the water in droplets into the air in order to speed up evaporative heat losses (Langford, 1983). About 150–200, 75-horsepower pumps are required to drive the spray devices for a 1000 MW power station (Clark and Brownell, 1973). The power required to operate the pumps represents only about 1% of the plant's output.

Cooling Towers

Effluent water may also be cooled with the use of so-called cooling towers. Cooling towers require much less space than cooling canals, but their height sometimes creates aesthetic problems. Dry cooling towers work on essentially the same principle as the radiator in an automobile. The warm water from the heat exchangers is run through a series of pipes around the interior of a tower, and heat from the water is transmitted through the pipes to a current of air rising through the tower. In a wet cooling tower, the heated water is sprayed into the tower as a mist and is allowed to fall in a thin film over a series of baffles. As the water moves downward, a rising current of air picks up heat from the water by way of evaporation and radiative heat loss. Figure 11.6 shows two basic types of wet cooling towers. The natural draft tower

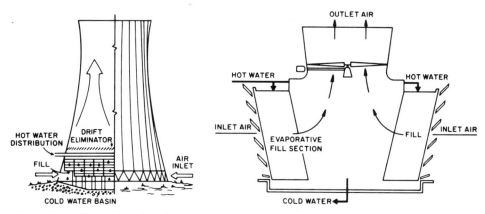

Figure 11.6   Natural draft (left) and mechanical draft (right) wet cooling towers. [Redrawn from Clark and Brownell (1973). Reproduced with permission of the American Littoral Society, Highlands, J. J. 07732.]

works much like a chimney. Air inside the tower rises as it is heated and is replaced by cooler air, which enters through openings in the bottom of the tower. The mechanical draft tower uses mechanical means to circulate air through the sides of the tower and out the top.

Problems

Unfortunately, none of the closed-cycle cooling schemes is without its problems. Cooling canals without spray devices require enormous areas of land. Such large areas of land are frequently not available near power plants, and even if they are, it is not always obvious that devoting so much land to cooling canals is the best option from an environmental standpoint.

The use of sprayers to speed up heat transfer between the water and atmosphere greatly reduces the land area requirements, but obviously increases the cost of running the cooling system. One of the chief drawbacks of cooling towers is their unsightliness. Natural draft cooling towers are typically 110–130 m tall. Mechanical draft towers are generally much shorter ($\sim$20 m high), but are costlier to run (Clark and Brownell, 1973). All wet cooling towers emit a plume of water vapor, which may cause problems under certain conditions. For example, fogging and/or icing may be caused or exacerbated by the plume of water vapor from a wet cooling tower. If saltwater is being used for cooling, salt fallout in the vicinity of cooling towers or near cooling canals using sprayers may damage vegetation. However, salt fallout from spray ponds is confined generally to the immediate vicinity of the plant, and salt fallout from mechanical draft wet cooling towers does not extend much further (Clark and Brownell, 1973). Although salt fallout from the much taller natural draft wet cooling towers can be expected to carry for some distance, the amount of salt fallout more than a short distance from the plant is not likely to be more than a few percent of the level that would be damaging to vegetation (Clark and Brownell, 1973). Langford (1983), for example, notes that saline water has been used in cooling towers at the Fleetwood Power Station in the United Kingdom for some years with few problems. Dry cooling towers obviously eliminate any problems associated with vapor plumes or salt fallout, but the capital cost of dry cooling towers is 2–3 times that of wet cooling towers, and corrosion problems are a major disadvantage.

Table 11.2 indicates the relative costs of various types of power plant cooling systems. Although the closed-cycle systems are several times more expensive than once-through cooling, Knighton (1973) concluded that use of a wet cooling tower would add no more than a few percent to a utility customer's electric bill. In other words, the cost of a closed cooling system is not large compared to the cost of the entire power plant. Closed cooling systems are almost mandatory at the largest power plants, since the amount of cooling water required for these installations using once-through cooling systems is prohibitive. In many cases the impact of power plants on the environment is much less severe when closed rather than open cooling systems are used.

INTERNAL PLANT KILLS

The quotation from *America the Raped* at the beginning of this chapter provides a good introduction to the problem of internal plant kills. These kills occur when organisms are drawn with the cooling water into the power plant. The larger organisms are impinged either on the trash rack or on the screen protecting the pumps. The smaller organisms are drawn through to the inner plant. Vigorous swimmers can of course avoid this fate, but planktonic organisms are largely at the mercy of the currents, and larval fish are often unable to resist the suction from a once-through cooling system's intake. According to Clark and Brownell (1973), the damage to aquatic biota from internal plant kills exceeds generally the damage due to external problems caused by the plant's effluent. The ecological significance of internal plant kills, however, is somewhat controversial (Langford, 1983). In the following discussion, we consider the effects of screen impingement and inner plant kills separately.

Screen Impingements

Some fish drawn in with the cooling water may be large enough to be impinged on the trash racks, but in general fish too large to pass through the 7–8 cm openings of the trash rack are vigorous enough swimmers to escape from the intake suction. As a

Table 11.2 Relative capital costs of various types of electric power plant cooling systems

Cooling system	Relative cost Fossil fuel plant	Nuclear plant
Once-through	1.0–1.5	1.5–2.5
Cooling canals or ponds	2.0–3.0	3.0–4.5
Wet cooling towers		
Mechanical draft	2.5–4.0	4.0–5.5
Natural draft	3.0–4.5	4.5–6.5
Dry cooling towers	8	12

Source. Knighton (1973) and Langford (1983).

result, trash rack kills are usually insignificant (Clark and Brownell, 1973). Smaller fish which pass through the trash rack may be impinged against the screen (Figure 11.2). Most of these fish remain pinned against the screen until they die of suffocation, exhaustion, or physical damage. The screen is generally cleaned once or twice a day by rotating it and dislodging the trapped fish and debris with a jet of water. The dead fish and debris are collected in a basket or similar device and carted off for disposal. Fish still lively enough to flop off the screen when it is rotated are immediately pinned against the screen again after they fall back into the water. The number of fish killed in this way is correlated positively both with the volume of water pumped through the plant and with the intake current velocity.

Table 11.3 lists some of the estimated fish kills due to impingement at electric power plants in the United States. In most cases the numbers for plants using once-through cooling systems are high, typically 10^5–10^7 fish per year. When these numbers are compared, however, to commercial fish catches or the standing stock of fish in the vicinity of the plant, the percentages are generally quite low. The 3×10^7 fish impinged each year at 17 Lake Michigan power plants, for example, represent only about 0.06% of the corresponding standing stocks in Lake Michigan (IAEA, 1980). On the other hand, the fact that impingement of yellow perch at the Monroe Power Plant on Lake Erie amounts to 2.7% of the commercial catch of this species seems noteworthy, since the Monroe Power Plant is by no means the only power plant located on the shores of Lake Erie.

Certainly measures can and have been taken to reduce impingement losses. Bubble curtains and mechanical diversion systems have been used with partial success to divert fish away from intakes, and velocity caps, which minimize the downward flow of water toward intake structures, have been used to minimize fish entrainment. Unfortunately, no such engineering modifications have proven generally satisfactory at estuarine sites, where the impingement problem is most severe because of the high densities of juvenile and larval fish. In such cases efforts to rescue organisms impinged on the intake screens may reduce losses significantly. At the Brunswick Steam Electric Plant, for example, a continuous rescue operation has resulted in 90% of the crabs, 80% of the shrimp, and 40% of the fish being returned to the estuary in live condition (IAEA, 1980). The most effective strategy, however, is to install a closed cooling system. This fact is illustrated dramatically in the comparison of impingement losses at the Davis-Besse, Bay Shore, and Acme power stations in Table 11.3. Although the Davis-Besse station generates more electricity than the other two stations combined, impingement losses at Davis-Besse are less than 0.1% of the losses at either of the other stations. The principal cause of the difference is the fact that Davis-Besse uses a closed cooling system that draws only 5–10% as much water as Acme or Bay Shore. The difference in impingement losses is particularly noteworthy, because the adult fish populations near Davis-Besse can be twice those near Bay Shore and 8 times those at Acme (Reutter and Herdendorf, 1984).

Inner Plant Kills

Organisms that pass through the 1 cm openings of the intake screens are drawn through the rest of the condenser cooling system. During this experience, they are subjected to a variety of stresses, including thermal shock, physical abrasion and buffeting, pressure shock, and possibly chlorination and nitrogen embolism. If an organism is not killed outright during its passage through the plant, it may be so physically damaged by the time it is discharged that it dies soon thereafter or easily falls victim to a predator.

Table 11.3   Fish Kills Due to Screen Impingement at Various Power Stations

Power plant	Impingement event	Period	Comments
Millstone Niantic Bay, Connecticut	Massive kill of small menhaden (more than 2 million), screens clogged.	1971	Occurring late summer, early fall, plant shut down on 8/21/71; cause unknown. Persistent low kill of 10 other species.
P. H. Robinson Galveston Bay, Texas	7.2 million fish impinged in 1 year.	1969–1970	Projected from sampling of operating plant; principal species were menhaden, anchovy, croaker; highest in March.
Indian Point No. 1, Hudson River, N.Y.	Yearly kill of 1–1.5 million fish.	1965–1972	Primarily white perch with 4–10% striped bass.
	Kill of 1.3 million.	1969–1970	10% striped bass. Plant closed February 8.
Indian Point No. 2	Massive kills. Maximum per day 120,000.	Jan. 1971	Testing cooling system of new plant (no heat); white perch and other species.
Port Jefferson Long Island, New York	2 truckloads (at least) of fish killed on screens in 3 days.	Jan. 26–28, 1966	Mostly small menhaden; also white perch.
Brayton Point Mount Hope Bay, Massachusetts	350,000 fish impinged in 1 year; mostly menhaden.	1971–1972	Heaviest from November to March, flounder, silverside, and others also impinged.
Oyster Creek Barnegat Bay, New Jersey	10,000 fish, 5000 crabs, destroyed per month in spring and summer.	1971	Estimated from 19 days of sampling; screen kill in cold season unknown.
Surrey Power Station, James River, Virginia	6 million river herring destroyed in 2–3 months.	Oct.–Dec., 1972	Estimated by AEC from screen sampling during partial power runs.
17 Lake Michigan plants	30 million fish impinged per year.	early 1970s	Mostly alewives and rainbow smelt. Biomass losses ~0.06% of standing stocks.

Table 11.3   (*Continued*)

Power plant	Impingement event	Period	Comments
4 Tennessee Valley Authority plants	22 thousand to 3.6 million fish impinged per year.	early 1970s	Mostly clupeoids and sciaenids. Losses 0.03 to 0.52% of total biomass. Losses of 8 major families < 3%.
Monroe Power plant, Lake Erie	122 thousand yellow perch impinged per year.	1970s	Loss equals 2.7% of Lake Erie commercial catch and 0.7% of perch population in western basin of lake.
Pilgrim New England	58 thousand fish impinged per year.	1970s	Mostly Atlantic herring and alewife.
Peach Bottom Pennsylvania	472 thousand fish impinged per year.	1973–1975	Mostly channel catfish, white crappie, and bluegill. Two units in operation.
Crystal River Florida	50–62 thousand fish impinged per year.	1969–1970	Mostly batfish, catfish, and perch.
Acme Power Plant Maumee River, Ohio	11.7 million fish impinged per year.	1976–1977	Mostly gizzard shad, emerald shiners, and alewives. Most between October and February.
Bay Shore Lake Erie at mouth of Maumee River	18.3 million fish impinged per year.	1976–1977	Mostly gizzard shad, emerald shiners, and alewives. Most between October and February.
Davis-Basse Lake Erie	4.4–6.6 thousand fish impinged per year.	1978–1979	Mostly goldfish and yellow perch. Most in colder months. Closed cooling system.

Source. Clark and Brownell (1973); Grimes (1975); Mather et al. (1977); IAEA (1980); Reutter and Herdendorf (1984).

Fish larvae and post larvae suffer particularly high mortalities due to inner plant stresses. For example, the EPA estimated that all larval menhaden were killed as a result of passage through the Brayton Point power plant on Mount Hope Bay, Massachusetts (EPA, 1972). Physical damage appeared to be the principal cause of death. Marcy (1971) found that no white perch survived passage through the Connecticut Yankee power plant at Haddam Point if the discharge temperature exceeded 28.3°C, and no larvae or juvenile fish of any species survived if the discharge temperature exceeded 32.8°C. Most of the dead fish were mangled, an indication that physical as well as thermal stress was a factor in their demise. At the Indian Point power plant on the Hudson River, 97.5% of larval and early juvenile fish died or were severely damaged by passage through the cooling system during the summer if the temperature rise across the condensers exceeded 15°C (Clark and Brownell, 1973).

Planktonic organisms generally survive inner plant stresses better than juvenile and larval fish. Zooplankton losses have been reported anywhere from 0 to 100%, the average loss being about 30–35% (Clark and Brownell, 1973; Langford, 1983). The percentage kill of zooplankton is quite dependent on effluent water temperature and degree of chlorination. For example, copepods suffered 100% mortality at the Northport power plant on Long Island Sound if the discharge temperature exceeded 34°C, a condition that exists from July to early October (Suchanek and Grossman, 1971). Copepod losses at the same plant, however, were only 33% in the fall and 4% in the winter (Clark and Brownell, 1973). At four power plants along the California coast, zooplankton losses were related linearly to temperature (Icanberry and Adams, 1974). Gammarid losses in power plant cooling systems are virtually 100% if the effluent temperature exceeds 32°C, and mysid mortality is 100% at temperatures above 27°C (Clark and Brownell, 1973). Davies and Jensen (1975) reported 100% mortality of zooplankton entrained in the Marshall power plant cooling system at chlorine concentrations of 1.5 mg/l, but no mortality at chlorine concentrations of 0.25–0.75 mg/l. These results suggest that the principal causes of inner plant zooplankton kills are thermal stress and chlorination, and that the effects of physical abrasion are of relatively minor importance. Obviously zooplankton mortalities can be expected to be highest during the summer months when ambient water temperatures are highest.

The impact of entrainment on phytoplankton is generally negative, and, as in the case of zooplankton, appears to result almost entirely from thermal stress and/or chlorination. Data summarized by Langford (1983) indicate an average reduction in photosynthetic rates of 40–50%, but in many cases it is unclear whether cells have actually been killed or just stressed temporarily. In temperate latitudes phytoplankton production may actually be stimulated in the warmer waters of the discharge plume (Morgan and Stross, 1969). There is no doubt that chlorination adversely affects phytoplankton. The data of Davis and Coughlan (1978), for example, show a very obvious correlation at chlorine concentrations above 0.1 mg/l (Figure 11.7). Fortunately the generation time of phytoplankton is relatively short, typically 12 hours to a few days, and natural physical processes can be expected to reseed effluent waters with healthy cells within a short time. It is therefore unlikely that entrainment has a significant negative impact on photosynthetic rates in waters surrounding electric power plants. In fact, the stimulation of undesirable cyanobacteria by the heated effluent may be more of a problem than any damage to phytoplankton caused by inner plant stresses.

Just how serious are inner plant kills to aquatic biota? Table 11.4 lists estimates of inner plant fish kills at several power plants. A comparison of Table 11.3 and 11.4 indicates that entrainment tends to kill far more fish than impingement, and the annual losses to entrainment are roughly 5% of the at-risk population. Intelligent siting and design of intake systems help to reduce such losses, but the most effective way to minimize entrainment losses is the use of a closed cooling system. The comparison of the Davis-Besse power plant with the Acme and Bay Shore facilities is again instructive. The entrainment of larval fish and fish eggs at Davis-Besse is only 3.6% and 0.01%, respectively, of the combined entrainment at the other plants, although Davis-Besse produces more power than Acme and Bay Shore combined. The difference reflects largely the use of a closed cooling system at Davis-Besse.

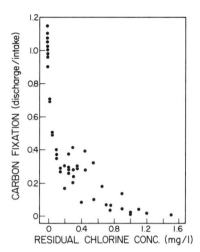

Figure 11.7   The effect of cooling-water chlorination on photosynthesis by marine phytoplankton at Fawley power station, U.K. [Redrawn from Davis and Coughlan (1976).]

Commentary

Judging from Tables 11.3 and 11.4, the number of fish killed each year from internal plant stresses in a once-through cooling system can run easily into the tens or hundreds of millions. The number of fish killed from thermal stress in late June, 1969 near the Turkey Point outfall was reportedly in the thousands, but not millions (FWPCA, 1970). The largest external plant fish kill on record involved 100–200 thousand menhaden, which were killed by cold shock when the Oyster Creek power plant on Barnegat Bay, New Jersey, shut down in late January, 1972 (Chapter 8). Other external plant kills have involved thousands and in some cases tens of thousands of fish (Clark and Brownell, 1973; Langford, 1983).

The greater magnitude of the internal plant kills reflects undoubtedly the fact that fish, including larval and juvenile fish, can usually avoid discharge plumes by simply swimming away. Small fish find it more difficult to escape the suction from a power plant's cooling water intake. One should keep in mind, however, that a large percentage of larval and juvenile fish never reach adulthood, since they are eaten by predators at an early age. Thus the killing of one million juvenile fish of a particular species may remove essentially the same percentage of the species and age group as the killing of 10^4 adults. On the other hand, the killing of juvenile and larval fish not only reduces the population and reproductive potential of the given species, but removes a soure of food for other organisms that prey on small fish. According to an Atomic Energy Commission staff analysis, entrainment of striped bass larvae by power plants along the Hudson River could result in over 60% mortality of each year's production of young striped bass (Goodyear, 1973). The impact of this loss on the adult population of striped bass has been estimated to be anywhere from 8–33%, depending on what assumptions are made about natural mortality and compensatory survival (Langford, 1983). Declines of 20–30% in the population of a valuable fish seem a bit much to simply ignore. What can be done to reduce the impact of internal plant kills?

Table 11.4   Fish Kills Due to Inner Plant Stresses at Various Power Stations

Power plant	Event	Period	Comments
Brayton Point Mount Hope Bay, Massachusetts	7–165 million menhaden (some river herring) killed per day.	Summer, 1971	Estimated from EPA's sampling techniques. 164.5 million killed on July 2. Fish mangled.
	50 million fish killed in 11 days.	August 10–21 1971	Estimated from net tows at discharge; menhaden and blueback herring; tests showed all fish died.
Millstone Niantic Bay, Connecticut	36 million fish killed in 16 days (probably menhaden and blueback herring).	November 2–18 1971	Estimated by sampling of vertebrae of dead fish in discharge canal.
	2.5 million flounder entrained.	April–June, 1972	Death rate not estimated.
Connecticut Yankee Connecticut River, CT	179 million fish larvae killed per year.	1969 and 1970	Primarily alewife and blueback herring. Loss equals 4% of transported population.
Cumberland Steam Cumberland River	57 million larvae entrained per year.	1970s	Entrainment equals 3.6% of larvae transported past plant.
Hanford Columbia River	142 thousand chinook salmon fry entrained per year.	1970s	Entrainment equals 8.8% of population.
17 Lake Michigan plants	alewife larvae	early 1970s	Entrainment equals 0.01% of population.
	smelt larvae	early 1970s	Entrainment equals 1% of population.
Acme Maumee River	80 million larval fish entrained per year.	1976–1977	Mostly gizzard shad and freshwater drum.
	178 million fish eggs entrained per year.	1976–1977	
Bay Shore Lake Erie	284 million larval fish entrained per year.	1976–1977	Mostly gizzard shad and white bass.
	426 million fish eggs entrained per year.	1976–1977	
Davis-Besse Lake Erie	13 million fish larvae entrained per year.	1976–1977	Mostly gizzard shad and emerald shiner.
	70 thousand fish eggs entrained per year.	1976–1977	

Source. Clark and Brownell (1973); IAEA (1980); Reutter and Herdendorf (1984).

Correctives

Engineering modifications such as bubble curtains, velocity caps, and mechanical diversion systems have had some success in reducing screen impingement kills, but have very little or no effect on entrainment losses. There are, however, several rather obvious ways of reducing internal plant kills at power plants that use once-through cooling systems.

1. If at all possible, power plants should not be located in estuarine areas, since these areas are known to serve as breeding and nursery grounds for many species of aquatic organisms, and the population density of juvenile and larval forms, which are most subject to entrainment, is likely to be highest in these estuarine regions.

2. According to Langford (1983, p. 192), "The greatest reductions in impingement mortalities have been at sites where water velocities have been significantly reduced." Clark and Brownell (1973) recommend that the speed of flow into the cooling system at the intake should be no more than about 8 cm s^{-1} and 15 cm s^{-1} for closed and open cooling systems, respectively. The idea is of course to maximize insofar as possible the likelihood that weakly swimming organisms will be able to resist being drawn into the cooling system. The higher limit of 15 cm s^{-1} for once-through cooling systems is made for practical rather than ecological reasons. Because of the large volumes of water needed for once-through cooling, Clark and Brownell (1973) felt that it would be impractical to expect the construction of a once-through cooling system intake structure large enough to keep flow speeds much below 15 cm s^{-1}, an opinion that is echoed by Langford (1983).

3. Impingement kills can be reduced significantly by designing a mechanism for rescuing the impinged organisms before they are injured seriously. As noted, the Brunswick Steam Electric Plant has reduced impingement losses by 40–90% with this approach.

4. The only effective way to reduce entrainment losses is to minimize the concentration of organisms in the cooling water or reduce the amount of water drawn from the environment. Judicious choice of the location of intake structure can help greatly in the former case. The intake at the Davis-Besse power plant, for example, is located offshore and near the bottom of Lake Erie. The Acme and Bay Shore plants both utilize open shoreline intake canals, which tend to attract schooling species that follow the shoreline. In general you can expect the concentration of organisms in the water to decrease with depth and with distance from shore.

   Reducing the amount of water drawn from the environment can be accomplished with a closed cooling system, but not without paying a price. The price includes the additional capital, operating, and maintenance costs associated with the closed cooling system, and reduced efficiency in electricity production. The latter loss of efficiency amounts to about 1.5–2.0% for fossil fuel plants and 3% for nuclear plants (Monn et al., 1979). The monetary costs, of course, are passed on to the consumer. The environmental costs, however, should not be overlooked. The reduction in plant efficiency translates into more carbon dioxide and oxides of nitrogen and sulfur released into the atmosphere at fossil fuel plants, and more radioactive wastes generated at nuclear plants. There may also be aesthetic considerations and in some cases significant additional environmental impacts associated with closed cooling systems, the destruction of mangroves at Turkey Point being an example of the latter. At large power stations, however, the availability of water alone may be sufficient to dictate the use of closed-cooling systems. Selection of the best system will require a full accounting of the pros and cons of the various alternatives.

POSSIBLE BENEFICIAL USES OF THERMAL DISCHARGES

The idea of putting the heated water discharged by power plants to some practical use has occurred to a number of people. About 50% and 58–68% of the energy released from the fuel is rejected as waste heat with the cooling water at fossil fuel and nuclear power plants, respectively (GESAMP, 1984). What can be done with all this wasted energy?

Unfortunately there are no uses to which a significant percentage of the heated effluent from power plants could possibly be put on a year-round basis in most parts of the world. Heated effluents have been and are being used in aquaculture and agriculture and for space and water heating, but demands for these uses drop off dramatically during certain times of the year. Furthermore, modifying present facilities to incorporate these uses will do little to help and may in fact aggravate water pollution problems. Nevertheless, such "beneficial" uses are worth examining to evaluate their pros and cons.

Cogeneration Power Plants

Electric power plants that produce a useful by-product in addition to electricity are called *cogeneration facilities*. In some cases the by-products include steam and both hot and cold water. In cases where both heating and cooling are possibilities, the network served by the plant is referred to as a *district heating/cooling (DHC) system*. Power plants that serve district heating or DHC systems are designed as cogeneration facilities when they are built. They are not the result of retrofitting conventional electric power plants.

Using heated power plant effluents for space and water heating is an old idea that has been developed extensively in the Soviet Union and some European countries. Much of this development occurred during the rebuilding following World War II, when it was convenient to lay the necessary pipes to set up district heating networks. About 54% of all space and water heating in the Soviet Union is provided by district heating, including 70% of all urban space and water heating (Karkheck, Powell, and Beardsworth, 1977). Table 11.5 indicates the percentages of space and water heating that is provided in some European countries by district heating networks.

Table 11.5   Percentage of Space and Water Heating Provided by District Heating in Some European Countries

Country	Percent space and water heating provided by district heating
Denmark	32
Sweden	20–25
Finland	14
West Germany	3

Source. Karkheck, Powell, and Beardsworth (1977).

District heating works in a very straightforward manner. The heated effluent from a power plant or similar source of heated water is pumped to buildings where it is used to warm rooms and heat water. In the process of heat exchange the effluent cools down until it is no longer useful for heating, at which point it is recycled to the power plant. The principal cost of district heating networks is the cost of distributing the hot water to the customers. The hot water is typically pumped through urethane-insulated pipes constructed of polymer concrete and lined with polymer (Karkheck, Powell, and Beardsworth, 1977). Obviously there is a limit as to how far hot water can be economically pumped for heating, but a study conducted in the United States has indicated that steam could be transmitted economically as much as 16 km for heating (Cook and Biswas, 1974), and hot water is easier to transmit over long distances than steam (Karkheck, Powell, and Beardsworth, 1977). Nevertheless, district heating is not a practical scheme in other than urban areas where the population density of customers is high.

To be useful for space and water heating, hot water must have a temperature in the 50–100°C range, and closer preferably to 100°C (Karkheck, Powell, and Beardsworth, 1977). In the United States, virtually all electric power plants operate with a steam exhaust pressure of about 0.05 atm, at which pressure the steam condenses at a temperature of about 33°C. As a result, the temperature of the condenser cooling water is usually only about 5–15°C above ambient, too low to be useful for district heating. However, if the exhaust pressure to the condensers is adjusted to 1 atm, the steam condenses at 100°C, and the temperature of the cooling water is raised correspondingly. Power plants to be used in a district heating network must obviously operate with a high steam exhaust pressure to produce sufficiently hot effluent water. This operation strategy sacrifices some efficiency in terms of electricity generation, but the loss of energy can be more than regained if the hot effluent water is utilized for district heating.

Although district heating is not as common in the United States as in the Soviet Union or Scandanavia, a number of DHC systems do exist (Collins, 1991a,b). The systems installed to date in the United States have been designed primarily to provide electricity and heating/cooling in high density urban situations suited ideally for the DHC concept. The DHC system in Hempstead, New York, for example, serves a 19,000-student junior college, a 17,000-seat indoor sports arena, a 483-room hotel/conference center, a 615-bed medical center, and a large county jail complex

(Collins, 1991b). A similar system in Hartford, Connecticut, serves 64 buildings, including 25% of the downtown area's high-rent office space (Collins, 1991a). A key feature of successful district heating and DHC systems is the ability to adjust the production of electricity and by-products to meet demands. In order to avoid pollution problems the system must have some way to recycle thermal effluent when the demand for by-products drops. Because of the high temperature of the thermal effluent required to make district heating and DHC systems practical, it would be unacceptable for excess hot water to be discharged to the environment. Some method of cooling and recycling the effluent is therefore necessary when the supply of hot water exceeds the demand. The plant must therefore have either a backup closed-cooling system to recycle unneeded hot water or a backup conventional electricity generating capacity.

Karkheck, Powell, and Beardsworth (1977) have estimated that 52% of the U.S. population could be served by district heating at a cost no greater than the present cost of heating using foreign oil. In the process, Karkheck, Powell, and Beardsworth (1977) argued that the United States could cut its imports of foreign oil by about 1.1 billion barrels per year, currently about 55% of oil imports (Flaven, 1985). The cost of setting up the district heating system, about $180 billion, would be recouped in about 14 years due to the anticipated savings of $13 billion per year in the balance of payments. One obvious flaw in this argument is that most fossil fuel power plants in the United States burn coal. A more likely impact of large-scale district heating would therefore be an extension of the useful lifetime of the United States' coal reserves. The number of U.S. cogeneration power plants will probably continue to grow, but because of the 30–40-year lifetime of existing conventional plants, it is doubtful that such facilities will account for a significant percentage of U.S. electricity production in the near future.

Agriculture

A variety of possible uses of thermal effluents in agriculture have been or are being tested. In so-called open-field thermal agriculture, heated water is used for frost protection during the spring and fall, for plant cooling during the summer, and for irrigation. In enclosed systems such as greenhouses, thermal effluents may be used in "controlled environment thermal agriculture" to regulate temperature and humidity levels to provide a controlled environment for plant and/or animal growth. Unfortunately many of the postulated uses of thermal effluents in agriculture would be highly seasonal and would come far short of using all the effluent from even a moderate-sized power plant. Nevertheless, Cook and Biswas (1974, p. 5) feel that, "In a relative sense, controlled-environment agriculture is probably the most practical (technologically and economically) of all beneficial uses for low grade waste heat." The following discussion indicates some of the specific applications for which thermal effluents have been tested.

A study at Springfield, Oregon, sponsored by the Eugene Water and Electric Board investigated the feasibility of using thermal effluents for frost protection in orchards. Warm (35°C) water was sprayed onto trees when there was danger of frost, and the heat released from the water as it cooled or froze warmed the twigs and buds, thus protecting them from frost damage. If too much water was applied, however, the limbs became overloaded with ice and broke. Therefore care had to be taken in regulating the amount of water applied (Miller, 1970).

At the same project site, heated water from the condenser of a pulp mill was piped to fields containing a variety of crops both to irrigate the crops and to maintain optimal soil temperatures. An analogous experiment conducted in Oregon by Boersma (1970) indicated that simply regulating soil temperature by supplying heat from underground pipes could substantially increase crop growth rates and might triple the yield of some crops. In the Springfield, Oregon, study the heated effluent used for frost protection and irrigation simply percolated through the soil and found its way ultimately to the Mackenzie River, where it entered the river at or below ambient river temperature.

The chief drawback to the foregoing type of open-field thermal agriculture is the large amount of money required to set up and maintain the system. Furthermore, water-pumping costs and heat losses with increasing distance from the power plant make this sort of scheme impractical except in the immediate vicinity of the thermal effluent. Finally, since this sort of water use would be highly seasonal, it would certainly not provide a satisfactory solution to the problem of thermal effluent disposal. On the other hand, the Springfield, Oregon, studies indicate that thermal effluents can be used to good advantage in open-field agriculture during certain times of the year and under appropriate climatological conditions.

Controlled-environment thermal agriculture has been tested in several desert areas as a possible means of growing crops in arid regions (Cook and Biswas, 1974). In studies conducted in Mexico and the middle east, crops were grown in greenhouses made of plastic and inflated with air. Heated effluent from power plants was used to desalinate seawater and produce distilled water, which in turn was used to irrigate the crops. The use of greenhouses reduced greatly irrigation water needs by cutting down on evaporation losses and provided protection from pests, disease, and sandstorms. Capital and maintenance costs for such operations are quite high, but if cheap labor is available, Cook and Biswas (1974) indicate that crops produced by this method would be cheaper than imported crops. Because of high capital and maintenance costs, however, such controlled-environment thermal agriculture would not be cost competitive with conventional arid-land farming techniques where the latter are applicable.

In New York, the feasibility of using thermal effluent to heat greenhouses has been studied (Bell et al., 1970), and although the cost of using thermal effluent for this purpose was found to be quite competitive with conventional heating costs, the scheme was rejected ultimately by a Con Ed and Westinghouse task force because of the limited potential of greenhouses to use thermal effluent. For example, a 1000 MW nuclear power plant could provide enough thermal effluent to heat 11.4 km^2 of greenhouses (Cook and Biswas, 1974). The idea, however, is not all that far-fetched. In Germany, for example, hot-houses with an area of 0.2 km^2 are heated by cooling water from power plants and produce crops throughout the year (Mitzinger, 1974).

Aquaculture

The idea of using thermal discharges to stimulate aquaculture production has been studied and proven economically practical in several countries, and there are presently at least 30 power stations throughout the world at which pilot scale or commercial aquaculture facilities have been developed (Langford, 1983). About two-thirds of those facilities use fresh water. The idea is to use warm discharge water either wholly or diluted with some natural water source to maintain a certain temperature for the organisms being cultured and in particular to extend the growing season into the colder months of the year when growth rates would normally decline due to cold water temperatures. In this way it may be possible to reduce greatly the time required to produce a marketable product and/or increase the number of generations produced per year.

The principal organisms cultured in freshwater systems have been catfish in the United States and carp in other countries (Table 11.6). In some cases it has proven advantageous to culture different species at different times of year to take advantage of natural variations in water temperature. For example, at the Mercer power station in New Jersey a pilot farm has grown rainbow trout in the winter and freshwater prawns in the summer (Langford, 1983). Similarly, at the Flevo power station in Holland, trout are grown in the winter; carp and eels are grown in the summer.

Japan has led the way in the culture of marine organisms in thermal effluent, the principal species of interest being shrimp, eel, yellowtail, seabream, ayu, and whitefish (Yee, 1972). According to Tanaka and Suzuki (1966), use of thermal discharges in conjunction with yellowtail culture has increased growth rates by almost 50% and has nearly doubled the average feeding efficiency. In another set of experiments conducted at a power plant near Matsuyama, Japan, the wintertime weight gain of shrimp grown using thermal discharges was 7 times that of control shrimp grown at ambient winter water temperatures.

In the United States, oysters are now being cultured year-round on a commercial scale by Long Island Oyster Farms, Inc., using the thermal effluent from the Long Island Lighting Company's power plant at Northport, Long Island (Yee, 1972). Old

Table 11.6   Organisms Cultured at Inland Power Stations using Freshwater in Various Countries

Country	Power station	Species cultured
U.S.	Gallatin (Tennessee)	Catfish
	Colorado City (Texas)	Catfish
	Hanford (Washington)	Catfish
	Fremont (Nebraska)	Catfish and *Tilapia*
	Lake Trinidad	Catfish
	Hutchinson	Catfish
	Mercer (New Jersey)	Trout and freshwater prawns
U.K.	Ratcliffe-on-Soar	Carp and eels
	Trawsfynydd	Trout
	Ironbridge	Carp
France	Cadarache	Eels
Netherlands	Flevo	Carp, grass carp, and trout
Germany	Finkenheerd	Carp
	Rheinberg	Carp, eels, and trout
Poland	Konin	Carp, grass carp, silver carp, and big-head carp
Hungary	Szazamlombatta	Carp, grass carp, silver carp, and big-head carp
Soviet Union	Klasson	Grass carp and carp
	Krivorozh	Grass carp and carp
	Shaktinsk	*Tilapia*
	Kirishi	Carp and trout

Source. Langford (1983).

oyster shells bearing oyster larvae are placed in mesh bags and suspended in the power plant's discharge basin. Ambient water temperatures range from -1 to $24°C$, and the thermal discharge is about $14.5°C$ above ambient. During the summer it is necessary to dilute the discharge with water of ambient temperature to avoid killing the oysters. The growth rate of the juvenile oysters in the discharge basin during the summer months is about double that of oysters growing at ambient temperatures, and growth continues during the winter in the discharge basin, whereas all growth comes to a stop in surrounding waters when the temperature drops below about $4.5°C$ (Cook and Biswas, 1974). Commercial sales amounted to $5 million in 1977 (Beall et al., 1977).

Although the use of thermal effluents has proven successful in some aquaculture operations, there are significant limitations to this method of utilizing waste heat. Plant shutdowns could prove disastrous to some aquaculture operations during the winter months unless there is some way to ameliorate the effect of thermal shock on the organisms being cultured. The aquaculture operation would either need an auxiliary source of heated water or the species being cultured would have to be rapidly adaptable to temperature changes. Carp and eels are rather tolerant to temperature changes, but trout, for example, are not.

Chlorine's extreme toxicity to many species of fish and invertebrates could weigh heavily against the use of thermal effluent for aquaculture at power plants which use chlorine to prevent fouling in the heat exchangers. Some method of dechlorinating the effluent or of bypassing the aquaculture operation when residuals became too high would be necessary. With respect to this issue the most desirable power stations for aquaculture are those that use mechanical or physical means to prevent fouling. The Northport plant, for example, uses brushes and sponge balls under pressure for condenser tube cleaning.

Pollutants that might accumulate in the flesh of cultured organisms include heavy metals and, in the case of nuclear power plants, radioisotopes. The metals of particular concern are copper, zinc, cadmium, iron, and nickel (Langford, 1983). There is no evidence that finfish cultured in power plant effluent have accumulated metals to any significant degree, perhaps because they are fed artificial diets that are essentially free of metal contaminants. Shellfish, on the other hand, filter plankton directly from the water and are well known for their tendencies to bioaccumulate metals and other pollutants (see Chapter 12). In some but not all cases, placing shellfish in clean water for a few days may reduce pollutant concentrations significantly. In general, however, careful monitoring for pollutants in the flesh of shellfish grown in power plant effluent will be advisable. From the standpoint of public acceptance, it is doubtful whether organisms cultured in the effluent from a nuclear power plant would be marketable, even if radiation levels in the tissue were below safety limits.

From the standpoint of thermal pollution, a little thought will convince you that aquaculture operations are not likely to significantly reduce stresses to the environment resulting from thermal discharges. Most aquaculture operations would probably be able to use only a small percentage of the effluent from a power plant anyway. For example, the catfish culturing system at the Tennessee Valley Authority's Gallatin, Tennessee, power plant uses less than 0.5% of the effluent from the plant (Anonymous, 1974). In any case, an aquaculture operation is obviously interested in using water that will stimulate growth. Such water could be discharged harmlessly into the environment in the first place. In fact, the waste products from the organisms growing in an aquaculture system may create pollution problems if the effluent is discharged. In such cases, it may be necessary to treat the effluent from the aquaculture operation before releasing it to the environment. Even if the effluent is recycled to the cooling system, some treatment may be necessary, since it would obviously be unwise to let the waste products accumulate continuously.

Other Uses

A variety of other possible uses for thermal discharges are discussed in a publication by Cook and Biswas (1974). These uses include desalination, melting of snow and ice, defogging airport runways, waste treatment, and warming bodies of water used for swimming. An analysis of these proposed schemes indicates that some are completely impractical,[4] and others would be practical under only rather special conditions. Power plant desalination units, for example, would be impractical unless there were a very high demand for freshwater ($\sim 5 \times 10^5$ m^3 d^{-1}) close to the power plant. On the other hand, the fact that a particular use is impractical in one part of the world need not detract from its use somewhere else. From the standpoint of thermal pollution, however, the most practical way to deal with many thermal discharges may simply be to use closed cooling systems that minimize the amount of water drawn from and discharged into the environment.

REFERENCES

Anonymous. 1974. 'Power plant' fish production. *Commercial Fish Farmer*, **1**(1), 10–13.

Bader, R. G., and M. A. Roessler. 1971. *An Ecological Study of South Biscayne Bay and Card Sound*. Progress Report to U.S. Atomic Energy Commission (AT(40-1)-3801-3) and Florida Power and Light Co.

Beall, S. E., C. C. Coutant, M. J. Olszewski, and J. S. Suffern. 1977. Energy from cooling water. *Industrial Water Eng.*, **14**(6), 8–14.

Bell, R. A., W. J. Cahill, A. S. Chiefetz, G. T. Cowherd, A. John, J. A. Nutant, and H. Wright. 1970. Combination urban-power systems utilizing waste heat. Paper presented at the conference on beneficial uses of thermal discharges. New York Dept. Environ. Conservation. Albany, N.Y. Sept. 17–18.

Boersma, L. 1970. Warm water utilization. Paper presented at the conference on beneficial uses of thermal discharges, New York Dept. Environ. Conservation. Albany, N.Y. Sept. 17–18.

Brungs, W. A. 1973. Effects of residual chlorine on aquatic life. *J. Water Pollut. Contr. Fed.*, **45**, 2180–2193.

Bush, R. M., et al. 1974. Potential effects of thermal discharges on aquatic systems. *Environ. Sci. Tech.*, **8**(6), 561–568.

Cairns, J., Jr., A. L. Buikema, Jr., A.G. Heath, and V. C. Parker. 1978. *Effects of temperature on aquatic organism sensitivity to selected chemicals*. Virginia Water Resources Res. Center. Bull. 106. Blacksburg, Va.

Clark, J. R. 1969. Thermal pollution and aquatic life. *Sci. Amer.*, **22**(3), 19–27.

---

[4]For example, using thermal discharges to keep the St. Lawrence Seaway free of ice.

Clark, J. and W. Brownell. 1973. *Electric Power Plants in the Coastal Zone: Environmental Issues.* American Littoral Society Special Publication No. 7, Highlands, NJ.

Collins, S. 1991a. The perfect match: combined cycles and district heating/cooling. *Power,* **135**(4), 117–120.

Collins, S. 1991b. Combined-cycle cogen plant cuts costs. *Power,* **135**(5), 63–64.

Cook, B., and A. K. Biswas. 1974. *Beneficial Uses of Thermal Discharges.* Environment Canada. Planning and Finance Service Report No. 2. Ottawa, Canada. 75 pp.

Davies, R. M., and L. D. Jensen. 1975. Zooplankton entrainment at three mid-Atlantic power plants. *J. Water Pollut. Contrl. Fed.,* **47**(8), 2130–2142.

Davis, M. H., and J. Coughlan. 1978. Response of entrained plankton to low-level chlorination at a coastal power station. In R. L. Jolley (Ed.), *Water Chlorination: Environmental Impacts and Health Effects.* Vol. II. Ann Arbor Science, MI. pp. 369–376.

DeMont, D. J., and R. W. Miller. 1971. First reported incidence of gas-bubble disease in the heated effluent of a steam generating station. Paper delivered at 25th Annual S. E. Assoc. of Game and Fish Comm. Charleston, SC, 13 pp.

Environmental Protection Agency. 1972. *Proceedings in the Matter of Pollution of Mount Hope Bay and Its Tributaries,* Washington, D.C. 2 Vols., 643 pp.

Environmental Protection Agency. 1976. *Quality Criteria for Water.* EPA-440/9-76-023. Washington, D.C. 501 pp.

Environmental Protection Agency. 1986. *Quality Criteria for Water.* EPA 440/5-86-001. Washington, D.C.

Fairbanks, R. B., W. S. Collings, and W. T. Sides. 1968. *Biological Investigations in the Cape Cod Canal prior to the Operation of a Steam Generating Plant.* Massachusetts Dept. Natural Resources. Boston, MA.

Flaven, C. 1985. *World Oil: Coping With the Dangers of Success.* Worldwatch Paper 66. Worldwatch Inst. Washington, D.C. 66 pp.

Florida State Board of Health. 1962. Survey, Biscayne Bay.

FWPCA. 1970. *Report on Thermal Pollution of Intrastate Waters of Biscayne Bay, Florida.* U.S. Dept. of the Interior. Southeast Water Laboratory. Federal Water Pollution Control Administration. Ft. Lauderdale, FL. 44 pp.

GESAMP. 1984. *Thermal Discharges in the Marine Environment.* Group of experts on the scientific aspects of marine pollution. Food and Agriculture Organization of the United Nations. Rome. 42 pp.

Goodyear, C. P. 1973. Probable reduction in striped bass in the Hudson. Testimony before AEC licensing board, Indian Point No. 2, Feb. 8.

Grimes, C. B. 1975. Entrapment of fishes on intake water screens at a steam electric generating station. *Chesapeake Science,* *16*(3), 172–177.

Hawkes, H. A. 1969. Ecological changes of applied significance from waste heat. In P. A. Krenkal and F. L. Parker (Eds.), *Biological Aspects of Thermal Pollution,* Vanderbuilt University Press. pp. 15–53.

IAEA. 1972. *Thermal Discharges at Nuclear Power Stations: Their Management and Environmental Impacts.* P. J. West (Ed.), International Atomic Energy Agency, Vienna.

IAEA. 1980. *Environmental Effects of Cooling Systems.* International Atomic Energy Agency, Vienna.

Icanberry, J., and J. R. Adams. 1974. Zooplankton survival in cooling water systems of four thermal power plants on the California coast. Interim report March 1971–Jan 1972. In J. W. Gibbons and R. R. Sharitz (Eds.), *Thermal Ecology,* Tech. Information Center, U.S. Atomic Energy Commission. Conference 730505. Washington, D.C.

Jokiel, P. L., and S. L. Coles. 1974. Effects of heated effluent on hermatypic corals at Kahe Point, Oahu. *Pac. Sci.*, **28**, 1–18.

Karkheck, J., J. Powell, and E. Beardsworth. 1977. Prospects for district heating in the United States. *Science*, **195**, 948–955.

Kaya, C. M. 1977. Reproductive biology of rainbow and brown trout in a geothermally heated stream, the Firehole River of Yellowstone National Park. *Trans. Am. Fish. Soc.*, **16**(4), 354–361.

Knighton, G. 1973. Supporting information for staff testimony on cooling towers. AEC Licensing Hearings, Indian Point No. 2. 6 pp.

Langford, T. E. 1983. *Electricity Generation and the Ecology of Natural Waters.* Liverpool University Press. Liverpool. 342 pp.

Lyerly, R. L., and C. E. Littlejohn. 1973. A summary report of the Turkey Point cooling canal system. R. L. Lyerly and Associates. Report prepared for Florida Power and Light Company. Dunedin, FL.

Marcy, B. C., Jr. 1971. Survival of young fish in the discharge canal of a nuclear power plant. *J. Fish. Res. Bd. Can.*, **28**, 1057–1060.

Marine, G. 1969. *America the Raped.* Avon Books, New York. 331 pp.

Mather, D., P. G. Heisey, and N. C. Magnusson. 1977. Impingement of fishes at Peach Bottom atomic power station, Pennsylvania. *Trans. Am. Fish. Soc.*, **106**(3), 258–267.

Miller, H. H. 1970. The thermal water horticultural demonstration project at Springfield, Oregon. Paper presented at the conference on beneficial uses of thermal discharges. New York Department Environmental Conservation. Albany, New York. Sept. 17–18.

Mitzinger, W. 1974. Energetics and landscaping, exemplified by the development of one district of the German Democratic Republic. Paper 2.3-2. Proc. 9th World Energy Conference, Detroit. London: W.E.C.

Monn, M. G., M. F. Rosenfeld, and N. A. Blum 1979. A survey of capital costs of closed-cycle cooling systems for stream electric power plants. Proc. 41st Annu. Mtg. American Power Conference. Chicago, IL.

Morgan, R. P., III, and R. G. Stross. 1969. Destruction of phytoplankton in the cooling water supply of a stream electric station. *Chesapeake Sci.*, **10**, 165–171.

NTAC. 1968. *Water Quality Criteria.* National Technical Advisory Committee, Federal Water Pollution Control Administration. Washington, D.C.

Reutter, J. M., and C. E. Herdendorf. 1984. *Fisheries and the design of electric power plants: the Lake Erie experience.* Ohio State University Sea Grant Tech. Bull. OHSU-TB-11. 14 pp.

Schneider, M. 1980. Gas bubble disease in fish. In *Environmental Effects of Cooling Systems.* International Atomic Energy Agency, Vienna. pp. 38–49.

Smithsonian Institution. 1973a. *Rocky Point menhaden kill.* Smithsonian Institution Center for Short-lived Phenomena, Report No. 1604.

Smithsonian Institution. 1973b. *Florida salinity induced fish kill.* Smithsonian Institution Center for Short-lived Phenomena. Report No. 1608.

Suchanek, T. H., Jr., and C. Grossman. 1971. Viability of zooplankton. In *Studies of the Effects of a Stream Electric Generating Plant on the Marine Environment at North Port, New York,* SUNY, Stony Brook, Marine Science Research Center, Tech. Rept. Serial No. 9: 61–74.

Tanaka, J., and S. Suzuki. 1966. High results of Serida culture by utilization of heated effluent water from fossil fuel power plants. *Fish. Cult.*, **3**(8), 13–16.

Thorhaug, A., N. Blake, and P. B. Schroeder. 1978. The effect of heated effluents from power plants on seagrass (*Thalassia*) communities quantitatively comparing estuaries in the subtropics to the tropics. *Mar. Pollut. Bull.*, **9**, 181–187.

Thorhaug, A., M. A. Roessler, S. D. Bach, R. Hixon, I. M. Brook, and M. N. Josselyn. 1979. Biological effects of power-plant thermal effluents in Card Sound, Florida. *Environ. Conserv.*, **6**(2), 127–137.

Truchan, J. G. 1978. Toxicity of residual chlorine to freshwater fish: Michigan's experience. In *Biofouling Control Perspectives: Technology and Ecological Effects. Pollution Eng. and Tech.*, **5**, 79–89. Marcel Dekker, New York.

Tucker, W. S. 1977. Personal communication to the author.

Yee, W. C. 1972. Thermal aquaculture: Engineering and economics. *Environ. Sci. Tech.*, **6**(3), 232–237.

Young, J. S., and C. I. Gibson. 1973. Effect of thermal effluent on migrating menhaden. *Mar. Poll. Bull.*, **4**(6), 94–95.

Zieman, J. C., and E. J. F. Wood. 1975. Effects of thermal pollution on tropical-type estuaries with emphasis on Biscayne Bay, Florida. In E. J. F. Wood and R. E. Johannes (Eds.), *Tropical Marine Pollution*, Elsevier, Amsterdam. pp. 75–98.

12

METALS

My name is Asae Fukuda, and I am a Minamata disease patient.

I am appearing in court on behalf of my daughter Itsuko, who died at the age of 11. My home is near a fishing port called *Hakariishi Port*, located ten odd kilometers north of the Chisso Plant.

I became pregnant in 1967. As this would be the birth of our first child, family and relatives were all very happy.

Thinking that it would be for the good of the child within me, I ate fresh fish every day. My happiness, however, was short lived.

Even seven months after birth, Itsuko could not support the weight of her own head, and could not see. Many times a day she suffered convulsions, contorted her infant features, and drooled, making it unbearable to watch.

When she became one year old, I was told at the Kumamato University Hospital that Minamata disease was suspected.

I did not think it could be so, and did not want to believe it, but the doctor at the University Hospital asked me if I was living near the sea, and if I ate a lot of fish. Everything he asked was so, and I could only accept that he was right. It was then that I began taking Itsuko to the hospital regularly.

Every time I heard of a good doctor, I took Itsuko to see him. Desperate to try anything, I also visited temples and shrines to pray.

Three or four times a month I took Itsuko to Kumamoto University Hospital for examinations, a trip that took two hours each way. On the days after examinations Itsuko was especially tired, and would cry all night. At such times I would carry Itsuko on my back to the nearby port to humor her. I cried with my child, who would not understand if I talked to her, as I thought of how great a relief it would be for us to throw ourselves in the sea and die.

When Itsuko was three, I wanted to dress her in fine clothing for the *Shichigosan* children's day, though she could not stand on her own. Even if I put the clothing on Itsuko, who was lying down, she would only be like an unmoving doll. It was so sad that I cried as I dressed her. I then put makeup on her as well. As I carried her on my back to the shrine I said, "Itsuko, when you're able to stand let's go once more to the shrine for *Shichigosan*."

At mealtime I gave her liquid food spoonful by spoonful, being very careful, as if praying, that she would not choke on it.

I was assailed by the thought that I made Itsuko this way because of the fish I ate during pregnancy. So I believed that I must care for her whatever should happen, and I nursed the child in her illness with all my heart.

However, during the winter when she became 11 years old, in February when it was still cold, Itsuko died while suffering respiratory difficulty . . .

I have been able to persevere this long because of my 11 years of suffering.

I cannot forgive the National Government, Kumamoto Prefecture, and Chisso, who robbed Itsuko of her life, and me of my health.

Whenever I think I am about to give up, I see Itsuko's face, continuing the struggle to live even as she fights to breathe, and I can hear her voice saying, "Mommy, don't give up the fight."

I shall continue this fight as long as I live.

Statement by Ms. Asae Fukuda to the International Forum on Minamata Disease held in Kumamoto City, November 7–8, 1988. (translated from Japanese by R. Davis) (Tsuru, et al., 1989, p. 338)

Metals are introduced into aquatic systems as a result of the weathering of soils and rocks, from volcanic eruptions, and from a variety of human activities involving the mining, processing, or use of metals and/or substances that contain metal contaminants. Although some metals such as manganese, iron, copper, and zinc are essential micronutrients, others such as mercury, cadmium, and lead are not required even in small amounts by any organism. Virtually all metals, including the essential metal micronutrients, are toxic to aquatic organisms as well as humans if exposure levels are sufficiently high. Table 12.1 gives some idea of the relative toxicity of various metals based on the maximum permissible concentrations in water recommended by the EPA for the protection of human health.

Despite the wide variation in maximum permissible concentrations for the metals listed in Table 12.1, there are definite similarities in the toxicology of certain of these metals. For example, virtually all heavy metals show a great affinity for sulfur, and metals such as mercury, lead, and cadmium appear to exert toxic effects largely by combining with sulfur-containing amino acids in proteins, thus interfering with enzyme-mediated processes and/or disrupting cellular structure. Although the particular proteins that are chiefly attacked by these metals may differ from one metal to another, the underlying biochemical interaction is common to all.

Most metals are rather insoluble in water with a neutral or basic pH, and rather than dissolving are adsorbed rapidly to particulate matter or assimilated by living organisms. In other than acidic waters, the concentration of metals dissolved in the water may therefore give a highly misleading picture of the degree of metal pollution

Table 12.1   Maximum Permissible Concentrations of Various Metals in Natural Waters for the Protection of Human Health   *TOTAL METALS*

Metal	Chemical Symbol	Maximum Permissible Concentration	
		mg m^{-3}	μmoles m^{-3}
Mercury	Hg	0.144	0.72
Lead	Pb	5	24
Cadmium	Cd	10	89
Selenium	Se	10	127
Thallium	Tl	13	64
Nickel	Ni	13.4	228
Silver	Ag	50	464
Manganese	Mn	50	910
Chromium	Cr	50	962
Iron	Fe	300	5372
Barium	Ba	1000	7281

uses atomic wt.

Source. EPA (1987); Federal Register *56* (110): 26460-26564 (1991).

Arsenic will

$mg/m^3 = ppb \approx$

and in some cases may underestimate significantly the total metal concentration in the water. Mercury concentrations measured by Andren (1974), for example, in the Mississippi River indicate that about 67% of the mercury in the water is associated with suspended sediments. Although it is doubtful that metals adhering to particulate materials exert a directly toxic effect on aquatic organisms, it is certainly possible that some of these metals could be desorbed in the acidic environment of a particle feeder's gut or absorbed by an organism in the process of pumping water over its gills. Lloyd (1961) found that suspended zinc was indeed toxic to rainbow trout, although apparently only dissolved zinc is toxic to Atlantic salmon (Sprague, 1964). There is unfortunately little information on the toxicity of suspended metals to most aquatic organisms.

Benthic organisms are likely to be most directly affected by sediment metal concentrations, because the benthos is the ultimate repository of the particulate materials that wash into aquatic systems. Sediments near sewer outfalls often contain unusually high metal concentrations, particularly if the sewer line is the recipient of discharges from industrial operations that utilize metals in one way or another and do not adequately recover these metals from their wastewater. Klein and Goldberg (1970), for example, found sediment mercury concentrations near the Hyperion sewage treatment plant outfall off the Palos Verdes peninsula in southern California to be as much as 50 times higher than sediment mercury concentrations in control areas. The effects of pollution on sediment metal concentrations and the difference between particulate and dissolved metal concentrations is clearly evident in Table 12.2, which lists concentrations of various metals in sediments at two locations around Oahu, Hawaii, in several unpolluted Australian bays, in river water, and in the open ocean. The enormous concentrations of metals in the sediments of the Ala Wai boat harbor undoubtedly reflect the use of certain metals such as lead, mercury, and copper in antifouling paints as well as the corrosion of metals from boats docked in the harbor. The high metal concentrations in the sediments of the West Loch of Pearl Harbor reflect the discharge of about 15 million

Table 12.2   Metal Concentrations in the Sediments at Two Locations on Oahu, Hawaii, and in Two Unpolluted Australian Bays,[a] and Dissolved Metal Concentrations in Average River Water and Seawater[b]

Location	Mercury	Arsenic	Copper	Zinc	Lead
Sediments					
Oahu					
Pearl Harbor, upper West Loch	0.25	11	57	80	50
Ala Wai boat harbor	58	1.6×10^2	7.0×10^3	1.4×10^3	5.0×10^3
Australia					
Halifax Bay	—	—	7.0	31	16
Cleveland Bay	—	—	6.2	30	15
Water					
Average river water	0.7	1.7	1.5	30	0.1
Average seawater	0.001	1.5	0.1	0.1	0.003

[a] Sediment values are μg/g dry weight.
[b] Dissolved concentrations are ng/g.
Source. City and County of Honolulu (1971), Knauer (1977), Martin and Whitfield (1983) and Mitra (1986).

liters of sewage per day into streams that flow into the loch,[1] the weathering of antifouling paints and corrosion of metals from naval ships, and the fact that the waters of West Loch are rather stagnant.[2]

The average concentrations of the same metals in river water and seawater provide an interesting comparison to the corresponding sediment values. Sediment metal concentrations even in the unpolluted Australian bays are between 10^3 and 10^7 times higher than the dissolved concentrations in river water and seawater, and in the polluted Ala Wai boat harbor the sediment metal concentrations are between 10^5 and 10^9 times greater than the river water and seawater concentrations. Metal concentrations as high as those found in the Ala Wai boat harbor sediments could seriously contaminate and stress benthic organisms. Furthermore, under appropriate conditions these metals could leach out of the sediments for many years after pollutant discharges had been stopped and hence continue to pollute the water column. Thus examination of metal concentrations in suspended particles, in aquatic organisms, and in sediments will usually provide a more sensitive and informative indicator of metal pollution than measurements of dissolved metal concentrations.

THE QUESTION OF BIOLOGICAL MAGNIFICATION

The fact that metal concentrations in aquatic organisms are typically several orders of magnitude higher than concentrations of the same metals in the water has led to some speculation that metals may become progressively concentrated at higher

[1] These discharges terminated in 1981.
[2] Hence the metals are flushed out inefficiently.

trophic levels in aquatic food chains due to food chain magnification. The high concentrations of mercury reported in tuna, swordfish, and marlin (Montague and Montague, 1971; DLNR, 1985) (top-level carnivores) is consistent with the food chain magnification hypothesis.

Table 12.3 lists concentrations of various metals in samples of phytoplankton and zooplankton collected at open ocean stations between California and Hawaii in 1973. Comparison with the seawater values in Table 12.2 shows that the metal concentrations in the phytoplankton are indeed about three orders of magnitude higher than the concentrations in the water, and for the first three metals at least, the concentrations in zooplankton are substantially higher than the phytoplankton concentrations. For the last four metals, however, there is little difference in concentration between the phytoplankton and zooplankton. Although the elevated metal concentrations in the zooplankton for the first three metals are suggestive of food chain magnification, we have already seen from the work of Hamelink, Waybrant, and Ball (1971) on DDT residues that such effects may be produced by entirely different mechanisms, namely by direct uptake of a pollutant from the water and by differences in pollutant equilibria between water and organism for different classes of organisms. Therefore it is impossible to say to what extent the increases in metal concentrations between phytoplankton and zooplankton for the first three metals in Table 12.3 reflect food chain magnification or simply differences in metal exchange equilibria between organisms and water.

In the case of mercury, Knauer and Martin (1972) have provided a thorough study of concentrations in a simple marine food chain consisting of phytoplankton, zooplankton, and anchovies in Monterey Bay, California. Some of their results are summarized in Table 12.4. For this metal at least, there is little evidence of magnification up the food chain, and in fact the variability in Hg concentration between some parts of the anchovies is greater than the difference in concentration between the anchovies and phytoplankton. The wide range of Hg concentrations in different parts of the anchovy reflects the importance of biochemical factors in affecting the relative tendency of different tissues to concentrate pollutants. Such

Table 12.3   Median Concentrations of Metals in Phytoplankton and Zooplankton Collected Between California and Hawaii

Metal	Phytoplankton[a]	Zooplankton[b]
Cu	3.2	13.9
Zn	19	260
Pb	<1.0	8.5
Fe	224	580
Mn	6.1	4.4
Ag	0.2	0.25
Cd	1.5	2.3
Hg	0.19	0.14

[a] Phytoplankton values are μg/g organic dry weight.
[b] Zooplankton values are μg/g dry weight.

Source. Martin and Knauer (1973).

biochemical or physiological differences may also play a major role in causing certain organisms to concentrate pollutants to a much higher level than other organisms, regardless of the relative position of the organisms in the aquatic food chain.

CASE STUDIES

It should be apparent from the previous discussion that some statements about metals and their effects on organisms are valid for most metals of major concern as pollutants. For example, it is true in general that the concentration of metals associated with particulate material, including living organisms, is at least several orders of magnitude higher than the concentration of metals dissolved in the water. It is also true in general that the concentration of metals may differ greatly between one organism and another, and between different parts of the same organism. In general one ascribes these concentration differences to differences in the tendency of metals to bind to the various molecular groups found within the cells of each organism, as well as to the degree of the organism's exposure to the metal as influenced by its metabolic characteristics and its position in the food chain.

Although such generalizations are useful in understanding some of the problems associated with metal pollution, it should be clear from an examination of Tables 12.1 and 12.3 that significant differences exist in the toxicity of various metals and in the mechanisms by which these metals are transferred through the biosphere. Thus more than a superficial understanding of metal pollution requires an in-depth examination of the characteristics and problems associated with each metal. The remainder of this chapter is therefore devoted to a detailed analysis of three of the most toxic metal pollutants, mercury, cadmium, and lead.

Mercury

Production and Uses
Mercury (Hg) is a highly unusual metal; it is a liquid at room temperature[3] and has a density even greater than that of lead.[4] Mercury is found in virtually all terrestrial rocks and soils, although the concentration of Hg in most soils and rocks is quite low, about 60 ppb. In some parts of the world, however, the concentration of Hg in the Earth's crust is considerably higher, particularly in regions of volcanic activity. The only ore that contains mercury in sufficient concentration to warrant extraction is cinnabar (HgS) or mercuric sulfide. World production of mercury from cinnabar peaked at about 10^4 tonnes per year around 1970 and has been declining ever since (Figure 12.1). Present production amounts to about 6×10^3 tonnes per year, of which about 15% is used in the United States. The principal producers are Spain and the Soviet Union.

What does the world do each year with 6×10^3 tonnes of mercury? The answer is that mercury is used for a great variety of purposes, including industrial and medical applications, as well as scientific research. In the United States over half of mercury use is accounted for by the manufacture of batteries and the production of

---

[3]The melting point of mercury is $-38.9°$C.
[4]A piece of lead will float on liquid mercury.

Table 12.4   Hg Levels in a Simple Food Chain

Trophic Level	Average Hg Concentration (ppb wet weight)
Phytoplankton	28
Zooplankton	12
Anchovy	
Skin	10
Gonads	15
Gills	30
Muscle	40
Liver	90

Source. Knauer and Martin (1972).

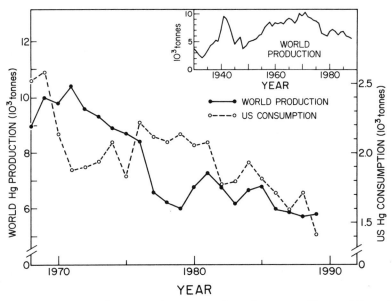

Figure 12.1 World production and U.S. consumption of mercury. (*Source*: Minerals Yearbook, U.S. Dept. of the Interior Bureau of Mines.)

chlorine and sodium hydroxide (Table 12.5). Mercury dry cell batteries are used in many products, including Geiger-Muller counters, radios, digital computers and calculators, hearing aids, military hardware, and scientific and communication devices (Mitra, 1986). The cathode of the battery consists of mercuric oxide mixed with graphite; the anode is usually metallic zinc. In the chlor-alkali industry, Hg is used as a cathode in an electrolytic process by which Cl_2 and $NaOH$ are produced from a solution of $NaCl$. For each tonne of Cl_2 produced, about 250–500 g of Hg are consumed (Montague and Montague, 1971). Prior to the 1970s, much of this "consumed" mercury was discharged as waste to the environment, although it is technically feasible to recover and recycle a large percentage of the Hg. Rapid expansion of the chlor-alkali industry in the United States began early in the twentieth century with the increase in demand for high quality NaOH used in the production of rayon. Since the 1940s, however, the principal users of mercury-grade NaOH have been the aluminum, glass, paper, petroleum, and detergent industries (Montague and Montague, 1971). The major user of the Cl_2 gas produced by chlor-alkali plants has been the plastics industry, for example in the production of polyvinyl chloride (PVC). Although it is possible to produce Cl_2 and NaOH by methods other than the so-called mercury-cell technique, for economic reasons this method has been utilized at virtually all chlor-alkali plants in the United States. The consumption of Hg by chlor-alkali plants in the United States has declined by about 45% from a peak of over 700 tonnes y^{-1} in the early 1960s, a result due in large part to recovery and recycling efforts designed to reduce mercury pollution. In Sweden, however, concern over Hg pollution is so great that chlor-alkali plants are no longer being built (Montague and Montague, 1971).

Because of its toxic qualities, mercury has been used for many years as a component both of antifouling paints and of mildew-proofing paints. The use of Hg in the former category was banned in 1972 under the aegis of the Federal Insecticides, Fungicides, and Rodenticides Act (FIFRA), and other components have been substituted for mercury to resist fouling. The use of Hg in paints still accounted for 16% of mercury consumption in the United States in 1989, but the Food and Drug Administration banned the use of Hg in latex paints for interior use as of August 20, 1990 (Emond, 1990).

Mercury is used in a variety of electrical apparatus, including neon lights, arc rectifiers, power control switches, oscillators, fluorescent lights, silent light switches, and in certain kinds of highway lights. In these applications the unusual properties of mercury often play an important role. Silent light switches, for example, consist of a hollow tube containing a puddle of Hg. Turning the switch to the "on" position inverts the tube, causing the Hg to flow to one end where it completes a circuit between two wires imbedded in the tube. Mercury vapor is the form of Hg used in neon lights, fluorescent lights, arc rectifiers, and in the bright, bluish highway lights found along many streets and roadways.

Table 12.5   U.S. Mercury Consumption Figures for 1989

Use	Consumption (tonnes)	Percent of Total Consumption
Chlor-alkali plants	381	31
Batteries	250	21
Paint	192	16
Wiring devices and switches	141	12
Measuring and control instruments	87	7
Pigments, pharmaceuticals, catalysts	40	3
Dental	39	3
Miscellaneous	35	3
Electric lights	31	3
Lab use	18	1
TOTAL	1214	

Source. Interior Dept. (1991).

The unusual properties of mercury are also important to its use in measuring and control instruments, including thermometers, manometers, barometers, diffusion pumps, Toepler pumps, and a variety of pressure gauges. When such devices are broken or discarded, there is a good chance that the mercury in them will be leaked to the environment. According to figures cited by Aaronson (1971), in Canada alone roughly 1.8 million mercury thermometers are broken each year in hospitals, and another 1.2 million are broken by the general public. If the mercury in these thermometers were simply discarded, the discharge would amount to over 6 tonnes of Hg per year.

Each year in the United States up to 100 million dental fillings containing mercury combined with silver, tin, and other metals in the form of an amalgam are used to restore decayed teeth (Anonymous, 1991). While this use of mercury accounts for only about 3% of mercury consumption in the United States, the practice has become somewhat controversial, largely because of a documentary aired by the television program 60 Minutes in December of 1990. The message conveyed by the 60 Minutes report was that mercury vapor released from the fillings could be causing serious health problems. As an example, 60 Minutes cited the case of a woman crippled by multiple sclerosis who went dancing the night after her amalgam fillings were removed.

The truth of the matter is that gold fillings are in fact preferable to amalgam fillings, but not because of any concern over mercury poisoning. Gold fillings last longer, are more resistant to wear, and are less likely to fracture than amalgam fillings (Anonymous, 1991). The problem with gold fillings is that they cost about 6 times as much as amalgam fillings. Studies have shown that when people chew, a small amount of Hg vapor is released from amalgam fillings, but current estimates indicate that even persons with moderate to large numbers of amalgam fillings receive only about 1% of the dose of mercury that might cause adverse health effects (Anonymous, 1991). Incidentally, once mercury accumulates in the human body, it is released at a rate of only about 1% per day. Thus, for example, removing one's amalgam fillings would not cause an overnight change in health due to mercury exposure. The fact that a woman crippled by multiple sclerosis went dancing the night after her fillings were removed is remarkable, but was almost certainly not the result of any change in her body burden of mercury. What 60 Minutes failed reprehensibly to report were the stories of numerous people who had their amalgam fillings removed to no avail. Another woman afflicted with multiple sclerosis, for example, who was unable to afford the cost of gold fillings, had all her teeth extracted and replaced with dentures. A month later her condition worsened and she was hospitalized (Anonymous, 1991).

A more legitimate concern about the use of mercury in fillings is that only about 50% of the total amount of mercury used in dental preparations is actually applied to the teeth (Mitra, 1986). When dentists prepare to fill a cavity, they invariably make up an excess of amalgam to ensure proper execution of the restorative work. The excess amalgam is either rinsed from the mouth or aspirated, and will be released to the environment unless caught in a strainer or trap in the waste drain. Although recovery of amalgam scrap is widespread in the United States, many dental offices do not contain appropriate solid waste traps.

In scientific research, mercury is used in measuring devices such as thermometers and manometers, as a catalyst in certain chemical reactions, and as a preservative to prevent deterioration of biological specimens due to microbial breakdown. Waste mercury from such use, as in a broken thermometer or once-used preservative, is typically discarded rather than recycled. Mercury consumption in general laboratory use has declined, however, by about 65% in the United States during the last 20 years. This trend most likely reflects the substitution of alternative methods and devices that do not involve the use of mercury.

Mercury is used as a catalyst to speed up a variety of chemical reactions. One of mercury's most important uses as a catalyst is in the production of several kinds of plastic, particularly urethane and polyvinylchloride (PVC). PVC is the most widely used plastic product in the world. It is used both in its unplasticized form, for example in bottles, pipe fittings, pipes, records, and valve parts, and in its plasticized form in rainwear, shower curtains, auto seat covers, baby pants, and so forth. Mercury has also been used as a catalyst in the production of synthetic acetate fiber. Discharges of spent mercury catalysts from industrial operations have been historically a significant cause of mercury pollution in the aquatic environment (Carter, 1977; Mitra, 1986). Consumption figures for Hg by the plastics industry are not tabulated specifically by the U.S. Bureau of Mines for proprietary reasons (Interior Dept., 1991), but it is obvious from Table 12.5 that such use accounts for no more than 3% of U.S. mercury consumption at the present time.

Several other uses of mercury are worth mentioning, in some cases largely because of their historical significance. Mercury in various forms has been used since 1914 as a coating on seeds to protect them from attack by fungi during storage and during the first few days after planting. Mercury-based fungicides were also used to control turf-grass disease on golf courses, parks, cemeteries, lawns, and so forth. The use of mercury to treat seed was greatly restricted in the United States under the aegis of the FIFRA in 1970 and 1972, and use as a fungicide was similarly restricted and ultimately banned by 1972 (Mitra, 1986). These actions were taken in part because of the impact of the mercury on seed-eating birds and because of concern over contamination of waterways due to runoff from areas of treated turf. Serious impacts on human health due to the practice of treating seed grain with mercury have occurred when the seed was consumed or fed to animals rather than planted. In 1969, for example, a farmer in New Mexico fed waste seed grain from a local grainery to his pigs. Subsequently the pigs were slaughtered, and both the farmer and his family consumed the meat. Three of the farmer's children suffered irreparable neurological damage as a result of consuming the meat, which was later found to contain 27 ppm mercury. Several years later, in 1972, a similar but more serious incident occurred in Iraq, when literally thousands of persons ate homemade bread prepared from wheat seed that had been treated with a mercurial fungicide. Four hundred fifty persons died of mercury poisoning (Mitra, 1986).

In the past mercury has been used in pulp and paper mills to prevent the growth of slime on paper-making equipment and to prevent the growth of mold and bacteria on pulp during storage. Such use declined precipitously after 1965 when the FDA ruled that paper products used as food containers should contain absolutely no mercury and was banned entirely in 1972 under the aegis of the FIFRA (Mitra, 1986).

The germ-killing properties of mercury have resulted in its use in various pharmaceutical products, dating back to at least 400 B.C. (Montague and Montague, 1971). For example, mercury came into widespread use for the treatment of siphilis as long ago as 1500. In pharmaceuticals today, mercury is used most commonly as a diuretic,[5] particularly in treating congenitive heart failure patients, who frequently accumulate body fluids.

Finally, mercury will combine with almost all other common metals except iron and platinum to form a fusion product called an *amalgam* (e.g., dental fillings). The tendency of mercury to form amalgams with other metals may be used to purify metals by separating them through amalgamation from their nonmetal contaminants. The purified metal is then recovered from the amalgam by electrolysis. This technique is used, for example, to produce high-purity zinc and aluminum.

Fluxes to the Environment

Of the mercury that is consumed each year, how much finds its way back into the environment in a manner and form that might cause pollution problems? In the United States at least, 25–30% of mercury consumption is accounted for by recycling (Interior Dept., 1991). In fact, it is very common in the United States to recover mercury from obsolete items and waste products such as amalgams, batteries, and industrial and control instruments. Nevertheless, the fact that 6×10^3 tonnes of Hg are extracted from ore deposits each year indicates that a comparable quantity is either accumulating in the form of supplies and manufactured products or being released back into the environment. Although there may be some net accumulation of supplies and manufactured products, it seems fair to assume that a quantity of mercury roughly equal to the amount that is mined each year is returned to the environment through emissions to the atmosphere, discharges to aquatic systems, or disposal in landfills. We conclude that about 6×10^3 tonnes of mercury are released to the environment each year as a direct result of human use. How does this figure compare to other fluxes of mercury through the environment?

To begin with, direct use of mercury is not the only cause of anthropogenic releases of mercury to the environment. Additional discharges result from the burning of fossil fuels and the processing of minerals and other metallic ores. Since mercury is a volatile substance, any process such as burning or smelting that elevates the temperature of a mercury-containing substance will tend to drive off Hg vapor into the atmosphere. Although the average concentration of mercury in crustal rocks and soils is only 60 ppb, the mercury concentration in fossil fuels such as coal and petroleum is much higher, typically 0.2–20 ppm (Mitra, 1986). Weiss, Koide, and Goldberg (1971) estimated that fossil fuel burning probably discharged about 1600 tonnes of Hg to the atmosphere each year. Based on trends in world oil production (Flavin, 1985), it is probably reasonable to scale up this figure to 2000 tonnes at the present time. Similar calculations by Weiss, Koide, and Goldberg (1971) showed that the smelting of sulfide ores might introduce as much as 2000 tonnes of Hg to the atmosphere annually. Based on trends in world production of iron ore since 1970, this figure is probably closer to 2700 tonnes at the present time.[6] Thus human

---

[5]A substance used to rid the body of salt and water.
[6]Statistics on world iron ore production can be found in annual editions of *Minerals Yearbook*, published by the U.S. Dept. of Interior Bureau of Mines.

activities appear to be responsible for the introduction of roughly 10,700 tonnes of Hg to the environment each year (Table 12.6), unless some major source of Hg release associated with human activities has been overlooked.

It is now instructive to compare this estimate of anthropogenic Hg fluxes with the natural flux of Hg from the atmosphere to the surface of the earth in rainfall. The concentration of mercury in ancient and presumably uncontaminated glacial ice is about 0.06 ppb (Weiss, Koide, and Goldberg, 1971). This figure is indicative of the mercury content of unpolluted rainfall. Taking the average annual global precipitation to be 4.96×10^5 km^3 (Berner and Berner, 1987), it is concluded that about 3×10^4 tonnes of Hg are naturally rained out on the surface of the earth each year. Since rain effectively washes Hg from the atmosphere, and since the average time between rains is about 10 days (Weiss et al., 1971), it is clear that some natural mechanism must recharge the atmosphere with Hg at the rate of about 3×10^4 tonnes per year. Our estimate of anthropogenic mercury fluxes to the environment is only about one-third of this figure. What natural process could be responsible for releasing so much mercury to the atmosphere?

Current thinking is that much of this natural release is due to the volatilization or sublimation of Hg from soils and rocks (Goldberg, 1976). Although direct quantitative evidence to support this "degassing" hypothesis is lacking, several related observations are at least consistent with the hypothesis. Goldberg (1976), for example, notes that the abundance of certain metals in atmospheric aerosols is remarkably consistent and correlated directly with the relative volatility of these elements in naturally occurring compounds, and Weiss, Koide, and Goldberg (1971) point out that Hg concentrations are markedly higher in sediments than in igneous rocks.[7] Furthermore, the mercury content of air is highly correlated with the season of the year and is highest during summer months (Mitra, 1986). Although the magnitude of crustal degassing is speculative, calculations by Goldberg (1976) indicate that the theory provides a reasonable explanation for the natural fluxes of Hg between the atmosphere and the surface of the earth.

Table 12.6 summarizes our present understanding of some of the more important mercury fluxes in the environment. Precipitation on the oceans amounts to about 3.86×10^5 km^3 y^{-1} (Berner and Berner, 1987), and the natural input of Hg to the oceans from precipitation is therefore estimated to be $(0.06)(3.86 \times 10^5) = 2.3 \times 10^4$ tonnes per year. Although the study of Weiss, Koide, and Goldberg (1971) suggested that human activities had roughly doubled the rate of mercury emissions to the atmosphere, there is now general agreement that anthropogenic fluxes are small compared to natural fluxes on a global scale (Dickson, 1972; Siegel, Siegel, and Thorarinsson, 1973). Deposition of mercury in polar ice fields, for example, appears to have changed very little since 800 BC (Mitra, 1986). The total input of mercury to the oceans therefore appears to be roughly 2.3×10^4 tonnes y^{-1} from rainfall + 3.8×10^3 tonnes y^{-1} from river runoff = 2.7×10^4 tonnes y^{-1}. This figure is about 4.5 times the present rate of mercury production from mines. Since mercury accumulates ultimately in the sea as a result of global recycling (Mitra, 1986), the implication of these calculations is that on a global scale anthropogenic inputs of Hg to the ocean are only about 20–25% of the total mercury flux.

---

[7] Igneous rocks, having once been very hot, would tend to have been degassed effectively, and indeed large amounts of Hg are known to be released in volcanic eruptions.

Table 12.6   Estimated Hg Fluxes in the Environment

Source of Hg	Estimated Annual Flux (tonnes)
Release from human use	6×10^3
Fossil fuel burning	2×10^3
Ore smelting	2.7×10^3
Rainfall (natural)	3×10^4
River runoff to oceans[a]	3.8×10^3

[a] *Source.* Weiss, Koide, and Goldberg (1971).

Localized Pollution.   Although this conclusion may be a source of some comfort, it is unfortunately a conclusion that applies only to global mercury fluxes. There has been very definite evidence of serious mercury pollution on a local scale in many parts of the world. The fact that these localized instances of pollution have involved discharges of Hg which were small compared to estimates of global fluxes has been no consolation to those who were affected by the discharges. A few examples will serve to illustrate the magnitude of local mercury pollution problems.

In Minamata, Japan, a factory operated by the Chisso Corporation discharged roughly 200–600 tonnes of Hg into Minamata Bay between 1932 and 1968 (Smith and Smith, 1975). The company had been using mercuric oxide as a catalyst in the production of acetaldehyde, a compound used in the manufacture of plastics, drugs, perfumes, and photographic chemicals (SCEP, 1970; Smith and Smith, 1975). Concentrations of Hg in the sediments of Minamata Bay were found to be ten to several hundred parts per million on a wet weight basis and ranged as high as 2000 ppm in the mud near the drainage channels from the Chisso factory. Assuming the water content of the sediments in Minamata Bay to be about 30–70% by weight, it is concluded that these concentrations would be 1.4–3.3 times higher if reported on a dry weight basis. For comparison, Andren and Harriss (1973) found that Hg levels in sediments from the Mississippi River, Mobile Bay, and the Florida Everglades ranged from 0.08 to 0.6 ppm (dry weight); Klein and Goldberg (1970) found Hg levels in marine sediments off La Jolla and Palos Verdes, California, to be 0.02–1.0 ppm (dry weight); and mercury concentrations in surface sediments near mid-Atlantic sewage sludge dump sites average about 0.04 ppm (dry weight). Based on these comparisons, Minamata Bay was grossly polluted with mercury.

In the United States, mercury concentrations in certain species of fish taken from Lake St. Clair (between Lake Huron and Lake Erie) increased by roughly a factor of 100 between 1935 and 1970. Concentrations in 1970 were as high as 3–7 ppm (Figure 4.11; Grant, 1971). This increase in Hg levels was almost certainly due to the development of the chlor-alkali industry in the area and in particular to the establishment of the Dow Chemical chlor-alkali plant at Sarnia, Ontario, on the St. Clair River between Lake Huron and Lake St. Clair. Between 1949 and 1970 this plant discharged an estimated 91 tonnes of Hg into the St. Clair River system (Wood, 1972). Studies in Canada showed that most of the Saskatchewan River system, including Lake Winnipeg, was polluted seriously with mercury, a condition due in part to the discharge of Hg by Saskatoon Cooperatives into the South Saskatchewan River. The company discharged about 30 tonnes of Hg between 1963 and 1970

(Wood, 1972). The Dryden chlorine plant at Lake Wabigoon in Manitoba was similarly implicated as contributing to the mercury pollution of most lakes and rivers in central Manitoba (Wood, 1972).

Back in the United States, two rivers in Virginia were found to be polluted heavily with mercury as a result of industrial discharges (Carter, 1977). In southwestern Virginia, the North Fork of the Holston River, which feeds into the Tennessee River system, was the unfortunate recipient of Hg discharges from the Mathieson Chemical Company[8] chlor-alkali plant at Saltville, Virginia, between 1950 and 1972. Although the plant was ultimately shut down, partly as a result of pollution control measures that were to be imposed, mercury has continued to leak into the river, both from the site of the plant itself, where the ground has been estimated by one Olin consultant to contain about 100 tonnes of Hg, and from two large "muck ponds" used for the disposal of waste from the plant (Carter, 1977). Mercury is believed to be present in the soil to depths of as much as 10 m at the old plant site, and wastes in the muck ponds extend to depths of about 25 m. Discharge of mercury to the North Fork of the Holston River from the muck ponds alone is estimated to be about 35 kg y^{-1}, and there is no satisfactory method, short of massive engineering efforts that might cost hundreds of millions of dollars, to greatly reduce the discharges from the old muck ponds (Carter, 1977). As of 1970 both the states of Virginia and Tennessee issued a ban on the eating of fish from the North Fork, but despite the closing of the Saltville plant in 1972, 75% of the fish sampled in 1976 as much as 110 km downstream from the plant were found to contain Hg in concentrations greater than 1 ppm (wet weight).

The other Virginia river found to be contaminated heavily with mercury is the South Fork of the Shenandoah River. The source of the mercury in this case appears to have been the mercuric sulfate used in the manufacture of acetate fiber at a plant operated by E.I. DuPont de Nemours and Company at Waynesboro, Virginia, between 1929 and 1950. Although the amount of Hg leaked to the stream by this plant was apparently quite small compared to the mercury discharges at Saltville, the mercury persisted in the sediments and biota of the stream long after use of Hg at the plant ceased in 1950 and despite major floods,[9] which might have been expected to flush the mercury from the system. On the basis of fish and sediment samples taken and analyzed by the Virginia State Water Control Board (SWCB), fish in the South River (which flows into the South Fork of the Shenandoah) below Waynesboro and in the entire South Fork were banned for eating purposes in June, 1977 (Carter, 1977). Sediment samples below the Waynesboro plant were found to contain as much as 240 ppm Hg. Sediments above the plant contained less than 1 ppm Hg. Bass taken as much as 124 km downstream from the plant were found to contain more than 1.0 ppm Hg. Remarkably, the SWCB failed to detect mercury pollution of the Shenandoah during sampling in 1970, because no analyses were made of sediments or biota. Only water was analyzed. This observation underscores a point noted earlier, namely that metals tend to become concentrated in sediments and in organisms and are usually found at relatively low levels in water. As a result the logical place to look for metal contamination is in sediments and organisms. Dissolved metal concentrations are not likely to be sensitive indicators of metal pollution.

[8]Later Olin Corporation.
[9]For example, the flooding after tropical storm Agnes in 1972.

The foregoing examples of mercury pollution, although far from inclusive, nevertheless serve to illustrate the point that mercury pollution can be a serious problem on a local scale, regardless of whether or not global fluxes of mercury have been affected significantly by human activities. It is impossible to appreciate the magnitude of the mercury pollution problem, however, solely from a knowledge of Hg levels in organisms and sediment, or of the amount of mercury discharged into a particular body of water. The great concern over mercury stems not from its mere presence, but from its effects on organisms, particularly humans. To understand these effects, and in order to make intelligent decisions about how to deal with mercury pollution, we must first understand something about the toxicology of mercury.

Toxicology

The toxicity of mercury depends very much on the chemical form in which the element is found. Some chemical forms of mercury are absorbed much more effectively by the body than others, and once inside the body, some forms of mercury are much more likely than others to produce serious damage to internal organs. Therefore some appreciation of the chemistry of mercury is necessary for an understanding of its toxicology.

The chemical forms of mercury may be broadly classified as being either organic or inorganic. The inorganic forms of mercury include metallic mercury such as the Hg used in thermometers, and various "salts" of mercury, in which the element is considered to be in the plus one (Hg^+) or plus two (Hg^{2+}) oxidation state. Salts of Hg^+ and Hg^{2+} are referred to as *mercurous* and *mercuric salts*, respectively, and are typified by mercurous chloride ($HgCl$) and mercuric chloride ($HgCl_2$) or mercuric oxide (HgO). At room temperature, metallic Hg may occur either as a gas or as a liquid.

If inorganic mercury is swallowed, over 98% is excreted rapidly in the urine and feces (Grant, 1971). Unless ingested repeatedly or in massive amounts, metallic Hg is relatively harmless. According to Goldwater (1971, p. 15) a person could swallow up to half a kilogram or more of metallic Hg "with no significant adverse effects." Montague and Montague (1971, p. 18) state that, "In an attempted suicide a person once actually injected two grams of metallic mercury directly into his veins without ill effect. The mercury merely formed a puddle in the right side of his heart." Ingestion of the more soluble salts of mercury, however, can cause serious problems. Mercuric chloride is corrosive to the intestinal tract, and like other forms of inorganic mercury, can seriously damage the liver and especially the kidneys if taken in sufficient amounts (Goldwater, 1971; Hammond, 1971). In fact, the liver and kidneys are the internal organs that generally contain the highest mercury concentrations in persons who have died from mercury poisoning (Smith and Smith, 1975). Once absorbed by the body, inorganic mercury may be transported via the blood to all parts of the body. Fortunately about 50% of the inorganic Hg transported in this way is present in the blood plasma (Grant, 1971) and in this form is readily excreted by the kidneys.

One of the most serious consequences of mercury poisoning can be brain damage, but unless ingested as a vapor, inorganic mercury does not readily penetrate the blood-brain barrier (Grant, 1971). When inhaled, inorganic Hg is of course initially deposited in the lungs, where in acute cases it may cause irritation and destruction of the lung tissue (Goldwater, 1971). From the lungs, metallic Hg may be transported by the blood to other parts of the body, including the brain. Unfortunately, any effects of mercury on the brain are likely to be permanent, since cells of the central nervous system, once damaged, do not recover. Most cases of mercury poisoning from inhalation are of the chronic sort. For many years chronic mercury poisoning was a serious problem in the felt hat industry, for example, because mercury nitrate was used to treat fur pelts at an early stage in the felting process. The hatters were constantly exposed to the mercury vapors emanating from the pelts, and often developed symptoms of chronic mercury poisoning. These symptoms include inflammation of the gums, metallic taste, diarrhea, mental instability, and tremors (Grant, 1969).[10] The practice of using Hg in felt making was banned in 1941. Inhalation of mercury remains a concern for persons who routinely work with mercury, especially those involved in the mining and extraction of mercury. From the standpoint of toxicology, one of the chief distinctions between inorganic mercury and organic mercury is the fact that most injuries due to inorganic Hg do not involve the central nervous system, but rather organs such as the kidneys, liver, and intestines. Consequently, the effects of inorganic Hg poisoning are often reversible, except in cases of severe poisoning or when inorganic Hg penetrates to the brain. Unfortunately the most common forms of organic mercury frequently cause irreversible damage by attacking the central nervous system.

The three major kinds of organic mercury compounds are phenyl mercury (e.g., phenylmercuric acetate or PMA), methoxy mercury (e.g., methoxyethyl mercury acetate), and alkyl mercury (e.g., methylmercuric acetate). Of these compounds, by far the most common as well as the most dangerous are certain members of the alkyl mercury group, the methyl mercury compounds. The ubiquity of methyl mercury can be accounted for by the fact that so-called methanogenic bacteria as well as certain molds in sediments are capable of converting virtually all other forms of mercury into methyl mercury (Grant, 1971; Goldwater, 1971; Wood, 1972). Relevant pathways are indicated in Figure 12.2. The final step involves methylation of Hg^{2+} with methylcobalamin, a vitamin B_{12} analogue, either by enzymatic means or by electrophilic attack of Hg^{2+} on methylcobalamin (Mitra, 1986). Wood (1972, p. 34) concluded that "All microorganisms capable of vitamin B_{12} synthesis are capable of methyl mercury synthesis." Methylation can take place under both aerobic and anaerobic conditions, but in the latter case formation of HgS and HgS_2^{2-} may prevent methylation (Mitra, 1986).

In addition to bacteria and molds found in sediments, there is evidence that microbes in the bodies of at least some animals may also methylate mercury (Hammond, 1971). Birds given nonmethylated mercury in their feed, for example, have been found to produce eggs containing methyl mercury (Grant, 1971), and mammalian liver contains a cobalamin-dependent transmethylase capable of methyl mercury synthesis (Mitra, 1986). It is not known whether fish themselves can convert Hg^{2+} to methyl mercury, but studies by Knauer and Martin (1972) on a simple marine food chain in Monterey Bay, California, showed that the percentage of mercury present in organisms as organic mercury was much higher in anchovies than in their zooplankton prey (Table 12.7). The higher percentage of organic Hg in the anchovies could have resulted from the methylation of inorganic

---

[10] The behavior displayed by Lewis Carroll's mad hatter in *Alice in Wonderland* is very typical of persons suffering from chronic mercury poisoning.

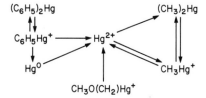

Figure 12.2   Transformations of mercury leading to methyl mercury. [Redrawn from Goldwater (1971).]

Hg within their bodies. Wood (1972), however, has noted that fish can take up at least some forms of methyl mercury from the water through diffusion across their gills, and such uptake could account in part for the high percentages of organic mercury usually found in fish. Wood (1972) indicates that about 95–100% of the mercury in fish is methyl mercury, although this generalization is not true in all cases (DLNR, 1986).

What makes methyl mercury so dangerous? First of all, if a person swallows food containing methyl mercury, about 90–95% of the methyl mercury is absorbed by the intestines (Grant, 1971). Once absorbed, methyl mercury is transported throughout the body by the circulatory system, primarily in the red blood cells. As with inorganic mercury, much of this methyl mercury is concentrated in the kidney and liver (Mitra, 1986). One factor contributing to the high rate of absorption of methyl mercury versus inorganic mercury may be the tendency of the liver to secrete absorbed methyl mercury along with bile into the small intestine, where the methyl mercury is once again absorbed through the intestinal wall back into the blood stream (Grant, 1971). Although some methyl mercury is excreted by the intestines and to a lesser degree by the kidneys, this excretory process is much less efficient than in the case of inorganic mercury excretion. In fact, some conversion of methyl mercury to inorganic Hg may occur in conjunction with methyl mercury excretion (Grant, 1971).

Although ingestion of methyl mercury may cause serious damage to organs such as the liver, kidneys, and pancreas, the most serious consequences of methyl mercury poisoning involve its effects on the central nervous system. Unlike inorganic mercury, methyl mercury effectively passes from the bloodstream into the brain, where it destroys cells, particularly in the cerebellum (leading to disturbances of equilibrium), and the frontal cortex (personality disturbances). In humans, about 10% of the total body burden of methyl mercury is found in the head, most of it presumably in the brain (Mitra, 1986). Even though the exact nature of the chemical

Table 12.7   Percentage Organic Mercury in a Simple Food Chain

Organisms	Percent Total Mercury as Organic Mercury
Phytoplankton	12–67
Zooplankton	17–78
Anchovies	76–100

Source. Knauer and Martin (1972) and R. Harriss (personal communication).

interaction that causes mercury, and particularly methyl mercury, to destroy cells is not fully understood, the underlying interaction is believed to be the attraction between mercury and sulfhydryl groups on proteins (Goldwater, 1971). Since proteins are essential constituents of cell membranes, binding of mercury to proteins could disrupt the cell membrane sufficiently to destroy the cell or seriously interfere with its functioning. This statement of course applies to cells in all parts of the body, not just in the central nervous system. Although central nervous system cells are not replaced when destroyed, it is possible in the case of a limited amount of cellular destruction for other central nervous system cells to functionally substitute for the dead cells, so that overtly there is no sign of trouble. Symptoms of mercury poisoning may not become apparent until a great deal of damage has already been done to the central nervous system, including parts of the brain as well as peripheral nerves (Smith and Smith, 1975).

Methyl mercury poisoning has also been found to be the cause of serious congenital disorders. Methyl mercury apparently passes readily through the placental barrier into the fetus, where concentrations of mercury may build up to levels several times higher than in the mother (Aaronson, 1971), reaching especially high levels in the fetal brain (Montague and Montague, 1971). Indeed, women showing no symptoms of mercury poisoning have given birth to children with serious congenital mental and/or physical defects caused by irreparable damage to the central nervous system of the fetus as a result of methyl mercury poisoning (Smith and Smith, 1975). In the case of methyl mercury poisoning in Minamata, Japan, mothers of children suffering from congenital effects generally showed lighter symptoms of mercury poisoning than did other members of the same family, as did women who had miscarriages or stillbirths (Smith and Smith, 1975). These results suggest that a pregnant woman may discharge a significant portion of the methyl mercury in her body through the placenta and into the fetus.

In addition to attacking central nervous system cells of the fetus, methyl mercury may interfere with reproduction in another way. Methyl mercury (as well as phenyl mercury) is known to interfere with the process of cell division, causing daughter cells to receive an unequal number of chromosomes. This phenomenon, known as nondisjunction, has been demonstrated in plant cells (onions), fruit flies, and tissue cultures from mice and humans (Grant, 1971; Montague and Montague, 1971; Mitra, 1986). Methyl mercury has been found to be the most potent causative agent (Grant, 1971). The mechanism of interaction is again presumably the attraction of Hg for sulfhydryl groups. Although there is no definite evidence that mercury has been responsible for creating genetic abnormalities in humans, mercury has been found to concentrate in the gonads of mice and trout (Montague and Montague, 1971), and considering the demonstrated mutagenic effects of mercury on human cell cultures, the possibility certainly exists that similar effects could be produced in living persons. Montague and Montague (1971, p. 62) note that, "Physicians performing autopsies on Japanese children from Minamata noted certain abnormalities that they could not attribute to direct mercury poisoning. It appeared as if the mercury eaten in fish at Minamata had entered the testes or ovaries of parents, damaged sex cells, and thus passed on genetic damage to the offspring." The other forms of organic mercury such as phenyl mercury and methoxy mercury can be toxic if ingested in sufficient amounts, but in general are of less concern than methyl mercury. In particular, if phenyl mercury is ingested, a large percentage is converted to inorganic mercury by the body, and thence effectively excreted by way of the kidneys and intestines (Grant, 1971). Neither phenyl mercury nor methoxy mercury is as effective in penetrating to the brain as methyl mercury, and hence neither class of mercury compound is as likely to cause permanent neurological damage. As far as humans are concerned, by far the most important mechanism of mercury intake is the ingestion of Hg-contaminated food (Mitra, 1986). Meat and especially fish products have in general been found to contain the highest levels of mercury (Goldwater, 1971), and at least in the case of fish, most of the mercury is usually in the form of methyl mercury (Wood, 1972). Thus from the standpoint of human health, methyl mercury is the most dangerous form of mercury for the following reasons:

1. It has the greatest tendency to cause irreparable neurological and perhaps genetic damage.
2. Insofar as we know it is more effectively retained if swallowed (more readily absorbed and more slowly excreted) than any other form of mercury.
3. It is probably the most abundant form of mercury in the food we eat.

Acceptable Levels of Exposure. What are safe levels of exposure to mercury? From the previous discussion, it should be obvious that the answer to this question depends on the chemical form of the mercury as well as the mechanism of ingestion. Furthermore, since Hg is not known to be required in any way by either plants or animals, any degree of exposure to mercury can hardly be considered beneficial. Since recent instances of large-scale mercury poisoning (Smith and Smith, 1975) have resulted from the eating of mercury-contaminated food, much attention has been directed toward establishing safe or acceptable mercury levels in food.

Assuming that virtually all the mercury we eat is methyl mercury (probably not far from the truth), it is possible to calculate the daily intake of mercury that would produce a given body burden of mercury in the steady state, if we know the excretion rate of mercury by the body. The excretion rate of methyl mercury by humans is about 1% per day (Mitra, 1986), so that in the steady state a person could consume each day about 1% as much mercury as his body contained. In other words, he would be excreting each day about as much Hg as he consumed. Based on studies of persons affected by mercury poisoning, it has been concluded that a body burden of about 25–30 mg of mercury is "dangerous" for an adult[11] in the sense that an adult having 25–30 mg of Hg in his body is likely to show overt signs of mercury poisoning (Smith and Smith, 1975). Assuming an excretion rate of 1% per day, it is concluded that a Hg intake of 0.25–0.30 mg/day would lead to a dangerous concentration of mercury in the body. Based on similar reasoning, the U.S. Environmental Protection Agency concluded that lowest observed effect levels of Hg poisoning in humans were associated with daily Hg intakes of 0.2–0.5 mg/day (EPA, 1980a). Applying the uncertainty factor of 10 appropriate to results from studies on prolonged ingestion by humans (Table 8.11), we conclude that the daily intake of methyl mercury should not exceed 0.02 mg = 20 μg (EPA, 1980a). Some persons have argued, however, that there really is no "safe" threshold limit for mercury. Smith and Smith (1975, p. 191), for example, have stated,

---

[11]Arbitrarily defined as a person weighing 70 kg.

"Although outward symptoms may not appear until a certain level (of mercury in the body) is reached, it can be assumed that damage is proportionate, to some extent, to the amount of mercury that is consumed and that passed through the body, whether or not the body's mercury content reaches a 'dangerous' level at any given moment. If so, there is no actual 'safety level.' The greater the methyl mercury intake, the greater the cell damage. The lower the intake, the less damage to cells. On the cellular level there is no 'threshold point.' "

The background level of organic mercury in food is roughly 0.02–0.05 ppm (WHO, 1973). Fruits, vegetables, grain, and milk, for example, typically contain Hg at concentrations less than 0.04 ppm (Goldwater, 1971; Mitra, 1986). Meat and freshwater fish tend to have a higher mercury content. A 1969 survey of U.S. pork and beef revealed mean Hg levels of 0.1 ppm, and the background level of Hg in freshwater fish is about 0.2 ppm (Montague and Montague, 1971). Marine fish typically contain Hg at levels of 0.01–0.08 ppm, but the average Hg concentrations in tuna and swordfish are 0.25 and 0.93 ppm, respectively (EPA, 1980a; Montague and Montague, 1971).

The high levels of mercury in tuna and swordfish have caused much concern, and on May 6, 1971 the FDA advised Americans to stop eating swordfish (Montague and Montague, 1971). As a result, the swordfish industry collapsed. Although no such recommendation was ever made with respect to tuna, it is instructive to note that consumption of a typical 6.5 ounce can of tuna containing 0.25 ppm Hg would result in the ingestion of 46 µg of mercury, over twice the maximum daily intake of methyl mercury recommended by the EPA. Consumption of three 6.5 ounce cans of tuna each week would lead to an average level of mercury consumption equal to the maximum level of methyl mercury intake recommended by the EPA, even if the rest of one's food contained no mercury at all. Looking at the problem another way, if one assumes an average daily rate of food consumption of 1.3 kg per person (OECD, 1975) and requires that the daily methyl mercury intake not exceed 20 µg, then the average methyl Hg concentration in food should not exceed 0.015 ppm. This figure is higher than the Hg concentration in many fruits and vegetables, but is lower than values found typically in meat and fish.

In July of 1969, the FDA established a safety level of mercury in food fish at 0.5 ppm wet weight. This figure was increased to 1.0 ppm in 1973 and was changed to 1.0 ppm methyl mercury in September, 1984. From the information in the previous paragraph, it should be clear that this recommendation was based in part on the assumption that Americans do not eat food containing close to 1.0 ppm Hg on a regular basis. Thus 1.0 ppm methyl mercury in food is a safe level only if such food is eaten infrequently, and if the rest of one's food averages less than 0.015 ppm Hg. According to studies conducted by the EPA, the single most important cause of methyl mercury intake for most Americans is the consumption of tuna. Methyl mercury accounts for more than 90% of the total mercury in tuna, and it is estimated that tuna account for about 21% of the fish consumed by Americans (EPA, 1980a). According to the FDA only 5% of Americans consume more than 27 g of fish per day. If the assumption is that all the mercury in tuna is methyl mercury, it follows that 95% of the U.S. population consumes less than 1.4 µg per day of methyl mercury in tuna. Contributions from other fish would probably increase this figure to about 4.2 µg d^{-1} (EPA, 1980a). Perhaps the principal consequence of changing the FDA action level for fish from 0.5 ppm total Hg to 1.0 ppm methyl mercury has been to give a boost to the swordfish industry. It is noteworthy, however, that over 50% of the swordfish tested by the EPA (1980a) contained more than 1.0 ppm total mercury.

The discussion up to this point has been concerned with the toxicity of mercury to humans, but there is no doubt that Hg pollution can have serious consequences for aquatic organisms as well. In Minamata Bay, which received mercury discharges from an acetaldehyde plant from 1932 to 1968, Smith and Smith (1975, p. 180) report that, "Unusual changes were detected in these waters [Minamata Bay] as long ago as 1950. Fish floated on the surface of the sea, shellfish frequently perished, and some of the seaweed died. In 1952 some birds such as the crow and the *amedori*—a type of sea bird—began to drop into the sea while flying. The area of the sea where dead fish could be seen floating spread throughout the bay and out into the Shiranui Sea. Sometimes octopus and cuttlefish floated so weakened that children could catch them with their bare hands."

This dismal picture describes an extreme case of mercury pollution. For those charged with the responsibility of setting water quality guidelines to protect aquatic life, it is obviously necessary to know not only the levels of mercury that can be expected to produce such consequences, but also the mercury levels that could be expected to produce chronic, sublethal stresses. The problem of setting water quality standards for mercury is compounded by the fact that aquatic organisms, like humans, are much more sensitive to methyl mercury than the other forms of mercury. Methylmercuric chloride, for example, is about 10 times more toxic to rainbow trout than mercuric chloride (EPA, 1980a).

When the present EPA water quality criteria for mercury were established in 1985, there was enough information to determine criterion maximum concentrations only for mercuric mercury. Given that most mercury currently being discharged is Hg^{2+} (EPA, 1985a), this limitation may not be a serious one. The final acute values in both freshwater and marine waters were calculated as outlined in Chapter 8, and the criterion maximum concentrations listed in Table 12.8 are 50% of those final acute values. The final residue values of 12 ng/l in freshwater and 25 ng/l in seawater were calculated by dividing the FDA action level of 1.0 ppm methyl mercury in fish and shellfish by bioconcentration factors (BCFs) for methyl mercury of 81,700 in freshwater and 40,000 in seawater (EPA, 1985a). The freshwater and seawater BCFs were obtained from studies with the fathead minnow and eastern oyster, respectively. For both freshwater and seawater, the final residue values were lower than both the final chronic value and the final plant value, and the final residue values therefore became the criterion continuous concentrations.

One interesting point about the criterion continuous concentrations for the protection of aquatic life is the fact that they are substantially lower than the criterion of 144 ng/l for the protection of human health (Table 12.1) and the drinking water standard of 2 µg/l (EPA, 1986). Although these differences may seem surprising, one should keep in mind that, unlike aquatic organisms, humans do not spend their lives bathed in water.

The figure of 144 ng/l assumes that people drink 2.0 l of water per day and consume 18.7 g of fish and shellfish.[12] The biological concentration factor in the fish

[12]The figure of 18. 7 g/day comes from a study by Cordle et al. (1978) and is almost 3 times higher than the figure of 6.5 g/day used by the EPA (1980b, 1985b). Since the water quality criteria with respect to mercury for the protection of human health were promulgated in 1980, the implication would seem to be that at the EPA the right hand does not always know what the left hand is doing.

Table 12.8   EPA Water Quality Criteria for Total Recoverable Mercury

Criterion Maximum Concentration		Criterion Continuous Concentration	
Marine waters	Freshwater	Marine waters	Freshwater
2.1 µg/l	2.4 µg/l	25 ng/l	12 ng/l

Source. EPA (1985a).

and shellfish is assumed to be 7337, which is substantially less than the BCFs used to derive the final residue values in freshwater and seawater. You should keep in mind, however, that final residue values must be chosen low enough to keep pollutant concentrations below FDA guidelines for any commercial species. The fish and shellfish consumed by Americans include many species for which the practical BCF is substantially less than $4-8 \times 10^4$. The mean value, based on the EPA (1980a) data, is 7337. The daily intake of mercury from the consumption of 2.01 of water containing 144 ng/l of Hg and the consumption of 18.7 g of fish and shellfish from water with a similar mercury concentration would therefore be $0.144(2 + 0.0187(7337)) = 20$ µg of mercury, the maximum acceptable daily intake rate of methyl mercury. Over 98% of this intake is estimated to come from the consumption of fish and shellfish. The validity of setting the criterion at 144 ng/l rests on the assumption that sources of exposure other than the consumption of fish, shellfish, and drinking water are negligible. This assumption is supported by the observation that nonfish eaters have the lowest blood concentrations of mercury (EPA, 1980a).

Most mercury in freshwater is believed to be in the form of mercuric mercury (EPA, 1980a), of which more than 98% is excreted in the feces and urine (Grant, 1971). The drinking water standard of 2 µg/l implies that it is acceptable to ingest as much as 4 µg/day of mercury in the water you drink, but if this were all present as Hg^{2+}, less than 80 ng would be retained by the body. For most Americans then and indeed for almost all persons, the dominant source of mercury exposure is the consumption of fish and shellfish.

Minamata Bay—A Case Study

Minamata Bay is undoubtedly the most serious and most heavily publicized case of mercury pollution anywhere in the world. Although the severity and extent of the damage caused by mercury pollution at Minamata makes the example atypical of most cases of mercury pollution, the scientific and medical problems associated with the mercury poisoning at Minamata are nevertheless relevant to virtually all instances of mercury pollution, regardless of the extent of mercury discharges or the degree of environmental damage or human suffering. Indeed, current safety guidelines with respect to mercury ingestion by humans are based in no small part on knowledge gained from the study of mercury poisoning victims at Minamata.

Minamata Bay is a semienclosed coastal indentation on the eastern side of the Shiranui Sea,[13] an inland sea on the western part of Japan's southern island of Kyushu (Figure 12.3). About 100,000 persons depend in part on fish taken from the Shiranui Sea as a source of food, and in this respect the people of Minamata City and its environs are typical. Historically Minamata has been a farming and fishing

[13]Sometimes called the Yatsushiro Sea.

Figure 12.3 Location of island of Kyushu (top), Minamata (center), and the Chisso factory and Minamata Bay (bottom).

area, and the population, which numbers about 36,000, consumes on the average between 286 g of fish in the winter and 410 g of fish in the summer per person per day (Smith and Smith, 1975).

In 1907 the parent company of the present Chisso Corporation built a factory in Minamata. Although the Chisso Corporation was initially involved primarily in the manufacture of fertilizer and carbide products,[14] the company soon expanded into

[14]Chisso in Japanese means nitrogen.

the manufacture of petrochemicals and plastics. That discharges from the Chisso plant were creating some water pollution problems in Minamata Bay during the first few decades of the plant's operation is evident from the fact that by 1925 Chisso was compensating Minamata fishermen for damage to their fishing areas (Smith and Smith, 1975). Serious pollution of the bay presumably did not begin until 1932, however, when the plant began using mercuric oxide (HgO) as a catalyst in the production of acetaldehyde and vinyl chloride. It is not clear how much mercury the plant discharged into Minamata Bay between 1932 and 1968, when discharges were terminated finally, but one estimate has put the total figure at 200–600 tonnes (Smith and Smith, 1975). Presumably the greatest mercury discharges occurred during the 1950s, when production at the plant was apparently at a maximum and the population of Minamata had grown to 50,000 persons (Smith and Smith, 1975).

The first reported signs of trouble involved animals rather than humans. Cats, which were fed in part on fish taken from the bay, began to go mad and die. By 1953 some dogs and pigs were dying in a similar manner. The sickness was characterized by salivating, loss of coordination, and convulsions. Effects on aquatic organisms also began to appear during the early 1950s. Shellfish and seaweeds were killed off in certain parts of the bay, and dead fish were found floating on the surface of the bay and even out into the Shiranui Sea (Smith and Smith, 1975). Despite these overt signs of trouble, people continued to fish in Minamata Bay and to eat fish taken from the bay.

The first reported case of a human affected by the disease occurred in April, 1956, when a 5-year-old girl was brought to the pediatrics department of the Chisso Corporation's Minamata factory. The girl was suffering from symptoms of brain damage, including delirium, disturbance of speech, and disturbance of gait. An investigation revealed that her younger sister as well as four neighbor children were suffering from the same symptoms, and "On May 1, 1956, Dr. Hajime Hosokawa, head of the Chisso factory hospital, reported to the Minamata Public Health Department: 'An unclarified disease of the central nervous system has broken out.' "(Smith and Smith, 1975, pp. 180–181). This date marks the official "discovery" of Minamata Disease.

During the summer of 1956, the occurrence of Minamata Disease reached "epidemic" proportions, although many of the persons found to be suffering from the disease at that time had in fact been ill since about 1953. In August, 1956, a research team from Kumamota University was appointed to investigate the cause of the sickness, and on October 4, 1956, this group reported that the disease was not infectious and in fact was caused by heavy metal poisoning associated with the eating of fish and shellfish from Minamata Bay (Smith and Smith, 1975). Although it would have seemed reasonable to ban the eating of fish and shellfish from Minamata Bay at that time, the government issued only a warning that eating fish from the bay was dangerous. Two years later, when the number of victims had exceeded 50, including 21 who had died, the Kumamoto Prefecture finally ordered a ban on the selling of fish from Minamata Bay, but there was still no restriction on catching fish (Smith and Smith, 1975).

Following the October 4, 1956 report of the Kumamoto University research team, an intensive effort was begun to identify the cause of the heavy metal poisoning. Although discharges from the Chisso factory were the obvious and logical source of the pollution, no cause-and-effect relationship had been proven, and the plant was therefore allowed to continue operation. The plant's effluent in fact contained a variety of metals, including manganese, thallium, arsenic, mercury, selenium, copper, and lead (Smith and Smith, 1975), and it was far from clear which metal or combination of metals was the cause of the poisoning. Initially manganese and then thallium was suspected as being the causative agent, but experimental feeding of these metals to cats failed to reproduce the symptoms of the disease. Finally, in September, 1958, Professor T. Takeuchi of Kumamoto University discovered that, "Clinical and pathological findings in cases of Minamata Disease coincided with certain cases of methyl mercury poisoning reported in England in 1940" (Smith and Smith, 1975, p. 182). Kumamoto University researchers then found that feeding methyl mercury to cats did reproduce the symptoms of Minamata disease, and in September, 1960, Professor M. Uchida of Kumamoto University extracted a methyl mercury compound (CH_3HgSCH) from Minamata Bay shellfish (Smith and Smith, 1975). Investigations of the bay following Professor Takeuchi's report revealed that Hg levels in the sediments of Minamata Bay were on the order of 10–100 ppm (wet weight), and concentrations in the mud of the Chisso drainage canal ranged as high as 2000 ppm. Mercury levels in fish and shellfish sampled from the bay ranged from 5 to 40 ppm wet weight (Smith and Smith, 1975). At that time the Chisso Corporation maintained that their Minamata plant could not be the source of the methyl mercury pollution, since the plant used only inorganic Hg in the production of acetaldehyde and vinyl chloride. Two pieces of evidence, however, indicated that the plant was indeed the cause of the poisoning.

1. Following the implication of methyl mercury as the disease-causing agent, Dr. Hajime Hosokawa, the head of the Chisso factory hospital, began experimental feeding of effluent from the acetaldehyde plant wastewater pipe to a cat. On October 7, 1959 the cat became ill, exhibiting the same symptoms—convulsions, salivating, disoriented movements—that had been observed in cats afflicted with Minamata disease. When Dr. Hosokawa reported this result to the Chisso management, he was told to discontinue his experiments, and that "There would be absolutely no more experiments connected with Minamata Disease" (Smith and Smith, 1975, p. 122). When Dr. Hosokawa's assistant went to get additional wastewater from the acetaldehyde plant, he was stopped by a guard. Being a good "company man," Dr. Hosokawa failed to communicate his finding to the public, and it was not until he testified on his deathbed in 1969 that the knowledge of his discovery became known to those outside the Chisso management.

2. Although Chisso denied scientists access to its wastewater shortly after the implication of methyl mercury as the Minamata poison in 1959, Dr. K. Irukayama of Kumamoto University discovered a bottle of sludge which had been obtained from the plant prior to that time sitting on a laboratory shelf in 1962. Analysis of the sludge revealed that it contained methyl mercury chloride, the implication being that the inorganic Hg used by the plant had been methylated prior to disposal from the plant. In fact, organic mercury was known to be a byproduct of the acetaldehyde production process (Kurland, 1989).

During this time several events occurred relevant to the mercury problem. In September, 1958 Chisso began discharging temporarily its acetaldehyde wastewater into the Minamata River, which flows into the Shiranui Sea north of Minamata

Bay. Within a few months, persons living in the vicinity of the river began to show symptoms of Minamata Disease (Smith and Smith, 1975).

In December, 1959 Chisso installed a wastewater treatment device called a *cyclator* at its Minamata plant. The cyclator was widely hailed as having made the factory wastewater safe, and at ceremonies marking the installation of the cyclator the president of Chisso drank water taken from the outflow of the treatment system. Later it was learned that the effluent from the cyclator on that day had not contained wastewater from the acetaldehyde process (Smith and Smith, 1975). Furthermore, testimony presented in court during the 1969–1973 trial of Chisso revealed that the cyclator had in fact not made the wastewater safe. Questioning of Eiichi Nishida, a Chisso engineer who had been head of the Minamata factory, resulted in the following exchange (Smith and Smith, 1975, p. 122):

LAWYER: *in other words, did you think that (the wastewater) which went through the Cyclator was safe?*
NISHIDA: *Well, that is . . .*
LAWYER: *That is?*
NISHIDA: *That is . . . at that point . . . I don't know.*
LAWYER: *I'm asking you whether you thought it was safe or not.*
NISHIDA: *Well, ah, at that time, you see, I thought that putting it through the cyclator—safe, well not . . .*
LAWYER: *Of course you didn't think it would be, is that correct?*
NISHIDA: *Ah, well, yes.*

Thus it is apparent that by the end of 1959 the Chisso Corporation had good reason to believe that the mercury in its wastewater was the cause of Minamata Disease,[15] and it is furthermore apparent that the company was aware or at least realized later that the cyclator installed in December, 1959, would not correct the problem.

On December 30, 1959 the Chisso Corporation negotiated what was, considering the information in the previous few paragraphs, a highly unethical contract with the Minamata Disease victims known at that time. The contract provided for payments to the victims, but specified that the payments were to be regarded as *mimai* (consolation) rather than as indemnity for damages. In fact, Chisso refused to accept responsibility for Minamata Disease, and the *mimai* contract stipulated that Chisso would not be liable for further compensation to the victims even if the company were later proven guilty (Smith and Smith, 1975). Furthermore, Chisso continued to discharge its wastewater into Minamata Bay. A stormy protest by fishermen in October of 1959, including demands for additional compensation and a cleanup of Minamata Bay, produced few tangible results. After receiving warnings from the government, the fishermen settled for meager payments. There was no cleanup (Smith and Smith, 1975).

After 1959 the furor over Minamata disease incredibly died down for several years. By 1962 there were 121 verified Minamata Disease victims, including 46 dead, but although Chisso continued to discharge its wastewater into the bay, the disease was considered "over," and people had begun to eat fish from the bay again (Smith and Smith, 1975).

[15]Based on Dr. Hosokawa's experiment and the report of the Kumamoto University research team.

Several years later, the history of Minamata was changed dramatically by events that occurred hundreds of kilometers away. In the city of Niigata, methyl mercury poisoning similar to Minamata Disease broke out in 1965. The source of pollution was found to be an acetaldehyde plant operated by the Showa Denko Company about 65 km upstream from Niigata on the Agano River. Whereas in Minamata the Chisso plant had provided about 1300 local jobs and hence enjoyed much sympathy, there were no such feelings in the case of Niigata. The Niigata victims, whose numbers eventually approached 500, began vigorous legal action against Showa Denko, resulting in a lawsuit and trial, which began on June 12, 1967 (Smith and Smith, 1975).

Aroused by the Niigata trial, some of the Minamata victims ultimately filed a lawsuit of their own against Chisso. Although the government had officially announced on September 26, 1968, that Chisso had been the cause of Minamata Disease (Smith and Smith, 1975), the Minamata plaintiffs' case was weakened by the existence of the 1959 *Mimai* contracts, which had stipulated that Chisso would not be held liable for further payments even if proven guilty. Furthermore, many Minamata victims declined to join in the lawsuit, but instead chose to seek compensation through a government-appointed Central Pollution Board, which was established in November, 1970. In fact, only about one-third of the 1959 *Mimai* signers were included among the 29 families (representing 45 victims) who sued Chisso on June 14, 1969 (Smith and Smith, 1975).

The Chisso trial lasted almost 4 years, with a verdict being reached finally on March 20, 1973. During the trial several events greatly strengthened the case of the plaintiffs. First, the Niigata trial of Showa Denko ended on September 29, 1971 with a decision in favor of the plaintiffs, and on November 13, 1971, Chisso president Kenichi Shimada signed a paper acknowledging moral (but not legal) responsibility on the part of Chisso for the Minamata poisoning. Finally, of course, there was the July 4, 1970 testimony of Dr. Hosokawa, which revealed that Chisso had known as early as 1959 that its sludge was the probable cause of Minamata Disease, but had concealed the evidence.

During the course of the trial, there were a number of clashes, both verbal and physical, between the Chisso Corporation and a so-called direct negotiations group, which insisted that Chisso deal directly with the Minamata victims rather than through the courts. The confrontation between Chisso and the direct negotiations group was highlighted by the beating of certain members of the direct negotiations group as well as American photographer W. Eugene Smith at a scheduled meeting at the Chisso plant at Goi on January 7, 1972 (Smith and Smith, 1975). A year later, on January 10, 1973, the direct negotiations group was instrumental in revealing that a number of Minamata victim's signatures had been forged on Central Pollution Board documents. At the time it was feared that the Central Pollution Board would announce its decision prior to the Minamata court decision, and it was felt that the Central Pollution Board was not disposed as favorably to the victims as was the court. In any case, the forgery scandal discredited the Central Pollution Board, forestalled its decision, and thus allowed the Minamata court to establish the precedent for compensation.

Finally, on March 20, 1973, the Minamata court announced its decision. Initial payments of $68,000 were to be made to the families of deceased victims and to severe cases. Chisso made no appeal and paid out compensation checks totaling $2.2 million within a few days (Smith and Smith, 1975). However, the trial decision and payments directly affected only the small group of victims who had sued Chisso. The decision did not directly affect those who were depending on the Central Pollution Board for compensation, nor did it apply to the many newly verified patients. As a result the direct negotiations group arranged a series of meetings with Chisso management, beginning on March 22, 1973, to discuss compensation for patients other than the trial group, as well as to negotiate additional and regular payments for the medical needs and care of all Minamata victims. Chisso, feeling that the number of victims might ultimately number in the thousands, was reluctant to agree to all the demands of the direct negotiations group. On April 1, 1973, however, a newly verified victim by the name of Kimito Iwamoto slashed his wrists with a broken ashtray in the presence of Chisso president Shimada, exclaiming that his life was worthless unless Chisso agreed to compensate him.[16] At his display of desperation, president Shimada gave in; Chisso agreed to compensate all victims fully. A few weeks later, on April 27, 1973, the Central Pollution Board finally announced its decision. The initial maximum compensation was to be $68,000, the minimum $60,000, identical to the Minamata court's previous ruling. In addition, Chisso was to make monthly payments to all patients. Since that decision, the number of verified victims has increased steadily. As of September, 1988, 2209 Minamata Disease victims had been verified, of whom 730 had died (Futatsuka, 1989; Katahira, 1989).

The cost in terms of human suffering from Minamata Disease is incalculable. The statement by Ms. Fukuda at the beginning of this chapter is a thought-provoking testimony to the frightening impact of the disease on its victims. The monetary cost of treating and caring for Minamata Disease victims has of course been staggering, and it is instructive to examine how the Chisso Corporation has dealt with this issue. Beginning in 1959, when the Kumamoto University research team concluded that the mercury in the effluent of Chisso's Minamata plant was the probable cause of Minamata Disease, the Chisso Corporation began to fragment into a large number of subsidiaries. While the formation of these subsidiaries could be rationalized to a certain extent on general principles as good business, one obvious consequence was the minimization and avoidance for the subsidiaries of the responsibility to pay compensation (Yamaguchi, 1989). In order to assist with the payments, Kumamoto Prefecture began making loans to Chisso in 1978, the loans being financed by the sale of prefectural bonds. A condition attached to the issue of the bonds was the following (Uzawa, 1989, p. 336): "Chisso, which has received this loan of public funds, will put its management on a firmer footing, and endeavor to repay the funds, as well as make itself able in the future to pay all compensation without any assistance; in addition, the Minamata Plant shall contribute to the stability of the local economy and society." There is no evidence, unfortunately, that the Kumamoto Prefecture is monitoring Chisso to ensure that these conditions are met, and the creation of numerous Chisso subsidiaries would seem to suggest that Chisso is not taking seriously the need to repay the loan, which by 1988 amounted to 60 billion yen.[17] Furthermore, the Minamata Chisso plant has been scaled back considerably, and by 1988 the work force, which once numbered 1300, had been reduced by more than a factor of 2 (Uzaza, 1989).

---

[16]He lived.
[17]Equivalent to about $450 million.

The decision to reduce the work force at the plant hardly seems consistent with the goal of having the plant contribute to the stability of the local economy and society. In discussing the attitude of Chisso management to the victims, Uzawa (1989, p. 337) made the following observation:

> Once I worked with Professor Susumu Nagai (Hosei University) in connection with the matter of compensation for Minamata disease patients to investigate Chisso's business operations. Even now I recall with a heavy heart the extremely insincere attitude of the Chisso main office during that investigation. During the investigation one of the company officers made this unforgettable statement to us: 'Most of the present Chisso employees joined the company after the Minamata disease issue had already become a thing of the past, and they can't understand why they should have to take the responsibility, or be criticized, for past events.' One could say that this statement reveals the true nature of Chisso, and is symbolic of the inhuman, unethical nature common to the many pollution issues in Japan, foremost among them Minamata disease.

Following the 1973 decision in the first Minamata Disease lawsuit, the Chisso Corporation, the Kumamoto Prefecture, and the Japanese national government were sued by Minamata Disease victims in subsequent legal actions. In 1979 the Kumamoto District Court found Chisso officers guilty of a criminal offense, and in 1987 recognized the responsibility of the Kumamoto Prefecture and the Japanese national government to pay compensation (Tsuru et al., 1989). The problem of identifying funds to pay the victims was solved to a certain extent by the latter decision, which identified two culpable entities with very deep pockets. The final appeal involving Chisso's criminal liability was heard in 1988; both the former Chisso president and the former plant manager were found guilty (Tsuru et al., 1989).

In reviewing what happened at Minamata, it seems obvious that discharges from the Chisso plant should have been terminated as early as 1956, when the Kumamoto University research team reported that heavy metal poisoning was the cause of Minamata Disease, since the plant was known to be discharging a variety of metals into the bay. The decision of Chisso to continue operating the plant even after 1959, when Dr. Hosokawa's experiments indicated that sludge from the plant could indeed cause Minamata Disease, was highly unethical and in fact immoral. Mercury discharges from the plant were finally terminated in May, 1968, because the mercury method of production had become outmoded (Smith and Smith, 1975). Why the Japanese government failed to halt the Chisso discharges earlier, at least temporarily while a thorough investigation could be conducted, is not clear, but such government inaction is not without precedent in the history of industrialized countries, including the United States.

Efforts to clean up Minamata Bay have not met with great success. The government decided to fill in or dredge all parts of the bay where sediment Hg levels exceeded 25 ppm (Smith and Smith, 1975), but 25 ppm mercury in sediments is not a very safe standard. Furthermore, dredging tends to stir up sediments and distribute them over a wider area. According to Wood (1972), mercury levels in fish were actually higher after the dredging of Minamata Bay than before. As of 1973 mercury levels in Minamata Bay shellfish averaged 0.47 ppm (Smith and Smith, 1975), a great improvement over the figures of 5–40 ppm reported earlier, but still higher than the provisional Japanese government standards of 0.4 ppm total mercury and 0.3 ppm methyl mercury (Tsuru et al., 1989, p. 415). In 1988 some fish in the bay still violated these provisional standards, yet at no time has the government enacted measures provided by law to prohibit fishing. For typical Minamata residents who eat 286–410 g of fish per day, the methyl mercury concentration in the fish would have to be less than 0.05–0.07 ppm in order to keep the intake of methyl mercury from fish consumption below the U.S. standard of 20 μg d^{-1}. It seems fair to say that there is still a serious mercury pollution problem in Minamata Bay.

How widespread has been the damage caused by mercury poisoning in Minamata no one can really say. Although obvious victims may continue to be verified for many years, mild symptoms associated with low-level mercury poisoning, particularly in congenital cases, may be extremely difficult to detect. In 1970, for example, an examination of junior high school students in the most heavily contaminated areas of Minamata revealed difficulty in articulating words among 18% of the students, sensory disturbance among 21%, and clumsy movements among 9% (Harada, 1989). For comparison, the average rate of mental deficiency among Japanese junior high school students is 9.7% (Smith and Smith, 1975). To what extent the high percentage of mental and physical deficiency among these students is due to low levels of mercury poisoning is anyone's guess.

Commentary

Was Minamata an isolated incident? Not really. The number of persons who actually died of mercury poisoning at Minamata exceeds the number known to have been killed by mercury in comparable incidents elsewhere, but the 6500 persons admitted to hospitals in Iraq as a result of methyl mercury poisoning in 1971 exceed by almost a factor of 3 the number of certified Minamata Disease victims, and the number of persons who died in the Iraq incident (450) is comparable to the number (730) who died at Minamata. Similar incidents, fortunately involving smaller numbers of victims, have been reported in other parts of the world. In Niigata, 690 persons have been certified to be victims of methyl mercury poisoning (Saito, 1989), and from 1958 to 1982 the Songhua River of China was seriously polluted with methyl mercury discharges, again from an acetaldehyde plant. The Chinese incident caused numerous cases of sublethal poisoning, documented symptoms including hearing and vision impairment (Pan, Jiang, and Wang, 1989). Fortunately there were no fatalities, but a "follow-up study showed . . . that there were many subclinical victims along the Songhua River who were easily overlooked" (Pan, Jiang, and Wang, 1989, p. 301).

High levels of mercury have been found in fish from several parts of North America, primarily in areas where chlor-alkali plants were discharging mercury-laden wastewater. Fish taken from the Canadian English-Wabigoon River system in 1970, for example, were found to contain as much as 27.8 ppm Hg, with many values between 10 and 20 ppm (Smith and Smith, 1975), and many fish taken from Lake St. Clair in the same year contained 5–7 ppm Hg (Grant, 1971). In both cases, the source of mercury was a chlor-alkali plant, the Dryden Paper Company plant at Lake Wabigoon in the case of the English-Wabigoon River and the Dow Chemical plant at Sarnia in the case of Lake St. Clair (Wood, 1972). It is noteworthy that the mercury levels in the fish sampled from these two systems were comparable to the levels of mercury found in fish and shellfish from Minamata (5–40 ppm) in 1959 and from Niigata (1–10 ppm) in the 1960s (Grant, 1971; Smith and Smith, 1975). Nevertheless, no widescale mercury poisoning incident has occurred either in the

United States or in Canada, presumably because most people in these countries eat fish much less frequently than do Japanese. There is evidence, however, that some Canadian Indians developed symptoms of methyl mercury intoxication as a result of eating mercury-contaminated fish (EPA, 1980a).

In the United States reduction of mercury pollution in aquatic systems has been accomplished primarily through enforcement of the 1899 Refuse Act, the Federal Insecticides, Fungicides, and Rodenticides Act (FIFRA) of 1947 and the Clean Water Act (CWA) of 1972. The FIFRA was used to eliminate the use of mercury compounds as antifouling agents and fungicides (Mitra, 1986). The Refuse Act was used to control discharges from chlor-alkali plants and other industrial operations prior to passage of the CWA. A chlor-alkali plant operated by Olin Matheson on the Niagra River, for example, was able to continue operation after being taken to court in 1970 because installation of pollution control devices reduced mercury discharges to about $220 \, \text{g d}^{-1}$ (Wood, 1972). Chlor-alkali plants operated by the Oxford Paper Company in Rumsford, Maine, and by Olin Matheson at Saltville, Virginia, simply closed down in 1970 and 1972, respectively, when faced with court orders to clean up their effluent (Wood, 1972; Carter, 1977). In Canada the Law on Fisheries was amended in 1972 to limit the discharge of mercury in wastewater by chlor-alkali plants to 1.82 g per tonne of chlorine. Discharges from U.S. plants are controlled indirectly by the criteria in Table 12.8, which apply to the receiving body of water. The impact of the Canadian and U.S. legislation on mercury levels in some fish populations has been quite dramatic (Figure 4.11), but further improvement would be desirable.

Even when mercury pollution is detected and halted, the problem of cleaning up the damage still remains. The lessons learned from Minamata Bay and U.S. and Canadian lakes and streams have made it clear that aquatic systems contaminated with mercury are generally very difficult to clean up. Once released to the environment, mercury may continue to cycle between the sediments, water, and biota for tens, hundreds, or even thousands of years before finally being flushed from the system. For example, it has been estimated that about 5000 years will be required for the mercury stored presently in the Lake St. Clair ecosystem to effectively flush out by natural processes (Wood, 1972), and a small amount of mercury (not much more than could be put in a Volkswagon gas tank) released to Virginia's South River and South Fork of the Shenandoah River between 1929 and 1950 has kept the system contaminated for decades, with sediment mercury levels exceeding 240 ppm in some places (Carter, 1977).

The following approaches for decontaminating mercury-polluted sediments have been suggested by Swedish workers (Mitra, 1986).

1. Dredging of sediments.
2. Increasing the pH of the sediments in order to favor demethylation and increase volatilization.
3. Introducing oxygen-consuming materials so as to create anaerobic conditions in the sediments and hence reduce mercury methylation.
4. Covering the sediments with fresh, finely divided, highly adsorptive materials such as clay or quartz.
5. Covering the sediments with any inorganic, inert material.

Unfortunately none of these methods is without drawbacks, either in terms of cost, permanence, effectiveness, or side effects. The Swedes, for example, dredged the sediments of polluted waterways on two occasions, and in each case found that mercury levels in fish were higher after dredging (Mitra, 1986). The Japanese experienced a similar, but temporary, problem in Minamata Bay. The burrowing activities of benthic infauna can frustrate efforts to seal off contaminated sediments with an overlayer of inert, adsorptive material, and Swedish scientists found that uptake of mercury by fish from sediments contaminated with phenylmercury was unaffected by ground silicates. Mercury uptake was reduced by a factor of two, however, when inorganic mercury was the contaminant (Mitra, 1986).

The problem of cleaning up mercury pollution is well illustrated by the case of the Olin Matheson chlor-alkali plant at Saltville, Virginia. So-called muck ponds at the site cover about 0.5 km^2 and extend along the Holston River for about 1 km. The ponds contain mercury to a depth of perhaps 24 m and leak mercury to the Holston River at an estimated rate of about 100 g d^{-1}. According to Carter (1977, p. 1017), ". . . it has become increasingly apparent that this problem can never be fully overcome; anything short of gargantuan engineering remedies, undertaken at costs that might run into the hundreds of millions, may bring nothing better than a modest, perhaps trifling, amelioration."

In summarizing mercury pollution, several points seem particularly worthy of keeping in mind:

1. There is no reason to believe that mercury fluxes to the environment have increased significantly on a global scale, and in fact mercury production has declined by about 40% since 1970. On a local scale, however, mercury pollution has been a serious problem, particularly in some areas affected by acetaldehyde and chlor-alkali plants.

2. For humans, the most important mechanism of mercury intake is the ingestion of mercury-containing fish. Tuna and swordfish contain naturally high concentrations of mercury, and consumption of the former is believed to be the single most important source of mercury ingestion for most Americans. The FDA standard of 1.0 ppm methyl mercury in fish is realistic only if one consumes less than 20 g of such fish per day. The average methyl mercury content of one's food should not exceed about 0.015 ppm.

3. There is a strong tendency for all forms of mercury to be converted to methyl mercury by microbes. Methyl mercury is the most toxic form of mercury, and damage done to the central nervous system by mercury poisoning is usually permanent.

4. Decontaminating an aquatic system polluted with mercury is likely to be difficult and costly, if possible at all. Mercury may continue to leach from contaminated sediments for hundreds or even thousands of years, and even in river systems natural flushing processes may take tens or hundreds of years to decontaminate the system.

5. Considering points 3–4, it is evident that prevention rather than cure is by far the most effective policy for dealing with mercury pollution.

Cadmium

Distribution, Production, and Uses

Cadmium is a metal chemically similar to zinc. Although it is widely distributed in the lithosphere, cadmium is usually found at quite low concentrations in crustal rocks and soils, typical concentrations being 100–300 ppb and 200–800 ppb, respectively (Nriagu and Sprague, 1987). Although Cd concentrations as high as 100 ppm are sometimes found in phosphatic rocks,[18] there are no ore deposits sufficiently rich in Cd to warrant extraction of the cadmium per se. Instead, cadmium is obtained invariably as a by-product of mining other metals, primarily the sulfide ores of zinc, and to a much lesser extent of lead and copper. Sphalerite and wurtzite, for example, both of which are forms of ZnS, may contain up to 5% Cd by weight, although median concentrations are only about 0.3% (Nriagu, 1980). Typically the ores are roasted to drive off gaseous sulfur products. Cadmium, being more volatile than either Zn, Pb, or Cu, is driven off in the roasting process. The cadmium is then collected in the flue dust, which may be recycled to further concentrate the cadmium. Flue dust cadmium levels are generally in the range of 20–40%. From the flue dust, cadmium and most other metal by-products are dissolved by acid oxidation. Final preparation of the Cd (>99.9% pure) may be effected by several different techniques,[19] the choice of method being determined by the nature of the other metals present and whether or not other metals are to be recovered (OECD, 1975).

World production of cadmium increased rather steadily from about 1000 tonnes in the early 1930s to over 15,000 tonnes by 1970, but has increased at an irregular rate of only about 1.3% per year since then (Figure 12.4). Since most cadmium is obtained as a by-product from zinc mines, the production of cadmium has little to do with the availability of or demand for cadmium. Instead, production is controlled largely by activities in the zinc industry. Since 1945 world cadmium production has averaged $30 \pm 2\%$ of world zinc production by weight and has accounted for about 5–10% of the income generated from zinc sales (Nriagu, 1980). Current production of Cd is about 21,000 tonnes y^{-1}. The major producing nations, in order of importance, are the Soviet Union, Japan, Belgium, Canada, and the United States (Llewellyn, 1991).

The U.S. Bureau of Mines does not keep a detailed inventory of Cd consumption in the United States, and the net consumption shown for the United States in Figure 12.4 is simply the difference between Cd imports, exports, and changes in stockpiles. Total consumption equals net consumption plus the consumption of recycled Cd, but historically recycling has accounted for less than 5% of Cd consumption (Nriagu, 1980). In contrast to world production, net consumption of Cd in the United States has been declining at a rate of about 2.5% per year for the last 20 years. Table 12.9 summarizes present use categories of Cd in the United States. The principal cause of the decline in U.S. consumption during the last 20 years has been a drop by about 60% in the use of Cd both for electroplating and in pigments. The use

---

[18] Due to the presence of fossilized fish teeth rich in cadmium.
[19] e.g., vacuum distillation, electrolysis.

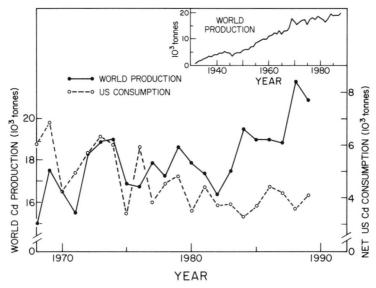

Figure 12.4 World production and net U.S. consumption of cadmium. (*Source*: Minerals Yearbook, U.S. Dept. of the Interior Bureau of Mines.)

of Cd in batteries, however, increased by more than a factor of 6 from 1968 to 1988.[20] The following comments briefly describe the nature of the principal uses of Cd:

Ni-Cd Batteries. Cadmium is used as the anode in nickel-cadmium batteries. These batteries are rechargeable, have a long lifetime of approximately 3000 cycles, low self-discharge rate, operate over a wide temperature range, and can deliver maximal currents with a low voltage drop. The chief disadvantages of Ni-Cd batteries are their low energy density and high cost. They are used as sealed cells in radios, alarm systems, emergency lighting, pacemakers, calculators, motor starters, walkie-talkies, portable appliances, and tools. Larger vented units are used in buses, diesel engines, aircraft, spacecraft, military applications, and standby power and lighting (Nriagu, 1980).

Electroplating. Almost all cadmium plating is done by electrodeposition, although some plating and coating is also done by vacuum deposition, dipping, or spraying. In electroplating, a thin layer of Cd is deposited by electrolysis on metal objects made of steel, iron, copper, brass, and other alloys to prevent corrosion. The Cd layer provides good solderability and conducts electricity well, is highly ductile,[21] and provides good corrosion resistance to tropical atmospheres, saltwater, and alkaline substances. Cadmium has the ability to protect steel, to which it is anodic, through sacrificial corrosion (Nriagu, 1980). Cadmium, however, does dissolve readily in acid solutions, including even weakly acidic solutions such as fruit juices. During the early 1940s hundreds of persons were poisoned from using Cd-plated utensils before the practice of plating utensils with cadmium was banned in the United States and Europe (McCaull, 1971).

---

[20] Data for 1968 came from OECD (1975). Data for 1988 from *Minerals Yearbook*, published by the Dept. of the Interior Bureau of Mines.
[21] So that plated parts can be stamped or shaped.

Table 12.9   Cadmium Use in the United States in 1989

Use	Tonnes	% of Total
Batteries	1434	35
Coating and plating	1229	30
Pigments	614	15
Plastics and synthetic products	410	10
Alloys and other uses	409	10
TOTAL	4096	

Source. Llewellyn (1991).

Pigments. Certain cadmium compounds are used as coloring agents in a variety of products, chiefly in plastics, but including coated fabrics, textiles, rubber, glass, paints, enamels, ceramic glazes, printing inks, and artists' colors. The colors, which range from yellow to red, are produced by mixtures of cadmium sulfide and cadmium selenide (reds), or of cadmium sulfide with zinc sulfide (yellow). Cadmium pigments have good hiding power and color intensity, do not bleed, are resistant to degradation by light, basic substances, and H_2S, and are heat stable up to about 600°C[22] (OECD, 1975).

Plastic Stabilizers. Mixtures of cadmium and barium combined with organic acid anions are used as heat stabilizers in plastics to retard degradation due to elevated temperatures. These stabilizers offer some protection against light-induced degradation as well. Compounds of lead and organotin account for the other major forms of plastic stabilizers. Because of the toxicity of Cd, the FDA has ruled that cadmium stabilizers cannot be incorporated into plastics used in food packaging (OECD, 1975).

Alloys. Alloys of cadmium have found use in a variety of applications. Metals such as bismuth, lead, and tin have been combined with cadmium to produce low-melting-point alloys that are used in fire-detection and firedoor release devices, as molds for casting plastics and as safety plugs in compressed-gas cylinders. Such alloys usually contain less than 20% Cd. Bearings made of cadmium ($\geq 98\%$) combined with nickel, copper, or silver have greater heat resistance and can run at higher speeds than tin or lead bearings, but use of such bearings has declined since World War II. Alloys consisting of cadmium (5–20%) combined with silver, copper, and zinc have been used as brazing compounds for soldering metals in cases where strong, leakproof, corrosion-resistant joints are required (OECD, 1975).

Miscellaneous. Cadmium is used in certain pesticides both in agriculture[23] and in nonagricultural applications.[24] Because Cd is effective in absorbing neutrons, it has been used to make control rods for nuclear reactors. Cadmium is also used in smoke detection devices, in solar cells, in photocells, and as a component of the phosphors in television tubes, X-ray screens, and luminescent dials. Small concentrations of Cd are found in virtually all fossil fuels. Concentrations in coal have been reported as high as 1–2 ppm, but typical levels in oil are more in the range 0.1–0.5 ppm.

---

[22] Hence their use in plastics, which are molded at high temperatures.
[23] For example, on broad beans, tomatoes, and wheat.
[24] For example, on golf courses.

Roughly 6 billion tonnes of fossil fuels are burned each year, and the concomitant release of Cd to the atmosphere amounts to about 65 tonnes (Nriagu, 1980).

Emissions to the Environment

Table 12.10 summarizes our present estimates of natural and anthropogenic fluxes of cadmium to the atmosphere and the ocean. There is little doubt that human activities have substantially increased emissions of cadmium to the atmosphere. The numbers in Table 12.10 indicate that the atmospheric flux has increased by almost an order of magnitude as a result of human activities. The chief cause of the increase has been the extraction and recycling of nonferrous metals, particularly zinc and copper. The incineration of waste, including sewage sludge, has also been an important contributor to the anthropogenic atmospheric emissions.

Table 12.10   Estimated Cd Emissions to the Environment

Flux	Tonnes per year
To the atmosphere	
Natural	
Volcanic eruptions	520
Vegetation	210
Windblown dusts	100
Forest fires	12
Seasalt sprays	1
Total	843
Anthropogenic	
Primary nonferrous metal production	4721
Waste incineration	1350
Secondary nonferrous metals	595
Wood combustion	200
Phosphate fertilizers	118
Iron and steel production	72
Fossil fuel combustion	65
Industrial applications	53
Rubber tire wear	10
Zinc mining	3
Total	7187
To the ocean	
Natural	
River runoff	5110
Atmospheric deposition	440
Total	5550
Anthropogenic	
River runoff	1700
Atmospheric deposition	2000
Total	3700

Source. Nriagu (1980) and Nriagu and Sprague (1987).

It is estimated that about half of the cadmium naturally emitted to the atmosphere is deposited in the oceans, but only 25–30% of the anthropogenic atmospheric emissions find their way directly into the sea. The explanation is that the industrial cadmium emissions are associated with larger particles than those involved in natural scavenging processes (Nriagu and Sprague, 1987). A major portion of the anthropogenic emissions is therefore deposited on the land. The data, however, suggest that cadmium inputs to the ocean have been increased by about 65% as a result of human activities. If the river fluxes can be taken as indicative of the pattern of freshwater inputs, then human activities have increased the input of cadmium to freshwater systems by 30–35%. One obvious fact about the data in Table 12.10 is that the net anthropogenic emissions to the ocean-atmosphere account for less than half of the present rate of Cd production. It is apparent, then, that a major fraction of the cadmium produced each year accumulates in the form of products and supplies or is discarded on land. With respect to land disposal, Nriagu (1980, p. 77) has commented, "The fate and ecological impacts of the cadmium so dissipated are essentially unknown." Cadmium-containing consumer products such as Ni-Cd batteries and various forms of plastic could remain stable in landfills for many years if they are not exposed to the atmosphere and do not come into contact with acidic water. Land disposal of Cd should not create environmental problems if reasonable precautions are taken, but experience has shown that many landfills are not operated in an environmentally safe manner (Chapter 16). The following discussion provides some additional information about the major fluxes listed in Table 12.10.

Natural Fluxes to the Atmosphere. The major input here is volcanic eruptions, which are also a significant source of natural mercury emissions into the atmosphere. The emissions result from the presence of the metals in volcanic rock and, of course, the volatility of the metals. At the temperatures of molten lava, metals such as cadmium are readily degassed into the atmosphere. Inputs from vegetation are the result of plant exudates, which Nriagu (1980) assumed to contain 2.75 ppm Cd. Here again, the volatility of Cd is a factor in the emissions. Windblown dust contains a small amount of Cd, and forest fires introduce Cd into the atmosphere because of the presence of cadmium in trees and other types of vegetation.

Anthropogenic Fluxes to the Atmosphere. Emissions resulting from the mining, extraction, and processing of zinc, copper, and lead account for over half the anthropogenic inputs of Cd to the atmosphere. Zinc operations are the principal culprit, despite the fact that cadmium is often obtained from the flue dust. However, some of the cadmium inevitably escapes from the exhaust system in the form of tiny aerosols, which may be dispersed over a wide area. Reprocessing of plated and galvanized metal also introduces significant amounts of Cd into the atmosphere because of the volatility of the Cd. For example, the melting point of iron, 1535°C, is much higher than the boiling point of Cd, 767°C. Obvious sources of Cd emissions associated with waste incineration include the burning of sewage sludge and wood painted with Cd-containing paint. As noted in Chapter 6, sewage sludge often contains elevated metal concentrations relative to natural soils, and simply applying sewage sludge to land can lead to Cd emissions due to volatilization. The U.S. FDA recommends that sewage sludge applied to the land not contain more than 29 ppm Cd (GESAMP, 1985). The high Cd content of some phosphate deposits, particularly those formed from fossilized fish teeth, accounts for the high Cd concentrations (up to 10 ppm) measured in phosphate fertilizers (OECD, 1975). The volatilization of Cd from fertilized agricultural lands can therefore introduce significant amounts of Cd to the atmosphere, and runoff from such lands can create water pollution problems as well. Horvarth et al. (1972), for example, found dissolved Cd concentrations to be 7 times higher in canal water adjacent to agricultural fields in South Florida than in canal water draining undeveloped land in the same region. A variety of metals, including Cd, had been applied to the agricultural fields in the form of pesticides as well as in fertilizer. The presence of Cd as an impurity in fossil fuels and tires accounts for atmospheric emissions of Cd due to fossil fuel combustion and tire wear. In the case of tires, the contamination results from the use of zinc oxide as an activator in the curing of rubber. Cadmium is invariably found as a contaminant in zinc compounds. As a result, Cd is present in tires at concentrations of 20–90 ppm (McCaull, 1971). Finally, the use of Cd by industry in the manufacture of finished products invariably results in some release to the environment. The sediments below one Ni-Cd battery plant located on a tributary of the Hudson River, for example, were found to contain 16.2% cadmium and 22.6% nickel (McCaull, 1971), and during the 1950s an investigation of water supplies on Long Island revealed that a groundwater recharge basin that received effluent from an aircraft manufacturing company contained Cd at 1.2 ppm (Lieber and Welsch, 1954). In the latter case, the company had been using Cd to anodize aluminum and other metals to retard corrosion. Seepage from the recharge basin was found to have contaminated local groundwater supplies to a depth of as much as 15 m for a distance of about 1 km from the recharge basin. Some test wells about 200 m from the recharge basin produced water containing 1–3 ppm Cd, 100–300 times the EPA limit for drinking water (EPA, 1987). Concern over cadmium pollution has led to serious efforts to minimize such pollution in recent years, but releases to the environment inevitably occur.

Toxicity

Human Health. With respect to toxicity, there are several notable similarities and dissimilarities between cadmium and mercury. Like mercury, cadmium is not required even in small amounts for the maintenance of life. Unlike mercury, only a single chemical species, the cadmium ion Cd^{2+}, is believed to exert a toxic effect. Transformation of Cd to more toxic compounds analogous to methyl mercury is not known to occur (OECD, 1975). The efficiencies of cadmium and inorganic mercury absorption by the intestines are comparable, about 5% and 2%, respectively. Low body iron stores or a calcium deficiency can increase intestinal absorption of Cd to 10–20% (GESAMP, 1985). Absorption of Cd vapor by the lungs is about 10–60% efficient, and about 0.1–0.2 µg of Cd is inhaled by smoking one cigarette (Friberg et al., 1986). For nonsmokers the consumption of Cd-contaminated food normally accounts for most of the total intake of Cd (GESAMP, 1985). An analysis of Cd levels in various foodstuffs, in air, and in water, for example, has indicated that persons in the United States ingest about 30 µg/day of cadmium from the food they eat, 2.6 µg/day from the water they drink, and 0.6 µg/day from the air they breathe (Table 8.12). Considerable variations exist, however, in the amount of Cd consumed with food. In Europe, for example, daily Cd intake averages only about 20 µg, but in Japan the average is 40–50 µg d^{-1} (GESAMP, 1985). Smokers and persons who are occupationally exposed to Cd may, of course, take in even more.[25]

Once ingested, cadmium is transported to all parts of the body by the bloodstream. Although almost all organs probably absorb some Cd, the highest concentrations are invariably found in the liver and kidneys. Roughly one-third and one-sixth of the body burden of Cd is stored in the kidneys and liver, respectively. After long-term, low-level exposure, most of the rest is found in the muscles (Friberg et al., 1986). Unlike mercury, Cd apparently does not effectively penetrate the placental barrier, so that poisoning of the fetus is of much less concern than is the case with methyl mercury (OECD, 1975). One of the major concerns with Cd poisoning is the fact that ingested Cd has a half-life of roughly 16–33 years in the human body. As a result, cadmium can be an extremely insidious poison, in the sense that ingestion of only small amounts over a period of many years may lead to the accumulation of toxic levels of Cd in the body. The body burden of Cd is only about 1 µg at birth, but steadily increases until roughly age 50, when it is typically 10–30 mg (GESAMP, 1985).

Once absorbed by the body, cadmium tends to be concentrated in the kidneys and liver by a low molecular weight protein called *thionein*. This protein contains large numbers of sulfhydryl groups, which attract Cd as well as other heavy metals such as mercury, zinc, and copper. The metal-protein complex is referred to as *metallothionein*, and is synthesized in the liver. The metallothionein is transported to the kidneys by the bloodstream, where it is filtered from the plasma by glomeruli and then reabsorbed in the tubuli along with other low molecular weight proteins (Friberg et al., 1986). A continuous catabolism of the cadmium metallothionein occurs after reabsorption, cadmium being split from the metallothionein and bound to newly formed metallothionein in the tubular cells (GESAMP, 1985). Binding heavy metals in metallothionein is believed to be a protective defense mechanism which prevents the metals from interacting with metabolically important proteins. In the case of Cd, kidney damage occurs when the amount of Cd in the kidneys overwhelms the ability of the kidneys to bind the cadmium in metallothionein. The unbound cadmium then damages the renal tubules, which lose their ability to reabsorb proteins. Symptoms of kidney damage include elevated urinary levels of protein (proteinuria) and cadmium. In more severe cases there may also be an increase of glucose in the urine (glucosuria) and of alkaline phosphatase in the blood (OECD, 1975). Extreme cases of cadmium intoxication are associated with osteomalacia (softening of the bones) and osteoporosis, apparently caused by disruption of the calcium-phosphorus balance in the renal tubules (OECD, 1975). The biochemical mechanism is believed to involve, at least in part, inhibition by Cd of the formation of a metabolite of vitamin D, referred to as 1,25-DHCC. This compound, which is formed in the kidney tubular cells, stimulates calcium absorption from the intestine and is necessary for normal bone mineralization (Friberg et al., 1986).

---

[25]McCaull (1971), for example, cites the example of construction workers who inhaled Cd vapor while cutting Cd-plated bolts with oxyacetylene torches in poorly ventilated places.

Inhalation of Cd vapor can lead to permanent or even fatal lung damage (OECD, 1975), and in a long-term study with rats has caused an increase in lung cancer. An excess of lung cancer has also been noted among previously heavily exposed cadmium workers, and it is possible that the Cd in cigarette smoke may contribute to the development of lung cancer in smokers (Friberg et al., 1986). At the present time, however, cadmium is not listed as a carcinogen by the EPA (EPA, 1987).

Cadmium has been linked to hypertension, largely as the result of experimental studies with animals (Yunice and Perry, 1961). There is no evidence, however, of an unusual incidence of hypertension or cardiovascular disease among workers exposed to cadmium (Friberg et al., 1986).

The derivation of the EPA water quality criteria for Cd with respect to human health has been discussed at some length in Chapter 8 and will be reviewed only briefly here. The criteria are based on kidney damage to persons who suffered long-term chronic exposure to Cd due to the consumption of Cd-contaminated food. Based on studies of such persons, the acceptable daily intake (ADI) of cadmium from food and water is 20–76 $\mu g \, d^{-1}$. The present intake of 30 $\mu g \, d^{-1}$ from food is due primarily to the consumption of vegetables and cereal crops, the concentrations in meat, fish, and fruits generally being quite low (GESAMP, 1985). If drinking water actually contained the EPA criterion of 10 ppb Cd, then consumption of 2 l of water per day would lead to the intake of 20 $\mu g \, d^{-1}$ of Cd and would increase overall intake to about 50 $\mu g \, d^{-1}$. Consumption of 6.5 g d^{-1} of fish and shellfish taken from such water would increase Cd intake by roughly 7 $\mu g \, d^{-1}$. Given an ADI of 20–76 $\mu g \, d^{-1}$, it seems doubtful whether intake should be allowed to increase from 30 to 57 $\mu g \, d^{-1}$. According to the EPA (1980c), only about 0.3% of U.S. water supplies contain more than 10 ppb Cd, and 37% contain less than 1.0 ppb.

The only extant guidelines with respect to Cd concentrations in food concern rice in Japan, the limits being 1.0 ppm in unpolished rice and 0.9 ppm in polished rice. However, daily ingestion of typically (in Japan) 335 μg of rice (Yamagata and Shigematsu, 1970) containing 1.0 ppm Cd would imply a daily intake of 335 μg of Cd, roughly 5–15 times the ADI. Realizing this fact, it is difficult to understand how the Japanese government could consider 0.9–1.0 ppm Cd in rice a tolerable concentration. In Japan the mean concentration of Cd in polished rice is only 0.07 ppm (Yamagata and Shigematsu, 1970). Consumption of 335 g d^{-1} of such rice would lead to a daily Cd intake of 23.5 μg, a figure which largely accounts for the difference in Cd intake between Japanese and Americans. The average Cd concentration of food consumed in the United States is about 23 ppb.

As is the case with mercury, some aquatic species are intoxicated by Cd at concentrations substantially lower than the present public water supply standard of 10 ppb. Derivation of the EPA freshwater criteria for the protection of aquatic organisms has been discussed in Chapter 8, and the results are summarized in Table 12.11. The criterion continuous concentration is based on the final chronic equation, the final plant value and the final residue value being less restrictive. The effect of water hardness on toxicity appears to be due to calcium, but not magnesium (EPA, 1985c). The calcium ion, Ca^{2+}, bears a charge identical to that of the cadmium ion, Cd^{2+}, and Ca^{2+} has virtually the same ionic radius, 0.99 angstroms, as does Cd^{2+}, 0.97 angstroms. Thus Ca^{2+} may compete effectively with Cd^{2+} for uptake by organisms.

Table 12.11   EPA Water Quality Criteria for Cadmium for the Protection of Aquatic Organisms in Freshwater and Saltwater

Water type	Criterion Continuous Concentration (ppb)	Criterion Maximum Concentration (ppb)
Marine	9.3	43
Freshwater		
50 mg/l $CaCO_3$	0.66	1.8
100 mg/l $CaCO_3$	1.1	3.9
200 mg/l $CaCO_3$	2.0	8.6

Source. EPA (1985c).

The marine final acute value was based on studies with 33 genera and was calculated according to the procedures in Chapter 8. The range of genus mean acute values was a factor of 2000 for the 33 genera studied. Several studies revealed a positive correlation between 96-hr TLm values and salinity (EPA, 1985c), a trend very likely reflecting the fact that Ca^{2+} is one of the principal ions in seawater. The criteria for marine waters are based almost entirely on studies conducted in full-strength seawater. The final chronic value was calculated by dividing the final acute value by an acute/chronic ratio of 9.1, the latter being derived from studies on two species of mysids whose mean acute values were close to the final acute value. This final chronic value was lower than both the final plant value and final residue value for marine waters, and therefore became the criterion continuous concentration for marine waters.

Itai-itai Disease—A Case Study

In 1955 two Japanese physicians reported the occurrence of a mysterious disease in the Jintsu River basin of Japan near the city of Toyama (Figure 12.5). The disease was characterized by severe pain in the back, joints, and lower abdomen, development of a waddling or ducklike gait, kidney lesions, proteinuria, glycosuria, and loss of calcium from the bones leading in some cases to multiple bone fractures (Yamagata and Shigematsu, 1970). In some bedridden patients the skeleton became so fragile that the slightest external pressure, for example from coughing, could cause bone fractures (Nogawa, 1980). Discussions with local doctors revealed that symptoms of the disease had been noted in the region as early as 1935. The disease was confined largely to postmenopausal women who had experienced multiple pregnancies in their earlier lives and who had lived in the region for all or most of their lives (Emmerson, 1970). For lack of a better name, the disease was called *Itai-itai* (loosely in Japanese "ouch-ouch") disease, because of the severe pain experienced by the more serious patients.

Initial efforts to identify the cause of Itai-itai disease after 1955 were unsuccessful. Some doctors suspected that the disease was caused by a local nutritional deficiency, as the victims were largely confined to a poor farming district. Others noted, however, that a mining and smelting operation had been operated since

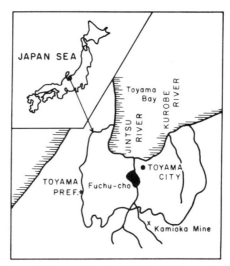

Figure 12.5   Toyama City and Jintsu River basin. Shaded area indicates region where disease occurred. [Redrawn from Yamagata and Shigematsu (1970).]

1924 by the Mitsui Mining Company at its Kamioka mining station upstream of the affected region on the Jintsu River (Figure 12.5). The mining operation produced zinc, lead, and cadmium, and until 1955 had routinely discharged its wastewater into the Jintsu River (Anonymous, 1971). It was also known that occasional accidents or flooding resulted in sludge's from the mine being washed into the river (Yamagata and Shigematsu, 1970). During World War II pollution control measures at the mine were relaxed, and the rice in downstream paddies was severely damaged by the resultant pollution (Anonymous, 1971). Cognizant of these facts, doctors began looking for a possible connection between Itai-itai disease and the mine wastewater.

Initially zinc was suspected of being the cause of the disease, but zinc-feeding experiments with animals failed to produce Itai-itai symptoms (Yamagata and Shigematsu, 1970). Analyses of metal concentrations in the bones of Itai-itai victims and in rice from the affected area revealed that Cd concentrations were about 10–100 times higher than in control samples. Zinc concentrations in rice from the affected area were little different from controls, but like cadmium, Zn levels in the bones of dead Itai-itai victims were elevated by roughly two orders of magnitude relative to controls. Subsequently it was discovered that feeding rats cadmium and zinc or a combination of cadmium, zinc, copper, and lead could produce bone degeneration similar to Itai-itai disease (Anonymous, 1971). Thus the cause of the disease was identified as being either cadmium or a combination of cadmium and other metals, particularly zinc. Although the implication of cadmium had been established by 1961, the Japanese government issued no official statement regarding the cause of Itai-itai disease until 1968, when an official press release stated: "The 'Itai-itai' disease is caused by chronic cadmium poisoning, on condition of the existence of such inducing factors as pregnancy, lactation, unbalanced internal secretion, aging, deficiency of calcium, etc. It sets up kidney troubles in the first stage and osteomalacia [softening of the bones] after that" (Yamagata and Shigematsu, 1970, p. 3).

The role of cadmium versus other factors in causing Itai-itai disease remains somewhat controversial. Cadmium itself had not been associated with a disruption of calcium metabolism prior to the occurrence of Itai-itai disease. However, two French reports and two British studies of persons occupationally exposed to cadmium have revealed bone changes similar to those seen in Itai-itai victims (Chang, Reuhl, and Wade, 1980). Neither zinc nor lead, the other metals released in large amounts by the Kamioka mine, is associated with a disruption of calcium metabolism. Studies of metal concentrations in the urine of inhabitants of the Toyama Prefecture during the early 1960s revealed higher concentrations of cadmium in the urine of persons from the endemic area, but no geographical differences in lead and zinc concentrations (Kjellstrom, 1986a). Studies of autopsied Itai-itai victims revealed that lead concentrations in the victims' bones were little different from controls, but Cd and Zn concentrations were orders-of-magnitude higher than in the bones of control corpses (Yamagata and Shigematsu, 1970).

There is rather convincing evidence that nutritional deficiency played a role in Itai-itai disease. Experimental studies with animals, for example, have shown that cadmium in the food does not lead to bone softening under normal dietary conditions unless the dose of cadmium is very high. Cadmium exposure, however, when coupled with a diet low in calcium and/or vitamin D, accelerates and potentiates osteoporotic and osteomalacic changes with a disturbance of bone salt metabolism (Chang, Reuhl, and Wade, 1980).

Perhaps the most convincing link with nutrition has been the results of treatment of Itai-itai disease patients. Patients with mild symptoms[26] were treated as outpatients, and were given vitamin D tablets and arrangements made for more milk, eggs, meat, and vegetables in their diet, and more sunlight. Pain decreased after about 2 months of treatment, and the patients were able to return to everyday life after about 4 months, by which time the ducklike gait had disappeared. Some impairment of mobility, however, remained for a long time (Kjellstrom, 1986a). Persons with moderate symptoms[27] and severe symptoms[28] were treated as inpatients. The inpatients received regular injections of vitamin D, ate regular hospital food, and were taken out in the sunlight as much as possible. After leaving the hospital, these persons received daily vitamin D tablets and were given vitamin D injections once or twice a month when they returned for checkups. Treatment was continued for 1–3 years. Pain began to decrease after about 3 weeks of treatment. The patients with medium symptoms were relieved of pain within about 2 months, could stand erect after 5 months, could walk inside the hospital after 8 months, and were able to return to everyday life after 13–14 months (Kjellstrom, 1986a). The persons with severe symptoms were able to sit up after 5 months, could stand after 7 months, were able to walk with a cane after 15 months, and could walk a few meters unaided after 20 months. In some cases pseudofractures of the bones persisted as long as 1–3 years (Kjellstrom, 1986a). Treatment with vitamin D and improved diet have obviously helped to alleviate the symptoms of Itai-itai disease, but recovery has been much slower than would be expected if the bone softening were due purely to a nutritional deficiency, and in many cases recovery has been incomplete.

[26]Ducklike gait, some limitation in movement, otherwise few bone symptoms.
[27]Able to crawl but barely walk.
[28]Cannot move at all.

Studies of cadmium concentrations in the Jintsu River system provide an interesting lesson in water quality monitoring. Most of the drainage water from the Kamioka mine was found to be weakly basic (Yamagata and Shigematsu, 1970), and in such water Cd is highly insoluble. As a result virtually all the Cd discharged from the mine was transported in particulate form. Filtered water samples from the Jintsu River were found to contain less than the U.S. drinking water standard of 10 ppb cadmium. Analyses of suspended materials in the river above and below the mine, however, revealed roughly an 80-fold increase in Cd concentration (Yamagata and Shigematsu, 1970). Evidently the cadmium was transported downstream in particulate form and ultimately settled out in the quiescent waters of rice paddies in the Fuchu-machi plain just south of Toyama city. Sediments from these rice fields contained about 3 ppm cadmium, versus less than 1 ppm in the sediments of control fields, and rice from the contaminated paddies averaged about 1 ppm Cd, versus a mean of 0.07 ppm Cd for Japanese rice in general (Yamagata and Shigematsu, 1970).

In 1955 the Kamioka mine built a dam to retain its wastewater in a lagoon. Since that time the incidence of Itai-itai disease has dropped sharply, and evidence of damage to crops has declined similarly (Anonymous, 1971). From the standpoint of public health, it is unfortunately not clear to what levels of cadmium Itai-itai victims were exposed, since the extensive analyses of water, sediment, and crops in the Fuchu-machi area were made after 1955. Based on these analyses, it has been estimated that inhabitants of the area were probably ingesting at least 600 μg of Cd per day prior to 1955 (Yamagata and Shigematsu, 1970). Examination of waste sludge piles near the mine, however, has indicated that pollution was probably more extreme in years past (Emmerson, 1970), and there is no doubt that pollutant discharges increased during World War II (Anonymous, 1971). Thus exposure of the Fuchu-machi population to cadmium may have been substantially greater than post-1955 analyses would indicate.

Between 1939 and 1954 approximately 200 persons were afflicted with Itai-itai disease, and of these nearly 100 died (Anonymous, 1971). Remarkably the Toyama Prefecture Department of Health did not begin mass screening of Jintsu River basin inhabitants for Itai-itai disease until 1967, and since that time an additional 132 victims have been diagnosed (Kjellstrom, 1986b). Only three cases, however, have been diagnosed since 1978, and of the 217 Itai-itai patients registered by the Toyama Prefecture in 1968, only 27 were less than 60 years old. Thus almost all Itai-itai victims seem to have developed the disease prior to roughly 1955, when the Kamioka mine initiated pollution control measures. It is fortunate and of considerable interest that similar outbreaks have not occurred in other parts of the world, since the number of zinc and cadmium mines is large. Some examples of what appeared to be the same disease have been reported near a zinc, lead, and cadmium mine on an island between Korea and Japan (Anonymous, 1971), and in 1969 the Japanese government took steps to halt cadmium pollution of Lake Suwa due to waste discharges from electroplating factories (Yamagata and Shigematsu, 1970). No Itai-itai cases, however, were associated with the Lake Suwa pollution, nor have any cases been reported in other areas of Japan where per capita ingestion of Cd is reported to be 200–240 μg d^{-1} (Yamagata and Shigematsu, 1970).

Correctives and Prospects for the Future

Although the occurrence of Itai-itai disease in the Jintsu River valley remains the only large-scale example of serious damage to human health from Cd pollution, the extremely long half-life of Cd in the human body and the fact that present levels of intake are comparable to ADIs has stimulated efforts to reduce cadmium emissions to the environment. Naively it might be assumed that recycling cadmium or finding substitutes for cadmium might do much to reduce the present rate of emissions. The fallacy in this logic is the fact that the mining of Cd-containing ore has almost nothing to do with the demand for cadmium, but instead is controlled largely by the demand for zinc. Substitutes for cadmium exist for most of its major uses, but ironically the metal mentioned most frequently as a possible substitute is zinc, and substitutions of other metals for cadmium in applications such as electroplating, pigments, and alloys yield inferior products (OECD, 1975). Plastic stabilizers made from tin are the principal alternatives to Cd stabilizers, and certain of these tin stabilizers have been approved by the FDA for use in food packaging. In fact organotin compounds are the most efficient stabilizers known for PVC, but they are more costly than Cd stabilizers (OECD, 1975). In the case of Ni-Cd batteries, recycling of Cd is a realistic option, and the European Economic Community took steps recently to initiate a systematic program of secondary Cd recovery from used batteries containing more than 0.025% Cd (Llewellyn, 1990). Recycling of Cd from batteries currently takes place in Sweden, France, Japan, and the Republic of Korea and is expected to become an integral part of Ni-Cd battery production in Europe and North America in the near future (Llewellyn, 1991). Some improvement in the cadmium pollution picture can therefore be expected from secondary recovery and substitutions in certain applications, but the fact remains that cadmium ores will continue to be mined as long as there is a demand for zinc and to a lesser extent lead and copper. As noted by Nriagu (1980, p. 57), "If the cadmium were not separated from the host zinc and lead ores, it would be dispersed in the environment on a much wider scale as flue particles and other industrial waste products." A significant improvement in the global Cd pollution picture therefore requires a serious effort to reduce emissions from some of the major anthropogenic sources listed in Table 12.10.

The technology for removing cadmium from industrial wastewater or from flue dust is well established. In wastewater, dissolved Cd can be precipitated with sodium sulfide, cemented by the addition of zinc, or separated out by ion exchange. If the Cd is incorporated into particulates, it can be dissolved by the addition of acid and then separated by one of the above techniques, or the solids can be settled out and the cadmium removed with the sludge. Cadmium released to the atmosphere by way of smoke stacks is primarily in the form of very fine particulate materials, which can be separated from the stack gases by wet scrubbers, fabric filters, or electrostatic precipitators (OECD, 1975).

Removal of Cd by any of the above techniques may lead to a solids disposal problem if the Cd is not to be recovered and recycled. Discharging Cd or Cd-containing compounds at sea is prohibited (OECD, 1975) and would certainly be unwise in any aquatic system. Land disposal, however, is acceptable as long as the soil is neutral or basic so that the cadmium is not mobilized by percolating groundwater. Efforts to reduce industrial emissions of Cd have been underway for some time and appear to have produced at least some positive effects. Concentrations of Cd deposited in snow and ice on Greenland, for example, declined by roughly a factor of 2.5 between 1970 and 1990 (Boutron et al., 1991). The realization that living organisms have no need for Cd, that the residence time of Cd in the human body is 16–33 years, that the accumulation of Cd in the kidneys and other organs can cause serious and sometimes permanent damage, and that present daily intake rates are comparable to ADIs, indicates that further reductions in anthropogenic Cd emissions would be desirable.

Lead

Production and Use

Lead (Pb) is widely distributed in the rocks and soils of the earth's crust, although the mean concentration is only 12–20 ppm (Settle and Patterson, 1980). Lead is mined primarily from deposits of the mineral galena, or lead sulfide (PbS). Since metallic lead can be separated from PbS by heating to low temperatures easily achieved by burning wood or charcoal,[29] it was not difficult for early civilization to extract lead (Figure 12.6). From written and archeological evidence, it is known that Pb was used by the Egyptians as long ago as 1500 BC. The Romans used lead extensively to line their aqueducts and water mains, and both the Greeks and Romans used Pb to line cooking vessels, since bronze pots tend to give food a bitter taste (Waldron and Stöfen, 1974). At least one author (Gilfillan, 1965) has suggested that endemic lead poisoning caused by the consumption of contaminated food and drink contributed significantly to the fall of the Roman empire. The use of lead or lead salts to sweeten wine was common in Europe after the ninth century, and associated outbreaks of lead poisoning were not infrequent. In colonial America lead poisoning was also a problem, mainly due to the use of lead condensing tubes for distilling rum and to the use of earthenware with a high Pb content (Waldron and Stöfen, 1974).

For obvious reasons the principal uses of lead today are of an industrial rather than culinary nature. Table 12.12 summarizes the major use categories in the United States as of 1989. The major use, accounting for almost 78% of total lead consumption in the United States, was the production of batteries. Other uses of significance from the standpoint of environmental pollution and human health were the use of lead in ammunition, paint pigments, and gasoline additives. The following comments provide additional information about the major use categories in Table 12.12.

1. The lead-acid battery is used extensively as a power source for starting motors in automobiles, power boats, lawn mowers, and a variety of other products. During 1989 over 80 million such batteries were sold in the United States. The design of the battery utilizes metallic lead as the negative electrode and lead dioxide (PbO_2) as the positive electrode of a voltaic cell. When the two electrodes are connected by an external electrical circuit, electrons flow from the negative to the positive electrode through the external circuit. As a result, the

---

[29]The overall reaction is $PbS + O_2 \rightarrow SO_2 + Pb$.

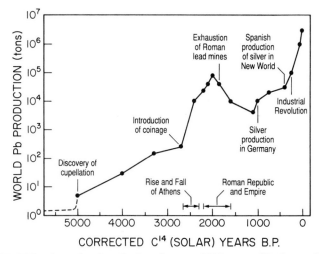

Figure 12.6 World lead production during the past 5500 years. [Redrawn from Settle and Patterson (1980).]

metallic lead (Pb) is oxidized to Pb^{2+}, while the lead in the PbO_2 is reduced, also to Pb^{2+}. The cell can be recharged many times by applying a voltage somewhat over 2 V across the terminals so as to reverse these reactions and convert Pb^{2+} back into Pb and PbO_2. The density of lead (11.3 g/cc) and hence the weight of lead-acid cells is an obvious drawback to their use, but for many purposes no better alternative has been found. In recent years use of lead in batteries has been increasing at a little over 1% per year in the United States.

Table 12.12   Use of Lead in the United States During 1989

Category	Tonnes
Metal products	
Batteries	1,012,155
Ammunition	62,940 ·
Cable covering: power and communication	22,605
Sheet lead	20,987
Solder	17,009
Casting metals	16,175
Pipes, traps, and other extruded products	9818
Brass and bronze: billets and ingots	9610
Type metal and other metal products	4564
Bearing metals	2586
Terne metal	2286
Calking lead: building construction	1831
Other oxides (paints, pigments, glass and ceramic products)	57,984
Miscellaneous (including gasoline additives)	42,684
Total	1,283, 234

Source. Woodbury (1991).

2. The use of lead in gasoline dates back to 1923, when tetraethyl lead was first introduced as an antiknock additive. This use was suspended during part of 1925 and 1926 pending the establishment of safety standards for its manufacture and handling, but after the spring of 1926 lead additives were commonly used in most gasoline sold in the United States and throughout the world. Beginning in 1960 organic alkyl lead compounds were blended into gasoline along with tetraethyl lead to further improve antiknock characteristics, and the additives were subsequently referred to as lead alkyls rather than just tetraethyl lead (Waldron and Stöfen, 1974). As of 1973, the lead content of gasoline sold in the United States averaged about 0.58 g/l.

   Because of concern over pollution of the environment with lead, in 1974 the EPA requested a phasedown in the use of lead alkyls in gasoline to 0.45 g/l on January 1, 1975, and to 0.13 g/l as of January 1, 1979. In December, 1974, however, the U.S. Circuit Court of Appeals in Washington ruled against the EPA, stating that evidence did not support the EPA contention that automobile emissions contributed significantly to blood lead levels, and that the EPA regulation was arbitrary and capricious.

   Since that time the legal status of gasoline lead additives has changed significantly. This change has resulted partly from the requirement for exhaust emission control devices on automobiles. The devices function poorly if at all when lead is present in the exhaust fumes, since the lead poisons the catalyst, which is designed to help oxidize unburned hydrocarbons. Furthermore, there is increasing awareness of the danger of environmental lead pollution due to the use of leaded gasoline. Since July 1, 1977 all new cars sold in the United States have been required to run on unleaded gas.[30] Furthermore, the gas filler on these cars must be of smaller diameter than the gas filler on older cars, so that it is impossible to insert a nozzle from a leaded gas pump into the gas tank filler on a new car. Finally, the government required that the lead content of leaded gas be reduced to 0.13 g/l by July 1, 1985 and to 0.10 g/l by January 1, 1986. The result of these regulations was a 10-fold decline in the use of lead in gasoline in the United States between the early 1970s and 1986 (Figure 12.7). Such use has presumably continued to decline since then, but the exact figures are proprietary (Woodbury, 1991). European countries were slow to follow suit, and in 1985 lead-free gasoline was still unavailable in Europe. At that time, however, the European Economic Community agreed to make lead-free gasoline available by 1989 and to reduce the lead content of leaded gas to 0.15 g/l (Dickson, 1985).

3. Lead has been used for many years in paints as a pigment, and in oil paints lead naphthenate is used as a drying agent. In 1918 40% of all painters were estimated to have symptoms of lead poisoning because of their contact with such paint (Craig and Berlin, 1971). The most serious problem associated with the use of Pb in paints, however, has been the poisoning of small children who eat paint chips or teethe on window sills or other surfaces that have been painted with lead-base paints. Dust from deteriorating paint may also contaminate windowsills, carpets, furniture, and toys and be ingested by children who touch the dust and subsequently put their fingers in their mouths.

---

[30]Defined as gasoline containing less than 0.013 g/l lead.

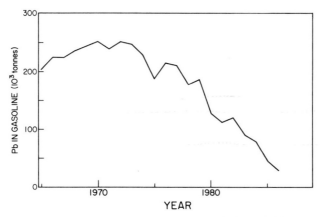

Figure 12.7   Use of lead in gasoline in the United States. (*Source*: Minerals Yearbook, U.S. Dept. of the Interior Bureau of Mines.)

Between 1954 and 1967, a total of 2038 children were treated for lead poisoning in New York City alone, and of these 128 died (Craig and Berlin, 1971). On January 1, 1973 the FDA banned the use of paint containing over 0.5% lead on residential surfaces accessible to children (Waldron and Stöfen, 1974), and, effective February 27, 1978, paint containing over 0.06% lead was banned as hazardous for use in residences, schools, hospitals, parks, playgrounds, and public buildings under the aegis of the Consumer Product Safety Act (Federal Register *42*: 44199, 1977). The only exceptions specifically written into the 1977 legislation were paints and coatings for motor vehicles and boats. The problem of poisoning from lead-based paints remains a serious public health problem, however, because many older buildings contain lead-based paints. A 1988 Public Health Service report revealed that 52% of American homes had layers of lead-based paint on their walls and woodwork (Blackman, 1991).

4. The use of lead in ammunition has created some serious environmental problems because of the tendency of scavengers and waterfowl to inadvertently ingest the lead. The decline in the population of California giant condors, for example, has been due in part to lead poisoning caused by feeding on dead animals shot by hunters (Crawford, 1985a,b). Similar poisoning of waterfowl such as ducks, geese, and coots is caused by the fact that the birds mistake lead shot in the sediments of marshes for seeds or grit. The sediments of marshes frequented by hunters may contain as many as 6–7 lead shots m^{-2} (EPA, 1972). Experiments with various waterfowl have shown that 4–5 number four shot are lethal to Canada Geese, and that 6 number six shot are lethal to Mallard, Pintail, and Redhead ducks (EPA, 1972). During the early 1970s, the U.S. Fish and Wildlife Service (FWS) estimated that about 2.4 million waterfowl out of a North American population of about 100 million died each year due to the ingestion of lead shot (Carter, 1977). As a result the FWS initiated a program to phase out the use of lead shot in certain parts of the country in 1976. The FWS action was opposed by the National Rifle Association (NRA) on the grounds that nontoxic steel shot was not as effective as lead shot, but the NRA's protest was rejected in court (Carter, 1977). The result of the FWS action is apparent in

the trend of lead use in ammunition in Figure 12.8. Federal and state action under the aegis of the Migratory Bird Treaty Act of 1918 and the Endangered Species Act of 1973 resulted ultimately in restrictions on the use of lead shot throughout much of the United States, and it is now estimated that about 92% of the total waterfowl harvest occurs in so-called nontoxic shot zones (Federal Register 55: 33626-33633, 1990).

As might be expected from this discussion, U.S. consumption of Pb declined during the 1970s (Figure 12.9). The decline was due primarily to the reduction in the use of lead alkyls in gasoline, but also reflected restrictions on the use of lead shot and lead-based paint. During the 1980s, however, U.S. consumption began to creep upward, as the demand for lead-acid batteries continued to increase.

World production of lead has been remarkably constant since 1970 and has averaged 3.41 ± 0.06 million tonnes y^{-1} (Figure 12.9). The constancy of production has not reflected a static situation, however, since the use patterns have changed significantly. Mine production, however, is affected not only by the demand for lead, but also by the extent of lead recycling. In the United States secondary lead recovery as a percentage of consumption increased from $44 \pm 2\%$ between 1965 and 1977 to $56 \pm 5\%$ from 1978 to 1989. This change obviously reduced the demand for mined lead. The net result of all the changes in demand and recycling has been a constant rate of world lead production for the past 20 years. While one should perhaps take consolation in the fact that world lead production has not increased recently, there is no question that the mining and use of such huge quantities of Pb has resulted in environmental pollution on a global scale. We now turn our attention to this issue.

Emissions

Several lines of evidence indicate that significant contamination of the environment with Pb has occurred on a global scale. Analyses of marine sediments, for example, have indicated that millions of years ago the flux of Pb to the oceans was about 13,000 tonnes y^{-1}, whereas river runoff alone contributes about 240,000 tonnes y^{-1} of Pb to the oceans at the present time (Waldron and Stöfen, 1974).

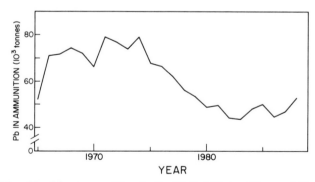

Figure 12.8   Use of lead in ammunition in the United States. (*Source*: Minerals Yearbook, U.S. Dept. of the Interior Bureau of Mines.)

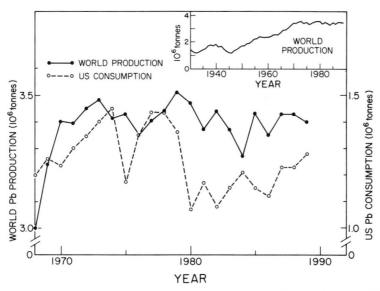

Figure 12.9   World production and U.S. consumption of lead. (*Source*: Minerals Yearbook, U.S. Dept. of the Interior Bureau of Mines.)

Analyses of Pb concentrations in seawater (Figure 12.10) have shown that surface waters are enriched with Pb relative to deep water, the cause of the enrichment presumably being atmospheric deposition. The lower concentrations in deep water reflect the fact that the residence time of bottom waters is approximately 500–1000 years. Had lead fluxes to the ocean been constant for thousands of years, we would expect to see uniform lead concentrations throughout the water column. However, surface water Pb concentrations in the North Pacific and North Atlantic are conservatively estimated to be 8–20 times greater and those in the South Pacific 2 times greater than natural concentrations due to industrial emissions of Pb to the atmosphere (Flegal and Patterson, 1983).

Lead concentrations in Greenland ice sheets sampled by Murozumi, Chow, and Patterson (1969) and Boutron et al. (1991) indicate that the Pb concentration of precipitation over the area increased by several orders of magnitude between roughly 1000 B.C. and 1950 (Figure 12.11). Much of the increase occurred after 1750 and accelerated greatly after about 1940. The increase between 1750 and 1940 presumably reflects the effect of the industrial revolution and associated emissions of lead to the atmosphere largely from lead smelters, whereas the more rapid increase after 1940 reflects the introduction of lead alkyls from vehicle exhaust emissions. Before the phaseout of leaded gasoline began, such emissions amounted to roughly 0.28 million tonnes y^{-1}, or a little over 8% of world Pb production (Settle and Patterson, 1980). The roughly 7.5-fold decline in the lead concentration of precipitation over Greenland from 1970 to 1990 presumably reflects the restrictions imposed on lead in gasoline (Boutron et al., 1991).

Table 12.13 provides a revealing summary of estimated global emissions of Pb to the atmosphere from natural and anthropogenic sources. The major natural source of emissions is windblown and volcanic dust, with a Pb concentration on the order of 10 ppm. By far the most important anthropogenic source has been the lead alkyls

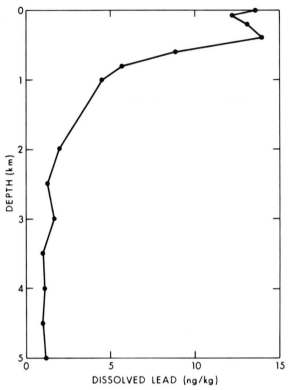

Figure 12.10 Dissolved Pb concentrations in the central N.E. Pacific (32°41'N, 145°W) in September, 1977. [Redrawn from Flegal and Patterson (1983).]

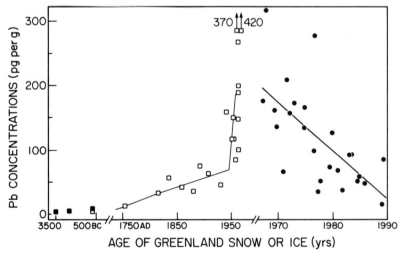

Figure 12.11 Lead concentrations in Greenland ice sheet. [Redrawn with permission from Boutron et al. (1991).]

in gasoline, and the estimate in Table 12.13 is based on the assumption that 70% of the Pb in gasoline is emitted to the atmosphere. Since world oil production has been rather constant since 1970 (Flavin, 1985) and since the use of lead alkyls in the United States has declined by at least 0.22 million tonnes in the last 20 years, it seems fair to say that global emissions of Pb from lead alkyls have declined by at least $(0.7)(0.22) = 0.15$ million tonnes during that time. It is noteworthy, however, that even if the lead content of all gasoline were reduced from the high of about 0.6 g/l that existed during the early 1970s to the "unleaded" limit of 0.013 g/l, emissions of lead to the atmosphere from the burning of gasoline would still amount to roughly 6000 tonnes per year, or several times the natural background rate. Under these conditions the principal anthropogenic emissions would come from ore smelting and to a lesser extent from coal burning. Reducing the lead content of gasoline has obviously done much to reduce anthropogenic emissions of Pb, but ore smelting and coal burning alone emit about 50 times as much lead into the atmosphere as do natural sources. To what extent have these emissions affected human exposure to Pb?

Human Exposure. Table 12.14 summarizes estimates of human intake of lead during prehistoric and modern times. In both cases the most important source of exposure was the consumption of food, but present rates of absorption are believed to be at least 150 times greater than during prehistoric times. A significant cause of contamination is the fallout of Pb from the atmosphere, and in modern times the chief source of Pb in the atmosphere has been the burning of leaded gasoline, ore smelting, and coal burning. Since the residence time of Pb in the atmosphere is on the order of a few days to as much as a month (Waldron and Stöfen, 1974), the impact of localized emission sources can obviously be spread by winds over a wide area,

Table 12.13   Global Lead Emissions to the Atmosphere from Natural and Anthropogenic Sources

Source	Emission (tonnes y^{-1})
Natural	
Windblown and volcanic dust	2000
Sea spray	<1000
Forest foliage	< 100
Volcanic sulfur	1
Total	<3100
Anthropogenic	
Lead alkyls	280,000
Iron smelting	47,000
Lead smelting	24,000
Zinc and copper smelting	42,000
Coal burning	15,000
Total	408,000

Source. Settle and Patterson (1980).

Table 12.14   Estimated Daily Amounts of Pb Absorbed into the Blood of Adult Humans in Prehistoric and Modern Times

Source	Intake	% Pb absorbed	Prehistoric (ng)	Contemporary Urban (ng)
Air	20 m^3	40%	0.3	6400
Water	2 kg	10%	<3	2250
Food	1.3 kg	7%	<182	18,200
Total			<185	26,850

Source. Modified from calculations in Settle and Patterson (1980).

including farmlands distant from urban and industrial centers. Significant additional contamination of food with lead may occur during processing, particularly if the food is dried and ground. Settle and Patterson (1980), for example, note that the Pb content of albacore tuna meat can increase by more than a factor of 50 during commercial drying and pulverizing. Other sources of lead contamination of foodstuffs include the use of lead arsenate as an insecticide and leaded ceramics and glazes in kitchenware. In areas such as Europe and the United States a significant additional source of lead contamination in the diet has been the use of lead-soldered cans for marketing food (Shea, 1973). Canned foods account for 10–15% of the food consumed by Americans, and in 1979 the FDA estimated that about 20% of the lead in the average daily diet of persons in the United States more than 1 year old came from canned food (FDA, 1979). About two-thirds of this Pb came from the solder. The implication was that lead solder accounted for about 13% of the lead ingested by Americans.

Lead is found in drinking water due both to its presence in raw water supplies and to the corrosion of plumbing materials in water distribution systems. The latter is usually the most important of the two sources. A recent survey by the EPA (1991), for example, revealed that less than 1% of public water systems in the United States had water entering the distribution system with Pb levels greater than 5 µg/l, the present standard for natural waters (Table 12.1). These systems serve less than 3% of Americans who receive their drinking water from public water systems.

The principal sources of lead in the distribution system are lead goosenecks or pigtails, lead service lines and interior household pipes, lead solders and fluxes used to connect copper pipes, and alloys containing lead, including some faucets made of brass or bronze. Lead solder and fluxes containing as much as 50% lead, for example, were commonly used to connect copper pipes throughout the United States prior to the 1986 amendments to the Safe Drinking Water Act (SDWA), which limited the lead content of solder and fluxes to 0.2%. EPA (1991) estimates indicate that there are about 10 million lead service lines/connections in the United States and that about 20% of all public water systems have some lead service lines/connections within their distribution.[31] Even if the distribution pipes contain little or no lead, brass and bronze in faucets and fixtures commonly contain lead and may be a major source of lead in drinking water that stands in faucets or fixtures (EPA, 1991). All water is corrosive to metal plumbing materials to some extent, and

[31]The 1986 amendments to the SDWA limited the lead content of new pipes, fittings, and patches in municipal water systems to less than 8%.

based on the foregoing it is not surprising to learn that lead concentrations of tap-water tend to exceed the lead concentrations of water entering distribution systems. For example, a random survey of 782 samples of tapwater in the United States, including 58 cities in 47 states, revealed average lead levels of 13 µg/l, with 10% of the values exceeding 33 µg/l (EPA, 1991). Lead is dissolved most readily by acidic water, and in communities where tapwater is soft and somewhat acidic, appreciable concentrations of Pb may leach into the water, particularly if the water is left standing overnight. Tapwater samples from homes served by lead water pipes in Glasgow, Scotland, for example, where the water is extremely soft, revealed mean Pb concentrations of 350 µg/l (Waldron and Stöfen, 1974), over 25 times the mean lead concentration of tapwater in the United States.

That the air people breathe is contaminated with lead should come as no surprise by this time. A person's exposure to lead through inhalation, however, depends very much on where he/she lives and whether or not he/she smokes cigarettes. The lead concentration of rural air, for example, is one to two orders of magnitude lower than the lead concentration of typical urban air (Settle and Patterson, 1980), and breathing rural air is therefore a very minor if not negligible cause of lead exposure compared to other sources at the present time. Cigarettes contain significant amounts of lead due to the use of lead arsenate as an insecticide on tobacco, and smoking a single cigarette results in the absorption of about 50–250 ng of lead (Moore, 1986). Heavy smokers may therefore absorb considerable amounts of lead due to their inhalation of cigarette smoke.

Absorption of inhaled Pb is about 40% efficient. Adults absorb about 10% of the Pb from the water they drink and about 7% from their food (Settle and Patterson, 1980). For children, gastrointestinal absorption of lead is more efficient, typically about 50% (EPA, 1980d). Once absorbed into the bloodstream, lead is transported to all parts of the body primarily by red blood cells, although its incorporation into tissues apparently occurs through exchange with the blood plasma (Waldron and Stöfen, 1974). Lead begins to appear in the liver and kidneys within a few hours of absorption, but ultimately 72–95% of the inorganic lead in the body is deposited in the bones, where it replaces calcium (EPA, 1980d). Once incorporated into the bone structure, lead is released back to the bloodstream at a steady but very slow rate of about 0.08–0.1%/day (Waldron and Stöfen, 1974). In contrast to inorganic lead, organic lead (e.g., lead alkyls) shows no special affinity for the bones, but instead tends to concentrate in lipid tissues, including those of the central nervous system. The highest concentrations of organic lead are often found in the brain and liver (Waldron and Stöfen, 1974).

Excretion of lead is accomplished primarily in the feces, although some lead is also excreted in urine and even in sweat. Absorbed lead is transferred to the intestines primarily in the bile fluid, although a small amount of release also seems to occur directly through the intestinal wall (Waldron and Stöfen, 1974). Because of the very slow release rate of lead from the bones, and because of the continual exposure of humans to lead, the excretion rate of lead is generally exceeded by the rate of lead absorption throughout the life of many persons. Hence the total body burden of lead tends to increase steadily with age (Figure 12.12). It is apparent from Figure 12.12 that the body burden of lead during the first few years of life is miniscule compared to the body burden in later years, and since humans are not known to require lead in even very small amounts, the accumulation apparent in Figure 12.12 can hardly be considered beneficial.

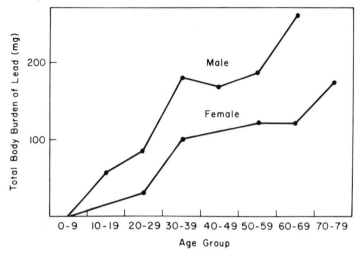

Figure 12.12   Total body burden of lead in humans as a function of age. [Data taken from Waldron and Stöfen (1974).]

Toxicology

Lead is a general metabolic poison. The pathological effects are varied, but in general reflect the tendency of lead to interact with proteins and hence to damage tissue and interfere with the proper functioning of enzymes (Waldron and Stöfen, 1974). Lead is known to inhibit active transport mechanisms involving ATP, to depress the activity of the enzyme cholinesterase, to suppress cellular oxidation-reduction reactions, and to inhibit protein synthesis (Waldron and Stöfen, 1974). Lead poisoning is also associated with the following problems.

Anemia. Lead is known to disrupt several enzymes involved in the production of heme, a constituent of hemoglobin and various other respiratory pigments, and has been shown to interfere with the uptake of iron by red blood cells. Anemia associated with lead poisoning is undoubtedly caused in part by these effects. There is also evidence, however, that lead causes a shortening of the life of red blood cells, probably due to disruption of the red cell membrane (Waldron and Stöfen, 1974).

Damage to the Central Nervous System. Damage to both the peripheral and central nervous system, including the brain, may be caused by a degeneration of nerve fibers and interference with the permeability of capillaries in the brain due to Pb intoxication. The exact mechanism responsible for these effects is not known, and indeed several factors, including enzyme inhibition and tissue damage, may be involved (Waldron and Stöfen, 1974).

Kidney Damage. Kidney damage characterized by atrophy of the renal tubules is a well-established effect of lead poisoning. The damage is associated with elevated levels of amino acids, sugar, and phosphate in the urine (Waldron and Stöfen, 1974). Gout caused by kidney damage has also been associated with lead poisoning, the consumption of illicitly distilled whiskey (moonshine) being a common cause. Concentrations of Pb in moonshine often exceed 10 ppm due to the use of lead in the distillation apparatus (EPA, 1980d). Waldron and Stöfen (1974) cite one example of a hospital in which 37 of 43 cases of gout admitted in 1967 were caused by the consumption of lead-contaminated liquor.

Effects on Children and Reproduction. From the standpoint of human health, the greatest danger of lead poisoning is undoubtedly to young children, particularly those living in urban areas. Neurological damage caused by the poisoning of such children may be permanent and can result in impaired physical as well as mental development. Even the unborn fetus is not protected from the effects of lead poisoning. There is evidence that the brain of the fetus is much more sensitive to lead poisoning than the brain of the infant or young child, and lead has been shown repeatedly to cause birth defects in experimental animals (EPA, 1980d). Although there is little information to indicate that Pb has a teratogenic effect on humans, exposure of pregnant women to lead can induce miscarriages and stillbirths (Waldron and Stöfen, 1974).

Although most of the body burden of lead is found in the bones, historically it has been much more practical to relate health effects on living persons to the level of lead in the blood, PbB.[32] Table 12.15 summarizes the lowest PbB levels associated with various health effects as summarized by the EPA (1980d) and CDC (1991). PbB levels as low as 100 ppb do not cause distinctive symptoms, but are associated with decreased intelligence and impaired neurobehavorial development, decreased stature or growth, decreased hearing acuity, and decreased ability to maintain a steady posture. The activity of δ-aminolaevulic acid dehydrase (ALA-d), an enzyme involved in heme metabolism, is also adversely affected at a PbB level of 100 ppb. Interference with vitamin D metabolism occurs at PbB levels of 100–150 ppb, and increased concentrations of protoporphyrin, a substrate needed for heme synthesis, and of ALA begin to appear in red blood cells and urine, respectively, at slightly higher PbB levels. Actual decreases in hemoglobin, somewhat loosely referred to in Table 12.15 as anemia, occur at PbB levels of about 400–500 ppb. Based on this evidence, the Centers for Disease Control have concluded that children's PbB levels should not exceed 100 ppb (CDC, 1991).

Unfortunately a great many children as well as adults in the United States have PbB levels greater than 100 ppb. For example, a study of urban children in 7 different United States cities between 1967 and 1970 revealed that about 29% had blood levels over 400 ppb, and almost 9% had blood lead levels over 500 ppb (Waldron and Stöfen, 1974). A blood lead level of 500-800 ppb is considered to be the threshold level of classical lead poisoning (Patterson, 1965). At the present time 1 in every 6 American children under 6 years of age is estimated to have lead poisoning (Blackman, 1991).

In 1975 the EPA promulgated a maximum contaminant level (MCL) of 50 ppb for lead in drinking water as an interim drinking water standard. In 1988, however, the EPA proposed to reduce the MCL to 5 ppb and to set a maximum contaminant level goal (MCLG) of zero for Pb in drinking water (EPA, 1991). The MCLG of zero was promulgated in 1991. According to the Safe Drinking Water Act, a MCL must be set as close to the MCLG as is feasible. The rationale for setting the MCLG at zero

---

[32] The Pb content of bones can now be assayed using noninvasive X-ray fluorescence techniques (Blackman, 1991).

Table 12.15   Lowest Blood Pb Levels Associated with Observed Biological Effects in Human

Lowest observed Effect level (ppb Pb in blood)	Effect	Population group
100	ALA-d inhibition	Children and adults
	Adverse effects on intelligence, hearing, and growth	Children
100–150	Interference with vitamin D metabolism	Children
150–200	Protoporphyrin elevation in red blood cells	Women and children
250–300	Protoporphyrin elevation in red blood cells	Adult males
400	Increased urinary ALA excretion	Children and adults
400	Anemia	Children
400	Coproporphyrin elevation	Adults and children
500	Anemia	Adults
500–600	Cognitive deficits	Children
500–600	Peripheral neuropathies	Adults and children
800–1000	Encephalopathic symptoms	Children
1000–1200	Encephalopathic symptoms	Adults

Source. EPA (1980d) and CDC (1991).

was in part the feeling that, "Drinking water should contribute minimal lead to total lead exposures because a substantial portion of the sensitive population already exceeds acceptable blood lead levels" (EPA, 1991, p. 26467). The EPA also considers lead to be a probable human carcinogen, but the evidence, based largely on the development of kidney tumors in rodents exposed to levels of lead far in excess of tolerable human doses, is of dubious relevance to humans for reasons discussed in Chapter 8. The EPA presently considers that the action level of Pb in drinking water has been exceeded if more than 10% of targeted tap samples contain more than 15 ppb lead (EPA, 1991).

Effects on Aquatic Organisms. The EPA water quality criteria for Pb for the protection of aquatic organisms are summarized in Table 12.16. As was the case with cadmium, the toxicity of lead in freshwater is negatively correlated with water hardness, and for three of the four species for which data were available to the EPA (1985d), the 96-hr TLm values were almost linearly related to water hardness. Final acute values for Pb in freshwater and saltwater were calculated by the procedures outlined in Chapter 8. In the former case the criteria are expressed in the form of an equation relating the criterion to water hardness. Final chronic values were calculated by dividing the final acute value by a mean acute/chronic ratio of 51, which was determined from studies on four different species. The final plant values were several orders of magnitude higher than the final chronic values and therefore did not enter into the calculation of the criterion continuous concentration. Final residue values were not calculated on the grounds that there were no maximum permissible tissue concentrations for lead. While it is technically true that there are no FDA action levels for lead in fish and shellfish in the United States, there are indications in FDA

Table 12.16   EPA Water Quality Criteria for Lead for the Protection of Aquatic Organisms in Freshwater and Saltwater

Water type	Criterion continuous Concentration (ppb)	Criterion maximum Concentration (ppb)
Marine	5.6	140
Freshwater		
50 mg/l CaCO₃	1.3	34
100 mg/l CaCO₃	3.2	82
200 mg/l CaCO₃	7.7	200

Source. EPA (1985d).

correspondence that after 1940 a limit of 7 ppm became the FDA guideline for the permissible lead content of canned goods (FDA, 1979). It seems reasonable to examine the implications of applying this standard to fish and shellfish.

The highest BCFs for edible freshwater and saltwater fish and shellfish reported by the EPA (1985d) are 45 and 370 for the bluegill fish and Eastern oyster, respectively. Dividing the FDA guideline of 7 ppm by these BCFs gives residue values of 156 ppb and 19 ppb for freshwater and saltwater, respectively. These figures are higher than the corresponding final chronic values and would therefore not have influenced the criterion continuous concentrations. It is noteworthy, however, that the average lead content of food is about 0.2 ppm (Settle and Patterson, 1980), and given the fact that consumption of such food has produced PbB levels above 300 ppb in many Americans, it is unclear why 7 ppm would be considered an acceptable Pb concentration in fish and shellfish. In the United Kingdom, for example, the lead content of fresh food must be less than 1 ppm (Moore, 1986). If 1 ppm were taken to be the criterion for fish and shellfish, the residue values in freshwater and saltwater would become 22 and 2.7 ppb, respectively. The latter figure is less than half the final chronic value for saltwater.

Commentary

In the case of lead pollution, there are no wide-scale disasters comparable to Minamata disease or Itai-itai disease. On the other hand, there is no doubt that many persons are today carrying blood lead levels that are within a factor of 2 or 3 of the levels associated with classical lead poisoning. About 15% of Americans, for example, have PbB levels greater than 200 ppb (Tackett, 1987). For comparison, Patterson (1965) has estimated that natural PbB levels in primitive humans were roughly 2.5 ppb. If an uncertainty factor of 10 (Chapter 8) is applied to the lowest observed effect levels in Table 12.15, then PbB levels should not exceed 10 ppb.

There are, to be sure, some encouraging aspects about the lead pollution situation. The phasing out of leaded gasoline is undoubtedly the most important, because during the peak of leaded gasoline use emissions of lead alkyls accounted for about 70% of the anthropogenic Pb input to the atmosphere. Fallout of lead onto crops and onto the surface waters of the ocean have undoubtedly declined in recent years as leaded gas has been phased out (Figure 12.11). The reduction in fallout onto crops and the decreased use of lead solder in cans has translated into lower lead levels in human diets (CDC, 1991).

Several other developments can be expected to reduce the level of human exposure to lead. One of the most important in the United States has been the sweeping restrictions passed by the Consumer Product Safety Commission in 1977 on the use of lead in paint. This legislation has been good news for infants and young children. The bad news is that there are a lot of old apartments and homes out there painted with lead-base paint. These buildings will continue to be a hazard to youngsters for many years unless the paint is removed. The Department of Health and Human Services (HHS) has called for the implementation of a four-point plan, including the elimination of leaded paint and contaminated dust in housing, to reduce the level of lead exposure to children. The problem is money. The HHS plan would cost about $10 billion over a period of 10 years (Blackman, 1991). The Office of Management and Budget raised allotments for lead-screening programs from $4 million in 1990 to a proposed $41 million in 1992, but has balked at further expenditures (Blackman, 1991).

Restrictions on the use of lead solder and pipes in water distribution systems mandated by the 1986 amendments to the SDWA can be expected to reduce the contribution to lead contamination from distribution networks. Although the FDA has not officially set a lead standard in food, it has made enough noise to cause food canners to take steps to reduce contamination problems from lead solder. The average lead level in evaporated milk, for example, dropped from 0.52 ppm to 0.10 ppm between 1972 and 1979, and the lead levels in canned juices similarly dropped from 0.30 ppm in 1973 to 0.05 ppm in 1979 (FDA, 1979). A major concern in this regard has been the lead content of infant foods. The average lead level in canned infant formula concentrate declined from 0.10 ppm to 0.06 ppm during the 1970s, and by 1979 all infant juice manufacturers had voluntarily switched from tin cans to glass jars (FDA, 1979). There are, incidentally, ways to make cans that do not involve the use of lead solder. Tin solder can be used instead of lead, but tin solder is more expensive. The cans can be welded, but welding is slower than soldering. A third alternative is a two-piece can constructed of tin plate or aluminum. The can is formed by forcing a "cup" of metal through a die, or by "redrawing" it (Moore, 1986). According to the FDA (1991), there is a definite movement away from lead-soldered cans by the food canning industry. This is good news for people who do not like to eat lead.

Despite the fact that anthropogenic emissions of lead have enriched the surface waters of the ocean with Pb by an order of magnitude in the northern hemisphere, the concentrations are still well below levels likely to exert a toxic stress. Surface water concentrations reported by Flegal and Patterson (1983), for example, are 0.010–0.035 ppb, more than two orders of magnitude smaller than the EPA's criterion continuous concentration. The best-documented effect of lead pollution in aquatic systems has been the poisoning of waterfowl that ingest lead shot in marshes. Additional inputs of lead to marshes in the United States have been largely eliminated by regulations requiring the use of nontoxic steel shot in most marsh areas, but there are still a lot of lead shot out there from previous years' hunting.

Although surface water concentrations of lead in the open ocean will undoubtedly decline as leaded gasoline is phased out, the same statement may not apply to some coastal areas. Recent sediments of the San Pedro, Santa Monica, and Santa Barbara basins off the coast of California, for example, have been found to contain about 30 ppm lead. Sediments deposited in the same areas prior to 1900 contain only 10 ppm lead (Chow et al., 1973). The recent increases in Pb concentration can be accounted for rather well from estimates of atmospheric fallout due to the burning of lead alkyls in gasoline (Chow et al., 1973). Recent sediment lead concentrations near White Point, California, however, are about 400 ppm, a figure that can hardly be accounted for by atmospheric fallout. These high lead concentrations presumably reflect the discharge of lead-containing industrial and domestic wastes from the White Point sewer outfall, and the magnitude of this discharge is not likely to be affected greatly by the switch to unleaded gasoline. Similarly, sediments of the Ala Wai boat harbor (Table 12.2), which contain 5000 ppm lead, have undoubtedly been polluted in part by the peeling and chipping of lead-based paints and the corrosion of lead-containing metal parts of boats. There is little prospect in the near future that such sources of pollution will be reduced significantly.

REFERENCES

Aaronson, T. 1971. Mercury in the environment. *Environment*, **13**(4), 16–23.

Andren, A. W. 1974. Ph.D. dissertation. Florida State Univeristy. Oceanography Department. Tallahassee, FL.

Andren, W. W., and R. C. Harriss. 1973. Methyl mercury in estuarine sediments. *Nature*, **245**, 256–257.

Anonymous. 1971. Cadmium pollution and Itai-itai disease. *Lancet*, **1**, 382–383.

Anonymous. 1991. The mercury in your mouth. *Consumer Reports*, **56**, 316–319.

Berner, E. K., and R. A. Berner. 1987. *The Global Water Cycle*. Simon and Schuster, Englewood Cliffs, NJ. 397 pp.

Blackman, A. 1991. Controlling a childhood menace. *Time*, **137**(8), 68–69.

Boutron, C. F., U. Gorfach, J. Candelone, M. A. Bolshov, and R. J. Delmas. 1991. Decrease in anthropogenic lead, cadmium and zinc in Greenland snows since the late 1960s. *Nature*, **353**, 153–156.

Carter, L. J. 1977. Chemical plants leave unexpected legacy for two Virginia rivers. *Science*, **198**, 1015–1020.

CDC. 1991. Preventing Lead Poisoning in Young Children. Centers for Disease Control, U.S. Dept. of Health and Human Services, Atlanta, GA. 108 pp.

Chang, L. W., K. R. Reuhl, and P. R. Wade. 1980. Pathological effects of cadmium poisoning. In J. O. Nriagu (Ed.), *Cadmium in the Environment*, Part II. Health Effects, Wiley, New York. pp. 783–839.

Chow, T. J., K. W. Bruland, K. Bertine, A. Soutar, M. Koide, and E. D. Goldberg. 1973. Lead pollution: records in southern California coastal sediments. *Science*, **181**, 551–552.

City and County of Honolulu. 1971. Department of Public Works. Water Quality Program for Oahu with Special Emphasis on Waste Disposal—Final Report Work Areas 6 and 7 Analysis of Water Quality Oceanographic Studies, Parts I and II. Honolulu, HI.

Cordle, F., P. Corneliussen, C. Jelinek, B. Hackley, R. Lehman, J. McLaughlin, R. Rhoden, and R. Shapiro. 1978. Human exposure to polychlorinated biphenyls and polybrominated biphenyls. *Environ. Health Perspect*, **24**, 157–172.

Craig, P. P., and E. Berlin. 1971. The air of poverty. *Environment*, **13**(5), 2–9.

Crawford, M. 1985a. The last days of the wild condor? *Science*, **229**, 844–845.

Crawford, M. 1985b. Condor agreement reached. *Science*, **229**, 1248.

Dickson, D. 1972. Mercury and lead in the Greenland ice sheet: A reexamination of the data. *Science*, **177**, 536–538.

Dickson, D. 1985. Europe to start removing lead from gas in 1989. *Science*, **228**, 37.

DLNR. 1985. *Hawaii Fisheries Plan 1985*. Dept. of Land and Natural Resources. Division of Aquatic Resources. State of Hawaii. Honolulu, HI. 163 pp.

Emmerson, B. T. 1970. 'Ouch-ouch' disease: The osteomalacia of cadmium nephropathy. *Annu. Intern. Med.*, **73**(5), 854–855.

Emond, S. 1990. Still a hazard after all these years. *Harvard Health Letter*, **16**(1), 7–8.

Environmental Protection Agency. 1972. *Water Quality Criteria*. EPA-R3-73-033. Washington, D.C. 594 pp.

Environmental Protection Agency. 1980a. *Ambient Water Quality Criteria for Mercury*. EPA 440/5-80-058. Washington, D.C.

Environmental Protection Agency. 1980b. *Seafood Consumption Data Analysis*. Stanford Research Institute International. Menlo Park, CA. Final Report, Task II. Contract No. 68-01-3887 to U.S. EPA.

Environmental Protection Agency. 1980c. *Ambient Water Quality Criteria for Cadmium*. EPA 440/5-80-025. Washington, D.C.

Environmental Protection Agency. 1980d. *Ambient Water Quality Criteria for Lead*. EPA 440/5-80-057. Washington, D.C.

Environmental Protection Agency. 1985a. *Ambient Water Quality Criteria for Mercury-1984*. EPA 440/5-84-026. Washington, D.C. 136 pp.

Environmental Protection Agency. 1985b. *Guidelines for Deriving Numerical National Water Quality Criteria for the Protection of Aquatic Organisms and Their Uses*. PB85-227049. Duluth, MN. 97 pp.

Environmental Protection Agency. 1985c. *Ambient Water Quality Criteria for Cadmium-1984*. EPA 440/5-84-032. Washington, D.C.

Environmental Protection Agency. 1985d. *Ambient Water Quality Criteria for Lead-1984*. EPA 440/5-84-027. Washington, D.C. 81 pp.

Environmental Protection Agency. 1986. *Quality Criteria for Water*. EPA 440/5-86-001. Washington, D.C.

Environmental Protection Agency. 1987. *Update #2 to Quality Criteria for Water 1986*. May 1, 1987. Office of Water Regulations and Standards. Criteria and Standards Division. Washington, D.C.

Environmental Protection Agency. 1991. Maximum contaminant level goals and national primary drinking water regulations for lead and copper; final rule. *Federal Register*, **56**(110), 26460–26564.

Food and Drug Administration. 1979. Lead in food; advance notice of proposed rulemaking: Request for data. *Federal Register*, **44**(171), 51233–51242.

Food and Drug Administration. 1991. Personal communication from John Thomas.

Flavin, C. 1985. *World Oil: Coping with the Dangers of Success*. Worldwatch Paper 66. Worldwatch Inst. Washington, D.C. 66 pp.

Flegal, A. R., and C. C. Patterson. 1983. Vertical concentration profiles of lead in the Central Pacific at 15°N and 20°S. *Earth and Planetary Sci. Letters*, **64**, 19–32.

Friberg, L., C. Elinder, T. Kjellstrom, and G. F. Nordberg. 1986. *Cadmium and Health: A Toxicological and Epidemiological Appraisal*. Vol. 2. CRC Press, Boca Raton, FL. 307 pp.

Futatsuka, M. 1989. Epidemiological aspects of methylmercury poisoning in Minamata. In S. Tsuru, T. Suzuki, H. Shiraki, K. Miyamoto, and M. Shimizu (Eds.), *For Truth and Justice in the Minamata Disease Case*. Proceedings of the International Forum on Minamata Disease 1988, Keiso Shobo, Tokyo. pp. 231–235.

GESAMP. 1985. *Cadmium, Lead, and Tin in the Marine Environment.* Joint Group of Experts on the Scientific Aspects of Marine Pollution. United Nations Environment Program Regional Seas Reports and Studies No. 56, Geneva. 90 pp.

Gilfillan, S. C. 1965. Lead poisoning and the fall of Rome. *J. Occup. Med.,* **7**, 53–60.

Goldberg, E. D. 1976. Rock volatility and aerosol composition. *Nature,* **260**(5547), 128–129.

Goldwater, L. J. 1971. Mercury in the environment. *Sci. Amer.,* **224**(5), 15–21.

Grant, N. 1969. Legacy of the mad hatter. *Environment,* **11**(4), 18–23, 43–44.

Grant, N. 1971. Mercury and man. *Environment,* **13**(4), 3–15.

Hamelink, J. L., R. C. Waybrant, and R. C. Ball. 1971. A proposal: Exchange equilibria control the degree chlorinated hydrocarbons are biologically magnified in lentic environments. *Trans. Am. Fish. Soc.,* **100**, 207–214.

Hammond, A. L. 1971. Mercury in the environment: Natural and human factors. *Science,* **171**, 788–789.

Harada, M. 1989. The intrauterine methylmercury poisoning known as 'congenital Minamata Disease'—a 20-year serial investigation and its recent problems. In S. Tsuru, T. Suzuki, H. Shiraki, K. Miyamoto, and M. Shimizu (Eds.), *For Truth and Justice in the Minamata Disease Case.* Proceedings of the International Forum on Minamata Disease 1988, Keiso Shobo, Tokyo. pp. 259–265.

Horvath, G., R. C. Harriss, and H. C. Mattraw. 1972. Land development and trace metal distribution in the Florida Everglades. *Mar. Pollution Bull.,* **3**, 182–183.

Interior Dept. 1991. Mercury. In *Minerals Yearbook.* Vol. 1. Metals and Minerals, U.S. Dept. of the Interior Bureau of Mines. pp. 705–708.

Katahira, K. 1989. Actual status of patients and urgency of redress. In S. Tsuru, T. Suzuki, H. Shiraki, K. Miyamoto, and M. Shimizu (Eds.), *For Truth and Justice in the Minamata Disease Case.* Proceedings of the International Forum on Minamata Disease 1988, Keiso Shobo, Tokyo. pp. 341–343.

Kjellstrom, T. 1986a. Itai-itai disease. In L. Friberg, C. Elinder, T. Kjellstrom, and G. F. Nordberg (Eds.), *Cadmium and Health: A Toxicological and Epidemiological Appraisal.* Vol. 2. Effects and Response. CRC Press, Boca Raton, FL. pp. 257–290.

Kjellstrom, T. 1986b. Effects on bone, on vitamin D, and calcium metabolism. In L. Friberg, C. Elinder, T. Kjellstrom, and G. F. Nordberg (Eds.), *Cadmium and Health: A Toxicological and Epidemiological Appraisal.* Vol. 2. Effects and Response, CRC Press, Boca Raton, FL. pp. 111–158.

Klein, D. H., and E. D. Goldberg. 1970. Mercury in the marine environment. *Environ. Sci. Tech.,* **4**(9), 765–768.

Knauer, G. A. 1977. Immediate industrial effects on sediment metals in a clean coastal environment. *Mar. Poll. Bull.,* **8**, 249–254.

Knauer, G. A., and J. H. Martin. 1972. Mercury in a marine pelagic food chain. *Limnol. Oceanogr.,* **17**, 868–876.

Kurland, L. T. 1989. An epidemiological overview of Minamata Disease and a review of earlier public health recommendations. In S. Tsuru, T. Suzuki, H. Shiraki, K. Miyamoto, and M. Shimizu (Eds.), *For Truth and Justice in the Minamata Disease Case.* Proceedings of the International Forum on Minamata Disease 1988, Keiso Shobo, Tokyo. pp. 240–249.

Lieber, M., and W. F. Welsch. 1954. Contamination of groundwater by cadmium. *J. Amer. Water Works Assoc.,* **46**, 541–547.

Llewellyn, T. O. 1990. Cadmium. In *Minerals Yearbook.* Vol. 1. Metals and Minerals, U.S. Dept. of the Interior Bureau of Mines. pp. 191–195.

Llewellyn, T. O. 1991. Cadmium. In *Minerals Yearbook.* Vol. 1. Metals and Minerals, U.S. Dept. of the Interior Bureau of Mines. pp. 207–210.

Lloyd, R. 1961. The toxicity of mixtures of zinc and copper sulphates to rainbow trout (*Salmo gairdnerii* Richardson). *Annu. Appl. Biol.*, **49**, 535–538.

Martin, J. H., and G. A. Knauer. 1973. The elemental composition of plankton. *Geochim. Cosmochim. Acta*, **37**, 1639–1653.

Martin, J. M., and M. Whitfield. 1983. The significance of the river input of chemical elements to the ocean. In C. S. Wong (Ed.), *Trace Metals in Sea Water*, Proceedings NATO Advanced Research Inst., March 30–April 1, 1981, Erice, Italy, Plenum, NY. pp. 265–296.

McCaull, J. 1971. Building a shorter life. *Environment*, **13**(7), 3–14, 38–41.

Mitra, S. 1986. *Mercury in the Ecosystem*, Trans. Tech. Publ., Switzerland. 327 pp.

Montague, K., and P. Montague. 1971. *Mercury*, Sierra Club, San Francisco. 158 pp.

Moore, M. R. 1986. Sources of lead exposure. In R. Lansdown and W. Y. Yule (Eds.), *The Lead Debate*. Croom Helm, London. pp. 131–189.

Murozumi, M., T. J. Chow, and C. Patterson. 1969. Chemical concentrations of pollutant lead aerosols, terrestrial dusts and sea salts in Greenland and Antarctic snow strata. *Geochim. Cosmochim. Acta*, **33**, 1247–1294.

Nogawa, K. 1980. Itai-itai disease and follow-up studies. In J. O. Nriagu (Ed.), *Cadmium in the Environment*. Part 2. Health Effects. Wiley, NY. pp. 2–37.

Nriagu, J. O. 1980. Production, uses, and properties of cadmium. In J. O. Nriagu (Ed.), *Cadmium in the Environment*. Vol. 1. Wiley-Interscience, NY. pp. 35–70.

Nriagu, J. O., and J. B. Sprague. 1987. *Cadmium in the Aquatic Environment*. Wiley-Interscience, NY. 272 pp.

OECD. 1975. *Cadmium and the Environment: Toxicity, Economy, Control*. Organization for Economic Co-operation and Development. Paris. 88 pp.

Pan, Y., X. Jiang, and S. Wang. 1989. The problem of methylmercury poisoning along the Songhua River in China. In S. Tsuru, T. Suzuki, H. Shiraki, K. Miyamoto, and M. Shimizu (Eds.), *For Truth and Justice in the Minamata Disease Case*. Proceedings of the International Forum on Minamata Disease 1988, Keiso Shobo, Tokyo. pp. 299–304.

Patterson, C. C. 1965. Contaminated and natural lead environments of man. *Arch. Environ. Health*, **11**, 344–360.

Saito, H. 1989. Niigata Minamata disease. In S. Tsuru, T. Suzuki, H. Shiraki, K. Miyamoto, and M. Shimizu (Eds.), *For Truth and Justice in the Minamata Disease Case*. Proceedings of the International Forum on Minamata Disease 1988, Keiso Shobo, Tokyo. pp. 297–298.

SCEP—Report of the Study of Critical Environmental Problems. 1970. *Man's Impact on the Global Environment*. MIT Press, Cambridge, MA. 319 pp.

Settle, D. M., and C. C. Patterson. 1980. Lead in albacore: Guide to lead pollution in Americans, *Science*, **207**, 1167–1176.

Shea, K. P. 1973. Canned milk. *Environment*, **15**(2), 6–11.

Siegel, B. Z., S. M. Siegel, and F. Thorarinsson. 1973. Icelandic geothermal activity and the mercury of the Greenland icecap, *Nature*, **241**, 526.

Smith, W. E., and A. M. Smith. 1975. *Minamata*. Holt, Rinehart, and Winston, NY. 192 pp.

Sprague, J. G. 1964. Lethal concentrations of copper and zinc for young Atlantic salmon, *J. Fish. Res. Bd. Can.*, **21**, 17–26.

Tackett, S. L. 1987. Lead in the environment: Effects of human exposure, *American Laboratory*, July. pp 32–41.

Tsuru, S., T. Suzuki, H. Shiraki, K. Miyamoto, and M. Shimizu (Eds.), 1989. *For Truth and Justice in the Minamata Disease Case*. Proceedings of the International Forum on Minamata Disease 1988, Keiso Shobo, Tokyo. 427 pp.

Uzawa, H. 1989. The responsibility of Chisso in the Minamata issue. In S. Tsuru, T. Suzuki, H. Shiraki, K. Miyamoto, and M. Shimizu (Eds.), *For Truth and Justice in the Minamata Disease Case*. Proceedings of the International Forum on Minamata Disease 1988, Keiso Shobo, Tokyo. pp. 335–337.

Waldron, H. A., and D. Stöfen. 1974. Sub-clinical Lead Poisoning. Academic Press, New York. 224 pp.

Weiss, H. V., M. Koide, and E. D. Goldberg. 1971. Mercury in a Greenland ice sheet: Evidence of recent input by man, *Science*, **174**, 692–694.

World Health Organization. 1973. *The Hazards to Health of Persistent Substances in Water*, World Health Organization, Regional Office for Europe of the WHO, Copenhagen.

Wood, J. M. 1972. A progress report on mercury, *Environment*, **14**(1), 33–39.

Woodbury, W. D. 1991. Lead. In *Minerals Yearbook*. Vol. 1. Metals and Minerals, U.S. Dept. of the Interior Bureau of Mines. pp. 627–654.

Yamagata, N., and I. Shigematsu. 1970. Cadmium pollution in perspective, *Bull. Inst. Publ. Health*, **19**, 1–27.

Yamaguchi, T. 1989. Chisso subsidiaries and liability. In S. Tsuru, T. Suzuki, H. Shiraki, K. Miyamoto, and M. Shimizu (Eds.), *For Truth and Justice in the Minamata Disease Case*. Proceedings of the International Forum on Minamata Disease 1988, Keiso Shobo, Tokyo. pp. 329–334.

Yunice, A., and H. M. Perry, Jr. 1961. The acute pressor effects of parental cadmium and mercury ions, *J. Lab. Clin. Med.*, **58**, 975.

13

OIL POLLUTION

Australia's Canberra Times reported on a group of Boy Scouts who were asked to write about the harmful effect to marine life of oil leakages from tankers, and at least one lad seemed to have a pretty good grip on the problem. "When my mum opened a can of sardines last night," he wrote, "it was full of oil and all the sardines were dead." (Associated Press story carried in the Honolulu Advertiser, August 24, 1978).

Petroleum in one form or another has been used by humans since at least biblical times. Natural oil seeps provided bitumen,[1] which was used as a building material in ancient cities, and in Iraq natural gas from the Kirkuk oil field at Baba Gurger has burned continuously since the time of Nebuchadnezzar. Over 3000 tonnes of oil per year were obtained from hand-dug pits at Baku near the Caspian Sea as early as the nineteenth century (Nelson-Smith, 1972).

In the United States, natural oil seeps such as those off Coal Oil Point, California, were noted as early as 1629, but the first commercial well, at Titusville, Pennsylvania, was not drilled until 1859. By 1900 world oil production from such wells had reached 20 million tonnes, and for the next 80 years doubled almost every 10 years, reaching 3 billion tonnes by 1978 (Figure 13.1). The earliest uses of crude oil were for caulking ships, embalming, treating leather, in medicine, and as already noted in building construction. With the advent of refineries in the midnineteenth century, however, it was possible to separate kerosene from the other components of crude oil, and kerosene obtained in this way rapidly replaced whale oil and vegetable oils as a fuel in lamps. For several decades oil was used primarily in lighting and to a minor extent as a lubricant, but the invention of the internal combustion engine and its widespread adoption in all forms of transportation during the last few decades of the nineteenth century rapidly changed this picture. The demand for oil as a fuel rose steadily, and today petroleum accounts for about 40% of world energy production (UN, 1991). Over half the oil is presently used in the transportation sector of the world economy (Table 13.1).

[1]A semisolid that remains after the more volatile components of crude oil have evaporated away.

417

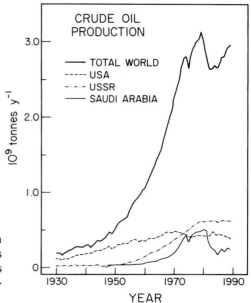

Figure 13.1 World crude oil production from 1930 to 1989. (*Source*: Various editions of the United Nations Energy Statistics Yearbook and the U.S. Dept. of the Interior's Minerals Yearbook.)

Table 13.1   Principal Uses of Oil in the World Economy in 1988

Sector of economy	% Use
Transportation	52
Industry	22
Residential/commercial	11
Electricity	6
Other	9

Source. OECD (1989).

The widespread use of petroleum and its products has inevitably resulted in the discharge of oil to the environment. With respect to aquatic systems, it is the marine environment that has received the greatest attention, since the majority of the more noteworthy oil spills have involved accidents at sea. The emphasis in this chapter on marine oil pollution is not meant to imply, however, that oil pollution of freshwater systems is not a serious problem. However, some of the issues associated with marine oil pollution, such as offshore oil drilling and use of supertankers to transport oil, are peculiar to the marine environment. Many of the other issues are equally relevant to freshwater and marine systems. In other words, a study of marine oil pollution encompasses most, if not all, of the problems relevant to freshwater oil pollution as well.

OIL DISCHARGES TO THE MARINE ENVIRONMENT

Table 13.2 summarizes the most recent estimates of oil inputs to the marine environment. In most cases the uncertainty of the estimates is rather large, and it seems reasonable to conclude that the numbers are accurate to no more than one significant figure. The following summary provides additional information on the various inputs.

Natural Sources

Marine seeps
There are presently known to be almost 200 significant marine oil seeps, most of which are near continental coastlines. Areas of significant seepage are found offshore of California and Alaska, off the northeast coast of South America, and in the Arabian Gulf, the Red Sea, and the South China Sea. Since most petroleum is produced from sedimentary rocks at depths greater than 1 km, it is unlikely that oil from underground deposits would reach the surface through seepage other than in tectonically active areas where petroleum source or reservoir beds have been uplifted.

TABLE 13.2   Estimates of Petroleum Inputs to the Marine Environment

Source	Input Rate (million tonnes per year)	
	Probable range	Best estimate
Natural sources		
Marine seeps	0.02 —2.0	0.2
Sediment erosion	0.005—0.5	0.05
Total natural sources	0.025—2.5	0.25
Offshore production	0.04 —0.06	0.05
Transportation		
Tanker operations	0.4 —1.5	0.7
Drydocking	0.02 —0.05	0.03
Marine terminals	0.01 —0.03	0.02
Bilge and fuel oils	0.2 —0.6	0.3
Tanker accidents	0.3 —0.4	0.4
Nontanker accidents	0.02 —0.04	0.02
Total transportation	0.95 —2.62	1.47
Atmosphere	0.05 —0.5	0.3
Municipal and industrial wastes and runoff		
Municipal wastes	0.4 —1.5	0.7
Refineries	0.06 —0.6	0.1
Nonrefinery industrial	0.1 —0.3	0.2
Urban runoff	0.01 —0.2	0.12
River runoff	0.01 —0.5	0.04
Ocean dumping	0.005—0.02	0.02
Total wastes and runoff	0.585—3.12	1.18
Total	2—9	3

Source. NRC (1985).

Based on this reasoning, one can divide the ocean margins into areas of high, medium, and low seepage probability. Estimates of seepage derived in this way vary widely, from about 0.02 million tonnes (Mt) y^{-1} to as much as 10 Mt y^{-1} (NRC, 1985). The seepage estimate can be narrowed down by considering the age of the oil and the probable amount of oil available for seepage. The age of oil deposits is on the order of 50 million years (Blumer, 1972). At a seepage rate of 0.02 Mt per year, 10^6 Mt of oil would have seeped to the surface in 50 million years. This figure is roughly 3–5 times the amount of oil believed to have been available for exploitation before the rapid increase in oil use began in the early twentieth century (Flavin, 1985). Allowing for unknowns with regard to the amount of petroleum that would have been available for seepage during geologic time and will become available in the future during the lifetimes of the seepage, an upper bound on the amount of oil available for seepage might be about 10^8 Mt (NRC, 1985). A seepage rate of 2.0 Mt y^{-1} would be required to deplete 10^8 Mt in 50 million years. Thus it seems reasonable to assume that natural seepage rates lie somewhere between 0.02 and 2.0 Mt y^{-1}. The NRC (1985) estimate of 0.2 Mt y^{-1} is simply the geometric mean of these two bounds. Given the probable range of seepage rates, 0.02–2.0 Mt y^{-1}, it is concluded that present underground oil deposits represent a small fraction of the quantity of oil that was formed underground over geologic time. Most of the latter has evidently seeped to the surface and been broken down by chemical and biological processes.

Sediment erosion

Estimates of the input of petroleum due to the erosion of continental rocks are based on indirect calculations, since there are only a few places where erosional inputs can be studied in detail. The calculations of the NRC (1985) are based on estimates of the amount of organic carbon transported to the ocean by rivers and the percentage of that carbon accounted for by petroleum. Fluvial inputs of organic carbon are estimated to be 400 Mt y^{-1}. Based on analyses of this carbon and of the carbon in ancient sedimentary rocks, petroleum is estimated to account for about 0.0125% of this figure, or 0.05 Mt y^{-1}. The uncertainty of this estimate is about a factor of 10.

Offshore Production

The estimate of 0.05 Mt y^{-1} is based on experience in the United States and extrapolation of this experience to other parts of the world, with allowance for differences in operating techniques. Offshore production discharges can of course vary greatly from one year to the next, depending on the occurrence of major offshore accidents. The Mexican Ixtoc I oil well blowout of 1979, for example, is estimated to have released about 0.48 Mt of crude oil into the Gulf of Mexico, almost 10 times the estimated annual average for the entire world. Of the 0.05 Mt y^{-1} average figure, about 19%, 6%, and 75% are estimated to come from offshore produced water discharges, minor spills, and major spills, respectively.[2] The water discharges associated with oil production can vary from virtually zero to about 80% of the volume of oil produced, depending on, for example, whether water injection is used to help

---

[2]Minor and major spills are defined as spills involving less than or more than, respectively, 50 barrels of oil.

extract the oil (NRC, 1985). Typical hydrocarbon concentrations in the discharged water are 35–70 ppm, depending on the type of equipment used and steps taken to separate the oil and water. Accidents include blowouts, the rupture of gathering lines, and other unpredictable occurrences.

Marine Transportation

Tanker operations

A significant portion of the world's oil production is transported by sea. About 80 ± 5% of the oil transported by tankers is crude, the remainder being refined products (Yamaguchi, 1990). The percentage of crude oil production transported by sea peaked at 48% in 1979, dropped to 33% in 1985, and increased to 38% in 1988. The percentage is expected to continue increasing as major consuming nations such as the United States are forced to rely more and more on foreign oil (Yamaguchi, 1990). Historically transportation losses have been a major cause of oil discharges to the ocean.

In the past a large percentage of the oil discharged to the ocean by tankers resulted from the practice of ballasting cargo bunkers with water after a tanker had offloaded its oil. Such ballasting was necessary to maintain the tanker in a sea-worthy condition for its return voyage. Although some ports maintain facilities to accept dirty ballast water, the capacity of such facilities is limited. They are not designed to handle more than a small percentage of the water needed to ballast a large oil tanker. To avoid discharging oil-laden ballast water in port while onload-ing its next shipment, a tanker would therefore normally flush its ballast tanks with clean water while at sea, discharging the oily water in the process. The amount of oil discharged by this procedure was about 0.35% of the tanker's carrying capacity (NAS, 1975). In the absence of any effort to prevent such losses, oil discharges due to cargo bunker ballasting would have amounted to 3.5–4.5 Mt y^{-1} in recent years.

Efforts to largely eliminate these losses occurred in two steps. The first was the introduction of the so-called Load On Top (LOT) procedure, depicted in Figure 13.2. Basically the idea is to transfer the oily wash water and ballast water to cargo holds, where the oil and water separate out.[3] The clean water at the bottom is then drawn off and discharged. Oil and water mixtures that remain are transferred to a slop tank, where further separation occurs. The clean water at the bottom of the slop tank is discharged, and at the next loading port, new cargo is loaded on top of the reclaimed oil in the slop tank. Further improvements were made with the introduc-tion of crude oil washing systems (COWs), which involve cleaning cargo tanks with high-pressure jets of crude oil while the ship is offloading. Residues that would otherwise remain in the tanks are pumped ashore with the cargo. The combination of LOT and COW is estimated by the NRC (1985) to have reduced oil discharges from ballasting by about a factor of 50.

The problem with LOT is that it may be impractical for tankers engaged in short-haul voyages, which account for about 15% of crude oil movement at sea (NRC, 1985). In order to circumvent this problem, regulations were put into effect requiring the use of segregated ballast tanks (SBTs) or clean ballast tanks (CBTs) on the

---

[3]Most components of oil are less dense than water.

majority of the tanker fleet. The use of SBTs or CBTs eliminates the need to pump ballast water into cargo tanks under normal circumstances. The only difference between SBTs and CBTs is that the latter do not require separate pipes and pumps for taking on and discharging ballast.

The international agreement that has been primarily responsible for effecting these changes is the International Convention for the Prevention of Pollution from Ships, often referred to as MARPOL, which initially resulted from a conference on marine pollution held in 1973 under the auspices of the International Maritime Consultative Organization (IMCO). The protocol for MARPOL was modified in 1978, the same year that the U.S. Congress passed the Port and Tanker Safety Act (public law 95-474). The provisions of MARPOL with respect to tanker safety and the Port and Tanker Safety Act of 1978 are very similar. The MARPOL protocol, which went into force in 1983, presently requires that all new crude oil tankers over 20,000 dead-weight tonnes (dwt)[4] have SBTs and COW, all new product tankers over 30,000 dwt have SBTs, all existing crude oil tankers over 40,000 dwt have SBTs or COW, and all existing product tankers over 40,000 dwt have SBTs or CBTs.

The obvious loopholes in the legislation are the exclusion of small tankers and the fact that existing crude oil tankers over 40,000 dwt are not required to have SBTs. The former consideration would appear to be of minor consequence. Tankers under 40,000 dwt account for about 40% of the number of oil tankers in the world fleet, but only 11% of the dead-weight tonnage (Drewry, 1991b). The U.S. Port and Tanker Safety Act, incidentally, applies to all tankers greater than 20,000 dwt, and these account for over 80% of the number of tankers and 97% of the dead-weight tonnage of the tanker fleet. The exclusion of existing crude oil tankers from the requirement for SBTs is of more consequence, since 70% by numbers and 77% by weight of the world's tanker fleet was built before 1982 (Drewry, 1991a). Thus most crude oil tankers can, at least for the time being, operate without SBTs. However, if they lack SBTs, they must have a COW system and should be equipped with LOT if they are to comply with the requirements of the International Convention for the Prevention of Pollution of the Sea (OILPOL), which has been ratified by nations representing almost all the world's merchant fleet (NRC, 1985). According to the provisions of OILPOL, which was last amended in 1969, tankers must not discharge more than 1/15,000 of their total cargo carrying capacity during any one ballast voyage (NRC, 1985).

The NRC's estimate of 0.7 Mt for the annual discharge of oil due to tanker operations is based on conditions that existed before the OILPOL protocols had been fully implemented, and before MARPOL-78 had come into force. The discharge is therefore presumably much smaller now. The OILPOL requirements, for example, are in many cases achievable with nothing more than LOT and COW (NRC, 1985). If the 1/15,000 OILPOL criterion is applied to the present crude and refined oil marine transport figure of 1.4×10^9 tonnes y^{-1}, the implication is that the discharge from tanker operations should be no more than 0.1 Mt y^{-1}.

Dry-docking

At the present time tankers go into dry dock about once every 2 years for maintenance and inspection (NRC, 1985). At such times they must be clean of oil and gas-free. Rigorous cleaning is required because of the danger of explosions. Some dry

---

[4]The dead weight is the weight of the ship without cargo.

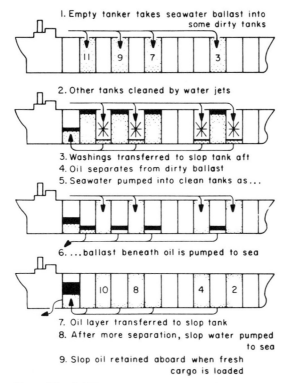

1. Empty tanker takes seawater ballast into
   some dirty tanks

2. Other tanks cleaned by water jets

3. Washings transferred to slop tank aft
4. Oil separates from dirty ballast
5. Seawater pumped into clean tanks as...

6. ...ballast beneath oil is pumped to sea

7. Oil layer transferred to slop tank
8. After more separation, slop water pumped
   to sea
9. Slop oil retained aboard when fresh
   cargo is loaded

Figure 13.2   An outline of the LOT procedure. [Redrawn from Nelson-Smith (1972) with permission of Elek Books Ltd.]

docks provide facilities for receiving the washings from this cleaning, but in the absence of such facilities tankers must discharge their washings at sea prior to entering dry dock. The use of COW systems has reduced the amount of clinage that must be removed, but some invariably remains. The NRC (1985) estimate of 0.03 Mt y^{-1} for dry-docking discharges is based on the assumption that 5% of tankers discharge oil equal to 0.4% of their dead-weight tonnage when they dry dock at intervals of 2 years. The present size of the tanker fleet, however, is about 30% smaller than the size assumed by the NRC (Drewry, 1991b), and the dry-docking discharges are presumably smaller by a similar percentage.

Marine terminals

Discharges at marine terminals are in part the result of human errors, for example overfilling tanks and disconnecting hoses without adequate drainage. Other causes of spillage include line or hose failures, submarine pipeline ruptures, and storage tank ruptures. The NRC estimate of 0.02 Mt y^{-1} for this category is based in part on U.S. Coast Guard records of spills at U.S. terminals and extrapolation of this information to the rest of the world. These more-or-less routine spills were estimated to amount to 0.01 Mt y^{-1}. Major spills, for example from submarine pipeline or storage tank ruptures, are relatively infrequent but not well quantified. The oil discharged in these large but infrequent spills was simply assumed to equal, on average, the oil discharged in the better-documented smaller spills.

Bilge and fuel oil

Both tankers and nontankers generate fuel oil sludge and bilge oil, and nontankers may also generate oily ballast water. The oil sludge comes from the use of heavy residual bunker fuel oil to power vessels and is estimated to account for about two-thirds of the total bilge and fuel oil discharge (NRC, 1985). Bunker fuel oil contains impurities such as sludge and water, and in the case of nontankers the lack of sufficient storage space in slop tanks may cause a large portion of the former to be discharged.

When at sea, tankers and nontankers generate 10–60 l of bilge oil per day. Most tankers retain the bilge oil in slop tanks for cargo oil or discharge it to shore reception facilities. Nontankers may either discharge the bilge oil at sea or at shore reception facilities. Use of the latter is facilitated if the ships are equipped with oily-water separators, but according to the NRC (1985) only about half of nontankers have such devices. About 30% of the estimated bilge and fuel oil discharge of 0.3 Mt y^{-1} comes from this bilge oil.

Nontankers, and particularly fishing vessels, may need to carry large quantities of ballast water for safety reasons, and because these vessels are not equipped with SBTs, the ballast water is carried in empty fuel tanks. When the ballast water is discharged, roughly 25% of the clingage is discharged as well. This input, however, is estimated to be only about 0.003 Mt y^{-1}.

Vessel Accidents

In recent years at least, a serious effort has been made to quantify oil discharges to the ocean from vessel accidents (Anonymous, 1991a). Not surprisingly, tanker accidents are by far the most important contributor in this category, the average discharge being about 0.4 Mt y^{-1}. The discharge from tanker accidents, however, can vary considerably from year to year, depending on the frequency of occurrence of supertanker[5] accidents. The largest tanker accident to date involved the *Atlantic Empress* and *Aegean Captain*, which crashed in a thunderstorm east of Tobago in 1979. About 0.3 Mt of oil was discharged by this single event. The *Castillo de Bellver* released about 0.25 Mt of oil when it caught fire and broke in half off the Cape of Good Hope in 1983, and the *Amoco Cadiz* discharged 0.22 Mt of oil when it grounded and split in half off the French coast in 1978 (Ware, 1989). For comparison, the wreck of the *Exxon Valdez*, the largest tanker accident in U.S. coastal waters, released only 0.035 Mt of oil into Prince William Sound in 1989. Tanker accidents are one of the most important causes of oil discharges to the ocean, both in terms of the quantity of oil and environmental impact. Because many tanker accidents occur near land, they tend to have a much greater impact on marine organisms than discharges from sources such as ballast water and bilge oil, which are normally released in the open ocean.

Atmosphere

The mechanisms here are vehicle exhaust and evaporation, followed by fallout in precipitation. Most of the fallout is believed to be the result of scavenging by rain of particulate material in the atmosphere. There is admittedly much uncertainty in the

---

[5]A *supertanker* is somewhat arbitrarily defined as a tanker of more than 200,000 dead-weight tonnes.

magnitude of the atmospheric input, the estimate of 0.3 Mt y^{-1} being little more than an educated guess. The principal problems in making an estimate are lack of data, the complex nature of atmospheric photochemistry, the large number of chemical compounds to be considered, and the difficulty of distinguishing hydrocarbons of petroleum origin from hydrocarbons derived from other sources. It is noteworthy, however, that about 68 Mt y^{-1} of petroleum are introduced into the atmosphere each year through the combustion of gasoline and other petroleum products (NAS, 1975). The NRC (1985) estimate that only 0.3 Mt y^{-1} enters the ocean from the atmosphere implies that almost all of the petroleum emitted into the atmosphere falls out on the land or is broken down photochemically in the atmosphere. While this conclusion is probably correct, the magnitude of the small percentage that reaches the ocean can be estimated only crudely. The NRC (1985) calculations, for example, are based almost entirely on estimates of normal alkane inputs, since information on atmospheric concentrations of other types of hydrocarbons is meagre.

Municipal and Industrial Wastes and Runoff

Municipal Wastes

At the present time municipal wastewaters probably account for more petroleum hydrocarbons discharged to the oceans than any other source. The estimate of 0.7 Mt y^{-1} made by the NRC (1985) is based on studies reported by Eganhouse and Kaplan (1981, 1982) on municipal wastewater in the Los Angeles area. Based on those studies, it was estimated that the per capita total hydrocarbon load in untreated wastewater was about 6.8 g d^{-1}. This figure was then normalized to petroleum consumption by multiplying by the number of persons estimated to be living within 80 km of the U.S. coastline and dividing by U.S. petroleum consumption in the year of the study. The conclusion was that the discharge of petroleum hydrocarbons in untreated wastewater amounts to about 0.032% of petroleum consumed. Primary and secondary sewage treatment were estimated to eliminate about 33% and 40%, respectively, of the petroleum hydrocarbons. The total discharge of 0.7 Mt y^{-1} of petroleum hydrocarbons in municipal wastewater was then calculated based on regional petroleum consumption figures and similar estimates of the percentage of petroleum hydrocarbons in wastewater removed by sewage treatment.

Industrial Wastes

Both oil refineries and other industrial operations may discharge oil into the ocean. Most nonrefinery industrial wastewater is believed to be discharged into municipal sewer systems and hence would be accounted for under the municipal wastes category. Some nonrefinery industrial waste is discharged directly into the marine environment, but there is little quantitative information on the magnitude of this discharge. The estimate of 0.2 Mt y^{-1} is admittedly little more than an educated guess. Some oil refineries also discharge wastewater into municipal sewer systems, but approximately 81% discharge directly into receiving waters (NRC, 1985). Studies have indicated that 5–75 g of petroleum hydrocarbons are discharged by refineries per tonne of operating capacity, the average value for the world being about 46 g t^{-1} (NRC, 1985). Since total refinery capacity is about 4200 Mt y^{-1}, the estimated discharge is $(46 \times 10^{-6})(4200) = 0.19$ Mt y^{-1}. However, only about half of this oil is discharged directly to the ocean, since only about half the refinery capacity is in coastal areas. The estimated discharge is therefore 0.1 Mt y^{-1}.

Runoff

Considered in this category are both urban runoff and river discharges. The estimate for urban runoff is based on studies in the United States and Sweden, which indicate a per capita contribution to urban runoff of roughly 1.0 g of petroleum hydrocarbons per day. Multiplying this figure by the estimated 120 million persons who live within 80 km of the U.S. coastline gives a discharge of 0.04 Mt y^{-1}. Given that the United States accounts for about 30% of the world's consumption of petroleum hydrocarbons, this figure was scaled up to 0.12 Mt y^{-1} for urban runoff worldwide. The river discharge figure was calculated using the per capita total hydrocarbon contribution to untreated wastewater of 6.8 g d^{-1}, which was then multiplied by the estimated 110 million persons whose wastewater enters interior rivers in the United States. On the assumption that only 5% of these hydrocarbons actually reach the ocean, the river discharge for the United States is estimated to be 0.014 Mt y^{-1}. Assuming this figure to be one-third of the world total, the total river discharge is estimated to be 0.04 Mt y^{-1}.

Ocean Dumping

The estimate of 0.02 Mt y^{-1} is based on the assumption that about 15 million tonnes of sewage sludge are discharged to the oceans each year and that this sludge contains about 0.1% petroleum hydrocarbons. The fact is, however, that ocean dumping of sewage sludge was prohibited by United States legislation in 1981 (Clark et al., 1984), and roughly half the 15 million tonne figure cited by the NRC (1985) was accounted for by U.S. sludge dumping which occurred prior to 1981. It therefore seems fair to say that oil inputs associated with ocean dumping of sludge are no more than 0.01 Mt y^{-1} at the present time.

Commentary

A certain amount of oil discharge is inevitably associated with oil use, and it is noteworthy that oil inputs to the ocean amount to only about 0.1% of current oil production. Furthermore, discharges from tanker ballasting operations, which accounted for over 20% of the NRC's (1985) estimated marine oil inputs, have almost certainly been reduced since data for the NRC report were assembled. On the other hand, the NRC (1985) report makes no mention of oil inputs caused by military hostilities. In recent years, wars in the Middle East have been a very significant cause of oil discharges to the Persian Gulf. In 1983, for example, a winter storm ruptured pipes around an oil platform in Iran's Nowruz offshore oil field, and thousands of barrels of oil per day began to flow into the northern Persian Gulf. Iraqi missles subsequently damaged two additional wells. Although capping the wells would probably have taken only about 3 weeks in the absence of hostilities (Begley, Carey, and Callcott, 1983), the Iran-Iraq war frustrated such efforts for more than 6 months. Estimates of the amount of oil that was discharged before the wells were finally capped range as high as 0.54 Mt (Ware, 1989). Eight years later the largest oil discharge to date occurred in almost the same location, when Iraqi forces deliberately released oil from the offshore tanker-loading station at Mina al-Ahmadi in Kuwait. Roughly 1.4 Mt of oil was discharged to the Gulf (Aldhous, 1991). These two incidents alone work out to an average discharge of 0.2 Mt y^{-1} from 1983 to 1991, and they are certainly not unique. During the first 10 months of 1990, for example, there were a total of 6 oil spills worldwide chalked up to either suspected sabotage or bombing by guerillas (Anonymous, 1991a), the average discharge rate being about 0.01 Mt y^{-1}.

One issue of particular concern about the Persian Gulf is the fact that it is shallow, with an average depth of only about 30 m (Aldhous, 1991). The ecosystem is therefore naturally productive, and there is great potential for an oil spill to damage living resources, including benthic organisms. This sensitivity is of course characteristic of virtually all coastal ecosystems, and it is unfortunately into coastal waters that much discharged oil goes. Geographically most of the remaining oil is discharged along the major tanker routes from the Middle East to Europe, the American continents, and the Far East.

The inputs of petroleum hydrocarbons to the ocean are therefore not distributed evenly, and much of the oil is released in coastal areas that are more susceptible to biological damage than the open ocean and more important to humans from the standpoint of fishery resources and recreation. What sort of impact are these oil discharges having on marine ecosystems, and what sort of changes can we expect with respect to oil pollution in the next few decades? To answer these questions, we must understand a little about the nature of oil; namely, how is it formed, what are its physical and chemical properties, what are its toxic characteristics, and what is its fate when released to the environment? The following three sections summarize our current understanding of these issues.

THE GENESIS OF OIL

Underground deposits of oil have been forming in one part of the world or another for hundreds of millions of years. The process of formation is undoubtedly very slow, probably requiring millions of years. Although no one is sure excatly how deposits of oil come to be formed, the following sequence of events describes the consensus opinion of many scientists.

Sedimentation

Organic detritus settles to the bottom of aquatic systems and is incorporated into the sediments. In most aquatic systems the rate of accumulation of organics in sediments is quite small compared to the primary production rate in the surface waters, since most of the organic carbon produced is ultimately respired. In highly productive systems, however, or in systems with estuarine type circulation patterns or stagnant bottom waters,[6] anoxic conditions may develop in bottom waters and hence prevent or retard the oxidation of detrital carbon. Such anoxic systems probably play an important role in the formation of oil, although the widespread nature of oil deposits has suggested to some that significant amounts of oil may have been formed from organics that accumulated in nominally oxidizing systems (Andreev et al., 1968). Such organics would necessarily have been rather refractory in nature. Since the oxygen level in sediments generally drops to zero below a depth of no more than a few centimeters, any organics that are preserved sufficiently long to be buried

---

[6]For example, the Black Sea and Cariaco Trench, respectively.

below a few centimeters of sediment would be effectively removed from the possibility of aerobic oxidation. Thus the first step in the formation of oil presumably requires one or more of the following conditions:

1. The existence of anoxic bottom waters.
2. The production of refractory carbon compounds.
3. Rapid sedimentation.

Under such circumstances organic detritus may accumulate at the bottom of the water column and ultimately become buried in the sediments.

Metamorphosis

Once buried in the sediments, organic carbon is ultimately incorporated into sedimentary rocks such as shales, sandstones, and carbonate rocks. Based on analyses of sediments, it seems likely that the organic carbon content of recently formed sedimentary rocks is on the order of no more than 0.1–1.0% (Andreev et al., 1968). Nevertheless, the organic carbon trapped in these rocks is apparently the source of the world's petroleum reserves. This carbon is slowly transformed under conditions of elevated temperature[7] and pressure found deep underground and perhaps through the mediation of catalysts such as aluminosilicate minerals. Humic acids, which are among the principal forms of organic carbon in sedimentary rocks, are probably the principal source of carbon for petroleum (Andreev et al., 1968). The presence of nitrogen and sulfur in virtually all crude oils, however, indicates that other types of organic substances are also involved. The sequence of transformations that convert sedimentary organic detritus into petroleum is a continuous process and undoubtedly highly complex. However, by correlating the composition of crude oil with its age, it is possible to get some idea of the sequence of transformations that organic compounds undergo while buried in sedimentary rocks. In general there is a tendency for the higher molecular weight compounds to be broken down with time, leading to the formation of paraffins and ultimately to the production of methane and perhaps graphite as end products (Andreev et al., 1968). Thus oil can be viewed as an intermediate stage in the breakdown of organic detritus in the absence of oxygen and under the influence of physical and chemical conditions (e.g., temperature, pressure, presence of catalysts) peculiar to deeply buried sedimentary rocks.

Migration

When contrasted to the low concentration of organic carbon in sedimentary rocks, the enormous deposits of oil in many parts of the world clearly imply the existence of some mechanism for concentrating or trapping the oil in a localized deposit from surrounding source rocks. In other words, the oil could not have been formed *in situ*, but instead formed over a much larger area and subsequently migrated to a localized deposit. The mode of migration is not known, but it seems reasonable to assume that liquid petroleum, once formed, could move upward through porous crustal materials until trapped by an impervious overlying substratum. The existence of natural oil seeps in various parts of the world testify to

[7]Probably 100–150°C.

the fact that such migrations can extend all the way to the surface if no impervious substratum traps the oil. Oil deposits are often overlaid by a pocket of gas and invariably underlain by a reservoir of water.[8] The existence of these two fluids in association with an oil deposit simply reflects the tendency of fluid substances to migrate upward through porous rocks until an impervious substratum is encountered. Thus the tendency of oil to migrate and form underground pools is by no means a characteristic peculiar to oil, but is a tendency common to all fluids. The migration and pooling of oil has been essential for its storage over millions of years and hence for its abundance today.

From the standpoint of modern industrial society, the most important characteristic of oil is the energy that can be derived from burning (oxidizing) it. This energy was of course originally fixed by way of photosynthesis millions of years ago and has been stored in the chemical bonds of the organic substances of which oil is composed. The anaerobic transformations that lead to the formation of oil release some of the original stored energy, but a sufficient amount remains in the chemical bonds of petroleum to provide a highly useful energy source. In summary, the genesis of petroleum deposits requires a) the formation by way of photosynthesis of energy-rich organic compounds; b) the storage of these compounds in an anaerobic environment or similar environment sufficiently reduced to retard or prevent their oxidation; c) the anaerobic transformation of these compounds into petroleum under conditions of high temperature and pressure; and d) the migration of the petroleum to underground traps or pools where it may remain for millions of years, during which time it undergoes additional transformations.

WHAT IS OIL

Natural crude oil is an exceedingly complex substance, composed of literally thousands of different kinds of organic molecules. Crude oil taken from different parts of the world may vary greatly in composition, depending on the age of the oil, the conditions of its formation, and so forth. Despite the complexity and variability of crude oil, some generalizations about its composition can be made.

Crude oil consists primarily of hydrocarbons, that is, compounds containing only carbon and hydrogen. Some crude oils contain as much as 98% hydrocarbons (Nelson-Smith, 1972). In addition to hydrocarbons, the organic substances in crude oil include compounds containing sulfur, nitrogen, and/or oxygen, with sulfur being more abundant than nitrogen and nitrogen more abundant than oxygen. In addition, there are small concentrations of metals such as nickel, vanadium, and iron (NAS, 1975).

Crude oils can be roughly characterized according to the relative amounts of the major kinds of hydrocarbons they contain. The three major types of hydrocarbons in crude oil are as follows.

Alkanes—Paraffins or Aliphatic Compounds

Alkanes or paraffins consist of one or more carbon atoms joined to four other atoms, either hydrogen or other carbon atoms. The simplest alkanes are methane, ethane, and propane, which contain one, two, and three carbon atoms, respectively

---

[8]The gas and water are less dense and more dense, respectively, than the oil.

(Figure 13.3). Larger alkanes, such as butane (C_4H_{10}) and pentane (C_5H_{12}), may exist in more than one configuration, as indicated in Figure 13.4. All such configurations, however, are classified as alkanes. Those configurations or isomers in which the carbon atoms are joined in a straight chain are referred to as *normal* or *straight chain alkanes*, whereas molecules consisting of more complex arrangements of the carbon atoms are referred to as *branched chain alkanes* (e.g., isobutane in Figure 13.4). Alkanes containing less than five carbon atoms are gases at room temperature. Alkanes containing five to about seventeen or eighteen carbon atoms[9] are primarily liquids at room temperature. Larger alkanes are solids at room temperature. Alkanes make up the major portion of the hydrocarbons in crude oil, accounting for anywhere from 60 to well over 90% of the hydrocarbon content (Fieser and Fieser, 1961).

Cycloalkanes or Naphthenes

Cycloalkanes or naphthenes are similar to alkanes, but the carbon chain is joined in a ring rather than being open ended. The simplest such cyclic compounds isolated from crude oil are cyclopentane and cyclohexane (Figure 13.5). Cyclopentane and cyclohexane are analogous to the normal alkanes pentane and hexane in the sense that the cyclic compounds have been formed by joining the two ends of the analogous normal alkanes while discarding two hydrogen atoms. Cyclic analogues of branched chain alkanes are also found in crude oil. In such cases the carbon ring contains alkyl substituents as well as hydrogen atoms (e.g., methylcyclohexane in Figure 13.5).

Aromatics

Aromatic compounds all contain one or more benzene rings, a 6-carbon ring structure in which the carbon atoms are joined by hybrid bonds intermediate in character between a single electron-pair bond and a double electron-pair bond. Some of the simpler aromatic compounds found in crude oil are indicated in Figure 13.6. These compounds may consist of a single aromatic ring (benzene) or of two or more rings joined together (e.g., napthalene). As was the case with the cycloalkanes, alkyl substituents may replace hydrogen atoms on the ring structure (e.g., toluene).

Refined Oil

Since oil pollution involves both refined petroleum products as well as crude oil, it is of some interest to know a little about the refining process and the types of substances that go into various petroleum products. The basic refinery technique takes advantage of the fact that the boiling point of hydrocarbons generally increases with increasing molecular size. Therefore one can separate hydrocarbons into various size fractions or approximate molecular weight groups by simply heating crude oil at a series of steadily increasing temperatures and collecting successively higher boiling fractions. This procedure, known as *fractional distillation*, is the basic oil refining technique. Table 13.3 indicates various fractions that may be collected in this way. The advent of the automobile created a demand for gasoline relative to the other components of crude oil which far exceeded the amount that could be obtained by simple fractional distillation. Therefore oil companies began to look for ways to convert the higher boiling fractions into gasoline by splitting or cracking the larger hydrocarbon molecules to obtain fragments in the C_4–C_{10} range.

---

[9]In other words, C_5 to C_{17} or C_{18} alkanes.

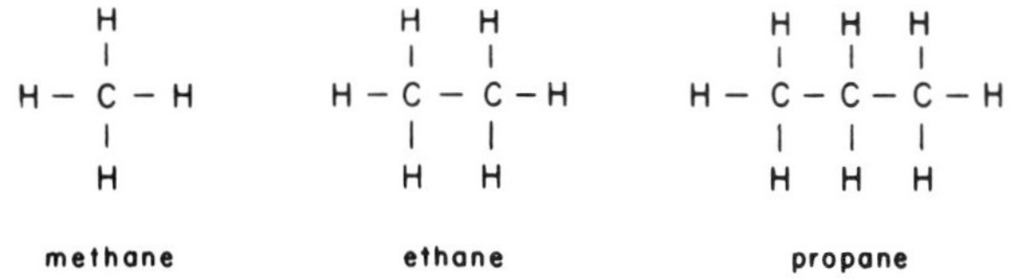

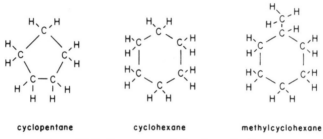

methane   ethane   propane

Figure 13.3 Structures of the simplest alkanes. Dashed lines indicate electron-pair bonds.

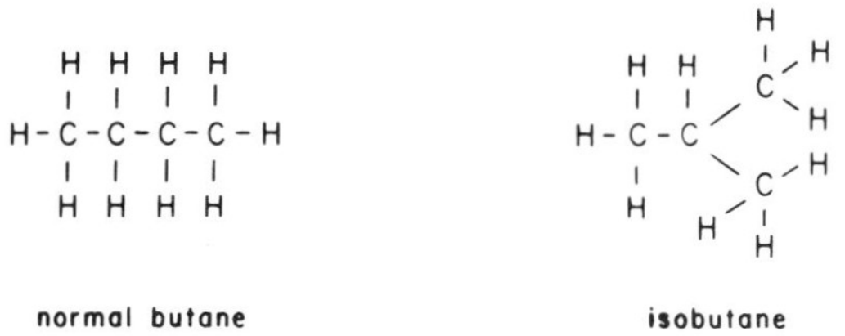

normal butane   isobutane

Figure 13.4 Structures of normal butane and isobutane.

cyclopentane   cyclohexane   methylcyclohexane

Figure 13.5 Structures of simple cycloalkanes.

Initially a thermal cracking technique was used. This method called for heating the heavier hydrocarbon components under pressure to effect their breakdown. Studies of the knocking characteristics of gasoline,[10] however, revealed that the hydrocarbons produced by thermally cracking crude oil were not of the sort best suited to retard knocking. Knocking is minimized in gasoline that contains primarily branched-chain alkanes, aromatics, and a class of hydrocarbons not found in crude oil, olefins. Olefins are analogous to alkanes and cycloalkanes, but contain at least some doubly bonded carbon atoms and correspondingly fewer hydrogen atoms. Two of the simplest olefins, ethylene and cyclohexene, are indicated in Figure 13.7. To produce gasoline with a more desirable mixture of hydrocarbons, a catalytic cracking procedure was developed. This technique yields a product rich in branched-chain compounds, olefins, cycloalkenes, and aromatics (Fieser and Fieser, 1961). Olefins may occur in concentrations as high as 30% in gasoline and account for about 1% of jet fuel. They are present in only minor amounts in other refined products (NRC, 1985).

With respect to aquatic pollution, it is important to note that the various hydrocarbons in crude or refined oil differ greatly in their toxicity to aquatic organisms, in

---

[10]Knocking occurs when gasoline burns explosively rather than smoothly in the cylinder of the engine.

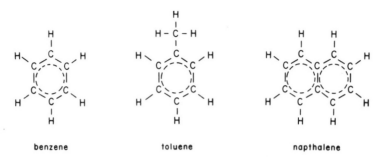

Figure 13.6 Simple aromatic compounds found in crude oil. The dashed circles inside the carbon hexagons indicate additional electron bonding.

Table 13.3   Petroleum Fractions Collected by Fractional Distillation Within Indicated Temperature Ranges

Fraction	Approximate boiling range	Approximate molecular size
Refinery gases	Up to 25°C	C_3-C_4
Gasoline	40–150°C	C_4-C_{10}
Naphta	150–200°C	C_{10}-C_{12}
Kerosene	200–300°C	C_{12}-C_{16}
Gas oils	300–400°C	C_{16}-C_{25}
Residual oil	Above 400°C	Above C_{25}

Source. Reprinted from Ryan (1977).

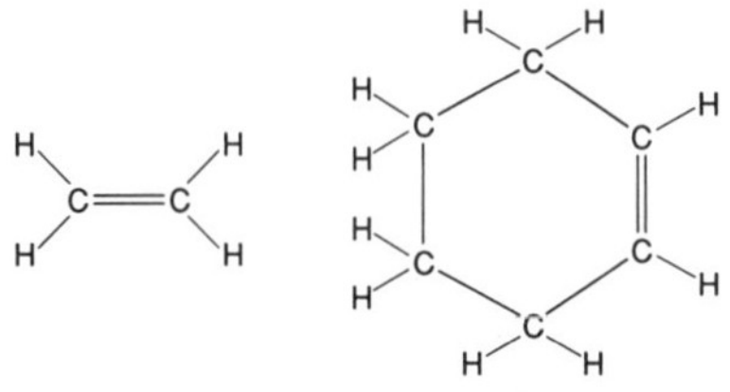

ethylene   cyclohexene

Figure 13.7   Structures of two simple olefins.

their resistance to degradation, and in their tendency to be incorporated into various components of the aquatic ecosystem. Thus the spilling of 100 tonnes of crude oil may have a very different effect on a coastal ecosystem than the spilling of 100 tonnes of gasoline or 100 tonnes of fuel oil. It is therefore appropriate at this time to examine the toxic characteristics of different petroleum fractions and products before attempting to evaluate the impact of oil pollution on aquatic systems.

TOXICOLOGY

Oiling and Ingestion

The toxic effects of oil fall into two general categories. The first category includes effects associated with coating or smothering of an organism with oil. Such effects are associated primarily with the higher-molecular-weight, water-insoluble hydrocarbons, the various tarry substances that coat the feathers on birds and cover intertidal organisms such as clams, oysters, and barnacles. Although some organisms such as tubeworms and barnacles are surprisingly little affected by such coatings (Nelson-Smith, 1972), the effect on organisms such as aquatic birds may be devastating. We will elaborate more on this point later on. The second category of toxic effects involves disruption of an organism's metabolism due to the ingestion of oil and the incorporation of hydrocarbons into lipid or other tissues in sufficient concentrations to upset the normal functioning of the organism. With respect to this second category of effects, it is generally agreed that aromatic hydrocarbons are the most toxic, followed by cycloalkanes, then olefins, and lastly alkanes. There is also a definite tendency for toxicity to be correlated positively with the molecular size of the hydrocarbons (Figure 13.8). Most toxic effects caused by ingestion of oil in water, however, are believed to be due to low-molecular-weight (C_{12}–C_{24}) alkanes and low-molecular-weight aromatics, since these compounds are the most soluble in water (NRC, 1985). Based on studies of the concentrations of hydrocarbons in water-soluble fractions of various oils, the NRC (1985) concluded that the contribution of compounds of higher molecular weight than alkylnaphthalenes was very small and

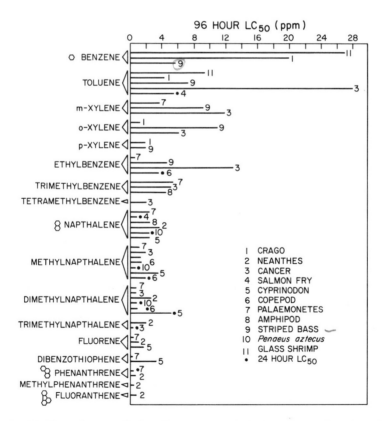

Figure 13.8 Median tolerance limits of some aromatic hydrocarbons for selected marine macroinvertebrates and fish. Duration of exposure was 96 hours unless otherwise noted. [Redrawn from NRC (1985).]

probably insignificant in terms of acutely toxic effects. Thus the compounds of greatest concern from the standpoint of ingestion are the low-molecular-weight aromatics such as benzene and toluene.

Toxic hydrocarbons apparently exert their effects in part by becoming incorporated into the fatty layer that makes up the interior of cell membranes (Nelson-Smith, 1972). As a result, the membrane is disrupted and ceases to properly regulate the exchange of substances between the interior and exterior of the cell. In extreme cases the cell membrane may lyse, allowing the contents of the cell to spill out and hence destroying the cell. There is also evidence that hydrocarbons interact with proteins in a variety of plants and animals (Nelson-Smith, 1972). Both enzymes and structural proteins appear to be affected. Once again, aromatics seem to be more toxic than other hydrocarbon classes with respect to this effect.

Based on the foregoing information, it is possible to understand a little about the relative toxicity of various types of oil. Refined petroleum products such as gasoline or kerosene contain virtually no high molecular weight hydrocarbons (Table 13.3), and hence exert very little in the way of a smothering or coating effect. However, because refined products do contain a higher percentage of low molecular weight hydrocarbons than crude oil, these products are more dangerous from the standpoint of ingestion than a comparable amount of crude oil (NAS, 1975; Tatem, Cox, and Anderson, 1978). With respect to this point, it is noteworthy that two of the more ecologically damaging oil spills, the grounding of the tanker *Tampico Maru* off Baja, California in 1957 and the grounding of the tanker *Florida* in Buzzards Bay, Massachusetts in 1969 involved spillage of refined petroleum, namely diesel oil and No. 2 fuel oil, respectively.

Crude oil on the other hand contains a significant fraction of high-molecular-weight hydrocarbons, which give it a viscous, sticky character. As a result, the greatest damage from crude oil discharges may be the coating of plants and animals with high-molecular-weight hydrocarbons. Undoubtedly the organisms most affected by oil coating or "oiling" are certain kinds of aquatic birds, namely auks, penguins, and diving sea ducks.[11] These birds are particularly susceptible to oiling for the following reasons:

1. They spend most of their lives on the surface of the sea.
2. They are poor fliers or are flightless.
3. They dive rather than fly in response to a disturbance.

Because of the third characteristic, such birds are unlikely to escape an oil slick, but instead tend to keep resurfacing in the slick after each dive. When oil is adsorbed to the feathers and down of these birds, their plumage becomes matted, and the air spaces which normally provide buoyancy and insulation become filled with water and oil. Because many of these birds are poor fliers at best, the additional load of oil and water in their feathers may make flight impossible, and some may literally drown due to the loss of buoyancy. Others presumably die due to their inability to

---

[11]Auks include murres, guillemots, razorbills, and puffins. Diving sea ducks include scoters, scaup, eiders, golden-eye, and long-tailed ducks. Of the penguins, only the Cape Penguins of South Africa and the Magellanic penguins of Argentina have been seriously involved. The distribution of other penguin species has tended to minimize the possibility of their being impacted by spilled oil.

maintain a proper body temperature without adequate insulation from their plumage. Invariably oiled birds attempt to clean themselves by preening their feathers, but in the process they may ingest as much as 50% of the oil in their plumage (Nelson-Smith, 1972) and die from the toxic effects of the ingested oil.

There is no doubt that large numbers of birds are killed each year from oiling. According to figures cited by Nelson-Smith (1972), about 150,000–450,000 birds die each year from oiling in the North Sea and North Atlantic. These estimates are based on counts of the number of dead and oiled birds that wash up on beaches, but exclude insofar as possible the effects of large oil spills. In other words, these losses are presumably due to small oil spills or the deliberate discharge of small amounts of oil from unknown sources. Large spills of crude oil can result in the killing of tens or even hundreds of thousands of birds. The breakup of the tanker *Torrey Canyon* off the coast of England in 1967, for example, is estimated to have resulted in the killing of 40,000–100,000 birds (mostly auks), and the stranding of the tanker *Gerd Maersk* in the Elbe estuary in 1955 is estimated to have killed 250,000–500,000 birds (Nelson-Smith, 1972). Only about 4500 dead and oiled birds were recovered after the breakup of the *Amoco Cadiz* off the French coast in 1978 (Mielke, 1990), but the grounding of the *Exxon Valdez* in Prince William Sound, Alaska, in 1989 is estimated to have killed 100,000 seabirds and 150 bald eagles (Hodgson, 1990), and as many as 40,000 seabirds may have died in the Severn Estuary in 1991 as the result of an oil spill caused by a fractured oil pipe at a steel plant in Wales (Anonymous, 1991b).

Weathering

The impact of a given oil discharge is determined very much by the nature of the hydrocarbons in the oil. Other factors, however, including weather conditions and the distance of the discharge from shore, also play an important role in determining the extent of ecological damage (Kerr, 1977; Mielke, 1990). This point requires some elaboration. Because most components of oil are less dense than water, oil discharges usually tend to float on the surface. Thus organisms that come into contact with the surface are most likely to be affected by oil pollution. However, hydrocarbons do have a finite, although small, solubility in water, and some of the oil will therefore dissolve in the water rather than simply floating on the surface. As already noted, the low-molecular-weight hydrocarbons are also the most soluble in water. On the other hand, the low-molecular-weight hydrocarbons are also the most volatile and therefore tend to evaporate from the surface most readily. Thus the characteristics of an oil slick are likely to change significantly within a day or so of the time the oil is released. Specifically, the percentage of dangerous low-molecular-weight hydrocarbons is likely to be greatly reduced due to evaporation and dissolution of these compounds, with evaporation accounting for most of the removal (Mielke, 1990). Most of the refinery gases, gasoline, and naphtha evaporate within a few days. Hence a slick that washes ashore a few days after discharge is probably much less toxic to intertidal organisms than a slick that is driven ashore immediately after discharge. Kerosene and gas oils may take several weeks or more to evaporate away. Indirect evidence from compositional changes and laboratory studies indicates that ultimately 20–40% of a typical crude oil slick may be removed by evaporation (Mielke, 1990). Gundlach et al. (1983), for example, estimated that 30% of the *Amoco Cadiz* spill was removed by evaporation.

Physical mixing of the water column can of course disperse a surface slick throughout the mixed layer if the turbulence is sufficiently great. By thus increasing the effective contact area between oil and water, mixing helps dissolve the oil as well as simply dispersing small droplets of oil within the water column. Because most components of the oil are rather insoluble in water, the dispersion process tends to produce an emulsion of oil-in-water or water-in-oil. The former consist of fine particles of oil dispersed in water. The latter contain up to 70% oil and are commonly referred to as *mousses*. In either case, mixing tends to expose subsurface organisms, including even benthic organisms if the water column is shallow enough, to the impact of an oil discharge, whereas in calm weather the same organisms would feel little or no direct impact from the discharge.

Ultimately most oil discharged into an aquatic system is degraded by a combination of biological, physical, and chemical processes, collectively referred to as *weathering*. Actual degradation rates depend on a great many factors, including the characteristics of the oil, temperature, availability of nutrients, degree of physical mixing, and so forth. About 90 species of bacteria and fungi are capable of subsisting on oil and hence degrading it (Mielke, 1990), but no single species is capable of metabolizing all the different components of crude oil. Under normal circumstances there are only about 10 bacteria capable of decomposing hydrocarbons in a liter of seawater, and growth of this bacterial population is normally slow in the period immediately following an oil spill (Mielke, 1990). As a result bacteria are of minor importance in removing the oil during the first few days after a spill. Bacterial degradation may, however, become significant during subsequent weeks, when the concentration of bacterial hydrocarbon decomposers may reach $5 \times 10^7/l$. The principal factors limiting bacterial degradation are the availability of inorganic nitrogen and phosphorus, temperature, oxygen concentration, and the degree of dispersion of the oil. Bacterial degradation was very ineffective in the open ocean following the Ixtoc I oil well blowout, for example, because of the low concentrations of inorganic nutrients in the surface waters (Mielke, 1990). In the case of the *Amoco Cadiz* tanker accident, however, bacteria consumed about 73,000 barrels of oil before the oil could reach the shoreline. Bacterial degradation rates tend to be positively correlated with temperature and according to Mielke (1990) are 4 times faster at 18°C than at 4°C. The result is that bacterial degradation takes a much longer time in polar climates than in temperate or tropical latitudes. Oxygen concentrations are not likely to be limiting in the open ocean, but may be a significant limiting factor in sediments and beach sands. The availability of oxygen, for example, probably accounts in part for the fact that degradation generally occurs more rapidly in coarse than in fine-grained sediment.

In the presence of oxygen and sunlight, the organic compounds in oil may be oxidized by photochemical mechanisms, the result being the production of oxygenated compounds such as carboxylic acids, alcohols, ketones, and phenols. These compounds are generally more soluble than their precursors, and in some cases are more toxic.

When the low-molecular-weight components of the oil have been removed through evaporation, dissolution, or degradation, there remain the tarry, viscous, residual oils. These high-molecular-weight compounds tend to form lumps of tar, which degrade much more slowly than the other components of the oil. Small tar balls have been noted floating on the surface of the ocean or washing up on beaches since as long ago as 1954 (Horn, Teal, and Backus, 1970), but undoubtedly have been components of some marine ecosystems for thousands if not millions of years. The lumps range in size from a few millimeters to about 10 cm in their longest dimension and are usually composed of paraffinic hydrocarbons containing as many as 40 carbon atoms (Mielke, 1990). The presence of barnacles and isopods growing on such tar lumps clearly indicates that the hydrocarbons are relatively nontoxic (Horn, Teal, and Backus, 1970). By determining the age of the attached organisms, Horn, Teal, and Backus (1970) inferred that at least some tar lumps float on the surface for as much as several weeks. It has been estimated, however, that 5–10% of the pelagic tar in the eastern Gulf of Mexico 4 years after the Ixtoc I blowout originated from that event (Mielke, 1990), a conclusion that obviously implies a residence time of several years for some pelagic tar. Ultimately some of the tar lumps wash up on beaches (NAS, 1975); others may become negatively buoyant due to weathering or other processes and sink to the bottom. Some tar lumps have been found in the stomachs of small fish (Horn, Teal, and Backus, 1970) and from that point may either be metabolized, excreted in feces, or transferred up the food chain. There are estimated to be about 0.3 Mt of pelagic tar in the oceans' surface waters at the present time (NRC, 1985).

In summary, oil discharged at sea is subject to a variety of degradative and removal processes, and if a discharge is sufficiently far from land most of the toxic components of the oil will have been removed or reduced to insignificant concentrations by the time the oil reaches shore. The principal ecological damage in such cases probably stems from the oiling of aquatic birds with the higher-molecular-weight hydrocarbons and from the aesthetic damage caused by the stranding of tar on recreational beaches.

When an oil discharge occurs near shore, the ecological damage is likely to be much greater. In such cases the oil may be washed ashore before weathering has had a chance to remove much of its toxicity. Intertidal and benthic organisms, many of which are sessile or have poor powers of locomotion, may then be destroyed in large numbers due to ingestion of the toxic hydrocarbons. The damage caused by the *Tampico Maru* and *Florida* tanker spills was intensified both by the fact that the tankers were carrying refined oil and by the fact that the spills occurred close to shore. Motile organisms such as fish can of course simply swim away to avoid an oil spill, but this defense mechanism is not infallible. Large numbers of dead fish washed ashore in Buzzards Bay after the grounding of the *Florida* in 1969 (NAS, 1975).

Lethal and Sublethal Effects

What concentrations of oil are toxic to marine organisms? Obviously the answer to this question depends on the type of oil or combination of hydrocarbons you are talking about, as well as the species involved. Adult fish are rather resistant to oil pollution, since their bodies, including the mouth and gill chambers, are coated with a slimy mucus that resists wetting by oil. Crude oil becomes toxic to adult and larval finfish at concentrations of roughly 100–500 ppm and 1–50 ppm, respectively (EPA, 1976; NRC, 1985), but individual aromatics such as benzene, toluene, and naphthalene are toxic at concentrations about 100 times lower (EPA, 1986). Whereas fish can swim away to avoid a spill, plankton cannot. Crude oil has been found to be toxic to zooplankton, including fish eggs, at concentrations ranging from 10 ppb to 10 ppm (NRC, 1985). Sublethal effects of oil have been studied on a variety of aquatic organisms. Effects on phytoplankton photosynthesis and growth have been reported at concentrations typically on the order of 0.1–10 ppm (NRC, 1985). Interference with larval development and growth occurs at crude oil concentrations ranging from 10 ppb to 100 ppm, depending on the organism and type of oil. Adults generally tend to be more tolerant, but some sublethal effects have been reported at concentrations as low as 10 ppb (EPA, 1976).

The general conclusion is that toxic effects of oil are not observed below concentrations of 10 ppb. There are, however, a few exceptions to this generalization. A reduction in the chemotactic perception of food by the snail *Nassarius obsoletus* has been reported at a kerosene concentration of 1 ppb (EPA, 1976), interference with the growth of microalgae was observed at an oil (unspecified type) concentration of 0.1 ppb (EPA, 1976), and changes in the rate of development of sea urchin eggs have been observed at a benzo(a)pyrene concentration of 0.1 ppb (NRC, 1985).

Some organisms that are particularly resistant to oil pollution may flourish in oil-polluted waters or for a period of time after an oil spill due to the absence of more sensitive predators or competitors. Both the bloodworm *Tubifex* and the annelid worm *Capitella capitata* appear to thrive in oil-polluted sediments (Nelson-Smith, 1972). Blumer et al. (1971), for example, noted a population explosion of *Capitella capitata* in some oil-polluted areas of Buzzards Bay following the *Florida* spill. Thompson et al. (1977) reported that scarlet prawns *Plesiopenaeus edwardsianus* were 3 times more abundant in oiled sediments off the northwest coast of Arabia than in unpolluted control areas. The prawns were apparently attracted by petroleum-derived compounds formed by bacterial metabolism of the oil. Cyanobacteria are apparently quite resistant to oil pollution, and patches of cyanobacteria have been noted on otherwise barren areas near refinery outfalls or on trickling filters designed to remove oil from effluents (Nelson-Smith, 1972).

Not a great deal is known about the effects of oil on macrophytes, most reported observations having been made following oil spills. There have been almost no laboratory studies. To a certain extent plants are surprisingly resistant to oil toxicity, and even if extensively damaged tend to recover more rapidly than many other organisms (NAS, 1975). A good example of the resiliency of plants is provided by events following the *Tampico Maru* spill. Giant kelps were seriously affected by the spill, but a population explosion of giant kelp occurred shortly thereafter (Nelson-Smith, 1972). The rapid recovery of the kelp was apparently due in part to the absence of the usual herbivore grazers, particularly sea urchins, which normally feed on the kelp (NAS, 1975). Evidently these herbivores were unable to reestablish themselves in the affected area as rapidly as the kelps were. Species of *Ulva* and *Enteromorpha* have also been observed to proliferate following oil spills, again presumably due to the absence of the usual herbivore grazers (NRC, 1985).

These observations should not, however, be taken to imply that macrophytes prefer oily water. Oil does adversely affect macrophytes. Intertidal macrophytes were seriously impacted, for example, by the oil spills resulting from the groundings of the tankers *Arrow* in 1970 and *Amoco Cadiz* in 1978. In most cases, however, macrophytes appear to recover rapidly following such spills, and because populations of their predators become reestablished slowly, a bloom of macrophytes may occur. Long-term adverse effects on macrophytes are evident, nevertheless, from some field studies. In the case of the *Arrow* spill, for example, one species of bladder wrack that disappeared during the spill had not reappeared 6 years later, and chronic adverse effects from oiling have been reported in certain sea grass communities (NRC, 1985). In the case of *Spartina*, a salt marsh macrophyte, adverse effects may not become apparent until a year or two after an oiling incident (NRC, 1985).

In summarizing this section, it is worthwhile to point out that the concentration of petroleum hydrocarbons in most parts of the ocean is less than 10 ppb (NRC, 1985). From the toxicity studies reviewed here, it would appear that most organisms are not likely to be affected by concentrations this low, particularly if these hydrocarbons consist predominantly of weathered oil from which many of the more toxic components have been removed. It is certainly true, however, that in some coastal areas oil pollution or various stresses or disruptions associated with oil production have significantly affected aquatic ecosystems. For example, the concentrations of petroleum hydrocarbons in polluted coastal sediments are often found to be as high as several thousand ppm and in extreme cases as high as 19% on a dry weight basis (NRC, 1985); natural hydrocarbon concentrations in unpolluted marine sediments are less than 250 ppm (NRC, 1985). Deep ocean sediment hydrocarbon concentrations are only 1–4 ppm (NAS, 1975).

An informative picture of the impact of oil production operations on a coastal ecosystem is provided by the case of coastal Louisiana, where more than 25,000 oil-producing wells are located in the same area as a large commercial fishery. From 1945 to 1975 it is estimated that at least 160,000 tonnes of oil were discharged into Louisiana's coastal waters, but the total production of the coastal fishery remained high (NAS, 1975). As was the case with Lake Erie, however, there were significant changes in the composition of the catch. The shrimp catch, for example, shifted from about 95% white shrimp to about a 50:50 mixture of white and brown shrimp. This shift in composition was apparently caused by destruction of the white shrimp's brackish water nursery grounds due to construction activities associated with the laying of pipelines and erection of drilling rigs. The original oyster beds were similarly destroyed, and although the total yield of oysters remained high, the yield per unit area on the new oyster grounds was only 11% of the pre-1945 figure (Howarth, 1981). Thus a significant disruption of the coastal ecosystem occurred not so much due to the discharge of oil, but rather to construction and drilling activities that accelerated erosion, altered tidal currents, and caused the silting up of some embayments (NAS, 1975).

Human Health

With respect to human health, there seems to be little cause for concern from oil pollution of the aquatic environment. In drinking water, petroleum compounds become organoleptically objectionable at concentrations far below levels associated with chronic toxicity (EPA, 1976). Thus consumption of oil-polluted water is unlikely to be a significant source of exposure for humans. A more probable route of exposure is the consumption of oil-contaminated fish or shellfish, and in this regard the chief concern is the fact that oil contains a number of carcinogenic compounds. Among these carcinogens are benzene and certain polynuclear aromatic hydrocarbons (PAHs). Benzene constitutes as much as 1.6% by volume of crude oil and up to 5% of refined products (Politzer et al., 1985). Because of benzene's volatility, however, the benzene content of other than recently discharged oil is likely to be quite small. Indeed most human exposure to benzene comes from inhalation, the principal sources being automobile exhausts and cigarette smoke. Urban residents in the United States, for example, inhale about 0.85 mg d^{-1} of benzene (Politzer et al., 1985).

The input of PAHs to the worldwide aquatic environment is estimated to be 0.23 million tonnes per year, about 75% of which is accounted for by oil spillage (Politzer et al., 1985). It is doubtful, however, whether the consumption of PAH-contaminated fish is a significant human health concern. There is no evidence, for example, that any of the known carcinogenic PAHs cause cancer in humans by oral intake (Politzer et al., 1985). Furthermore, when the concentration of petroleum hydrocarbons in marine organisms exceeds about 200–300 ppm, the organisms acquire a distinctly tainted taste, become unpalatable, and are unlikely to be eaten anyway. Experimental studies have shown that organisms contaminated with PAHs can be depurated by simply putting them in clean water for a period of time. For example, the half-life of benzo(a)pyrene in mollusks is about 15 days (Politzer et al., 1985). Finally, seafood is by no means the only foodstuff containing PAHs. Fruits, vegetables, and cooked or smoked meats often contain PAH concentrations on the order of a few ppb (Politzer et al., 1985). Fish and shellfish do not contain unusually high concentrations of PAHs relative to other foodstuffs, and for most persons the consumption of aquatic organisms is therefore a minor source of exposure to PAHs.

A CASE STUDY—BUZZARDS BAY

On September 16, 1969, the oil barge *Florida* went aground within a few hundred meters of land off Fassets Point in Buzzards Bay, Massachusetts (Figure 13.9). Damage to the barge resulted in the release of its cargo of 570 tonnes of No. 2 fuel oil. A storm the following day drove the oil ashore, mixing it into water and sediments (Teal and Howarth, 1984). The spill occurred within 10 km of the Woods Hole Oceanographic Institution, and therefore provided an excellent opportunity for scientists to intensively study the impact of the spill. Since the spill occurred in shallow water and consisted of a mixture of relatively low molecular weight hydrocarbons,[12] it was not surprising that the destruction of marine life was severe in the immediate vicinity of the spill. The oil penetrated into benthic sediments as much as 13 m below sea level and washed into marshes and tidal inlets, penetrating to 0.5 m or more in the inshore sediments. Sediment oil concentrations were as high as 400 ppm in sublittoral areas and over 2000 ppm in some portions of the intertidal zone (NRC, 1985). In the area of the initial spill, which covered about 2 km^2, virtually all marine organisms were killed. The dead organisms included fish and shellfish, worms, crabs, snails, clams, lobsters, and numerous other invertebrates. Ninety-five percent of the organisms taken in trawls in the affected area were found to be dead or dying (Blumer et al., 1971), and marsh grass (*Spartina alterniflora*) was completely killed in intertidal areas where oil concentrations exceeded 2000 ppm in the sediments.

The killing of benthic plants and animals resulted in the destabilization of the bottom sediments, a result similar to the effects of thermal pollution on the benthos

---

[12]No. 2 fuel oil is a type of diesel oil, which contains hydrocarbons that distill above 275°C. The hydrocarbon range is primarily C_{12}–C_{25}, with some higher molecular weight components (Politzer et al., 1985).

of Biscayne Bay. The oil-contaminated sediments were then easily eroded by currents and wave action and transported to other parts of the bay, carrying the oil with them. In this way the polluted area was enlarged rapidly and by the spring of 1970 covered about 22 km², roughly 10 times the area affected by the initial spill (Blumer et al., 1971).

Analyses of surviving shellfish revealed that they were contaminated with petroleum hydrocarbons for several years after the spill, and as a result shellfishing was banned in the area during that time. Placing shellfish in clean water for as much as 6 months failed to rid them of their hydrocarbon burden (Blumer et al., 1971). As expected, the hydrocarbons had been stored in the lipid tissue of the shellfish, where at least some of them remained, resistant to biological breakdown or excretion, much like chlorinated hydrocarbon pesticides. Damages paid by the owners of the oil for cleanup and as compensation for the destruction of fishing and shellfishing resources totaled about $500,000 (Blumer et al., 1971).

Although some bacterial breakdown of the petroleum hydrocarbons in the sediments was evident in all parts of the bay within a year of the spill (Blumer et al., 1971), a 7-year study of fiddler crab populations in the adjacent salt marshes by Krebs and Burns (1977) revealed that some toxic components of the oil persisted in these sediments for many years. Fiddler crab populations in heavily oiled areas of the marsh were greatly reduced following the spill, and living crabs found in these areas displayed aberrant behavior, including impaired locomotor and burrowing behavior, lengthy escape response time, increased molting, and the display of mating colors, although the time of the spill did not correspond to the mating season. Juvenile crabs were unable to overwinter successfully in areas of the marsh where sediment petroleum hydrocarbon concentrations exceeded 200 ppm, evidently due to the abnormally shallow burrows dug by crabs in these areas. As a result the juveniles evidently died from exposure to the cold. The abnormal burrow construction was attributed by Krebs and Burns (1977) to impairment of the crabs' locomotor abilities by ingested hydrocarbons. Some crabs were found to contain as much as 280 ppm hydrocarbons in body tissues. For at least 4 years after the spill, the percentage of female crabs in some areas of the marsh was abnormally low, presumably due to differential mortality or emigration from polluted areas. By the seventh year after the spill, crab populations in most areas of the marsh were returning to normal, although recovery in some parts of the marsh was still incomplete. The recovery pattern was found to be correlated highly with the loss of the naphthalene fraction of the oil in the sediments. Oil and some of its effects persisted in part of the area impacted by the spill for at least 12 years, and although most areas of the salt marsh were normal in gross appearance by that time, complete recovery had not yet occurred (Teal and Howarth, 1984).

In summarizing the effects of the Buzzards Bay spill, several points deserve emphasis. First, the persistence of the hydrocarbons in the ecosystem and their effects on organisms far exceeded any visible evidence of the spill which a casual observer might have noted. Because the spilled oil was relatively light and contained few of the high molecular weight residual oils found in crude oil, visible evidence of the spilled oil on the beaches and salt marshes was gone within a few days. The bodies of the many dead animals initially killed by the spill had decomposed or washed away within a similar time period. Thus to a casual observer any evidence of the spill or its effects had disappeared within a week. Only careful

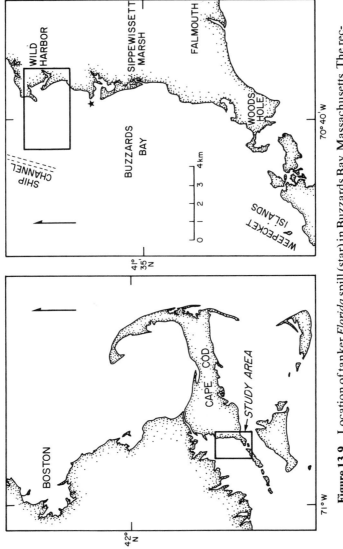

Figure 13.9 Location of tanker *Florida* spill (star) in Buzzards Bay, Massachusetts. The rectangle including Wild Harbor was intensively sampled by scientists. [Redrawn from Sanders et al. (1980).]

442

chemical analyses of the sediments and organisms revealed the persistent presence of petroleum hydrocarbons in the ecosystem, and a detailed and thorough study of benthic and salt marsh organisms was required to appreciate the extent and duration of the biological damage caused by the spill.

Second, the persistence of the hydrocarbons in the ecosystem greatly exceeded what might have been expected from the characteristics of the oil and the nature of mechanisms for detoxifying or removing spilled oil from the ocean. Although some of the lighter hydrocarbons undoubtedly evaporated away, the fact that other low molecular weight hydrocarbons evidently found their way into the sediments and continued to kill organisms for several months after the spill indicates that some of the acutely toxic, low molecular weight hydrocarbons persisted in these sediments for at least several months. Even though bacterial breakdown of the hydrocarbons was evident in all parts of the affected area within 1 year of the spill, it is generally the case that the least toxic hydrocarbons are metabolized first (Blumer et al., 1971), and it is apparent that some toxic hydrocarbons remained in the ecosystem for roughly 10 years (Krebs and Burns, 1977; Teal and Howarth, 1984). Once stored in the lipid tissues of organisms or in sediment, which is usually anoxic below the first few centimeters, petroleum hydrocarbons are subject to only slow degradation and thus mimic persistent poisons such as chlorinated pesticides in their longevity (Blumer et al., 1971). Metabolic studies revealed the induction of hydrocarbon-metabolizing enzymes in fish, and within 4 years of the spill hydrocarbon burdens in the fish were down to background levels (Burns and Teal, 1979). Induction of similar enzymes was also evident in fiddler crabs, but was evidently insufficient to deal with the body burden of hydrocarbons within the lifetime of these organisms (Burns, 1976).

Finally, it is appropriate to compare the effects of the Buzzards Bay spill with the impact of some of the other noteworthy oil spills that have occurred in the past 25 years. In terms of the amount of oil released, the Buzzards Bay spill was certainly not one of the larger spills. The Ixtoc I oil well blowout, the largest accidental oil spill of any kind, released about 840 times as much oil, and the grounding of the *Amoco Cadiz*, one of the largest tanker accidents to date, released about 400 times as much oil. The biological impact of the Ixtoc I blowout was not very thoroughly studied, but the damage to aquatic organisms appears to have been surprisingly small considering the amount of oil released. Shrimp landings during subsequent years in the affected area, for example, were as high or higher than in years preceding the blowout (Mielke, 1990). There was some loss of aquatic birds due to oiling, but along the Texas coast the percentage of oiled birds never exceeded 10%. The minimal impact on organisms along the U.S. coast undoubtedly reflected the extensive weathering of the oil at sea and the fact that only about 1% of the spilled oil actually reached U.S. beaches. There was a loss of revenue to the recreational and tourist industries along the Texas coast in the months following the accident, but a year later there was almost no visual evidence of the spill. Cleanup costs along the Texas coast amounted to about $4 million. However, the lost oil revenues and estimated total cost to the Mexicans of capping the well and containing the environmental damage amounted to about $220 million (Mielke, 1990).

The *Amoco Cadiz* spill was similar to the *Florida* grounding in the sense that the oil was discharged almost on the beach and that strong winds mixed the oil into the water column and drove it ashore. The quantity and type of oil discharged, however, were quite different. The *Amoco Cadiz* was carrying about 0.22 million tonnes of light Arabian crude oil, and virtually the entire cargo was lost. About 400 km of French coastline were oiled. The impact of the *Amoco Cadiz* spill on biota was thoroughly studied, but any analysis of long-term effects was compromised seriously by the fact that almost exactly 2 years after the *Amoco Cadiz* accident the tanker *Tanio* broke up about 65 km off the Brittany coast and released 7000 tonnes of oil into many of the same areas affected by the *Amoco Cadiz* spill.

Not surprisingly, there was almost a complete kill of the flora and fauna in exposed intertidal mud flats and marsh areas after the Amoco Cadiz spill (Teal and Howarth, 1984). In the water column there was considerable mortality of zooplankton, and an immediate kill of several tonnes of fish. About 4500 seabirds are known to have died from oiling. There were declines in recruitment, growth rates, and catches of several commercially important bottom fish, and an unusual incidence of fin rot in some species during the first year after the spill. Adverse effects on plaice continued to be found for several years in heavily impacted areas, and some oysters were still contaminated with hydrocarbons, although there was little evidence of histopathological or biochemical damage (Mielke, 1990). France recently won a court settlement of $120 million against Amoco for the damage caused by the spill (Joyce, 1990).

The general consensus that has emerged from analyses of oil spills is that circumstances at the time of the spill play a large role in determining their environmental impact. Spills that occur close to shore are likely to cause much more environmental damage than spills in the open ocean. In the former case, there is a high probability that benthic and intertidal organisms will be impacted, and there is less time for weathering to reduce the toxicity of the oil before significant impacts occur. This fact is dramatically illustrated in the comparison of the effects of the Ixtoc I and *Amoco Cadiz* spills. Unfortunately, the concentration of marine organisms, including commercially important species, is much greater in coastal areas than in the open ocean. Thus the potential for damage is far greater when spills occur in coastal areas. As already noted, much of the oil released to the ocean is unfortunately discharged in the nearshore environment.

Most scientists agree that in general the impact of a spill of refined petroleum is likely to cause more ecological damage than the release of a comparable amount of crude oil, because the latter contains a smaller percentage of the more dangerous low-molecular-weight components. This rationale presumably accounts for the extent of biological damage caused by the comparatively small amount of refined oil released by the tanker *Florida*. A large spill of crude oil, however, can expose organisms to the same concentrations of these toxic components as a small spill of refined product and also contaminates the system with the viscous residual oils not found in refined products. Thus, with the exception of the smothering and coating effects of the residual oils, "The same sorts of effects can be found in spills of all types of oils. The major difference in results from spills of different oil types would probably come primarily from the dilution of the more toxic compounds" (Teal and Howarth, 1984, p. 36). Given the environmental and monetary costs associated with these oil discharges, it is now appropriate to ask what can and has been done to reduce them and to ameliorate their effects when they occur.

CORRECTIVES

The wreck of the tanker *Torrey Canyon* off the coast of England in 1967 is a milestone in the history of oil pollution for several reasons. First, the spill was clearly the result of carelessness, an issue that has remained up to the present day (Hodgson, 1990). On a morning when the visibility was 13 km and his ship within range of a lightship, three lighthouses, and a radio beacon, an experienced captain ran his tanker at top speed onto Seven Stones Reef, a well-known navigational obstacle in a channel almost 10 km wide (Ware, 1989). Second, efforts to clean up the spill illustrated dramatically the inadequacy of understanding and techniques for minimizing the environmental damage caused by such spills. It is now generally agreed, for example, that detergents used to disperse the oil caused more damage to marine organisms than the oil itself (Nelson-Smith, 1967). Finally, the public reaction to the spill presaged an era of environmental awareness; similar reactions to subsequent spills have been influential in pressuring governments and the oil industry to take steps to minimize the amount of oil discharged to the environment and to ameliorate the adverse effects caused by the oil that is released.

Prevention

Certainly one area where significant progress has been made is in the routine operations of oil tankers. The use of LOT and COW and the requirements for clean or segregated ballast tanks mandated by MARPOL and the 1978 U.S. Port and Tanker Safety Act have done much to reduce routine discharges of oil from tankers due to ballasting operations, at one time a major source of oil discharges to the ocean. In addition, the Port and Tanker Safety Act mandates that tankers of 20,000 dwt or above be equipped with inert gas systems to prevent fires and explosions in the cargo tanks, and that tankers of 10,000 gross tons or above be equipped with a dual radar system and two remote steering gear control systems operable separately from the navigating bridge. The requirement for backup radar and steering systems was undoubtedly influenced by the *Amoco Cadiz* grounding, which was caused in part by a faulty steering system. It is noteworthy that the Port and Tanker Safety Act applies not just to vessels of U.S. registry, but to any vessel that operates on or enters the navigable waters of the United States, or which transfers oil or hazardous materials in any port or place subject to the jurisdiction of the United States.[13] Thus, for example, any foreign oil tanker which offloads oil at U.S. ports must abide by the stipulations of the Port and Tanker Safety Act.

Although such legal developments have helped to reduce the probability of oil spillage, the efficacy of the legislation depends in part on conscientious enforcement and human motivation. The Port and Tanker Safety Act, for example,

---

[13]The only noteworthy exception to this restriction is vessels engaged in so-called innocent passage through the territorial sea of the United States or in transit through the navigable waters of the United States that form a part of an international strait.

stipulates that no person may be issued a Federal license to pilot any steam vessel unless he is of sound health and has no physical limitations that would hinder or prevent the performance of a pilot's duties. Nevertheless, Joseph Hazelwood, the captain of the *Exxon Valdez*, had a history of alcoholism, having been arrested several times for drunken driving and treated in 1985 for alcohol abuse. The *Exxon Valdez* grounded on Bligh Reef because it left the normal outbound shipping lane in Prince William Sound to avoid icebergs and subsequently failed to resume a normal course. At the time of the grounding Hazelwood was absent from the bridge, having turned over the ship to an uncertified third mate. Hazelwood had an unacceptably high blood-alcohol level 9 hours after the accident (Lemonick, 1989). The fact that tanker accidents sometimes occur during inclement weather is understandable. However, accidents such as the wrecks of the *Torrey Canyon* and the *Exxon Valdez*, both of which occurred during fair weather, are inexcusable. Such events certainly raise the question of whether there is always a competent person on the bridge when tankers are underway. During the first 10 months of 1990 tanker collisions or groundings accounted for a total of 34 oil spills worldwide (Anonymous, 1991a). How many of these were at least in part the result of human error or incompetence is unclear, but it seems reasonable to speculate that better training and adherence to rules might reduce the frequency of these incidents.

Another area of concern is the condition and maintenance of the world's tanker fleet. This fleet currently numbers about 2750 vessels, 55% of which are more than 15 years old (Drewry, 1991a). Over 430 of the fleet are supertankers, and these account for almost 50% of the fleet's dead weight tonnage. About 70% of the supertankers are more than 15 years old, most of them having been built between 1972 and 1976 (Drewry, 1991a). Although tankers do go into dry dock at approximately 2-year intervals for routine maintenance and inspection, there is evidence that at least some of the tankers that put out to sea are structurally unsound. Carter (1978, p. 514), for example, citing the number of tanker accidents chalked up to structural failure, commented that in some cases, "an old rust bucket had simply broken up at sea and sunk." This problem still exists. On January 24, 1990, for example, a tanker split in two in Wakasa Bay, Japan, discharging 770 tonnes of oil. On August 4 of the same year a tanker sank during a typhoon near Japan, discharging over 800 tonnes of oil. At least five other oil discharges during 1990 were attributable to tankers' sinking (Anonymous, 1991a). It seems likely that the frequency of such incidents could be reduced by establishing a stricter program of tanker inspection and maintenance and/or by requiring that tankers be retired after reaching a certain age. Although some persons would undoubtedly object to the idea of arbitrarily mothballing tankers at a given age (say 20 years), perhaps feeling that from an economic standpoint it would be more logical to run tankers until they fall apart, it is clear from an environmental standpoint that there is a need to get tankers off the sea before they become structurally unsafe. The present system of periodic maintenance and inspection is evidently not getting the job done. The present tanker fleet includes over 200 ships more than 25 years old. It is a fair bet that such old tankers should not be in the oil transportation business.

Cleanup

Oil discharges result from human error, acts of God, sabotage, and routine operations associated with the transportation and use of oil. No matter how conscientious and careful we may be, discharges are bound to occur. Some of these discharges may be of little environmental consequence, but when a spill occurs with potentially serious environmental consequences, what can be done to ameliorate the damage? Unfortunately no cleanup method is without drawbacks or limitations, but it is worthwhile to review the state of the art to see what approaches are available.

Offloading

When a tanker goes aground, collides with another ship, or for any reason is in danger of discharging its cargo, an obvious move to minimize environmental damage is to offload the oil. Much has been made, for example, of the fact that the *Exxon Valdez* discharged 0.036 Mt of oil into Prince William Sound, but few people are aware of the fact that the tanker was carrying 5 times as much oil. Almost 80% of the cargo was successfully pumped into smaller vessels while the tanker lay atop Bligh Reef (Hodgson, 1990). Unfortunately tankers often go aground in rough weather, and under such conditions it is difficult if not impossible to attempt a reasonably safe offloading operation. For example, none of the 0.025 Mt of No. 6 fuel oil carried by the *Argo Merchant* was offloaded when the tanker grounded on the Nantucket Shoals in December, 1976. Instead, rough seas during the following week caused the tanker to break up, releasing its entire cargo. The *Amoco Cadiz* similarly lost its entire cargo when it broke up off the coast of Brittany during heavy seas in 1978. The U.S. Coast Guard has developed special offloading systems involving hydraulically operated submersible pumps for offloading oil in high-risk situations, but the fact remains that it is simply unsafe to bring a receiving vessel alongside a stricken tanker in rough seas.

Burning

Once oil has been spilled onto the water, there are basically three short-term approaches to cleanup, namely burning, chemical dispersal, and mechanical collection. The idea of burning an oil slick is logically appealing, at least if the slick is in a place where the fire itself is not likely to pose a threat, but in practice it has been found difficult to ignite and maintain a combustion of heavy oil, even with the aid of flame throwers. "In the case of the *Torrey Canyon* disaster, the addition of thousands of gallons of aviation fuel and Napalm combined with aerial bombing of the ship failed to produce any sustained burning" (Tully, 1969, p. 83). The difficulty in burning such slicks arises because many of the more volatile (and more flammable) hydrocarbons evaporate away; and because heat is rapidly transferred to the water beneath the oil, the temperature of the oil falls below the flash point. Oil can be burned if it is confined, but the tendency of spilled oil to spread out in a thin film on the surface of the water makes burning difficult. In general maintaining a burn of a surface slick requires that the oil be relatively fresh and at least 3 mm thick (Westermeyer, 1991). It is possible through the use of chemical additives to a slick to maintain a controlled burn (Tully, 1969), and some small-scale experiments have produced burn percentages greater than 90% (Westermeyer, 1991).

There are some examples in which unintentional fires consumed a significant percentage of spilled oil. When the tanker *Burmah Agate* collided with a freighter in 1979, much of the 0.034 Mt of oil discharged from the tanker was consumed in a fire caused by the collision that burned out of control for over 2 months. Similarly, when a series of explosions caused the supertanker *Mega Borg* to catch fire in the Gulf of

Mexico in June, 1990, most of the spilled oil either burned or evaporated (Yamaguchi, 1990). However, deliberately setting fire to a grounded tanker could prove futile, and at worst could lead to an explosion that might cause far more damage than would otherwise have occurred. Thus for several reasons oil burning is not currently an important oil spill response tool of any country (Westermeyer, 1991).

Chemical Dispersal

Detergents or similar surface active substances may be added to an oil slick to break up the film of oil into small droplets that become dispersed in the water column. The rationale for applying such dispersants is based in part on aesthetic considerations, since the dispersed oil is much less visible to an observer than a surface slick. It is also assumed, however, that dispersing the oil into small droplets hastens its break-down by microorganisms, since the contact area between oil and water is increased (Canevari, 1969). However, there a number of drawbacks to the use of chemical dispersants. Evaporation of the oil is reduced because less oil is concentrated at the surface; and because the oil is dispersed throughout the mixed layer, a greater number and variety of organisms are likely to come into contact with it. As noted by the EPA (1972, p. 262): "Because of the finer degree of dispersion, the soluble toxic fractions dissolve more rapidly and reach higher concentrations in sea water than would result from natural dispersal. The droplets themselves may be ingested by filter-feeding organisms and thus become an integral part of the marine food chain." One of the most serious problems with the early use of dispersants was the toxicity of the dispersants themselves, particularly if their solvent fractions consisted of a mixture of hydrocarbons. In the case of the *Torrey Canyon*, the dispersant applied to the oil slick contained a petroleum base solvent rich in low molecular weight aromatics (Nelson-Smith, 1972), and it is now generally agreed that this dispersant caused far more damage to marine life than the crude oil itself. As a result of such experiences, dispersants were reformulated. They are now more effective and far less toxic (Westermeyer, 1991).

At the present time it appears that dispersants probably do serve a useful purpose in the cleanup of some oil spills. For example, where there is danger that large numbers of aquatic birds may be killed due to coating from an oil slick, it may be more desirable to break up the slick with a dispersant and risk some damage to other aquatic species than to allow the birds to become coated with oil (Straughan, 1972). In fact the United Kingdom relies almost exclusively on dispersants to combat oil spills (Westermeyer, 1991), and the U.S. National Research Council recently recommended that they be considered a potential first response tool (NRC, 1989). As a general rule, however, dispersants should be applied to an oil slick while the slick is still far enough from shore that mixing processes rapidly dilute the oil-dispersant mixture to subtoxic concentrations. It is noteworthy, for example, that in the case of the *Torrey Canyon* dispersants were applied directly to oil that had come ashore, and as a result intertidal organisms were subjected to extremely high concentrations of the oil-dispersant mixture. According to the NAS (1975), the damage to organisms was perhaps due as much to the misuse of the dispersant as to its inherent toxicity.

Mechanical Containment and Cleanup

Mechanical means for cleaning up oil spills involve the use of booms or skimmers and absorbants. Because of concern over the effectiveness and toxicity of dispersants, this approach has been used almost exclusively to combat oil spills in the United States (Westermeyer, 1991). Unfortunately booms and skimmers are ineffective in currents greater than 1 knot, waves higher than about 2 m, or winds greater than 20 knots (Westermeyer, 1991). The devices are more effective in retrieving spills of crude oil than of refined products (NAS, 1975). If an oil spill is successfully contained by booms or similar devices, there is still the problem of somehow removing it. For example, an oil-absorbant material in the form of a continuous belt may be used to sop up the oil. Extraction of the oil from the belt is accomplished with the use of special rollers, which scrape and squeeze the oil into a recovery vessel (Hodgson, 1990). Oil may also be sopped up by distributing straw or polyurethane foam over the oil and then collecting the oil-soaked absorbant. In general the efficiency of absorbants in removing oil depends on the characteristics of the oil, such as viscosity and surface tension, but under appropriate conditions mechanical containment and absorption can be a highly effective means of cleaning up an oil slick. Unfortunately the most serious spills often occur in foul weather when containment devices are of little use.

Sinking

A fourth cleanup method is worthy of mention, not because it is recommended for use, but because it has been used in the past, and there is a need to point out why it should probably not be used in the future. The technique is to sink the oil by adding sand, talc, or chalk so as to cause the oil to agglutinate into globules more dense than water. While this method does remove the oil from sight, it should be clear that any benthic organisms beneath the slick are likely to be severely stressed if not killed by the blanket of oily globules. Furthermore, as Blumer et al. (1971) have noted, degradation of hydrocarbons in sediments is likely to proceed at a much slower rate than in the water column, and movements of oil-laden sediments may spread the pollution over an area many times larger than the initial spill. In this respect it is noteworthy that, "Oil from the blowout at Santa Barbara was carried to the sea bottom by clay minerals and . . . within four months after the accident the entire bottom of the Santa Barbara basin was covered with oil from the spill" (Blumer et al., 1971). Thus sinking oil onto the sediments does not appear to be a very desirable means of cleaning up a spill, and the EPA (1972, p. 263) has stated that, "Sinking of oil is not recommended." Nevertheless, one response suggested by a European scientist to the most recent Persian Gulf oil spill was to spray the slick with cement powder to bind the oil and make it sink *en masse* (Aldhous, 1991). Other scientists were skeptical of this idea (Aldhous, 1991).

Bioremediation

Bioremediation of oil spills involves the stimulation of microbial breakdown of the spilled oil. In theory bioremediation may take any or all of three forms: 1) Stimulation of indigenous microorganisms through addition of nutrients; 2) introduction of special assemblages of natural oil-degrading microbes; and 3) introduction of genetically engineered microorganisms with special oil-degrading properties (OTA, 1991). Bioremediation is not a first-response cleanup tactic, because even under ideal circumstances the time required for bioremediation to work is on the order of weeks or months. Thus, for example, spraying nutrients on an offshore oil slick is not going to keep the oil off the beaches. Although bioremediation is still very much an experimental tactic, the successful use of this technique to help clean up almost 180 km of beaches following the *Exxon Valdez* spill has stimulated considerable interest in its potential usefulness.

There are at least 70 microbial genera known to contain organisms capable of degrading components of petroleum (OTA, 1991). Most of these organisms are either bacteria, fungi, or yeasts, and they are widely distributed in aquatic systems. No one organism or even a combination of organisms is capable of degrading all the thousands of compounds found in crude oil, but given time there is no doubt that a large percentage of the compounds found in most natural oils can be degraded by microbes. Normal alkanes are the hydrocarbons most easily metabolized. Microorganisms are capable of utilizing n-alkanes with up to 44 carbon atoms, with C_{10}–C_{24} compounds usually being the easiest to metabolize. Branched chain alkanes are degraded more slowly than n-alkanes, and cycloalkanes are even more resistant, but all are subject to biodegradation. Low molecular weight aromatics may also be metabolized, but they are somewhat more resistant than the alkanes and cycloalkanes. By far the slowest degradation rates are observed for high-molecular-weight aromatics, asphaltenes, and resins.[14]

The factor that seems to most frequently limit the rate at which microorganisms can degrade spilled oil is the availability of inorganic nutrients, in particular nitrogen, phosphorus, and perhaps trace metals. This scenario is very reminiscent of nutrient limitation in microalgae and is the rationale for the first type of bioremediation. In the case of the *Exxon Valdez* cleanup, for example, simply applying fertilizer to oiled beaches increased oil biodegradation rates by as much as a factor of 2–4 for more than 30 days. A second application after 3–5 weeks stimulated microbial activity five- to ten-fold (OTA, 1991).

Once inorganic nutrients have been applied, the principal factors limiting oil biodegradation rates are the nature of the oil, the concentration and diversity of the microbial population, the availability of oxygen, and temperature. Natural weathering processes, and evaporation in particular, tend to remove some of the most easily metabolized components of the oil, and weathered crude oil is therefore much less subject to biodegradation than a fresh spill. The natural population of oil-metabolizing organisms can be expected to increase greatly during the first few days or weeks following a spill, and speeding up this process is of course the rationale for seeding a spill with natural oil-degrading microbes. The availability of oxygen can be a serious limiting factor on low-energy beaches and in fine-grained sediments, because anaerobic biodegradation proceeds several orders of magnitude more slowly than aerobic biodegradation. There is general agreement, for example, that the success of the *Exxon Valdez* bioremediation effort was due in part to the fact that the beaches treated were all very coarse, consisting mostly of cobbles and coarse sand (OTA, 1991). The slow rates of oil degradation noted in some low-energy French beaches after the *Amoco Cadiz* spill were undoubtedly the result of oxygen limitation (OTA, 1991). Slow anaerobic degradation can also be expected if oil sinks to the sea floor and is covered by sediments, one of the principal arguments against sinking oil. Finally, over the range of temperatures likely to be encountered in natural waters, microbial rates will tend to be positively correlated with tempera-

---

[14]Asphaltenes are nonvolatile solids with molecular weights exceeding 1000 and consist mainly of polycyclic and heterocyclic aromatic structures with alkyl and naphthenic side chains. Resins include polar and often heterocyclic compounds containing nitrogen, sulfur, and oxygen as constituents.

ture, reported Q_{10}s generally being in the 2–3 range. Thus, other factors being equal, bioremediation is likely to be more effective in tropical than polar climates. The cold temperatures in Prince William Sound, however, did not prevent bioremediation from greatly improving conditions on many of the beaches there.

Summary

One thought-provoking fact about oil spills is that in most cases very little of the spilled oil has been cleaned up. Evaporation alone may rapidly remove as much as 30–60% of the oil, and much of the remainder is slowly lost through natural weathering processes and biodegradation. Less than 10% of the oil discharged as a result of the *Amoco Cadiz* and Ixtoc I accidents and virtually none of the oil spilled by the *Florida* and *Argo Merchant* was ever cleaned up (Teal and Howarth, 1984).

This realization does not mean that we should simply stand back and let nature take its course. It does, however, mean that nature is on our side, and this fact is important to keep in mind during cleanup operations. Some well-intentioned actions taken in an effort to clean up spilled oil have probably done more harm than good, because they have adversely affected natural cleansing mechanisms. During the *Exxon Valdez* cleanup, for example, workers sprayed hot (60°C) water onto rocks using high-pressure hoses to remove adhering oil, in the process killing shoreline organisms (Hodgson, 1990). According to Joyce (1991), the hot water killed barnacles, mussels, clams, eelgrass, and rockweed, and effectively sterilized many parts of the beaches. National Oceanic and Atmospheric Administration official Sylvia Earle was quoted by Joyce (1991) as saying, "As far as Alaska's shoreline is concerned, the environment would have been better off if there had been less aggressive hot-water treatment and we had let nature take its course. Sometimes, the best, and ironically the most difficult, thing to do in the face of an ecological disaster is to do nothing."

It seems fair to say that our understanding of how to deal with spilled oil has improved considerably since the days of the *Torrey Canyon* disaster, when well-intentioned persons tried futilely to burn the slick with napalm and aerial bombing and then spread highly toxic chemicals on the beach in order to disperse the oil that had washed ashore. We now realize that the most effective arsenal for cleaning up oil spills includes a combination of mechanical containment methods, absorbants, dispersants, and bioremediation, and that the extent to which one or another method should be used depends on circumstances. Mechanical containment, absorption, and dispersion are first-response techniques, the latter being best suited for offshore spills and rough seas. Mechanical containment and absorption work well in calm weather and nearshore environments. Bioremediation is more of a finishing or polishing tool and appears best suited for high-energy beach cleanup.

Oil Fingerprinting

When oil is present in the environment, how can we tell whether the oil is derived from petroleum or some other source, and in the former case is there some way to identify the source of the petroleum if it is not obvious? The answer to both questions is a qualified yes.

Petroleum hydrocarbons differ in several significant ways from strictly biogenic hydrocarbons. For example:

1. In biogenic hydrocarbons there is a predominance of odd-number-carbon normal alkanes. In petroleum the ratio of odd- to even-number-carbon normal alkanes is near unity (NRC, 1985). The relative number of odd- and even-number-carbon normal alkanes is usually expressed as the *odd–even preference ratio (OEP)* or *carbon preference index (CPI)*.

2. Biogenic hydrocarbons generally contain a much smaller percentage of cycloalkanes and aromatics than does petroleum. Aromatics, for example, account for less than 1% of the hydrocarbons in most marine organisms, but up to 39% of some crude oils (NAS, 1975; Nelson-Smith, 1972).

3. Petroleum hydrocarbons contain a higher $^{13}C/^{12}C$ ratio and a much lower $^{14}C/^{12}C$ ratio than biogenic hydrocarbons.

The above list is by no means exhaustive, but it serves to make the point that there are characteristic differences between biogenic and petroleum hydrocarbons that can be used to determine the origin of the hydrocarbons, at least in a semiquantitative way. The problems with this approach are basically severalfold: First, many of the important differences can be expressed at best in a qualitative manner, and secondly, there are exceptions to some of the rules. For example, the predominance of odd-number-carbon normal alkanes in C_{20}-C_{30} biogenic hydrocarbons has not been observed in some marine bacteria and fish (NRC, 1985). Thus an OEP or CPI of approximately 1.0 could not, by itself, be taken to indicate that hydrocarbons were of petroleum origin. However, an OEP or CPI substantially greater than 1.0 would certainly imply a biogenic origin. A third issue is the fact that weathering can substantially change the chemical composition of oil, whether of petroleum or biogenic origin. However, weathering does not significantly change stable isotope ratios such as $^{13}C/^{12}C$ (NRC, 1985), and hence these may continue to serve as useful indicators of the source of the hydrocarbons.

Since hydrocarbons are produced naturally by aquatic organisms, the mere presence of hydrocarbons in the water cannot be taken to be a sign of petroleum pollution. Analyses based on the foregoing techniques indicate that concentrations of biogenic hydrocarbons in most aquatic systems are on the order of a few parts per billion. When there is evidence of significant petroleum inputs, a question naturally arises as to the source of the petroleum. The source could be natural (i.e., seeps), or it could be pollution. How can we determine the source of the petroleum if it is not obvious?

The answer is that the chemical composition of petroleum differs greatly from one region of the world to another, and these differences can be used in many cases to determine the source of spilled oil. Figure 13.10, for example, reveals dramatic compositional differences in oils from various sources based on a simple chromatographic analysis. The oils were first heated to drive off all the hydrocarbons having a boiling point less than 316°C to simulate the effects of weathering, and were then analyzed on a gas chromatograph, an instrument that partially separates the hydrocarbons in the oil by vaporizing the oil and passing the vapor through a fractionation column. The fractionation column acts as a series of tiny plates upon which the molecules in the oil vapor are successively adsorbed and desorbed. The time required for molecules to pass through the fractionation column is correlated positively with the molecules' molecular weight or boiling point. Thus the method separates effectively the hydrocarbons according to approximate molecular size or carbon content. In the chromatogram, peak height is proportional to the concentration of organics that boil at the indicated temperature. The unresolved envelope of peaks is caused primarily by the presence of cyclic and highly branched components of the oil, which are not effectively separated by the chromatography technique. Most of the chromatograms, however, show a series of distinct peaks that are associated with normal alkanes containing the number of carbon atoms indicated below each chromatogram. The differences in the chromatograms of the six oils in Figure 13.10 are obvious to even a casual observer and can be quantified for purposes of characterization. Other useful classification characteristics include the carbon and sulfur isotopic composition; the total content of sulfur, nitrogen, vanadium, and nickel; and the OEP or CPI. Fluorescence techniques that take advantage of differences in the characteristic fluorescence spectra of different oils have also shown considerable promise as a method of oil fingerprinting (Anonymous, 1986). These methods obviously work best when there are some candidate pollution sources whose fingerprints can be examined for comparison. The problem becomes more difficult if there are no obvious suspects.

COMMENTARY

According to the 1975 NAS panel report (p. 106): "The most damaging, indisputable adverse effects of petroleum are the oiling and tarring of beaches, the endangering of seabird species, and the modification of benthic communities along polluted coastlines where petroleum is heavily incorporated in the sediments." The first effect is largely an aesthetic problem, but a definite nuisance on recreational beaches, particularly in areas such as Bermuda that depend heavily on tourism (Butler, 1975). With respect to seabirds, Nelson-Smith (1972, p. 159) commented that, "There is no doubt that oil pollution has been the main cause of a dramatic decline in numbers of auks and sea-ducks throughout the North Atlantic and adjacent areas during the first half of this century." Auks have few natural enemies and a very low reproductive rate, and their populations therefore recover slowly from large-scale kills. It has been estimated that in the absence of oil pollution it would take a guillemot colony about 53 years to double in size (Nelson-Smith, 1972). After the wreck of the *Torrey Canyon* in 1967, the number of puffins on the Sept Isles on Brittany declined by a factor of 6, and the razorbill population declined by a factor of 9 (Nelson-Smith, 1972). Despite chronic oil pollution of the North Sea and adjacent waters, 25% of the colonies of common murres and razorbills in the British Isles maintained stable numbers during the 1970s, and 50% actually increased (NRC, 1985). With few exceptions, however, other auk populations in the Atlantic have continued to decline. Although some of this decline may have been due to overhunting and drowning in fishing nets, there is reason to believe that oil pollution has been a factor, particularly in Arctic populations (NRC, 1985). Sea ducks have higher natural reproductive rates than auks, but their habit of congregating in areas adjacent to shipping lanes and other areas particularly subject to oil pollution has resulted in kill-offs. Adult populations, however, appear to recover within a few years (NRC, 1985).

In summarizing its assessment of oil pollution effects, the NRC (1985, p. 490) commented: "In contrast to offshore situations where impact may be minimal and

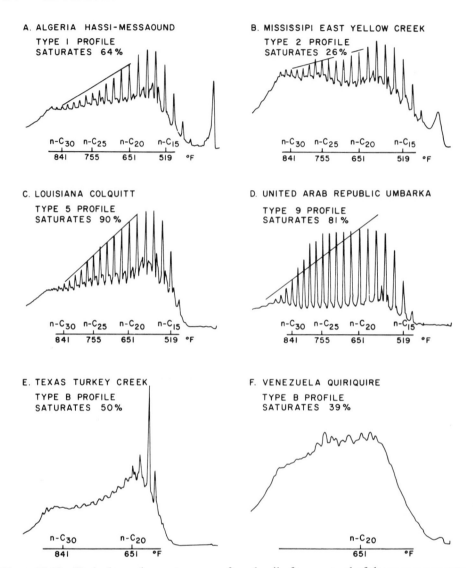

Figure 13.10 Typical gas chromatograms of crude oil after removal of those components having a boiling point less than 316°C. [Redrawn from Miller (1973).]

transient . . ., there are greater immediate concerns for coastal waters (for example, biologically productive estuaries) receiving on the average a greater proportion of discharged petroleum." Certainly it is now apparent from the work of Blumer et al. (1971), Krebs and Burns (1977), Sanders et al. (1980) and others that oil pollution damage to coastal benthic communities can be of much longer duration than was once supposed due to the low degradation rate of hydrocarbons that become incorporated in sediments or the lipid tissues of organisms. Indeed under certain conditions it is apparent that spilled petroleum may persist in the environment for several decades (NRC, 1985). Furthermore, construction and drilling activities in coastal areas may permanently disrupt intertidal and nearshore communities by accelerating erosion and siltation and by altering tidal current patterns.

There is no doubt that some of the damage caused by oil pollution in the coastal environment could be reduced. Tanker accidents such as the *Exxon Valdez* are inexcusable. More strict enforcement of legislation such as the Port and Tanker Safety Act would help. Exxon now requires that tanker crews be on board at least 4 hours before leaving port, a move intended to allow time for sobering up (Lemonick, 1989). Requirements that tankers be equipped with inert gas systems to prevent fires and explosions, segregated ballast tanks, and backup radar and steering systems have undoubtedly helped to reduce oil discharges from tankers. The U.S. Oil Pollution Act of 1990 has designated funding for research aimed at reducing the impact of oil pollution, including improved cleanup methods. One controversial step that is not presently being mandated is a requirement for double hulls or at least double bottoms on tankers. This idea was suggested by the United States at the MARPOL conference, but the proposal was not adopted (Timagenis, 1980). The rationale for double hulls is that breaching the outer hull would not lead to a loss of cargo. Opponents of the idea point out that double-hulled tankers would be more ponderous than conventional tankers with the same cargo capacity, and that flooding the outer hull might cause the tanker to sink (Yamaguchi, 1990).

Tanker accidents, however, were not perceived by the NRC (1985) as being the most serious cause of marine oil pollution. Tanker accidents, particularly involving supertankers, can have devastating effects on coastal ecosystems, and the persistence of petroleum hydrocarbons in some components of these systems is on the order of decades. Nevertheless, coastal ecosystems appear to recover remarkably fast from single incidents of oil spillage. There was a record harvest, for example, of 40 million pink salmon in Prince William Sound the year following the *Exxon Valdez* spill (Holden, 1990). The NRC (1985) report expressed more concern over chronic stresses associated with the long-term more-or-less continuous input of petroleum hydrocarbons to coastal waters. Regions associated with heavy tanker traffic and/or impacted by municipal and industrial wastewaters are the areas most likely to be affected by this sort of stress. However, impacts on biota from petroleum are in many such cases difficult to distinguish from effects caused by other mechanisms, including overfishing and other forms of pollution.

The global oil pollution picture will change dramatically during the twenty-first century for two reasons. In the short term marine transportation of oil will increase substantially as the oil production capacity of the major consuming nations, in particular the United States, declines. Yamaguchi (1990), for example, has projected almost a 50% increase in marine oil transportation during the 1990s. Much of the oil will be transported from the Middle East, which currently accounts for 56% of proven reserves (Flavin, 1985).

In the longer term, however, oil production, transportation, and use will certainly decline. Proven oil reserves have changed very little since 1970, and there is no reason to believe that substantial deposits of oil remain undiscovered. At current rates of oil consumption, the world's oil resources will be largely exhausted sometime during the latter half of the twenty-first century (Flavin, 1985). This scenario will undoubtedly translate into a reduction in oil pollution, but unless there is a major shift in the world's economy, a substitute energy source for oil will have to be found. Two of the most obvious substitutes, coal and nuclear energy, are beset with significant environmental problems of their own. The relevance of these problems to aquatic pollution are discussed in Chapters 14 and 15.

REFERENCES

Aldhous, P. 1991. Big test for bioremediation. *Nature*, **349**, 447.

Andreev, P. F., A. I. Bogomolov, A. F. Dobryanskii, and A. A. Kartsev. 1968. *Transformation of Petroleum in Nature*. Pergamon Press, New York. 466 pp.

Anonymous. 1986. Oil spills. *New Scientist*, **111**, 27.

Anonymous. 1991a. Another crude year. *Discover*, **12**(1), 70–71.

Anonymous. 1991b. Oil pollution. *Honolulu Star Bulletin and Advertiser*. Feb. 24.

Begley, S., J. Carey, and J. Callcott. 1983. Death of the Persian Gulf. *Newsweek*, **102**(4), 79.

Blumer, M. 1972. Submarine seeps: Are they a major source of open ocean oil pollution? *Science*, **176**, 1257–1258.

Blumer, M., H. L. Sanders, J. F. Grassie, and G. R. Hampson. 1971. A small oil spill. *Environment*, **13**(2), 2–12.

Burns, K. A. 1976. Microsomal mixed function oxidases in an estuarine fish, *Fundulus heteroclitus*, and their induction as a result of environmental contamination. *Comp. Biochem. Physiol.*, **53B**, 443–446.

Burns, K. A., and J. M. Teal. 1979. The West Falmouth oil spill; hydrocarbons in the saltmarsh ecosystem. *Estuarine Coastal Mar. Sci.*, **8**, 349–360.

Butler, V. N. 1975. Pelagic tar. *Sci. Amer.*, **232**(6), 90–97.

Canevari, G. P. 1969. The role of chemical dispersants in oil cleanup. In D. P. Hoult (Ed.), *Oil on the Sea*. Plenum Press, New York. pp. 29–51.

Carter, L. J. 1978. *Amoco Cadiz* points up the elusive goal of tanker safety. *Science*, **200**, 514–516.

Clark, C. S., H. S. Bjornson, C. C. Linnemann, Jr., and P. S. Gartside. 1984. *Evaluation of Health Risks Associated with Wastewater Treatment and Sludge Composting*. EPA-600/S1-84-014. Research Triangle Park, NC.

Drewry. 1991a. *Shipping Statistics and Economics*, No. 251 (Sept). Drewry Shipping Consultants, Ltd.

Drewry. 1991b. *Shipping Statistics and Economics*, No. 252 (Oct). Drewry Shipping Consultants, Ltd.

Eganhouse, R. P., and I. R. Kaplan. 1981. Transport dynamics and mass emission rates to the ocean. *Environ. Sci. Tech.*, **16**, 180–186.

Eganhouse, R. P., and I. R. Kaplan. 1982. Extractable organic matter in municipal wastewaters. 1. Petroleum hydrocarbons: Temporal variations and mass emission rates to the ocean. *Environ. Sci. Tech.*, **16**, 180–186.

Environmental Protection Agency. 1972. *Water Quality Criteria*. EPA-R3-73-033. Washington, D.C. 594 pp.

Environmental Protection Agency. 1976. *Quality Criteria for Water*. Washington, D.C. 256 pp.

Environmental Protection Agency. 1986. *Quality Criteria for Water*. EPA 440/5-86-001. Washington, D.C.

Fieser, L., and M. Fieser. 1961. *Advanced Organic Chemistry*. Reinhold, New York. 1157 pp.

Flavin, C. 1985. *World Oil: Coping with the Dangers of Success*. Worldwatch Paper 66. Worldwatch Inst., Washington, D.C. 66 pp.

Gundlach, E. R., P. D. Boehm, M. Marchand, R. M. Atlas, D. M. Wood, and D. A. Wolfe. 1983. The fate of *Amoco Cadiz* oil. *Science*, **221**, 122–129.

Hodgson, B. 1990. Alaska's big spill. Can the wilderness heal? *National Geographic*, **177**(1), 5–43.

Holden, C. 1990. Spilled oil looks worse on TV. *Science*, **250**, 371.

Horn, H. M., J. M. Teal, and R. H. Backus. 1970. Petroleum lumps on the surface of the sea. *Science*, **168**, 245–246.

Howarth, R. W. 1981. Fish versus fuel: a slippery quandary. *Tech. Rev.*, **83**(3), 68–77.

Joyce, C. 1990. French finally clean up after *Amoco Cadiz* disaster. *New Scientist*, **127**(1728), 24.

Joyce, C. 1991. Hot water sterilizes Alaska's oiled beaches. *New Scientist*, **130**, No. 1765, 14.

Kerr, R. A. 1977. Oil in the ocean: Circumstances control its impact. *Science*, **198**, 1134–1136.

Krebs, C. T., and K. A. Burns. 1977. Long-term effects of an oil spill on populations of the salt-marsh crab *Uca pugnax*. *Science*, **197**, 484–487.

Lemonick, M. D. 1989. The two Alaskas. *Time*, **133**(16), 56–66.

Mielke, J. E. 1990. *Oil in the Ocean: The Short- and Long-term Impacts of a Spill*. Congressional Research Service Report for Congress. 24 pp.

Miller, J. W. 1973. A multiparameter oil pollution source identification system. In *Prevention and Control of Oil Spills*. American Petroleum Inst. Environmental Protection Agency, and U.S. Coast Guard, Washington, D.C. pp. 195–203.

NAS. 1975. *Petroleum in the Marine Environment*. National Academy of Sciences, Washington, D.C. 107 pp.

Nelson-Smith, A. 1967. Oil emulsifiers and marine life. In *The Journal of the Devon Trust for Nature Conservation, Ltd*. Supplement. Conservation and The *Torrey Canyon*. July. pp. 29–33.

Nelson-Smith, A. 1972. *Oil Pollution and Marine Ecology*. Elek Science, London. 260 pp.

NRC. 1985. *Oil in the Sea—Inputs, Fates, and Effects*. National Research Council, National Academy Press, Washington, D.C.

NRC. 1989. *Using Oil Spill Dispersants on the Sea*. National Research Council Marine Board, National Academy Press, Washington, D.C.

OECD. 1989. *Annual Market Report 1989*. International Energy Agency. Paris. 84 pp.

OTA. 1991. *Biotechnology for Marine Oil Spills*. Congress of the United States Office of Technology Assessment. S/N 052-003-01240-5. Washington, D.C. 31 pp.

Politzer, I. R., I. R. DeLean, and J. L. Laseter. 1985. *Impact on Human Health of Petroleum in the Marine Environment*. Am. Petroleum Inst., Washington, D.C. 162 pp.

Ryan, P. R. 1977. The composition of oil—A guide for readers. *Oceanus*, **20**(4), 4.

Sanders, H. L., J. F. Grassle, G. R. Hampson, L. S. Morse, S. Garner-Price, and C. C. Jones. 1980. Anatomy of an oil spill: Long-term effects from the grounding of the barge *Florida* off West Falmouth, Massachusetts. *J. Mar. Res.*, **38**, 265–380.

Straughan, D. 1972. Factors causing environmental changes after an oil spill. *J. Pet. Tech.* (March), 250–254.

Tatem, H. E., B. A. Cox, and J. W. Anderson. 1978. The toxicity of oils and petroleum hydrocarbons to estuarine crustaceans. *Estuarine and Coastal Mar. Sci.*, **6**, 365–373.

Teal, J. M., and R. W. Howarth. 1984. Oil spill studies: A review of ecological effects. *Environ. Management*, **8**(1), 27–44.

Thompson, H. C., Jr., R. N. Farragut, and M. H. Thompson. 1977. Relationship of scarlet prawns (*Plesiopenaeus edwardsianus*) to a benthic oil deposit off the northwest coast of Arabia, Dutch West Indies. *Environ. Pollut.*, **13**(4), 239–253.

Timagenis, G. J. 1980. *International Control of Marine Pollution*. Vols. I and II. Oceana Publications, Inc. Dobbs Ferry, NY.

Tully, P. R. 1969. Removal of floating oil slicks by the controlled combustion technique. In D. P. Hoult (Ed.), *Oil on the Sea*. Plenum Press, New York. pp. 81–91.

U.N. 1991. *United Nations Energy Statistics Yearbook*. Department of International Economic and Social Affairs.

Ware, L. 1989. Oil in the sea: The big spills and blowouts. *Audubon*, **91**, 109.

Westermeyer, W. E. 1991. Oil spill response capabilities in the United States. *Environ. Sci. Tech.*, **25**(2), 196–200.

Yamaguchi, N. D. 1990. *The Changing Environment for Seaborne Oil Trade: An Overview of Fleet Developments and Forecast Tonnage Requirements Through the 1990s*. East-West Center Energy Program. Honolulu, HI. 48 pp.

14

RADIOACTIVITY

Once a bright hope, shared by all mankind, including myself, the rash proliferation of atomic power plants has become one of the ugliest clouds overhanging America. DAVID LILIENTHAL, first chairman of the Atomic Energy Commission. (*New York Times, July 20, 1969*)

PHYSICAL BACKGROUND

Before becoming immersed in a discussion of the problems associated with the use and disposal of radioactive substances, it is appropriate to first outline briefly just what radioactivity is and how it affects living organisms. This subject is one that requires some understanding of nuclear and atomic physics, topics that may be unfamiliar to many persons concerned with aquatic pollution. The following discussion provides the basic physical information relevant to the subject of radioactivity.

For our purposes, we may assume that atoms are composed of nothing more than protons, neutrons, and electrons. Protons and neutrons each have a rest mass of approximately one atomic mass unit (amu), or 1.660×10^{-24} g. Electrons are much smaller, having a rest mass of only about 5.5×10^{-4} amu. Hence the total mass of an atom is determined largely by the number of protons and neutrons it contains. For example, the common hydrogen atom, which contains 1 proton, 1 electron, and 0 neutrons, has a rest mass of about 1 amu. The common nitrogen atom, which contains 7 protons, 7 electrons, and 7 neutrons, has a rest mass of about 14 amu. In atomic physics the approximate mass of an atom is usually written as a superscript to the left of the chemical symbol for the element. Thus the symbols ^{1}H and ^{14}N would designate a hydrogen atom with a mass of 1 amu and a nitrogen atom with a mass of 14 amu, respectively.

The electrical charge of an atom is determined by the relative number of protons and electrons. Each proton bears a positive charge of one electrostatic unit (esu) or 1.6×10^{-19} coulomb; each electron bears a negative charge of 1 esu. Neutrons bear no electrical charge. Atoms containing equal numbers of protons and electrons are obviously electrostatically neutral; atoms that contain more protons than electrons bear a net positive charge and are called *cations*; atoms that contain more electrons than protons bear a net negative charge and are called *anions*. Atoms bearing either a positive or a negative charge are referred to as *ions*.

The nucleus of an atom consists of all the protons and neutrons bound by as yet poorly understood forces in a small region at the center of the atom. The electrons are distributed around the nucleus at distances large compared to the dimensions of the nucleus.[1] The identity of the atom is determined solely by the number of protons in the nucleus. An atom containing only 1 proton, for example, is a hydrogen atom, regardless of the number of electrons or neutrons it contains. An atom containing 7 protons is a nitrogen atom, and an atom containing 8 protons is an oxygen atom. The atomic number of an atom is the number of protons in its nucleus and is usually designated by a subscript to the left of the chemical symbol for the element. Thus $^{1}_{1}H$ indicates a hydrogen atom whose nucleus contains no neutrons $(1 - 1 = 0)$. The symbol $^{31}_{15}P$ indicates a phosphorus atom whose nucleus contains 16 neutrons $(31 - 15 = 16)$.

Atoms of the same element whose nuclei contain different numbers of neutrons are called *isotopes* of the element. For example, $^{1}_{1}H$, $^{2}_{1}H$, and $^{3}_{1}H$ are 3 isotopes of hydrogen containing 0, 1, and 2 neutrons, respectively. Collectively, isotopes are referred to as *nuclides*. There is more than one isotope of all known elements, with hydrogen having the fewest number of isotopes, three. Uranium, the naturally occurring element with the highest atomic number (92), has 14 isotopes, ranging in atomic weight from 227 to 240 amu.

For reasons that will become clear soon, the common isotopes of most elements are not radioactive. However, at least 1 isotope of all elements is unstable or radioactive. For example, of the 3 hydrogen isotopes, ^{1}H and ^{2}H are stable, but ^{3}H is radioactive. A *radioactive isotope*, or *radioisotope*, is an isotope that has a tendency to spontaneously decay or transform itself into some other isotope. Why some nuclides are radioactive is not well understood, but is undoubtedly determined by the nature of the forces that hold the nucleus together. Figure 14.1 shows the number of protons and neutrons in stable nuclides. It is clear from this figure that the relative number of protons and neutrons in stable nuclei follows a definite pattern. For the lightest elements, the number of protons and neutrons in stable nuclides is approximately the same, but in the heaviest stable nuclides there are about 50% more neutrons than protons. No nuclides with atomic number greater than 83 are stable. Nuclides that do not fall within the zone of stability indicated by the data points in Figure 14.1 are radioactive. These nuclei tend to spontaneously disintegrate in such a manner as to produce "daughter" nuclei that lie within the zone of stability or at least no further away. If the daughter nuclide is itself unstable, then it will also tend to decay, producing another nuclide, and so forth. Thus the decay of the original radionuclide can lead to a decay series involving a number of daughter nuclides. Ultimately the

---

[1]The diameter of an atomic nucleus is on the order of 10^{-12} cm. Electrons are found with maximum probability at distances on the order of 10^{-8} cm from the nucleus.

decay series leads to the production of a stable nuclide. For example, the radio-nuclide uranium-238 (^{238}U) decays through a series of 13 intermediate radio-nuclides to produce the stable nuclide lead-206 (^{206}Pb). On the other hand, ^{3}H decays into the stable nuclide helium-3 (^{3}He) in a single step.

Although there are quite a few mechanisms by which unstable radionuclides may decay, only three mechanisms are of direct importance insofar as the problem of radioactivity is concerned.

1. The radionuclide may decay by emitting an alpha particle (α), which consists of 2 protons and 2 neutrons. This mode of decay is common for elements of atomic number greater than 83, although some radionuclides of low atomic number also decay by this mechanism. Since the α particle contains 2 protons, it may be regarded as the nucleus of a helium atom and designated $^4_2$He. An example of α-particle decay is the decay of $^{238}_{92}$U as follows:

$$^{238}_{92}\text{U} \rightarrow {}^{234}_{90}\text{Th} + {}^4_2\text{He} \tag{14.1}$$

2. The radionuclide may decay by emitting a beta (β) particle. A beta particle has a mass identical to that of an electron and a charge of $+1$ esu (β^+) or -1 esu (β^-). If the beta particle is charged negatively, it is nothing more than an *elec-*

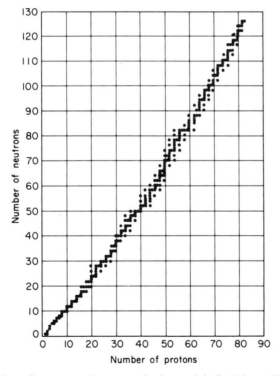

Figure 14.1 Number of protons and neutrons in the nuclei of stable nuclides. (From *General Chemistry*, 4th Ed., by P. W. Selwood. Copyright © 1950, 1959, 1965 by Holt, Rinehart and Winston, Inc. Reprinted by permission of Holt, Rinehart and Winston.)

tron. If it is charged positively, it is called a *positron.* An example of beta particle decay is the decay of ^3_1H as follows:

$$^3_1H \rightarrow\ ^3_2He + \beta^- \tag{14.2}$$

3. When radionuclides decay, they frequently emit gamma rays (γ). Gamma rays are nothing more than electromagnetic radiation, analogous to radio waves, visible light, and X rays. Gamma rays, however, differ from other types of electromagnetic radiation in that they originate from the nucleus of atoms and are usually of higher energy. Since gamma rays bear no charge, they are highly penetrating, whereas many alpha and beta particles can be stopped by a relatively thin layer of metal, soil, rock, or biological tissue. Returning to a previous example, the decay of $^{238}_{92}U$ by alpha particle emission is more correctly written:

$$^{238}_{92}U \rightarrow\ ^{234}_{90}Th + {}^4_2He + \gamma \tag{14.3}$$

to indicate that a gamma ray is emitted along with the alpha particle.

Radiation from radionuclides therefore consists largely of the alpha and beta particles and gamma rays which are emitted when radionuclides decay. It is also true that a few radionuclides spontaneously decay by emitting neutrons, but most environmental concern over neutron radiation is related to the high-energy neutrons produced in nuclear reactors. These high-energy neutrons influence the release of radioactive wastes from the reactors.

The decay of a radionuclide is a probabilistic phenomenon. In other words, although there is a certain probability that a given radionuclide will decay within a given time, the process of decay is by no means certain. The probability of decay can be characterized by the so-called half-life of the radionuclide. The half-life is defined as the amount of time during which half of the atoms of a given radionuclide can be expected to decay. For example, the half-life of ^{14}C is about 5770 years. Thus a sample containing 1000 atoms of ^{14}C at time zero would be expected to contain only 500 atoms of ^{14}C 5770 years later, since 50% of the ^{14}C would have decayed. After a second period of 5770 years, 50% of the 500 atoms would have decayed, leaving 250 atoms of ^{14}C. After a third period of 5770 years, 50% of the 250 atoms of ^{14}C would have decayed, leaving 125 atoms, and so forth. Since the amount of radioactivity released by a sample of a given radioisotope is directly proportional to the amount of radioisotope present, it follows from the foregoing discussion that the amount of radioactivity released from a radioactive sample steadily decreases with time. After 10 half-lives, the number of radioactive atoms, and hence the radioactivity, has been reduced by a factor of 1024, and after 20 half-lives by a factor of a little over one million, assuming of course that the radionuclide has decayed to a stable nuclide. The half-lives of radionuclides vary tremendously, from less than a second for some very short-lived radionuclides to 5×10^{15} years for $^{144}_{60}Nd$. The fact that some radionuclides have half-lives on the order of a day or less is convenient from the standpoint of radioactive waste disposal, since simply storing such waste for a period of a few months results in virtually all of these radionuclides decaying away. Unfortunately some radionuclides produced in nuclear reactors have half-lives on the order of tens to hundreds of thousands of years or even longer. To avoid contaminating the environment with these radionuclides, some more-or-less permanent and isolated repository must be found for their disposal.

RADIATION TOXICOLOGY

What sort of environmental damage do radionuclides cause? On the molecular level the effects of radiation on matter are understood in a semiquantitative way. In passing through material, radiation interacts with the atoms and molecules of which the material is composed, breaking chemical bonds to form positively and negatively charged ion pairs and free radicals.[2] These ions or radicals may then recombine in a variety of ways, leading to the production of chemical species that are not normally present in the material.

In the case of living organisms, which typically contain 70% or more water, many of the radiation interactions involve water. The most important initial products resulting from the interaction of radiation with water are H_2O^- and hydroxide and hydrogen free radicals (BEIR, 1990). Of these chemical species, the hydroxide radical is believed the most effective in causing damage because it is a strong oxidizing agent. Molecules such as the hydroxide radical may react with enzymes, nucleic acids, or other chemical species in the cell, perhaps leading to the destruction of the cell or causing the cell to function in an aberrant manner. What actually happens to a given cell, tissue, or organ following the initial production of ion pairs and other reactive species by the passage of radiation can be an exceedingly complex and drawn-out process, as suggested by the fact that some radiation-induced cancers may have a latency period of 30 years or more (Wilber, 1969). Recent studies have indicated that as many as ten distinct mutations may be necessary in a cell before it becomes malignant. For example, the loss of two or more suppressor genes, which normally inhibit cell growth, and the simultaneous activation of oncogenes have been shown to be necessary to produce some of the most common forms of cancer (Marx, 1989). Regardless of the detailed long-term effects, however, radiation that causes this general sort of damage is referred to as ionizing radiation, due to the initial production of ion pairs associated with the interaction between radiation and matter.

Of the types of radiation we have discussed, alpha particles penetrate the shortest distance in matter because of their relatively large size and charge. In other words, alpha particles have a high probability of interacting with the atoms and molecules in a given material, and through this interaction their movement is rapidly slowed and brought to a stop. Alpha particles generally travel only a few centimeters in air and can be stopped by a piece of paper or a thin layer of skin. Thus an alpha emitter is of relatively little concern as an external source of radiation, but may be of great concern if ingested (internal source). In the latter case, the fact that the alpha particle is stopped over a very short distance means that ionizing damage may be quite intense in the immediate vicinity of the radionuclide, although tissue a short distance away may be virtually unaffected.

Beta particles have a charge equal to one-half the charge of an alpha particle and a rest mass about 7200 times smaller than that of an alpha particle. Because of their smaller size and reduced charge, beta particles have a smaller probability of interacting with matter than do alpha particles and hence penetrate further into matter, spreading their ionizing damage over a longer distance. Beta particles may travel a meter or more in air before coming to a stop and up to several centimeters through tissue. Obviously beta emitters are of little concern to health unless one is standing in the immediate vicinity of the source, and therefore like alpha emitters

---

[2] Free radicals are highly reactive molecular fragments containing an unpaired electron.

are of primary concern as internal sources. The chief difference between alpha- and beta-particle damage is the fact that alpha-particle damage is relatively intense but confined to the immediate vicinity of the source. Beta-particle damage is less intense but extends over a greater distance.

Gamma rays have no electrostatic charge, but they do carry momentum and are capable of ionizing atoms and molecules just as alpha and beta particles do. Because of their chargeless character, gamma rays penetrate matter easily, and in fact may pass through an organism without causing any damage at all. When gamma rays do interact with matter, they generally cause ionization over a long path length, although the intensity of damage is likely to be much less than that caused by an alpha or beta particle. Because of their ability to penetrate matter, gamma rays are of concern from both external and internal sources of radiation.

As already noted, neutron damage is of concern primarily in the immediate vicinity of a nuclear reaction or an atomic explosion. Neutrons have a mass equal to one-fourth that of an alpha particle, but because they bear no charge, neutrons are not associated directly with ionization damage. However, they may, "like a bull in a china shop, . . . wreck local havoc by bumping atoms out of their stable states. Neutrons thus induce radioactivity in nonradioactive materials or tissues through which they pass" (Odum, 1971, p. 452).

The toxic effects of a particular radionuclide depend on a number of factors, including the intensity and energy of the emitted radiation, the location of the radionuclide, and the types of organisms affected. The intensity or activity of radiation is measured usually by simply noting the number of atoms that decay or disintegrate per unit time. The SI[3] measure of activity is the becquerel (Bq). A becquerel of radioactivity equals one disintegration per second. From the standpoint of radiation damage, however, the amount of radiation actually absorbed by an organism is of more importance than the number of disintegrations per second, because the number of ion pairs and other excited chemical species produced by the passage of radiation through matter is directly proportional to the absorbed energy of the radiation for a given type (α, β, γ) of radiation. The energy of the α, β and γ radiation emitted by radionuclides may vary greatly from one radionuclide to another. For example, ^{14}C emits beta particles with a maximum energy of 0.156 million electron volts (meV), whereas the beta particles emitted by ^{32}P have a maximum energy of 1.71 meV. Thus 1 Bq of ^{32}P would be expected to produce about $1.71/0.56 = 11$ times as many ion pairs in an organism as would 1 Bq of ^{14}C. The SI measure of absorbed radiation energy is the *gray* (*Gy*). One gray is defined as the absorbed dose of one joule of energy per kilogram of material. In radiation monitoring work, another unit of absorbed radiation, the roentgen, is often used. One roentgen corresponds to that amount of X or gamma radiation which, when interacting in 1 kg of dry air, will cause the liberation of electrons that, when completely stopped, will generate 2.58×10^{-4} coulomb of charge. Finally, in setting radiation standards, allowance is made for the fact that the same amount of absorbed radiation energy may cause different amounts of biological damage, depending on the intensity of the ionization caused by the absorbed energy. Recall, for example, that alpha particles are stopped over a very short distance and therefore produce more intense ionization per unit path length than do beta particles or gamma rays, which release their energy over a longer path length. Alpha particles therefore cause greater biological damage per

[3]International system of units.

unit energy deposited than do beta particles. To allow for this difference in damage intensity, a so-called quality factor (Q) is used to multiply the radiation-absorbed dose in grays to estimate the relative amount of biological damage caused by the absorption of various types of radiation by humans. The product of the dose in grays times the Q factor is called the *dose equivalent* in sieverts (Sv). Table 14.1 lists the Q factors for various types of radiation. Notice that alpha particles are estimated to cause 20 times as much biological damage per unit absorbed energy as are beta particles and gamma rays. Neutrons are intermediate in damage potential.

The No-Threshold and Linear Dose-Response Hypotheses

In setting radiation standards, it is now generally assumed that there is no threshold limit for radiation damage, and that the amount of damage caused is a linear function of the total amount of radiation absorbed (BEIR, 1990). The assumptions of linearity and no threshold are independent, as illustrated in Figure 14.2. The curves in Figure 14.2A and 14.2C are both linear, but in the latter there is a threshold dose below which there is no effect. Neither of the curves in Figure 14.2A and 14.2B has a threshold, but only the former is linear. The combined assumptions of linearity and no threshold imply that effect is directly proportional to dose, an assumption that seems to hold in many cases at low levels of radiation exposure. The assumption, for example, describes rather well the excess of mortality from all forms of cancer other than leukemia at doses less than 4 Sv (BEIR, 1990; Figure 14.3A). The leukemia data, however, are best described by a curvilinear function (Figure 14.3B).

The assumption of no threshold for radiation damage is a thought-provoking one, since it implies that any artificial release of radioactivity to the environment can be expected to increase the probability of occurrence of various disorders associated with radiation damage. These disorders include cancer, leukemia, genetic defects, fetal and neonatal deaths, and a general shortening of life (Gofman and Tamplin, 1970). In setting radiation standards, you must therefore presumably make some value judgment regarding the benefits derived from the use of radionuclides versus the cost in terms of disease, death, and deterioration of the environment. For example, a figure of $100,000 per person-sievert has sometimes been used in the United States as a figure of merit in evaluating trade-offs in nuclear waste management (e.g., Adams and Rogers, 1978). Person-sieverts are calculated by multiplying the total number of persons exposed by their average individual doses in sieverts. Considering the fact that the recent report of the Committee on the Biological Effects of Ionizing Radiation (BEIR-V report) estimates that a dose of 1000 person-Sv resulting from a single exposure incident would likely produce about 80

Table 14.1   Q Factors for Selected Types of Radiation

Type of Radiation	Q
X rays, γ rays, and electrons	1
High-energy protons and neutrons of unknown energy	10
α particles, multiply-charged particles, fission fragments, and heavy particles of unknown charge	20

Source. NRC (1991a).

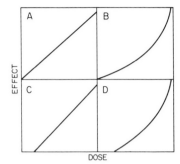

Figure 14.2 Hypothetical dose-effect curves describing the relationship between radiation dose and health effects.

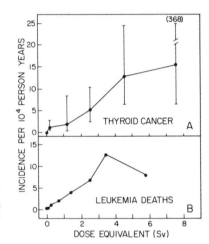

Figure 14.3 Actual dose-effect data describing incidence of thyroid cancer (A) and leukemia deaths (B) caused by ionizing radiation. [*Source*: ICRP (1991) and BEIR (1990).]

fatal cancers (BEIR, 1990), the foregoing figure of merit would put a price tag of no more than about $1.25 million on each fatal cancer. This price tag would obviously be reduced if the likely increase in genetic defects, fetal deaths, and so on resulting from each person-Sv were also considered. Although such a balancing of human misery against social and technological advances and benefits may seem repugnant to some, it is worth pointing out that roughly 48,000 people die annually on the highways in the United States (U.S. Dept. Health & Human Services, 1991), and automobiles contribute significantly to the pollution of both air and water. Even though most persons would undoubtedly like to see these "costs" of automobile use reduced, it is also likely that most persons would agree that the benefits from automobile use still outweigh the price in terms of human lives and environmental deterioration. How much of this sort of price are we willing to pay for nuclear power, nuclear weapons, and the use of radiation in science and medicine?

Health Effect Estimates

To answer this question, we must first have some idea of the relationship between radiation exposure and various related diseases and disorders. Most information regarding health effects on humans has come from follow-up studies on the survivors of the bombing of Hiroshima and Nagasaki at the end of World War II and on persons who were treated with X rays for medical purposes prior to 1955. The health effects estimates in the BEIR-V report, for example, were based primarily on the following studies:

1. A mortality study of 120,321 persons who were resident in Hiroshima and Nagasaki in 1950, 91,228 of whom were exposed to radiation from the atomic bombs dropped on August 6 and 9, 1945. Most of the radiation received by the exposed persons was in the form of γ rays, the estimated doses ranging from 10 mSv to more than 2 Sv (BEIR, 1990).

2. A study of 14,106 patients treated with radiotherapy to the spine for ankylosing spondylitis in the United Kingdom between 1935 and 1954.

3. A study of approximately 150,000 women in a number of countries treated for cervical cancer. About 70% of the women were treated with either radium implants or external radiotherapy.

4. A study of 31,710 women in Canadian tuberculosis sanatoria, a substantial number of whom were exposed to multiple fluoroscopies in conjunction with artificial pneumothorax treatments between 1930 and 1952.

5. A study of 601 women treated with radiotherapy for postpartem acute mastitis in New York State during the 1940s and 1950s as well as 1239 nonexposed women who also suffered from mastitis but were not treated with radiotherapy. Siblings of both groups of women were also included in the study.

6. A study of 1742 women first treated between 1930 and 1956 in two Massachusetts sanatoria, of whom 1044 were subjected to regular fluoroscopy in conjunction with treatment by artificial pneumothorax.

Several problems have complicated the interpretation of results from these and similar studies. First, it has sometimes been difficult to determine to how much radiation a person was exposed. A major revision of radiation health effects in the BEIR-V report, for example, resulted from recalculation of the radiation doses received by the survivors of the bombing of Hiroshima and Nagasaki. The shielding provided by air, humidity, windows, walls, and roofs was recalculated, and the doses received by the survivors, "were individually reconstructed, taking into account whether the person was facing or turned away from the blast, and, if sideways, which side of the body was exposed" (Marshall, 1990a, p. 23). Suffice it to say that such calculations are difficult and time-consuming and involve a certain amount of educated guesswork. Another problem is the long latency period in the development of certain types of cancer, an issue that has been overlooked in some short-term studies of dose-effect relationships (Gofman and Tamplin, 1970). By focusing on persons who had been exposed to artificial radiation prior to 1955, the BEIR-V study was able to follow rather thoroughly the time-course of various adverse health effects and to develop mathematical models to describe the observed results. An additional complicating factor is the tendency of rapidly dividing cells to be more sensitive to radiation damage than slowly dividing cells. As a result young children and especially babies *in utero* are much more susceptible to radiation damage than adults, and the most sensitive parts of the body are regions such as the reproductive organs and bone marrow where cell division is relatively rapid. For example, the excess cancer mortality expected from a single dose of radiation is about twice as high if exposure occurs at age 5 than at age 30 (BEIR, 1990).

Controlled experiments conducted on plants and animals have provided much valuable additional information on radiation-effect relationships. From such studies it has become clear that mammals are among the most sensitive organisms to acute doses of radiation and that microorganisms are the least sensitive. Figure 14.4 indicates the sensitivity range of mammals, insects, and bacteria to various acute doses of radiation. The left end of each bar indicates the acute dose that would severely affect reproduction in the most sensitive species of each group, and the right end of each bar roughly corresponds to the TLm for the more resistant species in the group. The arrows indicate doses that would kill or damage sensitive life-history stages such as embryos. The acute lethal dose for humans is about 3 Sv (EPA, 1973).

Studies of radiation damage to higher plants have shown that the acute lethal dose is directly proportional to the chromosome volume or DNA content of the cell, as indicated in Figure 14.5. These results suggest that in cases of acute lethal damage, the chromosomes in the cell represent a sort of target, the size of which is directly proportional to the probability of being hit by a particle or ray of radiation. This sort of relationship between lethal dose and chromosome size breaks down in higher animals, presumably because effects on specific organs are of critical importance and tend to mask effects due to differences in cellular structure.

Estimates of harm to the average person from current levels of radiation exposure require a certain extrapolation of experimental data, because most persons experience a chronic low-level exposure that is rather different from the acute doses received by bomb survivors and persons treated with radiation for medical reasons. It is now generally agreed that a given dose of radiation absorbed over a short time period is likely to cause more damage than the same dose of radiation received over a long time period. For example, Sparrow (1962) discovered that pine trees exposed to a radiation level of 1 roentgen per day for 10 years suffered about the same reduction in growth as pine trees exposed to a single dose of 60 roentgens. In other words, chronic doses of radiation administered over a long period of time had less effect per unit of radiation than a single acute dose. This effect would be expected if the single acute dose caused irreparable damage, whereas chronic doses caused only slight damage that could be repaired between exposures. The ratio of damage caused by acute and chronic doses of the same amount of radiation is referred to as the *dose-rate effectiveness factor* (*DREF*), and in the pine-tree example cited would be

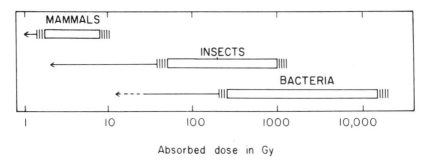

Absorbed dose in Gy

Figure 14.4 Relative sensitivity of three groups of organisms to acute doses of X or γ radiation. [Redrawn with permission from Odum (1971). Copyright © 1971 by W. B. Saunders Co.]

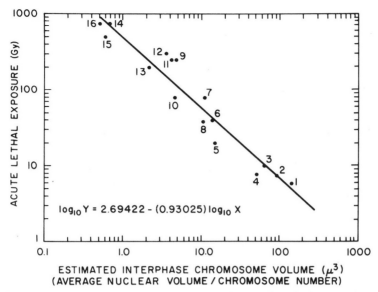

Figure 14.5   Relationship between chromosome volume of seed plants and acute lethal doses of radiation. [Redrawn with permission from Sparrow et al. (1963). Copyright © 1963 by the American Association for the Advancement of Science.]

(10)(365)/60 = 61. The DREFs for humans and animals appear to be substantially lower, with all reported values falling in the 2–10 range (BEIR, 1990). The BEIR-V report assumed a DREF of 2.0 for human health effects. This assumption is conservative in the sense that it would overestimate the adverse health effects caused by chronic exposure if the true DREF were greater than 2.0.

Current Levels of Exposure

With this introduction, it is now useful to examine current levels of radiation exposure and compare these with international guidelines. Tables 14.2 and 14.3 provide the relevant information. It is apparent from Table 14.2 that for the average person the major sources of radiation exposure are natural, the primary radioisotope of concern being radon-222, which is produced in the decay series of uranium-238. The importance of ^{222}Rn in natural radiation has been recognized only recently (BEIR, 1990). Actually it is not so much ^{222}Rn, but rather its daughter radionuclides, that are of concern in public health. Radon-222 is an inert gas and therefore has no tendency to be incorporated into biological tissue. Its daughter radionuclides, which include isotopes of polonium, lead, and bismuth, are not gases and tend to become incorporated into aerosols that lodge in the lungs. Both ^{222}Rn and its daughters are α particle emitters. The other natural radionuclide of principal concern from the standpoint of internal exposure is ^{40}K, which emits β particles. The primary natural sources of external exposure are cosmic rays and α particle-emitting radioisotopes in the decay series of ^{238}U, which includes uranium, thorium, and radium. These latter elements are associated with soils and rocks. The principal artificial source of radiation exposure in the United States is the use of radiation in medicine. The estimate of 0.53 mSv y^{-1}, however, is about 30% smaller than the cor-

Table 14.2   Estimates of Annual Average Radiation Exposure to a Member of the U.S. Population

Source	mSv	% of total
Natural		
Radon	2.0	55
Cosmic	0.27	8
Terrestrial	0.28	8
Internal	0.39	11
Total natural	3	82
Artificial		
X-ray diagnosis	0.39	11
Nuclear medicine	0.14	4
Consumer products	0.10	3
Other		
Occupational	<0.01	<0.3
Nuclear fuel cycle	<0.01	<0.03
Fallout	<0.01	<0.03
Miscellaneous	<0.01	<0.03
Total natural, artificial, and other	3.6	100

Source. BEIR (1990).

Table 14.3   Radiation Dose-Equivalent Limits for Protection of Individual Members of the General Public Due to Sources Other Than Natural Background and Exposures Received as a Patient for Medical Purposes

Exposed Area	Dose Limit	Weighting factor
Whole body		
Continuous or frequent exposure	1 mSv per year	
Infrequent exposure	5 mSv in 1 year	
Tissues and organs[a]	50 mSv per year	
Gonads		0.25
Breast		0.15
Red bone marrow		0.12
Lung		0.12
Thyroid		0.03
Bone surface		0.03
Remainder		0.30

[a]The dose equivalent limit for tissues and organs is calculated from a weighted mean of the doses to individual tissues and organs using the weighting factors indicated in the right-hand column.

Source. NCRP (1989) and NRC (1991a).

responding figures for 1970 and 1978 (Gillette, 1972; Marx, 1979). The difference presumably reflects efforts to reduce unnecessary exposure to radiation in medical diagnosis and therapy. The 1972 BEIR report, for example, noted that use of improved equipment, proper shielding of reproductive organs, and elimination of unnecessary X rays could reduce the "genetically significant dose" currently received by the general population by 50% (Gillette, 1972), and in 1979 Department of Health, Education, and Welfare secretary Joseph Califano directed the Food and

Drug Administration to accelerate steps to reduce unnecessary exposure to medical and dental radiation (Marx, 1979). The principal cause of exposure from consumer products is the radon in water supplies. Building materials, mining, agricultural products, and coal burning also contribute (BEIR, 1990). Cigarette smokers are exposed in addition to polonium-210 in tobacco, an exposure that contributes undoubtedly to the correlation between cigarette smoking and lung cancer.

It is apparent from Table 14.2 that nuclear power in the United States presently accounts for very little of the average American's radiation exposure. Most of this exposure is concentrated in the workers producing the nuclear fuels and running the power plants (Marx, 1979). However, when evaluating exposure levels associated with the use of nuclear power a number of potential sources must be considered, including radiation exposure to miners, risks in storage and reprocessing of spent fuel, the probability of exposure after final waste disposal, the risks of a major nuclear plant leak, and the hazards of terrorism. The nuclear power industry has existed for only about 30 years, and because of this relatively short track record it is difficult to say how frequently certain types of events are likely to occur. In 1975, for example, the U.S. Nuclear Regulatory Commission estimated that the probability of a person's dying from a nuclear reactor accident was about 1 in 5×10^{11} per year per nuclear reactor (NRC, 1975). At the present time the nuclear power industry has accumulated about 4000 reactors years of operation (Flavin, 1987). Given a world population of about 5×10^9 persons, the implication is that about 40 persons should have died by now from accidents at nuclear power plants. While it is true that close to this number of persons have been killed more-or-less outright by nuclear power plant accidents, estimates are that about 39,000 persons will ultimately die from radiation-induced cancers caused by the release of radionuclides from the 1986 Chernobyl accident (Marshall, 1987a). The dose of radiation to the human population from the Chernobyl accident is ultimately expected to be about 1.2 million person-Sv, or about 0.23 mSv per person. Although Chernobyl has been by far the most serious accident at a nuclear facility, it is by no means the only time a civilian population has received dangerous doses of radiation as a result of an accident. The U.S. government revealed recently that large amounts of radioactive iodine leaked from fuel processing tanks at its Hanford, Washington, facility from 1944 to 1947 (Marshall, 1990b). Over 13,000 persons may have received radiation doses to the thyroid gland greater than 330 mSv, and 1400 children may have received doses of 150–6500 mSv from the consumption of radioactive milk. How often such incidents can be expected to occur in the future is a major cause of uncertainty in the estimation of the costs associated with nuclear weapons and the nuclear power industry.

The health costs that the United States is supposedly willing to tolerate can be estimated from Table 14.3 and a knowledge of radiation dose-effect relationships. According to the BEIR-V report, for example, continuous exposure of the U.S. population to an additional 1 mSv of radiation would increase the frequency of cancer mortality by about 3% and reduce average life expectancy by about 1 month (BEIR, 1990). It is unrealistic, however, to think that all members of the U.S. population would experience a uniform increase in radiation exposure due to the operation of nuclear power plants. According to the ICRP (1984), dose limits for members of the general public are to be applied to the average effective dose equivalents to the members of the so-called critical group, which is defined as persons representative of those individuals in the population expected to receive the highest dose equivalent. The critical group is expected to include more than one but no more than a few tens of persons (ICRP, 1984). The NRC (1991a), however, stipulates that the total effective dose equivalent to the individual likely to receive the highest dose should not exceed the annual dose limit. Regardless of whether the dose limit is applied to the single member of the general public who receives the highest dose or the average dose received by a critical group of a few tens of persons, it seems fair to say that the average dose to the general public will be substantially less than the dose limit. The ICRP, for example, has estimated that the average dose to individual members of the public would be less than 10% of the dose equivalent limit, "provided that the practices exposing the public are few and cause little exposure outside the critical groups" (ICRP, 1977, p. 24). The implication is that present guidelines would allow the radiation dose to individual members of the public to increase by no more than 0.1 mSv y^{-1}. The resultant increase in cancer fatalities would amount to 0.3%, too small a change to be detectable with any degree of statistical confidence. The actual body count would amount to about 1200 additional cancer fatalities per year in the United States.

Importance of Certain Radionuclides

When guidelines with respect to the release of radioactivity by nuclear power plants are established, it is often the case that a small number of radionuclides turn out to be of special importance. These radionuclides are the ones that tend to be incorporated into living organisms, either because they are radioisotopes of essential elements or behave chemically very much like essential elements. In a few cases radionuclides are of special concern because they may be inhaled and concentrated in the lungs. Table 14.4 lists some of the radionuclides that are of particular concern with respect to human health. Tritium (^{3}H), carbon-14, phosphorus-32, and radioactive iodine (^{129}I and ^{131}I) of course behave chemically like stable hydrogen, carbon, phosphorus and iodine, respectively, and are utilized by the body in a similar manner. Tritium (^{3}H) may be incorporated into water molecules, and ^{14}C into organic molecules, and thus be distributed throughout the body. Both ^{3}H and ^{14}C are β emitters. Phosphorus-32 tends to concentrate in the bones, which consist of calcium phosphate, where its β radiation may induce damage to bone marrow, possibly leading to leukemia. ^{129}I and ^{131}I concentrate in the thyroid gland, where their β and γ radiation may cause thyroid cancer. Both radium and strontium are chemically similar to calcium, and therefore both tend to concentrate in bones. Cesium-137 is chemically similar to potassium, an element that is found in fluids throughout the body. Uranium and plutonium are not chemically similar to elements commonly found in the body, but may be inhaled on dust particles and lodge in the lungs, where their α radiation may of course lead to lung cancer. ^{222}Rn, as already noted, is of concern because its daughter radionuclides lodge in the lungs, where their α particle emissions create problems similar to those caused by ^{238}U and ^{239}Pu.

Effects on Aquatic Systems

Up to this point we have been concerned primarily with the effects of radiation on humans, but radiation may obviously affect aquatic organisms in similar ways. In the case of aquatic organisms, however, the fate of individual organisms is less of a concern than the health of the species or community in general. Thus while the

Table 14.4   Radionuclides that are of Special Concern as Health Hazards to Humans

Radionuclide	Half-life	Cause for Concern
^{3}H	12.3 years	Behaves like stable hydrogen; assimilated by body in water
^{14}C	5.8 10^4 years	Behaves like stable carbon; passed up food chain
^{32}P	14.3 days	Behaves like stable phosphorus; concentrated in bones
^{90}Sr	28 years	Behaves similarly to calcium; concentrated in bones
^{129}I	1.7×10^7 years	Behaves lile stable iodine; localized in thyroid gland
^{131}I	8.05 days	
^{137}Cs	30 years	Behaves similarly to potassium; found throughout the body
^{226}Ra	1.6×10^3 years	Behaves similarly to calcium; concentrated in the bones
^{238}U	4.5×10^9 years	Intake likely by way of inhalation of dust; concentrated in lungs and kidneys
^{239}Pu	2.4×10^4 years	Intake likely by way of inhalation of dust; concentrated in lungs
^{222}Rn	3.8 days	Radioactive daughters taken up by way of inhalation of dust and concentrated in lungs

deaths of 250,000 Americans out of a population of 250 million (i.e., 0.1%) might seem an alarming statistic to many persons, the loss of 0.1% of a population of crabs or tunicates would not be very likely to cause much public alarm.

Acute radiation doses that have been shown to be lethal to aquatic organisms lie in the 2–550 Gy range for invertebrates and adult fish, and as low as 0.16 Gy for fish embryos (Anderson and Harrison, 1986). Adverse effects on fish reproduction have been reported at chronic dose rates in the 6–120 mGy d^{-1} range (Anderson and Harrison, 1986), but there is unfortunately not much other reliable information concerning the effects of chronic low-level radiation exposure on aquatic organisms. Sublethal effects associated with genetic defects have been reported at acute doses as low as 0.5 Gy (Anderson and Harrison, 1986). In wild populations, however, "genetic damage may be removed by natural selection, and somatically weakened individuals are probably eaten by predators. Consequently aquatic organisms adversely affected by radiation are not readily recognized in the field" (EPA, 1973).

The aquatic organisms that have undoubtedly been subjected to the highest levels of long-term chronic stress from radiation exposure are those living in the vicinity of discharges from major nuclear power or nuclear research facilities. In White Oak Creek, a stream that receives radioactive wastes from the U.S. Oak Ridge National Laboratory, mosquito fish have been subjected to radiation doses of 0.6–3.6 mGy d^{-1} from contaminated sediments. In 1965, when the dose rate was about 3.6 mGy d^{-1}, the fish were found to be producing larger broods than usual, but with an increased incidence of dead and deformed embryos (Woodhead, 1984). The former response was assumed to be an adaptation to the radiation stress, and the latter was assumed to be a manifestation of lethal mutations in the genome. Chironomoid larvae living in the same stream were estimated to be receiving a radiation dose of about 6 mGy d^{-1} and were found to have an increased frequency of chromosomal aberrations. Although the observed aberrations were apparently lethal in overall effect, the abundance of the worms was unaffected (Woodhead,

1984). Columbia River salmon spawning near the outfalls from the U.S. nuclear installation at Hanford, Washington, were apparently unaffected by radiation doses of 1–2 mGy per week (EPA, 1973), and population and community studies of algae and invertebrates revealed no effects from radiation doses as high as 1.0 Gy d^{-1} in ponds and streams on the Hanford Reservation (Woodhead, 1984).

The examples just cited represent extreme cases of aquatic pollution, and although the list is not long, the implication is that at the population and community level aquatic systems are not being affected adversely by present levels of artificial radiation exposure, even in the most severe cases. This conclusion does not apply at the organismal and suborganismal level. Aquatic species appear to be of roughly comparable sensitivity to humans with respect to radiation effects. The highest dose rates associated with the release of low-level radioactive waste in the ocean are about 2.4 mGy d^{-1} at the end of the discharge pipe from the Windscale reprocessing plant in the United Kingdom (Woodhead, 1984). Most aquatic species obviously experience far lower doses of radiation. In the ocean at roughly 100 m below the surface, the radiation dose rate is only about 0.3 mGy per year. About 93% of this radiation comes from the decay of naturally occurring ^{40}K (Odum, 1971).

At the present time the chief concern over contamination of aquatic organisms with radionuclides involves the possible transfer of these radionuclides to human consumers. The transfer problem is compounded by the fact that certain radionuclides are greatly concentrated by some aquatic organisms. For example, the phytoplankter *Dunaliella* has been shown to concentrate ^{65}Zn by as much as a factor of 10^3–10^4 relative to the concentration in the water, and certain lamellibranchs have been found to concentrate ^{54}Mn by a factor of 10^4 (Wilber, 1969). Figure 14.6 indicates the concentration of ^{90}Sr in various components of a lake food web relative to the water. From studies such as this one, it has become obvious that concentrations of radionuclides dissolved in the water may give a very misleading picture of the degree of radioactive contamination in an aquatic system. This situation is reminiscent of many examples of heavy metal or pesticide pollution, where organisms and sediments often contain much higher concentrations of pollutants than the water. Permissible radionuclide discharge levels from nuclear installations have therefore been established only after making a careful analysis of the possible pathways of the radionuclides from the discharge into humans, considering not only direct contact between humans and water, but also various food chain pathways. It is then generally assumed that, "controls which are applied to the discharge of radioactive wastes into the marine environment to limit the potential exposure of human populations also provide quite adequate protection for populations of marine organisms" (Woodhead, 1984, p. 1266). Stating a similar feeling, Loutit (1956) commented, "If we take sufficient care radiobiologically to look after mankind, with few exceptions the rest of nature will take care of itself." Odum (1971, p. 457) has criticized this sort of reasoning as a "dangerous oversimplification," but such logic nevertheless remains the basis for setting radioisotope discharge limits. With this introduction to radiation toxicology, let us now examine the major applications to which nuclear energy has been put.

SR - 90 IN PERCH LAKE FOOD WEB

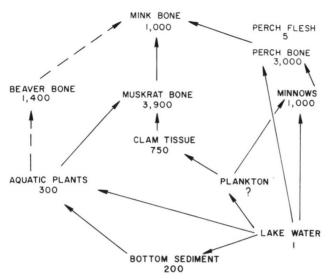

Figure 14.6 ^{90}Sr concentration factors in various components of a small Canadian lake receiving low-level atomic wastes. [Redrawn from Ophel (1963). Used with permission of Chalk River Nuclear Laboratories, Atomic Energy of Canada Limited, Chalk River, Ontario, Canada.]

NUCLEAR FISSION AND FISSION REACTORS

Energy can be generated from nuclear reactions by two basic mechanisms, fission and fusion. Nuclear fission involves the splitting of a heavy radionuclide to produce two daughter nuclides of lower atomic number. Fission is ordinarily induced by bombarding a heavy nuclide such as ^{235}U or ^{239}Pu with a neutron. For example, the fission of ^{235}U might occur according to the reaction

$$^{235}U + {}^{1}n \rightarrow {}^{140}Ba + {}^{93}Kr + 3{}^{1}n + \sim 200 \text{ meV of energy} \qquad (14.4)$$

The fragments formed from the splitting of ^{235}U are variable, since there are obviously a great many ways to split ^{235}U. However, the distribution of daughter nuclides produced by the fissioning of ^{235}U does follow a definite pattern. As indicated in Figure 14.7, the probability distribution function is bimodal, with the most abundant daughter nuclides having masses of about 140 amu and a little over 90 amu. Two other characteristics of ^{235}U fissioning are also highly predictable.

1. Large amounts of energy are released whenever ^{235}U fissions. One gram of ^{235}U can release energy equivalent to 20 million grams of TNT.

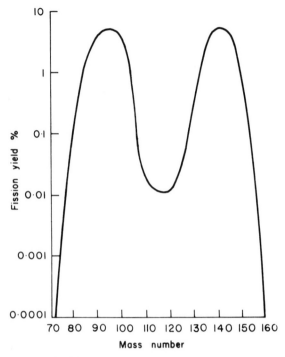

Figure 14.7 Probability distribution of daughter nuclides produced by the fissioning of ^{235}U. [Redrawn with permission from Hunt (1974). Copyright © 1974 by Pergamon Press.]

2. On the average, between 2.4 and 2.9 neutrons are produced for each neutron used to split the ^{235}U (Metz, 1976).

Because of the latter characteristic, it is potentially possible to set off a chain reaction of ^{235}U fissioning, assuming that the neutrons produced by the fissioning collide with other ^{235}U atoms and induce further fissioning. Establishing a chain reaction, however, is not trivial. The neutrons produced by the fissioning of ^{235}U are of rather high energy, and as a result they have a low probability of interacting with ^{235}U. In effect, they simply bounce off. However, if the speed of the neutrons is reduced, their probability of fissioning ^{235}U is increased greatly. To slow down the neutrons, a so-called moderator is introduced into the system. The moderator is nothing more than a substance containing low atomic number atoms such as carbon or beryllium. Water may also be used as a moderator. Atoms of low atomic number have a low probability of absorbing neutrons, but may effectively slow down fast neutrons by way of collisions.

Once the neutrons are slowed sufficiently, a second problem arises. Natural uranium consists of about 99.3% ^{238}U and only about 0.7% ^{235}U. When ^{238}U captures a slow neutron, the dominant reaction is not fission; the neutron simply combines with the ^{238}U nucleus, forming ^{239}U. Therefore natural uranium must be enriched artificially in ^{235}U to produce a fuel that can sustain a chain reaction for several years.[4] In practice, it is only necessary to produce fuel containing 2–3% ^{235}U to operate a fission reactor, because ^{235}U captures slow neutrons to produce fission 200 times more efficiently than does ^{238}U (Hunt, 1974). However, the process of ^{235}U enrichment is tedious and expensive, because there is no simple chemical means for separating ^{235}U from ^{238}U. Separation techniques involve gaseous diffusion of uranium hexafluoride[5] or passage of the uranium ions through a magnetic field (Hunt, 1974).

Finally there is the problem of the critical mass. If a piece of uranium fuel is too small, so many neutrons will escape from the fuel before colliding with ^{235}U that it will be impossible to sustain a chain reaction. If the size of the fuel element is increased, the surface/volume ratio declines, and a smaller percentage of the neutrons escape from the fuel before interacting with ^{235}U. The mass of fuel that is just large enough to sustain a chain reaction is called the *critical mass*. Obviously the size of the critical mass depends in part on the percentage of ^{235}U in the fuel; the larger the percentage of ^{235}U, the smaller the critical mass.

Plutonium may also be used as a fuel in a nuclear chain reaction, but since plutonium is not a naturally occurring element, it is necessary to use somewhat extraordinary means to generate plutonium fuel. If ^{238}U absorbs a neutron, it forms ^{239}U. Uranium-239 is radioactive, with a half-life of 23.5 minutes, and decays by way of β^- emission to produce ^{239}Np. Neptunium-239 has a half-life of 2.3 days and decays by way of β^- emission to form ^{239}Pu. Thus ^{239}Pu, a fissionable radionuclide, can be produced by bombarding ^{238}U with neutrons, and indeed some ^{239}Pu is produced by this mechanism in conventional nuclear reactors. The efficiency of Pu production is low, however, since ^{238}U does not effectively absorb slow neutrons.

The efficiency of Pu production can be increased by removing the neutron moderator. Under these conditions, two things happen:

1. The neutrons produced by fission reactions are not slowed and therefore have a lower probability of interacting with ^{239}Pu or ^{235}U.
2. The average number of neutrons produced per fission increases from about 2.4 to 2.9 (Metz, 1976).

To sustain a chain reaction under these conditions, the percentage of fissionable material, either ^{235}U or ^{239}Pu, in the fuel must be increased from about 2–3% to 15–30% (Metz, 1976). In so-called breeder reactors, a compact fuel core is surrounded by a blanket of ^{238}U. High-energy neutrons escaping from the fuel core combine with the ^{238}U to produce ^{239}Pu and thus generate additional fissile fuel. Because of the increased neutron yield per fission, such reactors can be designed to produce more fuel in the form of ^{239}Pu than they consume in the form of ^{235}U or ^{239}Pu. Hence the name, "breeders." A breeder reactor can also be designed using thorium-232 as a

---

[4]Natural uranium may be used as a fuel in reactors that use graphite or heavy water rather than ordinary water as a moderator, since graphite and heavy water absorb fewer neutrons. Heavy water is ^{2}H$_2$O; ordinary or light water is composed almost entirely of ^{1}H$_2$O.

[5]Uranium hexafluoride containing U-235 diffuses through a suitable membrane about 0.33% faster than ^{238}UF$_6$.

fuel. Bombarding ^{232}Th with slow neutrons produces ^{233}Th, which then decays by way of β^- emission with a half-life of 22.1 minutes to produce protactinium-233, which in turn decays by way of β^- emission with a half-life of 27.4 days to yield ^{233}U, a radionuclide, which, like ^{235}U and ^{239}Pu, can be bombarded with slow neutrons to sustain a fission chain reaction.

The first use of nuclear fission was of course for the production of bombs rather than for the generation of electricity. The United States began producing plutonium at special reactors near Hanford, Washington, in 1944 (Seymour, 1971), and plutonium from these reactors was used to build the bomb that was dropped on Nagasaki, Japan, in 1945. The bomb that was dropped on Hiroshima, Japan, was made of ^{235}U-enriched uranium from the government plant at Oak Ridge, Tennessee (Novick, 1969). In the case of a bomb, the goal is to release an enormous amount of energy in a short time. In a nuclear reactor the release of energy must of course occur at a slow and controllable rate.

At the present time all nuclear reactors in the United States use ^{235}U-enriched uranium as a fuel, and with a few exceptions all use water as a moderator. A diagram of the core region of a typical reactor is shown in Figure 14.8. The core of such a reactor consists of literally thousands of small uranium oxide fuel rods, each enclosed in a cladding of stainless steel or zirconium alloy. Each fuel rod is about 1 cm in diameter and several meters long. A typical 500 megawatt (MW) nuclear reactor would probably contain about 20,000 such fuel rods (Adler, 1973). Although no single fuel element contains a critical mass of fissile material, when the fuel elements are assembled together in the core of the reactor, a critical mass is achieved. To control the rate of the nuclear reaction, control rods made of cadmium are inserted into the reactor core. Since cadmium is a good absorber of neutrons, the control rods can effectively slow down or if necessary completely stop the chain reaction in the core. The ^{235}U-enriched fuel elements are normally used for a period of 2–3 years, during which time the percentage of ^{235}U decreases from about 3% to roughly 0.8%. At that time the rods must be replaced, since the percentage of ^{235}U is becoming marginal for sustaining a chain reaction (Cohen, 1977).

The water circulating through the core serves several purposes. As already noted, it slows down, by a factor of 1000 or more, the high-energy neutrons produced by the fissioning of ^{235}U (Novick, 1969). Second, it cools the reactor core. The reactor core is typically operated at a temperature of 540°C (Novick, 1969), but the intense heat produced by the chain reaction in the core would rapidly raise this temperature if water or a similar coolant were not constantly pumped through the core. If the cooling system were to fail and if the chain reaction were not quickly stopped, the rise in temperature could crack or melt the cladding and perhaps even melt the fuel, the result being an uncontrolled release of energy. Finally, heating the water is used to generate electricity. In a so-called boiling water reactor (BWR), the water is converted to steam by the temperature in the core, and the steam in turn used to drive a turbine to generate electricity as indicated in Figure 14.9. An alternative to the BWR is the pressurized water reactor (PWR), in which the water that circulates through the reactor core is kept under sufficient pressure to keep it from boiling and is cooled by way of a heat exchanger with a secondary cooling circuit (Figure 14.9). The water in the secondary circuit is converted to steam as it passes through the heat exchanger, and this steam is then used to drive a turbine to produce electricity. Most nuclear

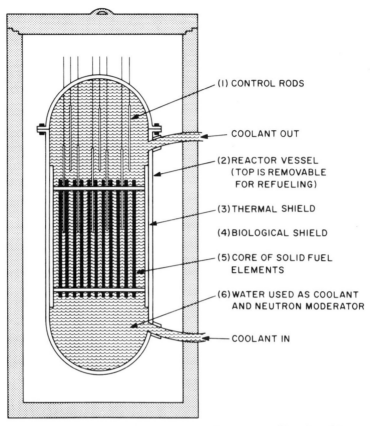

(1) CONTROL RODS

COOLANT OUT

(2) REACTOR VESSEL
(TOP IS REMOVABLE
FOR REFUELING)

(3) THERMAL SHIELD

(4) BIOLOGICAL SHIELD

(5) CORE OF SOLID FUEL
ELEMENTS

(6) WATER USED AS COOLANT
AND NEUTRON MODERATOR

COOLANT IN

Figure 14.8   Cross-section of a pressurized water nuclear reactor. (Reprinted from *Poisoned Power*. Copyright © 1971 by John W. Gofman and Arthur R. Tamplin. Permission granted by Rodale Press, Inc. Emmaus, PA. 18049.)

reactors in the United States are PWRs. Although BWRs and PWRs account for almost all the nuclear reactors in the United States, there are several other reactor designs that have been commonly used in other countries. In the United Kingdom and France, gas-cooled nuclear reactors are used (Joseph et al., 1973), and in Canada highly successful reactors have been designed using heavy water rather than ordinary water as the moderator (Robertson, 1978).

Although electricity was produced from the first breeder reactor as long ago as 1951, breeder reactor technology is still largely in the experimental stage. Small-scale prototypes have been built in the United States, the Soviet Union, Japan, the United Kingdom, West Germany, and France (Zaleski, 1980). Since the neutrons produced in these reactors must be of the fast variety, water is ruled out as a primary coolant. Instead, all breeder reactors now use liquid sodium as a primary coolant and are hence referred to as liquid metal fast breeder reactors (LMFBRs). Figure 14.10 shows a schematic of one possible design of a LMFBR. Based on experience to date, one of the principal problems with LMFBRs is the design of the heat exchangers. Sodium reacts explosively with both air and water, so that it is essential

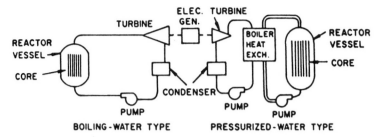

Figure 14.9 Schematic of two types of nuclear power reactors. [Redrawn from Joseph et al. (1973).]

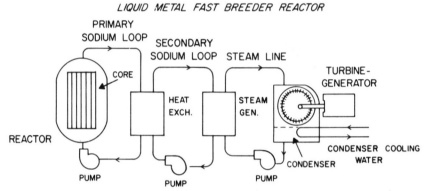

Figure 14.10 Diagram of a LMFBR. (Redrawn from *Poisoned Power*. Copyright © 1971 by John W. Gofman and Arthur R. Tamplin. Permission granted by Rodale Press, Inc. Emmaus, PA. 18049.)

to keep the sodium coolant in a completely enclosed system; a small leak in the sodium circuit could lead to a serious accident. However, since power is ultimately generated by the production of steam, a heat exchanger in which liquid sodium and water are in close proximity is a feature of all LMFBRs. Leaks in this system, usually associated with welding connections, led to at least three accidents in the Soviet Union's breeder reactors (Metz, 1976) and have been a problem in the French and British reactors as well. Because of this problem, a secondary sodium loop is utilized in all LMFBRs to minimize the possibility that the primary coolant will come into contact with water or air and react explosively.

A second problem with breeder reactors is the large amounts of plutonium involved. A large LMFBR would contain about a tonne of ^{239}Pu, only about 0.5% of which would represent a critical mass (Hammond, 1972). It is conceivable, although improbable, that terrorist groups could seize shipments of plutonium or spent fuel rods and use the ^{239}Pu to manufacture a bomb. Since plutonium is chemically different from uranium and other elements, separation of plutonium is accomplished much more easily than separation of ^{235}U from ^{238}U. Furthermore, "the information necessary to construct a crude nuclear bomb is readily available—pamphlets describing the required steps have been circulated in Great Britain . . ." (Hammond, 1972, p. 149).

Perhaps the most serious problem with breeder reactors, however, has been their cost. For example, the French Superphenix breeder reactor, the largest and most advanced breeder reactor in the world, has been producing electricity at about twice the cost of conventional PWRs (Dickson, 1987). While acknowledging that breeders are more expensive than conventional nuclear power plants over the roughly 30-year design lifetime of the plants, Weinberg (1986) has argued that the lifetime of a breeder reactor may be much longer than 30 years, perhaps as long as 100–150 years. If so, electricity production by breeders would become very cheap following their amortization time. However, the idea that breeder reactors might last as long as 100–150 years before being decommissioned is speculative at this time.

The principal motivation for breeder reactors is the fact that they consume only about 2% as much uranium as conventional fission reactors. Hence the lifetime of the fission nuclear power industry could be extended by about a factor of 50 with the use of breeder reactors. This fact is of great interest to countries such as France and Japan, which do not have much uranium. It is of less relevance on a global scale, since at current prices world uranium resources could sustain present rates of nuclear power production for about 100 years (Weinberg, 1986). The United States, in particular, has been reluctant to pursue the breeder option, because U.S. deposits of both uranium and coal are large. Hence the United States can afford to take a cautious approach to the development of technologies that might someday replace fossil fuel and conventional nuclear power production methods.

NUCLEAR FUSION

Nuclear fusion involves the combining of two elements of low atomic number. To be useful from the standpoint of energy generation, the fusion reaction must of course release large amounts of energy. That such high energy-yielding reactions can and do occur is well known. Nuclear fusion accounts for the energy output of the sun and stars, and so-called hydrogen bombs utilize nuclear fusion to produce an explosive release of energy. At the present time, however, nuclear fusion has not been used to generate electric power, because no satisfactory means has been found to control a nuclear fusion reaction. Nevertheless it is conceivable that within the next 50 years or so nuclear fusion reactors will become a reality, and it is therefore worthwhile to examine how such a reactor might work and later on to discuss its potential radiation problems.

Although a number of fusion reactions are potentially utilizable as a source of energy (Holdren, 1978), the most likely candidate for use in a nuclear reactor is the fusion of ^{2}H (deuterium) with ^{3}H (tritium) as follows:

$$^2\text{H} + {}^3\text{H} \rightarrow {}^4\text{He} + n \tag{14.8}$$

Although there is an abundance of deuterium (D) in the world's oceans, tritium (T), as already noted, is radioactive and virtually nonexistent in nature. The tritium necessary to fuel D-T fusion reactors, however, could presumably be supplied by breeding from lithium according to the reaction

$$^7\text{Li} + n\,(\text{fast}) \rightarrow {}^3\text{H} + {}^4\text{He} + n\,(\text{slow}) \tag{14.9}$$

or

$$^6\text{Li} + \text{n (slow)} \rightarrow {}^3\text{H} + {}^4\text{He} \tag{14.10}$$

Thus a D-T nuclear reactor would be a breeder reactor as well as a fusion reactor. Another possible reaction for use in a fusion reactor is the fusion of two deuterium atoms as follows:

$$^2\text{H} + {}^2\text{H} \rightarrow {}^3\text{He} + \text{n} \tag{14.11}$$

or

$$^2\text{H} + {}^2\text{H} \rightarrow {}^3\text{H} + {}^1\text{H} \tag{14.12}$$

If a D-D reaction were utilized, there would be no need to design a breeder system because of the great abundance of deuterium in the oceans.

Despite the fact that reactions such as the fusion of deuterium and tritium or deuterium and deuterium can be produced in a nuclear bomb, there are great obstacles to be overcome before these reactions can be used for constructive purposes. One important limitation, and undoubtedly the most controversial, is the fact that the temperature required to sustain these reactions is on the order of 100 million °C (Metz, 1972), about 10,000 times the temperature at the surface of the sun. This constraint was challenged, however, when in 1989 electrochemists Stanley Pons and Martin Fleischmann reported at a press conference that electrolysis of a lithium solution in heavy water, with palladium cathode and platinum anode, produced neutrons, tritium, and large amounts of heat, all from the fusion of deuterium nuclei (Garwin, 1991). The irreproducibility of these results, however, has cast considerable doubt on their authenticity. Even if so-called cold fusion can be achieved, there is still a legitimate question as to whether cold fusion will be a practical way to generate electricity. Although it seems premature to dismiss the possibility of cold fusion altogether, for the time being most of the emphasis in fusion research is focused on high-temperature reactions.

Unfortunately there is no substance known to man that could possibly contain a reaction at 100 million °C without melting. The most promising approach to solving this problem has been to ionize the atoms involved in the reaction and to contain the ionic soup or plasma with a magnetic field (Furth, 1990). The design of such a magnetic confinement reactor, however, is a formidable engineering task, since the magnets required to produce the confinement magnetic field must be of the superconducting variety, and hence will function only at temperatures less than roughly $-150°$C. Furthermore, the system must be designed to avoid damage to the magnets from the enormous flux of neutrons[6] emerging from the plasma in the center of the magnetic field, where the fusion reaction will occur. Since the magnets will be separated from the plasma by only a few meters, it is obvious that an enormous gradient in temperature as well as neutron flux must exist.

Even if magnetic containment devices are developed that can satisfactorily produce fusion reactions, there will still be formidable engineering obstacles to the development of fusion power reactors. Some method must be found to protect criti-

---

[6]About 10^{13} cm^{-2} s^{-1}.

cal components of the reactor from the enormous flux of high energy neutrons that will be produced in the fusion reaction. Much of the neutron energy could be absorbed by surrounding the inner chamber with a blanket of water about a meter thick, and indeed this absorption of neutron energy is likely to be the principal mechanism by which energy will be extracted from the reactor. High-energy neutron damage to reactor vessel material[7] could result in periodic replacement of the reactor vessel. The associated costs and shutdown time would probably make this an unacceptable operating requirement for utilities (Parkins, 1978).

The fact that no heat transfer surfaces can be located near the region of the reactor where fusion occurs and that heat must be extracted instead by absorbing high-energy neutrons at a distance from the central region of the reactor places a minimum size restriction on the reactor design. Because of this size restriction and the high cost of materials required to build a fusion reactor, the cost of conducting meaningful experiments has become a serious obstacle to fusion research. The United States, for example, recently chose to abandon plans to build a $1.9 billion fusion reactor known as the Burning Plasma Experiment (Holden, 1991). The cost of the reactor would have doubled the Energy Department's fusion budget over a 5-year period. One consequence of these high costs has been an international effort to pool resources. The United States, Japan, the Soviet Union, and Europe, for example, have begun the design phase of a so-called International Thermonuclear Experimental Reactor, which it is hoped will be capable of sustaining a fusion chain reaction (Holden, 1991). The design phase alone is expected to cost $1 billion (Hamilton, 1991a). Such high costs obviously raise the question of whether electricity derived from fusion would be cost competitive with alternative sources in the foreseeable future. For example, the cost of the stainless steel required to build a hypothetical fusion reactor designed by the University of Wisconsin Fusion Feasibility Study Group would alone have exceeded the total cost of a fossil-fueled or fission power plant of equivalent output (Parkins, 1978).

In summary, there are a number of difficult scientific as well as economic problems that will have to be overcome if fusion power is to become a viable energy option. Most scenarios do not envision fusion power becoming a significant factor in electricity production until roughly the middle of the twenty-first century (Holdren, 1978; Furth, 1990). The appeal of fusion power is the virtually inexhaustible supply of the fuel. Because of the abundance of deuterium and lithium in the ocean, a fusion-based electrical system would be essentially autarkic.

RADIATION RELEASES BY POWER PLANTS

There are basically two mechanisms by which radionuclides are created in power plants. In a fission reactor some of the fission products themselves may be radioactive, and these radionuclides may decay to produce daughter radionuclides. Important examples of such radioactive fission products are ^{137}Cs, ^{131}I, and ^{90}Sr. Tritium is the only radionuclide directly involved in a fusion reaction. Tritium is consumed in the D-T reaction (Eq. 14.8), but may be produced in a D-D reaction (Eq. 14.12). The second mechanism by which radionuclides are produced in a nuclear reactor is a process called *neutron activation*, in which a neutron produced by either a fission or

[7]Loss of ductility and swelling.

fusion reaction combines with a stable nuclide to produce a radionuclide. An example of such a neutron activation reaction would be the formation of the radionuclide iron-55 (^{55}Fe) from the stable nuclide ^{54}Fe according to the equation

$$^{54}\text{Fe} + \text{n} \rightarrow {}^{55}\text{Fe} \tag{14.13}$$

Important neutron activation products include ^{55}Fe, ^{32}P, ^{65}Zn, ^{3}H, and the transuranic elements.[8] All transuranics are radioactive, and none exists naturally. Obviously the nature of the neutron activation products produced by a nuclear reactor depends very much on the nature of the elements in the vicinity of the core. Tritium is the neutron activation product produced in greatest abundance because of the use of water as a coolant and/or moderator. Power plants attempt to minimize the production of other neutron activation products by removing impurities and corrosion products from the water and surrounding the reactor with nonactivating neutron absorbers.

Routine Radionuclide Releases

Once fission has been induced in the fuel elements of a conventional nuclear power reactor, radioactive fission fragments as well as neutron-activated radionuclides begin to accumulate inside the fuel elements. The amount of radioactive waste produced in this manner is enormous. A single 1000 MW nuclear reactor produces as much long-lived radioactive wastes in 1 year as a thousand Hiroshima bombs, and 10 such reactors operating for 2 years would generate as much radioactive waste as was produced by the combined U.S. and Soviet bomb tests prior to the 1963 partial test ban (Gofman and Tamplin, 1971). In fact the heat generated by the decay of radionuclides inside a fuel element after 2–3 years of use is so intense that the fuel element would literally melt unless artificially cooled (Novick, 1969).

Except for diffusion of some radionuclides through the fuel element cladding, little radioactivity escapes from the fuel elements unless a crack develops in the cladding. Radionuclides leaked from the fuel elements to the cooling water are normally discharged to the environment. If the radionuclides are gases they are released by way of stacks to the atmosphere. Nongaseous radionuclides are normally separated from the cooling water along with other impurities prior to recycling of the water. The impurities and nongaseous radionuclides are then disposed of in some convenient manner. In short, unless there has been a leak or malfunction in the reactor core, the level of radioactivity in these wastes is sufficiently low that they can be legally discharged to the atmosphere (gases), to the nearest convenient waterway (liquids), or simply buried in the ground in some remote and government-supervised location (solids).

There is no doubt that serious leaks in fuel element cladding have developed on occasion in some reactors. In June, 1965, unusually high discharges of radioactivity began to occur from the stacks of Pacific Gas and Electric's (PGE's) Humboldt Bay power plant at Eureka, California, a 68.5 MW BWR that had gone into operation in August 1963 (Novick, 1969; Gofman and Tamplin, 1971). The level of radioactivity in the stack gases continued to increase in the following months, and in August,

---

[8]Elements having atomic numbers greater than 92.

1965, the mean discharge rate of radioactivity from the stacks was 3 billion Bq. The Atomic Energy Commission's (AEC's) limit for the mean annual discharge rate for the Humboldt plant was 1.8 billion Bq. Fortunately most of the radioactivity was in the form of xenon and krypton, which are noble gases and highly inert chemically, but some radioactive release occurred in the form of ^{131}I. It was apparent at that time that the fuel rods' stainless steel cladding was developing leaks, and between September and December, 1965, 25% of the fuel elements were replaced with zircaloy-clad fuel elements. Although radioactive emissions dropped initially following this change, by August of 1966 emissions had risen to 1.5 billion Bq, presumably due to the failure of some of the remaining stainless steel fuel claddings. At that time a request by PGE for an increase in their radioactivity discharge rate limit to 7.8 billion Bq was denied by the AEC. Realizing that they could not continue to operate the Humboldt Bay plant with its leaky stainless steel cladding and still maintain their discharge under the 1.8 billion Bq annual average, PGE then replaced two-thirds of the remaining stainless steel-clad fuel elements with zircaloy-clad fuel elements. The plant was then run at 40% of its rated power "in order to extend the life of the remaining stainless steel-clad fuel in the core" (Stroube, 1966).[9] From an economic standpoint at least, the desire of PGE to use the stainless steel-clad fuel elements as much as possible, despite their tendency to leak, was understandable, since the initial fuel loading for the Humboldt reactor was worth about $4 million (Novick, 1969). Stainless steel rather than zircaloy was initially chosen as the cladding, because stainless steel was cheaper. The design of nonleaky cladding is obviously crucial to the successful operation of a fission power plant. Although the problem at the Humboldt plant appears to have been an unusual case, containing the radioactivity inside the fuel rods is no simple matter. Some radioactivity invariably escapes through the cladding by way of diffusion or through small cracks. This leakage contributes to the radioactivity routinely discharged by a nuclear power plant.

The second principal cause of radionuclides routinely discharged by a nuclear power plant is neutron activation of stable elements surrounding the fuel rods. Under normal circumstances these neutron-activated radionuclides constitute the principal source of radioactive release by a power plant (Joseph et al., 1973). Atoms in the primary coolant, including any corrosion products, the reactor vessel itself, and any other nearby materials are subject to neutron bombardment from the fuel elements. In the United States, where most reactors are water cooled, these activation products appear largely in the primary coolant. The principal gaseous activation products are nitrogen-13 and ^{41}Ar. The principal nongaseous activation products include tritium, ^{32}P, ^{51}Cr, ^{55}Fe, ^{54}Mn, ^{60}Co, and ^{65}Zn (Joseph et al., 1973). Many of the nongaseous neutron activation products are metals and result from activation of stable corrosion products. In U.S. Naval nuclear-powered ships, the principal source of radioactivity in liquid wastes is ^{60}Co, resulting from neutron activation of corrosion and wear products from reactor plant metal surfaces in contact with reactor cooling water (Miles et al., 1979).

Whether the nuclear reactor is a PWR or BWR, an attempt is made to keep the concentration of impurities in the primary coolant as low as possible by filtering

---

[9]The plant was shut down permanently in 1976 because it lies in a seismically active zone and was not structurally engineered to withstand earthquakes.

and demineralizing the water periodically. This procedure reduces corrosion rates as well as removing dissolved and particulate substances that otherwise would be subject to neutron activation. In a BWR the radioactive gases in the cooling water are removed by simply pumping large volumes of air through the steam turbine air ejector and venting the gases to the atmosphere, usually after a 20–30-minute delay to allow for decay of short-lived radionuclides (Joseph et al., 1973). In PWRs the radioactive gases are collected in special storage tanks and removed only periodically when the coolant is withdrawn.

The amount of radioisotopes that can be discharged by a nuclear power plant in the United States is determined by several pieces of legislation. In all cases the intent of the legislation is to protect members of the general public who might ingest, inhale, or come into contact with the radioisotopes, but the approaches taken to address this issue and the acceptable limits of exposure differ somewhat from one piece of legislation to the next. The NRC, for example, has established radioisotope concentration limits for wastewater based on the assumption that a person drinks 2 liters of the wastewater each day (NRC, 1991a). The maximum allowable dose is 0.5 mSv y^{-1}. The assumption of strictly additive effects is used to determine whether wastewater containing more than one radioisotope can be discharged. For example, the maximum effluent concentrations for tritium, ^{55}Fe, and ^{60}Co are 37,000, 3700, and 111 Bq/l, respectively. If wastewater from a nuclear power plant contained only these three radionuclides and if their concentrations were 7400, 1110, and 11.1 Bq l^{-1}, then the concentrations expressed as fractions of the maximum effluent concentrations would be 7400/37000 = 0.2, 1110/3700 = 0.3, and 11.1/111 = 0.1 for tritium, ^{55}Fe, and ^{60}Co, respectively. Since the sum of the fractions is less than 1.0, it would be acceptable to discharge this water according to the NRC (1991a).

In the foregoing example it was assumed that a person drank 2 l of the wastewater from the power plant each day, a rather improbable scenario. The other two pieces of legislation that limit radioisotope discharges from power plants take a more sophisticated approach, in the sense that they require a careful analysis of the pathways leading to human exposure and a quantitative assessment of the corresponding doses. The Environmental Protection Agency (1991a) requires that the annual dose equivalent to any member of the public resulting from planned discharges of radioisotopes associated with the uranium fuel cycle[10] not exceed 0.25 mSv to the whole body, 0.75 mSv to the thyroid gland, and 0.25 mSv to any other organ. Radiation exposure resulting from radon and its daughters, however, is excluded from this restriction. The NRC (1991b) requirement is that the amount of radioisotopes discharged with the wastewater from all nuclear power plants at a site to unrestricted areas will not result in an estimated annual dose from all pathways of exposure in excess of 0.05 mSv to the whole body of any individual in an unrestricted area.

Pathway analysis requires that a person be able to model the various routes by which radionuclides, once released to the environment, come into contact with humans. In applying this approach, a person must obviously be able to estimate with some accuracy the efficiency and rate at which radionuclides are transferred from one compartment in the model to another. Often only one or a few pathways are found to be much more important than any others in causing human exposure. These pathways are referred to as the *critical pathways.*

---

[10]The uranium fuel cycle means the operations of milling uranium ore, chemical conversion of uranium, isotopic enrichment of uranium, fabrication of uranium fuel, generation of electricity by nuclear power plants, and reprocessing of spent uranium fuel.

One of the best-known applications of critical pathway analysis was the determination of acceptable radioisotope discharge rates from the U.S. government's plutonium production reactors at Hanford, Washington. Although plutonium production and recovery at Hanford finally ceased in 1988 (Holden, 1990), for many years the Hanford reactors provided plutonium needed by the Department of Defense for the production of nuclear weapons. Prior to 1971 the cooling systems for at least some of the Hanford reactors were of the once-through variety (NAS, 1978), and the quantity of radionuclides discharged to the Columbia River far exceeded discharge rates from conventional electric power plants. Although the Hanford discharge occurred about 580 km from the mouth of the Columbia River, some radionuclides did reach the sea in significant amounts. A critical pathway analysis revealed that of these radionuclides, ^{32}P and ^{65}Zn posed the greatest danger to humans. Oysters grown in a bay several kilometers from the mouth of the river were found to have elevated concentrations of both radionuclides, and a person eating 230 g of these oysters per week for a year would have received about 0.3–0.6% of the acceptable yearly radiation dose at that time (Joseph et al., 1973). The figure of 230 g per week was estimated to be the average oyster consumption rate of the so-called critical group of persons who relied heavily on seafood for nutrition. In the Columbia River itself near the Hanford reactors, a critical pathway analysis indicated that human consumption of ^{32}P-containing fish was the most important mechanism for human exposure. Figure 14.11 depicts this critical pathway from the reactor wastewater to the river water to the fish and finally to humans.

Accidents

Much of the concern over nuclear power has centered around the issue of power plant accidents, in particular the sort of accidents that might lead to an uncontrolled nuclear chain reaction and perhaps release large amounts of radioactivity into populous areas. Obviously the best method of dealing with this problem is to try insofar

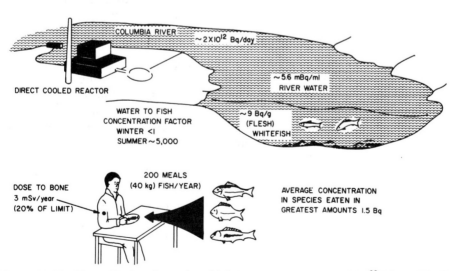

Figure 14.11   The critical pathway by which humans were exposed to ^{32}P from Hanford reactor wastes in the Columbia River. [Redrawn from *Radioactivity in the Marine Environment* (1973), p. 254, with the permission of the National Academy of Sciences, Washington, D.C.]

as possible to design a fail-proof, fool-proof power plant. However, since human endeavors are subject to errors and oversights, a person must anticipate that accidents will happen. Therefore power plants should be designed to minimize the danger both to plant personnel and the public when accidents occur. The following summaries of the most noteworthy accidents illustrate the nature of the problem.

The NRX Accident

On December 12, 1952, an operator at Canada's experimental nuclear reactor (NRX) at Chalk River, Ontario, mistakenly opened three or four (no one is sure) valves, causing three or four of the reactor's twelve shutoff rods to be withdrawn from the reactor core. At that time the reactor, which contained a natural-uranium-fueled, heavy-water-moderated core, was being operated on low power in conjunction with an experiment (Novick, 1969). The operator who mistakenly turned the valves was working in the basement of the reactor building and initially at least had no indication that he had done anything wrong. However, red lights quickly turned on at the control desk upstairs, indicating that the chain reaction in the core was out of control, and the supervisor at the desk immediately phoned the operator in the basement to tell him to stop whatever he was doing. The supervisor then went to the basement, realized the operator's mistake, and closed the valves. The red lights on the control desk then went out, indicating that the reaction in the core was again under control. Unfortunately subsequent events indicated that closing of the valves had caused the control rods to drop only part way into the core, so that although the reaction was technically under control, in fact there was only a small margin of safety. The cause of this apparent mechanical failure has never been known (Novick, 1969).

The supervisor, then thinking that he had fully corrected the problem, phoned an assistant in the control room, intending to have him return the reactor to normal operation by pushing buttons 3 and 4 on the control desk. Instead he inadvertently told the operator to push buttons 1 and 3. Although the supervisor quickly realized his mistake and attempted to correct his message, the assistant in the control room had already laid down the phone to carry out the order. Pushing button 1 caused four more shutoff rods to be withdrawn from the core, in addition to the three or four partially withdrawn rods that had evidently failed to fall back into place. As a result the chain reaction in the core again went out of control, the power output increased steadily, and the red lights on the control panel again flashed on (Novick, 1969).

About 20 seconds after pushing button 1, the assistant in the control room realized that something was wrong and pushed an emergency button, which should have caused all the control rods to drop back into the core. However, only one of the seven or eight shutoff rods dropped back into place, and that rod moved very slowly, requiring about 90 seconds to fall the necessary 3 m. Two red lights remained on at the control desk, indicating that the nuclear reaction was still out of control. Realizing the severity of the situation, the assistant supervisor ordered the heavy water drained from the reactor vessel, a move that would insure that the chain reaction came to a halt. The water required about 30 seconds to drain from the reactor, and at the end of that time instruments in the control room indicated that the reactor was shut down (Novick, 1969). The damage, however, had already been done.

The buildup of heat in the core during the roughly 70 seconds that the shutoff rods were withdrawn had melted some of the uranium fuel elements, and the molten uranium and aluminum then reacted violently with the water and steam in the core. The uranium-water reaction produced hydrogen gas, which, at the high temperature in the core, in turn reacted violently with inflowing air, causing an explosion that lifted the 4-tonne gas-holder dome over a meter into the air, jamming it among surrounding structures (Novick, 1969). The runaway nuclear fissions occurring in the core released energy equivalent to that produced by exploding about half a tonne of TNT, although fortunately the release of nuclear energy occurred over a sufficiently long time span (several seconds) that no violent nuclear explosion was produced. A cloud of radioactivity was released into the air, however, setting off automatic alarm systems in the vicinity of the reactor. Although no lives were lost and the exposure of personnel to radiation was relatively mild, the reactor core was destroyed by the accident (Novick, 1969).

Windscale

In 1950 and 1951 the British government began operating two plutonium-producing nuclear reactors at its facility at Windscale, Cumberland, in a sparsely populated area on the Irish Sea. The reactors were fueled with natural uranium fuel elements clad in steel and inserted in a 15-m cube of graphite, which served as a neutron moderator (Novick, 1969). The core of each reactor was cooled by means of air that was exhausted through two 125-m stacks to the atmosphere (Dunster et al., 1958).

It is a characteristic of a graphite moderator that some of the energy transmitted to the graphite by the high-energy neutrons emitted from the fuel elements is not immediately released as heat, but is instead stored as chemical energy within the graphite matrix of carbon atoms. In effect some of the carbon atoms in the graphite have been promoted to an energetically elevated or "excited" state. This excited state is thermodynamically unstable, but may persist for considerable time before the system reverts spontaneously to a lower-level energy state, at which time the stored energy is released as heat. This release of energy chemically stored in the graphite matrix is referred to as *Wigner release*, after a German physicist whose research contributed to the understanding of the process (Hunt, 1974).

Since unexpected releases of energy within a reactor core can lead to serious damage, some method of controlling Wigner release is necessary in a graphite-moderated reactor. Regular procedures to control Wigner release were first begun at Windscale, following a spontaneous Wigner release in 1952 (Novick, 1969). By accident the British discovered that heating the graphite slightly could trigger the release of the stored energy, so that by periodically heating the graphite, it was possible to prevent a large-scale buildup of stored energy. On October 7, 1957, such an energy release procedure was begun on Windscale Pile No. 1 while the reactor was in a shutdown state following a low-power run.

On October 8 operators noted that heating of the core was occurring at a faster rate than expected, and cadmium rods were inserted to dampen the chain reaction in the uranium (Novick, 1969). In retrospect, it is apparent that by this time one or more fuel elements in the core had in fact overheated, and the steel jackets melted or cracked. As the temperature in the core continued to rise, the uranium in the damaged fuel elements, now exposed to the air, began to burn, causing nearby fuel elements to overheat and crack, thus spreading the fire.

The first indication of a serious problem did not occur until noon on October 10, when radioactivity levels in the stack gases from Pile No. 1 began to rise sharply, and it became apparent to operators that some of the fuel elements had failed. In attempting to locate the damaged fuel elements, operators opened a plug on the side of the reactor and observed that the four fuel elements they could see inside the core were red hot. At this time the operator first realized that a fire had broken out in the core. Push rods subsequently inserted into the core came out dripping with molten uranium (Novick, 1969). Efforts to remove the damaged fuel elements failed, because the fuel cartridges were badly distorted. Fuel elements surrounding the fire were removed, however, so that at least the fire was contained. However, about 150 fuel elements were on fire by the time this action was taken. Efforts to cool the system by blowing carbon dioxide into the core failed, since "by this time the region involved was too hot to be effectively cooled in this way" (Dunster et al., 1958, p. 297).

During the early moring of October 11, a decision was made to attempt to put out the fire by pumping water into the core. This decision was reached with much difficulty, since it was known that water striking molten uranium could cause an explosion. Pumping of water into the core began at 9 A.M. on October 11. Fortunately no explosion occurred, and by October 12 the core was "quite cold" (Novick, 1969, p. 10). According to an official of the Atomic Energy Commission, "They will not guarantee that they could do it a second time without an explosion" (McCullough, 1958, p. 78).

The net result of the Windscale incident was the complete loss of the nuclear reactor and the release of an amount of radioactivity equal to about 10% of the radioactivity produced by the atomic bomb dropped on Hiroshima (Novick, 1969). The plume of radioactivity emitted from the Pile No. 1 stack from October 10 to October 12 moved in roughly a south–southeast direction over the English countryside and on into northern Europe. Examination of fallout indicated that the principal radionuclide emitted was ^{131}I, with a half-life of fortunately only 8 days. According to the report of Dunster et al. (1958), by far the most significant danger to human health from the incident was the contamination of cows' milk due to grazing in fields on which ^{131}I had been deposited. As a result sampling of cows' milk was immediately begun over a wide area, and milk was seized over roughly a 520-km^2 area for periods ranging from 20 to 40 days until radiation levels in the milk had dropped to acceptable levels (Dunster et al., 1958).

The Windscale accident occurred primarily because an unusually large amount of Wigner energy had been stored in the graphite moderator during the previous low-power run, and the release of this energy produced sufficient heat to melt the cladding on some of the fuel elements. It has since been learned that this problem only occurs after low-power runs, since at normal power the temperature reached in the graphite is high enough to give the excited carbon atoms sufficient "mobility" to return to their normal state during the operation of the reactor. Thus there is much less energy stored in the graphite after a run at normal power than after a low-power run (Hunt, 1974).

The SL-1 Incident
On August 11, 1958, a small, 3-MW nuclear BWR began producing power at the U.S. National Reactor Testing Station in Idaho. The reactor, called *SL-1*, was fueled with ^{235}U-enriched fuel elements clad in aluminum. Neutron absorbing "poison" strips consisting of thin plates of aluminum and boron had been tack welded to one or

both sides of each of the forty fuel elements in the core (Nelson et al., 1961). These strips were incorporated into the core design to serve as a burnable poison, "the depletion of which would compensate for the burning of fuel" (Nelson et al., 1961, p. 17). The neutrons produced by the fissioning of the fuel in the core were moderated by ordinary water, and the power output was regulated by means of five cadmium control rods. The reactor was designed to be a prototype of a low power BWR for use in geographically remote locations, a type of power plant that had been requested by the Defense Department in 1955. Part of the purpose of operating the prototype reactor was to gain plant-operating experience with military personnel, since it was envisioned that military personnel would one day operate such reactors in the field. Accordingly, military personnel were on the site from the time the reactor began producing power in 1958 and operated the reactor under the supervision of Combustion Engineering, the contractor responsible for overseeing the operation of the reactor (Nelson et al., 1961).

The first signs of trouble with the reactor appeared in 1959, when it was noted that the boron poison strips were bowed in the 76-mm sections between tack welds. During an August, 1960 inspection large amounts of the boron strips were found to be missing, and fuel elements in the center of the core were extremely difficult to remove by hand. Removal of fuel elements caused flaking of material and resulted in some plates' falling off. Because it was evident that further removal of fuel elements would cause additional loss of boron, no further inspections were conducted. At that time it was estimated that about 18% of the boron originally present in the core was missing. Because of this loss of boron, the reactivity of the core had increased, thus reducing the ability of the control rods to render the core subcritical. To compensate for this problem, strips of cadmium were inserted in two of the control rod shrouds on November 11, 1960 (Nelson et al., 1961).

The second major problem with the reactor was the tendency for control rods to stick in their shrouds. This problem was noted early in the operation of the reactor and became more severe with time. Some personnel felt that the bowing of the boron strips was producing a lateral force on the control rod shrouds sufficient to restrict free movement of the control rods, but this causative mechanism was never proven. In any case the problem of control rod sticking became sufficiently severe during the latter few months of 1960 that a program of "rod exercising" was initiated to maintain the rods in an operable status (Nelson et al., 1961).

On December 23, 1960 the reactor was shut down for maintenance on various components of the system. The only work on the core involved the insertion of 44 cobalt flux measuring assemblies into coolant channels between plates of the fuel elements. Removal of the control-rod drive assemblies was required to install these assemblies. The installation was accomplished by the day crew on January 3, 1961, and the night shift was assigned the job of reassembling the control-rod drives and preparing the reactor for startup.

At approximately 9 P.M. on the evening of January 3, 1961 the nuclear chain reaction in the SL-1 reactor went out of control, rapidly releasing energy roughly equivalent to that produced by detonating 10 kg of TNT. This rapid release of energy led to an explosion, probably caused by rapid vaporization of cooling water and/or to explosive reactions between oxygen and hydrogen gas in the reactor. The explosion killed three men on the night shift and released a cloud of radioactivity that was detected by radiation-monitoring devices around the reactor. Because of the damage to the reactor and because all three operators were killed, no one can say for sure what set off the explosion, but in the light of the foregoing discussion a probable mechanism is not difficult to formulate. In reassembling the control-rod drives, the operator on the night shift would have been required to lift the control rods. To produce an energy release of the observed magnitude, it would have been necessary only to jerk the central control rod upward by about 55–60 cm in about 1 second (Nelson et al., 1961). If additional boron had somehow been lost from the core during the 11-day shutdown, the uncontrolled chain reaction could have been set off more easily. That the military operators on the night shift would have deliberately jerked the central control rod in the manner described is highly unlikely, since the operators had been instructed never to raise the control rod more than 10 cm. However, the training procedure had not indicated the reason for this restriction. It is therefore conceivable that an operator, upon finding the central control rod stuck in its shroud, would have attempted to pull it out, unaware of the danger if the rod released suddenly. In testimony following the incident, it was stated that the central control rod never gave trouble, but records indicated that at least once shortly before the incident the central control rod "failed to fall freely when called upon to scram" (Nelson et al., 1961, p. VI). Although it is possible to postulate other mechanisms as the cause of the explosion, there is no evidence to support alternative hypotheses, and considering the history of the reactor, it seems most likely that the accident was caused by a rapid withdrawal of the central control rod, with the loss of the boron from the core as perhaps a contributing factor.

The SL-1 incident occurred rather early in the U.S. reactor program development and to a certain extent reflects the naivete and/or inexperience of authorities in the area of reactor safety. As Nelson et al. (1961, pp. V–VI) noted, ". . . the condition of the reactor core and the reactor control system had deteriorated to such an extent that a prudent operator would not have allowed operation of the reactor to continue without thorough analysis and review, and subsequent appropriate corrective action, with respect to the possible consequences or hazards resulting from the known deficiencies."

The Fermi Reactor Accident

The Enrico Fermi fast breeder reactor, one of only two commercial breeder reactors operated in the United States, was started up at Lagoona Beach, Michigan, 48 km from the city of Detroit, in 1963. The reactor had a maximum capacity of 200 MW, was cooled by liquid sodium, and contained about half a tonne of ^{235}U in its core, enough ^{235}U to make 40 Hiroshima-sized nuclear bombs (Novick, 1969). On the evening of October 4, 1966 operators at the plant were preparing to start up the reactor after a shut down period during which modifications had been made to correct a problem with the steam generators.

Between roughly 11 P.M. on October 4 and 8 A.M. the following morning, the reactor was run at a very low power level while the liquid sodium primary coolant heated to about 290°C. At 8 A.M. on October 5 the operators began to slowly increase the power output of the reactor, with the intention of testing the newly repaired steam generator and steam system. By that afternoon at 3 P.M. the reactor was producing 20 MW of heat energy, still only 10% of its maximum capacity.

At that time an erratic signal appeared on one of the control room instruments monitoring neutron production in the core, and although the instrument settled down again a short time later, the erratic signal reappeared when the power level had reached 34 MW. At about this time an operator noticed that instruments monitoring the temperature in the core indicated higher than normal readings in two regions of the core, and shortly thereafter radiation alarms sounded in the reactor building and elsewhere. Realizing that something had obviously gone wrong, the operators stopped the chain reaction at 3:20 P.M. by inserting six shutoff rods completely into the core. Subsequent sampling of the liquid sodium coolant revealed high concentrations of radioactive fission products, indicating that some of the fuel elements had melted. Had a sufficient amount of molten ^{235}U dropped to the bottom of the reactor and formed a critical mass, a much larger energy release could have been touched off, but fortunately the reactor was shut down before such a condition developed.

The cause of the partial meltdown was not determined until over a year after the accident, when a flexible periscope inserted into the core revealed and photographed a triangular piece of metal lying on the bottom of the reactor. The piece of metal was finally identified as a zirconium plate that had been welded to a cone about 30 cm high on the bottom of the reactor. The purpose of the cone was to direct the liquid sodium coolant upward into the core, and six zirconium strips had been welded to the cone as a last-minute modification in the design of the reactor,[11] ironically to disperse molten fuel falling on the cone in the event of a meltdown. On October 5 the zirconium strip which had come unwelded from the cone had evidently blocked one or two openings that normally admitted coolant to the core, thus causing the partial meltdown.

Although the Fermi reactor accident caused no loss of life and released little radioactivity to the environment, the accident is nevertheless noteworthy if the AEC's Hazards Summary Report for the Fermi reactor is compared with what actually took place on October 5, 1966. As noted by Novick (pp. 164–165):

> As is required by the AEC, the Hazards Summary Report contains a section describing the "maximum credible accident." This is an attempt to specify an accident which is not expected to occur, but which is the worst which the designers feel could occur. Page 603.15 of the Hazards Summary states:
>
>> The maximum credible accident in the Fermi reactor is the melting of some or all of the fuel in one core subassembly, due to either complete or partial plugging of the nozzle of that subassembly or to a flow restriction within the subassembly . . . .
>
> As a result of the October 5 accident, fuel in at least two subassemblies melted. Some damage was done to at least two other subassemblies.
>
>> The reactor would probably be shut down automatically as a result of the reactivity loss due to the melting of the fuel . . . .
>
> This did not occur on October 5; the reactor was shut down manually.
>
>> . . . . fission products . . . would have been released to the primary coolant system, and . . . the inert gas system. Since these systems are normally radioactive and sealed, there would be *no additional outward* effect caused by the addition of fission products from melted fuel [emphasis added].

In fact, there were outward effects, including high radiation levels in the reactor containment and fission product detection buildings. It is fortunate that there was leakage of radioactivity, for it was the sounding of radiation alarms that prompted the shut-

---

[11]The zirconium strips did not appear on the final construction "as built" drawings (Novick, 1969).

down of the reactor. SHELDON NOVICK (from *The Careless Atom*, copyright © 1969 by Sheldon Novick. Reprinted by permission of Houghton Mifflin Company).

The Three Mile Island Incident

At 4 A.M. on March 28, 1979 an auxiliary feedwater pump broke down on the Three Mile Island Unit 2 nuclear power plant near Harrisburg, Pennsylvania. Although the plant had officially been in operation for only 3 months, three of the four auxiliary feedwater pumps in the system had been taken out of commission 2 weeks before the accident and left out (Marshall, 1979a). As a result there was no margin of safety when the accident began at 4 A.M. Operation of the plant under these conditions was in fact a violation of federal regulations. In addition, operators had violated instructions by failing to close a pressure relief valve (PORV) in the primary coolant loop, which had been leaking for weeks prior to the accident (Ahearne, 1987b), and had inadvertently left closed a pair of block valves in the discharge lines of the emergency feedwater pumps (Lewis, 1980). The block valves were supposed to be open at all times during the normal operation of the plant.

Because of the feedwater pump failure, the core of the reactor began to overheat. The increase in temperature and pressure caused the reactor's control rods to automatically drop into the core, shutting down the reactor. Unfortunately the closed block valves caused a loss of feedwater to the steam generators, which soon boiled dry. In fact, for 13.5 hours "the reactor core was left partially exposed above the cooling water, while temperatures inside the reactor vessel climbed off the recording chart" (Marshall, 1979a, p. 280). The zirconium fuel element cladding began to melt and/or crack, and some portions of the core became so hot that the uranium oxide fuel melted (Booth, 1987). The failure of the fuel elements' cladding and the melting of part of the fuel caused large amounts of radioisotopes to escape. Following the initial loss of coolant, three additional mechanical failures intensified the problem (Marshall, 1979a).

1. The PORV in the primary coolant loop correctly opened so as to let out overheated water, but then failed to close. As a result there was a dangerous drop in pressure inside the primary coolant loop.
2. A water level indicator on the pressurizing system evidently malfunctioned, causing a technician to think the system was filled with water when in fact it was not. As a result the technician concluded incorrectly that the situation was under control.
3. When the emergency core cooling system automatically switched on, another automatic system designed to contain radioactive leaks failed to turn on. Both systems should have been automatically activated at the same time.

Finally, technicians in the control room turned off the emergency and primary cooling pumps because they did not realize that the PORV was stuck in an open position and were concerned that the system would lose the steam bubble at the top of the pressurizer (Silver, 1987). As a result they aggravated the overheating problem in the core. The core of the reactor was evidently in a highly unstable condition for about 16 hours.

The total amount of radioactivity released to the atmosphere as a result of the Three Mile Island accident has been estimated to be $1-5 \times 10^{17}$ Bq, but most was in the form of radioisotopes of xenon and krypton, which are inert gases and have half-lives of only a few days (Mynatt, 1982). Although small amounts of ^{131}I and ^{137}Cs were found in samples of milk and water taken near Three Mile Island, the plant's filters were apparently successful in preventing large-scale emissions of highly toxic radioisotopes. Cleaning up the damaged reactor, however, required more than 8 years and cost about $1 billion (Booth, 1987). The reactor is now mothballed and will ultimately be disposed of around 2010 when the Unit 1 reactor at Three Mile Island is decommissioned (Booth, 1987).

Chernobyl

At 1 A.M. on April 25, 1986 electrical engineers assumed control of the Soviet Union's Chernobyl No. 4 nuclear power reactor, one of four graphite-moderated, water-cooled, 1000-MW electrical power plants located in the Ukraine about 130 km from Kiev. The engineers wanted to test the generator's capacity to power emergency systems while coasting after a steam shut off. The Chernobyl No. 4 reactor was an example of a so-called boiling-water pressure tube reactor, or RBMK. At the time such reactors accounted for more than half of the nuclear-generated electrical capacity in the Soviet Union (Ahearne, 1987a). One disturbing feature of RBMKs is the fact that when voids are formed in the reactor core, for example if the coolant water boils and becomes less dense or if there is a substantial steam-air mixture, neutrons that would otherwise be captured by the water go into the graphite, where they are efficiently moderated to cause more fissions. Consequently the absence of water increases the rate of fissioning and the reactivity increases. In the jargon of the trade, RBMKs are said to have a positive void coefficient. RBMKs depend on a complex computer-run control system to handle this effect (Ahearne, 1987a). They are particularly sensitive to the positive void coefficient at low power output.

At 2 P.M. the emergency core cooling system, which would have drawn power and affected the test results, was shut off, the first of numerous safety violations. At 11:10 P.M. monitoring systems were adjusted to low power levels, but an operator failed to reprogram the computer to maintain power at the desired level, about 25% of capacity. As a result power fell to about 1% of capacity, a dangerously low level. At that point the majority of the control rods were withdrawn to increase power, but the accumulation in the fuel rods of xenon, a neutron absorber, frustrated the engineers' attempts to increase power.[12] Virtually all control rods were then withdrawn, an additional violation of safety standards. The power output then climbed to a little over 6% of capacity and stabilized. At 1:03 A.M., April 26, all eight feedwater pumps were activated to ensure adequate cooling after the test. The combination of low power output and high water flow necessitated many manual adjustments. The operators were having difficulty getting a stable flow of water, and because of this problem an operator turned off the emergency shutdown signals that would normally have reacted to instabilities in the water-steam separator by tripping the reactor and shutting everything down (Edwards, 1987). At 1:22 A.M. the computer indicated excess reactivity, but the operators chose to go ahead with the experiment anyway. Because of the instabilities of the reactor and the accumulation of xenon in the fuel rods, the operators realized that it would take a long time to start up the reactor if it shut down. Since they wanted to repeat the experiment immediately if it were done correctly, an operator manually disconnected the only remaining automatic trip safety system. The test began at 1:23 A.M., and power started to rise. Because of the instability of the reactor, any increase in power at this point triggered an even larger

---

[12]Xenon-135 is produced in the decay chain from tellurium, one of the direct products of ^{235}U fission.

increase. Facing catastrophe, the operators tried to insert the control rods, but the rods had 5 m of graphite at their ends, and the addition of this moderator further accelerated the reaction (Edwards, 1987). Within 4 seconds the power output increased to more than 100 times the reactor's capacity. The uranium fuel disintegrated, burst through its cladding, and reacted explosively with the cooling water. The resultant steam explosion sheared 1600 water pipes, flung the reactor's cap aside, blew through the concrete walls of the reactor hall, and threw burning blocks of graphite and fuel into the compound. Radioactive dust rose high into the atmosphere on a plume of intense heat (Edwards, 1987).

As a result of the accident, 31 persons were killed outright. Most of the immediate casualties were firemen who fought about 30 fires caused by the explosion, most ignited by graphite. The firemen were successful in preventing the spread of the flames from reactor 4 to an adjoining reactor, and with the exception of the fire in the reactor itself, all fires were extinguished by 5 A.M. The firemen, however, received large doses of radiation. Other immediate casualties included construction workers engaged in building a fifth reactor and a physician and paramedics who aided the fire fighters.

How many casualties may ultimately result from the Chernobyl accident is a matter of speculation and debate. The Soviets have estimated that about 3.7×10^{18} Bq of radioactivity were released as a result of the accident. About half of this amount was in the form of inert gases, but the remainder was in the form of biologically dangerous radionuclides, in particular ^{131}I. To a certain extent the impact of the accident on persons living near the reactor was mitigated by the fact that the accident occurred at night, that it was not raining, and that the plume rose very high, about 5 km into the sky. Nevertheless, some Soviet citizens did receive large radiation doses, in part because the Soviets did not evacuate them from the immediate area or take other appropriate measures quickly enough. Children living in a village about 9 km from the power plant, for example, were found to have received thyroid doses as high as 2.5 Sv as a result of ingesting ^{131}I in contaminated milk (Edwards, 1987). Ultimately much of the radioactivity released to the atmosphere fell out in the western Soviet Union, Europe, and parts of Asia.

In the short term the major concern from this fallout was the presence of ^{131}I, but in the longer term many of the health effects will likely result from the fallout of ^{137}Cs, which has a half-life of 30 years (Goldman, 1987). The total dose to the human population from the radioisotopes released at Chernobyl will probably amount to about 1.2 million person-sieverts, roughly half of which will be received over a period of many decades as ^{137}Cs levels gradually decline. The increase in fatal cancers associated with this dose is estimated to be about 39,000. Although this figure may seem large, it is trivial compared to the 630 million cancer fatalities expected over the same time frame for the affected population (Marshall, 1987a). The damaged reactor has been entombed in a sarcophagus of concrete and ideally should remain isolated from the environment for thousands of years while the radiation levels inside gradually decline. The monetary cost of the accident to the Soviet Union has been estimated at $3–5 billion (Table 14.5).

Summary
While the Chernobyl accident has been characterized correctly as the most serious accident ever to occur at a nuclear power plant, it is in some respects just another example of the sort of thing that has been happening at nuclear power plants from time to time for the past 40 years. From the foregoing discussion, one can see four general causes of accidents at nuclear power plants.

Table 14.5   Monetary Losses to the Soviet Union from the Chernobyl Accident, June 1986

Loss	Estimated cost (billions of dollars)
Replacement cost of plant	1.04–1.25
Lost agricultural output	1.00–1.90
Site cleanup	0.35–0.69
Health care for victims	0.28–0.56
Lost export earnings	0.22–0.66
Relocation of residents	0.07
Total	2.96–5.13

Source. Flavin (1987).

1. Human errors (e.g., an operator turning the wrong valves at the NRX reactor).
2. Violation of rules (e.g., turning off the emergency core cooling system at the Chernobyl reactor).
3. Equipment malfunctions (e.g., the zirconium plate that came unwelded and blocked the core cooling system of the Fermi reactor).
4. Inadequate technical knowledge (e.g., how to deal properly with Wigner energy release at the Windscale reactor).

Experience and better design of power plants can and have helped avoid accidents from the latter two causes; better training and selection of personnel would help reduce accidents from the first two. The fact that accidents and near-accidents continue to occur has suggested to many, however, that nuclear power is simply not worth the risks. As a result there has been growing disillusionment with the nuclear power industry (Figure 14.12) and a rapid acceleration in the cost of producing elec-

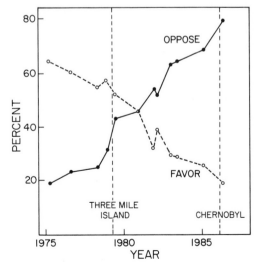

Figure 14.12 U.S. public opinion on building more nuclear power plants, 1975–1986. [*Source*: Flavin (1987).]

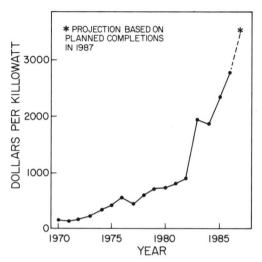

Figure 14.13   Average cost of new U.S. nuclear power plants entering operation, 1970–1987. [*Source*: Flavin (1987).]

tricity from nuclear fission. The latter development is particularly noteworthy, because it was in large part the apparent cheapness of nuclear power that caused utilities to buy nuclear plants in the first place. In the United States, for example, the cost of nuclear power plants, uncorrected for inflation, rose by a factor of 15–20 between 1970 and the late 1980s (Figure 14.13), or about a factor of 7–10 when corrected for inflation. The increased costs reflect the economic realities of designing and operating plants in a way that minimizes the possibility of an accident and the costs associated with proper handling and disposal of the radioactive wastes produced by the uranium fuel cycle. By 1987 the cost of producing electricity in the United States from new nuclear plants was at least twice the cost of electricity produced from new fossil fuel plants (Flavin, 1987). The result was a de facto moratorium on nuclear power plant construction in the United States.

The increase in the cost of nuclear power plants in the United States has been more rapid than in other countries, in part because the American system is dominated by private utilities. To a certain extent public authorities in other countries have been able to pass on the extra financial costs to the public at large (Flavin, 1987). An exception to this pattern has been in the area of power plant insurance, where the United States government has stepped in to bail out the U.S. nuclear power industry. Obviously nuclear power plants must be insured, since a private utility could certainly expect to be sued in the event that a nuclear accident resulted in loss of life or damage to health or property. In 1957 a committee appointed by the Atomic Energy Commission (AEC) at the request of the U.S. Congress released a report called *WASH-740*, in which monetary estimates were put on the amount of damage that might be caused by various accidents at a nuclear power plant. The worst accident postulated by the committee involved a core meltdown that resulted in a radiation release in which about half the fission products accumulated in the reactor fuel were emitted to the atmosphere. Because a 1000-MW reactor, after operating for only 1 year, contains more radioactive strontium, cesium, and iodine than were released in all the weapon tests prior to 1969 (Novick, 1969), it is obvious that such an accident could cause enormous damage and loss of life. The committee's estimate of the monetary cost of such an accident was put at $7 billion (Gofman and Tamplin, 1971). No insurance company or consortium of insurance companies would agree to insure nuclear power plants for such a large amount of money, and the utilities made it clear to Congress that without adequate insurance there would be no commercial nuclear power plants. To break this impasse, Senator Clinton Anderson and Congressman Melvin Price introduced a bill to provide government insurance for privately operated nuclear power plants. The bill, called the *Price-Anderson Act*, was passed by Congress in 1957 and became Public Law 85-256. The Price-Anderson Act modified the Atomic Energy Act so that the liability for a single reactor accident could not exceed $500 million plus whatever amount of private insurance was available, and then required that the AEC provide the $500 million (Novick, 1969). Private insurance companies were induced ultimately to put up an additional $74 million in insurance coverage and to agree to cover the liability for small accidents, but the total coverage of $574 million was still only about 8% of the maximum possible damage estimated by the 1957 AEC committee. Furthermore, as nuclear power plants became larger, the potential damage from an accident increased. In 1967 the largest commercial nuclear power plant in operation in the United States had a capacity of only 265 MW (Gofman and Tamplin, 1971). Most nuclear power plants built since 1970 have an electrical output of 800–1000 MW. It is thought provoking to realize that the Chernobyl accident released only 3–4% of the radioisotopes in the reactor's core (Flavin, 1987). If 50% of the fission products in the core had been released, as postulated in the WASH-740 worst-case scenario, the economic costs in terms of health care, lost agricultural output, and relocation costs could have been roughly 15 times as great, or as much as $38 billion. The most recent estimates indicate that as many as 100,000 immediate deaths and $150 billion of economic damage could result from a serious accident at a nuclear power plant located in a densely populated area of the United States (Flavin, 1987). By 1986 the nuclear power industry's maximum liability for any single accident under the Price-Anderson Act had been increased to only $665 million, and an effort to increase this liability to $6.5 billion failed in the final weeks of the 99th congress (Crawford, 1986). What all this means is that the United States government has in effect decreed that the cost of a major accident will not be reflected in the price tag of a nuclear power plant and will be borne largely by the victims of the accident and by the government. Assuming the president declared a national disaster, most of the monetary cost would ultimately be passed along to taxpayers. To some at least, the potential damage from such an accident is so great as to raise a serious moral question about whether nuclear power plants should be allowed to exist at all. As noted by Young (1957), "It is a reasonable question of public policy as to whether a hazard of this magnitude should be permitted, if it actually exists. . . . Even if insurance could be found, there is a serious question whether the amount of damage to persons and property would be worth the possible benefits accruing from atomic development."

Waste Disposal

The problem of deciding what to do with the enormous amount of radioactive waste that has been produced since the dawn of the nuclear age in 1942 has become something of a nightmare for the 25 countries presently committed to nuclear power (Flavin, 1987; Gibson, 1991). The problem is a complex one, because the various problems posed by radioisotopes vary greatly from one radioisotope to another. Thus a multifaceted approach is needed to deal with the problem.

Types of Radioactive Waste

The radioactive wastes of greatest concern are undoubtedly the so-called high-level wastes. These include irradiated reactor fuel and liquid or solid wastes resulting from the reprocessing of reactor fuel to recover fissionable isotopes. In some countries the term *intermediate level wastes* is used to designate the more highly radioactive wastes associated with maintenance of nuclear facilities, such as the spent ion exchange resins from cleanup of reactor coolant. These wastes may include transuranic elements. The United States, however, includes transuranic elements in a separate waste category, and regards as low-level waste what some other countries would regard as non-transuranic-containing intermediate level waste (Carter, 1987). The term *low-level waste* is a catchall classification that includes all radioactive waste not assigned to some other category. The designation implies such items as contaminated glassware, lab coats, and paper trash, which require little if any shielding. The radioisotopes have relatively short half-lives (Norman, 1984). This characterization, however, can be misleading. In the United States low-level wastes are radioactive wastes that are not high-level wastes, transuranic wastes, or uranium mill tailings (Robertson, 1984). As such, low-level wastes may be extremely radioactive and contain large quantities of fission products with half-lives longer than 25 years, such as ^{90}Sr and ^{137}Cs. It is also useful for purposes of this discussion to keep in mind that radioactive wastes have been generated from two different sources, the private sector and military operations. The military waste has resulted primarily from the production of nuclear bombs. The private sector waste has been generated mainly by the nuclear power industry, but also includes waste from the use of radioisotopes in medicine and scientific research.

History of Disposal

Low-level wastes. Low-level waste has been disposed of in several ways. From 1946 until a ban was placed on ocean dumping in 1983 by the so-called London Dumping Convention, more than 50 sites in the North Atlantic and Pacific oceans were used for disposal of low-level radioactive wastes (Gibson, 1991). Most of the U.S. waste discharged to the ocean was packaged in concrete-lined 55-gallon drums and disposed of at two sites, one near the Farallon Islands about 50–80 km off San Francisco and another about 210 km off Sandy Hook, New Jersey. The United States virtually ceased ocean dumping in the early 1960s, although a few barrels a year were dumped until 1970 (Norman, 1982a). The total amount of the U.S. discharge from 1946 to 1970 amounted to about 3.5×10^{15} Bq (Deese, 1977). In contrast, European countries were discharging about 3.7×10^{15} Bq of low-level waste per year at a site about 885 km off the tip of Land's End, England, just prior to the signing of the London Dumping Convention (Norman, 1982a).

   Low-level wastes have also been disposed of in shallow-land-burial sites. Until 1962 the United States utilized facilities at the Oak Ridge National Laboratory in Tennessee for this purpose, but between 1962 and 1967 five commercially operated shallow-land-burial sites were open for business (Figure 14.14). These included Beatty, Nevada; Maxey Flats, Kentucky; West Valley, New York; Richland, Washington; and Sheffield, Illinois. A sixth commercial site was opened at Barnwell, South Carolina, in 1971. The federal low-level waste has similarly been

DOE STORAGE AND DISPOSAL SITE
● COMMERCIAL DISPOSAL SITE
■ PROPOSED HIGH-LEVEL WASTE SITE
□ PROPOSED TRANSURANIC WASTE SITE

Figure 14.14   Location of radioactive waste sites in the U.S. [*Source*: Robertson (1984).]

disposed of at five major Department of Energy facilities and several minor sites (Figure 14.14).

The results of the low-level waste disposal program have been less than reassuring. Nearly all of the sites have leaked. Tritium was detected in groundwater surrounding the West Valley site, and plutonium turned up about 1.5 km off-site within 3 years of the opening of the Maxey Flats dump (Gibson, 1991). At the present time only three of the low-level waste sites are functional: Beatty, Nevada; Barnwell, South Carolina; and Hanford, Washington. It is expected, however, that about a dozen regional sites will begin receiving low-level wastes in 1993 as a result of the 1985 amendments to the National Low-Level Radiation Waste Policy Act.

According to Robertson (1984), the problems encountered at low-level waste disposal sites could have been avoided. The key to effective isolation of the wastes is minimizing contact of water with the waste and minimizing migration rates in groundwater. For example, in some cases at West Valley and the Oak Ridge National Laboratory, burial trenches were excavated below the water table. Proper site selection and careful attention to the design of burial trenches should allow effective isolation of low-level radioactivity over the time frame (no more than a few hundred years) that it remains potentially hazardous.

High-level and Transuranic Wastes.   The problem of satisfactorily disposing of high-level and transuranic wastes is much more difficult for several reasons. The intense radiation emitted by the waste poses a considerable health hazard and also produces an enormous amount of heat. Furthermore, some of the radioisotopes have very long half-lives and hence should be isolated from the environment for thousands and preferably tens of thousands of years.

In the early days of nuclear reactors, it was assumed that spent fuel elements would be recycled to recover fissionable isotopes, and indeed exactly this was done at several U.S. facilities. At Hanford, for example, spent fuel elements from plutonium-production reactors were initially stored under water for several months to allow most of the ^{131}I to decay away, and the cladding then removed with sodium hydroxide. The spent fuel inside was then dissolved in nitric acid. Several different schemes were used from one time to another to separate the uranium and pluton-ium from the nitric acid mixture, but the so-called PUREX (plutonium-uranium extraction) process, a solvent extraction method, was ultimately chosen and was used in the United States from 1955 until 1988 (Holden, 1990). The United States stopped recycling spent fuel elements in 1988.

After removal of most of the uranium and plutonium, the nitric acid solution containing the other radioactive wastes was neutralized with sodium hydroxide or sodium carbonate and stored in large underground tanks constructed of carbon steel with a concrete outer lining. This choice of storage container proved to be an unfortunate decision. The level of radioactivity in the waste solution stored in the tanks was sufficient to cause the contents of the tanks to boil for about 50 years (NAS, 1978). Although the boiling of the wastes had been anticipated and an appropriate cooling system provided on each tank to keep the boiling under control, the alkaline nature of the wastes combined with the high temperatures produced a highly corrosive solution that tended to eat through the carbon steel tanks. By 1978 leaks had been detected in 20 of the tanks, and a total of 5500 m^3 of highly radio-active liquid waste was estimated to have leaked into the ground (NAS, 1978; Schumacher, 1990). About 3800 m^3 of this radioactive water was cooling water that evidently leaked unnoticed from a single tank during the period 1968–1978, a dis-covery that was not reported publicly until 1990 (Schumacher, 1990). The Depart-ment of Energy has maintained that radioactivity from that leak remained in the soil beneath the tank and did not penetrate into the water table, 50 m below the sur-face. The previously reported worst spill occurred in 1973, when 435 m^3 of waste with a total activity of 10^{16} Bq escaped into the ground before the leak was detected and the tank pumped out. According to the NAS (1978, p. 38) report, the leak "was apparently the result of carelessness; a key employee was on vacation, and his sub-stitute failed to heed indications from the monitoring devices that something had gone wrong." In this case as well, the radioisotopes remained in the soil well above the groundwater, but such incidents did little to inspire confidence in the govern-ment's waste isolation program.

Despite a 1957 recommendation by a National Academy of Sciences committee that the military wastes be solidified and a deep geologic repository sought for their storage, the AEC, mindful of the cost of such a storage program, continued to opt for on-site storage of the wastes near the generation facilities at Hanford, Savannah River, and Idaho Falls (Carter, 1977). However, as leaks from the carbon steel tanks began to occur, it became apparent that some other form of storage would be necessary. In 1965 a program was initiated at Hanford and Savannah River to remove from the waste the principal heat-generating radionuclides, ^{90}Sr and ^{137}Cs, and then to evaporate the waste down to a damp salt cake or sludge that would no longer be much of a threat to escape from the tanks.

This program has been completed, but now what will be done with the solidified waste? Current plans call for incorporating the ^{90}Sr and ^{137}Cs into some sort of cap-sules that will presumably be buried at the U.S. high-level waste disposal site, probably Yucca Mountain, Nevada (Figure 14.14). Progress has been slow, but the Department of Energy is constructing a Defense Waste Processing Facility (DWPF) that will mix the highly radioactive sludge at Savannah River with borosilicate glass, which will then be poured into stainless steel containers about 3 m tall and 1 cm in diameter (Marshall, 1987b). Each canister will generate about 600 watts of heat, and the radiation field at the outside surface will be 63 Gy/h, enough to kill a person in 5 minutes. Until a permanent repository is available, the canisters will be stored in a bunker-like building at Savannah River, where they will be cooled by natural air circulation.

The salt cake at Savannah River will be stripped of most long-lived transuranic radionuclides, mixed with cement, and set in wide concrete vaults. The radioactive cement will be regarded as low-level waste. The separated transuranic isotopes, mainly ^{238}Pu, will be packaged in special drums and someday will presumably be stored at the Waste Isolation Pilot Project (WIPP) site near Carlsbad, New Mexico (Figure 14.14). The technology at Savannah River will also be used to package about 2300 m^3 of waste left by a bankrupt commercial nuclear reprocessing plant near West Valley, New York. If all goes well, the DWPF at Savannah River will probably become functional around 1995. A similar facility will probably be built at Hanford to process the salt cake and sludge there, but the Hanford facility is not expected to be finished before the year 2000.

A somewhat different procedure for handling high-level wastes has been used at the Idaho Falls facility. Instead of being neutralized with sodium hydroxide or sodium carbonate, the wastes were left in the acid state and stored in corrosion resistant stainless steel tanks. Subsequently the wastes were run through a relatively inexpensive high-temperature process to convert them to calcine, a granular material similar to fine sand. There is presently disagreement as to whether this procedure was better or worse than the approaches used at Hanford and Savannah River. By converting the waste to a granular form, the calcining process avoided the leakage problems that were experienced at Hanford. On the other hand, by not separating out the ^{90}Sr and ^{137}Cs from the other waste, the procedure used at Idaho Falls increased the volume of high-level waste by more than an order of magnitude (Marshall, 1987b).

The overall cost of cleaning up all the radioactive pollution at Department of Energy nuclear weapons facilities has been estimated at $100 billion over the next 30 years (Hamilton, 1991b). This price tag, however, does not include the cost of disposing of all the civilian spent fuel, which by the year 2020 may be 10 times as great as all the military high-level waste (Marshall, 1987b). Although some of the spent civilian fuel was reprocessed by Nuclear Fuel Services at West Valley from 1966 to 1972, the vast majority of spent civilian fuel elements remain at the power plants that used them. The old fuel elements are stored in cooling ponds for a period of 5–10 years and then transferred to air-cooled dry casks. Reprocessing of these fuel elements is an option, but due to the contamination of materials used in the procedure, reprocessing increases the waste volume by as much as a factor of 10 (Gibson, 1991). This fact combined with the general decline of the nuclear power industry and hence the demand for uranium and plutonium in the United States probably means that the spent fuel elements will be disposed of intact (Marshall, 1987b) along with the high-level waste from the military program. One of the main problems, as with the military high-level waste, is the extreme heat produced by the ^{90}Sr and ^{137}Cs in the fuel elements. However, given the delays in identifying a suitable permanent high-level waste disposal site, some of the civilian waste may be at least 50 years old before it is placed in a permanent depository. Given the half-lives of ^{90}Sr and ^{137}Cs, the heat output of this waste will decline by about a factor of 2 every 30 years for the next few hundred years. Therefore a workable solution to the long-term disposition of these fuel elements may be to store them on site for a time on the order of 100 years (Harrison, 1984) and then to bury them in appropriate caskets when their heat output has declined to an acceptable level (Krauskopf, 1990).

The Search for Long-term Disposal Sites

The identification of a suitable disposal site for high-level and transuranic radioactive wastes has become a very difficult problem. From a scientific and public health standpoint, the principal concern is the potential contamination of groundwater. From a political standpoint, the principal problem is the fact that no one wants these wastes in his/her backyard. In 1976 the Department of Energy authorized the construction of a facility for storage of military transuranic wastes at the WIPP site near Carlsbad, New Mexico (Figure 14.14), and work on this $2.2 billion facility is proceeding at the present time. The identification of a suitable high-level waste site has proven more difficult, and in 1982 Congress tried to get some movement on this issue by passing the Nuclear Waste Policy Act, which set out a detailed schedule of activities by the Department of Energy (DOE) and the NRC that would lead to the start of high-level waste disposal into a mined repository not later than 1998 (Krauskopf, 1990). The site tentatively chosen is Yucca Mountain, Nevada (Figure 14.14), but there has been considerable resistance to the designation of this site by environmentalists and the state of Nevada. Because the spent commercial reactor fuel to be deposited at Yucca Mountain will contain transuranic elements, both the WIPP and Yucca Mountain sites must be capable of isolating the waste from the environment for thousands and preferably tens of thousands of years. Indeed the EPA has mandated that the DOE design a repository to hold waste securely for at least 10,000 years (Pollock, 1986). Where could radioactive wastes be safely stored for such a long period of time?

Although aboveground storage systems could certainly be devised and would be most convenient from the standpoint of monitoring, the history and stability of sociopolitical institutions suggests that aboveground or nearground storage would not be a wise means of long-term disposal. A well-placed bomb, or even a small explosion or fire set by saboteurs, could result in significant releases of radioactivity from such a facility. Radioactive waste could of course be loaded aboard a spacecraft and simply fired off into outer space, but if something went wrong with the launch and the spacecraft crashed into some inconvenient place, the scattered radioactive debris might prove virtually impossible to clean up.

Current thinking therefore indicates that long-term waste disposal sites will be deep in the ground, either on continents or in the seabed. Ideally such a location would be in a geologically stable area where the probability of crustal movements or volcanic activity's redistributing the wastes would be nil, where groundwater movements would be minimal, and where the adsorption characteristics of the sediment or rocks would tend to trap radionuclides that had leached from containment canisters. The principal geological media being considered are granite and salt, but clay, basalt, tuff, shale, and diabase are also under consideration in some countries.

No single type of geological formation is clearly superior, and the choice of medium is limited by the options available in each country (Pollock, 1986).

The U.S. wastes will be packaged in corrosion-resistant canisters made of stainless steel, zirconium, or titanium. It would be optimistic and probably unrealistic to assume that the canisters would not sooner or later come into contact with groundwater. Even a dry substratum such as salt contains about 0.5% water (Cohen, 1977), and rock at a depth of a few hundred meters is nearly everywhere saturated with water (Krauskopf, 1990). Contact between groundwater and the metal canister would ultimately corrode the canister, exposing the material inside. How much time would be required for the metal canister to corrode would of course depend on how much groundwater came into contact with the container, as well as on the chemistry of the groundwater. If the groundwater were as corrosive as seawater, Heath (1977) indicates that confinement could probably not be guaranteed for more than a few thousand years, and de Marsily et al. (1977) state that the canister "is not expected to survive more than a few hundred years in the geologic environment." These time periods are obviously short compared to the tens of thousands of years during which some transuranics should be isolated.

Once the metal canister has corroded away, the material inside will be directly exposed to the leaching action of groundwater. The radionuclides will probably be present in a cylinder of glass or uranium oxide in the case of reprocessed fuel and spent fuel elements, respectively. How rapidly radionuclides will leach from these cylinders is a matter of some debate, particularly since certain properties of the glass or uranium oxide matrices may be altered by the radiation from the entrapped radionuclides. Obviously if most of the radionuclides remain trapped in the cylinders for hundreds of thousands of years, then the potential danger from even the long-lived transuranics will have been reduced considerably. However, if a substantial portion of the radionuclides leach out of the cylinder within a few thousand years, the long-lived transuranics will still pose a considerable hazard. Although certain types of glass show great resistance to leaching (Sales and Boatner, 1984), there is by no means a consensus of opinion that either glass or uranium oxide matrices will trap the radionuclides for tens of thousands of years. If the radionuclides leach out, what then?

A radionuclide that is not adsorbed by sediment particles or temporarily trapped by ion exchange with elements in the soil or rock will presumably migrate by advection with the groundwater. Since groundwater movement in repository sites is likely to be upward (de Marsily et al., 1977), the radionuclides would ultimately be transported to the surface. How long this movement might take would depend on the extent to which adsorption and ion exchange processes retarded the migration of the radionuclides. For example, movement of the strontium ion could be slowed by as much as a factor of 100 relative to groundwater flow due to exchange with calcium ions (Cohen, 1977).

Although it is obviously impractical now to set up an experiment to study the long-term movement of various radionuclides in the ground, such an experiment was set up in what is now the southeastern part of the Gabon Republic, near the equator on the coast of West Africa about 1.9 billion years ago. At that time natural uranium contained about 3% ^{235}U rather than the present 0.7%. In fact when the earth was formed about 4.5 billion years ago, uranium consisted of about 17% ^{235}U, but since the half-life of ^{235}U is about 640 times shorter than that of ^{238}U, the ratio of ^{235}U to ^{238}U has decreased steadily during the history of the earth. It is possible to sustain a nuclear chain reaction in uranium using ordinary water as a moderator if ^{235}U is present at greater than about 1% concentration, so that 1.9 billion years ago there was no theoretical reason why a chain reaction could not occur in a nearly pure deposit of uranium. Evidently such a chain reaction did occur in a large underground deposit of uranium at a place now called Oklo in southeastern Gabon. Studies there have revealed that some of the uranium deposits now contain only about 0.4% ^{235}U, a reflection of the fact that much of the ^{235}U was fissioned during the probably several hundred thousand years that the chain reaction went on. According to Cowan (1976) as much as 6 tonnes of ^{235}U were consumed by the natural reactor. From the standpoint of radioactive waste disposal, it is obviously of interest to know what became of the reactor's fission products. According to a study by the U.S. Energy Research and Development Administration, plutonium, the long-lived transuranic of probably greatest environmental concern, was efficiently confined within the reactor zone (Cowan, 1976). The principal radionuclides that escaped to the environment were probably ^{85}Kr, ^{90}Sr, and ^{137}Cs, but none of these has a half-life longer than 30 years.

Deep-seabed Disposal. An alternative to disposing of high-level and long-lived radioactive wastes in an on-land disposal site is deep-seabed disposal. This option was investigated by the United States and other nations for some time (Hollister et al., 1981), but U.S. participation in the international subseabed disposal research effort terminated in 1986 (Miles et al., 1987). At the present time Japan is the only nation actively exploring this option. The three possibilities being considered are bedrock beneath deep ocean basins and abyssal plains; ocean trenches where plate tectonics would be expected to carry the waste into the earth's mantle; and bedrock regions in areas of high sedimentation, such as river deltas (Gibson, 1991). The appeal of deep-seabed disposal is severalfold. First, for an earthquake-prone country such as Japan, there may be no suitable depository on land. Second, disposing of the wastes in the deep-seabed obviously eliminates any concern over groundwater contamination. Third, deep-sea sediments form a very effective barrier against upward migration of isotopes of concern. Relevant to this last point is the study of Colley and Thomson (1990) of the distribution of natural radionuclides in deep-sea turbidite sediments more than 0.5 million years old. By evaluating the relationship between the distribution of ^{238}U and its daughter radionuclides in the sediments, they discovered that, with the exception of ^{226}Ra, the rate of migration of the radionuclides was negligible and that none of the radionuclides, including ^{226}Ra, had escaped from the sediments into the overlying water. One of the major problems with deep-seabed disposal is the fact that an accident or miscalculation could create an international problem, whereas pollution from an on-land disposal site would presumably be localized. Furthermore, recovering leaky canisters from the deep-seabed would be very difficult compared to recovery from an on-land site. However, because of resistance to on-land disposal and the difficulty nations such as Japan will have in identifying a suitable on-land disposal site, the deep-seabed remains a possible option for long-term, high-level radioactive waste disposal.

Transmutation

The problem of high-level and transuranic waste disposal would be much more tractable if it were necessary to isolate the wastes for only a few hundred years rather than tens of thousands of years. The radioisotopes that create the long-term problem are actinides, namely isotopes of plutonium, uranium, neptunium, americium, and curium, and these can be separated from the other waste. Indeed the WIPP site is being designed specifically to handle long-lived transuranic radioactive waste. The fact is, however, that once the actinides have been separated, they can be destroyed by a process called *transmutation*. The idea is to bombard the long-lived radioisotopes with neutrons to transform them into stable isotopes or radioisotopes with much shorter half-lives. One approach would simply be to incorporate the actinides into nuclear fuel elements and transmute them in a reactor at only a modest loss in reactor efficiency (Angino, 1977; Steinberg, 1990). Theoretically the actinides could also be bombarded with neutrons using particle accelerators (Flam, 1991). Transmutation does not eliminate the radioactive waste disposal problem, but it does reduce greatly the length of time during which the wastes must be isolated. The problem with transmutation is that it requires reprocessing of spent fuel, a procedure that costs money and generates a much greater volume of radioactive waste. Thus there is a tradeoff. The need to guarantee long-term isolation of the waste can be eliminated, but in the short-term both the volume of waste and the overall cost of waste disposal increase. At the present time the United States is not pursuing vigorously the transmutation option, but more interest may develop if it becomes clear that no suitable long-term disposal site can be identified.

Uranium Mine Wastes

Natural uranium ores contain a large number of radionuclides in addition to the radioisotopes of uranium. These radionuclides consist primarily of the various daughter radionuclides produced in the long decay series of ^{238}U. Of the radionuclides in this decay series, particular concern is attached to radium-226 due to its tendency to concentrate in bones and to its daughter radon-222, whose daughter radionuclides may be concentrated in the lungs. A study of U.S. uranium miners reported in 1955 that levels of radioactivity in the air of most U.S. mines far exceeded international standards, with some of the air samples having 59 times the permissible level of radiation (Novick, 1969). Remarkably no significant action was taken by the U.S. government until 1967, when enforceable radiation standards were finally adopted for uranium miners.

Although the radiation problem associated with working conditions in the mines is primarily an air pollution problem due to ^{222}Rn, the disposal of uranium mine wastes has become a serious water pollution problem in certain parts of the southwestern United States. Only about 1 g of uranium oxide is produced from every 400 g of raw ore processed by a uranium mine, and the waste material, resembling sand and called *tailings*, has been dumped routinely in piles at a convenient location near the mine. These mine tailings contain ^{226}Ra as well as other radionuclides, and have unfortunately often been located near streams into which wind or rainwater runoff may transport them. The situation is disturbingly reminiscent of the case of cadmium pollution in the Jintsu River basin (Chapter 12). The radioactive tailings have contaminated water in a number of streams draining into the Colorado River and Lake Mead, a water system that provides both drinking and irrigation water to parts of seven states, California, Nevada, Utah, Wyoming, Colorado, New Mexico, and Arizona. Water sampled from Colorado's San Miguel River in 1955 was found to contain a radiation level of 3.2 Bq/l, over 4 times the maximum contaminant level proposed recently by the EPA for radium (EPA, 1991b). Unfortunately some of the areas where water has been contaminated by mine tailings were also exposed to considerable fallout from atmospheric bomb tests at the Nevada Test Site. According to Novick (1969), until 1959 people drinking water from the Animas River were receiving about 3 times the maximum permissible exposure to ^{226}Ra and ^{90}Sr.

In addition to creating water pollution problems, uranium mine tailings will, unless covered, continue to release radon gas for more than 100,000 years, and according to Victor Gilinsky, a nuclear physicist and member of the NRC, could become "the dominant contribution to radiation exposure from the nuclear fuel cycle" (Carter, 1978, p. 191). "In fact, according to the American Physical Society's 1977 report on waste management and the nuclear fuel cycle, the ingestion hazard from tailings becomes greater than that from high-level wastes within the first 1000 years" (Carter, 1978, p. 191). Incredibly, some of these tailings have been used as fill material beneath homes and other buildings. Carter (1978) reported that in Salt Lake City's Fire Station No. 1, which was built on such tailings around 1958, the exposure to radon daughters was 7 times greater than that allowed for uranium mine workers.

In response to this problem, Congress passed the Uranium Mill Tailings Radiation Control Act in November, 1978. The purpose of the act was to require the uranium milling industry to clean up the mill tailings. Simply stabilizing the piles requires covering them with gravel and earth and then establishing vegetation, and indeed such a stabilization program was begun in 1967. Proper cleanup, however, requires that the piles of tailings be graded to resist erosion and then covered with several meters of soil to reduce radon emissions to no more than twice background levels. Fortunately most of the roughly 200 million tonnes of tailings are in piles fairly remote from population centers, but one 1.8 million-tonne pile is located only 30 blocks from Utah's state capitol in Salt Lake City, and two piles in Colorado are located in the towns of Grand Junction and Durango. It is likely that these piles will have to be removed to another disposal site rather than merely covered over.

In 1978 Congress agreed to shoulder the cost of cleaning up 75 million tonnes of tailings generated prior to 1978 as a result of government contracts with the Atomic Energy Commission. The cost of that cleanup alone has been estimated between $700 and $900 million (Crawford, 1985). As might be expected, money has been a problem in cleaning up the remaining tailings. Since 1978 the financial position of the uranium mining industry in the United States has deteriorated seriously, the number of persons employed by the industry having declined by roughly an order of magnitude (Crawford, 1985). Because of a loophole in the 1978 legislation, uranium millers can delay stabilizing tailings until mills are permanently closed, and as a result mills can defer cleanup by simply postponing the retirement dates for facilities. Overall cleanup costs have been estimated to be $2–4 million (Crawford, 1986). In retrospect, the tailings problem seems to be one that could and should have been faced up to long ago; certainly there can be little excuse for failing to provide some sort of stabilization for the mine tailings instead of letting wind and rain transport them into the streams of a large and important watershed.

Decommissioning Nuclear Reactors
Nuclear power plants have a useful lifetime of roughly 30 years. The constraints radiation buildup places on routine maintenance and the inevitable embrittlement

of the reactor pressure vessel due to neutron bombardment make it difficult to extend the operational lifetime of the plant much beyond this time frame (Pollock, 1986). Since the first commercial reactor went into service in 1957, it is obvious that the problem of decommissioning nuclear power plants must now be faced by the industry, and indeed a few small reactors have already been decommissioned. Until the late 1970s it was felt that old reactors would simply be entombed in concrete until the radiation levels inside declined to safe levels. When reactors are initially shut down, most of the radioactivity in the pressure vessel and other components near the core is due to cobalt-60, which is formed by neutron activation of stable cobalt, a constituent of most steel. Because ^{60}Co has a half-life of only 5.3 years, it was felt that after a time on the order of decades radiation levels would have dropped to relatively harmless levels.

However, the scientists had overlooked two long-lived radioisotopes formed as the result of neutron activation of trace constituents in steel. The isotopes are nickel-59 and niobium-94, with half-lives of 80,000 years and 20,300 years, respectively. These radioisotopes contribute only a small percentage of the radioactivity when a plant is initially decommissioned, and for this reason may have been overlooked in early studies. Because of their long half-lives, however, these radioisotopes can emit radiation well above permitted levels long after ^{60}Co has decayed to insignificance (Norman, 1982b). This fact, incidentally, was brought to public attention largely as a result of calculations done by undergraduate students.

An alternative to entombment is to remove the reactor fuel and liquids, flush out the system, and put the plant under constant guard to prevent public access for 30–100 years. By that time the ^{60}Co and other short-lived radioisotopes will have decayed away to low levels, and the plant can be dismantled with relative ease. The contaminated parts would be shipped to a low-level burial site.

A third scenario is to dismantle the reactor within a few years of the time the plant shuts down. This procedure requires the use of remotely controlled cutting and blasting equipment to dismantle the plant because of the high radiation levels in the structural components.

After some study and a little experience, it has been concluded that dismantling a reactor shortly after shutdown is the most economical way to decommission the reactor. It is cheaper to dismantle the reactor after radiation levels have declined, but the monetary savings are more than outweighed by the cost of maintaining and guarding the facility for several decades (Norman, 1982b). Experience has come from the dismantling of two small reactors, the Elk River plant in Minnesota and the Sodium Reactor Experiment in Santa Susana, California. Both were shut down after only a few years of operation and were decommissioned at costs of $6 million and $10 million, respectively (Norman, 1982b). The largest reactor to be decommissioned to date has been the 72 MW reactor in Shippingport, Pennsylvania, the first commercial reactor in the United States. In that case, however, the DOE opted not to dismantle the reactor vessel, but instead encased it in concrete and barged it intact through the Panama Canal and up the Pacific coast to the Hanford reservation, where it was buried in an earthern trench (Pollock, 1986). This strategy saved money, but a valuable learning experience was lost. The Shippingport reactor had been in operation for 25 years, and the radiation levels in the reactor were undoubtedly far greater than in the Elk River and Sodium Reactor Experiment units, both of which were shut down after only a few years of operation and produced far less power. Transporting and burying reactor vessels intact does not appear to be an option for 1000-MW reactors, and it would have been useful to obtain some experience dismantling the reactor vessel of a smaller unit that had been in operation for a period of time more typical of the roughly 30-year lifetime of commercial nuclear power plants.

The cost of decommissioning a typical 1000-MW nuclear reactor will probably be about $150 million and will generate roughly 17,000 m^3 of contaminated steel and concrete (Pollock, 1986). The latter figure is about 25% of the volume of low-level wastes generated annually in the United States (Norman, 1982b). These figures are quite significant, because, by the year 2010, 65–70 large commercial nuclear units will have ceased operation in the United States (Pollock, 1986). The price tag of $150 million per reactor is a small percentage of the roughly $5 billion cost of the most recently completed plants (Stoler, 1984), but it is noteworthy that the nuclear power industry has made almost no effort during the operation of the plants to set aside funds for decommissioning (Pollock, 1986). The result is that the decommissioning costs will be passed along to the utilities' customers at the time of decommissioning. As noted by Pollock (1986, p. 36), this approach seems "unfair to utility customers if they are charged for decommissioning a reactor from which they did not receive power."

COMMENTARY

It is tempting to say that the nuclear power industry is dead, and indeed some persons have said exactly that. 1974 was the last year a nuclear power plant was ordered in the United States and the order not subsequently cancelled (Flavin, 1987). The global situation is not much different. In 1972 the International Atomic Energy Agency projected global nuclear power generating capacity in the year 2000 to be 3.5×10^{12} watts. Today it seems unlikely that nuclear generating capacity in the year 2000 will be more than 10–15% of that figure (Flavin, 1987).

Undoubtedly the most important factor in the decline of nuclear power has been economics. New nuclear power plants are no longer cost competitive with other means of generating electricity, and some would argue that nuclear power was never cost competitive. There is no doubt that the failure to take a full accounting of the costs of the uranium fuel cycle, including proper disposal of uranium mill tailings, disposal of radioactive wastes generated by power plants, and decommissioning of old plants, was a factor in making nuclear energy appear economically attractive early on. Because issues such as radioactive waste disposal and decommissioning have not yet been resolved, we still do not have a full accounting of the economic costs of nuclear power, but the calculations have gone far enough to make it clear that nuclear power, at least under present circumstances, is not viable economically.

Radioactive pollution and the economics of nuclear power are certainly not separate issues, but the public perception of the health hazards associated with nuclear power would be a serious problem for the industry at the present time even if the economics were favorable. It is thought provoking, however, to compare the environmental and health costs of nuclear power and conventional fossil fuel power. After all, fossil-fuel-burning power plants produce wastes, and the principal waste product by weight is carbon dioxide. A typical 1000-MW, coal-burning power plant emits about 270 kg of CO_2 per second from its exhaust stacks (Cohen, 1977).

Carbon dioxide is not considered a health hazard, but its accumulation in the atmosphere as the result of fossil fuel burning may be causing a gradual global warming as a result of the *greenhouse effect* (Schneider, 1989). Such a warming may have major environmental, economic, and social repercussions. From the standpoint of human health the most directly harmful waste emitted by fossil-fuel-burning power plants is sulfur dioxide (SO_2). In the absence of efforts to control SO_2 emissions, a 1000-MW, coal-burning power plant may emit about 4.5 kg of SO_2 per second, and such emissions could be expected to cause 25 fatalities, 60,000 cases of respiratory disease, and $12 million in property damage per year (Cohen, 1977). Fossil fuel burning also produces nitrogen oxides, and these together with sulfur oxides are the principal cause of acid rain (Chapter 15). Coal-burning plants also emit benzpyrene, the principal cancer-causing agent in cigarettes. A more complete accounting of the environmental costs of electricity derived from fossil fuel burning would include the thousands of persons who have suffered and died from black lung disease in coal mines and the various environmental costs associated with the use of oil (Chapter 13).

On the other side of the ledger, of course, are the tens of thousands of persons who will probably die from radiation-induced cancers caused by the accident at Chernobyl, and the uncertainty over how often accidents as serious or more serious than Chernobyl can be expected to occur. Certainly the nuclear power industry as a whole has learned from its mistakes, but safety remains a major issue. On October 11, 1991, less than 6 years after the explosion at the Chernobyl No. 4 reactor, an electrical fire broke out in the turbine hall next to the Chernobyl No. 2 reactor, blowing the roof off the building. Almost 50 firefighters worked for 6 hours to extinguish the blaze (Freeland, 1991).

In the near term at least, the nuclear power industry will probably remain in serious trouble until it becomes economically competitive with other forms of electricity and can come to grips with the issues of power plant accidents and the various forms of radioactive waste generated by the uranium fuel cycle. It is true that burning fossil fuels is not without its environmental and public-health costs, but body count comparisons will probably not convince the public that nuclear power is the way to go. The fact is that accidents such as Three Mile Island and Chernobyl should never have occurred. The nuclear power industry needs to get its act together.

If there is any appeal to nuclear power, it would seem to be in the long rather than the short term anyway. The reserves of ^{238}U, for example, could provide all the world's electrical power for thousands of years if used in breeder reactors; and if fusion reactors someday become operable, the world's electrical energy needs could be satisfied from lithium-deuterium fusion for literally millions of years. This prospect has some appeal. Sooner or later the wisdom of burning fossil fuels to generate electricity is going to be seriously questioned. That question has already been asked in countries such as France, and the response has and will depend on the alternatives available. Within 50–100 years, for example, it may be practical to generate electricity on a large scale from photovoltaics, but that time is not now. In the meantime, energy conservation and nuclear power appear to be the principal ways of reducing fossil fuel consumption. It now seems obvious that both governments and the public embraced the latter option too quickly and without asking enough questions. There is still time to examine the questions and to obtain informed answers. We should proceed with caution and continue to explore alternatives, including energy conservation.

REFERENCES

Adams, J. A., and V. L. Rogers. 1978. *A Classification System for Radioactive Waste Disposal. What Waste Goes Where.* NUREG-0456. FBDU-224-10. U.S. Nuclear Regulatory Commission.

Adler, C. 1973. *Ecological Fantasies.* Green Eagle Press, New York. 350 pp.

Ahearne, J. F. 1987a. Nuclear power after Chernobyl. *Science,* **236,** 673–679.

Ahearne, J. F. 1987b. TMI and Chernobyl. *Science,* **238,** 145.

Anderson, S. L., and F. L. Harrison, 1986. *Effects of Radiation on Aquatic Organisms and Radiobiological Methodologies for Effects Assessment.* U.S. EPA Office of Radiation Programs. Washington, D.C. 128 pp.

Angino, E. E. 1977. High-level and long-lived radioactive waste disposal. *Science,* **198,** 885–890.

BEIR. 1990. *Health Effects of Exposure to Low Levels of Ionizing Radiation.* Committee on the Biological Effects of Ionizing Radiation. BEIR-V. National Research Council. National Academy Press. Washington, DC. 421 pp.

Booth, W. 1987. Postmortem on three mile island. *Science,* **238,** 1342–1345.

Carter, L. J. 1977. Radioactive wastes: Some urgent unfinished business. *Science,* **195,** 661–666, 704.

Carter, L. J. 1978. Uranium mine tailings: Congress addresses a long-neglected problem. *Science,* **202,** 191–195.

Carter, L. J. 1987. *Nuclear Imperatives and Public Trust: Dealing with Radioactive Waste.* Resources for the Future. Washington, D.C. 473 pp.

Cohen, B. L. 1977. The disposal of radioactive wastes from fission reactors. *Sci. Amer.,* **236**(6), 21–31.

Colley, S., and J. Thomson. 1990. Limited diffusion of U-series radionuclides at depth in deep-sea sediments. *Nature,* **346,** 260–263.

Cowan, G. A. 1976. A natural uranium reactor. *Sci. Amer.,* **235**(7), 36–47.

Crawford, M. 1985. Mill tailings: A $4-billion problem. *Science,* **229,** 537–538.

Crawford, M. 1986. Toxic waste, energy bills clear congress. *Science,* **234,** 537–538.

Deese, D. A. 1977. Seabed emplacement and political reality. *Oceanus,* **20**(1), 47–63.

de Marsily, G., E. Ledoux, A. Barbreau, and J. Margat. 1977. Nuclear waste disposal: Can the geologist guarantee isolation? *Science,* **197,** 519–527.

Dickson, D. 1987. Slowdown for French fast breeders? *Science,* **238,** 472.

Dunster, H. J., H. Howells, and W. L. Templeton. 1958. District surveys following the Windscale incident, October 1957. In *Proceedings of the Second U.N. International Conference on the Peaceful Uses of Atomic Energy.* Vol. 18. U.N. Publications, New York. pp. 296–308.

Edwards, M. 1987. Chernobyl—one year after. *National Geographic,* **171,** 632–653.

Environmental Protection Agency. 1973. *Water Quality Criteria.* EPA-R3-73-033, Washington, D.C., 594 pp.

Environmental Protection Agency. 1991a. Environmental radiation protection standards for nuclear power operations. *Code of Fed. Reg.,* **40**(190), 6–7.

Environmental Protection Agency. 1991b. National primary drinking water regulations; radionuclides; proposed rule. *Fed. Register,* **56**(138), 33050–33127.

Flam, F. 1991. A nuclear cure for nuclear waste. *Science,* **252,** 1613.

Flavin, C. 1987. *Reassessing nuclear power: The fallout from Chernobyl.* Worldwatch Inst. Washington, D.C. 91 pp.

Freeland, K. 1991. Chernobyl fire revives Soviets' worst nightmare. *Honolulu Advertiser*, Oct. 14.

Furth, H. P. 1990. Magnetic confinement fusion. *Science*, **249**, 1522–1527.

Garwin, R. L. 1991. Fusion: The evidence reviewed. *Science*, **254**, 1394–1395.

Gibson, D. 1991. Confronting eternity: The long-lived problem of nuclear waste. *Calypso Log*. October. pp. 14–16.

Gillette, R. 1972. Radiation standards: The last word or at least a definitive one. *Science*, **178**, 966–967, 1012.

Gofman, J. W., and A. R. Tamplin. 1970. Radiation: The invisible casualities. *Environment*, **12**(3), 12–19, 49.

Gofman, J. W., and A. R. Tamplin. 1971. *Poisoned Power*. Rodale Press, Emmaus, PA. 368 pp.

Goldman, M. 1987. Chernobyl: A radiobiological perspective. *Science*, **238**, 622–623.

Hamilton, D. 1991a. Geographic fission on fusion. *Science*, **253**, 259.

Hamilton, D. 1991b. DOE priorities. *Science*, **253**, 1343.

Hammond, A. 1972. Fission: The pros and cons of nuclear power. *Science*, **178**, 147–150.

Harrison, J. M. 1984. Disposal of radioactive wastes. *Science*, **226**, 11–14.

Heath, G. R. 1977. Barriers to radioactive waste migration. *Oceanus*, **201**(1), 26–30.

Holden, C. 1990. Bailing out of the bomb business. *Science*, **250**, 753.

Holden, C. 1991. Fusion panel lowers its sights. *Science*, **254**, 193.

Holdren, J. P. 1978. Fusion energy in context: Its fitness for the long term. *Science*, **200**, 168–180.

Hollister, C. D., D. R. Anderson, and G. R. Heath. 1981. Subseabed disposal of nuclear wastes. *Science*, **213**, 1321–1325.

Hunt, S. E. 1974. *Fission, Fusion and the Energy Crisis*. Pergamon Press, New York. 164 pp.

International Commission on Radiobiological Protection. 1977. *Recommendations of the International Commission on Radiobiological Protection*. ICRP Publication No. 26. Pergamon Press, Oxford. 53 pp.

International Commission on Radiobiological Protection. 1984. A Compilation of the Major Concepts and Quantities in Use by the International Commission on Radiobiological Protection. *Annals of the International Commission on Radiobiological Protection*, **14**(4). Pergamon Press, Oxford.

International Commission on Radiobiological Protection. 1985. Quantitative Bases for Developing a Unified Index of Harm. *Annals of the International Commission on Radiobiological Protection*, **15**(3). Pergamon Press, Oxford.

International Commission on Radiobiological Protection. 1991. Risks Associated with Ionizing Radiations. *Annals of the International Commission on Radiobiological Protection*, **22**(1).

Joseph, A. B., P. F. Gustafson, I. R. Russell, E. A. Schuert, H. L. Volchok, and A. Tamplin. 1973. Sources of radioactivity and their characteristics. In *Radioactivity in the Marine Environment*. Committee on Oceanography, National Research Council, National Academy of Sciences, Washington, D.C. pp. 6–41.

Krauskopf, K. B. 1990. Disposal of high-level nuclear waste: Is it possible? *Science*, **249**, 1231–1232.

Lewis, H. W. 1980. The safety of fission reactors. *Sci. Amer.*, **242**, 53–65.

Loutit, J. F. 1956. The experimental animal for the study of biological effects of radiation. *Proc. Int. Conf. on Peaceful Uses of Atomic Energy*, **11**, 3–6.

Marshall, E. 1979. A preliminary report on Three Mile Island. *Science*, **204**, 280–281.

Marshall, E. 1987a. Recalculating the cost of Chernobyl. *Science*, **236**, 658–659.

Marshall, E. 1987b. Savannah River's $1-billion glassmaker. *Science*, **235**, 1314–1317.

Marshall, E. 1990a. Academy panel raises radiation risk estimate. *Science*, **247**, 22–23.

Marshall, E. 1990b. Radiation exposure: Hot legacy of the cold war. *Science*, **249**, 474.

Marx, J. 1979. Low-level radiation: just how bad is it? *Science*, **204**, 160–164.

Marx, J. 1989. Many gene changes found in cancer. *Science*, **246**, 1386–1388.

Metz, W. D. 1972. Magnetic containment fusion: What are the prospects. *Science*, **178**, 291–293.

Metz, W. D. 1976. European breeders (II): The nuclear parts are not the problem. *Science*, **191**, 368–372.

Miles, E. L., C. D. Hollister, and G. R. Heath. 1987. Subseabed waste disposal. *Science*, **238**, 144.

Miles, M. E., G. L. Sjoblom, and J. D. Eagles. 1979. *Environmental Monitoring and Disposal of Radioactive Wastes from U.S. Naval Nuclear-powered Ships and their Support Facilities.* Naval Sea Systems Command. Dept. of the Navy Report NT-79-1. 34 pp.

Mynatt. F. R. 1982. Nuclear reactor safety research since Three Mile Island. *Science*, **216**, 131–135.

National Academy of Sciences. 1978. *Radioactive Wastes at the Hanford Reservation. A Technical Review.* Panel on Hanford Wastes, Committee on Radioactive Waste Management, Commission on Natural Resources, National Research Council, National Academy of Sciences, Washington, D.C. 269 pp.

National Council on Radiation Protection and Measurements. 1989. *Radiation Protection for Medical and Allied Health Personnel.* Bethesda, MD.

Nelson, C. A., C. K. Beck, P. A. Morris, D. I. Walker, and F. Western. 1961. *Report on the SL-1 Incident, January 3, 1961.* Atomic Energy Commission Investigation Board Report, Joint Committee on Atomic Energy, Congress of the United States. U.S. Government Printing Office, Washington, D.C.

Norman, C. 1982a. U.S. considers ocean dumping of radwastes. *Science*, **215**, 1217–1219.

Norman, C. 1982b. A long-term problem for the nuclear industry. *Science*, **215**, 376–379.

Norman, C. 1984. Delay likely in high-level program. *Science*, **223**, 259.

Novick, S. 1969. *The Careless Atom.* Houghton Mifflin, New York. 225 pp.

Nuclear Regulatory Commission. 1975. *Reactor Safety Study. An Assessment of Accident Risks in U.S. Commercial Nuclear Power Plants.* NUREG-75/014. U.S. Nuclear Regulatory Commission.

Nuclear Regulatory Commission. 1991a. Standards for protection against radiation; final rule. *Fed. Register,* **56**(98), 23360–23474.

Nuclear Regulatory Commission. 1991b. Numerical guides for design objectives and limiting conditions for operation to meet the criterion "as low as is reasonably achievable" for radioactive material in light-water-cooled nuclear power reactor effluents. *Code of Fed. Regulations,* **10**(50). Appendix I.

Odum, E. P. 1971. *Fundamentals of Ecology.* Saunders, Philadelphia, 574 pp.

Ophel, I. L. 1963. The fate of radiostrontium in a freshwater community. In V. Schultz and W. Klement (Eds.), *Radioecology.* Reinhold, New York. pp. 213–216.

Parkings, W. E. 1978. Engineering limitations of fusion power plants. *Science*, **199**, 1403–1408.

Pollock, C. 1986. *Decommissioning: Nuclear Power's Missing Link.* Worldwatch Inst., Washington, D.C. 54 pp.

Robertson, J. A. L. 1978. The CANDU reactor system: An appropriate technology. *Science*, **199**, 657–664.

Robertson, J. B. 1984. Geologic problems at low-level radioactive waste-disposal sites. In J. Bredehoeft (Ed.), *Groundwater Contamination*. National Academy Press, Washington, D.C. pp. 104–108.

Sales, B. C., and L. A. Boatner. 1984. Lead-iron phosphate glass: A stable storage medium for high-level nuclear waste. *Science*, **226**, 45–48.

Schneider, S. H. 1989. The changing climate. *Sci. Amer.*, **261**, 70–79.

Schumacher, E. 1990. Million-gallon water leakage revealed. Honolulu Star-Bulletin. Sept. 30. p. A27.

Seymour, A. H. 1971. *Radioactivity in the Marine Environment*. Committee on Oceanography, National Research Council, National Academy of Sciences, Washington, D.C. pp. 1–5.

Silver, E. G. 1987. TMI and Chernobyl. *Science*, **238**, 144–145.

Singer, S. 1986. High-level nuclear waste disposal. *Science*, **234**, 127–128.

Sparrow, A. H. 1962. *The Role of the Cell Nucleus in Determining Radiosensitivity*. Brookhaven Lecture Series No. 17. Brookhaven National Laboratory Publication No. 766.

Sparrow, A. H., L. A. Schairer, and R. C. Sparrow. 1963. Relationship between nuclear volumes, chromosome numbers, and relative radiosensitivities. *Science*, **141**, 163–166.

Steinberg, M. 1990. Transmutation of high-level nuclear waste. *Science*, **250**, 887.

Stoler, P. 1984. Pulling the nuclear plug. *Time*, **123**(7), 34–42.

U.S. Dept. Health and Human Services. 1991. Vital statistics of the U.S. 1988. Vol. II—mortality, Part A.

Weinberg, A. M. 1986. Are breeder reactors still necessary? *Science*, **232**, 695–696.

Wilber, C. G. 1969. *The Biological Aspects of Water Pollution*. Charles C. Thomas, Springfield, IL. 296 pp.

Woodhead, D. S. 1984. Contamination due to radioactive materials. In O. Kinne (Ed.), *Marine Ecology V. Ocean Management. Part 3: Pollution and Protection of the Seas—Radioactive Materials, Heavy Metals and Oil*. Wiley-Interscience, New York. pp. 1111–1287.

Young, H. W. 1957. Vice President, Liberty Mutual Insurance Co., appearing on behalf of the American Mutual Alliance, *Hearings*, on Governmental Indemnity before the JCAE, 84th Congress, second session, pp. 248–250.

Zaleski, C. P. 1980. Breeder reactors in France. *Science*, **208**, 137–144.

15

ACID DEPOSITION

This most excellent canopy, the air, look you, this brave o'erhanging firmament, this majestical roof fretted with golden fire, why, it appeareth nothing to me but a foul and pestilent congregation of vapours. (William Shakespeare, Hamlet, II, ii, 299)

Acid deposition is a form of pollution that has drawn increasing attention during the last 40 years. The problem, however, is not new and can be traced back at least several hundred years. The term *acid deposition* refers to the transfer of acidic substances from the atmosphere to the surface of the earth and includes a variety of processes that can be broadly classified as either dry deposition or wet deposition. The former includes both absorption and adsorption of gases and impaction and gravitational settling of fine aerosols and coarse particles (Cowling, 1982). Wet deposition or precipitation includes deposition resulting from rain, snow, hail, dew, fog, or frost. The problems caused by acid deposition are often cited as being the result of acid rain, although it is obvious from the foregoing discussion that acid rain is only one form of acid deposition. However, it has been through studies of temporal and geographical trends in the composition and acidity of rainfall that much of our appreciation of the causes and magnitude of the acid deposition phenomenon has developed, and acid rain is certainly one of the more important forms of acid deposition. We therefore begin our discussion of acid deposition with an analysis of the nature and causes of acid rain.

ACID RAIN

According to the chemical definition of acidity, water is acidic if the concentration of hydrogen ions H^+ exceeds that of hydroxide ions OH^-. Water is classified as basic if the hydroxide ion concentration exceeds the hydrogen ion concentration. If the concentration of H^+ and OH^- are both expressed in moles per liter, then at $25°C$

$$(H^+)(OH^-) = 10^{-14} \tag{15.1}$$

where (H^+) and (OH^-) are the concentrations of H^+ and OH^-, respectively. The value of the product $(H^+)(OH^-)$ is slightly temperature dependent, but for most intents and purposes we may assume the product to equal 10^{-14}. In neutral water, it follows that $(H^+) = (OH^-) = 10^{-7}$. The pH of water is defined as the negative of the common logarithm of the H^+ activity. For our purposes we may assume that H^+ activity and H^+ concentration are synonymous. Therefore in neutral water $-\log_{10}(H^+) = \text{pH} = 7$. From the foregoing discussion, it follows that acidic water has a pH less than 7, and basic water has a pH greater than 7. For example, water with a pH of 10 would have a H^+ concentration of 10^{-10} and a OH^- concentration of 10^{-4}.

Although it might therefore seem logical to define acid rain as any rainwater having a pH < 7, acid rain is not defined in this way. The reason stems from the fact that natural waters invariably contain some dissolved gases, including in particular carbon dioxide, CO_2. When CO_2 dissolves in water, it combines with H_2O to form carbonic acid H_2CO_3 by the reaction

$$CO_2 + H_2O \rightarrow H_2CO_3 \tag{15.2}$$

H_2CO_3 may then dissociate to release H^+ ions by the reactions

$$H_2CO_3 \rightarrow H^+ + HCO_3^- \tag{15.3}$$

$$HCO_3^- \rightarrow H^+ + CO_3^= \tag{15.4}$$

Because H_2CO_3 is a source of H^+ ions, a sample of distilled water exposed to the air will actually contain an excess of H^+ ions over OH^- ions. The pH of such a water sample would be about 5.6 to 5.7. Because of this fact, acid rain is usually defined as rainwater having a pH less than 5.6 (Cowling, 1982). In other words, a pH of 5.6 is about what one would expect if the rainwater contained no dissolved substances other than atmospheric gases. In fact rainwater normally contains a variety of dissolved substances in addition to gases. The reason is that raindrops form on tiny atmospheric aerosols, which consist of particles of dust blown from the surface of the earth or even salt crystals injected into the atmosphere at the surface of the ocean. Because of the presence of these other dissolved substances, the pH of rainwater may vary widely, in some cases being greater than 7 and in some cases substantially lower. Charlson and Rodhe (1982), for example, have pointed out that pH values as low as 4.5 could result from removal by rainwater of naturally occurring acids (notably H_2SO_4) from the air. However, the pH of natural precipitation normally falls in the 5.0–5.6 range (Wellford et al., 1982), and for this reason acid rain can be defined as rainwater having a pH less than this range rather than simply rainwater with a pH less than 5.6.

Human activities have affected the pH of rainwater primarily by introducing into the atmosphere large quantities of sulfur oxides (SO_x) and various oxides of nitrogen, collectively referred to as NO_x. The sulfur oxides consist primarily of SO_2, while NO_x is a mixture of NO and NO_2. SO_x and NO_x are gases that can combine with water to form sulfuric acid (H_2SO_4) and nitric acid (HNO_3). Possible equations for such reactions are

$$3SO_2 + 3H_2O + O_3 \rightarrow 3H_2SO_4 \qquad (15.5)$$

$$SO_2 + 2OH\cdot \rightarrow H_2SO_4 \qquad (15.6)$$

$$2NO + H_2O + O_3 \rightarrow 2HNO_3 \qquad (15.7)$$

$$NO + O_3 \rightarrow NO_2 + O_2 \qquad (15.8a)$$

$$NO_2 + OH\cdot \rightarrow HNO_3 \qquad (15.8b)$$

Unfortunately H_2SO_4 and HNO_3 are much stronger acids than H_2CO_3. Even a dilute solution of H_2SO_4 or HNO_3 will have a pH of less than one. Hence the formation of these acids can greatly increase the acidity of rainwater. The atmospheric reactions which produce H_2SO_4 and HNO_3 from SO_x and NO_x are collectively favored by the presence of light, ozone (O_3), hydroxyl radicals ($OH\cdot$), and of course H_2O. As a result, summer rain tends to contain a higher concentration of H_2SO_4 than winter rain, because summer humidities are higher, and of course solar insolation is greatest during the summer (Galloway, 1979).

How do human activities introduce large quantities of SO_x and NO_x into the atmosphere? In the case of SO_x, we know that fossil fuels contain small but significant amounts of sulfur. When these fuels are burned, the sulfur is oxidized to form compounds such as SO_2. In general coal has a higher sulfur content than oil, which in turn has a higher sulfur content than natural gas. Burning of fossil fuels by humans has introduced large amounts of sulfur oxides into the atmosphere owing to the oxidation of the sulfur in the fuels. Fossil fuel burning also explains how humans have introduced nitrogen oxides into the atmosphere, but in the case of NO_x the principal source of the N is not the nitrogen in the fuels themselves. Instead, the NO_x compounds result from the oxidation of atmospheric N_2. Because the earth's atmosphere consists of almost 80% N_2 and since most fossil fuel burning occurs in a normal atmosphere, it is not surprising that some of the N_2 is oxidized to NO_x in the burning process. Note that while SO_x emissions can be controlled to a certain extent by switching from one fossil fuel to another, the choice of fossil fuel has very little to do with the rate of NO_x emissions.

Since what goes up usually comes back down, most of the SO_x and NO_x introduced into the atmosphere ultimately returns to earth. Sulfur dioxide, for example, has a life expectancy in the atmosphere of less than 5 days (Wellford et al., 1982). Some of the SO_x and NO_x return in rainfall, but a significant fraction may be returned as dry deposition. For example, Bischoff et al. (1984) estimate that in the eastern U.S. dry deposition accounts for 1.4 times as much SO_x deposition and 26% as much NO_x deposition as does wet deposition. Once sulfate or nitrate particles are deposited on the ground in dry form, they may later combine with surface water or groundwater to produce H_2SO_4 or HNO_3. Thus both wet and dry deposition can have serious impacts on the acidity of surface and ground water.

HISTORY OF THE ACID DEPOSITION PROBLEM

There is evidence to suggest that the phenomenon of acid deposition occurred as long ago as the seventeenth century. Sulfur pollution became a problem in London during that time (Brimblecombe, 1977, 1978), and Koide and Goldberg (1971) have reported that sulfate deposits in the Greenland icecap began increasing during the beginning of the industrial revolution. Acid rain was reported in Manchester, England in 1852 (Smith, 1852) and described more thoroughly 20 years later (Smith, 1872). Modern awareness and concern over acid deposition developed during the 1950s, when Barrett and Brodin (1955) discovered acid rain in Scandanavia, Houghton (1955) reported finding acid fog and cloud water in New England, and Gorham (1955) discovered that rain in the English Lake District was acidic when-ever the wind blew from urban/industrial areas. Recognition that the damage caused by acid rain crossed international boundaries came with the work of Oden (1968) in Sweden and Likens et al. (1972) in the United States.

The two major factors that have contributed to the acceleration of the acid deposition problem in recent years have been the increased use of fossil fuels and the increase in the height of exhaust stacks at power plants and various industrial operations. The increase in stack heights is noteworthy, since the change was made to reduce local air pollution problems associated with stack emissions. For exam-ple, a suffocating smog that hovered over London, England, for several days in 1952 resulted in 2500–4000 deaths in a week, 3 or more times the normal death rate (Hamilton, 1979). Such incidents led to serious attempts to clean up urban air in large, industrialized cities. One logical remedy was to build tall smokestacks so that the emissions could be sent up into the atmosphere where they were expected to dis-perse and be rendered harmless. While achieving this goal, the increase in stack heights created acid deposition problems for many kilometers downwind of the stacks. Thus a local fallout problem was transformed into an acid deposition prob-lem over a much wider area.

The first reports of the seriousness of acid deposition effects came from Scan-danavia. In Sweden, the pH of lakes dropped about two units between the 1930s and 1960. By the 1960s about 50% of Swedish lakes were found to have a pH less than 6, and 5000 lakes had a pH less than 5 (EPA, 1980). As a result of this change in pH, salmon populations were decimated in western Sweden, and other fish populations were seriously affected in central and eastern sections of the country. Norway likewise experienced a decline in salmon populations during the 1960s (EPA, 1980).

In North America there has been a serious increase in the acidity of lakes both in parts of Canada and the United States. In southern Ontario 33 of 150 lakes surveyed were found to have a pH of less than 4.5, and 32 lakes had pHs between 4.5 and 5.5 (EPA, 1980). Fish affected by the acidity included smallmouth bass, walleye, north-ern pike, lake trout, lake herring, perch, and rock bass. The largest single Canadian source of SO_x and NO_x has been the enormous smelters at Sudbury, Ontario, about 50 km north of Lake Huron. Several hundred lakes within an 80-km radius of Sud-bury contain few or no fish (EPA, 1980). A U.S. study during the 1960s and 1970s revealed that over 50% of remote Adirondack Mountain lakes at elevations greater than 600m had a pH less than 5, and 90% of these lakes contained no fish at all. For comparison, a study between 1929 and 1937 revealed that only 4% of the same lakes had a pH less than 5 and were devoid of fish (EPA, 1980). The mean pH of Adiron-dack lakes during the period 1930 to 1938 was 6.5, but between 1969 and 1975 the mean pH had dropped to 4.8 (EPA, 1980). More recent studies have shown that the number of fishless lakes in the Adirondacks has increased substantially since the early 1970s and that the phenomenon is no longer confined to lakes at high elevations (Environment Canada, 1981). Thus there is clear evidence of an increase in the acidity of U.S. Adirondack lakes and of a decline in their fish populations.

The cause of the increase in acidity of lakes in the eastern United States is not hard to determine. Figure 15.1 shows a plot of the pH of precipitation in the eastern United States in 1955–1956 and again in 1972–1973. During this time the minimum pH of rainfall in this area dropped from about 4.4 to 4.0 (Bischoff et al., 1984), and the area over which the mean pH was less than 5 (acid rain) increased from about 50% of the land area to almost 100%. The problem of acid deposition has been most acute in the northeastern United States, because this area lies directly downwind from major sources of SO_2 emissions, the numerous coal-burning power plants and smelters in the Ohio River valley. Figure 15.2, which shows the distribution of rainfall pH in a random month (June, 1966), dramatically illustrates the difference in rainfall acidification between the western and eastern United States. Western U.S. coal contains a significantly lower sulfur content than eastern U.S. coal, and in part because of this fact the pH of rain tends to be higher in the west than in the east. The SO_2 emission problem is particularly acute in the summer months (Fig. 15.2), because increased humidity and radiation during the summer result in more rapid oxidation of SO_2 to H_2SO_4 (Galloway, 1979). The result is apparent in Fig. 15.3, which shows the concentrations of NO_3^- and $SO_4^=$ in rainwater from the Hubbard Brook watershed in New Hampshire. Note that the concentration of NO_3^- shows little seasonal dependence, but the concentration of $SO_4^=$ is 2–3 times higher in the summer than in the winter.

SUSCEPTIBILITY OF LAKES TO ACID DEPOSITION EFFECTS

Natural bodies of water contain a variety of dissolved and particulate substances that buffer the pH. In other words, these substances tend to neutralize acids or bases and thereby tend to maintain the pH near a certain value. Although the natural pH of different lakes may vary greatly from one region of a country to another, the organisms that inhabit these lakes have presumably evolved so as to be able to carry out normal activities efficiently at the lake's normal pH.

The ability of a lake to resist pH changes due to acid deposition will depend on the buffering capacity of both the lake and its watershed. In most cases the principal buffer is the bicarbonate ion, HCO_3^- (McCormick, 1989). If the watershed consists of hard, insoluble bedrock (e.g., igneous or metamorphic rock) covered with thin, sandy, infertile soil, the buffering capacity of both the lake and watershed is likely to be low (Johnson, 1979). On the other hand, if the soils and rocks surrounding the lake are susceptible to chemical leaching (e.g., limestone), it is likely that a high concentration of bicarbonate and other buffers will have leached into the lake, and the watershed itself will probably have considerable capacity to absorb the acidity in acid deposition. We conclude that bodies of water most susceptible to acid deposition effects are

1. those that lie downwind from major sources of SO_x and NO_x emissions.

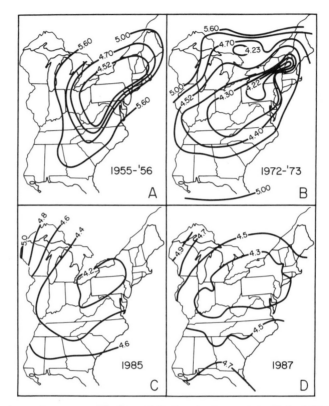

Figure 15.1   Annual average pH of precipitation in the eastern United States in selected years. [*Source*: Likens et al. (1976), McCormick (1989), and Li (1992).]

2. those whose watersheds are underlain with chemically unreactive bedrock such as granite and basalt and contain a thin cover of soil and vegetation.
3. those that receive runoff from low-order streams (i.e., near-headwater streams whose water has not been buffered significantly by reactions with soils and rocks).

Figure 15.4 illustrates the approximate distribution of land areas in the United States and Canada that are low in natural buffers and therefore particularly susceptible to acidification. As is apparent from Figure 15.4, most of the Canadian shield, an area of approximately 2.5 million km^2, is very susceptible to acid deposition. In the United States the principal sensitive areas are the Adirondack Mountains and northern New England in the East and Oregon, most of Washington and Idaho, and the mountainous portion of California in the West.

ACID DEPOSITION TOXICOLOGY

With this introduction to the subject of acid deposition, we examine here the effects of acidity on living organisms. With respect to fish, the most serious chronic effect of

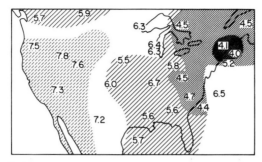

Figure 15.2 The acidification pattern of rainfall over the United States in June, 1966. [*Source*: Vermeulen (1980).]

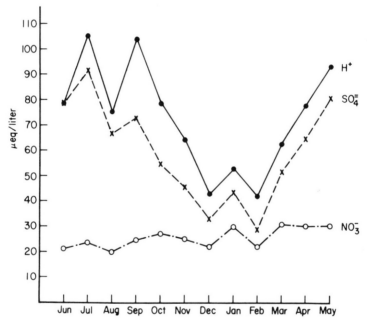

Figure 15.3 Mean monthly bulk precipitation values for selected ions in Hubbard Brook during 1963–1974. [*Source*: Galloway (1979).]

increased acidity in surface waters appears to be interference with the fish's reproductive cycle. Calcium levels in the female fish may be lowered to the point where she cannot produce eggs, or (a) the eggs fail to pass from the ovaries or (b) if fertilized, the eggs and/or larvae develop abnormally (EPA, 1980). Acutely toxic effects involving kills of adult fish may be due to a variety of causes. Failure of the fish to properly osmoregulate may result in improper body salt levels and/or failure of the fish to properly metabolize elements such as calcium and sodium. Increased acidity may also lead to the mobilization of certain metals that would otherwise remain in the soils and rocks of the watershed or in the sediments of the lake. Aluminum, for example, is derived primarily from the weathering of rocks, and the solubility of

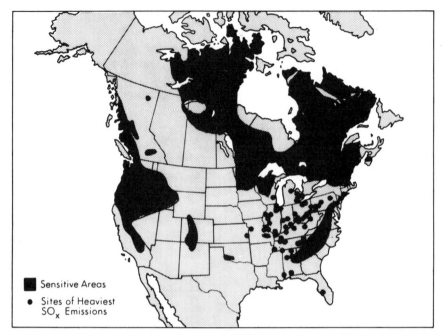

Figure 15.4   Regions in North America containing lakes that are sensitive to acidification by acid deposition. The shaded areas have igneous or metamorphic bedrock geology. Sources of major SO_x emissions are indicated by solid dots. [*Source*: Environment Canada (1981).]

aluminous soils increases dramatically as the pH is reduced from 7 to 5. Increases in dissolved Al concentrations due to acid deposition effects appear to be the chief cause of fish mortality in the Adirondacks (Cronan and Schofield, 1979; Schofield and Trojnar, 1980). Acutely toxic effects seem to result from severe necrosis of the fish's gill epithelium, and chronic effects such as reductions in growth have been detected at Al concentrations as low as 100 parts per billion (Cronan and Schofield, 1979). In Scandanavia on the other hand, the chief cause of fish mortalities has been attributed to disruption of the fishes' osmoregulatory mechanism controlling sodium balance due to low pH's and low calcium concentrations (Leivestad et al., 1976). There is no chart to indicate at exactly what pH a particular species of fish can no longer function, but when the pH drops to 5.5, most species are endangered, and when the pH reaches 4.5, almost all are dead (Environment Canada, 1981; McCormick, 1989).

The effects of acid deposition on aquatic systems are unfortunately not limited to the direct effects of increased acidity on fish populations. The reproduction of frogs and salamanders has been shown to be affected adversely by acid deposition. These amphibians lay their eggs in the meltwater pools of early spring. The snow that feeds these ponds may contain sulfuric and nitric acid in sufficient concentrations to prevent hatching of 80% or more of the eggs. Hatching failure under normal conditions is only about 10–15% (Environment Canada, 1981). Lake acidification generally results in a decline in algal species diversity, with a shift to more acid tolerant forms (EPA, 1980), and generally destabilizes the plankton community (Marmorek, 1984). Acidification of lakes may result in a decline in the biomass and activity of decomposer organisms such as bacteria and fungi, with the result that organic waste products tend to accumulate in the water column and sediments (EPA, 1980; Francis et al., 1984). Thus acid deposition may affect adversely a wide variety of organisms in aquatic systems, leading both to a serious imbalance in cycling processes involving materials within the system and to a severe decline in the biomass of living organisms.

Under appropriate conditions the fertilizing effects of acid deposition may actually be beneficial to the production of certain crops, but experiments conducted by the U.S. Environmental Protection Agency have shown that acid deposition can damage the foliage of crops such as radishes, beets, carrots, mustard greens and brocoli (Environment Canada, 1981). Acidification of soils may also render certain essential nutrients unavailable to terrestrial plants. For example, the recycling of critical elements such as nitrogen and phosphorus requires that these elements be released from decaying organic matter through the activities of soil decomposer microbes. As soil pH is reduced, the activities of these microbes is suppressed, just as the activities of aquatic microbes is suppressed by lake acidification (EPA, 1980). As a result dead organic matter tends to accumulate on the ground, and the essential nutrients bound in these dead organics are unavailable to stimulate new growth. In addition, essential elements such as calcium and magnesium, which are normally retained on soil particles, may be washed out of the soil by acid rain and hence become unavailable to plants. This process occurs because at low pH the chemical bond that is responsible for the binding of Ca and Mg to soil particles breaks down. Furthermore, acid deposition may actually stimulate the uptake of toxic heavy metals by vegetation. For example, lettuce exposed to acid rain has been found to retain elevated levels of cadmium (EPA, 1980). Although uptake of such heavy metals may not noticeably affect the growth of potentially marketable crops, the metal concentrations in the crop may make the product unfit for human consumption.

Acid rain has been shown to adversely affect photosynthesis by terrestrial plants. The production of carbohydrates, which are among the chief components of the edible portions of the plants (seeds, fruits, roots), is often suppressed when plants are grown in an acid rain environment (EPA, 1980). Furthermore, acid is known to cause bleaching of chlorophyll a, the principal light-absorbing pigment in photosynthesis. Whether such bleaching has occurred to a significant extent in the real world due to the effects of acid rain is not known, but such bleaching may in part explain the reduction in carbohydrate production observed in crops exposed to acid rain.

The direct impact of acid deposition on forests is difficult to quantify, because stresses caused by other pollutants and the indirect effects of SO_x and NO_x may be equally if not more damaging. For example, hydrocarbons, produced mainly by automobile engines, react with NO_x in the presence of sunlight to produce O_3. The O_3 may in turn react with SO_x and NO_x to produce sulfuric and nitric acid. Furthermore, ozone itself can be very damaging to trees and other vegetation and is believed to be the principal cause of damage to pine trees in the Appalachian and Blue Ridge Mountains (Postel, 1984). Acidification of forest soils can increase the concentration of soluble aluminum ions, which damage the fine roots of trees and prevent them from taking up adequate amounts of nutrients and water (McCormick, 1989). Deficiencies of calcium and magnesium frequently result.

Concern over the effects of acid deposition on human health center around the issue of heavy metal toxicity. As we already noted, acidification of water tends to mobilize heavy metals such as lead and mercury. Such metals tend to remain in particulate form and sink to the bottom of aquatic systems when the water is neutral or basic, but such particulate compounds may dissolve rapidly if the water becomes acidic. In New York State, water from the Hinckley Reservoir has become so acidic that it leaches lead from water conduit pipes, and dissolved lead concentrations in the tapwater exceed New York State public health guidelines (EPA, 1980). Other effects of acid rain on human health are suspected but not well documented. For example, how does exposure to acid rain or acid mists affect skin and hair? How does breathing in air containing mists of H_2SO_4 and HNO_3 affect our lungs? There is no doubt that SO_2 is a toxic gas (Park, 1987), and high concentrations have been shown to cause respiratory problems in humans (McCormick, 1989). Studies at Brookhaven laboratory in New York have suggested that increased concentrations of SO_2 in the air may be associated with an increase in mortality rates, and it has been hypothesized that increased SO_2 levels may cause chronic bronchitis and emphysema. However, not all scientists agree with these statements (Environment Canada, 1981).

One rather well documented effect of acid deposition is its effect on building components and stone statues. SO_2, for example, transforms calcite in stone into gypsum, which is much more soluble in water than calcite. A crust of gypsum and other compounds forms on the exterior of the stone, and then rapidly washes away in the next rain. The exposed fresh stone is then subject to further attack. Such deterioration causes a loss in the structural integrity of stone buildings and causes stone statues to lose their detail. SO_2 has also been implicated in the corrosion of steel (EPA, 1980). Even steel protected by zinc galvanizing is subject to corrosion, because the SO_2 reacts with the zinc in the protective covering. Paint has also been shown to be affected adversely by exposure to acid rain. Oil-based and automobile paints are evidently particularly vulnerable (EPA, 1980).

MAGNITUDE OF ANTHROPOGENIC EMISSIONS

Estimates of the amount of sulfur emitted to the atmosphere each year range from 140 to 220 million tonnes (EPA, 1980; Mackenzie et al., 1984). Some of these emissions are completely natural. Marine aerosols, for example, introduce $CaSO_4$ into the atmosphere, and volcanoes emit SO_2. The anthropogenic contribution to global sulfur emissions has been increasing steadily during the last century and is expected to reach 75 million tonnes by the year 2000 (Figure 15.5). Nevertheless, human activities account for only about 30–50% of total sulfur emissions. The anthropogenic contribution is much larger, however, in certain continental areas. In fact most of the anthropogenic sulfur emissions occur on only 5% of the earth's surface, primarily the industrial regions of Europe, eastern North America, and East Asia (Postel, 1984). Indeed Bischoff et al. (1984) note that, "Natural inputs of sulfur-bearing compounds into the atmosphere of the eastern United States are insignificant compared to the magnitudes of anthropogenic fluxes." Of the global anthropogenic sulfur emissions, about 70% are accounted for by the burning of coal, 16% by the combustion of petroleum, and 14% by the operations of oil refineries and

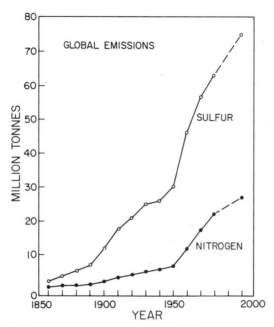

Figure 15.5   Historical global anthropogenic SO$_x$ and NO$_x$ emissions. [*Source*: Dignon and Hamseed (1989). Dashed lines are projections based on Postel (1984).]

nonferrous smelting (EPA, 1980). In the United States, anthropogenic SO$_2$ emissions peaked at about 27 million tonnes of sulfur in 1970 and have since declined to about 21 million tonnes (Figure 15.6, McCormick, 1989). About 75% of these emissions are accounted for by stationary sources east of the Mississippi, and about 92% of the latter are located near the industrialized Ohio River valley (EPA, 1980). Electric utilities account for about 66% of anthropogenic SO$_2$ emissions for the United States as a whole (Postel, 1984).

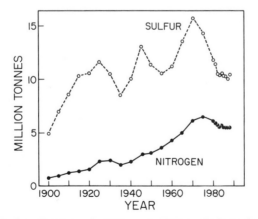

Figure 15.6   Historical anthropogenic SO$_x$ ands NO$_x$ emissions for the United States. [*Source*: Gschwandtner et al. (1988) and Kohout et al. (1990).]

Global emissions of NO_x are known only rather crudely, but it is estimated that about 50% of the NO_x in the earth's atmosphere is anthropogenic (McCormick, 1989). Natural sources of NO_x include processes such as chemical and bacterial nitrification in soil (McCormick, 1989). Anthropogenic emissions of NO_x in the United States currently amount to about 6.5 million tonnes of nitrogen per year, or roughly 27% of the world total (Figures 15.5 and 15.6). About 44% of U.S. anthropogenic emissions are contributed by transportation-related activities and 29% by the electric utility industry (Postel, 1984). There is general agreement that NO_x emissions have increased in importance relative to SO_x emissions since approximately 1950, primarily because control measures were developed more rapidly for stationary SO_x sources. In areas such as New England, for example, increases in the acidity of rain since 1964 have been accounted for largely by increases in rainwater nitrate concentrations. This fact is apparent in Fig. 15.7, which shows the relationship between H^+ and NO_3^- inputs to the Hubbard Brook Experimental Forest in New Hampshire over a 10-year period. Nevertheless, on a global scale SO_x still accounts for about 70% of the acidity of rainfall (Postel, 1984). In the Netherlands, for example, "there is a pronounced relationship between the total interior emission of SO_2 and the acidification of rainfall (Vermeulen, 1980). In the eastern United States, H_2SO_4, HNO_3, and HCl account for about 65%, 30%, and 5%, respectively, of the acidity in rainfall (Galloway, 1979). Near Denver, however, HNO_3 accounts for 80% of rainwater acidity and H_2SO_4 for only 10% (EPA, 1980). The difference reflects the relative importance of different NO_x and SO_x sources in the two areas and the effectiveness of various measures to control SO_x and NO_x emissions.

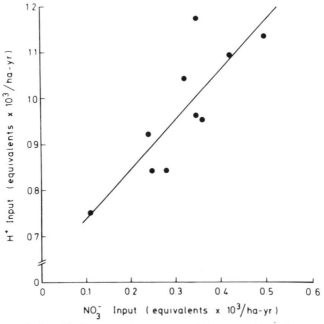

Figure 15.7 Relationship between the annual H^+ input and NO_3^- input during the period 1964–1965 to 1973–1974. [*Source*: Likens et al. (1976).]

CORRECTIVES

Methods for reducing anthropogenic emissions of NO_x and SO_x to the atmosphere may be grouped into three general categories. These categories are

(1) conservation of energy
(2) use of substitutes for fossil fuels
(3) removal of the N and S from the fossil fuels and/or combustion products before the NO_x and SO_x escape to the atmosphere.

Conservation of energy can be encouraged by either economic or political means. An obvious example has been the improved fuel efficiency of new automobiles in response to the rise in oil prices during the 1970s (Flavin and Durning, 1988). Possible substitutes for fossil fuel energy include nuclear energy, geothermal energy, wind, and solar energy. At the present time it does not appear likely that fossil fuel energy alternatives will become a major factor in the U.S. energy picture in the near future. Energy conservation has certainly helped to control the rise of fossil fuel combustion, but major reductions in emissions of SO_x and NO_x will require more than energy conservation. We conclude that some methods must be found to reduce the amount of SO_x and NO_x produced in the combustion of fossil fuels.

We will concentrate here on the problem of reducing SO_x and NO_x emissions from stationary sources such as steam-electric power plants and industrial boilers. The rationale for this approach is that only about 2–3% of anthropogenic SO_x emissions in the United States are accounted for by mobile sources due to the low S content of motor vehicle fuels (Bomberger and Phillips, 1979; Hendrey, 1983). Although historically 45–60% of NO_x emissions associated with fossil fuel burning have been due to mobile sources (Postel, 1984), this percentage has decreased in recent years due to the improved fuel efficiency of new automobiles, the use of NO_x control measures on new internal combustion engines, and the phasing out of older automobiles. Automobiles sold in the United States today have catalytic converters that reduce NO_x emissions by 76% (Postel, 1984), and the fuel efficiency of these cars is 92% higher than that of cars sold in 1973 (Flavin and Durning, 1988). As a result, stationary sources now account for more than 50% of anthropogenic NO_x emissions in countries such as the United States and West Germany (Postel, 1984).

There are basically four different stages in fossil fuel use where N and/or S may be removed. These stages are 1) fuel pretreatment, 2) fuel conversion, 3) fuel combustion and, 4) fuel postcombustion. We consider SO_x and NO_x removal separately, because both the origins of the S and N and removal processes are very different.

SO_x removal

Pretreatment

Coal Cleaning. Coal contains anywhere from 0.5 to 5.0% sulfur (McCormick, 1989). This sulfur may occur as pyritic sulfur (pyrite or marcasite), as sulfatic sulfur, or as organic sulfur. Sulfatic sulfur is unimportant because of its low concentration. The relative amounts of pyritic and organic sulfur vary widely between coals from different regions and even within a single vein. Mechanical coal cleaning takes advantage of the fact that pyrite has a specific gravity about twice that of coal and most of the minerals associated with coal. The approach in mechanical cleaning is to crush the coal into small particles and then to separate out the pyrite in a multistage cleaning process based on particle size and density. In effect the coal is given a sophisticated water bath. The pulverized coal floats on the surface, and impurities including pyrite sink to the bottom. As much as 90% removal of the pyrite is possible if the coal contains large, readily liberated pyrite (Bomberger and Phillips, 1979), but in practice pyritic S removal efficiencies rarely exceed 65% and average about 50% (McCandless et al., 1987). Because pyritic sulfur accounts for about 70% of the sulfur in coal (McCandless et al., 1987), mechanical cleaning can be expected to reduce the sulfur content of coal by about (70%)(50%) = 35%. Since the 1990 amendments to the Clean Air Act of 1970 are intended to reduce anthropogenic SO_x emissions in the United States by about 50% (Anonymous, 1991), it is evident that something more than mechanical coal cleaning will be needed to meet emission standards in the United States.

Chemical removal of sulfur from coal can be achieved in several ways. In some processes, the coal is pulverized and the pyrite is oxidized to free sulfur and/or sulfate. The sulfate dissolved in the leaching solution and the free sulfur are extracted with toluene. Pyrite removal efficiencies of 90–95% can be achieved with such methods (McCandless et al., 1987). In a process developed by Battelle Laboratories, both pyritic and organic sulfur are leached from the coal by treatment with NaOH and $Ca(OH)_2$ at elevated temperatures and pressures (Bomberger and Phillips, 1979). McCandless et al. (1987) report that up to 95% of the pyritic S and as much as 40% of the organic sulfur can be removed from coal by such methods. At the present time it appears that some form of coal cleaning will play a role in SO_x emission control at operations that burn other than low-sulfur coal, in part simply to control the variability in the sulfur content of the feed coal. Coal cleaning, however, produces its own forms of waste and generates emissions to the air, water, and land (McCandless et al., 1987).

Removal of S from Oil. The heavy fraction of crude oil known as *atmospheric residuum* has been used for years as a fuel by some power plants. This oil may contain over 4% sulfur (Bomberger and Phillips, 1979). To meet the New Source Performance Standards (NSPS) guidelines mandated by the Clean Air Act, this S content must be reduced to less than 0.8%. A standard procedure for achieving this reduction is catalytic hydrodesulfurization. The procedure consists of treating the oil with hydrogen at elevated temperatures in the presence of a suitable catalyst. The S is captured in the form of H_2S, a gas that can be removed by standard techniques. Such procedures are capable of reducing the fuel oil S content to as low as 0.2% (Bomberger and Phillips, 1979).

Conversion

Coal Gasification. Coal gasification involves treatment of pulverized coal with steam and oxygen to produce a gas containing CO_2, CO, and H_2. The sulfur is converted to H_2S, which is separated out by scrubbing with a solvent such as methanol.

The gas produced in the process is a low or medium grade fuel, depending on whether air or pure oxygen, respectively, is used as the oxygen source (Bomberger and Phillips, 1979). It is possible to upgrade the medium grade gas to synthetic natural gas by removing the CO_2 and converting the CO and H_2 to methane, but the process is costly.

Coal Liquefaction. Liquefaction of coal can be achieved by hydroliquefaction, solvent extraction, pyrolysis, or by synthesis from CO and H_2 derived from coal gasification. The net result of coal liquefaction is a product with a lower ratio of carbon to hydrogen (10 or less by weight) than is found in crude coal (12 to 20). In hydroliquefaction, hydrogen is added with the help of a catalyst to pulverized coal at high temperatures and pressure. Cleavage of the coal releases the sulfur as H_2S, the oxygen as H_2O, and the nitrogen as NH_3. All three compounds are gases that can be removed in a straightforward way. The liquid product consists of fuel oil and some lighter hydrocarbons. Solvent extraction is somewhat similar to hydroliquefaction, except that no synthetic catalysts are used, and the design of the reactor vessel is different. If hydrogen is added at about 1% of the weight of the coal, the product is primarily a solid known as solvent-refined coal (SRC) that melts at 148–177°C (Bomberger and Phillips, 1979). Since the SRC contains less than 1% sulfur and is low in ash, it can be used as a solid replacement for coal. If hydrogen is added at up to 4% of the coal weight, the primary solvent extraction products are liquids. However, the cost of producing liquid fuels by solvent extraction is much higher than the cost of producing SRC. Pyrolysis of coal has been practiced for a number of years in order to produce coke. In the process, some liquid and gaseous fractions are also produced, but the standard coking process is not designed to maximize the production of liquid fuels. By suitable modifications in the pyrolysis procedure, higher yields of liquid hydrocarbons can be achieved. However, the liquids must be catalytically hydrotreated to reduce their sulfur content. The solid residue (char) from the pyrolysis treatment also has too high a sulfur content to be used as a fuel, but gasification of the char may produce a suitably low sulfur fuel (Bomberger and Phillips, 1979).

Methanol Production. The CO and H_2O produced from coal gasification can be used to produce H_2 and CO_2 by the reaction $CO + H_2O \rightarrow H_2 + CO_2$. The H_2 produced by this reaction may then be used to produce methanol (CH_3OH) from either CO or CO_2 by the reactions

$$CO + 2H_2 \rightarrow CH_3OH \tag{15.9}$$

$$CO_2 + 3H_2 \rightarrow CH_3OH + H_2O \tag{15.10}$$

Sulfur is removed in the form of H_2S. Crude methanol from coal could be used as a clean utility fuel, particularly for peaking power (Bomberger and Phillips, 1979). Higher molecular weight hydrocarbons can be formed from H_2, CO, and CO_2 by Fischer-Tropsch synthesis.

Combustion

Fluidized Bed Combustion. Fluidized bed combustion of coal can be used to reduce emissions of both SO_x and NO_x. The combustion chamber in such a process consists of a perforated bed of crushed limestone or dolomite, ash, and partially burned coal. A distribution system in the bottom of the bed introduces powdered coal and combustion air. The air blown into the combustion chamber suspends the particles and causes them to float freely and behave like a liquid; hence the term *fluidized bed*. The two key features of the process are

1. Combustion takes place without visible flames at a temperature of about 850–900°C, much lower than the temperature of 1650°C needed in conventional coal-fired systems (McCormick, 1989). Because the oxidation of N_2 to NO_x requires a high activation energy, the lower combustion temperature greatly reduces NO_x formation from the air. It has less effect, however, on the formation of NO_x from nitrogen in the coal.

2. The SO_2 and SO_3 gases produced by the combustion process are in direct contact with the limestone. Absorption of these gases is about 90% complete if the bed contains adequate limestone.

From the standpoint of SO_x removal, the major drawback of the process is that limestone utilization efficiency is low. As a result limestone must be added at up to twice the stoichiometric requirements for SO_2 and SO_3 removal. Disposal of spent limestone therefore becomes a significant solid waste disposal problem. Research currently underway may identify a means for regenerating the limestone and thereby eliminating the solids disposal problem.

Lime Injection in Multistage Burners (LIMB). This technique is similar to fluidized bed combustion in the sense that NO_x emissions are controlled by reducing the combustion temperature, and SO_x gases are absorbed with powdered limestone. LIMB, however, does not involve the use of a fluidized bed. The SO_x absorbent is injected directly into the firebox, and the combustion temperature is lowered with the use of special burners. SO_x emissions are typically reduced by 50–70% (McCormick, 1989). The chief advantage of LIMB over fluidized bed combustion is that LIMB can be cost-effectively retrofitted into existing plants. Retrofitting plants to accommodate fluidized bed combustion is impractical in some cases due to lack of space, and in any case capital costs are far lower for LIMB than fluidized bed combustion.

Postcombustion

Stack Gas Scrubbing. Stack gas scrubbing is a demonstrated technology for removing SO_2 and SO_3 from stack gases at 90–95% efficiencies. More than 1000 plants, primarily in the United States and Japan, employ this technique (McCormick, 1989). There are basically two types of stack gas scrubbing, wet and dry. In wet scrubbing the SO_x is absorbed into water droplets, where sulfurous or sulfuric acids reacts with a base. The absorber is usually a countercurrent device in which the flue gases pass upwards through a spray or slurry containing the base.

Usually lime or limestone is used as the base, in which case the reaction product is a sludge of calcium sulfite and calcium sulfate with a consistency similar to that of toothpaste. The sludge must be disposed of at a landfill or in a pond. As is the case with fluidized bed systems, disposal of this sludge is a problem. In a typical 1000-

MW power plant burning coal containing 3.5% sulfur, wet scrubbing would produce 225,000 tonnes of sludge per year (McCormick, 1989). Processes can be used to recover sulfuric acid or pure sulfur from the sludge, but recovery is expensive and energy intensive. Air oxidation can be used to convert the calcium sulfate to gypsum, and in Japan, which has little natural gypsum, methods have been developed for purifying and marketing the converted gypsum.

Dry scrubbing methods have been developed primarily to reduce the sludge disposal problems associated with wet scrubbing. In the so-called Saarberg-Holter process, a slurry of lime or soda ash sprayed into the flue gases absorbs the SO$_x$ and is dried at the same time. In the Walther process, ammonia is sprayed into the gases, the end product being pelletized ammonium sulphate. The latter can be used as a fertilizer. In addition to producing less waste, dry scrubbing has several advantages compared to wet scrubbing. Dry scrubbers tend to have fewer operating problems, need fewer operators, have lower capital and operating and maintenance costs, and use much less water than wet scrubbers (McCormick, 1989). The principal disadvantage is that they are not as efficient in removing SO$_x$, the range of efficiencies being roughly 50–90% (McCormick, 1989). Consequently they have not been widely used.

Electron Beam Method. One of the most recently developed techniques for removing both SO$_x$ and NO$_x$ from stack gases is the electron beam method. The strategy is to spray the flue gas with water to reduce its temperature to 60–90°C and then to spray the cooled gases with ammonia. The gas mixture is then passed through an electron beam reactor, which converts the SO$_x$ and NO$_x$ to acids, and these in turn react with the ammonia and water vapor to produce ammonium compounds that can be used as fertilizers (McCormick, 1989). The technique is capable of removing 80–90% of the SO$_x$ from the stack gases (McCormick, 1989).

NO$_x$ REMOVAL

PreTreatment and Conversion
Neither pretreatment nor conversion of fossil fuels is a cost-effective way to reduce NO$_x$ emissions, because the N in the fuels is a minor source of NO$_x$.

Combustion
Since most of the NO$_x$ is produced from the oxidation of atmospheric N$_2$, the strategy in combustion control is to reduce the combustion temperature and/or the time the air stays in the combustion chamber. Both fluidized bed combustion and LIMB use this approach. In the former case the heat exchanger tubes are immersed in the fluidized bed, and heat transfer from the hot solids to the tubes is much more efficient than is the case in a conventional boiler; therefore, combustion temperatures can be lower. In the LIMB procedure the air is split into several streams and introduced into the flames in stages (McCormick, 1989). This approach allows the reduction of both the intensity and temperature of combustion. Unfortunately neither fluidized bed combustion nor LIMB is capable of reducing NO$_x$ emissions by more than 35–50% (Postel, 1984; McCormick, 1989).

Postcombustion

Both wet and dry methods exist for removing NO_x from stack gases, but the former suffer from the low solubility of NO and NO_2 in water. Hence very large absorbers and expensive catalysts are required. By far the most common dry technique for removing NO_x from stack gases is *selective catalytic reduction (SCR)*. In this procedure a catalyst and a reducing agent are used to convert the NO_x to N_2. The system must be designed to remove fly ash before the SCR treatment, because fly ash in the flue gases tends to clog the catalyst. Fly ash removal generally requires lower flue gas temperatures than are desirable for SCR, but NO_x removal efficiencies of 90% have been achieved with this method, and at a cost roughly half that of scrubbers (Postel, 1984). One attracive feature of several SCRs is that they also remove SO_x. The other dry removal technique is the *electron beam method*, which is capable of removing 50–90% of the NO_x as well as 80–90% of the SO_x (McCormick, 1989).

Integrated Gasification Combined Cycle

One of the most promising new technologies for reducing both SO_x and NO_x from the stack gases of coal-burning power plants is the *integrated gasification combined cycle (IGCC) process*. The idea is to convert crushed coal to a useful gas mixture by reacting it under pressure and high temperature in an almost pure stream of oxygen. The gaseous products consist mainly of CO and H_2. The gaseous sulfur products, mainly H_2S, are converted to elemental sulfur, which is sold as a byproduct. The synthetic gas is then burned at a reduced temperature to minimize NO_x production. In studies reported by Spencer et al. (1986) a 100-MW IGCC plant near Barstow, California, produced SO_x and NO_x emissions comparable or lower than those produced by a power plant burning natural gas and equal to only 10–11% of the new source performance standard limitations mandated by the Clean Air Act.

Comments

There is no question that the technology now exists to reduce SO_x and NO_x emissions from stationary sources to levels that would create virtually no adverse environmental impact. There are basically two problems with putting this technology to work. First, the most efficient methods, for example IGCC, require designing and building a power plant from scratch. The lifetime of a typical fossil fuel power plant is 30–40 years, and hence for the next few decades it can be expected that most electricity will be generated by old power plants. It is possible to retrofit old power plants to a certain extent to accommodate emission control technologies, the use of scrubbers on power plants in the United States and Japan being an obvious example. Retrofitting, however, is less effective in controlling SO_x and NO_x emissions than building a new plant with built-in control features. Effective technologies for reducing NO_x emissions have proven especially difficult to identify, both because little of the nitrogen originates from the fossil fuels themselves and because of the low solubility of NO and NO_2 in water. Hence it is fortunate that in most parts of the world where acid rain is a problem, SO_x is responsible for about 70% of the acidity and NO_x for only 30% (Postel, 1984).

The second major problem in implementing SO_x and NO_x emission controls is the need for motivation in the form of either political pressure or monetary incentives. It is this second problem to which we now turn our attention.

LEGAL ASPECTS

The legal aspects of acid deposition are a thorny issue, partly because of the long-range effects of SO_x and NO_x emissions. For example, more than 80% of the acid in rainwater falling on Sweden and Norway is generated in other European countries (Postel, 1984), and the United States contributes half the acid rain in Canada (Smith, 1980). Canada on the other hand is responsible for only about 10–15% of the acid rain in the United States (Environmental Canada, 1981).

The cost of controlling SO_x and NO_x emissions is thought provoking. Canada has estimated cleanup costs at \$400–500 million per year through the year 2000 (Smith, 1980), and in the United States, which emits about 5 times as much SO_x as Canada (Postel, 1984), the cost is likely to be much higher. The Congressional Office of Technology Assessment has estimated the cost of a proposed 9-million-tonne-per-year SO_2 reduction program at \$3.4 billion per year, and the National Commission on Air Quality estimated the cost of a 10-year reduction program of 6.9 million tonnes of SO_2 per year at \$2.2 billion per year (Wellford et al., 1982). In Japan, which has gone further than any other country to reduce SO_x and NO_x emissions, pollution control accounts for about 25% of the total cost of coal-generated power, and Japan's electricity costs are among the highest in the world (Postel, 1984). For comparison, U.S. SO_x and NO_x emission control is not expected to increase the cost of electricity by more than 2% in the eastern United States as a whole, although the increase may amount to 3–7% in states heavily dependent on coal (Wellford et al., 1982).

The benefits from acid deposition reduction are not as easy to estimate in monetary terms and are a subject of much controversy. The 1980 Crocker study performed at the University of Wyoming at the direction of the EPA estimated the value of reducing acid rain in the eastern United States at \$5 billion per year. This figure included "\$2 billion in effects on materials, \$1.75 billion in damage to forest ecosystems, \$1 billion in direct effects on agriculture, \$250 million in effects on aquatic systems, and \$100 million in other effects, including damage to water supply systems" (Wellford et al., 1982). If these figures are approximately correct, then the cost of control measures, though high, would still be outweighed by the benefits from a reduction in acid deposition. Undoubtedly the most thorough study of the costs of acid deposition in the United States was carried out by the National Acid Precipitation Assessment Program (NAPAP), a 10-year, \$570 million research effort funded by the U.S. government. The principal findings of the study, described in a massive 6000-page report, were as follows:

1. Acid deposition has adversely affected aquatic life in about 10% of streams and lakes in the eastern United States.

2. Acid deposition has contributed to the decline of red spruce at high elevations by reducing the spruce's cold tolerance.

3. Acid deposition has contributed to erosion and corrosion of buildings and materials.

4. Acid rain and related pollutants, especially fine sulfate particles, has reduced visibility throughout the northeastern and parts of the western United States. (Roberts, 1991)

How disturbing a person finds these conclusions seems to depend very much on who is making the assessment, and it is not hard to see why the conclusions of the report have been controversial. The fact that during the 1960s and 1970s roughly half of high-elevation Adirondack Mountain lakes had a pH less than 5 and were devoid of fish suggested that a serious problem was developing. The adversely affected high-elevation lakes were nevertheless a small percentage of eastern U.S. lakes, and the NAPAP report clearly indicates that lake and stream acidification caused by acid deposition has not become a widespread problem in the eastern United States. Does this conclusion mean that the threat of acid deposition was overestimated? At least some persons have concluded that this is indeed the case. The NAPAP report, for example, was the impetus for a December, 1990, *60 Minutes* television documentary that criticized severely the Bush administration and Congress for wasting the taxpayers' money by passing legislation aimed at controlling SO_x and NO_x emissions (Roberts, 1991). Many environmentalists, however, disagreed with the tone of the *60 Minutes* program. While acknowledging that the impact of acid deposition on lakes and streams had not become as serious as was once predicted, they argued that the problems caused by acid deposition were in fact much broader than had been realized or suspected when the NAPAP study began (Roberts, 1991). As noted by Mahoney (1991), the most recent director of the NAPAP, "Some of the effects are clearly less severe than the allegations, [but] there is abundant indication that we need [SO_x and NO_x] controls."

The U.S. government has in fact taken two important steps to reduce SO_x and NO_x emissions. The first step was taken in 1979, when the federal government put into effect its new source performance standards (NSPS). These regulations call for removal of 70–90% of the SO_x from stack gases in all new power plants, with the percentage removal depending on whether low or high sulfur fuel is burned. NO_x emissions must be no more than 38% and 50% of SO_x emissions on a weight basis for oil-fired plants and most coal-fired plants, respectively (EPA, 1991).

The NSPS were a landmark piece of legislation, because they were the first significant effort by the U.S. government to mandate a reduction in SO_x and NO_x emissions from stationary sources. However, they fell short of what environmentalists had hoped to see, because projections made in 1979 indicated that even with the NSPS in effect, SO_x emissions in the United States would increase by 10% from 1975 to 1995 (Carter, 1979). Although the calculated increase would have been 28% without the NSPS, merely reducing the increase from 28% to 10% was not what environmentalists had in mind. The problem was that the NSPS applied only to power plants on which construction had begun after August 17, 1971. Since fossil fuel plants typically remain in service for about 40 years (Carter, 1979), it was obvious that NO_x and SO_x emissions from old plants would continue to be a problem for several decades unless steps were taken to control their emissions. Indeed some of these older plants would have been in flagrant violation of NSPS requirements. Carter (1979), for example, noted that the Ohio Edison Company's Burger plant on the Ohio–West Virginia border emitted 10 times the amounts of SO_x permitted by the NSPS.

This problem was finally addressed in 1990, when amendments to the Clean Air Act were passed mandating reductions in SO_x and NO_x emissions from older power plants. Under the new Act, each power plant is given a reduction target, and utilities can earn credits that can be sold or traded if they exceed the targets (Anonymous, 1991). Thus utilities are given an incentive not only to meet but also to exceed the target reductions. It is hoped that this approach will encourage innovation and energy efficiency. The new Act is expected to reduce SO_x and NO_x emissions from stationary sources by almost 50% and 15–30%, respectively (Anonymous, 1991).

A CASE STUDY—THE NETHERLANDS

This chapter has tended to focus on conditions in the United States, which emits more SO_x and NO_x than any other nation on earth.[1] The problems caused by such emissions, however, are more severe in parts of Europe and the Soviet Union, which collectively emit 3–4 times as much SO_x as the United States (Postel, 1984). It is therefore of some interest to examine how conditions in these countries have evolved and how the problem of acid deposition is being addressed. We examine here the case of the Netherlands, which has the second highest SO_x deposition rate of any nation in western Europe.

Following World War II, about 85% of Dutch fuel consumption was accounted for by burning coal. However, during the next 15 years, oil gradually displaced coal as the major fuel source. In 1959, the largest natural gas field in the world was discovered in the province of Groninger in the Netherlands. Exploitation of this field began in 1963 (Vermeulen, 1980). The resultant pattern in energy use is illustrated in Figure 15.8.

Between 1946 and 1956, SO_x emissions in the Netherlands increased by about 125%, and between 1956 and 1967 by another 110% (Vermeulen, 1980). By 1967 SO_x concentrations in Dutch air were sometimes being reported as high as 500–2000 µg m^{-3}. These figures may be compared to the current EPA primary standards, which require that 24h average SO_2 concentrations not exceed 365 µg m^{-3} more than once per year and that average annual concentrations not exceed 80 µg m^{-3}. As noted by Vermeulen (1980), "In order to keep the Netherlands habitable, drastic measures were required immediately." In 1968 the Dutch Clean Air Act was passed. The act was aimed in part at reducing SO_x emissions. Although the act did not become effective until December 29, 1970, industries rapidly began cleaning up their emissions in anticipation of the act's implementation. Although some emission reductions were achieved by building desulfurization installations and by using low-sulfur oil, the principal mechanism for reducing SO_x emissions was to switch to natural gas as a fuel. Between 1967 and 1976 the percentage of the Netherlands fuel consumption accounted for by natural gas rose from 18% to 60% (Vermeulen, 1980), as illustrated in Figure 15.8. As a result SO_x emissions dropped by almost 60% between 1967 and 1977.

[1]This statement became true only after the breakup of the Soviet Union.

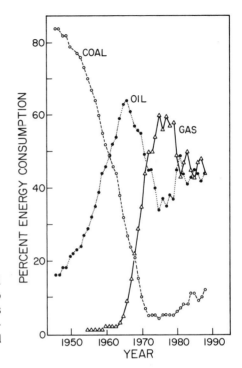

Figure 15.8   The share of coal, oil, and natural gas in Dutch fuel consumption from 1946 to 1988. [*Source*: Vermeulen (1979) and various editions of *United Nations Energy Statistics Yearbook*, Dept. of International Economic and Social Affairs.]

In response to the Arab oil boycott and the general instability in oil prices, the Dutch government decided that it would be wise to conserve the nation's natural gas as a strategic reserve. As a result of this shift in policy, Dutch power stations were required to decrease their natural gas consumption, and the percentage of Dutch energy consumption accounted for by gas rapidly dropped to about 45% (Figure 15.8). This decline was compensated for largely by an increase in the consumption of coal and oil, both of which have a higher sulfur content. According to Vermeulen (1980), in the absence of additional SO_x and NO_x emission controls, this switch back to oil and coal would have led to an increase of 230% in SO_x emissions by 1985, and an additional increase of 100% between 1985 and 2000. In summarizing this situation, Vermeulen (1980) commented, "With coal and oil as the most important fuels in the years to come it is obvious that a large and costly effort will have to take place in the field of desulfurization. . . . Because most desulfurization methods have not yet been technically perfected to a degree that the efficiency is sufficiently high and the costs are acceptable for large-scale application, in my opinion hardly any positive effect can be expected before 1985. After 1985, the acidification of precipitation will depend strongly on the level of technology of desulfurization and the velocity with which these installations will be built. Except for the costs, there must be the political will to handle the problem vigorously, and that is a separate problem. . . . In my opinion, the outlook for the Netherlands with respect to the acidification of precipitation is pretty grim."

Vermeulen's bleak assessment has unfortunately proven to be qualitatively correct. A serious effort was made to reduce SO_x emissions in the Netherlands, and the result was a 41% decline in Dutch SO_x emissions between 1980 and 1985 (Wollast, 1989). Almost 70% of the decrease was due to a drop in SO_x emissions by utilities.

During the same time interval Belgium, West Germany, France, and the United Kingdom, which collectively are responsible for more than half of the acid rain that falls in the Netherlands, achieved comparable reductions in SO_x emissions of 42%, 25%, 48%, and 24%, respectively (Wollast, 1989). Unfortunately there was very little reduction in NO_x emissions, either in the Netherlands or in neighboring countries, and as a result the overall deposition of acid in the Netherlands declined by only about 16% between 1980 and 1986. According to Wollast (1989) about a 75% reduction in acid deposition in the Netherlands is needed to prevent damage to sensitive ecosystems, particularly coniferous forests and heathland on poor sandy soils.

It is problematic at this time whether such a reduction in acid deposition can be achieved in the Netherlands. Certainly some effort must be made to reduce NO_x emissions, about 50% of which are accounted for by motor vehicles. According to Wollast (1989) modification of combustion conditions and the fitting of three-way catalytic converters on heavier passenger automobiles would reduce NO_x emissions by 14% in the year 2000. In the case of SO_x, a limited reduction of the sulfur content of oil, flue gas scrubbing, and a limited substitution of natural gas for heavy fuel oil would reduce SO_x emissions by 30% between 1985 and 2000. Even if neighboring countries responded with similar reductions, acid deposition in the Netherlands would still be over 3 times the target figure in the first decade of the twenty-first century. Further reductions in NO_x emissions could be achieved with the use of selective catalytic reduction for stack gases and the requirement of catalytic converters on all passenger automobiles, but these measures would still leave acid deposition at 2.5 times the target level by the year 2010. The implication, according to Wollast (1989) is that technical measures alone will not solve the acid deposition problem in the Netherlands and that some effort to reduce energy consumption will probably be required to bring deposition rates within desired limits. According to Wollast (1989, p. 95), "The picture presented by these scenarios does not give direct cause for optimistic expectations for the future development of acidification and its effects."

REFERENCES

Anonymous. 1991. New clean air act will reduce acid rain, then cap emissions. *Environmental Defense Fund Letter*, **23**(1), 1, 5.

Barrett, E. and G. Brodin. 1955. The acidity of Scandinavian precipitation. *Tellus*, **7**, 1–257.

Bischoff, W. D., V. L. Paterson, and F. T. Mackenzie. 1984. Geochemical mass balance for sulfur- and nitrogen-bearing acid components: Eastern United States. In O. P. Bricker (Ed.), *Geological Aspects of Acid Deposition*, Acid Precipitation Series. Vol. 7. Butterworth. pp. 1–21.

Bomberger, D. C. and R. C. Phillips. 1979. Technological options for mitigation of acid rain. In C. G. Gunnerson and B. E. Willard (Eds.), *Acid Rain*. Am. Soc. Civil Eng., New York. pp. 132–166.

Brimblecombe, P. 1977. London air pollution, 1500–1900. *Atmos. Envirn.*, **11**, 1157–1162.

Brimblecombe, P. 1978. Air pollution in industrializing England. *J. Air. Poll. Control Assoc.*, **28**, 115–118.

Canada Today. 1981. How many more lakes have to die? *Canada Today*, **12**(2). Canadian Embassy, 1771 N Street Room 300, Washington, D.C.

Carter, L. J. 1979. Uncontrolled SO_2 emissions bring acid rain. *Science*, **204**, 1179–1182.

Charlson, R. J. and H. Rodhe. 1982. Factors controlling the acidity of rainwater. *Nature*, **295**, 683–685.

Cowling, E. B. 1982. A status report on acid precipitation and its biological consequences as of April 1981. In F. M. D'Itri (Ed.), Acid Precipitation. Ann Arbor Science. pp. 3–20.

Cronan, C. S., and C. L. Schofield. 1979. Aluminum leaching response to acid precipitation: Effects on high-elevation watersheds in the northeast. *Science*, **204**, 304–305.

Dignon, J., and S. Hameed. 1989. Global emissions of nitrogen and sulfur oxides from 1860 to 1980. *J. Air Waste Management Assoc.*, **39**(2), 180–186.

Environment Canada. 1981. *Downwind. The Acid Rain Story*. Cat. No. En 56–561 1981E, Minister of Supply and Services Canada, Ottawa, Ontario. 20 pp.

Environmental Protection Agency. 1980. *Acid Rain*. Washington. 36 pp.

Environmental Protection Agency. 1991. Standard for sulfur dioxide. *Code of Fed. Reg.*, **40**(60), 250.

Flavin, C., and A. B. Durning. 1988. *Building on Success: The Age of Energy Efficiency*. Worldwatch Paper 82. Worldwatch Inst. Washington, D.C. 74 pp.

Francis, A. J., H. L. Quinby, and G. R. Hendrey. 1984. Effect of lake pH on microbial decomposition of allochthonous litter. In G. R. Hendrey (Ed.), *Early Biotic Responses to Advancing Lake Acidification*, Acid Precipitation series. Vol. 6. Butterworth. pp. 1–21.

Galloway, J. N. 1979. Acid precipitation: Spatial and temporal trends. In C. G. Gunnerson and B. E. Willard (Eds.), *Acid Rain*. Am. Soc. Civil Eng. New York. pp. 1–20.

Gorham, E. 1955. On the acidity and salinity of rain. *Geochim. Cosmochim. Acta*, **7**, 231–239.

Gschwandtner, G., J. K. Wagner, and R. B. Husar. 1988. *Comparison of Historic SO_2 and NO_x Emission Data Sets*. U.S. Environmental Protection Agency. EPA/600/S7-88/009. Research Triangle Park, NC.

Hamilton, L. D. 1979. Health effects of acid precipitation. In Proceedings, Action Seminar on Acid Precipitation (Toronto: A.S.A.P. organizing committee). pp. 117–134.

Hendrey, G. R. 1983. Automobiles and acid rain. *Science*, **222**, 8.

Houghton, H. 1955. On the chemical composition of fog and cloud water. *J. Meteorol.*, **12**, 355–357.

Johnson, N. M. 1979. Acid rain: Neutralization within the Hubbard Brook ecosystem and regional implications. *Science*, **204**, 497–499.

Kohout, E. J., D. J. Miller, L. A. Nieves, D. S. Rothman, D. L. Saricks, F. Stodolsky, and D. A. Hanson. 1990. *Current Emission Trends for Nitrogen Oxides, Sulfur Dioxide, and Volatile Organic Compounds by Month and State: Methodology and Results*. Policy and Economic Analyis Group, Environmental Assessment and Information Sciences Division, Argonne National Laboratory, IL. ANL/EAIS/TM-25.

Koide, M. and E. D. Goldberg. 1971. Atmospheric sulfur and fossile fuel combustion. *J. Geophys. Res.*, **76**, 6589–6596.

Leivestad, H., G. Hendry, I. Muniz, and E. Snekvik. 1976. Effects of acid precipitation on freshwater organisms. In F. H. Braekke (Ed.), SNSF Research Report 6/76. pp. 87–111.

Li, Y. 1992. Seasalt and pollution inputs over the continental United States. *Water, Air, and Soil Pollution*, **64**, 561–573.

Likens, G. E., H. Borman, J. S. Eaton, R. S. Pierce, and N. M. Johnson. 1976. Hydrogen ion input to the Hubbard Brook experimental forest, New Hampshire, during the last decade. *Water, Air, and Soil Pollution*, **6**, 435.

Likens, G. E., F. H. Bormann, and N. M. Johnson. 1972. Acid Rain. *Environment*, **14**, 33–40.

Likens, G. E., R. F. Wright, J. N. Galloway, and T. J. Butler. 1979. Acid rain. *Sci. Amer.*, **241**(4), 43–51.

Mackenzie, F. T., W. D. Bischoff, and V. B. Patterson. 1984. *Geochemical cycles and trends in estimates of inputs of anthropogenic chemical constituents to the environment.* Ecosystems Research Center, Cornell University, NY.

Mahoney, J. 1991. Quoted by Roberts (1991), p. 1303.

Marmorek, D. R. 1984. Changes in the temporal behavior and size structure of plankton systems in acid lakes. In G. R. Hendrey (Ed.), *Early Biotic Responses to Advancing Lake Acidification.* Acid Precipitation Series. Vol. 6. Butterworth. pp. 23–41.

Marshall, E. 1984. Canada goes it alone on acid rain controls. *Science*, **223**, 1275.

McCandless, L. C., A. B. Onursal, and J. M. Moore. 1987. *Assessment of Coal Cleaning Technology: Final Report.* Environmental Protection Agency, EPA/600/S7-86/037. Research Triangle Park, NC.

McCormick, J. 1989. *Acid Earth: The Global Threat of Acid Pollution.* 2nd ed. Earthscan Publ. LTD, London. 225 pp.

Oden, S. 1968. *The acidification of air and precipitation, and its consequences in the natural environment.* Ecol. Comm. Bull. No. 1, National Science Research Council, Stockholm (Arlington, Va: Translation Consultants, TR-1172).

Park, C. C. 1987. *Acid Rain: Rhetoric and Reality.* Methuen, London. 272 pp.

Postel, S. 1984. *Air Pollution, Acid Rain, and the Future of Forests.* Worldwatch Paper 58. Worldwatch Inst., Washington, D.C. 54 pp.

Roberts, L. 1991. Learning from an acid rain program. *Science*, **251**, 1302–1305.

Schofield, C. L. and J. R. Trojnar. 1980. Aluminum toxicity to fish in acidified waters. In T. Y. Toribara, M. W. Miller, and P. E. Morrow (Eds.), *Polluted Rain*, Plenum Press. New York. pp. 347–366.

Smith, R. A. 1852. On the air and rain of Manchester. *Mem. Manchester Lit. Phil. Soc.*, Ser. 2, **10**, 207–217.

Smith, R. A. 1872. *Air and Rain.* Longmans, Green.

Smith, R. J. 1980. Acid rain agreement. *Science*, **209**, 890.

Spencer, D. F., S. B. Alpert, and H. H. Gilman. 1986. Cool water: Demonstration of a clean and efficient new coal technology. *Science*, **232**, 609–612.

Vermeulen, A. J. 1980. The acidic precipitation phenomenon: A study of this phenomenon and the relationship between the acid content of precipitation and the emission of sulfur dioxide and nitrogen oxides in the Netherlands. In T. Y. Toribara, M. W. Miller, and P. E. Morrow (Eds.), *Polluted Rain.* Plenum Press. New York. pp. 7–60.

Wellford, Wegman, Krulwich, Gold, and Hoff. 1982. *Fact sheet on acid rain.* Canadian Embassy. 1771 N Street N.W. Washington, D.C. 20036. 8 pp.

Wollast, R. 1989. Acidification. In I. F. Langeweg (Ed.), *Concern for Tomorrow.* National Inst. of Public Health and Environmental Protection. Brussels. pp. 79–96.

16

GROUNDWATER POLLUTION

Groundwater pollution is an environmental problem that has attracted significant national attention only in recent years. The incident that most effectively focused public interest on the problem was the contamination of the Love Canal in the city of Niagara Falls, New York. About 20,000 tonnes of hazardous wastes were buried in 55-gal drums in the abandoned canal during the 1930s and 1940s by the Hooker Chemical Company. The accumulated waste was covered with soil, and the land sold to the Niagara Falls Board of Education for $1 in 1953. The Board of Education proceeded to build a school and playground over the old canal site, and the surrounding area was gradually developed for residential homes. Several years of heavy precipitation leading up to 1977 saturated the old canal site with water, and toxic chemicals began to leak out of the corroding drums buried in the canal. The resultant pollution caused the New York state health commissioner to declare a state of emergency, and President Carter shortly thereafter declared Love Canal a national disaster area. A $15 million fund was established to purchase 550 homes within a 30-sq block area surrounding the dump site (Epstein et al., 1982), and hundreds of residents were evacuated and resettled. The impact of this chemical waste dump on both the physical and emotional well-being of the Love Canal residents has been a subject of much controversy (Culliton, 1980) and will probably never be fully understood.

Incidents such as the pollution of Love Canal have made it clear how serious a problem the careless disposal of hazardous wastes can be. Federal regulations enacted during the 1970s have helped to restrict air and surface-water contamination, and as a result there has been an increase in subsurface disposal. When hazardous wastes are buried in the ground, there is an excellent chance that groundwater will become contaminated unless special precautions are taken to minimize this probability. Unfortunately it is often extremely difficult and expensive to clean up groundwater once it has become contaminated, and because dilution effects are minimal, "concentrations of contaminants are often much higher in groundwater than in surface water" (Patrick et al., 1987, p. 56). These realizations

543

have led scientists and environmental groups to begin an examination of the degree of groundwater contamination in the United States, to study the potential for future contamination, and to carefully assess our present use of and future needs for groundwater.

RELIANCE ON GROUNDWATER

Groundwater accounts for about 96% of all freshwater in the United States, the remainder being present in lakes and streams. It is estimated that $1.2–3.8 \times 10^{17}$ liters of groundwater lie within about 0.75 km of the surface in the United States (Patrick et al., 1987). Americans obtain about 50% of their drinking water from groundwater sources; community groundwater systems supply about 29% of drinking water demands, and domestic wells account for roughly 19%. About 95% of the rural population depends on groundwater for drinking water, and according to Pye et al. (1983, p. 38), "of the major cities in the United States, 75% depend upon well water for most of their supplies." As Figure 16.1 indicates, fresh groundwater withdrawals in the United States have been about 300 billion liters d^{-1} in recent years, which is roughly 10% of the effective recharge rate. The leveling off of withdrawals since 1975 has been due largely to more frugal and efficient use of water for irrigation, which has accounted for about 68% of all groundwater use in the United States since 1960 (Patrick et al., 1987). California and Texas, where groundwater is used extensively for irrigation, together account for 37% of all groundwater use in the United States (GRF, 1984).

General Aquifer Information

What is groundwater? According to Pye et al. (1983, p. 2) groundwater "is water that occurs in permeable saturated strata of rock, sand, or gravel called aquifers." The two main types of aquifers, confined and unconfined aquifers, are illustrated in Figure 16.2. An unconfined aquifer is not overlain by impermeable material and may be recharged by percolation of rainwater through the overlying soil and rock. The water in the aquifer is at atmospheric pressure. A so-called perched aquifer may occur where there is a limited layer of impermeable material above the water table. Confined or artesian aquifers are bounded on top and bottom by impermeable geologic formations called *aquitards*. The water in the artesian aquifer is under greater than atmospheric pressure. Such aquifers may have no recharge areas, or may have "discrete and variable recharge areas where the geologic material of the aquifer forms an outcrop at the surface" (Patrick et al., 1987, p. 2–3). Most aquifers occur within 0.75 km of the earth's surface (Burmaster, 1982) and in areal extent may vary greatly from very small to as large as several hundred thousand square kilometers. For example, the Ogallala aquifer in the High Plains of the United States covers an area of 3.9×10^5 km² and is as much as 0.30–0.45 km thick in parts of Nebraska. An important point about all aquifers is that recharging of the aquifer can occur only when precipitation exceeds evapotranspiration in the aquifer's recharge area. In the arid southwestern United States for example, aquifer recharge occurs mainly in wet or multiyear cycles (Patrick et al., 1987). If use of the aquifer is to be sustained over long periods of time, it is therefore critical that withdrawals

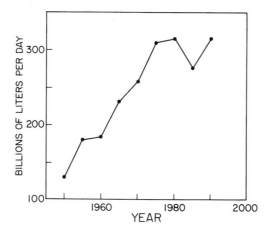

Figure 16.1 Groundwater withdrawals in the United States from 1950 to 1990. [*Source*: CEQ (1991) and W. B. Solley (U. S. Geological Survey), personal communication.]

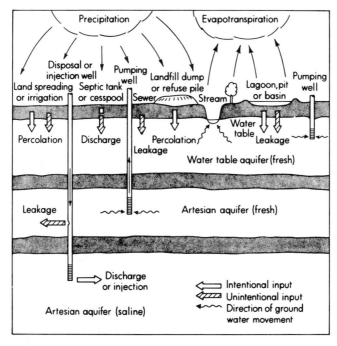

Figure 16.2 Diagram of confined and unconfined aquifers and disposal methods which can contaminate them. [Redrawn from Burmaster (1982).]

from the aquifer not exceed average recharge rates. The Ogallala aquifer, for example, is being overdrawn at the rate of more than 3.7 km^3 per year (Patrick et al., 1987). In Texas, where the aquifer is less than 30 m thick, it has been estimated that at the current rate of use and without an imported water supply, about 40% of the land now being irrigated in the High Plains of West Texas will have to be converted to dryland farming or be abandoned by the year 2000, and 60% by the year 2020. In fact, Gaines County, Texas, had completely dried up its groundwater resources by the early 1980s (Pye et al., 1983). Groundwater withdrawals in the southern High Plains have been averaging about 2.5×10^{10} liters d^{-1}, but the recharge rate is only 2% of this figure (Patrick et al., 1987). Burmaster (1982, p. 8) indicates that, "In 1975 approximately 25 percent of all groundwater withdrawals were overdrafts, mostly in agricultural regions." In some cases overdrafts result in the eventual elimination of the groundwater resource, as in the case of Gaines County, Texas; in other cases overdrafts result in contamination of fresh groundwater with saline groundwater, which may underlie lenses of freshwater, particularly in coastal areas.

THE EXTENT OF GROUNDWATER POLLUTION

Unfortunately saltwater encroachment associated with overdrafting of aquifers is only one mechanism by which groundwater may become contaminated. In some cases entirely natural processes may seriously contaminate groundwater. For example, in the southwestern and south central United States, natural leaching of chloride, sulfate, nitrate, fluoride, iron and arsenic ions, and radioactivity from naturally occurring deposits of uranium have created water quality problems in certain areas (GRF, 1984). However, most concern over groundwater contamination has centered on pollution associated with human activities. Sources of contamination associated with human activities but not directly related to waste disposal include accidents, certain agricultural activities, mining, highway deicing, acid rain, saltwater intrusion, and improper well construction and maintenance. Sources associated with waste disposal include private sewage disposal systems, land disposal of solid wastes, municipal wastewater, wastewater impoundments, land spreading of sludge, brine disposal from the petroleum industry, mine wastes, deep-well disposal of liquid wastes, animal feedlot wastes, and radioactive wastes (GRF, 1984). The following information gives some idea of the amount of the hazardous waste generated by these activities.

Septic tanks

Almost one-third of the U.S. population uses septic tanks for sewage treatment. These tanks handle 11–17 billion liters d^{-1} of sewage (Patrick et al., 1987). For septic tanks to work properly, it is important that a zone of unsaturated soil occur between the leach bed and the water table so that the effluent from the septic tank does not enter an unconfined aquifer. In a 1980 study, the EPA (1980a) concluded that a third of all septic tank installations in the United States were operating improperly. Based on studies conducted by the EPA during the 1970s, improperly operating septic tanks were judged to be a major cause of groundwater contamination in the northwestern, northeastern, and southeastern United States (GRF, 1984). Elevated concentrations of pathogens and nitrate are problems typically associated with septic tank pollution, but, "more recently, widespread use of septic tank cleaners containing degreasing agents such as trichloroethylene has resulted in groundwater contamination by synthetic organic chemicals" (GRF 1984, p. 27).

Saltwater Contamination

Contamination of fresh groundwater by saltwater can occur by several mechanisms. Overdrafting is one obvious mechanism. Most freshwater aquifers are underlain by brackish or saline groundwater, and in general salinity can be expected to increase with depth. A U.S. Geological Survey inventory reported 114 cases of saltwater intrusion in 1977 (EPA, 1977a). Use of salt for highway deicing is another source of potential contamination. During the winter of 1982–1983 at least 8.5 million tonnes of dry salts and abrasives and an additional 6.7 million liters of liquid salts were used for highway deicing in the United States (Patrick et al., 1987). When this salt washes off roads it may easily move with percolating water into underground aquifers. An additional problem is created by the fact that piles of salt to be used for deicing have frequently been stored uncovered along roads. Rain or snow melt can dissolve this salt and, through percolation, introduce it into aquifers. Groundwater contamination caused by the disposal of oil-field brines has been documented in at least 21 states (Patrick et al., 1987). At the present time the ratio of brine to crude oil recovered is on the order of 10:1, but the ratio may be much higher (e.g., 100:1) in older wells (Patrick et al., 1987). In the early days of oil and gas exploration these brines were disposed of in unlined pits. Unfortunately these so-called evaporation pits were often quite leaky, and it was easy for the brines to percolate into the ground and contaminate groundwater. Use of such pits has been wholly or at least partially eliminated in all parts of the United States. Texas, for example, banned unlined pits in 1969. Today brines are usually disposed of by injection into deep underground formations considered unsuitable for other purposes, or by reinjection into oil-producing formations to enhance recovery. However, as noted by Patrick et al. (1987, p. 73), "These injection practices, if not properly controlled, pose a potential threat to groundwater." The Safe Drinking Water Act standard for drinking water is 500 ppm of total dissolved solids (TDS), a figure that corresponds to a salinity less than 2% that of seawater.

Sewage

Improper disposal of municipal wastewater and both human and animal waste can pose a threat to groundwater. The total volume of sewage handled by municipal sewage treatment plants in the United States is about 70 billion liters d^{-1}. The principal components of domestic sewage that are regarded as a threat to groundwater are nitrate, heavy metals, and pathogens. Manure generated at feedlots is a similar threat to groundwater from the standpoint of nitrate and pathogen contamination. Between 1945 and 1984 there were 168 outbreaks of disease in the United States that could be attributed to contamination of groundwater by pathogens (Patrick et al., 1987). More than 31,000 cases of illness were associated with these outbreaks; in almost all cases either bacteria or viruses were implicated as the causative agents.

Mechanisms by which sewage may contaminate groundwater supplies include leaks in sewer lines due to age; disruption by tree roots, seismic activity, or poor construction; leakage from inadequately sealed sewage lagoons; and improper land disposal of treated wastewater.

Mining Activities

Mining activities can contaminate groundwater in several ways. Coal is often found in deposits that contain iron pyrites (FeS_2) and other sulfides. When these sulfides are exposed to oxygen as a result of mining activities, they become oxidized in the presence of water and form sulfuric acid, H_2SO_4. The result can be highly acidic groundwater and is a serious problem in the coal mining areas of Pennsylvania and Appalachia (Patrick et al., 1987). Drainage water and waste water from metal mines can seriously contaminate groundwater, because these waters typically contain high concentrations of toxic heavy metals. Uranium and copper mines are cited by Patrick et al. (1987) as producing drainage and waste water that pose the most serious threats to groundwater, because the polluted water contains high concentrations of both dissolved toxic substances (e.g. arsenic, sulfuric acid, copper) as well as radioisotopes. Metal mine leachate has increased the concentration of manganese in wells in Washington, caused arsenic poisoning of cattle in Idaho, and increased the radioactivity of groundwater in Wyoming (Patrick et al., 1987).

Toxic Chemicals

Contamination of groundwater by toxic organic chemicals is a problem that has only recently attracted much attention, primarily because concentrations of organic chemicals have not routinely been measured in groundwater. Recent groundwater monitoring has shown that toxic organic chemicals are an increasing problem, but according to Patrick et al. (1987, p. 254), "To date only about 10% of the organic chemicals contaminating drinking water in the United States have been identified." The combined effects of dilution, biological activity and chemical reactions are generally much less efficient in reducing the concentrations of toxic chemicals in groundwater than in surface water, and as a result groundwater concentrations of certain toxic organic chemicals in some areas are often considerably higher than the concentration of the same chemicals found in raw or treated drinking water drawn from the most contaminated surface supplies (Burmaster, 1982). The problem of monitoring the concentrations of toxic chemicals in water supplies is made difficult by the fact that approximately 50,000–75,000 chemicals are in use and being distributed through the environment at the present time, and an additional 700–800 are added each year (Patrick et al., 1987). Although not all of these chemicals are toxic, merely keeping up-to-date information on the possible toxicity of so many compounds is a prodigious task. Table 16.1 gives some idea of the variety and concentration of toxic organic chemicals that have been detected in drinking-water wells in the United States. Included where information is available are the water quality standards established by the EPA under the Clean Water Act, the evidence for carcinogenicity, and the principal uses of the various chemicals. It is obvious from the data in Table 16.1 that a number of toxic chemicals have somehow found their way into water supplies at concentrations that vastly exceed any reasonable safe level. How have certain groundwater supplies come to be so grossly contaminated with toxic chemicals?

Table 16.1   Toxic Organic Chemicals Found in Drinking-Water Wells[1]

Trichloroethylene (TCE)

Principal uses

Mainly as a solvent for vapor degreasing in metal industries. Also used to extract caffeine from coffee, as a dry cleaning agent, and as a chemical intermediate in the production of pesticides, waxes, gums, resins, tars, paints, varnishes, and specific chemicals such as chloroacetic acid.

Carcinogenicity: Animal carcinogen.

CWA recommended limit: 0.27–27 ppb.

Reported concentrations (ppb)	*State*
27,300	Pennsylvania
14,000	Pennsylvania
3800	New York
3200	Pennsylvania
1530	New Jersey
900	Massachusetts

Toluene

Principal uses

About 70% is converted to benzene. About 15% used to produce chemicals such as toluene diisocyanate, phenol, benzyl and benzoyl derivatives, benzoic acid, toluene sulfonates, nitrotoluenes, vinyltoluene, and saccharin. Remainder used as solvent for paints and coatings and as a component of automobile and aviation fuel.

Carcinogenicity: no.

CWA recommended limit: 14,300 ppb.

Reported concentrations (ppb)	*State*
6400	New Jersey
260	New Jersey
55	New Jersey

1,1,1-Trichloroethane

Principal uses

Widely used as a substitute for carbon tetrachloride. Also used in liquid form as a degreaser and for cold cleaning, dip-cleaning, and bucket cleaning of metals. Also used as a dry-cleaning agent, vapor degreasing agent and propellant.

Carcinogenicity: no.

CWA recommended limit: 18,400 ppb.

Reported concentrations (ppb)	*State*
5440	Maine
5100	New York
1600	Connecticut
965	New Jersey

Acetone

Principal uses

Used as a solvent, in the production of lubricating oils, and as an intermediate in the manufacture of chloroform and of various pharmaceuticals and pesticides.

Table 16.1 *Continued*

Carcinogenicity: no.

CWA recommended limit: no criteria.

Reported concentrations (ppb)	*State*
3000	New Jersey

Methylene Chloride

Principal uses

Mainly as a low-temperature extractant of substances that are affected adversely by high temperature. Can be used as a solvent for oil, fats, waxes, bitumen, cellulose acetate, and esters. Also used as paint remover and degreaser.

Carcinogenicity: potential carcinogen.

CWA recommended limit: 0.019–1.9 ppb.

Reported concentrations (ppb)	*State*
3000	New Jersey
47	New York

Dioxane

Principal uses

Primarily as a solvent for cellulose acetate, dyes, fats, greases, lacquers, mineral oil, paints, polyvinyl polymers, resins, varnishes, and waxes. Also used in paint and varnish strippers, as a wetting agent and dispersing agent in textile processing, dye baths, stain and printing compositions, and in preparation of histological slides.

Carcinogenicity: Animal carcinogen.

CWA recommended limit: No criteria, but 2480 ppb suggested goal (EPA, 1977b).

Reported concentrations (ppb)	*State*
2100	Massachusetts

Ethyl Benzene

Principal uses

Used in manufacture of cellulose acetate, styrene, and synthetic rubber. Also as a solvent or diluent and as a component of automotive and aviation fuel. Present in significant quantities in mixed xylenes, which are used as diluents in the paint industry, in insecticides, and in gasoline.

Carcinogenicity: no.

CWA recommended limit: 1400 ppb.

Reported concentrations	*State*
2000	New Jersey

Tetrachloroethylene

Principal uses

Solvent with particular use as a dry cleaning agent, degreaser, chemical inter-mediate, fumigant, and medically as an anthelmintic.

Table 16.1   *Continued*

Carcinogenicity: Animal carcinogen.

CWA recommended limit: 0.08–8 ppb.

Reported concentrations (ppb)	*State*
1500	New Jersey
740	Connecticut
717	New York

Cyclohexane

Principal uses

Used as a chemical intermediate, as a solvent for fats, oils, waxes, resins, and certain synthetic rubbers, and as an extractant of essential oils in the perfume industry.

Carcinogenicity: Not tested in animal bioassay.

CWA recommended limit: no criteria.

Reported concentrations (ppb)	*State*
540	New York

Chloroform

Principal uses

Widely used as a solvent (especially in lacquer industry) in the extraction and purification of penicillin and other pharmaceuticals, in the manufacture of artificial silk, plastics, floor polishes, and fluorocarbons, and in sterilization of catgut. Often found in drinking water supplies as a result of chlorination.

Carcinogenicity: Animal carcinogen.

CWA recommended limit: 0.019–1.9 ppb.

Reported concentrations (ppb)	*State*
490	New York
420	New Jersey
67	New York

Di-n-butyl phthalate

Principal uses

Used in plasticizing vinyl acetate emulsion systems and in plasticizing cellulose esters. Also used in insect repellent.

Carcinogenicity: no.

CWA recommended limit: 35,000 ppb.

Reported concentrations (ppb)	*State*
470	New York

Carbon tetrachloride

Principal uses

Solvent for oils, fats, lacquers, varnishes, rubber, waxes and resins. Synthesis of

Table 16.1 *Continued*

fluorocarbons. Used also as azeotropic drying agent for spark plugs, a dry-cleaning agent, a fire extinguishing agent, a fumigant, and as an anthelmintic.

Carcinogenicity: Animal carcinogen.

CWA recommended limit: 0.04–4 ppb

Reported concentrations (ppb)	*State*
400	New Jersey
135	New York

Benzene

Principal uses

Constituent of motor fuels, solvent for fats, inks, oils, paints, plastics, and rubber. Used in the extraction of oils from seeds and nuts and in photogravure printing. Also used as a chemical intermediate. By alkylation, chlorination, nitration, and sulfonation, chemicals such as styrene, phenols, and maleic anhydride are produced. Also used in manufacture of detergents, explosives, pharmaceuticals, and dye-stuffs.

Carcinogenicity: Human carcinogen.

CWA recommended limit: 0.066–6.6 ppb.

Reported concentrations (ppb)	*State*
330	New Jersey
230	New Jersey
70	Connecticut
30	New York

1,2-Dichloroethylene

Principal uses

Solvent for waxes, resins, and acetylcellulose. Used in extraction of rubber, as a refrigerant, in the manufacture of pharmaceuticals and artificial pearls, and in the extraction of oils and fats from fish and meat.

Carcinogenicity: potential carcinogen.

CWA recommended limit: 0.0033–0.33 ppb.

Reported concentrations (ppb)	*State*
323	Massachusetts
294	Massachusetts
91	New York

Ethylene dibromide (EDB)

Principal uses

Principally as a fumigant for ground pest control and as a constituent of ethyl gasoline. Also used in fire extinguisers, gauge fluids, and water proofing preparations. Also used as a solvent for celluloid, fats, oils, and waxes.

Carcinogenicity: Animal carcinogen.

CWA recommended limit: no criteria.

Table 16.1 *Continued*

Reported concentrations (ppb)	State
300	Hawaii
100	Hawaii
35	California

Xylene

Principal uses

Solvent, constituent of paint, lacquers, varnishes, inks, dyes, adhesives, cements, cleaning fluids, and aviation fuels. Used as a chemical feedstock for xylidines, benzoic acid, phthalic anhydride, isophthalic and terephthalic acids and their esters. Used in manufacture of quartz crystal oscillators, hydrogen peroxide, perfumes, insect repellents, epoxy resins, pharmaceuticals, and in the leather industry.

Carcinogenicity: Not tested.

CWA recommended limit: No criteria, but 6000 ppb suggested goal (EPA, 1977b).

Reported concentrations (ppb)	State
300	New Jersey
69	New York

1,1-Dichloroethylene

Principal uses

Chemical intermediate in synthesis of methylchloroform and in production of polyvinylidene chloride copolymers, which are components of Saran wrap and of polymer coatings of ship tanks, railroad cars, fuel storage tanks, and for coating steel pipes and structures.

Carcinogenicity: Suggested animal carcinogen.

CWA recommended limit: 0.0033–0.33 ppb.

Reported concentrations (ppb)	State
280	New Jersey
118	Massachusetts
70	Maine

1,2-Dichloroethane

Principal uses

Manufacture of ethylene glycol, diaminoethylene, polyvinyl chloride, nylon, viscose rayon, styrene-butadiene rubber, and various plastics. Solvent for resins, asphalt, bitumen, rubber, cellulose acetate, cellulose ester, and paint. Degreaser in engineering, textile, and petroleum industries. Extracting agent for soybean oil and caffeine. Antiknock agent in gasoline, a pickling agent, fumigant, and drycleaning agent. Used in photography, xerography, water softening, and in production of adhesives, pharmaceuticals, and varnishes.

Carcinogenicity: Animal carcinogen.

CWA recommended limit: 0.094–9.4 ppb.

Table 16.1 *Continued*

Reported concentrations (ppb)	*State*
250	New Jersey

Bis (2-ethylhexyl) phthalate

Principal uses

Plasticizer for resins and in the manufacture of organic pump fluids.

Carcinogenicity: Not tested in animal bioassay, but rated a carcinogen by the National Cancer Institute.

CWA recommended limit: 15,000 ppb.

Reported concentrations (ppb)	*State*
170	New York

Dibromochloromethane

Principal uses

Chemical intermediate in the manufacture of fire extinguishing agents, aerosol propellants, refrigerants, and pesticides. Presence in drinking water believed due to the haloform reaction that may occur during water chlorination.

Carcinogenicity: Halomethanes as a group considered carcinogenic.

CWA recommended limit: 0.019–1.9 ppb for total halomethanes.

Reported concentrations (ppb)	*State*
55	New York
20	Delaware

Vinyl chloride

Principal uses

Used as a vinyl monomer in the manufacture of polyvinyl chloride and synthetic rubber. Also used as chemical intermediate and solvent.

Carcinogenicity: Human and animal carcinogen.

CWA recommended limit: 0.2–20 ppb.

Reported concentrations (ppb)	*State*
50	New York

Chloromethane

Principal uses

Used as methylating agent and chlorinating agent in organic chemistry, as an extractant for greases, oils, and resins in petroleum refineries, as a solvent in the synthetic rubber industry, as a refrigerant, as a propellant in polystyrene foam production, and as an intermediate in drug manufacture.

Carcinogenicity: Halomethanes as a group considered carcinogenic.

CWA recommended limit: 0.019–1.9 ppb for total halomethanes.

Reported concentrations (ppb)	*State*
44	Massachusetts

Table 16.1   *Continued*

Lindane (gamma isomer of benzene hexachloride or BHC).

Principal uses

Insecticide used on seed and soil treatments, foliage application, wood protection.

Carcinogenicity: Animal carcinogen.

CWA recommended limit: 0.00186–0.186 ppb.

Reported concentrations (ppb)	*State*
22	New York

1,1,2-Trichloroethane

Principal uses

Chemical intermediate and solvent, but not used as widely as the 1,1,1 isomer.

Carcinogenicity: Animal carcinogen.

CWA recommended limit: 0.06–6 ppb.

Reported concentrations (ppb)	*State*
20	New York

Bromoform

Principal uses

Pharmaceutical manufacturing, as an ingredient in fire-resistant chemicals and gauge fluid, and as a solvent for waxes, greases, and oils.

Carcinogenicity: potential carcinogen.

CWA recommended limits: 0.019–1.9 ppb.

Reported concentrations (ppb)	*State*
20	Delaware

1,1-Dichloroethane

Principal uses

Solvent, cleaning and degreasing agent. Intermediate in organic syntheses.

Carcinogenicity: Suggested animal carcinogen.

CWA recommended limits: No criteria.

Reported concentrations (ppb)	*State*
7	Maine

Parathion

Principal uses

Insecticide.

Carcinogenicity: Suggested animal carcinogen.

CWA recommended limit: 0.013 ppb is Criterion Continuous Concentration.

Table 16.1   *Continued*

Reported concentrations (ppb)	State
4.6	California

[1]CWA recommended limits are concentration limits recommended under section 304(a)(1) of the Clean Water Act. Recommended limits for noncarcinogens represent levels at which exposure to a single chemical due to the consumption of water and contaminated fish is not anticipated to produce adverse effects in humans. Recommended limits for suspected or proven carcinogens are presented as concentrations in water associated with a range of incremental cancer risks from 10^{-7} to 10^{-5}, i.e., one additional case of cancer in populations ranging from 10 million to 100,000, respectively.

Source. Pye et al. (1983), Sittig (1980, 1985).

Disposal of toxic chemicals as well as many other hazardous wastes in improperly constructed and/or supervised landfills and surface impoundments appears to have been one major mechanism by which groundwater has become contaminated. As of 1980 the EPA estimated that there were a total of 200,000 landfills and dumps in the United States receiving 135 million tonnes per year of municipal solid wastes and 215 million tonnes per year of industrial solid wastes. In addition, about 176,000 surface impoundments located at 78,000 different sites were estimated to be receiving 38 trillion (i.e., 38×10^{12}) liters per year of liquid wastes.

Certainly not all wastes are hazardous. However the EPA (1980a) has estimated that about 55 million tonnes of hazardous waste are generated each year in the United States. These wastes include substances that are toxic (e.g., pesticides, PCBs; toxic organic chemicals, and heavy metals); reactive (e.g., wastes that have a tendency to explode); ignitible (e.g., benzene, toluene, paint, and varnish removers); corrosive (e.g., alkaline cleaners, acid liquids); infectious (e.g., improperly treated sewage sludges); or radioactive (Durso-Hughes and Lewis, 1982). According to the EPA (1980a) these hazardous wastes are generated at about 750,000 different sites in the United States. Larger industrial firms have at least in the past tended to dispose of wastes in landfills on their own property, and according to the EPA (1980a) about 50,000 such sites have been used for the disposal of hazardous wastes. Of these 50,000 sites, about 1200 are considered to pose a threat to the environment (Clay, 1990).

Further information on the quality of landfill and surface impoundments is revealing. A 1978 Waste Age survey identified 15,000 active municipal landfills in the United States. According to the EPA (1980b) only 31% of these landfills met state regulations. Of the industrial waste impoundments surveyed by the EPA (1980a), about one-third contained liquid wastes with potentially hazardous constituents, 70% were unlined, only 5% were known to be monitored for groundwater quality, and 30% were both unlined and located on permeable ground overlying usable aquifers. A study of over 80,000 surface impoundment sites of all types (e.g., industrial, municipal, agricultural, etc.) revealed that almost 40% were "potentially hazardous" (Pye et al., 1983, p. 60).

Roughly 10% of all industrial waste produced is hazardous (Durso-Hughes and Lewis, 1982). The industries responsible for the majority of the hazardous waste production are (in descending order by volume of hazardous waste produced) the organic chemical industry, the primary metals industry, the electroplating industry, the inorganic chemical industry, the textile industry, oil refineries, and the rubber and plastics industry (Durso-Hughes and Lewis, 1982). About 60% of industrial hazardous waste in the United States is generated by the chemical and allied products industry (EPA, 1980d). The type of hazardous waste generated will of course vary greatly between one industry and the next. For example cyanide wastes are associated with metallurgical operations; sulfite wastes are generated by paper and pulp manufacturing; mercury is a common waste product in the electrical industry; and the petrochemical industry produces a wide variety of toxic organic wastes ranging from pesticides to phenol-rich tar wastes (Patrick et al., 1987).

Illegal Disposal

Unfortunately disposal of wastes in improperly constructed and/or supervised land disposal sites is not the only mechanism by which hazardous wastes find their way into groundwater. As the costs associated with the proper disposal of hazardous wastes have increased, illegal dumping has become an increasingly attractive alternative. Some of the most widely publicized examples of illegal waste disposal operations have involved cases in New Jersey and Pennsylvania, but "illegal dumping operations have been identified in Kentucky, Texas, Ohio, Michigan, California, North Carolina, and many other states" (Epstein et al., 1982, p. 177). Patrick et al. (1987) document 44 case histories of illegal dumping in New Jersey, although only 5 of these 44 examples involved cases where groundwater was clearly affected. One of the best-known examples of illegal dumping in New Jersey involved a facility operated by Chemical Control, Inc., in Elizabeth, New Jersey. When state inspectors visited the facility in the winter of 1979, they found about 40,000 rusty and leaky 55-gal drums of hazardous industrial waste, in some places stacked 5-high and packed together like sardines in a can. "In a loft in the incinerator building on the site, investigators discovered about a hundred pounds of explosive dried picric acid; several pounds of radioactive material; several large bottles of a liquid labeled 'Nitro'; over twelve hundred 'lab pacs', packages of material wastes from research laboratories containing toxic agents; dozens of containers of explosive compressed gases; 'leaking containers of chromic acid, isopropanol, mineral spirits and petroleum naphtha, all designated hazardous substances'; and cylinders of mustard gas, 'a highly toxic nerve gas'. Also found inside the building were two other storage areas, labeled by the state as the 'pesticide room' and the 'boiler room', each crammed with rusted and leaking drums." (Epstein et al., 1982, p. 155)

The chemical wastes stored at the Chemical Control facility clearly constituted a hazard to the environment and to human health. The chemicals stored at the site included substances known to cause mucous membrane and respiratory tract irritation; cardiovascular, hepatotoxic, and neurotoxic effects; blood disorders; allergies; skin disoders; and to be carcinogenic, mutagenic, teratogenic, and toxic to unborn children. In addition, state experts indicated that there was a serious risk of a fire or explosion at the site (Epstein et al., 1982). A 5-m, high-pressure gas line was located less than 200 m from the site, and a large liquid natural gas tank, several bulk gasoline storage tanks, and ten propane gas tanks were all located within about 600 m of the site. Since a school and large residential area were located within 300 m of the facility, a serious fire or explosion at the site could have killed thousands of people. After several unsuccessful attempts to get the company to clean up the dump, the state finally took control of the site and initiated a cleanup. However, before the cleanup was completed, a serious fire did break out on April 11, 1980. Flames reached 100 m into the air, and smoke clouds rolled 24 km out to sea. Firemen battled the blaze for 10 hours before bringing it under control. Fortunately 10,000 drums of the most hazardous waste had been removed from the site before the fire, and the wind was blowing offshore at the time of the blaze. The operator of the Chemical Control dump was convicted in 1978 on three counts of illegal dumping and was sentenced to a 2-year prison term (Epstein et al., 1982).

This particular incident gives some idea of the magnitude of hazardous wastes which may accumulate at an illegally operated disposal site. Although the principal threats at the chemical control site were from fire, explosions, and air pollution, other illegal dumping operations have posed a very direct threat to water supplies. For example, ABM Disposal, a Philadelphia-based waste-handling operation, was found guilty of numerous illegal dumping operations that contaminated groundwater. In 1977 ABM haulers were convicted in three separate illegal dumping incidents, and in March 1977, "ABM personnel were convicted for illegally dumping dangerous pharmaceutical wastes into a groundwater well located in a garage in Montgomery County" (Epstein et al., 1982, p. 163). In another incident an ABM truck was discovered dumping cyanide waste containing "many, many times the lethal dose" of cyanide into Ridley Creek, a favorite swimming spot for children in the Chester, Pennsylvania, area. According to K. Welks, attorney for Pennsylvania's Department of Environmental Resources, "If anyone had been swimming at the time, it could have killed them" (cited by Epstein et al., 1982, p. 163). In another incident involving ABM, drums containing tonnes of waste and provided by ABM were stacked at a facility operated by Eastern Rubber Reclaiming, Inc., in downtown Chester, Pennsylvania. The operator of the facility "made more money on the drums, 'whenever he needed it', by opening them and pouring the contents into the ground or into lagoons on his property and selling the empty drums at $6.00 each to a local recycler. Tanker trucks filled with hazardous wastes were also allowed to discharge their contents on the grounds of Eastern Rubber Reclaiming" (Epstein et al., 1982, p. 164). The facility was raided in the fall of 1977, but before a proper inventory and cleanup could be completed, the site exploded in flames on February 2, 1978. The fire destroyed most of the drums on site, although about 4500 remained intact, and "when the smoke cleared, inspectors also found several tanker trucks filled with waste abandoned at the site" (Epstein et al., 1982, p. 165). Cleanup costs for the Eastern Rubber Reclaiming site were about $1.5 million.

These examples give some idea of the magnitude of the illegal dumping problem. Illegal dumping is big business, and there is a strong suggestion that organized crime may have become involved in parts of the country. Furthermore, "we are now seeing a trend away from midnight dumping of drums off the backs of trucks toward more complex sewering operations, in which it is extremely difficult to locate the underground disposal line carrying the waste away from the disposal site" (Epstein et al., 1982, p. 177). Major causes of illegal dumping have undoubtedly been industry irresponsibility or ignorance, but government laxity is also a factor. The Department of Justice during the Reagan administration indicted only about 30 entities and individuals per year for improper disposal of hazardous wastes. This figure doubled during the Bush administration (E. Boling, Dept. of Justice, personal communication).

Magnitude of the Problems

How serious a problem is groundwater pollution? Unfortunately no comprehensive national survey of groundwater contamination has been undertaken in the United States. Furthermore, only a small percentage of the contaminants now being found in groundwater have been tested for human health effects. Hence it is very difficult to make a quantitative statement about the national risk associated with drinking contaminated groundwater. The EPA did commission five regional groundwater assessment studies during the 1970s (GRF, 1984) and has conducted eight national drinking water surveys since 1975 (EPA, 1987). Unfortunately the regional studies and half of the drinking water surveys were completed before incidents such as Valley of the Drums[1] and Love Canal called national attention to the problems created by improper disposal of hazardous waste. There is little doubt that the conclusions of the surveys have been biased to some extent by the types of pollutants which investigators at the time of the studies considered most likely to be found in the waters that they tested. If a comprehensive national study were conducted today, it seems likely that a more thorough search would be made for the presence of a variety of toxic organic chemicals which recent studies have revealed in water supplies contaminated by improper disposal of chemical wastes.

What does clearly emerge from the reports and studies is that incidents of groundwater contamination have occurred in every state and that incidents are beginning to occur with increasing frequency (Patrick et al., 1987). Industrial impoundments, land disposal sites, and septic tanks and cesspools appear to be the most important sources of groundwater contamination on a national level. Municipal wastewater, petroleum exploration, and mining are of secondary importance as causes of contamination (EPA, 1980c). The four pollutants most commonly reported have been nitrates, heavy metals, microorganisms, and organic chemicals (Patrick et al., 1987). Human and animal wastes were judged to be of primary or secondary importance as causes of groundwater contamination in all five of the regional EPA surveys, and were considered to be among the top three contaminants in every state surveyed except California in a similar study conducted by the Environmental Assessment Council (GRF, 1984). Industrial wastes are also considered to be high-priority contaminants in most regions of the country and are judged to be the biggest source of groundwater contamination in the Northeast. In the southcentral and southwestern portions of the country, disposal of oil-field brines appear to be a major source of groundwater pollution and accounts for a high percentage of industry-related contamination. With the exception of industrial, human, and animal wastes, the important sources of contamination vary considerably from one region of the country to another. For example, some coastal states have serious problems with saltwater intrusion; chloride contamination from roadsalts is a serious problem in some snowbelt states; and high concentrations of dissolved solids are a pollution problem in areas having soluble aquifers (GRF, 1984).

---

[1] A 3-ha site in Kentucky in which 17,000 waste drums contaminated surface waters with some 200 organic chemicals and 30 metals.

What percentage of the aquifers in the United States are contaminated? Several estimates have been made of the areal extent of contamination of usable surface aquifers by surface impoundments and landfills (considered the most important sources) and by subsurface disposal systems, petroleum exploration, and mining (secondary sources). An EPA study (EPA, 1980c) put the percentage of contaminated aquifers at 0.5–1%; a similar analysis by Lehr (1982) indicated that up to 2% of the aquifers might be contaminated. As noted by GRF (1984, p. 4), "Although this might not seem to indicate a large problem, much of the contamination occurs in areas of heaviest reliance on groundwater." The following examples provide some indication of the extent of the problem:

1. In Nassau and Suffolk Counties on Long Island, more than 36 community wells were closed in 1980 due to contamination with tetrachloroethylene (TCE), trichloroethane, trichloroethene, and other volatile synthetic organic compounds. More than 3 million people in these counties are dependent on groundwater as their sole source of drinking water. The Council on Environmental Quality (CEQ, 1981) estimated that over 2 million persons were affected by these well closings.

3. In January 1980, California public health officials closed 39 public wells in 13 cities in the San Gabriel Valley because of TCE contamination. Over 400,000 people were affected by the well closings (Brumaster, 1982). Nineteen wells were closed in rural California east of Sacramento (Roberts, 1981).

2. In Jackson Township, New Jersey, about 100 drinking water wells were closed because of organic chemical contamination apparently caused by illegal disposal of chemicals at a landfill (Burmaster, 1982). Throughout the state about 400 municipal and 40 private wells have been closed because of groundwater contamination (Roberts, 1981).

4. In May 1978, four wells providing 80% of the drinking water for Bedford, Massachusetts, were closed. Toxic organic chemicals including up to 2100 ppb of dioxane and up to 500 ppb of TCE were found in the water (Burmaster, 1982). Throughout the state 22 public and private wells were closed because of groundwater contamination. A third of the commonwealth's 351 communities were affected by the closings (Roberts, 1981).

5. In Washington County, Illinois, 81% of 221 dug wells and 34% of drilled wells during the 1970s had a nitrate nitrogen concentration in excess of 10 ppm, the maximum allowable concentration recommended by the EPA for public water supplies (Pye and Patrick, 1983).

These examples indicate that groundwater pollution can and has affected the water supplies of large numbers of persons in some areas, despite the fact that current estimates of the areal extent of contamination are only in the 1–2% range for the United States as a whole. Previous discussion and Figure 16.2 indicate that there is a wide variety of mechanisms by which pollutants may be introduced into aquifers. However, both the regional studies conducted by the EPA during the 1970s and more recent information have indicated that improper disposal of hazardous waste, and particularly industrial hazardous waste, is the principal mechanism responsible for groundwater contamination on a national level. For example, a report to the Committee on Environmental and Public Works of the U.S. Senate, which reviewed a number of case histories of groundwater contamination that resulted in well closings, implicated the following types of substances in the indicated number of cases: metals, 619; organics, 242; insecticides, 201; chlorides, 26; nitrates, 23 (Pye and Patrick, 1983). The nature of the most frequently cited contaminants strongly suggests that industrial waste disposal was the cause of pollution in most cases. In this respect it is noteworthy that only about 10% of hazardous waste in the United States is disposed of in an environmentally sound method (Wood et al., 1984). Cost has been a major factor in determining the type of disposal method employed. As noted by the EPA (1980d, p. 15), "Environmentally sound technologies are available for treatment and disposal of hazardous waste. Costs vary widely, according to type and volume of waste handled, and are substantially in excess of unsound practices." The irony of this statement is that we are now beginning to realize that the true cost of improper hazardous waste disposal is usually far in excess of the cost of proper waste disposal when allowance is made for the cost of polluted water supplies, damage to the environment and human health, and cleanup (if cleanup is possible). The following case study provides a good illustration of these points.

A CASE STUDY—THE ROCKY MOUNTAIN ARSENAL

The Rocky Mountain Arsenal is a 70 km^2 facility located adjacent to the northeast sector of the city of Denver, Colorado (Figure 16.3). The Arsenal was operated for 40 years by the U.S. Army for the production of chemical warfare agents. It was a principal center for the production of nerve gas and its constituents and for emptying canisters of unused mustard gas. In 1947 the Shell Oil Company began leasing part of the land for the production of herbicides and associated chemicals, a practice that continued until 1982. From 1943 to 1956 liquid wastes were discharged into one or another of several hundred unlined ponds (Figure 16.4). The water contained a variety of complex organic and inorganic compounds, including proven or suspected carcinogens such as toluene, trichlorobenzene, vinyl chloride, xylene, carbon tetrachloride, chlordane, and aldrin (Anonymous, 1983a). The wastewater was rather easy to trace, since it contained a high chloride concentration, sometimes as high as 5 parts per thousand (5 ‰).

Groundwater flow in the region of the Arsenal is toward the northwest. Much of the land to the north of the Arsenal is irrigated farmland (Figure 16.4), and damage to crops from contamination of the irrigation water was apparent by 1951 (Konikow and Thompson, 1984). Similar damage was reported in 1952 and 1953. Particularly severe crop damage occurred in 1954, when precipitation was less than half the average value and use of groundwater for irrigation was therefore greater than usual. As a result of this repeated crop damage, several investigations were done to determine the extent of the groundwater contamination. A study by Petri and Smith (1956), for example, showed that contaminated groundwater extended over an area of several square kilometers to the north and northwest of the unlined disposal ponds.

As a result of these studies several steps were taken to alleviate the groundwater pollution problem. First, a 40-ha, asphalt-lined evaporation pond was constructed, and liquid wastes were discharged into that pond beginning in 1956 (Figure 16.4). Second, from 1968 or 1969 to about 1974, pond C (Figure 16.4) was maintained in a full condition most of the time by pumping in water from the freshwater reservoirs to the south. Water from pond C infiltrated into the ground at the rate of about 28 liters per second, and thus helped to dilute and flush contaminated groundwater.

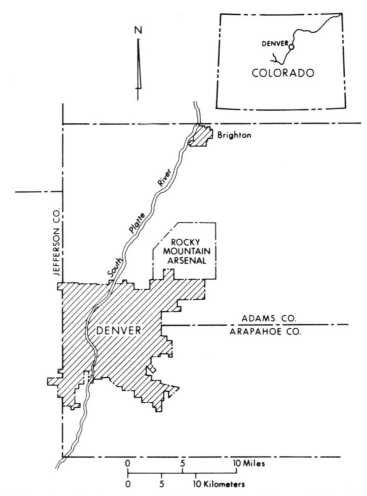

Figure 16.3 Location of Rocky Mountain Arsenal and city of Denver, Colorado. [Redrawn from Konikow and Thompson (1984).]

According to Konikow and Thompson (1984, p. 95), "By 1972 the areal extent and magnitude of contamination, as indicated by chloride concentration, had significantly diminished. Chloride concentrations were then above 1000 mg l^{-1} [1 ‰] in only two relatively small parts of the contaminated area and were almost at normal background levels in the middle of the affected area."

Unfortunately the asphalt-lined pond ultimately developed leaks, and there were new claims of crop damage in 1973 and 1974. In response to these new claims, the Colorado Department of Health conducted a study that revealed the presence of a variety of contaminants in wells downgradient from the disposal ponds. For example, diisopropylmethyl-phosphonate (DIMP), a nerve gas byproduct, was detected at a concentration of 0.57 ppb in a well located about 13 km downgradient from the disposal ponds and 1.6 km upgradient from two municipal water supply wells. A DIMP concentration of 48 ppm was measured in a groundwater sample collected near the disposal ponds. Other organic contaminants detected in wells or springs in the area included dicyclopentadiene (DCPD), endrin, aldrin, dieldrin, and several organo-sulfur compounds (Konikow and Thompson, 1984).

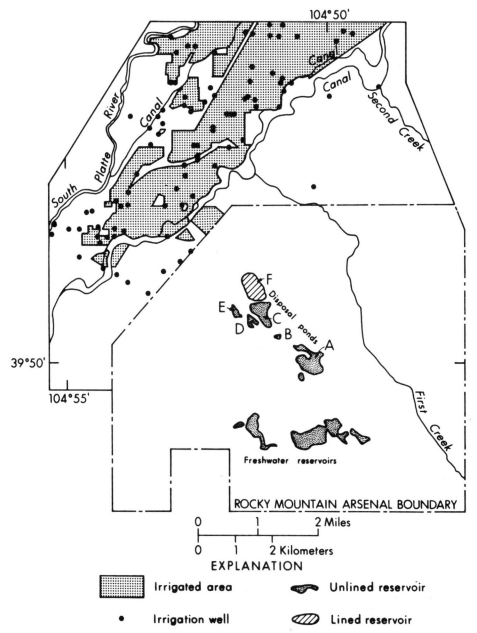

Figure 16.4 Location of evaporation ponds within the Rocky Mountain Arsenal and irrigated lands (stipled areas) to the north and west of the Arsenal. Dots indicate locations of irrigation wells. Evaporation pond F is the asphalt-lined pond. The other evaporation ponds are unlined. [Redrawn from Konikow and Thompson (1984).]

563

As a result of these discoveries the Colorado Department of Health issued cease and desist, cleanup, and monitoring orders in April 1975 to the Rocky Mountain Arsenal and Shell Oil Company. The cease and desist order required a halt to unauthorized discharges of contaminants into both surface and groundwater north of the Arsenal. Continued monitoring of groundwater revealed the presence of even more contaminants, including Nemagon (dibromochloro-propane) and several industrial solvents. Water samples taken from several hundred observation wells both within the Arsenal boundary and to the north and west provided a clear picture of the extent of contamination.

In response to the orders from the Colorado Department of Health, a computer model of groundwater flow was developed in order to predict the effect of possible remedial actions. The model predicted, for example, that it would probably take many decades for the contaminated aquifer to recover naturally. However, it was also predicted that certain water management policies could reduce recovery time to a matter of years. The solution ultimately adopted consisted of several components and was implemented in several phases. First, a dike was built to intercept the flow of contaminated surface water. Second, an impervious barrier consisting of a 1-m-wide, 7.6-m-deep, and 0.5-km-long trench filled with a mixture of soil and clay was built and anchored 0.6 km deep into the bedrock along the northern boundary of the Arsenal. Groundwater was then pumped from the south side (upgradient) of the barrier using six 20-cm-diameter wells spaced at equal intervals parallel to the barrier. The water was pumped at the rate of about 38 m^3h^{-1} through two columns of granular activated carbon to remove organics. Each column contained about 10 tons of activated carbon. Finally the treated water was reinjected by gravity flow into the ground on the north side (downgradient) of the barrier using twelve 0.5-m-diameter injection wells. This initial system is indicated schematically in Figure 16.5. The activated carbon columns were replaced whenever the concentration of DIMP in the treated water reached 50 ppb. In practice activated carbon use rates ranged from 100–150 mg of carbon per liter of water treated (Konikow and Thompson, 1984). The exhausted carbon was regenerated by a commercial vendor.

This system was operated for a total of 3 years, and preliminary results were sufficiently encouraging that an expanded containment system was constructed. The larger system consisted of a 2.1-km-long barrier of soil and clay, 7.6 to 15.2 m deep, with 54 withdrawal wells and 38 reinjection wells along the northern boundary of the Arsenal (Figure 16.6). Organics were removed from the water by three pulsed-bed adsorbers that contained 13.5 tonnes of activated carbon each. These new adsorbers were anticipated to be about 4 times more efficient than the cartridge filters used in the smaller pilot system. The new system became operational in 1983 and was capable of treating 136 m^3 of water per hour. The cost of the expanded system was $6 million.

Subsequently two other containment and treatment systems were built in addition to the initial north boundary system. A system built by Shell and completed in 1983 was located at one corner of the western boundary of the Arsenal (Figure 16.6). Like the original north boundary system, this so-called Irondale system consisted of a series of dewatering wells, a liquid treatment facility, and recharge wells. However, the Irondale system did not involve the use of a soil and clay slurry trench to block groundwater flow, but instead depended entirely on the dewatering wells to intercept groundwater movement. A system built by the U.S. Army and completed in

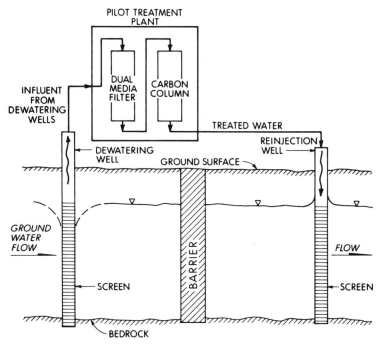

Figure 16.5 Schematic diagram of barrier and treatment system installed along northern boundary of the Rocky Mountain Arsenal. [Redrawn from Konikow and Thompson (1984).]

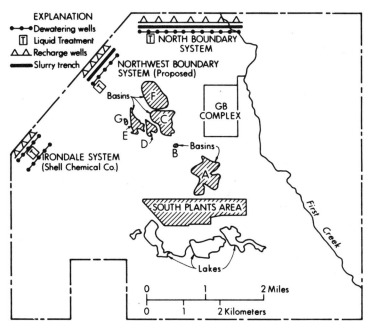

Figure 16.6 Location of various barrier and treatment systems at boundaries of the Rocky Mountain Arsenal. [Redrawn from Konikow and Thompson (1984).]

1985 was located along a portion of the northwest boundary of the Arsenal (Figure 16.6) and consisted of components similar to those used in the north boundary system. Both of these latter two systems were designed primarily to control the movement of Nemagon across the Arsenal boundary (Konikow and Thompson, 1984).

Although these containment measures worked reasonably well and despite the fact that the Rocky Mountain Arsenal was officially closed in 1982, considerable additional effort has been required to comply with the cease and desist, cleanup, and monitoring orders issued in 1975 by the Colorado Department of Health; and federal law has required that the site be returned to its original state (Anonymous, 1983a). Initial estimates of the cost of a complete cleanup were $2–$6 billion (Anonymous, 1983b), and in 1983 the U.S. Army sued Shell for $1.8 billion in damages, claiming that most of the wastes were the result of Shell's manufacturing and packaging of pesticides. Ultimately Shell and the Army reached an agreement whereby Shell would pay for 50% of the first $500 million in cleanup costs, 35% of the next $200 million, and 20% of any cleanup costs in excess of $700 million. The final price tag for restoration of the site is expected to be $1 billion (Anonymous, 1988).

As a part of the restoration, the Army transported some 76,000 drums of corrosive salts from the Arsenal to a landfill in Utah (Anonymous, 1988), and a special 2.5–5.0 km^2 landfill is being constructed to store contaminated soil excavated from the Arsenal (Anonymous, 1984). The liquid wastes in Basin F have proven to be a particularly difficult problem because of the tendency of the basin to leak. A temporary solution was to store the wastes in aboveground tanks, but the 32 million liters of toxic liquid were so corrosive that they threatened to eat through the tanks by mid-1993 (Anonymous, 1990). An agreement was therefore reached to burn the wastes in a 1050°C incinerator, the total cost of building the special incinerator and burning the wastes being about $35 million (Anonymous, 1990).

In considering the cost of containing the flow of contaminated groundwater from the Rocky Mountain Arsenal and ultimately restoring the site after manufacturing operations had ceased, one is reminded of a comment by Wood (1972), who noted, "The most satisfactory cure for groundwater pollution is prevention." Furthermore, regardless of whether groundwater pollution problems are solved by prevention or by cleanup, experience has shown that in many cases legal action is required in order to produce a satisfactory response to the problem. The case of the Rocky Mountain Arsenal provides a good illustration of this point. The cleanup did not begin until cease and desist, cleanup, and monitoring orders were issued by the Colorado Department of Health. With this point in mind, it is useful to ask how well-suited present laws are for dealing with groundwater pollution problems and how vigorously these laws are being enforced by the government.

LEGAL CONSIDERATIONS

At the present time there is no single agency responsible for the protection of groundwater, and as a result effective management of contamination problems has sometimes been difficult. Furthermore, there is no federal program specifically designed to protect groundwater. Most of the concern at the federal level has been with problems associated with industrial hazardous wastes. Nevertheless, despite the lack of agencies, programs, or legislation focused specifically on the problem of groundwater pollution, there are a number of laws which, if effectively enforced, would provide all of the legal muscle needed to deal with groundwater pollution. The following is a summary of the most important pieces of legislation.

RCRA

The Resource Conservation and Recovery Act (RCRA) was passed by Congress in 1976 to provide for careful planning and management practices in the treatment, storage, and disposal of both municipal solid wastes and hazardous wastes. As noted by Pye et al. (1983, p. 243), RCRA "is foremost among federal statutes aimed at minimizing groundwater contamination." The two key features of RCRA are the Subtitle C program, which governs hazardous waste management, and the Subtitle D program, which concerns municipal solid waste disposal. Subtitle C authorizes the EPA to regulate hazardous wastes "from cradle to grave." Subtitle C directs the EPA to establish standards for all hazardous waste management facilities and to incorporate these standards into RCRA permits for individual facilities. After November 19, 1980 it was unlawful to treat, store, or dispose of any hazardous waste without a RCRA permit.

While Subtitle C sounds like a potentially very effective piece of legislation, there have been some fairly serious problems in its implementation. There were more than 10,000 hazardous waste facilities in existence at the time of the effective date of the program, and it was impossible for the EPA to thoroughly review the permit applications of these facilities overnight. Hence interim status standards were established, and facilities that filed a timely application were eligible to continue operating under an interim status until such time as their permit applications were processed. The interim status standards included requirements for waste analysis and inspection, groundwater monitoring, and other "housekeeping" measures, and also established certain technical standards.[2] However, with Anne Burford running the EPA as of May, 1981, even these interim status standards were not enforced effectively. For example, in April of 1982 the EPA suspended the requirement that companies submit an annual report describing in detail how much and what kind of hazardous waste they handled, and suspended the requirement that industries file a yearly report on groundwater conditions near a waste site (Sun, 1982). Congress reacted in 1984 by amending the RCRA so as to cancel interim status for land disposal facilities that had failed to submit a final permit application by November 8, 1985 (Patrick et al., 1987). As a result of this requirement and in particular because of the difficulty in obtaining commercial pollution liability insurance, about two-thirds of the interim status land disposal facilities were forced to close down in 1985.

After much delay, the EPA did finally promulgate hazardous waste facility permitting standards on July 26, 1982 (the regulations were due in 1978). The criteria became effective in January, 1983. The permitting standards contain two key elements, a liquids management strategy and a groundwater monitoring and response program. The liquids management strategy requires the use of liners and leachate collection systems on new surface impoundments, waste piles, and landfills. Although existing facilities are not required to employ liners and leachate

---

[2]For example, containers holding hazardous waste must be in good condition and not leaking.

collection systems, both new and existing facilities are subject to groundwater monitoring requirements, "which can lead to the imposition of treatment or removal requirements when groundwater is contaminated" (Pye et al., 1983, p. 249).

Congress intended that administration of the Subtitle C program would ultimately pass to the states. Authorization of state programs was envisioned as taking place in two steps. The state first obtains interim authorization by submitting to the EPA a description of the program, including a description of the agency that will administer the program, the staff, costs, etc. If the program is approved, the state may apply for final authorization, but must demonstrate that the state program is equivalent to the federal program, consistent with the federal program and other state programs, and provides for adequate enforcement. As of this writing, almost all states had obtained final authorization to administer Subtitle C.

The Subtitle D program involving municipal solid waste is also envisioned as being administered largely by the states, although with guidance from the EPA. Under Subtitle D the states are required to develop solid waste management plans, and the EPA is required to provide technical and financial assistance in the development of those plans. The key element in the Subtitle D program is the establishment of guidelines for classifying facilities as either open dumps or sanitary landfills. A facility may be classified as a sanitary landfill "only if there is no reasonable probability of adverse effects on health or the environment from the disposal of solid waste at such facility." States with approved Subtitle D plans are to classify all municipal solid waste facilities as either open dumps or sanitary landfills. Any facilities that are classified as open dumps must be either upgraded or closed. Contamination of groundwater at a facility is considered to have occurred if the concentration of any pollutant at the "solid-waste boundary" exceeds the maximum contaminant levels (MCLs) specified by the Safe Drinking Water Act. An obvious flaw in the law is that MCLs have been set for only a small fraction of the contaminants now found in groundwater.

A potentially powerful feature of RCRA is the imminent-hazard provision, which authorizes the EPA administrator to bring suit in district court to restrain handling of solid or hazardous waste at a facility if such handling poses an "imminent and substantial endangerment to health or the environment." The EPA may also issue administrative orders "as may be necessary to protect public health and the environment." Clearly RCRA is a potentially powerful tool for dealing with groundwater pollution from hazardous and municipal solid waste.

SDWA

The Safe Drinking Water Act (SDWA) was enacted by Congress in 1974 to provide for sanitary drinking water supplies. Three features of SDWA are of particular importance.

First, the SDWA establishes a set of primary and secondary drinking water standards (i.e., maximum contaminant levels). The primary drinking water standards apply to contaminants that the EPA has determined to have an adverse effect on human health. The 1986 amendments to the SDWA and subsequent modifications mandated standards for a total of 83 such contaminants, including 33 organic pesticides, 21 volatile organics, and 18 inorganic chemicals. A serious flaw in the current primary standard list is that of all the organic and inorganic chemicals that have been found as groundwater contaminants, MCLs have been established only for total trihalomethanes, six organic pesticides, eight volatile organics, and ten inorganic chemicals (EPA, 1988). The secondary drinking water standards apply to contaminants or characteristics that might adversely affect the odor or appearance of drinking water to such an extent that a substantial number of persons would not drink the water. To date, the EPA has established a total of 13 secondary MCLs.

Second, the SDWA contains an underground-injection-control (UIC) program, which gives EPA direct authority over underground waste disposal wells. The UIC program was the first federal effort directly aimed at the control of groundwater pollution. Although the provisions in RCRA would be sufficient to regulate underground waste disposal wells, in practice the UIC program of SDWA has been used to regulate underground injection of wastes. Like the Subtitle C and D sections of RCRA, the UIC program is designed to be administered by the states. According to the provisions of UIC, underground injection is prohibited without an authorized state permit, and in order to obtain a state permit, an applicant must satisfy the state that drinking water sources will not be endangered by the injection of waste. The state in turn must adopt drinking water standards in conformance with the federal standards and implement procedures for the enforcement of those standards in order to obtain EPA authorization to enforce the drinking water standards and the provisions of UIC.

The third important feature of SDWA is the sole-source aquifer program. This program is specifically aimed at groundwater protection, and applies to areas that have only one aquifer as a principal source of drinking water. No new underground injection wells can be drilled in such areas without a permit, and if the EPA determines that a sole-source aquifer would, if contaminated, create "a significant hazard to public health," then "no commitment of federal financial assistance through grants, contracts, or loan guarantees may be given to any program which EPA determines may contaminate such an aquifer so as to create a significant hazard to human health" (Patrick et al., 1987, p. 383).

CWA

The Clean Water Act (CWA), which was passed by Congress in 1972, is the most comprehensive federal water pollution control program in existence, but is concerned primarily with the reduction and control of pollutant discharges into navigable waterways. The CWA can, however, affect groundwater pollution problems indirectly to the extent that certain instances of groundwater contamination result from leaching of polluted surface waters into an aquifer. Furthermore, the CWA does contain several provisions that are directly relevant to groundwater pollution. First, Section 104 of the CWA requires the EPA to "establish, equip and maintain a water quality surveillance system for the purpose of monitoring navigable waters and ground waters." In practice, however, Section 104 has been implemented primarily with respect to surface waters, with little attention given to groundwater. Second, Section 208 of the CWA calls for the development and implementation of area-wide plans for the management of waste treatment. These plans are to be worked out through a cooperative effort between the EPA and the states. Section 208 is potentially the most effective portion of the CWA for controlling groundwater pollution, but in practice "the programs implemented under Section 208 have, unfortunately, received lower priority and achieved much more limited results than have other parts of the Clean Water Act" (Patrick et al., 1987, p. 387). Furthermore, the area-wide planning authorized under Section 208 is no longer funded. Third, Section 402 of the CWA establishes a National Pollution Discharge Elimination System (NPDES), which allows the EPA or authorized states to issue permits for pollutant discharges if certain conditions are met. These conditions require compliance with specific water quality standards and "are the principal mechanism for enforcing measures to reduce and control the discharge of pollutants into surface waters" (Pye et al., 1983, p. 253). An important feature of Section 402 is a subsection that requires that a state have adequate authority to control the discharge of pollutants into wells as a condition for authorizing a state NPDES program. Thus a state with an authorized NPDES program could clearly control groundwater pollution under the CWA to the extent that the state could control the disposal of pollutants in wells. However, the CWA defines the term *pollutant* in a way that specifically excludes wastes associated with oil and gas production, and hence states could not, for example, regulate the disposal of oil field brines in injection wells under Section 402. Finally the CWA contains a provision that gives the EPA an imminent-hazard authority to restrain the discharge of pollutants where such discharge poses an imminent and substantial danger to the health or livelihood of human beings. This provision gives the EPA broad authority to protect against both enviromental and economic injury, and "could be effective in restraining activities threatening to contaminate groundwater supplies" (Patrick et al., 1987, p. 388).

CERCLA

The Comprehensive Environment Response, Compensation and Liability Act (CERCLA) was enacted by Congress in 1980. CERCLA was designed to allow the federal government to respond immediately to the release or threatened release of hazardous substances into the environment. Specifically CERCLA requires that remedial action be taken at inactive waste disposal sites that pose a threat to the environment or human health.[3] CERCLA authorizes the federal government to clean up the contamination at an inactive site or spill if it is impossible to obtain a satisfactory response from the private sector. Such federal cleanups are financed by a $1.6 billion Hazardous Substance Response Trust Fund, commonly referred to as *Superfund*, seven-eighths of which is provided by taxes on industries and the remaining one-eighth from general revenues. However, CERCLA specifically prohibits the government from using Superfund monies for cleanup if a responsible private party will conduct the work. Private parties such as owners or operators of dumps may be held responsible for cleanup under CERCLA without regard to fault or negligence, and may avoid liability only with a few narrowly drawn defenses. Subsequent to the enactment of CERCLA, Congress passed the National Contingency Plan (NCP) on July 16, 1982, to ensure effective response to both CERCLA and CWA. The NCP is basically a program to establish priorities for cleanup operations and to determine the most effective remedial response. A key component of the NCP is the Hazard Ranking System (HRS), which is used to establish cleanup priorities. Considerations taken into account include the population at risk, the

[3] Active waste disposal sites are covered by RCRA.

hazard potential of the pollutants, the potential for groundwater contamination, and so on. As of February, 1991 there were a total of 1183 sites throughout the country on the Superfund National Priorities List (Figure 16.7). It is clear from Figure 16.7 that hazardous waste sites that pose a threat to human health or the environment are widely distributed throughout the United States, but are most concentrated in the northeast.

It would be an understatement to say that CERCLA proved to be a controversial piece of legislation during the first decade of its existence. Although roughly 1200 abandoned hazardous waste sites were identified as priorities for cleanup, only 54 of those sites had been permanently cleaned up by the end of 1990 (Rubin and Setzer, 1990). In addition to the small number of sites that have been restored, a major criticism leveled at the Superfund program has been its reliance on pumping and treating as a means of aquifer restoration. Travis and Doty (1990, p. 1465), for example, have noted that, "A recent EPA study involving 19 sites where pumping and treating had been ongoing for up to 10 years concluded that although significant mass removal of contaminants had been achieved, there had been little success in reducing concentrations to the target levels. The typical experience is an initial drop in concentrations by a factor of 2–10, followed by a leveling with no further decline. To exacerbate the problem, once pumps are turned off, concentrations rise again." Such results have led some persons to question whether the Superfund program is not a waste of taxpayers' money. According to Stipp (1991), "The Superfund program has little success to show for the $7.5 billion in taxpayers money it has spent so far." In an interview in November, 1990 EPA Administrator William Reilly was more upbeat about the future of Superfund, but acknowledged that the program was plagued with problems during its early years (Rubin and Setzer, 1990).

Figure 16.7 Distribution of the 1183 Superfund priority sites as of February, 1991. [*Source*: Code of Federal Regulations **40**(300), subpart K, Appendix B.]

Other Legislation

RCRA, SDWA, CWA, and CERCLA are the principal pieces of federal legislation that can be used to control groundwater pollution. However, several other federal laws are relevant to groundwater pollution in one way or another and are worthy of mention at this time. The Toxic Substances Control Act (TSCA) was enacted in 1976 to regulate the manufacture, use, and disposal of hazardous chemicals and chemical mixtures. Although protection of groundwater supplies is not a specific objective of TSCA, the fact that TSCA gives the EPA authority to regulate the disposal of hazardous chemicals or substances containing hazardous chemicals is obviously relevant to groundwater pollution. Furthermore, since TSCA authorizes the EPA to limit the manufacture, processing, distribution, and use of chemicals, it is obvious that TSCA can have a very significant impact on groundwater pollution by limiting the quantities and kinds of hazardous chemicals which require disposal. The Uranium Mine Tailings Radiation Control Act (UMTRCA) of 1978 and the National Low-Level Radiation Waste Policy Act (NLRWPA) of 1980 are both relevant to the contamination of groundwater by radionuclides. As previously noted (Chapter 14), it had been customary prior to UMTRCA for uranium mining and milling operations to leave uranium-tailing piles uncovered and exposed to wind and rain (Hileman, 1982). A full assessment of the impact of this practice on radionuclide concentrations in groundwater as well as surface water began shortly after passage of the UMTRCA (Gallaher and Goad, 1981). Some degree of groundwater contamination was found at 75% of the active uranium mills licensed in the United States (Patrick et al., 1987). Under NLRWPA each state was to assume responsibility for the disposal of low-level radioactive wastes generated by commercial operations within its borders by 1986 (Pye et al., 1983), but the 1985 amendments to NLRWPA extended this deadline to 1993 and stipulated that it was the policy of the Federal Government that the responsibilities of the states for the disposal of low-level radioactive wastes could be most safely and effectively managed on a regional basis. In the past such wastes had been stored at a few commercial and defense sites around the country, and as noted in Chapter 14, there had been leakage of radioactivity at almost all of these sites. It is now anticipated that there will be a total of about a dozen regional low-level radioactive waste sites. Although "it is generally agreed that the technology exists for siting and safe packaging, handling, transport, and isolation of the wastes" (ibid, p. 69), the incident of plutonium migration from shallow trenches containing low-level radioactive wastes at Maxey Flats, Kentucky (Chap. 14), indicates that when the subject is radioactive wastes disposal, we do not always know as much as we think we know. Finally it is worth mentioning here the National Environmental Policy Act (NEPA) of 1969. The critical provision of NEPA is the requirement that an environmental impact statement be prepared for any proposed legislation or action that might affect significantly the quality of the environment. Thus under NEPA any major projects sponsored or permitted by the federal government may be evaluated for potential adverse effects on the environment, including groundwater.

Enforcement

It should be apparent from the foregoing discussion that there exist federal statutes that, if rigorously enforced, could be used to effectively deal with groundwater pollution problems. However, it should also be clear that at least some of the provisions of these laws have not been effectively enforced, either because of lack of manpower or lack of desire on the part of the government. This inability or unwillingness to take action was apparent during the Reagan administration, and particularly during the time that Anne Gorsuch Burford headed the EPA. For example, Rita Lavelle, who headed the CERCLA hazardous waste program and Superfund under Burford as of March 31, 1982 was appointed to her post after previously serving as director of communications for subsidiaries of Aerojet-General Corporation of California. Aerojet-General's liquid fuel plant had previously been cited by EPA as one of the 40 most hazardous chemical waste sites in the United States, and in 1979 California officials accused the company of discharging almost 80 m³ per day of hazardous waste into a swamp and pond (Sun, 1982). Lavelle was fired by EPA head Burford on February 7, 1983 and in December, 1983 was convicted by a U.S. district court in Washington of perjury and obstructing a congressional inquiry. Lavelle had denied under oath that she knew Aerojet-General was being investigated by the EPA in a hazardous waste disposal case, but the prosecution established that not only had Lavelle been aware of the investigation, but in fact had called Aerojet-General officials to warn them (Lowther, 1983). In March, 1983 Burford herself resigned, and ultimately 21 top EPA officials resigned in response to allegations of perjury, conflict of interest, and political manipulation.

On the state and local level lack of enforcement of aquifer protection laws has also been a problem. As noted by Pye et al. (1983, p. 272), "Ineffective enforcement appears to be a much more pervasive feature of groundwater regulation than of surface water regulation. Officials in state agencies regularly complain that monitoring and enforcement programs are understaffed and underfunded . . . Some regulations are enforced minimally, others not at all (Dawson 1979). Most violations are handled informally at the agency level, with only light sanctions imposed." The cost of establishing and maintaining an adequate groundwater monitoring program and the difficulty in obtaining accurate measurements of the wide variety of possible contaminants when concentrations are in the parts per billion range appear to be major factors which discourage vigorous enforcement of existing statutes. Maugh (1982), for example, cites a study conducted by the Centers for Disease Control in which 29 laboratories were sent samples containing a known concentration of PCBs. Only 3 of the 29 laboratories produced results that were within two standard deviations of the correct value. Given this sort of result, it is natural to wonder just how accurate measurements of low concentrations of toxic substances in water supplies really are, and how much time and expense would be required to establish a monitoring program which produced reliable results for the wide variety of hazardous substances which might be found in groundwater. Undoubtedly the same question has occurred to state and local officials.

CORRECTIVES

Under the category of correctives it is appropriate to distinguish between prevention and cleanup. It should be clear by now that prevention of groundwater pollution should be the preferred strategy for dealing with the problem, but we must also deal with aquifers that became contaminated in the past and be prepared to deal with incidents of contamination in the future. Do we now have at our disposal effective methods for cleaning up contaminated groundwater?

Cleanup

If the goal of cleanup is to restore aquifers to a pristine condition, then the answer to this question in many cases will apparently be no. We do, however, have methods for containing the contaminated plume of groundwater and reducing the mass of the contaminant. In some cases these measures may be sufficient. For example, the water need not be potable if it is to be used exclusively for irrigation. However, even if the goal is not to produce pristine water quality, cleanup costs are usually high. Current estimates are that it will cost about $64 billion to clean up the 2200 waste sites expected to be on the National Priorities List by the year 2000 (Rubin and Setzer, 1990), an estimate that works out to a little over $29 million per site.

There is no question that the high cost of cleaning up contaminated aquifers and the failure of groundwater pumping and treating efforts to restore potable water quality have led some scientists to seriously question the wisdom of large scale groundwater restoration programs. Muller (1982), for example, has stated that cleanup of an aquifer that has been contaminated by organic chemicals is almost never physically or economically feasible, and Pye and Patrick (1983, p. 717) noted that, "recent studies of remedial action have concluded that it is complicated, time-consuming, expensive, and often not feasible, and that the best solution to ground water contamination is prevention . . . Often it is more cost-effective to locate a new source of water than to attempt treatment." Unfortunately a new source of water is not always available, at least not at a reasonable cost. For example, an Army Corps of Engineers study of the feasibility of providing irrigation water to Great Plains farmland currently irrigated with water from the Ogallala aquifer by building a system of canals to import water from South Dakota, Missouri, and Arkansas indicated that the cost would be $3.6–$22.6 billion (Patrick et al., 1987). Hence it is sometimes necessary to consider the cleanup option.

There are potentially a number of methods available for restoring groundwater quality. These methods include: a) eliminating the source of contamination and allowing restoration to proceed by natural processes; b) accelerating the rate of removal of contaminants by the use of withdrawal wells, drains, or trenches; c) accelerating the rate of flushing by recharging with clean water; d) installing impermeable barriers to block the spread of contamination; e) inducing chemical or biological reactions to neutralize or immobilize the contaminant; and f) excavating and removing the contaminated portion of the aquifer (GRF, 1984). Note that options a–e have all been employed at one time or another at the Rocky Mountain Arsenal. Methods of treating contaminated water include reverse osmosis, ultra-filtration, ion-exchange resins, wet-air oxidation, ozonation and ultraviolet radiation, coagulation and precipitation, aerobic biological treatment, and activated carbon (Patrick et al., 1987). The nature of the aquifer and of the contaminants will obviously dictate to a large extent what sort of remedial measures are appropriate, and whether cleanup is feasible at all. In some cases, for example, grout curtains are used to block the flow of contaminated groundwater. The curtains are formed by injecting grout[4] under pressure through a number of closely spaced wells. However,

---

[4]Grout may be composed of a variety of substances including cement, fly ash, epoxy resins, etc.

there is always a question of whether the grout curtain will really form an impervious barrier, and the practical depth limit of the curtain is about 15–18 m (Patrick et al., 1987).

The remedial measures taken at Love Canal provide a good illustration of the application of cleanup techniques. The problems that developed at Love Canal are associated with a problem known as the *bathtub effect* (GRF, 1984). When landfills are located in low-permeability rocks and in humid areas, the trenches in which the wastes are buried tend to become filled with water from rain or snow. "This surface water seeps downward through the landfill cover, fills the trenches, and eventually overflows—the so-called 'bathtub' effect" (GRF, 1984, p. 8). In order to deal with this problem, a mounded clay cap was placed over the old canal. The cap was compacted to maximize its resistance to water seepage, and graded to divert stormwater into surface drains. The clay cap also helps to block the emission of fumes from volatile chemicals. About 3.5–6 m below ground a 1-m wide and 2.1 km-long barrier drain filled with crushed stone and sand was installed to isolate the canal from the surrounding environment. A pipe at the bottom of this trench was installed to carry leachate to a treatment plant (Wood et al., 1984).

The Rocky Mountain Arsenal and Love Canal cases illustrate that remedial action to eliminate or at least minimize the effects of groundwater pollution are possible, but the costs can be very high and the results may leave something to be desired. However, when the public health is threatened, a cleanup may be demanded. Such has been the case in "Silicon Valley" near San Jose, California, where organic solvents from the local electronics industry were detected in groundwater (GRF, 1984), and of course remedial action was required at the Rocky Mountain Arsenal, despite the $1 billion price tag. Clearly prevention of groundwater pollution in the first place is much more cost effective than trying to decontaminate a polluted aquifer.

Prevention

There are many methods for preventing groundwater contamination, and certainly one technique would be the use of improved waste disposal methods. According to the Geophysics Research Forum panel (GRF, 1984, p. 18), "Most wastes can safely be disposed of in the subsurface if repositories are selected, designed, and engineered on the basis of the nature of the wastes and adequate knowledge of the hydrology, geology and hydrogeochemistry of the site. There should be a more thorough search of disposal sites that can be used safely to isolate toxic wastes from the biosphere for long periods . . . A strategy should be developed that provides for the segregation, treatment, and disposal of wastes according to their hazards and their chemical affinities." The diversity of possible contaminants and disposal options means that each situation must be evaluated separately in its own particular context.

For example, one useful strategy that has been cited by both the U.S. military and the Department of Energy for the storage of radioactive wastes is the use of multiple barriers for waste containment. The waste should be stored in a form that is not readily soluble (e.g., incorporation of radioactive wastes into borosilicate glass); the containers holding the wastes should be highly resistant to development of leaks and cracks; the waste repositories should be backfilled with material that is chemically highly sorptive and of low permeability (e.g., clay); the storage facility should be located in a medium of low permeability; to the extent that groundwater is present, the natural flow should be away from the biosphere; the material surrounding the storage facility should contain minerals such as zeolites that would tend to sorb the contaminants or should consist of highly porous rocks in which diffusion into the matrix could retard flow (GRF, 1984).

Another point relevant to the siting of waste disposal facilities is the fact that a number of closed hydrologic basins having only internal drainage exist in parts of Nevada, Utah, and adjacent states. These basins are located in arid regions where leaching effects from rainfall are minimal, and even if the contaminants were moved as a result of groundwater flow, they would remain within the basin. Finally, Abelson (1984) has noted that most drinking water is drawn from wells that are less than 100 m deep, so that in some areas waste liquids with a density greater than that of water could be safely discharged through deep wells if injected at a depth greater than 100 m.

There are a number of more-or-less obvious and common sense techniques that could be used to prevent or at least reduce groundwater contamination from certain sources. For example, installation of sewer systems or the use of waterless toilets could, under appropriate conditions, be a feasible method for eliminating pollution from septic tanks. However, sewer systems are costly to build and maintain, and may not be cost effective in areas of low population density. Waterless toilets, which operate by acid incineration, gas incineration, or composting, would eliminate the organic solids and liquids generated by septic tanks, but would obviously not deal with domestic wastewater resulting from bathing, laundry, etc., and would produce an ash or compost, which would ultimately require disposal. Where overdrafting of an aquifer for irrigation purposes has produced groundwater pollution problems due to high salinity or has seriously depleted the groundwater resource, growing cool-season crops, crops with low water requirements, or use of subsurface irrigation systems may significantly reduce water withdrawal rates. Economic incentives may be used to encourage such practices. It is apparent, for example, that the solution to the current overdrawals from the Ogallala aquifer for irrigation will almost certainly be a return to dryland farming. Land spreading of sludge need not create groundwater pollution problems if the application site is carefully chosen and if the rate of sludge application is closely matched with the ability of the system to absorb organic waste. "Sludge disposal may best be sited on land where there is a mild gradient and a good soil cover that both allows slow percolation and has a high attenuation capacity, where there is a low water table, and where the vegetation is not to be used for food" (Patrick et al., 1987, p. 270). In fact, land farming of sludge has been used for 30 years by the petroleum industry, and the Solid Waste Management Committee of the American Petroleum Institute has concluded that such sludge application will not significantly contaminate the soil and groundwater so long as the area to be farmed is closely matched to the type and application rate of the sludge (Patrick et al., 1987). In fact, careful application of sludge can be a highly effective way to upgrade strip-mined areas and similar unproductive land (see Chapter 6).

Perhaps the most fundamental way to prevent groundwater pollution from hazardous waste disposal is to simply reduce the amount of hazardous waste. Modification in manufacturing procedures and/or recycling are the obvious mechanisms for reducing industrial hazardous wastes. Estimates of the degree of reduction that could be achieved by recovery and recycling vary widely. A study by Arthur D. Little

Co. concluded that only about 3% of industrial waste was potentially salable or reusable (Durso-Hughes and Lewis, 1982). However, the EPA has estimated that the total volume of hazardous waste could be reduced by 20% if available recycling and recovery technologies were used, and "Dr. Paul Palmer of Zero Waste Systems, Inc., a California-based commercial chemical recycling company, . . . claims that 80 percent of the chemical waste now buried at landfills or processed through other disposal methods could be recycled" (Durso-Hughes and Lewis, 1982, p. 16). An industrial toxics project known as the 33–50 program is designed to reduce toxic waste to a level 33% below 1988 levels by the end of 1992 and to 50% below these levels in 1995 (Koshland, 1991). Polsgrave (1982) has gone so far as to suggest the idea of a waste exchange, in which waste materials are transferred from seller to buyer through the hands of the exchange, which earns its income from commissions charged on completed transactions. According to Polsgrove (1982), waste raw materials such as iron and plastics would be the top priority waste, while processing residues such as solvents from pharmaceutical and paint processing and rejected lead plate from lead acid batteries would receive lower priority.

Regardless of the percentage of hazardous waste that could be economically recycled on a national level, there is no question that some companies have greatly reduced waste generation and at the same time saved considerable money by recycling. As noted in Chapter 9, for example, the "Pollution Prevention Pays" program begun by the 3M Corporation in 1975 has reduced its waste stream by a factor of 2 and in the process saved the company over $300 million (Pollock, 1987). In another case, Reliable Plating Co. of Milwaukee, Wisconsin, was forced by the EPA in 1980 to purchase a $30,000 ion-transfer system to recover chromium from the company's nickel/chrome plating operation. Although the company at first considered the device a regulatory burden, Reliable officials were pleasantly surprised to learn that in its first 8 months of operation the ion-transfer system reduced the company's purchases of chrome by a factor of 10 and cut water use by a factor of 70 (Durso-Hughes and Lewis, 1982).

The EPA Groundwater Protection Strategy

Certainly one important mechanism for preventing groundwater contamination would be a well-coordinated program involving the federal, state, and local governments to establish priorities and policies for groundwater use, to establish groundwater classification standards, to develop innovative approaches to groundwater protection, and to gather needed scientific and engineering information on groundwater contamination, assessment, enhancement, and protection. In fact just such a program was proposed by the EPA in November, 1980 and was implemented in August, 1984 when the EPA's Office of Ground-Water Protection published its Groundwater Protection Strategy. The goal of the program is the assessment, protection, and enhancement of "the quality of groundwater to the levels necessary for current and projected future uses and for the protection of public health and significant ecological systems" (EPA, 1980b). The strategy has the following four major objectives (EPA, 1984):

1. strengthen state groundwater programs;
2. address groundwater contamination problems of national concern;

3. create a policy framework for guiding EPA programs;
4. strengthen internal groundwater organization.

A key feature of the program has been the development of a groundwater classification system that recognizes different levels of protection for different aquifers. Aquifers are classified primarily on the basis of current or projected uses, and recognition is made of the facts that not all aquifers have high-quality yields nor do all uses require the same level of quality (EPA 1980b). Aquifers are classified in one of three categories as follows:

1. Class I aquifers are groundwaters of unusually high value that are in need of special protective measures because they have a relatively high potential for contaminants to enter and/or to be transported within the groundwater flow system (Patrick et al., 1987). They are also regarded either as irreplaceable in the sense that no alternative source of drinking water is available for a substantial population or as ecologically vital in the sense that they provide the base flow for a particularly sensitive ecological system that, if polluted, would destroy a unique habitat (Patrick et al., 1987). One might expect that groundwaters classified as sole source aquifers under the SDWA would be considered Class I groundwaters, but such is not automatically the case. The criteria for Class I groundwaters are more strict than those for sole source aquifers. The goal of EPA protection strategy for Class I aquifers is to keep pollutant concentrations below the maximum contaminant levels (MCLs) specified in the SDWA.

2. Class II groundwaters are all other aquifers that are currently used, or potentially available, for drinking water and other beneficial use. A Class II designation allows limited siting of hazardous waste facilities and, while ensuring quality suitable for drinking, permits exemptions to requirements under certain circumstances to allow less stringent standards when the protection of human health and the environment can be demonstrated (EPA, 1986). The EPA envisioned that most fresh groundwater would fall into Class II.

3. Class III groundwaters are not potential sources of drinking water and are of limited beneficial use, either because they are saline or because they are otherwise contaminated beyond levels that allow drinking or other beneficial use. Class III aquifers include groundwaters with a TDS greater than 10,000 ppm or "those that are so contaminated, naturally or by human activity, that cleanup using methods reasonably employed in public water system treatment is not possible" (Patrick et al., 1987, p. 299). Regulations for protection of Class III aquifers are more lenient and allow siting of hazardous waste facilities, but include technical requirements and monitoring rules to protect public health (Pye et al., 1983).

Perhaps the most controversial aspect of the EPA's Ground-Water Protection Strategy has been giving the states the primary role in groundwater protection. This policy actually went into effect before the Ground-Water Protection Strategy was formalized in 1984, because "with the confirmation of Administrator [Anne] Gorsuch at EPA in May 1981, the Reagan administration . . . quietly blocked all further development of the proposed [Ground-Water Protection] Strategy, and . . .

de facto turned over all responsibilities for groundwater-quality management to the states" (Burmaster 1982, p. 36). While the decision to shift responsibility for groundwater protection from the federal government to state and local authorities was consistent with other Reagan administration policies, "The states alone are incapable of supporting the research, policy development, and program implementation necessary to protect groundwater quality; the federal government must play a role to insure equity and efficiency among the states" (Burmaster 1982, p. 36). In summary, while it is undoubtedly reasonable to turn over a certain amount of regulatory responsibility to the states, the federal government and the EPA in particular must continue to play an active role in setting groundwater protection policies and working with the states to help implement those policies.

REFERENCES

Abelson, P. H. 1984. Groundwater contamination. *Science*, **224**, 673.

Anonymous. 1983a. Who will Shell-out to clean up nerve gas? *New Scientist*, 13 October, p. 74.

Anonymous. 1983b. Toxics suit asks $1.8 billion. *Engineering News-Record*, **211**, 24.

Anonymous. 1984. Colorado toxics fix. *Engineering News-Record*, **213**, 16.

Anonymous. 1988. Costs set for arsenal plan. *Engineering News-Record*, **220**, 11.

Anonymous. 1990. Superfund's worst site begins to come clean. *Engineering News-Record*, **225**, 39.

Burmaster, D.E. 1982. The new pollution. *Environment*, **24**(2), 6–13, 33–36.

Council on Environmental Quality. 1981. *Contamination of ground water by toxic organic chemicals*. Washington, D.C. 84 pp.

Council on Environmental Quality. 1991. *Environmental Quality*. 21st Annual Report. U. S. Government Printing Office, Washington, D.C. 388 pp.

Clay, D. R. 1990. Hazardous waste sites. *Science*, **247**, 1166.

Culliton, B. J. 1980. Continuing confusion over Love Canal. *Science*, **209**, 1002–1003.

Dawson, J. W. 1979. State Groundwater Protection Programs—Inadequate. In *Proceedings of the Fourth National Ground-water Quality Symposium*, Minneapolis, MN, 20–22 September, 1978. pp. 102–108.

Durso-Hughes, K. and J. Lewis. 1982. Recycling hazardous waste. *Environment*, **24**, 14–20.

Environmental Protection Agency. 1977a. Waste Disposal Practices and Their Effects on Ground Water. Report to Congress, prepared by the Office of Solid Waste Management programs, 512 pp.

Environmental Protection Agency. 1977b. *Multimedia Environmental Goals for Environmental Assessment*. EPA-600/7-77-136, Washington, D.C.

Environmental Protection Agency. 1980a. *Groundwater Protection*. Washington, D.C., 36 pp.

Environmental Protection Agency. 1980b. Proposed Ground Water Protection Strategy. Office of Drinking Water, Nov. 1980, 61 pp.

Environmental Protection Agency. 1980c. Planning workshops to develop recommendations for a groundwater protection strategy. Office of Drinking Water, Washington, D.C., 171 pp.

Environmental Protection Agency. 1980d. *Everybody's Problem, Hazardous Waste*. Office of Water and Waste Management, Washington, D.C., 36 pp.

Environmental Protection Agency. 1984. *Ground-water Protection Strategy*. Office of Ground-water Protection, Washington, D.C., 56 pp.

Environmental Protection Agency. 1986. *Guidelines for Ground-water Classification Under the EPA Ground-water Protection Strategy*, final draft. Office of Ground-water Protection, Washington, D.C., 137 pp.

Environmental Protection Agency. 1987. Drinking Water; Proposed Substitution of Con-taminants and Proposed List of Additional Substances Which may Require Regulation Under the Safe Drinking Water Act. *Federal Register*, **52**(130), 25719–25723.

Environmental Protection Agency. 1988. *The Safe Drinking Water Act*. San Francisco, 50 pp.

Epstein, S. S., L. O. Brown, and C. Pope. 1982. *Hazardous Waste in America*. Sierra Club Books, San Francisco, CA, 593 pp.

Gallaher, B. M., and M. S. Good. 1981. Water Quality Aspects of Uranium Mining and Mill-ing in New Mexico. Special Publication No. 10, New Mexico Geological Society, Santa Fe, NM. pp 85–91.

GRC. 1984. *Studies in Geophysics: Ground-water Contamination*. Geophysics Research Forum. National Academy Press, Washington, D.C. 179 pp.

Hileman, B. 1982. Nuclear waste disposal: A case of benign neglect? *Environmental Science Technology*, **16**(5), 271A–275A.

Konikow, L. F. and D. W. Thompson. 1984. Groundwater contamination and aquifer reclamation at the Rocky Mountain Arsenal, Colorado. In *Groundwater Contamination*, National Research Council Geophysics Study Committee, National Academy Press, Washington, D.C. pp. 93–103.

Koshland, D. E., Jr. 1991. Toxic chemicals and toxic laws. *Science*, **253**, 949.

Lehr, J. H. 1982. How much groundwater have we really polluted? *Ground Water Monitoring Review*, Winter, pp. 4–5.

Lowther, W. 1983. Turmoil at EPA. *Macleans*, **96**, 50.

Maugh, T. 1982. Just how hazardous are dumps? *Science*, **215**, 490–493.

Muller, D. W. 1982. cited by Maugh (1982), p. 491.

Patrick, R., E. Ford, and J. Quarles. 1987. *Groundwater Contamination in the United States*. 2nd ed. University of Pennsylvania Press, Philadelphia, PA. 513 pp.

Petri, L. R., and R. O. Smith. 1956. Investigation of the quality of groundwater in the vicinity of Derby, Colorado. U.S. Geol. Survey Open File Rept., 77 pp.

Pollock, C. 1987. *Mining Urban Wastes: The Potential for recycling*. Worldwatch Paper 76. Washington, D.C. 58 pp.

Polsgrove, C. 1982. The waste exchange option. *Environment*, **24**, 15, 37–41.

Pye, V. I., and R. Patrick. 1983. Ground water contamination in the United States. *Science*, **221**, 713–718.

Pye, V. I., R. Patrick, and J. Quarles. 1983. *Groundwater Contamination in the United States*. University of Pennsylvania Press, Philadelphia, PA. 315 pp.

Roberts, P. 1981. Keeping America's water drinkable. *Mainliner Magazine*, June, pp. 58, 86, 89, 112.

Rubin, D. K., and S. W. Setzer. 1990. The Superfund decade: Triumphs and troubles. *Engineering News-Record*, **225**, 38–44.

Sittig, M. 1980. *Priority Toxic Pollutants*. Noyes Data Corp., Park Ridge, NJ, 370 pp.

Sittig, M. 1985. *Handbook of Toxic and Hazardous Chemicals and Carcinogens*. 2nd edition. Noyes Publications, Park Ridge, NJ. 950 pp.

Solley, W. B. 1989. Reflections on water use in the United States. In *U. S. Geological Survey Yearbook for Fiscal Year 1988*. U. S. Geological Survey, Washington, D.C. pp. 28–30.

Stipp, D. 1991. Throwing good money at bad water yields scant improvement. *Wall Street Journal*. May 15.

Sun, M. 1982. EPA relaxes hazardous waste rules. *Science*, **216**, 275–276.

Travis, C. C., and C. B. Doty. 1990. Can contaminated aquifers at Superfund sites be remediated? *Environmental Science and Technology*, **24**(10), 1464–1466.

Wood, L. A. 1972. Groundwater degradation—causes and cures. In *Proceedings, 14th Water Quality Conference*, Urbana, IL. pp. 19–25.

Wood, E. F., R. A. Ferrara, W. G. Gray, and G. F. Pindar. 1984. *Ground Water Contamination from Hazardous Wastes*. Prentice-Hall, Englewood Cliffs, NJ. 163 pp.

17

PLASTICS IN THE SEA

I hopes ya swabs won't be throwing no plastics overboard. POPEYE THE SAILOR, on a poster aimed at educating the public about the adverse affects caused by disposal of plastic waste in the ocean.

THE NATURE OF THE PROBLEM

Plastics synthesized from petrochemicals were first produced more than a century ago, but the most significant growth of the commercial plastics industry did not begin until World War II, when shortages of rubber and other materials created a demand for substitute products. Plastics were used to produce items such as bugles, canteens, and dinnerware for the military, and in the private sector products such as nylon stockings and Saran Wrap became overnight sensations. The growth in the diversity of uses for plastic continued after the war with the development of products such as Tupperware, nylon zippers, and acrylic dentures. This rapid expansion of the plastics industry has continued more-or-less unabated to the present time (Figure 17.1). By 1988 plastics production in the United States had reached almost 27 million tonnes per year, or about 110 kg for every person in the United States (O'Hara et al., 1988).

The magnitude of plastics production and use has, not surprisingly, created a significant solid waste disposal problem. Until recently virtually all plastic products were nonbiodegradable. Although plastics do break down as a result of physical and chemical weathering, under most circumstances the breakdown process is slow. Plastics discarded in the aquatic environment may remain in largely unaltered form for years if not decades. In this respect plastics resemble certain other synthetic organics such as PCBs in terms of persistence.

Plastic debris takes basically two forms. So-called raw plastic consists of small resin pellets a few millimeters in diameter. These small pellets are convenient to ship and are ultimately melted down and molded into manufactured products. It

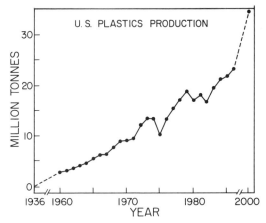

Figure 17.1   Plastics production in the United States during the twentieth century. [*Source*: O'Hara et al. (1988).]

has been quite common, however, for raw plastic pellets to be discharged with wastewater from manufacturing plants or to be lost in the process of shipment. The resemblance of these pellets to fish eggs has undoubtedly been partly responsible for their ingestion by certain organisms that feed on plankton. Plastic pellets have been reported in concentrations as high as 3500 and 34,000 km^{-2} in the surface waters of the Atlantic and Pacific Oceans, respectively (O'Hara et al., 1988).

The second major category of plastic debris is of course lost or discarded manufactured products. Only rough estimates can be made of the amount of this debris. Approximately 10 million tonnes of solid waste are discharged into the ocean each year, and of this figure roughly 10% is plastic (O'Hara et al., 1988). Hence about 1 million tonnes of plastic debris are introduced into the ocean each year, or about 100 tonnes per hour. The nature of this debris may vary greatly from one area to another. Certainly a major problem in some parts of the ocean has been lost or discarded fishing nets and traps. Lost sections of gill nets, for example, may drift in surface waters and continue to entangle fish for literally years until they eventually sink due to the weight of accumulated carcasses. Lost lobster and crab pots may also attract and catch organisms, perhaps over an even longer time frame. In areas of offshore oil production, it is common for plastic debris to consist largely of items discarded or lost from oil platforms, from oil drilling operations, and from ships engaged in seismic studies. There is a general consensus that at least historically most of the plastic debris in the ocean has come from ships, but in some areas much of the nearshore debris has a significant land-based origin. This condition may exist, for example, near coastal municipalities which have combined sewer systems.

EFFECTS

Aesthetics

Certainly one of the most obvious adverse effects associated plastic debris in the ocean is aesthetic. Probably more than any other factor, the appearance of plastic debris on recreational beaches has sensitized the public to the problem of plastic pollution. Annual beach cleanups sponsored by environmental groups have become common in all coastal states. The amount of debris picked up by volunteers during these cleanups is remarkable. In October, 1984, for example, volunteers picked up almost 24 tonnes of debris during a 3-hour period along 560 km of Oregon coastline. Included in the debris were 48,900 chunks of polystyrene larger than a baseball; 5300 food utensils; 4900 bags or sheets of plastic; 4800 plastic bottles; 2000 plastic strapping bands; 1500 six-pack rings; and 1100 pieces of fishing gear. Along a 240-km stretch of North Carolina beach, 8000 plastic bags were found in 3 hours, and more than 15,600 six-pack rings were picked up in 3 hours on a 480-km section of Texas coastline (O'Hara et al., 1988). Although certainly not all beach litter is plastic, national studies of the composition of beach debris indicate that 55–65% indeed consists of plastic items (O'Hara et al., 1988). Although only a small percentage of the items picked up in beach cleanups appear to have been left behind by beach users (CEE, 1987), beachgoers are not without fault. In Los Angeles County, for example, beach users leave behind about 68 tonnes of trash each week (O'Hara et al., 1988).

Perhaps the most famous incident of plastic pollution on recreational beaches was the "floatable episode" of June, 1976 when large amounts of debris washed up on beaches on Long Island, New York (Swanson et al., 1978). Plastic items accounted for most of the debris and consisted mainly of tampon applicators (more than 300/km of beach), condoms, sanitary napkin liners, and disposable diapers. Other plastic items included straws, pieces of styrofoam cups, plastic bottle caps, corks, plastic toys, and plastic cigar and cigarette tips. Within 9 days after the debris first began to appear, all beaches were closed to swimming, shellfishing was banned, and the Governor of New York declared most of Long Island a disaster area. The incident was apparently caused by a combination of factors, including combined sewer overflows, ocean dumping of sewage sludge, disposal of trash by commercial and recreational vessels, unusually heavy rainfall, and onshore winds (O'Hara et al., 1988). A similar incident, or more correctly series of incidents, occurred during the summer of 1988, when large amounts of garbage, including needles, syringes, and other medical wastes, washed up on Long Island and New Jersey beaches. Estimates of the economic costs associated with the 1988 incident, including impacts on tourism and recreational fishing, have ranged as high as $3 billion (Wagner, 1990).

Ingestion

Marine organisms sometimes consume plastic debris, either inadvertently in the process of feeding or deliberately because they mistake the debris for food. The extent to which marine organisms have been killed or stressed as a result of plastics ingestion is very poorly known, because an autopsy is necessary to confirm the problem. The best-documented cases of plastics ingestion have undoubtedly involved sea turtles, who presumably mistake plastic bags and sheeting for jellyfish, a common prey. At least five species of turtles are known to consume plastics (Balazs, 1985). In one case on Long Island, 11 of 15 dead leatherback turtles that washed ashore during a 2-week period had plastic bags blocking their stomach openings. Ten of the turtles had ingested four, 7.6-liter-sized bags, and one had eaten 15 bags (Anonymous, 1982). Another leatherback found in New York had ingested 46 m of monofilament fishing line (CEE, 1987).

In addition to sea turtles, there is documentation of ingestion of plastics by nine species of whales (CEE, 1987). As with the turtles, most of the plastic items have been bags and sheeting, although in one case about a liter of tightly packed trawl net was found in the stomach of a sperm whale. Most of the information concerning plastics consumption by whales has come from dead animals that had stranded, and in most cases it has not been clear that the ingested plastics were the cause of death. In 1985, however, "a young sperm whale was found dying on the shores of New Jersey as the result of a mylar balloon lodged in its stomach and three feet of purple ribbon wound through its intestines" (O'Hara et al., 1988, p. 22); in Hawaii a captive dolphin died from ingesting a piece of membrane plastic (CEE, 1987); and a pygmy sperm whale that was taken into captivity after stranding along the Texas coast later died due to ingestion of paper and plastic bags (Jones et al., 1986).

Consumption of plastics by sea birds has received much attention in recent years, and a comprehensive review of the subject has been reported by Day (1985). Plastic objects, most commonly raw polyethylene pellets, have been found in the digestive tracts of at least 50 of the world's 280 species of marine birds. Birds which feed primarily on crustaceans or squid tend to have the highest incidence of plastics ingestion, and the color, shape, and/or size of the ingested objects follow a pattern that suggests that the birds are selective consumers of plastics. In other words, the birds are not consuming the plastic debris by accident. The parakeet auklet, for example, which feeds mainly on planktonic crustaceans, selectively consumes light-brown plastic objects that are primarily cylindrical, spherical, box- or pill-shaped (CEE, 1987). Only a small percentage of marine birds ingest dark-colored plastic debris. Of particular concern is the fact that some birds feed plastic debris to their young (Fry and Fefer, 1986). For example, 90% of Laysan albatross chicks examined on Midway and Oahu Islands contained plastic objects. Ingested plastic debris, whether in adults or chicks, can fill a bird's stomach to such an extent that the bird no longer feeds properly due to false feelings of satiation. In some cases the debris may cause serious internal injuries.

Finfish are also known to consume plastic debris, although documentation is largely anecdotal. The problem seems to be most common with demersal fish, plastic pellets being the form of debris most frequently mentioned. Individual juvenile flounder, for example, have been reported to contain as many as thirty, 2–5-cm-diameter plastic pellets in their intestines (Kartar et al., 1973; 1976). The presence of such pellets may cause intestinal blockage in small fish (CEE, 1987).

Entanglement

Entanglement with plastic debris occurs when a marine organism accidentally or intentionally comes into contact with debris and becomes ensnared. Incidents of entanglement are the most easily documented of the biological impacts of plastic debris and undoubtedly the most publicized. Probably the most serious impacts have resulted from the accidental loss or in some cases intentional disposal of fishing nets or parts of fishing nets. Both gill nets and trawl nets have been involved. Much of the publicity concerning this problem came from studies in the North Pacific, where the total length of all gill nets used in the 15 major drift net fisheries during the 1980s was about 170,000 km, more than 4 times the circumference of the earth (CEE, 1987). Particular attention was focused on the North Pacific high seas drift net fisheries conducted by Japan, Taiwan, and South Korea. Every night during the fishing season each vessel involved in this fishery set a 9–27-km-long drift gill net (Eisenbud, 1985), which was picked up the following day. The total length of these drift nets was about 33,000 km.

The use of these nets to catch species of squid, salmon, and billfish proved to be highly controversial for a variety of reasons. Many of the fish that become entangled in the nets are not utilized in the catch, either because they are not recovered when the net is retrieved or because they are considered not worth saving and are simply thrown back into the ocean. Fisheries data furthermore indicate that the drift nets are so efficient at catching fish that some of the most important fishery resources in the North Pacific were being seriously depleted. These strictly fishery issues are not directly relevant to the problem of plastics pollution. However, when nets or pieces of nets are lost or discarded, they become marine debris.

The quantity of nets lost or discarded each year is not well known, most information on the subject being anecdotal or semiquantitative at best. Estimates of 0.05–0.06% per set have been made for the Japanese salmon pelagic gill net fishery, and total net losses from Japanese North Pacific squid and salmon fisheries have been put at 2500 km per year (Breen, 1990). In the New England groundfish gill net fishery, gill nets with a total length of typically 900–1100 m are anchored on the bottom to catch demersal fish (CEE, 1987). Since these nets remain submerged when lost, observations of lost nets must be made with a submersible vessel. A 1984 survey of prime demersal fishing sites in the Gulf of Maine by the submersible *Johnson Sea-Link II* revealed 10 lost nets in an area a little over 10 ha (Carr et al., 1985).

Trawl nets account for 3–4 times as many fish caught by weight as do gill nets on a worldwide basis[1] and, like gill nets, are subject to loss and breakage, particularly when used in bottom trawling operations. Roughly 5500 km of nets are used by 12 major North Pacific fisheries (CEE, 1987), and although this figure is quite small compared to the length of nets used in the gill net fisheries, studies of beach debris on Amchitka Island in the Aleutian Islands from 1972 to 1982 revealed that trawl net fragments accounted for 76–85% of the plastic litter collected (Merrell, 1985). Low et al. (1985) have estimated that during the early 1980s 50–70 trawl nets or large portions of trawl nets were lost each year by fishing vessels off Alaska.

In addition to their use in fishing nets, plastics have also become an important construction material in traps designed to catch crabs and lobsters. A standard size wooden lobster trap, for example, typically includes about 0.4 m^2 of plastic netting (CEE, 1987). In the New England lobster fishery, about 2.5 million traps are used annually, and of these at least 20% or 500,000 are lost each year. King crab fishermen in Alaska lose about 10% of their crab pots each year, and in the Gulf of Mexico annual trap losses in the stone crab fishery average 25% (CEE, 1987).

Ghost Fishing

Entanglement in lost or discarded fishing gear is usually referred to as ghost fishing. Plastic fishing nets, which are typically made of nylon, polypropylene, or polyethylene, are often difficult to discern in the water even during daylight hours, and in the case of gill nets may drift for literally years before washing ashore or sinking. Entanglement in most cases is undoubtedly accidental, but in some cases it appears that marine organisms are unwittingly attracted to net debris. Sea turtles, for exam-

---

[1] Purse seines, trawl nets, and gill nets account for roughly 45%, 22%, and 6% of the world fish catch. The remainder of the catch is taken by lines (9%) and by pound and other trap nets (9%).

ple, may become entangled in floating net fragments because of their natural attraction to sargassum mats and other natural floating masses that provide shelter and offer concentrated food sources (CEE, 1987). The impact of entanglement on marine organisms has undoubtedly been most thoroughly studied in the case of the northern fur seal population of the Pribilof Islands. The population was seriously depleted during the latter half of the nineteenth century due to overharvesting, but an international ban on harvesting in 1911 allowed the population to recover during the next 30 years. The population reached the apparent carrying capacity of the ecosystem of about 2 million individuals by 1940 and remained at that level through most of the 1950s. Some harvesting of females was permitted from 1956 to 1968. It was assumed that when the harvesting ceased the population would return to the carrying capacity. Instead the population declined at a rate of 4–8% per year (Figure 17.2). The decline has been attributed largely to entanglement, the estimated entanglement loss being at least 50,000 seals per year (CEE, 1987). The impact appears to have been heaviest on seals younger than 2 years (French and Reed, 1990). Current estimates indicate that the number of northern fur seals is no more than one-third the number that existed during the 1940s and 1950s.

Although much of the concern relative to ghost fishing has been focused on the northern fur seal population, almost half of the 34 extant pinniped species (seals, sea lions, and walruses) are known to have become entangled in marine debris (Henderson, 1990). Particularly noteworthy has been the entanglement of Hawaiian monk seals, an endangered species. Almost all documented incidents have involved lost or discarded fishing gear (Henderson, 1990). As is the case with the northern fur seals, juveniles appear to be more susceptible than adults. Other pinnipeds involved in entanglement incidents include California and Stellar sea lions, northern elephant seals, Cape and New Zealand fur seals, and harbor seals (CEE, 1987; Stewart and Yochem, 1990). Fortunately entanglement does not appear to be having a significant impact on pinniped demography and population dynamics in all cases (Stewart and Yochem, 1990).

Birds and cetaceans are also known to become entangled in fishing nets, but in both cases most losses are associated with active fishing gear rather than lost or discarded nets. During the 1980s, for example, the Japanese salmon gill net fishery was estimated to kill more than 250,000 seabirds each year in U.S. waters during the 2-

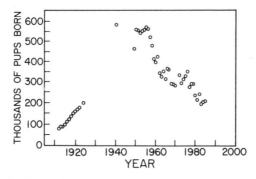

Figure 17.2 Reproduction of northern fur seals during the twentieth century. [*Source*: French and Reed (1990).]

month fishing season (King, 1985). The birds were apparently attracted to fish caught in the net. Ghost-fishing nets attract birds in a similar manner, and 100 or more dead birds have been found entangled in large pieces of gill net debris (Jones and Ferrero, 1985). A 1.5-km section of a 3.5-km long derelict gill net retrieved in the North Pacific contained 99 dead seabirds, 2 salmon sharks, 1 ragfish, and 75 recently entangled salmon (DeGange and Newby, 1980). The net was estimated to have been adrift for about a month. Cetaceans presumably become entangled in nets or pieces of netting as a result of unintentional collisions. DeGange and Newby (1980) found two dead Dall's porpoises in a 3.5 km derelict gill net off Amchitka Island.

Lost or discarded lobster, crab, and demersal fish traps are the other form of plastic debris principally associated with ghost fishing. In the simplest form of trap ghost fishing, captured animals die in lost traps, and their bodies act as bait (Breen, 1990). However, captured live animals may also attract other organisms to a lost trap, and in some cases organisms are apparently even attracted to empty traps, perhaps because they perceive the traps as a source of shelter (Breen, 1990). Death presumably results as a result of either starvation or cannibalism. Some estimates of trap ghost fishing loss rates are summarized in Table 17.1. It is apparent from this table that losses to trap ghost fishing can be quite substantial.

Other Causes of Entanglement

Other than lost or discarded nets and traps, the principal sorts of plastic associated with entanglement are monofilament fishing line, packing straps, and six-pack rings. Monofilament fishing line, for example, is the most common type of debris reported to entangle sea turtles (Balazs, 1985), the affected species including green, hawksbill, olive ridley and leatherback turtles, all of which are endangered or threatened species (CEE, 1987).

Brown pelicans, another protected species, become entangled in both monofilament fishing line and six-pack rings. Entanglement of these birds is a major problem in both California and Florida (Gress and Anderson, 1983). The pelicans actively collect pieces of nets and fishing line for nest material (Bourne, 1977) and apparently become entangled in six-pack rings when they try to dive or feed through the rings (CEE, 1987). Stewart and Yochem (1990) report that packing straps are the most common cause of entanglement of northern elephant seals in the Southern California Bight.

Table 17.1 Estimated Loss Rates Due to Trap Ghost Fishing

| Fishery | Annual loss rate | |
	Weight	Value
Newfoundland snow crab	10 tonnes	
U.S. American lobster	670 tonnes	$2.5 million
British Columbia Dungeness crab	7% of landings	$80,000
British Columbia sablefish	300 tonnes[1]	

[1]Commercial catch is 1000–4000 tonnes per year.

Source. Breen (1990).

Damage to Vessels

Damage to vessels from marine debris results from collisions with floating objects, entanglement of debris in propeller blades, and clogging of water intakes for engine cooling systems. The amount of such damage caused by plastic debris is not well known. Takehama (1990) has estimated the annual cost of damage to Japanese fishing vessels caused by floating debris to be roughly 4 billion yen, which is about 0.2% of the total cost of operating the vessels. The contribution of plastic debris to this damage is unknown. According to the U.S. Coast Guard about 10% of vessel accidents in U.S. waters are caused by debris (CEE, 1987). Unfortunately, plastic debris is not specifically coded for in the data analysis, and hence there is no way to judge the contribution of plastics to the problem. Anecdotal information, however, indicates that plastic garbage bags are considered to be the leading cause of engine damage of commercial and recreational vessels in New England waters, and the fact that some boating supply companies have built devices on propellers to combat entanglement indicates that the problem is of more than infrequent occurrence (CEE, 1987).

CORRECTIVES

Solutions to the problems caused by plastics in the sea can be broadly classified as legal, technological, and educational. The most important legal developments in recent years have been adoption of a resolution by the United Nations General Assembly requiring all nations conducting open ocean high-seas driftnet fishing to cease all driftnetting by December 31, 1992 (Anonymous, 1992) and the ratification of Annex V of the International Convention for the Prevention of Pollution from Ships (MARPOL) by the United States on December 31, 1987. The end of drift net fishing was the result primarily of fisheries considerations rather than concern over pollution. The nets took many nontarget species, and their extensive use, particularly in the North Pacific, posed a serious threat of overfishing. The disappearance of this method of fishing, however, also eliminated one of the important sources of plastic debris in the ocean, namely lost or discarded portions of drift nets. Unfortunately the lack of quantitative information concerning the impact of these derelict gill nets on marine organisms will make difficult an assessment of the effect of this development from a pollution standpoint.

MARPOL Annex V

Ratification of Annex V of MARPOL by the United States was a particularly noteworthy event, because Annex V is an optional component of MARPOL and could come into force only when ratified by countries representing 50% of the world's shipping tonnage.[2] The U.S. vote put the percentage of shipping tonnage associated with ratifying nations over the 50% mark. The legislation actually went into effect on December 31, 1988.

[2]Annexes I and II of MARPOL, which deal with oil discharges from ships and the transport of hazardous liquids, respectively, are automatically adopted by any country that signs onto MARPOL. Annexes III, IV, and V are all optional and concern the transport of hazardous materials in packaged form, the discharge of sewage from ships, and the discharge of garbage from ships, respectively.

As summarized in Table 17.2, Annex V concerns at-sea disposal of various forms of garbage from offshore platforms and privately owned vessels. Public vessels (e.g., military) are exempt from Annex V restrictions, but are expected to comply to the extent possible. The legislation actually permits disposal from vessels of most forms of garbage in most parts of the ocean, the restricted regions being primarily coastal regions and special areas designated by the Convention. Disposal of plastics, however, is prohibited by Annex V in all parts of the ocean.

While Annex V would superficially seem to be an important step in controlling plastic pollution of the ocean, it is one thing to legislate and another thing to implement. For years ships had been accustomed to disposing of their garbage, including plastics, at sea. How were vessels going to dispose of their plastic waste after December 31, 1988?

Recognizing this problem, the Marine Environment Protection Committee (MEPC) of the International Maritime Organization (IMO) developed guidelines for the implementation of Annex V at consecutive sessions in 1987 and 1988. The purpose of the guidelines was to assist governments that had ratified Annex V in developing and enacting domestic laws that would enforce and implement Annex V; assist vessel operators in complying with the requirements of Annex V and domestic laws; and assist port and terminal operators in determining the need for and providing adequate reception facilities for garbage generated by ships (Edwards and Rymarz, 1990). The guidelines fell into six general areas as follows.

Table 17.2 Summary of MARPOL Annex V at-sea Garbage Disposal Regulations

Garbage type	Private vessels Outside Special Areas	Private vessels Inside Special Areas[1]	Offshore Platforms
Plastics—including synthetic ropes, fishing nets, and plastic garbage bags	Disposal prohibited	Disposal prohibited	Disposal prohibited
Floating dunnage, lining and packing materials	>25 nautical miles offshore	Disposal prohibited	Disposal prohibited
Paper rags, metal, bottles, crockery, and similar refuse	>12 nautical miles offshore	Disposal prohibited	Disposal prohibited
All other garbage, including paper rags, glass, etc., comminuted or ground[2]	>3 nautical miles offshore	Disposal prohibited	Disposal prohibited
Food waste not comminuted or ground	>12 nautical miles offshore	>12 nautical miles offshore	Disposal prohibited
Food waste comminuted or ground	>3 nautical miles offshore	>12 nautical miles offshore	>12 nautical miles offshore
Mixed refuse types[3]	3	3	3

[1]Middle Eastern Gulf areas and the Mediterranean, Baltic, and Black Seas have been designated special areas.

[2]Comminuted or ground garbage must be able to pass through a screen with a mesh size no larger than one inch.

[3]When garbage is mixed with other harmful substances having different disposal or discharge requirements, the most stringent disposal requirements apply.

Source. Edwards and Rymarz (1990).

1. *Training, Education, and Information.* Educating both seafarers and port authorities about the requirements of Annex V, the environmental impact of at-sea garbage disposal, and the need for the development of better vessel and port waste management facilities was perceived as essential for the successful implementation of Annex V. Knowledge of national and international regulations governing garbage disposal at sea are now, for example, expected to be a part of maritime certification examinations and requirements. Ships are expected to post a summary declaration stating the restrictions for discharging garbage at sea and the penalties for failure to comply (Edwards and Rymarz, 1990). Curricula at maritime colleges and technical institutes are to include information on environmental impacts of garbage in the marine environment. Efforts to educate the general public about Annex V are to be made through radio and television broadcasts, articles in periodicals and trade journals, voluntary public projects such as beach cleanups, posters, brochures, and so forth. Figure 17.3, an educational sticker for boaters developed by the Center for Marine Conservation, is one example of the sort of educational material that has been developed as a part of this effort. An active exchange is being encouraged between governments and IMO of technical information on shipboard waste management, copies of current domestic laws and regulations related to Annex V, educational materials developed to raise the level of compliance with Annex V, and information on the nature and extent of marine debris.

2. *Minimizing Garbage Production.* One obvious way to facilitate a reduction in garbage disposal is to decrease the amount of garbage produced. In the case of plastics, this reduction may be accomplished to a certain extent by switching to reusable packaging and containers and by packaging provisions in something other than disposable plastic. Bulk packaging of consumable items may also help, at least if the product has a sufficiently long shelf life. Fishing vessel operators are encouraged to develop methods to minimize accidental encounters between ships and gear. Benthic traps, trawls, and gill nets should be designed to have degradable panels or sections made of natural fibers, wood, or wire. Efforts should be made to develop methods for recycling plastic waste returned to shore as garbage.

3. *Shipboard Garbage Handling and Storage Procedures.* Compliance with Annex V will obviously require that ships have facilities for collecting, sorting, processing, storing, and disposing of garbage. Waste management plans should be incorporated into crew- and vessel-operating manuals. Garbage collection devices must allow convenient separation of different forms of waste according to mandated disposal methods. There should be a clear separation, for example, between plastic waste and food waste. On-board processing facilities may include incinerators, compactors, and comminuters. Compactors are appealing because they reduce the volume of waste and hence facilitate storage. Compactors also facilitate sinking if garbage is discharged, and comminuters enhance assimilation of food wastes and other discharged garbage, but these latter considerations are irrelevant to plastics, which cannot be disposed of at sea. Incineration might seem an easy way to avoid collecting and storing plastic waste at sea, but there is a problem. According to the 1988 MEPC guidelines, "Ash from the combustion of some plastic products may contain heavy metal or other residues which can be toxic and should therefore not be discharged into the sea. Such ashes should be retained on board, where possible,

MARPOL Annex V
Sticker Produced for Boaters

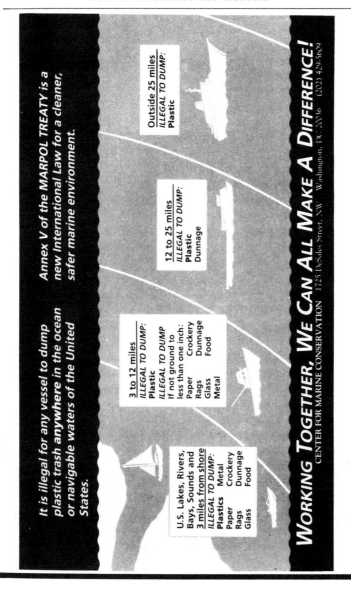

Figure 17.3 MARPOL Annex V educational poster produced by the Center for Marine Conservation.

and discharged at port reception facilities" (Bean, 1990, p. 993). Indeed plastics are believed to account for 70 and 90% of the lead and cadmium, respectively, in municipal incinerator ash due to the use of these metals in stabilizers and pigments, and the fly ash from such incinerators often qualifies as a hazardous waste under the Federal Resource Conservation and Recovery Act due to its high Pb and Cd content (Bean, 1990). In general, then, ship operators should plan on retaining the ash from plastics incineration.

4. *Shipboard Equipment for Processing Garbage.* Shipboard equipment that might be used in processing plastic waste include comminuters, compactors, and incinerators. Comminuters are relevant because it is usually impractical to compact uncomminuted plastics. Much manual labor is required to size the material for feed. Comminuted plastics, however, can be rapidly compacted to produce a medium density material, which requires minimal onboard storage space (Edwards and Rymarz, 1990). Incineration can reduce the volume of plastic waste by about 95%, but produces a potentially hazardous ash and of course eliminates any possibility of plastics recycling. In addition, combustion of plastics may produce dangerous gases such as hydrochloric acid vapor, and if adequate levels of oxygen are not supplied to the incinerator, high levels of soot will form in the exhaust stream (Edwards and Rymarz, 1990).

5. *Port Reception Facilities.* Upgrading port reception facilities to handle plastic waste is a key to the successful implementation of Annex V. It is critical that vessel operators know the capabilities of various ports to handle waste so that they can intelligently plan and direct on-board collection, processing, and storage. It would obviously be futile for a ship operator to carefully commute, compact, and store plastic waste if the next port of call could not accept the waste. With respect to this point it is noteworthy that of the special areas designated in MARPOL Annex V, restrictions were initially enforced only in the Baltic Sea. Dumping restrictions were not enforced in the other areas until the regions could demonstrate proper reception facilities and subsequently satisfy a 1-year waiting period (O'Hara et al., 1988).

6. *Ensuring Compliance.* Compliance with Annex V will to a large extent be voluntary, because it is obviously impractical to place an observer on every seagoing vessel. The degree of compliance will depend on the extent to which seafarers appreciate and understand the problems caused by improper disposal of garbage at sea. This perception and understanding will come through various forms of training and education. Also important will be the need to create economic incentives to comply with Annex V, to reduce the amount of vessel-generated garbage, and to recycle. A passenger ship operator, for example, should be rewarded for switching from disposable polystyrene cups to paper cups, but the latter presently cost 2–3 times as much (Martinez, 1990). Imaginative strategies for reducing the amount of waste generated at sea, minimizing the use of plastics, and recovering and recycling plastics where possible should be rewarded economically.

Other Legislation

At about the same time as the ratification of MARPOL Annex V, several other pieces of legislation relevant to the problem of plastic debris were passed in the United States. The Marine Plastic Pollution Research and Control Act (MPPRCA), Public Law 100-220, went into effect at the same time as Annex V and was intended to implement the provisions of Annex V. The MPPRCA prohibits the disposal of plastics anywhere at sea by U.S. vessels and prohibits the disposal of plastics by any vessel in U.S. waters, including bays, sounds, inland waterways, and the U.S. Exclusive Economic Zone (i.e., the coastal ocean within 200 nautical miles of the U.S. shoreline). The MPPRCA also directs all federal agencies, including the Navy and Coast Guard, to bring their vessels into full compliance with Annex V regulations by 1994.

Sometimes controversial legislation has been proposed or passed concerning the use of degradable plastics. Plastic products can be made either photodegradable or biodegradable, in the former case by changing the molecular structure of the plastic so that it decomposes in the presence of ultraviolet light and in the latter case by adding cornstarch or other vegetable additives that can be broken down by microorganisms. A number of states have passed laws requiring that six-pack rings be degradable (O'Hara et al., 1988), and similar federal legislation was passed by Congress in 1988. The federal law, however, has not to date been implemented. Other degradable plastic items, such as shopping bags, have become available in recent years.

The logic behind degradable plastics is that degradation will render the plastic harmless, or at least less of a threat to the environment. This wisdom has been questioned, however, by some environmentalists. They argue that the production of degradable plastic products encourages rather than discourages the use of plastics, and that under certain conditions degradable plastics are no more degradable than ordinary plastics. The deposition of plastic debris in landfills, for example, virtually precludes photodegradation, and biodegradation is minimal in well-operated landfills due to the absence of air and water in the compacted waste (Denison, 1990). Furthermore, it is only the cornstarch or vegetable additives that actually degrade in biodegradable plastics. The plastic dust that remains following biodegradation may be less visible and pose no threat of entanglement, but it may be more of a threat from the standpoint of ingestion than the undegraded product. On the other hand, a requirement that fishing nets and traps include degradable components might do much to reduce the loss of marine life caused by ghost fishing. The consensus would seem to be that degradable plastics may prove useful in reducing the adverse effects associated with some types of plastic debris, but they are no panacea. Indeed there are some uses of plastics for which degradability would clearly be undesirable. For example, hazardous liquids should certainly not be stored in a degradable plastic bottle.

Solutions Through Technology

An obvious way to reduce the amount of plastic debris is to recycle plastic products. This idea is not as easy to implement, however, as you might think. Plastic is a generic term covering hundreds of different types of products. Each type of plastic, for example, polypropylene, polyethylene, and polyvinyl chloride, has characteristics that make it best suited for certain applications. You cannot simply throw all sorts of plastic products into a pot, melt them down, and then mold them into useful products.

Nevertheless, the concept of plastic recycling is not new, and some forms of plastic have been recycled for more than a decade. Recycling of plastic milk bottles made from high-density polyethylene began during the mid-1970s. Recycling was facilitated by the facts that the bottles were easy to identify, they were all made from the same type of plastic, and there was a large enough supply to make recycling economical. Similar recycling of plastic soft-drink bottles, all of which are made from polyethylene terephthalate (PET), began in 1979. As of 1987 plastic milk and soft drink bottles accounted for about 40% of all plastic bottles sold in the United States, and in that year almost 70,000 tonnes of PET bottles, or 20% of all those sold, were recycled (O'Hara et al., 1988).

Plastic bottles are recycled in a variety of products, but because of concerns about purity and food contact, they are not recycled as beverage containers. Plastic milk bottles, for example, are recycled as plastic lumber, underground pipes, toys, pails, traffic barrier cones, garden furniture, golf bag liners, kitchen drain boards, milk bottle carriers, trash cans, and signs (O'Hara et al., 1988). PET is recycled primarily as fiberfill, which is used to stuff pillows, ski jackets, sleeping bags, and automobile seats (O'Hara et al., 1988). Obviously recycling of certain types of plastics has proven to be feasible. What is needed now is a significant increase in the quantity and variety of plastic products which are recycled. Among the obvious problems to be solved are the development of methods to separate different types of used plastic and the identification of suitable products that can be made from recycled plastics. The importance of solving these problems was underscored in 1984 with the formation of the Plastics Recycling Foundation in the United States. The Foundation has established a Center for Plastics Recycling Research at Rutgers University to carry out research and development of methods for recycling plastic products.

A somewhat controversial issue is the significance and source of plastic resin pellets. Such pellets are a minor component of the plastic debris in the ocean, accounting for only about 0.5%, for example, of the pieces of plastic collected in the North Pacific; but they account for about 70% of the plastic found in the stomach of seabirds (Bruner, 1990). Studies conducted during the 1970s made it clear that large numbers of pellets were escaping from plastics factories, and efforts to correct this problem were undertaken by the industry. Dow Chemical, for example, estimates that pellet reclamation procedures undertaken at its Louisiana plant recover more than 80 tonnes of pellets per year that would otherwise escape into marine areas (O'Hara et al., 1988). Nevertheless, pellets are still detected in coastal areas of the United States, and it is unclear whether most of the escapement occurs during transportation and handling or whether losses at the manufacturing level are still a major source of pollution. Identification of the principal sources of the pellets and development of techniques for reducing pellet escapement are needed.

Education

Education of both seafarers and the general public is perceived as being crucial to a successful resolution of the plastic debris problem. One of the best examples of the impact of public education efforts has been the proliferation of annual beach cleanup days and the number of persons involved (Figure 17.4). From 2100 persons who participated in the first beach cleanup in Oregon in 1984, beach cleanups have spread to all coastal states and involved about 100,000 persons in 1990.

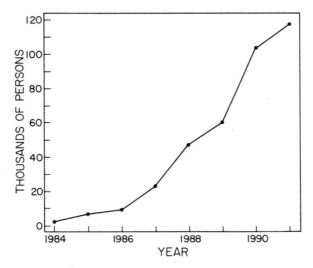

Figure 17.4   Volunteer participation in beach cleanups from 1984 to 1990. [*Source*: Debenham (1990) and Younger (1992).]

A number of public and private organizations, including in particular the Center for Marine Conservation, the National Oceanic and Atmospheric Administration, and the International Maritime Organization, have been actively involved in public awareness and education efforts. Since most of the derelict plastic in the ocean comes from vessels, it is vital that seafarers are aware of the adverse effects caused by plastics pollution and understand and respect laws designed to minimize this form of pollution. Although MARPOL Annex V has been an important step in reducing the input of plastics from vessels, there remain serious problems associated with ghost fishing by lost trawl nets and traps. To a certain extent these problems will probably always exist, but they can be minimized with the help of technology and a conscientious effort by fishermen to reduce net and trap losses. Training and education will do much to create the awareness and sustain the motivation needed to further reduce pollution of the oceans by plastics.

REFERENCES

Anonymous, 1982. Confused sea turtles said to be dying from diet of plastic trash. The Baltimore Sun, December 28.

Anonymous, 1992. Would anyone like a souvenir Japanese driftnet? *Mar. Cons. News*, **4**(1), 1.

Balazs, G. 1985. Impact of ocean debris on marine turtles: Entanglement and ingestion. In R. S. Shomura and H. O. Yoshida (Eds.), *Proceedings of the Workshop on the Fate and Impact of Marine Debris*, U. S. Dept. of Commerce, Honolulu, HI. pp. 387–429.

Bean, M. J. 1990. Redressing the problem of persistent marine debris through law and public policy: Opportunities and pitfalls. In R. S. Shomura and M. L. Godfrey (Eds.), *Proceedings of the Second International Conference on Marine Debris*. Vol. II. U. S. Dept. of Commerce, Honolulu, HI. pp. 989–997.

Bourne, W. R. P. 1977. Nylon netting as a hazard to birds. *Marine Pollution Bulletin*, **8**(4), 75–76.

Breen, P. A. 1990. A review of ghost fishing by traps and gillnets. In R. S. Shomura and M. L. Godfrey (Eds.), *Proceedings of the Second International Conference on Marine Debris*. Vol. I. U. S. Dept. of Commerce, Honolulu, HI. pp. 571–599.

Bruner, R. G. 1990. The plastics industry and marine debris: Solutions through education. In R. S. Shomura and M. L. Godfrey (Eds.), *Proceedings of the Second International Conference on Marine Debris*. Vol. II. U. S. Dept. of Commerce, Honolulu, HI. pp. 1077–1089.

Carr, H. A., A. W. Hulbert, and E. H. Amaral. 1985. Underwater survey of simulated lost demersal and lost commercial gill nets off New England. In R. S. Shomura and H. O. Yoshida (Eds.), *Proceedings of the Workshop on the Fate and Impact of Marine Debris*. U. S. Dept. of Commerce, Honolulu, HI. pp. 438–447.

Center for Environmental Education. 1987. *Plastics in the Ocean: More Than a Litter Problem*. Washington, D.C. 127 pp.

Day, R. H., D. H. S. Wehle, and F. C. Coleman. 1985. Ingestion of plastic pollutants by marine birds. In R. S. Shomura and H. O. Yoshida (Eds.), *Proceedings of the Workshop on the Fate and Impact of Marine Debris*. U. S. Dept. of Commerce, Honolulu, HI. pp. 344–386.

Debenham, P. 1990. Education and awareness: Keys to solving the marine debris problem. In R. S. Shomura and M. L. Godfrey (Eds.), *Proceedings of the Second International Conference on Marine Debris*. Vol. II. U. S. Dept. of Commerce, Honolulu, HI. pp. 1100–1114.

Degange, A. R., and T. C. Newby. 1980. Mortality of seabirds and fish in a lost salmon driftnet. *Marine Pollution Bulletin*, **11**, 322–323.

Denison, R. A. 1990. Degradable plastics: Right question, wrong answer. *Environmental Defense Fund Letter*, **21**(1), 4.

Edwards, D. T., and E. Rymarz. 1990. International regulations for the prevention and control of pollution by debris from ships. In R. S. Shomura and M. L. Godfrey (Eds.), *Proceedings of the Second International Conference on Marine Debris*. Vol. II. U. S. Dept. of Commerce, Honolulu, HI. pp. 956–988.

Eisenbud, R. 1985. Problems and prospects for the pelagic driftnet. *Boston College Environmental Affairs Law Review*, **12**(3), 473–490.

French, D. P., and M. Reed. 1990. Potential impact of entanglement in marine debris on the population dynamics of the northern fur seal, *Callorhinus ursinus*. In R. S. Shomura and M. L. Godfrey (Eds.), *Proceedings of the Second International Conference on Marine Debris*. Vol. I. U. S. Dept. of Commerce, Honolulu, HI. pp. 431–452.

Fry, D. M., and S. I. Fefer. 1986. Ingestion of floating plastic debris by seabirds in the Hawaiian Islands. Paper presented at the Sixth International Ocean Disposal Symposium, April 21–25, Pacific Grove, CA. In Program and Abstracts, pp. 73–74.

Gress, F., and D. W. Anderson. 1983. *California Brown Pelican Recovery Plan*. U. S. Fish and Wildlife Service, Portland, OR.

Henderson, J. R. 1990. Recent entanglements of Hawaiian monk seals in marine debris. In R. S. Shomura and M. L. Godfrey (Eds.), *Proceedings of the Second International Conference on Marine Debris*. Vol. I. U. S. Dept. of Commerce, Honolulu, HI. pp. 540–553.

Jones, L. L., and R. C. Ferrero. 1985. Observations of net debris and associated entanglements in the North Pacific Ocean and Bering Sea, 1978–1984. In R. S. Shomura and H. O. Yoshida (Eds.), *Proceedings of the Workshop on the Fate and Impact of Marine Debris*. U. S. Dept. of Commerce, Honolulu, HI. pp. 183–196.

Jones, S. C., R. J. Tarpley, and S. Fernandez. 1986. Cetacean strandings along the Texas coast, U. S. A. Paper presented at the 11th International Conference on Marine Mammals, April 2–6, Guaymas, Mexico.

Kartar, S., R. A. Milne, and M. Sainsbury. 1973. Polystyrene waste in the Severn Estuary. *Marine Pollution Bulletin*, **4**, 44.

Kartar, S., R. A. Milne, and M. Sainsbury. 1976. Polystyrene spherules in the Severn Estuary—a progress report. *Marine Pollution Bulletin*, **7**, 52.

King, B. 1985. Trash and debris on the beaches of Padre Island National Seashore. In Sixth Annual Minerals Management Service, Gulf of Mexico OCS Regional Office, Information Transfer Meeting. Session IV.E. Trash and Debris on Gulf of Mexico Waterfront Beaches. 23 October. Metairie, LA. (unpublished manuscript).

Low, L. L., R. E. Nelson, Jr., and R. E. Narita. 1985. Net loss from trawl fisheries off Alaska. In R. S. Shomura and H. O. Yoshida (Eds.), *Proceedings of the Workshop on the Fate and Impact of Marine Debris*. U. S. Dept. of Commerce, Honolulu, HI. pp. 130–153.

Martinez, L. A. 1990. Shipboard waste disposal: Taking out the trash under the new rules. In R. S. Shomura and M. L. Godfrey (Eds.), *Proceedings of the Second International Conference on Marine Debris*. Vol. II. U. S. Dept. of Commerce, Honolulu, HI. pp. 895–914.

Merrell, T. R., Jr. 1985. Fish nets and other plastic litter on Alaska beaches. In R. S. Shomura and H. O. Yoshida (Eds.), *Proceedings of the Workshop on the Fate and Impact of Marine Debris*. U. S. Dept. of Commerce, Honolulu, HI. pp. 160–182.

O'Hara, K. J., S. Iudicello, and R. Bierce. 1988. *A Citizen's Guide to Plastics in the Ocean: More Than a Litter Problem*. Center for Marine Conservation, Washington, D.C. 143 pp.

Stewart, R. S., and P. K. Yochem. 1990. Pinniped entanglement in synthetic materials in the Southern California Bight. In R. S. Shomura and M. L. Godfrey (Eds.), *Proceedings of the Second International Conference on Marine Debris*. Vol. I. U. S. Dept. of Commerce, Honolulu, HI. pp. 554–561.

Swanson, R. L., H. M. Stanford, J. S. O'Connor, S. Chanesman, C. A. Parker, P. A. Eisen, and G. F. Mayer. 1978. June 1976 pollution of Long Island ocean beaches. *Journal Environmental Engineering Division, ASCE*, **104**(EE6), 1067–1085.

Takehama, S. 1990. Estimation of damages to fishing vessels caused by marine debris, based on insurance statistics. In R. S. Shomura and M. L. Godfrey (Eds.), *Proceedings of the Second International Conference on Marine Debris*. Vol. II. U. S. Dept. of Commerce, Honolulu, HI. pp. 792–809.

Wagner, K. D. 1990. Medical wastes and the beach washups of 1988: Issues and impacts. In R. S. Shomura and M. L. Godfrey (Eds.), *Proceedings of the Second International Conference on Marine Debris*. Vol. II. U. S. Dept. of Commerce, Honolulu, HI. pp. 811–823.

Younger, L. K. 1992. Participation soars at 1991 international coastal cleanup. *Marine Cons. News*, **4**(1), 17.

INDEX

The DisplayMessage procedure shown in Figure 5-29 displays a message in a label control. As the Handles section in the procedure header indicates, the procedure is processed when the xSalesTextBox's TextChanged event occurs and also when the xClearButton's Click event occurs. When either of these events occurs, the memory address of the control associated with the event is sent to the DisplayMessage procedure's `sender` parameter. In this case, for example, the memory address of the xSalesTextBox is sent to the `sender` parameter when the text box's TextChanged event occurs. Likewise, the memory address of the xClearButton is sent to the `sender` parameter when the button's Click event occurs.

The Select Case statement in the procedure displays the appropriate message based on the type of control whose address is stored in the `sender` parameter. Notice that the Select Case statement uses the Boolean value True as the *selectorExpression*. The `TypeOf sender Is TextBox` expression in the first Case clause compares the `sender` parameter's type to the TextBox type. If both types are the same, the expression evaluates to True. In that case, because the value of the expression (True) matches the value of the *selectorExpression* (True), the procedure displays the "The text box invoked the procedure." message in the xMsgLabel. However, if both types are not the same, the `TypeOf sender Is TextBox` expression evaluates to False. Here, because the value of the expression (False) does not match the value of the *selectorExpression* (True), the procedure displays the "The Clear button invoked the procedure." message in the xMsgLabel.

To code and then test the DisplayMessage procedure:

1 Open the **TypeOf-Is Operator Solution** (TypeOf-Is Operator Solution.sln) file, which is contained in the VB2005\Chap05\TypeOf-Is Operator Solution folder. If necessary, open the designer window.

2 Open the Code Editor window. Replace the <your name> and <current date> text in the comments with your name and the current date.

3 In the DisplayMessage procedure, enter the code shown in Figure 5-29.

4 Close the Code Editor window. Save the solution, then start the application.

5 Enter **3** as the sales amount. Doing this invokes the xSalesTextBox's TextChanged event, which is associated with the DisplayMessage procedure. The procedure displays the "The text box invoked the procedure." message in the xMsgLabel, as shown in Figure 5-30.

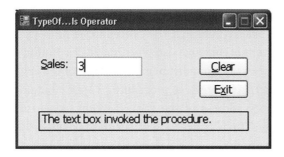

Figure 5-30: Message shown in the interface

6 Click the **Clear** button. Clicking the button invokes its Click event, which is associated with the DisplayMessage procedure. The procedure displays the "The Clear button invoked the procedure." message in the xMsgLabel.

7 Click the **Exit** button to end the application. You are returned to the designer window. Close the solution.

THE LIKE COMPARISON OPERATOR

The **Like operator** allows you to use pattern-matching characters to determine whether one string is equal to another string. Figure 5-31 shows the syntax of the Like operator and includes a listing of the operator's pattern-matching characters. The figure also shows several examples of using the Like operator.

Like operator	

Syntax

string **Like** *pattern*

The Like operator evaluates to True when the *string* matches the *pattern*; otherwise it evaluates to False.

Pattern-matching characters	Matches in *string*
?	any single character
*	zero or more characters
#	any single digit (0-9)
[*charlist*]	any single character in the *charlist* (for example, [a-z] matches any lowercase letter)
[!*charlist*]	any single character not in the *charlist* (for example, [!a-z] matches any character that is not a lowercase letter)

Examples

```
isEqual = firstName.ToUpper Like "B?LL"
```

assigns the Boolean value True to the `isEqual` variable when the string stored in the `firstName` variable begins with the letter B, followed by one character and then the two letters LL; otherwise, assigns the Boolean value False to the `isEqual` variable

```
If state Like "K*" Then
```

the condition evaluates to True when the string stored in the `state` variable begins with the letter K, followed by zero or more characters; otherwise, it evaluates to False

```
id Like "###*"
```

evaluates to True when the string stored in the `id` variable begins with three digits, followed by zero or more characters; otherwise, it evaluates to False

```
If firstName.ToUpper Like "T[OI]M" Then
```

the condition evaluates to True when the string stored in the `firstName` variable begins with the letter T, followed by either the letter O or the letter I, followed by the letter M; otherwise, it evaluates to False

```
isLowercase = letter Like "[a-z]"
```

assigns the Boolean value True to the `isLowercase` variable when the string stored in the `letter` variable is a lowercase letter; otherwise, it evaluates to False

```
letter Like "[!a-zA-Z]"
```

evaluates to True when the string stored in the `letter` variable is not a letter of the alphabet

Figure 5-31: Syntax and examples of the Like operator

In the Like operator's syntax, both *string* and *pattern* must be String expressions. However, *pattern* can contain one or more of the pattern-matching characters described in Figure 5-31. The Like operator evaluates to True when *string* matches *pattern*; otherwise it evaluates to False.

Study each example shown in Figure 5-31. The `isEqual = firstName.ToUpper Like "B?LL"` example contains the question mark (?) pattern-matching character, which is used to match one character in the *string*. Examples of *strings* that would make the `firstName.ToUpper Like "B?LL"` expression evaluate to True include "Bill", "Ball", "bell", and "bull". Examples of *strings* for which the expression would evaluate to False include "BPL", "BLL", and "billy".

The `state Like "K*"` condition in the second example uses the asterisk (*) pattern-matching character to match zero or more characters. Examples of *strings* that would make the condition evaluate to True include "KANSAS", "Ky", and "Kentucky". Examples of *strings* for which the condition would evaluate to False include "kansas" and "ky".

The `id Like "###*"` example contains two different pattern-matching characters: the number sign (#), which matches a digit, and the asterisk (*), which matches zero or more characters. Examples of *strings* that would make the example evaluate to True include "178" and "983Ab". Examples of *strings* for which the example would evaluate to False include "X34" and "34Z".

The `firstName.ToUpper Like "T[OI]M"` condition in the fourth example contains a *charlist* (character list)—in this case, the two letters O and I—enclosed in square brackets ([]). The condition evaluates to True when the string stored in the `firstName` variable is either "Tom" or "Tim" (entered in any case). When the `firstName` variable does not contain "Tom" or "Tim"—for example, when it contains "Tam" or "Tommy"—the condition evaluates to False.

The `isLowercase = letter Like "[a-z]"` example also contains a *charlist* enclosed in square brackets. However, the *charlist* represents a range of values; in this case, it represents the lowercase letters "a" through "z". Notice that you use a hyphen (-) to specify a range of values. The statement assigns the Boolean value True to the `isLowercase` variable when the string stored in the `letter` variable is a lowercase letter; otherwise, it assigns the Boolean value False.

The last example shown in Figure 5-31, `letter Like "[!a-zA-Z]"`, also contains a *charlist* that specifies a range of values. However, the *charlist* in this example is preceded by an exclamation point (!), which stands for "not". The example evaluates to True when the string stored in the `letter` variable is *not* a letter of the alphabet; otherwise, it evaluates to False.

In the next set of steps, you will code and then test a procedure that uses the Like operator to validate a part number.

To code and then test the xValidateButton's Click event procedure:

1 Open the **Like Operator Solution** (Like Operator Solution.sln) file, which is contained in the VB2005\Chap05\Like Operator Solution folder. If necessary, open the designer window.

2 Open the Code Editor window. Replace the <your name> and <current date> text in the comments with your name and the current date.

3 Locate the xValidateButton's Click event procedure. Position the insertion point in the blank line below the '`followed by a letter` comment. If necessary, press the **Tab** key twice to align the insertion point with the apostrophe in the previous line.

4 To be valid, a part number must begin with three digits followed by a letter of the alphabet. Type **if partnumber like "###[A-Z]" then** and press **Enter**, then type **me.xMsgLabel.text = "Valid part number"** and press **Enter**.

5 Type **else** and press **Enter**, then type **me.xMsgLabel.text = "Invalid part number"**. Figure 5-32 shows the completed Click event procedure for the xValidateButton.

Visual Basic code

```vb
Private Sub xValidateButton_Click(ByVal sender As Object, _
    ByVal e As System.EventArgs) Handles xValidateButton.Click
    ' validates a part number

    Dim partNumber As String
    partNumber = Me.xPartTextBox.Text.ToUpper

    ' the part number must contain 3 digits
    ' followed by a letter
    If partNumber Like "###[A-Z]" Then
        Me.xMsgLabel.Text = "Valid part number"
    Else
        Me.xMsgLabel.Text = "Invalid part number"
    End If

    ' set the focus and select the existing text
    Me.xPartTextBox.Focus()
    Me.xPartTextBox.SelectAll()
End Sub
```

Figure 5-32: The xValidateButton's Click event procedure

6 Close the Code Editor window. Save the solution, then start the application.

7 Enter **123a** as the part number, then press **Enter** to select the Validate button. The xValidateButton's Click event procedure displays the "Valid part number" message, as shown in Figure 5-33.

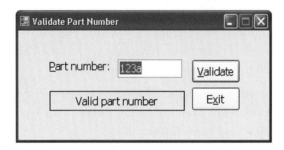

Figure 5-33: Message displayed in the xMsgLabel

8 Enter **45b** as the part number, then press **Enter**. The xValidateButton's Click event procedure displays the "Invalid part number" message in the xMsgLabel.

9 On your own, test the application using the following part numbers: **2233b**, **123ab**, **tr675**, **999Z**, **767%**, and **56**.

10 When you are finished testing, click the **Exit** button to end the application. You are returned to the designer window. Close the solution.

You have completed Lesson A. You can either take a break or complete the end-of-lesson questions and exercises before moving on to Lesson B.

SUMMARY

TO CREATE A SELECTION STRUCTURE THAT EVALUATES BOTH A PRIMARY AND A SECONDARY DECISION:

» Place (or nest) the selection structure for the secondary decision within either the true path or false path of the selection structure for the primary decision.

TO VERIFY THAT AN ALGORITHM WORKS CORRECTLY:

» Desk-check the algorithm. Desk-checking, also called hand-tracing, means that you use sample data to walk through each of the steps in the algorithm manually, just as though you were the computer.

TO CODE A MULTIPLE-PATH (OR EXTENDED) SELECTION STRUCTURE:

» Use either the If...Then...Else statement or the Select Case statement.

TO SPECIFY A RANGE OF VALUES IN A CASE CLAUSE CONTAINED IN A SELECT CASE STATEMENT:

» Use the To keyword when you know both the upper and lower bounds of the range. The syntax for using the To keyword is **Case** *smallest value in the range* **To** *largest value in the range.*

» Use the Is keyword when you know only one end of the range—either the upper end or the lower end. The Is keyword is used in combination with one of the following comparison operators: =, <, <=, >, >=, <>.

TO DETERMINE WHETHER TWO OBJECT REFERENCES REFER TO THE SAME OBJECT:

» Use the Is comparison operator. The syntax for using the Is operator is *objectReference1* **Is** *objectReference2.* The Is operator returns the Boolean value True when both *objectReferences* contain the same address; otherwise it returns the Boolean value False.

TO DETERMINE WHETHER AN OBJECT IS A SPECIFIED TYPE:

» Use the TypeOf...Is comparison operator. The syntax for using the TypeOf...Is operator is **TypeOf** *object* **Is** *objectType.* The TypeOf...Is operator returns the Boolean value True when the *object's* type is the same as the *objectType*; otherwise it returns the Boolean value False.

TO USE PATTERN MATCHING TO DETERMINE WHETHER ONE STRING IS EQUAL TO ANOTHER STRING:

» Use the Like comparison operator. The syntax for using the Like operator is *string* **Like** *pattern*, where *pattern* can contain one or more pattern-matching characters. Refer to Figure 5-31 for the pattern-matching characters. The Like operator returns the Boolean value True when the *string* matches the *pattern*; otherwise it returns the Boolean value False.

QUESTIONS

Use the following code to answer Questions 1 through 3:

```
If number <= 100 Then
        number = number * 2
ElseIf number > 500 Then
        number = number * 3
End If
```

1. If the number variable contains the number 90, what value will be in the number variable after the preceding code is processed?

   a. 0

   b. 90

   c. 180

   d. 270

2. If the number variable contains the number 1000, what value will be in the number variable after the preceding code is processed?

   a. 0

   b. 1000

   c. 2000

   d. 3000

3. If the number variable contains the number 200, what value will be in the number variable after the preceding code is processed?

   a. 0

   b. 200

   c. 400

   d. 600

Use the following code to answer Questions 4 through 7:

```
If id = 1 Then
      Me.xNameLabel.Text = "Janet"
ElseIf id = 2 OrElse id = 3 Then
      Me.xNameLabel.Text = "Mark"
ElseIf id = 4 Then
      Me.xNameLabel.Text = "Jerry"
Else
      Me.xNameLabel.Text = "Sue"
End If
```

4. What will the preceding code display when the id variable contains the number 2?

   a. Janet
   b. Jerry
   c. Mark
   d. Sue

5. What will the preceding code display when the id variable contains the number 4?

   a. Janet
   b. Jerry
   c. Mark
   d. Sue

6. What will the preceding code display when the id variable contains the number 3?

   a. Janet
   b. Jerry
   c. Mark
   d. Sue

7. What will the preceding code display when the id variable contains the number 8?

   a. Janet
   b. Jerry
   c. Mark
   d. Sue

8. A nested selection structure can appear in _____ of another selection structure.

   a. only the true path

   b. only the false path

   c. either the true path or the false path

9. If the *selectorExpression* in a Select Case statement is an Integer variable named code, which of the following Case clauses is valid?

   a. `Case Is > 7`
   b. `Case 3, 5`
   c. `Case 1 To 4`
   d. All of the above.

Use the following Case statement to answer Questions 10 through 12:

```
Select Case id
        Case 1
                Me.xNameLabel.Text = "Janet"
        Case 2 To 4
                Me.xNameLabel.Text = "Mark"
        Case 5, 7
                Me.xNameLabel.Text = "Jerry"
        Case Else
                Me.xNameLabel.Text = "Sue"
End Select
```

10. What will the preceding Case statement display when the `id` variable contains the number 2?

    a. Janet   b. Mark

    c. Jerry   d. Sue

11. What will the preceding Case statement display when the `id` variable contains the number 3?

    a. Janet   b. Mark

    c. Jerry   d. Sue

12. What will the preceding Case statement display when the `id` variable contains the number 6?

    a. Janet   b. Mark

    c. Jerry   d. Sue

13. Which of the following can be used to determine whether the `sender` parameter contains the address of the xNameTextBox?

    a. `If sender Is Me.xNameTextBox Then`

    b. `If sender = Me.xNameTextBox Then`

    c. `If sender Like Me.xNameTextBox Then`

    d. `If TypeOf sender Is xNameTextBox Then`

14. Which of the following can be used to determine whether the `sender` parameter contains the address of a label control?

    a. `If sender Is Label Then`

    b. `If sender = Label Then`

    c. `If sender Like Label Then`

    d. `If TypeOf sender Is Label Then`

15. Which of the following can be used to determine whether a String variable named `partNum` contains two characters followed by a digit?

    a. `If partNum Like "??#" Then`

    b. `If partNum Is = "**#" Then`

    c. `If partNum = "##?" Then`

    d. `If partNum Like "**[0-9]" Then`

16. Which of the following can be used to determine whether a String variable named `item` contains either the word "shirt" or the word "skirt"?

    a. `If item.ToUpper = "SHIRT" OrElse item.ToUpper = "SKIRT" Then`

    b. `If item.ToUpper Like "S[H-K]IRT" Then`

    c. `If item.ToUpper Is "S[HK]IRT" Then`

    d. both a and b

17. Which of the following can be used to determine whether the percent sign (%) is the last character entered in the xRateTextBox?

    a. `If "%" Like Me.xRateTextBox.Text Then`

    b. `If Me.xRateTextBox.Text Like "?%" Then`

    c. `If Me.xRateTextBox.Text Like "*%" Then`

    d. `If Me.xRateTextBox.Text Like [*%] Then`

18. A procedure needs to display the appropriate fee to charge a golfer. The fee is based on the following fee schedule:

Fee	Criteria
0	Club members
15	Non-members golfing on Monday through Thursday
25	Non-members golfing on Friday through Sunday

In this procedure, which is the primary decision and which is the secondary decision? Why?

19. List the three errors commonly made when writing selection structures. Which error makes the selection structure inefficient, but not incorrect?

20. Explain what the term "desk-checking" means.

EXERCISES

1. Write the Visual Basic code for the algorithm shown in Figure 5-10 in this lesson. The salesperson's code and sales amount are entered in the xCodeTextBox and xSalesTextBox controls, respectively. Store the text box values and bonus amount in variables. Display the appropriate bonus amount in the xMsgLabel.

2. Write the Visual Basic code that displays the message "Highest honors" when a student's test score is 90 or above. When the test score is 70 through 89, display the message "Good job". For all other test scores, display the message "Retake the test". Use the If/ElseIf/Else selection structure. The test score is stored in an Integer variable named score. Display the appropriate message in the xMsgLabel.

3. Write the Visual Basic code that compares the contents of an Integer variable named quantity with the number 10. When the quantity variable contains a number that is equal to 10, display the string "Equal" in the xMsgLabel. When the quantity variable contains a number that is greater than 10, display the string "Over 10". When the quantity variable contains a number that is less than 10, display the string "Not over 10". Use the If/ElseIf/Else selection structure.

4. Write the Visual Basic code that corresponds to the flowchart shown in Figure 5-34. Store the salesperson's code, which is entered in the xCodeTextBox, in an Integer variable named `code`. Store the sales amount, which is entered in the xSalesTextBox, in a Decimal variable named `sales`. Display the result of the calculation, or the error message, in the xMsgLabel.

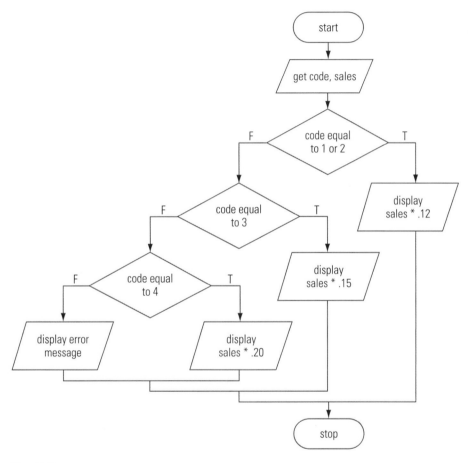

Figure 5-34

5. Write the Visual Basic code that corresponds to the flowchart shown in Figure 5-35. Store the salesperson's code, which is entered in the xCodeTextBox, in an Integer variable named `code`. Store the sales amount, which is entered in the xSalesTextBox, in a Decimal variable named `sales`. Display the result of the calculation, or the error message, in the xMsgLabel.

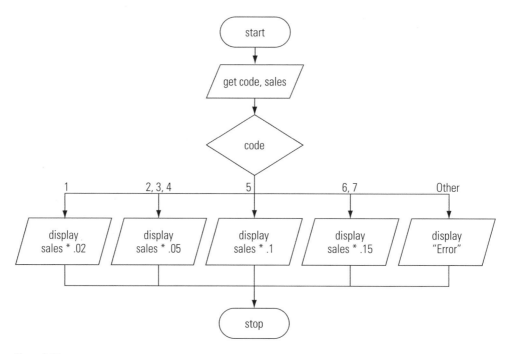

Figure 5-35

6. A procedure needs to display a shipping charge based on the state name stored in the state variable. The state name is stored using uppercase letters. Write a Select Case statement that assigns the shipping charge to a Double variable named shipCharge. Use the following chart to determine the appropriate shipping charge. Display an appropriate message in the xMsgLabel when the state variable contains a value that does not appear in the chart. Also assign the number 0 to the shipCharge variable.

State name	Shipping charge
HAWAII	$25.00
OREGON	$30.00
CALIFORNIA	$32.50

7. Rewrite the code from Exercise 6 using an If...Then...Else statement.

8. The price of a concert ticket depends on the seat location stored in the seatLoc variable. The seat location is stored using uppercase letters. Write a Select Case statement that displays the price in the xPriceLabel. Use the following chart to determine the appropriate price. Display an appropriate message in the xPriceLabel when the seatLoc variable contains a value that does not appear in the chart.

Seat location	Concert ticket price
BOX	$75.00
PAVILION	$30.00
LAWN	$21.00

9. Rewrite the code from Exercise 8 using an If...Then...Else statement.

10. A procedure needs to display a shipping charge based on the five-digit ZIP code stored in a String variable named `zipCode`. Write an If...Then...Else statement that assigns the shipping charge to a Double variable named `shipCharge`. All ZIP codes beginning with "605" have a $25 shipping charge. All ZIP codes beginning with "606" have a $30 shipping charge. All other ZIP codes are charged $35 for shipping.

11. Rewrite the code from Exercise 10 using a Select Case statement.

12. The DisplayCapital procedure is invoked when the Click event occurs for one of the following buttons: xAlabamaButton, xArizonaButton, xIllinoisButton, and xOregonButton. Write an If...Then...Else statement that displays the name of the appropriate state's capital in the xCapitalLabel. Use the following chart:

State	Capital
Alabama	Montgomery
Arizona	Phoenix
Illinois	Springfield
Oregon	Salem

13. Rewrite the code from Exercise 12 using a Select Case statement.

14. In this exercise, you complete two procedures that display a message based on a code entered by the user.

    a. If necessary, start Visual Studio 2005 or Visual Basic 2005 Express Edition. Open the Animal Solution (Animal Solution.sln) file, which is contained in the VB2005\Chap05\Animal Solution folder. If necessary, open the designer window.

    b. Open the Code Editor window. Complete the If...Then...Else button's Click event procedure by writing an If...Then...Else statement that displays the string "Dog" when the `animal` variable contains the number 1. Display the string "Cat" when the `animal` variable contains the number 2. Display the string "Bird" when the `animal` variable contains anything other than the number 1 or the number 2. Display the appropriate string in the xMsgLabel.

c. Save the solution, then start the application. Test the If...Then...Else button's code three times, using the numbers 1, 2, and 5.

d. Click the Exit button to end the application.

e. Complete the Select Case button's Click event procedure by writing a Select Case statement that displays the string "Dog" when the `animal` variable contains either the letter "D" or the letter "d". Display the string "Cat" when the `animal` variable contains either the letter "C" or the letter "c". Display the string "Bird" when the `animal` variable contains anything other than the letters "D", "d", "C", or "c". Display the appropriate string in the xMsgLabel.

f. Save the solution, then start the application. Test the Select Case button's code three times, using the letters D, c, and x.

g. Click the Exit button to end the application. Close the Code Editor window, then close the solution.

15. In this exercise, you complete two procedures that display the name of the month corresponding to a user's entry.

a. If necessary, start Visual Studio 2005 or Visual Basic 2005 Express Edition. Open the Month Solution (Month Solution.sln) file, which is contained in the VB2005\Chap05\Month Solution folder. If necessary, open the designer window.

b. Open the Code Editor window. Complete the If...Then...Else button's Click event procedure by writing an If...Then...Else statement that displays the name of the month corresponding to the number entered by the user. For example, if the user enters the number 1, the procedure should display the string "January". If the user enters an invalid number (one that is not in the range 1 through 12), display an appropriate message. Display the appropriate string in the xMsgLabel.

c. Save the solution, then start the application. Test the If...Then...Else button's code three times, using the numbers 3, 7, and 20.

d. Click the Exit button to end the application.

e. Now assume that the user will enter the first three characters of the month's name (rather than the month number) in the text box. Complete the Select Case button's Click event procedure by writing a Select Case statement that displays the name of the month corresponding to the characters entered by the user. For example, if the user enters the three characters "Jan" (in any case), the procedure should display the string "January". If the user enters "Jun", the procedure should display "June". If the three characters entered by the user do not match any of the expressions in the Case clauses, display an appropriate message. Display the appropriate string in the xMsgLabel.

    f. Save the solution, then start the application. Test the Select Case button's code three times, using the following data: jun, dec, xyz.

    g. Click the Exit button to end the application. Close the Code Editor window, then close the solution.

16. In this exercise, you complete a procedure that calculates and displays a bonus amount.

    a. If necessary, start Visual Studio 2005 or Visual Basic 2005 Express Edition. Open the Bonus Solution (Bonus Solution.sln) file, which is contained in the VB2005\Chap05\Bonus Solution folder. If necessary, open the designer window.

    b. Open the Code Editor window. Complete the Calculate button's Click event procedure by writing an If...Then...Else statement that assigns the number 25 to the bonus variable when the user enters a sales amount that is greater than or equal to $100, but less than or equal to $250. When the user enters a sales amount that is greater then $250, assign the number 50 to the bonus variable. When the user enters a sales amount that is less than $100, assign the number 0 as the bonus.

    c. Save the solution, then start the application. Test the Calculate button's code three times, using sales amounts of 100, 300, and 40.

    d. Click the Exit button to end the application. Close the Code Editor window, then close the solution.

17. In this exercise, you complete a procedure that calculates and displays the total amount owed by a company.

    a. If necessary, start Visual Studio 2005 or Visual Basic 2005 Express Edition. Open the Seminar Solution (Seminar Solution.sln) file, which is contained in the VB2005\Chap05\Seminar Solution folder. If necessary, open the designer window.

    b. Open the Code Editor window. Assume you offer programming seminars to companies. Your price per person depends on the number of people the company registers, as shown in the following chart. For example, if the company registers seven people, then the total amount owed is $560, which is calculated by multiplying the number 7 by the number 80. Display the total amount owed in the xTotalLabel. Use the Select Case statement to complete the Calculate button's

Click event procedure. Display an appropriate message when the number registered is not numeric.

Number of registrants	Price per person
1–4	$100
5–10	$80
11 or more	$60
Less than 1	$0

c. Save the solution, then start the application. Test the Calculate button's code five times, using the following data: 7, 4, 11, a, and –3.

d. Click the Exit button to end the application. Close the Code Editor window, then close the solution.

LESSON B
OBJECTIVES

AFTER STUDYING LESSON B, YOU SHOULD
BE ABLE TO:

» Include a group of radio buttons in an interface

» Designate a default radio button

» Include a check box in an interface

» Create and call an independent Sub procedure

» Generate random numbers

» Invoke a radio button's Click event procedure from code

THE MATH PRACTICE APPLICATION

COMPLETING THE USER INTERFACE

Recall that Susan Chen, the principal of a local primary school, wants an application that the first and second grade students can use to practice both adding and subtracting numbers. The problems displayed for the first grade students should use numbers from 1 through 10 only. The problems for the second grade students should use numbers from 10 through 99. Because the students have not learned about negative numbers yet, the subtraction problems should never ask them to subtract a larger number from a smaller one. The application should display the addition or subtraction problem on the screen, then allow the student to enter the answer, and then verify that the answer is correct. If the student's answer is not correct, the application should give him or her as many chances as necessary to answer the problem correctly. Recall that Ms. Chen wants the application to keep track of the number of correct and incorrect responses made by the student, and she wants the ability to control the display of that information. Figure 5-36 shows the sketch of the Math Practice application's user interface.

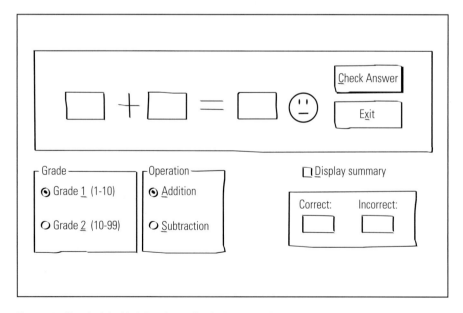

Figure 5-36: Sketch of the Math Practice application's user interface

The user interface contains one text box, four radio buttons, one check box, three picture boxes, four group boxes, and various labels. To save you time, the VB2005\Chap05\Math Solution folder contains a partially completed Math Practice application. When you open the application, you will notice that most of the user interface has already been created and the properties of the existing objects have been set. You will complete the user interface in this lesson.

To open the partially completed application:

1 Start Microsoft Visual Studio 2005 or Visual Basic 2005 Express Edition, if necessary, and close the Start Page window.

2 Open the **Math Solution** (Math Solution.sln) file, which is contained in the VB2005\Chap05\Math Solution folder. If necessary, open the designer window.

3 Auto-hide the Toolbox, Solution Explorer, and Properties windows, if necessary. Figure 5-37 shows the partially completed user interface for the Math Practice application.

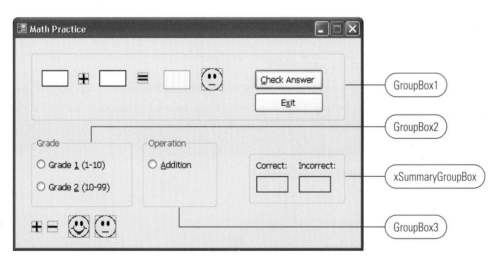

Figure 5-37: Partially completed user interface for the Math Practice application

As Figure 5-37 indicates, the names of the four group boxes are GroupBox1, GroupBox2, GroupBox3, and xSummaryGroupBox. The GroupBox1 control contains eight controls, which are named (from left to right) xNum1Label, xOperatorPictureBox, xNum2Label, PictureBox1, xAnswerTextBox, xFacePictureBox, xCheckAnswerButton, and xExitButton. The GroupBox2 control contains two radio buttons named xGrade1RadioButton and xGrade2RadioButton, and the GroupBox3 control contains one radio button named xAdditionRadioButton. The xSummaryGroupBox control contains four label controls named Label1, xCorrectLabel, Label2, and xIncorrectLabel. In addition to the controls already mentioned, the form also contains four picture boxes that are positioned at the bottom of the form. The names of the picture boxes are xPlusPictureBox, xMinusPictureBox, xHappyPictureBox, and xNeutralPictureBox. You will learn the purpose of these picture boxes later in this lesson. Only two controls are missing from the interface: the Subtraction radio button and the Display summary check box.

ADDING A RADIO BUTTON TO THE FORM

You use the **RadioButton tool** in the toolbox to add a radio button to an interface. A **radio button** allows you to limit the user to only one choice in a group of two or more related and mutually exclusive choices. You will use radio buttons in the Math Practice application to limit the user to one grade level selection (either Grade 1 or Grade 2) and one mathematical operation selection (either Addition or Subtraction).

Each radio button in an interface should be labeled so that the user knows its purpose. You enter the label using sentence capitalization in the radio button's Text property. Each radio button also should have a unique access key, which allows the user to select the button using the keyboard. In the next set of steps, you will add the missing Subtraction radio button to the interface.

To add a radio button to the interface:

1   Temporarily display the Toolbox window. Click the **RadioButton** tool in the toolbox, and then drag a radio button into the GroupBox3 control, placing it immediately below the xAdditionRadioButton. The RadioButton1 control appears in the interface.

2   Set the following properties for the RadioButton1 control:

Name:	**xSubtractionRadioButton**
Location:	**8, 59**
Size:	**91, 20**
Text:	**&Subtraction**

3   Click the form's **title bar**. Figure 5-38 shows the Subtraction radio button in the interface.

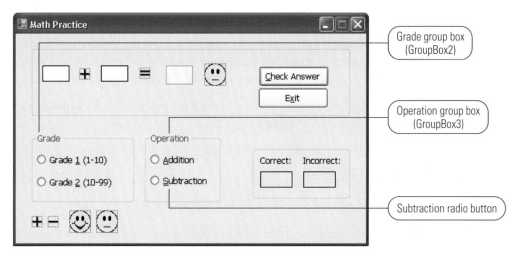

Figure 5-38: Subtraction radio button added to the interface

Radio buttons that are related should be grouped together. For example, two groups of radio buttons appear in the Math Practice interface: one group contains the two Grade radio buttons, and the other contains the two Operation radio buttons. To include two groups of radio buttons in an interface, at least one of the groups must be placed within a container, such as a group box, panel, or table layout panel; otherwise, the radio buttons are considered to be in the same group and only one can be selected at any one time. In this case, the radio buttons pertaining to the grade choice are contained in the Grade group box, and the radio buttons pertaining to the operation choice are contained in the Operation group box. Placing each group of radio buttons in a separate group box allows the user to select one button from each group.

The minimum number of radio buttons in a group is two, because the only way to deselect a radio button is to select another radio button. The recommended maximum number of radio buttons in a group is seven. Notice that each group of radio buttons in the Math Practice application contains the minimum number of required radio buttons, two.

It is customary in Windows applications to have one of the radio buttons in each group of radio buttons already selected when the user interface first appears. The selected button is called the **default radio button** and is either the radio button that represents the user's most likely choice or the first radio button in the group. You designate a radio button as the default radio button by setting the button's **Checked property** to the Boolean value True. In the Math Practice application, you will make the first radio button in each group the default radio button.

To designate the first radio button in each group as the default radio button:

1 Click the **Grade 1 (1-10)** radio button to select it, and then use the Properties window to set its **Checked** property to **True**. When you set the Checked property to True in the Properties window, a colored dot appears inside the circle in the radio button.

2 Click the **Addition** radio button, then set its **Checked** property to **True**. A colored dot appears inside the circle in the Addition radio button.

》GUI DESIGN TIP

Radio Button Standards

》 Use radio buttons when you want to limit the user to one choice in a group of related and mutually exclusive choices.

》 The minimum number of radio buttons in a group is two, and the recommended maximum number is seven.

》 The label in the radio button's Text property should be entered using sentence capitalization.

》 Assign a unique access key to each radio button in an interface.

》 Use a container (such as a group box, panel, or table layout panel) to create separate groups of radio buttons. Only one button in each group can be selected at any one time.

》 Designate a default radio button in each group of radio buttons.

ADDING A CHECK BOX CONTROL TO THE FORM

You use the **CheckBox tool** in the toolbox to add a check box to an interface. Check boxes work like radio buttons in that they are either selected or deselected only; but that is where the similarity ends. You use radio buttons when you want to limit the user to only one choice from a group of related and mutually exclusive choices. You use **check boxes**, on the other hand, to allow the user to select any number of choices from a group of one or more independent and nonexclusive choices. Unlike radio buttons, where only one button in a group can be selected at any one time, any number of check boxes on a form can be selected at the same time. As with radio buttons, each check box in an interface should be labeled so that the user knows its purpose. You enter the label using sentence capitalization in the check box's Text property. Each check box also should have a unique access key.

To add a check box to the interface:

1 Temporarily display the Toolbox window. Click the **CheckBox** tool in the toolbox, and then drag a check box onto the form, positioning it immediately above the xSummaryGroupBox. The CheckBox1 control appears in the interface.

2 Set the following properties for the CheckBox1 control:

Name:	**xSummaryCheckBox**
Location:	**352, 144**
Text:	**&Display summary**

3 Click the form's **title bar**. Figure 5-39 shows the Display summary check box in the interface.

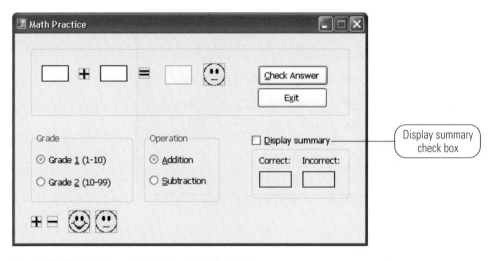

Figure 5-39: Display summary check box added to the interface

》GUI DESIGN TIP

Check Box Standards

》 Use check boxes when you want to allow the user to select any number of choices from a group of one or more independent and nonexclusive choices.

》 The label in the check box's Text property should be entered using sentence capitalization.

》 Assign a unique access key to each check box in an interface.

Now that you have completed the user interface, you can lock the controls in place and then set each control's TabIndex property.

LOCKING THE CONTROLS AND SETTING THE TABINDEX PROPERTY

Recall that when you have completed a user interface, you should lock the controls in place and then set each control's TabIndex property appropriately.

To lock the controls and then set each control's TabIndex property:

1 Right-click the **form**, and then click **Lock Controls** on the context menu.

2 Click **View** on the menu bar, and then click **Tab Order**. Use Figure 5-40 to set the TabIndex values for the controls on the form. (As you learned in Chapter 2, picture boxes do not have a TabIndex property.)

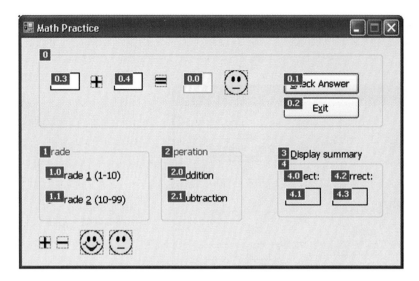

Figure 5-40: Correct TabIndex values

3 Press **Esc** to remove the TabIndex boxes from the form.

Next, you will start the application to observe how you select and deselect radio buttons and check boxes.

To observe how you select and deselect radio buttons and check boxes:

1 Save the solution, and then start the application. Notice that the Grade 1 and Addition radio buttons already are selected, as the colored dot inside each button's circle indicates. Also notice that the four picture boxes located at the bottom of the form, as well as the xSummaryGroupBox and its contents, do not appear in the interface when the application is started. This is because the Visible property of those controls is set to False in the Properties window. You will learn more about the Visible property of a control in Lesson C.

You can select a different radio button by clicking it. You can click either the circle or the text that appears inside the radio button.

2 Click the **Subtraction** radio button. The computer selects the Subtraction radio button as it deselects the Addition radio button. This is because both radio buttons belong to the same group and only one radio button in a group can be selected at any one time.

3 Click the **Grade 2 (10-99)** radio button. The computer selects the Grade 2 (10-99) radio button as it deselects the Grade 1 (1-10) radio button. Here again, the Grade 1 (1-10) and Grade 2 (10-99) radio buttons belong to the same group, so selecting one deselects the other.

4 After selecting a radio button in a group, you then can use the ↑ and ↓ keys on your keyboard to select another radio button in the group. Press ↑ to select the Grade 1 (1-10) radio button, and then press ↓ to select the Grade 2 (10-99) radio button.

5 Press **Tab**. Notice that the focus moves to the Subtraction radio button rather than to the Addition radio button. In Windows applications, only the selected radio button in a group of radio buttons receives the focus.

6 You can select a check box by clicking either the square or the text that appears inside the control. Click the **Display summary** check box to select it. A check mark appears inside the check box to indicate that the check box is selected.

7 Click the **Display summary** check box to deselect it. The computer removes the check mark from the check box.

8 When a check box has the focus, you can use the spacebar on your keyboard to select and deselect it. Press the **spacebar** to select the Display summary check box. A check mark appears inside the check box. Press the **spacebar** again to deselect the check box, which removes the check mark.

9 Click the **Exit** button to end the application. You are returned to the designer window.

CODING THE MATH PRACTICE APPLICATION

The TOE chart for the Math Practice application is shown in Figure 5-41. According to the TOE chart, the Click event procedures for seven of the controls, as well as the Load event procedure for the MainForm, need to be coded. In this lesson, you code all but the Click event procedures for the xExitButton (which already has been coded for you) and the xCheckAnswerButton and xSummaryCheckBox controls (which you code in Lesson C).

Task	Object	Event
1. Display the plus sign in the xOperatorPictureBox 2. Generate and display two random numbers in the xNum1Label and xNum2Label controls	xAdditionRadioButton	Click
Display either the happy face or the neutral face icon (from xCheckAnswerButton)	xFacePictureBox	None
Get and display the user's answer	xAnswerTextBox	None
1. Calculate the correct answer to the math problem 2. Compare the correct answer to the user's answer 3. Display appropriate icon in the xFacePictureBox 4. If the user's answer is correct, generate and display two random numbers in the xNum1Label and xNum2Label controls 5. If the user's answer is incorrect, display the "Try again!" message 6. Add 1 to the number of either correct or incorrect responses 7. Display the number of correct and incorrect responses in the xCorrectLabel and xIncorrectLabel controls, respectively	xCheckAnswerButton	Click
Display the number of correct responses (from xCheckAnswerButton)	xCorrectLabel	None
End the application	xExitButton	Click
Display an addition problem when the form first appears on the screen	MainForm	Load
Generate and display two random numbers in the xNum1Label and xNum2Label controls	xGrade1RadioButton, xGrade2RadioButton	Click
Display the number of incorrect responses (from xCheckAnswerButton)	xIncorrectLabel	None
Display two random numbers (from xGrade1RadioButton, xGrade2RadioButton, xAdditionRadioButton, xSubtractionRadioButton, xCheckAnswerButton)	xNum1Label, xNum2Label	None

Figure 5-41: TOE chart for the Math Practice application *(Continued)* ▶

Task	Object	Event
Display either the plus sign or the minus sign (from xAdditionRadioButton and xSubtractionRadioButton)	xOperatorPictureBox	None
1. Display the minus sign in the xOperatorPictureBox 2. Generate and display two random numbers in the xNum1Label and xNum2Label controls	xSubtractionRadioButton	Click
Display or hide the xSummaryGroupBox	xSummaryCheckBox	Click

Figure 5-41: TOE chart for the Math Practice application

Notice that the task of generating and displaying two random numbers in the xNum1Label and xNum2Label controls appears in the Task column for five of the controls. For example, the task is listed as Step 2 for the xAdditionRadioButton and xSubtractionRadioButton controls. It is listed as the only task for the xGrade1RadioButton and xGrade2RadioButton controls, and it also appears in Step 4 for the xCheckAnswerButton. Rather than entering the appropriate code in the Click event procedures for each of the five controls, you will enter the code in an independent Sub procedure. You then will have the five Click event procedures call (or invoke) the Sub procedure. First, you will learn how to create an independent Sub procedure.

CREATING AN INDEPENDENT SUB PROCEDURE

There are two types of Sub procedures in Visual Basic: event procedures and independent Sub procedures. The procedures that you coded in previous chapters were event procedures. An event procedure is a Sub procedure that is associated with a specific object and event, such as a button's Click event or a text box's TextChanged event. As you already know, the computer automatically processes an event procedure when the event occurs. An **independent Sub procedure**, on the other hand, is a collection of code that can be invoked from one or more places in an application. Unlike an event procedure, an independent Sub procedure is independent of any object and event, and is processed only when called, or invoked, from code.

Programmers use independent Sub procedures for two reasons. First, they allow the programmer to avoid duplicating code in different parts of a program. If different sections of a program need to perform the same task, it is more efficient to enter the appropriate code once, in a procedure, and then call the procedure to perform its task when needed. Second, procedures allow large and complex applications, which typically are written by a team of programmers, to be broken into small and manageable tasks. Each member of the team is assigned one or more tasks to code as a procedure. When each programmer completes his or her procedure, all of the procedures are gathered together into one application.

As do all procedures, independent Sub procedures have both a procedure header and procedure footer. In most cases, the procedure header begins with the `Private` keyword, which indicates that the procedure can be used only by the other procedures in the current form. Following the `Private` keyword in the procedure header is the `Sub` keyword, which identifies the procedure as a Sub procedure. After the `Sub` keyword is the procedure name. The rules for naming an independent Sub procedure are the same as those for naming variables and constants. Recall that the rules state that the name must begin with a letter or an underscore and can contain only letters, numbers, and the underscore character. No punctuation characters or spaces are allowed in the name, and the name cannot be a reserved word. In addition, the recommended maximum number of characters to include in a procedure name is 32.

Procedure names are entered using Pascal case. Recall that Pascal case means that you capitalize the first letter in the name and the first letter of each subsequent word in the name. You should select a descriptive name for the Sub procedure. The name should indicate the task the procedure performs. It is a common practice to begin the name with a verb. For example, a good name for a Sub procedure that generates and displays two random numbers is GenerateAndDisplayNumbers.

Following the procedure name in the procedure header is a set of parentheses that contains an optional *parameterlist*. The *parameterlist* lists the data type and name of one or more memory locations, called parameters. The parameters store the information passed to the procedure when it is invoked. You will learn about parameters in Chapter 7.

Unlike the procedure header, which varies with each procedure, the procedure footer for a Sub procedure is always `End Sub`. Between the procedure header and procedure footer, you enter the instructions you want the computer to process when the procedure is invoked.

To create an independent Sub procedure named GenerateAndDisplayNumbers:

1 Open the Code Editor window. Notice that the xExitButton's Click event procedure already contains the appropriate code.

2 Replace the <your name> and <current date> text in the comments with your name and the current date.

3 Click the blank line above the `End Class` statement, then press **Enter** to insert another blank line. In the new blank line, type **private sub GenerateAndDisplayNumbers()** and press **Enter**. When you press Enter, the Code Editor automatically enters the procedure footer for you, as shown in Figure 5-42.

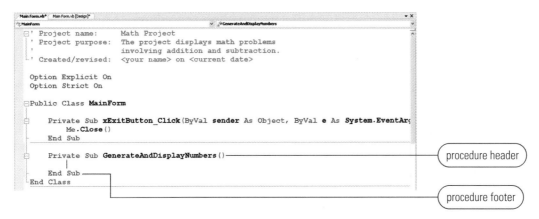

Figure 5-42: GenerateAndDisplayNumbers procedure header and footer

Figure 5-43 shows the pseudocode for the GenerateAndDisplayNumbers procedure.

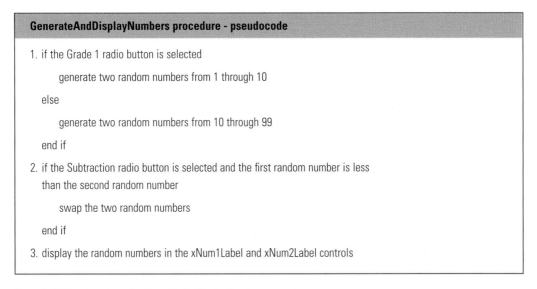

Figure 5-43: Pseudocode for the GenerateAndDisplayNumbers procedure

To begin coding the GenerateAndDisplayNumbers procedure:

1 Type **' generates and displays two random numbers** and press **Enter** twice.

Recall that before you begin coding a procedure, you first study the procedure's pseudocode to determine the variables and named constants (if any) the procedure will use. In this case, the procedure will not use any named constants. However, it will use two Integer variables to store the two random numbers it generates.

2 Type **dim randomNum1 as integer** and press **Enter**. Then type **dim randomNum2 as integer** and press **Enter** twice.

Step 1 in the pseudocode is to determine whether the Grade 1 radio button is selected in the interface. You can use the radio button's Checked property to make that determination. The Checked property will contain the Boolean value True if the radio button is selected; otherwise, it will contain the Boolean value False.

3 Type **' generate random numbers** and press **Enter**, then type **if me.xGrade1RadioButton.checked then** and press **Enter**.

If the Grade 1 radio button is selected, then the GenerateAndDisplayNumbers procedure should generate two random numbers from 1 through 10.

GENERATING RANDOM NUMBERS

Visual Basic provides a **pseudo-random number generator**, which is a device that produces a sequence of numbers that meet certain statistical requirements for randomness. Pseudo-random numbers are chosen with equal probability from a finite set of numbers. The chosen numbers are not completely random because a definite mathematical algorithm is used to select them, but they are sufficiently random for practical purposes. Figure 5-44 shows the syntax you use to generate random integers. The figure also includes several examples of generating random integers.

>> TIP

You also can write the condition in Step 3 as Me.xGrade1Radio Button.Checked = True.

>> TIP

In Exercise 5 at the end of this lesson, you learn how to use the Random.NextDouble method to generate a random floating-point number.

Using the pseudo-random number generator

Syntax

Dim *randomObjectName* **As New Random**

randomObjectName.**Next**(*minValue*, *maxValue*)

Examples

```
Dim number As Integer
Dim randomGenerator As New Random
number = randomGenerator.Next(0, 51)
```

creates a Random object named randomGenerator, then assigns (to the number variable) a random integer that is greater than or equal to 0, but less than 51

```
Dim number As Integer
Dim randomGenerator As New Random
number = randomGenerator.Next(50, 100)
```

creates a Random object named randomGenerator, then assigns (to the number variable) a random integer that is greater than or equal to 50, but less than 100

```
Dim number As Integer
Dim randomGenerator As New Random
number = randomGenerator.Next(-10, 0)
```

creates a Random object named randomGenerator, then assigns (to the number variable) a random integer that is greater than or equal to –10, but less than 0

Figure 5-44: Syntax and examples of generating random integers

To use the pseudo-random number generator in a procedure, you first create a Random object using the Dim statement, as shown in Figure 5-44. The Random object represents the pseudo-random number generator in the procedure. After creating a Random object, you can generate random integers using the **Random.Next method**. In the method's syntax (shown in Figure 5-44), *randomObjectName* is the name of the Random object. The *minValue* and *maxValue* arguments in the syntax must be integers, and *minValue* must be less than *maxValue*. The Random.Next method returns an integer that is greater than or equal to *minValue*, but less than *maxValue*.

In the first example shown in Figure 5-44, the number = randomGenerator.Next(0, 51) statement assigns (to the number variable) a random integer that is greater than or equal to 0, but less than 51. The number = randomGenerator.Next(50, 100) statement in the second example assigns a random integer that is greater than or equal to 50, but less than 100. In the last example, the number = randomGenerator.Next(-10, 0) statement assigns a random integer that is greater than or equal to –10, but less than 0.

According to the pseudocode shown earlier in Figure 5-43, if the Grade 1 radio button is selected, then the GenerateAndDisplayNumbers procedure should generate two random numbers from 1 through 10. To generate numbers within that range, you use the number 1 as the *minValue* and the number 11 as the *maxValue*. If the Grade 1 radio button is not selected, then the Grade 2 button must be selected. In that case, the GenerateAndDisplayNumbers procedure should generate two random numbers from 10 through 99. To generate numbers within that range, you use the numbers 10 and 100 as the *minValue* and *maxValue*, respectively.

To continue coding the GenerateAndDisplayNumbers procedure:

1 Enter the additional code shown in Figure 5-45, then position the insertion point as shown in the figure.

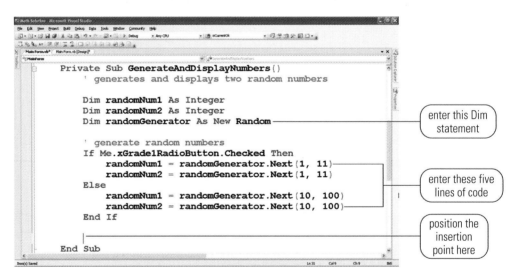

Figure 5-45: Random number generation code entered in the procedure

Step 2 in the pseudocode is to determine whether the Subtraction radio button is selected and whether the first random number is less than the second random number. If both conditions are true, then the procedure should swap (interchange) the two random numbers, because no subtraction problem should result in a negative number.

2 Enter the additional code shown in Figure 5-46, then position the insertion point as shown in the figure.

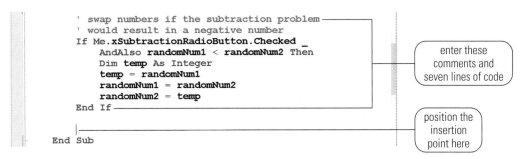

```
    ' swap numbers if the subtraction problem
    ' would result in a negative number
    If Me.xSubtractionRadioButton.Checked _
        AndAlso randomNum1 < randomNum2 Then
        Dim temp As Integer
        temp = randomNum1
        randomNum1 = randomNum2
        randomNum2 = temp
    End If

End Sub
```

enter these comments and seven lines of code

position the insertion point here

Figure 5-46: Additional code entered in the procedure

The last step in the pseudocode is to display the random numbers in the xNum1Label and xNum2Label controls.

3 Type **' display numbers** and press **Enter**. Type **me.xNum1Label.text = convert.tostring(randomnum1)** and press **Enter**, then type **me.xNum2Label.text = convert.tostring(randomnum2)** and press **Enter**. Figure 5-47 shows the completed GenerateAndDisplayNumbers procedure.

```
Private Sub GenerateAndDisplayNumbers()
    ' generates and displays two random numbers

    Dim randomNum1 As Integer
    Dim randomNum2 As Integer
    Dim randomGenerator As New Random

    ' generate random numbers
    If Me.xGrade1RadioButton.Checked Then
        randomNum1 = randomGenerator.Next(1, 11)
        randomNum2 = randomGenerator.Next(1, 11)
    Else
        randomNum1 = randomGenerator.Next(10, 100)
        randomNum2 = randomGenerator.Next(10, 100)
    End If

    ' swap numbers if the subtraction problem
    ' would result in a negative number
```

Figure 5-47: Completed GenerateAndDisplayNumbers procedure *(Continued)* ▶

```
        If Me.xSubtractionRadioButton.Checked _
            AndAlso randomNum1 < randomNum2 Then
            Dim temp As Integer
            temp = randomNum1
            randomNum1 = randomNum2
            randomNum2 = temp
        End If

        ' display numbers
        Me.xNum1Label.Text = Convert.ToString(randomNum1)
        Me.xNum2Label.Text = Convert.ToString(randomNum2)
    End Sub
```

Figure 5-47: Completed GenerateAndDisplayNumbers procedure

4 Save the solution.

Next, you will code the Click event procedures for the Grade 1 and Grade 2 radio buttons.

CODING THE XGRADE1RADIO BUTTON AND XGRADE2RADIO BUTTON CLICK EVENT PROCEDURES

According to the TOE chart shown earlier in Figure 5-41, the xGrade1RadioButton and xGrade2RadioButton controls should generate and display two random numbers when clicked. Recall that the code to generate and display the random numbers is entered in the GenerateAndDisplayNumbers procedure. The xGrade1RadioButton and xGrade2RadioButton controls can use the code entered in the GenerateAndDisplayNumbers procedure simply by invoking the procedure. You can invoke an independent Sub procedure using the **Call statement**, whose syntax is **Call** *procedurename*([*argumentlist*]). In the syntax, *procedurename* is the name of the procedure you are invoking (calling), and *argumentlist* (which is optional) is a comma-separated list of arguments you want passed to the procedure. If you have no information to pass to the procedure that you are calling, as is the case in the GenerateAndDisplayNumbers procedure, you simply include an empty set of parentheses after the *procedurename*, like this: `Call GenerateAndDisplayNumbers()`. (You will learn how to pass information to a procedure in Chapter 7.) Figure 5-48 shows two examples of including the `Call GenerateAndDisplayNumbers()` statement in the Click event procedures for the xGrade1RadioButton and xGrade2RadioButton controls.

»TIP

The word "Call" is optional when invoking a Sub procedure. In other words, you can call the GenerateAndDisplayNumbers procedure using either the statement `Call GenerateAndDisplayNumbers()` or the statement `GenerateAndDisplayNumbers()`.

Example 1

```
Private Sub xGrade1RadioButton_Click(ByVal sender As Object, _
        ByVal e As System.EventArgs) _
        Handles xGrade1RadioButton.Click
    Call GenerateAndDisplayNumbers()
End Sub

Private Sub xGrade2RadioButton_Click(ByVal sender As Object, _
        ByVal e As System.EventArgs) _
        Handles xGrade2RadioButton.Click
    Call GenerateAndDisplayNumbers()
End Sub
```

Example 2

```
Private Sub ProcessGradeRadioButtons(ByVal sender As Object, _
        ByVal e As System.EventArgs) _
        Handles xGrade1RadioButton.Click, _
        xGrade2RadioButton.Click
    Call GenerateAndDisplayNumbers()
End Sub
```

Figure 5-48: Two examples of including the Call statement in the Click event procedures for the grade radio buttons

In the first example shown in Figure 5-48, the Call statement is entered in both Click event procedures. In the second example, the Call statement is entered in a procedure named ProcessGradeRadioButtons. According to its Handles section, the ProcessGradeRadioButtons procedure is processed when the Click event occurs for either the xGrade1RadioButton or the xGrade2RadioButton. In this case, neither example is better than the other; both simply represent different ways of performing the same task.

To call the GenerateAndDisplayNumbers procedure when the Grade 1 and Grade 2 radio buttons are clicked:

1 Open the code template for the xGrade1RadioButton's Click event procedure.

2 Change `xGrade1RadioButton_Click`, which appears after `Private Sub` in the procedure header, to **ProcessGradeRadioButtons**.

3 Click immediately before the word `Handles` in the procedure header. Type _ (the underscore, which is the line continuation character) and press **Enter**, then press **Tab** twice to indent the line.

4 Position the insertion point at the end of the `Handles xGrade1RadioButton.Click` line. Type **, xGrade2RadioButton.Click**, then press **Enter**.

5 Type **call generateanddisplaynumbers()**, then click the blank line below the Call statement. Figure 5-49 shows the completed ProcessGradeRadioButtons procedure.

>>TIP

An easy way to enter the
Call GenerateAnd
DisplayNumbers()
statement is to type *call*,
press the Spacebar,
press Ctrl+Spacebar, type
ge, and then press Tab.

```
Private Sub ProcessGradeRadioButtons(ByVal sender As Object, ByVal e
        Handles xGrade1RadioButton.Click, xGrade2RadioButton.Click
    Call GenerateAndDisplayNumbers()

End Sub
```

Figure 5-49: Completed ProcessGradeRadioButtons procedure

When the user clicks either the Grade 1 radio button or the Grade 2 radio button, the computer processes the `Call GenerateAndDisplayNumbers()` statement contained in the ProcessGradeRadioButtons procedure. When the Call statement is processed, the computer leaves the ProcessGradeRadioButtons procedure, temporarily, to process the instructions contained in the GenerateAndDisplayNumbers procedure. When the GenerateAndDisplayNumbers procedure ends, which is when the computer processes the procedure's `End Sub` statement, the computer returns to the ProcessGradeRadioButtons procedure, to the line below the Call statement. In the ProcessGradeRadioButtons procedure, the line below the Call statement is the `End Sub` statement, which ends the procedure. In the next set of steps, you will test the code you have entered so far to verify that it is working correctly.

To test the code:

1 Save the solution, then start the application.

2 Click the **Grade 2 (10-99)** radio button. The computer leaves the ProcessGradeRadioButtons procedure, temporarily, to process the instructions in the GenerateAndDisplayNumbers procedure. The GenerateAndDisplayNumbers procedure generates and displays two random integers from 10 through 99, as shown in Figure 5-50. (Do not be concerned if the random numbers on your screen are different from the ones shown in the figure.)

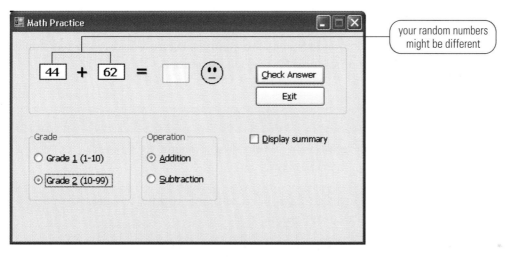

Figure 5-50: Two random numbers displayed in the interface

When the GenerateAndDisplayNumbers procedure ends, the computer returns to the ProcessGradeRadioButtons procedure, to the line immediately below the Call statement. That line is the End Sub statement, which ends the ProcessGradeRadioButtons procedure.

3 Click the **Grade 1 (1-10)** radio button. The computer leaves the ProcessGradeRadioButtons procedure, temporarily, to process the instructions in the GenerateAndDisplayNumbers procedure. The GenerateAndDisplayNumbers procedure generates and displays two random integers from 1 through 10. When the GenerateAndDisplayNumbers procedure ends, the computer returns to the ProcessGradeRadioButtons procedure, to the line immediately below the Call statement. That line is the End Sub statement, which ends the ProcessGradeRadioButtons procedure.

4 Click the **Exit** button. You are returned to the Code Editor window.

Next, you will code the Click event procedures for the Addition and Subtraction radio buttons.

CODING THE XADDITIONRADIO BUTTON AND XSUBTRACTION RADIOBUTTON CLICK EVENT PROCEDURES

According to the TOE chart shown earlier in Figure 5-41, when the user clicks either the xAdditionRadioButton or the xSubtractionRadioButton, the radio button's Click event procedure should display the appropriate mathematical operator (either a plus sign or a minus sign) in the xOperatorPictureBox. It also should generate and display two random numbers in the xNum1Label and xNum2Label controls. Figure 5-51 shows two examples of coding the Click event procedures for these radio buttons.

Example 1

```
Private Sub xAdditionRadioButton_Click(ByVal sender As Object, _
        ByVal e As System.EventArgs) _
        Handles xAdditionRadioButton.Click
    Me.xOperatorPictureBox.Image = Me.xPlusPictureBox.Image
    Call GenerateAndDisplayNumbers()
End Sub

Private Sub xSubtractionRadioButton_Click(ByVal sender As Object, _
        ByVal e As System.EventArgs) _
        Handles xSubtractionRadioButton.Click
    Me.xOperatorPictureBox.Image = Me.xMinusPictureBox.Image
    Call GenerateAndDisplayNumbers()
End Sub
```

Figure 5-51: Examples of coding the Click event procedures for the operation radio buttons *(Continued)* ▶

Example 2

```
Private Sub ProcessOperationRadioButtons(ByVal sender As Object, _
        ByVal e As System.EventArgs) _
        Handles xAdditionRadioButton.Click, _
        xSubtractionRadioButton.Click
    If sender Is Me.xAdditionRadioButton Then
        Me.xOperatorPictureBox.Image = _
            Me.xPlusPictureBox.Image
    Else 'xSubtractionRadioButton is sender
        Me.xOperatorPictureBox.Image = _
            Me.xMinusPictureBox.Image
    End If
    Call GenerateAndDisplayNumbers()
End Sub
```

Figure 5-51: Examples of coding the Click event procedures for the operation radio buttons

In Example 1 in Figure 5-51, both Click event procedures first display the appropriate operator in the xOperatorPictureBox. The Click event procedure for the xAdditionRadioButton displays the plus sign by assigning the Image property of the xPlusPictureBox, which is located at the bottom of the form, to the Image property of the xOperatorPictureBox. Likewise, the Click event procedure for the xSubtractionRadioButton displays the minus sign by assigning the Image property of the xMinusPictureBox, which also is located at the bottom of the form, to the Image property of the xOperatorPictureBox. After assigning the appropriate operator, the Click event procedures shown in Example 1 call the GenerateAndDisplayNumbers procedure to generate and display two random numbers in the xNum1Label and xNum2Label controls.

In the second example shown in Figure 5-51, the code to display the operator and random numbers is entered in the ProcessOperationRadioButtons procedure rather than in the individual Click event procedures. According to the Handles section, the ProcessOperationRadioButtons procedure is processed when either the xAdditionRadioButton's Click event or the xSubtractionRadioButton's Click event occurs. Notice that the procedure uses a selection structure to determine whether the sender parameter contains the address of the xAdditionRadioButton. If it does, then the procedure displays the plus sign in the xOperatorPictureBox; otherwise, it displays the minus sign in the control. Here again, neither example is better than the other; both simply represent two different ways of performing the same task.

» TIP

To remove a graphic from a picture box while a procedure is running, set the picture box's Image property to the keyword Nothing.

To code the xAdditionRadioButton and xSubtractionRadioButton Click event procedures, then test the code:

1 Open the code template for the xAdditionRadioButton's Click event procedure. Change `xAdditionRadioButton_Click`, which appears after `Private Sub` in the procedure header, to **ProcessOperationRadioButtons**.

2 Click immediately before the word `Handles` in the procedure header. Type _ (the underscore, which is the line continuation character) and press **Enter**, then press **Tab** twice to indent the line.

3 Position the insertion point at the end of the `Handles xAdditionRadioButon.Click` line, then type **, _** (a comma, space, and the underscore) and press **Enter**. Type **xSubtractionRadioButton.click** and press **Enter** twice.

4 Enter the comments and additional code shaded in Figure 5-52, which shows the completed ProcessOperationRadioButtons procedure.

```
Private Sub ProcessOperationRadioButtons(ByVal sender As Object, _
        ByVal e As System.EventArgs) _
        Handles xAdditionRadioButton.Click, _
        xSubtractionRadioButton.Click
    ' display the appropriate operator
    If sender Is Me.xAdditionRadioButton Then
        Me.xOperatorPictureBox.Image = Me.xPlusPictureBox.Image
    Else ' xSubtractionRadioButton is sender
        Me.xOperatorPictureBox.Image = Me.xMinusPictureBox.Image
    End If

    Call GenerateAndDisplayNumbers()
End Sub
```

Figure 5-52: Completed ProcessOperationRadioButtons procedure

5 Save the solution, then start the application. Notice that, even though the Grade 1 and Addition radio buttons are selected in the interface, an addition problem does not automatically appear in the interface. You will fix that problem in the next section.

6 Click the **Subtraction** radio button. A minus sign appears in the xOperatorPictureBox, and two random integers from 1 through 10 appear in the interface.

7 Click the **Addition** radio button. A plus sign appears in the xOperatorPictureBox, and two random integers from 1 through 10 appear in the interface.

8 Click the **Grade 2** radio button. Two random integers from 10 through 99 appear in the interface.

9 Click the **Exit** button. You are returned to the Code Editor window.

In the Math Practice application, you want an addition problem to be displayed automatically when the form first appears on the screen. You can accomplish this task in two ways: either you can use the Call statement to call the GenerateAndDisplayNumbers procedure, or you can use the PerformClick method to invoke the xAdditionRadioButton's Click event procedure. Whichever way you choose, the appropriate code must be entered in the form's Load event procedure, which is the last procedure you code in this lesson.

CODING THE FORM'S LOAD EVENT PROCEDURE

As you learned in Chapter 3, the instructions in a form's **Load event procedure** are processed when the application is started and the form is loaded into memory. The form is not displayed on the screen until all of the instructions in its Load event procedure are processed. To automatically display an addition problem when the Math Practice interface first appears, you can enter either the `Call GenerateAndDisplayNumbers()` statement or the `Me.xAdditionRadioButton.PerformClick()` statement in the MainForm's Load event procedure. The latter statement uses the **PerformClick method** to invoke the Addition radio button's Click event, which causes the computer to process the code contained in the Click event procedure. The PerformClick method's syntax is **[Me.]**_object_**.PerformClick()**, where _object_ is the name of the object whose Click event you want invoked.

To automatically display an addition problem when the Math Practice interface first appears:

1 Click the **Class Name** list arrow in the Code Editor window, then click (**MainForm Events**) in the list. Click the **Method Name** list arrow, and then click **Load** in the list. The template for the MainForm's Load event procedure appears in the Code Editor window.

2 Type **call generateanddisplaynumbers()** and press **Enter**.

3 Save the solution, then start the application. An addition problem appears in the interface, as shown in Figure 5-53.

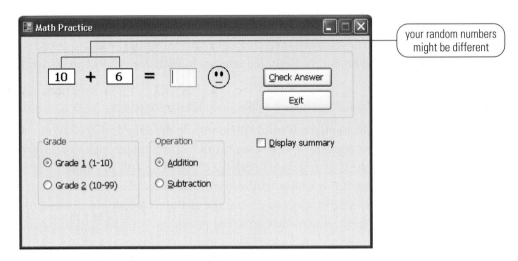

Figure 5-53: Addition problem displayed when the form first appears

4 Click the **Exit** button to end the application. Close the Code Editor window, then close the solution.

You have completed Lesson B. You can either take a break or complete the end-of-lesson questions and exercises before moving on to Lesson C.

SUMMARY

TO LIMIT THE USER TO ONLY ONE CHOICE IN A GROUP OF TWO OR MORE RELATED AND MUTUALLY EXCLUSIVE CHOICES:

» Use the RadioButton tool to add a radio button control to the interface.

» To include two groups of radio buttons in an interface, at least one of the groups must be placed within a container, such as a group box, panel, or table layout panel.

TO ALLOW THE USER TO SELECT ANY NUMBER OF CHOICES FROM A GROUP OF ONE OR MORE INDEPENDENT AND NONEXCLUSIVE CHOICES:

» Use the CheckBox tool to add a check box control to the interface.

TO CREATE A COLLECTION OF CODE THAT CAN BE INVOKED FROM ONE OR MORE PLACES IN A PROGRAM:

» Create an independent Sub procedure. The Sub procedure's name should begin with a verb and indicate the task performed by the procedure. The name should be entered using Pascal case.

TO GENERATE RANDOM INTEGERS:

» Create a Random object to represent the Visual Basic pseudo-random number generator. Typically, the syntax for creating a Random object is **Dim** *randomObjectName* **As New Random**.

» Use the Random.Next method to generate a random integer. The syntax of the Random.Next method is *randomObjectName*.**Next(***minValue***,** *maxValue***)**, where *randomObjectName* is the name of the Random object, and *minValue* and *maxValue* are integers. The Random.Next method returns an integer that is greater than or equal to *minValue*, but less than *maxValue*.

TO CALL (INVOKE) AN INDEPENDENT SUB PROCEDURE:

» Use the Call statement. The syntax of the Call statement is **Call** *procedurename* **(**[*argumentlist*]**)**, where *procedurename* is the name of the procedure you want to call, and *argumentlist* (which is optional) contains the information you want to send to the Sub procedure.

TO PROCESS CODE WHEN THE FORM IS LOADED INTO MEMORY:

» Enter the code in the form's Load event procedure.

TO INVOKE AN OBJECT'S CLICK EVENT FROM CODE:

» Use the PerformClick method. The syntax of the PerformClick method is **[Me.]***object*.**PerformClick()**, where *object* is the name of the object whose Click event you want invoked.

QUESTIONS

1. The minimum number of radio buttons in a group is _____.

   a. one

   b. two

   c. three

   d. There is no minimum number of radio buttons.

2. The text appearing in check boxes and radio buttons should be entered using _____.

   a. book title capitalization

   b. sentence capitalization

   c. either book title capitalization or sentence capitalization

3. It is customary in Windows applications to designate a default check box.

   a. True   b. False

4. A form contains two group boxes, each containing three radio buttons. How many radio buttons can be selected on the form?

   a. one   b. two

   c. three   d. six

5. A form contains two group boxes, each containing three check boxes. How many check boxes can be selected on the form?

   a. one   b. two

   c. three   d. six

6. If a radio button is selected, its _____ property contains the Boolean value True.

   a. Checked   b. On

   c. Selected   d. Selection

7. Which of the following statements declares an object that represents the pseudo-random number generator in a procedure?

   a. `Dim generator As New RandomNumber`

   b. `Dim numberGenerator As New Generator`

   c. `Dim numberRandom As New Random`

   d. `Dim number As New RandomObject`

8. Which of the following statements generates a random number from 1 to 25, inclusive? (The Random object's name is `randomGenerator`.)

    a. `number = randomGenerator.Next(1, 25)`

    b. `number = randomGenerator.Next(1, 26)`

    c. `number = randomGenerator(1, 25)`

    d. `number = randomGenerator.NextNumber(1, 26)`

9. You can use the _____ statement to invoke an independent Sub procedure.

    a. Call   b. Get

    c. Invoke   d. ProcedureCall

10. Which of the following statements invokes the xAlaskaRadioButton's Click event?

    a. `Me.xAlaskaRadioButton.Click()`

    b. `Me.xAlaskaRadioButton.ClickIt`

    c. `Me.Click.xAlaskaRadioButton()`

    d. `Me.xAlaskaRadioButton.PerformClick()`

EXERCISES

1. In this exercise, you use the PerformClick method to invoke a radio button's Click event.

    a. Use Windows to make a copy of the Math Solution folder, which is contained in the VB2005\Chap05 folder. Rename the folder Math Solution-PerformClick.

    b. If necessary, start Visual Studio 2005 or Visual Basic 2005 Express Edition. Open the Math Solution (Math Solution.sln) file contained in the VB2005\Chap05\Math Solution-PerformClick folder. Open the designer window.

    c. Modify the form's Load event procedure so that it uses the PerformClick method to invoke the Addition radio button's Click event.

    d. Save the solution, then start the application. An addition problem automatically appears in the interface.

    e. Click the Exit button to end the application. Close the Code Editor window, then close the solution.

2. In this exercise, you code an application for Woodland School. The application allows a student to select the name of a state and the name of a capital city. After making his or her selections, the student can click the Verify Answer button to verify that the selected city is the capital of the selected state.

   a. If necessary, start Visual Studio 2005 or Visual Basic 2005 Express Edition. Open the Capitals Solution (Capitals Solution.sln) file, which is contained in the VB2005\Chap05\Capitals Solution folder. If necessary, open the designer window.

   b. Designate the first radio button in each group as the default radio button for the group.

   c. Enter the code to invoke the Click event for the two default radio buttons when the form is read into the computer's internal memory.

   d. Declare two module-level String variables named `capital` and `choice`.

   e. Code the state radio buttons' Click event procedures so that each assigns the appropriate capital to the `capital` variable, and each removes the contents of the xMsgLabel.

   f. Code the capital radio buttons' Click event procedures so that each assigns the selected capital to the `choice` variable, and each removes the contents of the xMsgLabel.

   g. Code the Verify Answer button's Click event procedure so that it displays the word "Correct" in the xMsgLabel if the student selected the appropriate capital; otherwise, display the word "Incorrect".

   h. Save the solution, then start the application. Test the application by selecting Illinois from the state group and Salem from the capital group. Click the Verify Answer button. The word "Incorrect" appears in the xMsgLabel. Now select Wisconsin from the state group and Madison from the capital group. Click the Verify Answer button. The word "Correct" appears in the xMsgLabel.

   i. Click the Exit button to end the application. Close the Code Editor window, then close the solution.

3. In this exercise, you code an application for Professor Juarez. The application displays a letter grade based on the average of three test scores entered by the professor.

   a. If necessary, start Visual Studio 2005 or Visual Basic 2005 Express Edition. Open the Juarez Solution (Juarez Solution.sln) file, which is contained in the VB2005\Chap05\Juarez Solution folder. If necessary, open the designer window.

   b. Code the Display Grade button's Click event procedure so that it displays the appropriate letter grade based on the average of three test scores. Each test is worth

100 points. Display an appropriate message if any of the test scores are not numeric. Use the following chart to complete the procedure:

Test average	Grade
90–100	A
80–89	B
70–79	C
60–69	D
below 60	F

 c. When the user makes a change to the contents of a text box, the application should remove the contents of the xGradeLabel. Code the appropriate event procedures.

 d. The application should select a text box's existing text when the text box receives the focus. Code the appropriate event procedures.

 e. Save the solution, then start the application. Test the application three times. For the first test, use scores of 90, 95, and 100. For the second test, use scores of 83, 72, and 65. For the third test, use scores of 40, 30, and 20.

 f. Click the Exit button to end the application. Close the Code Editor window, then close the solution.

4. In this exercise, you create an application that allows the user to enter the total number of calories and grams of fat contained in a specific food. The application should calculate and display two values: the food's fat calories (the number of calories attributed to fat) and its fat percentage (the ratio of the food's fat calories to its total calories). You calculate the number of fat calories in a food by multiplying the number of fat grams contained in the food by the number nine, because each gram of fat contains nine calories. To calculate the fat percentage, you divide the food's fat calories by its total calories, and then multiply the result by 100. The application should display the message "This food is high in fat." when the fat percentage is over 30%; otherwise, it should display the message "This food is not high in fat."

 a. If necessary, start Visual Studio 2005 or Visual Basic 2005 Express Edition.

 b. Create a Visual Basic Windows-based application. Name the solution Fat Calculator Solution, and name the project Fat Calculator Project. Save the application in the VB2005\Chap05 folder.

 c. Assign the filename Main Form.vb to the form file object.

 d. Assign the name MainForm to the form.

e. Design an appropriate interface. Use the GUI design guidelines listed in Appendix A to verify that the interface you create adheres to the GUI standards outlined in this book.

f. Code the application.

g. Save the solution, then start the application. Test the application using both valid and invalid data.

h. Stop the application. Close the Code Editor window, then close the solution.

DISCOVERY EXERCISE

5. In this exercise, you generate and display random floating-point numbers.

a. If necessary, start Visual Studio 2005 or Visual Basic 2005 Express Edition. Open the Random Float Solution (Random Float Solution.sln) file, which is contained in the VB2005\Chap05\Random Float Solution folder. If necessary, open the designer window.

b. You can use the Random.NextDouble method to return a floating-point random number that is greater than or equal to 0.0, but less than 1.0. The syntax of the Random.NextDouble method is *randomObjectName*.**NextDouble**. Code the Display Random Number button's Click event procedure so that it displays a random floating-point number in the xNumberLabel.

c. Save the solution, then start the application. Click the Display Random Number button several times. Each time you click the button, a random number that is greater than or equal to 0.0, but less than 1.0, appears in the xNumberLabel.

d. Click the Exit button to end the application.

e. You can use the following formula to generate random floating-point numbers within a specified range: (*maxValue* – *minValue* + **1**) * *randomObjectName*.**NextDouble** + *minValue*. For example, assuming the Random object's name is `randomGenerator`, the formula `(10 - 1 + 1) * randomGenerator.NextDouble + 1` generates floating-point numbers that are greater than or equal to 1.0, but less than 11.0. Modify the Display Random Number button's Click event procedure so that it displays a random floating-point number that is greater than or equal to 25.0, but less than 51.0. Display two decimal places in the floating-point number.

f. Save the solution, then start the application. Click the Display Random Number button several times. Each time you click the button, a random number that is greater than or equal to 25.0, but less than 51.0, appears in the xNumberLabel.

g. Click the Exit button to end the application. Close the Code Editor window, then close the solution.

LESSON C
OBJECTIVES

AFTER STUDYING LESSON C, YOU SHOULD
BE ABLE TO:

» Code a check box's Click event procedure

» Display and hide a control

COMPLETING THE MATH PRACTICE APPLICATION

CODING THE XCHECKANSWERBUTTON CLICK EVENT PROCEDURE

To complete the Math Practice application, you still need to code the Click event procedures for the xCheckAnswerButton and the xSummaryCheckBox. Before you can code these procedures, you need to open the Math Practice application from Lesson B.

To open the Math Practice application from Lesson B:

1 Start Visual Studio 2005 or Visual Basic 2005 Express Edition, if necessary, and close the Start Page window.

2 Open the **Math Solution** (Math Solution.sln) file, which is contained in the VB2005\Chap05\Math Solution folder. If necessary, open the designer window. Figure 5-54 shows the user interface for the Math Practice application.

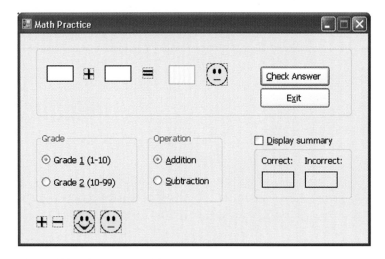

Figure 5-54: User interface for the Math Practice application

Figure 5-55 shows the pseudocode for the xCheckAnswerButton's Click event procedure.

xCheckAnswerButton's Click event procedure - pseudocode

1. if the Addition radio button is selected

       calculate the correct answer by adding the random number stored
       in the xNum1Label to the random number stored in the xNum2Label

  else

       calculate the correct answer by subtracting the random number stored
       in the xNum2Label from the random number stored in the xNum1Label

  end if

2. if the user's answer equals the correct answer

       display the happy face icon in the xFacePictureBox

       add 1 to the number of correct responses

       clear the contents of the xAnswerTextBox

       call the GenerateAndDisplayNumbers procedure to generate and display two random numbers

  else

       display the neutral face icon in the xFacePictureBox

       add 1 to the number of incorrect responses

       display the "Try again!" message in a message box

       select the existing text in the xAnswerTextBox

  end if

3. send the focus to the xAnswerTextBox

4. display the number of correct and incorrect responses in the xCorrectLabel and xIncorrectLabel controls

Figure 5-55: Pseudocode for the xCheckAnswerButton's Click event procedure

Study the procedure's pseudocode to determine the variables and named constants (if any) the procedure will use. In this case, the procedure will use a named constant for the "Try again!" message. It also will use the six Integer variables listed in Figure 5-56.

Name	Purpose
randomNum1	store the random number contained in the xNum1Label
randomNum2	store the random number contained in the xNum2Label
userAnswer	store the user's answer, which is contained in the xAnswerTextBox
correctAnswer	store the correct answer
correctResponses	store the number of correct responses made by the user; declare as a static variable
incorrectResponses	store the number of incorrect responses made by the user; declare as a static variable

Figure 5-56: Variables used by the xCheckAnswerButton's Click event procedure

Notice that the correctResponses and incorrectResponses variables must be declared as static variables. As you learned in Chapter 3, a static variable is a local variable that retains its value even when the procedure in which it is declared ends. In this case, the correctResponses and incorrectResponses variables need to be static variables because they must keep a running tally of the number of correct and incorrect responses.

To begin coding the xCheckAnswerButton's Click event procedure:

1 Open the Code Editor window, then open the code template for the xCheckAnswerButton's Click event procedure.

2 Type ' **calculates the correct answer and then compares** and press **Enter**, then type ' **the correct answer to the user's answer** and press **Enter**.

3 Type ' **keeps track of the number of correct** and press **Enter**. Type ' **and incorrect responses** and press **Enter** twice.

4 Type **const Message as string = "Try again!"** and press **Enter**.

5 Enter the following six statements. Press **Enter** twice after typing the last statement.

   dim randomNum1 as integer

   dim randomNum2 as integer

   dim userAnswer as integer

   dim correctAnswer as integer

   static correctResponses as integer

   static incorrectResponses as integer

6 Now you will assign the two random numbers and the user's answer to the appropriate variables. Enter the following comment and three assignment statements. Press **Enter** twice after typing the last assignment statement.

' assign random numbers and user's answer

integer.tryparse(me.xNum1Label.text, randomnum1)

integer.tryparse(me.xNum2Label.text, randomnum2)

integer.tryparse(me.xAnswerTextBox.text, useranswer)

Step 1 in the pseudocode shown in Figure 5-55 is to determine whether the Addition radio button is selected in the interface. If it is, the procedure should add the two random numbers together; otherwise, it should subtract the second random number from the first random number.

7 Enter the comments and selection structure shown in Figure 5-57, then position the insertion point as shown in the figure. (You also could use an If...Then...Else statement rather than a Select Case statement to determine whether the Addition radio button is selected.)

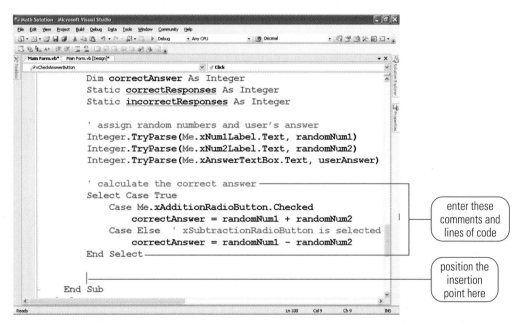

Figure 5-57: Comments and selection structure entered in the procedure

8 Save the solution.

Step 2 in the pseudocode shown in Figure 5-55 is to determine whether the user's answer is correct. You can do so by comparing the contents of the userAnswer variable to the contents of the correctAnswer variable.

To continue coding the xCheckAnswerButton's Click event procedure:

1 Type **' determine whether the user's answer is correct** and press **Enter**, then type **if useranswer = correctanswer then** and press **Enter**. (You also could use a Select Case statement to compare the contents of both variables.)

If the user's answer equals the correct answer, the procedure should perform the following four tasks: display the happy face icon in the xFacePictureBox, add the number 1 to the number of correct responses, clear the contents of the xAnswerTextBox, and call the GenerateAndDisplayNumbers procedure to generate and display two random numbers.

2 Enter the additional code shown in Figure 5-58, then position the insertion point as shown in the figure.

```
' determine whether the user's answer is correct
If userAnswer = correctAnswer Then
    Me.xFacePictureBox.Image = Me.xHappyPictureBox.Image ─── enter these
    correctResponses = correctResponses + 1                   four lines
    Me.xAnswerTextBox.Text = String.Empty                     of code
    Call GenerateAndDisplayNumbers()  ───────────────────┐
    |                                                    │
    End If                                               │── position the
End Sub  ──────────────────────────────────────────────┘    insertion
                                                             point here
```

Figure 5-58: Additional code entered in the procedure

If the user's answer is not correct, the procedure should perform the following four tasks: display the neutral face icon in the xFacePictureBox, add the number 1 to the number of incorrect responses, display the "Try again!" message in a message box, and select the existing text in the xAnswerTextBox.

3 Type **else** and press **Enter**. Complete the If...Then...Else statement by entering the additional code shown in Figure 5-59, then position the insertion point as shown in the figure.

```
' determine whether the user's answer is correct
If userAnswer = correctAnswer Then
    Me.xFacePictureBox.Image = Me.xHappyPictureBox.Image
    correctResponses = correctResponses + 1
    Me.xAnswerTextBox.Text = String.Empty
    Call GenerateAndDisplayNumbers()
Else
    Me.xFacePictureBox.Image = Me.xNeutralPictureBox.Image ─┐ enter these
    incorrectResponses = incorrectResponses + 1             │ five lines
    MessageBox.Show(Message, "Math Practice", _             │ of code
        MessageBoxButtons.OK, MessageBoxIcon.Information)    │
    Me.xAnswerTextBox.SelectAll()  ────────────────────────┘
    End If
                                                           ┌ position the
    |  ─────────────────────────────────────────────────── │ insertion
End Sub                                                    └ point here
```

Figure 5-59: Completed If...Then...Else statement shown in the procedure

The last two steps in the pseudocode shown in Figure 5-55 are to send the focus to the xAnswerTextBox and then display the number of correct and incorrect responses in the xCorrectLabel and xIncorrectLabel controls.

4 Type **me.xAnswerTextBox.focus()** and press **Enter**.

5 Type **me.xCorrectLabel.text = convert.tostring(correctresponses)** and press **Enter**.

6 Type **me.xIncorrectLabel.text = convert.tostring(incorrectresponses)** and press **Enter**.

7 Save the solution.

Before testing the code in the Check Answer button's Click event procedure, you will code the xSummaryCheckBox's Click event procedure.

CODING THE XSUMMARY CHECKBOX CLICK EVENT PROCEDURE

Recall that the four picture boxes located at the bottom of the form do not appear in the interface when the Math Practice application is started. This is because their Visible property is set to False in the Properties window. The Visible property of the xSummaryGroupBox is also set to False, which explains why you do not see the control and its contents when the form appears on the screen. As you learned in Chapter 4, the group box and the controls contained in the group box are treated as one unit. As a result, hiding the group box also hides the controls contained within the group box.

According to the TOE chart shown earlier in Figure 5-41, the xSummaryCheckBox's Click event procedure is responsible for both displaying and hiding the xSummaryGroupBox. The procedure should display the group box when the user selects the check box, and it should hide the group box when the user deselects the check box. You can use a check box's Checked property to determine whether the check box was selected or deselected. If the Checked property contains the Boolean value True, then the check box was selected. If it contains the Boolean value False, then the check box was deselected.

To code the xSummaryCheckBox's Click event procedure:

1 Open the code template for the xSummaryCheckBox's Click event procedure.

2 Type **if me.xSummaryCheckBox.checked then** and press **Enter**. (You also could use the Select Case statement rather than the If...Then...Else statement to determine the value stored in the Checked property.)

If the user selected the xSummaryCheckBox, then the procedure should display the xSummaryGroupBox. You can do so by setting the xSummaryGroupBox's Visible property to the Boolean value True.

3 Type **me.xSummaryGroupBox.visible = true** and press **Enter**.

If the user deselected the xSummaryCheckBox, then the procedure should hide the xSummaryGroupBox. You can do so by setting the xSummaryGroupBox's Visible property to the Boolean value False.

4 Type **else** and press **Enter**, then type **me.xSummaryGroupBox.visible = false**. Figure 5-60 shows the application's code.

> **» TIP**
> You also can write the condition in Step 2 as
> `If Me.xSummary`
> `CheckBox.Checked =`
> `True Then`.

```
' Project name: Math Project
' Project purpose: The project displays math problems
'                  involving addition and subtraction.
' Created/revised: <your name> on <current date>
Option Explicit On
Option Strict On

Public Class MainForm

    Private Sub xExitButton_Click(ByVal sender As Object, _
        ByVal e As System.EventArgs) Handles xExitButton.Click
        Me.Close()
    End Sub

    Private Sub GenerateAndDisplayNumbers()
        ' generates and displays two random numbers

        Dim randomNum1 As Integer
        Dim randomNum2 As Integer
        Dim randomGenerator As New Random
```

Figure 5-60: Math Practice application's code *(Continued)*

▶

```
                ' generate random numbers
            If Me.xGrade1RadioButton.Checked Then
                    randomNum1 = randomGenerator.Next(1, 11)
                    randomNum2 = randomGenerator.Next(1, 11)
            Else
                    randomNum1 = randomGenerator.Next(10, 100)
                    randomNum2 = randomGenerator.Next(10, 100)
            End If

                ' swap numbers if the subtraction problem
                ' would result in a negative number
            If Me.xSubtractionRadioButton.Checked _
                    AndAlso randomNum1 < randomNum2 Then
                    Dim temp As Integer
                    temp = randomNum1
                    randomNum1 = randomNum2
                    randomNum2 = temp
            End If

                ' display numbers
            Me.xNum1Label.Text = Convert.ToString(randomNum1)
            Me.xNum2Label.Text = Convert.ToString(randomNum2)
        End Sub

        Private Sub ProcessGradeRadioButtons(ByVal sender As Object, _
                ByVal e As System.EventArgs) _
                Handles xGrade1RadioButton.Click, xGrade2RadioButton.Click
            Call GenerateAndDisplayNumbers()
        End Sub

        Private Sub ProcessOperationRadioButtons(ByVal sender As Object, _
                ByVal e As System.EventArgs) _
                Handles xAdditionRadioButton.Click, _
                xSubtractionRadioButton.Click

            ' display the appropriate operator
            If sender Is Me.xAdditionRadioButton Then
                    Me.xOperatorPictureBox.Image = Me.xPlusPictureBox.Image
            Else      ' xSubtractionRadioButton is sender
                    Me.xOperatorPictureBox.Image = Me.xMinusPictureBox.Image
            End If

            Call GenerateAndDisplayNumbers()

        End Sub
```

Figure 5-60: Math Practice application's code *(Continued)* ▶

```vb
      Private Sub MainForm_Load(ByVal sender As Object, _
            ByVal e As System.EventArgs) Handles Me.Load
        Call GenerateAndDisplayNumbers()

   End Sub

   Private Sub xCheckAnswerButton_Click(ByVal sender As Object, _
            ByVal e As System.EventArgs) Handles
            xCheckAnswerButton.Click

     ' calculates the correct answer and then compares
     ' the correct answer to the user's answer
     ' keeps track of the number of correct
     ' and incorrect responses

     Const Message As String = "Try again!"
     Dim randomNum1 As Integer
     Dim randomNum2 As Integer
     Dim userAnswer As Integer
     Dim correctAnswer As Integer
     Static correctResponses As Integer
     Static incorrectResponses As Integer

     ' assign random numbers and user's answer
     Integer.TryParse(Me.xNum1Label.Text, randomNum1)
     Integer.TryParse(Me.xNum2Label.Text, randomNum2)
     Integer.TryParse(Me.xAnswerTextBox.Text, userAnswer)

     ' calculate the correct answer
     Select Case True
         Case Me.xAdditionRadioButton.Checked
             correctAnswer = randomNum1 + randomNum2
         Case Else ' xSubtractionRadioButton is selected
             correctAnswer = randomNum1 - randomNum2
     End Select

     ' determine whether the user's answer is correct

     If userAnswer = correctAnswer Then
         Me.xFacePictureBox.Image = Me.xHappyPictureBox.Image
         correctResponses = correctResponses + 1
         Me.xAnswerTextBox.Text = String.Empty
         Call GenerateAndDisplayNumbers()
```

Figure 5-60: Math Practice application's code *(Continued)*

▶

```
        Else
             Me.xFacePictureBox.Image = Me.xNeutralPictureBox.Image
             incorrectResponses = incorrectResponses + 1
             MessageBox.Show(Message, "Math Practice", _
                  MessageBoxButtons.OK, MessageBoxIcon.Information)
             Me.xAnswerTextBox.SelectAll()
        End If

        Me.xAnswerTextBox.Focus()
        Me.xCorrectLabel.Text = Convert.ToString(correctResponses)
        Me.xIncorrectLabel.Text = Convert.ToString(incorrectResponses)

    End Sub

    Private Sub xSummaryCheckBox_Click(ByVal sender As Object, _
             ByVal e As System.EventArgs) Handles xSummaryCheckBox.Click
        If Me.xSummaryCheckBox.Checked Then
             Me.xSummaryGroupBox.Visible = True
        Else
             Me.xSummaryGroupBox.Visible = False
        End If
    End Sub
End Class
```

Figure 5-60: Math Practice application's code

When coded, the Click event procedure for a check box will always contain a selection structure that determines whether the check box was selected or deselected. The selection structure is not necessary in a radio button's Click event procedure, because clicking a radio button always selects the button; the user cannot deselect a radio button by clicking it.

Now verify that the Click event procedures for the xCheckAnswerButton and xSummaryCheckBox controls are working correctly.

To test the application's code:

1 Save the solution, then start the application.

2 Type the correct answer to the addition problem appearing in the interface, then press **Enter** to select the Check Answer button, which is the default button on the form. The happy face icon and a new addition problem appear in the interface.

3 Click the **Display summary** check box to select it. A check mark appears in the check box, and the xSummaryGroupBox and its contents appear in the interface. Notice that the label controls within the group box indicate that you have made one correct response and zero incorrect responses.

4 Click inside the text box in which you enter the answer. Type an incorrect answer to the current addition problem, then press **Enter**. A neutral face icon appears in the interface, and a message box appears on the screen, as shown in Figure 5-61.

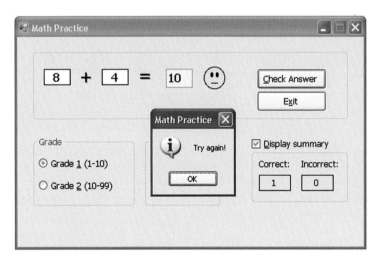

Figure 5-61: Result of entering an incorrect response to the addition problem

5 Click the **OK** button to close the message box. Notice that the number of incorrect responses changes from 0 to 1. Also notice that the incorrect answer is selected in the xAnswerTextBox. You can remove the incorrect answer simply by typing another answer in the text box.

6 Type the correct answer to the current addition problem, then press **Enter**. The number of correct responses changes from 1 to 2, and the happy face icon appears in the interface.

7 Click the **Display summary** check box to deselect it. The check mark is removed from the check box, and the xSummaryGroupBox and its contents disappear from the interface.

8 Click the **Exit** button to end the application. Close the Code Editor window, then close the solution.

You have completed Lesson C and Chapter 5. You can either take a break or complete the end-of-lesson questions and exercises.

SUMMARY

TO DISPLAY OR HIDE A CONTROL:

» Set the control's Visible property to the Boolean value True to display the control. Set the control's Visible property to the Boolean value False to hide the control.

TO CODE A CHECK BOX CONTROL'S CLICK EVENT PROCEDURE:

» Use a selection structure to determine whether the check box was either selected or deselected by the user.

QUESTIONS

1. Which of the following statements hides the xDivisionPictureBox?

    a. `Me.xDivisionPictureBox.Hide`

    b. `Me.xDivisionPictureBox.Hide = True`

    c. `Hide.xDivisionPictureBox = True`

    d. None of the above.

2. If a check box is deselected, its _____ property contains the Boolean value False.

    a. Checked   b. Deselected

    c. On   d. None of the above.

3. When coded, a check box's Click event procedure will always contain a selection structure that determines whether the check box is selected or deselected.

    a. True   b. False

4. Like a check box, a radio button can be deselected by clicking it.

    a. True   b. False

EXERCISES

1. In this exercise, you modify the selection structures contained in the Math Practice application.

   a. Use Windows to make a copy of the Math Solution folder, which is contained in the VB2005\Chap05 folder. Rename the folder Modified Math Solution.

   b. If necessary, start Visual Studio 2005 or Visual Basic 2005 Express Edition. Open the Math Solution (Math Solution.sln) file contained in the VB2005\Chap05\Modified Math Solution folder. Open the designer window.

   c. Change the If...Then...Else statement in the xSummaryCheckBox's Click event procedure to a Select Case statement.

   d. Change the first selection structure in the xCheckAnswerButton's Click event procedure to an If...Then...Else statement.

   e. Change the second selection structure in the xCheckAnswerButton's Click event procedure to a Select Case statement.

   f. Change the If...Then...Else statement in the ProcessOperationRadioButtons procedure to a Select Case statement.

   g. Save the solution, then start the application. Test the application to verify that it is working correctly.

   h. Click the Exit button to end the application. Close the Code Editor window, then close the solution.

2. In this exercise, you code an application for Ned's Health Club. The application calculates a member's monthly dues.

   a. If necessary, start Visual Studio 2005 or Visual Basic 2005 Express Edition. Open the Health Solution (Health Solution.sln) file, which is contained in the VB2005\Chap05\Health Solution folder. If necessary, open the designer window.

   b. Declare a module-level variable named `additionalCharges`.

   c. Code each check box's Click event procedure so that it adds the appropriate additional charge to the `additionalCharges` variable when the check box is selected, and subtracts the appropriate additional amount from the `additionalCharges` variable when the check box is deselected. The additional charges are $30 per month for tennis, $25 per month for golf, and $20 per month for racquetball. Each check box's Click event procedure should display the contents of the `additionalCharges` variable in the xAdditionalLabel, and also remove the contents of the xTotalLabel.

d. Code the Calculate button's Click event procedure so that it calculates the monthly dues. The dues are calculated by adding the basic fee to the total additional charge. Display the total due with a dollar sign and two decimal places.

e. When the user makes a change to the contents of the xBasicTextBox, the application should remove the contents of the xTotalLabel. Code the appropriate event procedure.

f. Save the solution, then start the application. Test the application by entering 80 in the text box, and then selecting the Golf check box. The number 25 appears in the xAdditionalLabel. Click the Calculate button. $105.00 appears in the xTotalLabel.

g. Now select the Tennis and Racquetball check boxes and deselect the Golf check box. The number 50 appears in the xAdditionalLabel. Click the Calculate button. $130.00 appears in the xTotalLabel.

h. Click the Exit button to end the application. Close the Code Editor window, then close the solution.

3. In this exercise, you create an application for Washington High School. The application displays a class rank, which is based on the code entered by the user. Use the following information to code the application:

Code	Rank
1	Freshman
2	Sophomore
3	Junior
4	Senior

a. If necessary, start Visual Studio 2005 or Visual Basic 2005 Express Edition. Open the Washington Solution (Washington Solution.sln) file, which is contained in the VB2005\Chap05\Washington Solution folder. If necessary, open the designer window.

b. Code the application appropriately. Allow the text box to accept only the numeric keys 1, 2, 3, and 4 and the Backspace key. Set the text box's MaxLength property to the number 1; this allows the user to enter only one character in the text box.

c. When a change is made to the code entered in the text box, clear the contents of the label control that displays the rank.

d. Center the rank in the label control.

    e. Save the solution, then start the application. Test the application using codes of 1, 2, 3, and 4. Also test the code using an empty text box (which should not display a rank in the label control). In addition, try to enter characters other than 1, 2, 3, or 4 in the text box.

    f. End the application. Close the Code Editor window, then close the solution.

4. In this exercise, you create an application for Barren Community Center. The application displays a seminar fee, which is based on the membership status and age entered by the user. Use the following information to code the application:

Seminar fee	Criteria
10	Club member younger than 65 years old
5	Club member at least 65 years old
20	Non-member

    a. If necessary, start Visual Studio 2005 or Visual Basic 2005 Express Edition.

    b. Create a Visual Basic Windows-based application. Name the solution Barren Solution, and name the project Barren Project. Save the application in the VB2005\Chap05 folder.

    c. Assign the filename Main Form.vb to the form file object.

    d. Assign the name MainForm to the form.

    e. Design an appropriate interface. Use the GUI design guidelines listed in Appendix A to verify that the interface you create adheres to the GUI standards outlined in this book. Use radio button controls for the status and age choices. Display the seminar fee in a label control. When the user clicks a radio button, clear the contents of the label control that displays the fee.

    f. Code the application.

    g. Save the solution, then start the application. Test the application appropriately.

    h. Stop the application. Close the Code Editor window, then close the solution.

5. In this exercise, you create an application for Golf Pro, a U.S. company that sells golf equipment both domestically and abroad. Each of Golf Pro's salespeople receives a commission based on the total of his or her domestic and international sales. The application you create should allow the user to enter the amount of domestic sales and the amount of international sales. It then should calculate and display the commission. Use the following information to code the application:

Sales	Commission
1–100,000	2% * sales
100,001–400,000	2,000 + 5% * sales over 100,000
400,001 and over	17,000 + 10% * sales over 400,000

a. If necessary, start Visual Studio 2005 or Visual Basic 2005 Express Edition.

b. Create a Visual Basic Windows-based application. Name the solution Golf Pro Solution, and name the project Golf Pro Project. Save the application in the VB2005\Chap05 folder.

c. Assign the filename Main Form.vb to the form file object.

d. Assign the name MainForm to the form.

e. Design an appropriate interface. Use the GUI design guidelines listed in Appendix A to verify that the interface you create adheres to the GUI standards outlined in this book.

f. Code the application. Keep in mind that the sales amounts may contain decimal places.

g. Save the solution, then start the application. Test the application appropriately.

h. Stop the application. Close the Code Editor window, then close the solution.

6. In this exercise, you create an application for Marshall Sales Corporation. Each of the company's salespeople receives a commission based on the amount of his or her sales. The application you create should allow the user to enter the sales amount. It then should calculate and display the commission. Use the following information to code the application:

Sales	Commission
1–100,000	2% * sales
100,001–200,000	4% * sales
200,001–300,000	6% * sales
300,001–400,000	8% * sales
400,001 and over	10% * sales

a. If necessary, start Visual Studio 2005 or Visual Basic 2005 Express Edition.

b. Create a Visual Basic Windows-based application. Name the solution Marshall Solution, and name the project Marshall Project. Save the application in the VB2005\Chap05 folder.

    c. Assign the filename Main Form.vb to the form file object.

    d. Assign the name MainForm to the form.

    e. Design an appropriate interface. Use the GUI design guidelines listed in Appendix A to verify that the interface you create adheres to the GUI standards outlined in this book.

    f. Code the application. Keep in mind that the sales amounts may contain decimal places.

    g. Save the solution, then start the application. Test the application appropriately.

    h. Stop the application. Close the Code Editor window, then close the solution.

7. In this exercise, you create an application for Willow Health Club. The application displays the number of daily calories needed to maintain your current weight. Use the following information to code the application:

Moderately active female: total calories per day = weight multiplied by 12 calories per pound

Relatively inactive female: total calories per day = weight multiplied by 10 calories per pound

Moderately active male: total calories per day = weight multiplied by 15 calories per pound

Relatively inactive male: total calories per day = weight multiplied by 13 calories per pound

    a. If necessary, start Visual Studio 2005 or Visual Basic 2005 Express Edition.

    b. Create a Visual Basic Windows-based application. Name the solution Willow Solution, and name the project Willow Project. Save the application in the VB2005\Chap05 folder.

    c. Assign the filename Main Form.vb to the form file object.

    d. Assign the name MainForm to the form.

    e. Design an appropriate interface. Use the GUI design guidelines listed in Appendix A to verify that the interface you create adheres to the GUI standards outlined in this book.

    f. Code the application.

    g. Save the solution, then start the application. Test the application appropriately.

    h. Stop the application. Close the Code Editor window, then close the solution.

8. In this exercise, you create an application for Johnson Products. The application calculates and displays the price of an order, based on the number of units ordered and the customer's status (either wholesaler or retailer). The price per unit is as follows:

Wholesaler		Retailer	
Number of units	Price per unit ($)	Number of units	Price per unit ($)
1–4	10	1–3	15
5 and over	9	4–8	14
		9 and over	12

a. If necessary, start Visual Studio 2005 or Visual Basic 2005 Express Edition.

b. Create a Visual Basic Windows-based application. Name the solution Johnson Solution, and name the project Johnson Project. Save the application in the VB2005\Chap05 folder.

c. Assign the filename Main Form.vb to the form file object.

d. Assign the name MainForm to the form.

e. Design an appropriate interface. Use the GUI design guidelines listed in Appendix A to verify that the interface you create adheres to the GUI standards outlined in this book.

f. Code the application. Use a Select Case statement to determine the customer's status. Use an If...Then...Else statement to determine the price per unit.

g. Save the solution, then start the application. Test the application appropriately.

h. Stop the application. Close the Code Editor window, then close the solution.

9. Jacques Cousard has been playing the lottery for four years and has yet to win any money. He wants an application that will select the six lottery numbers for him. Each lottery number can range from 1 to 54 only. (An example of six lottery numbers would be: 4, 8, 35, 15, 20, 3.)

a. If necessary, start Visual Studio 2005 or Visual Basic 2005 Express Edition.

b. Create a Visual Basic Windows-based application. Name the solution Lottery Solution, and name the project Lottery Project. Save the application in the VB2005\Chap05 folder.

c. Assign the filename Main Form.vb to the form file object.

d. Assign the name MainForm to the form.

e. Design an appropriate interface. Use the GUI design guidelines listed in Appendix A to verify that the interface you create adheres to the GUI standards outlined in this book.

f. Code the application. For now, do not worry if the lottery numbers are not unique. You will learn how to display unique numbers in Chapter 9.

g. Save the solution, then start the application. Test the application appropriately.

h. Stop the application. Close the Code Editor window, then close the solution.

10. Ferris Seminars offers computer seminars to various companies. The owner of Ferris Seminars wants an application that the registration clerks can use to calculate the registration fee for each customer. Many of Ferris Seminars' customers are companies that register more than one person for a seminar. The registration clerk will need to enter the number registered for the seminar, and then select either the Seminar 1 radio button or the Seminar 2 radio button. If a company is entitled to a 10% discount, the clerk will need to click the 10% discount check box. After the selections are made, the clerk will click the Calculate Total Due button to calculate the total registration fee. Seminar 1 is $100 per person, and Seminar 2 is $120 per person.

a. If necessary, start Visual Studio 2005 or Visual Basic 2005 Express Edition. Open the Ferris Solution (Ferris Solution.sln) file, which is contained in the VB2005\Chap05\Ferris Solution folder. If necessary, open the designer window.

b. Code the application appropriately.

c. Save the solution, then start the application. Test the application.

d. End the application. Close the Code Editor window, then close the solution.

DEBUGGING EXERCISE

11. In this exercise, you debug an existing application. The purpose of this exercise is to demonstrate the importance of testing an application thoroughly.

a. If necessary, start Visual Studio 2005 or Visual Basic 2005 Express Edition. Open the Debug Solution (Debug Solution.sln) file, which is contained in the VB2005\Chap05\Debug Solution folder. If necessary, open the designer window. The application displays a shipping charge, which is based on the total price entered by the user. If the total price is greater than or equal to $100 but less than $501, the shipping charge is $10. If the total price is greater than or equal to $501 but less than $1001, the shipping charge is $7. If the total price is greater than or equal to $1001, the shipping charge is $5. No shipping charge is due if the total price is less than $100.

b. Start the application. Test the application using the following total prices: 100, 501, 1500, 500.75, 30, 1000.33. You will notice that the application does not display the correct shipping charge for some of these total prices.

c. Click the Exit button to end the application.

d. Correct the application's code, then save the solution and start the application. Test the application using the total prices listed in Step b.

e. Click the Exit button to end the application. Close the Code Editor window, then close the solution.

THE REPETITION STRUCTURE

CREATING THE SHOPPERS HAVEN APPLICATION

The manager of Shoppers Haven wants an application that allows a store clerk to enter an item's original price and its discount rate. The discount rates range from 10% through 30% in increments of 5%. The application should calculate and display the amount of the discount and also the discounted price.

PREVIEWING THE COMPLETED APPLICATION

Before creating the Shoppers Haven application, you first preview the completed application.

To preview the completed application:

1 Use the Run command on the Windows Start menu to run the **Shoppers** (**Shoppers.exe**) file, which is contained in the VB2005\Chap06 folder. The user interface for the Shoppers Haven application appears on the screen. The interface contains a new control: a list box. You will learn about list boxes in Lesson C.

2 Type **56.99** in the Original price text box, then click **15** in the Discount rate list box.

3 Click the **Calculate** button. The item's discount and discounted price appear in the interface, as shown in Figure 6-1.

Figure 6-1: Interface showing the discount and discounted price amounts

4 Click the **Exit** button to end the application.

The Shoppers Haven application uses the repetition structure, which you will learn about in Lessons A and B. You will code the Shoppers Haven application in Lesson C.

LESSON A
OBJECTIVES

AFTER STUDYING LESSON A, YOU SHOULD
BE ABLE TO:

» Code the repetition structure using the For...Next and

   Do...Loop statements

» Include the repetition structure in pseudocode

» Include the repetition structure in a flowchart

» Initialize and update counters and accumulators

THE REPETITION STRUCTURE (LOOPING)

THE REPETITION STRUCTURE

As you learned in Chapter 1, the three programming structures are sequence, selection, and repetition. Every program contains the sequence structure, where the program instructions are processed one after another in the order they appear in the program. Most programs also contain the selection structure, which you learned about in Chapters 4 and 5. Recall that programmers use the selection structure when they need the computer to make a decision and then take the appropriate action based on the result of the decision.

In addition to including the sequence and selection structures, many programs also include the repetition structure. Programmers use the **repetition structure**, referred to more simply as a **loop**, when they need the computer to repeatedly process one or more program instructions until some condition is met, at which time the repetition structure ends. For example, you may want to process a set of instructions—such as the instructions to calculate net pay—for each employee in a company. Or, you may want to process a set of instructions until the user enters a negative sales amount, which indicates that he or she has no more sales amounts to enter. As with the sequence and selection structures, you already are familiar with the repetition structure. For example, shampoo bottles typically include a direction that tells you to repeat the "apply shampoo to hair," "lather," and "rinse" steps until your hair is clean.

A repetition structure can be either a pretest loop or a posttest loop. In both types of loops, the condition is evaluated with each repetition, or iteration, of the loop. In a **pretest loop**, the evaluation occurs *before* the instructions within the loop are processed. In a **posttest loop**, the evaluation occurs *after* the instructions within the loop are processed. Depending on the result of the evaluation, the instructions in a pretest loop may never be processed. The instructions in a posttest loop, however, always will be processed at least once. Of the two types of loops, the pretest loop is the most commonly used.

»TIP

Pretest and posttest loops also are called top-driven and bottom-driven loops, respectively.

You code a repetition structure (loop) in Visual Basic using one of the following statements: For...Next, Do...Loop, and For Each...Next. You will learn about the For...Next and Do...Loop statements in this lesson. The For Each...Next statement is covered in Chapter 9.

THE FOR...NEXT STATEMENT

You can use the **For**...**Next statement** to code a loop whose instructions you want processed a precise number of times. The loop created by the For...Next statement is a pretest loop, because the loop's condition is evaluated *before* the instructions in the loop are processed. Figure 6-2 shows the syntax of the For...Next statement and includes two examples of using the statement. (Notice that you can use either the Convert.ToString method or a numeric data type's ToString method to convert the contents of a numeric variable to a String.)

For...Next statement

Syntax

For *counter* [**As** *datatype*] = *startvalue* **To** *endvalue* [**Step** *stepvalue*]

    [*statements*]

Next *counter*

Examples

```
Dim numberSquared As Integer
For number As Integer = 1 To 3
    numberSquared = number * number
    MessageBox.Show(Convert.ToString(number) _
        & " squared is " _
        & Convert.ToString(numberSquared), _
        "Number Squared", MessageBoxButtons.OK, _
        MessageBoxIcon.Information)
Next number
```

displays the squares of the numbers 1, 2, and 3 in message boxes

```
Dim numberSquared As Integer
For number As Integer = 3 To 1 Step -1
    numberSquared = number * number
    MessageBox.Show(number.ToString & " squared is " _
        & numberSquared.ToString, "Number Squared", _
        MessageBoxButtons.OK, MessageBoxIcon.Information)
Next number
```

displays the squares of the numbers 3, 2, and 1 in message boxes

---

Figure 6-2: Syntax and examples of the For...Next statement *(Continued)* ▶

```
Dim rate As Decimal
For rate = .05D To .1D Step .01D
    Me.xRateLabel.Text = Me.xRateLabel.Text _
        & rate.ToString("P0") & ControlChars.NewLine
Next rate
displays the values 5 %, 6 %, 7 %, 8 %, 9 %, and 10 % in the xRateLabel
```

Figure 6-2: Syntax and examples of the For...Next statement

»TIP

You can use the `Exit For` statement to exit the For...Next statement prematurely—in other words, exit it before it has finished processing. You may need to do this if the computer encounters an error when processing the loop instructions.

»TIP

You can nest For...Next statements, which means that you can place one For...Next statement within another For...Next statement.

The For...Next statement begins with the For clause and ends with the Next clause. Between the two clauses, you enter the instructions you want the loop to repeat. In the syntax, *counter* is the name of a numeric variable that the computer can use to keep track of the number of times it processes the loop instructions. Notice that *counter* appears in both the For clause and the Next clause. Although, technically, you do not need to specify the name of the *counter* variable in the Next clause, doing so is highly recommended because it makes your code more self-documenting.

You can use the **As** *datatype* portion of the For clause to declare the *counter* variable, as shown in the first two examples in Figure 6-2. When you declare a variable in the For clause, the variable has block scope and can be used only by the For...Next loop. Alternatively, you can declare the *counter* variable in a Dim statement, as shown in the last example in Figure 6-2. As you know, when a variable is declared in a Dim statement at the beginning of a procedure, it has procedure scope and can be used by the entire procedure.

When deciding where to declare the *counter* variable, keep in mind that if the variable is needed only by the For...Next loop, then it is a better programming practice to declare it in the For clause. As you learned in Chapter 4, this is because fewer unintentional errors occur in applications when the variables are declared using the minimum scope needed. You should declare the *counter* variable in a Dim statement only when other statements in the procedure need to use its value.

The *startvalue*, *endvalue*, and *stepvalue* items control the number of times the loop instructions are processed. The *startvalue* tells the computer where to begin counting, and the *endvalue* tells the computer when to stop counting. The *stepvalue* tells the computer how much to count by—in other words, how much to add to (or subtract from if the *stepvalue* is a negative number) the *counter* variable each time the loop is processed. If you omit the *stepvalue*, a *stepvalue* of positive 1 is used. In the first example shown in Figure 6-2, the *startvalue* is 1, the *endvalue* is 3, and the *stepvalue* (which is omitted) is 1. Those values tell the computer to start counting at 1 and, counting by 1s, stop at 3—in other words, count 1, 2, and then 3. The computer will process the loop instructions shown in the first example three times.

The For clause's *startvalue*, *endvalue*, and *stepvalue* must be numeric and can be either positive or negative, integer or non-integer. If the *stepvalue* is positive, the *startvalue* must be less than or equal to the *endvalue* for the loop instructions to be processed. In other words, the instruction `For number As Integer = 1 To 3` is correct, but the instruction `For number As Integer = 3 To 1` is not correct because you cannot count from 3 (the *startvalue*) to 1 (the *endvalue*) by adding increments of 1 (the *stepvalue*). If, on the other hand, the *stepvalue* is negative, then the *startvalue* must be greater than or equal to the *endvalue* for the loop instructions to be processed. In other words, the instruction `For number As Integer = 3 To 1 Step -1` is correct, but the instruction `For number As Integer = 1 To 3 Step -1` is not correct because you cannot count from 1 to 3 by subtracting increments of 1. When processing the For...Next statement, the computer performs the three tasks listed in Figure 6-3.

Tasks performed when processing the For...Next statement

1. If the *counter* variable is declared in the For clause, the computer creates the variable and initializes it to the *startvalue*; otherwise, it just performs the initialization task. This is done only once, at the beginning of the loop.

2. If the *stepvalue* is positive, the computer checks whether the value in the *counter* variable is greater than the *endvalue*. (Or, if the *stepvalue* is negative, the computer checks whether the value in the *counter* variable is less than the *endvalue*.) If it is, the computer stops processing the loop, and processing continues with the statement following the Next clause. If it is not, the computer processes the instructions within the loop, and then the next task, task 3, is performed. Notice that the computer evaluates the loop condition *before* processing the instructions within the loop.

3. The computer adds the *stepvalue* to the contents of the *counter* variable. It then repeats tasks 2 and 3 until the *counter* variable's value is greater than (or less than, if the *stepvalue* is negative) the *endvalue*.

Figure 6-3: Processing tasks for the For...Next statement

Figure 6-4 describes how the computer processes the code shown in the first example in Figure 6-2. Notice that when the For...Next statement in that example ends, the value stored in the `number` variable is 4.

Processing steps for the first example in Figure 6-2

1. The computer creates the `numberSquared` variable and initializes it to the number 0.

2. The computer creates the `number` (*counter*) variable and initializes it to 1 (*startvalue*).

3. The computer checks whether the value in the `number` variable is greater than 3 (*endvalue*). It's not.

4. The assignment statement multiplies the value in the `number` variable by itself and assigns the result to the `numberSquared` variable.

5. The MessageBox.Show method displays the message "1 squared is 1".

6. The computer processes the Next clause, which adds 1 (*stepvalue*) to the contents of the `number` variable, giving 2.

7. The computer checks whether the value in the `number` variable is greater than 3 (*endvalue*). It's not.

8. The assignment statement multiplies the value in the `number` variable by itself and assigns the result to the `numberSquared` variable.

9. The MessageBox.Show method displays the message "2 squared is 4".

10. The computer processes the Next clause, which adds 1 (*stepvalue*) to the contents of the `number` variable, giving 3.

11. The computer checks whether the value in the `number` variable is greater than 3 (*endvalue*). It's not.

12. The assignment statement multiplies the value in the `number` variable by itself and assigns the result to the `numberSquared` variable.

13. The MessageBox.Show method displays the message "3 squared is 9".

14. The computer processes the Next clause, which adds 1 (*stepvalue*) to the contents of the `number` variable, giving 4.

15. The computer checks whether the value in the `number` variable is greater than 3 (*endvalue*). It is, so the computer stops processing the For...Next statement. Processing continues with the statement following the Next clause.

Figure 6-4: Processing steps for the code shown in the first example in Figure 6-2

Figures 6-5 and 6-6 show the pseudocode and flowchart, respectively, corresponding to the first example in Figure 6-2.

Pseudocode

repeat for number = 1 to 3
    calculate numberSquared by multiplying number by number
    display the message and the numberSquared
end repeat

Figure 6-5: Pseudocode for the first example shown in Figure 6-2

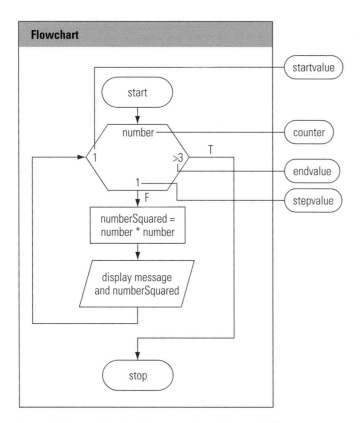

Figure 6-6: Flowchart for the first example shown in Figure 6-2

The For...Next loop is represented in a flowchart by a hexagon, which is a six-sided figure. Four values are recorded inside the hexagon: the name of the *counter* variable, the *startvalue*, the *stepvalue*, and the *endvalue*. Notice that a greater-than sign (>) precedes the *endvalue* in the hexagon shown in Figure 6-6. The greater-than sign indicates that the loop stops when the value in the *counter* variable is greater than the *endvalue*. When the *stepvalue* is a negative number, a less-than sign (<) should precede the *endvalue* in the hexagon, because a loop with a negative *stepvalue* stops when the value in the *counter* variable is less than the *endvalue*.

Next, you will view an application that uses the For...Next statement.

THE MONTHLY PAYMENT CALCULATOR APPLICATION

Figure 6-7 shows the Visual Basic code for the xCalcButton's Click event procedure, which is contained in the Monthly Payment Calculator application. The procedure calculates and displays the monthly car payments, using a term of five years and annual interest rates of 5%, 6%, 7%, 8%, 9%, and 10%. Figure 6-8 shows a sample run of the application.

Visual Basic code

```
Private Sub xCalcButton_Click(ByVal sender As Object, _
    ByVal e As System.EventArgs) Handles xCalcButton.Click
    ' calculates the monthly payments on a loan
    ' using a term of 5 years and interest rates
    ' of 5% through 10%

    Const Term As Double = 5.0
    Dim principal As Double
    Dim monthlyPayment As Double
    Dim isConverted As Boolean

    Me.xPaymentsLabel.Text = String.Empty
    isConverted = _
        Double.TryParse(Me.xPrincipalTextBox.Text, principal)

    If isConverted Then
        ' calculate and display payments
        For rate As Double = 0.05 To 0.1 Step 0.01
            monthlyPayment = _
            -Financial.Pmt(rate / 12.0, Term * 12.0, principal)
            Me.xPaymentsLabel.Text = Me.xPaymentsLabel.Text _
                & rate.ToString("P0") & " -> " _
                & monthlyPayment.ToString("C2") _
                & ControlChars.NewLine
        Next rate
    Else    ' principal cannot be converted to a number
        MessageBox.Show("Please re-enter the principal.", _
        "Payment Calculator", MessageBoxButtons.OK, _
        MessageBoxIcon.Information)
    End If

    Me.xPrincipalTextBox.Focus()
    Me.xPrincipalTextBox.SelectAll()
End Sub
```

For...Next statement

Figure 6-7: Code for the xCalcButton's Click event procedure

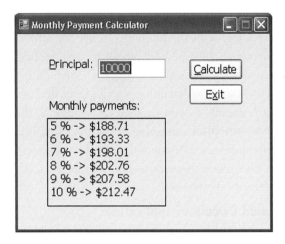

Figure 6-8: Sample run of the application that contains the procedure

The procedure shown in Figure 6-7 begins by declaring a named constant for the term and initializing it to the number five. The procedure also declares two variables to store the principal and payment amounts, and a variable to store the value returned by the TryParse method. After declaring the constant and variables, the procedure clears the contents of the xPaymentsLabel, and then uses the TryParse method to convert the contents of the xPrincipalTextBox to a Double number. If the contents of the text box cannot be converted to a number, the statement in the selection structure's false path displays an appropriate message in a message box. If the conversion is successful, on the other hand, the computer processes the For...Next statement located in the selection structure's true path.

The For clause in the For...Next statement tells the computer to create a Double *counter* variable named `rate` and initialize it to .05. It also tells the computer to repeat the instructions in the For...Next loop six times, using annual interest rates of .05, .06, .07, .08, .09, and .1. The first instruction in the For...Next loop uses the Financial.Pmt method, which you learned about in Chapter 4, to calculate the monthly payment. Notice that the instruction uses the negation operator to change the negative number returned by the method to a positive number. Also notice that the For...Next statement's *counter* variable (`rate`), which keeps track of the annual interest rates, is divided by 12 and used as the *Rate* argument in the Financial.Pmt method. It is necessary to divide the annual interest rate by 12 to get a monthly rate, because you want to display monthly payments rather than an annual payment. The `Term` constant, on the other hand, is multiplied by 12 to get the number of monthly payments; the result is used as the *Nper* argument in the Financial.Pmt method. Lastly, the `principal` variable, which stores the principal entered by the user, is used as the *PV* argument in the method.

The second instruction in the For...Next loop is an assignment statement that concatenates the current contents of the xPaymentsLabel with the following items: the annual interest rate converted to a string and formatted to a percentage with zero decimal places, the string " -> " (a space, a hyphen, a greater-than sign, and a space), the monthly payment amount converted to a string and formatted with a dollar sign and two decimal places, and the `ControlChars.NewLine` constant. As you learned in Chapter 3, the `ControlChars.NewLine` constant advances the insertion point to the next line in a control. The concatenated string is then assigned to the xPaymentsLabel's Text property. The last instructions in the xCalcButton's Click event procedure send the focus to the xPrincipalTextBox and also select the text box's existing text.

To code and then test the Monthly Payment Calculator application:

1 Start Visual Studio 2005 or Visual Basic 2005 Express Edition, if necessary, and close the Start Page window.

2 Open the **Payment Calculator Solution** (Payment Calculator Solution.sln) file, which is contained in the VB2005\Chap06\Payment Calculator Solution folder. If necessary, open the designer window.

3 Open the Code Editor window. Replace the <your name> and <current date> text in the comments with your name and the current date.

4 Locate the code template for the xCalcButton's Click event procedure, then enter the comments and code shown in Figure 6-7.

5 Close the Code Editor window. Save the solution, then start the application.

6 Enter **10000** as the principal, then press **Enter** to select the Calculate button, which is the default button on the form. The button's Click event procedure displays the monthly payments shown in Figure 6-8.

7 Click the **Exit** button to end the application. You are returned to the designer window. Close the solution.

As you learned earlier, you also can use the Do...Loop statement to code a repetition structure in Visual Basic.

THE DO...LOOP STATEMENT

Unlike the For...Next statement, the **Do...Loop statement** can be used to code both a pretest loop and a posttest loop. Figure 6-9 shows two slightly different versions of the Do...Loop statement's syntax. You use the first version to code a pretest loop, and the second version to code a posttest loop. Figure 6-9 also includes an example of using each syntax to display the numbers 1, 2, and 3 in message boxes.

Do...Loop Statement

Do...Loop syntax (pretest loop)

Do {While | Until} *condition*

  [*instructions to be processed either while the condition is true or until the condition becomes true*]

Loop

Do...Loop syntax (posttest loop)

Do

    [*instructions to be processed either while the condition is true or until the condition becomes true*]

Loop {While | Until} *condition*

Pretest loop example

```
Dim number As Integer = 1
Do While number <= 3
    MessageBox.Show(number.ToString, "Numbers", _
        MessageBoxButtons.OK, MessageBoxIcon.Information)
    number = number + 1
Loop
```

Posttest loop example

```
Dim number As Integer = 1
Do
    MessageBox.Show(number.ToString, "Numbers", _
        MessageBoxButtons.OK, MessageBoxIcon.Information)
    number = number + 1
Loop Until number > 3
```

Figure 6-9: Syntax and examples of the Do...Loop statement

»TIP

You can use the Exit Do statement to exit the Do...Loop statement prematurely—in other words, exit it before the loop has finished processing. You may need to do this if the computer encounters an error when processing the loop instructions.

»TIP

You can nest Do...Loop statements, which means that you can place one Do...Loop statement within another Do...Loop statement.

The Do...Loop statement begins with the Do clause and ends with the Loop clause. Between both clauses, you enter the instructions you want the computer to repeat. The **{While | Until}** portion of each syntax shown in Figure 6-9 indicates that you can select only one of the keywords appearing within the braces. In this case, you can choose either the While keyword or the Until keyword. As both examples shown in the figure indicate, you do not type the braces ({}) or the pipe symbol (|) when entering the Do...Loop statement. You follow the While or Until keyword with a *condition*, which can contain variables, constants, properties, methods, and operators. Like the condition used in the If...Then...Else statement, the condition used in the Do...Loop statement also must evaluate to a Boolean value—either True or False. The condition determines whether the computer processes the loop instructions. The `While` keyword indicates that the loop instructions should be processed *while* the condition is true. The `Until` keyword, on the other hand, indicates that the loop instructions should be processed *until* the condition becomes true. Notice that the keyword (either `While` or `Until`) and the condition appear in the Do clause in a pretest loop, but appear in the Loop clause in a posttest loop. Figures 6-10 and 6-11 describe how the computer processes the code shown in the examples in Figure 6-9.

Processing steps for the pretest loop example

1. The computer creates the `number` variable and initializes it to 1.

2. The computer processes the Do clause, which checks whether the value in the `number` variable is less than or equal to 3. It is.

3. The MessageBox.Show method displays 1 (the contents of the `number` variable).

4. The `number = number + 1` statement adds 1 to the contents of the `number` variable, giving 2.

5. The computer processes the Loop clause, which returns processing to the Do clause (the beginning of the loop).

6. The computer processes the Do clause, which checks whether the value in the `number` variable is less than or equal to 3. It is.

7. The MessageBox.Show method displays 2 (the contents of the `number` variable).

8. The `number = number + 1` statement adds 1 to the contents of the `number` variable, giving 3.

9. The computer processes the Loop clause, which returns processing to the Do clause (the beginning of the loop).

10. The computer processes the Do clause, which checks whether the value in the `number` variable is less than or equal to 3. It is.

11. The MessageBox.Show method displays 3 (the contents of the `number` variable).

12. The `number = number + 1` statement adds 1 to the contents of the `number` variable, giving 4.

13. The computer processes the Loop clause, which returns processing to the Do clause (the beginning of the loop).

14. The computer processes the Do clause, which checks whether the value in the `number` variable is less than or equal to 3. It isn't, so the computer stops processing the Do...Loop statement. Processing continues with the statement following the Loop clause.

Figure 6-10: Processing steps for the pretest loop example shown in Figure 6-9

Processing steps for the posttest loop example

1. The computer creates the `number` variable and initializes it to 1.

2. The computer processes the Do clause, which marks the beginning of the loop.

3. The MessageBox.Show method displays 1 (the contents of the `number` variable).

4. The `number = number + 1` statement adds 1 to the contents of the `number` variable, giving 2.

5. The computer processes the Loop clause, which checks whether the value in the `number` variable is greater than 3. It isn't, so processing returns to the Do clause (the beginning of the loop).

6. The MessageBox.Show method displays 2 (the contents of the `number` variable).

7. The `number = number + 1` statement adds 1 to the contents of the `number` variable, giving 3.

8. The computer processes the Loop clause, which checks whether the value in the `number` variable is greater than 3. It isn't, so processing returns to the Do clause (the beginning of the loop).

9. The MessageBox.Show method displays 3 (the contents of the `number` variable).

10. The `number = number + 1` statement adds 1 to the contents of the `number` variable, giving 4.

11. The computer processes the Loop clause, which checks whether the value in the `number` variable is greater than 3. It is, so the computer stops processing the Do...Loop statement. Processing continues with the statement following the Loop clause.

Figure 6-11: Processing steps for the posttest loop example shown in Figure 6-9

Figure 6-12 shows the pseudocode and flowchart associated with the pretest loop example in Figure 6-9. Figure 6-13 shows the pseudocode and flowchart for the posttest example.

Pseudocode for the pretest loop example

1. assign 1 to the number variable

2. repeat while the value in the number variable is less than or equal to 3
    display the contents of the number variable
    add 1 to the number variable
  end repeat

Figure 6-12: Pseudocode and flowchart for the pretest loop example shown in Figure 6-9 *(Continued)* ▶

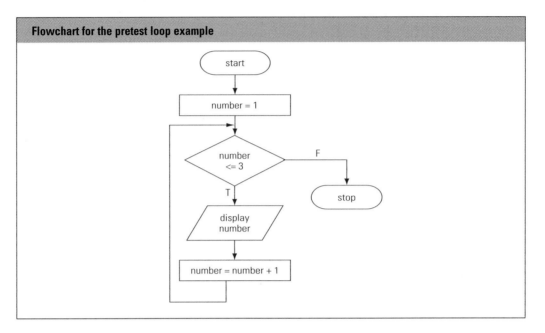

Figure 6-12: Pseudocode and flowchart for the pretest loop example shown in Figure 6-9

Pseudocode for the posttest loop example

1. assign 1 to the number variable

2. repeat
   display the contents of the number variable
   add 1 to the number variable
   end repeat until the value in the number variable is greater than 3

Figure 6-13: Pseudocode and flowchart for the posttest loop example shown in Figure 6-9 *(Continued)* ▶

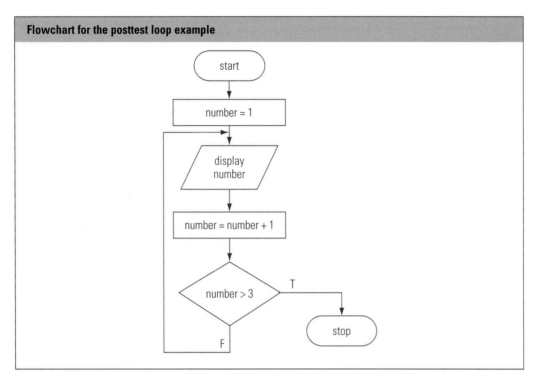

Flowchart for the posttest loop example

Figure 6-13: Pseudocode and flowchart for the posttest loop example shown in Figure 6-9

Notice that a diamond represents the loop condition in the flowcharts shown in Figures 6-12 and 6-13. As with the selection structure diamond, which you learned about in Chapter 4, the repetition structure diamond contains a comparison that evaluates to either True or False only. The result of the comparison determines whether the computer processes the instructions within the loop.

Like the selection diamond, the repetition diamond has one flowline entering the diamond and two flowlines leaving the diamond. The two flowlines leaving the diamond should be marked with a "T" (for True) and an "F" (for False). In the pretest loop's flowchart in Figure 6-12, the flowline entering the repetition diamond, as well as the symbols and flowlines within the True path, form a circle or loop. In the posttest loop's flowchart in Figure 6-13, the loop is formed by all of the symbols and flowlines in the False path. It is this loop, or circle, that distinguishes the repetition structure from the selection structure in a flowchart.

Although it appears that the pretest and posttest loops produce the same results—in this case, for instance, both examples in Figure 6-9 display the numbers 1 through 3—that will not always be the case. In other words, the two loops are not always interchangeable. The difference between both loops is demonstrated in the examples shown in Figure 6-14.

Examples and processing steps

Pretest loop example

```
Dim number As Integer = 10
Do While number <= 3
    MessageBox.Show(number.ToString, "Numbers", _
        MessageBoxButtons.OK, MessageBoxIcon.Information)
    number = number + 1
Loop
```

Processing steps for the pretest loop

1. The computer creates the `number` variable and initializes it to 10.

2. The computer processes the Do clause, which checks whether the value in the `number` variable is less than or equal to 3. It isn't, so the computer stops processing the Do...Loop statement. Processing continues with the statement following the Loop clause.

Posttest loop example

```
Dim number As Integer = 10
Do
    MessageBox.Show(number.ToString, "Numbers", _
        MessageBoxButtons.OK, MessageBoxIcon.Information)
    number = number + 1
Loop Until number > 3
```

Processing steps for the posttest loop

1. The computer creates the `number` variable and initializes it to 10.

2. The computer processes the Do clause, which marks the beginning of the loop.

3. The MessageBox.Show method displays 10 (the contents of the `number` variable).

4. The `number = number + 1` statement adds 1 to the contents of the `number` variable, giving 11.

5. The computer processes the Loop clause, which checks whether the value in the `number` variable is greater than 3. It is, so the computer stops processing the Do...Loop statement. Processing continues with the statement following the Loop clause.

Figure 6-14: Examples showing that the pretest and posttest loops do not always produce the same results

Comparing the processing steps shown in both examples in Figure 6-14, you will notice that the instructions in the pretest loop are not processed. This is because the `number <= 3` condition, which is evaluated *before* the instructions are processed, evaluates to False. The instructions in the posttest loop, on the other hand, are processed one time, because the `number > 3` condition is evaluated *after* (rather than *before*) the loop instructions are processed.

USING COUNTERS AND ACCUMULATORS

Many times an application will need to display a subtotal, a total, or an average. You calculate this information using a repetition structure that includes a counter, or an accumulator, or both. A **counter** is a numeric variable used for counting something, such as the number of employees paid in a week. An **accumulator** is a numeric variable used for accumulating (adding together) something, such as the total dollar amount of a week's payroll. Two tasks are associated with counters and accumulators: initializing and updating. **Initializing** means to assign a beginning value to the counter or accumulator. Typically, counters and accumulators are initialized to zero; however, they can be initialized to any number, depending on the value required by the application. The initialization task is performed before the loop is processed, because it needs to be performed only once. **Updating**, also called **incrementing**, means adding a number to the value stored in the counter or accumulator. The number can be either positive or negative, integer or non-integer. A counter is always incremented by a constant value—typically the number 1—whereas an accumulator is incremented by a value that varies. The assignment statement that updates a counter or an accumulator is placed within the loop in a procedure, because the update task must be performed each time the loop instructions are processed. The Sales Express application, which you view next, includes both a counter and an accumulator, as well as a repetition structure.

THE SALES EXPRESS APPLICATION

The sales manager at Sales Express wants an application that allows him to enter the amount of each salesperson's sales. The application then should display the average amount the company sold during the prior year. The application will use a counter to keep track of the number of sales amounts entered by the sales manager, and use an accumulator to total the sales amounts. After all of the sales amounts are entered, the application will calculate the average sales amount by dividing the value stored in the accumulator by the value stored in the counter. It then will display the average sales amount on the screen. Figure 6-15 shows the pseudocode for the xCalcButton's Click event procedure, which is contained in the Sales Express application.

---

xCalcButton Click event procedure - pseudocode

1. get a sales amount from the user

2. repeat while the user entered a sales amount
> if the sales amount can be converted to a number
>> add 1 to the counter variable
>> add the sales amount to the accumulator variable
>
> else
>> display an appropriate message in a message box
>
> end if
> get a sales amount from the user

> end repeat

3. if the counter variable contains a value that is greater than zero
> calculate the average sales amount by dividing the accumulator variable by the counter variable
> display the average sales amount in the xAverageLabel

else
> display the number 0 in the xAverageLabel

end if

Figure 6-15: Pseudocode for the xCalcButton's Click event procedure

Step 1 in the pseudocode is to get a sales amount from the user. Step 2 is a pretest loop whose instructions are processed as long as the user enters a sales amount. The first instruction in the loop is a selection structure that checks whether the sales amount can be converted to a number. If the sales amount can be converted to a number, the counter variable is incremented by one and the accumulator variable is incremented by the sales amount; otherwise, an appropriate message is displayed in a message box. After the selection structure within the loop is processed, the procedure requests another sales amount from the user. It then checks whether a sales amount was entered to determine whether the loop instructions should be processed again. When the user has finished entering sales amounts, the loop ends and processing continues with Step 3 in the pseudocode.

Step 3 in the pseudocode is a selection structure that checks whether the counter variable contains a value that is greater than zero. Before using a variable as the divisor in an expression, you always should verify that the variable does not contain the number zero because, as in math, division by zero is not mathematically possible. Dividing by zero in a procedure will cause the application to end abruptly with an error. As indicated in Step 3, if the counter variable contains a value that is greater than zero, the average sales amount is calculated and then displayed in the xAverageLabel; otherwise, the number zero is displayed in the xAverageLabel.

Notice that "get a sales amount from the user" appears twice in the pseudocode shown in Figure 6-15: immediately above the loop and also within the loop. The "get a sales amount from the user" entry that appears above the loop is referred to as the **priming read**, because it is used to prime (prepare or set up) the loop. In this case, the priming read gets only the first salesperson's sales amount from the user. Because the loop in Figure 6-15 is a pretest loop, the first value determines whether the loop instructions are processed at all. The "get a sales amount from the user" entry that appears within the loop gets the sales amounts for the remaining salespeople (if any) from the user. Figure 6-16 shows the Visual Basic code for the xCalcButton's Click event procedure.

Visual Basic code

```vb
Private Sub xCalcButton_Click(ByVal sender As Object, _
     ByVal e As System.EventArgs) Handles xCalcButton.Click
     ' calculates and displays the average sales amount

     Const Prompt As String = _
        "Enter a sales amount. Click Cancel to end."
     Const Title As String = "Sales Entry"
     Const Message As String = _
        "Please re-enter the sales amount."
     Dim inputSales As String
     Dim sales As Decimal
     Dim salesCount As Integer
     Dim salesAccum As Decimal
     Dim average As Decimal
     Dim isConverted As Boolean

     ' get first sales amount
     inputSales = InputBox(Prompt, Title, "0")

     ' repeat as long as the user enters a sales amount
     Do While inputSales <> String.Empty
        ' try to convert the sales amount to a number
        isConverted = Decimal.TryParse(inputSales, sales)

        ' if the sales amount can be converted to a
        ' number, update the counter and accumulator;
        ' otherwise, display a message
        If isConverted Then
           salesCount = salesCount + 1
           salesAccum = salesAccum + sales
```

> **» TIP**
> You also can write the loop condition shown in Figure 6-16 as `Do While inputSales <> ""`.

Figure 6-16: Code for the xCalcButton's Click event procedure *(Continued)* ▶

```
          Else
              MessageBox.Show(Message, _
              "Sales Express", MessageBoxButtons.OK, _
                 MessageBoxIcon.Information)
          End If

          ' get the next sales amount
          inputSales = InputBox(Prompt, Title, "0")
      Loop
      ' if counter is greater than 0, calculate
      ' and display the average sales amount; otherwise
      ' display 0 as the average sales amount
      If salesCount > 0 Then
          average = _
              salesAccum / Convert.ToDecimal (salesCount)
          Me.xAverageLabel.Text = average.ToString("C2")
      Else
          Me.xAverageLabel.Text = "0"
      End If
  End Sub
```

Figure 6-16: Code for the xCalcButton's Click event procedure

The xCalcButton's Click event procedure declares three named constants and six variables. The `Prompt` and `Title` constants are used as arguments in the InputBox function,
and the `Message` constant is used as an argument in the MessageBox.Show method. The
`inputSales` variable is used to store the sales amount returned by the InputBox function
as a String. The `sales` and `isConverted` variables are used by the TryParse method. The
`sales` variable will store the sales amount after it has been converted to Decimal, and the
`isConverted` variable will store a Boolean value that indicates whether the conversion
was successful. The `salesCount` variable is the counter variable that will keep track of
the number of valid sales amounts entered. The `salesAccum` variable is the accumulator
variable that the computer will use to total the sales amounts. The remaining variable,
`average`, will store the average sales amount after it has been calculated.

Recall that counters and accumulators must be initialized, or given a beginning value; typically, the beginning value is the number zero. Because the Dim statement automatically
assigns a zero to Integer and Decimal variables when the variables are created, you do not
need to enter any additional code to initialize the `salesCount` and `salesAccum` variables.
In cases where you need to initialize a counter or an accumulator to a value other than zero,
you can do so either in the Dim statement that declares the variable or in an assignment
statement. For example, to initialize the `salesCount` variable to the number one, you
could use either the declaration statement `Dim salesCount As Integer = 1` or the

assignment statement `salesCount = 1` in your code. (To use the assignment statement, the variable must already be declared.)

After the variables are declared, the InputBox function in the code displays a dialog box that prompts the user to either enter a sales amount or click the Cancel button. Clicking the Cancel button indicates that the user has no more sales amounts to enter. As you learned in Chapter 3, the value returned by the InputBox function depends on whether the user clicks the dialog box's OK button, Cancel button, or Close button. In this case, if the user enters a sales amount and then clicks the OK button in the dialog box, the InputBox function returns (as a string) the sales amount contained in the input area of the dialog box. However, if the user clicks either the dialog box's Cancel button or its Close button, the function returns a zero-length string (""). The assignment statement that contains the InputBox function assigns the function's return value to the `inputSales` variable, which has a String data type.

Next, the computer evaluates the loop condition in the Do...Loop statement to determine whether the loop instructions should be processed. In this case, the `inputSales <> String.Empty` condition compares the contents of the `inputSales` variable to the `String.Empty` value. As you learned in Chapter 2, the `String.Empty` value represents a zero-length, or empty, string. If the `inputSales` variable is not empty, the loop condition evaluates to True and the computer processes the loop instructions. If, on the other hand, the `inputSales` variable is empty, the loop condition evaluates to False and the computer skips over the loop instructions.

Now take a closer look at the instructions within the loop. The first instruction in the loop uses the TryParse method to convert the contents of the `inputSales` variable to a Decimal number. If the conversion is successful, the Decimal number is stored in the `sales` variable and the Boolean value True is assigned to the `isConverted` variable. Otherwise, the number zero and the Boolean value False are stored in the `sales` and `isConverted` variables, respectively.

The next instruction in the loop is a selection structure that uses the value stored in the `isConverted` variable to determine the next instruction to process. If the `isConverted` variable does not contain the Boolean value True, the MessageBox.Show method in the selection structure's false path displays an appropriate message. However, if the `isConverted` variable does contain the Boolean value True, the two instructions in the selection structure's true path are processed. The first instruction, `salesCount = salesCount + 1`, updates the counter variable by adding a constant value of one to it. Notice that the counter variable appears on both sides of the assignment operator. The statement tells the computer to add one to the contents of the `salesCount` variable, and then place the result back in the `salesCount` variable. The `salesCount` variable's value will be incremented by one each time the loop is processed. The second instruction in the true path, salesAccum = salesAccum + sales, updates the accumulator variable by adding a sales amount to it. Notice that the accumulator variable also appears on both sides of the assignment operator. The statement tells the computer to add the contents of the sales variable to the contents of the salesAccum variable, and then place the result back in the salesAccum variable. The salesAccum variable's value will be incremented by a sales amount, which will vary, each time the loop is processed.

The last instruction in the loop, inputSales = InputBox (Prompt, Title, "0"), displays a dialog box that prompts the user for another sales amount. Notice that the instruction appears twice in the code—before the Do...Loop statement and within the Do...Loop statement. Recall that the input instruction located above the loop is referred to as the priming read, and its task is to get only the first sales amount from the user. The input instruction located within the loop gets each of the remaining sales amounts (if any) from the user. If you forget to enter the input instruction within the loop, you will create an endless or infinite loop. You can stop an endless loop by clicking Debug on the menu bar, and then clicking Stop Debugging.

After the user enters another sales amount, the computer returns to the Do clause, where the loop condition is tested again. If the condition evaluates to True, the loop instructions are processed again. If the condition evaluates to False, the loop stops and the instruction after the Loop clause is processed. The instruction after the Loop clause is a selection structure that determines whether the salesCount variable contains a value that is greater than zero. Recall that before using a variable as the divisor in an expression, you first should verify that the variable does not contain the number zero, because division by zero is mathematically impossible and will cause the program to end with an error. In this case, if the salesCount variable contains a value that is greater than zero, the computer processes the instructions in the selection structure's true path. The first instruction in the true path calculates the average sales amount and assigns the result to the average variable. The second instruction in the true path displays the contents of the average variable in the xAverageLabel. However, if the value in the salesCount variable is not greater than zero, the computer processes the instruction in the selection structure's false path. The instruction in the false path displays the number zero in the xAverageLabel.

To code and then test the Sales Express application:

1 Open the **Sales Express Solution** (Sales Express Solution.sln) file, which is contained in the VB2005\Chap06\Sales Express Solution folder. If necessary, open the designer window.

2 Open the Code Editor window. Replace the <your name> and <current date> text in the comments with your name and the current date.

3 Open the code template for the xCalcButton's Click event procedure, then enter the comments and code shown earlier in Figure 6-16.

4 Close the Code Editor window. Save the solution, then start the application.

5 Click the **Calculate** button. The Sales Entry dialog box appears, as shown in Figure 6-17.

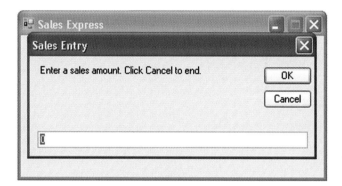

Figure 6-17: Sales Entry dialog box

6 Type **a** in the input area of the Sales Entry dialog box, then press **Enter** to select the OK button. The Sales Entry dialog box closes and the message box shown in Figure 6-18 appears.

Figure 6-18: Message box

7 Press **Enter** to select the OK button, which closes the message box. The Sales Entry dialog box appears.

8 Type **84** in the input area of the dialog box, then press **Enter** to select the OK button. Type **78** in the dialog box, then press **Enter**.

9 Click the **Cancel** button in the Sales Entry dialog box. The xCalcButton's Click event procedure calculates and displays the average sales amount. See Figure 6-19.

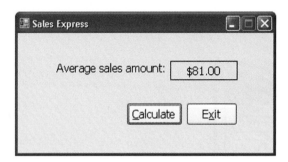

Figure 6-19: Average sales amount displayed in the interface

10 Click the **Exit** button to end the application. You are returned to the designer window. Close the solution.

You have completed Lesson A. You can either take a break or complete the end-of-lesson questions and exercises before moving on to Lesson B.

SUMMARY

TO HAVE THE COMPUTER REPEAT A SET OF INSTRUCTIONS UNTIL SOME CONDITION IS MET:

» Use a repetition structure (loop). You can code a repetition structure in Visual Basic using one of the following statements: For...Next, Do...Loop, and For Each...Next.

TO USE THE FOR...NEXT STATEMENT TO CODE A LOOP:

» Refer to Figure 6-2 for the syntax of the For...Next statement. The For...Next statement can be used to code pretest loops only. In the syntax, *counter* is the name of the numeric variable that keeps track of the number of times the loop instructions are processed. The *startvalue, endvalue,* and *stepvalue* items control the number of times the loop instructions are processed. The *startvalue, endvalue,* and *stepvalue* items must be numeric and can be positive or negative, integer or non-integer. If you omit the *stepvalue*, a *stepvalue* of positive 1 is used.

TO FLOWCHART A FOR...NEXT LOOP:

» Use a hexagon that shows the name of the *counter*, the *startvalue*, the *stepvalue*, and the *endvalue*.

TO USE THE DO...LOOP STATEMENT TO CODE A LOOP:

» Refer to Figure 6-9 for the two versions of the Do...Loop statement's syntax. The Do...Loop statement can be used to code both pretest and posttest loops. In a pretest loop, the loop condition appears in the Do clause; it appears in the Loop clause in a posttest loop. The loop condition must evaluate to a Boolean value.

TO FLOWCHART THE DO...LOOP STATEMENT:

» Use the selection/repetition diamond. The two flowlines leading out of the diamond should be marked with a "T" (for True) and an "F" (for False).

TO USE A COUNTER:

» Initialize the counter, if necessary. Update the counter using an assignment statement within a repetition structure. You update a counter by incrementing (or decrementing) its value by a constant amount.

TO USE AN ACCUMULATOR:

» Initialize the accumulator, if necessary. Update the accumulator using an assignment statement within a repetition structure. You update an accumulator by incrementing (or decrementing) its value by an amount that varies.

QUESTIONS

1. How many times will the MessageBox.Show method in the following code be processed?

```
For counter As Integer = 4 To 11 Step 2
      MessageBox.Show("Hello")
Next counter
```

    a. 3   b. 4

    c. 5   d. 8

2. What is the value stored in the counter variable when the loop in Question 1 stops?

    a. 10   b. 11

    c. 12   d. 13

3. Assume you know the precise number of times a loop's instructions should be processed. You can use the _____ statement to code this loop.

   a. Do...Loop   b. For...Next

   c. either a or b

4. The _____ loop processes the loop instructions at least once, whereas the _____ loop instructions might not be processed at all.

   a. posttest, pretest   b. pretest, posttest

5. Which of the following clauses stops the loop when the value in the age variable is less than the number zero?

   a. `Do While age >= 0`   b. `Do Until age < 0`

   c. `Loop While age >= 0`   d. All of the above.

6. How many times will the MessageBox.Show method in the following code be processed?

   ```
   Dim counter As Integer
   Do While counter > 3
        MessageBox.Show("Hello")
        counter = counter + 1
   Loop
   ```

   a. 0   b. 1

   c. 3   d. 4

7. How many times will the MessageBox.Show method in the following code be processed?

   ```
   Dim counter As Integer
   Do
     MessageBox.Show("Hello")
     counter = counter + 1
   Loop While counter > 3
   ```

   a. 0   b. 1

   c. 3   d. 4

   Refer to Figure 6-20 to answer Questions 8 through 11.

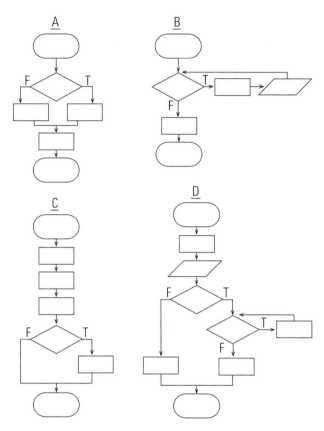

Figure 6-20

8. Which of the following programming structures are used in flowchart A in Figure 6-20? (Select all that apply.)

a. sequence   b. selection

c. repetition

9. Which of the following programming structures are used in flowchart B in Figure 6-20? (Select all that apply.)

a. sequence   b. selection

c. repetition

10. Which of the following programming structures are used in flowchart C in Figure 6-20? (Select all that apply.)

a. sequence   b. selection

c. repetition

11. Which of the following programming structures are used in flowchart D in Figure 6-20? (Select all that apply.)

   a. sequence   b. selection

   c. repetition

12. A procedure allows the user to enter one or more values. The first input instruction will get the first value only and is referred to as the _____ read.

   a. entering   b. initializer

   c. priming   d. starter

EXERCISES

1. Write a Visual Basic Do clause that processes the loop instructions as long as the value in the `quantity` variable is greater than the number 0. Use the `While` keyword.

2. Rewrite the Do clause from Exercise 1 using the `Until` keyword.

3. Write a Visual Basic Do clause that stops the loop when the value in the `inStock` variable is less than or equal to the value in the `reorder` variable. Use the `Until` keyword.

4. Rewrite the Do clause from Exercise 3 using the `While` keyword.

5. Write a Visual Basic Loop clause that processes the loop instructions as long as the value in the `letter` variable is either Y or y. Use the `While` keyword.

6. Rewrite the Loop clause from Exercise 5 using the `Until` keyword.

7. Write a Visual Basic Do clause that processes the loop instructions as long as the value in the `empName` variable is not "Done" (in any case). Use the `While` keyword.

8. Rewrite the Do clause from Exercise 7 using the `Until` keyword.

9. Write a Visual Basic assignment statement that updates the `quantity` counter variable by 2.

10. Write a Visual Basic assignment statement that updates the `total` counter variable by –3.

11. Write a Visual Basic assignment statement that updates the `totalPurchases` accumulator variable by the value stored in the `purchases` variable.

12. Write a Visual Basic assignment statement that subtracts the contents of the `salesReturns` variable from the `sales` accumulator variable.

13. Write the Visual Basic code for a pretest loop that uses an Integer variable named `evenNum` to display the even integers between 1 and 9 in the xNumbersLabel. Use the For...Next statement. Display each number on a separate line in the control.

14. Rewrite the pretest loop from Exercise 13 using the Do...Loop statement.

15. Change the pretest loop from Exercise 14 to a posttest loop.

16. Write the Visual Basic code that corresponds to the flowchart shown in Figure 6-21. (Display the calculated results on separate lines in the xCountLabel.)

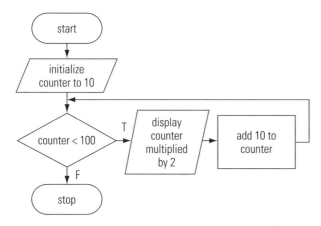

Figure 6-21

17. Write a For...Next statement that displays the numbers from 0 through 117, in increments of 9, in the xNumbersLabel. Display each number on a separate line in the control.

18. Write a For...Next statement that calculates and displays the squares of the even numbers from 2 through 12. Display the results in the xNumbersLabel. Display each number on a separate line in the control.

19. What will the following code display in message boxes?

```
Dim x As Integer
Do While x < 5
     MessageBox.Show(Convert.ToString(x))
     x = x + 1
Loop
```

20. What will the following code display in message boxes?

```
Dim x As Integer
Do
  MessageBox.Show(Convert.ToString(x))
  x = x + 1
Loop Until x > 5
```

21. An instruction is missing from the following code. What is the missing instruction, and where does it belong in the code?

```
Dim number As Integer = 1
Do While number < 5
     MessageBox.Show(number.ToString)
Loop
```

22. An instruction is missing from the following code. What is the missing instruction, and where does it belong in the code?

```
Dim number As Integer = 10
Do
     MessageBox.Show(number.ToString)
Loop Until number = 0
```

23. What will the following code display?

```
Dim totalEmp As Integer
Do While totalEmp <= 5
     MessageBox.Show(totalEmp.ToString)
     totalEmp = totalEmp + 2
Loop
```

24. What will the following code display?

```
Dim totalEmp As Integer = 1
Do
     MessageBox.Show(totalEmp.ToString)
     totalEmp = totalEmp + 2
Loop Until totalEmp >= 3
```

25. In this exercise, you modify the repetition structure contained in the Monthly Payment Calculator application.

   a. Use Windows to make a copy of the Payment Calculator Solution folder, which is contained in the VB2005\Chap06 folder. Rename the folder Modified Payment Calculator Solution.

   b. If necessary, start Visual Studio 2005 or Visual Basic 2005 Express Edition. Open the Payment Calculator Solution (Payment Calculator Solution.sln) file contained in the VB2005\Chap06\Modified Payment Calculator Solution folder. Open the designer window.

   c. Change the For...Next statement in the xCalcButton's Click event procedure to a Do...Loop statement.

   d. Save the solution, then start the application. Test the application to verify that it is working correctly.

   e. Click the Exit button to end the application. Close the Code Editor window, then close the solution.

26. In this exercise, you modify the repetition structure contained in the Sales Express application.

   a. Use Windows to make a copy of the Sales Express Solution folder, which is contained in the VB2005\Chap06 folder. Rename the folder Modified Sales Express Solution.

   b. If necessary, start Visual Studio 2005 or Visual Basic 2005 Express Edition. Open the Sales Express Solution (Sales Express Solution.sln) file contained in the VB2005\Chap06\Modified Sales Express Solution folder. Open the designer window.

c. Each time the application is started, the user will enter five sales amounts. Change the Do...Loop statement in the xCalcButton's Click event procedure to a For...Next statement. If the sales amount cannot be converted to a number, use the Exit For statement to exit the loop. Calculate the average only when the user enters five valid sales amounts.

d. Save the solution, then start the application. Test the application to verify that it is working correctly.

e. Click the Exit button to end the application. Close the Code Editor window, then close the solution.

DEBUGGING EXERCISE

27. The following code should display a 10% commission for each sales amount that is entered. The code is not working properly because an instruction is missing. What is the missing instruction, and where does it belong in the code?

```
Dim salesInput As String
Dim sales As Double
salesInput = InputBox("Enter a sales amount", "Sales")
Double.TryParse(salesInput, sales)
Do While sales > 0
    MessageBox.Show("Commission: " & sales * .1)
    Double.TryParse(salesInput, sales)
Loop
```

DEBUGGING EXERCISE

28. The following code should display a 10% commission for each sales amount that is entered. The code is not working properly. What is wrong with the code, and how can you fix it?

```
Dim salesInput As String
Dim sales As Double
salesInput = InputBox("Enter a sales amount", "Sales")
Double.TryParse(salesInput, sales)
Do
    salesInput = InputBox("Enter a sales amount", "Sales")
    Double.TryParse(salesInput, sales)
    MessageBox.Show("Commission: " & sales * .1)
Loop Until sales <= 0
```

LESSON B
OBJECTIVES

AFTER STUDYING LESSON B, YOU SHOULD
BE ABLE TO:

» Nest repetition structures

NESTED REPETITION STRUCTURES

NESTING REPETITION STRUCTURES

In Chapter 5, you learned how to nest selection structures. You also can nest repetition structures. In a nested repetition structure, one loop, referred to as the **inner loop**, is placed entirely within another loop, called the **outer loop**. Although the idea of nested loops may sound confusing, you already are familiar with the concept. A clock, for instance, uses nested loops to keep track of the time. For simplicity, consider a clock's second and minute hands only. You can think of the second hand as being the inner loop and the minute hand as being the outer loop. As you know, the second hand on a clock moves one position, clockwise, for every second that has elapsed. After the second hand moves 60 positions, the minute hand moves one position, clockwise. The second hand then begins its journey around the clock again. Figure 6-22 illustrates the logic used by a clock's second and minute hands. The outer loop in the figure corresponds to a clock's minute hand, and the inner loop corresponds to a clock's second hand. Notice that the entire inner loop is contained within the outer loop, which must be true for the loops to be nested and to work correctly.

>> **TIP**

Although both loops in Figure 6-22 are pretest loops, you also could write the logic using two posttest loops or a combination of a pretest and a posttest loop.

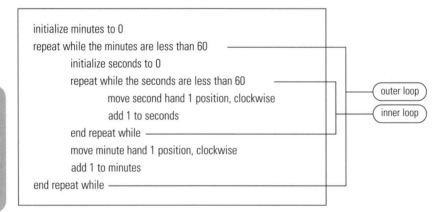

Figure 6-22: Nested loops used by a clock

In the next section, you view a different version of the Monthly Payment Calculator application that you viewed in Lesson A. This version uses two For...Next statements, one nested within the other.

MONTHLY PAYMENT CALCULATOR APPLICATION—NESTED FOR...NEXT STATEMENTS

Figure 6-23 shows the code for the xCalcButton's Click event procedure, which uses a nested repetition structure to calculate and display the monthly car payments. The outer For...Next statement controls the interest rates, which range from 5% to 10% in increments of 1%. The inner For...Next statement controls the terms, which are 3 years, 4 years, and 5 years.

Visual Basic

```vb
Private Sub xCalcButton_Click(ByVal sender As Object, _
    ByVal e As System.EventArgs) Handles xCalcButton.Click
    ' calculates the monthly payments on a loan
    ' using terms of 3, 4, and 5 years and interest
    ' rates of 5% through 10%

    Const TermHeading As String = _
        "           3 yrs        4 yrs        5 yrs"
    Dim principal As Double
    Dim monthlyPayment As Double
    Dim isConverted As Boolean

    Me.xPaymentsLabel.Text = String.Empty
    isConverted = _
        Double.TryParse(Me.xPrincipalTextBox.Text, principal)

    If isConverted Then
        ' display the term in the heading
        Me.xPaymentsLabel.Text = TermHeading _
            & ControlChars.NewLine
```

Figure 6-23: Nested repetition structure shown in the xCalcButton's Click event procedure *(Continued)* ▶

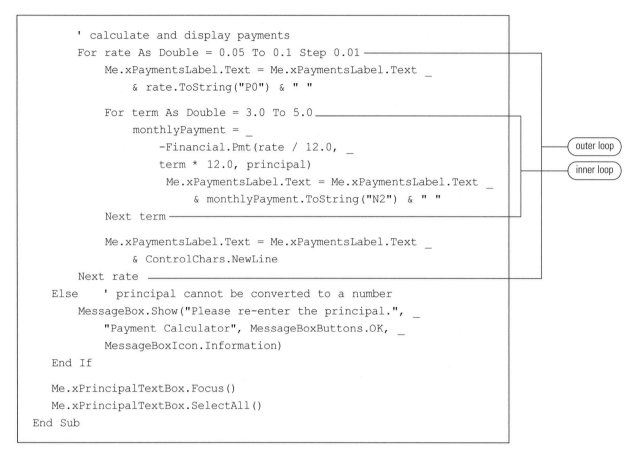

```
    ' calculate and display payments
    For rate As Double = 0.05 To 0.1 Step 0.01
        Me.xPaymentsLabel.Text = Me.xPaymentsLabel.Text _
            & rate.ToString("P0") & " "

        For term As Double = 3.0 To 5.0
            monthlyPayment = _
                -Financial.Pmt(rate / 12.0, _
                term * 12.0, principal)
                Me.xPaymentsLabel.Text = Me.xPaymentsLabel.Text _
                    & monthlyPayment.ToString("N2") & " "
        Next term

        Me.xPaymentsLabel.Text = Me.xPaymentsLabel.Text _
            & ControlChars.NewLine
    Next rate
Else    ' principal cannot be converted to a number
    MessageBox.Show("Please re-enter the principal.", _
        "Payment Calculator", MessageBoxButtons.OK, _
        MessageBoxIcon.Information)
End If

Me.xPrincipalTextBox.Focus()
Me.xPrincipalTextBox.SelectAll()
End Sub
```

outer loop

inner loop

Figure 6-23: Nested repetition structure shown in the xCalcButton's Click event procedure

When processing the `For rate As Double = 0.05 To 0.1 Step 0.01` clause, the computer creates the `rate` variable and assigns the number 0.05 to it. The computer then checks whether the `rate` variable contains a value that is greater than 0.1; at this point, it doesn't. Therefore, the computer processes the instructions contained in the outer loop. The first instruction in the outer loop displays the current rate in the xPaymentsLabel. When processing the second instruction in the outer loop, `For term As Double = 3.0 To 5.0`, the computer creates the `term` variable and initializes it to 3.0. The computer then checks whether the `term` variable contains a value that is greater than 5; at this point, it doesn't. Therefore, the computer processes the instructions contained in the inner loop.

The first instruction in the inner loop calculates the monthly car payment, using 0.05 as the rate and 3 as the term. The second instruction displays the monthly car payment in the xPaymentsLabel. The `Next term` clause increments the `term` variable by 1, giving 4. The computer then processes the `For term As Double = 3.0 To 5.0` clause again. This time, however, the computer simply checks whether the value in the `term` variable is greater than 5; it's not. Therefore, the computer again processes the instructions contained in the inner loop. Those instructions calculate the monthly car payment (using 0.05 as the rate and 4 as the term) and display the result in the xPaymentsLabel.

The `Next term` clause is processed next and increments the `term` variable by 1, giving 5. Then the computer processes the `For term As Double = 3.0 To 5.0` clause, which checks whether the value in the `term` variable is greater than 5; it's not. As a result, the computer processes the instructions contained in the inner loop. Those instructions calculate the monthly car payment (using 0.05 as the rate and 5 as the term) and display the result in the xPaymentsLabel.

The `Next term` clause is processed next and increments the `term` variable by 1, giving 6. Then the computer processes the `For term As Double = 3.0 To 5.0` clause, which checks whether the value in the `term` variable is greater than 5; it is. As a result, the inner For...Next statement ends and the computer processes the instruction located immediately after the `Next term` clause; that instruction moves the cursor to the next line in the xPaymentsLabel.

The computer then processes the `Next rate` clause, which increments the `rate` variable by 0.01, giving 0.06. It then processes the `For rate As Double = 0.05 To 0.1 Step 0.01` clause, which checks whether the `rate` variable contains a value that is greater than 0.1; it doesn't. Therefore, the computer again processes the instructions contained in the outer loop. Recall that the instructions in the outer loop display the rate in the xPaymentsLabel, and then use a For...Next statement to calculate and display the monthly car payments—this time using a rate of 0.06 and terms of three to five years. The outer For...Next statement ends when the value in the `rate` variable is 0.11.

To code and then test the Nested Payment application:

1 Open the **Nested Payment Solution** (Nested Payment Solution.sln) file, which is contained in the VB2005\Chap06\Nested Payment Solution folder. If necessary, open the designer window.

2 Open the Code Editor window. Replace the <your name> and <current date> text in the comments with your name and the current date.

3 Open the code template for the xCalcButton's Click event procedure, then enter the comments and code shown earlier in Figure 6-23.

4 Close the Code Editor window. Save the solution, then start the application.

5 Type **10000** in the Principal text box, then press **Enter** to select the Calculate button. The monthly payments shown in Figure 6-24 appear in the interface.

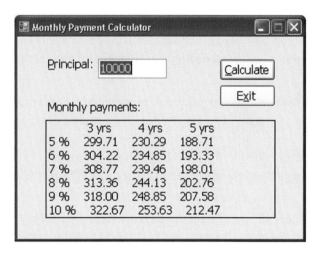

Figure 6-24: Monthly payments shown in the interface

6 Click the **Exit** button to end the application. You are returned to the designer window. Close the solution.

You have completed Lesson B. You can either take a break or complete the end-of-lesson questions and exercises before moving on to Lesson C.

SUMMARY

TO NEST A REPETITION STRUCTURE:

» Place the entire inner loop within the outer loop.

QUESTIONS

1. What appears in the xAsterisksLabel when the computer processes the following code?

```
For x As Integer = 1 To 2
   For y As Integer = 1 To 3
      Me.xAsterisksLabel.Text = _
         Me.xAsterisksLabel.Text & "*"
   Next y
   Me.xAsterisksLabel.Text = _
      Me.xAsterisksLabel.Text & _
      ControlChars.NewLine

Next x
```

a. ***
   ***

b. ***
   ***
   ***

c. **
   **
   **

d. ***
   ***
   ***
   ***

2. What number appears in the xSumLabel when the computer processes the following code?

```
Dim sum As Integer
Dim y As Integer
Do While y < 3
      For x As Integer = 1 To 4
            sum = sum + x
      Next x
      y = y + 1
Loop
Me.xSumLabel.Text = Convert.ToString(sum)
```

a. 5
b. 8

c. 15
d. 30

EXERCISES

1. In this exercise, you modify one of the repetition structures contained in the Monthly Payment Calculator application.

   a. Use Windows to make a copy of the Nested Payment Solution folder, which is contained in the VB2005\Chap06 folder. Rename the folder Nested Payment Solution-Ex1.

   b. If necessary, start Visual Studio 2005 or Visual Basic 2005 Express Edition. Open the Nested Payment Solution (Nested Payment Solution.sln) file contained in the VB2005\Chap06\Nested Payment Solution-Ex1 folder. Open the designer window.

   c. Change the For...Next statement that controls the term to a Do...Loop statement.

   d. Save the solution, then start the application. Test the application to verify that it is working correctly.

   e. Click the Exit button to end the application. Close the Code Editor window, then close the solution.

2. In this exercise, you modify both repetition structures contained in the Monthly Payment Calculator application.

   a. Use Windows to make a copy of the Nested Payment Solution folder, which is contained in the VB2005\Chap06 folder. Rename the folder Nested Payment Solution-Ex2.

   b. If necessary, start Visual Studio 2005 or Visual Basic Express Edition. Open the Nested Payment Solution (Nested Payment Solution.sln) file contained in the VB2005\Chap06\Nested Payment Solution-Ex2 folder. Open the designer window.

c. Change both For...Next statements to Do...Loop statements.

d. Save the solution, then start the application. Test the application to verify that it is working correctly.

e. Click the Exit button to end the application. Close the Code Editor window, then close the solution.

3. Professor Arkins needs an application that allows him to assign a grade to any number of students. Each student's grade is based on three test scores, with each test worth 100 points. The application should total the test scores and then assign the appropriate grade according to the following chart:

Total points earned	Grade
270–300	A
240–269	B
210–239	C
180–209	D
below 180	F

a. If necessary, start Visual Studio 2005 or Visual Basic 2005 Express Edition.

b. Create a Visual Basic Windows-based application. Name the solution Grade Calculator Solution, and name the project Grade Calculator Project. Save the application in the VB2005\Chap06 folder.

c. Assign the filename Main Form.vb to the form file object.

d. Assign the name MainForm to the form.

e. Create an appropriate interface. Code the application using a nested repetition structure.

f. Save the solution, then start the application. Test the application to verify that it is working correctly.

g. End the application. Close the Code Editor window, then close the solution.

LESSON C
OBJECTIVES

AFTER STUDYING LESSON C, YOU SHOULD
BE ABLE TO:

» Include a list box in an interface

» Select a list box item from code

» Determine the selected item in a list box

CODING THE SHOPPERS HAVEN APPLICATION

SHOPPERS HAVEN

Recall that the manager of Shoppers Haven wants an application that allows a store clerk to enter an item's original price and its discount rate. The discount rates range from 10% through 30% in increments of 5%. The application should calculate and display the amount of the discount and also the discounted price. Figure 6-25 shows the TOE chart for the application.

Task	Object	Event
1. Calculate the discount and discounted price 2. Display the discount and discounted price in the xDiscountLabel and xDiscountPriceLabel	xCalcButton	Click
End the application	xExitButton	Click
Display the discount (from xCalcButton)	xDiscountLabel	None
Display the discounted price (from xCalcButton)	xDiscountPriceLabel	None
Get and display the discount rates	xRateListBox	None
Clear the xDiscountLabel and xDiscountPriceLabel	xRateListBox xOrigPriceTextBox	SelectedValueChanged TextChanged
Get and display the original price Select the existing text	xOrigPriceTextBox	None Enter
1. Fill the xRateListBox with values 2. Select a default value in the xRateListBox	MainForm	Load

Figure 6-25: TOE chart for the Shoppers Haven application

To save you time, the VB2005\Chap06 folder contains a partially completed Shoppers Haven application.

To open the partially completed application:

1 Start Visual Studio 2005 or Visual Basic 2005 Express Edition, if necessary, and close the Start Page window.

2 Open the **Shoppers Haven Solution** (Shoppers Haven Solution.sln) file, which is contained in the VB2005\Chap06\Shoppers Haven Solution folder. Auto-hide the Toolbox, Solution Explorer, and Properties windows, if necessary. Figure 6-26 shows the partially completed user interface. Only the list box control is missing from the interface.

Figure 6-26: Partially completed user interface for the Shoppers Haven application

INCLUDING A LIST BOX IN AN INTERFACE

You use the **ListBox tool** in the toolbox to add a list box to an interface. You can use a **list box** to display a list of choices from which the user can select zero choices, one choice, or more than one choice. The number of choices the user is allowed to select is controlled by the list box's **SelectionMode property**. The default value for the property is One, which allows the user to select only one choice at a time in the list box. However, you also can set the property to None, MultiSimple, or MultiExtended. The None setting allows the user to scroll the list box, but not make any selection in it. The MultiSimple and MultiExtended settings allow the user to select more than one choice. (You will learn how to use the MultiSimple and MultiExtended settings in Discovery Exercise 9 at the end of this lesson.)

You can make a list box any size you want. If you have more items than fit into the list box, the control automatically displays a scroll bar that you can use to view the complete list of items. The Windows standard for list boxes is to display a minimum of three selections and a maximum of eight selections at a time. You should use a label control to provide keyboard access to the list box. For the access key to work correctly, you must set the TabIndex property of the label control to a value that is one less than the list box's TabIndex value.

To add a list box control to the form, then lock the controls and set the TabIndex values:

1 Click the **ListBox** tool in the toolbox, and then drag the mouse pointer to the form. Position the mouse pointer below the Discount rate label, then release the mouse button.

2 Set the list box's **(Name)** property to **xRateListBox**. (Do not be concerned that the list box's name appears inside the control. The name will not appear when the application is started.)

3 Set the list box's **Location** property to **188, 46**. Set its **Size** property to **66, 61**.

4 Lock the controls on the form, then use the information shown in Figure 6-27 to set the TabIndex values.

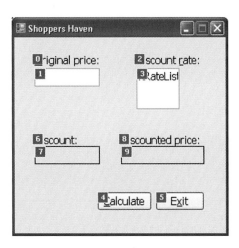

Figure 6-27: Correct TabIndex values

5 Press **Esc** to remove the TabIndex boxes from the form.

6 Save the solution.

ADDING ITEMS TO A LIST BOX

»TIP

In Discovery Exercise 10 at the end of this lesson, you will learn how to use the Items collection's Insert, Remove, RemoveAt, and Clear methods, as well as its Count property.

The items in a list box belong to a collection called the **Items collection**. A **collection** is a group of one or more individual objects treated as one unit. The first item in the Items collection appears as the first item in the list box, the second item appears as the second item in the list box, and so on. A unique number called an **index** identifies each item in a collection. The first item in the Items collection—and, therefore, the first item in the list box—has an index of zero. The second item has an index of one, and so on.

You use the Items collection's **Add method** to specify each item you want displayed in a list box. In most cases, you enter the Add methods in the form's Load event procedure. As you learned in Chapter 3, a form's Load event occurs before the form is displayed the first time. Figure 6-28 shows the syntax of the Add method and includes two examples of using the method to add items to a list box. In the syntax, *object* is the name of the control to which you want the item added, and *item* is the text you want displayed in the control. Each item in a list box is considered an object.

Add method

Syntax

object.**Items.Add(***item***)**

Examples

```
Me.xAnimalListBox.Items.Add("Dog")
Me.xAnimalListBox.Items.Add("Cat")
Me.xAnimalListBox.Items.Add("Horse")
```

displays Dog, Cat, and Horse in the xAnimalListBox

```
For code As Integer = 100 To 105
    Me.xCodeListBox.Items.Add(code.ToString)
Next code
```

displays 100, 101, 102, 103, 104, and 105 in the xCodeListBox

---

Figure 6-28: Syntax and examples of the Add method

»TIP

You also can use the String Collection Editor dialog box to add items to a list box. You will learn how to use the dialog box in Discovery Exercise 11 at the end of this lesson.

The Add methods shown in the first example in Figure 6-28 add the strings "Dog", "Cat", and "Horse" to the xAnimalListBox. In the second example, the `Me.xCodeListBox.Items.Add(code.ToString)` instruction contained in the For...Next statement displays numbers from 100 through 105, in increments of 1, in the xCodeListBox. Notice that the ToString method is used to convert each number to a String before adding it to the list box. Figure 6-29 shows the xAnimalListBox and xCodeListBox after the computer processes the Add methods.

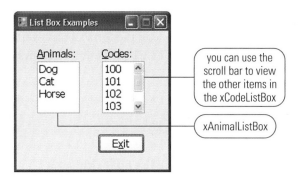

Figure 6-29: Items added to the list boxes

When you use the Add method to add an item to a list box, the position of the item in the list depends on the value stored in the list box's **Sorted property**. If the Sorted property contains its default value, False, the item is added at the end of the list. If the Sorted property is set to True, on the other hand, the item is sorted along with the existing items, and then placed in its proper position in the list. For example, if the Sorted property of the xAnimalListBox shown in Figure 6-29 were set to True, the list box items would appear in the following order: Cat, Dog, Horse.

Visual Basic sorts the list box items in dictionary order, which means that numbers are sorted before letters, and a lowercase letter is sorted before its uppercase equivalent. The items in a list box are sorted based on the leftmost characters in each item. Because of this, the items "Personnel", "Inventory", and "Payroll" will appear in the following order when the list box's Sorted property is set to True: Inventory, Payroll, Personnel. Similarly, the items 1, 2, 3, and 10 will appear in the following order when the list box's Sorted property is set to True: 1, 10, 2, 3.

The application determines whether you display the list box items in sorted order, or display them in the order in which they are added to the list box. If several list items are selected much more frequently than other items, you typically leave the list box's Sorted property set to False, and then add the frequently used items first so that the items appear at the beginning of the list. However, if the list box items are selected fairly equally, you typically set the list box's Sorted property to True, because it is easier to locate items when they appear in a sorted order.

»GUI DESIGN TIPS

List Box Standards

» A list box should contain a minimum of three selections.

» A list box should display a minimum of three selections and a maximum of eight selections at a time.

» Use a label control to provide keyboard access to the list box. Set the label's TabIndex property to a value that is one less than the list box's TabIndex value.

» List box items are either arranged by use, with the most used entries appearing first in the list, or sorted in ascending order.

You will use the Add method to add the discount rates to the xRateListBox in the Shoppers Haven interface. You will enter the appropriate code in the form's Load event procedure, because you want the list box to display its values when the interface first appears on the screen.

To add the discount rates to the xRateListBox:

1 Open the Code Editor window. Replace the `<your name>` and `<current date>` text in the comments with your name and the current date.

2 Open the MainForm's Load event procedure. Type **' fill the list box with values** and press **Enter**, then type **' and select a default value** and press **Enter** twice.

3 Type the For...Next loop shown in Figure 6-30, then position the insertion point as shown in the figure.

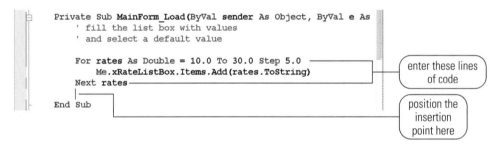

```
Private Sub MainForm_Load(ByVal sender As Object, ByVal e As
    ' fill the list box with values
    ' and select a default value

    For rates As Double = 10.0 To 30.0 Step 5.0
        Me.xRateListBox.Items.Add(rates.ToString)
    Next rates
    
End Sub
```

enter these lines of code

position the insertion point here

Figure 6-30: For...Next loop entered in the MainForm's Load event procedure

4 Save the solution, then start the application. See Figure 6-31.

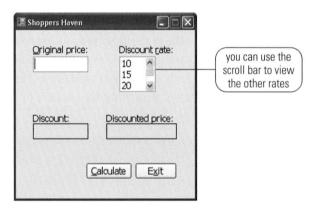

you can use the scroll bar to view the other rates

Figure 6-31: Discount rates shown in the xRateListBox

5 Scroll down the list box to verify that it also contains the numbers 25 and 30.

6 Scroll to the top of the list box, then click **15** in the list. See Figure 6-32.

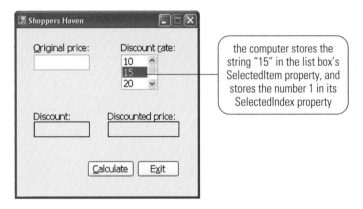

Figure 6-32: Second item selected in the xRateListBox

Notice that when you select an item in a list box, the item appears highlighted in the list. In addition, the item's value (in this case, the string "15") is stored in the list box's SelectedItem property, and the item's index (in this case, the number 1) is stored in the list box's SelectedIndex property. You will learn more about the SelectedItem and SelectedIndex properties in the next section.

7 Click the **Exit** button to end the application.

THE SELECTEDITEM AND SELECTEDINDEX PROPERTIES

You can use either the **SelectedItem property** or the **SelectedIndex property** to determine whether an item is selected in a list box. When no item is selected, the SelectedItem property contains the empty string, and the SelectedIndex property contains the number –1. Otherwise, the SelectedItem and SelectedIndex properties contain the value of the selected item and the item's index, respectively.

If a list box allows the user to make only one selection, it is customary in Windows applications to have one of the list box items already selected when the interface appears. The selected item, called the **default list box item**, should be either the item selected most frequently or, if all of the items are selected fairly equally, the first item in the list. You can use the SelectedItem and SelectedIndex properties to select the default list box item from code. Figure 6-33 shows examples of using the SelectedItem and SelectedIndex properties.

SelectedItem and SelectedIndex properties

Examples of the SelectedItem property

```
If Me.xRateListBox.SelectedItem.ToString = "15" Then
```
converts the selected item to String and then compares the result with the string "15"

```
If Convert.ToInt32(Me.xRateListBox.SelectedItem) = 15 Then
```
converts the selected item to Integer and then compares the result with the integer 15

```
Me.xRateListBox.SelectedItem = "20"
```
selects the 20 item in the list box

Examples of the SelectedIndex property

```
If Me.xRateListBox.SelectedIndex = 0 Then
```
determines whether the first item is selected in the list box

```
Me.xRateListBox.SelectedIndex = 2
```
selects the third item in the list box

Figure 6-33: Examples of the SelectedItem and SelectedIndex properties

The first three examples shown in Figure 6-33 use the SelectedItem property. The If clause in the first example converts the item selected in the xRateListBox to a String, and then compares the result with the string "15". The If clause in the second example converts the selected item to an Integer, and then compares the result with the integer 15. The statement in the third example uses the SelectedItem property to select the number 20 in the xRateListBox.

The last two examples shown in Figure 6-33 use the SelectedIndex property. The `If Me.xRateListBox.SelectedIndex = 0 Then` clause compares the contents of the xRateListBox's SelectedIndex property with the number zero to determine whether the first item is selected in the list box. You can use the `Me.xRateListBox.SelectedIndex = 2` statement shown in the last example to select the third item in the xRateListBox.

To select a default item in the xRateListBox:

1 The insertion point should be positioned below the `Next rates` line in the MainForm's Load event procedure. Type **me.xRateListBox.selectedindex = 0** and press **Enter**. Figure 6-34 shows the completed Load event procedure.

```
Private Sub MainForm_Load(ByVal sender As Object, ByVal e As
    ' fill the list box with values
    ' and select a default value

    For rates As Double = 10.0 To 30.0 Step 5.0
        Me.xRateListBox.Items.Add(rates.ToString)
    Next rates
    Me.xRateListBox.SelectedIndex = 0

End Sub
```

Figure 6-34: Completed MainForm's Load event procedure

2 Save the solution, then start the application. A default item—in this case, the first item—is selected in the xRateListBox, as shown in Figure 6-35.

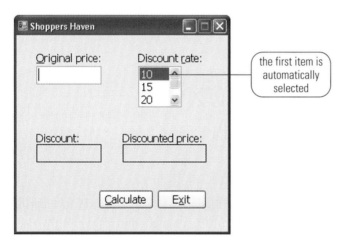

Figure 6-35: First item selected in the xRateListBox

3 Click the **Exit** button to end the application.

»GUI DESIGN TIP

Default List Box Item

» If a list box allows the user to make only one selection, then a default item should be selected in the list box when the interface first appears. The default item should be either the most used selection or the first selection in the list. However, if a list box allows more than one selection at a time, you do not select a default item.

THE SELECTEDVALUECHANGED AND SELECTEDINDEXCHANGED EVENTS

Each time either the user or a statement selects an item in a list box, the list box's **SelectedValueChanged** and **SelectedIndexChanged events** occur. You can use the procedures associated with these events to perform one or more tasks when the selected item has changed. In the Shoppers Haven application, you will code the xRateListBox's SelectedValueChanged procedure so that it clears both the discount and discounted price amounts whenever a change is made to the discount rate.

To code the list box's SelectedValueChanged event procedure:

1 Open the code template for the xRateListBox's SelectedValueChanged event. Type **' clear the calculated amounts** and press **Enter** twice.

2 Type **me.xDiscountLabel.text = string.empty** and press **Enter**, then type **me.xDiscountPriceLabel.text = string.empty** and press **Enter**.

3 Save the solution.

CODING THE TEXT BOX'S TEXTCHANGED EVENT PROCEDURE

According to the TOE chart (shown earlier in Figure 6-25), in addition to clearing the contents of the xDiscountLabel and xDiscountPriceLabel controls when the user selects a different discount rate, you also should clear the controls when the xOrigPriceTextBox's TextChanged event occurs. Recall that a text box control's TextChanged event occurs whenever a change is made to the contents of the control.

To code the text box's TextChanged event procedure:

1 In the xRateListBox's SelectedValueChanged procedure, replace `xRateListBox_SelectedValueChanged` in the procedure header with **ClearLabels**.

2 Position the insertion point immediately before the `Handles` keyword in the procedure header. Type _ (an underscore) and press **Enter**, then press **Tab** to indent the Handles clause.

3 Position the insertion point at the end of the `Handles xRateListBox.SelectedValueChanged` line. Type **, xOrigPriceTextBox.TextChanged**. Figure 6-36 shows the completed ClearLabels procedure.

```
Private Sub ClearLabels(ByVal sender As Object, ByVal e As System.EventArgs) _
    Handles xRateListBox.SelectedValueChanged, xOrigPriceTextBox.TextChanged
    ' clear the calculated amounts

    Me.xDiscountLabel.Text = String.Empty
    Me.xDiscountPriceLabel.Text = String.Empty

End Sub
```

line continuation character

enter this code

Figure 6-36: Completed ClearLabels procedure

4 Save the solution.

CODING THE XCALCBUTTON'S CLICK EVENT PROCEDURE

The xCalcButton's Click event procedure is the only procedure you still need to code. Figure 6-37 shows the procedure's pseudocode.

xCalcButton Click event procedure - pseudocode
1. convert the original price to a number
2. if the original price can be converted to a number
convert the current discount rate to a number
calculate the discount by multiplying the original price by the discount rate, and then dividing the result by 100
calculate the discounted price by subtracting the discount from the original price
end if
3. display the discount and discounted price in the xDiscountLabel and xDiscountPriceLabel controls

Figure 6-37: Pseudocode for the xCalcButton's Click event procedure

To code the xCalcButton's Click event procedure:

1 Open the code template for the xCalcButton's Click event procedure. Type '**calculate the discount and discounted price** and press **Enter** twice.

Recall that before you begin coding a procedure, you first study the procedure's pseudocode to determine the variables and named constants (if any) the procedure will use. In this case, the procedure will not use any named constants; however, it will use five variables.

2 Type **dim originalPrice as double** and press **Enter**. The procedure will use the `originalPrice` variable to store the original price entered in the xOrigPriceTextBox.

3 Type **dim rate as double** and press **Enter**. The procedure will use the `rate` variable to store the discount rate selected in the xRateListBox.

4 Type **dim discount as double** and press **Enter**, then type **dim discountPrice as double** and press **Enter**. The procedure will use these variables to store the calculated amounts.

5 Type **dim isPriceOk as boolean** and press **Enter** twice. The `isPriceOk` variable will store the value returned by the TryParse method when it attempts to convert the contents of the xOrigPriceTextBox to a Double number.

6 The first step in the pseudocode is to convert the original price to a number. Type **ispriceok** = _ and press **Enter**. Press **Tab**, then type **double.tryparse(me.xOrigPriceTextBox.text, originalprice)** and press **Enter**.

7 Next is a selection structure that determines whether the TryParse method succeeded in converting the original price to a number. Type **if ispriceok then** and press **Enter**.

8 If the TryParse method was successful, the first instruction in the selection structure's true path should convert the rate currently selected in the xRateListBox to a number. Type **rate = convert.todouble(me.xRateListBox.selecteditem)** and press **Enter**.

9 The next instruction in the true path should calculate the discount amount. Type **discount = originalprice * rate/100.0** and press **Enter**.

10 The last instruction in the true path should calculate the discounted price amount. Type **discountprice = originalprice – discount**.

11 Step 3 in the pseudocode is to display the discount and discounted price amounts in the appropriate label controls. Position the insertion point at the end of the `End If` clause, then press **Enter** twice. Type **me.xDiscountLabel.text = discount.tostring("C2")** and press **Enter**. Then type **me.xDiscountPriceLabel.text = discountprice.tostring("C2")** and press **Enter**. Figure 6-38 shows the code for the Shoppers Haven application.

```
' Project name:     Shoppers Haven Project
' Project purpose:  The project calculates the discount
'                   and discounted price.
' Created/revised:  <your name> on <current date>

Option Explicit On
Option Strict On

Public Class MainForm

    Private Sub xExitButton_Click(ByVal sender As Object, _
        ByVal e As System.EventArgs) Handles xExitButton.Click
        Me.Close()
    End Sub

    Private Sub xOrigPriceTextBox_Enter(ByVal sender As Object, _
        ByVal e As System.EventArgs) Handles xOrigPriceTextBox.Enter
        ' select the existing text

        Me.xOrigPriceTextBox.SelectAll()
    End Sub

    Private Sub MainForm_Load(ByVal sender As Object, _
        ByVal e As System.EventArgs) Handles Me.Load
        ' fill the list box with values
        ' and select a default value

        For rates As Double = 10.0 To 30.0 Step 5.0
            Me.xRateListBox.Items.Add(rates.ToString)
        Next rates
        Me.xRateListBox.SelectedIndex = 0

    End Sub

    Private Sub ClearLabels (ByVal sender As Object, ByVal e As System.EventArgs) _
        Handles xRateListBox.SelectedValueChanged, xOrigPriceTextBox.TextChanged
        ' clear the calculated amounts

        Me.xDiscountLabel.Text = String.Empty
        Me.xDiscountPriceLabel.Text = String.Empty

    End Sub
```

Figure 6-38: Shoppers Haven application's code *(Continued)* ▶

```
    Private Sub xCalcButton_Click(ByVal sender As Object, _
        ByVal e As System.EventArgs) Handles xCalcButton.Click
        ' calculate the discount and discounted price

        Dim originalPrice As Double
        Dim rate As Double
        Dim discount As Double
        Dim discountPrice As Double
        Dim isPriceOk As Boolean

        isPriceOk = _
            Double.TryParse(Me.xOrigPriceTextBox.Text, originalPrice)
        If isPriceOk Then
            rate = Convert.ToDouble(Me.xRateListBox.SelectedItem)
            discount = originalPrice * rate / 100.0
            discountPrice = originalPrice - discount
        End If

        Me.xDiscountLabel.Text = discount.ToString("C2")
        Me.xDiscountPriceLabel.Text = discountPrice.ToString("C2")

    End Sub
End Class
```

Figure 6-38: Shoppers Haven application's code

In the next set of steps, you will test the Shoppers Haven application to verify that it is working correctly.

To test the Shoppers Haven application:

1 Save the solution, then start the application.

2 Type **100** in the Original price text box, then click **20** in the Discount rate list box. Click the **Calculate** button. The xCalcButton's Click event procedure displays the correct discount and discounted price amounts, as shown in Figure 6-39.

Figure 6-39: Discount and discounted price amounts shown in the interface

3 Click **10** in the Discount rate list box. The ClearLabels procedure, which is associated with the xRateListBox's SelectedValueChanged event, removes the discount and discounted price amounts from the label controls.

4 Click the **Calculate** button. The xCalcButton's Click event procedure displays $10.00 and $90.00 as the discount and discounted price amounts, respectively.

5 Press **Tab** twice to move the focus to the xOrigPriceTextBox. The text box's Enter event procedure selects the existing text in the text box.

6 Type **9** in the text box. The ClearLabels procedure, which is associated with the xOrigPriceTextBox's TextChanged event, removes the discount and discounted price amounts from the label controls.

7 Click the **Calculate** button. The xCalcButton's Click event procedure displays $0.90 and $8.10 as the discount and discounted price amounts, respectively.

8 Click the **Exit** button to end the application. Close the Code Editor window, then close the solution.

You have completed Lesson C and Chapter 6. You can either take a break or complete the end-of-lesson questions and exercises.

SUMMARY

TO ADD A LIST BOX CONTROL TO A FORM:

» Use the ListBox tool in the toolbox.

TO SPECIFY WHETHER THE USER CAN SELECT ZERO CHOICES, ONE CHOICE, OR MORE THAN ONE CHOICE IN A LIST BOX:

» Set the list box's SelectionMode property to None, One, MultiSimple, or MultiExtended.

TO ADD ITEMS TO A LIST BOX:

» Use the Items collection's Add method, whose syntax is *object*.**Items.Add(***item***)**. In the syntax, *object* is the name of the control to which you want the item added, and *item* is the text you want displayed in the control.

TO AUTOMATICALLY SORT THE ITEMS IN A LIST BOX:

» Set the list box's Sorted property to True.

TO DETERMINE THE ITEM SELECTED IN A LIST BOX, OR TO SELECT A LIST BOX ITEM FROM CODE:

» Use either the list box's SelectedItem property or its SelectedIndex property.

QUESTIONS

1. You use the _____ method to add items to a list box.

    a. Add

    b. AddList

    c. Item

    d. ItemAdd

2. The items in a list box belong to the _____ collection.

    a. Items

    b. List

    c. ListItems

    d. Values

3. The _____ property stores the index of the item selected in a list box.

    a. Index

    b. SelectedIndex

    c. Selection

    d. SelectionIndex

4. Which of the following statements selects the "Cat" item, which appears third in the xAnimalListBox?

    a. `Me.xAnimalListBox.SelectedIndex = 2`

    b. `Me.xAnimalListBox.SelectedIndex = 3`

    c. `Me.xAnimalListBox.SelectedItem = "Cat"`

    d. both a and c

5. The _____ event occurs when the user selects a different item in a list box.

    a. SelectionChanged   b. SelectedItemChanged

    c. SelectedValueChanged   d. None of the above.

EXERCISES

1. In this exercise, you modify the repetition structure contained in the Shoppers Haven application.

    a. Use Windows to make a copy of the Shoppers Haven Solution folder, which is contained in the VB2005\Chap06 folder. Rename the folder Modified Shoppers Haven Solution.

    b. If necessary, start Visual Studio 2005 or Visual Basic 2005 Express Edition. Open the Shoppers Haven Solution (Shoppers Haven Solution.sln) file contained in the VB2005\Chap06\Modified Shoppers Haven Solution folder. Open the designer window.

    c. Change the For...Next statement in the MainForm's Load event procedure to a Do...Loop statement.

    d. Save the solution, then start the application. Test the application to verify that it is working correctly.

    e. Click the Exit button to end the application. Close the Code Editor window, then close the solution.

2. Powder Skating Rink holds a weekly ice-skating competition. Competing skaters must perform a two-minute program in front of a panel of judges. The number of judges varies from week to week. At the end of a skater's program, each judge assigns

a score of zero through 10 to the skater. The manager of the ice rink wants an application that calculates and displays a skater's average score. The application also should display the skater's total score and the number of scores entered.

a. If necessary, start Visual Studio 2005 or Visual Basic 2005 Express Edition.

b. Create a Visual Basic Windows-based application. Name the solution Powder Solution, and name the project Powder Project. Save the application in the VB2005\Chap06 folder.

c. Assign the filename Main Form.vb to the form file object.

d. Assign the name MainForm to the form.

e. Design an appropriate interface. Use the GUI design guidelines listed in Appendix A to verify that the interface you create adheres to the GUI standards outlined in this book.

f. Code the application.

g. Save the solution, then start the application. Test the application using your own sample data.

h. End the application. Close the Code Editor window, then close the solution.

3. In this exercise, you code an application that allows the user 10 chances to guess a random number generated by the computer. The random number should be an integer from 1 through 50, inclusive. Each time the user makes an incorrect guess, the application will display a message that tells the user either to guess a higher number or to guess a lower number. When the user guesses the random number, the application should display a "Congratulations!" message. However, if the user is not able to guess the random number after 10 tries, the application should display the random number in a message.

a. If necessary, start Visual Studio 2005 or Visual Basic 2005 Express Edition.

b. Create a Visual Basic Windows-based application. Name the solution Random Solution, and name the project Random Project. Save the application in the VB2005\Chap06 folder.

c. Assign the filename Main Form.vb to the form file object.

d. Assign the name MainForm to the form.

e. Design an appropriate interface. Use the GUI design guidelines listed in Appendix A to verify that the interface you create adheres to the GUI standards outlined in this book.

f. Code the application.

g. Save the solution, then start the application. Test the application appropriately.

h. End the application. Close the Code Editor window, then close the solution.

4. In this exercise, you code an application that displays the telephone extension corresponding to the name selected in a list box. The names and extensions are shown here:

Smith, Joe   3388

Jones, Mary   3356

Adkari, Joel   2487

Lin, Sue   1111

Li, Vicky   2222

a. If necessary, start Visual Studio 2005 or Visual Basic 2005 Express Edition. Open the Phone Solution (Phone Solution.sln) file, which is contained in the VB2005\Chap06\Phone Solution folder. If necessary, open the designer window.

b. The items in the list box should be sorted. Set the appropriate property.

c. Code the form's Load event procedure so that it adds the five names shown above to the xNamesListBox. Select the first name in the list.

d. Code the list box's SelectedValueChanged event procedure so that it assigns the item selected in the xNamesListBox to a variable. The procedure then should use the Select Case statement to display the telephone extension corresponding to the name stored in the variable.

e. Save the solution, then start the application. Test the application by clicking each name in the list box.

f. Click the Exit button to end the application. Close the Code Editor window, then close the solution.

5. In this exercise, you modify the application that you coded in Exercise 4. The application will now assign the index of the selected item, rather than the selected item itself, to a variable.

a. Use Windows to make a copy of the Phone Solution folder, which is contained in the VB2005\Chap06 folder. Change the name of the folder to Modified Phone Solution.

b. If necessary, start Visual Studio 2005 or Visual Basic 2005 Express Edition. Open the Phone Solution (Phone Solution.sln) file contained in the VB2005\Chap06\Modified Phone Solution folder. Open the designer window.

c. Modify the list box's SelectedValueChanged event procedure so that it assigns the index of the item selected in the xNamesListBox to a variable. Modify the Select Case statement so that it displays the telephone extension corresponding to the index stored in the variable.

d. Save the solution, then start and test the application.

e. Click the Exit button to end the application. Close the Code Editor window, then close the solution.

6. In this exercise, you create an application that displays the first 10 Fibonacci numbers (1, 1, 2, 3, 5, 8, 13, 21, 34, and 55).

a. If necessary, start Visual Studio 2005 or Visual Basic 2005 Express Edition. Open the Fibonacci Solution (Fibonacci Solution.sln) file, which is contained in the VB2005\Chap06\Fibonacci Solution folder. If necessary, open the designer window.

b. Code the application. Notice that, beginning with the third number in the series, each Fibonacci number is the sum of the prior two numbers. In other words, 2 is the sum of 1 plus 1, 3 is the sum of 1 plus 2, 5 is the sum of 2 plus 3, and so on. Display the numbers in the xNumbersLabel.

c. Save the solution, then start and test the application.

d. Click the Exit button to end the application. Close the Code Editor window, then close the solution.

7. In this exercise, you create an application that displays a multiplication table similar to the one shown in Figure 6-40, where x is a number entered by the user and y is the result of multiplying x by the numbers 1 through 9.

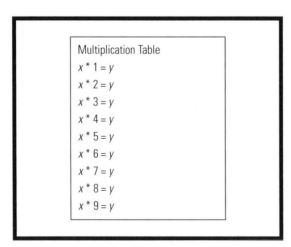

Figure 6-40

    a. If necessary, start Visual Studio 2005 or Visual Basic 2005 Express Edition.

    b. Create a Visual Basic Windows-based application. Name the solution Multiplication Solution, and name the project Multiplication Project. Save the application in the VB2005\Chap06 folder.

    c. Assign the filename Main Form.vb to the form file object.

    d. Assign the name MainForm to the form.

    e. Design an appropriate interface, then code the application.

    f. Save the solution, then start and test the application.

    g. End the application. Close the Code Editor window, then close the solution.

8. In this exercise, you create an application that allows the user to enter the gender (either F or M) and GPA for any number of students. The application should calculate the average GPA for all students, the average GPA for male students, and the average GPA for female students.

    a. If necessary, start Visual Studio 2005 or Visual Basic 2005 Express Edition.

    b. Create a Visual Basic Windows-based application. Name the solution GPA Solution, and name the project GPA Project. Save the application in the VB2005\Chap06 folder.

    c. Assign the filename Main Form.vb to the form file object.

    d. Assign the name MainForm to the form.

    e. Design an appropriate interface, then code the application. Do not use the InputBox function in the application.

    f. Save the solution, then start and test the application.

    g. End the application. Close the Code Editor window, then close the solution.

DISCOVERY EXERCISE

9. In this exercise, you learn how to create a list box that allows the user to select more than one item at a time.

    a. If necessary, start Visual Studio 2005 or Visual Basic 2005 Express Edition. Open the Multi Solution (Multi Solution.sln) file, which is contained in the VB2005\Chap06\Multi Solution folder. If necessary, open the designer window. The interface contains a list box named xNamesListBox. The list box's Sorted property is set to True, and its SelectionMode property is set to One.

    b. Open the Code Editor window. Notice that the form's Load event procedure adds five names to the list box.

c. Code the xSingleButton's Click event procedure so that it displays, in the xResultLabel, the item selected in the xNamesListbox. For example, if the user clicks Debbie in the list box and then clicks the Single Selection button, the name Debbie should appear in the xResultLabel.

d. Save the solution, then start the application. Click Debbie in the list box, then click Ahmad, and then click Bill. Notice that, when the list box's SelectionMode property is set to One, you can select only one item at a time in the list.

e. Click the Single Selection button. The name "Bill" appears in the xResultLabel.

f. Click the Exit button to end the application.

g. Change the list box's SelectionMode property to MultiSimple. Save the solution, then start the application. Click Debbie in the list box, then click Ahmad, then click Bill, and then click Ahmad. Notice that, when the list box's SelectionMode property is set to MultiSimple, you can select more than one item at a time in the list. Also notice that you click to both select and deselect an item. (You also can use Ctrl+click and Shift+click, as well as press the Spacebar, to select and deselect items when the list box's SelectionMode property is set to MultiSimple.)

h. Click the Exit button to end the application.

i. Change the list box's SelectionMode property to MultiExtended. Save the solution, then start the application.

j. Click Debbie in the list, then click Jim. Notice that, in this case, clicking Jim deselects Debbie. When a list box's SelectionMode property is set to MultiExtended, you use Ctrl+click to select multiple items in the list. You also use Ctrl+click to deselect items in the list. Click Debbie in the list, then Ctrl+click Ahmad, and then Ctrl+click Debbie.

k. Next, click Bill in the list, then Shift+click Jim; this selects all of the names from Bill through Jim.

l. Click the Exit button to end the application.

As you know, when a list box's SelectionMode property is set to One, the item selected in the list box is stored in the SelectedItem property, and the item's index is stored in the SelectedIndex property. However, when a list box's SelectionMode property is set to either MultiSimple or MultiExtended, the items selected in the list box are stored (as strings) in the SelectedItems property, and the indices of the items are stored (as integers) in the SelectedIndices property.

m. Code the xMultiButton's Click event procedure so that it first clears the contents of the xResultLabel. The procedure should then display the selected names (which are stored in the SelectedItems property) on separate lines in the xResultLabel.

7 Click the **Exit** button. A message box containing the "Do you want to exit?" message appears. Click the **No** button. Notice that the form remains on the screen. In Lesson C, you will learn how to prevent the computer from closing a form.

8 Click the **Exit** button, then click the **Yes** button in the message box. The application closes.

The Harvey Industries payroll application uses a combo box and a Function procedure. You learn about Function procedures in Lesson A, and about combo boxes in Lesson B. You begin coding the payroll application in Lesson B and complete it in Lesson C.

LESSON A

OBJECTIVES

AFTER STUDYING LESSON A, YOU SHOULD
BE ABLE TO:

» Explain the difference between a Sub procedure and a
Function procedure

» Create a procedure that receives information passed to it

» Explain the difference between passing data *by value* and
passing data *by reference*

» Create a Function procedure

CREATING SUB AND FUNCTION PROCEDURES

PROCEDURES

As you already know, a procedure is a block of program code that performs a specific task. Most procedures in Visual Basic are either Sub procedures or Function procedures. The difference between both types of procedures is that a **Function procedure** returns a value after performing its assigned task, whereas a **Sub procedure** does not return a value. Although you have been using Sub procedures since Chapter 1, this lesson provides a more in-depth look into their creation and use. After exploring the topic of Sub procedures, you then will learn how to create and use Function procedures.

» TIP
You also can create Property procedures in Visual Basic. Property procedures are covered in Chapter 11.

SUB PROCEDURES

As you learned in Chapter 5, there are two types of Sub procedures in Visual Basic: event procedures and independent Sub procedures. An event procedure is a Sub procedure that is associated with a specific object and event, such as a button's Click event or a text box's TextChanged event. As you learned in Chapter 1, the computer automatically processes an event procedure when the event occurs. An independent Sub procedure, on the other hand, is a procedure that is independent of any object and event, and is processed only when called, or invoked, from code. You learned how to create an independent Sub procedure in Chapter 5. You also learned how to use the Call statement to invoke the procedure. Figure 7-2 shows the syntax of an independent Sub procedure, as well as the syntax of the Call statement. The figure also includes the independent Sub procedure and Call statement from the Math Practice application that you created in Chapter 5.

Independent Sub procedure

<u>Syntax</u>

Private Sub *procedurename*(**[***parameterlist***]**) ————————————————— (procedure header)

    [*statements*]

End Sub

<u>Example</u>

```
Private Sub GenerateAndDisplayNumbers()
    ' generates and displays two random numbers

    Dim randomNum1 As Integer
    Dim randomNum2 As Integer
    Dim randomGenerator As New Random

    ' generate random numbers
    If Me.xGrade1RadioButton.Checked Then
        randomNum1 = randomGenerator.Next(1, 11)
        randomNum2 = randomGenerator.Next(1, 11)
    Else
        randomNum1 = randomGenerator.Next(10, 100)
        randomNum2 = randomGenerator.Next(10, 100)
    End If

    ' swap numbers if the subtraction problem
    ' would result in a negative number
    If Me.xSubtractionRadioButton.Checked _
        AndAlso randomNum1 < randomNum2 Then
        Dim temp As Integer
        temp = randomNum1
        randomNum1 = randomNum2
        randomNum2 = temp
    End If

    ' display numbers
    Me.xNum1Label.Text = Convert.ToString(randomNum1)
    Me.xNum2Label.Text = Convert.ToString(randomNum2)
End Sub
```

Call statement

<u>Syntax</u>

Call *procedurename*(**[***argumentlist***]**)

<u>Example</u>

```
Call GenerateAndDisplayNumbers()
```

Figure 7-2: Independent Sub procedure and Call statement

As Figure 7-2 indicates, following the *procedurename* in an independent Sub procedure header is a set of parentheses that contains an optional *parameterlist*. The *parameterlist* lists the data type and name of one or more memory locations, called **parameters**. The parameters store the information passed to the procedure when the procedure is invoked. You will learn more about the *parameterlist* in the next section. If the procedure does not require any information to be passed to it, as is the case with the GenerateAndDisplayNumbers procedure, an empty set of parentheses follows the *procedurename* in the procedure header.

In the Call statement's syntax, *procedurename* is the name of the procedure you are invoking (calling), and *argumentlist* (which is optional) is a comma-separated list of arguments you want passed to the procedure. If you do not need to pass any information to the procedure you are calling, you simply include an empty set of parentheses after the *procedurename* in the Call statement, as in the `Call GenerateAndDisplayNumbers()` statement.

> **» TIP**
>
> As you learned in Chapter 5, Sub procedure names should be entered using Pascal case. Also, it is a common practice to begin a procedure's name with a verb.

PASSING INFORMATION TO AN INDEPENDENT SUB PROCEDURE

As you learned in the previous section, an independent Sub procedure can contain one or more parameters in its procedure header. Each parameter stores data that is passed to the procedure when the procedure is invoked by the Call statement. The Call statement passes the information in its optional *argumentlist*. The number of arguments listed in the Call statement's *argumentlist* should agree with the number of parameters listed in the *parameterlist* in the procedure header. If the *argumentlist* includes one argument, then the procedure header should have one parameter in its *parameterlist*. Similarly, a procedure that is passed three arguments when called requires three parameters in its *parameterlist*. (Refer to the TIP on this page for an exception to this general rule.)

> **» TIP**
>
> Visual Basic allows you to specify that an argument in the Call statement is optional. You can learn more about optional arguments by completing Exercise 21 at the end of this lesson.

In addition to having the same number of parameters as arguments, the data type and position of each parameter in the *parameterlist* should agree with the data type and position of its corresponding argument in the *argumentlist*. For instance, if the argument is an integer, then the parameter that will store the integer should have a data type of Integer, Short, or Long, depending on the size of the integer. Likewise, if two arguments are passed to a procedure—the first one being a String variable and the second one being a Decimal variable—the first parameter should have a data type of String and the second parameter should have a data type of Decimal.

You can pass a literal constant, named constant, keyword, or variable to an independent Sub procedure. However, in most cases, you will pass a variable.

PASSING VARIABLES

Each variable you declare in an application has both a value and a unique address that represents the location of the variable in the computer's internal memory. Visual Basic allows you to pass either the variable's value or its address to the receiving procedure. Passing a variable's value is referred to as **passing by value**. Passing a variable's address is referred to as **passing by reference**. The method you choose—*by value* or *by reference*—depends on whether you want to give the receiving procedure access to the variable in memory. In other words, it depends on whether you want to allow the receiving procedure to change the contents of the variable.

>> **TIP**

The internal memory of a computer is like a large post office, where each memory cell, like each post office box, has a unique address.

Although the idea of passing information *by value* and *by reference* may sound confusing at first, it is a concept with which you already are familiar. To illustrate, assume that you have a savings account at a local bank. During a conversation with a friend, you mention the amount of money you have in the account. Telling someone the amount of money in your account is similar to passing a variable *by value*. Knowing the balance in your account does not give your friend access to your bank account. It merely gives your friend some information that he or she can use—perhaps to compare to the amount of money he or she has saved.

The savings account example also provides an illustration of passing information *by reference*. To deposit money to or withdraw money from your account, you must provide the bank teller with your account number. The account number represents the location of your account at the bank and allows the teller to change the account balance. Giving the teller your bank account number is similar to passing a variable *by reference*. The account number allows the teller to change the contents of your bank account, similar to the way the variable's address allows the receiving procedure to change the contents of the variable passed to the procedure.

PASSING BY VALUE

To pass a variable *by value* in Visual Basic, you include the keyword `ByVal` (which stands for "by value") before the variable's corresponding parameter in the *parameterlist*. When you pass a variable *by value*, the computer passes only the contents of the variable to the receiving procedure. When only the contents are passed, the receiving procedure is not given access to the variable in memory, so it cannot change the value stored inside the variable. You pass a variable *by value* when the receiving procedure needs to *know* the variable's contents, but the receiving procedure does not need to *change* the contents. Unless specified otherwise, variables are passed *by value* in Visual Basic.

Figure 7-3 shows a sample run of the Pet Information application, and Figure 7-4 shows the application's code, which passes two String variables *by value* to an independent Sub procedure.

Figure 7-3: Sample run of the Pet Information application

Visual Basic code

```
' Project name:     Pet Information Project
' Project purpose:  The project displays a message
'                   that contains a pet's name and age.
' Created/revised:  <your name> on <current date>

Option Explicit On
Option Strict On

Public Class MainForm

    Private Sub ShowMsg(ByVal pet As String, ByVal years As String)
        ' displays the pet information passed to it

        Me.xMessageLabel.Text = "Your pet " & pet & " is " _
            & years & " years old."
    End Sub

    Private Sub xGetInfoButton_Click(ByVal sender As Object, _
        ByVal e As System.EventArgs) Handles xGetInfoButton.Click
        ' gets the pet information, then displays the
        ' information in a message

        Dim petName As String
        Dim petAge As String

        petName = InputBox("Pet's name:", "Name Entry")
        petAge = InputBox("Pet's age (years):", "Age Entry")

        Call ShowMsg(petName, petAge)
    End Sub

    Private Sub xExitButton_Click(ByVal sender As Object, _
        ByVal e As System.EventArgs) Handles xExitButton.Click
        Me.Close()
    End Sub
End Class
```

parameterlist

argumentlist

Figure 7-4: Code for the Pet Information application

Notice that the number, data type, and sequence of the arguments in the Call statement match the number, data type, and sequence of the corresponding parameters in the procedure header. Also notice that the names of the parameters do not need to be identical to the names of the corresponding arguments. In fact, to avoid confusion, it usually is better to use different names for the arguments and parameters.

Study the code shown in Figure 7-4. The xGetInfoButton's Click event procedure first declares two String variables named petName and petAge. The next two statements in the procedure use the InputBox function to prompt the user to enter the name and age (in years) of his or her pet. If the user enters "Spot" as the name and "4" as the age, the computer stores the string "Spot" in the petName variable, and stores the string "4" in the petAge variable.

Next, the Call ShowMsg(petName, petAge) statement calls the ShowMsg procedure, passing it the petName and petAge variables *by value*, which means that only the contents of the variables—in this case, "Spot" and "4"—are passed to the procedure. You know that the variables are passed *by value* because the keyword ByVal appears before each variable's corresponding parameter in the ShowMsg procedure header. At this point, the computer temporarily leaves the xGetInfoButton's Click event procedure to process the code contained in the ShowMsg procedure.

The first instruction processed in the ShowMsg procedure is the procedure header. When processing the procedure header, the computer creates the variables listed in the *parameterlist*, and stores the information passed to the procedure in those variables. In this case, the computer creates the pet and years variables, storing the string "Spot" in the pet variable and the string "4" in the years variable. The variables that appear in a procedure header are procedure-level variables, which means they can be used only by the procedure in which they are declared. In this case, the pet and years variables can be used only by the ShowMsg procedure.

After processing the ShowMsg procedure header, the computer processes the assignment statement contained in the procedure. The assignment statement uses the values stored in the procedure's parameters (pet and years) to display the appropriate message in the xMessageLabel. In this case, the statement displays the message "Your pet Spot is 4 years old."

Next, the computer processes the ShowMsg procedure footer (End Sub), which ends the ShowMsg procedure. At this point, the pet and years variables are removed from the computer's internal memory. (Recall that a procedure-level variable is removed from the computer's memory when the procedure in which it is declared ends.) The computer then returns to the xGetInfoButton's Click event procedure to process the statement immediately following the Call ShowMsg(petName, petAge) statement. In this case, the statement following the Call statement is End Sub, which ends the xGetInfoButton's Click event procedure. The computer then removes the petName and petAge procedure-level variables from its internal memory.

> **»TIP**
>
> You cannot determine by looking at the Call statement whether a variable is being passed *by value* or *by reference*. You must look at the procedure header to make the determination.

To code and then test the Pet Information application:

1 Start Visual Studio 2005 or Visual Basic 2005 Express Edition, if necessary, and close the Start Page window.

2 Open the **Pet Information Solution** (Pet Information Solution.sln) file, which is contained in the VB2005\Chap07\Pet Information Solution folder. If necessary, open the designer window.

3 Open the Code Editor window. Replace the <your name> and <current date> text in the comments with your name and the current date.

4 Enter the comments and code shown in Figure 7-4.

5 Close the Code Editor window. Save the solution, then start the application.

6 Click the **Get Information** button. Type **Spot** in the Name Entry dialog box, then press **Enter** to select the OK button. Type **4** in the Age Entry dialog box, then press **Enter**. The application displays the message shown earlier in Figure 7-3.

7 Click the **Exit** button to end the application. You are returned to the designer window. Close the solution.

Next, you view another example of passing information *by value*. Figure 7-5 shows a sample run of the Bonus Calculator application, and Figure 7-6 shows the application's code. In the code, the xCalcButton's Click event procedure passes three values to the CalcBonus procedure: the values contained in two variables and the value contained in a named constant.

Figure 7-5: Sample run of the Bonus Calculator application

Visual Basic code

```vb
' Project name:      Bonus Calculator Project
' Project purpose:   The project calculates a 10% bonus.
' Created/revised:   <your name> on <current date>

Option Explicit On
Option Strict On

Public Class MainForm

    Private Sub CalcBonus(ByVal sales1 As Decimal, _
                          ByVal sales2 As Decimal, _
                          ByVal rate As Decimal)
        ' calculates and displays a bonus amount based
        ' on the sales amounts and bonus rate passed to it

        Dim bonus As Decimal

        bonus = (sales1 + sales2) * rate
        Me.xBonusLabel.Text = bonus.ToString("C2")
    End Sub

    Private Sub xExitButton_Click(ByVal sender As Object, _
        ByVal e As System.EventArgs) Handles xExitButton.Click
        Me.Close()
    End Sub

    Private Sub xCalcButton_Click(ByVal sender As Object, _
        ByVal e As System.EventArgs) Handles xCalcButton.Click
        ' calculates a 10% bonus amount

        Const BonusRate As Decimal = 0.1D
        Dim reg1Sales As Decimal
        Dim reg2Sales As Decimal
        Dim isConvertedReg1 As Boolean
        Dim isConvertedReg2 As Boolean

        isConvertedReg1 = _
            Decimal.TryParse(Me.xReg1TextBox.Text, reg1Sales)
        isConvertedReg2 = _
            Decimal.TryParse(Me.xReg2TextBox.Text, reg2Sales)
```

parameterlist

Figure 7-6: Code for the Bonus Calculator application *(Continued)* ▶

```
        If isConvertedReg1 AndAlso isConvertedReg2 Then
            Call CalcBonus(reg1Sales, reg2Sales, BonusRate)
        Else                                                        ──── argumentlist
            MessageBox.Show("The sales amounts must be numeric.", _
                "Bonus Calculator", MessageBoxButtons.OK, _
                MessageBoxIcon.Information)
        End If

        Me.xReg1TextBox.Focus()
    End Sub

    Private Sub xReg1TextBox_Enter(ByVal sender As Object, _
        ByVal e As System.EventArgs) Handles xReg1TextBox.Enter
        Me.xReg1TextBox.SelectAll()
    End Sub

    Private Sub xReg2TextBox_Enter(ByVal sender As Object, _
        ByVal e As System.EventArgs) Handles xReg2TextBox.Enter
        Me.xReg2TextBox.SelectAll()
    End Sub

    Private Sub xRegion1TextBox_TextChanged(ByVal sender As Object, _
        ByVal e As System.EventArgs) Handles xReg1TextBox.TextChanged
        Me.xBonusLabel.Text = String.Empty
    End Sub

    Private Sub region2TextBox_TextChanged(ByVal sender As Object, _
        ByVal e As System.EventArgs) Handles xReg2TextBox.TextChanged
        xBonusLabel.Text = String.Empty
    End Sub
End Class
```

Figure 7-6: Code for the Bonus Calculator application

Study the code shown in Figure 7-6. The xCalcButton's Click event procedure first declares a Decimal named constant, two Decimal variables, and two Boolean variables. The next two statements in the procedure use the TryParse method to convert the contents of the text boxes to Decimal numbers. The procedure then uses a selection structure to check whether the conversions were successful. If one or both conversions were not successful, the instruction in the selection structure's false path displays an appropriate message. However, if both conversions were successful, the Call CalcBonus(reg1Sales, reg2Sales, BonusRate) statement in the selection structure's true path calls the CalcBonus procedure, passing it two variables and a named constant *by value*, which means that only the contents of the variables and named constant are passed to the procedure. Here again, you know that the variables and named constant are passed *by value* because the keyword `ByVal` appears before each corresponding parameter in the CalcBonus procedure header. At this point, the computer temporarily leaves the xCalcButton's Click event procedure to process the code contained in the CalcBonus procedure.

The first instruction processed in the CalcBonus procedure is the procedure header. When processing the procedure header, the computer creates the three procedure-level variables listed in the *parameterlist* (`sales1`, `sales2`, and `rate`), and stores the information passed to the procedure in those variables. After processing the procedure header, the computer processes the statements contained in the procedure. The first statement declares an additional procedure-level variable named `bonus`. The next statement calculates the bonus by adding the value stored in the `sales1` variable to the value stored in the `sales2` variable, and then multiplying the sum by the value stored in the `rate` variable. The third statement in the procedure displays the bonus, formatted with a dollar sign and two decimal places, in the xBonusLabel. As Figure 7-5 shows, the procedure displays a bonus amount of $300.00 when the user enters the numbers 1000 and 2000 as the sales amounts for Region 1 and Region 2.

Next, the computer processes the CalcBonus procedure footer, which ends the CalcBonus procedure. At this point, the procedure's variables (`sales1`, `sales2`, `rate`, and `bonus`) are removed from the computer's internal memory. The computer then returns to the xCalcButton's Click event procedure and processes the `Me.xReg1TextBox.Focus()` statement. After setting the focus, the computer processes the `End Sub` statement, which ends the xCalcButton's Click event procedure. The computer then removes the `BonusRate` named constant and the `reg1Sales`, `reg2Sales`, `isConvertedReg1`, and `isConvertedReg2` variables from its internal memory.

To code and then test the Bonus Calculator application:

1 Open the **Bonus Calculator Solution** (Bonus Calculator Solution.sln) file, which is contained in the VB2005\Chap07\Bonus Calculator Solution folder. If necessary, open the designer window.

2 Open the Code Editor window. Replace the <your name> and <current date> text in the comments with your name and the current date.

3 Enter the comments and code shown in Figure 7-6.

4 Close the Code Editor window. Save the solution, then start the application.

5 Type **1000** in the Region 1 sales text box, then press **Tab**. Type **2000** in the Region 2 sales text box, then click the **Calculate** button. The application displays the bonus amount shown earlier in Figure 7-5.

6 Click the **Exit** button to end the application. You are returned to the designer window. Close the solution.

PASSING BY REFERENCE

In addition to passing a variable's value to a procedure, you also can pass a variable's address—in other words, its location in the computer's internal memory. As you learned earlier, passing a variable's address is referred to as passing *by reference*, and it gives the receiving procedure access to the variable being passed. You pass a variable *by reference* when you want the receiving procedure to change the contents of the variable. To pass a variable *by reference* in Visual Basic, you include the keyword `ByRef` (which stands for "by reference") before the variable's corresponding parameter in the receiving procedure's header. The `ByRef` keyword tells the computer to pass the variable's address rather than its contents.

Figure 7-7 shows a sample run of the Gross Pay application, and Figure 7-8 shows the application's code. In the code, the xCalcButton's Click event procedure passes three variables to the CalcGrossPay procedure: two *by value* and one *by reference*.

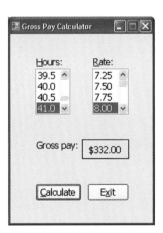

Figure 7-7: Sample run of the Gross Pay application

Visual Basic code

```
' Project name:     Gross Pay Project
' Project purpose:  The project calculates gross pay.
' Created/revised:  <your name> on <current date>

Option Explicit On
Option Strict On

Public Class MainForm

    Private Sub CalcGrossPay(ByVal hours As Decimal, _
                             ByVal rate As Decimal, _
                             ByRef gross As Decimal)

        ' calculates gross pay
        If hours <= 40D Then
            gross = hours * rate
        Else
            gross = hours * rate + (hours - 40D) * rate / 2D
        End If
    End Sub

    Private Sub MainForm_Load(ByVal sender As Object, _
        ByVal e As System.EventArgs) Handles Me.Load
        ' fills list boxes with values, then
        ' selects a default item

        For hours As Decimal = 0.5D To 50D Step 0.5D
            Me.xHoursListBox.Items.Add(hours.ToString)
        Next hours

        For rates As Decimal = 7.25D To 10.5D Step 0.25D
            Me.xRateListBox.Items.Add(rates.ToString)
        Next rates

        Me.xHoursListBox.SelectedItem = "40.0"
        Me.xRateListBox.SelectedIndex = 0
    End Sub

    Private Sub xCalcButton_Click(ByVal sender As Object, _
        ByVal e As System.EventArgs) Handles xCalcButton.Click
        ' calculates and displays a gross pay amount
```

parameterlist

Figure 7-8: Code for the Gross Pay application *(Continued)*   ▶

```
        Dim hoursWkd As Decimal
        Dim rateOfPay As Decimal
        Dim grossPay As Decimal

        hoursWkd = Convert.ToDecimal(Me.xHoursListBox.SelectedItem)
        rateOfPay = Convert.ToDecimal(Me.xRateListBox.SelectedItem)

        Call CalcGrossPay(hoursWkd, rateOfPay, grossPay)                        argumentlist

        Me.xGrossLabel.Text = grossPay.ToString("C2")
    End Sub

    Private Sub xExitButton_Click(ByVal sender As Object, _
        ByVal e As System.EventArgs) Handles xExitButton.Click
        Me.Close()
    End Sub

    Private Sub xHoursListBox_SelectedValueChanged(ByVal _
        sender As Object, ByVal e As System.EventArgs) _
        Handles xHoursListBox.SelectedValueChanged
        Me.xGrossLabel.Text = String.Empty
    End Sub

    Private Sub xRateListBox_SelectedValueChanged(ByVal _
        sender As Object, ByVal e As System.EventArgs) _
        Handles xRateListBox.SelectedValueChanged
        Me.xGrossLabel.Text = String.Empty
    End Sub
End Class
```

Figure 7-8: Code for the Gross Pay application

Notice that the number, data type, and sequence of the arguments in the Call statement match the number, data type, and sequence of the corresponding parameters in the procedure header. Also notice that the names of the parameters do not need to be identical to the names of the arguments to which they correspond.

Study the code shown in Figure 7-8. The xCalcButton's Click event procedure first declares three Decimal variables. The next two statements in the procedure convert the items selected in the list boxes to Decimal numbers. The Call CalcGrossPay(hoursWkd, rateOfPay, grossPay) statement then calls the CalcGrossPay procedure, passing it three variables. Two of the variables (hoursWkd and rateOfPay) are passed *by value*, which means that only the contents of the variables are passed to the procedure. The grossPay variable, however, is passed *by reference*, which means that the variable's address in memory, rather than its contents, is passed to the procedure. Here again, you know that the hoursWkd and rateOfPay variables are passed *by value* because the keyword ByVal appears before the corresponding parameters in the CalcGrossPay procedure header. You know that the grossPay variable is passed *by reference* because the keyword ByRef appears before the corresponding parameter in the CalcGrossPay procedure header. When processing the Call statement, the computer temporarily leaves the xCalcButton's Click event procedure to process the code contained in the CalcGrossPay procedure.

The first instruction processed in the CalcGrossPay procedure is the procedure header. The ByVal keyword that appears before the names of the first two parameters in the procedure header indicates that the procedure will be receiving the contents of the hoursWkd and rateOfPay variables. The ByRef keyword that appears before the last parameter in the procedure header indicates that the procedure will be receiving the address of the grossPay variable. When you pass a variable's address to a procedure, the computer uses the address to locate the variable in memory. It then assigns the name appearing in the procedure header to the memory location. In this case, for example, the computer first locates the grossPay variable in memory. It then assigns the name gross to the location. At this point, the memory location has two names: one assigned by the xCalcButton's Click event procedure, and the other assigned by the CalcGrossPay procedure.

After processing the CalcGrossPay procedure header, the computer processes the selection structure contained in the procedure. In this case, if the hours variable contains a value that is less than or equal to the number 40, the selection structure's true path calculates the gross pay by multiplying the contents of the hours variable by the contents of the rate variable, assigning the result to the gross variable. However, if the hours variable contains a value that is greater than the number 40, the selection structure's false path calculates the gross pay by multiplying the contents of the hours variable by the contents of the rate variable, and then adding to that an additional half-time for the hours worked over 40. The result is assigned to the gross variable. Figure 7-9 shows the contents of memory after the CalcGrossPay procedure header and selection structure are processed. The figure is based on the user selecting 41.0 in the Hours list box and 8.00 in the Rate list box. Notice that changing the contents of the gross variable also changes the contents of the grossPay variable. This is because the names refer to the same location in memory.

>> **TIP**

Although the grossPay and gross variables refer to the same location in memory, the grossPay variable is recognized only within the xCalcButton's Click event procedure. Similarly, the gross variable is recognized only within the CalcGrossPay procedure.

Memory locations belonging only to the xCalcButton's Click event procedure:	
<u>hoursWkd</u> 41.0	<u>rateOfPay</u> 8.00

Memory locations belonging only to the CalcGrossPay procedure:	
<u>hours</u> 41.0	<u>rate</u> 8.00

Memory location belonging to both procedures:
grossPay   (xCalcButton's Click event procedure) <u>gross</u   (CalcGrossPay procedure) 332.00

Figure 7-9: Contents of memory after the CalcGrossPay procedure header and selection structure are processed

As Figure 7-9 indicates, one memory location belongs to both the xCalcButton's Click event procedure and the CalcGrossPay procedure. Although both procedures can access the memory location, each procedure uses a different name to do so. The xCalcButton's Click event procedure uses the name grossPay to refer to the memory location, whereas the CalcGrossPay procedure uses the name gross.

The End Sub statement in the CalcGrossPay procedure is processed next and ends the procedure. At this point, the computer removes the memory locations that belong only to the CalcGrossPay procedure; those memory locations are the hours and rate variables. In addition, it removes the name gross, which is assigned to the memory location that belongs to both procedures. Figure 7-10 shows the contents of memory after the hours and rate variables, as well as the gross variable name, are removed from memory. Notice that the grossPay memory location now has only one name, which is the name assigned to it by the xCalcButton's Click event procedure.

Memory locations belonging only to the xCalcButton's Click event procedure:		
<u>hoursWkd</u> 41.0	<u>rateOfPay</u> 8.00	<u>grossPay</u> 332.00

Figure 7-10: Contents of memory after the appropriate variables and variable name are removed

The computer then returns to the xCalcButton's Click event procedure, to the statement located immediately below the Call statement. The statement displays the gross pay—in this case, $332.00—in the xGrossLabel. Next, the computer processes the End Sub statement in the xCalcButton's Click event procedure, which ends the procedure. The computer then removes the hoursWkd, rateOfPay, and grossPay variables from its internal memory.

To code and then test the Gross Pay application:

1 Open the **Gross Pay Solution** (Gross Pay Solution.sln) file, which is contained in the VB2005\Chap07\Gross Pay Solution folder. If necessary, open the designer window.

2 Open the Code Editor window. Replace the <your name> and <current date> text in the comments with your name and the current date.

3 Enter the comments and code shown in Figure 7-8.

4 Close the Code Editor window. Save the solution, then start the application.

5 Click **41.0** in the Hours list box, then click **8.00** in the Rate list box. Click the **Calculate** button. The application displays the gross pay amount shown earlier in Figure 7-7.

6 Click the **Exit** button to end the application. You are returned to the designer window. Close the solution.

As you learned earlier, in addition to creating Sub procedures, you also can create Function procedures in Visual Basic.

FUNCTION PROCEDURES

Like a Sub procedure, a Function procedure, which typically is referred to simply as a **function**, is a block of code that performs a specific task. However, unlike a Sub procedure, a function returns a value after completing its task. Some functions, such as the Val and InputBox functions, are built into Visual Basic. Recall that the Val function returns the numeric equivalent of a string, and the InputBox function returns the user's response to a prompt that appears in a dialog box.

You also can create your own functions in Visual Basic. After creating a function, you then can invoke it from one or more places in an application. You invoke a function that you create in exactly the same way as you invoke a built-in function: you simply include the function's name in a statement. Usually the statement will display the function's return value, or use the return value in a calculation, or assign the return value to a variable. As is true with Sub procedures, you also can pass (send) information to a function that you create, and the information can be passed either *by value* or *by reference*. Figure 7-11 shows the syntax you use to create a Function procedure. It also includes an example of a Function procedure.

Function procedure

Syntax

Private Function *procedurename*(*[parameterlist]*) **As** *datatype*

    [*statements*]

    Return *expression*

End Function

Example
```
Private Function CalcNew(ByVal price As Double) As Double
    ' calculates and returns a new price using the current
    ' price passed to it and a 5% price increase rate

    Return price * 1.05
End Function
```

Figure 7-11: Syntax and an example of a Function procedure

Like Sub procedures, Function procedures have both a procedure header and procedure footer. The procedure header for a Function procedure is almost identical to the procedure header for a Sub procedure, except it includes the `Function` keyword rather than the `Sub` keyword. The `Function` keyword identifies the procedure as a Function procedure, which returns a value after completing its task. Also different from a Sub procedure header, a Function procedure header includes the **As** *datatype* clause. You use the clause to specify the data type of the value returned by the function. For example, if the function returns a string, you include `As String` at the end of the procedure header. Similarly, if the function returns a number having a decimal place, you can include one of the following in the procedure header: As Decimal, As Double, or As Single. The *datatype* you use depends on the size of the number and whether you want the number stored with a fixed decimal point or a floating decimal point.

The procedure footer in a Function procedure is always End Function. Between the procedure header and procedure footer, you enter the instructions you want the computer to process when the function is invoked. In most cases, the last statement in a Function procedure is the Return statement. The syntax of the Return statement is **Return** *expression*, where *expression* represents the one and only value that will be returned to the statement that called the function. The data type of the *expression* in the Return statement must agree with the data type specified in the **As** *datatype* clause in the procedure header. The **Return statement** alerts the computer that the function has completed its task and ends the function after returning the value of its *expression*.

In the example shown in Figure 7-11, the CalcNew function is passed the current price of an item. The current price is passed *by value* and stored in the price variable. The Return price * 1.05 statement in the function first calculates the new price, and then returns the new price to the statement that called the function. In the next section, you view an application that creates and uses a function.

THE PINE LODGE APPLICATION

The owner of the Pine Lodge wants an application that calculates an employee's new hourly pay, given the employee's current hourly pay and raise rate. Figure 7-12 shows a sample run of the Pine Lodge application, and Figure 7-13 shows the application's code.

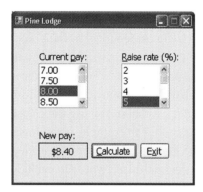

Figure 7-12: Sample run of the Pine Lodge application

Visual Basic code

```
' Project name:      Pine Lodge Project
' Project purpose:   The project calculates the amount
'                    of an employee's new pay.
' Created/revised:   <your name> on <current date>

Option Explicit On
Option Strict On

Public Class MainForm

    Private Function GetNewPay(ByVal current As Decimal, _
        ByVal rate As Decimal) As Decimal
        ' calculates and returns the new hourly pay
        ' based on the current hourly pay and raise rate
        ' passed to it

        Dim raise As Decimal
        Dim newPay As Decimal

        raise = current * rate
        newPay = current + raise

        Return newPay
    End Function

    Private Sub ClearLabel(ByVal sender As Object, _
        ByVal e As System.EventArgs) _
        Handles xCurrentListBox.SelectedValueChanged, _
        xRateListBox.SelectedValueChanged
        Me.xNewPayLabel.Text = String.Empty
    End Sub

    Private Sub xExitButton_Click(ByVal sender As Object, _
        ByVal e As System.EventArgs) Handles xExitButton.Click
        Me.Close()
    End Sub

    Private Sub MainForm_Load(ByVal sender As Object, _
        ByVal e As System.EventArgs) Handles Me.Load
        ' fills the list boxes with values, then selects
        ' the first value
```

GetNewPay function

Figure 7-13: Code for the Pine Lodge application *(Continued)* ▶

```
        For pay As Decimal = 7D To 12D Step 0.5D
            Me.xCurrentListBox.Items.Add(pay.ToString("N2"))
        Next pay

        For rate As Integer = 2 To 11
            Me.xRateListBox.Items.Add(rate.ToString)
        Next rate

        Me.xCurrentListBox.SelectedIndex = 0
        Me.xRateListBox.SelectedIndex = 0
    End Sub

    Private Sub xCalcButton_Click(ByVal sender As Object, _
        ByVal e As System.EventArgs) Handles xCalcButton.Click
        ' calls a function to calculate an employee's new
        ' hourly pay, then displays the new hourly pay

        Dim currentPay As Decimal
        Dim raiseRate As Decimal
        Dim newHourlyPay As Decimal

        currentPay = Convert.ToDecimal(Me.xCurrentListBox.SelectedItem)
        raiseRate = _
            Convert.ToDecimal(Me.xRateListBox.SelectedItem) / 100D

        newHourlyPay = GetNewPay(currentPay, raiseRate)

        Me.xNewPayLabel.Text = newHourlyPay.ToString("C2")
    End Sub
End Class
```

invokes the GetNewPay function and assigns the function's return value to the newHourlyPay variable

Figure 7-13: Code for the Pine Lodge application

Study the code shown in Figure 7-13. The MainForm's Load event procedure fills the two list boxes with values, and then selects the first value in each list. The ClearLabel Sub procedure removes the contents of the xNewPayLabel when the SelectedValueChanged event occurs for either list box. The xExitButton's Click event procedure ends the application. The xCalcButton's Click event procedure first declares three Decimal variables. It then converts the item selected in the xCurrentListBox to a Decimal number and assigns the result to the currentPay variable. The next statement in the procedure first converts the item selected in the xRateListBox to a Decimal number. It then divides the number by 100 and assigns the result to the raiseRate variable. (Because the values listed in the xRateListBox represent percentages, you need to divide the selected value by 100 to change the percentage to its decimal equivalent.) The newHourlyPay = GetNewPay(currentPay, raiseRate) statement calls the GetNewPay function, passing it the values stored in the currentPay and raiseRate variables. The computer stores the values in the current and rate variables that appear in the GetNewPay procedure header.

After processing the GetNewPay procedure header, the computer processes the statements contained in the function. The first two statements declare procedure-level variables named raise and newPay. The next statement calculates the employee's raise by multiplying the value stored in the current variable by the value stored in the rate variable; the statement assigns the raise amount to the raise variable. The next statement calculates the new pay by adding the value stored in the current variable to the value stored in the raise variable, assigning the result to the newPay variable. The Return newPay statement in the function returns the contents of the newPay variable to the statement that called the function. That statement is the newHourlyPay = GetNewPay(currentPay, raiseRate) statement in the xCalcButton's Click event procedure. After processing the Return statement, the GetNewPay function ends, and the computer removes the current, rate, raise, and newPay variables from its internal memory.

When the newHourlyPay = GetNewPay(currentPay, raiseRate) statement receives the value returned by the GetNewPay function, it assigns the value to the newHourlyPay variable. The Me.xNewPayLabel.Text = newHourlyPay. ToString("C2") statement then displays the new hourly pay in the xNewPayLabel. Finally, the computer processes the End Sub statement in the xCalcButton's Click event procedure, which ends the procedure. The computer then removes the currentPay, raiseRate, and newHourlyPay variables from its internal memory.

To code and then test the Pine Lodge application:

1 Open the **Pine Lodge Solution** (Pine Lodge Solution.sln) file, which is contained in the VB2005\Chap07\Pine Lodge Solution folder. If necessary, open the designer window.

2 Open the Code Editor window. Replace the <your name> and <current date> text in the comments with your name and the current date.

3 Enter the comments and code shown in Figure 7-13.

4 Close the Code Editor window. Save the solution, then start the application.

5 Click **8.00** in the Current pay list box, then click **5** in the Raise rate (%) list box. Click the **Calculate** button. The application displays the new pay amount shown earlier in Figure 7-12.

6 Click the **Exit** button to end the application. You are returned to the designer window. Close the solution.

You have completed Lesson A. You can either take a break or complete the end-of-lesson questions and exercises before moving on to Lesson B.

SUMMARY

TO CREATE AN INDEPENDENT SUB PROCEDURE:

» Refer to the syntax shown in Figure 7-2.

TO CALL AN INDEPENDENT SUB PROCEDURE:

» Refer to the syntax shown in Figure 7-2.

TO CREATE AN INDEPENDENT FUNCTION PROCEDURE:

» Refer to the syntax shown in Figure 7-11.

TO PASS INFORMATION TO A SUB OR FUNCTION PROCEDURE:

» Include the information in the Call statement's *argumentlist*. In the procedure header's *parameterlist*, include the names of memory locations that will store the information.

» The number, data type, and sequence of the arguments listed in the Call statement's *argumentlist* should agree with the number, data type, and sequence of the parameters listed in the *parameterlist* in the procedure header.

TO PASS A VARIABLE BY VALUE TO A PROCEDURE:

» Include the `ByVal` keyword before the parameter name in the procedure header's *parameterlist*. Because only the value stored in the variable is passed, the receiving procedure cannot access the variable.

TO PASS A VARIABLE BY REFERENCE:

» Include the `ByRef` keyword before the parameter name in the procedure header's *parameterlist*. Because the address of the variable is passed, the receiving procedure can change the contents of the variable.

QUESTIONS

1. Which of the following is false?

    a. A Function procedure can return one or more values to the statement that called it.

    b. A procedure can accept one or more items of data passed to it.

    c. The *parameterlist* in a procedure header is optional.

    d. At times, a memory location inside the computer's internal memory may have more than one name.

2. The items listed in the Call statement are referred to as _____.

    a. arguments   b. parameters

    c. passers   d. None of the above.

3. Each memory location listed in the *parameterlist* in the procedure header is referred to as _____.

    a. an address   b. a constraint

    c. a parameter   d. a value

4. To determine whether a variable is being passed to a procedure *by value* or *by reference*, you will need to examine _____.

    a. the Call statement

    b. the procedure header

    c. the statements entered in the procedure

    d. Either a or b.

5. Which of the following statements can be used to call the CalcArea Sub procedure, passing it two variables *by value*?

    a. `Call CalcArea(length, width)`

    b. `Call CalcArea(ByVal length, ByVal width)`

    c. `Call CalcArea ByVal(length, width)`

    d. `Call ByVal CalcArea(length, width)`

6. Which of the following procedure headers receives only the contents of a String variable?

    a. `Private Sub DisplayName(ByContents empName As String)`

    b. `Private Sub DisplayName(ByValue empName As String)`

    c. `Private Sub DisplayName ByVal(empName As String)`

    d. None of the above.

7. Which of the following is a valid procedure header for a procedure that receives an integer first and a number with a decimal place second?

    a. `Private Sub CalcFee(base As Integer, rate As Decimal)`

    b. `Private Sub CalcFee(ByRef base As Integer, ByRef rate As Decimal)`

    c. `Private Sub CalcFee(ByVal base As Integer, ByVal rate As Decimal)`

    d. None of the above.

8. A Function procedure can return _____.

    a. one value only

    b. one or more values

9. The procedure header specifies the procedure's _____.

    a. name   b. parameters

    c. type (either Sub or Function)   d. All of the above.

10. Which of the following is false?

    a. The sequence of the arguments listed in the Call statement should agree with the sequence of the parameters listed in the procedure header.

    b. The data type of each argument in the Call statement should match the data type of its corresponding parameter in the procedure header.

    c. The name of each argument in the Call statement should be identical to the name of its corresponding parameter in the procedure header.

    d. When you pass information to a procedure *by value*, the procedure stores the value of each item it receives in a separate memory location.

11. Which of the following instructs a function to return the contents of the `stateTax` variable to the statement that called the function?

    a. `Return stateTax`

    b. `Return stateTax ByVal`

    c. `Send stateTax`

    d. `SendBack stateTax`

12. Which of the following is a valid procedure header for a procedure that receives the value stored in an Integer variable first, and the address of a Decimal variable second?

    a. `Private Sub CalcFee(ByVal base As Integer, ByAdd rate As` ↵
    `Decimal)`

    b. `Private Sub CalcFee(base As Integer, rate As Decimal)`

    c. `Private Sub CalcFee(ByVal base As Integer, ByRef rate As` ↵
    `Decimal)`

    d. None of the above.

13. Which of the following is a valid procedure header for a procedure that receives the number 15?

    a. `Private Function CalcTax(ByVal rate As Integer) As Decimal`

    b. `Private Function CalcTax(ByAdd rate As Integer) As Decimal`

    c. `Private Sub CalcTax(ByVal rate As Integer)`

    d. Both a and c.

14. If the statement `Call CalcNet(netPay)` passes the address of the `netPay` variable to the CalcNet procedure, the variable is said to be passed _____.

    a. *by address*   b. *by content*

    c. *by reference*   d. *by value*

15. Which of the following is false?

    a. When you pass a variable *by reference*, the receiving procedure can change its contents.

    b. When you pass a variable *by value*, the receiving procedure creates a procedure-level variable that it uses to store the passed value.

    c. Unless specified otherwise, all variables in Visual Basic are passed *by value*.

    d. To pass a variable *by reference* in Visual Basic, you include the `ByRef` keyword before the variable's name in the Call statement.

16. A Sub procedure named CalcEndingInventory is passed four Integer variables named `begin`, `sales`, `purchases`, and `ending`. The procedure's task is to calculate the ending inventory, based on the beginning inventory, sales, and purchase amounts passed to the procedure. The procedure should store the result in the `ending` variable. Which of the following procedure headers is correct?

    a. `Private Sub CalcEndingInventory(ByVal b As Integer, ByVal s As Integer, ByVal p As Integer, ByRef final As Integer)`

    b. `Private Sub CalcEndingInventory(ByVal b As Integer, ByVal s As Integer, ByVal p As Integer, ByVal final As Integer)`

    c. `Private Sub CalcEndingInventory(ByRef b As Integer, ByRef s As Integer, ByRef p As Integer, ByVal final As Integer)`

    d. `Private Sub CalcEndingInventory(ByRef b As Integer, ByRef s As Integer, ByRef p As Integer, ByRef final As Integer)`

17. Which of the following statements should you use to call the CalcEndingInventory procedure described in Question 16?

    a. `Call CalcEndingInventory(begin, sales, purchases, ending)`

    b. `Call CalcEndingInventory(ByVal begin, ByVal sales, ByVal purchases, ByRef ending)`

    c. `Call CalcEndingInventory(ByRef begin, ByRef sales, ByRef purchases, ByRef ending)`

    d. `Call CalcEndingInventory(ByVal begin, ByVal sales, ByVal purchases, ByVal ending)`

18. The memory locations listed in the *parameterlist* in a procedure header are procedure-level and are removed from the computer's internal memory when the procedure ends.

    a. True   b. False

19. The CalcTax function's procedure header is `Private Function CalcTax(ByVal sales As Decimal) As Decimal`. Which of the following is a valid Return statement for the function? (The `tax` and `rate` variables are Decimal variables declared in the function.)

    a. `Return sales * .06D`   b. `Return tax`

    c. `Return sales * rate`   d. All of the above.

20. Which of the following statements invokes a Function procedure named Calculate, passing it the contents of two Decimal variables named hours and rate? The statement should assign the function's return value to a variable named grossPay.

    a. `grossPay = Call Calculate(hours, rate)`

    b. `Call Calculate(hours, rate, grossPay)`

    c. `grossPay = Calculate(hours, rate)`

    d. None of the above.

21. Explain the difference between a Sub procedure and a Function procedure.

22. Explain the difference between passing a variable *by value* and passing it *by reference*.

EXERCISES

1. Write the Visual Basic code for a Sub procedure that receives an integer passed to it. The procedure, named HalveNumber, should divide the integer by 2 and then display the result in the xNumLabel.

2. Write an appropriate statement to invoke the HalveNumber procedure created in Exercise 1. Pass the integer 87 to the procedure.

3. Write the Visual Basic code for a Sub procedure that prompts the user to enter the name of a city, and then stores the user's response in the String variable whose address is passed to the procedure. Name the procedure GetCity.

4. Write an appropriate statement to invoke the GetCity procedure created in Exercise 3. Pass a String variable named city to the procedure.

5. Write the Visual Basic code for a Sub procedure that receives four Integer variables: the first two *by value* and the last two *by reference*. The procedure should calculate the sum of and the difference between the two variables passed *by value*, and then store the results in the variables passed *by reference*. (When calculating the difference, subtract the contents of the second variable from the contents of the first variable.) Name the procedure CalcSumAndDiff.

6. Write an appropriate statement to invoke the CalcSumAndDiff procedure created in Exercise 5. Pass the Integer variables named `firstNum`, `secondNum`, `sum`, and `difference` to the procedure.

7. Write the Visual Basic code for a Sub procedure that receives three Decimal variables: the first two *by value* and the last one *by reference*. The procedure should divide the first variable by the second variable, and then store the result in the third variable. Name the procedure CalcQuotient.

8. Write the Visual Basic code for a Function procedure that receives the value stored in an Integer variable. Name the procedure DivideNumber. The procedure should divide the integer by 2 and then return the result (which may contain a decimal place).

9. Write an appropriate statement to invoke the DivideNumber procedure created in Exercise 8. The name of the Integer variable passed to the function is `number`. Assign the value returned by the function to the `answer` variable.

10. Write the Visual Basic code for a Function procedure that prompts the user to enter the name of a state, and then returns the user's response to the calling procedure. Name the procedure GetState.

11. Write an appropriate statement to invoke the GetState procedure created in Exercise 10. Display the function's return value in a message box.

12. Write the Visual Basic code for a Function procedure that receives the contents of four Integer variables. The procedure should calculate the average of the four integers and then return the result (which may contain a decimal place). Name the procedure CalcAverage.

13. Write an appropriate statement to invoke the CalcAverage procedure created in Exercise 12. The Integer variables passed to the function are named `num1`, `num2`, `num3`, and `num4`. Assign the function's return value to a Decimal variable named `average`.

14. In this exercise, you experiment with passing variables *by value* and *by reference*.

    a. If necessary, start Visual Studio 2005 or Visual Basic 2005 Express Edition. Open the Passing Solution (Passing Solution.sln) file, which is contained in the VB2005\Chap07\Passing Solution folder. If necessary, open the designer window.

    b. Open the Code Editor window. Study the application's existing code. Notice that the myName variable is passed *by value* to the GetName procedure.

    c. Start the application. Click the Display Name button. When prompted to enter a name, type your name and press Enter. Explain why the xDisplayButton's Click event procedure does not display your name in the xNameLabel. Click the Exit button to end the application.

    d. Modify the application's code so that it passes the myName variable *by reference* to the GetName procedure.

    e. Save the solution, then start the application. Click the Display Name button. When prompted to enter a name, type your name and press Enter. This time, your name appears in the xNameLabel. Explain why the xDisplayButton's Click event procedure now works correctly.

    f. Click the Exit button to end the application. Close the Code Editor window, then close the solution.

15. In this exercise, you create an application that allows the user to enter any number of monthly rainfall amounts. The application should calculate and display the total rainfall amount and the average rainfall amount.

    a. If necessary, start Visual Studio 2005 or Visual Basic 2005 Express Edition. Open the Rainfall Solution (Rainfall Solution.sln) file, which is contained in the VB2005\Chap07\Rainfall Solution folder. If necessary, open the designer window.

    b. Open the Code Editor window. The xCalcButton's Click event procedure should call the CalcTotalAndAverage Sub procedure, passing it the rainCounter, rainAccum, and avgRain variables. Complete the procedure by entering the appropriate Call statement.

    c. The CalcTotalAndAverage procedure should calculate the total and average rainfall amounts. Complete the Sub procedure by entering the appropriate code.

    d. Save the solution, then start the application. Test the application to verify that it is working correctly.

    e. Click the Exit button to end the application. Close the Code Editor window, then close the solution.

16. In this exercise, you modify the Gross Pay application that you completed in the lesson.

    a. Use Windows to make a copy of the Gross Pay Solution folder, which is contained in the VB2005\Chap07 folder. Rename the folder Modified Gross Pay Solution.

    b. If necessary, start Visual Studio 2005 or Visual Basic 2005 Express Edition. Open the Gross Pay Solution (Gross Pay Solution.sln) file contained in the VB2005\Chap07\Modified Gross Pay Solution folder. Open the designer window.

    c. Modify the code so that it uses a Function procedure rather than a Sub procedure to calculate and return the gross pay amount.

    d. Save the solution, then start and test the application.

    e. Click the Exit button to end the application. Close the Code Editor window, then close the solution.

17. In this exercise, you modify the Pine Lodge application that you completed in the lesson.

    a. Use Windows to make a copy of the Pine Lodge Solution folder, which is contained in the VB2005\Chap07 folder. Rename the folder Modified Pine Lodge Solution.

    b. If necessary, start Visual Studio 2005 or Visual Basic 2005 Express Edition. Open the Pine Lodge Solution (Pine Lodge Solution.sln) file contained in the VB2005\Chap07\Modified Pine Lodge Solution folder. Open the designer window.

    c. Modify the code so that it uses a Sub procedure rather than a Function procedure to calculate the new pay amount.

    d. Save the solution, then start and test the application.

    e. Click the Exit button to end the application. Close the Code Editor window, then close the solution.

18. In this exercise, you modify the Rainfall application from Exercise 15.

    a. Use Windows to make a copy of the Rainfall Solution folder, which is contained in the VB2005\Chap07 folder. Rename the folder Modified Rainfall Solution.

    b. If necessary, start Visual Studio 2005 or Visual Basic 2005 Express Edition. Open the Rainfall Solution (Rainfall Solution.sln) file contained in the VB2005\Chap07\Modified Rainfall Solution folder. Open the designer window.

    c. Open the Code Editor window. Modify the code so that it uses two Function procedures rather than a Sub procedure to calculate the total and average rainfall amounts.

d. Save the solution, then start the application. Test the application to verify that it is working correctly.

e. Click the Exit button to end the application. Close the Code Editor window, then close the solution.

19. In this exercise, you code an application that uses two independent Sub procedures: one to convert a temperature from Fahrenheit to Celsius, and the other to convert a temperature from Celsius to Fahrenheit.

    a. If necessary, start Visual Studio 2005 or Visual Basic 2005 Express Edition.

    b. Create a Visual Basic Windows-based application. Name the solution Temperature Solution, and name the project Temperature Project. Save the application in the VB2005\Chap07 folder.

    c. Assign the filename Main Form.vb to the form file object.

    d. Assign the name MainForm to the form.

    e. Create an appropriate interface. Code the application appropriately.

    f. Save the solution, then start and test the application.

    g. End the application. Close the Code Editor window, then close the solution.

20. In this exercise, you modify the application that you created in Exercise 19 so that it uses two functions rather than two Sub procedures.

    a. Use Windows to make a copy of the Temperature Solution folder, which is contained in the VB2005\Chap07 folder. Rename the folder Modified Temperature Solution.

    b. If necessary, start Visual Studio 2005 or Visual Basic 2005 Express Edition. Open the Temperature Solution (Temperature Solution.sln) file contained in the VB2005\Chap07\Modified Temperature Solution folder. Open the designer window.

    c. Modify the code so that it uses two functions rather than two Sub procedures to convert the temperatures.

    d. Save the solution, and then start and test the application.

    e. End the application. Close the Code Editor window, then close the solution.

DISCOVERY EXERCISE

21. In this exercise, you learn how to specify that one or more arguments are optional in a Call statement.

   a. If necessary, start Visual Studio 2005 or Visual Basic 2005 Express Edition. Open the Optional Solution (Optional Solution.sln) file, which is contained in the VB2005\Chap07\Optional Solution folder. If necessary, open the designer window.

   b. Open the Code Editor window. Study the application's existing code. Notice that the xCalcButton's Click event procedure contains two Call statements. The first Call statement passes three variables (`sales`, `bonus`, and `rate`) to the GetBonus procedure. The second Call statement, however, passes only two variables (`sales` and `bonus`) to the procedure. (Do not be concerned about the jagged line that appears below the second Call statement.) Notice that the `rate` variable is omitted from the second Call statement. You indicate that the `rate` variable is optional in the Call statement by including the keyword `Optional` before the variable's corresponding parameter in the procedure header; you enter the `Optional` keyword before the `ByVal` keyword. You also assign a default value that the procedure will use for the missing parameter when the procedure is called. You assign the default value by entering the assignment operator, followed by the default value, after the parameter in the procedure header. In this case, you will assign the number .1 as the default value for the `rate` variable. (Optional parameters must be listed at the end of the procedure header.)

   c. Change the `ByVal  bonusRate  As  Double` in the procedure header appropriately.

   d. Save the solution, then start the application. Calculate the bonus for a salesperson with an "A" code, $1000 in sales, and a rate of .05. The `Call GetBonus(sales, bonus, rate)` statement calls the GetBonus procedure, passing it the number 1000, the address of the `bonus` variable, and the number .05. The GetBonus procedure stores the number 1000 in the `totalSales` variable. It also assigns the name bonusAmount to the `bonus` variable, and stores the number .05 in the `bonusRate` variable. The procedure then multiplies the contents of the `totalSales` variable (1000) by the contents of the `bonusRate` variable (.05), assigning the result (50) to the `bonusAmount` variable. The `Me.xBonusLabel.Text = bonus.ToString("C2")` statement then displays $50.00 in the xBonusLabel.

e. Now calculate the bonus for a salesperson with a code of "B" and a sales amount of $2000. The `Call GetBonus(sales, bonus)` statement calls the GetBonus procedure, passing it the number 2000 and the address of the `bonus` variable. The GetBonus procedure stores the number 2000 in the `totalSales` variable, and assigns the name `bonusAmount` to the `bonus` variable. Because the Call statement did not supply a value for the `bonusRate` variable, the default value (.1) is assigned to the variable. The procedure then multiplies the contents of the `totalSales` variable (2000) by the contents of the `bonusRate` variable (.1), assigning the result (200) to the `bonusAmount` variable. The `Me.xBonusLabel.Text = bonus.ToString("C2")` statement then displays $200.00 in the xBonusLabel.

f. Click the Exit button to end the application. Close the Code Editor window, then close the solution.

LESSON B
OBJECTIVES

AFTER STUDYING LESSON B, YOU SHOULD
BE ABLE TO:

» Add a combo box to a form

» Add items to a combo box

» Sort the contents of a combo box

» Select a combo box item from code

» Determine the current item in a combo box

» Round a number

» Code a combo box's TextChanged event procedure

CODING THE HARVEY INDUSTRIES PAYROLL APPLICATION

HARVEY INDUSTRIES

Recall that the payroll manager at Harvey Industries wants an application that he can use to calculate an employee's weekly gross pay, Social Security and Medicare (FICA) tax, federal withholding tax (FWT), and net pay. Figure 7-14 shows the TOE chart for the application.

Task	Object	Event
1. Calculate the gross pay 2. Calculate the FWT 3. Calculate the FICA tax 4. Calculate the net pay 5. Display the gross pay, FWT, FICA, and net pay in xGrossLabel, xFwtLabel, xFicaLabel, and xNetLabel	xCalcButton	Click
End the application	xExitButton	Click
Display the gross pay, FWT, FICA, and net pay (from xCalcButton)	xGrossLabel, xFwtLabel, xFicaLabel, xNetLabel	None
Clear the xGrossLabel, xFwtLabel, xFicaLabel, and xNetLabel	xNameTextBox, xAllowComboBox	TextChanged
	xHoursListBox, xRateListBox	SelectedValueChanged
	xMarriedRadioButton, xSingleRadioButton	Click

Figure 7-14: TOE chart for the Harvey Industries application *(Continued)* ▶

Task	Object	Event
Select the existing text	xNameTextBox	Enter
Get and display the name, hours worked, pay rate, marital status, and number of withholding allowances	xNameTextBox, xHoursListBox, xRateListBox, xMarriedRadioButton, xSingleRadioButton, xAllowComboBox	None
1. Fill the xHoursListBox, xRateListBox, and xAllowComboBox with values 2. Select a default value in the xHoursListBox, xRateListBox, and xAllowComboBox	MainForm	Load
Verify that the user wants to close the application, then take the appropriate action based on the user's response		FormClosing

Figure 7-14: TOE chart for the Harvey Industries application

To save you time, the VB2005\Chap07\Harvey Industries Solution folder contains a partially completed Harvey Industries application.

To open the partially completed application:

1 Start Visual Studio 2005 or Visual Basic 2005 Express Edition, if necessary, and close the Start Page window.

2 Open the **Harvey Industries Solution** (Harvey Industries Solution.sln) file, which is contained in the VB2005\Chap07\Harvey Industries Solution folder. Auto-hide the Toolbox, Solution Explorer, and Properties windows, if necessary. Figure 7-15 shows the partially completed user interface. Only the combo box control is missing from the interface.

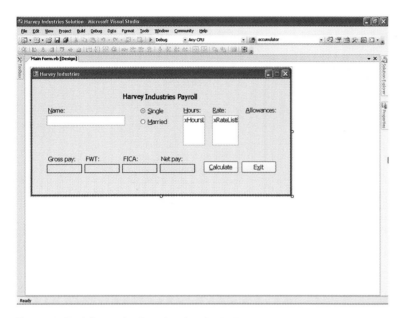

Figure 7-15: Partially completed user interface for the Harvey Industries application

INCLUDING A COMBO BOX IN AN INTERFACE

You use the **ComboBox tool** in the toolbox to add a combo box to an interface. A **combo box** is similar to a list box in that it allows the user to select from a list of choices. However, unlike a list box, a combo box also can contain a text field that allows the user to type an entry that is not on the list. In addition, a combo box can save space on a form because, unlike in a list box, the full list of choices can be hidden in a combo box.

Three styles of combo boxes are available in Visual Basic. The style is controlled by the combo box's **DropDownStyle property**, which can be set to Simple, DropDown (the default), or DropDownList. Each style of combo box contains a text portion and a list portion. When the DropDownStyle property is set to either Simple or DropDown, the text portion of the combo box is editable. However, in a Simple combo box the list portion is always displayed, while in a DropDown combo box the list portion appears only when the user clicks the combo box's list arrow. When the DropDownStyle property is set to the third style, DropDownList, the text portion of the combo box is not editable, and the user must click the combo box's list arrow to display the list of choices. When the user selects an item in the list portion of a combo box, or when he or she types a value in the text portion, the combo box's TextChanged event occurs. Figure 7-16 shows an example of each combo box style. You should use a label control to provide keyboard access to the combo box, as shown in the figure. For the access key to work correctly, you must set the TabIndex property of the label to a value that is one less than the combo box's TabIndex value.

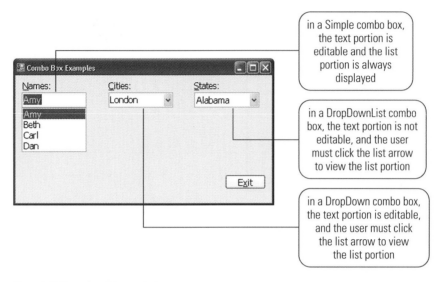

in a Simple combo box, the text portion is editable and the list portion is always displayed

in a DropDownList combo box, the text portion is not editable, and the user must click the list arrow to view the list portion

in a DropDown combo box, the text portion is editable, and the user must click the list arrow to view the list portion

Figure 7-16: Examples of the combo box styles

As you do with list boxes, you use the Items collection's Add method to add an item to a combo box. Like list boxes, combo boxes have a Sorted property that you can use to sort the items in the list. They also have a SelectedIndex property and a SelectedItem property, which you can use to either select an item in the list portion of the control or determine which item is selected in the list portion. To get or set the value that appears in the text portion of the combo box, you use the Text property. It is easy to confuse a combo box's Text property with its SelectedItem property. The SelectedItem property contains the value of the item selected in the list portion of the combo box, whereas the Text property contains the value that appears in the text portion. A value can appear in the text portion as a result of the user either selecting an item in the list portion of the control or typing an entry in the text portion itself. If the combo box is a DropDownList style, where the text portion is not editable, you can use the SelectedItem and Text properties interchangeably. However, if the combo box is either a Simple or DropDown style, where the user can type an entry in the text portion, you should use the Text property because it contains the value either selected or entered by the user. Figure 7-17 shows the code used to fill the combo boxes in Figure 7-16 with values. It also shows three different ways of specifying the default item in a combo box.

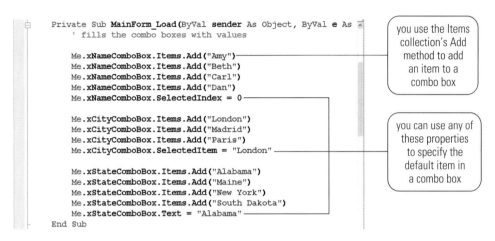

```
Private Sub MainForm_Load(ByVal sender As Object, ByVal e As
    ' fills the combo boxes with values

    Me.xNameComboBox.Items.Add("Amy")
    Me.xNameComboBox.Items.Add("Beth")
    Me.xNameComboBox.Items.Add("Carl")
    Me.xNameComboBox.Items.Add("Dan")
    Me.xNameComboBox.SelectedIndex = 0

    Me.xCityComboBox.Items.Add("London")
    Me.xCityComboBox.Items.Add("Madrid")
    Me.xCityComboBox.Items.Add("Paris")
    Me.xCityComboBox.SelectedItem = "London"

    Me.xStateComboBox.Items.Add("Alabama")
    Me.xStateComboBox.Items.Add("Maine")
    Me.xStateComboBox.Items.Add("New York")
    Me.xStateComboBox.Items.Add("South Dakota")
    Me.xStateComboBox.Text = "Alabama"
End Sub
```

you use the Items collection's Add method to add an item to a combo box

you can use any of these properties to specify the default item in a combo box

Figure 7-17: Code corresponding to the combo boxes shown in Figure 7-16

▶▶ GUI DESIGN TIPS

Combo Box Standards

» Use a label control to provide keyboard access to the combo box. Set the label's TabIndex property to a value that is one less than the combo box's TabIndex value.

» Combo box items are either arranged by use, with the most used entries appearing first in the list, or sorted in ascending order.

The Harvey Industries interface will provide a combo box for the user to specify the number of withholding allowances. The combo box will have the DropDown style, which will allow the user to either select the number from the list portion of the control, or enter the number in the text portion.

To add a DropDown style combo box to the form, then lock the controls and set the TabIndex values:

1 Click the **ComboBox** tool in the toolbox, and then drag the mouse pointer to the form. Position the mouse pointer below the Allowances label, then release the mouse button.

2 Set the combo box's (**Name**) property to **xAllowComboBox**.

3 Set the combo box's **Location** property to **588, 98**. Set its **Size** property to **51, 27**.

4 Lock the controls on the form, then use the information shown in Figure 7-18 to set the TabIndex values.

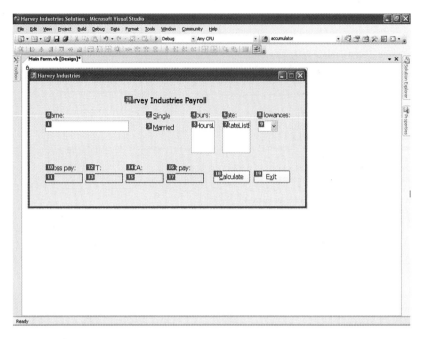

Figure 7-18: Correct TabIndex values

5 Press **Esc** to remove the TabIndex boxes from the form.

6 Save the solution.

In the next set of steps, you will code the MainForm's Load event procedure so that it displays the items 0 through 10 in the list portion of the xAllowComboBox. The code also will select the first item in the list portion.

To finish coding the MainForm's Load event procedure:

1 Open the Code Editor window. Notice that many of the events listed in the application's TOE chart (shown earlier in Figure 7-14) have already been coded for you.

2 Replace the <your name> and <current date> text in the comments with your name and the current date.

3 Locate the MainForm's Load event procedure, then position the insertion point in the blank line above the `Me.xHoursListBox.SelectedItem = "40.0"` statement. Press **Tab** twice, then type **for allowances as integer = 0 to 10** and press **Enter**.

4 Type **me.xAllowComboBox.items.add(allowances.tostring)**, then position the insertion point at the end of the `Next` clause. Press the **spacebar**, then type **allowances** and press **Enter**.

5 Position the insertion point in the blank line below the `Me.xRateListBox.SelectedItem = "9.50"` statement. Press **Tab** twice, then type **me.xAllowComboBox.selectedindex = 0** and press **Enter**. Figure 7-19 shows the completed Load event procedure.

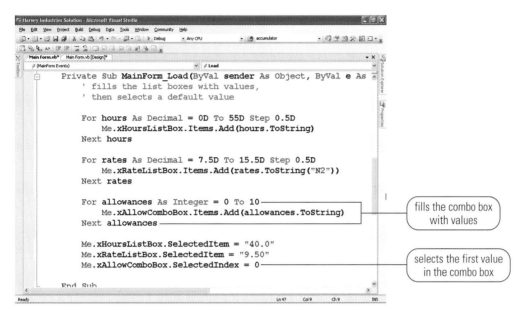

Figure 7-19: Completed Load event procedure

6 Save the solution, then start the application. The MainForm's Load event procedure fills both list boxes and the combo box with values. The procedure also selects the 40.0 and 9.50 items in the xHoursListBox and xRateListBox, respectively. The Me.xAllowComboBox.SelectedIndex = 0 statement in the procedure displays the first item, which is the number 0, in the text portion of the combo box, as shown in Figure 7-20.

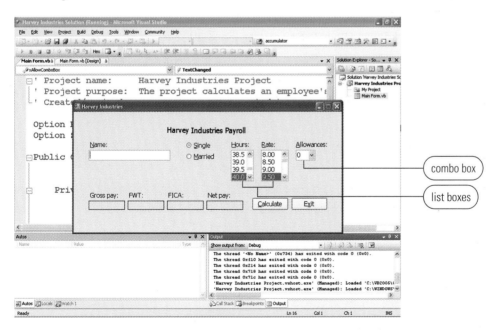

Figure 7-20: Completed user interface

7 The user can select the number of allowances from the list portion of the combo box. Click the **list arrow** button in the Allowances combo box to open the list portion of the control. Scroll down the list until you see the number 10, then click **10** in the list. Notice that the number 10 now appears in the text portion of the combo box.

8 The user also can specify the number of allowances by typing a number in the text portion of the combo box. Type **12** in the text portion of the combo box.

9 Click the **Exit** button to end the application. You are returned to the Code Editor window.

Next, you will associate the xAllowComboBox's TextChanged event with the ClearLabels procedure. Doing this will clear the calculated amounts from the interface when a change is made to the number of allowances.

To associate the combo box's TextChanged event with the ClearLabels procedure:

1 Locate the ClearLabels procedure in the Code Editor window.

2 Position the insertion point at the end of the procedure header. (The procedure header ends with the `xMarriedRadioButton.Click` line.) Type **,** **xAllowComboBox.textchanged**, then click in the blank line below the procedure header. Figure 7-21 shows the completed ClearLabels procedure.

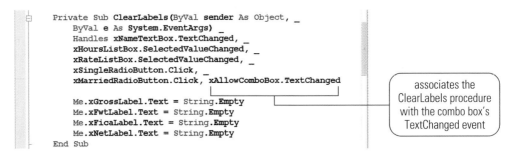

```
Private Sub ClearLabels(ByVal sender As Object, _
    ByVal e As System.EventArgs) _
    Handles xNameTextBox.TextChanged, _
    xHoursListBox.SelectedValueChanged, _
    xRateListBox.SelectedValueChanged, _
    xSingleRadioButton.Click, _
    xMarriedRadioButton.Click, xAllowComboBox.TextChanged

    Me.xGrossLabel.Text = String.Empty
    Me.xFwtLabel.Text = String.Empty
    Me.xFicaLabel.Text = String.Empty
    Me.xNetLabel.Text = String.Empty
End Sub
```

associates the ClearLabels procedure with the combo box's TextChanged event

Figure 7-21: Completed ClearLabels procedure

3 Save the solution.

CODING THE XCALCBUTTON'S CLICK EVENT PROCEDURE

According to the application's TOE chart, shown earlier in Figure 7-14, the xCalcButton's Click event procedure is responsible for calculating and displaying the gross pay, FWT (federal withholding tax), FICA tax, and net pay. Figure 7-22 shows the procedure's pseudocode.

xCalcButton Click event procedure - pseudocode

1. convert the Text property of the xAllowComboBox to a number

2. if the Text property can be converted to a number
    convert the item selected in the xHoursListBox to a number
    convert the item selected in the xRateListBox to a number

    if the Single radio button is selected
        assign "S" as the marital status
    else
        assign "M" as the marital status
    end if

    if the number of hours worked is less than or equal to 40
        calculate the gross pay = number of hours worked * pay rate
    else
        calculate the gross pay = 40 * pay rate + (number of hours worked − 40) * pay rate * 1.5
    end if

    call the GetFwt function to calculate and return the FWT
    calculate the FICA tax = gross pay * 7.65%
    round the gross pay, FWT, and FICA tax to two decimal places
    calculate the net pay = gross pay − FWT − FICA tax
    display the gross pay, FWT, FICA tax, and net pay in the xGrossLabel, xFwtLabel, xFicaLabel, and xNetLabel
else
    display the "Please re-enter the allowances" message in a message box
end if

Figure 7-22: Pseudocode for the xCalcButton's Click event procedure

To begin coding the xCalcButton's Click event procedure:

1 Open the code template for the xCalcButton's Click event procedure. Type ' **calculates and displays the gross pay,** and press **Enter**, then type ' **taxes, and net pay** and press **Enter** twice.

Recall that before you begin coding a procedure, you first study the procedure's pseudocode to determine the variables and named constants (if any) the procedure will use. In this case, the procedure will use a named constant for the FICA tax rate (7.65%), and two named constants for the message box's *prompt* and *title* arguments. The procedure also will use nine variables.

2 Type **const FicaRate as decimal = .0765d** and press **Enter**. Type **const Message as string =** _ and press **Enter**. Press **Tab**, then type **"Please re-enter the allowances"** and press **Enter**. Type **const Title as string = "Harvey Industries"** and press **Enter**.

3 Type **dim allowances as integer** and press **Enter**. The allowances variable will store the number of withholding allowances, which is selected in the xAllowComboBox.

4 Type **dim isAllowancesOk as boolean** and press **Enter**. The isAllowancesOk variable will store a Boolean value that indicates whether the number of allowances can be converted to a number.

5 Type **dim status as string** and press **Enter**. The status variable will store the letter "S" if the Single radio button is selected in the interface; otherwise, it will store the letter "M".

6 Type **dim hours as decimal** and press **Enter**. The hours variable will store the number of hours worked, which is selected in the xHoursListBox.

7 Type **dim rate as decimal** and press **Enter**. The rate variable will store the pay rate, which is selected in the xRateListBox.

8 Type **dim gross as decimal** and press **Enter**. The gross variable will store the gross pay amount.

9 Type **dim fwt as decimal** and press **Enter**. The fwt variable will store the federal withholding tax amount.

10 Type **dim fica as decimal** and press **Enter**. The fica variable will store the FICA tax amount.

11 Type **dim net as decimal** and press **Enter** twice. The net variable will store the net pay amount.

12 Save the solution.

Step 1 in the pseudocode is to convert the Text property of the xAllowComboBox to a number.

To continue coding the xCalcButton's Click event procedure:

1 Type **isallowancesok = _** and press **Enter**. Press **Tab**, then type **integer.tryparse(me.xAllowComboBox.text, allowances)** and press **Enter**.

Step 2 in the pseudocode is a selection structure that determines whether the TryParse method was able to convert the Text property of the combo box to an integer.

2 Type **if isallowancesok then** and press **Enter** twice, then type **else** and press **Enter**.

3 If the TryParse method could not convert the Text property to an integer, the selection structure's false path should display an appropriate message in a message box. Type the MessageBox.Show method shown in Figure 7-23, then position the insertion point as shown in the figure.

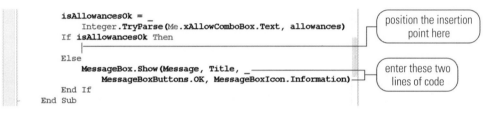

Figure 7-23: MessageBox.Show method entered in the procedure

4 If the TryParse method was able to convert the Text property of the combo box to an integer, the selection structure's true path should convert to numbers the items selected in the two list boxes. Type **hours = convert.todecimal(me.xHoursListBox.selecteditem)** and press **Enter**, then type **rate = convert.todecimal(me.xRateListBox.selecteditem)** and press **Enter** twice.

5 The next task is a nested selection structure that determines the employee's marital status. Type the nested selection structure shown in Figure 7-24, then position the insertion point as shown in the figure.

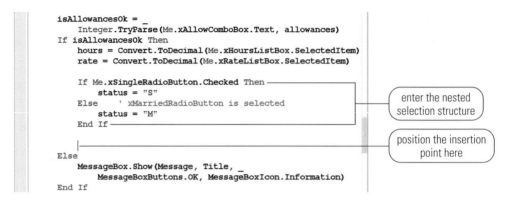

```
isAllowancesOk = _
    Integer.TryParse(Me.xAllowComboBox.Text, allowances)
If isAllowancesOk Then
    hours = Convert.ToDecimal(Me.xHoursListBox.SelectedItem)
    rate = Convert.ToDecimal(Me.xRateListBox.SelectedItem)

    If Me.xSingleRadioButton.Checked Then
        status = "S"
    Else    ' xMarriedRadioButton is selected
        status = "M"
    End If


Else
    MessageBox.Show(Message, Title, _
        MessageBoxButtons.OK, MessageBoxIcon.Information)
End If
```

enter the nested selection structure

position the insertion point here

Figure 7-24: Nested selection structure entered in the procedure

6 The next task is another nested selection structure that compares the number of hours worked with the number 40, and then calculates the gross pay based on the result. Type ' **calculate gross pay** and press **Enter**, then type **if hours <= 40 then** and press **Enter**.

7 If the number of hours worked is less than or equal to 40, the nested selection structure's true path should calculate the gross pay by multiplying the number of hours worked by the pay rate. Type **gross = hours * rate** and press **Enter**.

8 However, if the number of hours worked is greater than 40, the employee is entitled to his or her regular pay rate for the hours worked up to and including 40, and then time and one-half for the hours worked over 40. Type **else** and press **Enter**, then type **gross = 40d * rate + (hours – 40d) * rate * 1.5d**.

9 Save the solution.

According to the pseudocode (shown earlier in Figure 7-22), the xCalcButton's Click event procedure should now call the GetFwt function to calculate and return the FWT (federal withholding tax). Before entering the appropriate instruction, you will create the GetFwt function.

CODING THE GETFWT FUNCTION

The amount of federal withholding tax (FWT) to deduct from an employee's weekly gross pay is based on the employee's weekly taxable wages and his or her filing status, which is either single (including head of household) or married. You calculate the weekly taxable wages by first multiplying the number of withholding allowances by $63.46 (the value of one withholding allowance), and then subtracting the result from the weekly gross pay. For example, if your weekly gross pay is $400 and you have two withholding allowances, your weekly taxable wages are $273.08 (400 minus 126.92, which is the result of multiplying 63.46 by 2). You use the weekly taxable wages, along with the filing status and the weekly Federal Withholding Tax tables, to determine the amount of tax to withhold. Figure 7-25 shows the weekly FWT tables for the year 2006.

> **» TIP**
> The FWT tax tables and withholding allowance amount are for wages paid in the year 2006.

FWT Tables – Weekly Payroll Period

Single person (including head of household)

If the taxable wages are:	The amount of income tax to withhold is			
Over	But not over	Base amount	Percentage	Of excess over
	$ 51	0		
$ 51	$ 192	0	10%	$ 51
$ 192	$ 620	$ 14.10 plus	15%	$ 192
$ 620	$1,409	$ 78.30 plus	25%	$ 620
$1,409	$3,013	$ 275.55 plus	28%	$1,409
$3,013	$6,508	$ 724.67 plus	33%	$3,013
$6,508		$1,878.02 plus	35%	$6,508

Married person

If the taxable wages are:	The amount of income tax to withhold is			
Over	But not over	Base amount	Percentage	Of excess over
	$ 154	0		
$ 154	$ 440	0	10%	$ 154
$ 440	$1,308	$ 28.60 plus	15%	$ 440
$1,308	$2,440	$ 158.80 plus	25%	$1,308
$2,440	$3,759	$ 441.80 plus	28%	$2,440
$3,759	$6,607	$ 811.12 plus	33%	$3,759
$6,607		$1,750.96 plus	35%	$6,607

Figure 7-25: Weekly FWT tables

Notice that both tables shown in Figure 7-25 contain five columns of information. The first two columns list various ranges, also called brackets, of taxable wage amounts. The first column (Over) lists the amount that a taxable wage in that range must be over, and the second column (But not over) lists the maximum amount included in the range. The remaining three columns (Base amount, Percentage, and Of excess over) tell you how to calculate the tax for each range. For example, assume that you are married and your weekly taxable wages are $288.46. Before you can calculate the amount of your tax, you need to locate your taxable wages in the first two columns of the Married table. In this case, your taxable wages fall within the $154 through $440 range. After locating the range that contains your taxable wages, you then use the remaining three columns in the table to calculate your tax. According to the table, taxable wages in the $154 through $440 bracket have a tax of 10% of the amount over $154; therefore, your tax is $13.45, as shown in Figure 7-26.

Taxable wages	$ 288.46
Of excess over	−154.00
	134.46
Percentage	* .10
	13.45
Base amount	+ 0.00
Tax	$ 13.45

Figure 7-26: FWT calculation for a married taxpayer with taxable wages of $288.46

As Figure 7-26 indicates, you calculate the tax by first subtracting 154 (the amount shown in the Of excess over column) from your taxable wages of 288.46, giving 134.46. You then multiply 134.46 by 10% (the amount shown in the Percentage column), giving 13.45. You add the amount shown in the Base amount column—in this case, 0—to that result, giving $13.45 as your tax.

Now assume that your taxable wages are $600 per week and you are single. Figure 7-27 shows how the correct tax amount of $75.30 is calculated.

Taxable wages	$ 600.00
Of excess over	−192.00
	408.00
Percentage	* .15
	61.20
Base amount	+ 14.10
Tax	$ 75.30

Figure 7-27: FWT calculation for a single taxpayer with taxable wages of $600

To calculate the federal withholding tax, the GetFwt function needs to know the employee's gross pay amount, as well as the number of his or her withholding allowances and his or her marital status. The gross pay amount and number of withholding allowances are necessary to calculate the taxable wages, and the marital status indicates the appropriate FWT table to use when calculating the tax. The function will receive the necessary information from the xCalcButton's Click event procedure, which will pass the information when it calls the function. Recall that the information is stored in the procedure's `gross`, `allowances`, and `status` variables. Figure 7-28 shows the pseudocode for the GetFwt function.

GetFwt function - pseudocode

1. calculate the taxable wages = gross pay − number of withholding allowances * 63.46

2. if the marital status is Single
    taxable wages value:
        <= 51
            tax = 0
        <= 192
            calculate tax = 0.1 * (taxable wages - 51)
        <= 620
            calculate tax = 14.10 + 0.15 * (taxable wages - 192)
        <= 1409
            calculate tax = 78.30 + 0.25 * (taxable wages - 620)
        <= 3013
            calculate tax = 275.55 + .28 * (taxable wages - 1409)
        <= 6508
            calculate tax = 724.67 + .33 * (taxable wages - 3013)
        other
            calculate tax = 1878.02 + 0.35 * (taxable wages - 6508)
    else
    taxable wages value:
        <= 154
            tax = 0
        <= 440
            calculate tax = 0.1 * (taxable wages - 154)
        <= 1308

Figure 7-28: Pseudocode for the GetFwt function *(Continued)* ▶

```
                calculate tax = 28.60 + 0.15 * (taxable wages - 440)
            <= 2440
                calculate tax = 158.80 + 0.25 * (taxable wages - 1308)
            <= 3759
                calculate tax = 441.80 + 0.28 * (taxable wages - 2440)
            <= 6607
                calculate tax = 811.12 + .33 * (taxable wages - 3759)
            other
                calculate tax = 1750.96 + 0.35 * (taxable wages - 6607)
        end if
    3. return tax
```

Figure 7-28: Pseudocode for the GetFwt function

To code the GetFwt function:

1 Scroll to the top of the Code Editor window, then position the insertion point in the blank line above the ClearLabels procedure header.

When it calls the GetFwt function, the xCalcButton's Click event procedure will need to pass the values stored in its `status`, `allowances`, and `gross` variables, which have a data type of String, Integer, and Decimal, respectively. In this case, it is appropriate to pass each variable's value rather than its address, because you do not want the GetFwt function to change the contents of the variables. You will store the values passed to the function in three parameters: a String variable named `marital`, an Integer variable named `numAllow`, and a Decimal variable named `weekPay`. The GetFwt function will use the information it receives to calculate and return the FWT as a Decimal number.

2 Press **Tab**. Type the code and comment shown in Figure 7-29, then position the insertion point as shown in the figure. (Notice that the Code Editor automatically enters the `End Function` procedure footer for you.)

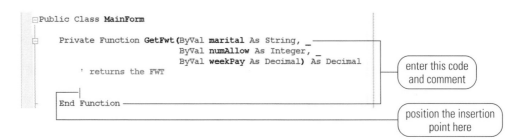

```
Public Class MainForm

    Private Function GetFwt(ByVal marital As String, _
                            ByVal numAllow As Integer, _
                            ByVal weekPay As Decimal) As Decimal
        ' returns the FWT

    End Function
```

enter this code and comment

position the insertion point here

Figure 7-29: GetFwt function header and footer

The GetFwt function will use a named constant for the withholding allowance amount (63.46). The function also will use two additional variables: one to store the taxable wages and the other to store the FWT.

3 Type **const WithAllow as decimal** = **63.46d** and press **Enter**.

4 Type **dim taxableWages as decimal** and press **Enter**, then type **dim tax as decimal** and press **Enter** twice.

Step 1 in the pseudocode shown in Figure 7-28 is to calculate the taxable wages.

5 Type **' calculate taxable wages** and press **Enter**, then type **taxablewages** = _ and press **Enter**. Press **Tab**, and then type **weekpay – convert.todecimal(numallow)** * **withallow** and press **Enter** twice.

6 Step 2 in the pseudocode is a selection structure that determines the marital status and then calculates the appropriate tax. Type **' determine marital status, then calculate FWT** and press **Enter**, then type **if marital = "S" then** and press **Enter**.

7 If the marital variable contains the letter "S", the selection structure's true path should calculate the federal withholding tax using the information from the Single tax table. Type the Select Case statement shown in Figure 7-30, then position the insertion point as shown in the figure.

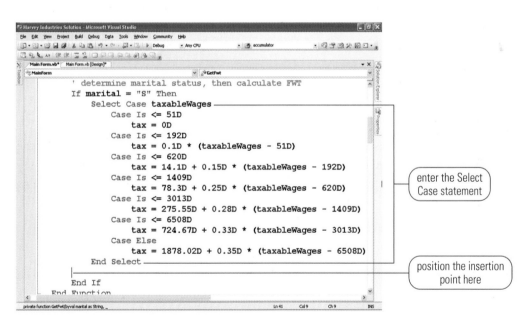

Figure 7-30: FWT calculations for Single taxpayers

If the `marital` variable does not contain the letter "S", the selection structure's false path should calculate the federal withholding tax using the information from the Married tax table.

8 Type **else** and press **Tab**, then type **' marital = "M"** and press **Enter**. Type the Select Case statement shown in Figure 7-31, then position the insertion point as shown in the figure.

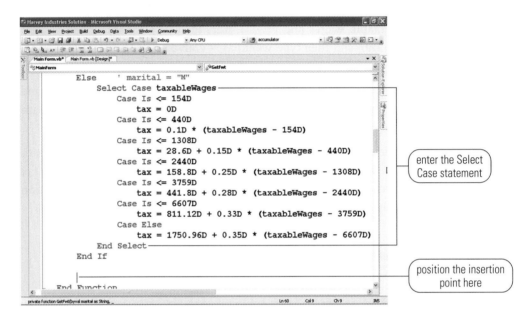

Figure 7-31: FWT calculations for Married taxpayers

The last step in the pseudocode is to return the federal withholding tax amount to the statement that invoked the function. The tax amount is stored in the `tax` variable.

9 Type **return tax**, then save the solution. Figure 7-32 shows the GetFwt function's code.

```
Private Function GetFwt(ByVal marital As String, _
                    ByVal numAllow As Integer, _
                    ByVal weekPay As Decimal) As Decimal

    ' returns the FWT

    Const WithAllow As Decimal = 63.46D
    Dim taxableWages As Decimal
    Dim tax As Decimal

    ' calculate taxable wages
    taxableWages = _
        weekPay - Convert.ToDecimal(numAllow) * WithAllow

    ' determine marital status, then calculate FWT
    If marital = "S" Then
        Select Case taxableWages
            Case Is <= 51D
                tax = 0D
            Case Is <= 192D
                tax = 0.1D * (taxableWages - 51D)
            Case Is <= 620D
                tax = 14.1D + 0.15D * (taxableWages - 192D)
            Case Is <= 1409D
                tax = 78.3D + 0.25D * (taxableWages - 620D)
            Case Is <= 3013D
                tax = 275.55D + 0.28D * (taxableWages - 1409D)
            Case Is <= 6508D
                tax = 724.67D + 0.33D * (taxableWages - 3013D)
            Case Else
                tax = 1878.02D + 0.35D * (taxableWages - 6508D)
        End Select

    Else ' marital = "M"
        Select Case taxableWages
            Case Is <= 154D
                tax = 0D
            Case Is <= 440D
                tax = 0.1D * (taxableWages - 154D)
            Case Is <= 1308D
                tax = 28.6D + 0.15D * (taxableWages - 440D)
```

Figure 7-32: GetFwt function's code *(Continued)* ▶

```
            Case Is <= 2440D
                tax = 158.8D + 0.25D * (taxableWages - 1308D)
            Case Is <= 3759D
                tax = 441.8D + 0.28D * (taxableWages - 2440D)
            Case Is <= 6607D
                tax = 811.12D + 0.33D * (taxableWages - 3759D)
            Case Else
                tax = 1750.96D + 0.35D * (taxableWages - 6607D)
        End Select

    End If

    Return tax
End Function
```

Figure 7-32: GetFwt function's code

Recall that you still need to complete the xCalcButton's Click event procedure.

COMPLETING THE XCALCBUTTON'S CLICK EVENT PROCEDURE

Now that you have created the GetFwt function, you can call the function from the xCalcButton's Click event procedure. Calling the GetFwt function is one of the tasks listed in the event procedure's pseudocode (shown earlier in Figure 7-22).

To complete the xCalcButton's Click event procedure:

1 Locate the xCalcButton's Click event procedure. Position the insertion point in the blank line above the last Else clause in the procedure. (The last Else clause appears above the MessageBox.Show method.) Press **Tab** three times to align the insertion point with the letter e in the Else clause.

Recall that the procedure needs to pass three values to the GetFwt function: the value stored in the status variable, the value stored in the allowances variable, and the value stored in the gross variable. The value returned by the function will be assigned to the fwt variable.

2 Type **' call a function to calculate the FWT** and press **Enter**, then type **fwt = getfwt(status, allowances, gross)** and press **Enter** twice.

3 The next task in the pseudocode for the xCalcButton's Click event procedure is to calculate the FICA tax by multiplying the gross pay amount by the FICA rate, which is 7.65%. Type **' calculate FICA tax** and press **Enter**, then type **fica = gross * ficarate** and press **Enter** twice.

Next, the procedure should round the gross pay, FWT, and FICA tax amounts to two decimal places. Rounding these amounts before making the net pay calculation will prevent the "penny off" error from occurring. (You can observe the "penny off" error by completing Exercise 3 at the end of this lesson.) You can use the **Math.Round function** to return a number rounded to a specific number of decimal places. The syntax of the Math.Round function is **Math.Round(**_value_[, _digits_]**)**, where _value_ is a numeric expression, and _digits_ (which is optional) is an integer indicating how many places to the right of the decimal point are included in the rounding. For example, `Math.Round(3.235, 2)` returns the number 3.24, but `Math.Round(3.234, 2)` returns the number 3.23. Notice that the Math.Round function rounds up a number only when the number to its right is at least 5; otherwise the Math.Round function truncates the excess digits. If the _digits_ argument is omitted, the Math.Round function returns an integer.

4 Type **' round gross pay, FWT, and FICA tax** and press **Enter**.

5 Type **gross = math.round(gross, 2)** and press **Enter**. Type **fwt = math.round(fwt, 2)** and press **Enter**, then type **fica = math.round(fica, 2)** and press **Enter** twice.

6 Next, the procedure should calculate the net pay by subtracting the FWT and FICA tax amounts from the gross pay amount. Type **' calculate net pay** and press **Enter**, then type **net = gross - fwt - fica** and press **Enter** twice.

7 The last task in the pseudocode is to display the gross pay, FWT, FICA tax, and net pay in the appropriate label controls in the interface. Type **' display calculated amounts** and press **Enter**, then enter the following four assignment statements.

Me.xGrossLabel.Text = gross.ToString("N2")

Me.xFwtLabel.Text = fwt.ToString("N2")

Me.xFicaLabel.Text = fica.ToString("N2")

Me.xNetLabel.Text = net.ToString("N2")

8 Save the solution. Figure 7-33 shows the code for the xCalcButton's Click event procedure.

```
Private Sub xCalcButton_Click(ByVal sender As Object, _
    ByVal e As System.EventArgs) Handles xCalcButton.Click
    ' calculates and displays the gross pay,
    ' taxes, and net pay

    Const FicaRate As Decimal = 0.0765D
    Const Message As String = _
        "Please re-enter the allowances"
    Const Title As String = "Harvey Industries"
    Dim allowances As Integer
    Dim isAllowancesOk As Boolean
    Dim status As String
    Dim hours As Decimal
    Dim rate As Decimal
    Dim gross As Decimal
    Dim fwt As Decimal
    Dim fica As Decimal
    Dim net As Decimal

    isAllowancesOk = _
        Integer.TryParse(Me.xAllowComboBox.Text, allowances)
    If isAllowancesOk Then
        hours = Convert.ToDecimal(Me.xHoursListBox.SelectedItem)
        rate = Convert.ToDecimal(Me.xRateListBox.SelectedItem)

        If Me.xSingleRadioButton.Checked Then
            status = "S"
        Else    ' xMarriedRadioButton is selected
            status = "M"
        End If

        ' calculate gross pay
        If hours <= 40 Then
            gross = hours * rate
        Else
            gross = 40D * rate + (hours - 40D) * rate * 1.5D
        End If
```

Figure 7-33: Completed xCalcButton's Click event procedure *(Continued)* ▶

```
            ' call a function to calculate the FWT
            fwt = GetFwt(status, allowances, gross)

            ' calculate FICA tax
            fica = gross * FicaRate

            ' round gross pay, FWT, and FICA tax
            gross = Math.Round(gross, 2)
            fwt = Math.Round(fwt, 2)
            fica = Math.Round(fica, 2)

            ' calculate net pay
            net = gross - fwt - fica

            ' display calculated amounts
            Me.xGrossLabel.Text = gross.ToString("N2")
            Me.xFwtLabel.Text = fwt.ToString("N2")
            Me.xFicaLabel.Text = fica.ToString("N2")
            Me.xNetLabel.Text = net.ToString("N2")

        Else

            MessageBox.Show(Message, Title, _
                MessageBoxButtons.OK, MessageBoxIcon.Information)

        End If
End Sub
```

Figure 7-33: Completed xCalcButton's Click event procedure

In the next set of steps, you will test both the xCalcButton's Click event procedure and the GetFwt function to verify that both are working correctly.

To test both the xCalcButton Click event procedure and GetFwt function:

1 Start the application.

First, you will calculate the weekly gross pay, taxes, and net pay for Karen Douglas. Last week, Karen worked 40 hours. She earns $10 per hour, and her marital status is Single. She claims one withholding allowance.

2 Type **Karen Douglas** in the Name text box. Scroll down the Rate list box, then click **10.00** in the list. Click the **list arrow** button in the Allowances combo box, then click **1** in the list. Click the **Calculate** button. The application calculates and displays Karen's gross pay, FWT, FICA tax, and net pay amounts, as shown in Figure 7-34.

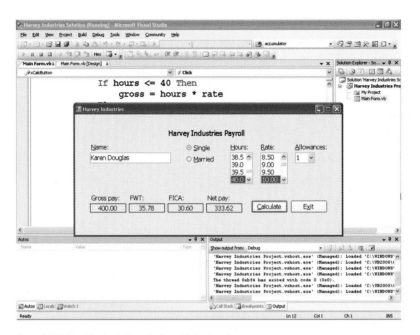

Figure 7-34: Payroll calculations displayed in the interface

Now calculate the weekly gross pay, taxes, and net pay for Carl Schmidt. Last week, Carl worked 39.5 hours. He earns $7.50 per hour, and his marital status is Married. He claims two withholding allowances.

3 Change the name entered in the Name text box to **Carl Schmidt**, then click the **Married** radio button. Click **39.5** in the Hours list box. Scroll up the Rate list box, then click **7.50** in the list. Type **2** in the text portion of the Allowances combo box. Click the **Calculate** button. The application displays 296.25, 1.53, 22.66, and 272.06 as Carl's gross pay, FWT, FICA tax, and net pay amounts, respectively.

4 Click the **Exit** button to end the application. Close the Code Editor window, then close the solution.

You have completed Lesson B. You can either take a break or complete the end-of-lesson questions and exercises before moving on to Lesson C.

SUMMARY

TO ADD A COMBO BOX TO A FORM:

» Use the ComboBox tool in the toolbox.

TO SPECIFY THE STYLE OF A COMBO BOX:

» Set the combo box's DropDownStyle property.

TO ADD ITEMS TO A COMBO BOX:

» Use the Items collection's Add method. The syntax of the method is *object*.**Items.Add(***item***),** where *object* is the name of the control to which you want the item added, and *item* is the text you want displayed in the control.

TO AUTOMATICALLY SORT THE ITEMS IN THE LIST PORTION OF A COMBO BOX:

» Set the combo box's Sorted property to True.

TO DETERMINE THE ITEM SELECTED IN THE LIST PORTION OF A COMBO BOX, OR TO SELECT AN ITEM FROM CODE:

» Use either the combo box's SelectedItem property or its SelectedIndex property.

TO DETERMINE THE VALUE THAT APPEARS IN THE TEXT POR-TION OF A COMBO BOX, OR TO SELECT AN ITEM FROM CODE:

» Use the combo box's Text property.

TO PROCESS CODE WHEN A DIFFERENT VALUE IS SELECTED IN A COMBO BOX:

» Enter the code in the combo box's TextChanged event procedure.

TO ROUND A NUMBER TO A SPECIFIC NUMBER OF DECIMAL PLACES:

» Use the Math.Round function. The function's syntax is **Math.Round(***value*[, *digits*]**),** where *value* is a numeric expression, and *digits* (which is optional) is an integer indicating how many places to the right of the decimal point are included in the rounding. If the *digits* argument is omitted, the Math.Round function returns an integer.

QUESTIONS

1. You use the _____ method to include items in a combo box.

    a. Add   b. AddList

    c. Item   d. ItemAdd

2. You use the _____ property to specify the style of a combo box.

    a. ComboBoxStyle   b. DropDownStyle

    c. DropStyle   d. Style

3. The items in a combo box belong to the _____ collection.

  a. Items

  b. List

  c. ListBox

  d. Values

4. Which of the following selects the "Cat" item, which appears third in the xAnimalComboBox?

  a. `Me.xAnimalComboBox.SelectedIndex = 2`

  b. `Me.xAnimalComboBox.SelectedItem = "Cat"`

  c. `Me.xAnimalComboBox.Text = "Cat"`

  d. All of the above.

5. The _____ property stores the item that appears in the text portion of a combo box.

  a. SelectedItem

  b. SelectedValue

  c. Text

  d. TextItem

6. Which of the following rounds the contents of the `number` variable to three decimal places?

  a. `Math.Round(3, number)`

  b. `Math.Round(number, 3)`

  c. `Round.Math(number, 3)`

  d. None of the above.

7. You use the _____ property to arrange the combo box items in ascending order.

  a. Alphabetical

  b. Arrange

  c. ListOrder

  d. Sorted

8. The _____ event occurs when the user either types a value in the text portion of a combo box, or selects a different value in the list portion.

  a. ChangedItem

  b. ChangedValue

  c. SelectedValueChanged

  d. TextChanged

EXERCISES

1. In this exercise, you modify the payroll application you created in the lesson. The modified code will use an independent Sub procedure rather than a Function procedure.

   a. Use Windows to make a copy of the Harvey Industries Solution folder, which is contained in the VB2005\Chap07 folder. Rename the folder Sub Harvey Industries Solution.

   b. If necessary, start Visual Studio 2005 or Visual Basic 2005 Express Edition. Open the Harvey Industries Solution (Harvey Industries Solution.sln) file contained in the VB2005\Chap07\Sub Harvey Industries Solution folder. Open the designer window.

   c. Change the GetFwt Function procedure to an independent Sub procedure, then modify the statement that calls the procedure.

   d. Save the solution, then start the application. Test the application by entering Karen Douglas in the Name text box. Then click 10.00 in the Rate list box, and click 1 in the Allowances combo box. Click the Calculate button. The calculated amounts should be identical to those shown in Lesson B's Figure 7-34.

   e. Click the Exit button to end the application. Close the Code Editor window, then close the solution.

2. In this exercise, you modify the payroll application you created in the lesson. In the modified code, the GetFwt function will be responsible for determining the employee's marital status.

   a. Use Windows to make a copy of the Harvey Industries Solution folder, which is contained in the VB2005\Chap07 folder. Rename the folder Modified Harvey Industries Solution.

   b. If necessary, start Visual Studio 2005 or Visual Basic 2005 Express Edition. Open the Harvey Industries Solution (Harvey Industries Solution.sln) file contained in the VB2005\Chap07\Modified Harvey Industries Solution folder. Open the designer window.

   c. The GetFwt function, rather than the xCalcButton's Click event procedure, should determine which radio button is selected. Make the appropriate modifications to the application's code.

d. Save the solution, then start the application. Test the application by entering Karen Douglas in the Name text box. Then click 10.00 in the Rate list box, and click 1 in the Allowances combo box. Click the Calculate button. The calculated amounts should be identical to those shown in Lesson B's Figure 7-34.

e. Click the Exit button to end the application. Close the Code Editor window, then close the solution.

3. In this exercise, you modify the payroll application you created in the lesson so that it no longer uses the Math.Round function. This will allow you to observe the "penny off" error.

a. Use Windows to make a copy of the Harvey Industries Solution folder, which is contained in the VB2005\Chap07 folder. Rename the folder No Rounding Harvey Industries Solution.

b. If necessary, start Visual Studio 2005 or Visual Basic 2005 Express Edition. Open the Harvey Industries Solution (Harvey Industries Solution.sln) file contained in the VB2005\Chap07\No Rounding Harvey Industries Solution folder. Open the designer window.

c. The Math.Round function appears in three statements in the xCalcButton's Click event procedure. Type an apostrophe at the beginning of each of the three statements, making them comments.

d. Save the solution, then start the application. Test the application by entering Carl Schmidt in the Name text box. Click the Married radio button, then click 39.5 in the Hours list box. Click 7.50 in the Rate list box, then click 2 in the Allowances combo box. Click the Calculate button. What is wrong with the calculated amounts?

e. Click the Exit button to end the application. Close the Code Editor window, then close the solution.

LESSON C
OBJECTIVES

AFTER STUDYING LESSON C, YOU SHOULD
BE ABLE TO:

» Prevent a form from closing

COMPLETING THE HARVEY INDUSTRIES PAYROLL APPLICATION

CODING THE MAINFORM'S FORMCLOSING EVENT PROCEDURE

To complete the Harvey Industries payroll application, you still need to code the MainForm's FormClosing event procedure. A form's **FormClosing event** occurs when a form is about to be closed. In most cases, this happens when the computer processes the Me.Close() statement in the form's code. However, it also occurs when the user clicks the Close button on the form's title bar. In the Harvey Industries application, the FormClosing event procedure is responsible for verifying that the user wants to close the application, and then taking the appropriate action based on the user's response. (Refer to the TOE chart shown in Figure 7-14 in Lesson B.) Figure 7-35 shows the pseudocode for this procedure.

MainForm FormClosing event procedure - pseudocode

1. use a message box to ask the user whether he or she wants to exit the application

2. if the user does not want to exit the application

       set the Cancel property of the procedure's e parameter to True

  end if

Figure 7-35: Pseudocode for the MainForm's FormClosing event procedure

To code the MainForm's FormClosing event procedure, then test the procedure:

1 Start Visual Studio 2005 or Visual Basic 2005 Express Edition, if necessary, and close the Start Page window.

2 Open the **Harvey Industries Solution** (Harvey Industries Solution.sln) file, which is contained in the VB2005\Chap07\Harvey Industries Solution folder. Auto-hide the Toolbox, Solution Explorer, and Properties windows, if necessary.

3 Open the Code Editor window. Click the **Class Name** list arrow, then click **(MainForm Events)** in the list. Click the **Method Name** list arrow, then click **FormClosing** in the list. The code template for the MainForm's FormClosing event procedure appears in the Code Editor window. (Notice that the code template appears above the MainForm's Load event procedure.)

4 Type ' **verify that the user wants to exit the application** and press **Enter** twice.

5 The procedure will use one variable, which will store the value returned by the MessageBox.Show method. Type **dim button as dialogresult** and press **Enter**.

Next, you will use the MessageBox.Show method to ask the user whether he or she wants to exit the application, and you will assign the user's response to the button variable. The message box will include Yes and No buttons and the Warning Message icon. The Yes button will be designated as the default button.

6 Type **button = messagebox.show("Do you want to exit?",** _ (be sure to type the line continuation character) and press **Enter**. Press **Tab**, then type **"Harvey Industries", messageboxbuttons.yesno,** _ and press **Enter**. Type **messageboxicon.exclamation,** _ and press **Enter**, then type **messageboxdefaultbutton.button1)** and press **Enter** twice.

If the user selects the No button in the message box, the FormClosing procedure should stop the computer from closing the form. You prevent the computer from closing a form by setting the **Cancel property** of the FormClosing procedure's e parameter to True.

7 Type ' **if the No button was selected, don't close the form** and press **Enter**.

8 Type **if button = windows.forms.dialogresult.no then** and press **Enter**, then type **e.cancel = true**.

9 Save the solution, then start the application. Click the **Close** button on the form's title bar. The computer processes the code contained in the FormClosing event procedure. The procedure displays the message box shown in Figure 7-36.

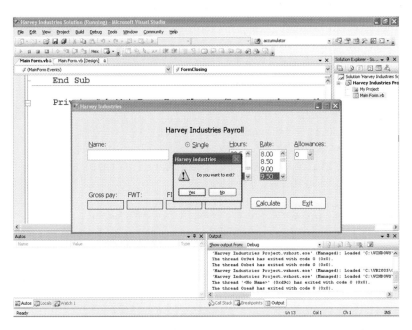

Figure 7-36: Message box displayed by the FormClosing event procedure

10 Click the **No** button. Notice that the form remains on the screen.

11 Click the **Exit** button, then click the **Yes** button to end the application. Close the Code Editor window, then close the solution. Figure 7-37 shows the code for the Harvey Industries application.

```
' Project name:      Harvey Industries Project
' Project purpose:   The project calculates an employee's weekly net pay.
' Created/revised:   <your name> on <current date>

Option Explicit On
Option Strict On

Public Class MainForm

    Private Function GetFwt(ByVal marital As String, _
                            ByVal numAllow As Integer, _
                            ByVal weekPay As Decimal) As Decimal
        ' returns the FWT

        Const WithAllow As Decimal = 63.46D
        Dim taxableWages As Decimal
        Dim tax As Decimal

        ' calculate taxable wages
        taxableWages = _
            weekPay - Convert.ToDecimal(numAllow) * WithAllow

        ' determine marital status, then calculate FWT
        If marital = "S" Then
            Select Case taxableWages
                Case Is <= 51D
                    tax = 0D
                Case Is <= 192D
                    tax = 0.1D * (taxableWages - 51D)
                Case Is <= 620D
                    tax = 14.1D + 0.15D * (taxableWages - 192D)
                Case Is <= 1409D
                    tax = 78.3D + 0.25D * (taxableWages - 620D)
                Case Is <= 3013D
                    tax = 275.55D + 0.28D * (taxableWages - 1409D)
                Case Is <= 6508D
                    tax = 724.67D + 0.33D * (taxableWages - 3013D)
                Case Else
                    tax = 1878.02D + 0.35D * (taxableWages - 6508D)
            End Select
```

Figure 7-37: Harvey Industries application's code *(Continued)* ▶

```
        Else    ' marital = "M"
            Select Case taxableWages
                Case Is <= 154D
                    tax = 0D
                Case Is <= 440D
                    tax = 0.1D * (taxableWages - 154D)
                Case Is <= 1308D
                    tax = 28.6D + 0.15D * (taxableWages - 440D)
                Case Is <= 2440D
                    tax = 158.8D + 0.25D * (taxableWages - 1308D)
                Case Is <= 3759D
                    tax = 441.8D + 0.28D * (taxableWages - 2440D)
                Case Is <= 6607D
                    tax = 811.12D + 0.33D * (taxableWages - 3759D)
                Case Else
                    tax = 1750.96D + 0.35D * (taxableWages - 6607D)
            End Select
        End If

        Return tax
End Function

Private Sub ClearLabels(ByVal sender As Object, _
    ByVal e As System.EventArgs) _
    Handles xNameTextBox.TextChanged, _
    xHoursListBox.SelectedValueChanged, _
    xRateListBox.SelectedValueChanged, _
    xSingleRadioButton.Click, _
    xMarriedRadioButton.Click, xAllowComboBox.TextChanged

    Me.xGrossLabel.Text = String.Empty
    Me.xFwtLabel.Text = String.Empty
    Me.xFicaLabel.Text = String.Empty
    Me.xNetLabel.Text = String.Empty
End Sub

Private Sub xNameTextBox_Enter(ByVal sender As Object, _
    ByVal e As System.EventArgs) Handles xNameTextBox.Enter
    Me.xNameTextBox.SelectAll()
End Sub
```

Figure 7-37: Harvey Industries application's code *(Continued)*   ▶

```vb
     Private Sub xExitButton_Click(ByVal sender As Object, _
         ByVal e As System.EventArgs) Handles xExitButton.Click
         Me.Close()
     End Sub

 Private Sub MainForm_FormClosing(ByVal sender As Object, _
         ByVal e As System.Windows.Forms.FormClosingEventArgs) _
         Handles Me.FormClosing
         ' verify that the user wants to exit the application

         Dim button As DialogResult
         button = MessageBox.Show("Do you want to exit?", _
             "Harvey Industries", MessageBoxButtons.YesNo, _
             MessageBoxIcon.Exclamation, _
             MessageBoxDefaultButton.Button1)

         ' if the No button was selected, don't close the form
         If button = Windows.Forms.DialogResult.No Then
             e.Cancel = True
         End If
     End Sub

     Private Sub MainForm_Load(ByVal sender As Object, _
         ByVal e As System.EventArgs) Handles Me.Load
         ' fills the list boxes with values,
         ' then selects a default value

         For hours As Decimal = 0D To 55D Step 0.5D
             Me.xHoursListBox.Items.Add(hours.ToString)
         Next hours

         For rates As Decimal = 7.5D To 15.5D Step 0.5D
             Me.xRateListBox.Items.Add(rates.ToString("N2"))
         Next rates

         For allowances As Integer = 0 To 10
             Me.xAllowComboBox.Items.Add(allowances.ToString)
         Next allowances

         Me.xHoursListBox.SelectedItem = "40.0"
         Me.xRateListBox.SelectedItem = "9.50"
         Me.xAllowComboBox.SelectedIndex = 0

     End Sub
```

Figure 7-37: Harvey Industries application's code *(Continued)*

▶

```
Private Sub xCalcButton_Click(ByVal sender As Object, _
    ByVal e As System.EventArgs) Handles xCalcButton.Click
    ' calculates and displays the gross pay,
    ' taxes, and net pay

    Const FicaRate As Decimal = 0.0765D
    Const Message As String = _
        "Please re-enter the allowances"
    Const Title As String = "Harvey Industries"
    Dim allowances As Integer
    Dim isAllowancesOk As Boolean
    Dim status As String
    Dim hours As Decimal
    Dim rate As Decimal
    Dim gross As Decimal
    Dim fwt As Decimal
    Dim fica As Decimal
    Dim net As Decimal

    isAllowancesOk = _
        Integer.TryParse(Me.xAllowComboBox.Text, allowances)
    If isAllowancesOk Then
        hours = Convert.ToDecimal(Me.xHoursListBox.SelectedItem)
        rate = Convert.ToDecimal(Me.xRateListBox.SelectedItem)

        If Me.xSingleRadioButton.Checked Then
            status = "S"
        Else    ' xMarriedRadioButton is selected
            status = "M"
        End If

        ' calculate gross pay
        If hours <= 40 Then
            gross = hours * rate
        Else
            gross = 40D * rate + (hours - 40D) * rate * 1.5D
        End If
```

Figure 7-37: Harvey Industries application's code *(Continued)* ▶

```
                  ' call a function to calculate the FWT
                  fwt = GetFwt(status, allowances, gross)

                  ' calculate FICA tax
                  fica = gross * FicaRate

                  ' round gross pay, FWT, and FICA tax
                  gross = Math.Round(gross, 2)
                  fwt = Math.Round(fwt, 2)
                  fica = Math.Round(fica, 2)

                  ' calculate net pay
                  net = gross - fwt - fica

                  ' display calculated amounts
                  Me.xGrossLabel.Text = gross.ToString("N2")
                  Me.xFwtLabel.Text = fwt.ToString("N2")
                  Me.xFicaLabel.Text = fica.ToString("N2")
                  Me.xNetLabel.Text = net.ToString("N2")
            Else
                  MessageBox.Show(Message, Title, _
                        MessageBoxButtons.OK, MessageBoxIcon.Information)
            End If
      End Sub
End Class
```

Figure 7-37: Harvey Industries application's code

You have completed Lesson C and Chapter 7. You can either take a break or complete the end-of-lesson questions and exercises.

SUMMARY

TO PROCESS CODE WHEN A FORM IS ABOUT TO BE CLOSED:

» Enter the code in the form's FormClosing event procedure. The FormClosing event occurs when the user clicks the Close button on a form's title bar. It also occurs when the computer processes the Me.Close() statement.

TO PREVENT A FORM FROM BEING CLOSED:

» Set the Cancel property of the FormClosing event procedure's e parameter to True.

QUESTIONS

1. The _____ event is triggered when you click a form's Close button.

   a. Close

   b. Closing

   c. FormClose

   d. FormClosing

2. The _____ event is triggered when the computer processes the `Me.Close()` statement.

   a. Close

   b. Closing

   c. FormClose

   d. FormClosing

3. Which of the following statements prevents a form from being closed?

   a. `e.Cancel = False`

   b. `e.Cancel = True`

   c. `e.Close = False`

   d. `sender.Cancel = True`

EXERCISES

1. In this exercise, you modify the Gross Pay application that you completed in Lesson A.

   a. Use Windows to make a copy of the Gross Pay Solution folder, which is contained in the VB2005\Chap07 folder. Rename the folder FormClosing Gross Pay Solution.

   b. If necessary, start Visual Studio 2005 or Visual Basic 2005 Express Edition. Open the Gross Pay Solution (Gross Pay Solution.sln) file contained in the VB2005\Chap07\FormClosing Gross Pay Solution folder. Open the designer window.

   c. Code the form's FormClosing event procedure so that it asks the user whether he or she wants to exit the application. Take the appropriate action based on the user's response.

   d. Save the solution, then start and test the application.

   e. Click the Exit button to end the application. Close the Code Editor window, then close the solution.

2. In this exercise, you modify the Pine Lodge application that you completed in Lesson A.

   a. Use Windows to make a copy of the Pine Lodge Solution folder, which is contained in the VB2005\Chap07 folder. Rename the folder FormClosing Pine Lodge Solution.

   b. If necessary, start Visual Studio 2005 or Visual Basic 2005 Express Edition. Open the Pine Lodge Solution (Pine Lodge Solution.sln) file contained in the VB2005\Chap07\FormClosing Pine Lodge Solution folder. Open the designer window.

   c. Code the form's FormClosing event procedure so that it asks the user whether he or she wants to exit the application. Take the appropriate action based on the user's response.

   d. Save the solution, then start and test the application.

   e. Click the Exit button to end the application. Close the Code Editor window, then close the solution.

DEBUGGING EXERCISE

3. In this exercise, you debug an existing application. The purpose of this exercise is to demonstrate a common error made when using functions.

   a. If necessary, start Visual Studio 2005 or Visual Basic 2005 Express Edition. Open the Debug Solution (Debug Solution.sln) file, which is contained in the VB2005\Chap07\Debug Solution folder. If necessary, open the designer window.

   b. Open the Code Editor window and study the existing code.

   c. Start the application. Click 20 in the Length list box, then click 30 in the Width list box. Click the Calculate Area button, which should display the area of a rectangle having a length of 20 feet and a width of 30 feet. Notice that the application is not working properly.

   d. Click the Exit button to end the application.

   e. Correct the application's code, then save the solution and start the application. Click 20 in the Length list box, then click 30 in the Width list box. Click the Calculate Area button, which should display the area of the rectangle.

   f. Click the Exit button to end the application. Close the Code Editor window, then close the solution.

8

MANIPULATING STRINGS

CREATING A HANGMAN GAME APPLICATION

Mr. Mitchell teaches second grade at Hinsbrook School. On days when the weather is bad and the students cannot go outside to play, he spends recess time playing a simplified version of the Hangman game with his class. The game requires two people to play. Currently, Mr. Mitchell thinks of a word that has five letters. He then draws five dashes on the chalkboard—one for each letter in the word. One student then is chosen to guess the word, letter by letter. When the student guesses a correct letter, Mr. Mitchell replaces the appropriate dash or dashes with the letter. For example, if the original word is *moose* and the student guesses the letter *o*, Mr. Mitchell changes the fives dashes on the chalkboard to *-oo--*. If the student's letter does not appear in the word, Mr. Mitchell begins drawing the Hangman image, which contains nine lines and one circle. The game is over when the student guesses all of the letters in the word, or when he or she makes 10 incorrect guesses, whichever comes first.

PREVIEWING THE COMPLETED APPLICATION

Before creating the Hangman Game application, you will preview the completed application.

To preview the completed application:

1 Use the Run command on the Windows Start menu to run the **Hangman (Hangman.exe)** file, which is contained in the VB2005\Chap08 folder. The Hangman Game application's user interface appears on the screen. As indicated in Figure 8-1, the interface contains a File menu. You will learn how to include a menu in an interface in Lesson B.

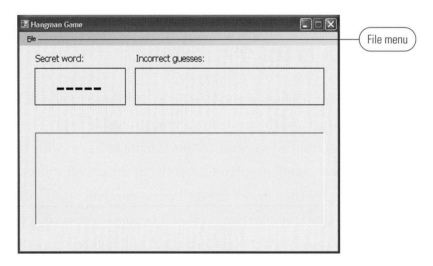

Figure 8-1: Hangman Game application's user interface

2 Click **File** on the menu bar, and then click **New Game**. The Hangman Game dialog box opens and prompts you to enter a 5-letter word.

3 Type **puppy** and press **Enter** to select the OK button. Five dashes (hyphens)—one for each letter in the word "puppy"—appear in the application's interface. In addition, the Letter dialog box opens and prompts you to enter a letter.

4 You will guess the letter y first. Type **y** and press **Enter** to select the OK button. The application replaces the last dash in the Secret word box with the letter Y. This indicates that the letter Y is the last letter in the word.

5 Now you will guess the letter x. Type **x** and press **Enter**. The letter x does not appear in the word "puppy", so the application displays the letter X in the Incorrect guesses

box. It also displays the bottom line of the Hangman image. Recall that the image contains nine lines and one circle.

6 Next, you will guess the letter a. Type **a** and press **Enter**. The letter a does not appear in the word "puppy", so the application displays the letter A in the Incorrect guesses box. It also displays another line in the Hangman image.

7 Now you will guess the letter u. Type **u** and press **Enter**. The application replaces the second dash in the Secret word box with the letter U.

8 Next, you will guess the letters d, g, and b. Type **d** and press **Enter**, then type **g** and press **Enter**, and then type **b** and press **Enter**. The letters you entered do not appear in the word "puppy", so the application displays the letters in the Incorrect guesses box. It also displays two additional lines and a circle in the Hangman image.

9 Now you will guess the letter p. Type **p** and press **Enter**. The application replaces the remaining dashes in the Secret word box with the letter P. It then displays the "Great guessing!" message in a message box, as shown in Figure 8-2.

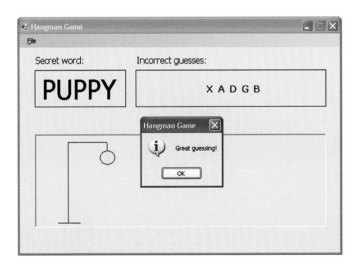

Figure 8-2: Result of guessing the word

10 Press **Enter** to close the message box. Click **File** on the menu bar, and then click **Exit** to end the application.

Before you can begin coding the Hangman Game application, you need to learn how to manipulate strings in Visual Basic and how to create a menu. You learn about string manipulation in Lesson A and about the MenuStrip tool in Lesson B. You complete the Hangman Game application in Lesson C.

LESSON A
OBJECTIVES

AFTER STUDYING LESSON A, YOU SHOULD
BE ABLE TO:

» Determine the number of characters contained in a string

» Remove characters from a string

» Replace one or more characters in a string

» Insert characters within a string

» Search a string for one or more characters

» Access characters contained in a string

» Compare strings

STRING MANIPULATION

MANIPULATING STRINGS IN VISUAL BASIC 2005

Many times, an application will need to manipulate (process) string data. For example, an application may need to verify that an inventory part number begins with a specific letter to determine the part's location in the warehouse. Or, it may need to determine whether the last three characters in an employee number are valid. In this lesson, you learn several ways of manipulating strings in Visual Basic. You begin by learning how to determine the number of characters contained in a string.

DETERMINING THE NUMBER OF CHARACTERS CONTAINED IN A STRING

If your application expects the user to enter a 10-digit phone number or a five-digit ZIP code, you should verify that the user entered the required number of characters. The number of characters contained in a string is stored in the string's **Length property**. Figure 8-3 shows the syntax and purpose of the Length property and includes several examples of using the property.

Length property

Syntax and purpose

*string.***Length**

stores the number of characters contained in a string

Examples

```
Dim numChars As Integer
numChars = Me.xZipTextBox.Text.Length
```

assigns to the numChars variable the number of characters contained in the xZipTextBox's Text property

```
Dim numChars As Integer
Dim fullName As String = "Paul Blackfeather"
numChars = fullName.Length
```

assigns the number 17 to the numChars variable

```
Dim phone As String
phone = InputBox("10-digit phone number", "Phone")
Do Until phone.Length = 10
      phone = InputBox("10-digit phone number", "Phone")
Loop
```

gets a phone number from the user until the number of characters contained in the phone number is equal to the number 10

```
Dim partNum As String
partNum = InputBox("Part number", "Part Number")
If partNum.Length >= 4 Then
      instructions to process when the condition evaluates to True
Else
      instructions to process when the condition evaluates to False
End If
```

gets a part number from the user, and then determines whether the part number contains at least four characters

Figure 8-3: Syntax, purpose, and examples of the Length property

In the first example in Figure 8-3, the numChars = Me.xZipTextBox.Text.Length statement assigns to the numChars variable the number of characters contained in the xZipTextBox's Text property. (Recall that the Text property of a control is treated as a string.) If the user enters the ZIP code 60111 in the xZipTextBox, the statement assigns the number five to the numChars variable. The Dim fullName As String = "Paul

`Blackfeather"` statement in the second example assigns the string "Paul Blackfeather" to a String variable named `fullName`. The `numChars = fullName.Length` statement then uses the `fullName` variable's Length property to determine the number of characters contained in the variable, assigning the result to the `numChars` variable. In this case, the number 17 will be assigned, because the `fullName` variable contains 17 characters. The pretest loop shown in the third example in Figure 8-3 processes the loop instruction, which prompts the user to enter a phone number, until the phone number entered contains 10 characters, at which time the loop ends. The code shown in the last example prompts the user to enter a part number, and stores the user's response in a String variable named `partNum`. The selection structure in the code then determines whether the `partNum` variable contains at least four characters.

REMOVING CHARACTERS FROM A STRING

At times, you may need to remove one or more characters from an item of data entered by the user. For example, you may need to remove a percent sign from the end of a tax rate so that you can use the tax rate in a calculation. You can use the **TrimStart method** to remove one or more characters from the beginning of a string, and the **TrimEnd method** to remove one or more characters from the end of a string. To remove one or more characters from both the beginning and end of a string, you use the **Trim method**. Each method returns a string with the appropriate characters removed (trimmed). Figure 8-4 shows the syntax and purpose of the TrimStart, TrimEnd, and Trim methods and includes several examples of using each method. In each syntax, *trimChars* is a comma-separated list of characters that you want removed (trimmed) from the *string*. The *trimChars* argument is optional in each syntax; if you omit the argument, the method removes one or more spaces from the beginning and/or end of the *string*. In other words, the default value for the *trimChars* argument is the space character (" ").

TrimStart, TrimEnd, and Trim methods

<u>Syntax and purpose of each method</u>

string.**TrimStart**[(*trimChars*)]

removes characters from the beginning of a string

string.**TrimEnd**[(*trimChars*)]

removes characters from the end of a string

string.**Trim**[(*trimChars*)]

removes characters from both the beginning and end of a string

---

<u>Examples</u>

```
Dim fullName As String
fullName = Me.xNameTextBox.Text.TrimStart
```

assigns to the `fullName` variable the contents of the xNameTextBox's Text property, excluding any leading spaces

---

```
Me.xNameTextBox.Text = Me.xNameTextBox.Text.TrimStart
```

removes any leading spaces from the xNameTextBox's Text property

---

```
Dim fullName As String
fullName = Me.xNameTextBox.Text.TrimEnd
```

assigns to the `fullName` variable the contents of the xNameTextBox's Text property, excluding any trailing spaces

---

```
Dim inputRate As String
Dim rate As String
inputRate = InputBox("Rate:", "Rate")
rate = inputRate.TrimEnd("%"c, " "c)
```

assigns to the `rate` variable the contents of the `inputRate` variable, excluding any trailing percent signs and spaces

---

```
Me.xNameTextBox.Text = Me.xNameTextBox.Text.Trim
```

removes any leading and trailing spaces from the xNameTextBox's Text property

---

```
Dim number As String
number = InputBox("Number:", "Number")
number = number.Trim("$"c, " "c, "%"c)
```

removes any leading and trailing dollar signs, spaces, and percent signs from the `number` variable

---

Figure 8-4: Syntax, purpose, and examples of the TrimStart, TrimEnd, and Trim methods

When processing the `fullName = Me.xNameTextBox.Text.TrimStart` statement in the first example in Figure 8-4, the computer first makes a temporary copy of the string stored in the xNameTextBox's Text property. It then removes any leading spaces from the temporary copy of the string, and assigns the resulting string to the `fullName` variable. If the user enters the string " Karen" (two spaces followed by the name Karen) in the xNameTextBox, the statement assigns the name "Karen" to the `fullName` variable; however, the xNameTextBox's Text property still contains " Karen" (two spaces followed by the name Karen). After the statement is processed, the computer removes the temporary copy of the string from its internal memory. Notice that the `fullName = Me.xNameTextBox.Text.TrimStart` statement does not remove the leading spaces from the xNameTextBox's Text property. To remove the leading spaces from the Text property, you use the statement shown in the second example in Figure 8-4: `Me.xNameTextBox.Text = Me.xNameTextBox.Text.TrimStart`.

When processing the `fullName = Me.xNameTextBox.Text.TrimEnd` statement shown in the third example, the computer first makes a temporary copy of the string stored in the xNameTextBox's Text property. It then removes any trailing spaces from the copied string, assigning the result to the `fullName` variable. After the statement is processed, the computer removes the copied string from its internal memory. If the user enters the string "Ned Yander   " (the name Ned Yander followed by four spaces) in the xNameTextBox, the statement assigns the name "Ned Yander" to the `fullName` variable; however, the statement does not change the contents of the xNameTextBox's Text property.

When processing the `rate = inputRate.TrimEnd("%"c, " "c)` statement shown in the fourth example, the computer first makes a copy of the string stored in the `inputRate` variable. It then removes any trailing percent signs and spaces from the copied string, and assigns the resulting string to the `rate` variable. For example, if the `inputRate` variable contains the string "3 %" (the number 3, a space, and a percent sign), the statement assigns the string "3" to the `rate` variable, but it does not change the value stored in the `inputRate` variable. Likewise, if the `inputRate` variable contains the string "15% " (the number 15, a percent sign, and two spaces), the statement assigns the string "15" to the `rate` variable, but it leaves the contents of the `inputRate` variable unchanged. The letter c that appears after each string in the *trimChars* argument is one of the literal type characters you learned about in Chapter 3. Recall that a literal type character forces a literal constant to assume a different data type. In this case, the c forces each string in the *trimChars* argument to assume the Char (character) data type.

>> TIP
The literal type characters are listed in Figure 3-7 in Chapter 3.

You can use the `Me.xNameTextBox.Text = Me.xNameTextBox.Text.Trim` statement shown in the fifth example in Figure 8-4 to remove any leading and trailing spaces from the xNameTextBox's Text property. Likewise, you can use the `number = number.Trim("$"c, " "c, "%"c)` statement shown in the last example to remove any leading and trailing dollar signs, spaces, and percent signs from the `number` variable.

THE REMOVE METHOD

Besides using the TrimStart, TrimEnd, and Trim methods, you also can use the Remove method to remove characters from a string. However, unlike the TrimStart, TrimEnd, and Trim methods, which remove characters from only the beginning and/or end of a string, the **Remove method** allows you to remove one or more characters located anywhere in a string. Figure 8-5 shows the syntax and purpose of the Remove method and includes several examples of using the method. Like the TrimStart, TrimEnd, and Trim methods, the Remove method returns a string with the appropriate characters removed.

Remove method

Syntax and purpose

string.**Remove**(*startIndex*, *count*)

removes characters from anywhere in a string

Examples

```
Dim fullName As String = "John Cober"
Me.xNameTextBox.Text = fullName.Remove(0, 5)
```

assigns the string "Cober" to the xNameTextBox's Text property

---

```
Dim fullName As String = "John"
Me.xNameTextBox.Text = fullName.Remove(2, 1)
```

assigns the string "Jon" to the xNameTextBox's Text property

```
Dim fullName As String = "Janis"
fullName = fullName.Remove(3, 2)
```

assigns the string "Jan" to the `fullName` variable

Figure 8-5: Syntax, purpose, and examples of the Remove method

Each character in a string is assigned a unique number that indicates its position in the string; the number is called an **index**. The first character in a string has an index of zero, the second character has an index of one, and so on. In the Remove method's syntax, *startIndex* is the index of the first character you want removed from the *string*, and *count* is the number of characters you want removed. For example, to remove only the first character from a string, you use the number zero as the *startIndex*, and the number one as the *count*. To remove the fourth through eighth characters, you use the number three as the *startIndex*, and the number five as the *count*.

When processing the `Me.xNameTextBox.Text = fullName.Remove(0, 5)` statement shown in the first example in Figure 8-5, the computer makes a copy of the string stored in the `fullName` variable. It then removes the first five characters from the copied string; in this case, the computer removes the letters J, o, h, and n, and the space character. The computer then assigns the resulting string ("Cober") to the xNameTextBox's Text property before removing the copied string from its internal memory. The contents of the `fullName` variable are not changed as a result of processing the `Me.xNameTextBox.Text = fullName.Remove(0, 5)` statement.

When processing the `Me.xNameTextBox.Text = fullName.Remove(2, 1)` statement in the second example, the computer makes a copy of the string stored in the `fullName` variable. It then removes one character, beginning with the character whose index is 2, from the copied string. The character with an index of 2 is the third character in the string—in this case, the letter h. The computer then assigns the resulting string ("Jon") to the xNameTextBox's Text property. Here again, the statement does not change the string stored in the `fullName` variable.

You can use the `fullName = fullName.Remove(3, 2)` statement shown in the last example in Figure 8-5 to remove two characters from the string stored in the `fullName` variable, beginning with the character whose index is 3. In this case, the letters i and s are removed, changing the contents of the `fullName` variable from "Janis" to "Jan".

At times, you may need to replace characters in a string, rather than just remove them.

REPLACING CHARACTERS IN A STRING

You can use the **Replace method** to replace a sequence of characters in a string with another sequence of characters. For example, you can use the Replace method to replace area code "800" with area code "877" in a phone number. Or, you can use it to replace the dashes in a Social Security number with the empty string. Figure 8-6 shows the syntax and purpose of the Replace method and includes several examples of using the method. In the syntax, *oldValue* is the sequence of characters that you want to replace in the *string*, and *newValue* is the replacement characters. The Replace method returns a string with all occurrences of *oldValue* replaced with *newValue*.

Replace method

<u>Syntax and purpose</u>

string.**Replace**(*oldValue, newValue*)

replaces all occurrences of a sequence of characters in a string with another sequence of characters

<u>Examples</u>

```
Dim name as String = "Mary James"
Dim newName as String
newName = name.Replace("James", "Smith")
```

assigns the string "Mary Smith" to the newName variable

```
Dim name as String = "James Jameston"
Dim newName as String
newName = name.Replace("James", "Smith")
```

assigns the string "Smith Smithton" to the newName variable

```
Dim socialNum As String = "000-11-9999"
socialNum = socialNum.Replace("-", "")
```

assigns the string "000119999" to the socialNum variable

```
Dim word As String = "latter"
word = word.Replace("t", "d")
```

assigns the string "ladder" to the word variable

Figure 8-6: Syntax, purpose, and examples of the Replace method

When processing the `newName = name.Replace("James", "Smith")` statement shown in the first example in Figure 8-6, the computer first makes a copy of the string stored in the name variable. It then replaces "James" with "Smith" in the copied string, and then assigns the result—in this case, "Mary Smith"—to the newName variable. Keep in mind that the Replace method replaces all occurrences of *oldValue* with *newValue*. Therefore, if the `name` variable contains "James Jameston", the string "Smith, Smithton" will be assigned to the `newName` variable, as indicated in the second example.

In the third example in Figure 8-6, the `socialNum = socialNum.Replace("-", "")` statement replaces each dash (hyphen) in the string stored in the `socialNum` variable with a zero-length (empty) string. Replacing the dashes with the empty string is the same as removing the dashes. After the statement is processed, the `socialNum` variable contains the string "000119999". As the last example in the figure indicates, the `word = word.Replace("t", "d")` statement replaces each letter "t" in the string stored in the `word` variable with the letter "d". The statement changes the contents of the `word` variable from "latter" to "ladder".

As mentioned earlier, the Replace method replaces all occurrences of *oldValue* with *newValue*. At times, however, you may need to replace only a specific occurrence of *oldValue* with *newValue*; for this, you use the Mid statement rather than the Replace method.

THE MID STATEMENT

You can use the **Mid statement** to replace a specified number of characters in a string with characters from another string. Figure 8-7 shows the syntax and purpose of the Mid statement and includes several examples of using the statement. In the syntax, *targetString* is the string in which you want characters replaced, and *replacementString* contains the replacement characters. *Start* is the character position of the first character you want replaced in the *targetString*. The first character in the *targetString* is in character position one, the second is in character position two, and so on. Notice that the character position is not the same as the index, which begins with zero. The optional *count* argument specifies the number of characters to replace in the *targetString*. If *count* is omitted, the Mid statement replaces the lesser of either the number of characters in the *replacementString* or the number of characters in the *targetString* from position *start* through the end of the *targetString*.

Mid statement

Syntax and purpose

Mid(_targetString_, _start_ [, _count_]**)** = _replacementString_

replaces a specific number of characters in a string with characters from another string

Examples

```
Dim fullName As String = "Rob Smith"
Mid(fullName, 7, 1) = "y"
```

changes the contents of the fullName variable to "Rob Smyth"

```
Dim fullName As String = "Rob Smith"
Mid(fullName, 7) = "y"
```

changes the contents of the fullName variable to "Rob Smyth"

```
Dim fullName As String = "Ann Johnson"
Mid(fullName, 5) = "Carl"
```

changes the contents of the fullName variable to "Ann Carlson"

```
Dim fullName As String = "Earl Cho"
Mid(fullName, 6) = "Liverpool"
```

changes the contents of the fullName variable to "Earl Liv"

Figure 8-7: Syntax, purpose, and examples of the Mid statement

The `Mid(fullName, 7, 1) = "y"` statement shown in the first example in Figure 8-7 replaces the letter "i", which is located in character position seven in the fullName variable, with the letter "y". After the statement is processed, the fullName variable contains the string "Rob Smyth". You also can omit the _count_ argument and use the `Mid(fullName, 7) = "y"` statement, which is shown in the second example, to replace the letter "i" in the fullName variable with the letter "y". Recall that when the _count_ argument is omitted from the Mid statement, the statement replaces the lesser of either the number of characters in the _replacementString_ (in this case, one) or the number of characters in the _targetString_ from position _start_ through the end of the _targetString_ (in this case, three).

The `Mid(fullName, 5) = "Carl"` statement in the third example replaces four characters in the fullName variable, beginning with the character located in character position five in the variable (the letter J). Here again, because the _count_ argument is omitted from the Mid statement, the statement replaces the lesser of either the number of characters in the *replacementString* (in this case, four) or the number of characters in the *targetString* from position *start* through the end of the *targetString* (in this case, seven). After the statement is processed, the `fullName` variable contains the string "Ann Carlson".

The `Mid(fullName, 6) = "Liverpool"` statement in the last example in Figure 8-7 replaces three characters in the `fullName` variable, beginning with the character located in character position six in the variable (the letter C). Here again, because the *count* argument is omitted from the Mid statement, the statement replaces the lesser of either the number of characters in the *replacementString* (in this case, nine) or the number of characters in the *targetString* from position *start* through the end of the *targetString* (in this case, three). After the statement is processed, the `fullName` variable contains the string "Earl Liv".

INSERTING CHARACTERS AT THE BEGINNING AND END OF A STRING

In addition to removing and replacing characters in a string, you also can insert characters within a string. To insert characters at either the beginning or end of a string, you can use the PadLeft and PadRight methods, respectively. Both methods pad the string with a character until the string is a specified length, then they return the padded string. The **PadLeft method** pads the string on the left; in other words, it inserts the padded characters at the beginning of the string, which right-aligns the characters within the string. The **PadRight method**, on the other hand, pads the string on the right, which inserts the padded characters at the end of the string and left-aligns the characters within the string. Figure 8-8 shows the syntax and purpose of the PadLeft and PadRight methods and includes examples of using the methods. In each syntax, *length* is an integer that represents the desired length of the *string*. In other words, *length* represents the total number of characters you want the *string* to contain. The *character* argument is the character that each method uses to pad the *string* until it reaches the desired *length*. Notice that the *character* argument is optional in each syntax; if omitted, the default *character* is the space character.

PadLeft and PadRight methods

Syntax and purpose of each method

string.**PadLeft**(*length*[, *character*])

pads the beginning of a string with a character until the string is a specified length

string.**PadRight**(*length*[, *character*])

pads the end of a string with a character until the string is a specified length

---

Examples

```
Dim number As Integer = 42
Dim outputNumber As String
outputNumber = Convert.ToString(number)
outputNumber = outputNumber.PadLeft(5)
```

assigns "   42" (three spaces and the string "42") to the `outputNumber` variable

---

```
Dim number As Integer = 42
Dim outputNumber As String
outputNumber = number.ToString.PadLeft(5)
```

assigns "   42" (three spaces and the string "42") to the `outputNumber` variable

---

```
Dim netPay As Decimal = 767.89D
Dim formattedNetPay As String
formattedNetPay = netPay.ToString("C2").PadLeft(15, "*"c)
```

assigns "********$767.89" to the `formattedNetPay` variable

```
Dim firstName As String = "Sue"
Dim leftAlignedName As String
leftAlignedName = firstName.PadRight(10)
```

assigns "Sue   " (the string "Sue" and seven spaces) to the `leftAlignedName` variable

---

```
Dim firstName As String = "Sue"
firstName = firstName.PadRight(10)
```

assigns "Sue   " (the string "Sue" and seven spaces) to the `firstName` variable

---

Figure 8-8: Syntax, purpose, and examples of the PadLeft and PadRight methods

The statements shown in the first two examples in Figure 8-8 produce the same results. In both examples, the code assigns five characters—three space characters, the character 4, and the character 2—to the `outputNumber` variable. However, the first example uses two assignment statements to accomplish the task, while the second example uses one assignment statement.

Study the two assignment statements shown in the first example. The first assignment statement, `outputNumber = Convert.ToString(number)`, converts the contents of the `number` variable to String, and then assigns the result to the `outputNumber` variable. When processing the second assignment statement, `outputNumber = outputNumber.PadLeft(5)`, the computer first makes a copy of the string stored in the `outputNumber` variable. It then pads the copied string with space characters until the string contains exactly five characters. In this case, the computer uses three space characters, which it inserts at the beginning of the string. The computer then assigns the resulting string—" 42"—to the `outputNumber` variable. Notice that the number 42 is right-aligned within the `outputNumber` variable.

Now study the assignment statement shown in the second example. When processing the `outputNumber = number.ToString.PadLeft(5)` statement, the computer first makes a copy of the value stored in the `number` variable; it then converts the copied value to a string. Next, it inserts space characters at the beginning of the string until the string has exactly five characters, and assigns the result to the `outputNumber` variable. Notice that when two methods appear in an expression, the computer processes the methods from left to right. In this case, the computer processes the ToString method before processing the PadLeft method.

When processing the `formattedNetPay = netPay.ToString("C2").PadLeft(15, "*"c)` statement shown in the third example in Figure 8-8, the computer first makes a copy of the number stored in the `netPay` variable. It then converts the number to a string and formats it with a dollar sign and two decimal places. The computer then pads the string with asterisks until the string contains exactly 15 characters. In this case, the computer inserts eight asterisks at the beginning of the string. The computer assigns the resulting string ("********$767.89") to the `formattedNetPay` variable. Many companies use this type of formatted net pay on paychecks given to their employees, because it makes it difficult for someone to change the amount.

When processing the `leftAlignedName = firstName.PadRight(10)` statement shown in the fourth example, the computer first makes a copy of the string stored in the `firstName` variable. It then pads the copied string with space characters until the string contains exactly 10 characters. In this case, the computer inserts seven space characters at the end of the string. The computer then assigns the resulting string—"Sue   "—to the `leftAlignedName` variable. Notice that the PadRight method left-aligns the name "Sue" within the `leftAlignedName` variable.

Keep in mind that the `leftAlignedName = firstName.PadRight(10)` statement does not change the contents of the `firstName` variable. To assign "Sue   " to the `firstName` variable, you would need to use the `firstName = firstName.PadRight(10)` statement shown in the last example in Figure 8-8.

The PadLeft and PadRight methods can be used to insert characters only at the beginning and end, respectively, of a string. To insert characters within a string, you use the Insert method.

THE INSERT METHOD

Visual Basic provides the **Insert method** for inserting characters anywhere within a string. Possible uses for the method include inserting an employee's middle initial within his or her name, and inserting parentheses around the area code in a phone number. Figure 8-9 shows the syntax and purpose of the Insert method and includes two examples of using the method. In the syntax, *startIndex* specifies where in the *string* you want the *value* inserted. To insert the *value* at the beginning of the *string*, you use the number zero as the *startIndex*. To insert the *value* as the second character in the *string*, you use the number one as the *startIndex*, and so on. The Insert method returns a string with the appropriate characters inserted.

Insert method

Syntax and purpose

*string.***Insert**(*startIndex, value*)

inserts characters within a string

Examples

```
Dim name As String = "Rob Smith"
Dim fullName As String
fullName = name.Insert(4, "T. ")
```

assigns the string "Rob T. Smith" to the fullName variable

```
Dim phone As String = "3120501111"
phone = phone.Insert(0, "(")
phone = phone.Insert(4, ")")
phone = phone.Insert(8, "-")
```

changes the contents of the phone variable to "(312)050-1111"

Figure 8-9: Syntax, purpose, and examples of the Insert method

When processing the `fullName = name.Insert(4, "T. ")` statement shown in the first example in Figure 8-9, the computer first makes a copy of the string stored in the `name` variable, and then inserts the *value* "T. " (the letter T, a period, and a space character) in the copied string. The letter T is inserted in *startIndex* position 4, which makes it the fifth character in the string. The period and space character are inserted in *startIndex* positions 5 and 6, making them the sixth and seventh characters in the string. After the statement is processed, the `fullName` variable contains the string "Rob T. Smith"; however, the `name` variable still contains "Rob Smith".

In the second example shown in Figure 8-9, the `phone = phone.Insert(0, "(")` statement changes the contents of the `phone` variable from "3120501111" to "(3120501111". The `phone = phone.Insert(4, ")")` statement then changes the contents of the variable from "(3120501111" to "(312)0501111", and the `phone = phone.Insert(8, "-")` statement changes the contents of the variable from "(312)0501111" to "(312)050-1111".

SEARCHING A STRING

In some applications, you might need to determine whether a string begins or ends with a specific character or characters. For example, you may need to determine whether a phone number entered by the user begins with area code "312", which indicates that the customer lives in Chicago. Or, you may need to determine whether a tax rate entered by the user ends with a percent sign, because the percent sign will need to be removed before the tax rate can be used in a calculation. In Visual Basic, you can use the **StartsWith method** to determine whether a specific sequence of characters occurs at the beginning of a string, and the **EndsWith method** to determine whether a specific sequence of characters occurs at the end of a string. Figure 8-10 shows the syntax and purpose of the StartsWith and EndsWith methods along with examples of using each method. In the syntax for both methods, *subString* is a string that represents the sequence of characters you want to search for either at the beginning or end of the *string*. The StartsWith method returns the Boolean value True when the *subString* is located at the beginning of the *string*; otherwise, it returns the Boolean value False. Similarly, the EndsWith method returns the Boolean value True when the *subString* is located at the end of the *string*; otherwise, it returns the Boolean value False. Both methods perform a case-sensitive search, which means that the case of the *subString* must match the case of the *string* for the methods to return the True value.

StartsWith and EndsWith methods

Syntax and purpose of each method

string.**StartsWith**(*subString*)

determines whether a specific sequence of characters occurs at the beginning of a string

string.**EndsWith**(*subString*)

determines whether a specific sequence of characters occurs at the end of a string

Examples

```
Dim pay As String
pay = InputBox("Pay rate", "Pay")
If pay.StartsWith("$") Then
     pay = pay.TrimStart("$"c)
End If
```

determines whether the string stored in the pay variable begins with the dollar sign; if it does, the dollar sign is removed from the string

```
Dim phone As String
phone = InputBox("10-digit phone number", "Phone")
Do While phone.StartsWith("312")
     Me.xPhoneListBox.Items.Add(phone)
     phone = InputBox("10-digit phone number", "Phone")
Loop
```

determines whether the string stored in the phone variable begins with "312"; if it does, the contents of the phone variable are added to the xPhoneListBox and the user is prompted to enter another phone number

```
Dim cityState As String
cityState = Me.xCityStateTextBox.Text.ToUpper
If cityState.EndsWith("CA") Then
     Me.xStateLabel.Text = "California customer"
End If
```

determines whether the string stored in the cityState variable ends with "CA"; if it does, the string "California customer" is assigned to the xStateLabel's Text property

```
Dim fullName As String
fullName = InputBox("Your name:", "Name")
fullName = fullName.ToUpper
If Not fullName.EndsWith("SMITH") Then
     Me.xNameLabel.Text = fullName
End If
```

determines whether the string stored in the fullName variable ends with "SMITH"; if it does not, the variable's value is assigned to the xNameLabel's Text property

Figure 8-10: Syntax, purpose, and examples of the StartsWith and EndsWith methods *(Continued)*

▶

```
Dim fullName As String
fullName = InputBox("Your name:", "Name")
If Not fullName.ToUpper.EndsWith("SMITH") Then
     Me.xNameLabel.Text = fullName
End If
```
determines whether the string stored in the fullName variable ends with "SMITH"; if it does not, the variable's value is assigned to the xNameLabel's Text property

Figure 8-10: Syntax, purpose, and examples of the StartsWith and EndsWith methods

In the first example shown in Figure 8-10, the If pay.StartsWith("$") Then clause determines whether the string stored in the pay variable begins with the dollar sign. If it does, the pay = pay.TrimStart("$"c) statement removes the dollar sign from the variable's contents. You also can write the If clause in this example as If pay. StartsWith("$") = True Then. Notice that the string "$" is used as the *subString* argument in the StartsWith method, but the character "$"c is used as the *trimChars* argument in the TrimStart method. This is because the *subString* argument in the StartsWith method must be a string, while the *trimChars* argument in the TrimStart method must be a listing of one or more characters.

In the second example shown in Figure 8-10, the Do While phone.StartsWith ("312") clause determines whether the string stored in the phone variable begins with "312". If it does, the contents of the variable are added to the xPhoneListBox and the user is prompted to enter another phone number. You also can write the Do clause in this example as Do While phone.StartsWith("312") = True. In the third example, the code assigns the string "California customer" to the xStateLabel's Text property when the string stored in the cityState variable ends with the uppercase string "CA". You also can write the If clause in this example as If cityState.EndsWith("CA") = True Then.

The fullName = fullName.ToUpper statement shown in the fourth example changes the contents of the fullName variable to uppercase. The If Not fullName.EndsWith("SMITH") Then clause then compares the contents of the fullName variable with the string "SMITH". If the fullName variable does not end with the string "SMITH", the Me.xNameLabel.Text = fullName statement assigns the contents of the fullName variable to the xNameLabel's Text property. You also can write the If clause in this example as If fullName.EndsWith("SMITH") = False Then. The code shown in the last example in Figure 8-10 is similar to the code shown in the fourth example, except the ToUpper method is included in the If clause rather than in an assignment statement. When processing the If Not fullName. ToUpper.EndsWith("SMITH") Then clause, the computer first makes a copy of the string stored in the fullName variable; it then converts the copied string to uppercase. The computer then determines whether the copied string ends with the string "SMITH".

Here again, notice that the computer processes the methods from left to right; in this case, it processes the ToUpper method before processing the EndsWith method. If the copied string does not end with the string "SMITH", the `Me.xNameLabel.Text = fullName` statement assigns the contents of the `fullName` variable to the xNameLabel's Text property. Unlike the code in the fourth example, the code in this example does not permanently change the contents of the `fullName` variable to uppercase. The If clause in this example also can be written as `If  fullName.ToUpper.EndsWith ("SMITH") = False Then`.

The StartsWith and EndsWith methods can be used only to determine whether a string begins or ends with a specific sequence of characters. To determine whether one or more characters appear anywhere in a string, you can use either the Contains method or the IndexOf method. You will learn about the Contains method first.

THE CONTAINS METHOD

You can use the **Contains method** to search a string to determine whether it contains a specific sequence of characters. For example, you can use the method to determine whether the area code "(312)" appears in a phone number, or whether the street name "Elm Street" appears in an address. Figure 8-11 shows the syntax and purpose of the Contains method and includes several examples of using the method. In the syntax, *subString* is a string that represents the sequence of characters for which you want to search within the *string*. The Contains method returns the Boolean value True when the *subString* is contained anywhere within the *string*; otherwise, it returns the Boolean value False. Like the StartsWith and EndsWith methods, the Contains method performs a case-sensitive search.

Contains method

Syntax and purpose

string.**Contains**(*subString*)

searches a string to determine whether it contains a specific sequence of characters, and then returns a Boolean value that indicates whether the characters appear within the string

Examples

```
Dim address As String = "345 Main Street, Glendale, CA"
Dim isContained As Boolean
isContained = address.Contains("Main Street")
```

assigns the Boolean value True to the `isContained` variable, because "Main Street" appears in the `address` variable

---

Figure 8-11: Syntax, purpose, and examples of the Contains method *(Continued)* ▶

```
Dim address As String = "345 Main Street, Glendale, CA"
Dim isContained As Boolean
isContained = address.Contains("Main street")
```

assigns the Boolean value False to the `isContained` variable, because "Main street" does not appear in the `address` variable

```
Dim address As String = "345 Main Street, Glendale, CA"
Dim isContained As Boolean
isContained = address.ToUpper.Contains("MAIN STREET")
```

assigns the Boolean value True to the `isContained` variable, because "MAIN STREET" appears in the `address` variable when its contents are converted to uppercase

```
Dim phone As String = "(312) 999-9999"
If phone.Contains("(312)") Then
```

the If...Then...Else statement's *condition* evaluates to True, because "(312)" appears in the `phone` variable

Figure 8-11: Syntax, purpose, and examples of the Contains method

In the first example shown in Figure 8.11, the `isContained = address.Contains("Main  Street")` statement assigns the Boolean value True to the `isContained` variable, because the *subString* "Main Street" is contained somewhere in the `address` variable. The `isContained = address.Contains("Main  street")` statement in the second example, however, assigns the Boolean value False to the `isContained` variable. This is because the string stored in the `address` variable ("Main Street") is not equivalent to the *subString* for which you are searching ("Main street"); the case of each string is different. (Recall that the Contains method performs a case-sensitive search.)

When processing the `isContained = address.ToUpper.Contains("MAIN STREET")` statement shown in the third example, the computer first makes a copy of the string stored in the `address` variable; it then converts the copied string to uppercase. The computer then determines whether the copied string contains the *subString* "MAIN STREET". In this case, the *subString* appears somewhere in the copied string, so the computer assigns the Boolean value True to the `isContained` variable.

In the first three examples shown in Figure 8-11, the Contains method appears in an assignment statement, where its return value is assigned to a Boolean variable. You also can use the Contains method in the *condition* of either a selection or repetition structure. For instance, in the last example shown in the figure, the Contains method appears in an If...Then...Else statement's *condition*. In the example, the Contains method returns the Boolean value True, because the phone variable contains the *subString* "(312)"; therefore, the *condition* evaluates to True. The If clause in the example also can be written as If phone.Contains("(312)") = True Then.

THE INDEXOF METHOD

In addition to using the Contains method to search for a *subString* anywhere within a *string*, you also can use the **IndexOf method**. Unlike the Contains method, which returns a Boolean value that indicates whether the *string* contains the *subString*, the IndexOf method returns an integer that represents the location of the *subString* within the *string*. Figure 8-12 shows the syntax and purpose of the IndexOf method and includes several examples of using the method. In the syntax, *subString* is the sequence of characters for which you are searching in the *string*, and *startIndex* is the index of the character at which the search should begin; in other words, *startIndex* specifies the starting position for the search. Recall that the first character in a string has an index of zero, the second character has an index of one, and so on. Notice that the *startIndex* argument is optional in the IndexOf method's syntax. If you omit the *startIndex* argument, the IndexOf method begins the search with the first character in the *string*. The IndexOf method searches for the *subString* within the *string*, beginning with the character whose index is *startIndex*. If the IndexOf method does not find the *subString*, it returns the number –1; otherwise, it returns the index of the starting position of the *subString* within the *string*.

IndexOf method

Syntax and purpose

string.**IndexOf**(*subString*[, *startIndex*])

searches a string to determine whether it contains a specific sequence of characters, and then returns an integer that indicates the starting position of the characters within the string

---

Examples

```
Dim message As String = "Have a nice day"
Dim indexNum As Integer
indexNum = message.IndexOf("nice", 0)
```

assigns the number 7 to the indexNum variable

Figure 8-12: Syntax, purpose, and examples of the IndexOf method *(Continued)* ▶

```
Dim message As String = "Have a nice day"
Dim indexNum As Integer
indexNum = message.IndexOf("nice")
```

assigns the number 7 to the indexNum variable

```
Dim message As String = "Have a nice day"
Dim indexNum As Integer
indexNum = message.IndexOf("Nice")
```

assigns the number –1 to the indexNum variable

```
Dim message As String = "Have a nice day"
Dim indexNum As Integer
indexNum = message.ToUpper.IndexOf("NICE")
```

assigns the number 7 to the indexNum variable

```
Dim message As String = "Have a nice day"
Dim indexNum As Integer
indexNum = message.IndexOf("nice", 5)
```

assigns the number 7 to the indexNum variable

```
Dim message As String = "Have a nice day"
Dim indexNum As Integer
indexNum = message.IndexOf("nice", 8)
```

assigns the number –1 to the indexNum variable

Figure 8-12: Syntax, purpose, and examples of the IndexOf method

You can use either the indexNum = message.IndexOf("nice",   0) statement shown in the first example in Figure 8-12, or the indexNum = message.IndexOf ("nice") statement shown in the second example, to search for the word "nice" in the message variable, beginning with the first character in the variable. In each case, the word "nice" begins with the eighth character in the variable. The eighth character has an index of seven, so both statements assign the number seven to the indexNum variable.

Like the StartsWith, EndsWith, and Contains methods, the IndexOf method performs a case-sensitive search, as the third example in Figure 8-12 indicates. In the example, the indexNum = message.IndexOf("Nice") statement assigns the number –1 to the indexNum variable, because the word "Nice" is not contained in the message variable. You can use the indexNum = message.ToUpper.IndexOf("NICE") statement shown in the fourth example to perform a case-insensitive search for the word "nice".

When processing the statement, the computer first makes a copy of the string stored in the message variable. The ToUpper method then converts the copied string to uppercase, and the IndexOf method searches the uppercase string for the word "NICE". The statement assigns the number seven to the indexNum variable because, ignoring case, the word "nice" begins with the character whose index is seven.

The indexNum = message.IndexOf("nice", 5) statement shown in the fifth example in Figure 8-12 searches for the word "nice" in the message variable, beginning with the character whose index is five; that character is the second letter "a". The statement assigns the number seven to the indexNum variable, because the word "nice" begins with the character whose index is seven. The indexNum = message. IndexOf("nice", 8) statement shown in the last example searches for the word "nice" in the message variable, beginning with the character whose index is eight; that character is the letter "i". The word "nice" does not appear anywhere in the "ice day" portion of the string stored in the message variable; therefore, the statement assigns the number –1 to the indexNum variable.

ACCESSING CHARACTERS CONTAINED IN A STRING

In some applications, it is necessary to access one or more characters contained in a string. For example, you may need to determine whether a specific letter appears as the third character in a string, because the letter identifies the department in which an employee works. Or, you may need to display only the string's first five characters, which identify an item's location in the warehouse. Visual Basic provides the **Substring method** for accessing any number of characters contained in a string. Figure 8-13 shows the syntax and purpose of the Substring method and includes several examples of using the method. Notice that the Substring method contains two arguments: *startIndex* and *count*. *StartIndex* is the index of the first character you want to access in the *string*. As you learned earlier, the first character in a string has an index of zero; the second character has an index of one, and so on. The *count* argument, which is optional, specifies the number of characters you want to access. The Substring method returns a string that contains *count* number of characters, beginning with the character whose index is *startIndex*. If you omit the *count* argument, the Substring method returns all characters from the *startIndex* position through the end of the string.

Substring method

Syntax and purpose

string.**Substring**(*startIndex*[, *count*])

accesses one or more characters contained in a string

Examples

```
Dim fullName As String = "Peggy Ryan"
firstName = fullName.Substring(0, 5)
lastName = fullName.Substring(6)
```

assigns "Peggy" to the firstName variable, and assigns "Ryan" to the lastName variable

```
Dim sales As String
sales = Me.xSalesTextBox.Text
If sales.StartsWith("$") Then
    sales = sales.Substring(1)
End If
```

determines whether the string stored in the sales variable begins with the dollar sign; if it does, assigns the contents of the variable, excluding the dollar sign, to the sales variable

```
Dim inputRate As String
inputRate = InputBox("Enter rate", "Tax Rate")
If inputRate.EndsWith("%") Then
    inputRate = inputRate.Substring(0, inputRate.Length - 1)
End If
```

determines whether the string stored in the inputRate variable ends with the percent sign; if it does, assigns the contents of the variable, excluding the percent sign, to the inputRate variable

Figure 8-13: Syntax, purpose, and examples of the Substring method

The firstName = fullName.Substring(0, 5) statement shown in the first example in Figure 8-13 assigns the first five characters contained in the fullName variable ("Peggy") to the firstName variable. The lastName = fullName.Substring(6) statement assigns to the lastName variable all of the characters contained in the fullName variable, beginning with the character whose index is 6. In this case, the statement assigns "Ryan" to the lastName variable.

The code shown in the second example uses the StartsWith method to determine whether the string stored in the sales variable begins with the dollar sign. If it does, the `sales = sales.Substring(1)` statement assigns all of the characters from the `sales` variable, beginning with the character whose index is 1, to the `sales` variable. In other words, it assigns all but the first character to the `sales` variable. The `sales = sales.Substring(1)` statement is equivalent to the statement `sales = sales.Remove(0, 1)`, as well as to the statement `sales = sales.TrimStart("$"c)`.

The code shown in the last example in Figure 8-13 uses the EndsWith method to determine whether the string stored in the `inputRate` variable ends with the percent sign. If it does, the `inputRate = inputRate.Substring(0, inputRate.Length - 1)` statement assigns to the `inputRate` variable all of the characters contained in the variable, excluding the last character (which is the percent sign). The `inputRate = inputRate.Substring(0, inputRate.Length - 1)` statement is equivalent to the statement `inputRate = inputRate.Remove(inputRate.Length - 1, 1)`, as well as to the statement `inputRate = inputRate.TrimEnd("%"c)`.

COMPARING STRINGS

In addition to using the =, <>, >, >=, <. <=, and Like comparison operators to compare two strings, you also can use the **String.Compare method**. Figure 8-14 shows the syntax and purpose of the String.Compare method and includes several examples of using the method. The *string1* and *string2* arguments in the syntax represent the two strings you want compared. The optional *ignoreCase* argument is a Boolean value that indicates whether you want to perform a case-insensitive or a case-sensitive comparison of both strings. When the *ignoreCase* argument is the Boolean value True, the String.Compare method performs a case-insensitive comparison. When the *ignoreCase* argument is the Boolean value False, or when the argument is omitted, the String.Compare method performs a case-sensitive comparison. The method returns an integer that indicates the result of comparing *string1* with *string2*. When both strings are equal, the method returns the number 0. When *string1* is greater than *string2*, the method returns the number 1. When *string1* is less than *string2*, the method returns the number –1. The String.Compare method uses rules called **word sort rules** when comparing the strings. Following these rules, numbers are considered less than lowercase letters, which are considered less than uppercase letters.

» TIP

You learned about the =, <>, <, <=, >, and >= operators in Chapter 4, and about the Like operator in Chapter 5.

String.Compare method

Syntax and purpose

String.Compare(*string1*, *string2*[, *ignoreCase*]**)**

compares two strings

Examples

```
Dim result As Integer
result = String.Compare("Dallas", "Dallas")
```
assigns the number 0 to the result variable

```
Dim result As Integer
result = String.Compare("Dallas", "DALLAS")
```
assigns the number –1 to the result variable

```
Dim result As Integer
result = String.Compare("Dallas", "DALLAS", True)
```
assigns the number 0 to the result variable

```
Dim result As Integer
result = String.Compare("Dallas", "Boston")
```
assigns the number 1 to the result variable

Figure 8-14: Syntax, purpose, and examples of the String.Compare method

Study the examples shown in Figure 8-14. The `result = String.Compare ("Dallas", "Dallas")` statement shown in the first example compares the string "Dallas" with the string "Dallas". Because both strings are equal, the String.Compare method returns the number 0, which the statement assigns to the `result` variable. The second example is identical to the first example, except the *string2* argument in the String.Compare method is capitalized. In this case, the String.Compare method returns the number –1, because the second character in *string1* (a) is less than the second character in *string2* (A). The third example is almost identical to the second example, except the String.Compare method contains the Boolean value True in its *ignoreCase* argument. In this case, the String.Compare method returns the number 0 because, ignoring case, both strings are equal. The `result = String.Compare("Dallas", "Boston")` statement shown in the last example compares the string "Dallas" with the string "Boston". In this example, the String.Compare method returns the number 1, because the first character in *string1* (D) is greater than the first character in *string2* (B).

You now have completed Lesson A. You can either take a break or complete the end-of-lesson questions and exercises before moving on to Lesson B.

SUMMARY

TO MANIPULATE STRINGS IN VISUAL BASIC:

» Use one of the techniques listed in Figure 8-15.

Technique	Syntax	Purpose
Length property	*string*.**Length**	determines the number of characters in a string
Trim method	*string*.**Trim**[(*trimChars*)]	removes characters from both the beginning and end of a string
TrimStart method	*string*.**TrimStart**[(*trimChars*)]	removes characters from the beginning of a string
TrimEnd method	*string*.**TrimEnd**[(*trimChars*)]	removes characters from the end of a string
Remove method	*string*.**Remove**(*startIndex*, *count*)	removes characters from anywhere in a string
Replace method	*string*.**Replace**(*oldValue*, *newValue*)	replaces all occurrences of a sequence of characters in a string with another sequence of characters
Mid statement	**Mid**(*targetString*, *start* [, *count*]) = *replacementString*	replaces a specific number of characters in a string with characters from another string
PadLeft method	*string*.**PadLeft**(*length*[, *character*])	pads the beginning of a string with a character until the string is a specified length
PadRight method	*string*.**PadRight**(*length*[, *character*])	pads the end of a string with a character until the string is a specified length
Insert method	*string*.**Insert**(*startIndex*, *value*)	inserts characters within a string

Figure 8-15: String manipulation techniques *(Continued)*   ▶

Technique	Syntax	Purpose
StartsWith method	*string*.**StartsWith**(*subString*)	determines whether a string begins with a specific sequence of characters
EndsWith method	*string*.**EndsWith**(*subString*)	determines whether a string ends with a specific sequence of characters
Contains method	*string*.**Contains**(*subString*)	determines whether a string contains a specific sequence of characters; returns a Boolean value
IndexOf method	*string*.**IndexOf**(*subString*[, *startIndex*])	determines whether a string contains a specific sequence of characters; returns an integer that indicates the starting position of the characters
Substring method	*string*.**Substring**(*startIndex*[, *count*])	accesses one or more characters contained in a string
String.Compare method	**String.Compare**(*string1*, *string2*[, *ignoreCase*])	compares two strings
Like operator	*string* **Like** *pattern*	compares two strings using pattern-matching characters
=, <>, <, <=, >, >=	*string1 operator string2*	compares two strings

Figure 8-15: String manipulation techniques

QUESTIONS

1. The `amount` variable contains the string "$56.55". Which of the following statements removes the dollar sign from the variable's contents?

   a. `amount = amount.Remove("$")`

   b. `amount = amount.Remove(0, 1)`

   c. `amount = amount.TrimStart("$"c)`

   d. Both b and c.

2. The `state` variable contains the string "MI   " (the letters M and I followed by three spaces). Which of the following statements removes the three spaces from the variable's contents?

   a. `state = state.Remove(2, 3)`

   b. `state = state.Remove(3, 3)`

   c. `state = state.TrimEnd(2, 3)`

   d. Both a and c.

3. Which of the following statements removes any dollar signs and percent signs from the beginning and end of the string stored in the `amount` variable?

   a. `amount = amount.Trim("$"c, "%"c)`

   b. `amount = amount.Trim("$, %"c)`

   c. `amount = amount.TrimAll("$"c, "%"c)`

   d. `amount = amount.TrimAll("$, %"c)`

4. Which of the following methods can be used to determine whether the string stored in the `partNum` variable begins with the letter A?

   a. `partNum.BeginsWith("A")`   b. `partNum.Starts("A")`

   c. `partNum.StartsWith("A")`   d. `partNum.StartsWith = "A"`

5. Which of the following methods can be used to determine whether the string stored in the `partNum` variable ends with either the letter B or the letter b?

   a. `partNum.Ends("B, b")`

   b. `partNum.Ends("B", "b")`

   c. `partNum.EndsWith("B", "b")`

   d. `partNum.ToUpper.EndsWith("B")`

6. Which of the following statements assigns the first three characters in the `partNum` variable to the `code` variable?

   a. `code = partNum.Assign(0, 3)`

   b. `code = partNum.Sub(0, 3)`

   c. `code = partNum.Substring(0, 3)`

   d. `code = partNum.Substring(1, 3)`

7. The word variable contains the string "Bells". Which of the following statements changes the contents of the word variable to "Bell"?

a. `word = word.Remove(word.Length - 1, 1)`

b. `word = word.Substring(0, word.Length - 1)`

c. `word = word.Replace("s", "")`

d. All of the above.

8. Which of the following statements changes the contents of the zip variable from "60121" to "60323"?

a. `Replace(zip, "1", "3")`

b. `zip.Replace("1", "3")`

c. `zip = zip.Replace("1", "3")`

d. `zip = zip.Replace("3", "1")`

9. Which of the following methods can be used to determine whether the amount variable contains the dollar sign?

a. `amount.Contains("$")`   b. `amount.IndexOf("$")`

c. `amount.IndexOf("$", 0)`   d. All of the above.

10. Which of the following statements changes the contents of the zip variable from "60537" to "60536"?

a. `Mid(zip, "7", "6")`   b. `Mid(zip, 4, "6")`

c. `zip = Mid(zip, 4, "6")`   d. None of the above.

11. Which of the following statements changes the contents of the word variable from "men" to "mean"?

a. `word = word.AddTo(2, "a")`

b. `word = word.Insert(2, "a")`

c. `word = word.Insert(3, "a")`

d. `word = word.Replace(2, "a")`

12. When the `message` variable contains the string "Happy holidays", the `message.IndexOf("day")` method returns _____.

    a. –1

    b. 0

    c. 10

    d. 11

13. Which of the following statements can be used to assign the fifth character in the `word` variable to the `letter` variable?

    a. `letter = word.Substring(4)`

    b. `letter = word.Substring(5, 1)`

    c. `letter = word(5).Substring`

    d. None of the above.

14. Which of the following statements assigns the nine characters contained in the `message` variable, followed by four exclamation points (!), to the `newMessage` variable?

    a. `newMessage = message.PadLeft(4, "!"c)`

    b. `newMessage = message.PadLeft(13, "!")`

    c. `newMessage = message.PadRight(4, "!")`

    d. `newMessage = message.PadRight(13, "!"c)`

15. The `String.Compare(city, "Paris", True)` method returns _____ when the `city` variable contains the string "PARIS".

    a. True

    b. False

    c. 0

    d. 1

EXERCISES

1. Write the Visual Basic statement that:

    a. displays in the xSizeLabel the number of characters contained in the `message` variable.

    b. removes the leading spaces from the `city` variable.

    c. removes the leading and trailing spaces from a String variable named `number`.

    d. removes any trailing spaces, commas, and periods from a String variable named `amount`.

e. uses the Remove method to remove the first two characters from the `fullName` variable.

f. uses the Mid statement to change the contents of the `word` variable from "mouse" to "mouth".

g. uses the Insert method to change the contents of the `word` variable from "mend" to "amend".

h. changes the contents of the `pay` variable from "235.67" to "****235.67".

2. Write the Visual Basic code that uses the EndsWith method to determine whether the string stored in the `inputRate` variable ends with the percent sign. If it does, the code should use the TrimEnd method to remove the percent sign from the variable's contents.

3. The `partNum` variable contains the string "ABCD34G". Write the Visual Basic statement that assigns the number 34 in the `partNum` variable to the `code` variable.

4. The `amount` variable contains the string "3,123,560". Write the Visual Basic statement that assigns the contents of the variable, excluding the commas, to the `amount` variable.

5. Write the Visual Basic statement that uses the IndexOf method to determine whether the `address` variable contains the street name "Elm Street" (entered in uppercase, lowercase, or a combination of uppercase and lowercase). Begin the search with the first character in the `address` variable, and assign the method's return value to an Integer variable named `indexNum`.

6. Write the Visual Basic statement that uses the Contains method to determine whether the `address` variable contains the street name "Elm Street" (entered in uppercase, lowercase, or a combination of uppercase and lowercase). Assign the method's return value to a Boolean variable named `isContained`.

7. Write the Visual Basic statement that uses the String. Compare method to determine whether the string stored in the `item` variable is equal to the string stored in the `itemOrdered` variable. Assign the method's return value to an Integer variable named `result`.

8. In this exercise, you complete an application that removes any leading and/or trailing spaces from a city name before adding the name to a combo box.

   a. If necessary, start Visual Studio 2005 or Visual Basic 2005 Express Edition. Open the City Names Solution (City Names Solution.sln) file, which is contained in the VB2005\Chap08\City Names Solution folder. If necessary, open the designer window. The interface allows the user to enter a city name. Code the Add Name button's Click event procedure so that it removes any leading and/or trailing spaces from the city name before adding the name to the combo box.

   b. Save the solution, then start the application. Test the application by entering spaces before and after the following city names: New York and Miami. Click the Exit button to end the application. Close the Code Editor window, then close the solution.

9. In this exercise, you complete an application that removes the dashes from a Social Security number.

   a. If necessary, start Visual Studio 2005 or Visual Basic 2005 Express Edition. Open the Social Security Solution (Social Security Solution.sln) file, which is contained in the VB2005\Chap08\Social Security Solution folder. If necessary, open the designer window. The interface allows the user to enter a Social Security number. Code the Remove Dashes button's Click event procedure so that it first verifies that the Social Security number contains exactly 11 characters. It then should remove the dashes from the Social Security number before displaying the number in the xNumberLabel. Use the Remove method to remove the dashes.

   b. Save the solution, then start the application. Test the application using the following data: 123-45 and 123-45-6789. Click the Exit button to end the application. Close the Code Editor window, then close the solution.

10. In this exercise, you modify the application from Exercise 9.

   a. Use Windows to make a copy of the Social Security Solution folder, which is contained in the VB2005\Chap08 folder. Rename the folder Modified Social Security Solution.

   b. If necessary, start Visual Studio 2005 or Visual Basic 2005 Express Edition. Open the Social Security Solution (Social Security Solution.sln) file contained in the VB2005\Chap08\Modified Social Security Solution folder. Open the designer window. Modify the Remove Dashes button's Click event procedure so that it removes the dashes using the Replace method rather than the Remove method.

c. Save the solution, then start the application. Test the application using the following data: 123-45 and 123-45-6789. Click the Exit button to end the application. Close the Code Editor window, then close the solution.

11. In this exercise, you complete an application that changes an area code.

a. If necessary, start Visual Studio 2005 or Visual Basic 2005 Express Edition. Open the Phone Solution (Phone Solution.sln) file, which is contained in the VB2005\Chap08\Phone Solution folder. If necessary, open the designer window. The interface allows the user to enter a phone number. Code the Change Area Code button's Click event procedure so that it uses the Mid statement to change the area code to 877 before displaying the phone number in the xNewLabel.

b. Save the solution, then start the application. Test the application using the following phone number: 123-456-7899. Click the Exit button to end the application. Close the Code Editor window, then close the solution.

12. In this exercise, you complete an application that right-aligns the numbers listed in a combo box.

a. If necessary, start Visual Studio 2005 or Visual Basic 2005 Express Edition. Open the Item Prices Solution (Item Prices Solution.sln) file, which is contained in the VB2005\Chap08\Item Prices Solution folder. If necessary, open the designer window. Modify the MainForm's Load event procedure so that it right-aligns the prices listed in the xRightComboBox.

b. Save the solution, then start the application. The prices listed in the xRightComboBox should be right-aligned. (The prices listed in the xLeftComboBox should still be left-aligned.) Click the Exit button to end the application. Close the Code Editor window, then close the solution.

13. In this exercise, you complete an application that changes the year number in a date from *yy* to 20*yy*.

a. If necessary, start Visual Studio 2005 or Visual Basic 2005 Express Edition. Open the Date Solution (Date Solution.sln) file, which is contained in the VB2005\Chap08\Date Solution folder. If necessary, open the designer window. The interface allows the user to enter a date. Code the Change Date button's Click event procedure so that it uses the Insert method to change the year number from *yy* to *20yy* before displaying the year number in the xDateLabel.

b. Save the solution, then start the application. Test the application using the following date: 11/15/07. Click the Exit button to end the application. Close the Code Editor window, then close the solution.

14. In this exercise, you complete an application that removes any percent sign from the end of a number.

a. If necessary, start Visual Studio 2005 or Visual Basic 2005 Express Edition. Open the Sales Tax Solution (Sales Tax Solution.sln) file, which is contained in the VB2005\Chap08\Sales Tax Solution folder. If necessary, open the designer window. The interface allows the user to enter a sales amount and a tax rate. Code the Calculate button's Click event procedure so that it uses the EndsWith method to determine whether the tax rate ends with a percent sign. If it does, the procedure should use the TrimEnd method to remove the percent sign from the rate.

b. Save the solution, then start the application. Test the application using the following data: a sales amount of 1000 and a tax rate of 5%, and then a sales amount of 5000 and a tax rate of 7%. Click the Exit button to end the application. Close the Code Editor window, then close the solution.

15. In this exercise, you complete an application that selects the appropriate delivery method from a list box.

a. If necessary, start Visual Studio 2005 or Visual Basic 2005 Express Edition. Open the Part Number Solution (Part Number Solution.sln) file, which is contained in the VB2005\Chap08\Part Number Solution folder. If necessary, open the designer window. The interface allows the user to enter a part number, which will consist of numbers and either one or two letters. The letter or letters represent the delivery method, as follows: "MS" represents "Mail – Standard", "MP" represents "Mail – Priority", "FS" represents "FedEx – Standard", "FO" represents "FedEx – Overnight", and "U" represents "UPS". Code the Select Delivery button's Click event procedure so that it uses the Contains method to select the appropriate delivery method in the list box. Display an appropriate message when the part number contains an invalid letter or letters.

b. Save the solution, then start the application. Test the application using the following data: 73mp6, 34fs56, 2u7, 78h89, 9FO8, and 1234ms. Click the Exit button to end the application. Close the Code Editor window, then close the solution.

16. In this exercise, you complete an application that determines the number of times a sequence of characters appears in a string.

a. If necessary, start Visual Studio 2005 or Visual Basic 2005 Express Edition. Open the Count Solution (Count Solution.sln) file, which is contained in the VB2005\Chap08\Count Solution folder. If necessary, open the designer window. The interface allows the user to enter a string. Code the Search button's Click event procedure so that it prompts the user to enter the sequence of characters for which he or she wants to search. The procedure should determine the number of times the sequence of characters appears in the string. Use the IndexOf method to search the string for the sequence of characters.

b. Save the solution, then start the application. Test the application by entering the string "The weather is beautiful!" (without the quotes), then clicking the Search button. Search for the two characters "ea" (without the quotes). The two characters appear twice in the string. Click the Exit button to end the application. Close the Code Editor window, then close the solution.

17. In this exercise, you complete an application that accesses only the first name from a person's full name.

a. If necessary, start Visual Studio 2005 or Visual Basic 2005 Express Edition. Open the Name Solution (Name Solution.sln) file, which is contained in the VB2005\Chap08\Name Solution folder. If necessary, open the designer window. The interface allows the user to enter a person's full name (first name followed by a space and his or her last name). Code the Display First Name button's Click event procedure so that it displays only the person's first name in the xFirstLabel.

b. Save the solution, then start the application. Test the application by entering your full name, and then clicking the Display First Name button. Your first name appears in the xFirstLabel. Click the Exit button to end the application. Close the Code Editor window, then close the solution.

18. In this exercise, you complete an application that displays a message.

a. If necessary, start Visual Studio 2005 or Visual Basic 2005 Express Edition. Open the Alphabet Solution (Alphabet Solution.sln) file, which is contained in the VB2005\Chap08\Alphabet Solution folder. If necessary, open the designer window. The interface allows the user to enter two letters of the alphabet. Code the Compare button's Click event procedure so that it compares the two letters and displays one of three messages in the xMessageLabel. When both letters are the same, display the "Both letters are the same." message. When the first letter comes after the second letter in the alphabet, display the "The letter *first* comes after the letter *second* in the alphabet." When the first letter comes before the second letter in the alphabet, display the "The letter *first* comes before the letter *second* in the alphabet." In the messages, *first* is the letter entered in the First letter text box, and *second* is the letter entered in the Second letter text box. Ignore the case when comparing the letters.

b. Save the solution, then start the application. Test the application by entering the letters n and B. The message "The letter n comes after the letter B in the alphabet." appears in the xMessageLabel. Test the application using different letters, then click the Exit button. Close the Code Editor window, then close the solution.

LESSON B
OBJECTIVES

AFTER STUDYING LESSON B, YOU SHOULD
BE ABLE TO:

» Include a MenuStrip control on a form

» Add elements to a menu

» Assign access keys to menu elements

» Assign shortcut keys to commonly used menu items

» Code a menu item's Click event procedure

ADDING A MENU TO A FORM

COMPLETING THE HANGMAN GAME APPLICATION'S USER INTERFACE

Recall that your task in this chapter is to create a simplified version of the Hangman game for Mr. Mitchell, who teaches second grade at Hinsbrook School. The VB2005\Chap08\Hangman Game Solution folder contains a partially completed Hangman game application. You complete the application's user interface in this lesson, and also begin coding the application.

To open the Hangman Game application:

1 Start Visual Studio 2005 or Visual Basic 2005 Express Edition, if necessary, and close the Start Page window.

2 Open the **Hangman Game Solution** (Hangman Game Solution.sln) file, which is contained in the VB2005\Chap08\Hangman Game Solution folder.

3 Permanently display the Toolbox window, if necessary.

4 Auto-hide the Solution Explorer and Properties windows, if necessary. Figure 8-16 shows the partially completed user interface for the Hangman Game application. The interface contains four label controls, 10 picture boxes, and a panel control. You create a panel control using the Panel tool, which is located in the Containers section of the toolbox. You will complete the interface by adding a menu to it.

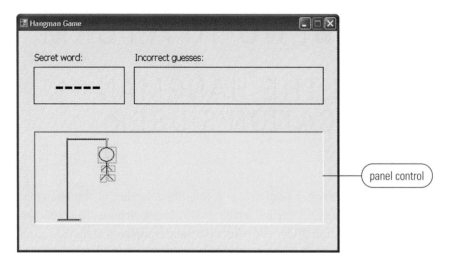

Figure 8-16: Partially completed interface for the Hangman Game application

CREATING MENUS

In Visual Basic 2005, you use a **MenuStrip control** to include one or more menus in an application. You instantiate a MenuStrip control using the **MenuStrip tool**, which is located in the Menus & Toolbars section of the toolbox. Each menu in an application contains a **menu title**, which appears on the menu bar at the top of a Windows form. When you click a menu title, its corresponding menu opens and displays a list of options, called **menu items**. The menu items can be commands (such as Open or Exit), separator bars, or submenu titles. As in all Windows applications, clicking a command on a menu executes the command, and clicking a submenu title opens an additional menu of options. Each of the options on a submenu is referred to as a **submenu item**. The purpose of a **separator bar** is to visually group together the related items on a menu or submenu. Figure 8-17 identifies the location of these menu elements.

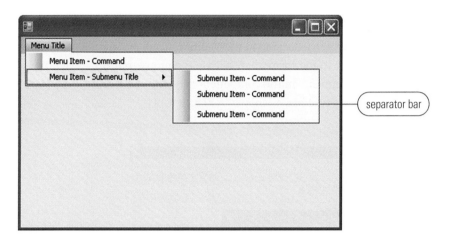

Figure 8-17: Location of menu elements

Each menu element is considered an object and has a set of properties associated with it. The most commonly used properties for a menu element are the Name and Text properties. The programmer uses the Name property to refer to the menu element in code. The Text property, on the other hand, stores the menu element's caption, which is the text that the user sees when he or she is working with the menu. The caption indicates the purpose of the menu element. Examples of familiar captions for menu elements include Edit, Save As, Copy, and Exit.

Menu title captions should be one word only, with the first letter capitalized. Each menu title should have a unique access key. The access key allows the user to open the menu by pressing the Alt key in combination with the access key.

Unlike the captions for menu titles, the captions for menu items can be from one to three words. Use book title capitalization for the menu item captions, and assign each menu item a unique access key. The access key allows the user to select the item simply by pressing the access key when the menu is open. If a menu item requires additional information from the user, the Windows standard is to place an ellipsis (...) at the end of the caption. The ellipsis alerts the user that the menu item requires more information before it can perform its task.

The menus included in your application should follow the standard Windows conventions. For example, if your application uses a File menu, it should be the first menu on the menu bar. File menus typically contain commands for opening, saving, and printing files, as well as for exiting the application. If your application requires Cut, Copy, and Paste commands, the commands should be placed on an Edit menu, which typically is the second menu on the menu bar.

To include a menu in the Hangman Game interface:

1 Click the **MenuStrip** tool, which is located in the Menus & Toolbars section of the toolbox. Drag the mouse pointer to the form, then release the mouse button. A MenuStrip control named MenuStrip1 appears in the component tray, and the words "Type Here" appear in a box on the form's title bar, as shown in Figure 8-18.

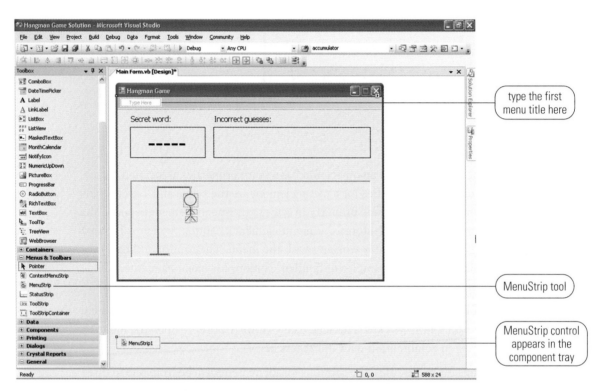

Figure 8-18: MenuStrip control added to the application

2 Auto-hide the Toolbox window, then temporarily display the Properties window. Click **(Name)** in the Properties list, and then type **xMainMenuStrip** and press **Enter**.

You will include only a File menu title on the form's menu bar. The File menu will contain three menu items: a New Game command, a separator bar, and an Exit command.

To add the File menu title and three menu items:

1 Click the **Type Here** box on the menu bar, then type **&File**. See Figure 8-19. Notice that a Type Here box appears below the File menu title and also to the right of the menu title. The Type Here box that appears below the menu title allows you to add a menu item to the File menu. The Type Here box that appears to the right of the menu title allows you to add another menu title to the menu bar.

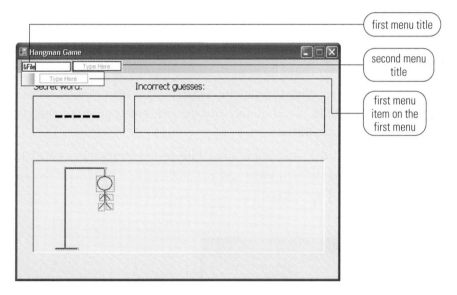

Figure 8-19: Menu title shown in the interface

2 Press **Enter**, then click the **File** menu title. Temporarily display the Properties window. Scroll the Properties window, if necessary, until you see the Text property. Notice that the property contains &File.

3 Scroll to the top of the Properties window, then click **(Name)**. Change the menu title's name to **xFileMenuTitle** and press **Enter**.

4 Click the **Type Here** box that appears in the row below the File menu title, then type **&New Game** and press **Enter**. Click the **New Game** menu item. Change the menu item's name to **xFileNewMenuItem** and press **Enter**.

5 Next, you will add a separator bar to the File menu. Place your mouse pointer on the Type Here box that appears below the New Game menu item, then click the **list arrow** that appears in the box. A drop-down list appears, as shown in Figure 8-20.

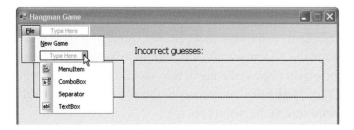

Figure 8-20: Drop-down list

6 Click **Separator** in the list. A separator bar appears below the New Game menu item. Click the **separator bar**, then change its name to **xFileSeparator** and press **Enter**.

7 Click the **Type Here** box that appears below the separator bar, then type **E&xit** and press **Enter**. Click the **Exit** menu item. Change the menu item's name to **xFileExitMenuItem** and press **Enter**.

8 Save the solution, then start the application. Click **File** on the form's menu bar. The File menu opens and displays the New Game and Exit options, which are separated by a separator bar, as shown in Figure 8-21.

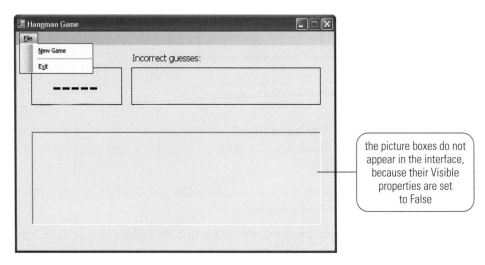

the picture boxes do not appear in the interface, because their Visible properties are set to False

Figure 8-21: Open File menu

9 Click the **Close** button on the form's title bar. The application closes and you are returned to the designer window.

ASSIGNING SHORTCUT KEYS TO MENU ITEMS

Shortcut keys appear to the right of a menu item and allow you to select an item without opening the menu. For example, in Windows applications you can select the Save command when the File menu is closed by pressing Ctrl + s. You should assign shortcut keys only to commonly used menu items. In the Hangman Game application, you will assign a shortcut key to the New Game option on the File menu.

To assign a shortcut key to the New Game menu item:

1 Click the **New Game** menu item on the File menu. Click **ShortcutKeys** in the Properties window, then click the **list arrow** in the Settings box. A box opens and allows you to specify the shortcut key's modifiers and key. You will assign Ctrl + n as the shortcut key. To do this, you need to specify Ctrl as the modifier and N as the key.

2 Click the **Ctrl** check box to select it. Click the **list arrow** that appears in the Key section. A drop-down list opens. Scroll the list until you see the letter N, then click **N** in the list. See Figure 8-22.

Figure 8-22: Shortcut key specified in the ShortcutKeys box

3 Press **Enter**. Ctrl + N appears in the ShortcutKeys property in the Properties window. It also appears to the right of the New Game menu item on the File menu.

4 Save the solution, then start the application. Click **File** on the Hangman Game application's menu bar. The Ctrl + N shortcut key appears to the right of the New Game menu item, as shown in Figure 8-23.

TIP
A menu item's access key can be used only when the menu is open, whereas the shortcut key can be used only when the menu is closed.

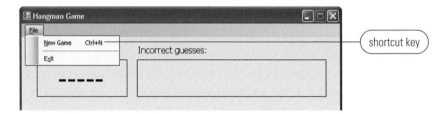

Figure 8-23: Shortcut key appears next to the New Game menu item

5 Click the **Close** button on the Hangman Game application's title bar. The application closes and you are returned to the designer window.

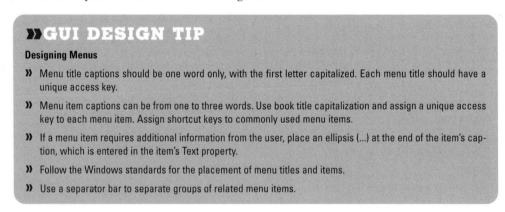

>> **GUI DESIGN TIP**

Designing Menus

» Menu title captions should be one word only, with the first letter capitalized. Each menu title should have a unique access key.

» Menu item captions can be from one to three words. Use book title capitalization and assign a unique access key to each menu item. Assign shortcut keys to commonly used menu items.

» If a menu item requires additional information from the user, place an ellipsis (...) at the end of the item's caption, which is entered in the item's Text property.

» Follow the Windows standards for the placement of menu titles and items.

» Use a separator bar to separate groups of related menu items.

Now that the interface is complete, you can begin coding the application. In this lesson, you will code only the Exit option's Click event procedure. You will code the New Game option's Click event procedure in Lesson C.

CODING THE EXIT OPTION'S CLICK EVENT PROCEDURE

When the user clicks the Exit option on the File menu, the option's Click event procedure should end the Hangman Game application.

To code the Exit option's Click event procedure, then test the procedure:

1 Open the Code Editor window. Replace the `<your name>` and `<current date>` text in the comments with your name and the current date.

2 Open the code template for the xFileExitMenuItem's Click event procedure. Type **me.close()** and press **Enter**. See Figure 8-24.

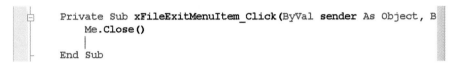

```
Private Sub xFileExitMenuItem_Click(ByVal sender As Object, B
    Me.Close()

End Sub
```

Figure 8-24: Completed xFileExitMenuItem's Click event procedure

3 Save the solution, then start the application. Click **File** on the Hangman Game application's menu bar, and then click **Exit** to end the application.

4 Close the Code Editor window, then close the solution.

You have completed Lesson B. You can either take a break or complete the end-of-lesson questions and exercises before moving on to Lesson C.

SUMMARY

TO ADD A MENUSTRIP CONTROL TO A FORM:
» Use the MenuStrip tool in the toolbox.

TO CREATE A MENU:
» Replace the words "Type Here" with the menu element's caption. Assign a meaningful name to each menu element. Assign a unique access key to each menu element, excluding menu items that are separator bars.

TO INCLUDE A SEPARATOR BAR ON A MENU:
» Place your mouse pointer on a Type Here box, then click the list arrow that appears in the box. Click Separator on the list.

TO ASSIGN A SHORTCUT KEY TO A MENU ITEM:
» Set the menu item's ShortcutKeys property.

QUESTIONS

1. The horizontal line in a menu is called _____.

    a. a menu bar   b. a separator bar

    c. an item separator   d. None of the above.

2. The underlined letter in a menu element's caption is called _____.

    a. an access key   b. a menu key

    c. a shortcut key   d. None of the above.

3. _____ allows you to access a menu item without opening the menu.

    a. An access key   b. A menu key

    c. A shortcut key   d. None of the above.

4. Which of the following is false?

    a. Menu titles should be one word only.

    b. Each menu title should have a unique access key.

    c. You should assign a shortcut key to commonly used menu titles.

    d. Menu items should be entered using book title capitalization.

5. Explain the difference between a menu item's access key and its shortcut key.

EXERCISES

1. In this exercise, you include a menu in an application.

    a. If necessary, start Visual Studio 2005 or Visual Basic 2005 Express Edition. Open the Commission Solution (Commission Solution.sln) file, which is contained in the VB2005\Chap08\Commission Solution folder. If necessary, open the designer window.

    b. Add a File menu and a Calculate menu to the application. Include an Exit menu item on the File menu. Include two menu items on the Calculate menu: 2% Commission and 5% Commission.

COMPLETING THE HANGMAN GAME APPLICATION

THE HANGMAN GAME APPLICATION

To complete the Hangman Game application, you need to code the New Game option's Click event procedure. The procedure should allow a student to enter a five-letter word, and then allow another student to guess the word, letter by letter. The game is over when the second student guesses all of the letters in the word, or when he or she makes 10 incorrect guesses, whichever comes first.

To open the Hangman Game application:

1 Start Visual Studio 2005 or Visual Basic 2005 Express Edition, if necessary, and close the Start Page window.

2 Open the **Hangman Game Solution** (Hangman Game Solution.sln) file, which is contained in the VB2005\Chap08\Hangman Game Solution folder.

3 Auto-hide the Toolbox, Solution Explorer, and Properties windows, if necessary. Figure 8-25 shows the TOE chart for the Hangman Game application.

Task	Object	Event
1. Hide the 10 picture boxes 2. Get a five-letter word from player 1 3. Display five dashes in the xWordLabel 4. Clear the xIncorrectLabel 5. Get a letter from player 2 6. Search the word for the letter 7. If the letter is contained in the word, replace the appropriate dash(es) 8. If the letter is not contained in the word, display the letter in the xIncorrectLabel, add 1 to the number of incorrect guesses, and then show the appropriate picture box 9. If all of the dashes have been replaced, the game is over, so display the message "Great guessing!" in a message box 10. If the user makes 10 incorrect guesses, the game is over, so display an appropriate message and the word in a message box	xFileNewMenuItem	Click
End the application	xFileExitMenuItem	Click
Display the Hangman images	xBottomPictureBox, xPostPictureBox, xTopPictureBox, xRopePictureBox, xHeadPictureBox, xBodyPictureBox, xRightArmPictureBox, xLeftArmPictureBox, xRightLegPictureBox, xLeftLegPictureBox	None
Display dashes and letters (from xFileNewMenuItem)	xWordLabel	None
Display the incorrect letters (from xFileNewMenuItem)	xIncorrectLabel	None

Figure 8-25: TOE chart for the Hangman Game application

CODING THE XFILENEWMENUITEM'S CLICK EVENT PROCEDURE

Each time the user wants to begin a new Hangman game, he or she will need to either press Ctrl + n or click File on the application's menu bar and then click New Game. Figure 8-26 shows the pseudocode for the xFileNewMenuItem's Click event procedure.

xFileNewMenuItem Click event procedure - pseudocode

1. hide the 10 picture boxes
2. repeat
      get a 5-letter word from player 1
  end repeat until the word contains 5 letters
3. convert the word to uppercase
4. display 5 dashes in the xWordLabel
5. clear the xIncorrectLabel
6. repeat until the game is over
      get a letter from player 2, then convert the letter to uppercase

      repeat for each letter in the word
          if the current letter is the same as the letter entered by player 2
              replace the appropriate dash in the xWordLabel
              assign True to the isDashReplaced variable
          end if
      end repeat

      if a dash was replaced in the xWordLabel
          if the xWordLabel does not contain any more dashes
              assign True to the isGameOver variable
              display the "Great guessing!" message in a message box
          else
              assign False to the isDashReplaced variable
          end if
      else
          display the incorrect letter in the xIncorrectLabel
          add 1 to the number of incorrect guesses counter

          value of the number of incorrect guesses counter:
              1   show the xBottomPictureBox
              2   show the xPostPictureBox
              3   show the xTopPictureBox
              4   show the xRopePictureBox
              5   show the xHeadPictureBox
              6   show the xBodyPictureBox
              7   show the xRightArmPictureBox
              8   show the xLeftArmPictureBox
              9   show the xRightLegPictureBox
             10 show the xLeftLegPictureBox
                 assign True to the isGameOver variable
                 display the "Sorry, the word is" message and the word in a message box
      end if
  end repeat

Figure 8-26: Pseudocode for the xFileNewMenuItem's Click event procedure

To code the xFileNewMenuItem's Click event procedure:

1 Open the Code Editor window, then open the code template for the xFileNewMenuItem's Click event procedure. Type **' simulates the Hangman game** and press **Enter** twice.

The procedure will use five variables. The `word` variable will store the word entered by the first player. The `letter` variable will store the current letter entered by the second player. The `isDashReplaced` variable will keep track of whether a dash was replaced in the word, and the `isGameOver` variable will indicate whether the game is over. The `incorrectCounter` variable will keep track of the number of incorrect guesses made by the second player.

2 Enter the following Dim statements. Press **Enter** twice after typing the last Dim statement.

dim word as string

dim letter as string

dim isDashReplaced as boolean

dim isGameOver as boolean

dim incorrectCounter as integer

3 The first step in the pseudocode is to hide the 10 picture boxes. Enter the following comment and assignment statements. Press **Enter** twice after typing the last assignment statement.

' hide the picture boxes

me.xBottomPictureBox.Visible = false

me.xPostPictureBox.Visible = false

me.xTopPictureBox.Visible = false

me.xRopePictureBox.Visible = false

me.xHeadPictureBox.Visible = false

me.xBodyPictureBox.Visible = false

me.xRightArmPictureBox.Visible = false

me.xLeftArmPictureBox.Visible = false

me.xRightLegPictureBox.Visible = False

me.xLeftLegPictureBox.Visible = false

4 The second step in the pseudocode is a posttest repetition structure that repeats its one instruction until the user enters a word that contains five characters. Type the comment and repetition structure shown in Figure 8-27, then position the insertion point as shown in the figure.

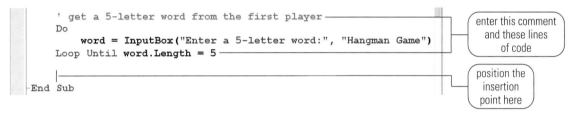

```
' get a 5-letter word from the first player
Do
    word = InputBox("Enter a 5-letter word:", "Hangman Game")
Loop Until word.Length = 5

|
End Sub
```

enter this comment and these lines of code

position the insertion point here

Figure 8-27: Comment and posttest repetition structure entered in the procedure

5 The third step in the pseudocode is to convert to uppercase the word entered by the first player. Type **' convert the word to uppercase** and press **Enter**. Type **word = word.toupper** and press **Enter** twice.

6 Steps 4 and 5 in the pseudocode are to display five dashes in the xWordLabel and then clear the xIncorrectLabel. Enter the following comments and assignment statements. (The first assignment statement assigns five hyphens to the xWordLabel.) Press **Enter** twice after typing the last assignment statement.

' display five dashes in the xWordLabel

' and clear the xIncorrectLabel

me.xWordLabel.text = "-----"

me.xIncorrectLabel.text = string.empty

7 Step 6 in the pseudocode is a pretest repetition structure that repeats its instructions until the game is over. Enter the following comments and Do clause.

' allow the second player to guess a letter

' the game is over when the second player

' either guesses the word or makes 10

' incorrect guesses

do until isgameover

8 The first task in the repetition structure is to get a letter from the second player, and then convert the letter to uppercase. Enter the following comments and assignment statements. Press **Enter** twice after typing the last assignment statement.

' get a letter from the second player, then

' convert the letter to uppercase

letter = inputBox("Enter a letter:", _

   "Letter", " ", 450, 400)

letter = letter.toupper

9 The next task in the repetition structure is a nested repetition structure that repeats its instructions for each letter in the word. Type **' search the word for the letter** and press **Enter**, then type **for indexnum as integer = 0 to 4** and press **Enter**. Change the Next clause to **Next indexNum**.

10 The nested repetition structure contains a selection structure that compares the current letter in the word with the letter entered by the second player. If both letters are the same, the selection structure's true path replaces the appropriate dash in the word. It also assigns the Boolean value True to the isDashReplaced variable to indicate that a dash was replaced in the word. Type the comments and selection structure shown in Figure 8-28, then position the insertion point as shown in the figure.

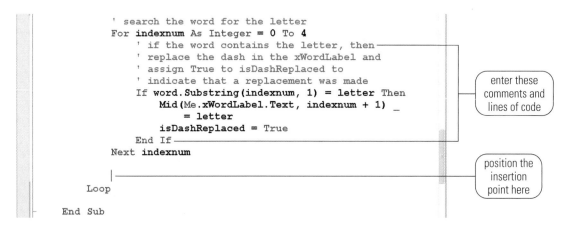

```
' search the word for the letter
For indexnum As Integer = 0 To 4
    ' if the word contains the letter, then
    ' replace the dash in the xWordLabel and
    ' assign True to isDashReplaced to
    ' indicate that a replacement was made
    If word.Substring(indexnum, 1) = letter Then
        Mid(Me.xWordLabel.Text, indexnum + 1) _
            = letter
        isDashReplaced = True
    End If
Next indexnum

    |
Loop
End Sub
```

enter these comments and lines of code

position the insertion point here

Figure 8-28: Comments and selection structure entered in the procedure

11 The next task in the outer repetition structure is a selection structure that determines whether a dash was replaced in the word. Type **' determine whether a dash was replaced** and press **Enter**, then type **if isdashreplaced then** and press **Enter**.

If a dash was replaced in the word, the selection structure's true path uses a nested selection structure to determine whether the word contains any more dashes. If all of the dashes have been replaced in the word, the game is over and the nested selection structure's true path assigns the Boolean value True to the isGameOver variable. It also displays the "Great guessing!" message in a message box. Otherwise, the nested selection structure's false path assigns the Boolean value False to the isDashReplaced variable; this resets the variable for the next search.

12 Type the comments and selection structure shown in Figure 8-29, then position the insertion point as shown in the figure.

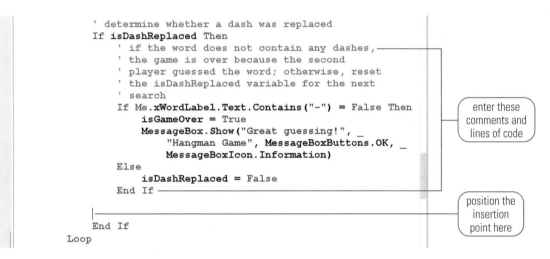

```
' determine whether a dash was replaced
If isDashReplaced Then
    ' if the word does not contain any dashes,
    ' the game is over because the second
    ' player guessed the word; otherwise, reset
    ' the isDashReplaced variable for the next
    ' search
    If Me.xWordLabel.Text.Contains("-") = False Then
        isGameOver = True
        MessageBox.Show("Great guessing!", _
            "Hangman Game", MessageBoxButtons.OK, _
            MessageBoxIcon.Information)
    Else
        isDashReplaced = False
    End If

End If
Loop
```

enter these comments and lines of code

position the insertion point here

Figure 8-29: Comments and nested selection structure entered in the procedure

If a dash was not replaced in the word, it means that the letter entered by the second player does not appear in the word. In that case, the outer selection structure's false path displays the incorrect letter in the xIncorrectLabel, and also adds the number one to the contents of the incorrectCounter variable. In addition, it uses a selection structure and the value stored in the incorrectCounter variable to determine which of the 10 picture boxes to display. If the incorrectCounter variable contains the number 10, the game is over because the second player reached the maximum number of incorrect guesses allowed. In that case, the selection structure assigns the Boolean value True to the isGameOver variable and also displays (in a message box) an appropriate message along with the word.

13 Type the comments and additional lines of code shaded in Figure 8-30, which shows the code for the Hangman Game application.

```
' Project name:      Hangman Game Project
' Project purpose:   The project simulates the Hangman game.
' Created/revised:   <your name> on <current date>

Option Explicit On
Option Strict On

Public Class MainForm
    Private Sub xFileExitMenuItem_Click(ByVal sender As Object, _
        ByVal e As System.EventArgs) Handles xFileExitMenuItem.Click
        Me.Close()
    End Sub

    Private Sub xFileNewMenuItem_Click(ByVal sender As Object, _
        ByVal e As System.EventArgs) Handles xFileNewMenuItem.Click
        ' simulates the Hangman game

        Dim word As String
        Dim letter As String
        Dim isDashReplaced As Boolean
        Dim isGameOver As Boolean
        Dim incorrectCounter As Integer

        ' hide the picture boxes
        Me.xBottomPictureBox.Visible = False
        Me.xPostPictureBox.Visible = False
        Me.xTopPictureBox.Visible = False
        Me.xRopePictureBox.Visible = False
        Me.xHeadPictureBox.Visible = False
        Me.xBodyPictureBox.Visible = False
```

Figure 8-30: Hangman Game application's code *(Continued)*   ▶

```
Me.xRightArmPictureBox.Visible = False
Me.xLeftArmPictureBox.Visible = False
Me.xRightLegPictureBox.Visible = False
Me.xLeftLegPictureBox.Visible = False
' get a 5-letter word from the first player
Do
    word = InputBox("Enter a 5-letter word:", "Hangman Game")
Loop Until word.Length = 5

' convert the word to uppercase
word = word.ToUpper

' display five dashes in the xWordLabel
' and clear the xIncorrectLabel
Me.xWordLabel.Text = "-----"
Me.xIncorrectLabel.Text = String.Empty

' allow the second player to guess a letter
' the game is over when the second player
' either guesses the word or makes 10
' incorrect guesses
Do Until isGameOver
    ' get a letter from the second player, then
    ' convert the letter to uppercase
    letter = InputBox("Enter a letter:", _
     "Letter", "", 450, 400)
    letter = letter.ToUpper

' search the word for the letter
For indexnum As Integer = 0 To 4
    ' if the word contains the letter, then
    ' replace the dash in the xWordLabel and
    ' assign True to isDashReplaced to
    ' indicate that a replacement was made
    If word.Substring(indexnum, 1) = letter Then
        Mid(Me.xWordLabel.Text, indexnum + 1) _
            = letter
        isDashReplaced = True
    End If
Next indexnum
```

Figure 8-30: Hangman Game application's code *(Continued)*

▶

```
            ' determine whether a dash was replaced
        If isDashReplaced Then
            ' if the word does not contain any dashes,
            ' the game is over because the second
            ' player guessed the word; otherwise, reset
            ' the isDashReplaced variable for the next
            ' search
            If Me.xWordLabel.Text.Contains("-") = False Then
                isGameOver = True
                MessageBox.Show("Great guessing!", _
                    "Hangman Game", MessageBoxButtons.OK, _
                    MessageBoxIcon.Information)
            Else
                isDashReplaced = False
            End If
        Else    ' processed when no dash was replaced
            ' display the incorrect letter, then update
            ' the incorrectCounter variable, then show
            ' the appropriate picture box
            Me.xIncorrectLabel.Text = _
                Me.xIncorrectLabel.Text & " " & letter
            incorrectCounter = incorrectCounter + 1
            Select Case incorrectCounter
                Case 1
                    Me.xBottomPictureBox.Visible = True
                Case 2
                    Me.xPostPictureBox.Visible = True
                Case 3
                    Me.xTopPictureBox.Visible = True
                Case 4
                    Me.xRopePictureBox.Visible = True
                Case 5
                    Me.xHeadPictureBox.Visible = True
```

Figure 8-30: Hangman Game application's code *(Continued)*

```
        Case 6
            Me.xBodyPictureBox.Visible = True
        Case 7
            Me.xRightArmPictureBox.Visible = True
        Case 8
            Me.xLeftArmPictureBox.Visible = True
        Case 9
            Me.xRightLegPictureBox.Visible = True
        Case 10
            Me.xLeftLegPictureBox.Visible = True
            isGameOver = True
            MessageBox.Show("Sorry, the word is" _
                & word & ".", "Hangman Game", _
                MessageBoxButtons.OK, _
                MessageBoxIcon.Information)
        End Select
    End If
Loop
End Sub
End Class
```

Figure 8-30: Hangman Game application's code

Now that you have finished coding the application, you will test it to verify that it is working correctly.

To test the Hangman Game application:

1 Save the solution, then start the application. Click **File** on the application's menu bar, then click **New Game**. The Hangman Game dialog box opens.

2 Type **cat** in the Hangman Game dialog box, then press **Enter** to select the OK button. Because the word does not contain five characters, the dialog box appears again and prompts you to enter a 5-letter word.

3 Type **puppy** in the Hangman Game dialog box, then press **Enter**. The Letter dialog box opens.

4 Type **p** in the Letter dialog box, then press **Enter** to select the OK button. The xFileNewMenuItem's Click event procedure replaces, with the letter P, three of the dashes in the xWordLabel, as shown in Figure 8-31.

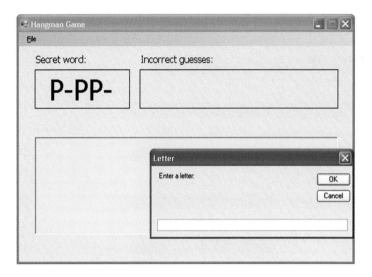

Figure 8-31: Dashes replaced with the letter P

5 Type **a** in the Letter dialog box, then press **Enter**. The xFileNewMenuItem's Click event procedure displays the letter A in the xIncorrectLabel and also makes the xBottomPictureBox visible, as shown in Figure 8-32.

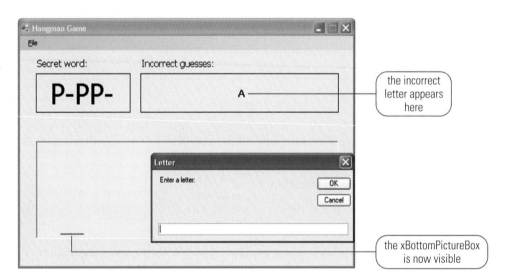

Figure 8-32: Result of entering an incorrect letter

6 Type **b** in the Letter dialog box, then press **Enter**. The xFileNewMenuItem's Click event procedure adds the letter B to the contents of the xIncorrectLabel and also makes the xPostPictureBox visible.

7 Type **u** in the Letter dialog box, then press **Enter**. Type **y** in the Letter dialog box, then press **Enter**. The xFileNewMenuItem's Click event procedure displays the "Great guessing!" message in a message box, as shown in Figure 8-33.

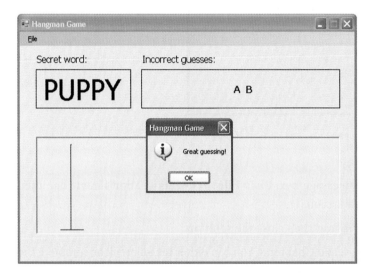

Figure 8-33: Result of guessing the word

8 Press **Enter** to close the message box.

Next, you will use the New Game option's shortcut key, which is Ctrl+N, to start a new game. You then will enter a word followed by two correct guesses and 10 incorrect guesses.

9 Press **Ctrl+n** to start a new game. Type **basic** and press **Enter**.

10 Type **c** and press **Enter**, then type **a** and press **Enter**.

11 Type the following 10 letters, pressing **Enter** after typing each letter: **d**, **e**, **f**, **g**, **h**, **j**, **k**, **x**, **y**, **z**. The xFileNewMenuItem's Click event procedure displays the 10 incorrect letters in the xIncorrectLabel. It also displays the 10 picture boxes and a message box, as shown in Figure 8-34.

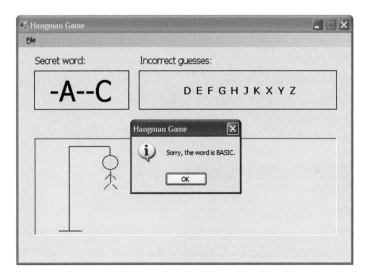

Figure 8-34: Result of making 10 incorrect guesses

12 Press **Enter** to close the message box. Click **File** on the application's menu bar, then click **Exit** to end the application.

13 Close the Code Editor window, then close the solution.

You have completed Lesson C and Chapter 8. You can either take a break or complete the end-of-lesson questions and exercises.

SUMMARY

TO ACCESS EACH CHARACTER IN A STRING:
» Use the Substring method.

TO REPLACE A CHARACTER IN A STRING WITH ANOTHER CHARACTER:
» Use the Mid statement.

TO DETERMINE WHETHER A SPECIFIC CHARACTER IS CONTAINED IN A STRING:
» Use the Contains method.

QUESTIONS

1. Which of the following For clauses can be used to access each character contained in the name variable, character by character? The variable contains 10 characters.

    a. `For indexNum As Integer = 0 To 10`

    b. `For indexNum As Integer = 0 To name.Length - 1`

    c. `For indexNum As Integer = 1 To 10`

    d. `For indexNum As Integer = 1 To name.Length - 1`

2. Which of the following can be used to change the first letter in the name variable from "K" to "C"?

    a. `Mid(name, 0, "C")`

    b. `Mid(name, 0) = "C"`

    c. `Mid(name, 1, "C")`

    d. `Mid(name, 1) = "C"`

3. If the word variable contains the string "Irene Turner", the `word.Contains("r")` method returns _____.

    a. True

    b. False

    c. 1

    d. 2

EXERCISES

1. In this exercise, you modify the Hangman Game application you created in the chapter.

    a. Use Windows to make a copy of the Hangman Game Solution folder, which is contained in the VB2005\Chap08 folder. Rename the copy Hangman Game Solution-Ex1.

    b. If necessary, start Visual Studio 2005 or Visual Basic 2005 Express Edition. Open the Hangman Game Solution (Hangman Game Solution.sln) file contained in the VB2005\Chap08\Hangman Game Solution-Ex1 folder. Open the designer window. Modify the application so that it verifies that the word entered by the first player contains only letters of the alphabet. Also verify that the character entered by the second player is a letter of the alphabet.

c. Save the solution, then start and test the application. Click Exit on the File menu to end the application. Close the Code Editor window, then close the solution.

2. In this exercise, you modify the Hangman Game application you created in the chapter.

a. Use Windows to make a copy of the Hangman Game Solution folder, which is contained in the VB2005\Chap08 folder. Rename the copy Hangman Game Solution-Ex2.

b. If necessary, start Visual Studio 2005 or Visual Basic 2005 Express Edition. Open the Hangman Game Solution (Hangman Game Solution.sln) file contained in the VB2005\Chap08\Hangman Game Solution-Ex2 folder. Open the designer window. Modify the application so it allows the first player to enter a word that contains any number of characters, up to a maximum of 10 characters.

c. Save the solution, then start and test the application. Click Exit on the File menu to end the application, then close the solution.

3. In this exercise, you create an application that allows the user to enter a word. The application should display the word in pig latin form. The rules for converting a word into pig latin form are as follows:

» If the word begins with a vowel (A, E, I, O, or U), then add the string "-way" (a dash followed by the letters w, a, and y) to the end of the word. For example, the pig latin form of the word "ant" is "ant-way".

» If the word does not begin with a vowel, first add a dash to the end of the word. Then continue moving the first character in the word to the end of the word until the first character is the letter A, E, I, O, U, or Y. Then add the string "ay" to the end of the word. For example, the pig latin form of the word "Chair" is "air-Chay".

» If the word does not contain the letter A, E, I, O, U, or Y, then add the string "-way" to the end of the word. For example, the pig latin form of "56" is "56-way".

a. Build an appropriate interface. Name the solution Pig Latin Solution. Name the project Pig Latin Project. Save the application in the VB2005\Chap08 folder.

b. Code the application. Save the solution, then start and test the application.

c. Stop the application. Close the Code Editor window, then close the solution.

4. Credit card companies typically assign a special digit, called a check digit, to the end of each customer's credit card number. Many methods for creating the check digit have been developed. One simple method is to multiply every other number in the credit card number by two, then add the products to the remaining numbers to get the total. You then take the last digit in the total and append it to the end of the number, as illustrated in Figure 8-35.

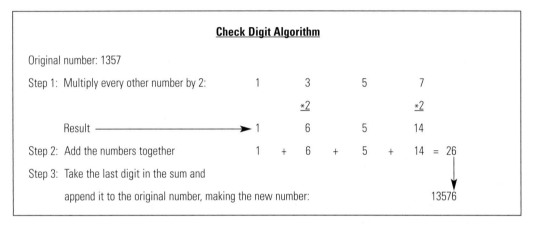

Figure: 8-35

a. If necessary, start Visual Studio 2005 or Visual Basic 2005 Express Edition. Open the Georgetown Solution (Georgetown Solution.sln) file, which is contained in the VB2005\Chap08\Georgetown Solution folder. If necessary, open the designer window.

b. Code the application so that it allows the user to enter a five-digit credit card number, with the fifth digit being the check digit. The application should use the method illustrated in Figure 8-35 to verify that the credit card number is valid. Display appropriate messages indicating whether the credit card number is valid or invalid.

c. Save the solution, then start and test the application. Stop the application. Close the Code Editor window, then close the solution.

5. In this exercise, you code an application that creates a new password.

a. If necessary, start Visual Studio 2005 or Visual Basic 2005 Express Edition. Open the Jacobson Solution (Jacobson Solution.sln) file, which is contained in the VB2005\Chap08\Jacobson Solution folder. If necessary, open the designer window.

   b. Code the application so that it allows the user to enter a password that contains five, six, or seven characters. The application then should create and display a new password as follows:

     1. Replace all vowels (A, E, I, O, and U) with the letter X.

     2. Replace all numbers with the letter Z.

     3. Reverse the characters in the password.

   c. Save the solution, then start and test the application. Stop the application. Close the Code Editor window, then close the solution.

6. Each salesperson at BobCat Motors is assigned an ID number, which consists of four characters. The first character is either the letter F or the letter P. The letter F indicates that the salesperson is a full-time employee. The letter P indicates that he or she is a part-time employee. The middle two characters are the salesperson's initials, and the last character is either a 1 or a 2. A 1 indicates that the salesperson sells new cars, and a 2 indicates that the salesperson sells used cars.

   a. If necessary, start Visual Studio 2005 or Visual Basic 2005 Express Edition. Open the BobCat Motors Solution (BobCat Motors Solution.sln) file, which is contained in the VB2005\Chap08\BobCat Motors Solution folder. If necessary, open the designer window.

   b. Code the application so that it allows the sales manager to enter a salesperson's ID and the number of cars the salesperson sold during the month. The application should allow the sales manager to enter this information for as many salespeople as needed. The application should calculate and display the total number of cars sold by each of the following four categories of employees: full-time employees, part-time employees, employees selling new cars, and employees selling used cars.

   c. Save the solution, then start and test the application. Stop the application. Close the Code Editor window, and then close the solution.

DEBUGGING EXERCISE

7. In this exercise, you find and correct an error in an application. The process of finding and correcting errors is called debugging.

   a. If necessary, start Visual Studio 2005 or Visual Basic 2005 Express Edition. Open the Debug Solution (Debug Solution.sln) file, which is contained in the VB2005\Chap08\Debug Solution folder. If necessary, open the designer window.

   b. Open the Code Editor window and review the existing code.

c. Start the application. Enter Tampa, Florida in the Address text box, then click the Display City button. The button displays the letter T in a message box, which is incorrect; it should display the word Tampa. Click the OK button to close the dialog box, then click the Exit button to end the application.

d. Correct the application's code, then save the solution and start the application.

e. Enter Tampa, Florida in the Address text box, then click the Display City button. The button should display the word Tampa in a message box. Close the dialog box.

f. Click the Exit button to end the application. Close the Code Editor window, then close the solution.

CHAPTER NINE

ARRAYS

CREATING THE PERRYTOWN GIFT SHOP APPLICATION

 John Blackfeather is the owner and manager of the Perrytown Gift Shop. Every Friday afternoon, Mr. Blackfeather calculates the weekly pay for his six employees. The most time-consuming part of this task, and the one prone to the most errors, is the calculation of the federal withholding tax (FWT). Mr. Blackfeather wants an application that he can use to quickly and accurately calculate the FWT.

PREVIEWING THE PERRYTOWN GIFT SHOP APPLICATION

Before creating the Perrytown Gift Shop application, you first preview the completed application.

To preview the completed application:

1 Use the Run command on the Windows Start menu to run the **Perrytown** (**Perrytown.exe**) file, which is contained in the VB2005\Chap09 folder.

2 First, you will use the application to calculate the FWT for a married employee with taxable wages of $288.46. Type **288.46** in the Taxable wages text box, then click the **Married** radio button. Click the **Calculate** button. The application calculates and displays $13.45 as the FWT, as shown in Figure 9-1.

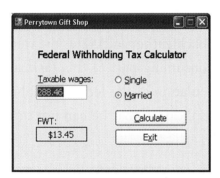

Figure 9-1: FWT for a married employee with taxable wages of $288.46

3 Next, you will calculate the FWT for a single employee with taxable wages of $600. Type **600** in the Taxable wages text box, then click the **Single** radio button. Click the **Calculate** button. The application calculates and displays $75.30 as the FWT.

4 Click the **Exit** button to end the application.

Before you can begin coding the Perrytown Gift Shop application, you need to learn how to create and use an array. You learn about one-dimensional arrays in Lessons A and B, and about two-dimensional arrays in Lesson C. You code the Perrytown Gift Shop application in Lesson C.

LESSON A
OBJECTIVES

AFTER STUDYING LESSON A, YOU SHOULD
BE ABLE TO:

» Declare and initialize a one-dimensional array

» Store data in a one-dimensional array

» Display the contents of a one-dimensional array

» Code a loop using the For Each...Next statement

» Access an element in a one-dimensional array

» Search a one-dimensional array

» Compute the average of a one-dimensional array's
  contents

» Find the highest entry in a one-dimensional array

» Update the contents of a one-dimensional array

» Sort a one-dimensional array

USING A ONE-DIMENSIONAL ARRAY

USING ARRAYS

All of the variables you have used so far have been simple variables. A **simple variable**, also called a **scalar variable**, is one that is unrelated to any other variable in memory. In many applications, however, you may need to reserve a block of variables, referred to as an array. An **array** is a group of variables that have the same name and data type and are related in some way. For example, each variable in the array might contain an inventory quantity, or each might contain a state name, or each might contain an employee record (name, Social Security number, pay rate, and so on). It may be helpful to picture an array as a group of small, adjacent boxes inside the computer's memory. You can write information to the boxes and you can read information from the boxes; you just cannot *see* the boxes.

Programmers use arrays to temporarily store related data in the internal memory of the computer. Examples of data stored in an array would include the federal withholding tax tables in a payroll program, and a price list in an order entry program. Storing data in an array increases the efficiency of a program, because data can be both written to and read from internal memory much faster than it can be written to and read from a file on a disk. In addition, after the data is entered into an array, which typically is done at the beginning of the program, the program can use the data as many times as desired. A payroll program, for example, can use the federal withholding tax tables stored in an array to calculate the amount of each employee's federal withholding tax.

The most commonly used arrays are one-dimensional and two-dimensional. You will learn about one-dimensional arrays in Lessons A and B, and about two-dimensional arrays in Lesson C. Arrays having more than two dimensions, which are used in scientific and engineering applications, are beyond the scope of this book.

ONE-DIMENSIONAL ARRAYS

You can visualize a **one-dimensional array** as a column of variables. A unique number called a **subscript** identifies each variable in a one-dimensional array. The computer assigns the subscripts to the array variables when the array is created. The subscript indicates the variable's position in the array. The first variable in a one-dimensional array is assigned a subscript of 0 (zero), the second a subscript of 1 (one), and so on. You refer to each variable in an array by the array's name and the variable's subscript, which is specified in a set of parentheses immediately following the array name. For example, to refer to the first variable in a one-dimensional array named states, you use states(0)—read "states sub zero." Similarly, to refer to the third variable in the states array, you use states(2). Figure 9-2 illustrates this naming convention.

»TIP

You also can visualize a one-dimensional array as a row of variables rather than as a column of variables.

»TIP

A subscript is also called an index.

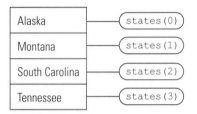

Alaska	────── states(0)
Montana	────── states(1)
South Carolina	────── states(2)
Tennessee	────── states(3)

Figure 9-2: Names of the variables in a one-dimensional array named states

Before you can use an array, you first must declare (create) it. Figure 9-3 shows two versions of the syntax you use to declare a one-dimensional array in Visual Basic. The figure also includes examples of using each syntax.

Declaring a one-dimensional array

Syntax - Version 1

{**Dim** | **Private**} *arrayname*(*highestSubscript*) **As** *datatype*

Syntax - Version 2

{**Dim** | **Private**} *arrayname*() **As** *datatype* = {*initialValues*}

Examples

```
Dim cities(3) As String
```

declares a four-element procedure-level array named cities; each element is automatically initialized using the keyword Nothing

```
Private numbers(5) As Integer
```

declares a six-element module-level array named numbers; each element is automatically initialized to the number zero

```
Private states() As String = {"Hawaii", "Alaska", "Maine"}
```

declares and initializes a three-element module-level array named states

```
Dim sales() As Decimal = {75.33D, 9.65D, 23.55D, 6.89D}
```

declares and initializes a four-element procedure-level array named sales

---

Figure 9-3: Syntax versions and examples of declaring a one-dimensional array

The {**Dim** | **Private**} portion in each syntax version indicates that you can select only one of the keywords appearing within the braces. In this case, you can select either the Dim keyword or the Private keyword. The appropriate keyword to use depends on whether you are creating a procedure-level array or a module-level array. *Arrayname* is the name of the array, and *datatype* is the type of data the array variables, referred to as **elements**, will store. Recall that each of the elements (variables) in an array has the same data type.

In Version 1 of the syntax, *highestSubscript* is an integer that specifies the highest subscript in the array. When the array is created, it will contain one element more than the number specified in the *highestSubscript* argument; this is because the first element in a one-dimensional array has a subscript of zero. For instance, the Dim cities(3) As String statement, which is shown in the first example in Figure 9-3, creates a procedure-level one-dimensional array named cities. The cities array contains four elements with subscripts of 0, 1, 2, and 3; each element can store a string. Similarly, the Private numbers(5) As Integer statement shown in the second example creates a module-level one-dimensional array named numbers. The numbers array contains six elements with subscripts of 0, 1, 2, 3, 4, and 5; in this case, each element can store an integer. As is true of module-level variables, module-level arrays are declared in the form's Declarations section, which begins with the Public Class statement and ends with the End Class statement.

When you use the syntax shown in Version 1 to declare a one-dimensional array, the computer automatically initializes each element in the array when the array is created. If the array's data type is String, each element in the array is initialized using the keyword Nothing. As you learned in Chapter 3, variables initialized to Nothing do not actually contain the word "Nothing"; rather, they contain no data at all. Elements in a numeric array are initialized to the number zero, and elements in a Boolean array are initialized to the Boolean value False. Date array elements are initialized to 12:00 AM January 1, 0001.

You use the syntax shown in Version 2 in Figure 9-3 to declare a one-dimensional array and, at the same time, specify each element's initial value. Assigning initial values to an array is often referred to as **populating the array**. You list the initial values in the *initialValues* section of the syntax, using commas to separate the values, and you enclose the list of values in braces ({}). Notice that the syntax shown in Version 2 does not include the *highestSubscript* argument; instead, an empty set of parentheses follows the array name. The computer automatically calculates the highest subscript based on the number of values listed in the *initialValues* section. If the *initialValues* section contains five values, the highest subscript in the array is 4. Likewise, if the *initialValues* section contains 100 values, the highest subscript in the array is 99. Notice that the highest subscript is always one number less than the number of values listed in the *initialValues* section; this is because the first subscript in a one-dimensional array is the number zero.

In the third example shown in Figure 9-3, the Private states() As String = {"Hawaii", "Alaska", "Maine"} statement declares a module-level one-dimensional String array named states. The states array contains three elements with subscripts of

0, 1, and 2. When the array is created, the computer assigns the string "Hawaii" to the `states(0)` element, "Alaska" to the `states(1)` element, and "Maine" to the `states(2)` element. Similarly, the `Dim sales() As Decimal = {75.33D, 9.65D, 23.55D, 6.89D}` statement shown in the last example declares a procedure-level one-dimensional Decimal array named `sales`. The `sales` array contains four elements with subscripts of 0, 1, 2, and 3. The computer assigns the number 75.33 to the `sales(0)` element, 9.65 to the `sales(1)` element, 23.55 to the `sales(2)` element, and 6.89 to the `sales(3)` element.

STORING DATA IN A ONE-DIMENSIONAL ARRAY

In most cases, you use an assignment statement to enter data into an existing array. Figure 9-4 shows the syntax of such an assignment statement and includes several examples of using the syntax to enter data into the arrays declared in Figure 9-3. In the syntax, *arrayname*(*subscript*) is the name and subscript of the array variable to which you want the *value* (data) assigned.

Storing data in a one-dimensional array

Syntax

arrayname(*subscript*) = *value*

Examples

```
cities(0) = "Madrid"
cities(1) = "Paris"
cities(2) = "Rome"
```

assigns the strings "Madrid", "Paris", and "Rome" to the `cities` array

---

```
For x As Integer = 1 To 6
    numbers(x - 1) = x * x
Next x
```

assigns the squares of the numbers from one through six to the `numbers` array

---

```
states(1) = "Virginia"
```

assigns the string "Virginia" to the second element in the `states` array

---

```
isConverted = Decimal.TryParse(Me.xSalesTextBox.Text, sales(0))
```

assigns either the value entered in the xSalesTextBox (converted to Decimal) or the number zero to the first element in the `sales` array

---

Figure 9-4: Syntax and examples of assignment statements used to enter data into a one-dimensional array

The three assignment statements shown in the first example in Figure 9-4 assign the strings "Madrid", "Paris", and "Rome" to the first three elements in the `cities` array, replacing the values stored in the array elements when the array was created. The code shown in the second example assigns the squares of the numbers from one through six to the `numbers` array, writing over the array's initial values. Notice that the number one must be subtracted from the value stored in the `x` variable when assigning the squares to the array; this is because the first array element has a subscript of zero rather than one. The `states(1) = "Virginia"` statement shown in the third example assigns the string "Virginia" to the second element in the `states` array, replacing the string "Alaska" that was stored in the element when the array was created. In the last example, the `isConverted = Decimal.TryParse(Me.xSalesTextBox.Text, sales(0))` statement converts the value entered in the xSalesTextBox to Decimal. If the conversion is successful, the statement assigns the Decimal result to the first element in the `sales` array; otherwise, it assigns the number zero to the first element in the array.

MANIPULATING ONE-DIMENSIONAL ARRAYS

The variables (elements) in an array can be used just like any other variables. For example, you can assign values to them, use them in calculations, display their contents, and so on. In the next several sections, you view sample procedures that demonstrate how one-dimensional arrays are used in an application. More specifically, the procedures will show you how to perform the following tasks using a one-dimensional array:

1. Display the contents of an array.

2. Access an array element using its subscript.

3. Search an array.

4. Calculate the average of the data stored in a numeric array.

5. Find the highest value stored in an array.

6. Update the array elements.

7. Sort the array elements.

In most applications, the values stored in an array come from a file on the computer's disk and are assigned to the array after it is declared. However, so that you can follow each procedure's code and its results more easily, most of the procedures you view in this chapter use the Dim statement to assign the appropriate values to the array. The first procedure you view displays the contents of a one-dimensional array.

DISPLAYING THE CONTENTS OF A ONE-DIMENSIONAL ARRAY

Figure 9-5 shows a sample run of the Months application, which assigns to a label control the item selected in a list box. Figure 9-6 shows the pseudocode for the MainForm's Load event procedure contained in the application, and Figure 9-7 shows the corresponding Visual Basic code. The Load event procedure demonstrates how you can display the contents of an array in a list box.

Figure 9-5: Sample run of the Months application

MainForm Load event procedure - pseudocode
1. declare and initialize a one-dimensional String array named months
2. repeat for each element in the months array
add the contents of the current array element to the xMonthListBox
end repeat
3. select the first item in the xMonthListBox

Figure 9-6: Pseudocode for the MainForm's Load event procedure

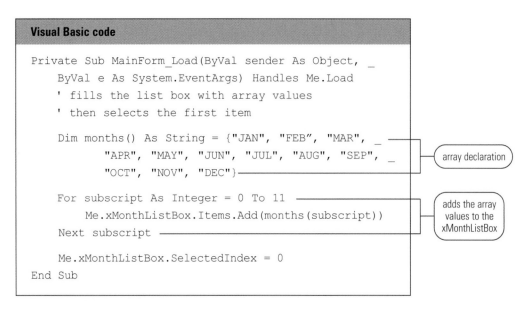

```
Visual Basic code

Private Sub MainForm_Load(ByVal sender As Object, _
    ByVal e As System.EventArgs) Handles Me.Load
    ' fills the list box with array values
    ' then selects the first item

    Dim months() As String = {"JAN", "FEB", "MAR", _
        "APR", "MAY", "JUN", "JUL", "AUG", "SEP", _
        "OCT", "NOV", "DEC"}

    For subscript As Integer = 0 To 11
        Me.xMonthListBox.Items.Add(months(subscript))
    Next subscript

    Me.xMonthListBox.SelectedIndex = 0
End Sub
```

array declaration

adds the array values to the xMonthListBox

Figure 9-7: Code for the MainForm's Load event procedure

The MainForm's Load event procedure declares a 12-element one-dimensional String array named months, using the names of the 12 months to initialize the array. The procedure then uses a For...Next loop to add the contents of each array element to the xMonthListBox. The first time the loop is processed, the subscript variable contains the number zero and the Me.xMonthListBox.Items.Add(months(subscript)) statement adds the contents of the months(0) element—JAN—to the xMonthListBox. The Next subscript statement then adds the number one to the value stored in the subscript variable, giving one. When the loop is processed the second time, the Me.xMonthListBox.Items.Add(months(subscript)) statement adds the contents of the months(1) element—FEB—to the xMonthListBox, and so on. The computer repeats the loop instructions for each element in the months array, beginning with the element whose subscript is zero (JAN) and ending with the element whose subscript is 11 (DEC). The computer stops processing the loop when the value contained in the subscript variable is 12, which is one number more than the highest subscript in the array.

To code and then test the Months application:

1 Start Visual Studio 2005 or Visual Basic 2005 Express Edition, if necessary, and close the Start Page window. Open the **Months Solution** (Months Solution.sln) file, which is contained in the VB2005\Chap09\Months Solution folder. If necessary, open the designer window.

2 Open the Code Editor window. Replace the <your name> and <current date> text in the comments with your name and the current date. Open the code template for the MainForm's Load event procedure, then enter the comments and code shown in Figure 9-7.

3 Close the Code Editor window. Save the solution, then start the application. The application displays the names of the 12 months in the xMonthListBox.

4 Click **AUG** in the list box, then click the **Display Selection** button. The application displays AUG in the interface, as shown earlier in Figure 9-5.

5 Click the **Exit** button to end the application, then close the solution.

The MainForm's Load event procedure in Figure 9-7 uses the For...Next statement to add each array element to the list box. You also could use the Do...Loop statement (which you learned about in Chapter 6) or the For Each...Next statement (which you learn about next).

>> TIP

As you learned in Chapter 6, you can code a repetition structure (loop) in Visual Basic using the For...Next, Do...Loop, or For Each...Next statement.

THE FOR EACH...NEXT STATEMENT

You can use the **For Each...Next statement** to code a loop whose instructions you want processed for each element in a group—for example, for each array variable in an array. Figure 9-8 shows the syntax of the For Each...Next statement. It also shows two ways of writing the loop from Figure 9-7 using the For Each...Next statement.

For Each...Next statement

Syntax

For Each *element* [**As** *datatype*] **In** *group*
    [*statements*]
Next *element*

Examples

```
For Each monthName As String In months
    Me.xMonthListBox.Items.Add(monthName)
Next monthName
```
adds the contents of the months array to the xMonthListBox

```
Dim monthName As String
For Each monthName In months
    Me.xMonthListBox.Items.Add(monthName)
Next monthName
```
same as the previous example

Figure 9-8: Syntax and examples of the For Each...Next statement

>> TIP

You can use the Exit For statement to exit the For Each...Next statement before it has finished processing. You may need to do this if the computer encounters an error when processing the loop instructions.

»TIP

You can nest For Each...Next statements, which means that you can place one For Each...Next statement within another For Each...Next statement.

The For Each...Next statement begins with the For Each clause and ends with the Next clause. Between the two clauses, you enter the instructions you want the loop to repeat for each *element* in the *group*. When using the For Each...Next statement to process an array, *element* is the name of a variable that the computer can use to keep track of each array variable, and *group* is the name of the array. You can use the **As** *datatype* portion of the For Each clause to declare the *element* variable, as shown in the first example in Figure 9-8. When you declare a variable in the For Each clause, the variable has block scope and can be used only by the For Each...Next loop. Alternatively, you can declare the *element* variable in a Dim statement, as shown in the last example in Figure 9-8. As you know, when a variable is declared in a Dim statement at the beginning of a procedure, it has procedure scope and can be used by the entire procedure. The data type of the *element* must match the data type of the *group*. For example, if the *group* is an Integer array, then the *element's* data type must be Integer. Likewise, if the *group* is a String array, then the *element's* data type must be String. In the examples shown in Figure 9-8, *group* is a String array named months, and *element* is a String variable named monthName. The Me.xMonthListBox.Items.Add(monthName) statement in both examples in Figure 9-8 adds the current array element to the xMonthListBox and will be processed for each element in the months array. The code shown in both examples will display the same result as shown earlier in Figure 9-5.

To modify and then test the code in the Months application:

1 Use Windows to make a copy of the Months Solution folder, which is contained in the VB2005\Chap09 folder. Rename the folder **Months Solution-ForEach**.

2 Open the **Months Solution** (Months Solution.sln) file contained in the VB2005\Chap09\Months Solution-ForEach folder. Open the designer window.

3 Open the Code Editor window. Locate the MainForm's Load event procedure. Replace the For...Next statement with the For Each...Next statement shown in the first example in Figure 9-8.

4 Close the Code Editor window. Save the solution, then start the application. Click **AUG** in the list box, then click the **Display Selection** button. The application displays AUG in the interface, as shown earlier in Figure 9-5.

5 Click the **Exit** button to end the application, then close the solution.

USING THE SUBSCRIPT TO ACCESS AN ELEMENT IN A ONE-DIMENSIONAL ARRAY

XYZ Corporation pays its managers based on six different salary codes, 1 through 6. Each code corresponds to a different salary amount. Figure 9-9 shows a sample run of the Salary Code application, which displays the salary amount associated with the salary code entered by the user. Figure 9-10 shows the pseudocode for the xDisplayButton's Click event procedure contained in the application, and Figure 9-11 shows the corresponding Visual Basic code. The Click event procedure uses an array element's subscript to access the element in the array.

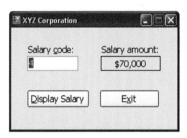

Figure 9-9: Sample run of the Salary Code application

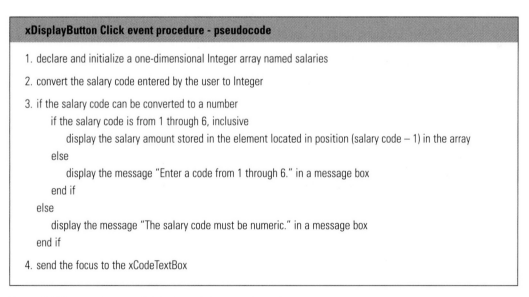

Figure 9-10: Pseudocode for the xDisplayButton's Click event procedure

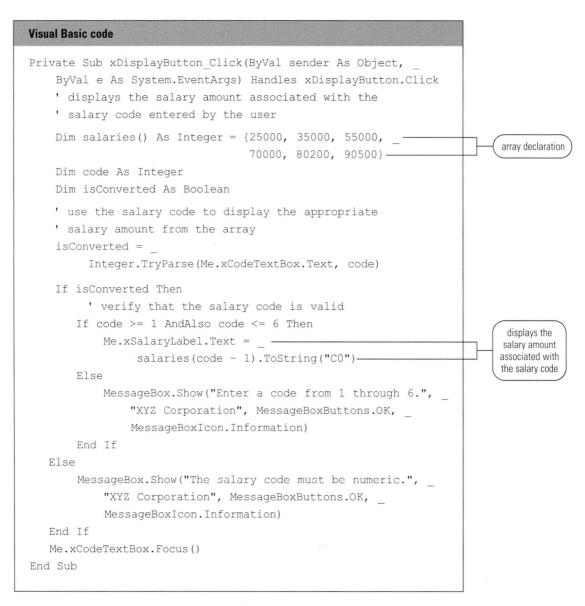

Visual Basic code

```
Private Sub xDisplayButton_Click(ByVal sender As Object, _
    ByVal e As System.EventArgs) Handles xDisplayButton.Click
    ' displays the salary amount associated with the
    ' salary code entered by the user
    Dim salaries() As Integer = {25000, 35000, 55000, _
                                70000, 80200, 90500}
    Dim code As Integer
    Dim isConverted As Boolean

    ' use the salary code to display the appropriate
    ' salary amount from the array
    isConverted = _
        Integer.TryParse(Me.xCodeTextBox.Text, code)

    If isConverted Then
        ' verify that the salary code is valid
        If code >= 1 AndAlso code <= 6 Then
            Me.xSalaryLabel.Text = _
                salaries(code - 1).ToString("C0")
        Else
            MessageBox.Show("Enter a code from 1 through 6.", _
                "XYZ Corporation", MessageBoxButtons.OK, _
                MessageBoxIcon.Information)
        End If
    Else
        MessageBox.Show("The salary code must be numeric.", _
            "XYZ Corporation", MessageBoxButtons.OK, _
            MessageBoxIcon.Information)
    End If
    Me.xCodeTextBox.Focus()
End Sub
```

array declaration

displays the salary amount associated with the salary code

Figure 9-11: Code for the xDisplayButton's Click event procedure

The procedure shown in Figure 9-11 declares a one-dimensional Integer array named salaries, using six salary amounts to initialize the array. The salary amount for salary code 1 is stored in salaries(0). Salary code 2's salary amount is stored in salaries(1), and so on. Notice that the salary code is one number more than the subscript of its corresponding salary amount in the array. After creating and initializing the array, the procedure declares an Integer variable named code and a Boolean variable named isConverted. It then uses the TryParse method to convert the salary code entered in the xCodeTextBox to Integer. If the conversion is not successful, an appropriate message is displayed in a message box. However, if the conversion is successful, the procedure verifies that the salary code is valid. To be valid, the salary code must be greater than or equal to the number one and, at the same time, less than or equal to the number six. Before accessing an array element, a procedure always should verify that the subscript is valid—in other words, that it is in the appropriate range. If the procedure uses a subscript that is not in the appropriate range, Visual Basic displays an error message and the procedure ends abruptly.

If the salary code is not valid, the xDisplayButton's Click event procedure displays an appropriate message in a message box. Otherwise, it uses the salary code, which is stored in the code variable, to access the appropriate salary amount from the array. Notice that, to access the correct element in the salaries array, the number one must be subtracted from the contents of the code variable. This is because the salary code is one number more than the subscript of its associated salary amount in the array. As Figure 9-9 indicates, the procedure displays $70,000 when the user enters the number four as the salary code.

To code and then test the Salary Code application:

1 Open the **Salary Code Solution** (Salary Code Solution.sln) file, which is contained in the VB2005\Chap09\Salary Code Solution folder. If necessary, open the designer window.

2 Open the Code Editor window. Replace the <your name> and <current date> text in the comments with your name and the current date. Open the code template for the xDisplayButton's Click event procedure, then enter the comments and code shown in Figure 9-11.

3 Close the Code Editor window. Save the solution, then start the application. Type **4** in the Salary code text box, then click the **Display Salary** button. The application displays $70,000 as the salary amount, as shown earlier in Figure 9-9.

4 Click the **Exit** button to end the application, then close the solution.

SEARCHING A ONE-DIMENSIONAL ARRAY

The sales manager at Jacobsen Motors wants an application that allows him to determine the number of salespeople selling above a specific amount. To accomplish this task, the application will search an array that contains the amount sold by each salesperson. The application will look for array values that are greater than the amount provided by the sales manager. Figure 9-12 shows a sample run of the Sales application. Figure 9-13 shows the pseudocode for the xSearchButton's Click event procedure contained in the application, and Figure 9-14 shows the corresponding Visual Basic code.

Figure 9-12: Sample run of the Sales application

xSearchButton Click event procedure - pseudocode

1. declare and initialize a one-dimensional Integer array named sales

2. convert the sales amount entered by the user to Integer

3. repeat for each element in the sales array
    if the value in the current array element is greater than the value entered by the user
        add 1 to the counter variable
    end if
  end repeat

4. display the contents of the counter variable in the xCountLabel

5. send the focus to the xSalesTextBox

Figure 9-13: Pseudocode for the xSearchButton's Click event procedure

Visual Basic code

```vb
Private Sub xSearchButton_Click(ByVal sender As Object, _
    ByVal e As System.EventArgs) Handles xSearchButton.Click
    ' searches the array, looking for values that are
    ' greater than the value entered by the user

    Dim sales() As Integer = {45000, 35000, 25000, 60000, 23000}    ← array declaration
    Dim counter As Integer
    Dim searchFor As Integer

    Integer.TryParse(Me.xSalesTextBox.Text, searchFor)

    ' search the array, updating the counter
    ' when the array value is greater than
    ' the searchFor value
    For Each salesAmount As Integer In sales
        If salesAmount > searchFor Then
            counter = counter + 1
        End If
    Next salesAmount

    ' display the counter value
    Me.xCountLabel.Text = Convert.ToString(counter)

    Me.xSalesTextBox.Focus()
End Sub
```

searches the array for values greater than a specific amount

Figure 9-14: Code for the xSearchButton's Click event procedure

The xSearchButton's Click event procedure declares a one-dimensional Integer array named `sales`, using five sales amounts to initialize the array. The procedure also declares two Integer variables named `counter` and `searchFor`. After declaring the array and variables, the procedure uses the TryParse method to convert the sales amount entered in the xSalesTextBox to Integer. Notice that the procedure does not verify that the conversion was successful. This is because the xSalesTextBox's KeyPress event procedure is coded to allow the text box to accept only numbers, the period, and the Backspace key. After converting the sales amount to Integer, the xSearchButton's Click event procedure uses a loop that repeats its instructions for each element in the `sales` array. Notice that the loop uses the `salesAmount` variable to represent each element in the array. The selection structure in the loop compares the contents of the current array element with the contents of the `searchFor` variable. If the array element contains a number that is greater than the number stored in the `searchFor` variable, the selection structure's true path adds the number one to the value stored in the `counter` variable. In the xSearchButton's Click event procedure, the `counter` variable is used as a counter to keep track of the number of salespeople selling over the amount entered by the sales manager. When the loop ends, which is when there are no more array elements to search, the procedure displays the contents of the `counter` variable in the xCountLabel. As Figure 9-12 indicates, the procedure displays the number two when the sales manager enters 35000 as the sales amount.

To code and then test the Sales application:

1 Open the **Sales Solution** (Sales Solution.sln) file, which is contained in the VB2005\Chap09\Sales Solution folder. If necessary, open the designer window.

2 Open the Code Editor window. Replace the <your name> and <current date> text in the comments with your name and the current date. Open the code template for the xSearchButton's Click event procedure, then enter the comments and code shown in Figure 9-14.

3 Close the Code Editor window. Save the solution, then start the application. Type **35000** in the Sales over text box, then click the **Search** button. The application displays the number 2 in the interface, as shown earlier in Figure 9-12.

4 Click the **Exit** button to end the application, then close the solution.

CALCULATING THE AVERAGE AMOUNT STORED IN A ONE-DIMENSIONAL NUMERIC ARRAY

Professor Jeremiah wants an application that calculates and displays the average test score earned by his students on the final exam. To accomplish this task, the application will add together the test scores stored in an array, and then divide the sum by the number of array elements. Figure 9-15 shows a sample run of the Average application. Figure 9-16 shows the pseudocode for the xCalcButton's Click event procedure contained in the application, and Figure 9-17 shows the corresponding Visual Basic code.

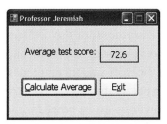

Figure 9-15: Sample run of the Average application

xCalcButton Click event procedure - pseudocode
1. declare and initialize a one-dimensional Integer array named scores

2. repeat for each element in the scores array
     add the contents of the current array element to an accumulator variable
  end repeat

3. calculate the average score by dividing the contents of the accumulator variable by the number of array elements
4. display the average score in the xAverageLabel

Figure 9-16: Pseudocode for the xCalcButton's Click event procedure

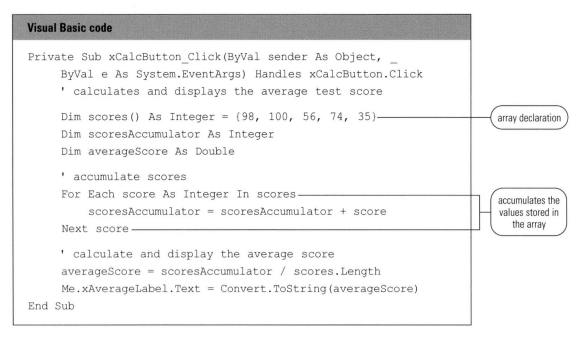

Figure 9-17: Code for the xCalcButton's Click event procedure

The xCalcButton's Click event procedure declares a one-dimensional Integer array named `scores`, using five test scores to initialize the array. It also declares an Integer variable named `scoresAccumulator` and a Double variable named `averageScore`. The loop in the procedure repeats its instruction for each element in the `scores` array. The instruction in the loop adds the score contained in the current array element to the `scoresAccumulator` variable. When the loop ends, the `averageScore = scoresAccumulator / scores.Length` statement uses the `scoresAccumulator` variable and the array's Length property to calculate the average test score. An array's **Length property**, whose syntax is *arrayname*.**Length**, stores an integer that represents the number of elements in the array. In this case, the Length property stores the number five, because the `scores` array contains five elements. The xCalcButton's Click event procedure then displays the average test score in the xAverageLabel. As Figure 9-15 indicates, the procedure displays the number 72.6.

To code and then test the Average application:

1 Open the **Average Solution** (Average Solution.sln) file, which is contained in the VB2005\Chap09\Average Solution folder. If necessary, open the designer window.

2 Open the Code Editor window. Replace the <your name> and <current date> text in the comments with your name and the current date. Open the code template for the xCalcButton's Click event procedure, then enter the comments and code shown in Figure 9-17.

3 Close the Code Editor window. Save the solution, then start the application. Click the **Calculate Average** button. The application displays the number 72.6 in the interface, as shown earlier in Figure 9-15.

4 Click the **Exit** button to end the application, then close the solution.

DETERMINING THE HIGHEST VALUE STORED IN A ONE-DIMENSIONAL ARRAY

Sharon Johnson keeps track of the amount of money she earns each week. She would like an application that displays the highest amount earned in a week. The application will store the pay amounts in a one-dimensional array. Similar to the Sales application, which you viewed earlier in Figures 9-12 through 9-14, the Pay application will need to search the array. However, rather than looking in the array for values that are greater than a specific amount, the Pay application will look for the highest amount in the array. Figure 9-18 shows a sample run of the Pay application. Figure 9-19 shows the pseudocode for the xHighButton's Click event procedure contained in the application, and Figure 9-20 shows the corresponding Visual Basic code.

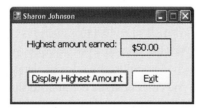

Figure 9-18: Sample run of the Pay application

xHighButton Click event procedure - pseudocode

1. declare and initialize a one-dimensional Decimal array named pays

2. declare a Decimal variable named highestPay and initialize it using the contents of the first element in the pays array

3. repeat for the second through the last array element
      if the value in the current array element is greater than the value in the highestPay variable
          assign the value in the current array element to the highestPay variable
    end if
  end repeat

4. display the contents of the highestPay variable in the xHighestLabel

Figure 9-19: Pseudocode for the xHighButton's Click event procedure

Visual Basic code

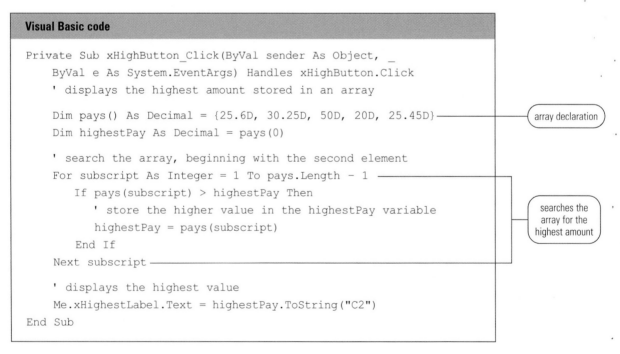

```
Private Sub xHighButton_Click(ByVal sender As Object, _
    ByVal e As System.EventArgs) Handles xHighButton.Click
    ' displays the highest amount stored in an array

    Dim pays() As Decimal = {25.6D, 30.25D, 50D, 20D, 25.45D}          array declaration
    Dim highestPay As Decimal = pays(0)

    ' search the array, beginning with the second element
    For subscript As Integer = 1 To pays.Length - 1
        If pays(subscript) > highestPay Then
            ' store the higher value in the highestPay variable        searches the
            highestPay = pays(subscript)                               array for the
        End If                                                         highest amount
    Next subscript

    ' displays the highest value
    Me.xHighestLabel.Text = highestPay.ToString("C2")
End Sub
```

Figure 9-20: Code for the xHighButton's Click event procedure

The xHighButton's Click event procedure declares a one-dimensional Decimal array named `pays`, and it initializes the array to the amounts that Sharon earned during the last five weeks. The procedure also declares a Decimal variable named `highestPay`, which will keep track of the highest value stored in the `pays` array. Notice that the `highestPay` variable is initialized using the value stored in the first array element. The first time the loop in the procedure is processed, the selection structure within the loop compares the value stored in the second array element—`pays(1)`—with the value stored in the `highestPay` variable. At this point, the `highestPay` variable contains the same value as the first array element. If the value stored in the second array element is greater than the value stored in the `highestPay` variable, then the `highestPay = pays(subscript)` statement assigns the array element value to the `highestPay` variable. The `Next subscript` clause then adds the number one to the `subscript` variable, giving 2. The next time the loop is processed, the selection structure compares the value stored in the third array element—`pays(2)`—with the value stored in the `highestPay` variable, and so on. When the loop ends, which is when the `subscript` variable contains the number five, the procedure displays the contents of the `highestPay` variable in the xHighestLabel. In this case, the procedure displays $50.00, as shown earlier in Figure 9-18.

To code and then test the Pay application:

1 Open the **Pay Solution** (Pay Solution.sln) file, which is contained in the VB2005\Chap09\Pay Solution folder. If necessary, open the designer window.

2 Open the Code Editor window. Replace the <your name> and <current date> text in the comments with your name and the current date. Open the code template for the xHighButton's Click event procedure, then enter the comments and code shown in Figure 9-20.

3 Close the Code Editor window. Save the solution, then start the application. Click the **Display Highest Amount** button. The application displays $50.00 in the interface, as shown earlier in Figure 9-18.

4 Click the **Exit** button to end the application, then close the solution.

UPDATING THE VALUES STORED IN A ONE-DIMENSIONAL ARRAY

The sales manager at Jillian Company wants an application that increases the price of each item the company sells. The application will store the prices in a one-dimensional array, and then increase the value stored in each array element. In addition, the sales manager wants the application to display each item's new price in the xNewPricesLabel. Figure 9-21 shows a sample run of the Prices application. Figure 9-22 shows the pseudocode for the xUpdateButton's Click event procedure contained in the application, and Figure 9-23 shows the corresponding Visual Basic code.

> **» TIP**
>
> The loop in Figure 9-20 searches the second through the last element in the `pays` array. The first element is not included in the search because its value is stored in the `highestPay` variable when the variable is created.

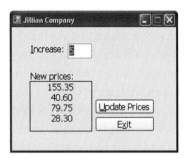

Figure 9-21: Sample run of the Prices application

xUpdateButton Click event procedure - pseudocode

1. declare a one-dimensional Double array named prices

2. convert the increase amount entered in the xIncreaseTextBox to Double

3. repeat for each element in the prices array
     add the increase amount to the value stored in the current array element
     display the contents of the current array element in the xNewPricesLabel
  end repeat

4. send the focus to the xIncreaseTextBox

Figure 9-22: Pseudocode for the xUpdateButton's Click event procedure

Visual Basic code

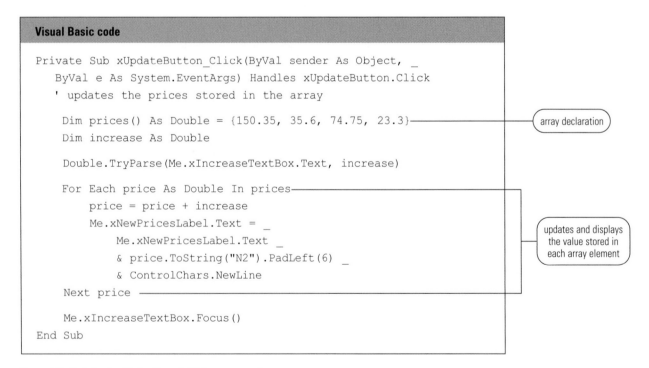

```
Private Sub xUpdateButton_Click(ByVal sender As Object, _
    ByVal e As System.EventArgs) Handles xUpdateButton.Click
    ' updates the prices stored in the array

    Dim prices() As Double = {150.35, 35.6, 74.75, 23.3}
    Dim increase As Double

    Double.TryParse(Me.xIncreaseTextBox.Text, increase)

    For Each price As Double In prices
        price = price + increase
        Me.xNewPricesLabel.Text = _
            Me.xNewPricesLabel.Text _
            & price.ToString("N2").PadLeft(6) _
            & ControlChars.NewLine
    Next price

    Me.xIncreaseTextBox.Focus()
End Sub
```

array declaration

updates and displays the value stored in each array element

Figure 9-23: Code for the xUpdateButton's Click event procedure

The xUpdateButton's Click event procedure declares a one-dimensional Double array named `prices`, using four values to initialize the array; it also declares a Double variable named `increase`. The procedure uses the TryParse method to convert the increase amount entered in the xIncreaseTextBox to Double. Notice that the procedure does not verify that the conversion was successful. This is because the xIncreaseTextBox's KeyPress event procedure is coded to allow the text box to accept only numbers, the period, and the Backspace key. After converting the increase amount to Double, the xUpdateButton's Click event procedure uses a loop to access each element in the `prices` array. Notice that the procedure uses a Double variable named `price` to represent each array element. The first instruction in the loop, `price = price + increase`, updates the contents of the current array element by adding the increase amount to it. The second instruction in the loop then displays the updated contents in the xNewPricesLabel. The loop ends when all of the array elements have been updated. Figure 9-21 shows the results when the user enters the number five as the increase amount.

To code and then test the Prices application:

1 Open the **Prices Solution** (Prices Solution.sln) file, which is contained in the VB2005\Chap09\Prices Solution folder. If necessary, open the designer window.

2 Open the Code Editor window. Replace the <your name> and <current date> text in the comments with your name and the current date. Open the code template for the xUpdateButton's Click event procedure, then enter the comment and code shown in Figure 9-23.

3 Close the Code Editor window. Save the solution, then start the application. Type **5** in the Increase text box, then click the **Update Prices** button. The application displays the new prices in the interface, as shown earlier in Figure 9-21. Notice that each new price is five dollars more than its corresponding original price.

4 Click the **Exit** button to end the application, then close the solution.

SORTING THE DATA STORED IN A ONE-DIMENSIONAL ARRAY

In some applications, you might need to arrange the contents of an array in either ascending or descending order. Arranging data in a specific order is called **sorting**. When an array is sorted in ascending order, the first element in the array contains the smallest value, and the last element contains the largest value. When an array is sorted in descending order, on the other hand, the first element contains the largest value, and the last element contains the smallest value. You use the **Array.Sort method** to sort the elements in a one-dimensional array in ascending order. The method's syntax is **Array.Sort**(*arrayname*), where *arrayname* is the name of the one-dimensional array to be sorted. To sort a one-dimensional array in descending order, you first use the Array.Sort method to sort the array in ascending order, and then use the **Array.Reverse method** to reverse the array elements. The syntax of the Array.Reverse method is **Array.Reverse**(*arrayname*), where *arrayname* is the name of the one-dimensional array whose elements you want reversed.

The State application that you view in this section uses both the Array.Sort and Array.Reverse methods. The application allows the user to enter the names of five states. It stores the state names in a module-level one-dimensional String array named `states`. The user then can choose to display the state names in either ascending or descending order. Figure 9-24 shows a sample run of the State application when the user enters the following state names and then clicks the Ascending Sort button: Illinois, Texas, Alaska, New York, and Idaho. Figure 9-25 shows the application's code.

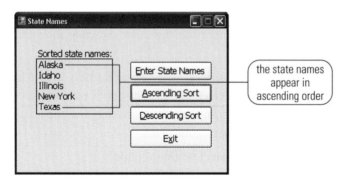

Figure 9-24: Sample run of the State application

Visual Basic code

```
' Project name:      State Project
' Project purpose:   The project allows the user to enter
'                    state names. It then sorts the names
'                    in ascending or descending order and
'                    displays them in the xStateLabel.
' Created/revised:   <your name> on <current date>

Option Explicit On
Option Strict On

Public Class MainForm

    ' module-level array
    Private stateNames(4) As String ─────────────────────

    Private Sub xExitButton_Click...

    Private Sub xEnterButton_Click(ByVal sender As Object, _
        ByVal e As System.EventArgs) Handles xEnterButton.Click
        ' allows the user to enter five state names
        ' stores the state names in the module-level
        ' stateNames array

        For subscript As Integer = 0 To stateNames.Length - 1
            stateNames(subscript) = _
                InputBox("State name", "State Names")
        Next subscript
    End Sub

    Private Sub xAscendButton_Click(ByVal sender As Object, _
        ByVal e As System.EventArgs) Handles xAscendButton.Click
        ' sorts the array values in ascending order, then
        ' displays them in the xStateLabel

        Array.Sort(stateNames) ──────────────────────────

        Me.xStateLabel.Text = String.Empty
        For Each name As String In stateNames
            Me.xStateLabel.Text = _
                Me.xStateLabel.Text & name _
                & ControlChars.NewLine
        Next name
    End Sub
```

module-level
array declaration

sorts the array values
in ascending order

Figure 9-25: Code for the State application *(Continued)* ▶

```
    Private Sub xDescendButton_Click(ByVal sender As Object, _
        ByVal e As System.EventArgs) Handles xDescendButton.Click
        ' sorts the array values in descending order, then
        ' displays them in the xStateLabel

        Array.Sort(stateNames)
        Array.Reverse(stateNames)

        Me.xStateLabel.Text = String.Empty
        For Each name As String In stateNames
            Me.xStateLabel.Text = _
                Me.xStateLabel.Text & name _
                & ControlChars.NewLine
        Next name
    End Sub
End Class
```

> sorts the array values in descending order

Figure 9-25: Code for the State application

» TIP

Recall that a one-dimensional array's Length property stores the number of elements in the array and is always one number more than the highest subscript in the array.

The `Private stateNames(4) As String` statement in Figure 9-25 declares a module-level one-dimensional String array named `stateNames`. The array contains five elements having subscripts of 0, 1, 2, 3, and 4. You typically use a module-level array when you need more than one procedure in the *same* form to use the *same* array, because a module-level array can be used by all of the procedures in the form, including the procedures associated with the controls contained on the form. In this application, the `stateNames` array will be used by the Click event procedures for the xEnterButton, xAscendButton, and xDescendButton.

The xEnterButton's Click event procedure uses a For...Next statement and the InputBox function to prompt the user for five state names. The procedure stores the names in the `stateNames` array. The xAscendButton's Click event procedure uses the Array.Sort method to sort in ascending order the values stored in the `stateNames` array. It uses a For Each...Next statement to display the sorted values in the xStateLabel. The xDescendButton's Click event procedure uses both the Array.Sort and Array.Reverse methods to sort in descending order the values stored in the `stateNames` array. It also uses a For Each...Next statement to display the sorted values in the xStateLabel.

To code and then test the State application:

1 Open the **State Solution** (State Solution.sln) file, which is contained in the VB2005\Chap09\State Solution folder. If necessary, open the designer window.

2 Open the Code Editor window. Replace the <your name> and <current date> text in the comments with your name and the current date. Open the code templates for both

the xAscendButton's Click event procedure and the xDescendButton's Click event procedure. In both procedures, enter the appropriate comments and code shown in Figure 9-25.

3 Close the Code Editor window. Save the solution, then start the application. Click the **Enter State Names** button. Type the following five state names, one at a time, in the State Names dialog box: **Illinois, Texas, Alaska, New York**, and **Idaho**. Press **Enter** after typing each name.

4 Click the **Ascending Sort** button. The application displays the state names in ascending order, as shown earlier in Figure 9-24. Click the **Descending Sort** button. The application displays the state names in descending order.

5 Click the **Exit** button to end the application, then close the solution.

You have completed Lesson A. You can either take a break or complete the end-of-lesson questions and exercises before moving on to Lesson B.

SUMMARY

TO DECLARE A ONE-DIMENSIONAL ARRAY:

» Use either of the following two syntax versions:

Version 1: {**Dim** | **Private**} *arrayname*(*highestSubscript*) **As** *datatype*

Version 2: {**Dim** | **Private**} *arrayname*() **As** *datatype* = {*initialValues*}

» Use the Dim keyword to declare a procedure-level array. Use the Private keyword to declare a module-level array. *Arrayname* is the name of the array, and *datatype* is the type of data the array variables will store.

» The *highestSubscript* argument that appears in Version 1 of the syntax is an integer that specifies the highest subscript in the array. Using Version 1's syntax, the computer automatically initializes the elements (variables) in the array.

» The *initialValues* section that appears in Version 2 of the syntax is a list of values separated by commas and enclosed in braces. The values are used to initialize each element in the array.

TO REFER TO A VARIABLE INCLUDED IN AN ARRAY:

» Use the array's name followed by the variable's subscript. Enclose the subscript in a set of parentheses following the array name.

TO DETERMINE THE NUMBER OF ELEMENTS (VARIABLES) IN A ONE-DIMENSIONAL ARRAY:

» Use the array's Length property in the following syntax: *arrayname*.**Length**.

TO SORT IN ASCENDING ORDER THE ELEMENTS STORED IN A ONE-DIMENSIONAL ARRAY:

» Use the Array.Sort method. The method's syntax is **Array.Sort(***arrayname***)**.

TO REVERSE THE ORDER OF THE ELEMENTS INCLUDED IN A ONE-DIMENSIONAL ARRAY:

» Use the Array.Reverse method. The method's syntax is **Array.Reverse(***arrayname***)**.

TO PROCESS INSTRUCTIONS FOR EACH ELEMENT IN A GROUP:

» Use the For Each...Next statement, whose syntax is shown in Figure 9-8.

QUESTIONS

1. Which of the following statements declares a one-dimensional array named `amounts` that contains five elements?

    a. `Dim amounts(4) As Double`

    b. `Dim amounts(5) As Double`

    c. `Dim amounts(4) As Double = {3.55, 6.70, 8, 4, 2.34}`

    d. Both a and c.

2. The `items` array is declared using the `Dim items(20) As String` statement. The `x` variable keeps track of the array subscripts and is initialized to the number zero. Which of the following Do clauses will process the loop instructions for each element in the array?

    a. `Do While x > 20`   b. `Do While x < 20`

    c. `Do While x >= 20`   d. `Do While x <= 20`

    Use the `sales` array to answer Questions 3 through 6. The array was declared using the following statement: `Dim sales() As Integer = {10000, 12000, 900, 500, 20000}`.

3. The statement `sales(3) = sales(3) + 10` will _____.

    a. replace the 500 amount with 10

    b. replace the 500 amount with 510

    c. replace the 900 amount with 10

    d. replace the 900 amount with 910

4. Which of the following If clauses can be used to verify that the array subscript, named x, is valid for the `sales` array?

   a. `If sales(x) >= 0 AndAlso sales(x) < 4 Then`

   b. `If sales(x) >= 0 AndAlso sales(x) <= 4 Then`

   c. `If x >= 0 AndAlso x < 4 Then`

   d. `If x >= 0 AndAlso x <= 4 Then`

5. Which of the following loops will correctly add 100 to each variable in the `sales` array? The x variable was declared as an Integer variable and was initialized to the number zero.

   a.
   ```
   Do While x <= 4
       x = x + 100
   Loop
   ```

   b.
   ```
   Do While x <= 4
       sales = sales + 100
   Loop
   ```

   c.
   ```
   Do While x < 5
       sales(x) = sales(x) + 100
   Loop
   ```

   d. None of the above.

6. Which of the following statements sorts the `sales` array in ascending order?

   a. `Array.Sort(sales)`   b. `sales.Sort`

   c. `Sort(sales)`   d. `SortArray(sales)`

Use the `numbers` array to answer Questions 7 through 12. The array was declared using the following statement: `Dim numbers() As Integer = {10, 5, 7, 2}`. The `total` and x variables were declared as Integer variables, and the `avg` variable was declared as a Decimal variable. The variables were initialized to the number zero.

7. Which of the following will correctly calculate the average of the elements included in the `numbers` array?

   a.
   ```
   Do While x < 4
       numbers(x) = total + total
       x = x + 1
   Loop
   avg = Convert.ToDecimal(total / x)
   ```

b. ```
Do While x < 4
   total = total + numbers(x)
   x = x + 1
Loop
avg = Convert.ToDecimal(total / x)
```

c. ```
Do While x < 4
   total = total + numbers(x)
   x = x + 1
Loop
avg = Convert.ToDecimal(total / x - 1)
```

d. ```
Do While x < 4
   total = total + numbers(x)
   x = x + 1
Loop
avg = Convert.ToDecimal(total / (x - 1))
```

8. The code in Question 7's answer a will assign _____ to the avg variable.

   a. 0                          b. 5

   c. 6                          d. 8

9. The code in Question 7's answer b will assign _____ to the avg variable.

   a. 0                          b. 5

   c. 6                          d. 8

10. The code in Question 7's answer c will assign _____ to the avg variable.

    a. 0                          b. 5

    c. 6                          d. 8

11. The code in Question 7's answer d will assign _____ to the avg variable.

    a. 0                          b. 5

    c. 6                          d. 8
```

12. Which of the following statements determines the number of elements included in the `numbers` array, and then assigns the result to an Integer variable named `elements`?

    a. `elements = Len(numbers)`

    b. `elements = Length(numbers)`

    c. `elements = numbers.Len`

    d. `elements = numbers.Length`

EXERCISES

1. Write the statement to declare a procedure-level one-dimensional Integer array named `numbers` that can store 20 integers. Then write the statement to store the number seven in the second element contained in the `numbers` array.

2. Write the statement to declare a module-level one-dimensional String array named `products` that has 10 elements. Then write the statement to store the string "Paper" in the third element contained in the `products` array.

3. Write the statement to declare and initialize a procedure-level one-dimensional Double array named `rates` that has five elements. Use the following numbers to initialize the array: 6.5, 8.3, 4, 2, 10.5.

4. Write three versions of the code to display, in the xRatesLabel, the contents of the `rates` array from Exercise 3. First use the For...Next statement. Then rewrite the code using the Do...Loop statement, and then rewrite the code using the For Each...Next statement.

5. Write the statement to sort the `rates` array in ascending order.

6. Write the statement to reverse the contents of the `rates` array.

7. Write three versions of the code to calculate the average of the elements stored in the `rates` array from Exercise 3. Display the average in the xAverageLabel. First use the For...Next statement. Then rewrite the code using the Do...Loop statement, and then rewrite the code using the For Each...Next statement.

8. Write three versions of the code to subtract the number one from each element in the `rates` array from Exercise 3. First use the Do...Loop statement. Then rewrite the code using the For...Next statement, and then rewrite the code using the For Each...Next statement.

9. In this exercise, you code an application that displays the lowest value stored in an array.

   a. If necessary, start Visual Studio 2005 or Visual Basic 2005 Express Edition. Open the Lowest Solution (Lowest Solution.sln) file, which is contained in the VB2005\Chap09\Lowest Solution-ForNext folder. If necessary, open the designer window.

   b. Open the Code Editor window. Locate the xDisplayButton's Click event procedure. The procedure declares and initializes a 20-element one-dimensional Integer array named `scores`. Code the procedure so that it displays (in the xLowestLabel) the lowest score stored in the array. Use the For...Next statement.

   c. Save the solution, then start the application. Click the Display Lowest button. The number 13 should appear in the xLowestLabel.

   d. Click the Exit button to end the application. Close the Code Editor window, then close the solution.

   e. Use Windows to make a copy of the Lowest Solution-ForNext folder. Rename the folder Lowest Solution-ForEachNext.

   f. Open the Lowest Solution (Lowest Solution.sln) file contained in the VB2005\Chap09\Lowest Solution-ForEachNext folder. Open the designer window.

   g. Open the Code Editor window. Modify the xDisplayButton's Click event procedure so that it uses the For Each...Next statement.

   h. Save the solution, then start the application. Click the Display Lowest button. The number 13 should appear in the xLowestLabel.

   i. Click the Exit button to end the application. Close the Code Editor window, then close the solution.

   j. Use Windows to make a copy of the Lowest Solution-ForNext folder. Rename the folder Lowest Solution-DoLoop.

   k. Open the Lowest Solution (Lowest Solution.sln) file contained in the VB2005\Chap09\Lowest Solution-DoLoop folder. Open the designer window.

   l. Open the Code Editor window. Modify the xDisplayButton's Click event procedure so that it uses the Do...Loop statement.

   m. Save the solution, then start the application. Click the Display Lowest button. The number 13 should appear in the xLowestLabel.

   n. Click the Exit button to end the application. Close the Code Editor window, then close the solution.

10. In this exercise, you code an application that updates each value stored in an array.

   a. If necessary, start Visual Studio 2005 or Visual Basic 2005 Express Edition. Open the Update Prices Solution (Update Prices Solution.sln) file, which is contained in the VB2005\Chap09\Update Prices Solution folder. If necessary, open the designer window.

   b. Open the Increase button's Click event procedure. Declare a one-dimensional array named `prices`. The array should contain the following 10 elements: 6.75, 12.50, 33.50, 10, 9.50, 25.50, 7.65, 8.35, 9.75, and 3.50.

   c. The procedure should ask the user for a percentage amount by which each price should be increased. It then should increase each price by that amount, and then display the increased prices in the interface.

   d. Save the solution, then start the application. Click the Increase button. Increase each price by 5%.

   e. Click the Exit button to end the application. Close the Code Editor window, then close the solution.

11. In this exercise, you modify the application from Exercise 10. The modified application allows the user to update a specific price.

   a. Use Windows to make a copy of the Update Prices Solution folder, which is contained in the VB2005\Chap09 folder. Rename the folder Modified Update Prices Solution.

   b. If necessary, start Visual Studio 2005 or Visual Basic 2005 Express Edition. Open the Update Prices Solution (Update Prices Solution.sln) file contained in the VB2005\Chap09\Modified Update Prices Solution folder. Open the designer window.

   c. Open the Increase button's Click event procedure. Modify the procedure so that it also asks the user to enter a number from one through 10. If the user enters the number one, the procedure should update the first price in the array. If the user enters the number two, the procedure should update the second price in the array, and so on.

d. Save the solution, then start the application. Click the Increase button. Increase the second price by 10%. Click the Increase button again. This time, increase the tenth price by 2%.

e. Click the Exit button to end the application. Close the Code Editor window, then close the solution.

12. In this exercise, you code an application that displays the number of students earning a specific score.

a. If necessary, start Visual Studio 2005 or Visual Basic 2005 Express Edition. Open the Scores Solution (Scores Solution.sln) file, which is contained in the VB2005\Chap09\Scores Solution folder. If necessary, open the designer window.

b. Open the Display button's Click event procedure. Declare a 20-element one-dimensional Integer array named `scores`. Assign the following 20 numbers to the array: 88, 72, 99, 20, 66, 95, 99, 100, 72, 88, 78, 45, 57, 89, 85, 78, 75, 88, 72, and 88.

c. Code the procedure so that it prompts the user to enter a score from zero through 100. The procedure then should display (in a message box) the number of students who earned that score.

d. Save the solution, then start the application. Use the application to answer the following questions:

How many students earned a score of 72?

How many students earned a score of 88?

How many students earned a score of 20?

How many students earned a score of 99?

e. Click the Exit button to end the application. Close the Code Editor window, then close the solution.

13. In this exercise, you modify the application from Exercise 12. The modified application allows the user to display the number of students earning a score within a specific range.

a. Use Windows to make a copy of the Scores Solution folder, which is contained in the VB2005\Chap09 folder. Rename the folder Modified Scores Solution.

b. If necessary, start Visual Studio 2005 or Visual Basic 2005 Express Edition. Open the Scores Solution (Scores Solution.sln) file contained in the VB2005\Chap09\Modified Scores Solution folder. Open the designer window.

c. Open the Display button's Click event procedure. Modify the procedure so that it prompts the user to enter both a minimum score and a maximum score. The procedure then should display (in a message box) the number of students who earned a score within that range.

d. Save the solution, then start the application. Use the application to answer the following questions.

How many students earned a score from 70 through 79?

How many students earned a score from 65 through 85?

How many students earned a score from 0 through 50?

e. Click the Exit button to end the application. Close the Code Editor window, then close the solution.

14. In this exercise, you create an application that generates and displays six unique random numbers for a lottery game. Each lottery number can range from 1 through 54 only.

a. If necessary, start Visual Studio 2005 or Visual Basic 2005 Express Edition. Open the Lottery Game Solution (Lottery Game Solution.sln) file, which is contained in the VB2005\Chap09\Lottery Game Solution folder. If necessary, open the designer window.

b. Open the Display Numbers button's Click event procedure. Code the procedure so that it displays six unique random numbers in the interface. (*Hint*: Store the numbers in a one-dimensional array.)

c. Save the solution, then start the application. Click the Display Numbers button several times. Each time you click the button, six unique random numbers between 1 and 54 (inclusive) should appear in the interface.

d. Click the Exit button to end the application. Close the Code Editor window, then close the solution.

DISCOVERY EXERCISE

15. In this exercise, you learn about the ReDim statement.

a. Display the Help screen for the ReDim statement. What is the purpose of the statement? What is the purpose of the `Preserve` keyword?

b. If necessary, start Visual Studio 2005 or Visual Basic 2005 Express Edition. Open the ReDim Solution (ReDim Solution.sln) file, which is contained in the VB2005\Chap09\ReDim Solution folder.

c. Open the Code Editor window and view the xDisplayButton's Click event procedure. Study the existing code, then modify the procedure so that it stores any number of sales amounts in the `sales` array.

d. Save the solution, then start the application. Click the Display Sales button, then enter the following sales amounts: 700, 550, and 800. The button's Click event procedure should display each sales amount in a separate message box.

e. Click the Display Sales button again, then enter the following sales amounts: 5, 9, 45, 67, 8, and 0. The button's Click event procedure should display each sales amount in a separate message box.

f. Click the Exit button to end the application. Close the Code Editor window, then close the solution.

LESSON B
OBJECTIVES

AFTER STUDYING LESSON B, YOU SHOULD
BE ABLE TO:

» Create parallel one-dimensional arrays

» Locate information in two parallel one-dimensional
arrays

PARALLEL ONE-DIMENSIONAL ARRAYS

USING PARALLEL ONE-DIMENSIONAL ARRAYS

Takoda Tapahe owns a small gift shop named Treasures. She wants an application that displays the price of the item whose product ID she enters. Figure 9-26 shows a portion of the gift shop's price list.

Product ID	Price
BX35	13
CR20	10
FE15	12
KW10	24
MM67	4

Figure 9-26: A portion of the gift shop's price list

Recall that all of the variables in an array have the same data type. So how can you store a price list, which includes a string (the product ID) and a number (the price), in an array? One solution is to use two one-dimensional arrays: a String array to store the product IDs and an Integer array to store the prices. Both arrays are illustrated in Figure 9-27.

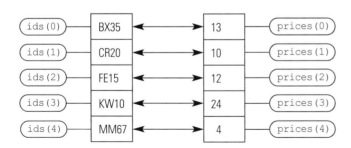

Figure 9-27: Illustration of a price list stored in two one-dimensional arrays

The arrays shown in Figure 9-27 are referred to as parallel arrays. **Parallel arrays** are two or more arrays whose elements are related by their position in the arrays. In other words, they are related by their subscript. The `ids` and `prices` arrays shown in Figure 9-27 are parallel because each element in the `ids` array corresponds to the element located in the same position in the `prices` array. For example, the first element in the `ids` array corresponds to the first element in the `prices` array. In other words, the item whose product ID is BX35 [`ids(0)`] has a price of $13 [`prices(0)`]. Likewise, the second elements in both arrays—the elements with a subscript of 1—also are related; the item whose product ID is CR20 has a price of $10. The same relationship is true for the remaining elements in both arrays. To determine an item's price, you simply locate the item's ID in the `ids` array and then view its corresponding element in the `prices` array. Figure 9-28 shows a sample run of the Price List application. Figure 9-29 shows the pseudocode for the xDisplayButton's Click event procedure contained in the application, and Figure 9-30 shows the corresponding Visual Basic code.

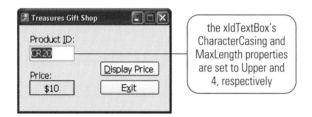

the xIdTextBox's CharacterCasing and MaxLength properties are set to Upper and 4, respectively

Figure 9-28: Sample run of the Price List application

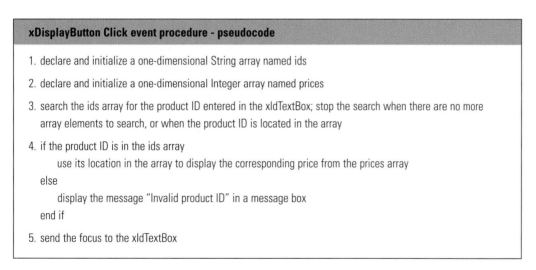

xDisplayButton Click event procedure - pseudocode

1. declare and initialize a one-dimensional String array named ids

2. declare and initialize a one-dimensional Integer array named prices

3. search the ids array for the product ID entered in the xIdTextBox; stop the search when there are no more array elements to search, or when the product ID is located in the array

4. if the product ID is in the ids array
       use its location in the array to display the corresponding price from the prices array
   else
       display the message "Invalid product ID" in a message box
   end if

5. send the focus to the xIdTextBox

Figure 9-29: Pseudocode for the xDisplayButton's Click event procedure

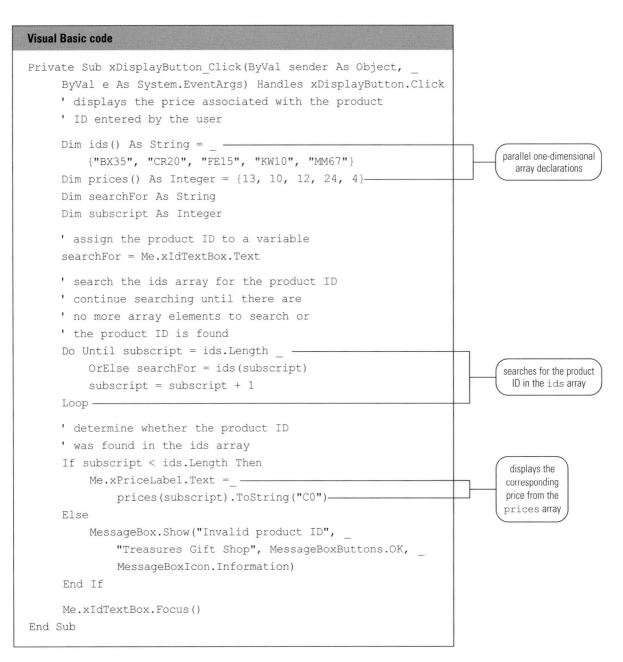

Visual Basic code

```vb
Private Sub xDisplayButton_Click(ByVal sender As Object, _
    ByVal e As System.EventArgs) Handles xDisplayButton.Click
    ' displays the price associated with the product
    ' ID entered by the user

    Dim ids() As String = _
        {"BX35", "CR20", "FE15", "KW10", "MM67"}
    Dim prices() As Integer = {13, 10, 12, 24, 4}
    Dim searchFor As String
    Dim subscript As Integer

    ' assign the product ID to a variable
    searchFor = Me.xIdTextBox.Text

    ' search the ids array for the product ID
    ' continue searching until there are
    ' no more array elements to search or
    ' the product ID is found
    Do Until subscript = ids.Length _
        OrElse searchFor = ids(subscript)
        subscript = subscript + 1
    Loop

    ' determine whether the product ID
    ' was found in the ids array
    If subscript < ids.Length Then
        Me.xPriceLabel.Text =_
            prices(subscript).ToString("C0")
    Else
        MessageBox.Show("Invalid product ID", _
            "Treasures Gift Shop", MessageBoxButtons.OK, _
            MessageBoxIcon.Information)
    End If

    Me.xIdTextBox.Focus()
End Sub
```

parallel one-dimensional array declarations

searches for the product ID in the ids array

displays the corresponding price from the prices array

Figure 9-30: Code for the xDisplayButton's Click event procedure using parallel one-dimensional arrays

The xDisplayButton's Click event procedure declares and initializes two parallel one-dimensional arrays: a five-element String array named ids and a five-element Integer array named prices. Notice that each product ID is stored in the ids array, and its price is stored in the corresponding location in the prices array. The procedure also declares a String variable named searchFor and an Integer variable named subscript. After declaring the arrays and variables, the procedure assigns the contents of the xIdTextBox to the searchFor variable. (As Figure 9-28 indicates, the xIdTextBox's CharacterCasing and MaxLength properties are set to Upper and 4, respectively.) The loop in the procedure searches for the product ID in each element in the ids array, stopping when there are no more array elements to search or when the product ID is located in the array. After the loop completes its processing, the selection structure in the procedure compares the number stored in the subscript variable with the value stored in the ids array's Length property. In this case, the Length property contains the number 5, because there are five elements in the ids array. If the subscript variable contains a number that is less than the number of elements in the ids array, it indicates that the loop stopped processing because the product ID was located in the array. In that case, the procedure displays the price located in the same position in the prices array. However, if the subscript variable's value is not less than the number of elements in the ids array, it indicates that the loop stopped processing because it reached the end of the array without finding the product ID. In that case, the message "Invalid product ID" is displayed in a message box. As Figure 9-28 indicates, the procedure displays a price of $10 when the user enters CR20 as the product ID.

To code and then test the Price List application:

1 Start Visual Studio 2005 or Visual Basic 2005 Express Edition, if necessary, and close the Start Page window. Open the **Price List Solution** (Price List Solution.sln) file, which is contained in the VB2005\Chap09\Price List Solution-Parallel folder. If necessary, open the designer window.

2 Open the Code Editor window. Replace the <your name> and <current date> text in the comments with your name and the current date. Open the code template for the xDisplayButton's Click event procedure, then enter the comments and code shown in Figure 9-30.

3 Close the Code Editor window. Save the solution, then start the application. Type **cr20** in the Product ID box, then click the **Display Price** button. The application displays $10 in the Price box, as shown earlier in Figure 9-28.

4 Click the **Exit** button to end the application, then close the solution.

You have completed Lesson B. You can either take a break or complete the end-of-lesson questions and exercises before moving on to Lesson C.

SUMMARY

TO CREATE TWO PARALLEL ONE-DIMENSIONAL ARRAYS:
» Create two one-dimensional arrays. When assigning values to both arrays, be sure that the value stored in each element in one array corresponds to the value stored in the same element in the other array.

QUESTIONS

1. Parallel arrays are related by their subscripts.

   a. True   b. False

2. The `state` and `capital` arrays are parallel arrays. If Illinois is stored in the second element in the `state` array, then Springfield is stored in the _____ element.

   a. `capital(1)`   b. `capital(2)`

EXERCISES

1. In this exercise, you code an application that allows Professor Carver to display a grade based on the number of points he enters. The grading scale is shown in Figure 9-31.

Minimum points	Maximum points	Grade
0	299	F
300	349	D
350	399	C
400	449	B
450	500	A

Figure 9-31

   a. If necessary, start Visual Studio 2005 or Visual Basic 2005 Express Edition. Open the Carver Solution (Carver Solution.sln) file, which is contained in the VB2005\Chap09\Carver Solution folder. If necessary, open the designer window.

b. Store the minimum points in a five-element one-dimensional Integer array named points. Store the grades in a five-element one-dimensional String array named grades. The arrays should be parallel arrays.

c. Code the Display Grade button's Click event procedure so that it searches the points array for the number of points entered by the user, and then displays the corresponding grade from the grades array.

d. Save the solution, and then start and test the application. Click the Exit button to end the application. Close the Code Editor window, and then close the solution.

2. In this exercise, you modify the application from Exercise 1. The modified application allows the user to change the grading scale when the application is started.

a. Use Windows to make a copy of the Carver Solution folder, which is contained in the VB2005\Chap09 folder. Rename the folder Modified Carver Solution.

b. If necessary, start Visual Studio 2005 or Visual Basic 2005 Express Edition. Open the Carver Solution (Carver Solution.sln) file contained in the VB2005\Chap09\ Modified Carver Solution folder. Open the designer window.

c. When the form is loaded into the computer's memory, the application should use the InputBox function to prompt the user to enter the total number of possible points—in other words, the total number of points a student can earn in the course. Modify the application's code to perform this task.

d. Modify the application's code so that it uses the grading scale shown in Figure 9-32. For example, if the user enters the number 500 as the total number of possible points, the code should enter 450, which is 90% of 500, as the minimum number of points for an A. If the user enters the number 300, the code should enter 270, which is 90% of 300, as the minimum number of points for an A.

Minimum points	Grade
Less than 60% of the possible points	F
60% of the possible points	D
70% of the possible points	C
80% of the possible points	B
90% of the possible points	A

Figure 9-32

e. Save the solution, then start the application. Enter 300 as the number of possible points, then enter 185 in the Points text box. Click the Display Grade button. A grade of D appears in the interface. Click the Exit button to end the application.

f. Start the application again. Enter 500 as the number of possible points, then enter 363 in the Points text box. Click the Display Grade button. A grade of C appears in the interface. Click the Exit button to end the application. Close the Code Editor window, then close the solution.

3. In this exercise, you code an application that allows Ms. Laury to display a shipping charge based on the number of items ordered by a customer. The shipping charge scale is shown in Figure 9-33.

Minimum order	Maximum order	Shipping charge
1	10	15
11	50	10
51	100	5
101	99999	0

Figure 9-33

a. If necessary, start Visual Studio 2005 or Visual Basic 2005 Express Edition. Open the Laury Solution (Laury Solution.sln) file, which is contained in the VB2005\Chap09\Laury Solution folder. If necessary, open the designer window.

b. Store the maximum order amounts in a four-element one-dimensional Integer array named order. Store the shipping charge amounts in a four-element one-dimensional Integer array named ship. The arrays should be parallel arrays.

c. Code the Display Shipping Charge button's Click event procedure so that it searches the order array for the number of items ordered by the user, and then displays the corresponding shipping charge from the ship array. Display the shipping charge with a dollar sign and two decimal places.

d. Save the solution, and then start and test the application. Click the Exit button to end the application. Close the Code Editor window, then close the solution.

LESSON C
OBJECTIVES

AFTER STUDYING LESSON C, YOU SHOULD BE ABLE TO:

» Create and initialize a two-dimensional array

» Store data in a two-dimensional array

» Search a two-dimensional array

» Determine the highest and lowest subscript in a two-dimensional array

TWO-DIMENSIONAL ARRAYS

USING TWO-DIMENSIONAL ARRAYS

As mentioned in Lesson A, you can visualize a one-dimensional array as a column of variables. A **two-dimensional array**, on the other hand, resembles a table in that the variables are in rows and columns. Each variable (element) in a two-dimensional array is identified by a unique combination of two subscripts, which the computer assigns to the variable when the array is created. The subscripts specify the variable's row and column position in the array. Variables located in the first row in a two-dimensional array are assigned a row subscript of 0 (zero). Variables located in the second row are assigned a row subscript of 1 (one), and so on. Similarly, variables located in the first column in a two-dimensional array are assigned a column subscript of 0 (zero). Variables located in the second column are assigned a column subscript of 1 (one), and so on. You refer to each variable in a two-dimensional array by the array's name and the variable's row and column subscripts, which are separated by a comma and specified in a set of parentheses immediately following the array name. For example, to refer to the variable located in the first row, first column in a two-dimensional array named `products`, you use `products(0, 0)`—read "`products` sub zero comma zero." Similarly, you use `products(1, 2)` to refer to the variable located in the second row, third column in the `products` array. Figure 9-34 illustrates this naming convention. Notice that the row subscript is listed first within the parentheses.

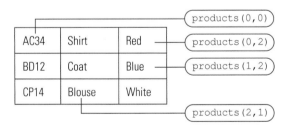

Figure 9-34: Names of some of the variables contained in the `products` array

Recall that you first must declare (create) an array before you can use it. Figure 9-35 shows two versions of the syntax you use to declare a two-dimensional array in Visual Basic. The figure also includes an example of using each syntax.

Declaring a two-dimensional array

Syntax - Version 1

{**Dim** | **Private**} *arrayname*(*highestRowSubscript, highestColumnSubscript*) **As** *datatype*

Syntax - Version 2

{**Dim** | **Private**} *arrayname*(,) **As** *datatype* = {{*initialValues*}, ...{*initialValues*}}

Examples

```
Dim cities(5, 3) As String
```

declares a six-row, four-column array named `cities`; each element is automatically initialized using the keyword `Nothing`

```
Dim scores(,) As Integer = {{75, 90}, _
                            {9, 25}, _
                            {23, 56}, _
                            {6, 12}}
```

declares and initializes a four-row, two-column array named `scores`

Figure 9-35: Syntax versions and examples of declaring a two-dimensional array

You use the `Dim` keyword to declare a procedure-level array and use the `Private` keyword to declare a module-level array. In each syntax version, *arrayname* is the name of the array, and *datatype* is the type of data the array variables will store. Recall that each of the variables in an array has the same data type. In Version 1 of the syntax, *highestRowSubscript* and *highestColumnSubscript* are integers that specify the highest row and column subscripts, respectively, in the array. When the array is created, it will contain one row more than the number specified in the *highestRowSubscript* argument, and one column more than the number specified in the *highestColumnSubscript* argument. This is because the first row subscript in a two-dimensional array is zero, and the first column subscript also is zero. When you declare a two-dimensional array using the syntax shown in Version 1 in Figure 9-35, the computer automatically initializes each element in the array when the array is created.

You use the syntax shown in Version 2 in Figure 9-35 to declare a two-dimensional array and, at the same time, specify each variable's initial value. Using Version 2's syntax, you include a separate *initialValues* section, enclosed in braces, for each row in the array. If the array has two rows, then the statement that declares and initializes the array should have two *initialValues* sections. If the array has five rows, then the declaration statement should have five *initialValues* sections. Within the individual *initialValues* sections, you enter one or more values separated by commas. The number of values to enter corresponds to the number of columns in the array. If the array contains 10 columns, then each individual *initialValues* section should contain 10 values. In addition to the set of braces that surrounds each individual *initialValues* section, notice in the syntax that a set of braces also surrounds all of the *initialValues* sections. Also notice that a comma appears within the parentheses that follow the array name. The comma indicates that the array is a two-dimensional array. (Recall that a comma is used to separate the row subscript from the column subscript in a two-dimensional array.)

Study the two examples shown in Figure 9-35. The statement shown in the first example creates a two-dimensional String array named `cities`; the array has six rows and four columns. The computer automatically initializes each variable in the `cities` array using the keyword `Nothing`. The statement shown in the second example creates a two-dimensional Integer array named `scores`; the array has four rows and two columns. The statement initializes the `scores(0, 0)` variable to the number 75, and initializes the `scores(0, 1)` variable to the number 90. The `scores(1, 0)` and `scores(1, 1)` variables are initialized to the numbers 9 and 25, respectively. The `scores(2, 0)` and `scores(2, 1)` variables are initialized to the numbers 23 and 56, respectively, and the `scores(3, 0)` and `scores(3, 1)` variables are initialized to the numbers 6 and 12, respectively.

STORING DATA IN A TWO-DIMENSIONAL ARRAY

As with one-dimensional arrays, you generally use an assignment statement to enter data into a two-dimensional array. Figure 9-36 shows the syntax of such an assignment statement and includes several examples of using the syntax to enter data into the arrays declared in Figure 9-35. In the syntax, *arrayname*(*rowSubscript*, *columnSubscript*) is the name and subscripts of the array variable to which you want the *value* (data) assigned.

Storing data in a two-dimensional array

Syntax

arrayname(*rowSubscript*, *columnSubscript*) = *value*

Examples

```
cities(0, 0) = "Madrid"
cities(0, 1) = "Paris"
cities(0, 2) = "Rome"
cities(0, 3) = "London"
```

assigns the strings "Madrid", "Paris", "Rome", and "London" to the variables contained in the first row in the `cities` array

```
For row As Integer = 1 To 4
      For col As Integer = 1 To 2
            scores(row - 1, col - 1)
      Next col
Next row
```

assigns the number zero to each variable in the `scores` array

```
For Each number As Integer In scores
      number = 0
Next number
```

assigns the number zero to each variable in the `scores` array

Figure 9-36: Syntax and examples of assignment statements used to enter data into a two-dimensional array

The code shown in the first example in Figure 9-36 uses four assignment statements to assign values to the elements contained in the first row in the `cities` array. The code shown in the second and third examples assigns the number zero to each element contained in the `scores` array. The second and third examples use a nested For...Next loop to make the assignments.

SEARCHING A TWO-DIMENSIONAL ARRAY

In Lesson B, you viewed an application created for Takoda Tapahe, the owner of a small gift shop named Treasures. As you may remember, the application allows Takoda to display the price of the item whose product ID she enters. Recall that the xDisplayButton's Click event procedure (shown earlier in Figure 9-30) uses two parallel one-dimensional arrays to store the gift shop's price list. In this lesson, the xDisplayButton's Click event procedure will use a two-dimensional array to store the price list. Figure 9-37 shows a sample run of the modified Price List application. Figure 9-38 shows the modified pseudocode for the xDisplayButton's Click event procedure, and Figure 9-39 shows the corresponding Visual Basic code.

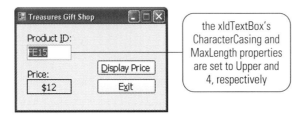

Treasures Gift Shop
Product ID:
FE15
Display Price
Price:
$12
Exit
the xIdTextBox's CharacterCasing and MaxLength properties are set to Upper and 4, respectively

Figure 9-37: Sample run of the modified Price List application

xDisplayButton Click event procedure - pseudocode

1. declare and initialize a two-dimensional String array named products
2. search the first column in the products array, looking for the product ID entered in the xIdTextBox; stop the search when there are no more array elements to search, or when the product ID is located in the array
3. if the product ID is in the first column of the products array
    use its location in the array to display the corresponding price from the second column in the array
else
    display the message "Invalid product ID" in a message box
end if
4. send the focus to the xIdTextBox

Figure 9-38: Modified pseudocode for the xDisplayButton's Click event procedure

Visual Basic code

```
Private Sub xDisplayButton_Click(ByVal sender As Object, _
     ByVal e As System.EventArgs) Handles xDisplayButton.Click
     ' displays the price associated with the product
     ' ID entered by the user
     Dim products(,) As String = {{"BX35", "13"}, _
                                  {"CR20", "10"}, _
                                  {"FE15", "12"}, _
                                  {"KW10", "24"}, _
                                  {"MM67", "4"}}
     Dim searchFor As String
     Dim row As Integer

     ' assign the product ID to a variable
     searchFor = Me.xIdTextBox.Text

     ' search for the product ID in the first
     ' column of the products array; continue
     ' searching until there are no more array
     ' elements to search or the product ID is found
     Do Until row = 5 _
        OrElse searchFor = products(row, 0)
        row = row + 1
     Loop

     ' determine whether the product ID
     ' was found in the products array
     If row < 5 Then
         Me.xPriceLabel.Text = "$" & products(row, 1)
     Else
         MessageBox.Show("Invalid product ID", _
             "Treasures Gift Shop", MessageBoxButtons.OK, _
             MessageBoxIcon.Information)
     End If

     Me.xIdTextBox.Focus()
End Sub
```

two-dimensional array declaration

searches for the product ID in the first column of the `products` array

displays the corresponding price from the second column of the `products` array

Figure 9-39: Code for the xDisplayButton's Click event procedure using a two-dimensional array

The procedure shown in Figure 9-39 declares and initializes a two-dimensional array named products; the array has five rows and two columns. Notice that each product ID is stored in the first column of the products array, and its price is stored in the corresponding row in the second column. The procedure also declares a String variable named searchFor and an Integer variable named row. After declaring the array and variables, the procedure assigns the contents of the xIdTextBox to the searchFor variable. (As Figure 9-37 indicates, the xIdTextBox's CharacterCasing and MaxLength properties are set to Upper and 4, respectively.) The loop in the procedure searches for the product ID in the first column in the products array, stopping when there are no more array elements to search or when the product ID is located in the array. After the loop completes its processing, the selection structure in the procedure compares the number stored in the row variable with the number five, which is the number of rows contained in the array. If the row variable contains a number that is less than five, it indicates that the loop stopped processing because the product ID was located in the first column in the array. In that case, the procedure displays the price located in the same row as the product ID, but in the second column of the array. However, if the row variable's value is not less than five, it indicates that the loop stopped processing because it reached the end of the array's first column without finding the product ID. In that case, the message "Invalid product ID" is displayed in a message box. As Figure 9-37 indicates, the procedure displays a price of $12 when the user enters FE15 as the product ID.

To code and then test this version of the Price List application:

1 Start Visual Studio 2005 or Visual Basic 2005 Express Edition, if necessary, and close the Start Page window. Open the **Price List Solution** (Price List Solution.sln) file, which is contained in the VB2005\Chap09\Price List Solution-Two Dimensional folder. If necessary, open the designer window.

2 Open the Code Editor window. Replace the <your name> and <current date> text in the comments with your name and the current date. Open the code template for the xDisplayButton's Click event procedure, then enter the comments and code shown in Figure 9-39.

3 Close the Code Editor window. Save the solution, then start the application. Type **fe15** in the Product ID box, then click the **Display Price** button. The application displays $12 in the Price box, as shown earlier in Figure 9-37.

4 Click the **Exit** button to end the application, then close the solution.

Now that you know how to use a two-dimensional array, you can begin coding the Perrytown Gift Shop application that you previewed at the beginning of this chapter.

THE PERRYTOWN GIFT SHOP APPLICATION

Recall that your task in this chapter is to create an application that John Blackfeather, the owner and manager of the Perrytown Gift Shop, can use to calculate the weekly federal withholding tax for his employees. To save you time, the VB2005\Chap09\Perrytown Solution folder contains a partially completed Perrytown Gift Shop application.

To open the Perrytown Gift Shop application:

1 Open the **Perrytown Solution** (Perrytown Solution.sln) file, which is contained in the VB2005\Chap09\Perrytown Solution folder. Figure 9-40 shows the user interface for the Perrytown Gift Shop application.

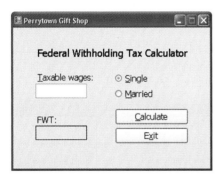

Figure 9-40: Interface for the Perrytown Gift Shop application

To calculate the federal withholding tax, Mr. Blackfeather will need to provide the employee's taxable wages and marital status, and then click the Calculate button. The Perrytown Gift Shop application will use the tax tables shown in Figure 9-41 to calculate the appropriate federal withholding tax (FWT).

FWT Tables – Weekly Payroll Period

Single person (including head of household)

If the taxable wages are:		The amount of income tax to withhold is		
Over	But not over	Base amount	Percentage	Of excess over
	$ 51	0		
$ 51	$ 192	0	10%	$ 51
$ 192	$ 620	$ 14.10 plus	15%	$ 192
$ 620	$1,409	$ 78.30 plus	25%	$ 620
$1,409	$3,013	$ 275.55 plus	28%	$1,409
$3,013	$6,508	$ 724.67 plus	33%	$3,013
$6,508		$1,878.02 plus	35%	$6,508

Married person

If the taxable wages are:		The amount of income tax to withhold is		
Over	But not over	Base amount	Percentage	Of excess over
	$ 154	0		
$ 154	$ 440	0	10%	$ 154
$ 440	$1,308	$ 28.60 plus	15%	$ 440
$1,308	$2,440	$ 158.80 plus	25%	$1,308
$2,440	$3,759	$ 441.80 plus	28%	$2,440
$3,759	$6,607	$ 811.12 plus	33%	$3,759
$6,607		$1,750.96 plus	35%	$6,607

Figure 9-41: Weekly FWT tables

Notice that both tables shown in Figure 9-41 contain five columns of information. The first two columns list various ranges, also called brackets, of taxable wage amounts. The first column (Over) lists the amount that a taxable wage in that range must be over, and the second column (But not over) lists the maximum amount included in the range. The remaining three columns (Base amount, Percentage, and Of excess over) tell you how to calculate the tax for each range. For example, assume that you are married and your weekly taxable wages are $288.46. Before you can calculate the amount of your tax, you need to locate your taxable wages in the first two columns of the Married table. In this case, your taxable wages fall within the $154 through $440 range. After locating the range that contains your taxable wages, you then use the remaining three columns in the table to calculate your tax. According to the table, taxable wages in the $154 through $440 bracket have a tax of 10% of the amount over $154; therefore, your tax is $13.45. In the Perrytown Gift Shop application, you will store each tax table in a separate two-dimensional array. To save you time, the application already contains the code to declare and initialize the arrays.

To view the code that declares and initializes the two-dimensional arrays:

1 Open the Code Editor window. Replace the <your name> and <current date> text in the comments with your name and the current date. The code to declare and initialize the arrays is located in the form's Declarations section, as shown in Figure 9-42. Notice that each array contains seven rows and four columns. The four columns in each array correspond to the "But not over", "Base amount", "Percentage", and "Of excess over" columns in each table. The last two zeroes in the first row in each array correspond to the empty blocks in the "Percentage" and "Of excess over" columns in the tax tables. The 99999 in the last row in each array represents the empty block in the "But not over" column in the tax tables. You can use any large number to represent the empty block in the "But not over" column, as long as the number is greater than the largest weekly taxable wage you expect the user to enter.

```
Public Class MainForm

    Private singleTable(,) As Double = {{51, 0, 0, 0}, _
                                        {192, 0, 0.1, 51}, _
                                        {620, 14.1, 0.15, 192}, _
                                        {1409, 78.3, 0.25, 620}, _
                                        {3013, 275.55, 0.28, 1409}, _
                                        {6508, 724.67, 0.33, 3013}, _
                                        {99999, 1878.02, 0.35, 6508}}

    Private marriedTable(,) As Double = {{154, 0, 0, 0}, _
                                         {440, 0, 0.1, 154}, _
                                         {1308, 28.6, 0.15, 440}, _
                                         {2440, 158.8, 0.25, 1308}, _
                                         {3759, 441.8, 0.28, 2440}, _
                                         {6607, 811.12, 0.33, 3759}, _
                                         {99999, 1750.96, 0.35, 6607}}
```

each array contains seven rows and four columns

Figure 9-42: Code to declare and initialize the two-dimensional arrays

Figure 9-43 shows the TOE chart for the Perrytown Gift Shop application. Almost all of the event procedures listed in the figure have been coded for you. You just need to code the xCalcButton's Click event procedure.

Task	Object	Event
1. Calculate the federal withholding tax 2. Display the federal withholding tax in the xFwtLabel	xCalcButton	Click
End the application	xExitButton	Click
Display the federal withholding tax (from xCalcButton)	xFwtLabel	None
Get and display the taxable wages Select the existing text Accept numbers, the period, and the Backspace Clear the contents of the xFwtLabel	xTaxableTextBox	None Enter Keypress TextChanged
Get the marital status Clear the contents of the xFwtLabel	xMarriedRadioButton, xSingleRadioButton	None Click

Figure 9-43: TOE chart for the Perrytown Gift Shop application

CODING THE XCALCBUTTON CLICK EVENT PROCEDURE

According to the TOE chart, the xCalcButton's Click event procedure is responsible for calculating the federal withholding tax (FWT) and displaying the calculated amount in the xFwtLabel. Figure 9-44 shows the pseudocode for the procedure.

xCalcButton Click event procedure - pseudocode

1. convert the taxable wages stored in the xTaxableTextBox to Double

2. if the xSingleRadioButton is selected
    assign the Single tax table, which is stored in the singleTable array, to the taxTable array
else
    assign the Married tax table, which is stored in the marriedTable array, to the taxTable array
end if

3. repeat until either there are no more rows to search in the taxTable array or the taxable wages have been
found
    if the taxable wages are less than or equal to the value stored in the first
    column of the current row in the taxTable array
        use the information stored in the second, third, and fourth columns in the
        taxTable array to calculate the federal withholding tax

        indicate that the taxable wages were found by assigning the value True
        to the isFound variable
    else
        add 1 to the contents of the row variable to continue the search in the next
        row in the taxTable array
    end if
end repeat

4. display the federal withholding tax in the xFwtLabel

5. send the focus to the xTaxableTextBox

Figure 9-44: Pseudocode for the xCalcButton's Click event procedure

To code the xCalcButton's Click event procedure:

1 Open the code template for the xCalcButton's Click event procedure. Type ' **calculates and displays the FWT** and press **Enter** twice.

Recall that before you begin coding a procedure, you first study the procedure's pseudocode to determine the variables and named constants (if any) the procedure will use. In this case, the procedure will use a two-dimensional array and four variables.

2 Type **dim taxTable(5, 3) as double** and press **Enter**. The procedure will use the `taxTable` array to store the appropriate tax table, which is based on the employee's marital status.

3 Type **dim taxableWages as double** and press **Enter**. The procedure will convert to Double the taxable wage amount entered by the user, and then store the result in the `taxableWages` variable.

4 Type **dim fwt as double** and press **Enter**. The fwt variable will store the employee's tax amount.

5 Type **dim row as integer** and press **Enter**. The procedure will use the row variable in a loop that searches for the taxable wages in the tax table.

6 Type **dim isFound as boolean** and press **Enter** twice. The isFound variable will store a Boolean value that indicates whether the taxable wages have been found in the tax table.

7 The first step in the pseudocode is to convert the taxable wages stored in the xTaxableTextBox to Double. Type **double.tryparse(me.xTaxableTextBox.text, taxablewages)** and press **Enter** twice.

8 Step 2 is to determine the selected radio button, and then assign the appropriate module-level array to the procedure-level taxTable array. Type the comments and selection structure shown in Figure 9-45, then position the insertion point as shown in the figure.

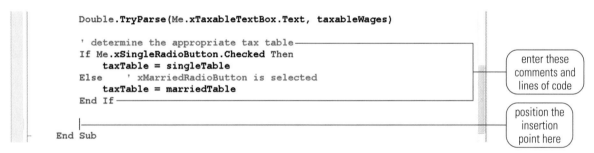

```
Double.TryParse(Me.xTaxableTextBox.Text, taxableWages)

' determine the appropriate tax table
If Me.xSingleRadioButton.Checked Then
    taxTable = singleTable
Else     ' xMarriedRadioButton is selected
    taxTable = marriedTable
End If

|
End Sub
```

enter these comments and lines of code

position the insertion point here

Figure 9-45: Comments and selection structure entered in the procedure

Before coding Step 3 in the pseudocode, you will learn about an array's GetUpperBound and GetLowerBound methods.

THE GETUPPERBOUND AND GETLOWERBOUND METHODS

Both an array's **GetUpperBound method** and its **GetLowerBound method** return an integer that indicates the highest subscript and lowest subscript, respectively, in the specified dimension in the array. Figure 9-46 shows the syntax of both methods and includes examples of using the methods. In each syntax, *arrayname* is the name of the array, and *dimension* is an integer that specifies the dimension whose upper or lower bound you want to retrieve. In a one-dimensional array, the *dimension* argument will always be 0 (zero). In a two-dimensional array, however, the *dimension* argument will be either 0 or 1. The zero represents the row dimension, and the 1 represents the column dimension.

GetUpperBound and GetLowerBound methods

Syntax

*arrayname.***GetUpperBound**(*dimension*)
*arrayname.***GetLowerBound**(*dimension*)

Examples

```
Dim upperRowSub As Integer
Dim lowerColSub As Integer
Dim products(20, 3) As String
upperRowSub = products.GetUpperBound(0)
lowerColSub = products.GetLowerBound(1)
```

assigns the number 20 to the `upperRowSub` variable, and assigns the number 0 to the `lowerColSub` variable

---

```
Dim lowerSub As Integer
Dim upperSub As Integer
Dim cities(5) As String
lowerSub = cities.GetLowerBound(0)
upperSub = cities.GetUpperBound(0)
```

assigns the number 0 to the `lowerSub` variable, and assigns the number 5 to the `upperSub` variable

Figure 9-46: Syntax and examples of the GetUpperBound and GetLowerBound methods

In the first example in Figure 9-46, the `upperRowSub = products.GetUpperBound(0)` statement assigns the number 20, which is the highest row subscript in the two-dimensional `products` array, to the `upperRowSub` variable. The `lowerColSub = products.GetLowerBound(1)` statement assigns the number 0, which is the lowest column subscript in the `products` array, to the `lowerColSub` variable. In the second example, the `lowerSub = cities.GetLowerBound(0)` statement assigns the number 0, which is the lowest subscript in the one-dimensional `cities` array, to the `lowerSub` variable. The `upperSub = cities.GetUpperBound(0)` statement assigns the number 5, which is the highest subscript in the `cities` array, to the `upperSub` variable. You will use the GetUpperBound method when coding the xCalcButton's Click event procedure.

To continue coding the xCalcButton's Click event procedure:

1 Step 3 in the pseudocode shown in Figure 9-44 is a loop that repeats its instructions either until there are no more rows to search in the `taxTable` array or until the taxable wages have been found. Type **' search for the taxable wages in the first column** and press **Enter**, then type **' in each row in the array** and press **Enter**. Type **do until row = taxtable.getupperbound(0) orelse isfound** and press **Enter**.

2 The first instruction in the loop is a selection structure that determines whether the taxable wages are less than or equal to the value stored in the first column of the current row in the `taxTable` array. Type **if taxablewages <= taxtable(row, 0) then** and press **Enter**.

3 The first instruction in the selection structure's true path should calculate the federal withholding tax, using the information stored in the second, third, and fourth columns in the taxTable array. Type **' calculate the FWT** and press **Enter**. Type **fwt = taxtable(row, 1) + taxtable(row, 2)** _ and press **Enter**. Press **Tab**, then type * **(taxablewages – taxtable(row, 3))** and press **Enter**.

4 The last instruction in the true path should assign the Boolean value True to the `isFound` variable to indicate that the taxable wages were located in the `taxTable` array. Type **isfound = true** and press **Enter**.

5 If the taxable wages are not less than or equal to the value stored in the first column of the current row in the taxTable array, the selection structure's false path should search the next row in the array. Type **else** and press **Enter**. Type **' continue searching for the taxable wages** and press **Enter**, then type **row = row + 1**.

6 The last two instructions in the pseudocode are to display the federal withholding tax in the xFwtLabel and then send the focus to the xTaxableTextBox. Position the insertion point in the blank line below the `Loop` clause, then press **Enter** to insert another blank line. Type **me.xFwtLabel.text = fwt.tostring("C2")** and press **Enter**, then type **me.xTaxableTextBox.focus()**.

7 Save the solution. Figure 9-47 shows the Perrytown Gift Shop application's code. (The following procedures are collapsed in the figure: xExitButton_Click, xTaxableTextBox_Enter, and ClearLabel.)

```
' Project name:      Perrytown Project
' Project purpose:   The project calculates the Federal
'                    Withholding Tax (FWT) based on the
'                    taxable wages and marital status
'                    entered by the user.
' Created/revised:   <your name> on <current date>

Option Explicit On
Option Strict On

Public Class MainForm
    Private singleTable(,) As Double = {{51, 0, 0, 0}, _
                                        {192, 0, 0.1, 51}, _
                                        {620, 14.1, 0.15, 192}, _
                                        {1409, 78.3, 0.25, 620}, _
                                        {3013, 275.55, 0.28, 1409}, _
                                        {6508, 724.67, 0.33, 3013}, _
                                        {99999, 1878.02, 0.35, 6508}}

    Private marriedTable(,) As Double = {{154, 0, 0, 0}, _
                                        {440, 0, 0.1, 154}, _
                                        {1308, 28.6, 0.15, 440}, _
                                        {2440, 158.8, 0.25, 1308}, _
                                        {3759, 441.8, 0.28, 2440}, _
                                        {6607, 811.12, 0.33, 3759}, _
                                        {99999, 1750.96, 0.35, 6607}}

    Private Sub xExitButton_Click...

    Private Sub xTaxableTextBox_Enter...

    Private Sub xTaxableTextBox_KeyPress(ByVal sender As Object, _
        ByVal e As System.Windows.Forms.KeyPressEventArgs) _
        Handles xTaxableTextBox.KeyPress
        ' accept only numbers, the period, and the Backspace key

        If (e.KeyChar < "0" OrElse e.KeyChar > "9") _
            AndAlso e.KeyChar <> "." _
            AndAlso e.KeyChar <> ControlChars.Back Then
            e.Handled = True
        End If
    End Sub
```

Figure 9-47: Perrytown Gift Shop application's code *(Continued)* ▶

```
    Private Sub ClearLabel...
    Private Sub xCalcButton_Click(ByVal sender As Object, _
        ByVal e As System.EventArgs) Handles xCalcButton.Click
        ' calculates and displays the FWT

        Dim taxTable(5, 3) As Double
        Dim taxableWages As Double
        Dim fwt As Double
        Dim row As Integer
        Dim isFound As Boolean

        Double.TryParse(Me.xTaxableTextBox.Text, taxableWages)

        ' determine the appropriate tax table
        If Me.xSingleRadioButton.Checked Then
            taxTable = singleTable
        Else    ' xMarriedRadioButton is selected
            taxTable = marriedTable
        End If

        ' search for the taxable wages in the first column
        ' in each row in the array
        Do Until row = taxTable.GetUpperBound(0) OrElse isFound
            If taxableWages <= taxTable(row, 0) Then
                ' calculate the FWT
                fwt = taxTable(row, 1) + taxTable(row, 2) _
                    * (taxableWages - taxTable(row, 3))
                isFound = True
            Else
                ' continue searching for the taxable wages
                row = row + 1
            End If
        Loop

        Me.xFwtLabel.Text = fwt.ToString("C2")
        Me.xTaxableTextBox.Focus()
    End Sub
End Class
```

Figure 9-47: Perrytown Gift Shop application's code

In the next set of steps, you will test the application to verify that it is working correctly.

To test the Perrytown Gift Shop application:

1 Start the application. First, you will calculate the FWT for a married taxpayer with taxable wages of $288.46. Type **288.46** in the Taxable wages text box, then click the **Married** radio button. Click the **Calculate** button. The application calculates and displays a tax of $13.45, as shown in Figure 9-48.

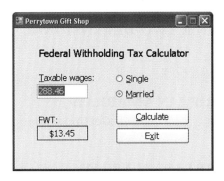

Figure 9-48: FWT displayed in the interface

2 Next, you will calculate the FWT for a single taxpayer with taxable wages of $600. Type **600** in the Taxable wages text box, then click the **Single** radio button. Click the **Calculate** button. The application calculates and displays a tax of $75.30.

3 Click the **Exit** button to end the application. Close the Code Editor window, then close the solution.

You have completed Lesson C and Chapter 9. You can either take a break or complete the end-of-lesson questions and exercises.

SUMMARY

TO DECLARE A TWO-DIMENSIONAL ARRAY:

» Use either of the following two syntax versions:

Version 1: {**Dim | Private**} *arrayname*(*highestRowSubscript*, *highestColumnSubscript*) **As** *datatype*

Version 2: {**Dim | Private**} *arrayname*(**,**) **As** *datatype* = {{*initialValues*}, {*initialValues*}, . . .{*initialValues*}}

» *Arrayname* is the name of the array, and *datatype* is the type of data the array variables will store.

» The *highestRowSubscript* and *highestColumnSubscript* arguments that appear in Version 1 of the syntax are integers that specify the highest row and column subscripts, respectively, in the array. Using Version 1's syntax, the computer automatically initializes the elements (variables) in the array.

» The *initialValues* section in Version 2 of the syntax allows you to specify the initial values for the array. You include a separate *initialValues* section for each row in the array. Each *initialValues* section should contain the same number of values as there are columns in the array.

TO REFER TO A VARIABLE INCLUDED IN A TWO-DIMENSIONAL ARRAY:

» Use the syntax *arrayname*(*rowSubscript*, *columnSubscript*).

TO DETERMINE THE HIGHEST SUBSCRIPT IN AN ARRAY:

» Use the GetUpperBound method, whose syntax is *arrayname*.**GetUpperBound** (*dimension*).

TO DETERMINE THE LOWEST SUBSCRIPT IN AN ARRAY:

» Use the GetLowerBound method, whose syntax is *arrayname*.**GetLowerBound** (*dimension*).

QUESTIONS

1. Which of the following statements creates a two-dimensional String array named `letters` that contains four rows and two columns?

   a. `Dim letters(1, 3) As String`

   b. `Dim letters(3, 1) As String`

   c. `Dim letters(2, 4) As String`

   d. `Dim letters(4, 2) As String`

   Use the `sales` array to answer Questions 2 through 4. The array was declared using the following statement:

   ```
   Dim sales(,) As Integer = {{1000, 1200, 900, 500, 2000}, _
                              {350, 600, 700, 800, 100}}
   ```

2. The `sales(1, 3) = sales(1, 3) + 10` statement will _____.

   a. replace the 900 amount with 910

   b. replace the 500 amount with 510

   c. replace the 700 amount with 710

   d. replace the 800 amount with 810

3. The `sales(0, 4) = sales(0, 4 - 2)` statement will _____.

   a. replace the 500 amount with 1200

   b. replace the 2000 amount with 900

   c. replace the 2000 amount with 19980

   d. result in an error

4. Which of the following If clauses can be used to verify that the array subscripts named `row` and `col` are valid for the `sales` array?

   a. `If sales(row, col) >= 0 AndAlso sales(row, col) < 5 Then`

   b. `If sales(row, col) >= 0 AndAlso sales(row, col) <= 5 Then`

   c. `If row >= 0 AndAlso row < 3 AndAlso col >= 0 AndAlso col < 6 Then`

   d. `If row >= 0 AndAlso row < 2 AndAlso col >= 0 AndAlso col < 5 Then`

5. Which of the following statements assigns the string "California" to the variable located in the third column, fifth row of a two-dimensional array named `states`?

   a. `states(3, 5) = "California"`

   b. `states(5, 3) = "California"`

   c. `states(2, 4) = "California"`

   d. `states(4, 2) = "California"`

6. Which of the following assigns the number zero to each element in a two-dimensional Integer array named `sums`? The `sums` array contains two rows and four columns.

   a.
```
For row As Integer = 0 To 1
      For column As Integer = 0 To 3
            sums(row, column) = 0
      Next column
 Next row
```

b.
```
Dim row As Integer
Dim column As Integer
Do While row < 2
    column = 0
    Do While column < 4
        sums(row, column) = 0
        column = column + 1
    Loop
    row = row + 1
Loop
```

c.
```
For x As Integer = 1 To 2
    For y As Integer = 1 To 4
        sums(x - 1, y - 1) = 0
    Next y
Next x
```

d. All of the above.

EXERCISES

1. Write the statement to declare a procedure-level two-dimensional Decimal array named `balances`. The array should have four rows and six columns. Then write a For...Next loop that stores the number 10 in each element in the array.

2. Rewrite the loop from Exercise 1 using a Do...Loop statement.

3. Write the statement to assign the Boolean value True to the variable located in the third row, first column of a Boolean array named `answers`.

4. In this exercise, you code an application that sums the values stored in a two-dimensional array.

   a. If necessary, start Visual Studio 2005 or Visual Basic 2005 Express Edition. Open the Inventory Solution (Inventory Solution.sln) file, which is contained in the VB2005\Chap09\Inventory Solution folder. If necessary, open the designer window.

   b. Code the Display Total button's Click event procedure so that it adds together the values stored in the `inventory` array. Display the sum in the xTotalLabel.

   c. Save the solution, and then start the application. Click the Display Total button to display the sum of the array values. Click the Exit button to end the application. Close the Code Editor window, then close the solution.

5. In this exercise, you code an application that sums the values stored in a two-dimensional array.

   a. If necessary, start Visual Studio 2005 or Visual Basic 2005 Express Edition. Open the Conway Solution (Conway Solution.sln) file, which is contained in the VB2005\Chap09\Conway Solution folder. If necessary, open the designer window.

   b. Code the Display Totals button's Click event procedure so that it displays the total domestic sales, total international sales, and total company sales in the appropriate label controls.

   c. Save the solution, and then start the application. Click the Display Totals button. The button's Click event procedure should display domestic sales of $235,000, international sales of $177,000, and company sales of $412,000. Click the Exit button to end the application. Close the Code Editor window, and then close the solution.

6. In this exercise, you code an application that displays the number of times a value appears in a two-dimensional array.

   a. If necessary, start Visual Studio 2005 or Visual Basic 2005 Express Edition. Open the Count Solution (Count Solution.sln) file, which is contained in the VB2005\Chap09\Count Solution folder. If necessary, open the designer window.

   b. Code the Display Count button's Click event procedure so that it displays the number of times each of the numbers from one through nine appears in the `numbers` array. (*Hint*: Store the counts in a one-dimensional array.)

   c. Save the solution, and then start the application. Click the Display Count button to display the nine counts. Click the Exit button to end the application. Close the Code Editor window, then close the solution.

7. In this exercise, you code an application that displays the highest score earned on the midterm exam and the highest score earned on the final exam.

   a. If necessary, start Visual Studio 2005 or Visual Basic 2005 Express Edition. Open the Highest Solution (Highest Solution.sln) file, which is contained in the VB2005\Chap09\Highest Solution folder. If necessary, open the designer window.

   b. Code the Display Highest button's Click event procedure so that it displays (in the appropriate label controls) the highest score earned on the midterm exam and the highest score earned on the final exam.

   c. Save the solution, and then start the application. Click the Display Highest button to display the highest scores earned on the midterm and final exams. Click the Exit button to end the application. Close the Code Editor window, then close the solution.

DEBUGGING EXERCISE

8. In this exercise, you debug an existing application.

a. If necessary, start Visual Studio 2005 or Visual Basic 2005 Express Edition. Open the Debug Solution (Debug Solution.sln) file, which is contained in the VB2005\Chap09\Debug Solution folder. If necessary, open the designer window.

b. Open the Code Editor window and review the existing code. Notice that the names array contains five rows and two columns. Column one contains five first names, and column two contains five last names. The xDisplayButton's Click event procedure should display the first and last names in the xFirstLabel and xLastLabel controls, respectively.

c. Notice that a jagged line appears below some of the lines of code in the Code Editor window. Correct the code to remove the jagged lines. Save the solution, then start the application. Click the Display button. If the "IndexOutOfRangeException was unhandled" error message appears in a box, read the troubleshooting tips, then click the Close button. Click Debug on the menu bar, and then click Stop Debugging.

d. Correct the errors in the application's code, then save the solution and start the application. Click the Display button to display the first and last names in the appropriate labels. Click the Exit button to end the application. Close the Code Editor window, then close the solution.

10

STRUCTURES AND SEQUENTIAL ACCESS FILES

CREATING THE FRIENDS APPLICATION

In this chapter, you code an application that allows you to add the names of your friends to a combo box, and also delete one or more names from the combo box. The application will store the names in a sequential access file. You will learn about sequential access files in Lesson B. You will code the Friends application in Lesson C.

PREVIEWING THE COMPLETED APPLICATION

Before creating the Friends application, you will preview the completed application.

To preview the completed application:

1 Use the Run command on the Windows Start menu to run the **Friends** (**Friends.exe**) file, which is contained in the VB2005\Chap10 folder. The Friends application's user interface appears on the screen.

2 First, you will add a name to the Friends combo box. Type **Jameston, Carol** in the text portion of the combo box, then click the **Add** button to add the name to the list portion of the combo box. See Figure 10-1.

Figure 10-1: Name added to the combo box

3 Now you will add two more names. Type **Adair, Mark** in the text portion of the combo box, then click the **Add** button. Type **Staney, Jane**, then click the **Add** button. The combo box now contains three names.

4 Next, you will remove a name from the combo box. Click **Jameston, Carol** in the combo box, then click the **Remove** button. The combo box now contains only two names.

5 Click the **Exit** button to end the application. The application saves the remaining two names in a sequential access file named friends.txt. The file is located in the VB2005\ Chap10 folder.

6 Run the **Friends** (**Friends.exe**) file again. The application reads the two names stored in the friends.txt file and displays the names in the combo box, as shown in Figure 10-2.

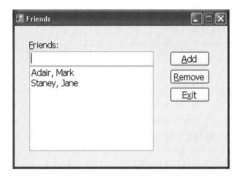

Figure 10-2: The combo box contains the names stored in the file

7 Click the **Exit** button to close the application.

LESSON A
OBJECTIVES

AFTER STUDYING LESSON A, YOU SHOULD
BE ABLE TO:

» Create a structure

» Declare a structure variable

» Pass a structure variable to a procedure

» Create and manipulate a one-dimensional array of
structures

STRUCTURES

CREATING A STRUCTURE

In previous chapters, you used only the data types built into Visual Basic, such as the Integer, Decimal, and String data types. You also can create your own data types in Visual Basic using the **Structure statement**. Data types created using the Structure statement are referred to as **user-defined data types** or **structures**. Figure 10-3 shows the syntax of the Structure statement and includes an example of using the statement to create a structure (user-defined data type) named Employee.

Structure statement

<u>Syntax</u>

Structure *structureName*
    Public *memberVariable1* **As** *datatype*
    [**Public** *memberVariableN* **As** *datatype*]
End Structure

<u>Example</u>

```
Structure Employee
     Public number As String
     Public firstName As String
     Public lastName As String
     Public salary As Decimal
End Structure
```

Figure 10-3: Syntax and an example of the Structure statement

 TIP
Most programmers use the Class statement, rather than the Structure statement, to create data types that contain procedures. You will learn about the Class statement in Chapter 11.

The Structure statement begins with the Structure clause, which contains the keyword `Structure` followed by the name of the structure. In the example shown in Figure 10-3, the name of the structure is Employee. The Structure statement ends with the End Structure clause, which contains the keywords `End Structure`. Between the Structure and End Structure clauses, you define the members included in the structure. The members can be variables, constants, or procedures. However, in most cases, the members will be variables; such variables are referred to as **member variables**. In this book, you learn how to include only member variables in a structure.

TIP
The Structure statement merely defines the structure. It does not actually create a structure variable.

As the syntax shown in Figure 10-3 indicates, each member variable's definition contains the keyword `Public` followed by the name of the variable, the keyword `As`, and the variable's *datatype*. The *datatype* identifies the type of data the member variable will store and can be any of the standard data types available in Visual Basic; it also can be another structure (user-defined data type). The Employee structure shown in Figure 10-3 contains four member variables: three are String variables and one is a Decimal variable.

In most applications, you enter the Structure statement in the form's Declarations section in the Code Editor window. Recall that the form's Declarations section begins with the Public Class clause and ends with the End Class clause. After entering the Structure statement, you then can use the structure to declare a variable.

USING A STRUCTURE TO DECLARE A VARIABLE

As you can with the standard data types built into Visual Basic, you also can use a structure (user-defined data type) to declare a variable. Variables declared using a structure are often referred to as **structure variables**. Figure 10-4 shows the syntax for creating a structure variable. The figure also includes examples of declaring structure variables using the Employee structure from Figure 10-3. In the syntax, *structureVariableName* is the name of the structure variable you are declaring and *structureName* is the name of the structure (user-defined data type).

Declaring a structure variable

<u>Syntax</u>

{**Dim** | **Private**} *structureVariableName* **As** *structureName*

Examples

```
Dim manager As Employee
```
declares a procedure-level Employee variable named manager

```
Private salaried As Employee
```
declares a module-level Employee variable named salaried

Figure 10-4: Syntax and examples of declaring a structure variable

Similar to the way the `Dim age As Integer` instruction declares an Integer variable named age, the `Dim manager As Employee` instruction, which is shown in the first example in Figure 10-4, declares an Employee structure variable named manager. However, unlike the age variable, the manager variable contains four member variables.

In code, you refer to the entire structure variable by its name; in this case, you refer to it using the name manager. To refer to an individual member variable within a structure variable, you precede the member variable's name with the name of the structure variable in which it is defined. You use the dot member access operator (a period) to separate the structure variable's name from the member variable's name. For instance, the names of the member variables within the manager structure variable are manager.number, manager.firstName, manager.lastName, and manager.salary. The Private salaried As Employee instruction shown in the second example in Figure 10-4 declares a module-level Employee structure variable named salaried. The names of the members within the salaried variable are salaried.number, salaried. firstName, salaried.lastName, and salaried.salary.

The member variables contained in a structure variable can be used just like any other variables. For example, you can assign values to them, use them in calculations, display their contents, and so on. Figure 10-5 shows various ways of manipulating the member variables created by the statements shown in Figure 10-4.

> **» TIP**
> The dot member access operator indicates that number, firstName, lastName, and salary are members of the manager and salaried variables.

Examples of manipulating member variables

```
manager.number = "0477"
manager.firstName = "Janice"
manager.lastName = "Lopenski"
manager.salary = 34500D
```

assigns data to the member variables contained in the manager variable

```
manager.salary = manager.salary * 1.05D
```

multiplies the contents of the manager.salary member variable by 1.05, and then assigns the result to the member variable

```
salaried.firstName = Me.xFirstTextBox.Text
```

assigns the value entered in the xFirstTextBox to the salaried.firstName member variable

```
Me.xSalaryTextBox.Text = salaried.salary.ToString("C2")
```

formats the value contained in the salaried.salary member variable to Currency with two decimal places, and then displays the result in the xSalaryTextBox

Figure 10-5: Examples of manipulating member variables

In the first example shown in Figure 10-5, the first three assignment statements assign the strings "0477", "Janice", and "Lopenski" to the String members of the `manager` variable. The fourth assignment statement in the example assigns a Decimal number to the Decimal member of the `manager` variable. The assignment statement in the second example multiplies the contents of the `manager.salary` member variable by 1.05, and then assigns the result to the member variable. The assignment statement in the third example assigns the value entered in the xFirstTextBox to the `salaried.firstName` member variable. The assignment statement in the last example displays the value stored in the `salaried.salary` member variable, formatted with a dollar sign and two decimal places, in the xSalaryTextBox.

Programmers use structures (user-defined data types) to group related items into one unit. The advantages of doing this will become more apparent as you read through the next two sections.

PASSING A STRUCTURE VARIABLE TO A PROCEDURE

The sales manager at Willow Pools wants an application that allows the salespeople to enter the length, width, and depth of a rectangular pool. The application should determine the amount of water required to fill the pool. You can make this determination by calculating the volume of the pool. To calculate the volume, you multiply the pool's length by its width, and then multiply the result by the pool's depth. Figure 10-6 shows a sample run of the Willow Pools application, and Figure 10-7 shows how you can code the application without using a structure.

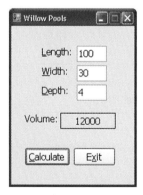

Figure 10-6: Sample run of the Willow Pools application

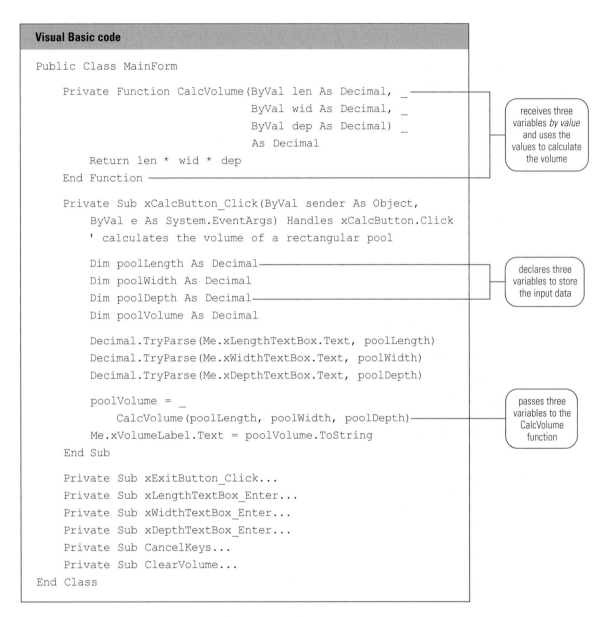

Visual Basic code

```
Public Class MainForm

    Private Function CalcVolume(ByVal len As Decimal, _
                          ByVal wid As Decimal, _
                          ByVal dep As Decimal) _
                          As Decimal
        Return len * wid * dep
    End Function

    Private Sub xCalcButton_Click(ByVal sender As Object, _
        ByVal e As System.EventArgs) Handles xCalcButton.Click
        ' calculates the volume of a rectangular pool

        Dim poolLength As Decimal
        Dim poolWidth As Decimal
        Dim poolDepth As Decimal
        Dim poolVolume As Decimal

        Decimal.TryParse(Me.xLengthTextBox.Text, poolLength)
        Decimal.TryParse(Me.xWidthTextBox.Text, poolWidth)
        Decimal.TryParse(Me.xDepthTextBox.Text, poolDepth)

        poolVolume = _
            CalcVolume(poolLength, poolWidth, poolDepth)
        Me.xVolumeLabel.Text = poolVolume.ToString
    End Sub

    Private Sub xExitButton_Click...
    Private Sub xLengthTextBox_Enter...
    Private Sub xWidthTextBox_Enter...
    Private Sub xDepthTextBox_Enter...
    Private Sub CancelKeys...
    Private Sub ClearVolume...
End Class
```

receives three variables *by value* and uses the values to calculate the volume

declares three variables to store the input data

passes three variables to the CalcVolume function

Figure 10-7: Code for the Willow Pools application (without a structure)

When the user clicks the Calculate button in the interface, the xCalcButton's Click event procedure declares four Decimal variables; it then assigns the user's input to three of the variables. Next, the poolVolume = CalcVolume(poolLength, poolWidth, poolDepth) statement calls the CalcVolume function, passing it three variables *by value*. The CalcVolume function uses the values stored in the variables to calculate the volume of the pool, which it returns as a Decimal number. The statement assigns the function's return value to the `poolVolume` variable. Finally, the procedure converts the contents of the `poolVolume` variable to String and displays the result in the xVolumeLabel.

Figure 10-8 shows a more convenient way of writing the code for the Willow Pools application. In this version of the code, a structure named Dimensions is used to group together the input data. Notice that the Structure statement that defines the Dimensions structure is entered in the MainForm's Declarations section. The Dimensions structure contains three member variables named `length`, `width`, and `depth`.

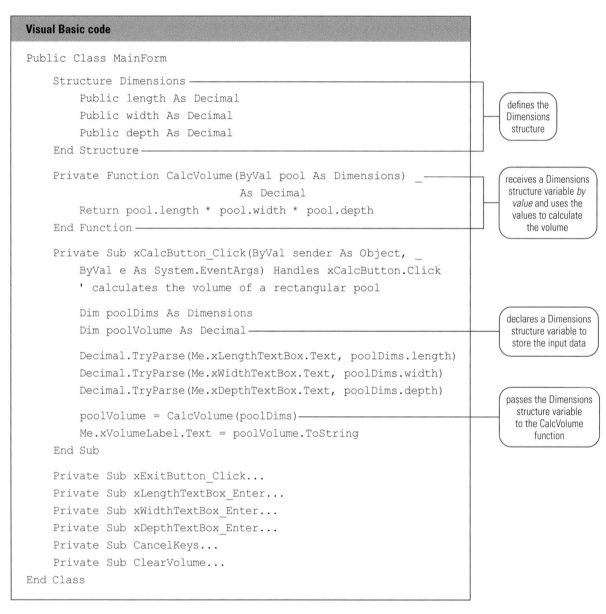

Visual Basic code

```vb
Public Class MainForm

    Structure Dimensions          ── defines the Dimensions structure
        Public length As Decimal
        Public width As Decimal
        Public depth As Decimal
    End Structure

    Private Function CalcVolume(ByVal pool As Dimensions) _    ── receives a Dimensions structure variable by value and uses the values to calculate the volume
                        As Decimal
        Return pool.length * pool.width * pool.depth
    End Function

    Private Sub xCalcButton_Click(ByVal sender As Object, _
        ByVal e As System.EventArgs) Handles xCalcButton.Click
        ' calculates the volume of a rectangular pool

        Dim poolDims As Dimensions
        Dim poolVolume As Decimal          ── declares a Dimensions structure variable to store the input data

        Decimal.TryParse(Me.xLengthTextBox.Text, poolDims.length)
        Decimal.TryParse(Me.xWidthTextBox.Text, poolDims.width)
        Decimal.TryParse(Me.xDepthTextBox.Text, poolDims.depth)

        poolVolume = CalcVolume(poolDims)          ── passes the Dimensions structure variable to the CalcVolume function
        Me.xVolumeLabel.Text = poolVolume.ToString
    End Sub

    Private Sub xExitButton_Click...
    Private Sub xLengthTextBox_Enter...
    Private Sub xWidthTextBox_Enter...
    Private Sub xDepthTextBox_Enter...
    Private Sub CancelKeys...
    Private Sub ClearVolume...
End Class
```

Figure 10-8: Code for the Willow Pools application (with a structure)

When the user clicks the Calculate button in the Willow Pools interface, the xCalcButton's Click event procedure declares a Dimensions structure variable named poolDims and a Decimal variable named poolVolume. It then assigns the user's input to the three member variables contained in the structure variable. The poolVolume = CalcVolume(poolDims) statement calls the CalcVolume function, passing it the poolDims structure variable *by value*. When you pass a structure variable to a Sub or Function procedure, all of its member variables are automatically passed. The CalcVolume function uses the values stored in the structure's member variables to calculate the volume of the pool, which it returns as a Decimal number. The statement assigns the function's return value to the poolVolume variable. Finally, the procedure converts the contents of the poolVolume variable to String and displays the result in the xVolumeLabel.

Notice that the xCalcButton's Click event procedure shown earlier in Figure 10-7 uses three scalar variables to store the input data. However, in Figure 10-8's code, the procedure uses only one structure variable for this purpose. The xCalcButton's Click event procedure in Figure 10-7 also must pass three scalar variables (rather than one structure variable) to the CalcVolume function, which must use three scalar variables (rather than one structure variable) to accept the data. Imagine if the input data consisted of 20 items rather than just three items! Passing a structure variable would be much less work than passing 20 individual scalar variables.

>> **TIP**

Recall from Chapter 9 that a scalar variable (also called a simple variable) is one that is unrelated to any other variable in memory.

To code and then test the Willow Pools application:

1 Start Visual Studio 2005 or Visual Basic 2005 Express Edition, if necessary, and close the Start Page window. Open the **Pool Solution** (Pool Solution.sln) file, which is contained in the VB2005\Chap10\Pool Solution-with structure folder. If necessary, open the designer window.

2 Open the Code Editor window. Replace the <your name> and <current date> text in the comments with your name and the current date.

3 Enter the Structure statement and CalcVolume function shown in Figure 10-8.

4 Locate the xCalcButton's Click event procedure, then enter the code shown in Figure 10-8.

5 Close the Code Editor window. Save the solution, then start the application. Type **100** in the Length box, then type **30** in the Width box. Type **4** in the Depth box, then click the **Calculate** button. The application displays the number 12000 as the volume, as shown earlier in Figure 10-6.

6 Click the **Exit** button to end the application, then close the solution.

As you will learn in the next section, another advantage of grouping related data into one unit is that the unit then can be stored in an array.

CREATING AN ARRAY OF STRUCTURE VARIABLES

In Chapter 9, you learned how to use two parallel one-dimensional arrays to store a price list for Takoda Tapahe, the owner of a small gift shop named Treasures. (The code is shown in Figure 9-30 in Chapter 9.) As you may remember, you stored each product's ID in a one-dimensional String array, and stored each product's price in the corresponding location in a one-dimensional Integer array. Also in Chapter 9, you learned how to store the Treasures price list in a two-dimensional String array. (The code is shown in Figure 9-39 in Chapter 9.) In addition to using parallel one-dimensional arrays or a two-dimensional array, you also can use a one-dimensional array of structure variables. Figure 10-9 shows a sample run of the Price List application, and Figure 10-10 shows the application's code using an array of structure variables.

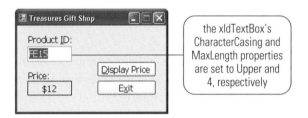

the xIdTextBox's CharacterCasing and MaxLength properties are set to Upper and 4, respectively

Figure 10-9: Sample run of the Price List application

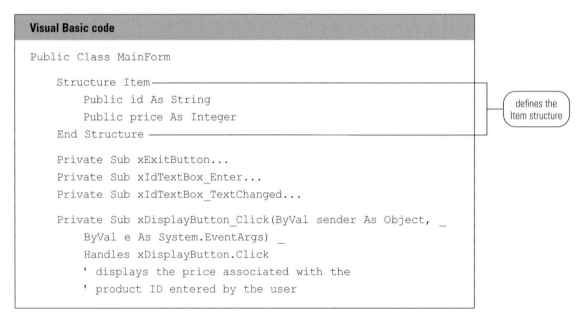

```
Visual Basic code

Public Class MainForm

    Structure Item
        Public id As String
        Public price As Integer
    End Structure

    Private Sub xExitButton...
    Private Sub xIdTextBox_Enter...
    Private Sub xIdTextBox_TextChanged...

    Private Sub xDisplayButton_Click(ByVal sender As Object, _
        ByVal e As System.EventArgs) _
        Handles xDisplayButton.Click
        ' displays the price associated with the
        ' product ID entered by the user
```

defines the Item structure

Figure 10-10: Code for the Price List application using an array of structure variables *(Continued)* ▶

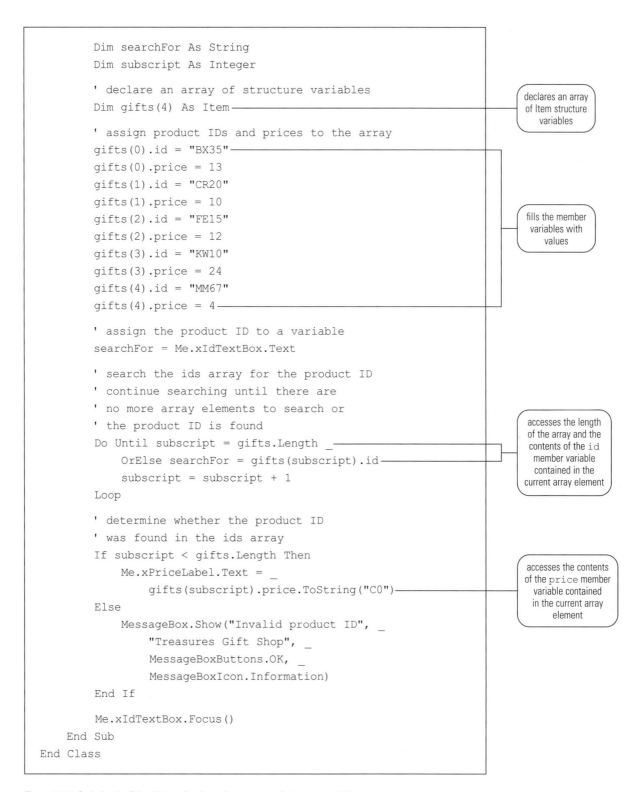

```
            Dim searchFor As String
            Dim subscript As Integer

            ' declare an array of structure variables
            Dim gifts(4) As Item

            ' assign product IDs and prices to the array
            gifts(0).id = "BX35"
            gifts(0).price = 13
            gifts(1).id = "CR20"
            gifts(1).price = 10
            gifts(2).id = "FE15"
            gifts(2).price = 12
            gifts(3).id = "KW10"
            gifts(3).price = 24
            gifts(4).id = "MM67"
            gifts(4).price = 4

            ' assign the product ID to a variable
            searchFor = Me.xIdTextBox.Text

            ' search the ids array for the product ID
            ' continue searching until there are
            ' no more array elements to search or
            ' the product ID is found
            Do Until subscript = gifts.Length _
                OrElse searchFor = gifts(subscript).id
                subscript = subscript + 1
            Loop

            ' determine whether the product ID
            ' was found in the ids array
            If subscript < gifts.Length Then
                Me.xPriceLabel.Text = _
                    gifts(subscript).price.ToString("C0")
            Else
                MessageBox.Show("Invalid product ID", _
                    "Treasures Gift Shop", _
                    MessageBoxButtons.OK, _
                    MessageBoxIcon.Information)
            End If

            Me.xIdTextBox.Focus()
        End Sub
    End Class
```

declares an array of Item structure variables

fills the member variables with values

accesses the length of the array and the contents of the id member variable contained in the current array element

accesses the contents of the price member variable contained in the current array element

Figure 10-10: Code for the Price List application using an array of structure variables

The xDisplayButton's Click event procedure declares a String variable named searchFor and an Integer variable named subscript. The searchFor variable will store the product ID entered by the user, and the subscript variable will keep track of the array subscripts. The procedure also declares a five-element one-dimensional array named gifts, using the Item structure defined in the MainForm's Declarations section as the array's data type. Each element in the gifts array is a structure variable that contains two member variables: a String variable named id and an Integer variable named price. After declaring the array, the procedure populates the array by assigning the appropriate IDs and prices to it. Notice that you refer to a member variable in an array element using the syntax *arrayname*(*subscript*).*memberVariableName*. For example, you use gifts(0).id to refer to the id member contained in the first element in the gifts array. Likewise, you use gifts(4).price to refer to the price member contained in the last element in the gifts array. Figure 10-11 illustrates this naming convention.

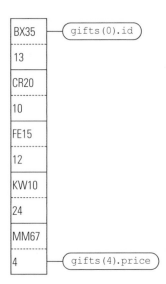

Figure 10-11: Names of some of the member variables in the gifts array

After populating the gifts array, the xDisplayButton's Click event procedure assigns the contents of the xIdTextBox to the searchFor variable. The loop in the procedure then searches for the product ID in each id member in the gifts array, stopping when there are no more id member elements to search or when the product ID is located. After the loop completes its processing, the selection structure in the procedure compares the number stored in the subscript variable with the value stored in the gifts array's Length property. In this case, the Length property contains the number five, because there are five elements in the gifts array. If the subscript variable contains a number that is less than five, it indicates that the loop stopped processing because the product ID was located in an `id` member in the array. In that case, the procedure displays the price from the corresponding `price` member in the array. However, if the `subscript` variable's value is not less than five, it indicates that the loop stopped processing because it reached the end of the array without finding the product ID. In that case, the message "Invalid product ID" is displayed in a message box. As Figure 10-9 indicates, the procedure displays a price of $12 if the user enters FE15 as the product ID.

To code and then test the Price List application:

1 Open the **Price List Solution** (Price List Solution.sln) file, which is contained in the VB2005\Chap10\Price List Solution-Structure folder. If necessary, open the designer window.

2 Open the Code Editor window. Replace the <your name> and <current date> text in the comments with your name and the current date. Enter the Structure statement and xDisplayButton's Click event procedure shown in Figure 10-10.

3 Close the Code Editor window. Save the solution, then start the application. Type **fe15** in the Product ID box, then click the **Display Price** button. The application displays $12 as the price, as shown earlier in Figure 10-9. Click the **Exit** button to end the application, then close the solution.

You have completed Lesson A. You can either take a break or complete the end-of-lesson questions and exercises before moving on to Lesson B.

> **» TIP**
> Recall that the value stored in an array's Length property is always one number more than the highest subscript in the array. You also could have written the first expression in the Do loop's condition as `subscript > gifts.GetUpperBound(0)`.

SUMMARY

TO CREATE A STRUCTURE (USER-DEFINED DATA TYPE):

» Use the Structure statement, whose syntax is shown in Figure 10-3. In most applications, you enter the Structure statement in the form's Declarations section.

TO DECLARE A STRUCTURE VARIABLE:

» Use the following syntax: {**Dim** | **Private**} *structureVariableName* **As** *structureName*.

TO REFER TO A MEMBER VARIABLE WITHIN A STRUCTURE VARIABLE:

» Use the syntax *structureVariableName.memberVariableName*.

TO CREATE AN ARRAY OF STRUCTURE VARIABLES:

» Use the Structure statement to create a structure. Declare the array using the structure's name as the data type.

TO REFER TO A MEMBER VARIABLE WITHIN A STRUCTURE VARIABLE STORED IN AN ARRAY:

» Use the syntax *arrayname***(***subscript***).***memberVariableName*.

QUESTIONS

1. You use the _____ statement to create a user-defined data type.

   a. Declare

   b. Define

   c. Structure

   d. UserType

2. In most applications, the code to define a user-defined data type is entered in the form's _____.

   a. Declarations section

   b. Definition section

   c. Load event procedure

   d. User-defined section

3. Which of the following statements assigns the string "Maple" to the `street` member variable within a structure variable named `address`?

   a. `address&street = "Maple"`

   b. `address.street = "Maple"`

   c. `street.address = "Maple"`

   d. None of the above.

4. An array is declared using the following statement: `Dim inventory(4) As Product`. Which of the following statements assigns the number 100 to the `quantity` member variable contained in the last array element?

   a. `inventory.quantity(4) = 100`

   b. `inventory(4).Product.quantity = 100`

   c. `inventory(3).quantity = 100`

   d. None of the above.

5. An application uses a structure named Employee. Which of the following statements creates a five-element one-dimensional array of Employee structure variables?

a. `Dim workers(4) As Employee`

b. `Dim workers(5) As Employee`

c. `Dim workers As Employee(4)`

d. `Dim workers As Employee(5)`

EXERCISES

1. Write a Structure statement that defines a structure named Book. The structure contains two String member variables named `title` and `author`, and a Decimal member variable named `cost`. Then write a Private statement that declares a Book variable named `fiction`.

2. Write a Structure statement that defines a structure named Tape. The structure contains three String member variables named `name`, `artist`, and `songLength`, and an Integer member variable named `songNum`. Then write a Dim statement that declares a Tape variable named `blues`.

3. An application contains the following Structure statement:

```
Structure Computer
        Public model As String
        Public cost As Decimal
End Structure
```

a. Write a Dim statement that declares a Computer variable named `homeUse`.

b. Write an assignment statement that assigns the string "IB-50" to the `model` member variable.

c. Write an assignment statement that assigns the number 2400 to the `cost` member variable.

d. Write a Private statement that declares a 10-element one-dimensional array of Computer variables. Name the array `business`.

e. Write an assignment statement that assigns the string "HPP405" to the `model` member variable contained in the first array element.

f. Write an assignment statement that assigns the number 3600 to the `cost` member variable contained in the first array element.

4. An application contains the following Structure statement:

```
Structure Friend
        Public last As String
        Public first As String
End Structure
```

a. Write a Dim statement that declares a Friend variable named school.

b. Write an assignment statement that assigns the value in the xFirstTextBox to the first member variable.

c. Write an assignment statement that assigns the value in the xLastTextBox to the last member variable.

d. Write an assignment statement that assigns the value in the last member variable to the xLastLabel.

e. Write an assignment statement that assigns the value in the first member variable to the xFirstLabel.

f. Write a Private statement that declares a five-element one-dimensional array of Friend variables. Name the array home.

g. Write an assignment statement that assigns the value in the xFirstTextBox to the first member variable contained in the last array element.

h. Write an assignment statement that assigns the value in the xLastTextBox to the last member variable contained in the last array element.

5. In this exercise, you modify the Price List application that you completed in the lesson.

a. Use Windows to make a copy of the Price List Solution-Structure folder, which is contained in the VB2005\Chap10 folder. Rename the folder Modified Price List Solution-Structure.

b. If necessary, start Visual Studio 2005 or Visual Basic 2005 Express Edition. Open the Price List Solution (Price List Solution.sln) file contained in the VB2005\Chap10\Modified Price List Solution-Structure folder. Open the designer window.

c. Modify the application so that it displays both the name and price corresponding to the product ID entered by the user. You will need to modify the interface and the code (including the Structure statement). The names of the products are shown in the following chart.

Product ID	Name
BX35	Necklace
CR20	Bracelet
FE15	Jewelry box
KW10	Doll
MM67	Ring

   d. Save the solution, then start and test the application. Click the Exit button to end the application. Close the Code Editor window, then close the solution.

6. In this exercise, you code an application that allows Professor Carver to display a grade based on the number of points he enters. The grading scale is shown in Figure 10-12.

Minimum points	Maximum points	Grade
0	299	F
300	349	D
350	399	C
400	449	B
450	500	A

Figure 10-12

   a. If necessary, start Visual Studio 2005 or Visual Basic 2005 Express Edition. Open the Carver Solution (Carver Solution.sln) file, which is contained in the VB2005\Chap10\Carver Solution folder. If necessary, open the designer window.

   b. Create a structure that contains two fields: an Integer field for the minimum points and a String field for the grades.

   c. Use the structure to declare a five-element one-dimensional array named gradeScale. Store the minimum points and grades in the gradeScale array.

   d. The application should search the gradeScale array for the number of points earned, and then display the appropriate grade from the array.

   e. Save the solution, then start and test the application. Click the Exit button to end the application. Close the Code Editor window, then close the solution.

LESSON B
OBJECTIVES

AFTER STUDYING LESSON B, YOU SHOULD
BE ABLE TO:

» Write information to a sequential access file

» Align the text written to a sequential access file

» Read information from a sequential access file

» Determine whether a file exists

SEQUENTIAL ACCESS FILES

FILE TYPES

In addition to getting information from the keyboard and sending information to the computer screen, an application also can get information from and send information to a file on a disk. Getting information from a file is referred to as "reading the file," and sending information to a file is referred to as "writing to the file." Files to which information is written are called **output files**, because the files store the output produced by an application. Files that are read by the computer are called **input files**, because an application uses the information in these files as input.

You can create three different types of files in Visual Basic: sequential, random, and binary. The file type refers to the way the information in the file is accessed. The information in a sequential access file is always accessed sequentially. In other words, it is always accessed in consecutive order from the beginning of the file through the end of the file. The information stored in a random access file can be accessed either in consecutive order or in random order. The information in a binary access file can be accessed by its byte location in the file. You learn about sequential access files in this chapter. Random access and binary access files are used less often in programs, so these file types are not covered in this book.

USING SEQUENTIAL ACCESS FILES

A **sequential access file** is often referred to as a **text file**, because it is composed of lines of text. The text might represent an employee list, as shown in Example 1 in Figure 10-13. It also might be a memo or a report, as shown in Examples 2 and 3 in the figure.

Examples

Example 1 — employee list

```
Bonnel, Jacob
Carlisle, Donald
Eberg, Jack
Hou, Chang
```

Example 2 — memo

```
To all employees:

Effective January 1, 2008, the cost of dependent coverage will
increase from $35 to $38.50 per month.

Jefferson Williams
Insurance Manager
```

Example 3 — report

```
ABC Industries Sales Report

State           Sales
California       15000
Montana          10000
Wyoming           7000
                 ------
Total sales:     $32000
```

Figure 10-13: Examples of text stored in a sequential access file

Sequential access files are similar to cassette tapes in that each line in the file, like each song on a cassette tape, is both stored and retrieved in consecutive order (sequentially). In other words, before you can record (store) the fourth song on a cassette tape, you first must record songs one through three. Likewise, before you can write (store) the fourth line in a sequential access file, you first must write lines one through three. The same holds true for retrieving a song from a cassette tape and a line of text from a sequential access file. To listen to the fourth song on a cassette tape, you must play (or fast-forward through) the first three songs. Likewise, to read the fourth line in a sequential access file, you first must read the three lines that precede it.

WRITING INFORMATION TO A SEQUENTIAL ACCESS FILE

You can use the **WriteAllText method** to write information to a sequential access file. Figure 10-14 shows the syntax of the WriteAllText method and includes examples of using the method to write text to a sequential access file.

Writing text to a sequential access file

Syntax

My.Computer.FileSystem.WriteAllText(*file*, *text*, *append***)**

Example 1

```
My.Computer.FileSystem.WriteAllText("msg.txt", "Hi", False)
```

Result

Hi|

Example 2

```
My.Computer.FileSystem.WriteAllText("msg.txt", _
    "Hi" & ControlChars.NewLine, False)
```

Result

Hi
|

Example 3

```
Const File As String = "top.txt"
Const Line As String = "The top salesperson is "
Dim name As String = "Carolyn"

My.Computer.FileSystem.WriteAllText(File, Line, False)
My.Computer.FileSystem.WriteAllText(File, name & ".", True)
My.Computer.FileSystem.WriteAllText(File, _
    ControlChars.NewLine & ControlChars.NewLine, True)
My.Computer.FileSystem.WriteAllText(File, "ABC Sales", True)
```

Result

The top salesperson is Carolyn.

ABC Sales|

Example 4

```
Dim file As String = "C:\VB2005\Chap10\report.txt"
Dim price As Decimal = 5.6D
My.Computer.FileSystem.WriteAllText(file, _
    "Total price: ", False)
My.Computer.FileSystem.WriteAllText(file, _
    price.ToString("C2") & ControlChars.NewLine, True)
```

Result

Total price: $5.60
    |

Figure 10-14: Syntax and examples of writing text to a sequential access file *(Continued)*   ▶

Example 5

```
Dim file As String = "C:\VB2005\Chap10\letters.txt"
My.Computer.FileSystem.WriteAllText(file, _
    Strings.Space(10) & "A" & Strings.Space(5) _
    & "B", False)
```

Result

```
          A     B|
```

Figure 10-14: Syntax and examples of writing text to a sequential access file

The "My" in the syntax shown in Figure 10-14 is a new feature in Visual Basic 2005. The **My feature** exposes a set of commonly used objects to the programmer. One of the objects exposed by the My feature is the My.Computer object. Using the **My.Computer object**, you can easily access other objects and methods used to manipulate files. For example, to write information to a file on your computer's disk, you use the WriteAllText method, which is available through the FileSystem object exposed by the My.Computer object. In the WriteAllText method's syntax (shown in Figure 10-14), the *file* argument is a string that contains the name of the sequential access file to which you want to write information. The *file* argument can contain an optional folder path. If the folder path is not specified, the computer looks for the file in the current project's bin\Debug folder. For example, if the current project is stored in the VB2005\Chap10\Payroll Solution\ Payroll Project folder, the computer will look for the file in the VB2005\Chap10\Payroll Solution\Payroll Project\bin\Debug folder. If the file whose name is specified in the *file* argument does not exist, the computer creates the file before writing the information to the file. The information to write to the sequential access file is specified in the *text* argument in the WriteAllText method's syntax. Like the *file* argument, the *text* argument must be a string. The *append* argument in the syntax is a Boolean value, either True or False. An *append* value of True indicates that you want the information contained in the *text* argument to be appended (or added) to the end of any existing information in the file. An *append* value of False, on the other hand, indicates that you do not want to *append* the *text* information to the existing information in the file. When the *append* argument is False, any existing information in the file is erased before the information in the *text* argument is written to the file.

The My.Computer.FileSystem.WriteAllText("msg.txt", "Hi", False) statement shown in Example 1 in Figure 10-14 tells the computer to write the string "Hi" to a file named msg.txt. If the msg.txt file does not exist, the computer creates it before writing the string "Hi" to it. However, if the msg.txt file exists, the False value that appears in the *append* argument tells the computer to erase the file's contents before writing the string to the file. The computer uses a file pointer to keep track of the next character to either write to a file or read from a file. After processing the `My.Computer.FileSystem.WriteAllText("msg.txt", "Hi", False)` statement in Example 1, the computer positions the file pointer immediately after the last letter in the string "Hi", as indicated by the | symbol in the example. To position the file pointer on the next line in the file, you concatenate the `ControlChars.NewLine` constant with the *text* argument, as shown in the second example in Figure 10-14.

The `My.Computer.FileSystem.WriteAllText(File, Line, False)` statement shown in Example 3 tells the computer to write the string "The top salesperson is " to the top.txt file. If the file does not exist, the computer creates it before writing the string to it. If the file exists, the False value that appears in the *append* argument tells the computer to erase the file's contents before writing the string to the file. After processing the statement, the computer positions the file pointer after the last character in the string (in this case, after the space character). The next statement in Example 3, `My.Computer.FileSystem.WriteAllText(File, name & ".", True)`, tells the computer to concatenate the contents of the `name` variable with a period, and then write the result to the top.txt file. The True value that appears in the *append* argument tells the computer to append (add) the concatenated string to the file. In this case, the concatenated string will be written on the same line as the previous string, immediately after the last space character in the string. The next statement, `My.Computer.FileSystem.WriteAllText(File, ControlChars.NewLine & ControlChars.NewLine, True)` writes two new line characters to the file, which positions the file pointer two lines below the previously written text in the file. The last statement in Example 3, `My.Computer.FileSystem.WriteAllText(File, "ABC Sales", True)`, writes the string "ABC Sales" to the file, adding the string immediately below the two new line characters. After the statement is processed, the file pointer is located after the last character in the string, as indicated in the example.

The WriteAllText methods shown in Example 4 write the string "Total price: " and the contents of the `price` variable (formatted with a dollar sign and two decimal places) on the same line in the file. The file pointer is then positioned at the beginning of the next line in the file. Example 5 shows how you can use the **Strings.Space method** to write a specific number of spaces to a file. The syntax of the Strings.Space method is **Strings. Space(***number***)**, where *number* represents the number of spaces you want to write. The `My.Computer.FileSystem.WriteAllText(file, Strings.Space(10) & "A" & Strings.Space(5) & "B", False)` statement writes 10 spaces, the letter "A", five spaces, and the letter "B" to the file. After the statement is processed, the file pointer is positioned immediately after the letter B in the file.

> **»TIP**
> Although it is not a requirement, the "txt" (short for "text") filename extension is typically used when naming sequential access files. This is because sequential access files are text files.

To code and test the Top Salesperson application, and then view the top.txt file:

1 Start Visual Studio 2005 or Visual Basic 2005 Express Edition, if necessary, and close the Start Page window. Open the **Top Salesperson Solution** (Top Salesperson Solution.sln) file, which is contained in the VB2005\Chap10\Top Salesperson Solution folder. If necessary, open the designer window.

2 Open the Code Editor window. Replace the <your name> and <current date> text in the comments with your name and the current date. Locate the code template for the xWriteButton's Click event procedure, then enter the code shown in Example 3 in Figure 10-14, except replace the string "Carolyn" with **Me.xNameTextBox.Text**.

3 Close the Code Editor window. Save the solution, then start the application. Type **Pierre Johnson** in the Name box, then click the **Write** button to write the appropriate information to the top.txt file. Click the **Exit** button to end the application.

4 Now you will view the contents of the top.txt file in a separate window. Click **File**. Either point to **Open** and then click **File**, or click **Open File**. Open the **top.txt** file, which is contained in the VB2005\Chap10\Top Salesperson Solution\Top Salesperson Project\bin\Debug folder. Press the ↓ key on your keyboard three times to position the insertion point at the end of the file. Figure 10-15 shows the contents of the top.txt file.

the top.txt window's Close button

the insertion point appears at the end of the file

Figure 10-15: Contents of the top.txt file

5 Click the **Close** button on the top.txt window to close the window, then close the solution.

ALIGNING COLUMNS OF INFORMATION IN A SEQUENTIAL ACCESS FILE

In Chapter 8, you learned how to use the PadLeft and PadRight methods to pad a string with a character until the string is a specified length. Recall that the syntax of the PadLeft method is *string*.**PadLeft**(*length*[, *character*]), and the syntax of the PadRight method is *string*.**PadRight**(*length*[, *character*]). In each syntax, *length* is an integer that represents the desired length of the *string*, and *character* (which is optional) is the character that each method uses to pad the *string* until it reaches the desired *length*. If the *character* argument is omitted, the default *character* is the space character. Figure 10-16 shows examples of using the PadLeft and PadRight methods to align columns of information written to a sequential access file.

Aligning columns of information in a sequential access file

Example 1

```
Dim formatPrice As String
For price As Double = 1.0 To 3.0 Step 0.5
    formatPrice = price.ToString("N2")
    My.Computer.FileSystem.WriteAllText("prices.txt", _
        formatPrice.PadLeft(4), True)
    My.Computer.FileSystem.WriteAllText("prices.txt", _
        ControlChars.NewLine, True)
Next price
```

Result

```
1.00
1.50
2.00
2.50
3.00
```

Example 2

```
Const Heading As String  = _
    "Name" & Strings.Space(11) & "Age"
Const Path As String = "C:\VB2005\Chap10\"
Dim name As String
Dim age As String
My.Computer.FileSystem.WriteAllText(Path & "info.txt", _
    Heading & ControlChars.NewLine, True)
name = InputBox("Enter name:", "Name")
Do While name <> String.Empty
    age = InputBox("Enter age:", "Age")
    My.Computer.FileSystem.WriteAllText(Path & "info.txt", _
        name.PadRight(15) & age _
        & ControlChars.NewLine, True)
    name = InputBox("Enter name:", "Name")
Loop
```

Result (when the user enters the following names and ages: Janice, 23, Sue, 67)

```
Name           Age
Janice         23
Sue            67
```

Figure 10-16: Examples of aligning columns of information in a sequential access file

The code in Example 1 shows how you can align a column of numbers by the decimal point. First, you format each number in the column to ensure that each has the same number of digits to the right of the decimal point. You then use the PadLeft method to insert spaces at the beginning of the number; this right-aligns the number within the column. Because each number has the same number of digits to the right of the decimal point, aligning each number on the right will, in effect, align each by its decimal point.

The code in Example 2 in Figure 10-16 shows how you can align the second column of information when the first column contains strings whose lengths vary. To align the second column, you first use either the PadRight or PadLeft method to ensure that each string in the first column contains the same number of characters. You then concatenate the padded string to the information in the second column before writing the concatenated string to the file. The code shown in Example 2, for instance, uses the PadRight method to ensure that each name in the first column contains exactly 15 characters. It then concatenates the 15 characters with the age stored in the age variable, and then writes the concatenated string to the file. Because each name has 15 characters, each age will automatically appear beginning in character position 16 in the file.

To code and test the Align Prices application, and then view the prices.txt file:

1 Open the **Align Prices Solution** (Align Prices Solution.sln) file, which is contained in the VB2005\Chap10\Align Prices Solution folder. If necessary, open the designer window.

2 Open the Code Editor window. Replace the <your name> and <current date> text in the comments with your name and the current date. Locate the code template for the xWriteButton's Click event procedure, then enter the code shown in Example 1 in Figure 10-16.

3 Close the Code Editor window. Save the solution, then start the application. Click the **Write Prices** button to write the appropriate information to the prices.txt file, then click the **Exit** button to end the application.

4 Open the **prices.txt** file, which is contained in the VB2005\Chap10\Align Prices Solution\Align Prices Project\bin\Debug folder. Press the ↓ key on your keyboard five times to position the insertion point at the end of the file. Figure 10-17 shows the contents of the prices.txt file.

Figure 10-17: Contents of the prices.txt file

5 Close the prices.txt window, then close the solution.

READING INFORMATION FROM A SEQUENTIAL ACCESS FILE

You can use the **ReadAllText method** to read the information stored in a sequential access file. Figure 10-18 shows the syntax of the ReadAllText method and includes examples of using the method to read information from a sequential access file. The *file* argument in the syntax is a string that contains the name of the sequential access file from which you want to read information. The *file* argument can contain an optional folder path. If the folder path is not specified, the computer looks for the *file* in the current project's bin\Debug folder. If the file whose name is specified in the *file* argument does not exist, an error occurs and your application will end abruptly. You will learn how to handle this situation in the next section.

Reading information from a sequential access file

Syntax

My.Computer.FileSystem.ReadAllText(*file***)**

Example 1

```
Dim contents As String
contents = My.Computer.FileSystem.ReadAllText("prices.txt")
```

reads the text contained in the prices.txt file, and then assigns the text to the `contents` variable

Example 2

```
Const Path As String = "C:\VB2005\Chap10\"
Me.xReportTextBox.Text = _
    My.Computer.FileSystem.ReadAllText(Path & "info.txt")
```

reads the text contained in the info.txt file, and then displays the text in the xReportTextBox

---

Figure 10-18: Syntax and examples of reading text from a sequential access file

The `contents = My.Computer.FileSystem.ReadAllText("prices.txt")` statement in the first example reads the text contained in a sequential access file named prices.txt, and it assigns the text to a String variable named `contents`. Because no folder path is specified in the *file* argument, the computer will look for the prices.txt file in the current project's bin\Debug folder. The `Me.xReportTextBox.Text = My.Computer.FileSystem.ReadAllText(Path & "info.txt")` statement in the second example reads the text contained in the info.txt file and assigns the text to the xReportTextBox's Text property. In this case, the computer will look for the info.txt file in the VB2005\Chap10 folder.

DETERMINING WHETHER A FILE EXISTS

As you learned earlier, an error occurs when the computer attempts to read a sequential access file that does not exist. You can prevent the error from occurring by first determining whether the file exists. You can make the determination using the **FileExists method**. Figure 10-19 shows the syntax of the FileExists method and includes two examples of using the method. The *file* argument in the syntax is the name of the file whose existence you want to verify. If you do not include a folder path in the *file* argument, the computer searches for the file in the current project's bin\Debug folder. The FileExists method returns the Boolean value True when the file exists; otherwise, it returns the Boolean value False.

Determining whether a file exists

Syntax

My.Computer.FileSystem.FileExists(*file* **)**

Example 1

```
If My.Computer.FileSystem.FileExists("prices.txt") Then
    Me.xPricesLabel.Text = _
        My.Computer.FileSystem.ReadAllText("prices.txt")
Else
    MessageBox.Show("File does not exist", "Prices", _
        MessageBoxButtons.OK, MessageBoxIcon.Information)
End If
```

reads and displays the prices.txt file if the file exists; otherwise, displays the "File does not exist" message in a message box

---

Example 2

```
Const Path As String = "C:\VB2005\Chap10\"
Dim answer As DialogResult
If My.Computer.FileSystem.FileExists(Path & "info.txt") Then
    answer = MessageBox.Show("Erase the file?", _
        "Information", MessageBoxButtons.YesNo, _
        MessageBoxIcon.Exclamation, _
        MessageBoxDefaultButton.Button2)
    If answer = Windows.Forms.DialogResult.Yes Then
        My.Computer.FileSystem.WriteAllText(Path & "info.txt", _
        String.Empty, False)
    End If
End If
```

if the info.txt file exists, displays the "Erase the file?" message in a message box; the file is erased if the user selects the Yes button in the message box

---

Figure 10-19: Syntax and examples of determining whether a file exists

The `If My.Computer.FileSystem.FileExists("prices.txt") Then` clause in the first example determines whether the prices.txt file exists in the current project's bin\Debug folder. If the file exists, the `Me.xPricesLabel.Text = My.Computer.FileSystem.ReadAllText("prices.txt")` statement assigns the contents of the file to the Text property of the xPricesLabel. If the file does not exist, an appropriate message is displayed in a message box. The `If My.Computer.FileSystem.FileExists(Path & "info.txt") Then` clause in the last example determines whether the info.txt file exists in the VB2005\Chap10 folder. If the file exists, the "Erase the file?" message is displayed in a message box, along with Yes and No buttons. If the user selects the Yes button, the `My.Computer.FileSystem.WriteAllText(Path & "info.txt", String.Empty, False)` statement erases the contents of the file by writing the empty string to it.

To code and then test the Read Prices application:

1 Open the **Read Prices Solution** (Read Prices Solution.sln) file, which is contained in the VB2005\Chap10\Read Prices Solution folder. If necessary, open the designer window.

2 Open the Code Editor window. Replace the <your name> and <current date> text in the comments with your name and the current date. Locate the code template for the xDisplayButton's Click event procedure, then enter the code shown in Example 1 in Figure 10-19.

3 Close the Code Editor window. Save the solution, then start the application. Click the **Display Prices** button. The button's Click event procedure first determines whether the prices.txt file exists in the current project's bin\Debug folder. In this case, the file exists, so the procedure reads the contents of the prices.txt file and assigns the contents to the Text property of the xPricesLabel. See Figure 10-20.

Figure 10-20: Prices shown in the xPricesLabel

4 Click the **Exit** button to end the application, then close the solution.

You have completed Lesson B. You can either take a break or complete the end-of-lesson questions and exercises before moving on to Lesson C.

SUMMARY

TO WRITE INFORMATION TO A SEQUENTIAL ACCESS FILE:

» Use the WriteAllText method. The method's syntax is **My.Computer.FileSystem.WriteAllText(***file***, ***text***, ***append***)**. In the syntax, *file* is the name of the file (including an optional folder path) to which you want to write information. If the folder path is not specified, the computer looks for the file in the current project's bin\Debug folder. If the file does not exist, the computer creates the file. The *text* argument in the syntax is the information to write to the file. The *append* argument is either the Boolean value True or the Boolean value False. You use True to append (add) the *text* to the end of the existing information in the file. You use False to erase the existing information in the file before the *text* is written to the file.

TO WRITE SPACES TO A SEQUENTIAL ACCESS FILE:

» Use the Strings.Space method. The method's syntax is **Strings.Space(***number***)**, where *number* represents the number of spaces you want to write.

TO ALIGN COLUMNS OF INFORMATION IN A SEQUENTIAL ACCESS FILE:

» Use the PadLeft and PadRight methods.

TO READ INFORMATION FROM A SEQUENTIAL ACCESS FILE:

» Use the ReadAllText method. The method's syntax is **My.Computer.FileSystem.ReadAllText(***file***)**, where *file* is the name of the file (including an optional folder path) from which you want to read information. If the folder path is not specified, the computer looks for the file in the current project's bin\Debug folder. If the file does not exist, an error occurs.

TO DETERMINE WHETHER A FILE EXISTS:

» Use the FileExists method. The method's syntax is **My.Computer.FileSystem.FileExists(***file***)**, where *file* is the name of the file (including an optional folder path) whose existence you want to verify. If the folder path is not specified, the computer looks for the file in the current project's bin\Debug folder. The FileExists method returns the Boolean value True if the file exists; otherwise, it returns the Boolean value False.

QUESTIONS

1. Which of the following statements writes the contents of the xAddressTextBox to a sequential access file named address.txt, replacing the file's existing text?

   a. `My.Computer.FileSystem.WriteAll("address.txt",`
      `Me.xAddressTextBox.Text, False)`

   b. `My.Computer.FileSystem.WriteAllText("address.txt",`
      `Me.xAddressTextBox.Text, False)`

   c. `My.Computer.FileSystem.WriteAllText("address.txt",`
      `Me.xAddressTextBox.Text, True)`

   d. `My.Computer.FileSystem.WriteText("address.txt",`
      `Me.xAddressTextBox.Text, Replace)`

2. Which of the following statements reads the contents of a sequential access file named address.txt and assigns the contents to a String variable named `fileContents`?

   a. `fileContents = My.Computer.File.Read("address.txt")`

   b. `fileContents = My.Computer.File.ReadAll("address.txt")`

   c. `fileContents = My.Computer.FileSystem.ReadAll("address.txt")`

   d. `fileContents = My.Computer.FileSystem.`
      `ReadAllText("address.txt")`

3. Which of the following clauses verifies the existence of a sequential access file named address.txt?

   a. `If My.FileExists("address.txt") Then`

   b. `If My.Computer.FileExists("address.txt") Then`

   c. `If My.Computer.FileSystem.FileExists("address.txt") Then`

   d. None of the above.

4. Which of the following can be used to write the string "Your pay is $56" to a sequential access file? The file's name is stored in the `file` variable. The `pay` variable contains the number 56.

   a. `My.Computer.FileSystem.WriteAllText(file, "Your pay is $",`
      `False)`
      `My.Computer.FileSystem.WriteAllText(file, pay.ToString, True)`

   b. `My.Computer.FileSystem.WriteAllText(file, "Your pay is $" &`
      `pay.ToString, False)`

   c. `My.Computer.FileSystem.WriteAllText(file, "Your ", False)`
      `My.Computer.FileSystem.WriteAllText(file, "pay is $", True)`
      `My.Computer.FileSystem.WriteAllText(Convert.ToString(pay), True)`

   d. All of the above.

5. Which of the following statements can be used to write 15 space characters to a sequential access file named msg.txt?

   a. `My.Computer.FileSystem.WriteAllText("msg.txt", Blank(15, " "),`
      `True)`

   b. `My.Computer.FileSystem.WriteAllText("msg.txt", Chars(15, " "),`
      `True)`

   c. `My.Computer.FileSystem.WriteAllText("msg.txt",`
      `Strings.Space(15), True)`

   d. `My.Computer.FileSystem.WriteAllText("msg.txt",`
      `Strings.Space(15, " "), True)`

EXERCISES

1. Write the string "Employee" and the string "Name" to a sequential access file named report.txt. Each string should appear on a separate line in the file.

2. Write the contents of a String variable named `capital`, followed by 20 spaces, the contents of a String variable named `state`, and the ControlChars.NewLine constant to a sequential access file named geography.txt.

3. Write the statement to assign the contents of the report.txt file to a String variable named `text`.

4. Write the statement to assign the contents of the report.txt file to the xReportLabel.

5. Write the code to determine whether the jansales.txt file exists. If it does, the code should display the string "File exists" in the xMessageLabel; otherwise, it should display the string "File does not exist" in the xMessageLabel.

6. In this exercise, you code an application that writes information to a sequential access file.

   a. If necessary, start Visual Studio 2005 or Visual Basic 2005 Express Edition. Open the Employee List Solution (Employee List Solution.sln) file, which is contained in the VB2005\Chap10\Employee List Solution folder. If necessary, open the designer window.

   b. The Write button should write the contents of the xNameTextBox to a sequential access file named names.txt. Each name should appear on a separate line in the file. Save the names.txt file in the VB2005\Chap10\Employee List Solution\Employee List Project folder.

   c. Save the solution, then start the application. Test the application by writing five names to the file, then click the Exit button to end the application.

   d. Open the names.txt file to verify that it contains five names, then close the names.txt window. Close the Code Editor window, then close the solution.

7. In this exercise, you code an application that reads and writes information to a sequential access file.

   a. If necessary, start Visual Studio 2005 or Visual Basic 2005 Express Edition. Open the Memo Solution (Memo Solution.sln) file, which is contained in the VB2005\Chap10\Memo Solution folder. If necessary, open the designer window.

   b. The Write button should write the contents of the xMemoTextBox to a sequential access file named memo.txt. Save the memo.txt file in the VB2005\Chap10\Memo Solution\Memo Project folder. The Read button should read the contents of the file, displaying the contents in the xMemoTextBox.

   c. Save the solution, then start the application. Test the application by writing the following memo to the file:

   To all employees:

   The annual picnic will be held at Rogers Park on Saturday, July 13. Bring your family for a day full of fun!

   Carolyn Meyer
   Personnel Manager

   d. Click the Exit button to end the application. Open the memo.txt file to verify that it contains the memo, then close the memo.txt window.

   e. Start the application again. Click the Read button to display the contents of the memo.txt file.

   f. Click the Exit button to end the application. Close the Code Editor window, then close the solution.

8. In this exercise, you code an application that writes information to a sequential access file.

   a. If necessary, start Visual Studio 2005 or Visual Basic 2005 Express Edition. Open the Report Solution (Report Solution.sln) file, which is contained in the VB2005\Chap10\Report Solution folder. If necessary, open the designer window.

   b. Open the Code Editor window. The application uses an array to store three state names and their corresponding sales. Code the application so that it creates the report shown in Example 3 in Figure 10-13. Save the report in a sequential access file named report.txt. Save the report.txt file in the VB2005\Chap10\Report Solution\Report Project folder.

   c. Save the solution, then start and test the application. Click the Exit button to end the application.

   d. Open the report.txt file to verify that it contains the report, then close the report.txt window. Close the Code Editor window, then close the solution.

DISCOVERY EXERCISE

9. In this exercise, you update the contents of a sequential access file.

   a. If necessary, start Visual Studio 2005 or Visual Basic 2005 Express Edition. Open the Pay Solution (Pay Solution.sln) file, which is contained in the VB2005\Chap10\Pay Solution folder. If necessary, open the designer window.

   b. Open the payrates.txt file, which is contained in the VB2005\Chap10\Pay Solution\Pay Project\bin\Debug folder. View the file's contents, then close the payrates.txt window.

   c. Code the Increase button's Click event procedure so that it increases each pay rate by 10%. Save the increased prices in a sequential access file named updated.txt. Save the file in the VB2005\Chap10\Pay Solution\Pay Project\bin\Debug folder.

   d. Save the solution, then start the application. Click the Increase button to update the pay rates, then click the Exit button to end the application.

   e. Open the payrates.txt file. Also open the updated.txt file. The pay rates contained in the updated.txt file should be 10% more than the pay rates contained in the payrates.txt file. Close the updated.txt and payrates.txt windows. Close the Code Editor window, then close the solution.

LESSON C
OBJECTIVES

AFTER STUDYING LESSON C, YOU SHOULD
BE ABLE TO:

» Determine whether a combo box contains a specific value

» Remove an item from a combo box while an application is running

» Delete a sequential access file while an application is running

» Save combo box items to a sequential access file

» Fill a combo box with values from a sequential access file

SEQUENTIAL ACCESS FILES AND COMBO BOXES

CODING THE FRIENDS APPLICATION

Recall that your task in this chapter is to create an application that allows you to add names to and remove names from a combo box. The application will store the names in a sequential access file. To save you time, the VB2005\Chap10\Friends Solution folder contains a partially completed Friends application.

To open the partially completed Friends application:

1 Start Visual Studio 2005 or Visual Basic 2005 Express Edition, if necessary, and close the Start Page window. Open the **Friends Solution** (Friends Solution.sln) file, which is contained in the VB2005\Chap10\Friends Solution folder. If necessary, open the designer window. Figure 10-21 shows the user interface for the Friends application.

Figure 10-21: Interface for the Friends application

Figure 10-22 shows the TOE chart for the Friends application.

Task	Object	Event
Read the friends.txt file and assign its contents to the list portion of the xFriendComboBox	MainForm	Load
Save the list portion of the xFriendComboBox in a sequential access file named friends.txt		FormClosing
End the application	xExitButton	Click
Get the name to add or remove	xFriendComboBox	None
Add the name entered in the text portion of the xFriendComboBox to the list portion, but only if it is not a duplicate name	xAddButton	Click
Remove (from the list portion of the xFriendComboBox) the name either entered or selected in the combo box	xRemoveButton	Click

Figure 10-22: TOE chart for the Friends application

CODING THE XADDBUTTON CLICK EVENT PROCEDURE

According to the TOE chart shown in Figure 10-22, the xAddButton's Click event procedure should add the name entered in the text portion of the xFriendComboBox to the list portion of the combo box, but only if the name is not already contained in the combo box. Figure 10-23 shows the pseudocode for the xAddButton's Click event procedure.

---

xAddButton Click event procedure - pseudocode

1. remove any leading and/or trailing spaces from the name entered in the text portion of the combo box

2. if the text portion of the combo box is empty
    display "Please enter a name." in a message box
  else
    if the combo box already contains the name entered in its text portion
      display "Duplicate name" in a message box
    else
      add the name to the combo box
    end if
  end if

3. send the focus to the combo box

Figure 10-23: Pseudocode for the xAddButton's Click event procedure

To code the xAddButton's Click event procedure:

1 Open the Code Editor window. Replace the <your name> and <current date> text in the comments with your name and the current date.

2 Open the code template for the xAddButton's Click event procedure. Type **' adds a name to the combo box** and press **Enter** twice.

Recall that before you begin coding a procedure, you first study the procedure's pseudocode to determine the variables and named constants (if any) the procedure will use. In this case, the procedure will use two named constants and one variable.

3 Type **const EnterMsg as string = "Please enter a name."** and press **Enter**, then type **const DuplicateMsg as string = "Duplicate name"** and press **Enter**. The named constants will be used to display the appropriate message in a message box.

4 Type **dim friendName as string** and press **Enter** twice. The friendName variable will store the name entered in the text portion of the xFriendComboBox.

5 The first step in the pseudocode shown in Figure 10-23 is to remove any leading and/or trailing spaces from the name entered in the text portion of the combo box. You can do this using the Trim method, which you learned about in Chapter 8. Type **friendname = me.xFriendComboBox.text.trim** and press **Enter** twice.

6 The next step is a selection structure that determines whether the text portion of the combo box is empty. Type **if friendname = string.empty then** and press **Enter**.

7 If the text portion is empty, the selection structure's true path should display an appropriate message. Type **messagebox.show(entermsg, "Friends", _** and press **Enter**. Press **Tab**, then type **messageboxbuttons.ok, messageboxicon.information)** and press **Enter**.

8 If the text portion of the combo box is not empty, the selection structure's false path should determine whether the name to be added is already contained in the combo box. You can do this using a nested selection structure along with the Contains method (which you learned about in Chapter 8). Type **else** and press **Enter**, then type **' check for a duplicate name** and press **Enter**. Type **if me.xFriendComboBox.items.contains(friendname) then** and press **Enter**.

9 If the Contains method returns the Boolean value True, the nested selection structure's true path should display an appropriate message. Type **messagebox.show(duplicatemsg, "Friends", _** and press **Enter**. Press **Tab**, then type **messageboxbuttons.ok, messageboxicon.information)** and press **Enter**.

10 If, on the other hand, the Contains method returns the Boolean value False, the nested selection structure's false path should add the name to the list portion of the combo box. Type **else** and press **Enter**, then type **me.xFriendComboBox.items.add(friendname)**.

11 The last step in the pseudocode is to send the focus to the xFriendComboBox. Type the additional code shown in Figure 10-24.

```
Private Sub xAddButton_Click(ByVal sender As Object, ByVal e As System.E'
    ' adds a name to the combo box

    Const EnterMsg As String = "Please enter a name."
    Const DuplicateMsg As String = "Duplicate name"
    Dim friendName As String

    friendName = Me.xFriendComboBox.Text.Trim

    If friendName = String.Empty Then
        MessageBox.Show(EnterMsg, "Friends", _
            MessageBoxButtons.OK, MessageBoxIcon.Information)
    Else
        ' check for a duplicate name
        If Me.xFriendComboBox.Items.Contains(friendName) Then
            MessageBox.Show(DuplicateMsg, "Friends", _
                MessageBoxButtons.OK, MessageBoxIcon.Information)
        Else
            Me.xFriendComboBox.Items.Add(friendName)
        End If
    End If
    Me.xFriendComboBox.Focus()                          enter this
End Sub                                                 line of code
```

Figure 10-24: Completed xAddButton's Click event procedure

Next, you will test the xAddButton's Click event procedure to verify that it is working correctly.

To test the xAddButton's Click event procedure:

1 Save the solution, then start the application. Type **Sue** in the text portion of the combo box, then click the **Add** button. The name Sue appears in the list portion of the combo box. See Figure 10-25.

Figure 10-25: Name entered in the xFriendComboBox

2 Type **Mary** in the text portion of the combo box, then click the **Add** button. The name Mary appears before the name Sue in the list portion of the combo box, because the combo box's Sorted property is set to True.

3 Type **Sue** in the text portion of the combo box, then click the **Add** button. The "Duplicate name" message appears in a message box. Click the **OK** button to close the message box.

4 Click the **Exit** button to end the application.

CODING THE XREMOVEBUTTON CLICK EVENT PROCEDURE

According to the TOE chart shown earlier in Figure 10-22, the xRemoveButton's Click event procedure will allow the user to remove a name from the list portion of the combo box. Figure 10-26 shows the pseudocode for the xRemoveButton's Click event procedure.

---

xRemoveButton Click event procedure - pseudocode

1. remove any leading and/or trailing spaces from the name entered in the text portion of the combo box

2. remove the name from the list portion of the combo box

3. send the focus to the combo box

Figure 10-26: Pseudocode for the xRemoveButton's Click event procedure

To code the xRemoveButton's Click event procedure, then test the procedure:

1 Open the code template for the xRemoveButton's Click event procedure. Type **' removes a name from the combo box** and press **Enter** twice.

2 The procedure will use a variable to store the name appearing in the text portion of the combo box. Type **dim friendName as string** and press **Enter**.

3 The first step in the pseudocode is to remove any leading and/or trailing spaces from the name entered in the text portion of the combo box. Type **friendname = me.xFriendComboBox.text.trim** and press **Enter**.

4 The next step is to remove the name from the list portion of the combo box. You can remove an item from a combo box using the Items collection's Remove method. The method's syntax is *object***.Items.Remove(***item***)**, where *object* is the name of the combo box and *item* is the item's value. In this case, *object* is Me.xFriendComboBox and *item* is friendName. Type **me.xFriendComboBox.items.remove(friendname)** and press **Enter**.

»» TIP

You also can use the Items collection's Remove method to remove an item from a list box.

5 The last step in the pseudocode is to send the focus to the xFriendComboBox. Type **me.xFriendComboBox.focus()**.

6 Save the solution, then start the application. Notice that the combo box does not contain the two names you entered in the previous set of steps. You will fix that problem later in this lesson.

7 First, you will add three names to the combo box. Type **Sue** in the combo box, then click the **Add** button. Type **Mary** in the combo box, then click the **Add** button. Type **John** in the combo box, then click the **Add** button.

8 Now you will remove two names from the combo box. Click **Sue** in the combo box, then click the **Remove** button. The xRemoveButton's Click event procedure removes the name Sue from the list portion of the combo box. Type **Mary** in the combo box, then click the **Remove** button to remove the name Mary from the list portion of the combo box.

9 Click the **Exit** button to end the application.

CODING THE MAINFORM FORMCLOSING EVENT PROCEDURE

According to the TOE chart shown earlier in Figure 10-22, the MainForm's FormClosing event procedure should save the list portion of the xFriendComboBox in a sequential access file named friends.txt. Figure 10-27 shows the pseudocode for the MainForm's FormClosing event procedure.

MainForm FormClosing event procedure - pseudocode
1. if the friends.txt file exists
delete the file
end if
2. repeat for each item in the list portion of the combo box
write the item and the ControlChars.NewLine constant to the file
end repeat

Figure 10-27: Pseudocode for the MainForm's FormClosing event procedure

To code the MainForm's FormClosing event procedure, then test the procedure:

1 Open the code template for the MainForm's FormClosing event procedure. Type **' saves the list portion of the combo box** and press **Enter**. Type **' in a sequential access file** and press **Enter** twice.

2 The procedure will store the filename, friends.txt, in a named constant. Type **const FriendsFile as string = "friends.txt"** and press **Enter** twice.

Rather than deleting the file, you also can simply erase the file's contents.

3 The first step in the pseudocode is a selection structure that determines whether the friends.txt file exists. If it does exist, the selection structure's true path will delete the file. You can use the DeleteFile method to delete a file while an application is running. The method's syntax is **My.Computer.FileSystem.DeleteFile(** *file***)**, where *file* is the name of the file to delete. Type **if my.computer.filesystem.fileexists(friendsfile) then** and press **Enter**, then type **my.computer.filesystem.deletefile(friendsfile)**.

4 The last step in the pseudocode is a repetition structure that repeats its instruction for each item in the list portion of the combo box. The instruction writes the current item followed by the ControlChars.NewLine constant to the file. As you learned in Chapter 3, the ControlChars.NewLine constant instructs the computer to issue a carriage return followed by a line feed, which advances the insertion point to the next line in a control, or in a file, or on the printer. Type the repetition structure shown in Figure 10-28.

```
Private Sub MainForm_FormClosing(ByVal sender As Object, ByVal e
        ' saves the list portion of the combo box
        ' in a sequential access file

        Const FriendsFile As String = "friends.txt"

        If My.Computer.FileSystem.FileExists(FriendsFile) Then
                My.Computer.FileSystem.DeleteFile(FriendsFile)
        End If

        For x As Integer = 0 To Me.xFriendComboBox.Items.Count - 1
                My.Computer.FileSystem.WriteAllText(FriendsFile, _
                        Me.xFriendComboBox.Items(x).ToString _
                        & ControlChars.NewLine, True)
        Next x
End Sub
```

enter these five lines of code

writes a carriage return character and a line feed character

Figure 10-28: Completed MainForm's FormClosing event procedure

5 Save the solution, then start the application. Type **Babbett, Jeremy** and click the **Add** button, then type **Williams, Carolyn** and click the **Add** button. Click the **Exit** button to end the application.

6 Now you will view the contents of the friends.txt file. Open the **friends.txt** file, which is contained in the VB2005\Chap10\Friends Solution\Friends Project\bin\Debug folder. Press the ↓ key on your keyboard two times to position the insertion point at the end of the file. Figure 10-29 shows the contents of the friends.txt file. As indicated in the figure, the carriage return and line feed characters at the end of each line are not visible in the file.

the carriage return and line feed characters at the end of each line are not visible in the file

Figure 10-29: Contents of the friends.txt file

7 Close the friends.txt window.

The last procedure you need to code is the MainForm's Load event procedure.

CODING THE MAINFORM LOAD EVENT PROCEDURE

According to the TOE chart shown earlier in Figure 10-22, the MainForm's Load event procedure will read the friends.txt file and assign its contents to the list portion of the xFriendComboBox. Figure 10-30 shows the pseudocode for the MainForm's Load event procedure.

MainForm Load event procedure - pseudocode
if the friends.txt file exists
read the file and assign its contents to a String variable named friends
search the friends variable for the ControlChars.NewLine constant, beginning with the first character in the variable
repeat until no ControlChars.NewLine constants can be found in the friends variable
access the name, which appears on the same line as the ControlChars.NewLine constant, but immediately before it
add the name to the list portion of the combo box
search the friends variable for the next ControlChars.NewLine constant, beginning two characters after the previous ControlChars.NewLine constant
end repeat
end if

Figure 10-30: Pseudocode for the MainForm's Load event procedure

To code the MainForm's Load event procedure:

1 Open the code template for the MainForm's Load event procedure. Type ' **fills the combo box with the names** and press **Enter**, then type ' **from a sequential access file** and press **Enter** twice.

2 The procedure will store the filename, friends.txt, in a named constant. Type **const FriendFile as string = "friends.txt"** and press **Enter**.

3 Type **dim friends as string** and press **Enter**. After reading the friends.txt file, the procedure will store the file's contents in the `friends` variable.

4 Type **dim name as string** and press **Enter**. The procedure will store a friend's name in the `name` variable before adding the name to the combo box.

5 Type **dim newLineIndex as integer** and press **Enter**. When the procedure locates the ControlChars.NewLine constant in the `friends` variable, it will store the constant's location in the `newLineIndex` variable.

6 Type **dim nameIndex as integer** and press **Enter** twice. The `nameIndex` variable will store the location of the first character in each name.

7 The first step in the pseudocode is a selection structure that determines whether the friends.txt file exists. Type **if my.computer.filesystem.fileexists(friendfile) then** and press **Enter**.

8 If the friends.txt file exists, the selection structure's true path should read the file and assign its contents to the `friends` variable. Type **friends = my.computer. filesystem.readalltext(friendfile)** and press **Enter**.

The next instruction in the true path should search the `friends` variable for the `ControlChars.NewLine` constant, beginning with the first character in the variable. You can perform the search using the IndexOf method, which you learned about in Chapter 8. As you may remember, the method's syntax is *string*.**IndexOf**(*subString*[*,* *startIndex*]**)**. Recall that *subString* is the sequence of characters for which you are searching in the *string*, and the optional *startIndex* specifies the starting position for the search. If the *startIndex* is omitted, the search begins with the first character in the *string*. The IndexOf method returns the number -1 when the *subString* cannot be located in the *string*.

9 In this case, *subString* will be the `ControlChars.NewLine` constant. The search should begin with the first character in the `friends` variable, so you can either use the number 0 as the *startIndex* or simply omit the *startIndex*. You will assign the IndexOf method's return value to the `newLineIndex` variable, because the value indicates the location of the `ControlChars.NewLine` constant. Type **newlineindex = friends. indexof(controlchars.newline)** and press **Enter**.

10 The next instruction in the selection structure's true path is a repetition structure that repeats its instructions until no `ControlChars.NewLine` constants can be found in the `friends` variable. Type **do until newlineindex = -1** and press **Enter**.

The first instruction in the repetition structure is to access the name, which appears on the same line as the `ControlChars.NewLine` constant, but immediately before it. Figure 10-31 shows the contents of the friends.txt file; it also shows the index of each character in the file. Notice that the `ControlChars.NewLine` constant is associated with two characters: the carriage return (CR) character and the line feed (LF) character.

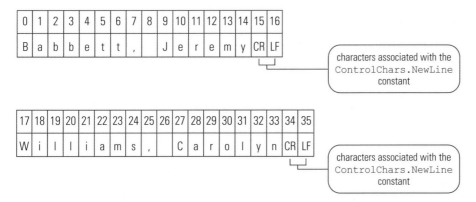

Figure 10-31: Index of each character

The number of characters in each name can be calculated by subtracting the index of the first character in the name from the index of the first character in the `ControlChars.NewLine` constant; the first character in the `ControlChars.NewLine` constant is the carriage return. For example, the name "Babbett, Jeremy" contains 15 characters, which is the difference between the letter B's index (0) and the carriage return's index (15). Similarly, the name "Williams, Carolyn" contains 17 characters, which is the difference between the letter W's index (17) and the carriage return's index (34). You can use this information, along with the Substring method (which you learned about in Chapter 8), to access each name. Recall that the Substring method's syntax is string.**Substring(**startIndex[, count]**)**. The startIndex argument is the index of the first character you want to access in the string. In this case, you will use the `nameIndex` variable as the startIndex argument, because the variable will always contain the index of the first character in each name. The count argument specifies the number of characters you want to access. You will use the expression `newLineIndex - nameIndex` as the count argument. The expression subtracts the index of the first character in the name from the index of the first character in the `ControlChars.NewLine` constant, resulting in the number of characters you need to access in the name. You will assign the Substring method's return value to the `name` variable.

To continue coding the MainForm's Load event procedure, then test the procedure:

1 Type **name = _** and press **Enter**. Press **Tab**, then type **friends.substring(nameindex, newlineindex - nameindex)** and press **Enter**.

2 According to the pseudocode shown earlier in Figure 10-30, the next instruction in the repetition structure should add the name to the list portion of the combo box; the name is stored in the name variable. Type **me.xFriendComboBox.items.add(name)** and press **Enter**.

The last instruction in the pseudocode is to search the friends variable for the next ControlChars.NewLine constant. You should begin the search two characters after the previous ControlChars.NewLine constant; if you don't, the Load event procedure will not work correctly because the IndexOf method will continue to find the same constant.

3 Type **nameindex = newlineindex + 2** and press **Enter** to update the nameIndex variable, which stores the index of the first character in each name. Recall that the nameIndex variable is used in the Substring method. It also is used in the second IndexOf method, which you will enter next. Type **newlineindex = _** and press **Enter**. Press **Tab**, then type **friends. indexof(controlchars.newline, nameindex)**. Figure 10-32 shows the completed Load event procedure.

```
Private Sub MainForm_Load(ByVal sender As Object, ByVal e As System.Even
    ' fills the combo box with the names
    ' from a sequential access file

    Const FriendFile As String = "friends.txt"
    Dim friends As String
    Dim name As String
    Dim newLineIndex As Integer
    Dim nameIndex As Integer

    If My.Computer.FileSystem.FileExists(FriendFile) Then
        friends = My.Computer.FileSystem.ReadAllText(FriendFile)
        newLineIndex = friends.IndexOf(ControlChars.NewLine)
        Do Until newLineIndex = -1
            name = _
                friends.Substring(nameIndex, newLineIndex - nameIndex)
            Me.xFriendComboBox.Items.Add(name)
            nameIndex = newLineIndex + 2
            newLineIndex = _
                friends.IndexOf(ControlChars.NewLine, nameIndex)
        Loop
    End If
End Sub
```

Figure 10-32: Completed Load event procedure

4 Save the solution, then start the application. The MainForm's Load event procedure reads the friends.txt file and displays the names in the combo box, as shown in Figure 10-33.

Figure 10-33: Names from the file appear in the combo box

5 Click the **Exit** button to end the application. Close the Code Editor window, then close the solution. Figure 10-34 shows the Friends application's code.

```
' Project name:      Friends Project
' Project purpose:   The project allows the user to add
'                    and delete names from a combo box.
'                    It also reads the names from and
'                    writes the names to a sequential
'                    access file.
' Created/revised:   <your name> on <current date>

Option Explicit On
Option Strict On

Public Class MainForm

    Private Sub xExitButton_Click...

    Private Sub xAddButton_Click(ByVal sender As Object, _
        ByVal e As System.EventArgs) Handles xAddButton.Click
        ' adds a name to the combo box

        Const EnterMsg As String = "Please enter a name."
        Const DuplicateMsg As String = "Duplicate name"
        Dim friendName As String
        friendName = Me.xFriendComboBox.Text.Trim
```

Figure 10-34: Friends application's code *(Continued)* ▶

```
            If friendName  =  String.Empty Then
                MessageBox.Show(EnterMsg, "Friends", _
                    MessageBoxButtons.OK, MessageBoxIcon.Information)
            Else
                ' check for a duplicate name
                If Me.xFriendComboBox.Items.Contains(friendName) Then
                    MessageBox.Show(DuplicateMsg, "Friends", _
                        MessageBoxButtons.OK, MessageBoxIcon.Information)
                Else
                    Me.xFriendComboBox.Items.Add(friendName)
                End If
            End If
            Me.xFriendComboBox.Focus()
        End Sub

        Private Sub xRemoveButton_Click(ByVal sender As Object, _
            ByVal e As System.EventArgs) Handles xRemoveButton.Click
            ' removes a name from the combo box

            Dim friendName As String
            friendName = Me.xFriendComboBox.Text.Trim
            Me.xFriendComboBox.Items.Remove(friendName)
            Me.xFriendComboBox.Focus()
        End Sub

        Private Sub MainForm_FormClosing(ByVal sender As Object, _
            ByVal e As System.Windows.Forms.FormClosingEventArgs) _
            Handles Me.FormClosing
            ' saves the list portion of the combo box
            ' in a sequential access file

            Const FriendsFile As String = "friends.txt"

            If My.Computer.FileSystem.FileExists(FriendsFile) Then
                My.Computer.FileSystem.DeleteFile(FriendsFile)
            End If

            For x As Integer = 0 To Me.xFriendComboBox.Items.Count - 1
                My.Computer.FileSystem.WriteAllText(FriendsFile, _
                    Me.xFriendComboBox.Items(x).ToString _
                    & ControlChars.NewLine, True)
            Next x
        End Sub
```

Figure 10-34: Friends application's code *(Continued)*

▶

```
    Private Sub MainForm_Load(ByVal sender As Object, _
        ByVal e As System.EventArgs) Handles Me.Load
        ' fills the combo box with the names
        ' from a sequential access file

        Const FriendFile As String = "friends.txt"
        Dim friends As String
        Dim name As String
        Dim newLineIndex As Integer
        Dim nameIndex As Integer

        If My.Computer.FileSystem.FileExists(FriendFile) Then
            friends = My.Computer.FileSystem.ReadAllText(FriendFile)
            newLineIndex = friends.IndexOf(ControlChars.NewLine)
            Do Until newLineIndex = -1
                name = _
                    friends.Substring(nameIndex, newLineIndex - nameIndex)
                Me.xFriendComboBox.Items.Add(name)
                nameIndex = newLineIndex + 2
                newLineIndex = _
                    friends.IndexOf(ControlChars.NewLine, nameIndex)
            Loop
        End If
    End Sub
End Class
```

Figure 10-34: Friends application's code

You have completed Lesson C and Chapter 10. You can either take a break or complete the end-of-lesson questions and exercises.

SUMMARY

TO REMOVE AN ITEM FROM THE LIST PORTION OF A COMBO BOX (OR LIST BOX) WHILE AN APPLICATION IS RUNNING:

» Use the Items collection's Remove method. The method's syntax is *object*.**Items. Remove(***item***)**, where *object* is the name of the combo box (or list box) and *item* is the item's value.

TO DELETE A FILE WHILE AN APPLICATION IS RUNNING:

» Use the DeleteFile method. The method's syntax is **My.Computer.FileSystem. DeleteFile(** *file***)**, where *file* is the name of the file to delete.

QUESTIONS

1. Which of the following statements writes the contents of the `city` variable on a separate line in the cities.txt file?

    a. `My.Computer.FileSystem.WriteAllText("cities.txt", city &`
       `ControlChars.NewLine, True)`

    b. `My.Computer.FileSystem.WriteAll("cities.txt", city, True)`

    c. `My.Computer.WriteAllText("cities.txt", city &`
       `ControlChars.NewLine, True)`

    d. `My.Computer.WriteAll("cities.txt", city & ControlChars.NewLine,`
       `True)`

2. Which of the following statements removes the string "Paris" from the list portion of the xCitiesComboBox?

    a. `Me.xCitiesComboBox.Items.Delete("Paris")`

    b. `Me.xCitiesComboBox.Items.Remove("Paris")`

    c. `Me.xCitiesComboBox.Remove.Item("Paris")`

    d. None of the above.

3. Which of the following statements deletes the cities.txt file?

    a. `My.Computer.FileSystem.Delete("cities.txt")`

    b. `My.Computer.FileSystem.DeleteFile("cities.txt")`

    c. `My.Computer.FileSystem.FileDelete("cities.txt")`

    d. None of the above.

EXERCISES

1. In this exercise, you modify the Friends application that you coded in this lesson.

    a. Use Windows to make a copy of the Friends Solution folder, which is contained in the VB2005\Chap10 folder. Rename the folder Friends Solution-Erase.

    b. If necessary, start Visual Studio 2005 or Visual Basic 2005 Express Edition. Open the Friends Solution (Friends Solution.sln) file contained in the VB2005\Chap10\ Friends Solution-Erase folder. Open the designer window.

    c. Open the Code Editor window. Modify the FormClosing event procedure so that it erases the contents of the friends.txt file, rather than deleting the file.

    d. Save the solution, then start and test the application. Click the Exit button to end the application. Close the Code Editor window, then close the solution.

2. In this exercise, you modify the Friends application that you coded in this lesson.

    a. Use Windows to make a copy of the Friends Solution folder, which is contained in the VB2005\Chap10 folder. Rename the folder Friends Solution-ListBox.

    b. If necessary, start Visual Studio 2005 or Visual Basic 2005 Express Edition. Open the Friends Solution (Friends Solution.sln) file contained in the VB2005\Chap10\ Friends Solution-ListBox folder. Open the designer window.

    c. Remove the combo box from the interface, then add a text box and a list box. The interface will use the text box to get the name, and use the list box to display the names.

    d. Open the Code Editor window. Modify the code appropriately. To remove a name, the user will need to select the name in the list box before clicking the Remove button.

    e. Save the solution, then start and test the application. Click the Exit button to end the application. Close the Code Editor window, then close the solution.

3. Glovers Industries stores, in a sequential access file named itemInfo.txt, the item numbers and prices of the items it sells. The company's sales manager wants an application that allows her to display the price corresponding to the item selected in a list box.

    a. If necessary, start Visual Studio 2005 or Visual Basic 2005 Express Edition. Open the Glovers Solution (Glovers Solution.sln) file, which is contained in the VB2005\Chap10\Glovers Solution folder. If necessary, open the designer window.

    b. Open the itemInfo.txt file, which is contained in the VB2005\Chap10\Glovers Solution\Glovers Project\bin\Debug folder. Notice that the item number and price appear on separate lines in the file. Close the itemInfo.txt file.

    c. Define a structure named Product. The structure should contain two member variables: a String variable to store the item number and a Decimal variable to store the price.

    d. Declare a module-level array that contains five Product structure variables.

    e. Code the MainForm's Load event procedure so that it reads the item numbers and prices from the itemInfo.txt file. The procedure should store the item numbers and prices in the module-level array. It also should add the item numbers to the list box.

    f. When the user selects an item in the list box, the item's price should appear in the xPriceLabel. Code the appropriate procedure.

    g. Save the solution, then start and test the application. Click the Exit button to end the application. Close the Code Editor window, then close the solution.

4. In this exercise, you modify the Friends application that you coded in this lesson.

a. Use Windows to make a copy of the Friends Solution folder, which is contained in the VB2005\Chap10 folder. Rename the folder Friends Solution-Ignore Case.

b. If necessary, start Visual Studio 2005 or Visual Basic 2005 Express Edition. Open the Friends Solution (Friends Solution.sln) file contained in the VB2005\ Chap10\Friends Solution-Ignore Case folder. Open the designer window.

c. Start the application. Enter Janice in the combo box, then click the Add button. Now enter janice in the combo box, then click the Add button. Because the case of the entries is different, the Add button's Click event procedure adds both names to the combo box. Remove Janice and janice from the list portion of the combo box, then click the Exit button to end the application.

d. Modify the Add button's Click event procedure so that it ignores the case of the entry when checking for duplicate names. In other words, rather than adding both Janice and janice to the list portion of the combo box, the procedure should display the "Duplicate name" message.

e. Save the solution, then start the application. Enter Janice in the combo box, then click the Add button. Now enter janice in the combo box, then click the Add button. The procedure should display the "Duplicate name" message. Click the OK button to close the message box. Remove Janice from the list box.

f. Click the Exit button to end the application. Close the Code Editor window, then close the solution.

5. In this exercise, you modify the Friends application that you coded in this lesson.

a. Use Windows to make a copy of the Friends Solution folder, which is contained in the VB2005\Chap10 folder. Rename the folder Friends Solution-Modified Add.

b. If necessary, start Visual Studio 2005 or Visual Basic 2005 Express Edition. Open the Friends Solution (Friends Solution.sln) file contained in the VB2005\Chap10\ Friends Solution-Modified Add folder. Open the designer window.

c. Modify the Add button's Click event procedure so that it ensures that the first and last names begin with an uppercase letter. The remaining letters in the names should be lowercase.

d. Save the solution, then start the application. Remove the names from the list portion of the combo box. Then enter the following names: smith, mary; CARTER, CAROL; rEISE, JeFF; and Tenny, Sam. Click the Exit button to end the application.

e. Open the friends.txt file, which is contained in the VB2005\Chap10\Friends Solution-Modified Add\Friends Project\bin\Debug folder. Each first and last name in the file should begin with an uppercase letter; the remaining letters in the name should be lowercase.

f. Close the friends.txt window. Close the Code Editor window, then close the solution.

6. Each year, WKRK-Radio polls its audience to determine the best Super Bowl commercial. The choices are as follows: Budweiser, FedEx, MasterCard, and Pepsi. The station manager wants an application that allows him to enter a caller's choice. The choice should be saved in a sequential access file. The application also should allow the station manager to display the number of votes for each commercial.

a. If necessary, start Visual Studio 2005 or Visual Basic 2005 Express Edition. Create a Visual Basic Windows-based application. Name the solution WKRK Solution, and name the project WKRK Project. Save the application in the VB2005\Chap10 folder.

b. Assign the filename Main Form.vb to the form file object. Assign the name MainForm to the form.

c. Design an appropriate interface, then code the application.

d. Save the solution, then start and test the application. End the application. Close the Code Editor window, then close the solution.

DISCOVERY EXERCISE

7. Glovers Industries stores, in a sequential access file named itemInfo.txt, the item numbers and prices of the items it sells. The company's sales manager wants an application that allows her to display the price corresponding to the item selected in a list box.

a. If necessary, start Visual Studio 2005 or Visual Basic 2005 Express Edition. Open the Glovers Solution (Glovers Solution.sln) file, which is contained in the VB2005\Chap10\Glovers Solution-Discovery folder. If necessary, open the designer window.

b. Open the itemInfo.txt file, which is contained in the VB2005\Chap10\Glovers Solution\Glovers Project\bin\Debug folder. Notice that each item's number and price appear on a separate line in the file; the number is first, followed by a comma and the price. Close the itemInfo.txt file.

c. Define a structure named Product. The structure should contain two member variables: a String variable to store the item number and a Decimal variable to store the price.

d. Declare a module-level array that contains five Product structure variables.

e. Code the MainForm's Load event procedure so that it reads the item numbers and prices from the itemInfo.txt file. The procedure should store the item numbers and prices in the module-level array. It also should add the item numbers to the list box.

f. When the user selects an item in the list box, the item's price should appear in the xPriceLabel. Code the appropriate procedure.

g. Save the solution, then start and test the application. Click the Exit button to end the application. Close the Code Editor window, then close the solution.

DEBUGGING EXERCISE

8. In this exercise, you find and correct an error in an application. The process of finding and correcting errors is called debugging.

a. If necessary, start Visual Studio 2005 or Visual Basic 2005 Express Edition. Open the Debug Solution (Debug Solution.sln) file, which is contained in the VB2005\Chap10\Debug Solution folder.

b. Open the Code Editor window. Review the existing code, then start and test the application. Notice that the application is not working correctly.

c. Correct the application's code, then save the solution. Start and test the application. When the application is working correctly, close the Code Editor window, then close the solution.

11

CLASSES AND OBJECTS

CREATING THE KESSLER LANDSCAPING APPLICATION

Monica Kessler is the owner of a landscaping company named Kessler Landscaping. She wants an application that she can use to estimate the cost of laying sod. According to Monica, the sod typically is laid on a rectangular piece of land.

PREVIEWING THE KESSLER LANDSCAPING APPLICATION

Before creating the Kessler Landscaping application, you first preview the completed application.

To preview the completed application:

1 Use the Run command on the Windows Start menu to run the **Kessler** (**Kessler.exe**) file, which is contained in the VB2005\Chap11 folder. The Kessler Landscaping application's user interface appears on the screen.

2 Type **75** in the Length (feet) box, then type **100** in the Width (feet) box. Type **1.25** in the Sod price (square yard) box, then click the **Calculate Total Price** button. The button's Click event procedure displays the total price of the sod, as shown in Figure 11-1.

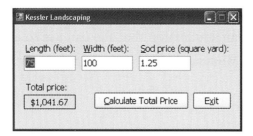

Figure 11-1: Interface showing the total price of the sod

3 Click the **Exit** button to end the application.

In this chapter, you learn how to create a class and then instantiate objects from the class. You will use a class to code the Kessler Landscaping application in Lesson C.

LESSON A
OBJECTIVES

AFTER STUDYING LESSON A, YOU SHOULD BE ABLE TO:

» Define a class

» Instantiate an object from a class that you define

» Add Property procedures to a class

» Create a default constructor

» Create a parameterized constructor

» Include methods other than constructors in a class

CLASSES AND OBJECTS

USING CLASSES AND OBJECTS

As you learned in the Overview, object-oriented programs are based on objects that are instantiated (created) from classes. Each object is an instance of the class from which it is created. A text box, for example, is an instance of the TextBox class. Similarly, buttons and labels are instances of the Button class and Label class, respectively. Recall that a class encapsulates (contains) the attributes (properties) that describe the object it creates, and the behaviors (methods and events) that allow the object to perform tasks or respond to actions. The TextBox class, for instance, contains the Name, CharacterCasing, and Text properties, as well as the Focus and SelectAll methods. Examples of events contained in the TextBox class include Click and TextChanged.

In previous chapters, you instantiated objects using classes that are built into Visual Basic, such as the TextBox and Label classes. You used the instantiated objects in a variety of ways in many different applications. For example, in some applications you used a text box to enter a name, while in other applications you used it to enter a sales tax rate. Similarly, you used label controls to identify text boxes and also to display the result of calculations. The ability to use an object for more than one purpose saves programming time and money—an advantage that contributes to the popularity of object-oriented programming.

»TIP

Recall that each tool in the toolbox represents a class. When you drag a tool from the toolbox to the form, the computer uses the class to instantiate the appropriate object.

DEFINING A CLASS

»TIP

Although you can code a class in a matter of minutes, the objects produced by such a class probably will not be of much use. The creation of a good class—one whose objects can be used in a variety of ways by many different applications—requires a lot of time, patience, and planning.

In addition to using the classes built into Visual Basic, you also can define your own classes and then create instances (objects) from those classes. The classes that you define can represent something encountered in real life, such as a credit card receipt, a check, and an employee. Like the Visual Basic classes, your classes must specify the attributes and behaviors of the objects they create. The attributes describe the characteristics of the objects, and the behaviors specify the tasks that the objects can perform or the events to which the objects can respond. You use the **Class statement** to define a class in Visual Basic. Figure 11-2 shows the syntax of the Class statement and includes an example of using the statement to create a class named TimeCard.

Class statement

Syntax

Public Class *classname*
   attributes section
   behaviors section
End Class

Example

```
Public Class TimeCard
    variables and Property procedures appear in the attributes section
    Sub and Function procedures appear in the behaviors section
End Class
```

Figure 11-2: Syntax and an example of the Class statement

The Class statement begins with the keywords `Public Class`, followed by the name of the class; it ends with the keywords `End Class`. Although it is not required by the syntax, the convention is to enter the class name using Pascal case, which means that you capitalize the first letter in the name and the first letter in any subsequent words in the name. The names of Visual Basic classes (for example, String and TextBox) also follow this naming convention. Within the Class statement, you define the attributes and behaviors of the class. The attributes are represented by variables and Property procedures, and the behaviors are represented by Sub and Function procedures, more commonly referred to as methods. You enter the Class statement in a class file. Figure 11-3 shows how to add a class file to the current project, and Figure 11-4 shows an example of a completed Add New Item – *projectname* dialog box.

>> **TIP**

You also can include Event declarations in a Class statement. However, that topic is beyond the scope of this book.

Adding a class file to a project

1. Click Project on the menu bar.

2. Click Add Class. The Add New Item – *projectname* dialog box opens with Class selected in the Visual Studio installed templates box.

3. Type the name of the class followed by a period and the letters vb in the Name box.

4. Click the Add button.

Figure 11-3: Procedure for adding a class file to a project

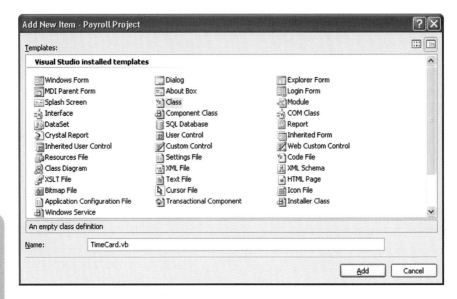

Figure 11-4: Completed Add New Item – *projectname* dialog box

When you click the Add button in the Add New Item – Payroll Project dialog box shown in Figure 11-4, the computer adds a file named TimeCard.vb to the current project. It also opens the file in the Code Editor window, as shown in Figure 11-5.

Figure 11-5: TimeCard.vb file opened in the Code Editor window

The Code Editor automatically enters the Class statement in the class file. Within the Class statement, you enter the attributes and behaviors of the class. After you define the class, you then can use it to instantiate one or more objects. Figure 11-6 shows two versions of the syntax you use to instantiate an object from a class; it also includes an example of using each version. In both versions, *class* is the name of the class the computer will use to instantiate the object, and *objectVariable* is the name of a variable that will store the object's address. You use the Dim keyword to create a procedure-level *objectVariable*. To create a module-level *objectVariable*, you use the Private keyword.

Instantiating an object from a class

Syntax - version 1

{**Dim** | **Private**} *objectVariable* **As** *class*
objectVariable = **New** *class*

Syntax - version 2

{**Dim** | **Private**} *objectVariable* **As New** *class*

Examples

```
Private employeeTimeCard As TimeCard
employeeTimeCard = New TimeCard
```

the first instruction creates a TimeCard variable named employeeTimeCard; the second instruction instantiates a TimeCard object and assigns its address to the employeeTimeCard variable

```
Dim employeeTimeCard As New TimeCard
```

the instruction creates a TimeCard variable named employeeTimeCard and also instantiates a TimeCard object; it assigns the object's address to the variable

---

Figure 11-6: Syntax and examples of instantiating an object from a class

The first example in Figure 11-6 uses Version 1 of the syntax. The Private employeeTimeCard As TimeCard statement creates a module-level variable named employeeTimeCard that can store the address of a TimeCard object. The employeeTimeCard = New TimeCard statement in the example uses the TimeCard class to instantiate a TimeCard object, and it assigns the object's address to the employeeTimeCard variable. The second example in the figure uses Version 2 of the syntax. In the example, the Dim employeeTimeCard As New TimeCard statement creates a variable named employeeTimeCard. It also uses the TimeCard class to instantiate a TimeCard object, assigning the object's address to the variable. Notice that the difference between both versions of the syntax used to instantiate an object relates to when the object is actually created. In Visual Basic, the computer creates the object when it processes the statement containing the New keyword.

An easy way to learn how to define classes and instantiate objects is to view examples of doing so. You will begin with a simple example of a class that contains attributes only. In the example, each of the attributes is represented by a Public variable.

EXAMPLE 1—USING A CLASS THAT CONTAINS PUBLIC VARIABLES ONLY

The sales manager at Sweets Unlimited wants an application that allows him to save each salesperson's name, sales amount, and bonus amount in a sequential access file named sales.txt. Notice that each salesperson is associated with three items of information: a name, a sales amount, and a bonus amount. In the context of object-oriented programming (OOP), each salesperson can be treated as an object having three attributes. By including the attributes in a Class statement, you can create a pattern that any application can use to instantiate a salesperson object. In this case, the Sweets Unlimited application will use the Class statement to create a class named Salesperson. It then will use the Salesperson class to instantiate a Salesperson object, which it will use to store each salesperson's information before writing the information to the sequential access file. Figure 11-7 shows the Salesperson class defined in a class file named Salesperson.vb. Notice that the class file contains the `Option Explicit On` and `Option Strict On` statements. As is true when coding a form, it is a good programming practice to enter both Option statements when coding a class; both statements have the same meaning in a class file as they do in a form file.

»TIP

The Class statement in Figure 11-7 groups the salesperson's information into one unit. You also can group the information using the Structure statement, which you learned about in Chapter 10.

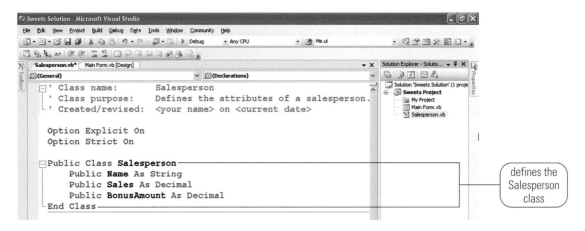

Figure 11-7: Salesperson class defined in the Salesperson.vb file

The Salesperson class shown in Figure 11-7 contains three attributes; each attribute is represented by a Public variable. When a variable in a class is declared using the Public keyword, it can be accessed by any application that contains an instance of the class. In this case, the variables included in the Salesperson class can be used by any application that instantiates a Salesperson object. Notice that, unlike most variable names, the variable names in the Salesperson class are not entered using camel case. This is because the convention is to use Pascal case for the names of the Public variables in a class.

Figure 11-8 shows the interface for the Sweets Unlimited application, and Figure 11-9 shows a sample of the sequential access file created by the application. Figure 11-10 shows the code for the xSaveButton's Click event procedure, which is contained in the application.

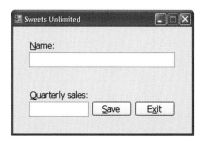

Figure 11-8: Interface for the Sweets Unlimited application

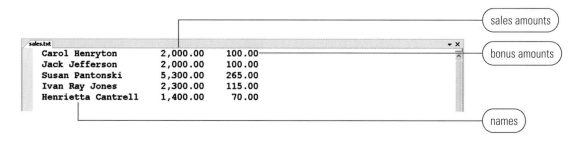

Figure 11-9: Sample sequential access file created by the application

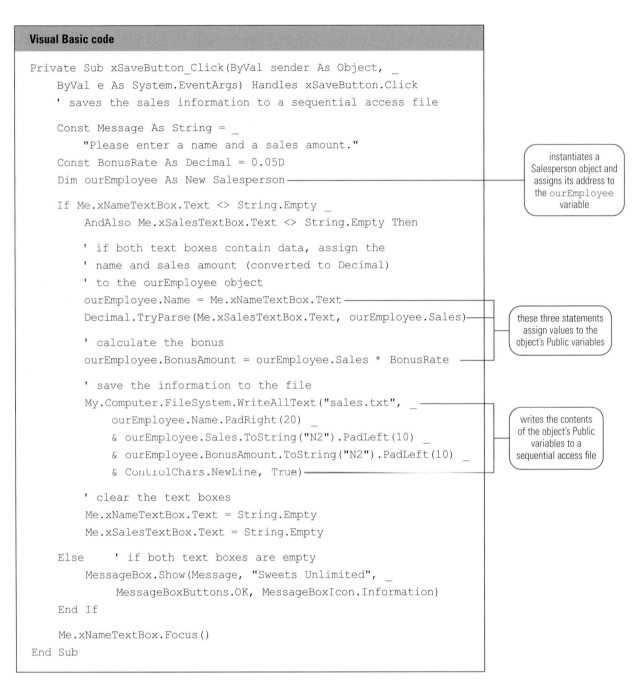

Visual Basic code

```
Private Sub xSaveButton_Click(ByVal sender As Object, _
    ByVal e As System.EventArgs) Handles xSaveButton.Click
    ' saves the sales information to a sequential access file

    Const Message As String = _
        "Please enter a name and a sales amount."
    Const BonusRate As Decimal = 0.05D
    Dim ourEmployee As New Salesperson

    If Me.xNameTextBox.Text <> String.Empty _
        AndAlso Me.xSalesTextBox.Text <> String.Empty Then

        ' if both text boxes contain data, assign the
        ' name and sales amount (converted to Decimal)
        ' to the ourEmployee object
        ourEmployee.Name = Me.xNameTextBox.Text
        Decimal.TryParse(Me.xSalesTextBox.Text, ourEmployee.Sales)

        ' calculate the bonus
        ourEmployee.BonusAmount = ourEmployee.Sales * BonusRate

        ' save the information to the file
        My.Computer.FileSystem.WriteAllText("sales.txt", _
            ourEmployee.Name.PadRight(20) _
            & ourEmployee.Sales.ToString("N2").PadLeft(10) _
            & ourEmployee.BonusAmount.ToString("N2").PadLeft(10) _
            & ControlChars.NewLine, True)

        ' clear the text boxes
        Me.xNameTextBox.Text = String.Empty
        Me.xSalesTextBox.Text = String.Empty

    Else    ' if both text boxes are empty
        MessageBox.Show(Message, "Sweets Unlimited", _
            MessageBoxButtons.OK, MessageBoxIcon.Information)
    End If

    Me.xNameTextBox.Focus()
End Sub
```

instantiates a Salesperson object and assigns its address to the ourEmployee variable

these three statements assign values to the object's Public variables

writes the contents of the object's Public variables to a sequential access file

Figure 11-10: Code for the xSaveButton's Click event procedure

The `Dim ourEmployee As New Salesperson` statement in the procedure first instantiates a Salesperson object, and then assigns the object's address to the `ourEmployee` variable. After the statement is processed, you can access the object's attributes using the syntax *objectVariable.attribute*, where *objectVariable* is the name of the variable that stores the object's address, and *attribute* is the name of the attribute you want to access. For example, you use `ourEmployee.Name` to access the Name attribute of the Salesperson object created in Figure 11-10. Likewise, you use `ourEmployee.Sales` and `ourEmployee.BonusAmount` to access the Sales and BonusAmount attributes, respectively.

Notice that the xSaveButton's Click event procedure uses assignment statements to assign values to the Name and BonusAmount attributes of the Salesperson object. The `ourEmployee.Name = Me.xNameTextBox.Text` statement assigns the contents of the xNameTextBox to the object's Name attribute. Similarly, the `ourEmployee.BonusAmount = ourEmployee.Sales * BonusRate` statement multiplies the contents of the object's Sales attribute by the value stored in the `BonusRate` constant (.05), and then stores the result in the object's BonusAmount attribute. The object's Sales attribute receives its value from the TryParse method contained in the `Decimal.TryParse(Me.xSalesTextBox.Text, ourEmployee.Sales)` statement in the procedure. The procedure uses the WriteAllText method, which you learned about in Chapter 10, to write the contents of the object's three attributes to a sequential access file named sales.txt file.

To code and then test the Sweets Unlimited application:

1 Start Visual Studio 2005 or Visual Basic 2005 Express Edition, if necessary, and close the Start Page window. Open the **Sweets Solution** (Sweets Solution.sln) file, which is contained in the VB2005\Chap11\Sweets Solution folder. If necessary, open the designer window.

2 Click **Project** on the menu bar, then click **Add Class**. The Add New Item – Sweets Project dialog box opens with Class selected in the Visual Studio installed templates box. Change the name entered in the Name box to **Salesperson.vb**, then click the **Add** button.

3 Type the comments and code shown in Figure 11-7, replacing the <your name> and <current date> text in the comments with your name and the current date.

4 Save the solution, then close the Salesperson.vb window.

5 Open the form's Code Editor window. Replace the <your name> and <current date> text in the comments with your name and the current date. Open the xSaveButton's Click event procedure, then enter the comments and code shown in Figure 11-10.

6 Close the Code Editor window. Save the solution, then start the application. Type **Carol Henryton** in the Name box, then type **2000** in the Quarterly sales box. Click the **Save** button. The xSaveButton's Click event procedure calculates the salesperson's bonus amount and then saves the salesperson's information to the sales.txt file.

7 Now enter the following four names and sales amounts:

Jack Jefferson	**2000**
Susan Pantonski	**5300**
Ivan Ray Jones	**2300**
Henrietta Cantrell	**1400**

8 Click the **Exit** button to end the application.

9 Click **File**, point to **Open**, then click **File**. (*If you are using the Express Edition*, click **File**, then click **Open File**.) Open the **sales.txt** file, which is contained in the VB2005\ Chap11\Sweets Solution\Sweets Project\bin\Debug folder. The file contains the same information shown earlier in Figure 11-9.

10 Close the sales.txt window, then close the solution.

Although you can define a class that contains only attributes represented by Public variables—like the Salesperson class shown in Figure 11-7—that is rarely done. The disadvantage of using Public variables in a class is that a class cannot control the values assigned to its Public variables. For example, the class cannot validate the values to ensure that they are appropriate for the variables. Besides, most classes contain not only attributes, but behaviors as well. This is because the purpose of a class in OOP is to encapsulate the properties that describe an object, the methods that allow the object to perform tasks, and the events that allow the object to respond to actions. The class used in the next example will show you how to include data validation code and methods in a class. (The creation of class events is beyond the scope of this book.)

EXAMPLE 2—USING A CLASS THAT CONTAINS A PRIVATE VARIABLE, A PROPERTY PROCEDURE, AND TWO METHODS

In this example, you view the code for a class named Square, which can be used to instantiate a Square object. Square objects have one attribute: the length of a side. Square objects also have two behaviors: they can initialize their side measurement when they are created, and they can calculate their area. Figure 11-11 shows the Square class defined in the Square.vb file.

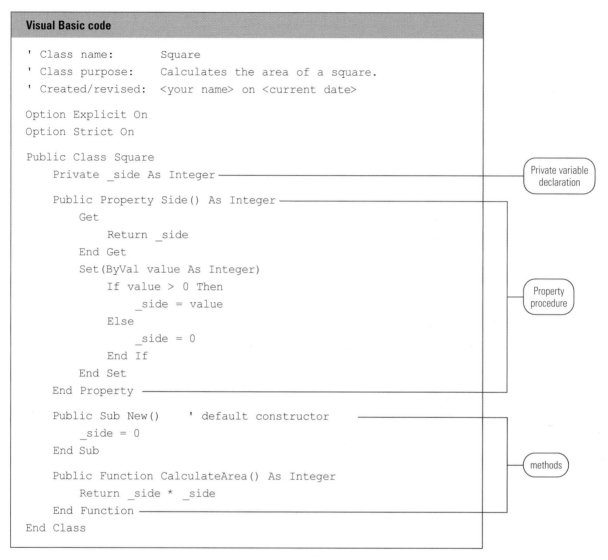

Visual Basic code

```vb
' Class name:       Square
' Class purpose:    Calculates the area of a square.
' Created/revised:  <your name> on <current date>

Option Explicit On
Option Strict On

Public Class Square
    Private _side As Integer

    Public Property Side() As Integer
        Get
            Return _side
        End Get
        Set(ByVal value As Integer)
            If value > 0 Then
                _side = value
            Else
                _side = 0
            End If
        End Set
    End Property

    Public Sub New()    ' default constructor
        _side = 0
    End Sub

    Public Function CalculateArea() As Integer
        Return _side * _side
    End Function
End Class
```

Private variable declaration

Property procedure

methods

Figure 11-11: Square class defined in the Square.vb file

The Square class contains the `Private _side As Integer` statement, which declares a Private variable named `_side`. When naming the Private variables in a class, the convention is to use the underscore as the first character and then use camel case for the remainder of the name. The `Private` keyword in the statement indicates that only the class in which it is defined can use the `_side` variable. In this case, the `_side` variable can be used only by the code entered in the Square class. The code uses the variable to store the side measurement of the square whose area is to be calculated.

When you use a class to instantiate an object in an application, only the Public members of the class are exposed (made available) to the application; the Private members of the class are hidden from the application. When an application needs to assign data to or retrieve data from a Private variable in a class, it must use a Public property to do so. In other words, an application cannot directly refer to a Private variable in a class. Rather, it must refer to the variable indirectly, through the use of a Public property. You create a Public property using a **Property procedure.** Figure 11-12 shows the syntax of a Property procedure. The figure also includes examples of Property procedures.

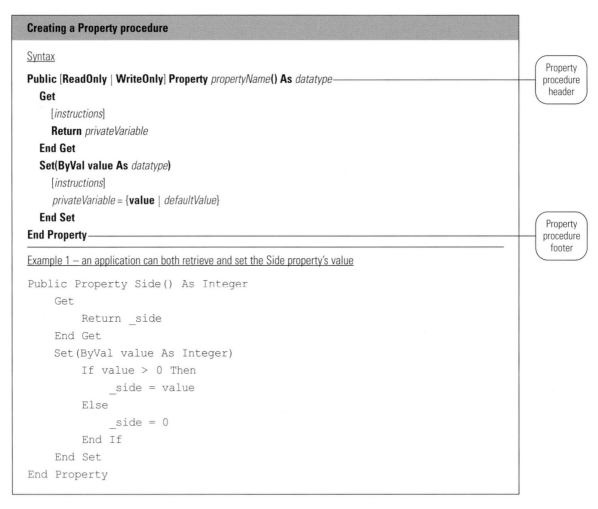

Creating a Property procedure

Syntax

Public [ReadOnly | WriteOnly] Property *propertyName***() As** *datatype* ——————— Property procedure header
  Get
    [*instructions*]
    Return *privateVariable*
  End Get
  Set(ByVal value As *datatype***)**
    [*instructions*]
    privateVariable = {**value** | *defaultValue*}
  End Set
End Property —————————————————————— Property procedure footer

Example 1 – an application can both retrieve and set the Side property's value

```
Public Property Side() As Integer
    Get
        Return _side
    End Get
    Set(ByVal value As Integer)
        If value > 0 Then
            _side = value
        Else
            _side = 0
        End If
    End Set
End Property
```

Figure 11-12: Syntax and examples of creating a Property procedure *(Continued)* ▶

Example 2 – an application can retrieve the Bonus property's value, but not set it

```
Public ReadOnly Property Bonus() As Decimal
    Get
        Return _bonus
    End Get
End Property
```

Example 3 – an application can set the AnnualSale property's value, but not retrieve it

```
Public WriteOnly Property AnnualSale() As Integer
    Set(ByVal value As Integer)
        _annualSale = value
    End Set
End Property
```

Figure 11-12: Syntax and examples of creating a Property procedure

In most cases, a Property procedure header begins with the keywords `Public Property`. However, as the syntax indicates, the header also can include either the `ReadOnly` keyword or the `WriteOnly` keyword. The `ReadOnly` keyword indicates that the property's value can be retrieved (read) by an application, but the application cannot set (write to) the property. The `WriteOnly` keyword indicates that an application can set the property's value, but it cannot retrieve the value. Following the `Property` keyword in the header is the name of the property. You should use nouns and adjectives to name a property, as in Side, Bonus, and AnnualSale. Property names should be entered using Pascal case. Notice that a set of parentheses, the keyword `As`, and the property's *datatype* follow the property name in the Property procedure header. The data type of the property must match the data type of the Private variable associated with the Property procedure. A Public Property procedure creates a property that is visible to any application that contains an instance of the class.

A Property procedure ends with the procedure footer, which contains the keywords `End Property`. Between the procedure header and procedure footer, you include a Get block of code, or a Set block of code, or both Get and Set blocks of code. The appropriate block or blocks of code to include depends on the keywords contained in the Property procedure header. If the procedure header contains the `ReadOnly` keyword, you include only a Get block of code in the Property procedure. The code contained in the **Get block** allows an application to retrieve the contents of the Private variable associated with the property. In the Property procedure shown in Example 2 in Figure 11-12, the `ReadOnly` keyword indicates that an application can retrieve the contents of the Bonus property, but it cannot set the property's value. However, if the Property procedure header contains the WriteOnly keyword, you include only a Set block of code in the procedure. The code in the **Set block** allows an application to assign a value to the Private variable associated with the property. In the Property procedure shown in Example 3 in Figure 11-12, the WriteOnly keyword indicates that an application can assign a value to the AnnualSale property, but it cannot retrieve the property's contents. If the Property procedure header does not contain the ReadOnly or WriteOnly keywords, you include both a Get block of code and a Set block of code in the procedure, as shown in Example 1 in Figure 11-12. In this case, an application can both retrieve and set the Side property's value.

The Get block uses the **Get statement**, which begins with the keyword Get and ends with the keywords End Get. Most times, you will enter only the **Return** *privateVariable* instruction within the Get statement. The instruction directs the computer to return the contents of the Private variable associated with the property. For instance, in the first example shown in Figure 11-12, the Return _side statement tells the computer to return the contents of the _side variable, which is the Private variable associated with the Side property. Similarly, the Return _bonus statement in the second example tells the computer to return the contents of the _bonus variable, which is the Private variable associated with the Bonus property. Notice that Example 3 in Figure 11-12 does not contain a Get statement. This is because the AnnualSale property is designated as a WriteOnly property.

The Set block uses the **Set statement**, which begins with the keyword Set and ends with the keywords End Set. As shown in Figure 11-12, the Set keyword is followed by a parameter enclosed in parentheses. The parameter begins with the keywords ByVal value As, followed by a *datatype*, which must match the data type of the Private variable associated with the Property procedure. The value parameter temporarily stores the value that is passed to the property by the application. You can enter one or more instructions within the Set statement. One of the instructions should assign the contents of the value parameter to the Private variable associated with the property. For instance, in the AnnualSale Property procedure shown in Example 3, the _annualSale = value statement assigns the contents of the procedure's value parameter to the Private _annualSale variable.

In the Set statement, you often will include instructions to validate the value received from the application before assigning it to the Private variable. For instance, the Set statement shown in Example 1 in Figure 11-12 includes a selection structure that determines whether the side measurement received from the application is valid. In this case, a valid side measurement is an integer that is greater than zero. If the side measurement is valid, the _side = value instruction in the selection structure's true path assigns the integer stored in the value parameter to the Private _side variable. However, if the side measurement is not valid, the _side = 0 instruction in the selection structure's false path assigns a default value—in this case, the number zero—to the Private _side variable. Notice that the Property procedure shown in Example 2 in Figure 11-12 does not contain a Set statement. This is because the Bonus property is designated as a ReadOnly property.

As shown earlier in Figure 11-11, the Square class contains two methods named New and CalculateArea. The New method is the default constructor for the class.

CONSTRUCTORS

A **constructor** is a method whose instructions the computer automatically processes each time an object is instantiated from the class. The sole purpose of a constructor is to initialize the class's Private variables. Figure 11-13 shows the syntax of a constructor. It also includes the constructor entered in the Square class.

Creating a constructor

Syntax

Public Sub New([*parameterlist*]**)**
  instructions to initialize the class's Private variables
End Sub

Example

```
Public Sub New()
     _side = 0
End Sub
```

Figure 11-13: Syntax and an example of a constructor

A constructor begins with the keywords Public Sub New, followed by a set of parentheses that contains an optional *parameterlist*. A constructor ends with the keywords End Sub. Constructors never return a value, so they are always Sub procedures rather than Function procedures. Within the constructor you enter the code to initialize the class's Private variables. The variables will be initialized when the class is used to instantiate an object. A class can have more than one constructor. Each constructor has the same name, New, but its parameters (if any) must be different from any other constructor in the class. A constructor that has no parameters is called the **default constructor**. A class can have only one default constructor.

The Square class defined earlier in Figure 11-11 contains one constructor, which is shown in the example in Figure 11-13. Notice that the constructor initializes the class's Private variable, named _side, to the number zero. The constructor is the Square class's default constructor; you can tell this because it has no parameters. The computer automatically processes the default constructor when you use the Square class to instantiate an object. Examples of statements that you can use to instantiate a Square object include `Dim mySquare As New Square` and `mySquare = New Square`.

A class also can contain methods other than constructors. The Square class defined in Figure 11-11, for example, contains a method named CalculateArea.

METHODS OTHER THAN CONSTRUCTORS

Except for constructors, which must be Sub procedures, the methods included in a class can be either Sub procedures or Function procedures. As you learned in Chapter 7, the difference between both types of procedures is that a Function procedure returns a value after performing its assigned task, whereas a Sub procedure does not return a value. Figure 11-14 shows the syntax of a method that is not a constructor. The figure also includes the CalculateArea method entered in the Square class.

Creating a method that is not a constructor

Syntax

Public {Sub | Function} *methodname*(**[***parameterlist***]**) **[As** *datatype***]**
    instructions
End {Sub | Function}

Example

```
Public Function CalculateArea() As Integer
     Return _side * _side
End Function
```

Figure 11-14: Syntax and an example of a method that is not a constructor

The {**Sub | Function**} in the syntax shown in Figure 11-14 indicates that you can select only one of the keywords appearing within the braces. In this case, you can choose either the Sub keyword or the Function keyword. The rules for naming methods are similar to the rules for naming properties. Like property names, method names should be entered using Pascal case. However, unlike property names, the first word in a method name should be a verb; any subsequent words should be nouns and adjectives. The name CalculateArea follows this naming convention.

The CalculateArea method in the Square class is represented by a Function procedure. The `Return _side * _side` statement within the procedure uses the contents of the class's Private variable, `_side`, to calculate the area of a square. The statement then returns the area to the application that called the procedure. The Area application, which you view next, uses the Square class to first instantiate and then manipulate a Square object. Figure 11-15 shows a sample run of the application, and Figure 11-16 shows the code for the xCalcButton's Click event procedure, which is contained in the application.

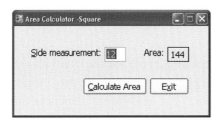

Figure 11-15: Sample run of the Area application

Visual Basic code

```
Private Sub xCalcButton_Click(ByVal sender As Object, _
    ByVal e As System.EventArgs) Handles xCalcButton.Click
    ' calculates and displays the area of a square

    Dim mySquare As New Square                                        creates a Square
    Dim area As Integer                                              object and assigns its
                                                                     address to the
                                                                     mySquare variable
    ' assign side measurement to the
    ' Square object's property
    Integer.TryParse(Me.xSideTextBox.Text, mySquare.Side)           assigns a value to the
                                                                     object's Side property
    ' calculate and display the area
    area = mySquare.CalculateArea()                                 uses the object's
    Me.xAreaLabel.Text = area.ToString                              CalculateArea method
                                                                    to calculate the area
    Me.xSideTextBox.Focus()
End Sub
```

Figure 11-16: Code for the xCalcButton's Click event procedure

The `Dim mySquare As New Square` statement in the xCalcButton's Click event procedure tells the computer to instantiate a Square object, and then assign the object's address to the `mySquare` variable. When creating the Square object, the computer uses the class's default constructor to initialize the class's Private variable (named `_side`). The `Integer.TryParse(Me.xSideTextBox.Text, mySquare.Side)` statement in the procedure converts the contents of the xSideTextBox to Integer and assigns the result to the Square object's Side property. Next, the `area = mySquare.CalculateArea()` statement uses the Square object's CalculateArea method to calculate and return the area; the result is assigned to the `area` variable. The `Me.xAreaLabel.Text = area.ToString` statement then displays the contents of the `area` variable in the xAreaLabel.

To code and then test the Area application:

1 Open the **Area Solution** (Area Solution.sln) file, which is contained in the VB2005\Chap11\Area Solution folder. If necessary, open the designer window.

2 Click **Project** on the menu bar, then click **Add Class**. Change the name entered in the Name box to **Square.vb**, then click the **Add** button.

3 Type the comments and code shown earlier in Figure 11-11, replacing the <your name> and <current date> text in the comments with your name and the current date. Notice that when you type the `Public Property Side() As Integer` statement in the Code Editor window and then press the Enter key, the Code Editor automatically enters the `Get, End Get, Set(ByVal value As Integer), End Set,` and `End Property` statements for you.

4 Save the solution, then close the Square.vb window.

5 Open the form's Code Editor window. Replace the <your name> and <current date> text in the comments with your name and the current date. Open the xCalcButton's Click event procedure, then enter the comments and code shown in Figure 11-16.

6 Close the Code Editor window. Save the solution, then start the application. Type **12** in the Side measurement box, then click the **Calculate Area** button. The xCalcButton's Click event procedure calculates the area of the square and displays the result (144) in the Area box, as shown earlier in Figure 11-15.

7 Click the **Exit** button to end the application, then close the solution.

EXAMPLE 3—USING A CLASS THAT CONTAINS A READONLY PROPERTY

In this example, you view the code for a class named CourseGrade, which can be used to instantiate a CourseGrade object. CourseGrade objects have three attributes: two test scores and a letter grade. CourseGrade objects also have two behaviors: they can initialize their attributes and they can calculate their letter grade. Figure 11-17 shows the CourseGrade class defined in the CourseGrade.vb file.

Visual Basic code

```vb
' Class name:       CourseGrade
' Class purpose:    Calculates a grade.
' Created/revised:  <your name> on <current date>

Option Explicit On
Option Strict On

Public Class CourseGrade
    Private _score1 As Integer
    Private _score2 As Integer
    Private _letterGrade As String

    Public Property Score1() As Integer
        Get
            Return _score1
        End Get
        Set(ByVal value As Integer)
            _score1 = value
        End Set
    End Property

    Public Property Score2() As Integer
        Get
            Return _score2
        End Get
        Set(ByVal value As Integer)
            _score2 = value
        End Set
    End Property

    Public ReadOnly Property LetterGrade() As String        ← ReadOnly property
        Get
            Return _letterGrade
        End Get
    End Property

    Public Sub New()
        _score1 = 0
        _score2 = 0
        _letterGrade = "?"
    End Sub

    Public Sub CalculateGrade()
        Dim total As Integer
        total = _score1 + _score2
```

Figure 11-17: CourseGrade class defined in the CourseGrade.vb file *(Continued)* ▶

```
      Select Case total
          Case Is >= 180
              _letterGrade = "A"
          Case Is >= 160
              _letterGrade = "B"
          Case Is >= 140
              _letterGrade = "C"
          Case Is >= 120
              _letterGrade = "D"
          Case Else
              _letterGrade = "F"
      End Select
    End Sub
End Class
```

Figure 11-17: CourseGrade class defined in the CourseGrade.vb file

The CourseGrade class contains three Private variables named _score1, _score2, and _letterGrade. It also contains three Property procedures named Score1, Score2, and LetterGrade. Notice that the LetterGrade property is designated as a ReadOnly property. As you learned earlier, an application can access the contents of a ReadOnly property, but it cannot set the property's value. In addition to the variables and Property procedures, the CourseGrade class also contains two methods named New and CalculateGrade. The New method is the class's default constructor and simply assigns initial values to the class's Private variables. The CalculateGrade method is responsible for assigning the appropriate letter grade to the class's Private _letterGrade variable. The CourseGrade class is used in the Grade application, which you view next. Figure 11-18 shows a sample run of the application, and Figure 11-19 shows the code for the xCalcButton's Click event procedure contained in the application.

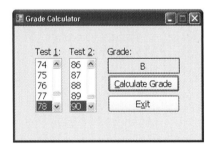

Figure 11-18: Sample run of the Grade application

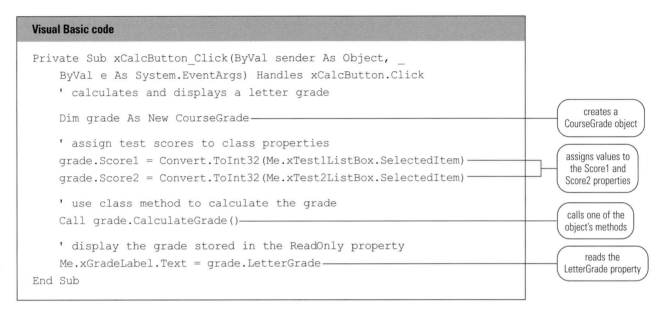

```
Visual Basic code

Private Sub xCalcButton_Click(ByVal sender As Object, _
    ByVal e As System.EventArgs) Handles xCalcButton.Click
    ' calculates and displays a letter grade

    Dim grade As New CourseGrade                                    ──── creates a
                                                                        CourseGrade object

    ' assign test scores to class properties
    grade.Score1 = Convert.ToInt32(Me.xTest1ListBox.SelectedItem) ──┐ assigns values to
    grade.Score2 = Convert.ToInt32(Me.xTest2ListBox.SelectedItem) ──┘ the Score1 and
                                                                       Score2 properties

    ' use class method to calculate the grade
    Call grade.CalculateGrade()                                    ──── calls one of the
                                                                        object's methods

    ' display the grade stored in the ReadOnly property
    Me.xGradeLabel.Text = grade.LetterGrade                        ──── reads the
End Sub                                                                LetterGrade property
```

Figure 11-19: Code for the xCalcButton's Click event procedure

The `Dim grade As New CourseGrade` statement in the xCalcButton's Click event procedure tells the computer to instantiate a CourseGrade object, and then assign the object's address to the `grade` variable. When creating the CourseGrade object, the computer uses the class's default constructor to initialize the three Private variables. The next two statements in the procedure assign the items selected in the list boxes to the CourseGrade object's Score1 and Score2 properties. The `Call grade.CalculateGrade()` statement invokes the CourseGrade object's CalculateGrade procedure, which assigns the appropriate letter grade to the `_letterGrade` variable. The `Me.xGradeLabel.Text = grade.LetterGrade` statement uses the CourseGrade object's LetterGrade property to access the letter grade. The statement displays the letter grade in the xGradeLabel.

To code and then test the Grade application:

1 Open the **Grade Solution** (Grade Solution.sln) file, which is contained in the VB2005\ Chap11\Grade Solution folder. If necessary, open the designer window.

2 Add a class file named **CourseGrade.vb** to the project. Type the comments and code shown in Figure 11-17, replacing the <your name> and <current date> text in the comments with your name and the current date.

3 Save the solution, then close the CourseGrade.vb window.

4 Open the form's Code Editor window. Replace the <your name> and <current date> text in the comments with your name and the current date. Open the xCalcButton's Click event procedure, then enter the comments and code shown in Figure 11-19.

5 Position your mouse pointer on the word `LetterGrade` in the last assignment statement, as shown in Figure 11-20. The box that appears indicates that LetterGrade is a Public ReadOnly property whose data type is String.

```
Private Sub xCalcButton_Click(ByVal sender As Object, ByVal e As System.
    ' calculates and displays a letter grade

    Dim grade As New CourseGrade

    ' assign test scores to class properties
    grade.Score1 = Convert.ToInt32(Me.xTest1ListBox.SelectedItem)
    grade.Score2 = Convert.ToInt32(Me.xTest2ListBox.SelectedItem)

    ' use class method to calculate the grade
    Call grade.CalculateGrade()

    ' display the grade stored in the ReadOnly property
    Me.xGradeLabel.Text = grade.LetterGrade
End Sub
    Public ReadOnly Property LetterGrade() As String
End Class
```

location of mouse pointer

Figure 11-20: Information pertaining to the LetterGrade property

6 Now you will observe what happens when you attempt to assign a value to a ReadOnly property. In the blank line below the Call statement, type **grade.lettergrade = "B"** and press **Enter**. A jagged line appears below the assignment statement, indicating that the statement contains an error. Position your mouse pointer on the jagged line, as shown in Figure 11-21. The box indicates that the LetterGrade property is ReadOnly.

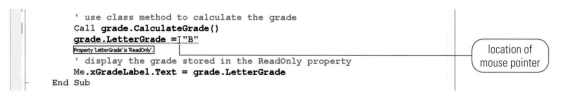

```
    ' use class method to calculate the grade
    Call grade.CalculateGrade()
    grade.LetterGrade = "B"
    Property 'LetterGrade' is 'ReadOnly'.
    ' display the grade stored in the ReadOnly property
    Me.xGradeLabel.Text = grade.LetterGrade
End Sub
```

location of mouse pointer

Figure 11-21: Result of attempting to assign a value to the LetterGrade property

7 Delete the `grade.LetterGrade = "B"` statement as well as the blank line immediately below it.

8 Position your mouse pointer on the word `CalculateGrade` in the Call statement. The box that appears indicates that CalculateGrade is a Public Sub procedure.

9 Position your mouse pointer on the word `Score1` in the first assignment statement. The box that appears indicates that Score1 is a Public property whose data type is Integer.

10 Close the Code Editor window. Save the solution, then start the application. Click **78** in the Test 1 list box, then click **90** in the Test 2 list box. Click the **Calculate Grade** button. The xCalcButton's Click event procedure displays the appropriate grade (B) in the Grade box, as shown earlier in Figure 11-18.

11 Click the **Exit** button to end the application, then close the solution.

EXAMPLE 4—USING A CLASS THAT CONTAINS TWO CONSTRUCTORS

In this example, you view the code for a class named FormattedDate. You can use the class to instantiate a FormattedDate object. FormattedDate objects have two attributes: a month number and a day number. FormattedDate objects also have three behaviors. First, they can initialize their attributes using values provided by the class. Second, they can initialize their attributes using values provided by the application in which they are instantiated. Third, they can return their month number and day number attributes, separated by a slash. Figure 11-22 shows the FormattedDate class defined in the FormattedDate.vb file.

Visual Basic code

```
' Class name:        FormattedDate
' Class purpose:     Inserts a slash between the month
'                    number and day number.
' Created/revised:   <your name> on <current date>

Option Explicit On
Option Strict On

Public Class FormattedDate
    Private _month As String
    Private _day As String

    Public Property Month() As String
        Get
            Return _month
        End Get
        Set(ByVal value As String)
            _month = value
        End Set
    End Property
```

Figure 11-22: FormattedDate class defined in the FormattedDate.vb file *(Continued)*

```
    Public Property Day() As String
        Get
            Return _day
        End Get
        Set(ByVal value As String)
            _day = value
        End Set
    End Property

    Public Sub New()     ' default constructor
        _month = String.Empty
        _day = String.Empty
    End Sub

    Public Sub New(ByVal monthNum As String, ByVal dayNum As String)
        Month = monthNum
        Day = dayNum
    End Sub

    Public Function GetNewDate() As String
        Dim newDate As String
        newDate = _month & "/" & _day
        Return newDate
    End Function
End Class
```

> accesses the Private _month variable directly

> uses the Month property to access the Private _month variable indirectly

> accesses the Private variables directly

Figure 11-22: FormattedDate class defined in the FormattedDate.vb file

The FormattedDate class contains two Private variables named _month and _day. It also contains two Property procedures named Month and Day. The Month Property procedure is associated with the _month variable, and the Day Property procedure is associated with the _day variable. In addition to the Private variables and Property procedures, the FormattedDate class also contains three methods: two are named New and one is named GetNewDate. The two New methods are the class's constructors. The first constructor is the default constructor, because it does not have any parameters. The computer processes the default constructor when you use a statement such as Dim payDate As New FormattedDate to instantiate a FormattedDate object. It also processes the default constructor when you use the payDate = New FormattedDate statement to create an instance of the FormattedDate class. The second constructor in the class allows you to specify the initial values for a FormattedDate object when the object is created. In this case, the initial values must be strings, because the constructor's *parameterlist* contains two String variables. You include the initial values, enclosed in a set of parentheses, in the statement that instantiates the object. For example, the `Dim payDate As New FormattedDate(monthNumber, dayNumber)` statement creates a FormattedDate object and passes two String variables (arguments) to the class's New procedure; recall that all procedures named New are class constructors. The computer determines which class constructor to use by matching the number, data type, and position of the arguments with the number, data type, and position of the parameters listed in each constructor's *parameterlist*. In this case, the computer will use the second constructor in the FormattedDate class, because that is the constructor that contains two String variables in its *parameterlist*. Constructors that contain parameters are called **parameterized constructors**. The method name combined with its optional *parameterlist* is called the method's **signature**.

The third method in the FormattedDate class is a Function procedure named GetNewDate. The function's purpose is to return the month and day numbers, separated by a slash. The month and day numbers are stored in the class's Private variables. The methods in a class can access the class's Private variables either directly (by name) or indirectly (through the Public properties). In the FormattedDate class shown in Figure 11-22, both the default constructor and the GetNewDate method use the names of the Private variables to access the variables directly. The default constructor assigns values to the Private variables, and the GetNewDate method retrieves the values stored in the Private variables. The parameterized New constructor, on the other hand, uses the Public properties to access the Private variables indirectly. This is because the values passed to the parameterized constructor come from the application rather than from the class itself. Values that originate outside of the class should always be assigned to the Private variables indirectly, through the Public properties. Doing this ensures that the Property procedure's Set block, which typically contains validation code, is processed. Currently, the Month and Day Property procedures in the FormattedDate class do not contain any data validation code. However, if validation code is added at a later date, the code in the parameterized constructor will not need to be modified. The Personnel application that you view next uses the FormattedDate class to instantiate and manipulate a FormattedDate object. Figure 11-23 shows a sample run of the application, and Figure 11-24 shows the code for two event procedures in the application: the xParameterizedButton's Click event procedure and the xDefaultButton's Click event procedure.

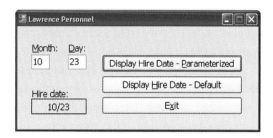

Figure 11-23: Sample run of the Personnel application

Visual Basic code

```vb
Private Sub xParameterizedButton_Click(ByVal sender As Object, _
    ByVal e As System.EventArgs) _
    Handles xParameterizedButton.Click
    ' displays a formatted date using the
    ' parameterized constructor

    Dim hireDate As New FormattedDate(Me.xMonthTextBox.Text, _
                            Me.xDayTextBox.Text)

    Me.xHiredLabel.Text = hireDate.GetNewDate()
End Sub

Private Sub xDefaultButton_Click(ByVal sender As Object, _
    ByVal e As System.EventArgs) Handles xDefaultButton.Click
    ' displays a formatted date using the
    ' default constructor

    Dim hireDate As New FormattedDate

    hireDate.Month = Me.xMonthTextBox.Text
    hireDate.Day = Me.xDayTextBox.Text

    Me.xHiredLabel.Text = hireDate.GetNewDate()
End Sub
```

- instantiates a FormattedDate object and provides the initial values
- uses the GetNewDate method to get the formatted date
- instantiates a FormattedDate object
- assigns values to the Month and Day properties
- uses the GetNewDate method to get the formatted date

Figure 11-24: Click event procedures for the xParameterizedButton and xDefaultButton

First, study the code contained in the xParameterizedButton's Click event procedure. The Dim hireDate As New FormattedDate(Me.xMonthTextBox.Text, Me.xDayTextBox.Text) statement tells the computer to create a FormattedDate object and assign its address to the hireDate variable. The computer will use the parameterized constructor to initialize the object. This is because the Dim statement contains two String arguments, which agrees with the *parameterlist* in the second constructor. The computer passes the two String arguments, *by value*, to the parameterized constructor. The constructor receives the values and stores them in the `monthNum` and `dayNum` variables listed in its procedure header. The `Month = monthNum` instruction in the parameterized constructor assigns the value stored in the `monthNum` variable to the Month property. When you assign a value to a property, the computer passes the value to the property's Set statement, where it is stored in the Set statement's `value` parameter. The `_month = value` instruction in the Set statement assigns the contents of the `value` parameter to the Private `_month` variable. Next, the `Day = dayNum` instruction in the parameterized constructor assigns the value stored in the `dayNum` variable to the Day property. In this case, the computer passes the value to the Day property's Set statement, where it is stored in the statement's `value` parameter. The `_day = value` instruction in the Set statement then assigns the contents of the `value` parameter to the Private `_day` variable. Finally, the `Me.xHiredLabel.Text = hireDate.GetNewDate()` statement in the xParameterizedButton's Click event procedure uses the FormattedDate object's GetNewDate method to return the month and day numbers, separated by a slash. The statement displays the formatted date in the xHiredLabel.

Compare the code contained in the xParameterizedButton's Click event procedure with the code contained in the xDefaultButton's Click event procedure (also shown in Figure 11-24). Notice that the `Dim hireDate As New FormattedDate (Me.xMonthTextBox.Text, Me.xDayTextBox.Text)` statement in the xParameterizedButton's Click event procedure was replaced with the following three lines of code in the xDefaultButton's Click event procedure:

```
Dim hireDate As New FormattedDate
hireDate.Month = Me.xMonthTextBox.Text
hireDate.Day = Me.xDayTextBox.Text
```

When processing the `Dim hireDate As New FormattedDate` statement, the computer instantiates a FormattedDate object, and it assigns the object's address to the `hireDate` variable. In this case, the computer uses the default constructor to initialize the class's Private variables. This is because the Dim statement does not contain any arguments, which agrees with the empty *parameterlist* in the default constructor. The `hireDate.Month = Me.xMonthTextBox.Text` statement then assigns the contents of the xMonthTextBox to the object's Month property. Recall that the Month property is a Public member of the class, and all Public members are exposed to any application that uses an instance of the class. Similarly, the `hireDate.Day = Me.xDayTextBox.Text` statement assigns the contents of the xDayTextBox to the object's Day property. Finally, the `Me.xHiredLabel.Text = hireDate.GetNewDate()` statement in the xDefaultButton's Click event procedure uses the FormattedDate object's GetNewDate method to return the month and day numbers, separated by a slash. The statement displays the formatted date in the xHiredLabel.

To code and then test the Personnel application:

1 Open the **Personnel Solution** (Personnel Solution.sln) file, which is contained in the VB2005\Chap11\Personnel Solution folder. If necessary, open the designer window.

2 Add a class file named **FormattedDate.vb** to the project. Type the comments and code shown earlier in Figure 11-22, replacing the <your name> and <current date> text in the comments with your name and the current date.

3 Save the solution, then close the FormattedDate.vb window.

4 Open the form's Code Editor window. Replace the <your name> and <current date> text in the comments with your name and the current date.

5 First, you will enter the appropriate code in the xParameterizedButton's Click event procedure, which is shown in Figure 11-24. Locate the xParameterizedButton's Click event procedure. In the blank line above the End Sub statement, type **dim hireDate as new formatteddate(**. The IntelliSense feature in the Code Editor window displays a box that allows you to view a method's signature, one signature at a time. The box shown in Figure 11-25, for instance, displays the first of the New method's two signatures.

Figure 11-25: First signature for the New method

6 You can use the down arrow in the box to display the second signature. Click the **down arrow** in the box to display the New method's second signature.

Figure 11-26: Second signature for the New method

7 To use the second signature, the application will need to provide two items of information: the month number and day number. Both items should have a data type of String. Type **me.xMonthTextBox.text, _** and press **Enter**. Press **Tab** eight times, then press the **spacebar** twice. Type **me.xDayTextBox.text)** and press **Enter** twice.

8 Type **me.xHiredLabel.text = hiredate.getnewdate()** and press **Enter**.

9 Now you will enter the appropriate code in the xDefaultButton's Click event procedure. Locate the xDefaultButton's Click event procedure, then enter the procedure's code from Figure 11-24.

10 Close the Code Editor window. Save the solution, then start the application. Type **10** in the Month box, then type **23** in the Day box. Click the **Display Hire Date - Parameterized** button. The button's Click event procedure displays the formatted date in the Hire date box, as shown earlier in Figure 11-23.

11 Type **7** in the Month box, then type **10** in the Day box. Click the **Display Hire Date – Default** button. The button's Click event procedure displays 7/10 in the Hire date box.

12 Click the **Exit** button to end the application, then close the solution.

You have completed Lesson A. You can either take a break or complete the end-of-lesson questions and exercises before moving on to Lesson B.

SUMMARY

TO DEFINE A CLASS:
» Use the syntax shown in Figure 11-2.

TO ADD A CLASS FILE TO A PROJECT:
» Click Project on the menu bar, and then click Add Class. In the Add New Item – *projectname* dialog box, verify that Class is selected in the Visual Studio installed templates box. In the Name box, type the name of the class followed by .vb, then click the Add button.

TO INSTANTIATE (CREATE) AN OBJECT FROM A CLASS:
» Use either of the syntax versions shown in Figure 11-6.

TO CREATE A PROPERTY PROCEDURE:
» Use the syntax shown in Figure 11-12. The Get block allows an application to retrieve the contents of the Private variable associated with the Property procedure. The Set block allows an application to assign a value to the Private variable associated with the Property procedure.

TO ACCESS THE PROPERTIES OF AN OBJECT:
» Use the syntax *objectVariable.property*, where *objectVariable* is the name of the variable that stores the object's address, and *property* is the name of the property you want to access.

TO ACCESS AN OBJECT'S METHODS:
» Use the syntax *objectVariable.method*, where *objectVariable* is the name of the variable that stores the object's address, and *method* is the name of the method you want to access.

TO CREATE A CONSTRUCTOR:

» Use the syntax shown in Figure 11-13. A constructor that has no parameters is called the default constructor.

TO CREATE A METHOD OTHER THAN A CONSTRUCTOR:

» Use the syntax shown in Figure 11-14.

QUESTIONS

1. Which of the following statements is false?

    a. An example of an attribute is the `_minutes` variable in a Time class.

    b. An example of a behavior is the SetTime method in a Time class.

    c. An object created from a class is referred to as an instance of the class.

    d. A class is considered an object.

2. In Visual Basic, you enter the Class statement in a _____.

    a. class file whose filename ends with .vb

    b. class file whose filename ends with .cls

    c. form file whose filename ends with .cla

    d. form file whose filename ends with .cls

3. The attributes of an object are represented by _____ in a class.

    a. constants   b. methods

    c. functions   d. variables

4. If a variable in a class is declared using the `Public` keyword, the variable can be accessed by any application that uses an instance of the class.

    a. True   b. False

5. Some constructors return a value.

    a. True   b. False

6. Following the naming convention discussed in the lesson, which of the following would be considered a good name for a method contained in a class?

   a. Bonus

   b. SalesIncome

   c. SetDate

   d. Expenses

7. Following the naming convention discussed in the lesson, which of the following would be considered a good name for a property contained in a class?

   a. CalcBonus

   b. sales

   c. FirstName

   d. Both b and c.

8. A constructor that has no parameters is called the _____ constructor.

   a. default

   b. empty

   c. parameterless

   d. parameter-free

9. The Product class contains a Private variable named _price and a Public method named CalculateNewPrice. The _price variable is associated with a Public property named Price. The method is a Function procedure. An application instantiates a Product object and assigns its address to a variable named `item`. Which of the following statements can the application use to assign the number 45 to the _price variable?

   a. `_price = 45`

   b. `Price = 45`

   c. `_price.item = 45`

   d. `item.Price = 45`

10. Which of the following statements can be used by the application in Question 9 to call the CalculateNewPrice method?

   a. `newPrice = Call CalculateNewPrice()`

   b. `newPrice = Price.CalculateNewPrice()`

   c. `newPrice = item.CalculateNewPrice()`

   d. `newPrice = CalculateNewPrice(item)`

11. A Private variable in a class can be accessed directly by a Public method in the same class.

   a. True

   b. False

12. An application can access the Private variables in a class _____.

    a. directly

    b. using properties created by Property procedures

    c. through Private methods contained in the class

    d. None of the above.

13. To expose a variable or method contained in a class, you declare the variable or method using the _____ keyword.

    a. `Exposed`   b. `Private`

    c. `Public`   d. `Viewable`

14. The name of the default constructor for a class named Animal is _____.

    a. Animal   b. AnimalConstructor

    c. Constructor   d. None of the above.

15. The method name combined with the method's optional *parameterlist* is called the method's _____.

    a. autograph   b. inscription

    c. signature   d. None of the above.

16. A constructor is _____.

    a. a Function procedure

    b. a Property procedure

    c. a Sub procedure

    d. either a Function procedure or a Sub procedure

17. Which of the following statements creates an Animal object and assigns the object's address to the dog variable?

    a. `Dim dog As Animal`

    b. `Dim dog As New Animal`

    c. `Dim dog As Animal`

       `dog = New Animal`

    d. Both b and c.

18. An application creates an Animal object and assigns its address to the dog variable. Which of the following statements calls the DisplayBreed method contained in the Animal class?

   a. `Call Animal.DisplayBreed()`

   b. `Call DisplayBreed.Animal()`

   c. `Call DisplayBreed().Dog`

   d. `Call dog.DisplayBreed()`

19. If you need to validate a value before assigning it to a Private variable, you enter the validation code in the _____ block in a Property procedure.

   a. Assign

   b. Get

   c. Set

   d. Validate

20. The Return statement is entered in the _____ block in a Property procedure.

   a. Get   b. Set

EXERCISES

1. Write a Class statement that defines a class named Book. The class contains three Public properties named `Title`, `Author`, and `Cost`. The `Title` and `Author` variables are String variables. The `Cost` variable is a Decimal variable.

2. Write a Class statement that defines a class named Tape. The class contains four Public String variables named `Name`, `Artist`, `SongNumber`, and `Length`.

3. Use the syntax shown in Version 1 in Figure 11-6 to declare a variable named `fiction` that can store the address of a Book object. Create the Book object and assign its address to the `fiction` variable.

4. Use the syntax shown in Version 2 in Figure 11-6 to declare a Tape object and assign its address to a variable named `blues`.

5. An application contains the class definition shown in Figure 11-27.

```
Public Class Computer
    Private _model As String
    Private _cost As Decimal

    Public Property Model() As String
        Get
            Return _model
        End Get
        Set(ByVal value As String)
            _model = value
        End Set
    End Property

    Public Property Cost() As Decimal
        Get
            Return _cost
        End Get
        Set(ByVal value As String)
            _cost = value
        End Set
    End Property

    Public Sub New()
        _model = String.Empty
        _cost = 0D
    End Sub

    Public Sub New(ByVal comType As String, ByVal price As Decimal)
        Model = comType
        Cost = price
    End Sub

    Public Function IncreasePrice() As Decimal
        Return _cost * 1.2D
    End Function
End Class
```

Figure 11-27

a. Write a Dim statement that creates a Computer object and initializes it using the default constructor. Assign the object's address to a variable named homeUse.

b. Write an assignment statement that uses the Computer object created in Step a to assign the string "IB-50" to the Model property.

   c. Write an assignment statement that uses the Computer object created in Step a to assign the number 2400 to the Cost property.

   d. Write an assignment statement that uses the Computer object created in Step a to call the IncreasePrice function. Assign the function's return value to a variable named `newPrice`.

   e. Write a Dim statement that creates a Computer object and initializes it using the parameterized constructor. Assign the object's address to a variable named `companyUse`. Use the following values to initialize the object: "IBM" and 1236.99.

6. Write the class definition for a class named Employee. The class should include Private variables and Property procedures for an Employee object's name and salary. The salary may contain a decimal place. The class also should contain two constructors: the default constructor and a constructor that allows an application to assign values to the Private variables.

7. Add the CalculateNewSalary method to the Employee class you defined in Exercise 6. The method should calculate an Employee object's new salary, based on a raise percentage provided by the application using the object. Before calculating the new salary, the method should verify that the raise percentage is greater than or equal to zero. If the raise percentage is less than zero, the method should assign the number 0 as the new salary.

8. Write the Property procedure for a ReadOnly property named BonusRate. The property's data type is Decimal.

9. In this exercise, you modify the Area application that you completed in this lesson.

   a. Use Windows to make a copy of the Area Solution folder, which is contained in the VB2005\Chap11 folder. Rename the folder Modified Area Solution.

   b. If necessary, start Visual Studio 2005 or Visual Basic 2005 Express Edition. Open the Area Solution (Area Solution.sln) file contained in the VB2005\Chap11\ Modified Area Solution folder. Open the designer window.

   c. Make the following modifications to the Square class:

     1) Add a Private variable named `_area`.

     2) Associate the `_area` variable with a ReadOnly Property procedure named Area.

3) Change the CalculateArea method to a Sub procedure. The method should calculate the area and then assign the result to the _area variable.

4) Include a parameterized constructor in the class. The constructor should accept one argument: the side measurement. After initializing the Private variable, the constructor should automatically call the CalculateArea method. (Use the Public property to initialize the Private variable.)

d. Modify the xCalcButton's Click event procedure so that it uses the parameterized constructor to instantiate the Square object. The parameterized constructor will automatically calculate the area of the square; therefore, you can delete the line of code that calls the CalculateArea method in the event procedure.

e. Save the solution, and then start and test the application. Click the Exit button to end the application, then close the solution.

10. In this exercise, you modify the Grade application that you completed in this lesson.

a. Use Windows to make a copy of the Grade Solution folder, which is contained in the VB2005\Chap11 folder. Rename the folder Modified Grade Solution.

b. If necessary, start Visual Studio 2005 or Visual Basic 2005 Express Edition. Open the Grade Solution (Grade Solution.sln) file contained in the VB2005\Chap11\ Modified Grade Solution folder. Open the designer window.

c. Include a parameterized constructor in the CourseGrade class. The constructor should accept two arguments: the two test scores.

d. Modify the xCalcButton's Click event procedure so that it uses the parameterized constructor to instantiate the CourseGrade object.

e. Save the solution, and then start and test the application. Click the Exit button to end the application, then close the solution.

11. In this exercise, you modify the Personnel application that you completed in this lesson.

a. Use Windows to make a copy of the Personnel Solution folder, which is contained in the VB2005\Chap11 folder. Rename the folder Modified Personnel Solution.

b. If necessary, start Visual Studio 2005 or Visual Basic 2005 Express Edition. Open the Personnel Solution (Personnel Solution.sln) file contained in the VB2005\ Chap11\Modified Personnel Solution folder. Open the designer window.

c. Modify the interface to allow the user to enter the year number.

d. The FormattedDate class should create an object that returns the month number, followed by a slash, the day number, a slash, and the year number. Make the appropriate modifications to the class.

e. Modify the FormattedDate class to validate the month number, which should be from 1 through 12. If the month number is invalid, assign the number –1 to the _month variable.

f. Make the necessary modifications to the Personnel application's code.

g. Save the solution, and then start and test the application. Click the Exit button to end the application, then close the solution.

12. In this exercise, you use the Employee class from Exercise 7 to create an object in an application.

a. If necessary, start Visual Studio 2005 or Visual Basic 2005 Express Edition. Open the Salary Solution (Salary Solution.sln) file, which is contained in the VB2005\Chap11\Salary Solution folder. If necessary, open the designer window.

b. Open the Employee.vb class file in the Code Editor window. Enter the class definition from Exercise 7, then close the Employee.vb window.

c. Open the form's Code Editor window. Use the comments that appear in the code to enter the missing instructions.

d. Save the solution, then start the application. Test the application by entering your name, a current salary amount of 54000, and a raise percentage of 10 (for 10%). The application should display the number $59,400.

e. Click the Exit button to end the application. Close the Code Editor window, then close the solution.

13. In this exercise, you modify the Sweets Unlimited application that you completed in this lesson.

a. Use Windows to make a copy of the Sweets Solution folder, which is contained in the VB2005\Chap11 folder. Rename the folder Modified Sweets Solution.

b. Delete the sales.txt file contained in the VB2005\Chap11\Modified Sweets Solution\Sweets Project\bin\Debug folder.

c. If necessary, start Visual Studio 2005 or Visual Basic 2005 Express Edition. Open the Sweets Solution (Sweets Solution.sln) file contained in the VB2005\ Chap11\Modified Sweets Solution folder. Open the designer window.

d. Modify the Salesperson class so that it uses Private variables and Public Property procedures, rather than Public variables. Include two constructors in the class: the default constructor and a parameterized constructor. Include a method that calculates the bonus amount, using the bonus rate provided by the application.

e. Make the necessary modifications to the Sweets Unlimited application's code.

f. Save the solution, and then start and test the application. Click the Exit button to end the application. View the contents of the sales.txt file, then close the sales.txt window. Close the form's Code Editor window, then close the solution.

LESSON B
OBJECTIVES

AFTER STUDYING LESSON B, YOU SHOULD
BE ABLE TO:

» Overload the methods in a class

» Create a derived class using inheritance

» Override a method in the base class

» Refer to the base class using the `MyBase` keyword

MORE ON CLASSES AND OBJECTS

EXAMPLE 5—USING A CLASS THAT CONTAINS OVERLOADED METHODS

In this example, you view the code for a class named Employee, which you can use to instantiate an Employee object. Employee objects have two attributes: an employee number and an employee name. Employee objects also have the following four behaviors:

1. They can initialize their attributes using values provided by the class.

2. They can initialize their attributes using values provided by the application in which they are instantiated.

3. They can calculate and return the gross pay for salaried employees. The gross pay is calculated by dividing the salaried employee's annual salary by 24, because salaried employees are paid twice per month.

4. They can calculate and return the gross pay for hourly employees. The gross pay is calculated by multiplying the number of hours the employee worked during the week by his or her pay rate.

Figure 11-28 shows the Employee class defined in the Employee.vb file.

Visual Basic code

```
' Class name:        Employee
' Class purpose:     Calculates the gross pay for salaried
'                    and hourly employees.
' Created/revised:   <your name> on <current date>

Option Explicit On
Option Strict On

Public Class Employee
    Private _number As String
    Private _empName As String
```

Figure 11-28: Employee class defined in the Employee.vb file *(Continued)* ▶

```
    Public Property Number() As String
        Get
            Return _number
        End Get
        Set(ByVal value As String)
            _number = value
        End Set
    End Property

    Public Property EmpName() As String
        Get
            Return _empName
        End Get
        Set(ByVal value As String)
            _empName = value
        End Set
    End Property

    Public Sub New()
        _number = String.Empty
        _empName = String.Empty
    End Sub

    Public Sub New(ByVal num As String, ByVal name As String)
        Number = num
        EmpName = name
    End Sub

    Public Overloads Function CalculateGross(ByVal salary As Decimal) _
        As Decimal
        ' calculates the gross pay for salaried
        ' employees, who are paid twice per month

        Return salary / 24D
    End Function

    Public Overloads Function CalculateGross(ByVal hours As Decimal, _
        ByVal rate As Decimal) As Decimal
        ' calculates the weekly pay for
        ' hourly employees

        Return hours * rate
    End Function
End Class
```

overloaded constructors

overloaded CalculateGross methods

Figure 11-28: Employee class defined in the Employee.vb file

The Employee class contains two Private variables named _number and _empName. It also contains two Property procedures named Number and EmpName. The Number Property procedure is associated with the _number variable, and the EmpName Property procedure is associated with the _empName variable. In addition to the Private variables and Property procedures, the Employee class also contains four methods. The two methods named New are the class's constructors. The first New method is the default constructor, and the second New method is a parameterized constructor. When two or more methods have the same name but different parameters, the methods are referred to as **overloaded methods**. The two constructors contained in the Employee class are overloaded methods, because each is named New and each has a different *parameterlist*. Keep in mind that you can overload any of the methods contained in a class, not just constructors. For example, the two CalculateGross methods in the Employee class also are overloaded methods, because they have the same name but a different *parameterlist*. However, if the methods being overloaded are not constructors, you must use the Overloads keyword in the procedure header, as shown in both CalculateGross methods in Figure 11-28. The Overloads keyword is not used when overloading constructors.

Overloading is useful when two or more methods require different parameters to perform the same task. For example, both overloaded constructors in the Employee class initialize the class's Private variables. However, the default constructor does not need to be passed any information to perform the task, while the parameterized constructor requires two items of information (the employee number and name). Similarly, both CalculateGross methods in the Employee class calculate and return a gross pay amount. However, the first CalculateGross method, which calculates and returns the gross pay amount for salaried employees, requires an application to pass it one item of information: the employee's annual salary. The second CalculateGross method, on the other hand, calculates and returns the gross pay amount for hourly employees and requires two items of information: the number of hours the employee worked and his or her rate of pay. Rather than using two overloaded CalculateGross methods in the Employee class, you could use two methods having different names. For example, you could use a method named CalcSalariedGross to calculate and return the gross pay amount for salaried employees, and a method named CalcHourlyGross to calculate and return the gross pay amount for hourly employees. The advantage of overloading the CalculateGross method is that you need to remember the name of only one method.

» TIP

The New methods in the FormattedDate class shown in Figure 11-22 in Lesson A also are overloaded methods. As you observed in Lesson A, the IntelliSense feature displays the signatures, one signature at a time, in the Code Editor window.

Next, you view the ABC Company application, which uses the Employee class to instantiate and manipulate an Employee object. Figures 11-29 and 11-30 show sample runs of the application, and Figure 11-31 shows the code for the xCalcButton's Click event procedure, which is contained in the application.

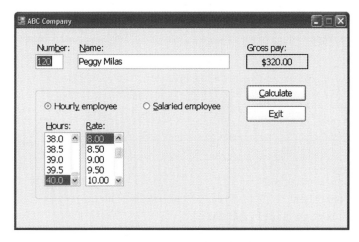

Figure 11-29: Sample run showing the gross pay for an hourly worker

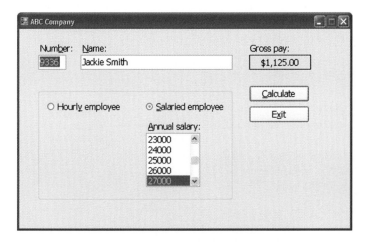

Figure 11-30: Sample run showing the gross pay for a salaried worker

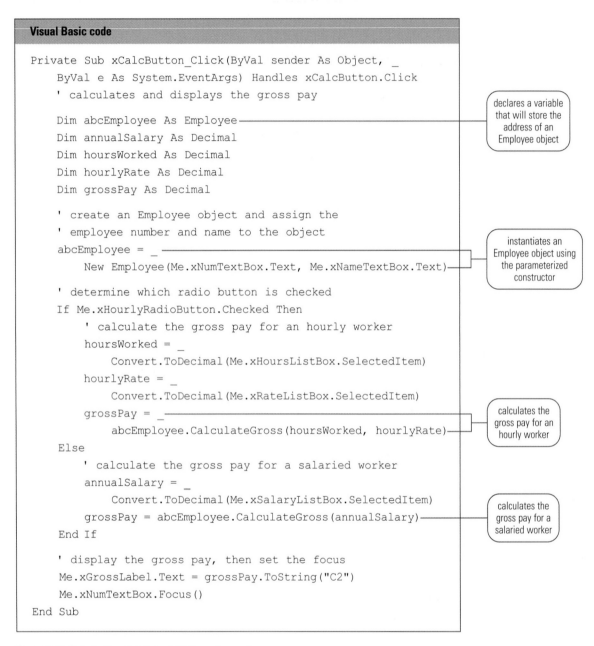

Visual Basic code

```
Private Sub xCalcButton_Click(ByVal sender As Object, _
    ByVal e As System.EventArgs) Handles xCalcButton.Click
    ' calculates and displays the gross pay

    Dim abcEmployee As Employee
    Dim annualSalary As Decimal
    Dim hoursWorked As Decimal
    Dim hourlyRate As Decimal
    Dim grossPay As Decimal

    ' create an Employee object and assign the
    ' employee number and name to the object
    abcEmployee = _
        New Employee(Me.xNumTextBox.Text, Me.xNameTextBox.Text)

    ' determine which radio button is checked
    If Me.xHourlyRadioButton.Checked Then
        ' calculate the gross pay for an hourly worker
        hoursWorked = _
            Convert.ToDecimal(Me.xHoursListBox.SelectedItem)
        hourlyRate = _
            Convert.ToDecimal(Me.xRateListBox.SelectedItem)
        grossPay = _
            abcEmployee.CalculateGross(hoursWorked, hourlyRate)
    Else
        ' calculate the gross pay for a salaried worker
        annualSalary = _
            Convert.ToDecimal(Me.xSalaryListBox.SelectedItem)
        grossPay = abcEmployee.CalculateGross(annualSalary)
    End If

    ' display the gross pay, then set the focus
    Me.xGrossLabel.Text = grossPay.ToString("C2")
    Me.xNumTextBox.Focus()
End Sub
```

declares a variable that will store the address of an Employee object

instantiates an Employee object using the parameterized constructor

calculates the gross pay for an hourly worker

calculates the gross pay for a salaried worker

Figure 11-31: Code for the xCalcButton's Click event procedure

The `Dim abcEmployee As Employee` statement in the xCalcButton's Click event procedure declares a variable that will store the address of an Employee object when the object is created. The Employee object is instantiated using the `abcEmployee = New Employee(Me.xNumTextBox.Text, Me.xNameTextBox.Text)` statement. When instantiating the object, the computer will use the class's parameterized constructor to initialize the object's Private variables.

Notice that the CalculateGross method appears in two statements in the xCalcButton's Click event procedure. The computer uses the signature of the CalculateGross method in each statement to determine which of the class's CalculateGross methods to process. In this case, the `grossPay = abcEmployee.CalculateGross(hoursWorked, hourlyRate)` statement tells the computer to process the CalculateGross method that contains two parameters, and then assign the return value to the `grossPay` variable. In the Employee class, the CalculateGross method that contains two parameters calculates and returns the gross pay amount for an hourly worker. The `grossPay = abcEmployee.CalculateGross(annualSalary)` statement, on the other hand, tells the computer to process the CalculateGross method that contains one parameter, and then assign the return value to the `grossPay` variable. In the Employee class, the CalculateGross method that contains one parameter calculates and returns the gross pay amount for a salaried worker. The xCalcButton's Click event procedure displays the contents of the `grossPay` variable, formatted with a dollar sign and two decimal places, in the xGrossLabel.

To code and then test the ABC Company application:

1 Start Visual Studio 2005 or Visual Basic 2005 Express Edition, if necessary, and close the Start Page window. Open the **ABC Solution** (ABC Solution.sln) file, which is contained in the VB2005\Chap11\ABC Solution-Overloads folder. If necessary, open the designer window.

2 Add a class file named **Employee.vb** to the project. Type the comments and code shown earlier in Figure 11-28, replacing the <your name> and <current date> text in the comments with your name and the current date.

3 Save the solution, then close the Employee.vb window.

4 Open the form's Code Editor window. Replace the <your name> and <current date> text in the comments with your name and the current date. Open the xCalcButton's Click event procedure, then enter the comments and code shown in Figure 11-31.

5 Close the Code Editor window. Save the solution, then start the application. Type **120** in the Number box, then type **Peggy Milas** in the Name box. Click **8.00** in the Rate list box, then click the **Calculate** button. The xCalcButton's Click event procedure displays the gross pay amount ($320.00) in the Gross pay box, as shown earlier in Figure 11-29.

6 Type **9336** in the Number box. Press **Tab**, then type **Jackie Smith** in the Name box. Click the **Salaried employee** radio button, then click **27000** in the Annual salary list box. Click the **Calculate** button. The xCalcButton's Click event procedure displays the gross pay amount ($1,125.00) in the Gross pay box, as shown earlier in Figure 11-30.

7 Click the **Exit** button to end the application, then close the solution.

EXAMPLE 6—USING A BASE CLASS AND A DERIVED CLASS

As you learned in the Overview, you can create one class from another class. In OOP, this is referred to as inheritance. The new class is called the derived class and it inherits the attributes and behaviors of the original class, called the base class. You indicate that a class is a derived class by including the Inherits clause in the derived class's Class statement. The syntax of the Inherits clause is **Inherits** *base*, where *base* is the name of the base class whose attributes and behaviors the derived class will inherit. You enter the Inherits clause below the Public Class header in the derived class. Figure 11-32 shows the code for a base class named Employee and a derived class named Salaried. Both class definitions are contained in the CompanyClass.vb file. (In Exercise 1 at the end of this lesson, you will add another derived class, named Hourly, to the file.)

Visual Basic code

```
' The CompanyClass.vb file contains the
' definitions for the Employee base class,
' and the Salaried and Hourly derived classes.

' Class name:      Employee
' Class purpose:   Base class.
' Created/revised: <your name> on <current date>

Option Explicit On
Option Strict On

Public Class Employee                             ─── beginning of base
    Private _number As String                          Employee class
    Private _empName As String

    Public Property Number() As String
        Get
            Return _number
        End Get
        Set(ByVal value As String)
            _number = value
        End Set
    End Property
```

Figure 11-32: Base Employee class and derived Salaried class defined in the CompanyClass.vb file *(Continued)* ▶

```
    Public Property EmpName() As String
        Get
            Return _empName
        End Get
        Set(ByVal value As String)
            _empName = value
        End Set
    End Property

    Public Sub New()
        _number = String.Empty
        _empName = String.Empty
    End Sub

    Public Sub New(ByVal num As String, ByVal name As String)
        Number = num
        EmpName = name
    End Sub

    Public Overridable Function CalculateGross() As Decimal
        ' this function will be overriden
        ' in the derived classes
    End Function
End Class

' Class name:      Salaried
' Class purpose:   Calculates the gross pay
'                  for salaried workers.
' Created/revised: <your name> on <current date>

Public Class Salaried
    Inherits Employee

    Private _salary As Decimal

    Public Property Salary() As Decimal
        Get
            Return _salary
        End Get
        Set(ByVal value As Decimal)
            _salary = value
        End Set
    End Property
```

indicates that the method can be overridden in the derived class

beginning of the derived Salaried class

the derived class inherits the base class's attributes and behaviors

Figure 11-32: Base Employee class and derived Salaried class defined in the CompanyClass.vb file *(Continued)* ▶

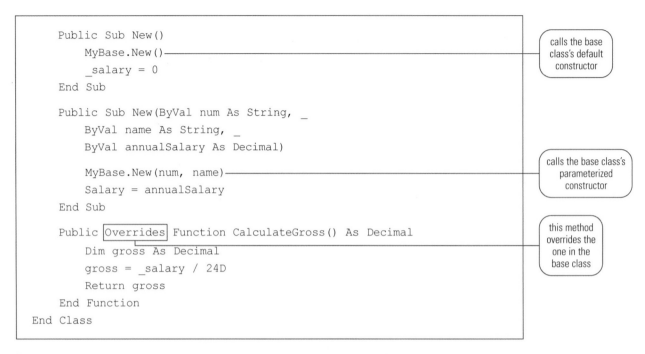

```
    Public Sub New()
        MyBase.New()                                          calls the base
        _salary = 0                                           class's default
    End Sub                                                   constructor

    Public Sub New(ByVal num As String, _
        ByVal name As String, _
        ByVal annualSalary As Decimal)
                                                              calls the base class's
        MyBase.New(num, name)                                 parameterized
        Salary = annualSalary                                 constructor
    End Sub
                                                              this method
    Public Overrides Function CalculateGross() As Decimal     overrides the
        Dim gross As Decimal                                  one in the
        gross = _salary / 24D                                 base class
        Return gross
    End Function
End Class
```

Figure 11-32: Base Employee class and derived Salaried class defined in the CompanyClass.vb file

» TIP

The three statements in the Salaried class's CalculateGross method are equivalent to the statement `return _salary / 24D`.

The Employee class definition shown in Figure 11-32 is almost identical to the Employee class definition shown in Figure 11-28; only the CalculateGross method in both class definitions differs. For instance, the CalculateGross method in Figure 11-28 is an overloaded method that can calculate the gross pay for either a salaried employee or an hourly employee. The CalculateGross method in the Employee class shown in Figure 11-32, on the other hand, does not perform any calculations; it doesn't even contain any code. In addition, its procedure header contains the `Overridable` keyword, which indicates that the method can be overridden by any class that is derived from the Employee class. In other words, classes derived from the Employee class will provide their own CalculateGross method.

Now study the code contained in the Salaried class shown in Figure 11-32. The `Inherits Employee` clause that appears below the `Public Class Salaried` clause indicates that the Salaried class is derived from the Employee class. This means that the Salaried class includes all of the attributes and behaviors of the Employee class. The Salaried class also contains an attribute of its own: salary. The salary attribute is represented by the Private `_salary` variable and the Public Salary Property procedure in the class. The salary attribute belongs to the Salaried class only.

Both constructors in the Salaried class contain the **MyBase.New(**[*parameterlist*]**)** statement. This statement tells the computer to process the code contained in the appropriate constructor in the base class. You refer to the base class using the `MyBase` keyword. For example, the `MyBase.New()` statement in the default constructor tells the computer to process the code contained in the base class's default constructor. The

`MyBase.New(num, name)` statement in the Salaried class's parameterized constructor indicates that the code in the base class's parameterized constructor should be processed.

The CalculateGross method in the Salaried class calculates the gross pay for a salaried employee. The method's header contains the `Overrides` keyword, which indicates that the method overrides (replaces) the CalculateGross method contained in the base Employee class.

In Figure 11-31, you viewed the code for the xCalcButton's Click event procedure, which used the overloaded methods defined in the Employee class shown in Figure 11-28. Figure 11-33 shows a different version of the procedure. In this version, the procedure uses the derived Salaried class shown in Figure 11-32. The code pertaining to the Salaried class is shaded in the figure. (You will complete the xCalcButton's Click event procedure in Exercise 1 at the end of this lesson.)

Visual Basic code

```vb
Private Sub xCalcButton_Click(ByVal sender As Object, _
    ByVal e As System.EventArgs) Handles xCalcButton.Click
    ' calculates and displays the gross pay

    Dim salEmployee As Salaried
    Dim annualSalary As Decimal
    Dim hoursWorked As Decimal
    Dim hourlyRate As Decimal
    Dim grossPay As Decimal

    ' determine which radio button is checked
    If Me.xHourlyRadioButton.Checked Then
        ' calculate the gross pay for an hourly worker

    Else
        ' calculate the gross pay for a salaried worker
        annualSalary = _
            Convert.ToDecimal(Me.xSalaryListBox.SelectedItem)
        salEmployee = _
            New Salaried(Me.xNumTextBox.Text, _
            Me.xNameTextBox.Text, annualSalary)
        grossPay = salEmployee.CalculateGross()
    End If

    ' display the gross pay, then set the focus
    Me.xGrossLabel.Text = grossPay.ToString("C2")
    Me.xNumTextBox.Focus()
End Sub
```

> you will complete the selection structure's true path in Exercise 1 at the end of this lesson

Figure 11-33: Salaried class used in the xCalcButton's Click event procedure

The `Dim  salEmployee  As  Salaried` statement declares a variable that can store the address of a Salaried object. The `salEmployee = New  Salaried (Me.xNumTextBox.Text, Me.xNameTextBox.Text, annualSalary)` statement instantiates a Salaried object, and it assigns the object's address to the `salEmployee` variable. The statement passes three items of information to the Salaried class's parameterized constructor, which uses the items to initialize the object when it is instantiated. The `grossPay = salEmployee.CalculateGross()` statement uses the Salaried class's CalculateGross method to calculate and return the gross pay, which the statement assigns to the `grossPay` variable.

To code and then test this version of the ABC Company application:

1 Open the **ABC Solution** (ABC Solution.sln) file, which is contained in the VB2005\ Chap11\ABC Solution-Inheritance folder. If necessary, open the designer window.

2 Add a class file named **CompanyClass.vb** to the project. Type the comments and code shown earlier in Figure 11-32, replacing the <your name> and <current date> text in the comments with your name and the current date.

3 Save the solution, then close the CompanyClass.vb window.

4 Open the form's Code Editor window. Replace the <your name> and <current date> text in the comments with your name and the current date. Open the xCalcButton's Click event procedure, then enter the comments and code shown earlier in Figure 11-33. (Do not be concerned about the jagged lines that appear below the `Dim hoursWorked As Decimal` and `Dim hourlyRate As Decimal` statements in the Code Editor window. The jagged lines will disappear when you complete the procedure in Exercise 1 at the end of this lesson.)

5 Close the Code Editor window. Save the solution, then start the application. Type **9336** in the Number box, then type **Jackie Smith** in the Name box. Click the **Salaried employee** radio button, then click **27000** in the Annual salary list box. Click the **Calculate** button. The xCalcButton's Click event procedure displays the gross pay amount ($1,125.00) in the Gross pay box, as shown earlier in Figure 11-30.

6 Click the **Exit** button to end the application, then close the solution.

You have completed Lesson B. You can either take a break or complete the end-of-lesson questions and exercises before moving on to Lesson C.

SUMMARY

TO CREATE TWO OR MORE METHODS THAT PERFORM THE SAME TASK BUT REQUIRE DIFFERENT PARAMETERS:

» Overload the methods by giving them the same name but different *parameterlists*. If the methods are not constructors, include the `Overloads` keyword in their procedure header.

TO HAVE A DERIVED CLASS INHERIT THE ATTRIBUTES AND BEHAVIORS OF A BASE CLASS:

» Include the **Inherits** *base* clause below the Public Class header in the derived class. In the clause, *base* is the name of the base class whose attributes and behaviors you want the derived class to inherit.

TO REFER TO THE BASE CLASS:

» Use the `MyBase` keyword.

TO ALLOW A DERIVED CLASS TO OVERRIDE A METHOD IN THE BASE CLASS:

» Include the `Overridable` keyword in the base class method's procedure header.

TO OVERRIDE A METHOD IN THE BASE CLASS:

» Include the `Overrides` keyword in the derived class method's procedure header.

QUESTIONS

1. When two methods have the same name but different *parameterlists*, the methods are referred to as _____.

   a. loaded

   b. overloaded

   c. overridden

   d. parallel

2. An overloaded method that is not a constructor must contain the _____ keyword in its procedure header.

   a. `Loaded`

   b. `Overloaded`

   c. `Overloads`

   d. `Overridden`

3. Which of the following clauses allows a derived class named Dog to have the same attributes and behaviors as its base class, which is named Animal?

a. `Inherits Animal`   b. `Inherits Dog`

c. `Overloads Animal`   d. `Overloads Dog`

4. A base class contains a method named CalcBonus. Which of the following procedure headers can be used in the base class to indicate that a derived class can provide its own code for the method?

a. `Public Inherits Sub CalcBonus()`

b. `Public Overloads Sub CalcBonus()`

c. `Public Overridable Sub CalcBonus()`

d. `Public Overrides Sub CalcBonus()`

5. A base class contains a method named CalcBonus. Which of the following procedure headers can be used in the derived class to indicate that it is providing its own code for the method?

a. `Public Inherits Sub CalcBonus()`

b. `Public Overloads Sub CalcBonus()`

c. `Public Overridable Sub CalcBonus()`

d. `Public Overrides Sub CalcBonus()`

EXERCISES

1. In this exercise, you modify the second version of the ABC Company application that you coded in this lesson.

a. Use Windows to make a copy of the ABC Solution-Inheritance folder, which is contained in the VB2005\Chap11 folder. Rename the folder Modified ABC Solution-Inheritance.

b. If necessary, start Visual Studio 2005 or Visual Basic 2005 Express Edition. Open the ABC Solution (ABC Solution.sln) file contained in the VB2005\Chap11\ Modified ABC Solution-Inheritance folder.

c. Open the CompanyClass.vb class file. Include another derived class definition in the file. Name the derived class Hourly. The Hourly class should contain a CalculateGross method that calculates the gross pay for an hourly employee. Calculate the gross pay by multiplying the hours worked by the pay rate.

  d. Close the CompanyClass.vb window. Open the form's Code Editor window, then complete the selection structure's true path in the xCalcButton's Click event procedure.

  e. Save the solution, and then start the application. Test the application using the data shown earlier in Figures 11–29 and 11–30. Click the Exit button to end the application. Close the Code Editor window, then close the solution.

2. In this exercise, you modify the Area application that you coded in Lesson A.

  a. Use Windows to make a copy of the Area Solution folder, which is contained in the VB2005\Chap11 folder. Rename the folder Area Solution-Overloads.

  b. If necessary, start Visual Studio 2005 or Visual Basic 2005 Express Edition. Open the Area Solution (Area Solution.sln) file contained in the VB2005\Chap11\Area Solution-Overloads folder.

  c. Open the Square.vb class file. Include another CalculateArea method in the class. The method should accept the side measurement provided by the application. Close the Square.vb window.

  d. Open the form's Code Editor window. Modify the xCalcButton's Click event procedure so that it uses the CalculateArea method you defined in Step c.

  e. Save the solution, and then start and test the application. Click the Exit button to end the application. Close the Code Editor window, then close the solution.

LESSON C
OBJECTIVES

AFTER STUDYING LESSON C, YOU SHOULD
BE ABLE TO:

» Create a class that contains more than one constructor

» Create a class that contains an overloaded method

» Use the class you created in an application

THE KESSLER LANDSCAPING APPLICATION

CODING THE KESSLER LANDSCAPING APPLICATION

Recall that your task in this chapter is to create an application that Monica Kessler, the owner of Kessler Landscaping, can use to estimate the cost of laying sod. Monica informs you that the sod typically is laid on a rectangular piece of land. Therefore, you will create a MyRectangle class that can be used to instantiate a MyRectangle object in the application. The MyRectangle object will have two attributes: a length measurement and a width measurement. It also will have three behaviors. First, it will be able to initialize its attributes using values provided by the class. Second, it will be able to initialize its attributes using values provided by the application. And third, it will be able to calculate its area. To save you time, the VB2005\Chap11\Kessler Solution folder contains a partially completed Kessler Landscaping application.

To open the Kessler Landscaping application:

1 Start Visual Studio 2005 or Visual Basic 2005 Express Edition, if necessary, and close the Start Page window.

2 Open the **Kessler Solution** (Kessler Solution.sln) file, which is contained in the VB2005\Chap11\Kessler Solution folder. If necessary, open the designer window. Figure 11-34 shows the user interface for the Kessler Landscaping application, and Figure 11-35 shows the application's TOE chart.

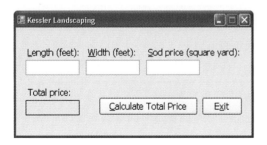

Figure 11-34: Interface for the Kessler Landscaping application

Task	Object	Event
1. Calculate the area of the rectangle 2. Calculate the total price 3. Display the total price in the xTotalPriceLabel	xCalcButton	Click
End the application	xExitButton	Click
Display the total price (from xCalcButton)	xTotalPriceLabel	None
Get the length in feet, width in feet, and price of the sod per square yard	xLengthTextBox, xWidthTextBox, xPriceTextBox	None
Clear the contents of the xTotalPriceLabel		TextChanged
Accept only numbers, the period, and the Backspace		KeyPress

Figure 11-35: TOE chart for the Kessler Landscaping application

Most of the Kessler Landscaping application has been coded for you. You will need to code only the MyRectangle class and the xCalcButton's Click event procedure. First you will define the MyRectangle class.

DEFINING THE MYRECTANGLE CLASS

As mentioned earlier, the MyRectangle class will contain two attributes and three behaviors. The attributes will be defined using two Private variables (named _length and _width) along with their associated Property procedures (named Length and Width). The behaviors will be defined using three methods. Two of the methods will be constructors. The third method, named CalculateArea, will be a Function procedure. Figure 11-36 shows the pseudocode for the methods contained in the MyRectangle class.

MyRectangle class methods - pseudocode
New method (default constructor) initialize the _length and _width variables to the number 0
New method (parameterized constructor) initialize the _length and _width variables using the object's Length and Width properties and the values provided by the application
CalculateArea method calculate the area by multiplying the contents of the _length variable by the contents of the _width variable, and then return the result

Figure 11-36: Pseudocode for the methods contained in the MyRectangle class

To define the MyRectangle class:

1 Click **Project** on the menu bar, then click **Add Class**. The Add New Item – Kessler Project dialog box opens with Class selected in the Visual Studio installed templates box. Change the name in the Name box to **MyRectangle.vb**, then click the **Add** button.

2 Insert a blank line above the `Public Class MyRectangle` clause. Position the insertion point in the blank line, then type the following comments, replacing the <your name> and <current date> text with your name and the current date. Press **Enter** twice after typing the last comment.

' **Class name:** **MyRectangle**

' **Class purpose:** **Calculates the area of a rectangle.**

' **Created/revised:** <your name> **on** <current date>

3 Type **option explicit on** and press **Enter**, then type **option strict on** and press **Enter**.

4 First you will declare the class's Private variables. Position the insertion point in the blank line below the `Public Class MyRectangle` clause. Type **private _length as decimal** and press **Enter**, then type **private _width as decimal** and press **Enter** twice.

Recall that when an application needs to assign data to or retrieve data from a Private variable in a class, it must use a Public property to do so. In other words, an application cannot refer, directly, to a Private variable in a class. Rather, it must refer to the variable indirectly, through the use of a Public property. Recall that you create a Public property using a Property procedure.

5 Type **public property Length() as decimal** and press **Enter**. The Property procedure's template appears in the Code Editor window, as shown in Figure 11-37.

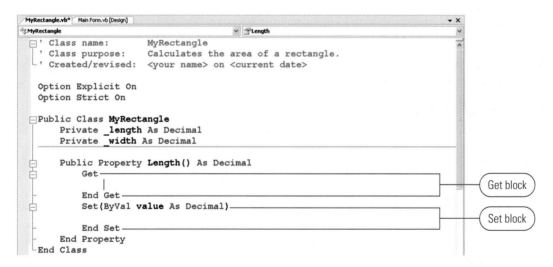

Figure 11-37: Template for the Length property procedure

»TIP

Recall that the Property procedure's data type must match the data type of the Private variable with which it is associated.

6 Recall that the code in the Get block allows an application to retrieve the contents of the Private variable associated with the Property procedure. Most times, you will enter only the Return *privateVariable* instruction in the Get block. Type **return _length**.

7 The code in the Set block, on the other hand, allows the application to assign a value to the Private variable associated with the Property procedure. Type the additional statement shown in Figure 11-38, then position the insertion point as shown in the figure.

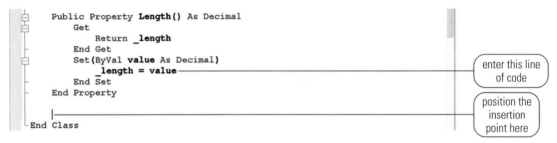

```
Public Property Length() As Decimal
    Get
        Return _length
    End Get
    Set(ByVal value As Decimal)
        _length = value                    ──── enter this line
    End Set                                        of code
End Property

|                                              ──── position the
                                                     insertion
End Class                                            point here
```

Figure 11-38: Property procedure associated with the `_length` variable

8 Create a Property procedure for the `_width` variable. Name the property **Width**. Complete the Get and Set blocks appropriately.

According to the pseudocode shown earlier in Figure 11-36, the MyRectangle class will use two constructors: the default constructor and a parameterized constructor. The default constructor will initialize the Private `_length` and `_width` variables to the number 0. The parameterized constructor, on the other hand, will use the object's Public Length and Width properties to initialize the Private variables to the values provided by the application. As you learned in Lesson A, values that originate outside the class should always be assigned to the Private variables indirectly, through the Public properties. Doing this ensures that the computer processes the Property procedure's Set block, which typically contains validation code.

9 Click the **Method Name** list arrow in the Code Editor window, then click **New** in the list. The template for the default constructor appears in the Code Editor window. Type **_length = 0D** and press **Enter**, then type **_width = 0D**.

10 Position the insertion point two lines below the End Sub clause, but before the End Class clause. Type **public sub new(byval len as decimal, byval wid as decimal)** and press **Enter**. The template for the parameterized constructor appears in the Code Editor window. Type **length = len** and press **Enter**, then type **width = wid**.

11 According to the pseudocode shown earlier in Figure 11-36, the MyRectangle class will contain a method named CalculateArea. The method is responsible for calculating and returning the area of the rectangle. Position the insertion point two lines below the last End Sub clause, but before the End Class clause. Type **public function CalculateArea() as decimal** and press **Enter**. The method's template appears in the Code Editor window. Type **return _length * _width**.

12 Save the solution. Figure 11-39 shows the completed MyRectangle class. Verify that your code agrees with the code shown in the figure, then close the MyRectangle.vb window.

Visual Basic code

```vb
' Class name:        MyRectangle
' Class purpose:     Calculates the area of a rectangle.
' Created/revised:   <your name> on <current date>
Option Explicit On
Option Strict On
Public Class MyRectangle
    Private _length As Decimal
    Private _width As Decimal

    Public Property Length() As Decimal
        Get
            Return _length
        End Get
        Set(ByVal value As Decimal)
            _length = value
        End Set
    End Property

    Public Property Width() As Decimal
        Get
            Return _width
        End Get
        Set(ByVal value As Decimal)
            _width = value
        End Set
    End Property

    Public Sub New()
        _length = 0D
        _width = 0D
    End Sub

    Public Sub New(ByVal len As Decimal, ByVal wid As Decimal)
        Length = len
        Width = wid
    End Sub

    Public Function CalculateArea() As Decimal
        Return _length * _width
    End Function
End Class
```

Figure 11-39: MyRectangle class defined in the MyRectangle.vb file

CODING THE XCALCBUTTON'S CLICK EVENT PROCEDURE

According to the TOE chart shown earlier in Figure 11-35, the xCalcButton's Click event procedure needs to calculate both the area of the rectangle and the total price, and then display the total price in the xTotalPricelabel. Figure 11-40 shows the pseudocode for the xCalcButton's Click event procedure.

---

xCalcButton Click event procedure - pseudocode

1. calculate the area of the rectangle whose measurements are entered by the user

2. calculate the total price of the sod, using the area of the rectangle and the price per square yard entered by the user

3. display the total price of the sod in the xTotalPriceLabel

4. send the focus to the xLengthTextBox

Figure 11-40: Pseudocode for the xCalcButton's Click event procedure

To code the xCalcButton's Click event procedure:

1 Close the MyRectangle.vb window, then open the form's Code Editor window. Replace the <your name> and <current date> text with your name and the current date.

2 Open the code template for the xCalcButton's Click event procedure. Type **' calculates the cost of laying sod** and press **Enter** twice.

3 The procedure will use four variables. Type **dim lawn as new myrectangle** and press **Enter**. The `lawn` variable will store the address of a MyRectangle object.

4 Type **dim sodPrice as decimal** and press **Enter**. The procedure will use the `sodPrice` variable to store the price of a square yard of sod.

5 Type **dim area as decimal** and press **Enter**. The procedure will use the `area` variable to store the area of the MyRectangle object.

6 Type **dim totalPrice as decimal** and press **Enter** twice. The procedure will use the `totalPrice` variable to store the total price of the sod.

7 Next, you will assign the length and width values, which are entered by the user, to the MyRectangle object's Length and Width properties. Type **decimal.tryparse (me.xLengthTextBox.text, lawn.length)** and press **Enter**, then type **decimal.tryparse(me.xWidthTextBox.text, lawn.width)** and press **Enter**.

8 Now assign the price of a square yard of sod, which is entered by the user, to the `sodPrice` variable. Type **decimal.tryparse(me.xPriceTextBox.text, sodprice)** and press **Enter** twice.

9 The first step in the pseudocode shown in Figure 11-40 is to calculate the area of the rectangle whose measurements are entered by the user. Type **' calculate the area (in square yards)** and press **Enter**, then type **area = lawn.calculatearea/9d** and press **Enter** twice.

10 The second step is to calculate the total price of the sod, using the area of the rectangle and the price per square yard entered by the user. Type **' calculate and display the total price** and press **Enter**, then type **totalprice = area * sodprice** and press **Enter**.

11 The third step is to display the total price of the sod in the xTotalPriceLabel. Type **me.xTotalPriceLabel.text = totalprice.tostring("C2")** and press **Enter** twice.

12 The last step in the pseudocode shown in Figure 11-40 is to send the focus to the xLengthTextBox. Type **' set the focus** and press **Enter**, then type **me.xLengthTextBox.focus()** and press **Enter**. Figure 11-41 shows the completed xCalcButton's Click event procedure.

```
Private Sub xCalcButton_Click(ByVal sender As Object, ByVal e As System.
    ' calculates the cost of laying sod

    Dim lawn As New MyRectangle
    Dim sodPrice As Decimal
    Dim area As Decimal
    Dim totalPrice As Decimal

    Decimal.TryParse(Me.xLengthTextBox.Text, lawn.Length)
    Decimal.TryParse(Me.xWidthTextBox.Text, lawn.Width)
    Decimal.TryParse(Me.xPriceTextBox.Text, sodPrice)

    ' calculate the area (in square yards)
    area = lawn.CalculateArea / 9D

    ' calculate and display the total price
    totalprice = area * sodPrice
    Me.xTotalPriceLabel.Text = totalPrice.ToString("C2")

    ' set the focus
    Me.xLengthTextBox.Focus()

End Sub
```

Figure 11-41: Completed xCalcButton's Click event procedure

Now that you have finished coding the application, you can test the application to verify that the code is working correctly.

To test the Kessler Landscaping application:

1 Save the solution, then start the application. Type **20** in the Length (feet) box, **35** in the Width (feet) box, and **2.56** in the Sod price (square yard) box. Click the **Calculate Total Price** button. The button's Click event procedure displays the total price in the Total price box, as shown in Figure 11-42.

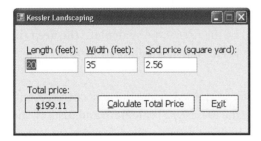

Figure 11-42: Total price displayed in the interface

2 Click the **Exit** button to end the application. Close the Code Editor window, then close the solution.

You have completed Lesson C and Chapter 11. You can either take a break or complete the end-of lesson questions and exercises.

SUMMARY

TO INCLUDE MORE THAN ONE CONSTRUCTOR IN A CLASS:

» Each constructor's parameters (if any) must be different from any other constructor in the class.

TO ASSIGN A VALUE TO A PRIVATE VARIABLE IN A CLASS:

» If the class is providing the value, the class's methods can assign the value, directly. However, if the value is provided by an application, you should assign the value using the Public property associated with the Private variable. Doing this ensures that the computer processes the variable's Set block, which many times contains validation code.

QUESTIONS

1. A class can contain only one constructor.

   a. True   b. False

2. A class contains a Private variable named `_location`. The variable is associated with a Public property named Location. Which of the following is the correct way for the default constructor to assign the value "Unknown" to the variable?

   a. `_location = "Unknown"`

   b. `_location.Location = "Unknown"`

c. `Location._location = "Unknown"`

d. None of the above.

3. A class contains a Private variable named `_capital`. The variable is associated with a Public property named Capital. Which of the following is the best way for a parameterized constructor to assign the value stored in its `capName` parameter to the variable?

a. `_capital = capName`

b. `Capital = capName`

c. `_capital.Capital = capName`

d. None of the above.

EXERCISES

1. In this exercise, you modify the Kessler Landscaping application that you completed in this lesson.

a. Use Windows to make a copy of the Kessler Solution folder, which is contained in the VB2005\Chap11 folder. Rename the folder Modified Kessler Solution.

b. If necessary, start Visual Studio 2005 or Visual Basic 2005 Express Edition. Open the Kessler Solution (Kessler Solution.sln) file contained in the VB2005\Chap11\ Modified Kessler Solution folder. Open the designer window.

c. Open the MyRectangle.vb file in the Code Editor window. The class should verify that the length and width values are valid. To be valid, the values must be greater than zero. If a value is not valid, assign the number –1 to the corresponding Private variable in the class.

d. Close the MyRectangle.vb file window, then open the form's Code Editor window. Modify the xCalcButton's Click event procedure so that it uses the MyRectangle class's parameterized constructor rather than its default constructor.

e. Save the solution, then start and test the application. Click the Exit button to end the application. Close the Code Editor window, then close the solution.

2. Jack Sysmanski, the owner of All-Around Fence Company, wants an application that he can use to calculate the cost of installing a fence. In this exercise, you modify the MyRectangle class that you coded in this lesson, and then use the class in the All-Around Fence Company application.

   a. If necessary, start Visual Studio 2005 or Visual Basic 2005 Express Edition. Open the Fence Solution (Fence Solution.sln) file, which is contained in the VB2005\Chap11\Fence Solution folder.

   b. Use Windows to copy the MyRectangle.vb file from the VB2005\Chap11\Kessler Solution\Kessler Project folder to the VB2005\Chap11\Fence Solution\Fence Project folder.

   c. Use the Project menu to add the existing MyRectangle.vb class file to the Fence project. Open the MyRectangle.vb file in the Code Editor window. Modify the MyRectangle class so that it calculates the perimeter of a rectangle. To calculate the perimeter, the class will need to add together the length and width measurements, and then multiply the sum by two.

   d. Close the MyRectangle.vb window, then open the form's Code Editor window. Code the Fence application so that it displays the cost of installing the fence.

   e. Save the solution, then start the application. Test the application using 120 feet as the length, 75 feet as the width, and 10 as the cost per linear foot. The application should display $3,900.00 as the installation cost.

   f. Click the Exit button to end the application. Close the Code Editor window, then close the solution.

3. The manager of Pool-Time, which sells in-ground pools, wants an application that the salespeople can use to determine the number of gallons of water required to fill an in-ground pool—a question commonly asked by customers. (*Hint*: To calculate the number of gallons, you need to find the volume of the pool. You can do so using the formula *length * width * depth*.) In this exercise, you modify the MyRectangle class that you created in this lesson, and then use the class in the Pool-Time application.

   a. If necessary, start Visual Studio 2005 or Visual Basic 2005 Express Edition. Open the Pool Solution (Pool Solution.sln) file, which is contained in the VB2005\Chap11 folder.

   b. Use Windows to copy the MyRectangle.vb file from the VB2005\Chap11\Kessler Solution\Kessler Project folder to the VB2005\Chap11\Pool Solution\Pool Project folder.

   c. Use the Project menu to add the existing MyRectangle.vb class file to the Pool project. Open the MyRectangle.vb file in the Code Editor window. Modify the MyRectangle class so that it also calculates and returns the volume, then close the MyRectangle.vb window.

d. Code the Pool application so that it displays the number of gallons. To calculate the number of gallons, you divide the volume by .13368.

e. Save the solution, then start the application. Test the application using 25 feet as the length, 15 feet as the width, and 6.5 feet as the depth. The application should display 18,233.84 as the number of gallons.

f. Click the Exit button to end the application. Close the Code Editor window, then close the solution.

4. In this exercise, you define a Triangle class. You also create an application that uses the Triangle class to create a Triangle object.

a. If necessary, start Visual Studio 2005 or Visual Basic 2005 Express Edition. Create an interface that allows the user to display either the area of a triangle or the perimeter of a triangle. (*Hint*: The formula for calculating the area of a triangle is *1/2 * base * height*. The formula for calculating the perimeter of a triangle is *a + b + c*, where *a*, *b*, and *c* are the lengths of the sides.) Name the solution Math Solution. Name the project Math Project. Save the application in the VB2005\Chap11 folder.

b. Add a class file to the project. Name the class file Triangle.vb. The Triangle class should verify that the dimensions are greater than zero before assigning the values to the Private variables. The class also should include a method to calculate the area of a triangle and a method to calculate the perimeter of a triangle.

c. Save the solution, and then start and test the application. End the application, then close the Code Editor window and the solution.

5. Maria Jacobsen, a professor at Mayflower College, wants an application that allows her to enter each student's name and three test scores. The application should calculate and display each student's average test score.

a. If necessary, start Visual Studio 2005 or Visual Basic 2005 Express Edition. Open the Mayflower Solution (Mayflower Solution.sln) file, which is contained in the VB2005\Chap11\Mayflower Solution folder. If necessary, open the designer window.

b. Add a class file to the project. Name the class file Student.vb. The Student class should contain four properties and two methods. The Mayflower application will use the class to create a Student object. It will store the user input in the object's properties, and use the object's methods to initialize the Private variables and calculate and return the average test score.

c. Save the solution, and then start and test the application. Click the Exit button to end the application. Close the Code Editor window, then close the solution.

6. In this exercise, you modify the Student class you created in Exercise 5. The modified class will include an additional method named ValidateScores. The ValidateScores method will verify that each test score is greater than or equal to a minimum value, but less than or equal to a maximum value. The minimum and maximum values will be passed to the method by the procedure that invokes the method.

a. Use Windows to make a copy of the Mayflower Solution folder, which is contained in the VB2005\Chap11 folder. Rename the folder Modified Mayflower Solution.

b. If necessary, start Visual Studio 2005 or Visual Basic 2005 Express Edition. Open the Mayflower Solution (Mayflower Solution.sln) file contained in the VB2005\ Chap11\Modified Mayflower Solution folder. Open the designer window.

c. Make the appropriate modifications to the Student class.

d. Modify the xCalcButton's Click event procedure so that it invokes the ValidateScores method, passing it the minimum and maximum values for a test score. In this case, the minimum value will be zero and the maximum value will be 100.

e. Save the solution, and then start and test the application. Click the Exit button to end the application. Close the Code Editor window, then close the solution.

7. Jeremiah Carter, the manager of the Accounts Payable department at Franklin Calendars, wants an application that he can use to keep track of the checks written by his department. More specifically, he wants to record (in a sequential access file) the check number, date, payee, and amount of each check.

a. If necessary, start Visual Studio 2005 or Visual Basic 2005 Express Edition. Create an appropriate interface and class.

b. Save the solution, and then start and test the application. End the application, then close the Code Editor window and the solution.

DISCOVERY EXERCISE

8. In this exercise, you modify the MyRectangle class from Exercise 3 so that it uses inheritance.

a. Use Windows to make a copy of the Pool Solution folder, which is contained in the VB2005\Chap11 folder. Rename the folder Modified Pool Solution.

b. If necessary, start Visual Studio 2005 or Visual Basic 2005 Express Edition. Open the Pool Solution (Pool Solution.sln) file contained in the VB2005\Chap11\Modified Pool Solution folder. Open the designer window.

c. Open the MyRectangle.vb file in the Code Editor window. Define a derived class named MyPool. The MyPool class should inherit the attributes and behaviors of the MyRectangle class, which is the base class. Remove the code pertaining to the depth measurement and volume calculation from the MyRectangle class. The derived MyPool class will now be responsible for the depth measurement and volume calculation.

d. Close the MyRectangle.vb window, then open the form's Code Editor window. Modify the Pool application appropriately.

e. Save the solution, then start the application. Test the application using 25 feet as the length, 15 feet as the width, and 6.5 feet as the depth. The application should display 18,233.84 as the number of gallons.

f. Click the Exit button to end the application. Close the Code Editor window, then close the solution.

DISCOVERY EXERCISE

9. Shelly Jones, the manager of Pennington Book Store, wants an application that she can use to calculate and display the total amount a customer owes.

a. If necessary, start Visual Studio 2005 or Visual Basic 2005 Express Edition. Open the Pennington Solution (Pennington Solution.sln) file contained in the VB2005\Chap11\Pennington Solution folder. Open the designer window.

b. A customer can purchase one or more books at either the same price or different prices. The application should keep a running total of the amount the customer owes, and display the total in the Total due box. For example, a customer might purchase two books at $6 and three books at $10. To calculate the total due, Shelly will need to enter 2 in the Quantity box and enter 6 in the Price box, and then click the Add to Sale button. At this point, the button's Click event procedure should display $12.00 in the Total due box. To complete the order, Shelly will need to enter 3 in the Quantity box and enter 10 in the Price box, and then click the Add to Sale button. The button's Click event procedure should display $42.00 as the total amount due. Before calculating the next customer's order, Shelly will need to click the New Order button. Code the application appropriately, using a class.

c. Save the solution, then start and test the application. Click the Exit button to end the application. Close the Code Editor window, then close the solution.

DEBUGGING EXERCISE

10. In this exercise, you find and correct an error in an application. The process of finding and correcting errors is called debugging.

   a. If necessary, start Visual Studio 2005 or Visual Basic 2005 Express Edition. Open the Debug Solution (Debug Solution.sln) file, which is contained in the VB2005\ Chap11\Debug Solution folder. If necessary, open the designer window.

   b. Open the Code Editor window. Review the existing code in the Main Form.vb and Computer.vb files. Notice that a jagged line appears below some of the lines of code. Correct the code to remove the jagged lines.

   c. Save the solution, then start the application. Correct any errors in the application's code, then save the solution and start and test the application again.

   d. Click the Exit button to end the application. Close the Code Editor window, then close the solution.

12

USING ADO.NET 2.0

CREATING THE TRIVIA GAME APPLICATION

Nancy Jacoby plans to audition for the *Jeopardy* show in a few months. Her friends are helping her prepare for the audition by sending trivia questions and answers, which Nancy records in a database named Trivia. Nancy wants an application that displays the questions and answers, and then allows her to select the correct answer. She also wants the application to keep track of the number of incorrect responses she makes.

PREVIEWING THE TRIVIA GAME APPLICATION

Before creating the Trivia Game application, you first preview the completed application.

To preview the completed application:

1 Use the Run command on the Windows Start menu to run the **Trivia Game** (**Trivia Game.exe**) file, which is contained in the VB2005\Chap12 folder. The Trivia Game application appears on the screen. See Figure 12-1.

Figure 12-1: Trivia Game application

2 Answer the first question correctly by clicking the **B** radio button, and then clicking the **Submit Answer** button.

3 Answer the second question incorrectly by clicking the **A** radio button, and then clicking the **Submit Answer** button.

4 Answer the remaining seven questions on your own. When you have submitted the answer for the last question, the application displays the number of incorrect responses in a message box.

5 Click the **OK** button to close the message box, then click the **Exit** button to end the application.

In this chapter, you will learn how to use ADO.NET 2.0 to access the data stored in a Microsoft SQL Server database. (A Microsoft Access version of Chapter 12 is available online.) You will code the Trivia Game application in Lesson C.

LESSON A
OBJECTIVES

AFTER STUDYING LESSON A, YOU SHOULD
BE ABLE TO:

» Define the terms used when talking about databases

» Connect an application to a database

» Bind table and field objects to controls

» Explain the purpose of the DataSet, BindingSource,

TableAdapter, and BindingNavigator objects

» Access the records in a dataset

» Move the record pointer

DATABASES

DATABASE TERMINOLOGY

In order to maintain accurate records, most businesses store information about their employees, customers, and inventory in files called databases. In general, a **database** is simply an organized collection of related information stored in a file. Many computer products exist for creating databases; some of the most popular are Microsoft SQL Server, Oracle, and Microsoft Access. You can use Visual Basic to access the data stored in databases created by these products. This allows a company to create a standard interface in Visual Basic that employees can use to access database information stored in a variety of formats. Instead of learning each product's user interface, the employee needs to know only one interface. The actual format of the database is unimportant and will be transparent to the user.

In this lesson, you learn how to access the data stored in a Microsoft SQL Server database. (To learn how to access the data stored in a Microsoft Access database, refer to the Microsoft Access version of Chapter 12. You can obtain the Microsoft Access version of Chapter 12 and the Access database files electronically from the Course Technology Web site by connecting to *www.course.com* and then navigating to the page for this book.) Databases created by Microsoft SQL Server are relational databases. A **relational database** is one that stores information in tables composed of columns and rows, similar to the format used in a spreadsheet. Each column in a table represents a field, and each row represents a record. A **field** is a single item of information about a person, place, or thing—for example, a name, a salary amount, a Social Security number, or a price. A **record** is a group of related fields that contain all of the necessary data about a specific person, place, or thing. The college you are attending keeps a student record on you. Examples of fields contained in your student record include your Social Security number, name, address, phone number, credits earned, and grades earned. The place where you are employed also keeps a record on you. Your employee record contains your Social Security number, name, address, phone number, starting date, salary or hourly wage, and so on. A group of related records is called a **table**. Each record in a table pertains to the same topic, and each contains the same type of information. In other words, each record in a table contains the same fields.

A relational database can contain one or more tables. A one-table database would be a good choice for storing the information regarding the college courses you have taken. An example of such a table is shown in Figure 12-2. Notice that each record in the table contains four fields: an ID field that indicates the department name and course number, a course title field, a number of credit hours field, and a grade field. In most tables, one of the fields uniquely identifies each record and is called the **primary key**. In the table shown in Figure 12-2, you could use either the ID field or the Title field as the primary key, because the data in those fields will be unique for each record.

»TIP
You do not have to be a business to make use of a database. Many people use databases to keep track of their medical records, their compact disc collections, and even their golf scores.

»TIP
Appendix C contains the steps for creating a SQL Server database.

»TIP
The databases are called relational because the information in the tables can be related in different ways. The databases created using Microsoft Access and Oracle also are relational databases.

ID	Title	Hours	Grade
CIS100	Intro to Computers	3	A
Eng100	English Composition	3	B
Phil105	Philosophy Seminar	2	C
CIS201	Visual Basic 2005	3	A

Figure 12-2: Example of a one-table relational database

To store information about your CD (compact disc) collection, you typically use a two-table database: one table to store the general information about each CD (such as the CD's name and the artist's name) and the other table to store the information about the songs on each CD (such as their title and track number). You then use a common field— for example, a CD number—to relate the records contained in both tables. Figure 12-3 shows an example of a two-table database that stores CD information. The first table is often referred to as the **parent table**, and the second table is referred to as the **child table**. In the parent table in Figure 12-3, the Number field is the primary key, because it uniquely identifies each record in that table. In the child table, the Number field is used solely to link the song title and track information to the appropriate CD in the parent table. In the child table, the Number field is called the **foreign key**.

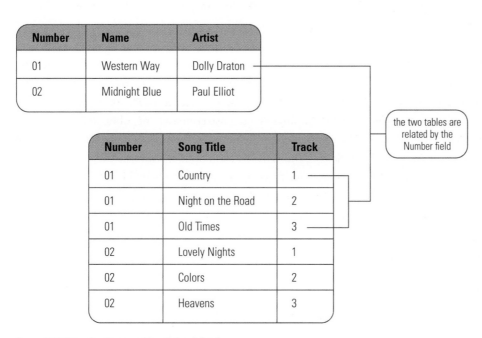

Figure 12-3: Example of a two-table relational database

Storing data in a relational database offers many advantages. The computer can retrieve data stored in a relational format both quickly and easily, and the data can be displayed in any order. For example, the information in the CD database shown in Figure 12-3 can be arranged by artist name, song title, and so on. A relational database also allows you to control the amount of information you want to view at a time. You can view all of the information in the CD database, or you can view only the information pertaining to a certain artist, or only the names of the songs contained on a specific CD.

ADO.NET 2.0

When a Visual Basic 2005 application needs to access the information stored in a database, the computer uses a technology called ADO.NET 2.0 to connect the application to the database. (ADO stands for ActiveX Data Objects.) The connection allows the application to read information from and write information to the database. After creating the connection between the database and the application, the computer opens the database and then makes a copy of the fields and records the application wants to access. The computer stores the copy, called a **dataset**, in its internal memory. The computer then closes both the database and the connection to the database. Notice that the connection is a temporary one. The computer reconnects the application to the database only when the application requests further information from the database, or when changes made to the dataset (which is in internal memory) need to be saved. After retrieving the additional information or saving any changes, the computer again closes the database and the connection to the database.

The applications you view in the following sections use the data contained in a Microsoft SQL Server database named Employees. The Employees database is stored in the Employees.mdf file, which is located in the VB2005\Chap12\SQL Databases folder. The .mdf filename extension stands for Master Database File and indicates that the file contains a SQL Server database. The Employees database contains one table, which is named tblEmploy. (It is not necessary to begin a table name with "tbl".) Figure 12-4 shows the table data displayed in a window in the IDE. You can open a database table in a window in the IDE by connecting the database to an application, right-clicking the table's name in the Server Explorer window, and then clicking Show Table Data.

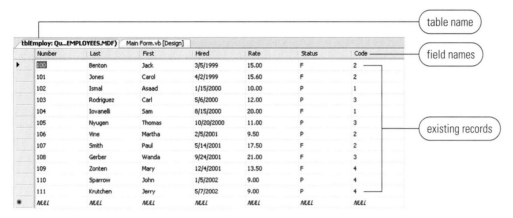

Figure 12-4: Data contained in the tblEmploy table

The tblEmploy table contains seven fields and 12 records. The Number, Last, First, Hired, and Rate fields store employee numbers, last names, first names, hire dates, and rates of pay, respectively. The Status field contains the employment status, which is either the letter F (for full-time) or the letter P (for part-time). The Code field identifies the employee's department: 1 for Accounting, 2 for Advertising, 3 for Personnel, and 4 for Inventory. In the tblEmploy table, the Number field is the primary key, because it uniquely identifies each record.

CONNECTING AN APPLICATION TO A DATABASE

Before an application can access the data stored in a database, you need to connect the application to the database. You can use the Data Source Configuration Wizard to make the connection. Figure 12-5 shows the procedure you follow when using the Data Source Configuration Wizard to connect an application to a database.

<table>
<tr><td>Connecting an application to a database</td></tr>
</table>

1. Open the application's solution file. Click Data on the menu bar, then click Show Data Sources to open the Data Sources window.

2. Click the Add New Data Source link in the Data Sources window. This opens the Data Source Configuration Wizard dialog box and displays the Choose a Data Source Type screen. If necessary, click Database.

3. Click the Next button, then continue using the Data Source Configuration Wizard to specify the data source. The Wizard will add several files to the project. The names of the files will appear in the Solution Explorer window. When the Wizard is finished, it adds a dataset to the Data Sources window.

Figure 12-5: Procedure for connecting an application to a database

To connect the Employees database to the Morgan Industries application:

1 Start Visual Studio 2005 or Visual Basic 2005 Express Edition, if necessary, and close the Start Page window. Open the **Morgan Industries Solution** (Morgan Industries Solution.sln) file, which is contained in the VB2005\Chap12\Morgan Industries Solution-DataGrid folder. If necessary, open the designer window and auto-hide the Toolbox, Solution Explorer, and Properties windows.

2 Click **Data** on the menu bar, then click **Show Data Sources** to open the Data Sources window.

3 Click **Add New Data Source** in the Data Sources window to start the Data Source Configuration Wizard, which displays the Choose a Data Source Type screen. If necessary, click **Database**. The Database option allows you to connect to a database and choose the database objects (such as tables and fields) for your application. The option also creates a dataset that contains the chosen objects. The Choose a Data Source Type screen is shown in Figure 12-6.

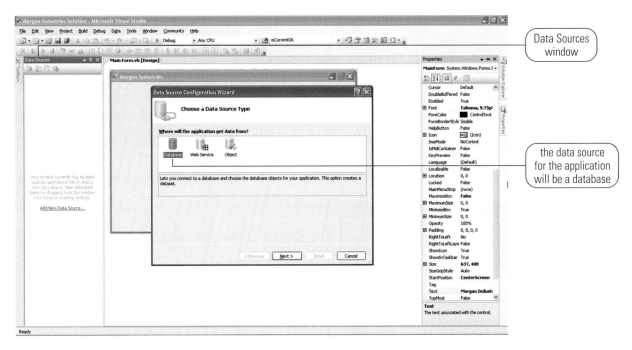

Figure 12-6: Choose a Data Source Type screen

4 Click the **Next** button to display the Choose Your Data Connection screen. See Figure 12-7.

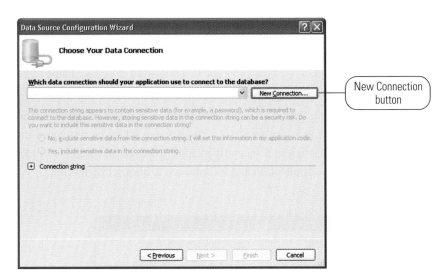

Figure 12-7: Choose Your Data Connection screen

5 Click the **New Connection** button to display the Add Connection dialog box. If Microsoft SQL Server Database File (SqlClient) does not appear in the Data source box, click the **Change** button to open the Change Data Source dialog box, click **Microsoft SQL Server Database File**, and then click the **OK** button to return to the Add Connection dialog box.

6 Click the **Browse** button in the Add Connection dialog box. Open the VB2005\ Chap12\SQL Databases folder, then click **Employees.mdf** in the list of filenames. Click the **Open** button. Figure 12-8 shows the completed Add Connection dialog box.

Figure 12-8: Completed Add Connection dialog box

7 Click the **Test Connection** button in the Add Connection dialog box. The "Test connection succeeded." message appears in a dialog box. Click the **OK** button to close the dialog box.

8 Click the **OK** button to close the Add Connection dialog box. Employees.mdf appears in the Choose Your Data Connection screen. Click the **Next** button. A message similar to the one shown in Figure 12-9 appears in a message box. The message asks whether you want to include the database file in the current project. By including the file in the current project, you can more easily copy the application and its database to another computer.

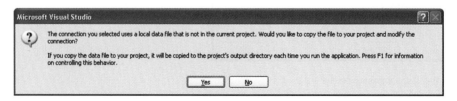

Figure 12-9: Message pertaining to copying the database file

9 Click the **Yes** button to add the Employees.mdf file to the current project. The file is added to the VB2005\Chap12\Morgan Industries Solution-DataGrid\Morgan Industries Project folder. The Save the Connection String to the Application Configuration File screen appears next. The name of the connection string, EmployeesConnectionString, appears on the screen, as shown in Figure 12-10. If necessary, select the check box.

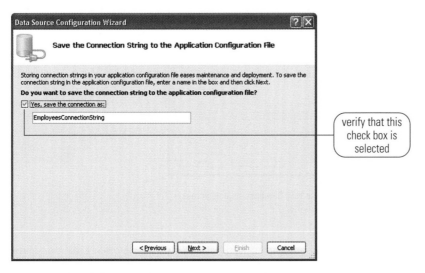

Figure 12-10: Save the Connection String to the Application Configuration File screen

10 Click the **Next** button to display the Choose Your Database Objects screen. Click the **plus box** that appears next to Tables, then click the **plus box** that appears next to tblEmploy. Notice that the database contains one table (named tblEmploy) and seven fields. You can use this screen to select the table and/or field objects to include in the dataset, which is automatically named EmployeesDataSet.

11 In this application, you will add all of the fields to the EmployeesDataSet. Click the **empty box** that appears before tblEmploy. Doing this selects the table and field check boxes, as shown in Figure 12-11.

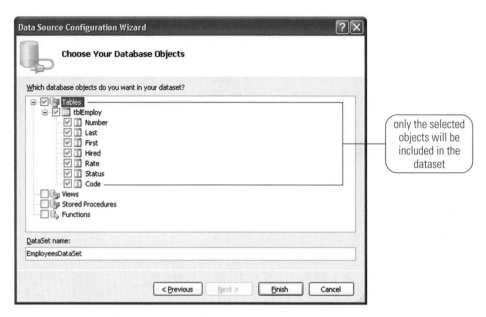

Figure 12-11: Database objects selected in the Choose Your Database Objects screen

12 Click the **Finish** button. The computer adds the EmployeesDataSet to the Data Sources window. Click the **plus box** that appears next to tblEmploy in the Data Sources window. As Figure 12-12 indicates, the EmployeesDataSet contains one table object and seven field objects.

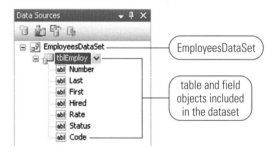

Figure 12-12: EmployeesDataSet added to the Data Sources window

PREVIEWING THE DATA CONTAINED IN A DATASET

You can use the procedure shown in Figure 12-13 to preview the data (fields and records) contained in a dataset.

Previewing the contents of a dataset

1. Right-click the Data Sources window, and then click Preview Data to open the Preview Data dialog box. You also can click the form, then click Data on the menu bar, and then click Preview Data on the menu.

2. Select the object to preview, and then click the Preview button.

3. When you are finished previewing the data, click the Close button in the dialog box.

Figure 12-13: Procedure for previewing the contents of a dataset

To preview the data contained in the EmployeesDataSet:

1 Right-click the **Data Sources window**, click **Preview Data**, and then click the **Preview** button in the Preview Data dialog box. The EmployeesDataSet contains 12 records (rows) and seven fields (columns), as shown in Figure 12-14. You can use the scroll bars that appear on the Results box to view the fields and records that are not currently visible.

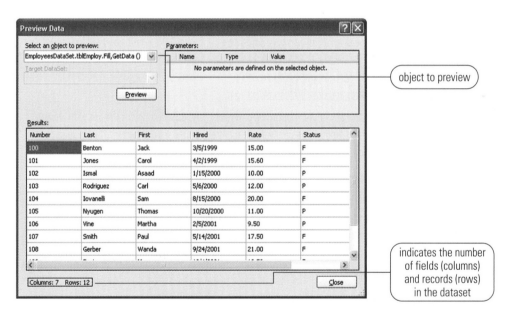

Figure 12-14: Data displayed in the Preview Data dialog box

Notice that `EmployeesDataSet.tblEmploy.Fill, GetData()` appears in the Select an object to preview box. EmployeesDataSet is the name of the dataset, and tblEmploy is the name of the table included in the dataset. Fill and GetData are methods. The Fill method populates an existing table with data, while the GetData method creates a new table and populates it with data.

2 Click the **Close** button in the Preview Data dialog box, then save the solution.

BINDING THE OBJECTS IN A DATASET

» TIP

Bound controls are also referred to as data-aware controls.

For the user to view the contents of a dataset while an application is running, you need to connect one or more objects in the dataset to one or more controls in the interface. Connecting an object to a control is called **binding**, and the connected controls are called **bound controls**. Figure 12-15 lists various ways of binding the objects in a dataset. Notice that you can bind an object either to a control that the computer creates for you, or to an existing control in the interface.

Binding the objects in a dataset
To have the computer create a control and then bind an object to it:
In the Data Sources window, click the object you want to bind. If necessary, use the object's list arrow to change the control type. Drag the object to an empty area on the form, then release the mouse button. The computer creates the appropriate control and binds the object to it.
To bind an object to an existing control:
In the Data Sources window, click the object you want to bind, then drag the object to the control on the form, and then release the mouse button. Alternatively, you can click the control on the form, then use the Properties window to set the appropriate property or properties. (Refer to the BINDING TO AN EXISTING CONTROL section in this lesson.)

Figure 12-15: Ways to bind the objects in a dataset

HAVING THE COMPUTER CREATE A BOUND CONTROL

As indicated in Figure 12-15, one way to bind an object from a dataset is to drag the object to an empty area on the form. When you do this, the computer creates the necessary control and automatically binds the object to it. The icon that appears before the object's name in the

Data Sources window indicates the type of control the computer will create. For example, the 🔲 icon shown in Figure 12-16 indicates that the computer will create a DataGridView control when you drag the tblEmploy table object to the form. A **DataGridView control** displays the table data in a row and columnar format, similar to a spreadsheet. Each row in the control represents a record, and each column represents a field. The abl icon, on the other hand, indicates that the computer will create a text box when you drag a field object to the form.

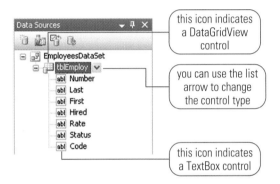

Figure 12-16: Icons displayed in the Data Sources window

You can use the list arrow that appears next to an object's name to change the type of control the computer creates. For example, to display the tblEmploy data in separate text box controls rather than in a DataGridView control, you first click tblEmploy in the Data Sources window. You then click its list arrow, as shown in Figure 12-17, and then click Details in the list. The Details option tells the computer to create a separate control for each field in the table.

Figure 12-17: Result of clicking the tblEmploy table object's list arrow

Similarly, to display the Last field data in a label control rather than in a text box, you first click Last in the Data Sources window. You then click its list arrow, as shown in Figure 12-18, and then click Label in the list.

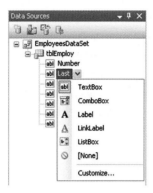

Figure 12-18: Result of clicking the Last field object's list arrow

To bind the tblEmploy object to a DataGridView control:

1 If necessary, click **tblEmploy** in the Data Sources window to select the tblEmploy object.

2 Drag the tblEmploy object from the Data Sources window to the form, then release the mouse button. The computer adds a DataGridView control to the form, and it binds the tblEmploy object to the control. See Figure 12-19.

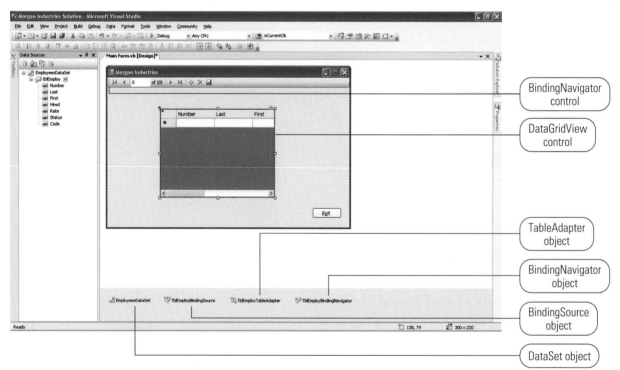

Figure 12-19: Result of dragging the tblEmploy object to the form

3 Temporarily display the Properties window. Set the TblEmployDataGridView control's **AutoSizeColumnsMode** property to **Fill**. The Fill setting automatically adjusts the column widths so that all of the columns exactly fill the display area of the control.

4 Click **Dock** in the Properties list, then click the **list arrow** that appears in the Settings box. Click the **rectangle that appears at the top of the box**. Doing this sets the Dock property to Top and anchors the TblEmployDataGridView control to the top of the form.

5 Set the TblEmployDataGridView control's **Size** property to **632, 250**.

6 Next, you will set the tab order for the controls. Set the TblEmployDataGridView control's **TabIndex** property to **0**. Set the TblEmployBindingNavigator control's **TabIndex** property to **1**, and set the xExitButton's **TabIndex** property to **2**.

7 Lock the controls on the form. Auto-hide the Data Sources window, then save the solution.

As Figure 12-19 indicates, besides adding a DataGridView control to the form, the computer also adds a BindingNavigator control. When an application is running, the **BindingNavigator control** allows you to move from one record to the next in the dataset. It also contains buttons that allow you to add and delete a record, as well as to save any changes made to the dataset. The computer also places four objects in the component tray: a DataSet, a BindingSource, a TableAdapter, and a BindingNavigator. As you learned in Chapter 1, the component tray stores objects that do not appear in the user interface when an application is running. An exception to this is the BindingNavigator object, which appears as the BindingNavigator control during both design time and run time. Figure 12-20 illustrates the relationships among the database, the objects in the component tray, and the controls on the form.

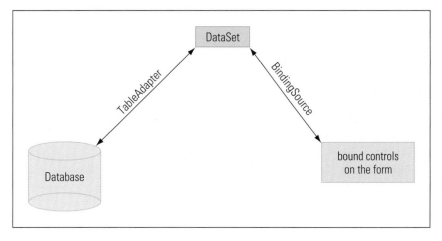

Figure 12-20: Illustration of the relationships among the database, the objects in the component tray, and the controls on the form

The **TableAdapter object** connects the database to the **DataSet object**, which stores the information you want to access from the database. The TableAdapter is responsible for retrieving the appropriate information from the database and storing it in the DataSet. It also is responsible for saving to the database any changes made to the data contained in the DataSet. The **BindingSource object** provides the connection between the DataSet object and the bound controls on the form. For example, the TblEmployBindingSource object (shown earlier in Figure 12-19) connects the EmployeesDataSet object to two bound controls: the DataGridView control and the BindingNavigator control. The BindingSource object allows the DataGridView control to display the data contained in the dataset. It also allows the BindingNavigator control to access the records stored in the dataset.

If a table object's control type is changed from DataGridView to Details, the computer automatically provides the appropriate controls (such as text boxes, labels, and so on) when you drag the table object to the form. It also adds the BindingNavigator control to the form, and adds the DataSet, BindingSource, TableAdapter, and BindingNavigator objects to the component tray. The appropriate controls and objects are also automatically included when you drag a field object to an empty area on the form. In addition to adding the controls and objects to the application, the computer also enters some code in the Code Editor window.

To view the code contained in the Code Editor window:

1 Open the Code Editor window. The code shown in Figure 12-21 was automatically entered when the tblEmploy object was dragged to the form.

Visual Basic code

```
Private Sub TblEmployBindingNavigatorSaveItem_Click_
  (ByVal sender As System.Object, _
  ByVal e As System.EventArgs) _
  Handles TblEmployBindingNavigatorSaveItem.Click
    Me.Validate()
    Me.TblEmployBindingSource.EndEdit()
    Me.TblEmployTableAdapter.Update(Me.EmployeesDataSet.tblEmploy)
End Sub

Private Sub MainForm_Load(ByVal sender As System.Object, _
    ByVal e As System.EventArgs) Handles MyBase.Load
    ' TODO: This line of code loads data into the 'EmployeesDataSet.tblEmploy'
  table. You can move, or remove it, as needed.
    Me.TblEmployTableAdapter.Fill(Me.EmployeesDataSet.tblEmploy)

    End Sub
End Class
```

Figure 12-21: Code automatically entered in the Code Editor window

The code contained in the TblEmployBindingNavigatorSaveItem's Click event procedure is processed when you click the Save button on the TblEmployBindingNavigator control. First, the code uses the Validate method to validate the changes you want saved to the dataset. It then uses the BindingSource object's **EndEdit method** to apply any pending changes (such as new records, deleted records, or changed records) to the dataset. The syntax of the EndEdit method is **Me.**_bindingSourceName_**.EndEdit( )**, where _bindingSourceName_ is the name of the BindingSource object in the application. Lastly, the code uses the TableAdapter object's **Update method** to commit, to the database, the changes made to the dataset. The syntax of the Update method is **Me.**_tableAdapterName_**. Update(Me.**_dataSetName_**.**_tableName_**)**. In the syntax, _tableAdapterName_ is the name of the TableAdapter object in the application. _DataSetName_ is the name of the DataSet object, and _tableName_ is the name of the table object contained in the dataset.

The code contained in the MainForm's Load event procedure, on the other hand, is processed when the application is started and the form is loaded into the computer's internal memory. The Load event procedure shown in Figure 12-21 contains one statement: `Me.TblEmployTableAdapter.Fill(Me.EmployeesDataSet.tblEmploy)`. The statement uses the TableAdapter object's **Fill method** to retrieve data from the database and store it in the dataset. The Fill method's syntax is **Me.**_tableAdapterName_**. Fill(Me.**_dataSetName_**.**_tableName_**)**. In the syntax, _tableAdapterName_ is the name of the TableAdapter object in the application. _DataSetName_ is the name of the DataSet object, and _tableName_ is the name of the table object contained in the dataset. In most applications, the statement to fill a dataset with data belongs in the form's Load event procedure, as shown in Figure 12-21. However, as the comments in the Load event procedure indicate, you can move the statement to another procedure; or, you can remove the statement from the Code Editor window.

To start and then test the Morgan Industries application:

1 Close the Code Editor window, then start the application. When the application is started, the statement in the MainForm's Load event procedure retrieves the appropriate data from the Employees database and loads the data into the EmployeesDataSet object. The data appears in the TblEmployDataGridView control, which is bound to the tblEmploy table contained in the EmployeesDataSet. Notice that scroll bars appear when there is more data than will fit into the control. See Figure 12-22.

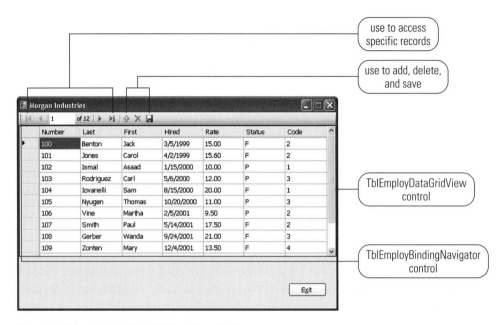

use to access specific records

use to add, delete, and save

TblEmployDataGridView control

TblEmployBindingNavigator control

Figure 12-22: Data displayed in the TblEmployDataGridView control

2 You can use the arrow keys on your keyboard to access a different field and/or record in the TblEmployDataGridView control. Press the **down arrow** on your keyboard to move the highlight to the next record, then press the **right arrow** to move it to the next field. Press the **left arrow**, then press the **up arrow** to return the highlight to its original location.

3 You can use the buttons on the TblEmployBindingNavigator control to access the first and last records in the EmployeesDataSet, as well as to access the previous and next records. You also can use the control to access a record by its record number. The first record in a dataset has a record number of one, the second has a record number of two, and so on. Click the **box that contains the number 1**, which is the current record number. Change the 1 to a **5**, and then press **Enter**. The highlight moves to the first field in the fifth record.

4 The TblEmployBindingNavigator control also contains buttons that allow you to add a record to the dataset, delete a record from the dataset, and save the changes made to the dataset. You can make a change to an existing record by modifying the data in the appropriate cell in the TblEmployDataGridView control. A **cell** is an intersection of a row and column in the control. For example, to change Jack Benton's status from full-time to part-time, click the letter **F** that appears in the cell located in the first row, sixth column. Type **P** and then press **Enter**.

5 Click the **Save Data** 💾 button that appears on the TblEmployBindingNavigator control, then click the **Exit** button to end the application.

6 Start the application again. Notice that the change you made to Jack Benton's Status field was not saved. This is due to the way Visual Basic manages a local database file. The method Visual Basic uses is controlled by the database file's Copy to Output Directory property. You will learn about the Copy to Output Directory property in the next section.

7 Click the **Exit** button to end the application.

THE COPY TO OUTPUT DIRECTORY PROPERTY

As you may remember, when you added the Employees.mdf database file to the Morgan Industries solution, the file was saved in the solution's project folder; in this case, it was saved in the Morgan Industries Project folder. A database file's **Copy to Output Directory property** determines the way the file is handled in a Visual Basic Windows-based application. When the property is set to its default setting, Copy always, the database file is copied from the project folder to the project folder's bin\Debug folder each time you start the application. In this case, for example, it is copied from the Morgan Industries Project folder to the Morgan Industries Project\bin\Debug folder. As a result, the Employees.mdf file will appear in two different folders in the solution. When you make a change to the data displayed in the DataGridView control and then click the Save Data button on the BindingNavigator control, the changes are recorded only in the database file stored in the bin\Debug folder; the database file stored in the project folder is not changed. The next time you start the application, the database file in the project folder is copied to the bin\Debug folder, overwriting the file that contains the changes. One way to fix this problem is to set the database file's Copy to Output Directory property to "Copy if newer." The "Copy if newer" setting tells the computer to compare the dates on both database files to determine which file has the newer (more current) date. The computer should copy the project folder's database file to the bin\Debug folder only when the file has the newer date. The file should not be copied when its date is older (less current) than the date of the file stored in the bin\Debug folder. (In Exercise 5 at the end of this lesson, you will learn a different way to fix the problem.)

To change the Copy to Output Directory property, then save and start the solution:

1 Temporarily display the Solution Explorer window, then click **Employees.mdf**. Temporarily display the Properties window, then set the Employees.mdf file's **Copy to Output Directory** property to **Copy if newer**.

2 Save the solution, then start the application. Change Jack Benton's Status field from F to **P**, then click the **Save Data** button on the TblEmployBindingNavigator control.

3 Click the **Exit** button to end the application, then start the application again. Jack Benton's Status field now contains P.

4 Change Jack Benton's Status field from P to **F**, then click the **Save Data** button.

5 Click the **Exit** button to end the application, then close the solution.

BINDING TO AN EXISTING CONTROL

As indicated earlier in Figure 12-15, you also can bind an object in a dataset to an existing control on the form. The easiest way to do this is by dragging the object from the Data Sources window to the control. However, you also can click the control and then set one or more properties in the Properties window. The appropriate property (or properties) to set depends on the control you are binding. For example, you use the DataSource and DataMember properties to bind a DataGridView control. However, you use the DataSource and DisplayMember properties to bind a ListBox control. To bind label and text box controls, you use the DataBindings/Text property.

When you drag an object from the Data Sources window to an existing control, the computer does not create a new control; rather, it merely binds the object to the existing control. Because a new control does not need to be created, the computer ignores the control type specified for the object in the Data Sources window. As a result, it is not necessary to change the control type in the Data Sources window to match the existing control's type. In other words, you can drag an object that is associated with a text box in the Data Sources window to a label control on the form. The computer will bind the object to the label, but it will not change the label to a text box.

To bind the dataset objects to existing controls:

1 Open the **Morgan Industries Solution** (Morgan Industries Solution.sln) file, which is contained in the VB2005\Chap12\Morgan Industries Solution-Labels folder. If necessary, open the designer window.

2 Click **Data** on the menu bar, then click **Show Data Sources** to open the Data Sources window. Click **Add New Data Source** to start the Data Source Configuration Wizard. If necessary, click **Database**.

3 Click the **Next** button to display the Choose Your Data Connection screen. Click the **New Connection** button to display the Add Connection dialog box. If Microsoft SQL Server Database File (SqlClient) does not appear in the Data source box, click the **Change** button to open the Change Data Source dialog box, then click **Microsoft SQL Server Database File**, and then click the **OK** button to return to the Add Connection dialog box.

4 Click the **Browse** button in the Add Connection dialog box. Open the VB2005\ Chap12\SQL Databases folder, then click **Employees.mdf** in the list of filenames. Click the **Open** button. Click the **Test Connection** button in the Add Connection dialog box. The "Test connection succeeded." message appears in a dialog box. Click the **OK** button to close the dialog box.

5 Click the **OK** button to close the Add Connection dialog box. Click the **Next** button, then click the **Yes** button to add the Employees.mdf file to the current project. The file is added to the VB2005\Chap12\Morgan Industries Solution-Labels\Morgan Industries Project folder. If necessary, select the check box in the Save the Connection String to the Application File screen.

6 Click the **Next** button to display the Choose Your Database Objects screen. Click the **plus box** that appears next to Tables, then click the **plus box** that appears next to tblEmploy. In this application, you will include the Number, Last, Status, and Code fields in the dataset. Click the **empty boxes** that appear next to the Number, Last, Status, and Code field names, then click the **Finish** button. The computer adds the EmployeesDataSet to the Data Sources window. Click the **plus box** that appears next to tblEmploy in the Data Sources window. The EmployeesDataSet contains one table object and four field objects.

7 Permanently display the Data Sources window. Drag the **Number field object** to the xNumberLabel control on the form, then release the mouse button. Drag the **Last field object** to the xLastNameLabel control, then release the mouse button. Drag the **Status field object** to the xStatusLabel control, then release the mouse button. Drag the **Code field object** to the xCodeLabel control, then release the mouse button. The computer binds the Number, Last, Status, and Code field objects to the xNumberLabel, xLastNameLabel, xStatusLabel, and xCodeLabel controls, respectively. In addition to binding the field objects to the appropriate controls, the computer also adds the DataSet, BindingSource, and TableAdapter objects to the component tray, as shown in Figure 12-23. However, notice that it does not include the BindingNavigator object and BindingNavigator control in the application. (You can use the BindingNavigator tool to add a BindingNavigator control to the form. You then would set the control's DataSource property to the name of the BindingSource object in the application. In this case, for example, you would set the control's DataSource property to TblEmployBindingSource.)

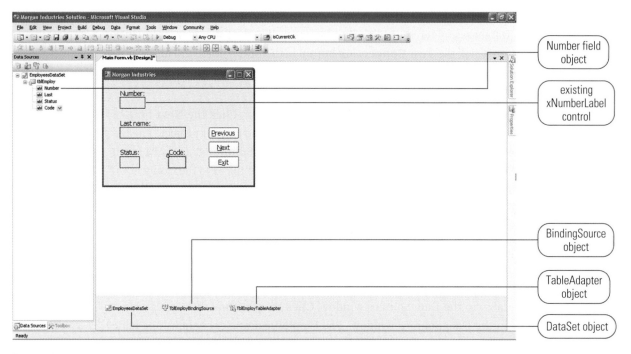

Figure 12-23: Result of dragging the field objects to the existing label controls

To test the Morgan Industries application:

1 Save the solution, then start the application. Notice that only the first record in the dataset appears in the interface. In this application, you will access the other records using the Next and Previous buttons on the form, rather than a BindingNavigator control.

2 Click the **Exit** button to end the application.

ACCESSING THE RECORDS IN A DATASET

The BindingSource object uses an invisible record pointer to keep track of the current record. It stores the position of the record pointer in its **Position property**. The first record is in position zero, the second is in position one, and so on. For example, if the record pointer is pointing to the third record, the Position property contains the integer 2. Figure 12-24 shows the Position property's syntax and includes examples of using the property.

Using the BindingSource object's Position property

Syntax

bindingSourceName.Position

Examples

```
recordNum = Me.TblEmployBindingSource.Position
```
assigns the position of the current record to the `recordNum` variable

```
Me.TblEmployBindingSource.Position = 4
```
moves the record pointer to the fifth record in the dataset

```
Me.TblEmployBindingSource.Position = _
    Me.TblEmployBindingSource.Position + 1
```
moves the record pointer to the next record in the dataset

Figure 12-24: Syntax and examples of the BindingSource object's Position property

In the first example shown in Figure 12-24, the `recordNum = Me.TblEmployBindingSource.Position` statement assigns the record pointer's position to the `recordNum` variable. The `Me.TblEmployBindingSource.Position = 4` statement in the second example moves the record pointer to the fifth record in the dataset. (Recall that the first record is in position zero.) In the last example, the `Me.TblEmployBindingSource.Position = Me.TblEmployBinding Source.Position + 1` statement moves the record pointer to the next record in the dataset.

In addition to using the Position property to move the record pointer in a dataset, you also can use the Move methods of the BindingSource object. Figure 12-25 shows each Move method's syntax and includes an example of using each method. Notice that you can use the Move methods to move the record pointer to the first, last, next, and previous record in the dataset associated with the BindingSource object.

Using the BindingSource object's Move methods

Syntax

bindingSourceName.**MoveFirst()**
bindingSourceName.**MoveLast()**
bindingSourceName.**MoveNext()**
bindingSourceName.**MovePrevious()**

Examples

```
Me.TblEmployBindingSource.MoveFirst()
```
moves the record pointer to the first record in the dataset

```
Me.TblEmployBindingSource.MoveLast()
```
moves the record pointer to the last record in the dataset

```
Me.TblEmployBindingSource.MoveNext()
```
moves the record pointer to the next record in the dataset

```
Me.TblEmployBindingSource.MovePrevious()
```
moves the record pointer to the previous record in the dataset

Figure 12-25: Syntax and examples of the BindingSource object's Move methods

To code and then test the Next and Previous buttons:

1 Auto-hide the Data Sources window, then open the Code Editor window.

2 When the Next button is clicked, its Click event procedure should move the record pointer to the next record in the dataset. You can use the TblEmployBindingSource object's MoveNext method to code the procedure. Open the code template for the xNextButton's Click event procedure. Type **' moves the record pointer to the next record** and press **Enter** twice. Type **me.TblEmployBindingSource.movenext()** and press **Enter**.

3 When the Previous button is clicked, its Click event procedure should move the record pointer to the previous record in the dataset. You can use the TblEmployBindingSource object's MovePrevious method to code the procedure. Open the code template for the xPreviousButton's Click event procedure. Type **' moves the record pointer to the previous record** and press **Enter** twice. Type **me.TblEmployBindingSource. moveprevious()** and press **Enter**.

4 Close the Code Editor window. Save the solution, then start the application. The first record appears in the interface.

5 Click the **Next** button to display the second record, then continue clicking the **Next** button until the record for employee number 111 appears in the interface.

6 Click the **Previous** button until the first record (record number 100) appears in the interface.

7 Click the **Exit** button to end the application, then close the solution.

You have completed Lesson A. You can either take a break or complete the end-of-lesson questions and exercises before moving on to Lesson B.

SUMMARY

TO CONNECT AN APPLICATION TO A DATABASE:

» Use the procedure shown in Figure 12-5.

TO PREVIEW THE DATA CONTAINED IN A DATASET:

» Use the procedure shown in Figure 12-13.

TO BIND AN OBJECT IN A DATASET:

» Use either of the ways shown in Figure 12-15.

TO HAVE THE COLUMNS EXACTLY FILL THE DISPLAY AREA IN A DATAGRIDVIEW CONTROL:

» Set the DataGridView control's AutoSizeColumnsMode property to Fill.

TO ANCHOR A CONTROL TO A BORDER OF THE FORM:

» Set the control's Dock property.

TO MOVE THE RECORD POINTER WHILE AN APPLICATION IS RUNNING:

» You can use a BindingNavigator control. Or, you can use the BindingSource object's Position property, or its Move methods (MoveFirst, MoveLast, MoveNext, or MovePrevious).

QUESTIONS

1. The field that links a child table to a parent table is called the _____.

   a. foreign key in the child table   b. foreign key in the parent table

   c. link key in the parent table   d. primary key in the child table

2. The process of connecting a control to a dataset is called _____.

   a. assigning   b. binding

   c. joining   d. None of the above.

3. The _____ object connects a database to a DataSet object.

   a. BindingSource   b. DataBase

   c. DataGridView   d. TableAdapter

4. The _____ property stores an integer that represents the location of the record pointer in a dataset.

   a. BindingNavigator object's Position

   b. BindingSource object's Position

   c. TableAdapter object's Position

   d. None of the above.

5. If the record pointer is positioned on record number five in a dataset, which of the following methods can be used to move the record pointer to record number four?

   a. GoPrevious()   b. Move(4)

   c. MovePrevious()   d. PositionPrevious

6. Which of the following statements is true about a relational database?

   a. Data stored in a relational database can be retrieved both quickly and easily by the computer.

   b. Data stored in a relational database can be displayed in any order.

   c. A relational database stores data in a column and row format.

   d. All of the above are true.

7. The _____ object provides the connection between a DataSet object and a control.

   a. Binding   b. BindingSource

   c. Connecting   d. None of the above.

8. If the current record is the second record in the dataset, which of the following statements will position the record pointer on the first record?

   a. `Me.TblEmployBindingSource.Position = 0`

   b. `Me.TblEmployBindingSource.Position =`
     `Me.TblEmployBindingSource.Position - 1`

   c. `Me.TblEmployBindingSource.MoveFirst()`

   d. All of the above.

EXERCISES

1. In this exercise, you learn how to open a database table in a window in the IDE. You also modify one of the Morgan Industries applications that you completed in the lesson.

   a. Use Windows to make a copy of the Morgan Industries Solution-Labels folder, which is contained in the VB2005\Chap12 folder. Rename the copy Modified Morgan Industries Solution-Labels.

   b. If necessary, start Visual Studio 2005 or Visual Basic 2005 Express Edition. Open the Morgan Industries Solution (Morgan Industries Solution.sln) file contained in the VB2005\Chap12\Modified Morgan Industries Solution-Labels folder. Open the designer window.

   c. Use the View menu to open the Server Explorer window. If necessary, click the plus box that appears next to Employees.mdf, then click the plus box that appears next to Tables. Right-click tblEmploy, then click Show Table Data. The table data appears in a window in the IDE, as shown earlier in Figure 12-4.

   d. Close the window that contains the table data, then close the Server Explorer window.

   e. Open the Code Editor window. Modify the code in the Click event procedures for the xNextButton and xPreviousButton. Rather than using the MoveNext and MovePrevious methods, use the Position property to move the record pointer.

   f. Save the solution, then start and test the application. Click the Exit button to end the application, then close the solution.

2. Carl Simons, the sales manager at Cartwright Industries, records the item number, name, and price of each product the company sells in a SQL Server database named Items. The Items database is stored in the Items.mdf file contained in the VB2005\Chap12\SQL Databases folder. Mr. Simons wants an application that allows the sales clerks to view the numbers, names, and prices of the items in the database.

a. If necessary, start Visual Studio 2005 or Visual Basic 2005 Express Edition. Open the Cartwright Solution (Cartwright Solution.sln) file contained in the VB2005\Chap12\ Cartwright Solution folder. If necessary, open the designer window.

b. Connect the Items database to the application. Display the tblItems data in label controls. Code the Next and Previous buttons appropriately.

c. Save the solution, then start and test the application. Click the Exit button to end the application, then close the solution.

3. Addison Grace, the owner of Addison Playhouse, records each patron's seat number, name, and phone number in a SQL Server database named Play. The Play database is stored in the Play.mdf file contained in the VB2005\Chap12\SQL Databases folder. Ms. Grace wants an application that allows her to view the contents of the database. The application also should allow her to add, delete, and save changes to the database.

a. If necessary, start Visual Studio 2005 or Visual Basic 2005 Express Edition. Open the Addison Playhouse Solution (Addison Playhouse Solution.sln) file contained in the VB2005\Chap12\Addison Playhouse Solution folder. If necessary, open the designer window.

b. Connect the Play database to the application. Display the tblReservations data in a DataGridView control.

c. Save the solution, then start and test the application. Click the Exit button to end the application, then close the solution.

4. In this exercise, you modify one of the Morgan Industries applications that you completed in the lesson so that it uses a list box.

a. Use Windows to make a copy of the Morgan Industries Solution-Labels folder, which is contained in the VB2005\Chap12 folder. Rename the copy Morgan Industries Solution-ListBox.

b. If necessary, start Visual Studio 2005 or Visual Basic 2005 Express Edition. Open the Morgan Industries Solution (Morgan Industries Solution.sln) file contained in the VB2005\Chap12\Morgan Industries Solution-ListBox folder. Open the designer window.

c. Unlock the controls, then delete the xNumberLabel from the form. Add a list box to the form. Name the list box xNumberListBox. Lock the controls, then set the tab order appropriately.

d. Set the xNumberListBox's DataSource property to TblEmployBindingSource, and set its DisplayMember property to Number.

e. Save the solution, then start and test the application. Click the Exit button to end the application, then close the solution.

DISCOVERY EXERCISE

5. As you learned in this lesson, you can save the changes made to a dataset by setting the database file's Copy to Output Directory property to "Copy if newer." In this exercise, you learn another way to save the changes made to a dataset.

   a. Use Windows to open the VB2005\Chap12\Morgan Industries Solution-NoCopy\Morgan Industries Project folder. Select the Employees.mdf and Employees_log.ldf files, then copy the files to the bin\Debug folder. (Hint: Press Ctrl+c to copy the files from the project folder. Open the bin\Debug folder, then press Ctrl+v to paste the files in that folder.)

   b. If necessary, start Visual Studio 2005 or Visual Basic 2005 Express Edition. Open the Morgan Industries Solution (Morgan Industries Solution.sln) file contained in the VB2005\Chap12\Morgan Industries Solution-NoCopy folder. If necessary, open the designer window.

   c. Click Employees.mdf in the Solution Explorer window, then set the file's Copy to Output Directory to "Do not copy."

   d. Use the View menu to display the Server Explorer window. Right-click Employees.mdf, then click Modify Connection to display the Modify Connection dialog box. Click immediately before the Employees.mdf filename in the Database filename (new or existing) box, then type \bin\Debug\ and press Enter. The dialog box closes.

   e. Close the Server Explorer window. Save the solution, then start and test the application by adding, deleting, and changing records. Click the Exit button to end the application, then start the application again. Verify that your changes were saved to the database. Click the Exit button to end the application, then close the solution.

   f. Use Windows to delete the Employees.mdf and Employees_log.ldf files from the VB2005\Chap12\Morgan Industries Solution-NoCopy\Morgan Industries Project folder.

   g. Open the Morgan Industries Solution (Morgan Industries Solution.sln) file contained in the VB2005\Chap12\Morgan Industries Solution-NoCopy folder. Right-click Employees.mdf in the Solution Explorer window, then click Exclude From Project. Save the solution. Start and test the application again. Click the Exit button to end the application, then close the solution.

LESSON B

OBJECTIVES

AFTER STUDYING LESSON B, YOU SHOULD
BE ABLE TO:

» Write SQL SELECT statements

» Create a query using the Query Configuration Wizard

» Associate a ToolStrip control with a query

CREATING QUERIES

THE DATASET DESIGNER

Before you begin this lesson, you will open one of the Morgan Industries applications that you completed in Lesson A.

To open the Morgan Industries application, then start the application:

1 Use Windows to make a copy of the Morgan Industries Solution-DataGrid folder, which is contained in the VB2005\Chap12 folder. Rename the copy Morgan Industries Solution-DataGrid-Query.

2 Start Visual Studio 2005 or Visual Basic 2005 Express Edition, if necessary, and close the Start Page window. Open the **Morgan Industries Solution** (Morgan Industries Solution.sln) file contained in the VB2005\Chap12\Morgan Industries Solution-DataGrid-Query folder. Open the designer window.

3 If necessary, auto-hide the Toolbox, Data Sources, Server Explorer, and Properties windows. If necessary, permanently display the Solution Explorer window.

4 Start the application. The data contained in the EmployeesDataSet appears in the TblEmployDataGridView control, as shown in Figure 12-26. Only 10 of the 12 records are displayed in the control; however, you can use the scroll bar to view the remaining two records.

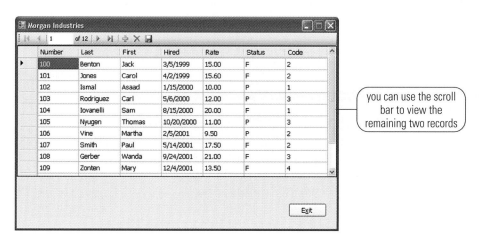

Figure 12-26: Data displayed in the TblEmployDataGridView control

5 Click the **Exit** button to end the application.

As you learned in Lesson A, data stored in a relational database can be displayed in any order. For example, you can arrange the data contained in the Employees database by employee number, pay rate, status, and so on. A relational database also allows you to control the amount of information you want to view at a time. For example, you can view all of the information in the Employees database, or you can view only the employee numbers and corresponding pay rates. You also can choose to view only the records for the part-time employees. You can use the DataSet Designer to indicate the order in which you want data displayed, and also to specify the fields and records you want to view from the database. Figure 12-27 shows three ways to open the DataSet Designer.

Opening the DataSet Designer

» Right-click the TableAdapter object in the component tray, then click Edit Queries in DataSet Designer.
» In the Solution Explorer window, right-click the name of the schema file, which has an .xsd extension, then click View Designer.
» Right-click the Data Sources window, then click Edit DataSet with Designer.

Figure 12-27: Three ways to open the DataSet Designer

To open the DataSet Designer:

1 Right-click the **TblEmployTableAdapter** object in the component tray, then click **Edit Queries in DataSet Designer**. Figure 12-28 shows the DataSet Designer open in the IDE. The key ⚷ symbol in the figure indicates that the Number field is the primary field in the tblEmploy table.

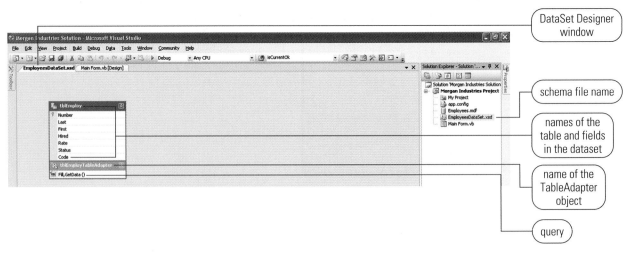

Figure 12-28: DataSet Designer

The Solution Explorer window shown in Figure 12-28 contains a file named EmployeesDataSet.xsd. The filename also appears on the tab in the DataSet Designer. The .xsd extension on the filename indicates that the file is an XML schema definition file. **XML**, which stands for **Extensible Markup Language**, is a text-based language used to store and share data between applications and across networks and the Internet. An **XML schema definition file** defines the tables and fields that make up the dataset. The .xsd file is added to the Solution Explorer window when you connect an application to a database.

The DataSet Designer in Figure 12-28 shows the name of the table object included in the EmployeesDataSet. It also lists the names of the seven field objects included in the dataset. In addition, it shows the name of the application's TableAdapter object, which connects the dataset to its underlying database. Every TableAdapter object contains one or more queries, which are listed below the TableAdapter object's name in the DataSet Designer. A **query** specifies the fields and records to retrieve from the database, as well as the order in which to arrange the fields and records. As Figure 12-28 indicates, the TblEmployTableAdapter in the Morgan Industries application contains one query. The query is associated with two methods named Fill and GetData. As you learned earlier, the Fill method populates an existing table with data, and the GetData method creates a new table and populates it with data. You can view the information contained in a query by right-clicking the query in the DataSet Designer window, and then clicking Configure. Doing this opens the TableAdapter Configuration Wizard.

To view the information contained in a query:

1 Right-click the **Fill, GetData()** query in the DataSet Designer window, then click **Configure**. The TableAdapter Configuration Wizard opens and displays the Enter a SQL Statement screen. The statement associated with the current query appears in the "What data should be loaded into the table?" box, as shown in Figure 12-29. As the screen indicates, the TableAdapter uses the data returned by the statement to fill its DataTable. You will learn about SQL (Structured Query Language) in the next section.

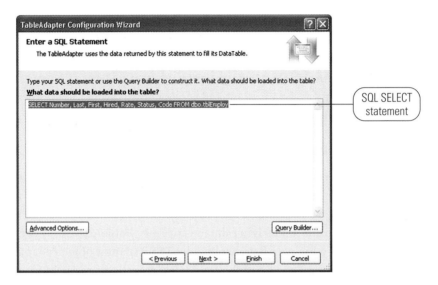

SQL SELECT statement

Figure 12-29: Enter a SQL Statement screen in the TableAdapter Configuration Wizard

2 Click the **Cancel** button to close the TableAdapter Configuration Wizard.

STRUCTURED QUERY LANGUAGE

Queries are created using a special language called **Structured Query Language** or, more simply, SQL. **SQL**, which is pronounced *ess-cue-el,* is a set of commands that allows you to access and manipulate the data stored in many database management systems on computers of all sizes, from large mainframes to small microcomputers. (SQL also can be pronounced like the word *sequel.*) You can use SQL commands to perform database tasks such as storing, retrieving, updating, deleting, and sorting data. The SELECT statement is the most commonly used command in SQL. The **SELECT statement** allows you to specify the fields and records you want to view, as well as to control the order in which the fields and records appear when displayed. Figure 12-30 shows the basic syntax of the SELECT statement. In the syntax, *fields* is one or more field names (separated by commas), and *table* is the name of the table containing the fields. Notice that the syntax contains two clauses that are optional: the WHERE clause and the ORDER BY clause. The **WHERE clause** allows you to limit the records that will be selected, and the **ORDER BY clause** allows you to control the order in which the records appear when displayed. Although you do not have to capitalize the keywords SELECT, FROM, WHERE, and ORDER BY in a SELECT statement, many programmers do so for clarity. Figure 12-30 also includes several examples of using the SELECT statement to access the data stored in the tblEmploy table in the Employees database. (Figure 12-4 in Lesson A shows the data contained in the tblEmploy table.) As you may remember, the tblEmploy table contains seven fields. The Number, Rate, and Code fields contain numeric data. The Last, First, and Hired fields contain text data, and the Status field contains one character.

» TIP

The SQL syntax, which refers to the rules you must follow to use the language, was accepted by the American National Standards Institute (ANSI) in 1986. You can use SQL in many database management systems and programming languages.

Using the SELECT statement

TIP

The full syntax of the SELECT statement contains other clauses and options that are beyond the scope of this book.

Syntax

SELECT *fields* **FROM** *table* [**WHERE** *condition*] [**ORDER BY** *field*]

Example 1

```
SELECT Number, Last, First, Hired, Rate, Status, Code FROM dbo.tblEmploy
```

selects all of the fields and records in the table

Example 2

```
SELECT Number, First, Last FROM dbo.tblEmploy
```

selects the Number, First, and Last fields from each record in the table

Example 3

```
SELECT Number, Last, First, Hired, Rate, Status, Code FROM dbo.tblEmploy
WHERE Status = 'F'
```

selects the records for full-time employees

Example 4

```
SELECT Number, Rate FROM dbo.tblEmploy WHERE Code = 3
```

selects the Number and Rate fields for employees in the Personnel department

Example 5

```
SELECT Number FROM dbo.tblEmploy WHERE Last LIKE 'Smith'
```

selects the Number field for records having a last name of Smith

Example 6

```
SELECT Number FROM dbo.tblEmploy WHERE Last LIKE 'S%'
```

selects the Number field for records having a last name that begins with the letter S

Example 7

```
SELECT Number, Last, First, Hired, Rate, Status, Code FROM dbo.tblEmploy
ORDER BY Code
```

selects all of the fields and records in the table, and sorts the records in ascending order by the Code field

Example 8

```
SELECT Number, Last, First, Hired, Rate, Status, Code FROM dbo.tblEmploy
WHERE Status = 'p' ORDER BY Code
```

selects the records for part-time employees, and sorts the records in ascending order by the Code field

Figure 12-30: Syntax and examples of the SELECT statement

Study the examples shown in Figure 12-30. Example 1's SELECT statement is the same as the SELECT statement shown in Figure 12-29. The statement tells the computer to select all of the fields and records from the tblEmploy table. Notice that the three letters "dbo" and a period precede the table name in the example. The letters "dbo" stand for "database object."

The SELECT statement in Example 2 selects only three of the fields from each record in the table. The SELECT statement in Example 3 uses the WHERE clause to limit the records that will be selected. In this case, the statement indicates that only records for full-time employees should be selected. Notice that, when comparing the contents of the Status field with a character, you enclose the character in single quotation marks.

The SELECT statement in Example 4 selects only the Number and Rate fields for employees working in the Personnel department. The SELECT statement in Example 5 shows how you can compare the contents of a text field with a string. You do so using the Like operator in the WHERE clause, and you enclose the string in single quotation marks. Example 5's SELECT statement selects the Number field for all records that have Smith as the last name. The SELECT statement in Example 6 shows how you can use the Like operator along with the % (percent sign) wildcard in the WHERE clause. As the example indicates, the `SELECT Number FROM dbo.tblEmploy WHERE Last LIKE 'S%'` statement selects the Number field for all records with a last name that begins with the letter S. Example 7's SELECT statement selects all of the fields and records from the tblEmploy table, and then uses the ORDER BY clause to sort the records in ascending order by the Code field. To sort the records in descending order, you use `SELECT Number, Last, First, Hired, Rate, Status, Code FROM dbo.tblEmploy ORDER BY Code DESC`. The DESC keyword stands for "descending". The last SELECT statement shown in Figure 12-30 selects the records for part-time employees, and it sorts the records in ascending order by the Code field. Notice that the statement compares the contents of the Status field (which contains uppercase letters) with a lowercase letter 'p'. The statement works correctly because SQL commands are not case-sensitive.

CREATING A NEW QUERY

As mentioned earlier, a TableAdapter object can contain more than one query. Figure 12-31 shows the procedure you follow when using the TableAdapter Query Configuration Wizard to create a new query.

Creating a query using the TableAdapter Query Configuration Wizard

1. Click the name of the TableAdapter in the DataSet Designer to select it. Right-click the name, then click Add Query to open the TableAdapter Query Configuration Wizard. (Alternatively, you can right-click the DataSet Designer, point to Add, and then click Query. Or, you can click the DataSet Designer, click Data on the menu bar, point to Add, and then click Query.)

2. Depending on the way you opened the TableAdapter Query Configuration Wizard, you may need to specify the data connection. If necessary, select the appropriate data connection, then click the Next button.

3. In the Choose a Command Type screen, select the Use SQL statements radio button, if necessary, then click the Next button.

4. In the Choose a Query Type screen, select the SELECT which returns rows radio button, if necessary, then click the Next button.

5. When the Specify a SQL SELECT statement screen appears, you can type the SELECT statement yourself, or you can use the Query Builder dialog box to construct the statement for you. When you have completed the SELECT statement, click the Next button in the Specify a SQL SELECT statement screen.

6. In the Choose Methods to Generate screen, select the Fill a DataTable and Return a DataTable check boxes, if necessary. Enter appropriate names for the query's FillBy and GetDataBy methods, then click the Next button.

7. Click the Finish button in the Wizard Results screen to add the query to the DataSet Designer.

Figure 12-31: Procedure for creating a query using the TableAdapter Query Configuration Wizard

To use the TableAdapter Query Configuration Wizard to create a query:

1 Click **tblEmployTableAdapter** in the DataSet Designer to select it, if necessary, then right-click **tblEmployTableAdapter**. Click **Add Query** on the shortcut menu to open the TableAdapter Query Configuration Wizard, which displays the Choose a Command Type screen. (If Add Query does not appear on the shortcut menu, point to Add and then click Query.) See Figure 12-32.

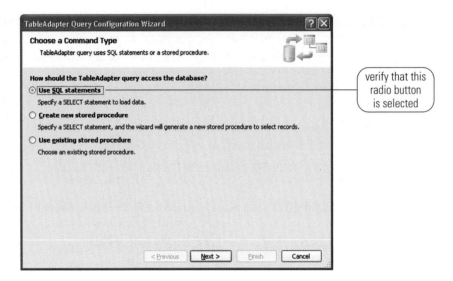

Figure 12-32: Choose a Command Type screen in the TableAdapter Query Configuration Wizard

2 Select the **Use SQL statements** radio button, if necessary, then click the **Next** button to display the Choose a Query Type screen. See Figure 12-33.

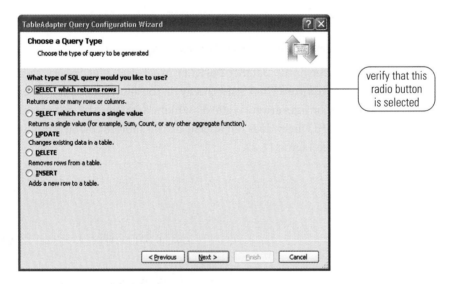

Figure 12-33: Choose a Query Type screen

3 Select the **SELECT which returns rows** radio button, if necessary, then click the **Next** button to display the Specify a SQL SELECT statement screen. See Figure 12-34.

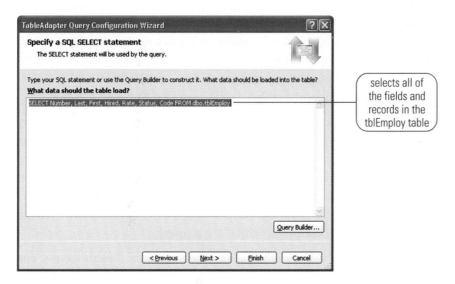

Figure 12-34: Specify a SQL SELECT statement screen

4 You can type the SELECT statement yourself, or you can use the Query Builder dialog box to construct the statement for you. In this case, you will use the Query Builder dialog box. Click the **Query Builder** button to display the Query Builder dialog box. See Figure 12-35.

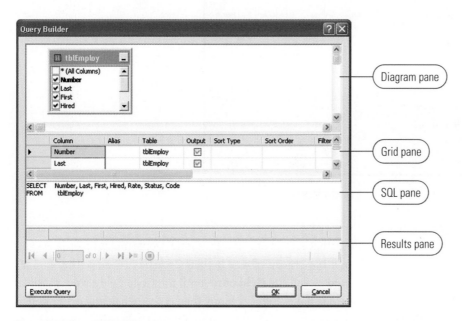

Figure 12-35: Query Builder dialog box

5 The SELECT statement that appears in the SQL pane tells the computer to select all of the fields and records contained in the tblEmploy table. Click the **Execute Query** button to run the query. The query results appear in the Results pane. Scroll the **Results pane**. The records appear in ascending order according to the Number field, which is the order in which they appear in the table.

6 Now you will modify the query so that it displays the records in descending order by the Number field. In the Grid pane, click the **blank cell** that appears below Sort Type in the first row. Click the **list arrow** in the cell, then click **Descending** in the list, and then click **any other cell**. The word Descending appears as the Number field's Sort Type, and the number 1 appears as its Sort Order. The number 1 indicates that the Number field is the primary field in the sort. As a result, the Query Builder adds the ORDER BY Number DESC clause to the SELECT statement.

7 Click the **Execute Query** button to run the query, then scroll the **Results pane**. The records appear in descending order according to the Number field. See Figure 12-36.

▶▶TIP

You also can sort records based on more than one field. For example, to sort the records by last name, with records having the same last name sorted by first name, you specify the Last field as the primary field (Sort Order = 1) and the First field as the secondary field (Sort Order = 2).

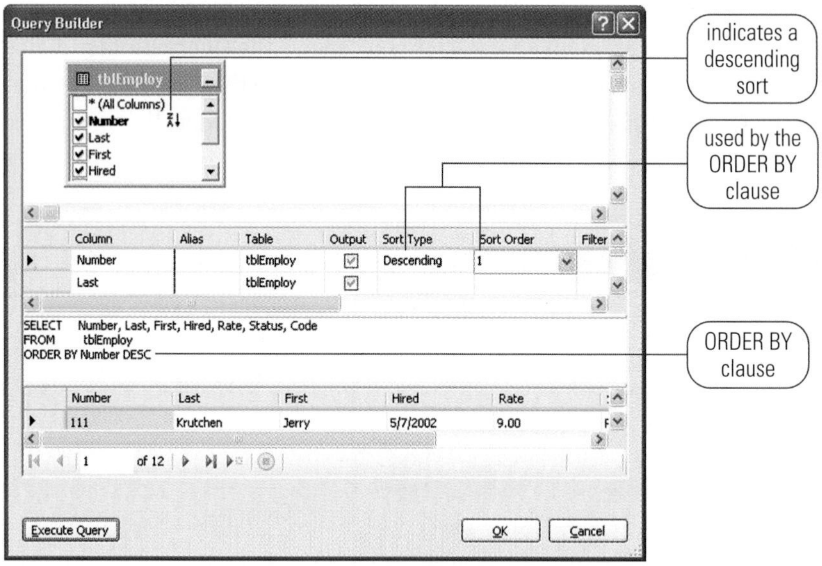

Figure 12-36: SELECT statement containing the ORDER BY clause

8 Next, you will create a query that selects only the Personnel Department records. As you may remember, those records have the number 3 in their Code field. Change the Number field's Sort Type to **Unsorted**. Scroll the **Grid pane** until the Code field's row is visible. Click the **blank cell** in the Code field's Filter column, then type **3**. Click **any other cell** in the Grid pane. The = 3 that appears in the Filter column for the Code field tells the Query Builder to include the WHERE (Code = 3) clause in the SELECT statement.

9 Scroll the **tblEmploy window**, which is contained in the Diagram pane, until the Code field is visible. The funnel ∇ symbol that appears to the right of the Code field name indicates that the field is used to filter the records. Click the **Execute Query** button to run the query, then scroll the **Results pane** to verify that it contains only the three Personnel Department records. See Figure 12-37.

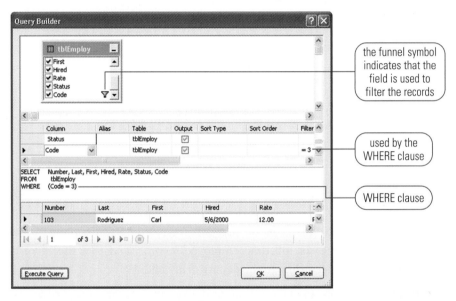

Figure 12-37: SELECT statement containing the WHERE clause

10 Lastly, you will create a query that selects only the records for full-time employees. You will display the records in ascending order according to the Code field. Select (highlight) the **= 3** that appears in the Code field's Filter column, then press **Delete** to remove the entry from the cell. Click the **blank cell** that appears in the Filter column for the Status field, then type **F**. Click the **blank cell** that appears in the Sort Type column for the Code field, then click **Ascending** in the list. Click the **SQL pane**. See Figure 12-38. Notice the Query Changed message that appears below the Results pane in the dialog box. The message alerts you that the information displayed in the Results pane is not from the current query.

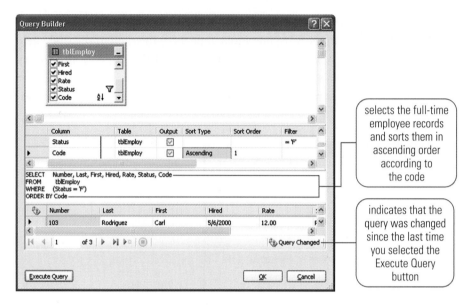

selects the full-time employee records and sorts them in ascending order according to the code

indicates that the query was changed since the last time you selected the Execute Query button

Figure 12-38: SELECT statement containing the WHERE and ORDER BY clauses

11 Click the **Execute Query** button to run the query, then scroll the **Results pane** to verify that it contains only the six full-time employee records, and that the records appear in ascending order according to the department code.

12 Click the **OK** button to close the Query Builder dialog box. You are returned to the Specify a SQL SELECT statement screen in the TableAdapter Query Configuration Wizard. The last SELECT statement you entered in the Query Builder appears in the "What data should the table load?" box. Click the **Next** button to display the Choose Methods to Generate screen.

13 If necessary, select the **Fill a DataTable** and **Return a DataTable** check boxes. Change the Fill a DataTable method's name from FillBy to **FillByFullTime**. Change the Return a DataTable method's name from GetDataBy to **GetDataByFullTime**. Figure 12-39 shows the completed Choose Methods to Generate screen.

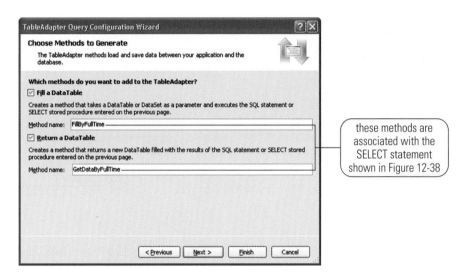

Figure 12-39: Completed Choose Methods to Generate screen

14 Click the **Next** button to display the Wizard Results screen. See Figure 12-40.

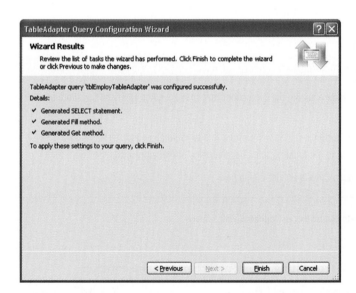

Figure 12-40: Wizard Results screen

15 Click the **Finish** button. The FillByFullTime, GetDataByFullTime query is added to the DataSet Designer, as shown in Figure 12-41.

Figure 12-41: Query names shown in the DataSet Designer

16 Save the solution, then close the EmployeesDataSet.xsd window. Auto-hide the Solution Explorer window.

ALLOWING THE USER TO RUN A QUERY

You can provide a ToolStrip control that allows the user to run a query after an application has been started. Figure 12-42 shows the steps you follow to include a ToolStrip control on a form, and associate the control with a query.

Adding a ToolStrip control that is associated with a query

1. Right-click the name of the TableAdapter object in the component tray, then click Add Query on the shortcut menu.

2. Select the Existing query name radio button in the Search Criteria Builder dialog box, then click the down arrow in the list box that appears next to the radio button. Click the name of the query in the list.

3. Click the OK button to close the Search Criteria Builder dialog box.

Figure 12-42: Procedure for adding a ToolStrip control that is associated with a query

To associate a ToolStrip control with a query, and then test the control:

1 Right-click **TblEmployTableAdapter** in the component tray, then click **Add Query** to open the Search Criteria Builder dialog box.

2 Click the **Existing query name** radio button to select it. Click the **down arrow** in the list box that appears next to the radio button, then click **FillByFullTime()** in the list. Figure 12-43 shows the completed Search Criteria Builder dialog box.

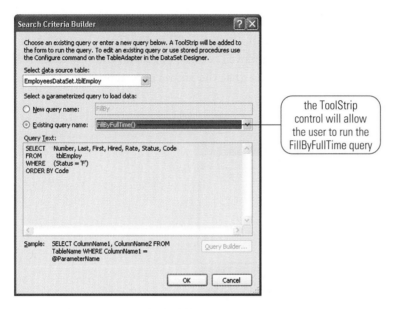

Figure 12-43: Completed Search Criteria Builder dialog box

3 Click the **OK** button to close the Search Criteria Builder dialog box. The computer adds a ToolStrip control to the form and a ToolStrip object to the component tray, as shown in Figure 12-44. The control and object are associated with the FillByFullTime query.

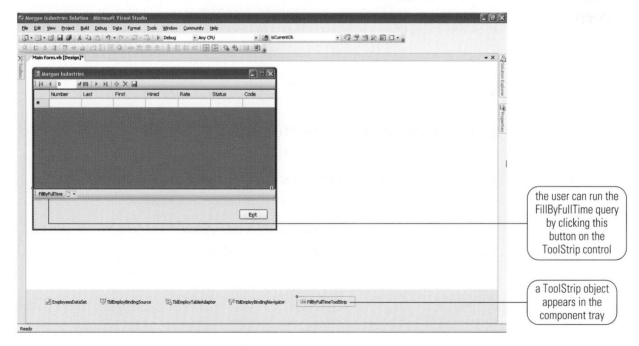

Figure 12-44: ToolStrip control and object added to the application

4 Save the solution, then start the application. The 12 records contained in the tblEmploy table appear in the DataGridView control. Click the **FillByFullTime** button on the ToolStrip control. Only the full-time employee records appear in the DataGridView control, as shown in Figure 12-45. Notice that the six records appear in ascending order according to the Code field.

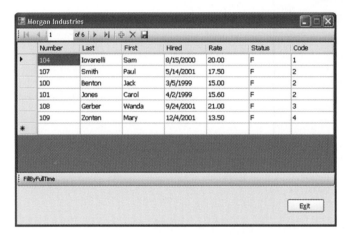

Figure 12-45: Result of clicking the FillByFullTime button

5 Click the **Exit** button to end the application, then close the solution.

You have completed Lesson B. You can either take a break or complete the end-of-lesson questions and exercises before moving on to Lesson C.

SUMMARY

TO OPEN THE DATASET DESIGNER:
» Use one of the ways listed in Figure 12-27.

TO VIEW THE INFORMATION CONTAINED IN A QUERY:
» Right-click the query in the DataSet Designer, then click Configure.

TO SPECIFY THE INFORMATION YOU WANT TO VIEW, AS WELL
AS THE ORDER OF THE INFORMATION:
» Use the SELECT statement. The statement's syntax is shown in Figure 12-30.

TO CREATE A QUERY:

» Use the procedure shown in Figure 12-31.

TO ADD A TOOLSTRIP CONTROL THAT IS ASSOCIATED WITH A QUERY:

» Use the procedure shown in Figure 12-42.

QUESTIONS

1. You can use the DataSet Designer to indicate the _____.

   a. fields to display

   b. records to display

   c. order of the displayed data

   d. All of the above.

2. SQL stands for _____.

   a. Select Query Language

   b. Semi-Quick Language

   c. Structured Quick Language

   d. Structured Query Language

3. Which of the following SELECT statements will select the First, Middle, and Last fields from a table named tblNames?

   a. `SELECT First, Middle, AND Last FROM dbo.tblNames`

   b. `SELECT First, Middle, OR Last FROM dbo.tblNames`

   c. `SELECT First, Middle, Last FROM dbo.tblNames`

   d. None of the above.

4. Which of the following SELECT statements will display the SocialNum field data in descending order? The field is contained in a table named tblPensionInfo.

   a. `SELECT SocialNum FROM dbo.tblPensionInfo DESC`

   b. `SELECT SocialNum FROM dbo.tblPensionInfo ORDER BY SocialNum DESC`

   c. `SELECT SocialNum FROM dbo.tblPensionInfo WHERE SocialNum DESC`

   d. None of the above.

5. Which of the following SELECT statements will select only records that have the letter A in their Status field? The field is contained in the Worker table.

   a. `SELECT Id, Name, Status FROM dbo.Worker WHERE Status = 'A'`

   b. `SELECT Id, Name, Status FROM dbo.Worker ORDDER BY Status = 'A'`

   c. `SELECT Id, Name, Status FROM dbo.Worker FOR Status = "A"`

   d. None of the above.

Use the following database table, named tblState, to answer Questions 6 and 7.

Field name	Data type
State	Text
Capital	Text
Population	Numeric

6. Which of the following statements allows you to view all of the records in the table?

   a. `SELECT ALL records FROM dbo.tblState`

   b. `SELECT State, Capital, Population FROM dbo.tblState`

   c. `VIEW ALL records FROM dbo.tblState`

   d. `VIEW State, Capital, Population FROM dbo.tblState`

7. Which of the following statements will retrieve all records having a population that exceeds 5,000,000?

   a. `SELECT ALL FROM dbo.tblState FOR Population > 5000000`

   b. `SELECT State, Capital, Population FROM dbo.tblState WHERE Population > "5000000"`

   c. `SELECT State, Capital, Population FROM dbo.tblState WHERE Population > '5000000'`

   d. None of the above.

8. In a SELECT statement, the _____ clause limits the records that will be selected.

   a. LIMIT   b. ORDER BY

   c. SET   d. None of the above.

9. If a funnel symbol appears next to a field's name in the Query Builder dialog box, it indicates that the field is _____.

a. used in an ORDER BY clause in a SELECT statement

b. used in a WHERE clause in a SELECT statement

c. the primary key

d. the foreign key

10. The SQL SELECT statement performs case sensitive comparisons.

a. True

b. False

EXERCISES

1. In this exercise, you modify the Morgan Industries application that you completed in this lesson.

   a. Use Windows to make a copy of the Morgan Industries Solution-DataGrid-Query folder, which is contained in the VB2005\Chap12 folder. Rename the folder Modified Morgan Industries Solution-DataGrid-Query.

   b. If necessary, start Visual Studio 2005 or Visual Basic 2005 Express Edition. Open the Morgan Industries Solution (Morgan Industries Solution.sln) file contained in the VB2005\Chap12\Modified Morgan Industries Solution-DataGrid-Query folder. Open the designer window.

   c. Create another query. The query should select only the records for part-time employees. Arrange the records in order by the Rate field. Use FillByPartTime and GetDataByPartTime as the names for the query's methods.

   d. Associate the FillByPartTime query with a ToolStrip control. Also associate the Fill query with a ToolStrip control.

   e. Save the solution, then start the application. Test the FillByFullTime, FillByPartTime, and Fill buttons on the ToolStrip controls. Click the Exit button to end the application, then close the solution.

2. In this exercise, you modify the application that you created in Exercise 2 in Lesson A.

   a. Use Windows to make a copy of the Cartwright Solution folder, which is contained in the VB2005\Chap12 folder. Rename the folder Cartwright Solution-Query.

   b. If necessary, start Visual Studio 2005 or Visual Basic 2005 Express Edition. Open the Cartwright Solution (Cartwright Solution.sln) file contained in the VB2005\Chap12\Cartwright Solution-Query folder. Open the designer window.

c. Create a query that displays the data in ascending order by the item number. Also create a query that displays the data in descending order by the price.

d. Associate the two queries with two ToolStrip controls. Save the solution, then start and test the application. Click the Exit button to end the application, then close the solution.

3. In this exercise, you use a SQL Server database named Courses. The database is stored in the Courses.mdf file, which is located in the VB2005\Chap12\SQL Databases folder. The Courses database contains one table named tblCourses. The table contains 10 records, each having four fields: course ID, course title, credit hours, and grade. The credit hours field is numeric; the other fields contain text.

a. If necessary, start Visual Studio 2005 or Visual Basic 2005 Express Edition. Create an appropriate Visual Basic Windows-based application. Name the solution College Courses Solution, and name the project College Courses Project. Save the application in the VB2005\Chap12 folder. The application should contain six queries. Five of the queries should allow you to display the data for a specific grade (A, B, C, D, F). The sixth query should display all of the records in their original order.

b. Associate the queries with ToolStrip controls. Save the solution, then start and test the application. Stop the application, then close the solution.

DISCOVERY EXERCISE

4. In this exercise, you include a ToolStrip control on a BindingNavigator control.

a. Use Windows to make a copy of the Morgan Industries Solution-DataGrid-Query folder, which is contained in the VB2005\Chap12 folder. Rename the folder Morgan Industries Solution-DataGrid-ToolStrip.

b. If necessary, start Visual Studio 2005 or Visual Basic 2005 Express Edition. Open the Morgan Industries Solution (Morgan Industries Solution.sln) file contained in the VB2005\Chap12\Morgan Industries Solution-DataGrid-ToolStrip folder. Open the designer window.

c. Unlock the controls on the form. Change the Dock property of the FillByFullTimeToolStrip control to None. Drag the FillByFullTimeToolStrip control to the TblEmployBindingNavigator control, which appears above the DataGridView control. The ToolStrip control appears as a button on the TblEmployBindingNavigator control. Lock the controls on the form.

d. Save the solution, then start and test the application. Click the Exit button to end the application, then close the solution.

LESSON C
OBJECTIVES

AFTER STUDYING LESSON C, YOU SHOULD BE ABLE TO:

» Connect a SQL Server database to an application

» Bind field objects to controls in an interface

» Position the record pointer in a dataset

» Determine the number of records in a dataset

THE TRIVIA GAME APPLICATION

CODING THE TRIVIA GAME APPLICATION

Recall that your task in this chapter is to create a Trivia Game application that displays questions along with answers from which the user can select. The application should keep track of the number of incorrect selections made by the user, and then display this information at the end of the game. To save you time, the VB2005\Chap12\Trivia Game Solution folder contains a partially completed Trivia Game application.

To open the partially completed Trivia Game application:

1 Start Visual Studio 2005 or Visual Basic 2005 Express Edition, if necessary, and close the Start Page window.

2 Open the **Trivia Game Solution** (Trivia Game Solution.sln) file, which is contained in the VB2005\Chap12\Trivia Game Solution folder. If necessary, open the designer window and auto-hide the Toolbox, Solution Explorer, and Properties windows. The Trivia Game application's user interface is shown in Figure 12-46.

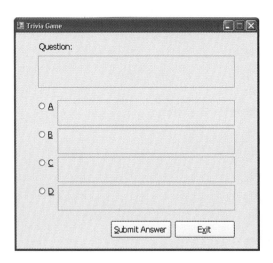

Figure 12-46: Trivia Game application's user interface

The Trivia Game questions and answers are stored in the Trivia database, which is contained in a SQL Server database file named Trivia.mdf. You will need to connect the Trivia database to the application.

To connect the Trivia database to the application, then preview the data:

1 Click **Data** on the menu bar, and then click **Show Data Sources** to display the Data Sources window. Click **Add New Data Source** in the Data Sources window to start the Data Source Configuration Wizard. If necessary, click **Database**.

2 Click the **Next** button to display the Choose Your Data Connection screen, then click the **New Connection** button to display the Add Connection dialog box. If Microsoft SQL Server Database File (SqlClient) does not appear in the Data source box, click the **Change** button to open the Change Data Source dialog box, click **Microsoft SQL Server Database File**, and then click the **OK** button to return to the Add Connection dialog box.

3 Click the **Browse** button in the Add Connection dialog box. Open the VB2005\ Chap12\SQL Databases folder, then click **Trivia.mdf** in the list of filenames. Click the **Open** button. Figure 12-47 shows the completed Add Connection dialog box.

Figure 12-47: Completed Add Connection dialog box

4 Click the **Test Connection** button in the Add Connection dialog box. The "Test connection succeeded." message appears in a dialog box. Click the **OK** button to close the dialog box.

5   Click the **OK** button to close the Add Connection dialog box. The Trivia.mdf database file-name now appears in the Choose Your Data Connection screen. Click the **Next** button, then click the **Yes** button to add the Trivia.mdf file to the current project. The Save the Connection String to the Application Configuration File screen appears next. If necessary, select the **check box**. Notice that the connection string's name is TriviaConnectionString.

6   Click the **Next** button to display the Choose Your Database Objects screen. Click the **plus box** that appears next to Tables, then click the **plus box** that appears next to Game1. Notice that the Game1 table contains six fields. You can use this screen to select the table and/or field objects to include in the dataset.

7   In this application, you need to include all of the fields in the dataset. Click the **empty box** that appears next to Game1. Doing this selects the table and field check boxes, as shown in Figure 12-48.

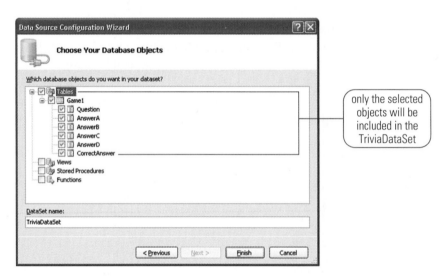

Figure 12-48: Database objects selected in the Choose Your Database Objects screen

8   Click the **Finish** button. The computer adds a dataset to the Data Sources window. Click the **plus box** that appears next to Game1 in the Data Sources window. See Figure 12-49.

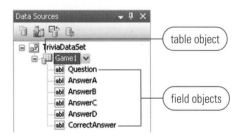

Figure 12-49: TriviaDataSet added to the Data Sources window

9 Now you will preview the data contained in the dataset. Right-click the **Data Sources window**, click **Preview Data**, and then click the **Preview** button in the Preview Data dialog box. See Figure 12-50.

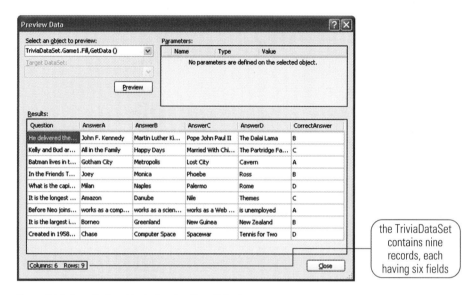

Figure 12-50: Data displayed in the Preview Data dialog box

10 Click the **Close** button in the Preview Data dialog box, then save the solution.

In the next set of steps, you will bind the field objects in the dataset to the appropriate text boxes on the form.

To bind the field objects to the text boxes, then test the application:

1 Click the **Question** field object in the Data Sources window, then drag the Question field object to the xQuestionTextBox, as shown in Figure 12-51.

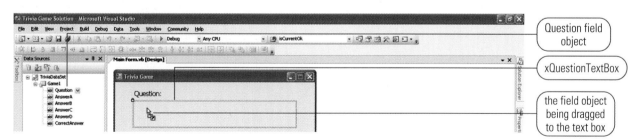

Figure 12-51: Question field object being dragged to the xQuestionTextBox

2 Release the mouse button. The computer binds the Question field object to the xQuestionTextBox. It also adds the TriviaDataSet, Game1BindingSource, and Game1TableAdapter objects to the component tray, as shown in Figure 12-52.

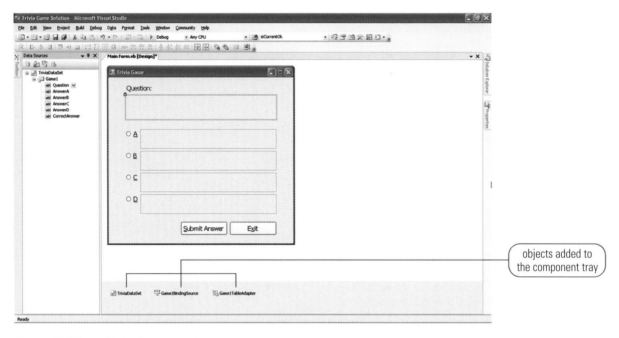

Figure 12-52: Objects added to the component tray

3 Drag the **AnswerA** field object to the xATextBox, then drag the **AnswerB** field object to the xBTextBox. Drag the **AnswerC** object to the xCTextBox, then drag the **AnswerD** object to the xDTextBox.

4 Save the solution, then start the application. The first record in the dataset appears in the interface, as shown in Figure 12-53.

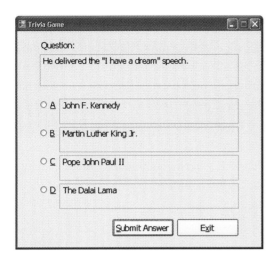

Figure 12-53: First record appears in the interface

5 Click the **Exit** button to end the application, then close the Data Sources window.

Now you can begin coding the Trivia Game application. Figure 12-54 shows the application's TOE chart. According to the TOE chart, only three event procedures need to be coded: the xExitButton's Click event procedure, the MainForm's Load event procedure, and the xSubmitButton's Click event procedure.

Task	Object	Event
End the application	xExitButton	Click
Fill the dataset with data	MainForm	Load
1. Compare the user's answer to the correct answer 2. Keep track of the number of incorrect answers 3. Display the next question and answers from the dataset 4. Display the number of incorrect answers	xSubmitButton	Click
Display questions from the dataset	xQuestionTextBox	None
Display answers from the dataset	xATextBox, xBTextBox, xCTextBox, xDTextBox	None
Allow the user to select an answer	xARadioButton, xBRadioButton, xCRadioButton, xDRadioButton	None

Figure 12-54: Trivia Game application's TOE chart

To begin coding the application:

1 Open the Code Editor window. Notice that the xExitButton's Click event procedure and the MainForm's Load event procedure are already coded for you. The only procedure you need to code is the xSubmitButton's Click event procedure. (Recall that the computer automatically enters the `Me.Game1TableAdapter.Fill(Me.TriviaDataSet.Game1)` statement when you drag an object from the Data Sources window to the form.)

2 Replace the <your name> and <current date> text in the comments with your name and the current date.

Figure 12-55 shows the pseudocode for the xSubmitButton's Click event procedure.

---

xSubmitButton Click event procedure - pseudocode

1. store the position of the record pointer in a variable

2. determine which radio button is selected, and assign its Text property (without the ampersand that designates the access key) to a variable named userAnswer

3. if the value contained in the userAnswer variable does not match the correct answer from the dataset
      add 1 to a counter variable named numIncorrect
  end if

4. if the record pointer is not pointing to the last record
      move the record pointer to the next record in the dataset
  else
      display the number of incorrect answers in a message box
  end if

Figure 12-55: Pseudocode for the xSubmitButton's Click event procedure

To code the xSubmitButton's Click event procedure:

1 Open the code template for the xSubmitButton's Click event procedure. Type ' **determines whether the user's answer is correct** and press **Enter**, then type ' **keeps track of the number of incorrect answers** and press **Enter**. Type ' **displays the next record or the number of** and press **Enter**, then type ' **incorrect answers** and press **Enter** twice.

2 Type **dim recPtrPos as integer** and press **Enter**, then type **dim userAnswer as string = string.empty** and press **Enter**. The `recPtrPos` variable will keep track of the position of the record pointer, and the `userAnswer` variable will store the user's answer to the question displayed in the xQuestionTextBox (either A, B, C, or D).

3 Type **static numIncorrect as integer** and press **Enter** twice. The numIncorrect variable will be used as a counter to keep track of the number of incorrect answers made by the user.

4 The first step in the pseudocode is to store the position of the record pointer in a variable. You can use the BindingSource object's Position property, which you learned about in Lesson A, to determine the current position of the record pointer in a dataset. Type **' store position of record pointer** and press **Enter**, then type **recptrpos = me.Game1BindingSource.position** and press **Enter** twice.

5 Next, you will determine which radio button is selected and assign its Text property (without the ampersand that designates the access key) to the userAnswer variable. Type the comments and Select Case statement shown in Figure 12-56.

```
' determine selected radio button, which
' represents the user's answer
Select Case True
    Case me.xARadioButton.Checked
        userAnswer = me.xARadioButton.Text.Replace("&", "")
    Case me.xBRadioButton.Checked
        userAnswer = me.xBRadioButton.Text.Replace("&", "")
    Case me.xCRadioButton.Checked
        userAnswer = me.xCRadioButton.Text.Replace("&", "")
    Case me.xDRadioButton.Checked
        userAnswer = me.xDRadioButton.Text.Replace("&", "")
End Select
```

Figure 12-56: Comments and Select Case statement

The next step in the pseudocode is to compare the contents of the userAnswer variable with the contents of the CorrectAnswer field in the current record. You can access the value stored in a field in the current record using the syntax **Me.***dataSetName.**tableName***(*recordNumber*)*.*fieldname*. If the userAnswer variable and CorrectAnswer field do not contain the same value, the procedure should increase (by one) the contents of the numIncorrect counter variable.

6 Position the insertion point two blank lines below the End Select clause, then type **' if necessary, update number of incorrect answers** and press **Enter**. Then type **if useranswer <> _** and press **Enter**. Press **Tab**, then type **me.triviadataset. game1(recptrpos).correctanswer then** and press **Enter**. Type **numincorrect = numincorrect + 1**.

The next step in the pseudocode is to determine whether the record pointer is pointing to the last record in the table. Although the Game1 table contains nine records, the last record is in record position eight; this is because the first record in a table is in record position zero. Therefore, you can determine whether the record pointer is on the last record by comparing the value stored in the recPtrPos variable with the number eight. Although this will work, it has one disadvantage: if records are added to or deleted from the Game1 table, the code in the xSubmitButton's Click event procedure will need to be changed. A better approach is to use the table object's Count property, whose syntax is **Me.**dataSetName.tableName.**Count**. The **Count property** returns an integer that represents the number of records in the table and is always one number more than the highest record position. Therefore, you can determine the highest record position in a table by subtracting the number one from the value stored in the table's Count property. In this case, rather than comparing the value stored in the recPtrPos variable with the number eight, you will compare it with the expression Me.TriviaDataSet.Game1.Count - 1.

7 Position the insertion point two blank lines below the End If clause, then type **' determine position of record pointer** and press **Enter**. Type **if recptrpos < me.TriviaDataSet.Game1.count – 1 then** and press **Enter**.

8 As you learned in Lesson A, you can use the BindingSource object's MoveNext method to move the record pointer to the next record. Type **me.Game1BindingSource. movenext()** and press **Enter**.

9 However, if the record pointer is pointing to the last record, it means that there are no more questions and answers to display. In that case, the procedure should display the number of incorrect answers made by the user. Type **else** and press **Enter**. Type **messagebox.show("Number incorrect: "** _ and press **Enter**. Press **Tab**, then type **& numincorrect.tostring, "Trivia Game",** _ and press **Enter**. Type **messagebox buttons.ok, messageboxicon.information)**.

10 Save the solution. Figure 12-57 shows the code for the Trivia Game application.

```
' Project name:     Trivia Game Project
' Project purpose:  The project displays trivia
'                   questions and answers from
'                   a database. It also keeps track
'                   of the number of incorrect
'                   responses made by the user.
' Created/revised:  <your name> on <current date>

Option Explicit On
Option Strict On

Public Class MainForm

    Private Sub xExitButton_Click(ByVal sender As Object, _
        ByVal e As System.EventArgs) Handles xExitButton.Click
        Me.Close()
    End Sub

    Private Sub MainForm_Load(ByVal sender As System.Object, _
        ByVal e As System.EventArgs) Handles MyBase.Load
        'TODO: This line of code loads data into the
        'TriviaDataSet.Game1' table. You can move,
        'or remove it, as needed.
        Me.Game1TableAdapter.Fill(Me.TriviaDataSet.Game1)

    End Sub

    Private Sub xSubmitButton_Click(ByVal sender As Object, _
        ByVal e As System.EventArgs) Handles xSubmitButton.Click
        ' determines whether the user's answer is correct
        ' keeps track of the number of incorrect answers
        ' displays the next record or the number of
        ' incorrect answers

        Dim recPtrPos As Integer
        Dim userAnswer As String = String.Empty
        Static numIncorrect As Integer

        ' store position of record pointer
        recPtrPos = Me.Game1BindingSource.Position

        ' determine selected radio button, which
        ' represents the user's answer
```

Figure 12-57: Trivia Game application's code *(Continued)*   ▶

```
        Select Case True
            Case Me.xARadioButton.Checked
                userAnswer = Me.xARadioButton.Text.Replace("&", "")
            Case Me.xBRadioButton.Checked
                userAnswer = Me.xBRadioButton.Text.Replace("&", "")
            Case Me.xCRadioButton.Checked
                userAnswer = Me.xCRadioButton.Text.Replace("&", "")
            Case Me.xDRadioButton.Checked
                userAnswer = Me.xDRadioButton.Text.Replace("&", "")
        End Select

        ' if necessary, update number of incorrect answers
        If userAnswer <> _
            Me.TriviaDataSet.Game1(recPtrPos).CorrectAnswer Then
            numIncorrect = numIncorrect + 1
        End If

        ' determine position of record pointer
        If recPtrPos < Me.TriviaDataSet.Game1.Count - 1 Then
            Me.Game1BindingSource.MoveNext()
        Else
            MessageBox.Show("Number incorrect: " _
                & numIncorrect.ToString, "Trivia Game", _
                MessageBoxButtons.OK, MessageBoxIcon.Information)
        End If
    End Sub
End Class
```

Figure 12-57: Trivia Game application's code

To test the application:

1 Start the application. When the first question appears on the screen, answer the question correctly by clicking the **B** radio button, and then clicking the **Submit Answer** button.

2 When the second question appears on the screen, answer the question incorrectly by clicking the **D** radio button, and then clicking the **Submit Answer** button.

3 Answer the remaining seven questions on your own. When you have submitted the answer for the last question, the xSubmitButton's Click event procedure displays the number of incorrect responses in a message box.

4 Click the **OK** button to close the message box, then click the **Exit** button to end the application.

5 Close the Code Editor window, then close the solution.

You have completed Lesson C and Chapter 12. You can either take a break or complete the end-of-lesson questions and exercises.

SUMMARY

TO DETERMINE THE NUMBER OF RECORDS IN A TABLE:

» Use the Count property, whose syntax is **Me.***dataSetName***.***tableName***.Count**.

QUESTIONS

1. Which of the following statements assigns the current position of the record pointer to a variable named recordPtr?

   a. recordPtr = Me.tblStateBindingSource.Position

   b. recordPtr = Me.tblStateBindingTable.Position

   c. recordPtr = Me.tblStateRecordSource.Position

   d. recordPtr = Me.tblStateTableSource.Position

2. Which of the following statements assigns the number of records contained in the Products table to the numRecords variable? The Products table is contained in the ProductDataSet.

   a. numRecords = Me.ProductDataSet.Products.Count

   b. numRecords = Me.ProductDataSet.Items.Count

   c. numRecords = Me.Product.ProductDataSet.Count

   d. numRecords = Me.Count(ProductDataSet.Product)

EXERCISES

1. In this exercise, you modify the Trivia Game application from the lesson.

   a. Use Windows to make a copy of the Trivia Game Solution folder, which is contained in the VB2005\Chap12 folder. Rename the folder Trivia Game Solution-Ex1.

   b. If necessary, start Visual Studio 2005 or Visual Basic 2005 Express Edition. Open the Trivia Game Solution (Trivia Game Solution.sln) file contained in the VB2005\Chap12\Trivia Game Solution-Ex1 folder. Open the designer window.

c. Display the question number (from 1 through 9) along with the word "Question" in the interface.

d. Add another button to the interface. The button should allow the user to start a new game. Allow the user to click the New Game button only after he or she has answered all nine questions.

e. Save the solution, and then start and test the application. End the application. Close the Code Editor window, then close the solution.

2. In this exercise, you modify the Trivia Game application from the lesson.

a. Use Windows to make a copy of the Trivia Game Solution folder, which is contained in the VB2005\Chap12 folder. Rename the folder Trivia Game Solution-Ex2.

b. If necessary, start Visual Studio 2005 or Visual Basic 2005 Express Edition. Open the Trivia Game Solution (Trivia Game Solution.sln) file contained in the VB2005\Chap12\Trivia Game Solution-Ex2 folder. Open the designer window.

c. When the user submits an incorrect answer, the application should display the correct answer in a message box before moving to the next record. Modify the application's code accordingly.

d. Save the solution, and then start and test the application. End the application. Close the Code Editor window, then close the solution.

3. In this exercise, you use a SQL Server database named Sports. The database is stored in the Sports.mdf file, which is located in the VB2005\Chap12\SQL Databases folder. The Sports database contains one table named tblScores. The table contains five records, each having four fields that store the following information: name of the opposing team, the date of the game, your favorite team's score, and the opposing team's score.

a. If necessary, start Visual Studio 2005 or Visual Basic 2005 Express Edition. Create an appropriate Visual Basic Windows-based application. Name the solution Sports Action Solution, and name the project Sports Action Project. Save the application in the VB2005\Chap12 folder. The application should allow the user to view each record, and also to add, delete, and save records.

b. Save the solution, and then start and test the application. End the application. Close the Code Editor window, then close the solution.

DISCOVERY EXERCISE

4. In this exercise, you modify the Trivia Game application from the lesson.

   a. Use Windows to make a copy of the Trivia Game Solution folder, which is contained in the VB2005\Chap12 folder. Rename the copy Trivia Game Solution-Disc.

   b. If necessary, start Visual Studio 2005 or Visual Basic 2005 Express Edition. Open the Trivia Game Solution (Trivia Game Solution.sln) file contained in the VB2005\Chap12\Trivia Game Solution-Disc folder. Open the designer window.

   c. Allow the user to answer the questions in any order, and also change his or her answers. (You will probably need to modify the interface by including additional buttons.) Only display the number of incorrect answers when the user requests that information.

   d. Save the solution, and then start and test the application. End the application. Close the Code Editor window, then close the solution.

DISCOVERY EXERCISE

5. In this exercise, you learn how to create a parameterized query. Jerry Schmidt, the manager of the Fiction Bookstore, uses a SQL Server database named Books to keep track of the books in his store. The database is stored in the Books.mdf file, which is contained in the VB2005\Chap12\SQL Databases folder. The database has one table named tblBooks. The table has three fields: a numeric field named BookNumber and two text fields named Title and Author. Mr. Schmidt wants an application that he can use to enter an author's name and then display only the titles of books written by the author. In this application, you need to allow the user to specify the records he or she wants to select while the application is running. You can use a parameterized query to accomplish this task. A parameterized query is simply a SELECT statement that contains an @ symbol followed by a *parameter* name in the WHERE clause. The *@parameterName* acts as a placeholder for the missing data. For example, you can create a parameterized query using `@Author` in the WHERE clause. Associate the query with a ToolStrip control in the interface.

   a. If necessary, start Visual Studio 2005 or Visual Basic 2005 Express Edition. Open the Fiction Bookstore Solution (Fiction Bookstore Solution.sln) file, which is contained in the VB2005\Chap12\Fiction Bookstore Solution folder.

   b. Code the application appropriately.

   c. Save the solution, and then start and test the application. End the application. Close the Code Editor window, then close the solution.

DEBUGGING EXERCISE

6. In this exercise, you find and correct an error in an application. The process of finding and correcting errors is called debugging.

   a. Open the Debug Solution (Debug Solution.sln) file, which is contained in the VB2005\Chap12\Debug Solution folder.

   b. Open the Code Editor window. Review the existing code. Notice that a jagged line appears below one of the lines of code in the Code Editor window. Correct the code to remove the jagged line.

   c. Save the solution, then start and test the application. Notice that the application is not working correctly. Click the Exit button to end the application.

   d. Correct the errors in the application's code, then save the solution. Start and test the application.

   e. Click the Exit button to end the application. Close the code Editor window, then close the solution.

GUI DESIGN GUIDELINES

CHAPTER 1—LESSON C

FormBorderStyle, ControlBox, MaximizeBox, MinimizeBox, and StartPosition properties:

» A splash screen should not have Minimize, Maximize, or Close buttons, and its borders should not be sizable. In most cases, a splash screen's FormBorderStyle property is set to either None or FixedSingle. Its StartPosition property is set to CenterScreen.

» A form that is not a splash screen should always have a Minimize button and a Close button, but you can choose to disable the Maximize button. Typically, the FormBorderStyle property is set to Sizable; however, it also can be set to FixedSingle. Most times, the form's StartPosition property is set to CenterScreen.

CHAPTER 2—LESSON A

Layout and Organization of the User Interface

» Organize the user interface so that the information flows either vertically or horizontally, with the most important information always located in the upper-left corner of the screen. When positioning the controls, maintain a consistent margin from the edge of the form.

» Group together related controls using either white (empty) space or a control instantiated from one of the tools contained in the Containers section of the toolbox.

» Use a meaningful caption in each button. Place the caption on one line and use from one to three words only. Use book title capitalization for button captions.

» Use a label to identify each text box in the user interface. Also use a label to identify other label controls that display program output. The label text should be meaningful. It also should be from one to three words only and appear on one line. Left-align the text within the label, and position the label either above or to the left of the control it identifies. Follow the label text with a colon (:) and use sentence capitalization.

» Size the buttons in a group of buttons relative to each other, and place the most commonly used button first in the group.

» Align the borders of the controls wherever possible to minimize the number of different margins used in the interface.

CHAPTER 2—LESSON B

Adding Graphics

» Include a graphic in an interface only if it is necessary to do so. If the graphic is used solely for aesthetics, use a small graphic and place it in a location that will not distract the user.

Selecting Appropriate Font Types, Styles, and Sizes

» Use only one font type for all of the text in the interface. Use a sans serif font, preferably the Tahoma font. If the Tahoma font is not available, use either Microsoft Sans Serif or Arial.

» Use an 8-, 9-, 10-, 11-, or 12-point font for the text in an interface.

» Limit the number of font sizes used to either one or two.

» Avoid using italics and underlining, because these font styles make text difficult to read.

» Limit the use of bold text to titles, headings, and key items that you want to emphasize.

Selecting Appropriate Colors

» Build the interface using black, white, and gray first, then add color only if you have a good reason to do so.

» Use white, off-white, or light gray for an application's background, and use black for the text.

» Never use a dark color for the background or a light color for the text. A dark background is hard on the eyes, and light-colored text can appear blurry.

» Limit the number of colors in an interface to three, not including white, black, and gray. The colors you choose should complement each other.

» Never use color as the only means of identification for an element in the user interface.

Setting the BorderStyle Property of a Text Box and Label

» Leave the BorderStyle property of text boxes at the default value, Fixed3D.

» Leave the BorderStyle property of labels that identify other controls at the default value, None.

» Set to FixedSingle the BorderStyle property of labels that display program output, such as those that display the result of a calculation.

» In Windows applications, a control that contains data that the user is not allowed to edit does not usually appear three-dimensional. Therefore, you should avoid setting a label control's BorderStyle property to Fixed3D.

Locking the Controls

» Lock the controls in place on the form.

Rules for Assigning Access Keys and Controlling the Focus

» Assign a unique access key to each control (in the interface) that can receive user input (for example, text boxes, buttons, and so on).

» When assigning an access key to a control, use the first letter of the caption or identifying label, unless another letter provides a more meaningful association. If you can't use the first letter and no other letter provides a more meaningful association, then use a distinctive consonant. Lastly, use a vowel or a number.

» Assign a TabIndex value (begin with 0) to each control in the interface, except for controls that do not have a TabIndex property. The TabIndex values should reflect the order in which the user will want to access the controls.

» To give users keyboard access to a text box, assign an access key to the text box's identifying label. Set the identifying label's TabIndex property to a value that is one number less than the value stored in the text box's TabIndex property. (In other words, the TabIndex value of the text box should be one number greater than the TabIndex value of its identifying label control.)

CHAPTER 3—LESSON B

InputBox Function's Prompt and Title Capitalization

» In the InputBox function, use sentence capitalization for the *prompt*, and book title capitalization for the *title*.

Rules for Assigning the Default Button

» The default button should be the button that is most often selected by the user, except in cases where the tasks performed by the button are both destructive and irreversible. The default button typically is the first button.

CHAPTER 4—LESSON B

Labeling a Group Box

» Use sentence capitalization for the optional identifying label, which is entered in the group box's Text property.

MessageBox.Show Method Standards

» Use sentence capitalization for the *text* argument, but book title capitalization for the *caption* argument. The name of the application typically appears in the *caption* argument.

» Avoid using the words "error," "warning," or "mistake" in the message, because these words imply that the user has done something wrong.

» Display the Warning Message icon in a message box that alerts the user that he or she must make a decision before the application can continue. You can phrase the message as a question.

» Display the Information Message icon in a message box that displays an informational message along with an OK button only.

» Display the Stop Message icon when you want to alert the user of a serious problem that must be corrected before the application can continue.

» The default button in the dialog box should be the one that represents the user's most likely action, as long as that action is not destructive.

CHAPTER 5—LESSON B

Radio Button Standards

» Use radio buttons when you want to limit the user to one choice in a group of related and mutually exclusive choices.

» The minimum number of radio buttons in a group is two, and the recommended maximum number is seven.

» The label in the radio button's Text property should be entered using sentence capitalization.

» Assign a unique access key to each radio button in an interface.

» Use a container (such as a group box, panel, or table layout panel) to create separate groups of radio buttons.

» Only one button in each group can be selected at any one time.

» Designate a default radio button in each group of radio buttons.

Check Box Standards

» Use check boxes when you want to allow the user to select any number of choices from a group of one or more independent and nonexclusive choices.

» The label in the check box's Text property should be entered using sentence capitalization.

» Assign a unique access key to each check box in an interface.

CHAPTER 6—LESSON C

List Box Standards

» A list box should contain a minimum of three selections.

» A list box should display a minimum of three selections and a maximum of eight selections at a time.

» Use a label control to provide keyboard access to the list box. Set the label's TabIndex property to a value that is one less than the list box's TabIndex value.

» List box items are either arranged by use, with the most used entries appearing first in the list, or sorted in ascending order.

Default List Box Item

» If a list box allows the user to make only one selection, then a default item should be selected in the list box when the interface first appears. The default item should be either the most used selection or the first selection in the list. However, if a list box allows more than one selection at a time, you do not select a default item.

CHAPTER 7—LESSON B
Combo Box Standards

» Use a label control to provide keyboard access to the combo box. Set the label's TabIndex property to a value that is one less than the combo box's TabIndex value.

» Combo box items are either arranged by use, with the most used entries appearing first in the list, or sorted in ascending order.

CHAPTER 8—LESSON B
Designing Menus

» Menu title captions should be one word only, with the first letter capitalized. Each menu title should have a unique access key.

» Menu item captions can be from one to three words. Use book title capitalization and assign a unique access key to each menu item. Assign shortcut keys to commonly used menu items.

» If a menu item requires additional information from the user, place an ellipsis (...) at the end of the item's caption, which is entered in the item's Text property.

» Follow the Windows standards for the placement of menu titles and items.

» Use a separator bar to separate groups of related menu items.

B

VISUAL BASIC TYPE CONVERSION FUNCTIONS

Figure B-1 lists the Visual Basic type conversion functions. As you learned in Chapter 3, you can use the conversion functions (rather than the Convert methods) to convert an expression from one data type to another.

Syntax	Return data type	Range for *expression*
CBool(*expression*)	Boolean	Any valid String or numeric expression
CByte(*expression*)	Byte	0 through 255 (unsigned)
CChar(*expression*)	Char	Any valid String expression; value can be 0 through 65535 (unsigned); only first character is converted
CDate(*expression*)	Date	Any valid representation of a date and time
CDbl(*expression*)	Double	−1.79769313486231570E+308 through −4.94065645841246544E-324 for negative values; 4.94065645841246544E-324 through 1.79769313486231570E+308 for positive values
CDec(*expression*)	Decimal	+/−79,228,162,514,264,337,593,543,950,335 for zero-scaled numbers, that is, numbers with no decimal places. For numbers with 28 decimal places, the range is +/−7.9228162514264337593543950335. The smallest possible non-zero number is 0.0000000000000000000000000001 (+/−1E-28)
CInt(*expression*)	Integer	−2,147,483,648 through 2,147,483,647; fractional parts are rounded
CLng(*expression*)	Long	−9,223,372,036,854,775,808 through 9,223,372,036,854,775,807; fractional parts are rounded
CObj(*expression*)	Object	Any valid expression
CSByte(*expression*)	SByte (signed Byte)	−128 through 127; fractional parts are rounded
CShort(*expression*)	Short	−32,768 through 32,767; fractional parts are rounded
CSng(*expression*)	Single	−3.402823E+38 through −1.401298E-45 for negative values; 1.401298E-45 through 3.402823E+38 for positive values
CStr(*expression*)	String	Depends on the *expression*
CUInt(*expression*)	UInt	0 through 4,294,967,295 (unsigned)
CULng(*expression*)	ULng	0 through 18,446,744,073,709,551,615 (unsigned)
CUShort(*expression*)	UShort	0 through 65,535 (unsigned)

Figure B-1: Visual Basic type conversion functions

C

CREATING A SQL SERVER DATABASE

In this appendix, you learn how to use Visual Studio 2005 Professional Edition to create a SQL Server database.

To create a SQL Server database:

1 Start Visual Studio 2005, if necessary, and close the Start Page window.

2 Use the View menu to open the Server Explorer window. Right-click the **Server Explorer** window, then click **Create New SQL Server Database** to open the Create New SQL Server Database dialog box.

3 In the Create New SQL Server Database dialog box, you enter the server name (which typically is .\SQLEXPRESS), your security preference, and the database name. Type **.\SQLEXPRESS** in the Server name box, then type **FirstDatabase** in the New database name box. Also verify that the Use Windows Authentication radio button is selected. See Figure C-1.

Figure C-1: Completed Create New SQL Server Database dialog box

4 Click the **OK** button to close the Create New SQL Server Database dialog box. Visual Studio creates two files named FirstDatabase.mdf and FirstDatabase_log.LDF. The files are located in the Program Files\Microsoft SQL Server\MSSQL.1\MSSQL\Data folder.

5 If necessary, widen the Server Explorer window so that you can view each line in its entirety. Click the **plus box** that appears next to the FirstDatabase.dbo entry in the Server Explorer window. Doing this displays the objects that can be included in a database. See Figure C-2.

Figure C-2: Server Explorer window showing the database name and objects

6 Now you will include a table in the database. Right-click **Tables** in the Server Explorer window, and then click **Add New Table**. Doing this displays a window in which you define the table. See Figure C-3.

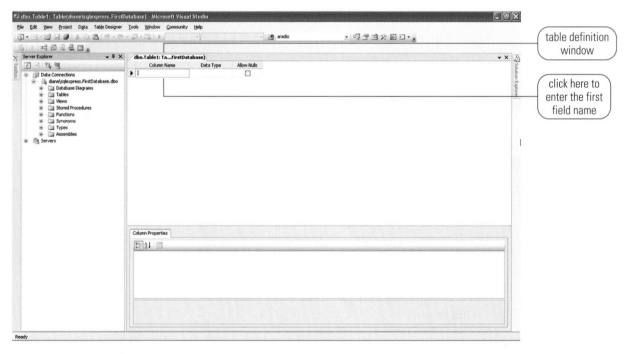

Figure C-3: Table definition window

7 The FirstDatabase will contain one table named tblEmpInfo. The tblEmpInfo table will contain three fields named EmpNum, FirstName, and LastName. In the box below the Column Name heading, type **EmpNum**, then press **Tab**. The EmpNum field will contain integers that represent employee numbers. Click the **down arrow** in the Data Type column. Scroll up the list, and then click **int**. Click the **check box** in the Allow Nulls column to deselect it. The 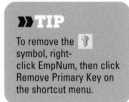 symbol that appears next to the EmpNum field indicates that the field is the table's primary key. (If the symbol does not appear next to the EmpNum field, right-click EmpNum, then click Set Primary Key on the shortcut menu.)

8 Use Figure C-4 to define the FirstName and LastName fields, which will contain the first and last names of the employees.

TIP

To remove the symbol, right-click EmpNum, then click Remove Primary Key on the shortcut menu.

Figure C-4: Completed table definition window

9 Click **File** on the menu bar, and then click **Save Table1** to open the Choose Name dialog box. Type **tblEmpInfo** in the Enter a name for the table box, then click the **OK** button.

10 Click the **plus box** that appears next to Tables in the Server Explorer window, if necessary, then click the **plus box** that appears next to tblEmpInfo. See Figure C-5.

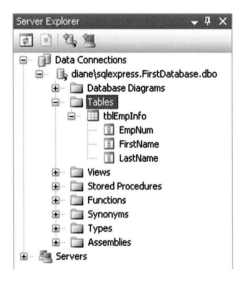

Figure C-5: Server Explorer window showing table and field names

11 After defining a table, you can fill it with data. Right-click **tblEmpInfo** in the Server Explorer window, then click **Show Table Data**. See Figure C-6.

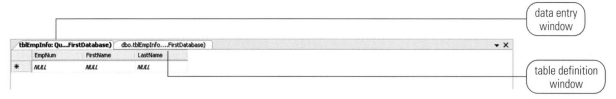

Figure C-6: Data entry window

12 Click the **NULL** that appears below the EmpNum column heading, then type **12** as the employee number. Press **Tab**, then type **Harold** as the first name and press **Tab**. Type **Jones** as the last name and press **Tab**.

13 Use the information shown in Figure C-7 to enter the remaining records.

Figure C-7: Records entered in the tblEmpInfo table

14 Close the data entry window, then close the table definition window.

INDEX

& (ampersand), 129, 130, 236
' (apostrophe), 154
* (asterisk), 37, 412
\ (backslash), 36
: (colon), 112
, (comma), 197, 450
{ } (curly braces), 493, 704
. (decimal point), 164–165, 194, 196, 197, 492
$ (dollar sign), 197, 250, 492, 647, 653–654
" (double quotes), 633
. . . (ellipsis), 669, 674
= (equal sign), 149, 193
! (exclamation point), 412
> (greater-than sign), 489
- (hyphen), 412
< (less-than sign), 489
( ) (parentheses), 69, 156, 157, 196, 197, 200, 703, 837, 839
% (percent sign), 162, 164, 197
. (period), 149, 196, 237, 356
| (pipe symbol), 493
+ (plus sign), 161
? (question mark), 412
[ ] (square brackets), 149, 412
_ (underscore), 69, 200, 450

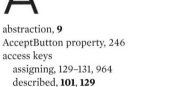

abstraction, **9**
AcceptButton property, 246
access keys
   assigning, 129–131, 964
   described, **101**, **129**
   rules for, 136
accumulators
   described, **499**
   repetition structures and, 499–506
Add method, **528**, 530, 592
Addition and Subtraction application, 297–299
ADO.NET (Microsoft)
   connecting applications to databases with, 899–906
   described, **898–899**
   overview, 893–960

algorithms
   described, **386**
   selection structures and, 386–396
ampersand (&), 129, 130, 236
apostrophe ('), 154
application(s). *See also specific applications*
   described, **7**
   designing, 95–192
   ending, 63–66
   exiting, 101
   previewing, 184–185, 282–283
   programmers, **3–4**
   running, 11–12
   starting, 63–66
   tasks, identifying, 104–106
APR (annual percentage rate), 228–229
Area Calculator application, 211–216
arguments
   described, **196**
   InputBox function and, 240
   variables and, 197
arithmetic expressions
   described, **156–158**
   variables and, 199–201
array(s)
   calculating average amounts stored in, 715–717
   declaring, 703
   described, **699–766**
   displaying the contents of, 707–709
   elements, **704**, 711–713
   highest value stored in, 717–719
   manipulating, 706–725
   numeric, 715–717
   of structure variables, 778–781
   one-dimensional, 702–734, 736–740
   parallel, **736–740**
   populating, **704**
   searching, 713–715, 748–750
   sorting data in, 705–706, 721–725, 746–747
   two-dimensional, **744–766**
   values stored in, updating, 719–721
Array.Reverse method, **722–724**
Array.Sort method, **722–724**
assembler, **5**

N

O